# CONTRIBUTIONS to AUTOMORPHIC FORMS, GEOMETRY, and NUMBER THEORY

# CONTRIBUTIONS to AUTOMORPHIC FORMS, GEOMETRY, and NUMBER THEORY

Edited by

Haruzo Hida, Dinakar Ramakrishnan,

and Freydoon Shahidi

The Johns Hopkins University Press
Baltimore and London

Printed in the United States of America on acid-free paper
9 8 7 6 5 4 3 2 1

The Johns Hopkins University Press
2715 North Charles Street
Baltimore, Maryland 21218-4363
www.press.jhu.edu

Library of Congress Cataloging-in-Publication Data

Contributions to automorphic forms, geometry, and number theory / Haruzo Hida, Dinakar Ramakrishnan, and Freydoon Shahidi, [editors].
p. cm.
Includes bibliographical references.
ISBN 0-8018-7860-8 (alk. paper)
1. Automorphic forms. 2. Geometry. 3. Number theory. I. Hida, Haruzo. II. Ramakrishnan, Dinakar. III. Shahidi, Freydoon.

QA353.A9C66 2004
515′9—dc22 2003060396

A catalog record for this book is available from the British Library.

# CONTENTS

*Preface* vii

*Joe Shalika and the Fine Hall Days, 1968–1971,* by Stephen Gelbart ix

1 Joseph A. Shalika, *Representation of the two by two unimodular group over local fields* 1

2 Jeffrey Adams, *Theta-10* 39

3 Jeffrey D. Adler, Lawrence Corwin, and Paul J. Sally, Jr., *Discrete series characters of division algebras and* $\mathbf{GL}_n$ *over a* $p$*-adic field* 57

4 James Arthur, *Automorphic representations of GSp(4)* 65

5 Don Blasius, *Elliptic curves, Hilbert modular forms, and the Hodge conjecture* 83

6 Siegfried Böcherer, Masaaki Furusawa, and Rainer Schulze-Pillot, *On the global Gross-Prasad conjecture for Yoshida liftings* 105

7 Daniel Bump, Solomon Friedberg, and Jeffrey Hoffstein, *Sums of twisted* $GL(3)$ *automorphic L-functions* 131

8 Bill Casselman, *Harmonic analysis of the Schwartz space of* $\Gamma\backslash\mathbf{SL}_2(\mathbb{R})$ 163

9 Laurent Clozel and Emmanuel Ullmo, *Equidistribution des points de Hecke* 193

10 James W. Cogdell and Ilya I. Piatetski-Shapiro, *Remarks on Rankin-Selberg convolutions* 255

11 Benedict H. Gross and Joe Harris, *On some geometric constructions related to theta characteristics* 279

12 Thomas C. Hales, *Can* $p$*-adic integrals be computed?* 313

13 Michael Harris, *Occult period invariants and critical values of the degree four L-function of GSp(4)* 331

14 Michael Harris and Stephen S. Kudla, *On a conjecture of Jacquet* 355

15 Hervé Jacquet, *Integral representation of Whittaker functions* 373

16 Hervé Jacquet, Erez Lapid, and Stephen Rallis, *A spectral identity for skew symmetric matrices* 421

17 Dihua Jiang and David Soudry, *Generic representations and local Langlands reciprocity law for p-adic* $SO_{2n+1}$ 457

18 Nicholas M. Katz, *Larsen's alternative, moments, and the monodromy of Lefschetz pencils* 521

19 Henry H. Kim and Freydoon Shahidi, *On the holomorphy of certain L-functions* 561

20 Victor Kreiman and V. Lakshmibai, *Richardson varieties in the Grassmannian* 573

21 Philip Kutzko, *Types and covers for SL(2)* 599

22 Robert P. Langlands, *Beyond endoscopy* 611

23 Dipendra Prasad, *An analogue of a conjecture of Mazur: A question in Diophantine approximation on tori* 699

24 Dinakar Ramakrishnan, *Existence of Ramanujan primes for GL*(3) 711

25 Peter Sarnak, *Nonvanishing of L-functions on* $\Re(s) = 1$ 719

26 Joseph A. Shalika, Ramin Takloo-Bighash, and Yuri Tschinkel, *Rational points and automorphic forms* 733

27 Joseph A. Shalika and Yuri Tschinkel, *Height zeta functions of equivariant compactifications of the Heisenberg group* 743

28 Marie-France Vignéras, *On highest Whittaker models and integral structures* 773

29 Jean-Loup Waldspurger, *Représentations de réduction unipotente pour SO(2n+1): quelques conséquences d'un article de Lusztig* 803

# PREFACE

Joseph Shalika is the author of some of the most important basic results in the modern theory of automorphic forms and representations. To comprehend the extent of his contributions, one can do no better than to peruse his articles, and also read his thesis, published for the first time here.

A sixtieth birthday conference for Shalika was held at the Johns Hopkins University during May 14–17, 2002. Among the participants were Jeffrey Adams, William Casselman, James Cogdell, Sol Friedberg, Masaki Furusawa, Stephen Gelbart, Thomas Hales, Hervé Jacquet, Dihua Jiang, Nicholas Katz, Henry Kim, Stephen Kudla, Philip Kutzko, V. Lakshmibai, Erez Lapid, Stephen Rallis, Dinakar Ramakrishnan, Paul Sally, Peter Sarnak, Freydoon Shahidi, Ramin Takloo-Bighash, and Yuri Tschinkel. We thank the Johns Hopkins math department, the NSF, and the Clay Math Foundation for providing financial support for the conference.

But this volume is more than a collection of articles by the participants of the conference. It contains, in addition, papers (written alone or jointly) by the following mathematicians: Jeffrey Adler, James Arthur, Don Blasius, Siegfried Böcherer, Laurent Clozel, Lawrence Corwin, Benedict Gross, Joe Harris, Michael Harris, Viktor Kreiman, Robert Langlands, Ilya Piatetski-Shapiro, Dipendra Prasad, Rainer Schulze-Pillot, Joseph Shalika, David Soudry, Emmanuel Ullmo, Marie-France Vignéras, and Jean-Loup Waldspurger.

All the articles in this volume were thoroughly refereed, and we thank the unnamed referees for their careful and timely job. All articles, except for the one by S. Gelbart, are research articles. We thank all the authors, first for contributing to the volume, and then for putting up with our demands for revision. We also thank Joseph Shalika for allowing us to publish his thesis here.

Finally, we thank the Johns Hopkins math department, especially Steve Zelditch and Christina Stanger, several people at the *American Journal of Mathematics,* including Bernard Shiffman and Mike Smith, and the Johns Hopkins University Press, for without their help this volume would not have seen the light of day.

# JOE SHALIKA AND THE FINE HALL DAYS, 1968–1971

By STEPHEN GELBART

I remember first meeting Joe in 1968. He was an assistant professor at Princeton, and I was a graduate student there working under Eli Stein. Below, I shall tell about how, in three years, I slowly moved from Eli Stein's field of harmonic analysis to the subject of automorphic forms in the style of Joe Shalika.

In 1968, being a graduate student was a complicated business. During my first year of graduate work, in 1967–68, graduate students received student deferments from the draft. But as the war in Vietnam reached its peak, it was announced in 1968 that graduate students would no longer be considered deferable. Rather than go and fight, most of the students in my year at Princeton had left the university to find acceptably deferable employment elsewhere. I was lucky enough to become an Instructor at Rutgers–Newark, where I taught three undergraduate courses a week, Monday, Wednesday, and Friday. That left me able to return to Princeton on Tuesdays and Thursdays and to finish up my Ph.D thesis as soon as possible.

My thesis was to carry out for the real numbers some of the analysis of the additive Fourier transform on matrix space. In general, that was to be similar to the analysis over the complex numbers that Stein had developed in his important paper on the missing unitary irreducible representations of $GL(n, \mathbf{C})$. For $GL(n, \mathbf{R})$, there would not be any new complementary series produced. But once again, the slightly modified Fourier operator $F^*$ led to a central operator which on the representation side was multiplication by a function that I had just been able to compute. That is, if $\Lambda$ is the parameterizing space of irreducible unitary representations of $G = GL(n, \mathbf{R})$, $f$ is in $L^2(G)$, and $T(f)(\lambda) = \int_G f(g)T^\lambda(g)\,d^*g$ for $\lambda$ in $\Lambda$, then

$$T(F^* f)(\lambda) = m(\lambda)T(f)(\lambda),$$

with $m(\lambda)$ explicitly computable.

Eli Stein was as helpful to me as an advisor could be. The completion of my thesis also owed a great deal to Joe Shalika. Let me explain how this happened. One prerequisite for this work was the infinite dimensional representation theory of real reductive *Lie* groups, a subject I liked very much. But as the thesis progressed and the problem was solved, Stein decided that I could work on certain aspects of the theory over $p$-adic fields. These were completely new fields of numbers for an analyst to deal with. As a matter of fact, until 1968, I would scan *Mathematical Reviews* by looking over the relevant sections on

Manuscript received October 18, 2002.

Research supported by the Minerva Foundation, Germany.

(real) Fourier and Functional Analysis: after 1968, my attention was focused uniformly on *Algebraic Number Theory*.

This transition took place in three steps. First, Stein had shown me a letter he received from Godement in 1968, just after Stein's paper was published in the *Annals of Mathematics*. The letter proposed a plan for $p$-adic groups which would do precisely what some of Stein's paper had done for the complex linear groups. Second, 1969 marked the publication of the English translation of the *Group Representations and Automorphic Forms*, by Gelfand, Graev, and Piatetski-Shapiro. (I have now admired and enjoyed this work for over thirty years.) I first looked at the book with Stein in the old Fine Hall common room. Together, we read the introduction in which the three authors thanked Godement for his help and mentioned Langlands for his new theory of Eisenstein series. Although I had been in graduate school for two years, this was the first time I had heard of Langlands. So I was impressed that Stein took him very seriously. What I learned quickly was that Joe had had an intense interest in Langlands' work for several years, as I shall now explain.

The combination of the two events above probably made me jump into the $p$-adic theory of my thesis. But the third factor was even more important: talking with (or rather listening to) Joe on the contents of the soon to be completed "Automorphic Forms On GL(2)," by Herve Jacquet and Robert Langlands. This monumental work was finished in 1969, and Joe had been learning of its contents from Jacquet, with the help of talking it over with Bill Casselman. Joe had already published several outstanding papers on $p$-adic and adelic harmonic analysis, and was one of the few masters who could readily appreciate the novelties of "the Book." I, on the other hand, was a complete novice, even being ill-prepared for the number theory involved.

In fact, I had never even taken a course in number theory. My original motivation for reading Jacquet and Langlands was purely local representation theoretic. I saw that a small part of their (local) representation theory was more or less equivalent to the major problem in my thesis for $GL(2, \mathbf{R})$. For the sake of definiteness, take $G = GL(2, \mathbf{R})$, and suppose $\pi$ is an irreducible unitary representation in the discrete series. For nice $f$ on matrix space, the function $f(x)|det(x)|^s$ belongs to $L^1(G)$ whenever $Re(s) > 1$, and the zeta-integral

$$\zeta(f, \pi, s) = \int_G f(x)\pi(x)|det(x)|^s dx$$

defines a holomorphic function of $s$ in this range; the same analysis can be given for the integrals $\zeta(f^*, \pi^*, s)$, where $f^*$ is essentially the modified Fourier transform $F^*$ mentioned above, and $\pi^*$ denotes the representation $\pi(x^{-1})$. So using the arguments in my book "Automorphic Forms on Adele Groups" (chapter 4, Further Notes), one indeed shows that the analytic continuation and functional equations for $\zeta(f, \pi, s)$ (just as in chapter 13, say, of Jacquet and Langlands' book) follow from the computation of $m$ of the modified Fourier operator $F^*$.

But Joe was interested in all of the Jacquet-Langlands GL (2) program, especially the adelic analysis, and it didn't take much of his lecturing to convince me of its importance. So I went into the adelic area by really studying the $GL(1)$ analysis due to Tate in his thesis. I remember that I followed the treatment of this in the book by Gelfand, Graev, and

Piatetski-Shapiro, which I thought was particularly beautiful. Joe taught me to consider the local and global theories as equally important: in this case, the global theory handed us an abstract functional equation, whereas the local theory led to knowledge of the concrete Dirichlet L-functions. (Once, at about this time, I was taking the elevator down to the third floor of [new] Fine Hall with Bernie Dwork, and he bluntly put it this way: "Tate's main theorem was the adelic, i.e., proved by global means; he included the local computations so that his global theory wouldn't be trivial!")

Anyway, after this, Joe many times used the blackboard of Princeton's Colloquium room to lecture me on Jacquet-Langlands' $GL(2)$ mysteries.

I was particularly lucky to have Joe's paper "Class Field Theory," which he had prepared for the 1969 Stony Brook Conference (Proceedings of Symposia in Pure Mathematics, AMS **20** [1971]). In it, Joe discusses some evidence for certain conjectures concerning class field theory that are further considered in Jacquet-Langlands' book. In particular, he surveys his work with Tannaka explicitly concerning the construction of dihedral forms on $GL(2)$.

I got my Ph.D degree in 1970 and enjoyed a year as an instructor at Princeton University, with more time to talk with Joe, while I also finished several projects with Eli Stein in real analysis.

Looking back on these years, the fact is that the number of people who really understood Jacquet-Langlands or even grasped the general concept of the Langlands program was incredibly small. And Joe was one of the few that really counted; some idea of the greatness of his work, particularly with Jacquet in the 70s and 80s, but also alone, shows itself in the many articles of this 60th "Anniversary" volume.

# CONTRIBUTIONS to AUTOMORPHIC FORMS, GEOMETRY, and NUMBER THEORY

# CHAPTER 1

# REPRESENTATION OF THE TWO BY TWO UNIMODULAR GROUP OVER LOCAL FIELDS

By Joseph A. Shalika

*Abstract.* Let $\Omega$ be a local field, i.e., a finite extension of the $p$-adic rational $\mathbf{Q}_p$. We assume throughout that $p$ is different from two. Let $\mathcal{O}_\mathfrak{p}$ denote the ring of integers in $\Omega$. The field $\Omega$ has the topology induced by the essentially unique extension of the valuation from $\mathbf{Q}_p$ to $\Omega$. With that topology $\Omega$ is a locally compact field. If $V$ is any finite-dimensional vector space over $\Omega$, then $V$ has a unique topology for which it is a topological vector space, and with that topology $V$ is locally compact. If $M_2(\Omega)$ denotes the algebra of two by two matrices over $\Omega$, then with the above topology $M_2(\Omega)$ becomes a topological ring. The closed subgroup $G = \mathrm{SL}(2, \Omega)$ of the group of units in $M_2(\Omega)$ then becomes a locally compact group. $K = \mathrm{SL}(2, \mathcal{O}_\mathfrak{p})$ is a maximal compact subgroup which is open in $G$ and every maximal compact subgroup of $G$ is the image of $K$ under some automorphism of $G$.

In this paper I will consider three problems basic to Fourier analysis on $G$. They are:

i. Construction of irreducible unitary representations $D$ of $G$
ii. Construction of irreducible unitary representations $\mathcal{D}$ of $K$
iii. Comparison of (i) and (ii), namely the decomposition of the representation, $\mathrm{ind}_{K\uparrow G}\mathcal{D}$, induced on $G$ by the representation $\mathcal{D}$ of $K$ into irreducible components

I will begin by studying a certain family of projective representations defined by Weil in [16]. By taking tensor products I will obtain a family of unitary representations of $G$ that I will decompose into irreducible components $D$. In certain cases I will pick out from the restriction of $D$ to $K$ irreducible representations $\mathcal{D}$ of $K$ and show that we can recover $D$ by inducing $\mathcal{D}$ from $K$.

The purpose of the fourth paragraph is to obtain all irreducible unitary representations of $\mathrm{SL}(2, \mathcal{O}_\mathfrak{p})$. (In §4.1 and §4.2, $\Omega$ may also be taken as the completion of a function field.) In §4.1 and §4.2 of this paper, I will construct a series of representations of $\mathrm{SL}(2, \mathcal{O}_\mathfrak{p})$ as monomial representations. In §4.3, I show that these together with certain explicit matrix representations considered by Hecke [6], Kloosterman [8], Maass [9] and also in the first three sections yield all irreducible unitary representations of $\mathrm{SL}(2, \mathcal{O}_\mathfrak{p})$.

I remark that some of the results of §4 are general and may by used to construct some irreducible representations of the compact $p$-adic groups (e.g., integral forms of non-compact semi-simple algebraic groups) provided they satisfy a certain criterion of semi-simplicity (namely the Lie-algebra of the integral form reduced mod $\mathfrak{p}^n$ has a nondegenerate Killing form) and also an analogue of Hensel's Lemma (the analogue of Lemma 4.2.3). Finally the combined results of the paragraph and §3 show that some of the discrete series of $\mathrm{SL}(2, \Omega)$ are monomial representations, i.e., induced from one-dimensional representations of open compact subgroups. It seems possible that one can construct some of the discrete series in higher dimensions in an analogous way.

**1. Introduction.** The discovery that Weil representation [16] (or the oscillator representation) of $\mathrm{SL}(2, \Omega)$ has the group $O(n)$ as a group of symmetries was essentially discovered by Kloosterman, who actually considered the analogous groups over $Z/p^n$, in his work on theta-series [8]. The fact that $O(n)$ is exactly the group of symmetries, in the situation of local fields, appears, I believe, for the first time in this paper. This observation was, soon after, extensively developed by R. Howe [7] in his theory of dual reductive pairs. Howe's theory was to prove fundamental in the modern theory of theta-functions.

This paper also owes much to the book of Gelfand, Graev, and Piatetskii-Shapiro [4].

F. I. Mautner surely deserves recognition as one of the founders of the theory of representations of $p$-adic groups. We note in particular his paper [11], in which the existence of what later came to be called supercuspidal representations, by Harish-Chandra, [5] appear for the first time.

Finally, I would like to say that the appearance of my paper led to a fruitful collaboration and lifelong friendship with P. J. Sally.

*Acknowledgment.* I would like to acknowledge my appreciation to Professors Mautner and Yoshizawa for the many hours of discussions we had together.

**1.1. Projective representations.** The purpose of this paragraph is to make the formal process of passing from projective representations to representations of a given group clear, hence I will ignore all continuity considerations.

Let $G$ be an abstract group, $U$ the group of unitary operators on a Hilbert space $\mathfrak{h}$, and $\mathbf{C}^1$ the complex numbers of absolute value one.

By a *projective representation* of $G$ is meant a homomorphism $P$ of $G$ into $U/\mathbf{C}$. For each $g$ in $G$, let $U_g$ be an element of $U$ belonging to the class of $P(g)$, then the operators $U_g$ satisfy

$$U_{g_0g_1} = \tilde{\omega}(g_0, g_1)U_{g_0}U_{g_1} \tag{1}$$

for all $g_0, g_1$ in $G$, where $\tilde{\omega}$ is a map of $G \times G$ to $\mathbf{C}^1$. From the associative law of $G$, one sees immediately that $\tilde{\omega}$ is a two-cocycle on $G$, i.e.,

$$\tilde{\omega}(g_0g_1, g_2)\tilde{\omega}(g_0, g_1) = \tilde{\omega}(g_0, g_1g_2)\tilde{\omega}(g_1, g_2). \tag{2}$$

A map of $G$ into $U$ satisfying (1) for some map $\tilde{\omega}$ from $G \times G$ to $\mathbf{C}^1$ (which is then necessarily a two-cocycle) will be called a *multi-valued representation* of $G$ *with associated two-cocycle* $\tilde{\omega}$.

Two multi-valued representations on $G$ in the same Hilbert space, defined respectively by $g \longmapsto U_g$ and $g \longmapsto U'_g$, are called equivalent if they come from the same projective representation of $G$, i.e., if there exists a map $\phi$ of $G$ to $\mathbf{C}^1$ such that

$$U_g = \phi(g)U'_g. \tag{3}$$

If $\tilde{\omega}$ and $\tilde{\omega}'$ are the two-cocycles associated respectively to these multi-valued representations, then (3) implies

$$\tilde{\omega}(g_0, g_1) = \tilde{\omega}'(g_0, g_1)\phi(g_0)\phi(g_1)/\phi(g_0g_1), \tag{4}$$

i.e., $\tilde{\omega}$ and $\tilde{\omega}'$ differ by the coboundary of the one cochain $\phi$, or $\tilde{\omega}$ and $\tilde{\omega}'$ are cohomologous.

Hence to each projective representation $P$ of $G$ is associated a unique cohomology class and $P$ may be factored by a representation of $G$ (i.e., a homomorphism from $G$ to $U$) if and only if that class is trivial.

Suppose $g \longmapsto U_{0,g}$ and $g \longmapsto U_{1,g}$ are two multi-valued representations of $G$ with associated two cocycles given respectively $\tilde{\omega}_0$ and $\tilde{\omega}_1$, then the map $g \longmapsto U_{0,g} \otimes U_{1,g}$ is another multi-valued representation of $G$ with associated two cocycle $\tilde{\omega}_0 \cdot \tilde{\omega}_1$.

In the following I will construct a family of multi-valued representations of SL(2, $\Omega$), all with the same associated cohomology class, and will prove that the square of that class is trivial (in fact these multi-valued representations will have the same associated two-cocycle whose square will be trivial). Hence, any $2n$-fold tensor product of these multi-valued representations will have an associated cohomology class which will be trivial, and consequently that tensor product will be equivalent to a representation of $G$.

**1.2. Generating relations for SL(2, $\Omega$).** Let $\Omega^*$ denote the multiplicative group of non-zero elements in $\Omega$.

In the construction of multi-valued representations of SL(2, $\Omega$), I will need the following theorem.

THEOREM 1.2.1. *$G = \mathrm{SL}(2, \Omega)$ is generated by elements of the form $u(b) = \left(\begin{smallmatrix} 1 & b \\ 0 & 1 \end{smallmatrix}\right)$, for $b \in \Omega$, and the element $w = \left(\begin{smallmatrix} 0 & 1 \\ -1 & 0 \end{smallmatrix}\right)$. Further, for $b \in \Omega^*$, if we put*

$$s(b) = \begin{pmatrix} b & 0 \\ 0 & b^{-1} \end{pmatrix} = wu(b^{-1})wu(b)wu(b^{-1}),$$

*then the relations between $u(b)$ and $w$ are given by*

(i) $w^2 = s(-1)$

(ii) $u(b)u(b') = u(b + b')$ *for $b$ and $b'$ in $\Omega$*

(iii) $s(a)s(a') = s(aa')$ *for $a$ and $a'$ in $\Omega^*$*

(iv) $s(a)u(b)s(a^{-1}) = u(ba^2)$ *for $b \in \Omega$ and $a \in \Omega^*$.*

*Proof.* One easily checks that these four relations are satisfied by $w$, $u(b)$ and $s(a)$.

I claim that SL(2, $\Omega$) is isomorphic to the group, which I will denote by $G'$, generated by the symbols $U'(b)$, for $b \in \Omega$, and $w'$, subject to the above four relations (with $u', w', s'$ instead of $u, w, s$).

For $\left(\begin{smallmatrix} a & b \\ c & d \end{smallmatrix}\right) \in G$, put $\psi\left(\begin{smallmatrix} a & b \\ c & d \end{smallmatrix}\right) = s'(-c^{-1})\,u'(ac)\,w'\,u'(dc^{-1})$, if $c$ belongs to $\Omega^*$, and $\psi\left(\begin{smallmatrix} a & b \\ 0 & d \end{smallmatrix}\right) = s'(a)\,u'(ba^{-1})$. Since indeed

$$\begin{pmatrix} a & b \\ c & d \end{pmatrix} = \begin{pmatrix} -c^{-1} & 0 \\ 0 & -c \end{pmatrix}\begin{pmatrix} 1 & ac \\ 0 & 1 \end{pmatrix}\begin{pmatrix} 0 & 1 \\ -1 & 0 \end{pmatrix}\begin{pmatrix} 1 & dc^{-1} \\ 0 & 1 \end{pmatrix},$$

for $c \in \Omega^*$, and

$$\begin{pmatrix} a & b \\ 0 & a^{-1} \end{pmatrix} \begin{pmatrix} a & 0 \\ 0 & a^{-1} \end{pmatrix} \begin{pmatrix} 1 & ba^{-1} \\ 0 & 1 \end{pmatrix},$$

it follows that $G$ is generated by $\left(\begin{smallmatrix} 1 & b \\ 0 & 1 \end{smallmatrix}\right)$, for $b \in \Omega$, and $\left(\begin{smallmatrix} 0 & 1 \\ -1 & 0 \end{smallmatrix}\right)$. Hence in order to prove that $\psi$ defines an isomorphism of $G$ into $G'$ it suffices to prove the two statements:

- $\psi\left(\begin{smallmatrix} a & b \\ c & d \end{smallmatrix}\right)\psi\left(\begin{smallmatrix} 1 & x \\ 0 & a \end{smallmatrix}\right) = \psi\left(\begin{smallmatrix} a & ax+b \\ c & cx+d \end{smallmatrix}\right)$
- $\psi\left(\begin{smallmatrix} a & b \\ c & d \end{smallmatrix}\right)\psi\left(\begin{smallmatrix} 0 & 1 \\ -1 & 0 \end{smallmatrix}\right) = \psi\left(\begin{smallmatrix} -b & a \\ -d & c \end{smallmatrix}\right)$,

for all $x \in \Omega$ and $\left(\begin{smallmatrix} a & b \\ c & d \end{smallmatrix}\right) \in G$. The first statement follows immediately from (ii). The second follows easily when either $c$ or $d$ is zero. Hence it remains to prove

$$s'(-c^{-1})\,u'(ac)\,w'\,u'(dc^{-1})\,w' = s'(d^{-1})\,u'(bd)\,w'\,u'(-cd^{-1}).$$

Using the definition of $s'(d)$ and the fact that $s'(-1)$ commutes with $w'$ by (i) and with $u'(b)$ by (ii) and hence lies in the center of $G'$, we have

$$w'\,u'(dc^{-1})\,w' = s'(-d^{-1}c)\,u'(-dc^{-1})\,w'\,u'(-d^{-1}c).$$

Hence it suffices to prove

$$s'(-c^{-1})\,u'(ac)\,s'(-d^{-1}c)\,u'(-dc^{-1}) = s'(d^{-1})\,u'(bd),$$

a relation which follows easily from (ii) to (iv). This proves the theorem. □

N.B. The previous proof makes no assumption on the field $\Omega$ and Theorem 1.2.1 is true for the group SL(2, $k$) for any field $k$.

**1.3. Computation of Gaussian sums.** As before let $\mathcal{O}_\mathfrak{p}$ denote the ring of integers in $\Omega$. Let $\mathfrak{p}$ be the maximal ideal of $\mathcal{O}_\mathfrak{p}$ and $\mathcal{O}_\mathfrak{p}^*$ the group of units in $\mathcal{O}_\mathfrak{p}$. Let $\pi$ be a prime element, i.e., a generator of the principal ideal $\mathfrak{p}$. Every non-zero element $x \in \Omega^*$ may be written uniquely in the form

$$x = \pi^{\omega(x)}\eta,$$

where $\omega(x)$ is an integer and $\eta$ belongs to $\mathcal{O}_\mathfrak{p}^*$. It follows that every (fractional) ideal (a non-zero proper $\mathcal{O}_\mathfrak{p}$-submodule of $\Omega$) of $\Omega$ has the form $\mathfrak{p}^\nu$ for some integer $\nu$.

Let $\Phi$ be a fixed non-trivial character of the additive group $\Omega^+$ of $\Omega$. For any ideal $\mathfrak{a}$, let $\mathfrak{a}^*$ denote the set of all $x \in \Omega$ satisfying $\Phi(xy) = 1$ for all $y \in \mathfrak{a}$. $\mathfrak{a}^*$ is again an ideal. Let $\mathfrak{p}^{\omega(\Phi)}$ be the conductor of $\Phi$, i.e., the largest ideal $\mathfrak{a}$ for which $\Phi$ restricted to $\mathfrak{a}$ is trivial.

LEMMA 1.3.1. *Let* $\mathfrak{a} = \mathfrak{p}^\nu$, $\omega = \omega(\Phi)$, *then* $\mathfrak{a}^* = \mathfrak{p}^{\omega-\nu}$.

*Proof.* An element $x \in \Omega$ belongs to $\mathfrak{a}^*$ if and only if $\pi^\nu x$ belongs to $\mathcal{O}_\mathfrak{p}^*$. Hence $\mathfrak{a}^* = \mathfrak{p}^{-\nu}\mathcal{O}_\mathfrak{p}^*$. By definition of the conductor of $\Phi$, $\mathfrak{p}^\omega$ is the largest ideal contained in $\mathcal{O}_\mathfrak{p}^*$. □

Let $\mu$ be a Haar measure on $\Omega^+$. If we put $N(\mathfrak{p}^\nu) = q^\nu$, where $q$ is the number of elements in the residue class field $\mathcal{O}_\mathfrak{p}/\mathfrak{p}$, it is well known that every ideal $\mathfrak{a}$ has finite measure given by

$$\mu(\mathfrak{a}) = N(\mathfrak{a})^{-1}\mu(\mathcal{O}_\mathfrak{p}).$$

*The additive Fourier transform.* The group $\Omega^+$ is self-dual, i.e., for any $x \in \Omega$ and any non-trivial character $\Phi$ of $\Omega^+$, let $\Phi_x$ denote the character defined by

$$\Phi_x(y) = \Phi(xy).$$

Then the map $x \longmapsto \Phi_x$ defines an isomorphism of $\Omega^+$ with its dual group. Hence we may define the Fourier transform of an element $h$ in $L^2(\Omega^+)$ by the integral

$$\hat{h}(x) = \int h(y)\Phi(2xy)\,dy,$$

which is defined for continuous functions $h$ of compact support on $\Omega^+$ and extends by continuity to $L^2(\Omega^+)$.

If we normalize the measure $\mu$ so that $\mu(\mathcal{O}_\mathfrak{p}) = q^{\omega(\Phi)/2}$, then it is easily verified that

$$\hat{\hat{f}}(x) = f(-x).$$

Denote this normalized measure by $d_\Phi x$. Finally we recall that for $a \in \Omega^*$,

$$d_\Phi(ax) = |a|d_\Phi x,$$

where $|a| = q^{-\omega(a)}$.

Lemma 1.3.2. *If $\Phi$ is a non-trivial character of $\Omega^+$, then the limit*

$$\mathcal{H}(\Phi) = \lim_{m\to-\infty}\int_{\mathfrak{p}^m}\Phi(x^2)\,d_\Phi x,$$

*which we will denote by $\int\Phi(x^2)\,d_\Phi x$, exists and is equal to 1 if $\omega = \omega(\Phi)$ is even and is equal to*

$$G(\Phi) = (\sqrt{q})^{-1}\sum_{t \bmod \mathfrak{p}}\Phi(\pi^{\omega-1}t^2),$$

*if $\omega = \omega(\Phi)$ is odd.*

*Proof.* For any subset $E$ of $\Omega$, let $\chi(E)$ denote the characteristic function of $E$. First, suppose $\omega \geq 0$. Choose $m \leq \omega$, then

$$I = \int_{\mathfrak{p}^m} \Phi(x^2)\, d_\Phi x = \sum_{x_0 \in \mathfrak{p}^m/\mathfrak{p}^\omega} \int \Phi(x^2)\chi(x_0 + \mathfrak{p}^\omega)(x)\, d_\Phi x,$$

or, writing $x = x_0 + \pi^\omega \lambda$, where $\lambda$ belongs to $\mathcal{O}_\mathfrak{p}$,

$$I = \frac{d\pi^\omega \lambda}{d\lambda} \sum_{x_0 \in \mathfrak{p}^m/\mathfrak{p}^\omega} \int_{\mathcal{O}_\mathfrak{p}} \Phi(2x_0\pi^\omega\lambda)\, d\lambda.$$

Since the residual characteristic $p$ is not two, we have

$$\int_{\mathcal{O}_\mathfrak{p}} \Phi(2x_0\pi^\omega\lambda)\, d\lambda = \chi((\mathfrak{p}^\omega)^*).$$

Hence

$$I = |\pi^\omega|\mu(\mathcal{O}_\mathfrak{p}) \sum_{x_0 \in \mathfrak{p}^m/\mathfrak{p}^\omega} \Phi(x^2)\chi(\mathcal{O}_\mathfrak{p})(x_0) = q^{-\omega/2} \sum_{x_0 \in \mathcal{O}_\mathfrak{p}/\mathfrak{p}^\omega} \Phi\left(x_0^2\right).$$

Hence the lemma is true if $\omega = 0$ or 1.

Now suppose $\omega$ is odd, put

$$\xi(x) = \Phi(\pi^{\omega-1}x).$$

Then $\xi$ is a character of $\Omega^+$ with conductor $\mathfrak{p}$, hence by the above $\mathcal{H}(\xi) = G(\xi)$. On the other hand

$$\int_{\mathfrak{p}^m} \xi(x^2)\, d_\xi x = \int_{\mathfrak{p}^m} \Phi(\pi^{\omega-1}x^2)\, d_\xi(x) = \left|x^{-(\omega-1)/2}\right| \int_{\mathfrak{p}^{m+(\omega-1)/2}} \Phi(x^2)\, d_\xi x,$$

and, since $d_\xi x\, q^{(\omega-1)/2} = d_\Phi x$, the limit $\lim_{m\to-\infty} \int_m \Phi(x^2)\, d_\Phi x$ exists and is equal to $(\sqrt{q})^{-1} \sum_{t \bmod \mathfrak{p}} \Phi(\pi^{\omega-1}t^2)$.

If $\omega$ is even put $\xi(x) = \Phi(\pi^\omega x)$—a similar computation then proves the lemma. □

An immediate corollary to this lemma is the fact that, for $b \in \Omega^*$ of the form $\pi^{\omega(b)}\eta$, the limit

$$\mathcal{H}(\Phi, b) = \lim_{m\to-\infty} \int_{\mathfrak{p}^m} \Phi(bx^2)\, d_\Phi x$$

exists and is equal to $q^{\omega(b)/2}\mathcal{H}(\Phi_b)$. It now follows that $\mathcal{H}(\Phi_b)$ is 1, if $\omega(\Phi) - \omega(b)$ is even and, is equal to $G(\Phi_b) = G(\Phi_\eta)$, if $\omega(\Phi) - \omega(b)$ is odd.

For $\eta \in \mathcal{O}_\mathfrak{p}^*$, put $(\frac{\eta}{\mathfrak{p}}) = 1$, if $\eta$ is the square of an element in $\Omega$, and $(\frac{\eta}{\mathfrak{p}}) = -1$ otherwise. The following facts are well-known:

i. $\mathcal{H}(\Phi)^2 = (\frac{-1}{\mathfrak{p}})^{\omega(\Phi)}$,

ii. $G(\Phi_\eta) = (\frac{\eta}{\mathfrak{p}})\, G(\Phi)$ for $\eta \in \mathcal{O}_\mathfrak{p}^*$.

*A transformation law.* Let $\mathcal{S}$ be the Schwartz-Bruhat space attached to $\Omega$. $\mathcal{S}$ is by definition the space of locally constant continuous complex-valued functions on $\Omega$ of compact support.

$\mathcal{S}$ may be also characterized as the space of continuous complex-valued functions $h$ on $\Omega$ such that $h$ and $\hat{h}$ are of compact support. Finally $\mathcal{S}$ with the $L^2$ topology is dense in $L^2(\Omega^+)$.

THEOREM 1.3.3. *Let $h$ belong to $\mathcal{S}$. For $b \in \Omega^*$, put $f_b(x) = \Phi(bx^2)$. Then $f_b \cdot h$ and $f_b * h$ also belong to $\mathcal{S}$. The Fourier transform of $f_b \cdot h$ is given explicitly by*

$$\widehat{f_b \cdot h} = \mathcal{H}(\Phi, b) f_{-1/b} * \hat{h}.$$

*Proof.* Formally we have

$$\widehat{f_b \cdot h} = \hat{f}_b * \hat{h},$$

and

$$\begin{aligned}\hat{f}_b(x) &= \int \Phi(by^2 + 2xy)\,dy = \int \Phi\left(b\left(\left(y + \frac{x}{b}\right)^2 - \frac{x^2}{b^2}\right)\right) dy \\ &= \Phi\left(-\frac{x^2}{b}\right) \int \Phi(by^2)\,dy.\end{aligned}$$

This formal argument can be made precise as follows:

Given $h$ and $b$ an element of $\Omega^*$, choose $m$ so that the support of $h$ lies in $\mathfrak{p}^m$ and the support of $\hat{h}$ lies in $\mathfrak{p}^m b$. Then we have

$$\widehat{f_b \cdot h} = (f_b \chi(\mathfrak{p}^m) \cdot h)^\wedge = (f_b \cdot \chi(\mathfrak{p}^m))^\wedge * \hat{h},$$

or the same, $$\widehat{f_b \cdot h}(x) = \int (f_b \cdot \chi(\mathfrak{p}^m))^\wedge(y) \hat{h}(x-y)\,dy.$$

If we restrict $x$ to lie in $\mathfrak{p}^m b$, since the support of $\hat{h}$ is contained in $\mathfrak{p}^m b$, this last integral becomes

$$\int_{\mathfrak{p}^m b} (f_b \cdot \chi(\mathfrak{p}^m))^\wedge(y) \hat{h}(x-y)\,dy.$$

However, for $y \in \mathfrak{p}^m b$, our initial argument shows that

$$(f_b \cdot \chi(\mathfrak{p}^m))^\wedge(y) = f_{-1/b}(y) \int_{\mathfrak{p}^m} \Phi(yz^2)\,dz.$$

Hence, for $x \in \mathfrak{p}^m b$,

$$\widehat{f_b \cdot h}(x) = \int_{\mathfrak{p}^m} \Phi(bz^2)\,dz \int_{\mathfrak{p}^m b} f_{-1/b}(y)\hat{h}(x-y)\,dy$$
$$= \int_{\mathfrak{p}^m} \Phi(bz^2)\,dz \int f_{-1/b}(y)\hat{h}(x-y)\,dy.$$

Taking the limit as $m \to -\infty$, proves the theorem. □

**1.4. Construction of multi-valued representations of SL(2, Ω).** Let $\mathfrak{h}$ denote the Hilbert space $L^2(\Omega^+)$. For $h \in \mathfrak{h}$ and for each non-trivial character $\Phi$ of $\Omega^+$, let

$$T_w h = \mathcal{H}(\Phi)\hat{h},$$

and, for $b \in \Omega$, let

$$M_b h = f_b \cdot h.$$

Since $|f_b| = |\mathcal{H}(\Phi)| = 1$, the operators $T_w$ and $M_b$ are unitary.

THEOREM 1.4.1. *The maps* $\left(\begin{smallmatrix} 0 & 1 \\ -1 & 0 \end{smallmatrix}\right) \longmapsto T_w$ *and* $\left(\begin{smallmatrix} 1 & b \\ 0 & 1 \end{smallmatrix}\right) \longmapsto M_b$ *may be extended to a multi-valued representation of* SL(2, Ω) *on the Hilbert space* $\mathfrak{h}$. *Let* $\tilde{\omega}(\Phi)$ *denote the associated two-cocycle. Then*

(i) $\tilde{\omega}(\Phi)$ *is independent of* $\Phi$.
(ii) $[\tilde{\omega}(\Phi)]^2 = 1$.

*Proof.* For $b \in \Omega^*$, put $T_w M_{b^{-1}} T_w M_b T_w M_{b^{-1}} = D_b$. We shall prove in the following that there exists a two-cocycle $\zeta$ and a one-cocycle $\varphi$ defined on $\mathcal{O}_{\mathfrak{p}}^*$, both independent of $\Phi$ and satisfying $\zeta^2 = \varphi^2 = 1$, such that

i. $T_w^2 = D_{-1}$
ii. $M_b\, M_{b'} = M_{b+b'}$ for $b, b'$ in $\Omega$
iii. $D_a\, D_{a'} = \zeta(a, a')\, D_{aa'}$ for $a, a'$ in $\Omega^*$
iv. $D_a\, M_b\, D_{a^{-1}} = \varphi(a)\, M_{ba^2}$ for $b \in \Omega$ and $a \in \Omega^*$.

This together with §1.2 will prove the theorem.

First of all, in the proof, it suffices to consider the restriction of the operators $T_w$ and $M_b$ to $\mathcal{S}$, which by previous lemmas is left stable by these unitary operators. An easy computation using theorem 1.3.3 then shows that, for $h \in \mathcal{S}$ and $a \in \Omega^*$,

$$(D_a h)(x) = |a|[\mathcal{H}(\Phi)]^{-1}\mathcal{H}(\Phi, a)h(ax).$$

Hence $a \longmapsto D_a$ is a projective representation of $\Omega^*$. If we put

$$X(a) = \mathcal{H}(\Phi, a)[\mathcal{H}(\Phi)]^{-1}|a|^{1/2},$$

then the two-cocycle $\zeta$ (on $\Omega^*$) associated to this projective representation is given by

$$\zeta(a, a') = X(aa')/X(a)X(a').$$

Writing $a = \pi^{\omega(a)}\eta$, $a' = \pi^{\omega(a')}\eta'$, we have by §1.3

$$X(a) = \mathcal{H}(\Phi_a)[\mathcal{H}(\Phi)]^{-1} = \left(\frac{\eta}{\mathfrak{p}}\right)^{\omega(\Phi)-\omega(a)} \theta(a),$$

where $\theta(a)$ is equal to one if $\omega(\Phi)$ and $\omega(\Phi) - \omega(a)$ have the same parity, to $G(\Phi)$ if $\omega(\Phi)$ is even and $\omega(\Phi) - \omega(a)$ is odd, and to $[G(\Phi)]^{-1}$ if $\omega(\Phi)$ is odd and $\omega(\Phi) - \omega(a)$ is even. Hence, since $\eta \longmapsto (\frac{\eta}{\mathfrak{p}})$ is a multiplicative character,

$$\zeta(a, a') = \left(\frac{\eta}{\mathfrak{p}}\right)^{\omega(a')} \left(\frac{\eta'}{\mathfrak{p}}\right)^{\omega(a)} \theta(a, a')/\theta(a)\theta(a').$$

Since $\theta(a, a')/\theta(a)\theta(a')$ is equal to one if $\omega(a)$ and $\omega(a')$ are even or of opposite parity, and to $[G(\Phi)]^2 = (\frac{-1}{\mathfrak{p}})$ if $\omega(a)$ and $\omega(a')$ are odd, $\zeta$ has the required property. Hence

$$\left(T_w^2 h\right)(x) = [\mathcal{H}(\Phi)]^2 \hat{\hat{h}}(x) = \left(\frac{-1}{\mathfrak{p}}\right)^{\omega(\Phi)} h(-x) = X(-1)h(-x) = (D_{-1}h)(x).$$

Clearly, $D_1 = M_0 = 1$. Finally

$$D_a\, M_b\, D_{a^{-1}} h = X(x)X(a^{-1}) \cdot M_{ba^2} h = [\theta(a)]^2\, M_{ba^2} h = \left(\frac{-1}{\mathfrak{p}}\right)^{\omega(a)} M_{ba^2} h.$$

□

**1.5. Construction of unitary representations of SL(2, Ω).** In this section I will obtain unitary representations of SL(2, Ω) by taking tensor products of the projective representations defined in §1.4.

Let $V$ be a metric vector space over $\Omega$, i.e., a finite-dimensional topological vector space over $\Omega$ together with a nondegenerate quadratic form $Q$. As in the case of $\Omega^+$, the additive group $V^+$ of $V$ is self-dual; in fact every character of $V^+$ has the form

$$\mathbf{x} \longmapsto \Phi(B(\mathbf{x}, \mathbf{y})),$$

where $\mathbf{y}$ is a fixed element of $V$ and $B$ is the bilinear form associated with $Q$:

$$B(\mathbf{x}, \mathbf{y}) = Q(\mathbf{x} + \mathbf{y}) - Q(\mathbf{x}) - Q(\mathbf{y}).$$

For $h$ in the Hilbert space $L^2(V^+) = \mathfrak{h}_V$, the Fourier transform is defined by

$$\hat{h}(\mathbf{x}) = \int_V h(\mathbf{y})\Phi(B(\mathbf{x}, \mathbf{y}))\, d\mathbf{y}.$$

The Haar measure $d\mathbf{y}$ on $V^+$ may again be normalized so that

$$\hat{\hat{h}}(\mathbf{x}) = h(-x).$$

Denote the normalized measure by $d_\Phi \mathbf{y}$.

By a lattice $\Lambda$ in $V$ is meant an open compact $\mathcal{O}_\mathfrak{p}$ submodule of $V$.

LEMMA 1.5.1. *Let $V$ be a metric vector space over $\Omega$ and $\Lambda$ a lattice in $V$. Then $\Lambda$ is an orthogonal direct sum of cyclic $\mathcal{O}_\mathfrak{p}$ submodules, i.e., if* $\dim_\Omega V = n$, *there exists an orthogonal basis $\mathbf{x}_1, \ldots, \mathbf{x}_n$ of $V$ such that $\Lambda$ is the $O_\mathfrak{p}$-submodule of $V$ generated by $\mathbf{x}_1, \ldots, \mathbf{x}_n$.*

*Proof.* Let $\mathfrak{a}$ be the subset of $\Omega$ consisting of all elements of the form $B(\mathbf{x}, \mathbf{y})$ where $\mathbf{x}$ and $\mathbf{y}$ belong to $\Lambda$. Since $B$ is continuous, $\mathfrak{a}$ is compact. Let $\mathfrak{a}_\mathbf{y}$ be the subset of $\Omega$ consisting of all elements of the form $B(\mathbf{x}, \mathbf{y})$, where $\mathbf{x}$ belongs to $\Lambda$ and $\mathbf{y}$ is a fixed element of $\Lambda$. Since $\Lambda$ is open in $V$, $\Lambda$ generates $V$ over $\Omega$, hence $\mathfrak{a}_\mathbf{y}$ and consequently $\mathfrak{a}$ are proper non-zero ideals in $\Omega$. Let $\mathfrak{a} = \mathfrak{p}^\nu$. Then there exists an $\mathbf{x}_1 \in \Lambda$ such that

$$B(\mathbf{x}_1, \mathbf{x}_1) = \pi^\nu \eta,$$

where $\eta$ belongs to $\mathcal{O}_\mathfrak{p}^*$; since otherwise

$$B(\mathbf{x}, \mathbf{y}) = \frac{1}{2}[B(\mathbf{x}+\mathbf{y}, \mathbf{x}+\mathbf{y}) - B(\mathbf{x}, \mathbf{x}) - B(\mathbf{y}, \mathbf{y})]$$

would belong to $\mathfrak{p}^{\nu+1}$ for all $\mathbf{x}, \mathbf{y}$ in $\Lambda$.

If $\langle \mathbf{x}_1 \rangle^\perp$ is the orthogonal complement of the space generated by $\mathbf{x}_1$, let $\Lambda' = \langle \mathbf{x}_1 \rangle^\perp \cap \Lambda$. Then $\Lambda'$ is an open compact $\mathcal{O}_\mathfrak{p}$ submodule (lattice) in $\langle \mathbf{x}_1 \rangle^\perp$, and for $\mathbf{x} \in \Lambda$, we have

$$\mathbf{x} = \frac{B(\mathbf{x}_1, \mathbf{x})}{B(\mathbf{x}_1, \mathbf{x}_1)}\mathbf{x}_1 + \left(\mathbf{x} - \frac{B(\mathbf{x}_1, \mathbf{x})}{B(\mathbf{x}_1, \mathbf{x}_1)}\mathbf{x}_1\right),$$

i.e., $\Lambda = (\langle \mathbf{x}_1 \rangle \cap \Lambda) + (\langle \mathbf{x}_1 \rangle^\perp \cap \Lambda)$ and, since $B$ is nondegenerate, this decomposition is unique. Obviously $\langle \mathbf{x}_1 \rangle \cap \Lambda$ is cyclic. The lemma now follows by induction on the dimension of $V$. □

Introducing a metric vector space of dimension $n$ over $\Omega$, we may describe any $n$-fold tensor product of the projective representations described in §1.4 as follow:

Let $\mathbf{x}_1, \ldots, \mathbf{x}_n$ be any orthogonal basis of $V$ of the type described in the previous lemma. If $l_i = B(\mathbf{x}_i, \mathbf{x}_i)/2$, then we have

$$Q(\mathbf{x}) = l_1 x_1^2 + \cdots l_n x_n^2,$$

for any $\mathbf{x} = x_1\mathbf{x}_1 + \cdots x_n\mathbf{x}_n$ in $V$. If $d\mathbf{x}$ denotes the unique Haar measure on $V^+$ whose restriction to $\langle \mathbf{x}_i \rangle$ induces $d_{\Phi_{l_i}} x$ on $\Omega^+$, then

$$L^2(V^+, d\mathbf{x}) \cong \bigotimes_{i=1}^{n} L^2(\Omega^+, d_{\Phi_{l_i}} x),$$

the isomorphism being uniquely determined by the relation

$$(h_1 \otimes \cdots \otimes h_n)(x_1, \ldots, x_n) = h_1(x_1) \cdots h_n(x_n),$$

for $h_i \in \mathcal{S}$.

Also, we have

$$(h_1 \otimes \cdots \otimes h_n)^\wedge = \hat{h}_1 \otimes \cdots \otimes \hat{h}_n,$$

the Fourier transform on the left being with respect to $d\mathbf{x}$ and the Fourier transform of $h_i$ being with respect to $d_{\Phi_{l_i}} x$. Hence $d\mathbf{x} = d_\Phi \mathbf{x}$.

Further, if $\mathfrak{p}^m \Lambda$ denotes the lattice in $V$ consisting of elements of the form $x\lambda$ for $x \in \mathfrak{p}^m$ and $\lambda \in \Lambda$, then

$$\int_{\mathfrak{p}^m \Lambda} \Phi(Q(\mathbf{x}))\, d_\Phi \mathbf{x} = \prod_{i=1}^{n} \int_{\mathfrak{p}^m} \Phi_{l_i}(x^2)\, d_{\Phi_{l_i}} x.$$

Hence the limit

$$\mathcal{H}(\Phi, Q) = \lim_{m \to -\infty} \int_{\mathfrak{p}^m \Lambda} \Phi(Q(\mathbf{x}))\, d_\Phi \mathbf{x}$$

exists and is equal to

$$\lim_{m \to -\infty} \prod_{i=1}^{n} \int_{\mathfrak{p}^m} \Phi_{l_i}(x^2)\, d_{\Phi_{l_i}} x = \prod_{i=1}^{n} \mathcal{H}(\Phi_{l_i}).$$

LEMMA 1.5.2. *The constant* $\mathcal{H}(\Phi, Q)$ *is independent of the choice of the lattice* $\Lambda$.

*Proof.* Let $\mathcal{S}(V)$ denote the Schwartz-Bruhat space for $V$; namely, the space of complex-valued functions $h$ on $V$ such that $h$ and $\hat{h}$ are of compact support.

Put $f_b(\mathbf{x}) = \Phi(bA(\mathbf{x}))$.

Then proceeding as in §1.4, one can show that, for $h \in \mathcal{S}(V)$,

$$\widehat{f_b \cdot h} = \mathcal{H}(\Phi, b, Q) f_{-1/b} * \hat{h},$$

where $\mathcal{H}(\Phi, b, Q) = \lim_{m \to -\infty} \int_{\mathfrak{p}^m \Lambda} \Phi(bQ(\mathbf{x}))\, d_\Phi \mathbf{x}$. Choosing a non-zero $h$ proves the lemma. □

The preceding results of this section and §1.4 demonstrate the following theorem.

THEOREM 1.5.3. *Let $V$ be a metric vector space over $\Omega$. For $h \in \mathfrak{h}_V$ and for each non-trivial character $\Phi$ of $\Omega^+$, let*

$$T_w h = \mathcal{H}(\Phi, Q)\hat{h},$$

*and for $b \in \Omega$,*

$$M_b h = f_b \cdot h.$$

*The maps*

$$\begin{pmatrix} 0 & 1 \\ -1 & 0 \end{pmatrix} \longmapsto T_w, \qquad \text{and} \qquad \begin{pmatrix} 1 & b \\ 0 & 1 \end{pmatrix} \longmapsto M_b$$

*may be extended to a projective representation of* $\mathrm{SL}(2, \Omega)$. *If the dimension of $V$ over $\Omega$ is even that representation is single-valued.*

Denote this representation defined by theorem 1.5.3 by $D(\Phi, V)$.

**1.6. Explicit kernel for $D(\Phi, V)$ on $\mathcal{S}(V)$.** Let $V$ be an even-dimensional metric vector space over $\Omega$: $\dim_\Omega V = n = 2m$. For $h \in \mathfrak{h}_V$ and $a \in \Omega^*$ put

$$(U_a h)(\mathbf{x}) = |a|^m h(a\mathbf{x}).$$

Then, $a \longmapsto U_a$ is a unitary representation of $\Omega^*$. From §1.4 and §1.5 we know that, if

$$(U'_a h)(\mathbf{x}) = \mathcal{H}(\Phi, a, Q)[\mathcal{H}(\Phi, Q)]^{-1}|a|^n h(a\mathbf{x}),$$

then $a \longmapsto U'_a$ is also a unitary representation of $\Omega^*$. Hence

$$a \longmapsto \mathcal{H}(\Phi, a, Q)[\mathcal{H}(\Phi, Q)]^{-1}|a|^m$$

defines a one-dimensional unitary representation on a character of $\Omega^*$. Denote this character by *sign*.

If $U_g$ denotes the unitary operator on $\mathfrak{h}_V$ corresponding to the element $g = \left(\begin{smallmatrix} a & b \\ c & d \end{smallmatrix}\right) \in \mathrm{SL}(2, \Omega)$ under the representation $D(\Phi, V)$, then one may easily derive from §1.1, §1.4 and §1.5 that $U_g$ is given by the following kernel on $\mathcal{S}(V)$:

$$(U_g h)(\mathbf{x}) = \int K(g; \mathbf{x}, \mathbf{y}) h(\mathbf{y})\, d\mathbf{y},$$

where for $c \in \Omega^*$

$$K(g; \mathbf{x}, \mathbf{y}) = \mathrm{sign}(-1)\mathcal{H}(\Phi, Q)\frac{\mathrm{sign}(c)}{|c|^{\mathrm{m}}}\,\Phi\left(\frac{aQ(\mathbf{x}) + dQ(\mathbf{y}) - B(\mathbf{x}, \mathbf{y})}{c}\right),$$

and for $c = 0$,

$$K(g; \mathbf{x}, \mathbf{y}) = |a|^m\, \mathrm{sign}(a)\Phi(baQ(\mathbf{x}))\Delta(a\mathbf{x} - \mathbf{y}),$$

where $\Delta$ denote the Dirac delta "function."

**1.7. Continuity of $D(\Phi, V)$.**

THEOREM 1.7.1 (CARTIER). *The representation $D(\Phi, V)$ is continuous.*

*Proof.* To prove the continuity of $g \longmapsto V_g$, it suffices to prove that $g_n$ is a sequence of elements of $G$ converging to the identity, then $U_{g_n}$ converges to the identity operator on $\mathfrak{h}_V$ (in the strong topology).

For $g = \left(\begin{smallmatrix} a & b \\ c & d \end{smallmatrix}\right)$ with $a \neq 0$, we have the identity

$$U_g = T_w M_{a^{-1}-ca^{-1}} T_w M_a T_w M_{a^{-1}+a^{-1}b}.$$

Hence it suffices to show that, for $b \in \Omega$, $b \longmapsto M_b$ is continuous.

Since $\mathcal{S}(V)$ is stable under $M_b$ and is dense in $\mathfrak{h}_V$, it is enough to show that $b \longmapsto M_b h$ is continuous at zero for $h \in \mathcal{S}(V)$. However we can choose $b$ so small that $\Phi(bQ(\mathbf{x})) = 1$ for all $\mathbf{x}$ in the support of $h$. For such $b$, $M_b h = h$. □

**1.8. Decomposition of $D(\Phi, V)$ into irreducible components.** Let $V$ be an even-dimensional metric vector space over $\Omega$ and $D(\Phi, V)$ the corresponding representation of SL(2, $\Omega$). Let $A = A(V)$ denote the group of automorphisms of $V$, i.e., the orthogonal group of the quadratic form $Q$.

THEOREM 1.8.1. *Let $C$ denote the algebra (with the weak topology) of all bounded operators on $\mathfrak{h}_V$ which commute with $D(\Phi, V)$. Then there is a continuous homomorphism of the group algebra $L^1(A)$ of $A$ into $C$. (A completely analogous procedure is used by Kloosterman in [8].)*

*Proof.* For $a \in A$ and $h \in \mathfrak{h}_V$, put $(L_a h)(\mathbf{x}) = h(a\mathbf{x})$. I claim that $L_a$ is a unitary operator on $\mathfrak{h}_V$ belonging to $C$.

Let $\mu$ be a Haar measure on $V^+$. For $E$ a measurable subset of $V^+$, put $\mu_a(E) = \mu(aE)$. Then $\mu_a$ is also a Haar measure on $V^+$; hence $\mu_a = c(a)\mu$ where $c(a)$ is a positive real-valued function on $A$ satisfying $c(a_1 a_2) = c(a_1)c(a_2)$. Since $A$ is generated by elements of order two, $c(a) = 1$ for all $a \in A$. This proves that $L_a$ is unitary. On the other hand, it follows easily from the definitions that

$$L_a T_w = T_w L_a$$
$$L_a M_b = M_b L_a, \qquad \text{for} \quad b \in \Omega,$$

i.e., $L_a$ belongs to $C$.

For $F \in L^1(A)$, put

$$L_F = \int F(a) L_a \, da.$$

Clearly $L_F$ commutes with $D(\Phi, V)$ and $F \longmapsto L_F$ is continuous in the weak topology. □

In the remainder of this section we restrict ourselves to two-dimensional metric vector spaces $V$. If the quadratic form $Q$ on $V$ has a non-trivial zero, then, as in the case of SL(2, $R$) [1], the representation $D(\Phi, V)$ is unitary equivalent to the usual action of SL(2, $\Omega$) on $L^2(\Omega \oplus \Omega)$, which (as in the case of SL(2, $R$)) is a direct integral of the Principal Series of representations.

Hence we may assume that $Q$ is a non-zero form in which case $V$ may be realized as a quadratic extension field of $\Omega$—the quadratic form becoming the norm $N = N_{V/\Omega}$ from that field to $\Omega$.

*Structure of the orthogonal group $A(V)$ of $N_{V/\Omega}$.* Let $V$ be a quadratic extension field of $\Omega$. Let $N^1$ denote the elements of relative norm one in $V^*$ and $\Sigma$ the Galois group of $V/\Omega$. An element $\sigma \in \Sigma$ operates on $N^1$ by sending $\epsilon \in N^1$ to $\epsilon^\sigma$.

Let $\mathcal{O}_{\mathfrak{P}}$ denote the ring of integers in $V$ with unique maximal ideal $\mathfrak{P}$. The following facts about $A(V)$ and $N^1$ are well-known.

(i) With the above action of $\Sigma$ on $N^1$, $A(V)$ is the semi-direct product of $\Sigma$ and $N^1$.

(ii) $N^1$ is the direct product of a finite subgroup $G_0$ (isomorphic to the elements of relative norm one in the residue class field of $V$) with $N^1 \cap (1 + \mathfrak{P})$.

(iii) $N^1 \cap (1 + \mathfrak{P})$ has a natural filtration:

$$N^1 \cap (1 + \mathfrak{P}) = N_1 \supsetneq N_2 \supsetneq \cdots \supsetneq N_m \supsetneq \cdots, \quad \text{where} \quad N_m = N^1 \cap (1 + \mathfrak{P}^m).$$

(iv) There is a unique character $\rho_0$ of $N^1$ of order two. $\rho_0$ is determined by its restriction to $G_0$.

Let $\delta = \delta_{V/\Omega}$ denote the different of $V/\Omega$. If $\rho$ belongs to $\hat{N}^1$ and if $m$ is the smallest integer for which $\rho | N_m = 1$, we define the *conductor* $\rho$ to be the ideal $\mathfrak{p}^m \delta$ of $\mathcal{O}_{\mathfrak{P}}$.

*Definition and elementary facts about Kloosterman sums.* We say that an ideal $\mathfrak{B}$ in $\mathcal{O}_{\mathfrak{P}}$ lies over an ideal $\mathfrak{b}$ of $\mathcal{O}_{\mathfrak{p}}$ if $\mathfrak{B} \cap \mathcal{O}_{\mathfrak{p}} = \mathfrak{b}$.

The norm $N = N_{V/\Omega}$ (respectively trace $\mathrm{Tr} = \mathrm{Tr}_{V/\Omega}$) from $V$ to $\Omega$ maps $\mathcal{O}_{\mathfrak{P}}$ to $\mathcal{O}_{\mathfrak{p}}$ and induces a norm $\mathfrak{N}$ (respectively trace Sp) from $\mathcal{O}_{\mathfrak{P}}/\mathfrak{B}$ to $\mathcal{O}_{\mathfrak{p}}/\mathfrak{b}$ such that the following diagram is commutative:

$$\begin{array}{ccc} \mathcal{O}_{\mathfrak{P}} & \longrightarrow & \mathcal{O}_{\mathfrak{P}}/\mathfrak{B} \\ \downarrow {\scriptstyle N,\mathrm{Tr}} & & \downarrow {\scriptstyle \mathfrak{N},\mathrm{Sp}} \\ \mathcal{O}_{\mathfrak{p}} & \longrightarrow & \mathcal{O}_{\mathfrak{p}}/\mathfrak{b}. \end{array}$$

We will denote the image of an element $\mathbf{x} \in \mathcal{O}_{\mathfrak{P}}$ in $\mathcal{O}_{\mathfrak{P}}/\mathfrak{B}$ by $\tilde{\mathbf{x}}$.

Given a proper ideal $\mathfrak{b}$ in $\mathcal{O}_{\mathfrak{p}}$, there exists a unique ideal $\mathfrak{B}$ lying over $\mathfrak{b}$ for which the bilinear form on $\mathcal{O}_{\mathfrak{P}}/\mathfrak{B}$ defined by

$$(\tilde{\mathbf{x}}, \tilde{\mathbf{y}}) \longmapsto \mathrm{Sp}(\tilde{\mathbf{x}} \cdot \tilde{\mathbf{y}})$$

is nondegenerate, namely, $\mathfrak{B} = \delta^{-1}\mathfrak{b}$. From now on given $\mathfrak{b}$ in $\mathcal{O}_{\mathfrak{p}}$, $\mathfrak{B}$ will denote this unique ideal of $\mathcal{O}_{\mathfrak{P}}$. If $\Phi$ is a character of $\mathcal{O}_{\mathfrak{p}}$ with conductor $\mathfrak{b}$, we may consider $\Phi$ as a character of $\mathcal{O}_{\mathfrak{p}}/\mathfrak{b}$. In that case every character of $\mathcal{O}_{\mathfrak{P}}/\mathfrak{B}$ has the form $\Phi_{\tilde{\mathbf{y}}}$, where

$$\Phi_{\tilde{\mathbf{y}}}(\tilde{\mathbf{x}}) = \Phi(\mathrm{Sp}(\tilde{\mathbf{x}} \cdot \tilde{\mathbf{y}})).$$

Let $\rho$ be a character of $N^1$ of conductor $\mathfrak{B}$. This is possible since $\mathfrak{B}$ is the product of an ideal in $\mathcal{O}_{\mathfrak{p}}$ by $\delta$. One easily sees that the map from $N^1$ to $\mathfrak{N}^1$, the elements of norm one in $\mathcal{O}_{\mathfrak{P}}/\mathfrak{B}$, is surjective; hence $\rho$ may be considered as a character of $\mathfrak{N}^1$.

*Definition.* A *Kloosterman sum* is a map of $\mathcal{O}_{\mathfrak{P}}$ to the complex numbers of the form

$$K(\Phi, \rho)(\mathbf{x}) = \sum_{\substack{\epsilon \in N^1 \\ \epsilon \bmod \mathfrak{B}}} \rho(\epsilon)\, \Phi(\mathrm{Tr}(\epsilon \mathbf{x})),$$

where $\Phi$ is a character of $\mathcal{O}_{\mathfrak{P}}$ with conductor $\mathfrak{b}$ and $\rho$ is a character of $N^1$ with conductor $\mathfrak{B} = \delta^{-1}\mathfrak{b}$. If $\mathfrak{b} = \mathfrak{p}$ ($\mathfrak{B} = \mathfrak{P}$), we also include in the definition the case $\rho = 1$.

Since $K(\Phi, \rho)(\mathbf{x})$ depends only on $\mathbf{x} \bmod \mathfrak{B}$, we may view $K(\Phi, \rho)$ as a function on $\mathcal{O}_{\mathfrak{P}}/\mathfrak{B}$:

$$K(\Phi, \rho)(\tilde{\mathbf{x}}) = \sum_{\epsilon \in \mathfrak{N}^1} \rho(\epsilon)\Phi(\mathrm{Sp}(\epsilon \cdot \tilde{\mathbf{x}})).$$

Let $e$ denote the ramification index of $V/\Omega$.

LEMMA 1.8.2. *If $\rho$ is not identically one, $K(\Phi, \rho)$ vanishes on the ideal $\mathfrak{p} = \mathfrak{P}^e$ of $\mathcal{O}_{\mathfrak{P}}$.*

*Proof.* The group $N^1$ operates on $\mathcal{O}_{\mathfrak{P}}$ by multiplication and hence operates on $\mathcal{O}_{\mathfrak{P}}/\mathfrak{B}$ by reduction mod $\mathfrak{B}$. If $\mathbf{x}$ belongs to $\mathfrak{p}$, it is readily seen that the stabilizer of the action of $N^1$ at $\tilde{\mathbf{x}}$ is a subgroup of $N^1$ of the form $N_m$ which contains $N^1 \cap (1 + \delta^{-2}\mathfrak{b})$ as a proper subgroup. Breaking up the sum over $\mathfrak{N}^1$ in the definition of $K(\Phi, \rho)$ into cosets mod $N_m$ and using the fact that $\rho$ has conductor $\mathfrak{B}$ gives the proof. □

Let $N(V^*)$ denote the subgroup of $\Omega^*$ consisting of norms from $V^*$. One has $(\Omega^*)^2 \subset N(V^*)$ and it is well-known that $[N(V^*) : (\Omega^*)^2] = 2$. Hence $N(V^*)$ has a unique character, denoted by $(\frac{\ }{\mathfrak{p}})$, of order two. Let $S^+ = S^+(V)$ (respectively, $S^- = S^-(V)$) denote those $\mathbf{x}$ in $V^*$ for which $(\frac{N(\mathbf{x})}{\mathfrak{p}}) = 1$ (respectively, $(\frac{N(\mathbf{x})}{\mathfrak{p}}) = -1$). Clearly, $V^*$ is the disjoint union of the (open) sets $S^+$ and $S^-$.

I will need the following lemma to prove the irreducibility of the representations that I will discuss in the next section.

LEMMA 1.8.3. *Let $\rho_0$ be the unique character of order 2 of $N^1$.*
(i) *If $\rho \neq \rho_0$, then there exists an $\mathbf{x} \in S^- \cap \mathcal{O}_{\mathfrak{P}}$ such that $K(\Phi, \rho)(\mathbf{x}) \neq 0$.*
(ii) *$K(\Phi, \rho_0)(\mathbf{x}) = 0$ for all $\mathbf{x} \in S^- \cap \mathcal{O}_{\mathfrak{P}}$.*

*Proof.* Suppose that $\rho \neq \rho_0$ and that $K(\Phi, \rho)(\mathbf{x}) = 0$ for all $\mathbf{x} \in S^-$.

With this assumption I claim that $K(\Phi, \rho)(\mathbf{x})$ is zero unless $\mathbf{x}$ belongs to $\mathcal{O}_{\mathfrak{P}}^*$, the unit group of $\mathcal{O}_{\mathfrak{P}}$. In the unramified case ($e = 1$), this is obvious from lemma 1.8.2. If $\Pi$ is a prime element for $V$, in the ramified case ($e = 2$), we know, again by lemma 1.8.2, that $K(\Phi, \rho)(\mathbf{x}) = 0$ unless $\mathbf{x}$ belongs to $\mathcal{O}_{\mathfrak{P}}^* \cup \Pi\mathcal{O}_{\mathfrak{P}}^*$. However, for $\mathbf{x} \in \Pi\mathcal{O}_{\mathfrak{P}}^*$, we have $(\frac{N(\mathbf{x})}{\mathfrak{p}}) = -1$. Hence the claim.

For $\tilde{\mathbf{x}} \in \mathcal{O}_{\mathfrak{P}}/\mathfrak{B}$, put $\tilde{\rho}(\tilde{\mathbf{x}}) = \rho(\tilde{\mathbf{x}})$, if $\tilde{\mathbf{x}}$ belongs to $\mathfrak{N}^1$, and $\tilde{\rho}(\tilde{\mathbf{x}}) = 0$ otherwise. Since Sp is nondegenerate, we have a Fourier expansion for $\tilde{\rho}$:

$$\tilde{\rho} = \sum_{\tilde{\mathbf{y}} \in \mathcal{O}_{\mathfrak{P}}/\mathfrak{B}} a_{\tilde{\mathbf{y}}} \, \Phi_{\tilde{\mathbf{y}}},$$

where

$$a_{\tilde{\mathbf{y}}} = \frac{1}{N(\mathfrak{B})} \sum_{\epsilon \in \mathfrak{N}^1} \rho(\epsilon)\overline{\Phi(\mathrm{Sp}(\epsilon\tilde{\mathbf{y}}))} = \frac{1}{N(\mathfrak{B})} K(\Phi, \rho)(-\tilde{\mathbf{y}}).$$

Hence by the preceding $a_{\tilde{\mathbf{y}}} = 0$ unless $\tilde{\mathbf{y}}$ belongs to $S^+ \cap \mathcal{O}_{\mathfrak{P}}^*$. For such $\tilde{\mathbf{y}}$ we may write uniquely $\tilde{\mathbf{y}} = \lambda\epsilon_0$, where $\lambda$ belongs to the unit group $(\mathcal{O}_{\mathfrak{p}}/\mathfrak{b})^*$ of $\mathcal{O}_{\mathfrak{p}}/\mathfrak{b}$ and $\epsilon_0$ is an element of $\mathfrak{N}^1$, taken mod $\pm 1$. Using lemma 1.8.2, an easy calculation then gives

$$\tilde{\rho}(\tilde{\mathbf{x}}) = -\Delta_{\rho,1}\frac{n^2}{2N(\mathfrak{B})} + \frac{1}{N(\mathfrak{B})} \sum_{\substack{\epsilon \in \mathfrak{N}^1 \\ \lambda \in \mathcal{O}_{\mathfrak{p}}/\mathfrak{b} \\ \epsilon_0 \in \mathfrak{N}^1 \bmod \pm 1}} \rho(\epsilon)\, \Phi(\lambda \mathrm{Sp}(\epsilon_0(\tilde{\mathbf{x}} - \epsilon))),$$

where $\Delta_{\rho,1} = 1$ if $\rho = 1$ and is zero otherwise, and where $n$ is the order of $\mathfrak{N}^1$.

Assume $\tilde{\mathbf{x}}$ is in $\mathfrak{N}^1$. Then

$$\sum_{\lambda \in \mathcal{O}_{\mathfrak{p}}/\mathfrak{b}} \Phi(\lambda \mathrm{Sp}(\epsilon_0(\tilde{\mathbf{x}} - \epsilon))) = N(\mathfrak{b}),$$

if $\mathrm{Sp}(\epsilon_0\tilde{\mathbf{x}}) = \mathrm{Sp}(\epsilon_0\epsilon)$ and is zero otherwise. Since $\mathrm{Sp}(\epsilon_0\tilde{\mathbf{x}}) = \mathrm{Sp}(\epsilon_0\epsilon)$ if and only if either $\epsilon = \tilde{\mathbf{x}}$ or $\epsilon = \epsilon_0^{-2}\tilde{\mathbf{x}}^{-1}$, we obtain

$$
\begin{aligned}
N(\mathfrak{B})\tilde{\rho}(\tilde{\mathbf{x}}) &= -\Delta_{\rho,1}\frac{n^2}{2} + \sum_{\substack{\epsilon\in\mathfrak{N}^1 \\ \lambda\in\mathcal{O}_{\mathfrak{p}}/\mathfrak{b}}} \rho(\epsilon)(\lambda \mathrm{Sp}(1 - \tilde{\mathbf{x}}^{-1}\epsilon)) \\
&\quad + \sum_{\substack{\epsilon\in\mathfrak{N}^1 \\ \lambda\in\mathcal{O}_{\mathfrak{p}}/\mathfrak{b} \\ \epsilon_0\in\mathfrak{N}^1 \bmod \pm 1 \\ \epsilon_0\neq\tilde{\mathbf{x}}^{-1}}} \rho(\epsilon)\Phi(\lambda\mathrm{Sp}(\epsilon_0(\tilde{\mathbf{x}} - \epsilon))) \\
&= -\Delta_{\rho,1}\frac{n^2}{2} + N(\mathfrak{b})\rho(\tilde{\mathbf{x}}) \\
&\quad + \sum_{\substack{\epsilon_0\in\mathfrak{N}^1 \bmod \pm 1 \\ \epsilon_0\neq\tilde{\mathbf{x}}^{-1}}} N(\mathfrak{b})\sum_{\epsilon}\rho(\epsilon)\left(\Delta(\epsilon - \tilde{\mathbf{x}}) + \Delta\left(\epsilon - \epsilon_0^{-2}\tilde{\mathbf{x}}^{-1}\right)\right) \\
&= -\Delta_{\rho,1}\frac{n^2}{2} + N(\mathfrak{b})\frac{n}{2}\rho(\tilde{\mathbf{x}}) + N(\mathfrak{b})\sum_{\substack{\epsilon_0\in\mathfrak{N}^1 \\ \epsilon_0\neq\tilde{\mathbf{x}}^{-1}}} \rho\left(\epsilon_0^{-2}\tilde{\mathbf{x}}^{-1}\right).
\end{aligned}
$$

This readily leads to a contradiction involving the orders of the various groups if $\rho \neq \rho_0$.

If $\rho = \rho_0$, the above expression reduces to the identity

$$N(\mathfrak{B}) = N(\mathfrak{b})(n - 1).$$

Reversing the steps and using the uniqueness of the Fourier expansion proves the lemma. □

We are now in a position to prove

THEOREM 1.8.4. *Let $V$ be a quadratic extension field of $\Omega$. Let $A$ denote the orthogonal group of the quadratic form $N_{V/\Omega}$. Then the image of the group algebra, $L^1(A)$, in the commuting algebra $C$ of the representation $D(\Phi, V)$ is dense. The kernel of the homomorphism $L^1(A) \longrightarrow C$ is one-dimensional. (An analogous theorem appears in Kloosterman [8].)*

*Proof.* The proof will be divided into two parts:

(A) Decomposition of $D(\Phi, V)$ into irreducible components

(B) Classification of those components according to unitary equivalence

For $h \in \mathfrak{h}_V$ and $\rho \in \hat{N}^1$, let $(L_\rho h)(\mathbf{x}) = \int_{N^1} \rho^{-1}(\epsilon)h(\epsilon\mathbf{x})\,d\epsilon$. The operators $L_\rho$ are orthogonal projections on $\mathfrak{h}_V$ (with a suitable normalization of the Haar measure

$d\epsilon$ on the compact group $N^1$) which commute with the action of $D(\Phi, V)$. Since $\sum_{\rho\in\hat{N}^1} L_\rho$ is the identity operator on $\mathfrak{h}_V$, we obtain an orthogonal direct sum of $\mathfrak{h}_V$ into stable subspaces:

$$\mathfrak{h}_V = \sum_{\rho\in\hat{N}^1} \mathfrak{h}_\rho,$$

where $\mathfrak{h}_\rho$ consists of those $h \in \mathfrak{h}_V$ satisfying $h(\epsilon\mathbf{x}) = \rho(\epsilon)h(\mathbf{x})$. Denote the restriction of $D(\Phi, V)$ to $\mathfrak{h}_\rho$ by $D(\rho)$. Then I claim:

(i) $D(\rho)$ is irreducible if $\rho$ is not of order two.

(ii) $D(\rho_0)$ splits into two inequivalent representations, which I will denote by $D^+$ and $D^-$; $D(\rho_0) = D^+ \oplus D^-$.

(iii) $D(\rho)$ and $D(\rho')$ interwine if and only if there exists $\sigma \in \Sigma$ such that $\rho' = \rho^\sigma$, i.e., $\rho'(\epsilon) = \rho(\epsilon^\sigma)$ for all $\epsilon \in N^1$.

*Proof of (i).* Let $\mathcal{T}$ be a bounded operator on $\mathfrak{h}_\rho$ commuting with $D(\rho)$. Then $\mathcal{T}$ must commute with the operators $M_b$ on the subspace $\mathfrak{h}_\rho$. Since the representation of $\Omega$ defined by $b \longmapsto M_b$ on the Hilbert space $\mathfrak{h}_\rho$ has a simple spectrum, $\mathcal{T}$ must be given by multiplication by a function $f_\rho$ on that space:

$$\mathcal{T}h = f_\rho h,$$

for $h \in \mathfrak{h}_\rho$, where $f_\rho$ is a function on $V$ satisfying $f_\rho(\epsilon x) > f_\rho(u)$ for all $\epsilon \in \mathbf{N}^1$. Since $D_a\mathcal{T} = \mathcal{T}D_a$ for all $a \in \Omega^*$, we must also have $f_\rho(a\mathbf{x}) = f_\rho(\mathbf{x})$. Hence $f_\rho$ may be regarded as a function $f'_\rho$ on $N(V^*)$:

$$f_\rho(\mathbf{x}) = f'_\rho(N(\mathbf{x})),$$

where $f'_\rho$ satisfies $f'_\rho(a^2\mathbf{x}) = f'_\rho(\mathbf{x})$, i.e., $f_\rho$ is constant on $S^+$ and $S^-$.

Since $\mathcal{T}$ commutes with $T_w$, we must also have

$$f_\rho \cdot T_w(L_\rho h) = T_w(f_\rho \cdot L_\rho h),$$

for $h \in \mathfrak{h}_V$. This condition is easily seen to be equivalent to

$$\int_V f_\rho(\mathbf{y})K_\rho(\mathbf{x}, \mathbf{y})\, h(\mathbf{y})\, d\mathbf{y} = f_\rho(\mathbf{x}) \int_V K_\rho(\mathbf{x}, \mathbf{y})h(\mathbf{y})\, d\mathbf{y},$$

where $K_\rho(\mathbf{x}, \mathbf{y}) = \int_{N^1} \rho^{-1}(\epsilon)\Phi(\mathrm{Tr}(\mathbf{x}\epsilon\mathbf{y}))\, d\epsilon$. Hence $f_\rho(\mathbf{y})K(\mathbf{x}, \mathbf{y}) = f_\rho(\mathbf{x})K(\mathbf{x}, \mathbf{y})$ for $\mathbf{x}, \mathbf{y}$ outside a set of measure zero in $V \times V$. (A similar argument is used in Gelfand and Graev in [3].)

On the other hand, if $\mathfrak{B}$ denotes the conductor of $\rho$, we may choose $\mathbf{y} \in \Omega^*$ so that the conductor of $\Phi_\mathbf{y}$ is $\mathfrak{B} \cap \mathcal{O}_\mathfrak{p}$. In that case,

$$K_\rho(\mathbf{x}, \mathbf{y}) = \int_{N^1} \rho^{-1}(\epsilon)\Phi_\mathbf{y}(\mathrm{Tr}(\epsilon\mathbf{x}))\, d\epsilon = c \cdot K(\Phi_\mathbf{y}, \rho)(\mathbf{x}).$$

Hence, if $\rho \neq \rho_0$, from Lemma 1.8.3 we have the existence of a pair $(\mathbf{x}, \mathbf{y})$ in $S^- \times S^+$ such that $K_\rho(\mathbf{x}, \mathbf{y}) \neq 0$. Hence, in this case, the commuting algebra $C_\rho$ of $D(\rho)$ is one-dimensional. Since $D(\rho)$ is unitary this proves (i).

In any case we have shown that $\dim C_\rho \leq 2$. □

*Proof of (ii).* For $\sigma \in \Sigma$ and $h \in \mathfrak{h}_V$, let $L_\sigma$ denote the unitary operator on $\mathfrak{h}_V$ defined by $(L_\sigma h)(\mathbf{x}) = h(\mathbf{x}^\sigma)$. If $h$ belongs to $\mathfrak{h}_\rho$,

$$(L_\sigma h)(\epsilon \mathbf{x}) = h(\epsilon^\sigma \mathbf{x}^\sigma) = \rho_0(\epsilon)(L_\sigma h)(\mathbf{x}),$$

because $\rho_0(\epsilon^\sigma) = \rho_0(\epsilon)$, i.e., $L_\sigma$ leaves $\mathfrak{h}_{\rho_0}$ stable. On the other hand, for $\mathbf{x} \in V^*$,

$$h(\mathbf{x}^\sigma) = h(\mathbf{x}^\sigma \mathbf{x}^{-1} \mathbf{x}) = \rho_0(\mathbf{x}^\sigma \mathbf{x}^{-1}) h(\mathbf{x}).$$

Hence, by Hilbert's Theorem 90, for $\sigma$ non-trivial, $L_\sigma | \mathfrak{h}_{\rho_o}$ is not a scalar multiple of the identity. Thus $\dim C_{\rho_0} = 2$. Hence $\mathfrak{h}_{\rho_0}$ decomposes into two stable subspaces

$$\mathfrak{h}_{\rho_0} = \mathfrak{h}^+ \oplus \mathfrak{h}^-,$$

and correspondingly $D(\rho_0)$ splits into two inequivalent representations:

$$D(\rho_0) = D^+ \oplus D^-.$$

$\mathfrak{h}^+$ (resp. $\mathfrak{h}^-$) consists of those $h \in \mathfrak{h}_V$ whose support is contained in $S^+$ (resp. $S^-$). □

*Proof of (iii).* Clearly $L_\sigma$ maps $\mathfrak{h}_\rho$ to $\mathfrak{h}_{\rho_0}$. Since $L_\sigma$ belongs to $C$, the representations $D(\rho)$ and $D(\rho^\sigma)$ are unitary equivalent. The converse may be proven as in the proof of (i).

Hence the operators $L_\sigma$ and $L_\rho$ generate (in the topological sense) C. This proves that the image of $L^1(A)$ in $C$ is dense.

Denote the restriction of $D(\Phi, V)$ to $\mathfrak{h}_\rho$ by $D(\Phi, \rho, V)$. □

*Computation of the kernel of* $L^1(A) \longrightarrow C$. Let $\sigma_0$ denote the non-trivial element of $\Sigma$. $A$ is the disjoint union of the cosets $N^1$ and $N^1\sigma_0$.

Let $F$ belong to $L^1(A)$. Suppose $L_F$ is identically zero, i.e.,

$$\int_A F(a) L_a h \, da = 0,$$

for all $h \in \mathfrak{h}_V$. We may write this integral as

$$\int_{N^1} F_1(\epsilon) h(\epsilon \mathbf{x}) \, d\epsilon + \int_{N^1} F_2(\epsilon) h(\epsilon \mathbf{x}^{\sigma_0}) \, d\epsilon,$$

where $F_1(\epsilon) = F(\epsilon)$ and $F_2(\epsilon) = F(\epsilon\sigma_0)$. Taking $h \in \mathfrak{h}_\rho$ we obtain

$$h(\mathbf{x}) \int_{N^1} F_1(\epsilon)\rho(\epsilon) \, d\epsilon + h(\mathbf{x}^{\sigma_0}) \int_{N^1} F_2(\epsilon)\rho(\epsilon) \, d\epsilon = 0,$$

for all $\mathbf{x} \in V$. For $\mathbf{x} \in V^*$, we have as before $h(\mathbf{x}^\sigma) = h(\mathbf{x})\rho(\mathbf{x}^\sigma \mathbf{x}^{-1})$. Hence, by Hilbert's Theorem 90,

$$\int_{N^1} F_1(\epsilon)\rho(\epsilon)\, d\epsilon = \int_{N^1} F_2(\epsilon)\rho(\epsilon)\, d\epsilon = 0,$$

provided $\rho$ is non-trivial. Therefore, $F_1 = c_1$, $F_2 = c_2$ are constant functions on $N^1$ and putting $\rho = 1$ we obtain $c_1 + c_2 = 0$. Conversely, if $F_1 = -F_2$ is constant, $L_F$ is zero. This completes the proof of the theorem. □

**1.9. Further classification of the representations $D(\Phi, V)$.** Let $G$ be a locally compact group, and $D$ and $D'$ be unitary representations of $G$ on a Hilbert space $\mathfrak{h}$. For $g \in G$, Let $U_g$ and $U'_g$ be the corresponding unitary operators on $\mathfrak{h}$. If $\varphi$ is a (continuous) automorphism of $G$, we say that $D$ and $D'$ are *conjugate with respect to* $\varphi$ if $U'_g = U_{\varphi(g)}$ for all $g \in G$.

THEOREM 1.9.1. *Let $\varphi$ denote the automorphism of* $\mathrm{SL}(2, \Omega)$ *defined by*

$$g \longmapsto \begin{pmatrix} a & 0 \\ 0 & d \end{pmatrix} g \begin{pmatrix} a^{-1} & 0 \\ 0 & d^{-1} \end{pmatrix},$$

*where $ad^{-1} = r$ belongs to $\Omega^*$. Then the representations $D(\Phi, V)$ and $D(\Phi_r, V)$ are conjugate with respect to $\varphi$.*

*Proof.* Let $M_b$ and $T_w$ be the operators on $\mathfrak{h}_V$ corresponding to $\left(\begin{smallmatrix} 1 & b \\ 0 & 1 \end{smallmatrix}\right)$ and $\left(\begin{smallmatrix} 0 & 1 \\ -1 & 0 \end{smallmatrix}\right)$ with respect to $D(\Phi, V)$. Then for $h \in \mathfrak{h}_V$, by an explicit computation, we have

$$(M_{\varphi(b)}h)(\mathbf{x}) = \Phi_r(bN(\mathbf{x}))h(\mathbf{x}) \qquad \text{and}$$

$$\begin{aligned} (T_{\varphi(w)}h)(\mathbf{x}) &= (D_r T_w h)(\mathbf{x}) \\ &= |r|^2 \mathcal{H}(\Phi, r)\mathcal{H}(\Phi)^{-1}(T_w h)(r\mathbf{x}) \\ &= \mathcal{H}(\Phi_r) \int h(\mathbf{y})\Phi_r(\mathrm{Tr}(\mathbf{x}\mathbf{y}^{\sigma_0}))\, d_{\Phi_r}\mathbf{y}. \end{aligned}$$

Hence $M_{\varphi(b)}$ and $T_{\varphi(w)}$ are precisely the operators on $\mathfrak{h}_V$ corresponding to $\left(\begin{smallmatrix} 1 & b \\ 0 & 1 \end{smallmatrix}\right)$ and $\left(\begin{smallmatrix} 0 & 1 \\ -1 & 0 \end{smallmatrix}\right)$ with respect to $D(\Phi_r, V)$. □

For any subset $S$ of $V^*$, let $N(S)$ denote its image in $\Omega^*$ under the map $N_{V/\Omega}$. We will denote the relation of unitary equivalence by $\sim$.

THEOREM 1.9.2. *Let $r$ belong to $\Omega^*$, then*

(i) *For $\rho \neq \rho_0$, one has $D(\Phi, \rho, V) \sim D(\Phi_r, \rho, V)$ if and only if $r \in N(V^*)$.*

(ii) *$D^+(\Phi, V) \sim D^+(\Phi_r, V)$ if and only if $r \in N(S^+)$, i.e., $r$ is a square in $\Omega^*$.*

(iii) *$D^+(\Phi, V) \sim D^-(\Phi_r, V)$ if and only if $r \in N(S^-)$, i.e., $r$ is in $N(V^*)$ but not a square.*

*Proof.* For $h \in \mathfrak{h}_V$ put $(L_{\mathbf{x}}h)(\mathbf{y}) = h(\mathbf{xy})$. Then $L_{\mathbf{x}}$ is an isometry from $L^2(V, d_\Phi \mathbf{x})$ to $L^2(V, d_{\Phi_{N(\mathbf{x})}}\mathbf{x})$. An explicit computation shows that

$$\left(L_{\mathbf{x}} M_b L_{\mathbf{x}}^{-1} h\right)(\mathbf{y}) = \Phi_{N(\mathbf{x})}(bN(\mathbf{y}))h(\mathbf{y}),$$

$$\left(L_{\mathbf{x}} T_w L_{\mathbf{x}}^{-1} h\right)(\mathbf{y}) = \mathcal{H}(\Phi_{N(\mathbf{x})}) \times \int h(\mathfrak{z})\Phi_{N(\mathbf{x})}(\mathrm{Tr}(\mathfrak{z}\mathbf{y}^{\sigma_0}))\, d_{\Phi_{N(\mathbf{x})}}\mathfrak{z}.$$

$L_{\mathbf{x}}$ maps to $\mathfrak{h}_\rho$ to $\mathfrak{h}_\rho$; it maps to $\mathfrak{h}^+$ to $\mathfrak{h}^+$ if and only if $\mathbf{x} \in S^+$ and $\mathfrak{h}^+$ to $\mathfrak{h}^-$ if and only if $\mathbf{x} \in S^-$. This proves the "if" part of the theorem.

Conversely, if $D(\Phi, \rho, V) \sim D(\Phi_r, \rho, V)$, then their restrictions to $B$, the subgroup of $G$ consisting of elements the form $\left(\begin{smallmatrix} 1 & b \\ 0 & 1 \end{smallmatrix}\right)$ for $b \in \Omega$, are also unitary equivalent:

$$D(\Phi, \rho, V)\Big| \begin{pmatrix} 1 & b \\ 0 & 1 \end{pmatrix} = \int_{V^*/N^1} \Phi(bN(\mathbf{x}))\, d\mathbf{x}.$$

Hence, if $\rho \neq \rho_0$, since these restrictions have a simple spectrum, the open sets $N(V^*)$ and $rN(V^*)$ must coincide up to a set of measure zero (the measure being the Haar measure of $\Omega^+$). Hence $r$ must belong to $N(V^*)$. Similarly in the other cases. □

THEOREM.1.9.3 (i) *For $\rho \neq \rho_0$, we have $D(\Phi, \rho, V) \sim D(\Phi, \rho, V')$ if and only if $V = V'$.*

(ii) *$D^+(\Phi, V) \sim D^+(\Phi, V')$ for any pair of quadratic extension fields $V$ and $V'$ of $\Omega$.*

*Proof.* (i) follows as in the proof of Theorem 1.9.2 using the fact that $N(V^*) = N(V'^*)$ if and only if $V = V'$. I will postpone the proof of (ii) until the conclusion of §3. □

**2.1. Representations of SL(2, $\mathcal{O}_\mathfrak{p}$).** In this section I will obtain irreducible representations of $K = \mathrm{SL}(2, \mathcal{O}_\mathfrak{p})$ by restricting the representations of $G = \mathrm{SL}(2, \Omega)$ to that subgroup. Some of these representations appear in a series of papers of Kloosterman [8].

*$K$-stable subspaces of $D(\Phi, V)$.* For the moment, I return to the case when $V$ is an even-dimensional metric vector space over $\Omega$. For a lattice $\Lambda$ in $V$ and a sublattice $\Lambda'$ of $\Lambda$, let $H_{\Lambda,\Lambda'}$ denote the Hilbert subspace of $\mathfrak{h}_V$ consisting of those complex-valued functions $h$ on $V$ satisfying

(i) the support of $h$ is contained in $\Lambda$,

(ii) $h(\mathbf{x} + \lambda') = h(\mathbf{x})$ for all $\lambda' \in \Lambda$.

For any lattice $\Lambda$ in $V$, let $\delta(\Lambda)$ denote the set of $\mathbf{x} \in V$ such that $B(\mathbf{x}, \mathfrak{z}) \in \mathcal{O}_\mathfrak{P}$ for all $\mathfrak{z} \in \Lambda$. $\delta(\Lambda)$ is again a lattice in $V$.

Let $D(\Phi, V)|K$ denote the restriction of $D(\Phi, V)$ to $K$ and as before let $\mathfrak{b}$ denote the conductor of $\Phi$.

THEOREM 2.1.1. *$H_{\Lambda,\Lambda'}$ is stable under the operators of $D(\Phi, V)|K$ if and only if $\Lambda = \delta(\Lambda') \cdot \mathfrak{b}$.*

*Proof.* For fixed $\Phi$, and any lattice $\Lambda$ in $V$, let $\Lambda^*$ denote the set of $\mathbf{x}$ in $V$ such that $\Phi(B(\mathbf{x}, \mathfrak{z})) = 1$ for all $\mathfrak{z} \in \Lambda$. If $\chi(\Lambda)$ denotes the characteristic function of $\Lambda$, we have as in §1.3.

(i) $\hat{\chi}(\Lambda) = \mu(\Lambda)\chi(\Lambda^*)$,

(ii) $\Lambda^* = \mathfrak{b} \cdot \delta(\Lambda)$.

Since $K$ is generated by $w = \left(\begin{smallmatrix} 0 & 1 \\ -1 & 0 \end{smallmatrix}\right)$ and $B \cap K$, $H_{\Lambda,\Lambda'}$ is stable under $D(\Phi, V)|K$ if and only if it is stable under $T_w$ and $M_b$ for $b \in \mathcal{O}_{\mathfrak{p}}$.

*$T_w$ condition.* For $\xi \in \Lambda$, let $\chi_\xi$ denote the characteristic function of $\xi + \Lambda'$. Clearly $H_{\Lambda,\Lambda'}$ is spanned by the $\chi_\xi$. We have

$$\hat{\chi}_\xi(\mathbf{x}) = \psi(B(\mathbf{x}, \xi))\, \mu(\Lambda')\chi(\Lambda'^*).$$

Assume $H_{\Lambda,\Lambda'}$ is stable under $T_w$, then we must have

(A) $\Lambda'^* \subset \Lambda$,

(B) $\Phi(B(\mathbf{x} + \lambda', \xi))\chi(\Lambda'^*)(\mathbf{x} + \lambda') = \Phi(B(\mathbf{x}, \xi))\chi(\Lambda'^*)(\mathbf{x})$ for $\mathbf{x} \in V$, $\lambda' \in \Lambda'$, $\xi \in \Lambda$.

Condition (B) is easily seen to be equivalent to the two conditions

(B1) $\Lambda' \subset \Lambda'^*$,

(B2) $\Phi(B(\lambda', \xi)) = 1$ for all $\lambda' \in \Lambda'$, $\xi \in \Lambda$.

Hence, by (B2), $\Lambda \subset \Lambda'^*$ and hence $\Lambda = \Lambda'^*$. Conversely if $\Lambda = \Lambda'^*$, $T_w$ leaves $H_{\Lambda,\Lambda'}$ stable.

*$M_b$ condition.* I claim that if $\Lambda = \Lambda'^*$, then for $b \in \mathcal{O}_{\mathfrak{p}}$, $M_b$ leaves $H_{\Lambda,\Lambda'}$ stable: For $\mathbf{x} \in V$, $\lambda' \in \Lambda'$, and $\xi \in \Lambda$,

$$\begin{aligned}(M_b\chi_\xi)(\mathbf{x} + \lambda') &= \Phi(b(Q(\mathbf{x} + \lambda')))\, \chi_\xi(\mathbf{x} + \lambda') \\ &= \Phi(b(Q(\mathbf{x}) + Q(\lambda') + B(\mathbf{x}, \lambda')))\, \chi_\xi(\mathbf{x}),\end{aligned}$$

which by (B2) is equal to $\Phi(bQ(\mathbf{x}))\, \chi_\xi(\mathbf{x})$, since $Q(\lambda') = (1/2)B(\lambda', \lambda')$. □

I will now study the representation of $K$ on the space $H_{\Lambda,\Lambda'}$ in more detail when $V$ is quadratic extension field of $\Omega$. In that case I take $\Lambda = \mathcal{O}_{\mathfrak{P}}$. Then one sees that the unique sublattice $\Lambda'$ of $\Lambda$ defined by Theorem 2.1.1 becomes $\delta^{-1}\mathfrak{b}$, the ideal in $\mathcal{O}_{\mathfrak{P}}$ which I have denoted by $\mathfrak{B}$. In this case I will denote the Hilbert space $H_{\Lambda,\Lambda'}$ by $H(\Phi, V)$.

Since $N^1 \subset \mathcal{O}^*_{\mathfrak{P}}$, the operators $L_\rho$ defined in §1.8 leave $H(\Phi, V)$ stable, (see also [8]) and thus we have an orthogonal direct sum decomposition of $H(\Phi, V)$ into $K$ stable subspaces:

$$H(\Phi, V) = \bigoplus_{\rho \in \hat{N}^1} H(\Phi, \rho, V),$$

where $H(\Phi, \rho, V)$ consists of those $h$ in $H(\Phi, V)$ satisfying $h(\epsilon \mathbf{x}) = \rho(\epsilon) h(\mathbf{x})$, for all $\epsilon \in N^1$. Denote the representation of $K$ on $H(\Phi, \rho, V)$ by $\mathcal{D}(\Phi, \rho, V)$. In what follows I restrict $\rho$ to be either the trivial character on $N^1$, or a character of $N^1$ with conductor $\mathfrak{B}$.

By an explicit computation (as in the proof of Lemma 1.8.2) one may conclude that for $h \in H(\Phi, \rho, V)$:

i. If $V/\Omega$ is unramified and if $\rho \neq 1$, then $h$ vanishes outside $\mathcal{O}^*_{\mathfrak{P}}$ and the spectrum of $\mathcal{D}(\Phi, \rho, V)$ restricted to $B \cap K$ is given by:

$$\mathcal{D}(\Phi, \rho, V)\Big| \begin{pmatrix} 1 & b \\ 0 & 1 \end{pmatrix} = \bigoplus_{\substack{\tilde{\mathbf{x}} \in (\mathcal{O}_{\mathfrak{P}}/\mathfrak{B})^* \\ \tilde{\mathbf{x}} \bmod \mathcal{N}^1}} \Phi(b\mathcal{N}(\tilde{\mathbf{x}})) = \bigoplus_{x \in (\mathcal{O}_{\mathfrak{p}}/\mathfrak{b})^*} \Phi(bx).$$

If $\rho = 1$ and if $\mathfrak{b} = \mathfrak{p}$, then

$$\mathcal{D}(\Phi, \rho, V)\Big| \begin{pmatrix} 1 & b \\ 0 & 1 \end{pmatrix} = \bigoplus_{\substack{\tilde{\mathbf{x}} \in \mathcal{O}_{\mathfrak{P}}/\mathfrak{P} \\ \tilde{\mathbf{x}} \bmod \mathcal{N}^1}} \Phi(b\mathcal{N}(\tilde{\mathbf{x}})) = \bigoplus_{x \in \mathcal{O}_{\mathfrak{p}}/\mathfrak{p}} \Phi(bx).$$

ii. If $V/\Omega$ is ramified, $\rho^2 \neq 1$ and $\Pi$ is a prime element in $V$ with $N_{V/\Omega}(\Pi) = \pi$ a prime element in $\Omega$, then $h$ vanishes outside $\mathcal{O}^*_{\mathfrak{P}} \cup \Pi\mathcal{O}^*_{\mathfrak{P}}$ and

$$\mathcal{D}(\Phi, \rho, V)\Big| \begin{pmatrix} 1 & b \\ 0 & 1 \end{pmatrix} = \bigoplus_{\substack{\tilde{\mathbf{x}} \in (\mathcal{O}_{\mathfrak{P}}/\mathfrak{B})^* \cup \Pi(\mathcal{O}_{\mathfrak{P}}/\mathfrak{B})^* \\ \tilde{\mathbf{x}} \bmod \mathcal{N}^1}} \Phi(b\mathcal{N}(\tilde{\mathbf{x}})) = \bigoplus_{x \in [(\mathcal{O}_{\mathfrak{p}}/\mathfrak{b})^*]^2 \cup \pi[(\mathcal{O}_{\mathfrak{p}}/\mathfrak{b})^*]^2} \Phi(bx).$$

If $\mathfrak{b} = \mathfrak{p}$, $\rho = \rho_0$, then $h$ vanishes outside $\mathcal{O}^*_{\mathfrak{P}}$ and

$$\mathcal{D}(\Phi, \rho, V)\Big| \begin{pmatrix} 1 & b \\ 0 & 1 \end{pmatrix} = \bigoplus_{\tilde{\mathbf{x}} \in (\mathcal{O}_{\mathfrak{P}}/\mathfrak{P})^*} \Phi(b\mathcal{N}(\tilde{\mathbf{x}})) = \bigoplus_{x \in [(\mathcal{O}_{\mathfrak{p}}/\mathfrak{p})^*]^2} \Phi(bx).$$

If $\mathfrak{b} = \mathfrak{p}$, $\rho = 1$, then

$$\mathcal{D}(\Phi, \rho, V)\Big| \begin{pmatrix} 1 & b \\ 0 & 1 \end{pmatrix} = \bigoplus_{\substack{\tilde{\mathbf{x}} \in \mathcal{O}_{\mathfrak{P}}/\mathfrak{P} \\ \tilde{\mathbf{x}} \bmod \mathcal{N}^1}} \Phi(b\mathcal{N}(\tilde{\mathbf{x}})) = 1 \oplus \bigoplus_{x \in [(\mathcal{O}_{\mathfrak{p}}/\mathfrak{p})^*]^2} \Phi(bx).$$

In each case, the spectrum of $\mathcal{D}(\Phi, \rho, V)\big|B \cap K$ is simple and the commuting algebra of $\mathcal{D}(\Phi, \rho, V)$ may be computed as in §1.8. This yields

THEOREM 2.1.2. (i) *For $\rho \neq \rho_0$, $\mathcal{D}(\Phi, \rho, V)$ is irreducible.*

(ii) *For $\rho = \rho_0$ and $V/\Omega$ ramified, $\mathcal{D}(\Phi, \rho, V)$ is irreducible.*

(iii) *For $\rho = \rho_0$ and $V/\Omega$ unramified, $H(\Phi, \rho_0, V)$ splits into two subspaces stable under $\mathcal{D}(\Phi, \rho_0, V)$:*

$$H(\Phi, \rho_0, V) = H^+(\Phi, V) \oplus H^-(\Phi, V),$$

*where $H^+(\Phi, V)$ (resp. $H^-(\Phi, V)$) consists of those $h \in H(\Phi, \rho_0, V)$ whose support is contained in $S^+$ (resp. in $S^-$).*

Denote the corresponding representations respectively by $\mathcal{D}^+(\Phi, V)$ and $\mathcal{D}^-(\Phi, V)$. Then

$$\mathcal{D}^+(\Phi, V)\Big| \begin{pmatrix} 1 & b \\ 0 & 1 \end{pmatrix} = \bigoplus_{x \in [(\mathcal{O}_{\mathfrak{p}}/\mathfrak{p})^*]^2} \Phi(bx),$$

$$\mathcal{D}^-(\Phi, V)\Big| \begin{pmatrix} 1 & b \\ 0 & 1 \end{pmatrix} = \bigoplus_{x \in \zeta[(\mathcal{O}_{\mathfrak{p}}/\mathfrak{p})^*]^2} \Phi(bx),$$

where $\zeta$ is a non-square unit in $\mathcal{O}_{\mathfrak{p}}/\mathfrak{p}$.

*Degree of $\mathcal{D}(\Phi, \rho, V)$.* Let $N(\mathfrak{b}) = q^m$.

| | | |
|---|---|---|
| $\rho^2 \neq 1$ | $q^{m-1}(q-1)$ | $q^{m-2} \cdot (q^2-1)/2$ |
| $\rho = \rho_0$ | $q-1$ | $(q-1)/2$ |
| $\rho = 1$ | $q$ | $(q+1)/2$ |

$$\text{Degree of } D^+(\Phi, V) = \text{Degree of } D^-(\Phi, V) = \frac{q-1}{2}.$$

THEOREM 2.1.3. *If $V'/\Omega$ is ramified, then $\mathcal{D}^+(\Phi, V) \sim \mathcal{D}(\Phi, \rho_0, V')$.*

*Proof.* For $\mathfrak{b} = \mathfrak{p}$, $\rho$ of conductor $\mathfrak{P}$, one easily sees that the representations $\mathcal{D}(\Phi, \rho, V)$ are identically one on the subgroup of $K$ consisting of elements congruent to $\left(\begin{smallmatrix} 1 & 0 \\ 0 & 1 \end{smallmatrix}\right)$ mod $\mathfrak{p}$.

Hence these representations may be considered as representations of $\mathrm{SL}(2, \mathcal{O}_\mathfrak{p}/\mathfrak{p})$. It is well known [8] that there are exactly two inequivalent representations of $\mathrm{SL}(2, \mathcal{O}_\mathfrak{p}/\mathfrak{p})$ of degree $(q-1)/2$. Since $\mathcal{D}^+(\Phi, V)$ and $\mathcal{D}^-(\Phi, V)$ have inequivalent restriction to $B \cap K$, whereas $\mathcal{D}^+(\Phi, V)$ and $\mathcal{D}(\Phi, \rho_0, V')$ have the same restrictions, the theorem is proved. □

*Inclusions.*

i. For conductor of $\Phi = \mathfrak{b}$, conductor of $\rho = \delta^{-1}\mathfrak{b}$; or for conductor of $\Phi = \mathfrak{p}$, $\rho = 1$, we have

$$H(\Phi, \rho, V) \subset \mathfrak{h}(\Phi, \rho, V).$$

ii. For $V/\Omega$ unramified, conductor of $\Phi = \mathfrak{p}$, we have

$$H^+(\Phi, V) \subset \mathfrak{h}^+(\Phi, V),$$

$$H^-(\Phi, V) \subset \mathfrak{h}^-(\Phi, V).$$

iii. For $V/\Omega$ ramified, conductor of $\Phi = \mathfrak{p}$, we have

$$H(\Phi, \rho_0, V) \subset \mathfrak{h}^+(\Phi, V),$$

since in the ramified case $\mathcal{O}^*_\mathfrak{P} \subset S^+$.

I remark (see also the above-cited work of Kloosterman) that it follows from [14], that the representations $\mathcal{D}(\Phi, \rho, V)$, $\mathcal{D}^+(\Phi, V)$, $\mathcal{D}^-(\Phi, V)$ for $\rho$ of conductor $\mathfrak{P}$ and $\Phi$ of conductor $\mathfrak{p}$ considered as representations of $\mathrm{SL}(2, \mathcal{O}_\mathfrak{p}/\mathfrak{p})$, together with the principal series of representations of that group [11] exhaust (up to unitary equivalence) all irreducible representations of $\mathrm{SL}(2, \mathcal{O}_\mathfrak{p}/\mathfrak{p})$.

**3.1. Induced representations.** Let $\mathcal{D}$ be a unitary representation of $K$ acting on a finite-dimensional Hilbert space $H$. Let $M(H)$ denote the complex vector space of linear transformations of $H$. Let $D$ denote the representation $\mathrm{ind}_{K\uparrow G}\mathcal{D}$ which $\mathcal{D}$ induces on $G$. Let $S_\mathcal{D}$ denote the complex vector space of continuous functions of compact support on $G$ with values in $M(H)$ satisfying

(i) $\mathcal{F}(kgk') = U_k\mathcal{F}(g)U_{k'}$, where $U_k$ is the unitary operator on $H$ corresponding to $k$ under the representation $\mathcal{D}$.

For $\mathcal{F}_1$, $\mathcal{F}_2$ in $S_\mathcal{D}$, put

(ii) $(\mathcal{F}_1 * \mathcal{F}_2)(g_0) = \int_G \mathcal{F}_1(g)\mathcal{F}_2(g^{-1}g_0)\,dg.$

$S_{\mathcal{D}}$ with the product defined by (ii) is a complex algebra. It is known that $S_{\mathcal{D}}$ is dense in the commuting algebra of the induced representation $D$.

Now let $\mathcal{D}$ be one of the following irreducible unitary representations of $K$: $\mathcal{D}(\Phi, \rho, V)$ for $\rho^2 \neq 1$, $\mathcal{D}^+(\Phi, V)$, or $\mathcal{D}^-(\Phi, V)$ when $V/\Omega$ is unramified, and $\mathcal{D}(\Phi, \rho_0, V)$ when $V/\Omega$ is ramified. With this restriction we have:

THEOREM 3.1.1. *The algebras $S_{\mathcal{D}}$ have complex dimension one. (This theorem is a generalization of a theorem of Mautner [10].)*

*Proof.* Let $\mathcal{F}$ belong to $S_{\mathcal{D}}$. Since (say, by elementary divisors)

$$G = \bigcup_{m \geq 0} K\tau^m K, \quad \text{where} \quad \tau = \begin{pmatrix} \pi & 0 \\ 0 & \pi^{-1} \end{pmatrix},$$

$\mathcal{F}$ is determined by its value on $\tau^m$ for $m \geq 0$. Choose a basis in $H$ for which the operators $M_b$ for $b \in \mathcal{O}_{\mathfrak{p}}$ are represented by diagonal matrices:

$$M_b \sim \begin{pmatrix} \ddots & & 0 \\ & \Phi_x(b) & \\ 0 & & \ddots \end{pmatrix},$$

where in each case $x$ runs over a certain subset, say, $B(\mathcal{D})$, of $\mathcal{O}_{\mathfrak{p}}$ as described in §2. Let $(\mathcal{F}^m_{x,y})$ denote the matrix of $\mathcal{F}(\tau^m)$ in this basis. Then, since

$$\begin{pmatrix} \pi^m & 0 \\ 0 & \pi^{-m} \end{pmatrix}\begin{pmatrix} 1 & b \\ 0 & 1 \end{pmatrix} = \begin{pmatrix} 1 & b\pi^{2m} \\ 0 & 1 \end{pmatrix}\begin{pmatrix} \pi^m & 0 \\ 0 & \pi^{-m} \end{pmatrix},$$

we have $\mathcal{F}(\tau^m)M_b = M_{b\pi^{2m}}\mathcal{F}(\tau^m)$, or in matrix form

$$\mathcal{F}^m_{x,z}\Phi(zb) = \Phi(xb\pi^{2m})\mathcal{F}^m_{x,z}.$$

Suppose that there exists an $\mathcal{F}$ in $S_{\mathcal{D}}$ such that $\mathcal{F}(\tau^m)$ is non-zero, then we have $z \equiv x\pi^{2m} \bmod \mathfrak{b}$ for all $z$ and $x$ in $B(\mathcal{D})$. Since any element in $B(\mathcal{D})$ is either a unit or $\pi$ times a unit, we derive a contradiction if $m \geq 1$. Hence $\mathcal{F}$ must have its support on $K$. In any case

$$\mathcal{F}(1 \cdot k) = \mathcal{F}(1)U_k = U_k\mathcal{F}(1);$$

since $\mathcal{D}$ is irreducible, $\mathcal{F}(1)$ must be a scalar multiple of the identity. □

It follows from Theorem 3.1.1 that each of the induced representations $\mathrm{ind}_{K\uparrow G}\mathcal{D}$ is irreducible. Combining this with the inclusion relations of §2 and Theorem 1.8.4 proves the following statements:

i. For $\Phi$ of conductor $\mathfrak{b}$, $\rho$ of conductor $\delta^{-1}\mathfrak{b}$, $\rho^2 \neq 1$,

$$\mathrm{ind}_{K\uparrow G}\mathcal{D}(\Phi, \rho, V) \sim D(\Phi, \rho, V).$$

ii. For $V/\Omega$ unramified,

$$\mathrm{ind}_{K\uparrow G}\mathcal{D}^{+}(\Phi, V) \sim D^{+}(\Phi, V),$$

$$\mathrm{ind}_{K\uparrow G}\mathcal{D}^{-}(\Phi, V) \sim D^{-}(\Phi, V).$$

iii. For $V/\Omega$ unramified

$$\mathrm{ind}_{K\uparrow G}\mathcal{D}(\Phi, \rho_0, V) \sim D(\Phi, V).$$

Since $\Omega$ is self-dual, given any ideal $\mathfrak{b}$ in $\mathcal{O}_{\mathfrak{p}}$ and any non-trivial character $\Phi$ of $\Omega^{+}$, there exists an $r \in \Omega^{*}$ such that $\Phi_r$ has conductor $\mathfrak{b}$, combining the above statements with Theorem 1.9.1 yields:

THEOREM 3.1.2. *For $V$ a quadratic extension field of $\Omega$, each of the representations $D(\Phi, \rho, V)$ for $\rho^2 \neq 1$, $D^{+}(\Phi, V)$ and $D^{-}(\Phi, V)$ is induced from an irreducible representation of some maximal compact subgroup of $G$.*

It follows from [10] that there exists an orthonormal basis in the Hilbert space corresponding respectively to each of the representations of Theorem 3.1.2 for which the corresponding matrix coefficients are of compact support on $G$.

Finally it now follows from Theorem 2.1.3 that for $\Phi$ of conductor $\mathfrak{p}$, we have $D^{+}(\Phi, V) \sim D^{-}(\Phi, V')$ for any pair of quadratic extension fields $V$ and $V'$. Hence by Theorem 1.9.1 this equivalence holds for any non-trivial $\Phi$. This completes the proof of Theorem 1.9.

**4.** The purpose of the present paragraph is to obtain all irreducible unitary representations of $\mathrm{SL}(2, \mathcal{O}_{\mathfrak{p}})$. (In §4.1 and §4.2, $\Omega$ may also be taken as the completion of a function field.) In sections §4.1 and §4.2 I will construct a series of representations of $\mathrm{SL}(2, \mathcal{O}_{\mathfrak{p}})$ as monomial representations. In section §4.3 I show that these, together with certain explicit matrix representations considered by Hecke [6], Kloosterman [8], Maass [9] and also in the preceding, yield all irreducible unitary representations of $\mathrm{SL}(2, \mathcal{O}_{\mathfrak{p}})$.

**4.1. General facts about induced representations.** In this paragraph I will prove a number of lemmas which I will use to obtain and classify irreducible representations of $\mathrm{SL}(2, \mathcal{O}_{\mathfrak{p}})$.

Let $G$ be a finite group, $H$ a subgroup, $u$ and $u'$ complex-valued irreducible representations of $H$ operating respectively on the complex vector spaces $V$ and $V'$. Let $\mathrm{ind}_{H\uparrow G}u$ and $\mathrm{ind}_{H\uparrow G}u'$ respectively denote the representations which $u$ and $u'$ induce on $G$. Let $S_{u,u'}$ be the space of maps $\mathcal{F}$ of $G$ into Hom(V′, V) satisfying

$$\mathcal{F}(hgh') = u(h)\mathcal{F}(g)u'(h'),$$

for all $g \in G$ and $h, h' \in H$.

It is well known that $S_{u,u'}$ is isomorphic to the space of intertwining operators (i.e., $G$-module homomorphisms) from $\mathrm{ind}_{H\uparrow G}u$ to $\mathrm{ind}_{H\uparrow G}u'$. Further if $V = V'$, $u = u'$, and for $\mathcal{F}_1, \mathcal{F}_2$ in $S_{u,u} = S_u$, we put

$$(\mathcal{F}_1 * \mathcal{F}_2)(g_0) = \sum_{g\in G} \mathcal{F}_1(g)\mathcal{F}_2(g^{-1}g_0), \tag{5}$$

then, with product defined by (5), $S_u$ is a complex algebra isomorphic to the commuting algebra of $\mathrm{ind}_{H\uparrow G}u$.

In what follows if $U$ is a representation of $G$, I will denote by $U|H$ its restriction to $H$, and if $U_1$ is a component of $U$, I will write $U_1 \subset U$.

I shall now apply these facts to the case when $H = N$ is normal in $G$. Let $N^*$ denote the set of equivalence classes of irreducible representations of $N$. $G$ operates on $N^*$ in an obvious fashion, namely for $u \in N^*$, $g \in G$, $n \in N$, put $u^g(n) = u(g^{-1}ng)$ then $u^g$ is an irreducible representation of $N$,

$$u^{g_1 \cdot g_2} = (u^{g_1})^{g_2},$$

and the map $u \longmapsto u^g$ obviously preserves equivalence.

For $u \in N^*$, let $N(u)$ be the subgroup of $G$ fixing the class of $u$, i.e., if $\sim$ denotes equivalence, $N(u)$ consists of those elements $g$ in $G$ for which $u \sim u^g$.

LEMMA 4.1.1. *Let $U_i$ be any irreducible representation of $N(u)$ such that $U_i|N \supset u$. Then the representations* $\mathrm{ind}_{N(u)\uparrow G}U_i$ *are irreducible, and* $\mathrm{ind}_{N(u)\uparrow G}U_i \sim \mathrm{ind}_{N(u)\uparrow G}U_j$ *if and only if $U_i \sim U_j$.*

*Proof.* As above, let $S_u$ be the commuting algebra of $\mathrm{ind}_{N\uparrow G}u$. Let $\mathcal{F}$ belong to $S_u$. Since $N$ is normal in $G$, we have

$$u(n)\mathcal{F}(g) = \mathcal{F}(ng) = \mathcal{F}(gg^{-1}ng) = \mathcal{F}(g)u^g(n).$$

Hence if $\mathcal{F}(g)$ is not zero, then $u$ and $u^g$ interwine; or, since $u$ is irreducible, we have $u \sim u^g$, i.e., $g$ belongs to $N(u)$. Hence if $\mathcal{F}$ belongs to $S_u$, then $\mathcal{F}$ vanishes outside $N(u)$. This immediately implies that $S_u$ is isomorphic to the commuting algebra of the representation $\mathrm{ind}_{N\uparrow N(u)}u$ of $N(u)$.

Decompose $\mathrm{ind}_{N\uparrow N(u)}u$ into irreducible representations:

$$\mathrm{ind}_{N\uparrow N(u)}u = \bigoplus_{i=1}^{r} n_i U_i,$$

where $n_i > 0$, $U_i$ is irreducible and $U_i \sim U_j$ if and only if $i = j$. Then by the preceding,

$$\dim S_u = \sum_{i=1}^{r} n_i^2.$$

On the other hand, by transitivity of inducing representations, we have

$$\operatorname{ind}_{N\uparrow G}u = \bigoplus_{i=1}^{r} n_i \operatorname{ind}_{N(u)\uparrow G}U_i.$$

If $\chi$ is the character of $\operatorname{ind}_{N\uparrow G}u$ and $\chi_i$ is the character of $\operatorname{ind}_{N(u)\uparrow G}U_i$, then $\chi = \sum_{i=1}^{r} n_i \chi_i$. Define the scalar product $\langle\ ,\ \rangle$ as usual: If $f_1$ and $f_2$ are complex-valued functions on $G$, we put $\langle f_1, f_2\rangle = (1/[G:1])\sum_{g\in G} f_1(g)\overline{f_2(g)}$. We have then

$$\begin{aligned}\dim S_u = \langle\chi,\chi\rangle &= \sum_{\substack{1\le i\le r\\ 1\le j\le r}} n_i n_j \langle\chi_i,\chi_j\rangle \\ &= \sum_{i=1}^{r} n_i^2 \langle\chi_i,\chi_j\rangle + \sum_{\substack{1\le i\le r\\ 1\le j\le r\\ i\ne j}} n_i n_j \langle\chi_i,\chi_j\rangle.\end{aligned}$$

Therefore we must have $\langle\chi_i,\chi_j\rangle = \delta_{ij}$. Since the $U_i$ are exactly those representations of $N(u)$ whose restriction to $N$ contains $u$ (Frobenius reciprocity), the theorem is proved. □

LEMMA 4.1.2. *Let $u$ and $u'$ belong to $N^*$. Then the following conditions are equivalent:*

(i) *$\operatorname{ind}_{N\uparrow G}u$ and $\operatorname{ind}_{N\uparrow G}u'$ interwine*

(ii) *$\operatorname{ind}_{N\uparrow G}u \sim \operatorname{ind}_{N\uparrow G}u'$*

(iii) *$u$ and $u'$ lie in the same orbit of $G$ in $N^*$, i.e., there exists $g\in G$ such that $u^g = u'$.*

*Proof.* $\operatorname{ind}_{N\uparrow G}u$ and $\operatorname{ind}_{N\uparrow G}u'$ interwine if and only if $\dim S_{u,u'} > 0$. Suppose that $\mathcal{F}$ belongs to $S_{u,u'}$ and $\mathcal{F}(g_0)\ne 0$ for some $g\in G$. Then

$$\begin{aligned}u(n)\mathcal{F}(g_0) = \mathcal{F}(ng_0) &= \mathcal{F}(g_0 g_0^{-1} n g_0) \\ &= \mathcal{F}(g_0)u'^{g_0}(n),\end{aligned}$$

for $n\in N$. Hence, the irreducible representations $u$ and $u'^{g_0}$ interwine, which implies $u\sim u'^{g_0}$. If $u\sim u'^{g_0}$, then it follows easily from the definitions that

$$\operatorname{ind}_{N\uparrow G}u \sim \operatorname{ind}_{N\uparrow G}u'^{g} \sim \operatorname{ind}_{N\uparrow G}u'.$$

Clearly (ii) implies (i). □

In the application (§4.2) I will obtain irreducible representations of $\mathrm{SL}(2,\mathcal{O}_{\mathfrak{p}}) = K$ be decomposing representations of $K_n = \mathrm{SL}(2,\mathcal{O}_{\mathfrak{p}}/\mathfrak{p}^n)$ of the form $\operatorname{ind}_{N\uparrow K_n}\Phi$, where $N$ is normal in $K_n$ and $\Phi$ is a one-dimensional representation of $N$. For suitable $\Phi$, $N(\Phi)$ is generated by $N$ and an auxiliary subgroup $A$:

$$N(\Phi) = A \cdot N.$$

The irreducible representations $U$ of $N(\Phi)$ satisfying $U|N \supset u$ are then obtained by the following Lemma.

LEMMA 4.1.3. *Let $G$ be a finite group generated by a subgroup $A$ and a normal subgroup $N$: $G = AN$, $\Phi$ a one-dimensional representation of $N$ invariant under $G$: $\Phi^g = \Phi$ for all $g \in G$, then*

(i) *the irreducible representations $U$ of $G$ satisfying $U|N \supset \Phi$ are in one-to-one correspondence with the irreducible representations $u$ of $A$ satisfying $u|(A \cap N) \supset \Phi|(A \cap N)$.*

(ii) *the multiplicity of $U$ in* $\mathrm{ind}_{N\uparrow G}\Phi$ = *the degree of $U$.*

*Proof.* Take $U \in G^*$ satisfying $U|N \supset \Phi$. Let $U|N = \bigoplus_{i=1}^r n_i u_i$, where $n_i > 0$, $u_i \in N^*$, $u_i \sim u_j$ if and only if $i = j$ and, say, $u_1 = \Phi$. Then $n_i$ is the multiplicity of $U$ in $\mathrm{ind}_{N\uparrow G} u_i$, hence $\mathrm{ind}_{N\uparrow G} u_i$ and $\mathrm{ind}_{N\uparrow G}\Phi$ interwine; thus by Lemma 4.1.2 there exists $g \in G$ such that $u_i \sim \Phi^g = \Phi$. Hence $U|N = (\deg U)\Phi$, which proves (ii). Let $U|A = u$, then $U(an) = u(a)\Phi(n)$. Clearly $a \longmapsto u(a)$ is an irreducible representation of $A$ and $u|(A \cap N) = \Phi|(A \cap N)$. Conversely, the same reasoning shows that if $u \in A^*$ satisfies $u|(A \cap N) \supset \Phi|(A \cap N)$, then $u|(A \cap N) = (\deg u)\Phi|(A \cap N)$. Hence, if we put $U(g) = U(an) = u(a)\Phi(n)$, then $U$ is well defined, and since $\Phi^g = \Phi$ for $g \in G$, $g \longmapsto U(g)$ is a representation of $G$, which is clearly irreducible. □

**4.2. Construction of the irreducible unitary representations of SL(2, $\mathcal{O}_\mathfrak{p}$).** Let $K_n = \mathrm{SL}(2, \mathcal{O}_\mathfrak{p}/\mathfrak{p}^n)$. It is easily verified that the natural map from $K$ to $K_n$ (namely reduction mod $\mathfrak{p}^n$) is surjective. Thus $K$ is totally disconnected every finite-dimensional (unitary) representation of $K$ factors through a representation of $K_n$. An irreducible representation of $K_n$ will be called *primitive* if it does not factor through a representation of $K_{n-1}$.

I will begin by describing the primitive representations of $K_n$ for $n$ even. For $n$ odd, $n \neq 1$ a similar method yields at present part of the primitive representations of $K_n$; these however together with certain representations of $K_n$ constructed in the earlier part of this paper and, also constructed by Kloosterman in [8] give all primitive representations of $K_n$. The method does not apply to the group $K_1$; however the representations of this group are well known.

Let $N_\nu$ denote the subgroup of $K_n$ consisting of all $n \in K_n$ satisfying $n \equiv 1 \bmod \mathfrak{p}^\nu$. Suppose first that $n = 2k$ is even. Let $\Phi$ be a one-dimensional representation of $N_k$. Since $N_k$ is abelian, we obtain all irreducible representations of $K_n$ by decomposing $\mathrm{ind}_{N_k\uparrow K_n}\Phi$, as $\Phi$ varies over $(N_k)^*$.

Call $\Phi$ *primitive* if $\Phi$ is non-trivial on $N_{n-1}$. It is easy to see that, if $\Phi$ is primitive, then each irreducible representation $\mathcal{D}$ of $G$ occurring in $\operatorname{ind}_{N_k \uparrow K_n} \Phi$ is primitive and that otherwise each such $\mathcal{D}$ is not primitive.

In what follows I will show that, for $\Phi$ primitive, ($n$ even) the group $N_k(\Phi)$ is generated by an abelian group $T(\Phi)$ and $N_k$:

$$N_k(\Phi) = T(\Phi)N_k.$$

Assuming this for the moment, by Lemma 4.1.3, the irreducible representations $U$ of $N_k(\Phi)$ satisfying $U|N_k \supset \Phi$ are of the form $\Psi_{\rho,\Phi}$:

$$\Psi_{\rho,\Phi}(tn) = \rho(t)\Phi(n),$$

for $t \in T(\Phi)$, $n \in N_k$ where $\rho \in T(\Phi)^*$ and satisfies $\rho|T(\Phi) \cap N_k = \Phi|T(\Phi) \cap N_k$. Thus we have

THEOREM 4.2.1. *For $n = 2k$ even,*

(i) *the primitive representations of $K_n$ are exactly of the form*

$$\operatorname{ind}_{N_k(\Phi)\uparrow K_n} \Psi_{\rho,\Phi} = \mathcal{D}(\rho, \Phi).$$

(ii) $\mathcal{D}(\rho, \Phi) \sim \mathcal{D}(\rho', \Phi')$ *implies* $\Phi' = \Phi^g$, *for some* $g \in K_n$ *in which case* $\operatorname{ind}_{N_k \uparrow K_n} \Phi$ *and* $\operatorname{ind}_{N_k \uparrow K_n} \Phi'$ *are equivalent.*

(iii) $\mathcal{D}(\rho, \Phi) \sim \mathcal{D}(\rho', \Phi)$ *if and only if* $\rho = \rho'$.

*Proof.* For $\Phi$ primitive, we have

$$\begin{aligned}\operatorname{ind}_{N_k \uparrow K_n} \Phi &= \operatorname{ind}_{N_k(\Phi)\uparrow K_n} \left(\operatorname{ind}_{N_k \uparrow N_k(\Phi)} \Phi\right) \\ &= \bigoplus_{\substack{\rho \in T(\Phi)^* \\ \rho|T(\Phi)\cap N_k = \Phi|T(\Phi)\cap N_k}} \operatorname{ind}_{N_k(\Phi)\uparrow K} \Psi_{\rho,\Phi};\end{aligned}$$

(i) and (iii) follow from Lemma 4.1.1. If $\mathcal{D}(\rho, \Phi) \sim \mathcal{D}(\rho', \Phi')$ then, by the Frobenius reciprocity, $\operatorname{ind}_{N_k \uparrow K_n} \Phi$ and $\operatorname{ind}_{N_k \uparrow K_n} \Phi'$ interwine. Thus (ii) follows from Lemmas 4.1.2. □

We will return to the case $n$ odd below.

*Structure of $N_k(\Phi)$.* Let $\mathfrak{g}$ denote the Lie algebra over $\mathcal{O}_\mathfrak{p}$ consisting of all $2 \times 2$ matrices $X$ of trace zero with coefficients in $\mathcal{O}_\mathfrak{p}$. We will call an element $X$ of $\mathfrak{g}$ *primitive* if $X \notin \mathfrak{p}\mathfrak{g}$. $K$ operates on $\mathfrak{g}$ by the "adjoint representation"

$$k \circ X = kXk^{-1} \text{ for } k \in K \text{ and } X \in \mathfrak{g},$$

and clearly leaves invariant the primitive elements. In order to find the structure of $N_k(\Phi)$, I need the following lemmas:

LEMMA 4.2.2. *Each primitive element of $\mathfrak{g}$ is in the $K$-orbit of one and only one of the following elements:*

(i) $\left(\begin{smallmatrix}\lambda & 0\\ 0 & -\lambda\end{smallmatrix}\right)$; $\lambda \in \mathcal{O}_{\mathfrak{p}}^*$, $\quad\lambda \bmod \pm 1$.

(ii) $\left(\begin{smallmatrix}0 & 1\\ \tilde{\omega} & 0\end{smallmatrix}\right)$; $\tilde{\omega} \in \mathfrak{p}$,

$\left(\begin{smallmatrix}0 & \zeta\\ \tilde{\omega} & 0\end{smallmatrix}\right)$; *$\zeta$ a fixed element of $\mathcal{O}_{\mathfrak{p}}^*$ satisfying $\left(\frac{\zeta}{\mathfrak{p}}\right) = -1$.*

(iii) $\left(\begin{smallmatrix}0 & 1\\ \tilde{\omega} & 0\end{smallmatrix}\right)$; $\tilde{\omega} \in \mathcal{O}_{\mathfrak{p}}^*$, $\left(\frac{\omega}{\mathfrak{p}}\right) = -1$.

*Proof.* Let $X = \left(\begin{smallmatrix}\mu & \sigma\\ \tau & -\mu\end{smallmatrix}\right)$. There are several cases.

(A) If $\sigma \in \mathcal{O}_{\mathfrak{p}}^*$, then

$$\begin{pmatrix}1 & 0\\ \mu\sigma^{-1} & 1\end{pmatrix}\begin{pmatrix}\mu & \sigma\\ \tau & -\mu\end{pmatrix}\begin{pmatrix}1 & 0\\ -\mu\sigma^{-1} & 1\end{pmatrix} = \begin{pmatrix}0 & \sigma\\ \tau + \mu^2\sigma^{-1} & 0\end{pmatrix}.$$

(B) If $\tau \in \mathcal{O}_{\mathfrak{p}}^*$, then

$$\begin{pmatrix}1 & -\mu\tau^{-1}\\ 0 & 1\end{pmatrix}\begin{pmatrix}\mu & \sigma\\ \tau & -\mu\end{pmatrix}\begin{pmatrix}1 & \mu\tau^{-1}\\ 0 & 1\end{pmatrix} = \begin{pmatrix}0 & \mu^2\tau^{-1} + \sigma\\ \tau & 0\end{pmatrix}.$$

(C) If $\sigma \in \mathfrak{p}$, $\tau \in \mathfrak{p}$, $\mu \in \mathcal{O}_{\mathfrak{p}}^*$, choose $\lambda$ so that $\lambda^2 = \mu^2 + \sigma$. If $\mu + \lambda \in \mathcal{O}_{\mathfrak{p}}^*$, then

$$\begin{pmatrix}1 & \sigma/(\lambda+\mu)\\ -\tau/2\lambda & (\lambda+\mu)/2\lambda\end{pmatrix}\begin{pmatrix}\mu & \sigma\\ \tau & -\mu\end{pmatrix}\begin{pmatrix}(\lambda+\mu)/2\lambda & -\sigma/(\lambda+\mu)\\ \tau/2\lambda & 1\end{pmatrix} = \begin{pmatrix}\lambda & 0\\ 0 & -\lambda\end{pmatrix}.$$

If $\mu - \lambda \in \mathcal{O}_{\mathfrak{p}}^*$, then

$$\begin{pmatrix}1 & \sigma/(\mu-\lambda)\\ -\tau/2\lambda & (\mu-\lambda)/2\lambda\end{pmatrix}\begin{pmatrix}\mu & \sigma\\ \tau & -\mu\end{pmatrix}\begin{pmatrix}(\mu-\lambda)/2\lambda & -\sigma/(\mu-\lambda)\\ \tau/2\lambda & 1\end{pmatrix} = \begin{pmatrix}\lambda & 0\\ 0 & -\lambda\end{pmatrix}.$$

(D) If $X = \left(\begin{smallmatrix}0 & \sigma\\ \tau & 0\end{smallmatrix}\right)$ where $\sigma \in \mathcal{O}_{\mathfrak{p}}^*$, $\tau \in \mathcal{O}_{\mathfrak{p}}^*$ and $\left(\frac{\sigma\tau}{\mathfrak{p}}\right) = 1$, choose $\lambda \in \mathcal{O}_{\mathfrak{p}}^*$ such that $\lambda^2 = \sigma\tau$, then

$$\begin{pmatrix}1/2 & \lambda/2\tau\\ -\lambda/\sigma & 1\end{pmatrix}\begin{pmatrix}0 & \sigma\\ \tau & 0\end{pmatrix}\begin{pmatrix}1 & -\lambda/2\tau\\ \lambda/\sigma & 1/2\end{pmatrix} = \begin{pmatrix}\lambda & 0\\ 0 & -\lambda\end{pmatrix}.$$

(E) If $X = \left(\begin{smallmatrix}0 & \sigma\\ \tau & 0\end{smallmatrix}\right)$ with $\sigma \in \mathcal{O}_{\mathfrak{p}}^*$, $\tau \in \mathcal{O}_{\mathfrak{p}}^*$ and $\left(\frac{\sigma\tau}{\mathfrak{p}}\right) = -1$, choose $a, b \in \Omega$ so that $a^2 - b^2\tau/\sigma = \sigma^{-1}$. Since in that case $a, b \in \mathcal{O}_{\mathfrak{p}}$ and $(a, b) = 1$ we may choose $c, d \in \mathcal{O}_{\mathfrak{p}}$ so that $ad - bc = 1$, in that case

$$\begin{pmatrix}a & b\\ b\tau & a\sigma\end{pmatrix}\begin{pmatrix}0 & \sigma\\ \tau & 0\end{pmatrix}\begin{pmatrix}a\sigma & -b\\ -b\tau & a\end{pmatrix} = \begin{pmatrix}0 & 1\\ \sigma\tau & 0\end{pmatrix}.$$

Finally, by using $k = \left(\begin{smallmatrix}0 & 1\\ -1 & 0\end{smallmatrix}\right)$ and $k = \left(\begin{smallmatrix}a & 0\\ 0 & a^{-1}\end{smallmatrix}\right)$, where $a \in \mathcal{O}_{\mathfrak{p}}^*$, it follows from (A)–(E) that each primitive $X \in \mathfrak{g}$ lies in the $K$-orbit of one of the elements of the lemma. The fact that these elements lie in distinct $K$-orbits follows easily by computation. □

Let $\mathfrak{g}_n$ denote the Lie algebra over $\mathcal{O}_\mathfrak{p}/\mathfrak{p}^n$ consisting of $2 \times 2$ matrices of trace zero with coefficients in $\mathcal{O}_\mathfrak{p}/\mathfrak{p}^n$. $K_n$ operates on $\mathfrak{g}_n$ as before: $k \circ \mathbf{x} = k\mathbf{x}k^{-1}$ for $k \in K_n$ and $\mathbf{x} \in \mathfrak{g}_n$; and, as before, we call an element $\mathbf{x} \in \mathfrak{g}_n$ primitive if $\mathbf{x} \notin \mathfrak{p}\mathfrak{g}_n$. The natural map (namely reduction mod $\mathfrak{p}^n$) from $\mathfrak{g}$ to $\mathfrak{g}_n$ is surjective. For $X \in \mathfrak{g}$, let $\tilde{X}$ denote its image in $\mathfrak{g}_n$.

Let $T_X$ denote the subgroup of $K$ consisting of those $g \in K$ such that $gXg^{-1} = X$, and $T_{\tilde{X},n}$ the subgroup of $K_n$ consisting of those $g \in K_n$ such that $g\tilde{X}g^{-1} = X$. Then we have

Lemma 4.2.3. *For $X$ primitive, the natural map from $K$ to $K_n$ induces a surjection from $T_X$ to $T_{\tilde{X},n}$.*

*Proof.* It suffices to prove the lemma for the special representatives of Lemma 4.2.2 (since $T_{k\circ X} = kT_Xk^{-1}\ldots$). For $X = \left(\begin{smallmatrix}\lambda & 0\\ 0 & -\lambda\end{smallmatrix}\right)$:

$$T_X = \left\{\begin{pmatrix} a & 0\\ 0 & a^{-1}\end{pmatrix}; a \in \mathcal{O}_\mathfrak{p}^*\right\},$$

$$T_{\tilde{X},n} = \left\{\begin{pmatrix} a & 0\\ 0 & a^{-1}\end{pmatrix}; a \in (\mathcal{O}_\mathfrak{p}/\mathfrak{p}^n)^*\right\}.$$

For $X = \left(\begin{smallmatrix}0 & \sigma\\ \tau & 0\end{smallmatrix}\right)$:

$$T_X = \left\{\begin{pmatrix} a & b\\ b\tau\sigma^{-1} & a\end{pmatrix}; a, b \in \mathcal{O}_\mathfrak{p}, \quad a^2 - b^2\tau\sigma^{-1} = 1\right\},$$

$$T_{\tilde{X},n} = \left\{\begin{pmatrix} a & b\\ b\tau\sigma^{-1} & a\end{pmatrix}; a, b \in \mathcal{O}_\mathfrak{p}/\mathfrak{p}^n, \quad a^2 - b^2\tau\sigma^{-1} = 1\right\}.$$

The lemma now follows from the usual inductive and compactness arguments of Hensel's lemma. We remark that in each case $T_X$ is abelian. □

Let $B$ denote the "Killing form" of $\mathfrak{g}_n$, i.e., $B(\mathbf{x}, \mathbf{y}) = \mathrm{Tr}(\mathbf{x} \cdot \mathbf{y})$ for $\mathbf{x}, \mathbf{y}$ in $\mathfrak{g}_n$. Let $\eta$ be a primitive character of $\mathcal{O}_\mathfrak{p}/\mathfrak{p}^n$, i.e., $\eta \in (\mathcal{O}_\mathfrak{p}/\mathfrak{p}^n)^*$ and $\eta$ is non-trivial on $\mathfrak{p}^{n-1}$, then

Lemma 4.2.4. (i) *$B$ is nondegenerate $\mathcal{O}_\mathfrak{p}/\mathfrak{p}^n$-valued bilinear form on $\mathfrak{g}_n \times \mathfrak{g}_n$ i.e., $B(\mathbf{x}, \mathbf{y}) = 0$ for all $\mathbf{x} \in \mathfrak{g}_n$ implies $\mathbf{y} = 0$.*

(ii) *If we put $\eta_\mathbf{y}(\mathbf{x}) = \eta(B(\mathbf{x}, \mathbf{y}))$, then $\mathbf{y} \longmapsto \eta_\mathbf{y}$ is a $K_n$-module isomorphism of $\mathfrak{g}_n$ to $\mathfrak{g}_n^*$.*

(iii) *$\eta_\mathbf{y}$ is primitive (i.e., non-trivial on $\mathfrak{p}^{n-1}\mathfrak{g}_n$) if and only if $\mathbf{y}$ is primitive.*

*Proof.* A simple computation shows that $B$ is nondegenerate. As $\mathbf{x}$ runs over $\mathfrak{g}_n$, the elements of $\mathcal{O}_\mathfrak{p}/\mathfrak{p}^n$ of the form $B(\mathbf{x}, \mathbf{y})$ ($\mathbf{y}$ fixed) form an ideal in $\mathcal{O}_\mathfrak{p}/\mathfrak{p}^n$;

hence $\mathbf{y} \longmapsto \eta_{\mathbf{y}}$ is an injection and hence an isomorphism (of abelian groups) since $\mathfrak{g}_n$ and $\mathfrak{g}_n^*$ have the same number of elements. The fact that this isomorphism is an isomorphism of $K_n$-modules follows from the invariance of $B$, i.e.,

$$B(k \circ \mathbf{x}, k \circ \mathbf{y}) = B(\mathbf{x}, \mathbf{y}) \text{ for } k \in K_n,$$

where $\mathbf{x}, \mathbf{y} \in \mathfrak{g}_n$. (iii) follows from (i) and the definitions. □

For $n$ even, $n = 2k$, each $n \in N_k$ may be written uniquely in the form $1 + \Lambda\pi^k$ where $\Lambda$ is taken mod $\mathfrak{p}^k$ and $\mathrm{Tr}(\Lambda) \equiv 0 \bmod \mathfrak{p}^k$. The map $n \longmapsto \Lambda \bmod \mathfrak{p}^k$ is an isomorphism of $N_k$ and $\mathfrak{g}_k$ considered as $K_n$-modules. Hence, by Lemma 4.2.4, we may identify $(N_k)^*$ with $\mathfrak{g}_k$. Thus, if $\mathbf{x} \in \mathfrak{g}_k$ and $\eta_{\mathbf{x}}$ is the corresponding element in $(N_k)^*$ we have $N_k(\eta_{\mathbf{x}}) = \{g \in K_n; g\mathbf{x}g^{-1} \equiv \mathbf{x} \bmod \mathfrak{p}^k\}$. Let $X$ be a pre-image of $\mathbf{x}$ in $\mathfrak{g}$, then by Lemma 4.2.3 we have

$$N_k(\eta_{\mathbf{x}}) = T_{\tilde{X},n} N_k.$$

This completes the result for even $n$.

For $n$ odd, $n = 2k + 1$ we may write $n \in N_{k+1}$ uniquely in the form $n = 1 + \Lambda\pi^{k+1}$ (again, $\Lambda$ is taken mod $\mathfrak{p}^k$), $\mathrm{Tr}(\Lambda) \equiv 0 \bmod \mathfrak{p}^k$. As before, $N_{k+1}$, $\mathfrak{g}_k$ and $\mathfrak{g}_k^*$ are isomorphic as $K_n$-modules. In this case however for $\mathbf{x} \in \mathfrak{g}_k$, $\mathbf{x}$ primitive, $X$ a pre-image in $\mathfrak{g}$,

$$N_{k+1}(\eta_{\mathbf{x}}) = T_{\tilde{X},n} N_k.$$

In what follows $X$ will denote an element of $\mathfrak{g}$ of the form $\left(\begin{smallmatrix} \lambda & 0 \\ 0 & -\lambda \end{smallmatrix}\right)$, where $\lambda \in \mathcal{O}_{\mathfrak{p}}^*$, or one of the form $\left(\begin{smallmatrix} 0 & \zeta \\ \tilde{\omega} & 0 \end{smallmatrix}\right)$, where $\zeta \in \mathcal{O}_{\mathfrak{p}}^*$ and $\tilde{\omega} \in \mathfrak{p}$. In particular $X$ is primitive. Let $\eta_X$ be the corresponding character of $N_{k+1}$: if $\xi$ is the character of $\mathfrak{p}^{k+1}$ in $\mathcal{O}_{\mathfrak{p}}/\mathfrak{p}^n$ defined by $\xi(\pi^{k+1}\lambda) = \eta(\lambda)$ ($\lambda \bmod \mathfrak{p}^k$), then $\eta_X(n) = \xi(\pi^{k+1}\mathrm{Tr}(\tilde{X}\Lambda)) = \eta(\mathrm{Tr}(X\Lambda))(\Lambda \bmod \mathfrak{p}^k)$. Let $B_n$ denote the subgroup of $K_n$ consisting of elements of the form

$$b = \begin{pmatrix} 1 + \mu\pi^k & \pi^k\sigma \\ \pi^{k+1}\tau & 1 + \mu'\pi^k \end{pmatrix} \qquad (\mu, \mu', \sigma, \tau \in \mathcal{O}_{\mathfrak{p}}/\mathfrak{p}^n).$$

By using the explicit form of $T_X$, one sees that $B_n$ is a normal subgroup of $T_{\tilde{X},n}N_k$. Let $\tilde{\xi}$ be any extension of $\xi$ to a character of $\mathfrak{p}^k$.

For $b \in B_n$, put

$$\eta_{X,\tilde{\xi}}(b) = \tilde{\xi}(\mathrm{Tr}(\tilde{X} \cdot \log b)),$$

i.e., if $X = \left(\begin{smallmatrix} \lambda & 0 \\ 0 & -\lambda \end{smallmatrix}\right)$,

$$\eta_{X,\tilde{\xi}} \begin{pmatrix} 1 + \mu\pi^k & \pi^k\sigma \\ \pi^{k+1}\tau & 1 + \mu'\pi^k \end{pmatrix} = \tilde{\xi}(2\lambda\pi^k(\mu - \pi^k\mu^2/2)),$$

and, if $X = \left(\begin{smallmatrix} 0 & \zeta \\ \tilde{\omega} & 0 \end{smallmatrix}\right)$,

$$\eta_{X,\tilde{\xi}} \begin{pmatrix} 1+\mu\pi^k & \pi^k\sigma \\ \pi^{k+1}\tau & 1+\mu'\pi^k \end{pmatrix} = \tilde{\xi}(\pi^k(\sigma\tilde{\omega}+\zeta\pi\tau)).$$

One checks immediately in each case that $\eta_{X,\tilde{\xi}}$ defines a one-dimensional representation of $B_n$. It is clear that $\eta_{X,\tilde{\xi}}$ is invariant under the action of $T_{\tilde{X},n}$.

By Lemma 4.1.3 the irreducible representations $U$ of $T_{\tilde{X},n}B_n$, such that $U|B_n \supset \eta_{X,\tilde{\xi}}$, are of the form $\Psi_{\rho,X,\tilde{\xi}}$:

$$\Psi_{\rho,X,\tilde{\xi}}(t\cdot b) = \rho(t)\eta_{X,\tilde{\xi}}(b),$$

for $t \in T_{\tilde{X},n}$, $b \in B_n$, where $\rho \in (T_{\tilde{X},n})^*$ satisfies $\rho|T_{X,n} \cap B_n = \eta_{X,\tilde{\xi}}|T_{X,n} \cap B_n$.

THEOREM 4.2.5. (i) *The representations* $\operatorname{ind}_{T_{\tilde{X},n}B_n \uparrow K_n} \Psi_{\rho,X,\tilde{\xi}} = \mathcal{D}(\rho, X, \tilde{\xi})$ *are primitive representations of* $K_n$.

(ii) $\mathcal{D}(\rho, X, \tilde{\xi}) \sim \mathcal{D}(\rho', X', \tilde{\xi}')$ *implies* $\tilde{X}$ *and* $\tilde{X}'$ *lie in the same orbit in* $\mathfrak{g}_k$, *i.e., there exists* $k \in K_k$ *such that* $k \circ \tilde{X} = \tilde{X}'$; *in that case* $\operatorname{ind}_{N_{k+1}\uparrow K_n}\eta_X$ *and* $\operatorname{ind}_{N_{k+1}\uparrow K_n}\eta_{X'}$ *are equivalent.*

(iii) $\mathcal{D}(\rho, X, \tilde{\xi}) \sim \mathcal{D}(\rho', X, \tilde{\xi}')$ *implies* $\tilde{\xi} = \tilde{\xi}'$.

(iv) $\mathcal{D}(\rho, X, \tilde{\xi}) \sim \mathcal{D}(\rho', X, \tilde{\xi})$ *if and only if* $\rho = \rho'$.

*Proof.* Let $M$ denote the subset of $K_n$ consisting of all elements $k \in T_{\tilde{X},n}N_k$ such that $\eta^k_{X,\tilde{\xi}} = \eta_{X,\tilde{\xi}}$.

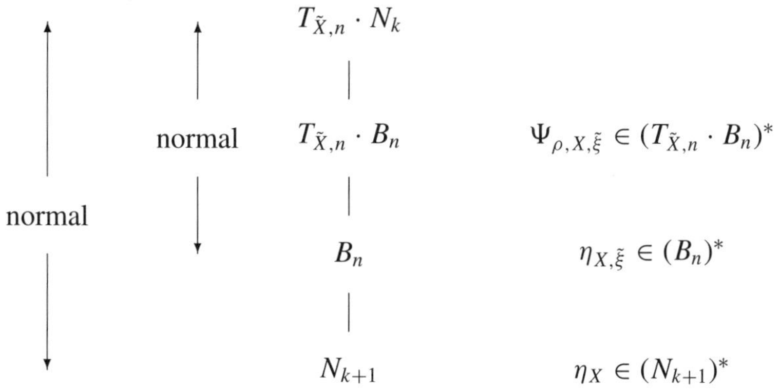


I claim that $M = T_{\tilde{X},n}B_n$ if $\xi = \xi'$ and that otherwise $M$ is empty. In any case the characters $\eta_{X,\tilde{\xi}}$ and $\eta'_{X,\tilde{\xi}}$ are fixed by the action of $T_{\tilde{X},n}B_n$. Hence $M$ is closed under (right and left) multiplication by elements of $T_{\tilde{X},n}B_n$. Hence it suffices to show that $M \cap N_k = B_n$, and for this it suffices to consider a system of representatives of $N_k/B_n$. Such a system of representatives may be chosen from elements of the form $\left(\begin{smallmatrix} 1 & 0 \\ \pi^k z & 1 \end{smallmatrix}\right) = Z$.

In the first case ($X = \left(\begin{smallmatrix}\lambda & 0\\ 0 & \lambda\end{smallmatrix}\right)$; $\lambda \in \mathcal{O}_{\mathfrak{p}}^*$), if $Z \in M$, we must have

$$\eta_{X,\xi}\left(\begin{pmatrix}1 & \pi^k\sigma\\ \pi^{k+1}\tau & 1\end{pmatrix}\right) = \eta_{X,\xi}\left(Z^{-1}\begin{pmatrix}1 & \pi^k\sigma\\ \pi^{k+1}\tau & 1\end{pmatrix}Z\right)$$
$$= \eta_{X,\xi}\left(\begin{pmatrix}1+\pi^{2k}\sigma z & \pi^k\sigma\\ \pi^{k+1}\tau & 1-\pi^{2k}\sigma z\end{pmatrix}\right),$$

i.e., $\tilde{\xi}(2\lambda\pi^{2k}\sigma z) = 1$ for all $\sigma \in \mathcal{O}_{\mathfrak{p}}/\mathfrak{p}^n$. Hence $z \in \mathfrak{p}$. In the second case, ($X = \left(\begin{smallmatrix}0 & \zeta\\ \tilde{\omega} & 0\end{smallmatrix}\right)$; $\zeta \in \mathcal{O}_{\mathfrak{p}}^*$, $\tilde{\omega} \in \mathfrak{p}$) if $Z \in M$, we must have, for $A \equiv 1 \bmod \mathfrak{p}^k$,

$$\eta_{X,\xi}\left(\begin{pmatrix}A & 0\\ 0 & A^{-1}\end{pmatrix}\right) = \eta_{X,\xi}\left(Z^{-1}\begin{pmatrix}A & 0\\ 0 & A^{-1}\end{pmatrix}Z\right)$$
$$= \eta_{X,\xi}\left(\begin{pmatrix}A & 0\\ \pi^k z\,(A^{-1}-A) & A^{-1}\end{pmatrix}\right),$$

i.e., $\tilde{\xi}(\zeta(A^{-1}-A)\pi^k z) = 1$ for all $A \equiv 1 \bmod \mathfrak{p}^k$ which implies $z \in \mathfrak{p}$. Hence in either case $Z \in B_n$. This shows that $M \subset T_{\tilde{X},n}B_n$. Hence the claim. Hence, by Lemma 4.1.2 for $\tilde{\xi} \neq \tilde{\xi}'$, $\mathrm{ind}_{B_n\uparrow T_{X,n}N_k}\eta_{X,\tilde{\xi}}$ and $\mathrm{ind}_{B_n\uparrow T_{X,n}N_k}\eta_{X,\tilde{\xi}'}$ have no irreducible components in common; and by Lemma 4.1.1

$$\mathrm{ind}_{B_n\uparrow T_{\tilde{X},n}N_k}\eta_{X,\tilde{\xi}} = \bigoplus_{\substack{\rho\in(T_{\tilde{X},n})^*\\ \rho|T_{X,n}\cap B_n=\eta_{X,\tilde{\xi}}|T_{X,n}\cap B_n}} \mathrm{ind}_{T_{\tilde{X},n}B_n\uparrow T_{\tilde{X},n}N_k}\Psi_{\rho,X,\tilde{\xi}}$$

is a decomposition of $\mathrm{ind}_{B_n\uparrow T_{\tilde{X},n}N_k}\eta_{X,\tilde{\xi}}$ into irreducible representations of $T_{\tilde{X},n}N_k$ which are inequivalent for varying $\rho$ and $\tilde{\xi}$. Since, for $\rho \in (T_{\tilde{X},n})^*$ satisfying $\rho|T_{\tilde{X},n} \cap B_n = \eta_{X,\tilde{\xi}}|T_{\tilde{X},n} \cap B_n$, the representations $\mathrm{ind}_{T_{\tilde{X},n}B_n\uparrow T_{\tilde{X},n}N_k}\Psi_{\rho,X,\tilde{\xi}}N_{k+1} \supset \eta_X$, the representations $\mathrm{ind}_{T_{\tilde{X},n}B_n\uparrow K_n}\Psi_{\rho,X,\tilde{\xi}}$ are primitive, and again by Lemma 4.1.1 and the above, they are irreducible and inequivalent for varying $\rho$ and $\tilde{\xi}$. This proves (i), (iii), and (iv). (ii) follows from the fact that $\mathcal{D}(\rho, X, \xi)|N_{k+1} \supset \eta_X$ and a similar argument. □

**4.3. Counting.** Let $X$ be a regular element in $\mathfrak{g}$, i.e., $X$ is primitive and has distinct eigenvalues. Let $T$ be a torus in $K$, i.e., the centralizer in $K$ of a regular element in $\mathfrak{g}$. By Lemma 4.2.2, each torus of $K$ is conjugate to one of the form

(i) $\left\{\begin{pmatrix}a & 0\\ 0 & a^{-1}\end{pmatrix}; a \in \mathcal{O}_{\mathfrak{p}}^*\right\}$,

(ii) $\left\{\begin{pmatrix}a & b\\ \tilde{\omega}b & a\end{pmatrix}; \tilde{\omega} \in \mathcal{O}_{\mathfrak{p}}^*,\ \left(\frac{\omega}{\mathfrak{p}}\right) = -1\right\}$,

(iii) $\left\{\begin{pmatrix}a & b\\ \tilde{\omega}b & a\end{pmatrix}; \tilde{\omega} \in \mathfrak{p},\ \tilde{\omega} = \pi^\nu\zeta,\ \pi > 0,\ \zeta \bmod \mathfrak{p},\quad \zeta \in \mathcal{O}_{\mathfrak{p}}^*/(\mathcal{O}_{\mathfrak{p}}^*)^2\right\}$.

I shall refer to these as the split, unramified, and ramified cases respectively. I have shown in §4 and in the preceding sections, that to each $T$ is associated a series of representations of $K$ parametrized by $T^*$ together with certain additive characters. In this paragraph, I will show that these representations exhaust all representations of $\mathrm{SL}(2, \mathcal{O}_{\mathfrak{p}})$. First I recall the pertinent results of §2.1.

In the notation of §2.1, if $V/\Omega$ is *unramified*, $\Phi \in \hat{\Omega}$ of conductor $\mathfrak{p}^n$, $\rho \in \hat{N}^1$ of conductor $\mathfrak{p}^n$ ($N^1$ may of course be identified with a torus of $K$), then, by Theorem 4.2.2, $\mathcal{D}(\Phi, \rho, V)$ is irreducible and (analogous to the proof of Theorem 1.8.4) $\mathcal{D}(\Phi, \rho, V) \sim \mathcal{D}(\Phi, \rho', V)$ if and only if $\rho' = \rho^{\pm 1}$. Further, by (i) of page 23 the representations $\mathcal{D}(\Phi, \rho, V)$ are primitive representations of $K_n$ having *degree* $q^{n-1}(q-1)$, $(q = N(\mathfrak{p}))$. Hence the *number* of inequivalent irreducible representations of $K_n$ of the form $\mathcal{D}(\Phi, \rho, V)$ ($V/\Omega$ unramified, $\Phi$ and $\rho$ of conductor $\mathfrak{p}^n$) is

$$\frac{1}{2}\left([N^1 : N_n] - [N^1 : N_{n-1}]\right) = \frac{1}{2}q^{n-2}(q^2-1).$$

Returning to §4.1 and §4.2, an easy computation, using Theorems 4.2.1 and 4.2.5 and the degrees of the various groups involved, shows that in the split case ($X = \left(\begin{smallmatrix}\lambda & 0\\ 0 & -\lambda\end{smallmatrix}\right)$; $\lambda \in \mathcal{O}_{\mathfrak{p}}^*$) the *number* of primitive representations of $K_n$ $(n \geq 2)$ constructed there $= (1/2)q^{n-2}(q-1)^2$ and that they have degree $= q^{n-1}(q+1)$, in the *unramified* case ($n$ even and $X = \left(\begin{smallmatrix}0 & 1\\ \tilde{\omega} & 0\end{smallmatrix}\right)$; $\tilde{\omega} \in \mathcal{O}_{\mathfrak{p}}^*$, $\left(\frac{\tilde{\omega}}{\mathfrak{p}}\right) = -1$) the *number* is $(1/2)q^{n-2}(q^2-1)$ and their *degree* is $q^{n-1}(q-1)$, and in the *ramified* case ($n \geq 2$, $X = \left(\begin{smallmatrix}0 & \zeta\\ \tilde{\omega} & 0\end{smallmatrix}\right)$; $\zeta \in \mathcal{O}_{\mathfrak{p}}^*$, $\tilde{\omega} \in \mathfrak{p}$) the *number* is $4q^{n-1}$ and their *degree* is $q^{n-2}(q^2-1)/2$. Thus (comparing degrees) I have constructed

$$\frac{1}{2}q^{n-2}(q-1)^2 + \frac{1}{2}q^{n-2}(q^2-1) + 4q^{n-1} = q^n + 3q^{n-1}$$

inequivalent primitive representations of $K_n$. However, by [12] the number of conjugacy classes of $K_n$ minus the number of conjugacy classes of $K_{n-1}$ $(n \geq 2) = q^n + 3q^{n-1}$. Finally the representations of $K_1$ are well known. See [2], [14]. Thus I have obtained all irreducible (unitary) representations of $\mathrm{SL}(2, \mathcal{O}_{\mathfrak{p}})$.

---

## REFERENCES


[1] P. Cartier, Representations of $\mathrm{SL}_2(r)$ and Quadratic forms, In *Seminar on representations of Lie groups*, vol. 1. Institute for Advanced Study, Princeton, NJ, 1965.

[2] G. F. Frobenius, Über gruppencharactere, *Berliner Sitzungsberichte,* 1896, pp. 985–1021.

[3] I. M. Gelfand and M. I. Graev, The group of matrices of second order with coefficients in a locally compact fields, *Uspehi Mati. Nauk,* 1963, pp. 29–99.

[4] I. M. Gelfand, M. I. Graev and I. I. Pyatetskii-Shapiro, *Representation theory and automorphic functions*, Translated from the Russian by K. A. Hirsch, W. B. Saunders Co., Philadelphia, Pa., 1969.

[5] Harish-Chandra, *Harmonic analysis on reductive p-adic groups*, Springer-Verlag, Berlin, 1970, Notes by G. van Dijk, Lecture Notes in Mathematics, vol. 162.

[6] E. Hecke, Zur Theorie der elliptischen Modulfunktionen, *Math. Ann.* **97** (1926), 210–242.

[7] R. Howe, $\theta$-series and invariant theory *Automorphic forms, representations and L-functions (Proc. Sympos. Pure Math., Oregon State Univ., Corvallis, Ore., 1977), Part 1*, Proc. Sympos. Pure Math., XXXIII, 275–285, Amer. Math. Soc., Providence, R.I., 1979.

[8] H. D. Kloosterman, The behaviour of general theta functions under the modular group and the characters of binary modular congruence groups, i–ii, *Annals of Mathematics*, **47** (1949), 317–447.

[9] H. Maass, Über eine neue Art von nicht analytischen automorphen Funktionen und die bestimmung von dirichletscher reihen durch funktional gleichunger, *Math. Ann.*, **121** (1949), 141–183.

[10] F. I. Mautner, Spherical functions over $p$-adic fields. II, *Amer. J. Math.*, **86** (1964), 171–200.

[11] ———,The $p$-adic principal series and its relation to hecke operators, In *Seminar on representations of Lie groups*, vol. 4. Princeton, NJ, 1965.

[12] H. W. Praetorius, Die charactere der modulargruppen der stufe $q^2$, *Abh. a. d. Math. Seminar d. hamburgischen Univ.* **9** (1933), 365–394.

[13] H. Rohrbach, Die character der binären konguence gruppen mod $p^2$, *Schriften d. math. Seminars d. Univ. Berlin* **1** (1932), 33–94.

[14] J. Schur, Undtersuchungen über die darstellung der endlichen gruppen durch gebrochene lineare substitution, *J. Reine Angew. Math.* **132** (1907), 85–137.

[15] H. Spies, Die darstellung der inhomogen modulargruppen mod $q^n$ durch die ganzen modulformen gerader dimension, *Math. Ann.* **111** (1935), 329–354.

[16] A. Weil, Sur certains groupes d'opérateurs unitaires, *Acta Math.* **111** (1964), 143–211.

## CHAPTER 2

# THETA-10

By Jeffrey Adams

---

*It is a pleasure to dedicate this paper to Joe Shalika, my freshman number theory professor.*

**1. Introduction.** The symplectic group $Sp(4, F)$ has a particular interesting unitary representation for $F$ a finite or local field. When $F$ is finite this representation was denoted $\Theta_{10}$ [30], and somehow the name has stuck. We discuss this representation in some detail in the case of $\mathbb{R}$.

This material is mostly well known, at least to the experts, and this paper is intended as a reference for non-specialists. It also serves as an introduction to some of the machinery of the Arthur conjectures as discussed in [5]. It has existed since 1990 as an informal set of notes.

**2. Background.** Here are some references for the basic material under discussion: admissible representations of real reductive groups, Harish-Chandra modules, reductive dual pairs, the Langlands classification, and $L$ and Arthur packets. The references are chosen for their accessibility rather than being the primary sources.

A good introduction to representation theory of real groups is A. Knapp's *Representation Theory of Semisimple Groups, An Overview Based on Examples* [21]. A quicker guide to the subject is a set of lecture notes by A. Knapp and P. Trapa from the 1998 Park City Conference [24]. The proceedings of the 1996 Edinburgh Conference [8] include a number of expository articles, including one on the Langlands program by A. Knapp [22].

We make repeated use of the Langlands classification (sometimes referred to as the Langlands–Knapp–Zuckerman classification) and its equivalent form, the Vogan classification. A summary of the statements, with references to more details, may be found in Sections 3 and 4 of D. Vogan's "The unitary dual of $G_2$" [34]. Some of the methods used here are discussed in "The Kazhdan-Lusztig conjecture for real reductive groups" by D. Vogan [36].

For more details on some of the technology used, the basic references are *Representations of Real Reductive Groups* by D. Vogan [32] and *Cohomological Induction and Unitary Representations* by A. Knapp and D. Vogan [25]. The introduction to [25] has an overview of the cohomological construction.

---

Manuscript received October 9, 2002

For the Langlands program, in addition to [22] cited above, see the article by A. Borel [9, Volume 2] in the proceedings of the 1977 Corvallis Conference [10]. The real case is explained in Section 11, pages 46–48. For Arthur's conjectures see *The Langlands Classification and Irreducible Characters for Real Reductive Groups* by J. Adams, D. Barbasch, and D. Vogan [5], especially the Introduction.

For basics of reductive dual pairs see "$\theta$-series and Invariant Theory" by R. Howe [15] and "Examples of dual reductive pairs" by S. Gelbart [13]. For the relationship with $\theta_{10}$ see "A counterexample to the 'generalized Ramanujan conjecture' for (quasi-) split groups" by R. Howe and I. Piatetski–Shapiro [16]. Another good reference is "A brief survey on the theta correspondence" by D. Prasad [28]. For the connection between dual pairs and $L$ and Arthur packets see "L-Functoriality for Dual Pairs" by J. Adams [1].

The representation $\theta_{10}$ over a finite field is discussed in [30]. It is one of the first examples of a cuspidal unipotent representation [27]. By a standard construction this also gives $\theta_{10}$ over a $p$-adic field and a corresponding cuspidal automorphic representation. Alternatively $\theta_{10}$ over any local field may be constructed via the dual pair correspondence with an anisotropic orthogonal group of rank 2. All of this is discussed in [16, §1].

**3. Notation.** In this section we establish some notation and conventions to be used throughout. Much of this is standard and the reader may want to skip ahead to Section 4 and refer back to this when necessary.

Let $V$ be a four-dimensional real vector space with a non-degenerate symplectic form $<\,,\,>$. Let $G = Sp(V) = Sp(V, <\,,\,>)$ be the isometry group of $<\,,\,>$. Choose a basis of $V$ such that $<\,,\,>$ is given by $J = \left(\begin{smallmatrix} 0 & I_2 \\ -I_2 & 0 \end{smallmatrix}\right)$. Then $Sp(V)$ is isomorphic to $G = Sp(4, \mathbb{R})$, the group of matrices $g$ satisfying $gJ\,{}^tg = J$. Thus $G$ consists of the matrices $\left(\begin{smallmatrix} A & B \\ C & D \end{smallmatrix}\right)$ where

$$A\,{}^tB = {}^t(A\,{}^tB) \quad C\,{}^tD = {}^t(C\,{}^tD) \quad A\,{}^tD - B\,{}^tC = I_2.$$

Let $\mathfrak{g}_0$ be the Lie algebra of $G$. Thus $\mathfrak{g}_0$ consists of matrices $X$ satisfying $XJ + J\,{}^tX = 0$, i.e., $X = \left(\begin{smallmatrix} A & B \\ C & -{}^tA \end{smallmatrix}\right)$ where $B$ and $C$ are symmetric.

We will make repeated use of the Cartan subgroups of $G$. There are four conjugacy classes of Cartan subgroups, isomorphic to $\mathbb{R}^* \times \mathbb{R}^*$, $\mathbb{R}^* \times S^1$, $\mathbb{C}^*$ and $S^1 \times S^1$ respectively. We choose explicit representatives $H^s$, $H^\ell$, $H^{sh}$ and $H^c$ as follows ($\ell$ and $sh$ stand for long and short Cayley transforms respectively [21, page 417]).

Let $H^s(x, y) = diag(x, y, \frac{1}{x}, \frac{1}{y})$ $(x, y \in \mathbb{R}^*)$. This gives the Cartan subgroup $H^s$.

For $2 \times 2$ matrices $A$, $B$ and $\theta \in \mathbb{R}$ let

$$M(A,B) = \begin{pmatrix} A_{1,1} & & A_{1,2} & \\ & B_{1,1} & & B_{1,2} \\ A_{2,1} & & A_{2,2} & \\ & B_{2,1} & & B_{2,2} \end{pmatrix} \tag{3.1}$$

$$t(\theta) = \begin{pmatrix} \cos(\theta) & \sin(\theta) \\ -\sin(\theta) & \cos(\theta) \end{pmatrix}. \tag{3.2}$$

For $x \in \mathbb{R}^*, \theta \in \mathbb{R}$ let $h^\ell(x,\theta) = M(diag(x, \frac{1}{x}), t(\theta))$ and $h^{sh}(x,\theta) = diag(xt(\theta), \frac{1}{x}t(\theta))$. This gives $H^\ell \simeq \mathbb{R}^* \times S^1$ and $H^{sh} \simeq \mathbb{C}^*$ respectively.

We also need the Lie algebras. Let $X^s(x, y) = diag(x, y, -x-y)$; this gives the Lie algebra $\mathfrak{h}_0^s$ of $H^s$.

Let $T(y) = \left(\begin{smallmatrix} & y \\ -y & \end{smallmatrix}\right)$. Let $X^\ell(x, y) = diag(xI_2 + T(y), -xI_2 + T(y))$ and $X^{sh}(x, y) = M(diag(x, -x), T(y))$. Finally let $X^c(x, y) = M(T(x), T(y))$. This gives the Lie algebras $\mathfrak{h}_0^\ell$, $\mathfrak{h}_0^{sh}$, and $\mathfrak{h}_0^c$ respectively.

Write $(x, y)$ for the element of $\mathrm{Hom}(\mathfrak{h}_0^c, \mathbb{C})$ taking $h^c(x', y')$ to $i(xx' + yy')$. Then $\alpha = (0, 2)$, $\beta = (1, -1)$, $\gamma = (1, 1)$, and $\mu = (2, 0)$ are a set $\Delta^+(\mathfrak{g}, \mathfrak{h}^c)$ of positive roots of $\mathfrak{h}^c$ in $\mathfrak{g}$, and $\rho$ (one-half the sum of the positive roots) is $(2, 1)$.

Let $\theta(g) = {}^t g^{-1}$; this is a Cartan involution of $G$, and let $K = G^\theta \simeq U(2)$ be a maximal compact subgroup of $G$. Let $\theta(X) = -{}^t X$ be the corresponding Cartan involution of $\mathfrak{g}_0$, and let $\mathfrak{k}_0 = \mathfrak{g}_0^\theta$, $\mathfrak{p}_0 = \mathfrak{g}_0^{-\theta}$. For $\star = c, \ell, sh$, or $s$ write $H^\star = T^\star A^\star$ as usual, with $T^\star = H^\star \cap K$. Let $\Delta^+(\mathfrak{k}, \mathfrak{h}^c) = \Delta^+(\mathfrak{g}, \mathfrak{h}^c) \cap \Delta(\mathfrak{k}, \mathfrak{h}^c) = \{\beta\}$, $\Delta^+(\mathfrak{p}, \mathfrak{h}^c) = \Delta^+(\mathfrak{g}, \mathfrak{h}^c) \cap \Delta(\mathfrak{p}, \mathfrak{h}^c) = \{\alpha, \gamma, \delta\}$. Let $\mathfrak{p}^\pm$ be the abelian subalgebra of $\mathfrak{p}$ corresponding to $\pm\Delta^+(\mathfrak{p}, \mathfrak{h}^c)$.

The irreducible representations of $K$ are parametrized by highest weights with respect to $\Delta^+(\mathfrak{k}, \mathfrak{h}^c)$. For $x \geq y, x, y \in \mathbb{Z}$ let $\mu(x, y)$ be the irreducible finite dimensional representation with highest weight $(x, y)$. By a highest weight module for $\mathfrak{g}$ we mean a module with a vector annihilated by $\mathfrak{p}^-$. For more on highest weight modules in general see [31].

We parametrize infinitesimal characters for $\mathfrak{g}$ by elements of $\mathrm{Hom}(\mathfrak{h}^c, \mathbb{C})$ via the Harish-Chandra homomorphism. See [12] or [24, Lecture 5] for the definitions of infinitesimal character and Harish-Chandra homomorphism. Write $\chi(x, y)$ for the infinitesimal character corresponding to $(x, y)$ $(x, y \in \mathbb{C})$.

Let $\lambda_0 = (1, 0)$, and let $\mathfrak{q} = \mathfrak{q}(\lambda_0)$ be the associated $\theta$-stable parabolic [32, Definition 5.2.1]. Thus $\mathfrak{q} = \mathfrak{l} \oplus \mathfrak{u}$ with $\mathfrak{l} \simeq \mathbb{C} \oplus \mathfrak{sl}(2, \mathbb{C})$. With the usual notation we have $\Delta(\mathfrak{l}, h^c) = \{\pm\alpha\}$ and $\Delta(\mathfrak{u}, h^c) = \{\beta, \gamma, \mu\}$, $\rho_{\mathfrak{l}} = (0, 1)$ and $\rho_{\mathfrak{u}} = (2, 0)$.

Let $L$ be the normalizer of $\mathfrak{q}$ as usual; $L \simeq U(1) \times SL(2, \mathbb{R})$. Note that $L \supset H^\ell$; in fact $L$ is the centralizer of $T^\ell$. Suppose $\lambda \in \mathrm{Hom}(\mathfrak{t}^\ell, \mathbb{C})$ is the differential of a character of $T^\ell$; equivalently $\lambda$ is the restriction of $(k, 0)$ for some $k \in \mathbb{Z}$. We normalize the derived functor construction as in [32, 6.3.1] and [25, 5.3b]; let $A_{\mathfrak{q}}(\lambda)$ be the $(\mathfrak{g}, K)$-module $R^1_{\mathfrak{q}}(\lambda - \rho(\mathfrak{u}))$. As usual we view $\lambda - \rho(\mathfrak{u})$ as an $(\mathfrak{l}, L \cap K)$-module. This has infinitesimal character $\chi(\lambda + \rho_{\mathfrak{l}}) = (k, 1)$.

This is an irreducible non-tempered unitary representation if $k \geq 1$, i.e., the matrix coefficents are in $L^{2+\epsilon}(G)$ [21, page 198] or [24, Lecture 8]. It has lowest $K$-type $\mu(k+1, 1)$ and has non-zero $(\mathfrak{g}, K)$-cohomology if $k \geq 2$. See [32, Definition 1.2.10, page 52] for the notion of lowest $K$-type. See [37] for a discussion of representations with $(\mathfrak{g}, K)$-cohomology, and in particular Theorem 4.6, p. 232 for the classification of these representations.

Let $\mathcal{O}$ be the nilpotent orbit in $\mathfrak{g}_0$ through the element $(\begin{smallmatrix} 0 & B \\ 0 & 0 \end{smallmatrix})$ with $B = diag(1, 0)$. This is a 4-dimensional orbit. Its closure consists of itself and the 0-orbit.

*Remark* 2.1. We take this oppportunity to clarify some notation. The algebraic group $Sp(2n)$ is simply connected, and is a two-fold cover of its adjoint group $PSp(2n)$. The group $PSp(2n, \mathbb{R})$ usually refers to the real points of the algebraic group $PSp(2n)$, or equivalently the real points of the complex group $PSp(2n, \mathbb{C}) = Sp(2n, \mathbb{C})/\pm I$. This group is disconnected as a real Lie group, and has identity component $Sp(2n, \mathbb{R})/\pm I$. Occasionally (for example in [36], pp. 251–254) $PSp(2n, \mathbb{R})$ refers to $Sp(2n, \mathbb{R})/\pm I$.

Also consider the symplectic similitude group $GSp(2n)$. This has center $Z \simeq G_m$ and the adjoint group $GSp(2n)/Z$ is isomorphic to $PSp(2n)$. Furthermore the set of real points of the adjoint group is isomorphic to $GSp(2n, \mathbb{R})/Z(\mathbb{R})$. So the notation $PGSp(2n, \mathbb{R})$ is unambiguous, and perhaps for this reason some authors prefer to use this group.

Simliar remarks hold over any local field.

**4. $\Theta_{10}$.** We give a number of descriptions of $\Theta_{10}$. For unexplained notation see Section 3.

THEOREM 4.1. *There is a unique irreducible representation $\Theta_{10}$ of $Sp(4, \mathbb{R})$ satisfying the following equivalent conditions:*

1. *$\Theta_{10}$ is a highest weight module for $\Delta^+(\mathfrak{k}, \mathfrak{h}^c) \cup \Delta(\mathfrak{p}_0^-, \mathfrak{h}^c)$, with highest weight $(2, 2)$;*
2. *$\Theta_{10}$ is the endpoint of the continuous spectrum in the parametrization [38] of unitary highest weight modules of $G$ with one-dimensional lowest $K$-types;*
3. *In the Langlands-Knapp-Zuckerman classification, $\Theta_{10}$ is a limit of holomorphic discrete series [21, Chapter 12, §7, page 460], with infinitesimal character $\chi(1, 0)$. Thus $\Theta_{10} = \Theta^G(\lambda, C, \chi)$ in the notation of [23, Theorem 1.1], with $\lambda = (1, 0)$, $C$ the holomorphic discrete series chamber (given by simple roots $\alpha, \beta$), and $\chi$ the trivial character of the center of $G$;*
4. *$\Theta_{10} = \psi_\alpha(\pi')$ where $\pi'$ is the holomorphic discrete series with infinitesimal character $\chi(\rho)$, and $\psi_\alpha$ is the translation functor to the $\alpha$-wall (cf. [32, 8.2.6]);*
5. *$\Theta_{10}$ has infinitesimal character $\chi(1, 0)$ and contains the $K$-type $\mu(2, 2)$;*
6. *In the Vogan-Zuckerman classification, $\Theta_{10} = \overline{X}(\mathfrak{q}, \delta \otimes \nu)(\mu(2, 2))$, where $\delta \otimes \nu \in H^{\ell^*}$ is given by $\delta \otimes \nu(h^\ell(\theta, x)) = e^{-i\theta} sgn(x)$ ([32], Definition 6.5.11);*
7. *$\Theta_{10}$ is the direct summand of $Ind_{MAN}^G(\pi_2 \otimes 1 \otimes 1)$ containing $\mu(2, 2)$, where $MA \simeq SL(2, \mathbb{R}) \times \mathbb{R}^*$ is the centralizer of $A^\ell$ and $\pi_2$ is the lowest holomorphic discrete series representation of $SL(2, \mathbb{R})$;*
8. *$\Theta_{10}$ has $K$-spectrum $\Sigma_{1 \le m \le n} n\, \mu(2m, 2n)$;*
9. *$\Theta_{10}$ corresponds to the $sgn$ representation of $O(2)$ in the dual pair correspondence for $(O(2), Sp(4, \mathbb{R}))$;*

10. $\Theta_{10}$ *is the direct summand of* $A_{\mathfrak{q}}(0)$ *containing* $\mu(2,2)$*; equivalently it is the tempered summand of this representation;*

11. $\Theta_{10}$ *is tempered, has infinitesimal character* $\chi(1,0)$ *and wave-front set* $\overline{\mathcal{O}}$ *[17].*

$\Theta_{10}$ *is unitary, tempered, and has Gelfand-Kirillov dimension* 2 *(cf. [35]).*

*Remark* 4.2. There is an outer automorphism of $Sp(4,\mathbb{R})$ that takes $\Theta_{10}$ to its contragredient $\Theta_{10}^*$. There is no intrinsic way to choose one of these two representations. The explict choices of the previous section enable us to make such a choice and denote it $\Theta_{10}$.

If we replace $Sp(4,\mathbb{R})$ with $G = PSp(4,\mathbb{R})$ (cf. Remark 2.1) there is a unique representation, whose restriction to the identity component $G^0$ of $G$ is $\Theta_{10} \oplus \Theta_{10}^*$ (pushed down to $G^0 = Sp(4,\mathbb{R})/\pm I$).

*Proof.* Most of these facts may be found in the literature, generally as special cases. We sketch the arguments.

The equivalence of (1) and (2) is a matter of reading the definitions of [38]. Thus the endpoint of the continuous spectrum in (2) is a limit of discrete series with highest weight (2, 2). The relation between the infinitesimal character $\chi(x,y)$ and highest weight $\tau = (x', y')$ is given by $(x', y') = (x+1, y+2)$, so $\chi = \chi(1,0)$. Note for later use that this implies the only highest weight modules with this infinitesimal character have highest weight (1, 1) or (2, 2). Also note that the series [38] of unitary highest weight modules for the universal cover of $G$ has four isolated points, two of which correspond to representations which factor to $G$.

The data $(\lambda, C, \chi)$ [23] giving the limit of discrete series described by (2) are the infinitesimal character, the "type" of discrete series (holomorphic) and the central character ($\mu(2,2)$ is trivial on the center of $G$). Thus (2) and (3) are equivalent. Item (4) is simply the definition of limit of discrete series in terms of translation functors [23, Section 1].

Let $\pi$ be any representation with lowest $K$-type $\mu(2,2)$ or $\mu(2,0)$ and infinitesimal character $\chi(1,0)$. The $\theta$-stable data attached to this data [32, Definition 5.4.8 and Corollary 5.4.9] are computed as follows. The element $\lambda$ of [32, Definition 5.3.22] is $(1,0)$, so the $\theta$-stable parabolic is $\mathfrak{q}$ (cf. Section 3). The $L \cap K$-type $\pi^0$ has highest weight $(-1,1)$, and so $\delta$ is this weight restricted to $H^c$ as stated. Then $\nu$ is determined in this case by the infinitesimal character and is trivial.

Thus $\pi$ is a subquotient of the standard module $X = X(\mathfrak{q}, \delta \otimes \nu)$ [32, Definition 6.5.2, page 392]. Now $X$ is computed as follows [32, Definition 6.5.1]. We have $L \simeq U(1) \times SL(2,\mathbb{R})$ and $X_L$ is a principal series of $L$ with odd K-types (because $\delta_{T^\ell \cap A^\ell}$ is the *sgn* character) and infinitesimal character 0 (because $d\nu = 0$) on $SL(2,\mathbb{R})$. Therefore $X_L$ is the sum of the limits of discrete series of $SL(2,\mathbb{R})$ tensored with $\delta|_{U(1)}$, which is $e^{-i\theta}$ on $U(1)$. Thus by [32, 6.5.10] $X(\mathfrak{q}, \delta \otimes \nu)$ is the direct sum of two pieces, each with a unique irreducible summand. By [32,

6.5.10] again this standard module has two lowest K-types $\mu(2,2)$ and $\mu(2,0)$ each of which is the unique lowest K-type of a summand. Thus $\Theta_{10}$ is the summand of $X$ containing $\mu(2,2)$, i.e., $\overline{X}(\mathfrak{q},\delta\otimes\nu)(\mu(2,2))$.

Thus there are unique irreducible representations with infinitesimal character $\chi(1,0)$ and lowest K-types $\mu(2,2)$ and $\mu(2,0)$ respectively, and they are distinct. Now let $\pi$ be any irreducible representation with this infinitesimal character and containing the K-type $\mu(2,2)$ (not necessarily lowest). By definition of the ordering of K-types and elementary K-type considerations, the only possible lower K-types are $\mu(2,0)$, which is ruled out by the above discussion, or trivial. The spherical principal series with infinitesimal character $\rho$ contains the holomorphic discrete series as a constituent (see the Appendix). Translated to infinitesimal character $\chi(1,0)$ we obtain the standard module $X$, so by (3) $X$ contains $\Theta_{10}$. Since $X$ contains $\mu(2,2)$ with multiplicity one we see the irreducible spherical constituent does not contain $\mu(2,2)$. This proves the representation defined by (5) is unique and equal to that defined by (1–4) and (6).

Item (7) is equivalent to (6) by [32, 6.6.2, 6.6.12–14]. The representation defined by (3) has the K-spectrum indicated in (8) by the Blattner formula [14]. Conversely if an irreducible representation $\pi$ has this K-spectrum it is clearly a highest weight module (since operators from $\mathfrak{p}_0^-$ lower weights), so this representation is $\Theta_{10}$ by (1).

For (9), the $sgn$ representation of $O(2)$ corresponds to an irreducible representation of $Sp(4,\mathbb{R})$ with highest weight $(2,2)$ [20], which is isomorphic to $\Theta_{10}$ by (1). See Section 5.

For (10) we compute the K-spectrum of $A_{\mathfrak{q}}(0)$ by the Blattner formula. We see it has K-types $\mu(1,1), \mu(2,2), \mu(3,1),\dots$, and as in (8) any constituent of this representation has a highest weight. Considering infinitesimal characters as in (2) we immediately see $A_{\mathfrak{q}}(0)$ has two constituents, with highest weights $(1,1)$ and $(2,2)$. By (2) the term with highest weight $(1,1)$ is not tempered. On the other hand $A_{\mathfrak{q}}(0)$ is completely reducible. This proves (10) is equivalent to (1).

For (11) $\Theta_{10}$ has the indicated wave-front set, and any representation with this wave-front set is a highest weight module. Using the infinitesimal character and the discussion in (2) we see the only other possible highest weight is $(1,1)$. This representation has the same wave-front set, but is not tempered. This proves (11).

This completes the proof of the theorem. □

**5. Reductive dual pairs.** We now list some representations of $Sp(4,\mathbb{R})$ coming from dual pairs. Fix a non-trivial unitary character of $\mathbb{R}$, and let $\omega_n$ be the corresponding oscillator representation of $Sp(2n,\mathbb{R})$. Fix $n$, and let $<\,,\,>_i$ be a set of representatives of the isomorphism classes of symmetric bilinear forms of dimension $2n$. There are $2n+1$ such forms, one each of signature $(p,q)$ $(p+q=2n)$. Let $G_i$ be isometry group of $<\,,\,>_i$. We write $G_i = G_{p,q}$ if $<\,,\,>_i$ has signature $(p,q)$. Then for all $p,q$ $(G_{p,q}, Sp(4,\mathbb{R}))$ is a reductive dual pair in $Sp(8n,\mathbb{R})$. Even though $G_{p,q}\simeq G_{q,p}$ this notation keeps track not just of $G$ but of the form and the embedding (which it is necessary to choose consistently). See [3] for details. Let $\iota : G_{p,q} \times Sp(4, \mathbb{R}) \to Sp(8n, \mathbb{R})$ be the corresponding map.

Let $\widetilde{Sp}(2n, \mathbb{R})$ be the non-trivial two-fold cover of $Sp(2n, \mathbb{R})$. If $H$ is a subgroup of $Sp(2n, \mathbb{R})$, let $\tilde{H}$ denote its inverse image in $\widetilde{Sp}(2n, \mathbb{R})$. In particular $\tilde{U}(n)$ is isomorphic to the $det^{\frac{1}{2}}$ cover of $U(n)$:

$$\tilde{U}(n) \simeq \{(g, z) \mid g \in U(n), z \in \mathbb{C}^*, det(g) = z^2\}.$$

Consequently if $(O(p, q), Sp(2n, \mathbb{R}))$ is a dual pair in $Sp(2n(p+q), \mathbb{R})$, then $\tilde{O}(p, q) \simeq \{(g, z) \mid det(g)^n = z^2\}$. If $n$ is even this covering splits over $O(p, q)$ by the map $g \to (g, det(g)^{n/2})$.

The covering of $Sp(2n(p+q), \mathbb{R})$ splits over $Sp(2n, \mathbb{R})$ if and only if $p+q$ is even, in which case the splitting is unique. Thus if $p+q$ and $n$ are even we obtain a map $\gamma : G_{p,q} \times Sp(2n\ R) \to \widetilde{Sp}(2n(p+q), \mathbb{R})$.

The oscillator representation $\omega$ of $Sp(8n, \mathbb{R})$ now yields a bijection between subsets of the admissible duals of $\tilde{G}_{p,q}$ $(p+q=2)$ and $\tilde{S}p(4, \mathbb{R})$. We are in the setting of the previous paragraph and via $\gamma$ we obtain a representation correspondence between representations of $G_{p,q}$ and $Sp(4, \mathbb{R})$.

Let $\pi^+[p, q]$ be the irreducible representation of $Sp(4, \mathbb{R})$ corresponding (via the embedding coming from the form of signature $(p, q)$) to the trivial representation of $G_{p,q}$. Similary $\pi^-[p, q]$ corresponds to the $sgn$ representation. The following identifications are not difficult to deduce from the literature. We use the notation of the Appendix. All representations have infinitesimal character $(1, 0)$ and are unitary.

(1) $\pi^+[2, 0]$: highest weight module, with lowest $K$-type $\mu(1, 1)$; non-tempered; $\overline{C}_\alpha$,

(2) $\pi^+[0, 2] = \pi^+[2, 0]^*$; LKT $\mu(-1, -1)$; $\overline{D}_\alpha$,

(3) $\pi^-[2, 0] = \Theta_{10}$; LKT $\mu(2, 2)$; $\overline{I}_\alpha$,

(4) $\pi^-[0, 2] = \Theta_{10}^*$; LKT $\mu(-2, -2)$; $\overline{L}_\alpha$,

(5) $\pi^+[1, 1]$: spherical; $\overline{B}_\alpha$,

(6) $\pi^-[1, 1]$: LKT $(1, -1)$; $\overline{H}_\alpha$,

(7) $\pi^+[4, 0] = \Theta_{10}$,

(8) $\pi^+[0, 4] = \Theta_{10}^*$,

(9) $\pi^+[2, 2] = \pi^+[1, 1]$.

Here is a brief justification for this table. In each example the lowest $K$-type of the representation of $Sp(4, \mathbb{R})$ is known by Howe's theory of joint harmonics [19, Proposition 3.4; see 4, Proposition 1.4, page 7]. Also the infinitesimal character is determined; see [29] or [1, page 107].

In (1–4) the orthogonal groups $O(2, 0)$ and $O(0, 2)$ are compact, and by [18] the corresponding representations of $Sp(4, \mathbb{R})$ are highest weight modules. These are determined by their lowest $K$-type and infinitesimal character. Similar comments apply to cases (7) and (8). In cases (5) and (9) the representation of $Sp(4,\mathbb{R})$ is spherical by the preceding paragraph and is determined by its infinitesimal character, which determines the representation. Finally the representation in (6) is determined by its lowest $K$-type and infinitesimal character.

**6. Packets.** We now describe some L and Arthur packets for $G$.

Let ${}^LG$ be the L-group for $G$ [26]. The identity component ${}^\vee G$, sometimes denoted ${}^LG^0$, is a complex connected group with root system of type $B_2 \simeq C_2$, and is adjoint since $G$ is simply connected. Therefore ${}^\vee G$ is isomorphic to the special orthogonal group on a complex 5-dimensional space. We define ${}^\vee G = SO(5,\mathbb{C})$ with respect to the standard inner product. Then ${}^LG$ is a trivial extension of ${}^\vee G$ by $\Gamma = Gal(\mathbb{C}/\mathbb{R})$, i.e., ${}^LG \simeq {}^\vee G \times \Gamma$. As usual we may drop the extension since it is trivial, and write ${}^LG = {}^\vee G$.

Let $W_\mathbb{R}$ be the Weil group of $\mathbb{R}$ [9]. This is the unique non-split central extension of $Gal(\mathbb{C}/\mathbb{R})$ by $\mathbb{C}^*$. It is given by generators and relations as $W_\mathbb{R} =< \mathbb{C}^*, j >$, where $j^2 = -1$ and $jzj^{-1} = \bar{z}$.

**6.1. *L*-packets.** We now define the L-homomorphism $\phi : W_\mathbb{R} \to {}^LG$ whose corresponding L-packet $\Pi_\phi$ contains $\Theta_{10}$. We begin by describing the more general L-homomorphism giving a general L-packet of discrete series. For $k, \ell \in \mathbb{Z}$ let $\phi(re^{i\theta}) = diag(t(k\theta), t(\ell\theta), 1)$ (cf. 3.2). To extend this to a homomorphism of $W_\mathbb{R}$ we need to choose $x = \phi(j)$ such that

$$x^2 = diag((-1)^k, (-1)^k, (-1)^\ell, (-1)^\ell, 1)$$

and such that conjugation by $x$ acts by inverse on $\phi(\mathbb{C}^*)$. This forces $x = diag(1,-1,1,-1,1)t$ with $t \in \phi(\mathbb{C}^*)$, which in turn implies $k, \ell$ are even. Then $\phi$ is an admissible homomorphism [26].

*Definition* 6.1. For $k, \ell \in \mathbb{Z}$ let $\phi_{k,\ell}$ be the admissible homomorphism from $W_\mathbb{R}$ to ${}^LG$ defined by:

$$\phi_{k,\ell}(re^{i\theta}) = diag(t(2k\theta), t(2\ell\theta), 1), \qquad \phi_{k,\ell}(j) = J.$$

For $k \neq \ell$ both non-zero the image of $\phi_{k,\ell}$ is contained in no proper Levi subgroup of ${}^LG$, and the corresponding L-packet consists of four discrete series representations with infinitesimal character $\chi(k,\ell)$. Up to conjugation we may assume $k > \ell > 0$.

The $L$-homomorphism giving $\Theta_{10}$ is $\phi_{1,0}$:

*Definition* 6.2.

$$\phi(re^{i\theta}) = diag(t(2\theta), 1, 1, 1), \qquad \phi(j) = diag(1,-1,1,-1,1)$$

The next proposition follows immediately.

PROPOSITION 6.3. *The L-packet $\Pi_\phi$ contains $\Theta_{10}$, and consists of four limits of discrete series representations.*

We describe this $L$-packet in more detail. The L-packet $\Pi_{\phi_{2,1}}$ contains four discrete representations, with Harish-Chandra parameters $(2, \pm 1)$, $(\pm 1, -2)$. Each of these representations is non-zero when we translate to the $\alpha$-wall, with infinitesimal character $\chi(1, 0)$, and these are the four limits of discrete series in $\Pi_\phi$. They have lowest $K$-types $\mu(2, 2)$, $\mu(2, 0)$, $\mu(0, -2)$, and $\mu(-2, -2)$ respectively. We can think of them as having Harish-Chandra parameters $(1, 0^+)$, $(1, 0^-)$, $(0^+, -1)$, and $(0^-, -1)$, respectively. Here $0^\pm$ indicates the limit; for example $(1, 0^+)$ is the translation to the wall of the discrete series representation with Harish-Chandra parameter $(2, 1)$.

Write $\gamma$ for the limit of discrete series representation with lowest $K$-type $\mu(2, 0)$. Then

$$\prod_\phi = \{\Theta_{10}, \gamma, \gamma^*, \Theta_{10}^*\}.$$

The centralizer $S_\phi$ of $\phi$ in ${}^\vee G$ is computed as follows. The centralizer of $\phi(\mathbb{C}^*)$ is isomorphic to $\mathbb{C}^* \times SO(3, \mathbb{C})$, and $\phi(j)$ acts on the centralizer as an involution. The fixed point set of this action is isomorphic to $\mathbb{Z}/2\mathbb{Z} \times O(2)$. The component group $\mathbf{S}_\phi$ is $\mathbb{Z}/2\mathbb{Z} \times \mathbb{Z}/2\mathbb{Z}$. The four characters of this group correspond to the four representations in the L-packet $\Pi_\phi$, with the trivial representation corresponding to $\gamma$. Note that in [2] and [5] we discussed the larger super-packet containing $\Theta_{10}$, which contains representations of the groups $Sp(p, q)$. In this case this is not necessary: we have obtained a bijection $\Pi_\phi \to \widetilde{S_\phi}\hat{}$.

**6.2. Arthur packets.** We describe some Arthur packets of unipotent representations of $Sp(4, \mathbb{R})$ [5]. There is some overlap with [7, Examples 1.4.2–3] and [5, Example 27.14].

We describe unipotent orbits by their Jordan form [11]: $SO(5, \mathbb{C})$ has four unipotent orbits: $\mathcal{O}(5)$, $\mathcal{O}(1, 1, 1, 1, 1)$, $\mathcal{O}(3, 1, 1)$, and $\mathcal{O}(2, 2, 1)$. That is if $\lambda$ is a partition of 5 then $\lambda$ determines a nilpotent orbit in $GL(5, \mathbb{C})$. If the multiplicity of every even entry of $\lambda$ is even, this orbit interesects $SO(5, \mathbb{C})$ and determines a unipotent orbit of $SO(5, \mathbb{C})$. We consider parameters $\psi$ corresponding to the first three cases in turn.

(1) $\mathcal{O}(5)$.

This is the principal orbit for the dual group, and it follows that $\psi(j) = I_5$. The corresponding representations have infinitesimal character $\rho$ and trivial associated variety: $\Pi_\psi = \{\textit{trivial}\}$. The centralizer of the image of $\psi$ is trivial, so there is no endoscopy for this packet. (There would be if we consider packets for all real forms of $Sp(2n,\mathbb{R})$ as in [5].) See [5, Theorem 27.18].

(2) $\mathcal{O}(1,1,1,1,1)$.

This is the 0-orbit for the dual group, with dual orbit the principal orbit of $Sp(4,\mathbb{R})$, and infinitesimal character 0. We take $\psi(z)=1$ $(z\in\mathbb{C}^*)$. Up to conjugacy there are three possibilities for $\psi(j)$. We write $\psi^\dagger$ with $\dagger=a,b,c$, where:

$$\psi^\dagger(j)=\begin{cases} I_5 & \dagger=a\\ diag(-1,-1,1,1,1) & \dagger=b\\ diag(-1,-1,-1,-1,1) & \dagger=c\end{cases}$$

Let $y=\psi(j)$. Then $y$ acts as an involution on ${}^\vee G$, and thereby defines a real form of ${}^\vee G$. These real forms are $SO(5,0)$, $SO(3,2)$, and $SO(4,1)$, respectively. The representations in each packet have infinitesimal character 0. There are three (minimal) principal series representations of $Sp(4,\mathbb{R})$ with infinitesimal character 0, the spherical one which is irreducible, and the two others each having two irreducible components.

In terms of the tables in the Appendix we see this as follows. After translation we are considering representations with infinitesimal character $\rho$, and both simple roots not in the $\tau$-invariant. See [32, Definition 7.3.8, p. 472] for the definition of the $\tau$-invariant. These are dual to representations of the real forms of ${}^\vee G$ with both roots in the $\tau$-invariant, i.e., one-dimensional representations. These are parametrized by ${}^\vee G(\mathbb{R})/{}^\vee G(\mathbb{R})^0$; there are 1, 2 and 2 of them respectively. In the first case $\Pi$ consists of the single irreducible spherical representation $\pi_{sph}$ with infinitesimal character 0 (this representation is itself a block). In the second case $\Pi$ consists of the two large discrete series representations $\overline{J},\overline{K}$ translated to infinitesimal character 0. We denote these $\overline{J}_0,\overline{K}_0$. The final case consists of the two representations $\overline{Y},\overline{Z}$ in the other block for $Sp(4,\mathbb{R})$, translated to $\overline{Y}_0,\overline{Z}_0$ at infinitesimal character 0.

It is immediate that $\pi_{sph}$, $\overline{K}_0+\overline{L}_0$, and $\overline{Y}_0+\overline{Z}_0$ are stable.

We turn now to endoscopy. See [5, Chapters 22 and 26] for details. Computing centralizers, we see $\mathbf{S}_\psi$ consists of 1, 2 and 2 elements respectively. Given $s\in\mathbf{S}_\psi$ let ${}^\vee H$ be the identity component of the centralizer of $s$ in ${}^\vee G$. Associated to ${}^\vee H$ and $\psi$ is an endoscopic group $H$, stable Arthur packet of unipotent representations of $H$, and virtual character of $Sp(4,\mathbb{R})$ obtained by lifting. The identity element corresponds to the stable sums above.

We consider endoscopy coming from the non-trivial elements. For $\psi^a$ we have $H=GL(1,\mathbb{R})\times SL(2,\mathbb{R})$; $\overline{K}_0-\overline{L}_0$ is the lift from $H$ of $sgn$ on $GL(1,\mathbb{R})$ and the irreducible (spherical) principal series of infinitesimal character 0 on the $SL(2,\mathbb{R})$ factor. In this case lifting is induction from a real parabolic subgroup. The corresponding construction at infinitesimal character $\rho$ yields $E-F=\overline{E}-\overline{F}+\overline{I}+\overline{J}-\overline{K}-\overline{L}$, and translating to infinitesimal character 0 all terms vanish except $\overline{K}_0-\overline{L}_0$.

Similarly we obtain $\overline{Y}_0 - \overline{Z}_0$ as the lift from $GL(1,\mathbb{R}) \times SL(2,\mathbb{R})$, of the trivial representation on $GL(1,\mathbb{R})$ times the reducible principal series of infinitesimal character 0 on $SL(2,\mathbb{R})$.

We summarize this as follows.

PROPOSITION 6.4. *(1)* $\Pi_{\psi^a} = \{\pi_{sph}\}$

*(2)* $\Pi_{\psi^b} = \{\overline{J}_0, \overline{K}_0\}$

*(2a)* $\overline{J}_0 - \overline{K}_0$ *is the lift of an irreducible principal series representation of* $GL(1,\mathbb{R}) \times SL(2,\mathbb{R})$,

*(3)* $\Pi_{\psi^c} = \{\overline{Y}_0, \overline{Z}_0\}$

*(4)* $\overline{Y}_0 - \overline{Z}_0$ *is the lift of a reducible principal series representation of* $GL(1,\mathbb{R}) \times SL(2,\mathbb{R})$.

(3) $\mathcal{O}(3,1,1)$

This is the most interesting case. To be concrete, let $\iota$ denote the embedding of $SO(3,\mathbb{C})$ given by $\iota(g) = diag(I_2, g)$. Let $\pi : SL(2,\mathbb{C}) \to SO(3,\mathbb{C})$ be the covering map and let $\psi = \iota \circ \pi : SL(2,\mathbb{C}) \to \mathrm{SO}(5,\mathbb{C})$.

We take $\psi(z) = 1$ ($z \in \mathbb{C}^*$). The centralizer of $\psi(SL(2,\mathbb{C}))$ is isomorphic to $O(2)$. Up to conjugation $O(2)$ has three elements of order two. Hence there are three Arthur parameters $\psi$ for this orbit, written $\psi^\dagger$ with $\dagger = a, b, c$, where

$$\psi^\dagger(j) = \begin{cases} I_5 & \dagger = a \\ diag(-1,-1,1,1,1) & \dagger = b \\ diag(1,-1,-1,-1,-1) & \dagger = c \end{cases}$$

As in [5, 22.8] we obtain an element $y \in {}^\vee G$ of order 2, defining a Cartan involution $\theta_y$. For $\dagger = a, b, c$, we have $y = diag(1,1,-1,1,-1), diag(-1-1$ $-,1,-1)$, and $diag(1,-1,1,-1,1)$, respectively. The Cartan involutions $\theta_y$ define the real forms $SO(3,2)$, $SO(4,1)$, and $SO(3,2)$, respectively. We obtain a block for this real form, and the representations in $\Pi_{\psi^\dagger}$ are dual in the sense of [33] to these.

Using some facts about associated varieties for representations of $SO(3,2)$ and $SO(4,1)$ we conclude (notation as in Section 5 and the Appendix):

THEOREM 6.5. *The Arthur packets defined by* $\psi^a$, $\psi^b$, *and* $\psi^c$ *are:*

*1.* $\Pi_{\psi^a} = \{\overline{B}_\alpha, \overline{H}_\alpha\} = \{\pi^+[1,1], \pi^-[1,1]\}$.

2. $\Pi_{\psi^b} = \{\overline{W}_\alpha, \overline{X}_\alpha\}$.

*3.* $\Pi_{\psi^c} = \{\overline{C}_\alpha, \overline{D}_\alpha, \overline{I}_\alpha, \overline{L}_\alpha\} = \{\pi^+[2,0], \pi^+[0,2], \pi^-[2,0] = \Theta_{10}, \pi^-[0,2] = \Theta_{10}^*\}$.

Now we compute the centralizer of the image of $\psi$. As noted the centralizer of $\psi(SL(2,\mathbb{C}))$ is isomorphic to $O(2)$. To compute the centralizer of the image of $\psi$ we compute the fixed points of $\psi(j)$ on this group.

LEMMA 6.6. *The centralizer* $S_\psi$ *and its component group* $\mathbf{S}_\psi$ *are as follows:*

*(1) If* $\psi = \psi^a$ *or* $\psi^b$ *then* $S_\psi \simeq O(2)$, *and* $\mathbf{S}_\psi = \mathbb{Z}/2\mathbb{Z}$, *with* $s = diag(1,-1,-1,-1,-1)$.

*(2)* $S_{\psi^c} = \mathbf{S}_{\psi^c} = S[O(1)\times O(1)\times O(1)] \simeq \mathbb{Z}/2\mathbb{Z}\times\mathbb{Z}/2\mathbb{Z}$. *Explicitly* $\mathbf{S}_{\psi^c} = \{s_1 = I_5, s_2 = diag(-1,-1,1,1,1), s_3 = diag(1,-1,-1,-1,-1), s_4 = s_2 s_3\}$.

In case (1) the identity component ${}^\vee H$ of the centralizer of the non-trivial element of $\mathbf{S}$ is $SO(4,\mathbb{C})$. In case a (respectively b) $\theta_y$ gives the real form $SO(2,2)$ (resp. $SO(3,1)$). The corresponding endoscopic groups are $SO(2,2)$ and $SO(3,1)$, respectively.

In case (2) the identity components of the centralizers of the elements of $\mathbf{S}$ are $SO(5)$, $SO(3)\times SO(2)$, $SO(4)$, and $SO(4)$, respectively. The corresponding real forms defined are $SO(3,2)$, $\mathbb{R}^*\times SO(2,1)$, $SO(2,2)$, and $SO(3,1)$, respectively. The corresponding endoscopic groups are isomorphic to $Sp(4,\mathbb{R})$, $U(1)\times SL(2,\mathbb{R})$, $SO(2,2)$, and $SO(3,1)$, respectively.

Let $sgn$ be the non-trivial one-dimensional representation of $SO(2,2)$ or $SO(3,1)$.

PROPOSITION 6.7. *Each Arthur packet* $\Pi_{\psi^\dagger}$ *is in bijection with* $\mathbf{S}_{\psi^\dagger}$. *The lifted characters in the Arthur packets defined by* $\psi^a$, $\psi^b$, *and* $\psi^c$ *are the following.*

*(1)* $\psi^a$:

*(1a)* $\overline{B}_\alpha + \overline{H}_\alpha$ *is stable,*

*(1b)* $\overline{B}_\alpha - \overline{H}_\alpha$ *is the lift from* $SO(2,2)$ *of the* sgn *representation,*

*(2)* $\psi^b$:

*(2a)* $\overline{W}_\alpha + \overline{X}_\alpha$ *is stable,*

*(2b)* $\overline{W}_\alpha - \overline{X}_\alpha$ *is the lift from* $SO(3,1)$ *of the* sgn *representation,*

*(3)* $\psi^c$:

*(3a)* $\overline{C}_\alpha + \overline{D}_\alpha + \overline{I}_\alpha + \overline{L}_\alpha$ *is stable,*

*(3b)* $\overline{C}_\alpha - \overline{D}_\alpha - \overline{I}_\alpha + \overline{L}_\alpha$ *is the lift of the trivial representation from* $SL(2,\mathbb{R})\times U(1)$,

*(3c)* $\overline{C}_\alpha - \overline{D}_\alpha + \overline{I}_\alpha - \overline{L}_\alpha$ *is the lift of the trivial representation from* $SO(3,1)$.

*(3d)* $\overline{C}_\alpha + \overline{D}_\alpha - \overline{I}_\alpha - \overline{L}_\alpha$ *is the lift of the trivial representation from* $SO(2,2)$.

*Remark* 6.8. There is a further choice required to define the lifting [5, Definition 26.15(iii)]. The affect of this choice is to interchange the trivial and *sgn* representations of the endoscopic group $H$. We have made a particular such choice above.

*Remark* 6.9. The packets given by $\psi^b$, $\psi^c$, are those of [7, 1.4.3] and [5, 27.27(a–b)]. The case of $\psi^c$ is [7, 1.4.2] and [5, 27.17(c)].

*Proof.* The proofs of these facts are all similar, based on the character table in the Appendix. The basic technique is that standard representations of $H$ at infinitesimal character $\rho$ for $G$ lift to standard representations in a simple way. To compute the lift of an irreducible representation, in particular the trivial representation, we write it as a linear combination of standard representations. We then compute the corresponding lift at $\rho$, and then translate to a wall (cf. [32, 8.2.6]). The information which we need is either in the Appendix or is readily obtained from smaller groups such as $SL(2,\mathbb{R})$.

We give a few examples.

$\psi^a$:

$$\begin{aligned}\overline{B}+\overline{H} &= (B-G-I-L)+(H-J-K)\\ &= B+H+G-(I+J+K+L),\end{aligned}$$

which is stable. Translating to the $\alpha$ wall we conclude that $\overline{B}_\alpha+\overline{H}_\alpha$ is stable. On the other hand the lift of the $sgn$ representation of $SO(2,2)$ is the translation to the $\alpha$ wall of

$$\begin{aligned}B-G-G+(-I+J+K-L) &= (\overline{B}+\overline{E}+\overline{F}+\overline{G}+\overline{H}+\overline{I}+\overline{J}+\overline{K}+\overline{L})\\ &\quad -\overline{2}(G+\overline{E}+\overline{F}+\overline{H}+\overline{J}+\overline{K})\\ &\quad -\overline{I}+J+K-\overline{L}\\ &= \overline{B}-\overline{E}-\overline{F}-\overline{G}-\overline{H}.\end{aligned}$$

Translating to the $\alpha$ wall we have

$$\mathit{Lift}(sgn)=\overline{B}_\alpha-\overline{E}_\alpha-\overline{F}_\alpha-\overline{G}_\alpha-\overline{H}_\alpha=\overline{B}_\alpha-\overline{H}_\alpha,$$

since $\alpha$ is in the $\tau$-invariant of $\overline{E}$, $\overline{F}$, and $\overline{G}$.

The case of $\psi^b$ is similar.

$\psi^c$:

$$\begin{aligned}\overline{C}+\overline{D}+\overline{I}+\overline{L} &= (C-E-H+I+J+K)\\ &\quad +(D-H-F+J+K+L)+I+L\\ &= (C+D)-(E+F)-2H+2(I+J+K+L).\end{aligned}$$

This is stable, and remains so upon passing to the $\alpha$ wall.

The lift of the trivial representation of $SO(2,2)$ is

$$\begin{aligned}A-H-H+(-I+J+K-L) &= (\overline{A}+\overline{C}+\overline{D}+\overline{E}+\overline{F}+\overline{G}+2\overline{H}+\overline{J}+\overline{K})\\ &\quad -2(\overline{H}+\overline{J}+\overline{K})-\overline{I}+\overline{J}+\overline{K}-\overline{L}\\ &= \overline{A}+\overline{C}+\overline{D}+\overline{E}+\overline{F}-\overline{I}-\overline{L}.\end{aligned}$$

Translating to the $\alpha$ wall $A$, $E$, $F$ vanish to give

$$\mathit{Lift}(\mathbb{C})=\overline{C}_\alpha+\overline{D}_\alpha-\overline{I}_\alpha-\overline{L}_\alpha.$$

We leave verification of the remaining cases to the reader. □

Table I. Block of the Trivial Representation.

| Representation | Description | Length | Composition Series | Formal Character of Standard Module | $\tau$-invariant |
|---|---|---|---|---|---|
| $\overline{A}$ | trivial | 3 | $\overline{A}+\overline{C}+\overline{D}+\overline{E}+\overline{F}+\overline{G}+2\overline{H}+\overline{J}+\overline{K}$ | A − C − D − G + E + F + H − I − J − K − L | $\alpha, \beta$ |
| $\overline{B}$ | non-spherical | 3 | $\overline{B}+\overline{E}+\overline{F}+\overline{G}+\overline{H}+\overline{I}+\overline{J}+\overline{K}+\overline{L}$ | B − G − I − L | $\beta$ |
| $\overline{C}$ | | 2 | $\overline{C}+\overline{E}+\overline{H}+\overline{J}$ | C − E − H + I + J + K | $\beta$ |
| $\overline{D}$ | | 2 | $\overline{D}+\overline{F}+\overline{H}+\overline{K}$ | D − H − F + J + K + L | $\beta$ |
| $\overline{E}$ | $A(2, 0)$ – highest weight | 1 | $\overline{E}+\overline{I}+\overline{J}$ | E − I − J | $\alpha$ |
| $\overline{F}$ | $A(0, 2)$ – lowest weight | 1 | $\overline{F}+\overline{K}+\overline{L}$ | F − K − L | $\alpha$ |
| $\overline{G}$ | | 2 | $\overline{G}+\overline{E}+\overline{F}+\overline{H}+\overline{J}+\overline{K}$ | G − E − F − H − I + J + K + L | $\alpha$ |
| $\overline{H}$ | A(1/2, 1/2) | 1 | $\overline{H}+\overline{J}+\overline{K}$ | H − J − K | $\beta$ |
| $\overline{I}$ | holomorphic discrete series | 3 | $\overline{I}$ | I | $\beta$ |
| $\overline{J}$ | discrete series | 3 | $\overline{J}$ | J | $*$ |
| $\overline{K}$ | discrete series | 3 | $\overline{K}$ | K | $*$ |
| $\overline{L}$ | anti-holomorphic discrete series | 3 | $\overline{L}$ | L | $\beta$ |

Table II. Block of $Sp(4, \mathbb{R})$ Dual to Block of the Trivial Representation of $SO(4, 1)$.

| Representation | Description | Length | Composition Series | Formal Character of Standard Module | $\tau$-invariant |
|---|---|---|---|---|---|
| $\overline{V}$ | dual to discrete series | 3 | $\overline{V} + \overline{W} + \overline{X} + \overline{Y} + \overline{Z}$ | V − W − X | $\beta$ |
| $\overline{W}$ | dual to $A_{\mathfrak{q}}(\lambda)$ | 2 | $\overline{W} + \overline{Y}$ | W − Y | $\alpha$ |
| $\overline{X}$ | dual to $A_{\mathfrak{q}}(\lambda)'$ | 2 | $\overline{X} + \overline{Z}$ | X − Z | $\alpha$ |
| $\overline{Y}$ | dual to one-dimensional | 1 | $\overline{Y}$ | Y | $\alpha, \beta$ |
| $\overline{Z}$ | dual to trivial | 1 | $\overline{Z}$ | Z | $\alpha, \beta$ |

**6.3. Relation with the theta correspondence.** Note that by 6.2(3) $\Pi_{\psi^c}$ consists of the representations corresponding to the trivial and $sgn$ representations of $O(2,0)$ and $O(0,2)$. This is an example of the philosophy of [1].

This can also be described in terms of derived functors. With notation as in Section 3 $A_{\mathfrak{q}}(k,0)$ is an irreducible unitary representation if $k \geq 1$. We define $A_{\mathfrak{q}'}(0,-k)$ analogously; $A_{\mathfrak{q}'}(0,-k) \simeq A_{\mathfrak{q}}(k,0)^*$. If $k \geq 1$ then $\{A_{\mathfrak{q}}((k,0)), A_{\mathfrak{q}}((0,-k))\}$ is an Arthur packet of a particularly simple type as in [6].

Now take $k = 0$. Then $A_{\mathfrak{q}}(0)$ is unitary and reducible; in fact

$$A_{\mathfrak{q}}(0) = \Theta_{10} \oplus \pi^+[2,0], \qquad A_{\mathfrak{q}'}(0) = \Theta_{10}^* \oplus \pi^+[0,2].$$

Therefore $\Pi_{\psi^c}$ consists of the four consitituents of $A_{\mathfrak{q}}(0)$ and $A_{\overline{\mathfrak{q}}}(0)$, and this is an Arthur packet of the previous type at singular infinitesimal character.

**7. Appendix: Character tables.** We give some information about the representations of $Sp(4,\mathbb{R})$ with infinitesimal character $\rho$. This information is reasonably well known, if not necessarily readily accessible. Each table lists standard modules $A, B, \ldots$ with their irreducible quotients $\overline{A}, \overline{B}, \ldots$. The composition series of the standard modules and the expressions of the irreducible modules (in the Grothendieck group) in terms of standard modules are given. The final column shows the $\tau$-invariant of each irreducible representation, with $\alpha$ (respectively $\beta$) a long (respectively short) simple root. For general information see [32] and [33].

There are three blocks for $Sp(4,\mathbb{R})$ with infinitesimal character $\rho$. One of these is the singleton consisting of the irreducible principle series module, which is dual to the trivial representation of $SO(5,0)$. The other two are the block of the trivial representation (dual to a block for $SO(3,2)$) and a block dual to the block of the trivial representation of $SO(3,1)$.

Table I is from [36]. The information in Table II may all be read off via duality from the corresonding dual block of the trivial representation of $SO(4,1)$. We note that $SO(4,1)$ has two one-dimensional representations *trivial* and $\chi$, and two representations with $(\mathfrak{g}, K)$-cohomology. We denote the latter $A_{\mathfrak{q}}(\lambda)$ and $A(\mathfrak{q},\lambda)' = A_{\mathfrak{q}}(\lambda) \otimes \chi$.

If $\alpha$ is not contained in the $\tau$-invariant of a representation $\overline{X}$ occuring in this list, we let $\psi_\alpha(\overline{X}) \neq 0$ be the translate of $\overline{X}$ to the $\alpha$ wall.

---

REFERENCES


[1] J. Adams, $L$-functoriality for dual pairs, *Astérisque,* 171–172 (1989), 85–129. Orbites unipotentes et représentations, II.

[2] ———, *Lifting of characters,* vol. 101 of *Progress in mathematics,* Birkhäuser, Boston, 1991.

[3] J. Adams and D. Barbasch, Genuine representations of the metaplectic group, preprint.

[4] ———, Genuine representations of the metaplectic group, *Compositio Math.* **113** (1998), 23–66.

[5] Jeffrey Adams, Dan Barbasch, and David A. Vogan, Jr., *The Langlands classification and irreducible characters for real reductive groups,* vol. 104 of *Progress in Mathematics*. Birkhäuser, Boston, 1992.

[6] Jeffrey Adams and Joseph F. Johnson, Endoscopic groups and packets of nontempered representations, *Compositio Math.* **64** (1987), 271–309.

[7] J. Arthur, *On Some Problems Suggested By the Trace Formula,* vol. 1024, *Lecture Notes in Mathematics,* Springer-Verlag, Berlin, 1983.

[8] T. N. Bailey and A. W. Knapp (eds.), *Representation theory and automorphic forms,* vol. 61 of *Proceedings of Symposia in Pure Mathematics*. American Mathematical Society, Providence, R.I., 1997. Papers from the Instructional Conference held in Edinburgh, March 17–29, 1996.

[9] A. Borel, *Automorphic L-Functions,* vol. 33 of *Proc. Symp. Pure Math.* American Math. Soc., Providence, R.I., 1979.

[10] A. Borel and H. Jacquet, Automorphic forms and automorphic representations. In *Automorphic forms, representations and L-functions (Proc. Sympos. Pure Math., Oregon State Univ., Corvallis, Ore., 1977), Part 1,* Proc. Sympos. Pure Math., XXXIII, pages 189–207. Amer. Math. Soc., Providence, R.I., 1979. With a supplement "On the notion of an automorphic representation" by R. P. Langlands.

[11] David H. Collingwood and William M. McGovern, *Nilpotent orbits in semisimple Lie algebras.* Van Nostrand Reinhold Co., New York, 1993.

[12] Patrick Delorme. Infinitesimal character and distribution character of representations of reductive Lie groups. In *Representation theory and automorphic forms (Edinburgh, 1996),* vol. 61 of *Proc. Sympos. Pure Math.,* pp. 73–81. Amer. Math. Soc., Providence, R.I., 1997.

[13] Stephen Gelbart, Examples of dual reductive pairs. In *Automorphic forms, representations and L-functions (Proc. Sympos. Pure Math., Oregon State Univ., Corvallis, Ore., 1977), Part 1,* Proc. Sympos. Pure Math., XXXIII, pages 287–296. Amer. Math. Soc., Providence, R.I., 1979.

[14] Henryk Hecht and Wilfried Schmid, A proof of Blattner's conjecture. *Invent. Math.* **31** (1975), 129–154.

[15] R. Howe, *$\theta$-series and Invariant Theory,* vol. 33 of *Proc. Symp. Pure Math.* American Math. Soc., Providence, R.I., 1979.

[16] R. Howe and I. I. Piatetski-Shapiro, A counterexample to the "generalized Ramanujan conjecture" for (quasi-) split groups. In *Automorphic forms, representations and L-functions (Proc. Sympos. Pure Math., Oregon State Univ., Corvallis, Ore., 1977), Part 1,* Proc. Sympos. Pure Math., XXXIII, pages 315–322. Amer. Math. Soc., Providence, R.I., 1979.

[17] Roger Howe. Wave front sets of representations of Lie groups. In *Automorphic forms, representation theory and arithmetic (Bombay, 1979),* vol. 10 of *Tata Inst. Fund. Res. Studies in Math.,* pages 117–140. Tata Inst. Fundamental Res., Bombay, 1981.

[18] ———, Remarks on classical invariant theory. *Trans. Amer. Math. Soc.* **313** (1989), 539–570.

[19] ———, Transcending classical invariant theory, *J. Amer. Math. Soc.* **2** (1989), 535–552.

[20] M. Kashiwara and M. Vergne, On the Segal-Shale-Weil representations and harmonic polynomials, *Invent. Math.* **44** (1978), 1–47.

[21] A. Knapp, *Representation Theory of Semisimple Groups,* Princeton University Press, Princeton, NJ, 1986.

[22] A. W. Knapp, Introduction to the Langlands program. In *Representation theory and automorphic forms (Edinburgh, 1996),* vol. 61 of *Proc. Sympos. Pure Math.,* pages 245–302. Amer. Math. Soc., Providence, R.I., 1997.

[23] A. W. Knapp and Gregg J. Zuckerman, Classification of irreducible tempered representations of semisimple groups. *Ann. of Math.* **116** (1982), 389–455.

[24] Anthony W. Knapp and Peter E. Trapa, Representations of semisimple Lie groups. In *Representation theory of Lie groups (Park City, UT, 1998),* vol. 8 of *IAS/Park City Math. Ser.,* pages 7–87. Amer. Math. Soc., Providence, R.I., 2000.

[25] Anthony W. Knapp and David A. Vogan, Jr., *Cohomological induction and unitary representations,* vol. 45 of *Princeton Mathematical Series*. Princeton University Press, Princeton, NJ, 1995.

[26] R. Langlands, *On the Classification of Irreducible Representations of Real Algebraic Groups*. Number 31 in Mathematical Surveys and Monographs. American Mathematical Society, Providence, 1989.

[27] George Lusztig, *Characters of reductive groups over a finite field,* vol. 107 of *Annals of Mathematics Studies*. Princeton University Press, Princeton, NJ, 1984.

[28] Dipendra Prasad, A brief survey on the theta correspondence. In *Number theory (Tiruchirapalli, 1996),* volume 210 of *Contemp. Math.,* pages 171–193. Amer. Math. Soc., Providence, R.I., 1998.

[29] Tomasz Przebinda, The duality correspondence of infinitesimal characters. *Colloq. Math.* **70** (1996), 93–102.

[30] Bhama Srinivasan, The characters of the finite symplectic group sp(4, $q$). *Trans. Amer. Math. Soc.* **131** (1968), 488–525.

[31] N. Wallach, T. Enright and R. Howe, *A Classification of Unitary Highest Weight Modules,* Birkhäuser, Boston, 1983.

[32] D. Vogan, *Representations of Real Reductive Lie Groups,* vol. 15 of *Progress in mathematics,* Birkhäuser, Boston, 1981.

[33] D. Vogan, Irreducible characters of semisimple Lie groups IV, Character-multiplicity duality. *Duke Math. J.* **49** (1982), 943–1073.

[34] D. Vogan, The unitary dual of $g_2$, *Invent. Math.* **116** (1994), 677–791.

[35] David A. Vogan, Jr., Gelfand-Kirillov dimension for Harish-Chandra modules. *Invent. Math.* **48** (1978), 75–98.

[36] David A. Vogan, Jr., The Kazhdan-Lusztig conjecture for real reductive groups. In *Representation theory of reductive groups (Park City, Utah, 1982),* pages 223–264. Birkhäuser Boston, Boston, MA, 1983.

[37] David A. Vogan, Jr., Cohomology and group representations. In *Representation theory and automorphic forms (Edinburgh, 1996),* vol. 61 of *Proc. Sympos. Pure Math.,* pages 219–243. Amer. Math. Soc., Providence, R.I., 1997.

[38] N. Wallach, The analytic continutation of the discrete series I, *Trans. Amer. Math. Soc.* **251** (1979), 1–17.

CHAPTER 3

# DISCRETE SERIES CHARACTERS OF DIVISION ALGEBRAS AND $\mathrm{GL}_n$ OVER A $p$-ADIC FIELD

By JEFFREY D. ADLER, LAWRENCE CORWIN, and PAUL J. SALLY, JR.

---

**1. Introduction.** Let $F$ be a $p$-adic field (with $p$ odd) and $D$ a central division algebra of degree $n$ over $F$. We assume throughout this paper that $(n, p) = 1$, the *tame case*. In this situation, the construction of the irreducible unitary representations of $D^\times$ has been known for some time; see [1], [2], [10] and [18]. (There is also a construction available when $p \mid n$. See [3] and [13]). The characters of these representations have been computed for $n = 2$ in [12]. More generally, some qualitative information about the nature of characters of $D^\times$ is available in [4].

One of our main goals in the current paper is to give explicit character formulas for the irreducible representations of $D^\times$. In the tame case, these representations are parametrized by admissible characters $\theta$ of the multiplicative groups of field extensions $E/F$ of degree dividing $n$. We indicate the dependence by writing $\theta \rightsquigarrow \pi_\theta$. Fix $\pi = \pi_\theta$ and write $\chi_\pi$ for its trace character. We describe a collection of elements of $D^\times$, called *normal* elements, which meets every conjugacy class in $D^\times$, and, in the theorem below, give a formula for $\chi_\pi(y)$ when $y$ is normal. This formula depends only on $\theta$ and $y$, and on certain algebraic data attached to $y$. While complicated, it can be used for explicit computations in many situations. We give several examples in Section 5.

The representation theory of $D^\times$ is related to the representation theory of $\mathrm{GL}_n(F)$ via the Matching Theorem (see [8], [16]). This provides a canonical bijection between the set of irreducible representations of $D^\times$ and the set of discrete series representations of $\mathrm{GL}_n(F)$. Among other things, this bijection preserves characters up to a sign. It was shown in [5] that, under this matching, supercuspidal representations of $\mathrm{GL}_n(F)$ correspond to representations $\pi_\theta$ where $\theta$ is an admissible character of an extension $E/F$ of degree $n$; and generalized special representations of $\mathrm{GL}_n(F)$ correspond to representations $\pi_\theta$ where $\theta$ is an admissible character of an extension $E/F$ of degree $m < n$. Using this and our formulas here, we can get interesting qualitative information about the difference between trace characters of supercuspidal and generalized special representations of $\mathrm{GL}_n(F)$. A special example of the character formulas in this paper appears in [6] in the case where $n = \ell$ is prime and $E/F$ is totally ramified. All of the details are present in that paper. Here, our general character formula includes the case when $n = \ell$ is prime and $E/F$ is

---

Jeffrey D. Adler was partially supported by the National Security Agency (#MDA904-02-1-0020).

unramified. Both the totally ramified and unramified cases are computed directly in [7] for $\mathrm{GL}_\ell(F)$ and, as DeBacker has observed, upon replacing $m-1$ by $\ell-1$ in [6, Theorem 4.2(c)], these results agree. (This is a typographical error in [6], not a disagreement.) The construction of $\pi_\theta$ from $\theta$ is inductive in two senses. First, $\pi_\theta$ is induced from a representation $\pi_0$ of some subgroup of $D^\times$. Second, the main ingredient in the inducing representation $\pi_0$ is an irreducible representation of the multiplicative group $D_1^\times$ of a division algebra of smaller degree (over an extension of $F$). Thus, $\chi_{\pi_\theta}$ may be described in terms of characters of representations of division algebras of smaller degree. A qualitative description of the relationship between these two characters appeared in [4]. One can use this information and an inductive argument to find an explicit formula for $\chi_{\pi_\theta}$.

*Acknowledgments.* The authors thank Allen Moy and Loren Spice for their assistance in preparing this manuscript.

*Comments by the senior author.* Joseph Shalika is a remarkable mathematician and a terrific colleague. The present paper is a natural continuation of our first collaboration, [17]. I salute Joe on his sixtieth birthday.

**2. Notation.** As in the introduction, let $F$ be a $p$-adic field of characteristic 0 and odd residual characteristic $p$ and $D$ a central division algebra over $F$ of degree $n$ (so that the center of $D$ is $F$ and $[D:F]=n^2$). Assume $(n,p)=1$. Write $R_D$ for the ring of integers in $D$ and $\wp_D$ for the prime ideal in $R_D$.

We choose a prime element $\varpi \in D$ so that $\varpi^n \in F$. Then $\varpi$ normalizes the maximal unramified extension $F_n$ of $F$ in $D$, and $\sigma : x \mapsto \varpi x \varpi^{-1}$ generates $\mathrm{Gal}(F_n/F)$. (In fact, given a generator $\sigma \in \mathrm{Gal}(F_n/F)$, there is a unique central division algebra $D/F$ so that $\sigma$ arises in this way.) We normalize the valuation and absolute value on $D$ so that $v(\varpi)=1$ and $|\varpi|=q^{-n}$.

An element of the form $\alpha\varpi^j$, where $\alpha$ is either zero or a $(q^n-1)$st root of unity in $F_n$, is called a *monomial*. If $y \in D^\times$, we may write $y$ uniquely as a sum of monomials,

$$y = \sum_{j=j_0}^{\infty} \alpha_j \varpi^j, \qquad \alpha_{j_0} \neq 0.$$

Then $v(y)=j_0$ and $|y|=q^{-nj_0}$. We say that $y$ is *normal* if all the monomials $\alpha_j\varpi^j$ appearing in the sum above commute. The following lemma illustrates the importance of normal elements.

LEMMA. *Every element $y \in D^\times$ is conjugate to a normal element. In fact, for any extension $E/F$ of degree dividing $n$, $E^\times$ embeds into the set of normal elements in $D^\times$.*

We introduce a filtration on $D^\times$ by putting $K_0 = R_D^\times$ and $K_m = 1 + \wp_D^m$ for $m \geq 1$. For any closed subgroup $S \subseteq D^\times$ and any irreducible smooth representation $\pi$ of $S$, we define the *depth* of $\pi$ to be the smallest nonnegative integer $d$ such that $\pi$ is trivial on $S \cap K_{d+1}$. (The notion of depth becomes more complicated when one replaces $D^\times$ by an arbitrary reductive $p$-adic group. The general definition is due to Moy and Prasad [15]. Note that, because we normalize our valuation differently from the way they do, our depths are not exactly the same as theirs. For example, if a representation $\pi$ of $D^\times$ has depth $d$ in our normalization, then it has depth $d/n$ in theirs.)

**3. Representations.** The irreducible representations of $D^\times$ are parametrized by admissible characters $\theta$ of extensions $E/F$ of degree dividing $n$. Given $\theta$, there exist a tower of fields $F = E_0 \subsetneq E_1 \subsetneq \cdots \subsetneq E_t = E$ and characters $\phi_0, \phi_1, \ldots, \phi_t$ of $E_0^\times, E_1^\times, \ldots, E_t^\times$ such that:

(1) $\theta = \prod_{i=0}^t (\phi_i \circ N_{E/E_i})$.

(2) For $1 \leq i \leq t$, $\phi_i$ is a "generic" character of $E_i^\times$ over $E_{i-1}$.

Let $d$ be the depth of $\theta$ and let $d_i$ be the depth of $\phi_i$, $0 \leq i \leq t$. The tower of fields is uniquely determined by $\theta$, as are the depths $d_1 > d_2 > \cdots > d_t$. Moreover, $d_0 \geq d_1$ or we may take $\phi_0 \equiv 1$. A detailed discussion of admissible characters and their properties may be found in [11] or [14].

We can (and will) embed each field $E_i$ in $D$ by identifying $E_i$ with some $L[m]$, where $L$ is a subfield of $F_n$, and $m$ is a monomial. Let $D_i$ be the centralizer of $E_i$ in $D$. The irreducible representation $\pi_\theta$ corresponding to $\theta$ is induced from a representation $\pi_0$ of the subgroup $D_t^\times(D_{t-1}^\times \cap K_{\lfloor d_t/2 \rfloor}) \cdots (D_1^\times \cap K_{\lfloor d_2/2 \rfloor})K_{\lfloor d_1/2 \rfloor}$. All irreducible representations of $D^\times$ are obtained in this way. Full details appear in the references mentioned in the introduction.

**4. Character formulas.** Let $\theta$ be an admissible character of an extension $E/F$ of degree dividing $n$, and let $\pi = \pi_\theta$ be the corresponding irreducible representation of $D^\times$. We compute the character values $\chi_\pi(y) = \chi_{\pi_\theta}(y)$, $y \in D^\times$. We can make some simplifying assumptions before starting:

(1) The depth of $\pi$ is minimal under twisting by characters of $D^\times$. This implies that $d = d_1$, and we may take the character $\phi_0$ arising in the Howe factorization to be trivial.

(2) The element $y$ is normal.

The following notation will be used in the character formulas below. Write

$$y = \gamma_0 \varpi^{j_0}(1 + \gamma_1 \varpi + \gamma_2 \varpi^2 + \cdots)$$

- Define a tower of fields $F \subseteq E^{(0)} \subseteq E^{(1)} \subseteq \cdots$ as follows: $E^{(0)} = F[\gamma_0 \varpi^{j_0}]$ and, for $j \geq 1$, $E^{(j)} = F[\gamma_0 \varpi^{j_0}, \gamma_1 \varpi, \ldots, \gamma_j \varpi^j]$.
- For $1 \leq i \leq t$, let $s(i)$ be the smallest index $j$ such that $E^{(j)} = E^{(d_i - 1)}$.
- For $0 \leq j < s(1)$, let $h(j)$ be the largest index $1 \leq i \leq t$ such that there is some index $j' < d_i$ with $E^{(j')} = E^{(j)}$.
- For $y \in D_t^\times K_d$, let $\mathrm{Gal}'(E_*/F)y$ denote the set of all normal $\tilde{y}$ in the group $D_t^\times (D_{t-1}^\times \cap K_{d_t}) \cdots (D_1^\times \cap K_{d_2}) K_{d_1}$ such that $\tilde{y}$ is conjugate to $y$ in $D^\times$, but not via an element of $D_t^\times$.
- Let $E_{i,j}$ denote the center of the division algebra of elements that commute with $E_i$ and $E^{(j)}$. (Note that when $y \in D_t^\times K_d$ and $j < d$, $E_i$ and $E^{(j)}$ commute, so this is simply the compositum $E_i E^{(j)}$.) Let $n(i, j)$, $e(i, j)$ and $f(i, j)$ denote the degree, ramification index and residual degree of $E_{i,j}/F$, respectively.

Note that when $y \in D_t^\times K_{d+1}$, then $\mathrm{Gal}'(E_*/F)y$ has a simpler interpretation: We can (and do) replace $y$ by a normal element of $D_t^\times$. In this case, $\mathrm{Gal}'(E_*/F)y$ is just the orbit of $y$ under the action of the group of $F$-automorphisms of $E$ that preserve every $E_i$ (a group that acts on $y$ via conjugation in $D^\times$). Moreover, $E_{i,j}$ can be defined as a compositum for all $j \leq d$.

In the particular case when $y \in D_t^\times K_d$, but $y$ is *not* conjugate to a normal element in $D_t^\times K_{d+1}$, the more complicated definitions above are necessary. It is this situation which creates the so-called "bad shell" for the character formulas (see, for example, [6, Theorem 4.2(d)]).

Take $\tilde{y}$ normal in $D^\times$. To each pair of indices $(i, i')$ with $1 \leq i \leq t$ and $1 \leq i' \leq s(i)$, we associate an eighth root of unity $H_{i,i'}(\tilde{y})$ (depending on $\theta$). This is the only term appearing in the character formula which we do not discuss in complete detail. The computation of these factors is somewhat long and complicated, and will appear in a future paper with full details of the computation of the character formulas listed in the Theorem below.

THEOREM. *Let $\pi = \pi_\theta$ be an irreducible representation of $D^\times$, and let $\chi_\pi = \chi_{\pi_\theta}$ be its character.*

1. *If $y$ is not conjugate to a normal element of $D_t^\times K_d$, then $\chi_\pi(y) = 0$.*
2. *If $y$ is a normal element in $D_t^\times K_d$, then*

$$\chi_\pi(y) = \sum_{\tilde{y} \in \mathrm{Gal}'(E_*/F)y} (\theta \circ N_{D_t/E_t})(\tilde{y}) \Big( \prod_{\substack{1 \leq i \leq t \\ i' \leq s(i)}} H_{i,i'}(\tilde{y}) \Big)$$

$$\times \left( \prod_{i=1}^{t} \frac{f(i, d_i)}{f(i-1, d_i)} \, \frac{q^{n/e(i-1,d_i)} - 1}{q^{n/e(i,d_i)} - 1} \right) q^{\alpha(y,\pi)/2},$$

*where*

$$\alpha(y,\pi) = \sum_{0\le j<s(1)} \left(\frac{n}{n(0,j)} - \frac{n}{n(h(j),j)}\right) + \sum_{i=1}^{t}(d_i - s(i))\left(\frac{n}{n(i-1,d_i-1)} - \frac{n}{n(i,d_i-1)}\right) - \left(\frac{n}{e(i-1,d_i-1)} - \frac{n}{e(i,d_i-1)}\right).$$

COROLLARY. *If $\theta$ is generic and $y$ is a normal element of $D_t^\times K_d$, then*

$$\chi_\pi(y) = \sum_{\tilde y\in \mathrm{Gal}(E_1/F)y} (\theta\circ N_{D_1/E_1})(\tilde y)\Big(\prod_{i'\le s(1)} H_{1,i'}(\tilde y)\Big) \times \left(\frac{f(1,d)}{f(0,d)}\frac{q^{n/e(0,d)}-1}{q^{n/e(1,d)}-1}\right) q^{\alpha(y,\pi)/2},$$

*where*

$$\alpha(y,\pi) = \sum_{0\le j<s(1)} \left(\frac{n}{n(0,j)} - \frac{n}{n(1,j)}\right) + (d-s(1))\left(\frac{n}{n(0,d-1)} - \frac{n}{n(1,d-1)}\right) - \left(\frac{n}{e(0,d-1)} - \frac{n}{e(1,d-1)}\right).$$

*Proof of the Corollary.* In this case, $t = 1$ and every $h(j) = 1$. □

*Remark.* Note a distinction between the cases $[E : F] = n$ and $[E : F] < n$. In each case, one may ask which Cartan subgroups (up to conjugacy) contain very regular elements (see §5.2) far from the identity on which the character $\chi_\pi$ does not vanish. In the former case, which corresponds to supercuspidal characters of $\mathrm{GL}_n(F)$, the answer is the single Cartan subgroup $E^\times$. In the latter case, which corresponds to generalized special characters of $\mathrm{GL}_n(F)$, $D_t$ is a division algebra which contains every extension of $E$ having degree $n$ over $F$. Thus, the answer is: every Cartan subgroup that contains $E^\times$.

## 5. Applications of the character formula.

**5.1. Degrees.** The degree of $\pi = \pi_\theta$ is $\chi_\pi(1)$. The character formula gives

$$\chi_\pi(1) = f\frac{q^n-1}{q^{n/e}-1}q^{\alpha(1,\pi)/2}.$$

Here, $e = e(E/F)$, $f = f(E/F)$, $n_i = [E_i : F]$ and

$$\alpha(1,\pi) = \frac{n}{e} - n + \sum_{i=1}^{t} d_i\left(\frac{n}{n_{i-1}} - \frac{n}{n_i}\right).$$

This follows from the facts that $\mathrm{Gal}'(E_*/F)\cdot 1 = \{1\}$, $s(i) = 0$ for all $i$, $h(j) = 1$ for all $j$, and $H_{i,i'}(1) = 1$. This agrees with [5, Theorem 3.25].

**5.2. Character values at very regular points.** Following [9], let us call an element $y \in D^\times$ *very regular* if the associated fields $E^{(j)}$ all have dimension $n$ (and thus are equal). Let $\theta$ be a generic character of a field $E/F$ of degree dividing $n$. (In this case, $E = E_1$ and $\theta = \phi_1$.) We will compute the character of $\pi = \pi_\theta$ at all very regular elements $y \in D_1^\times$. In fact, our assumption on $y$ is slightly weaker than this. We assume only that the fields $E^{(j)}$ are all equal and contain $E$. Then

$$\chi_\pi(y) = \sum_{\tilde{y}\in \mathrm{Gal}(E_1/F)y} (\theta \circ N_{D_1/E_1})(\tilde{y}) \cdot H_{1,0}(\tilde{y}).$$

This follows from the Corollary above and the facts that $s(1) = 0$ and $\alpha(y,\pi) = 0$. By the Matching Theorem, we have now partially computed the characters of all very supercuspidal representations and some generalized special representations of $\mathrm{GL}_n(F)$ (all generalized special representations, in case $n$ is the product of two primes).

**5.3. *n* prime.** In this situation, $\theta$ is always generic (under the assumption that $\chi = 1$), and so the Corollary above applies. For ease of computation, we compute the character values only at those elements $y$ which are conjugate to a normal element in $E^\times \cdot K_{d+1}$. Without loss of generality, we may in fact assume that $y$ is a normal element in $E^\times$. Then the character formula takes several different forms, depending on the value of $s(1)$. If $s(1) = 0$, then

$$\chi_\pi(y) = \sum_{\tilde{y}\in \mathrm{Gal}(E/F)y} \theta(\tilde{y})H_{1,0}(\tilde{y}).$$

If $0 < s(1) < d$, then

$$\chi_\pi(y) = \sum_{\tilde{y}\in \mathrm{Gal}(E/F)y} \theta(\tilde{y})\Big(\prod_{i'\le s(1)} H_{1,i'}(\tilde{y})\Big)q^{(n-1)s(1)/2}.$$

If $s(1) = d$, then

$$\chi_\pi(y) = \sum_{\tilde{y}\in \mathrm{Gal}(E/F)y} \theta(\tilde{y})\Big(\prod_{i'\le d} H_{1,i'}(\tilde{y})\Big)q^{(n-1)d/2}$$

if $E/F$ is unramified and

$$\chi_\pi(y) = \sum_{\tilde{y}\in \mathrm{Gal}(E/F)y} \theta(\tilde{y})\Big(\prod_{i'\le d} H_{1,i'}(\tilde{y})\Big)q^{(n-1)(d-1)/2}$$

if $E/F$ is ramified. If $s(1) > d$, then

$$\chi_\pi(y) = \sum_{\tilde{y} \in \mathrm{Gal}(E/F)y} \theta(\tilde{y}) \Big( \prod_{i' \leq s(1)} H_{1,i'}(\tilde{y}) \Big) n q^{(n-1)d/2}$$

if $E/F$ is unramified, and

$$\chi_\pi(y) = \sum_{\tilde{y} \in \mathrm{Gal}(E/F)y} \theta(\tilde{y}) \Big( \prod_{i' \leq s(1)} H_{1,i'}(\tilde{y}) \Big) \frac{q^n - 1}{q - 1} q^{(n-1)(d-1)/2}$$

if $E/F$ is ramified. In the notation of [6], $s(1) = \nu_0(y)$.

UNIVERSITY OF AKRON, AKRON, OH 44325-4002
*E-mail address:* adler@uakron.edu

RUTGERS UNIVERSITY, PISCATAWAY, NJ 08854

UNIVERSITY OF CHICAGO, CHICAGO, IL 60637-1514
*E-mail address:* sally@math.uchicago.edu

---

## REFERENCES

[1] L. Corwin, Representations of division algebras over local fields, *Adv. in Math.* **13** (1974), 259–267.
[2] ———, Representations of division algebras over local fields. II, *Pacific J. Math.* **101** (1982), no. 1, 49–70.
[3] ———, The unitary dual for the multiplicative group of arbitrary division algebras over local fields, *J. Amer. Math. Soc.* **2** (1989), no. 3, 565–598.
[4] L. Corwin and R. Howe, Computing characters of tamely ramified $p$-adic division algebras, *Pacific J. Math.* **73** (1977), no. 2, 461–477.
[5] L. Corwin, A. Moy and P. J. Sally, Jr., Degrees and formal degrees for division algebras and $\mathrm{GL}_n$ over a $p$-adic field, *Pacific J. Math.* **141** (1990), no. 1, 21–45.
[6] ———, Supercuspidal character formulas for $\mathrm{GL}_\ell$, Representation theory and harmonic analysis (Cincinnati, OH, 1994), *Contemp. Math.*, vol. 191, Amer. Math. Soc., Providence, RI, 1995, pp. 1–11.
[7] S. DeBacker, On supercuspidal characters of $\mathrm{GL}_\ell$, $\ell$ a prime, Ph.D. thesis, University of Chicago, 1997.
[8] P. Deligne, D. Kazhdan and M.-F. Vignéras, Représentations des algèbres centrales simples $p$-adiques, Representations of reductive groups over a local field, Hermann, Paris, 1984, pp. 33–117.
[9] G. Henniart, Correspondance de Jacquet-Langlands explicite. I. Le cas modéré de degré premier, Séminaire de Théorie des Nombres, Paris, 1990–91, *Progr. Math.*, vol. 108, Birkhäuser Boston, Boston, MA, 1993, pp. 85–114.
[10] R. Howe, Representation theory for division algebras over local fields (tamely ramified case), *Bull. Amer. Math. Soc.* **77** (1971), 1063–1066.
[11] ———, Tamely ramified supercuspidal representations of $\mathrm{GL}_n$, *Pacific J. Math.* **73** (1977), no. 2, 437–460.
[12] H. Jacquet and R. P. Langlands, Automorphic forms on $GL_2$, *Lecture Notes in Math.*, no. 114, Springer-Verlag, Berlin, 1970.
[13] H. Koch, Eisensteinsche Polynomfolgen und Arithmetik in Divisionsalgebren über lokalen Körpern, *Math. Nachr.* **104** (1981), 229–251.

[14] A. Moy, Local constants and the tame Langlands correspondence, *Amer. J. Math.* **108** (1986), 863–930.
[15] A. Moy and G. Prasad, Unrefined minimal $K$-types for $p$-adic groups, *Invent. Math.* **116** (1994), 393–408.
[16] J. D. Rogawski, Representations of GL($n$) and division algebras over a $p$-adic field, *Duke Math. J.* **50** (1983), no. 1, 161–196.
[17] P. J. Sally, Jr. and J. Shalika, Characters of the discrete series of representations of SL(2) over a local field, *Proc. Nat. Acad. Sci. U. S. A.* **61** (1968), 1231–1237.
[18] J.-K. Yu, Construction of tame supercuspidal representations, *J. Amer. Math. Soc.* **14** (2001), no. 3, 579–622.

## CHAPTER 4

# AUTOMORPHIC REPRESENTATIONS OF GSp(4)

By James Arthur

---

**1.** In this note, we shall describe a classification for automorphic representations of GSp(4), the group of similitudes of four-dimensional symplectic space. The results are part of a project [A3] on the automorphic representations of general classical groups. The monograph [A3] is still in preparation. When complete, it will contain a larger classification of representations, subject to a general condition on the fundamental lemma.

In the case of GSp(4), the standard fundamental lemma for invariant orbital integrals has been established [Ha], [W]. However, a natural variant of the standard fundamental lemma is also needed. To be specific, the theorem we will announce here is contingent upon a fundamental lemma for twisted, weighted, orbital integrals on the group $GL(4) \times GL(1)$ (relative to a certain outer automorphism). This has not been established. However, it seems likely that by methods of descent, perhaps in combination with other means, one could reduce the problem to known cases of the standard fundamental lemma. (The papers [BWW], [F], and [Sc] apply such methods to the twisted analogue of the fundamental lemma, but not its generalization to weighted orbital integrals.)

The general results of [A3] are proved by a comparison of spectral terms in the stabilized trace formula. It is for the existence of the stabilized trace formula [A2] (and its twisted analogues) that the fundamental lemma is required. However, any discussion of such methods would be outside the scope of this paper. We shall be content simply to state the classification for GSp(4) in reasonably elementary terms. The paper will in fact be somewhat expository. We shall try to motivate the classification by examining the relevant mappings from a Galois group (or some extension thereof) to the appropriate $L$-groups.

Representations of the group GSp(4) have been widely studied. The papers [HP], [Ku], [Y], [So], and [Ro] contain results that were established directly for GSp(4). Results for groups of higher rank in [CKPS] and [GRS] could also be applied (either now or in the near future) to the special case of GSp(4).

**2.** Let $F$ be a local or global field of characteristic zero. If $N$ is any positive integer, the general linear group $GL(N)$ has an outer automorphism

$$g \to g^{\vee} = {}^{t}g^{-1}, \qquad g \in GL(N),$$

over $F$. Standard classical groups arise as fixed point groups of automorphisms in the associated inner class. In this paper, we shall be concerned with classical groups of similitudes. We therefore take the slightly larger group

$$\widetilde{G} = \mathrm{GL}(N) \times \mathrm{GL}(1)$$

over $F$, equipped with the outer automorphism

$$\alpha : (x, y) \to (x^{\vee}, \det(x)y), \qquad x \in \mathrm{GL}(N), \qquad y \in \mathrm{GL}(1).$$

The corresponding complex dual group

$$\widehat{\widetilde{G}} = \mathrm{GL}(N, \mathbb{C}) \times \mathbb{C}^*$$

comes with the dual outer automorphism

$$\widehat{\alpha} : (g, z) \to (g^{\vee} z, z), \qquad g \in \mathrm{GL}(N, \mathbb{C}), \qquad z \in \mathbb{C}^*.$$

Motivated by Langlands's conjectural parametrization of representations, we consider homomorphisms

$$\tilde{\psi} : \Gamma_F \to \widehat{\widetilde{G}},$$

from the Galois group $\Gamma_F = \mathrm{Gal}(\bar{F}/F)$ into $\widehat{\widetilde{G}}$. Each $\tilde{\psi}$ is required to be continuous, which is to say that it factors through a finite quotient $\Gamma_{E/F} = \mathrm{Gal}(E/F)$ of $\Gamma_F$, and is to be taken up to conjugacy in $\widehat{\widetilde{G}}$. Any $\tilde{\psi}$ may therefore be decomposed according to the representation theory of finite groups. We first write

$$\tilde{\psi} = \psi \oplus \chi : \sigma \to \psi(\sigma) \oplus \chi(\sigma), \qquad \sigma \in \Gamma_F,$$

where $\psi$ is a (continuous) $N$-dimensional representation of $\Gamma_F$, and $\chi$ is a (continuous) 1-dimensional character on $\Gamma_F$. We then break $\psi$ into a direct sum

$$\psi = \ell_1 \psi_1 \oplus \cdots \oplus \ell_r \psi_r,$$

for inequivalent irreducible representations

$$\psi_i : \Gamma_F \to \mathrm{GL}(N_i, \mathbb{C}), \qquad 1 \leq i \leq r,$$

and multiplicities $\ell_i$ such that

$$N = \ell_1 N_1 + \cdots + \ell_r N_r.$$

We shall be interested in maps $\tilde{\psi}$ that are $\widehat{\alpha}$*-stable*, in the sense that the homomorphism $\widehat{\alpha} \circ \tilde{\psi}$ is conjugate to $\tilde{\psi}$. It is clear that $\tilde{\psi}$ is $\widehat{\alpha}$-stable if and only if the $N$-dimensional representation

$$\psi^{\vee} \otimes \chi : \sigma \to \psi(\sigma)^{\vee} \chi(\sigma), \qquad \sigma \in \Gamma_F,$$

is equivalent to $\psi$. This in turn is true if and only if there is an involution $i \leftrightarrow i^{\vee}$ on the indices such that for any $i$, the representation $\psi_i^{\vee} \otimes \chi$ is equivalent to $\psi_{i^{\vee}}$, and $\ell_i$ equals $\ell_{i^{\vee}}$. We shall say that $\tilde{\psi}$ is $\widehat{\alpha}$*-discrete* if it satisfies the further constraint that for each $i$, $i^{\vee} = i$ and $\ell_i = 1$.

Suppose that $\tilde{\psi}$ is $\widehat{\alpha}$-discrete. Then

$$\psi = \psi_1 \oplus \cdots \oplus \psi_r,$$

for distinct irreducible representations $\psi_i$ of degree $N_i$ that are $\chi$-self dual, in the sense that $\psi_i$ is equivalent to $\psi_i^\vee \otimes \chi$. We write

$$\psi_i(\sigma)^\vee \chi(\sigma) = A_i^{-1}\psi_i(\sigma)A_i, \qquad \sigma \in \Gamma_F, \qquad 1 \leq i \leq r,$$

for fixed intertwining operators $A_i \in \mathrm{GL}(N_i, \mathbb{C})$. Applying the automorphism $g \to g^\vee$ to each side of the last equation, we deduce from Schur's lemma that

$${}^tA_i = c_i A_i,$$

for some complex number $c_i$ with $c_i^2 = 1$. The operator $A_i$ can thus be identified with a bilinear form on $\mathbb{C}^n$ that is symmetric if $c_i = 1$ and skew-symmetric if $c_i = -1$. We are of course free to replace any $\psi_i$ by a conjugate

$$B_i^{-1}\psi_i(w)B_i, \qquad B_i \in \mathrm{GL}(N_i, \mathbb{C}).$$

This has the effect of replacing $A_i$ by the matrix

$$B_i A_i {}^t B_i.$$

We can therefore assume that the intertwining operator takes a standard form

$$A_i = \begin{pmatrix} 0 & & & 1 \\ & & \cdot & \\ & \cdot & & \\ 1 & & & 0 \end{pmatrix}, \qquad \text{if } c_i = 1,$$

and

$$A_i = \begin{pmatrix} 0 & & & & & 1 \\ & & & & -1 & \\ & & & \cdot & & \\ & & \cdot & & & \\ & \cdot & & & & \\ & 1 & & & & \\ -1 & & & & & 0 \end{pmatrix}, \qquad \text{if } c_i = -1,$$

We shall say that $\psi_i$ is *orthogonal* or *symplectic* according to whether $c_i$ equals 1 or $-1$.

We have shown that the image of the homomorphism $\tilde{\psi}_i = \psi_i \oplus \chi$ is contained in the subgroup

$$\{(g, z) \in \mathrm{GL}(N_i, \mathbb{C}) \times \mathbb{C}^* : \ g^\vee z = A_i^{-1} g A_i\}$$

of $\mathrm{GL}(N_i, \mathbb{C}) \times \mathbb{C}^*$. If $(g, z)$ belongs to this subgroup, $z$ is the image

$$g \to z = \Lambda(g)$$

of $g$ under a rational character $\Lambda$. The subgroup of $\mathrm{GL}(N_i, \mathbb{C}) \times \mathbb{C}^*$ therefore projects isomorphically onto the subgroup

$$\{g \in \mathrm{GL}(N_i, \mathbb{C}) : A_i{}^t g A_i^{-1} g = \Lambda(g) I\}$$

of $\mathrm{GL}(N_i, \mathbb{C})$. This subgroup is, by definition, the group $\mathrm{GO}(N_i, \mathbb{C})$ of orthogonal similitudes if $c_i = 1$, and the group $\mathrm{GSp}(N_i, \mathbb{C})$ of symplectic similitudes if $c_i = -1$. In each case, the rational character $\Lambda$ is called the *similitude character* of the group. Set

$$N_{\pm} = \sum\nolimits_{i \in I_{\pm}} N_i, \qquad I_{\pm} = \{i : c_i = \pm 1\}.$$

We then obtain a decomposition

$$\psi = \psi_+ \oplus \psi_-,$$

where $\psi_+$ takes values in a subgroup of $\mathrm{GL}(N_+, \mathbb{C})$ that is isomorphic to $\mathrm{GO}(N_+, \mathbb{C})$, while $\psi_-$ takes values in a subgroup of $\mathrm{GL}(N_-, \mathbb{C})$ that is isomorphic to $\mathrm{GSp}(N_-, \mathbb{C})$. The original representation takes a form

$$\tilde{\psi} = \psi_+ \oplus \psi_- \oplus \chi,$$

in which the two similitude characters satisfy

$$\Lambda(\psi_+(\sigma)) = \Lambda(\psi_-(\sigma)) = \chi(\sigma), \qquad \sigma \in \Gamma_F.$$

The complex group $\mathrm{GSp}(N_-, \mathbb{C})$ is connected. It is therefore isomorphic to the complex dual group $\widehat{G}_-$ of a split group $G_-$ over $F$. If $N_+$ is odd, $\mathrm{GO}(N_+, \mathbb{C})$ is also connected. It is again isomorphic to a complex dual group $\widehat{G}_+$, for a split group $G_+$ over $F$. If $N_+ = 2n_+$ is even, however, the mapping

$$\nu : g \to \Lambda(g)^{-n_+} \det(g), \qquad g \in \mathrm{GO}(N_+, \mathbb{C}),$$

is a nontrivial homomorphism from $\mathrm{GO}(N_+, \mathbb{C})$ to the group $\{\pm 1\}$, whose kernel $\mathrm{SGO}(N_+, \mathbb{C})$ is connected. (See [Ra, §2].) The composition of $\psi_+$ with $\nu$ then provides a homomorphism from $\Gamma_F$ to a group of outer automorphisms of $\mathrm{SGO}(N_+, \mathbb{C})$. In this case, we take $G_+$ to be the quasisplit group over $F$ whose dual group is isomorphic to the group $\mathrm{SGO}(N_+, \mathbb{C})$, equipped with the given action of $\Gamma_F$. Having defined $G_+$ and $G_-$ in all cases, we write $G$ for the quotient of $G_+ \times G_-$ whose dual group is isomorphic to the subgroup

$$\widehat{G} = \{g_+, g_-, z) \in \widehat{G}_+ \times \widehat{G}_- \times \mathbb{C}^* : \ \Lambda(g_+) = \Lambda(g_-) = z\}$$

of $\widehat{G}_+ \times \widehat{G}_- \times \mathbb{C}^*$.

The quasisplit groups $G$ over $F$, obtained from $\widehat{\alpha}$-discrete homomorphisms $\tilde{\psi}$ as above, are called the (elliptic, $\alpha$-twisted) endoscopic groups for $\widetilde{G}$. Any such $G$ is determined up to isomorphism by a partition $N = N_+ + N_-$, and an extension $E$ of $F$ of degree at most two (with $E = F$ unless $N_+$ is even). One sees easily that there is a natural $L$-embedding

$$^L G = \widehat{G} \rtimes \Gamma_F \hookrightarrow {}^L\widetilde{G} = (\mathrm{GL}(n, \mathbb{C}) \times \mathbb{C}^*) \times \Gamma_F$$

of $L$-groups. Given the $\widehat{\alpha}$-discrete parameter $\tilde{\psi}$, we conclude that the mapping $\sigma \to \tilde{\psi}(\sigma) \times \sigma$ from $\Gamma_F$ to ${}^L\widetilde{G}$ factors through a subgroup ${}^LG$, for a unique endoscopic group $G$. The discussion above can also be carried out for $\widehat{\alpha}$-stable maps $\tilde{\psi}$ that are not $\widehat{\alpha}$-discrete. In this case, however, the mapping $\sigma \to \tilde{\psi}(\sigma) \times \sigma$ could factor through several subgroups ${}^LG$ of ${}^L\widetilde{G}$.

The classification of $\widehat{\alpha}$-discrete maps $\tilde{\psi}$ has been a simple exercise in elementary representation theory. The group $\Gamma_F$ plays no special role, apart from the property that its quotients of order two parametrize quadratic extensions of $F$. The discussion would still make sense if $\Gamma_F$ were replaced by a product of the group SL(2, $\mathbb{C}$) with the Weil group $W_F$, or more generally, the Langlands group $L_F$ of $F$. We recall [Ko, §12] that $L_F$ equals $W_F$ in the case that $F$ is local archimedean, and equals the product of $W_F$ with the group SU(2) if $F$ is local nonarchimedean. If $F$ is global, $L_F$ is a hypothetical group, which is believed to be an extension of $W_F$ by a product of compact, semisimple, simply connected groups. We assume its existence in what follows. Then in all cases, $L_F$ comes with a projection $w \to \sigma(w)$ onto a dense subgroup of $\Gamma_F$.

Having granted the existence of $L_F$, we consider continuous homomorphisms

$$\tilde{\psi} = \psi \oplus \chi : L_F \times \mathrm{SL}(2, \mathbb{C}) \to \widehat{\widetilde{G}} = \mathrm{GL}(n, \mathbb{C}) \times \mathbb{C}^*.$$

In this context, we also impose the condition that the restriction of $\tilde{\psi}$ to $L_F$ be unitary, or equivalently, that the image of $L_F$ in $\widehat{\widetilde{G}}$ be relatively compact. Assume that $\tilde{\psi}$ is $\widehat{\alpha}$-stable and $\widehat{\alpha}$-discrete. The discussion above then carries over verbatim. We obtain a decomposition

$$\psi = \psi_1 \oplus \cdots \oplus \psi_r,$$

for distinct irreducible representations

$$\psi_i : L_F \times \mathrm{SL}(2, \mathbb{C}) \to \mathrm{GL}(N_i, \mathbb{C}),$$

such that $\psi_i$ is equivalent to $\psi_i{}^\vee \otimes \chi$. We shall again say that $\psi_i$ is *symplectic* or *orthogonal*, according to whether its image is contained in the subgroup GSp($N_i$, $\mathbb{C}$) or GO($N_i$, $\mathbb{C}$) of GL($N_i$, $\mathbb{C}$). Combining the symplectic and orthogonal components as before, we see that $\tilde{\psi}$ factors through a subgroup ${}^LG$ of ${}^L\widetilde{G}$, for a unique (elliptic, $\alpha$-twisted) endoscopic group $G$ for $\widetilde{G}$.

We are working with a product $L_F \times$ SL(2, $\mathbb{C}$), in place of the original group $\Gamma_F$. This means that the irreducible components of $\psi$ decompose into tensor products

$$\psi_i = \mu_i \otimes \nu_i, \qquad 1 \leq i \leq r,$$

for irreducible representations $\mu_i\colon\ L_F \to \mathrm{GL}(m_i, \mathbb{C})$ and $\nu_i\colon\ \mathrm{SL}(2, \mathbb{C}) \to \mathrm{GL}(n_i, \mathbb{C})$ such that $N_i = m_i n_i$. Any irreducible representation of SL(2, $\mathbb{C}$) is automatically self dual. This means that for any $i$, $\mu_i$ is equivalent to $\mu_i{}^\vee \otimes \chi$. Moreover, the representation $\nu_i$ of SL(2, $\mathbb{C}$) is symplectic or orthogonal according to whether it is even or odd dimensional. It follows that $\psi_i$ is symplectic if and only if either $\mu_i$ is symplectic and $\nu_i$ is odd dimensional, or $\mu_i$ is orthogonal and $\nu_i$ is even dimensional.

**3.** We have classified the $\widehat{\alpha}$-discrete representations of the group $L_F \times \mathrm{SL}(2, \mathbb{C})$ in order to motivate a classification of symplectic automorphic representations. At this point, we may as well specialize to the case that $\psi$ is purely symplectic, which is to say that the corresponding group $\widehat{G}$ is purely symplectic. We assume henceforth that $N$ is even, and that $G$ is the split group over $F$ such that $\widehat{G}$ is isomorphic to GSp($N$, $\mathbb{C}$). Then $G$ is isomorphic to the general spin group

$$\mathrm{GSpin}(N+1) = (\mathrm{Spin}(N+1) \times \mathbb{C}^*)/\{\pm 1\}$$

over $F$. Our ultimate concern will in fact be the special case that $N$ equals 4. In this case, there is an exceptional isomorphism between GSpin($N+1$) and GSp($N$), so that $G$ is isomorphic to the group GSp(4) of the title.

For the given group $G \cong \mathrm{GSpin}(N+1)$, we write $\Psi(G) = \Psi(G/F)$ for the set of continuous homomorphisms $\psi$ from $L_F \times \mathrm{SL}(2, \mathbb{C})$ to $\widehat{G}$, taken up to conjugacy in $\widehat{G}$, such that the image of $L_F$ is relatively compact. For any such $\psi$, we set

$$\chi(w) = \Lambda(\psi(w, u)), \qquad w \in L_F, \qquad u \in \mathrm{SL}(2, \mathbb{C}),$$

where $\Lambda \colon \widehat{G} \to \mathbb{C}^*$ is the similitude character on $\widehat{G}$. Then $\chi = \chi_\psi$ is a one-dimensional unitary character on $L_F$. The correspondence $\psi \to \widetilde{\psi} = \psi \oplus \chi_\psi$ gives a bijection from $\Psi(G)$ to the subset of (equivalence classes of) $\widehat{\alpha}$-stable representations $\widetilde{\psi}$ such that the mapping

$$(w, u) \to \widetilde{\psi}(w, u) \times \sigma(w), \qquad w \in L_F, \qquad u \in \mathrm{SL}(2, \mathbb{C}),$$

factors through the subgroup ${}^LG$ of ${}^L\widetilde{G}$. We shall write $\Psi_2(G)$ for the subset of elements $\psi \in \Psi(G)$ such that $\widetilde{\psi}$ is $\widehat{\alpha}$-discrete. For any unitary 1-dimensional character $\chi$ on $L_F$, we also write $\Psi(G, \chi)$ and $\Psi_2(G, \chi)$ for the subsets of elements $\psi$ in $\Psi(G)$ and $\Psi_2(G)$, respectively, such that $\chi_\psi = \chi$.

If $\psi$ belongs to $\Psi(G)$, we set

$$S_\psi = \mathrm{Cent}_{\widehat{G}}(\mathrm{Im}(\psi)) = \{s \in \widehat{G} \colon s\psi(w, u) = \psi(w, u)s, (w, u) \in L_F \times \mathrm{SL}(2, \mathbb{C})\},$$

and also

$$\mathcal{S}_\psi = S_\psi / S_\psi^0 Z(\widehat{G}),$$

where $Z(\widehat{G}) \cong \mathbb{C}^*$ is the center of $\widehat{G}$. Then $\psi$ belongs to $\Psi_2(G)$ if and only if the connected group $S_\psi^0$ equals $Z(\widehat{G})$, which is to say that the group $S_\psi$ is finite modulo $Z(\widehat{G})$. It is not hard to compute $S_\psi$ directly in terms of the irreducible components $\psi_i$ of $\psi$. For example, if $\psi$ belongs to the subset $\Psi_2(G)$, and has $r$ components, then $S_\psi$ is isomorphic to $(\mathbb{Z}/2\mathbb{Z})^{r-1} \times \mathbb{C}^*$, while $\mathcal{S}_\psi$ is isomorphic to $(\mathbb{Z}/2\mathbb{Z})^{r-1}$.

We assume from now on that $F$ is global. We write $V_F$ for the set of valuations of $F$, and $V_{F,\infty}$ for the finite subset of archimedean valuations in $V_F$. The Langlands group $L_F$ is supposed to come with an embedding $L_{F_v} \hookrightarrow L_F$ for each $v \in V_F$. This embedding is determined up to conjugacy in $L_F$, and extends the usual conjugacy classes of embeddings $W_{F_v} \hookrightarrow W_F$ and $\Gamma_{F_v} \hookrightarrow \Gamma_F$. It gives rise to a restriction mapping

$$\psi \to \psi_v = \psi|_{L_{F_v} \times \mathrm{SL}(2,\mathbb{C})}, \qquad \psi \in \Psi(G),$$

from $\Psi(G) = \Psi(G/F)$ to $\Psi(G/F_v)$, which in turn provides an injection $S_{\psi_v} \to S_\psi$, and a homomorphism $\mathcal{S}_{\psi_v} \to \mathcal{S}_\psi$. Consider the special case that $\psi$ is unramified at $v$. This means that $v$ lies in the complement of $V_{F,\infty}$, and that for each $u \in \mathrm{SL}(2, \mathbb{C})$, the function

$$w_v \to \psi_v(w_v, u), \qquad w_v \in L_{F_v},$$

depends only on the image of $w_v$ in the quotient

$$W_{F_v}/I_{F_v} \cong F_v^*/\mathcal{O}_v^*$$

of $W_{F_v} = L_{F_v}$. Following standard notation, we write $\varpi_v$ for a fixed uniformizing element in $F_v^*$. Then $\varpi_v$ maps to a generator of the cyclic group $W_{F_v}/I_{F_v}$, and can also be mapped to the element

$$\begin{pmatrix} |\varpi_v|^{\frac{1}{2}} & 0 \\ 0 & |\varpi_v|^{-\frac{1}{2}} \end{pmatrix}$$

in $\mathrm{SL}(2, \mathbb{C})$. Composed with $\psi_v$, these mappings yield a semisimple conjugacy class

$$c_v(\psi) = c(\psi_v) = \psi_v\left(\varpi_v, \begin{pmatrix} |\varpi_v|^{\frac{1}{2}} & 0 \\ 0 & |\varpi_v|^{-\frac{1}{2}} \end{pmatrix}\right)$$

in $\widehat{G}$.

The Langlands group is assumed to have the property that its finite dimensional representations are unramified almost everywhere. It follows that any $\psi \in \Psi(G)$ determines a family

$$c(\psi) = \{c_v(\psi) = c(\psi_v) \colon v \notin V_\psi\}$$

of semisimple conjugacy classes in $\widehat{G}$, indexed by the complement of a finite subset $V_\psi \supset V_{F,\infty}$ of $V_F$. We note that if $\psi$ belongs to a subset $\Psi_2(G, \chi)$ of $\Psi(G)$, the family of complex numbers $c(\chi) = \{c_v(\chi)\}$ is equal to the image $\Lambda(c(\psi)) = \{\Lambda(c_v(\psi))\}$ of $c(\psi)$ under the similitude character. In general the relationships among the different elements in any family $c(\psi)$ convey much of the arithmetic information that is wrapped up in the homomorphism $\psi$.

There is of course another source of semisimple conjugacy classes in $\widehat{G}$, namely automorphic representations. If $\pi$ is an automorphic representation of $G$, the Frobenius–Hecke conjugacy classes provide a family

$$c(\pi) = \{c_v(\pi) = c(\pi_v) \colon\ v \notin V_\pi\}$$

of semisimple conjugacy classes in $\widehat{G}$, indexed by the complement of a finite subset $V_\pi \supset V_{F,\infty}$ of $V_F$. The elements in $c(\pi)$ are constructed in a simple way from the inducing data attached to unramified constituents $\pi_v$ of $\pi$. (See [B], for example.) Now a one dimensional character $\chi$ of $L_F$ amounts to an idèle class character of $F$, and this in turn can be identified with a character on the center of $G(\mathbb{A})$. Let $L^2_{\mathrm{disc}}(G(F)\backslash G(\mathbb{A}), \chi)$ be the space of $\chi$-equivariant, square integrable functions on $G(F)\backslash G(\mathbb{A})$ that decompose discretely under the action of $G(\mathbb{A})$. We write $\Pi_2(G, \chi)$ for the set of equivalence classes of automorphic representations of $G$ that are constituents of $L^2_{\mathrm{disc}}(G(F)\backslash G(\mathbb{A}), \chi)$. If $\pi$ belongs to $\Pi_2(G, \chi)$, the family $c(\chi)$ is equal to the image $\Lambda(c(\pi)) = \{\Lambda(c_v(\pi))\}$ of $c(\pi)$ under $\Lambda$. In general, the relationships among the different elements in any family $c(\pi)$ convey much of the arithmetic information that is wrapped up in the automorphic representation $\pi$.

The following conjecture was an outgrowth of Langlands's conjectural theory of endoscopy. We have stated it here somewhat informally. A more precise assertion, which applies to any group, is given in [A1] and [AG].

*Conjecture. (i) For any $\psi$, there is a canonical mapping $\pi \to \psi$ from $\Pi_2(G, \chi)$ to $\Psi_2(G, \chi)$ such that*

$$c(\pi) = c(\psi), \qquad \pi \in \Pi_2(G, \chi).$$

*(ii) Any fiber of the mapping is of the form*

$$\{\pi \in \Pi_2(G, \chi)\colon\ c(\pi) = c(\psi)\}, \qquad \psi \in \Psi_2(G, \chi),$$

*and can be characterized explicitly in terms of the groups $\mathcal{S}_{\psi_v}$, the diagonal image of the map*

$$\mathcal{S}_\psi \to \prod_v \mathcal{S}_{\psi_v},$$

*and a character*

$$\varepsilon_\psi : \mathcal{S}_\psi \to \{\pm 1\}$$

*attached to certain symplectic root numbers* [A1, §8].

**4.** The conjecture describes a classification of the automorphic $\chi$-discrete spectrum of $G$ in terms of mappings

$$\psi = \psi_1 \oplus \cdots \oplus \psi_r = (\mu_1 \otimes \nu_1) \oplus \cdots \oplus (\mu_r \otimes \nu_r)$$

in $\Psi_2(G, \chi)$. It is the simplest way to motivate what one might try to prove. The conjecture would actually be very difficult to establish in the form stated above (and in [A1] and [AG]). Indeed, one would first have to establish the existence and fundamental properties of the global Langlands group $L_F$. However, there is a natural way to reformulate the conjecture as a classification of automorphic representations of $G$ in terms of those of general linear groups. The idea is to reinterpret the "semisimple" constituents $\mu_i$ of $\psi$.

We may as well keep the idèle class character $\chi$ fixed from this point on. In the formulation above, $\mu_i$ stands for an irreducible unitary representation of $L_F$ of dimension $m_i$ that is $\chi$-self dual. The main hypothetical property of $L_F$ is that its irreducible unitary representations of dimension $m$ should be in canonical bijection with the unitary cuspidal automorphic representations of GL($m$). We could therefore bypass $L_F$ altogether by interpreting $\mu_i$ as an automorphic representation. From now on, $\mu_i$ will stand for a unitary, cuspidal automorphic representation of GL($m_i$) that is $\chi$-self dual, in the sense that the representation

$$x \to \mu_i(x^\vee)\chi(\det x), \qquad x \in \mathrm{GL}(m_i, \mathbb{A}),$$

is equivalent to $\mu_i$. As an automorphic representation of GL($m_i$), $\mu_i$ comes with a family

$$c(\mu_i) = \{c_v(\mu_i)\colon\ v \notin V_{\mu_i}\}$$

of Frobenius–Hecke conjugacy classes in GL($m_i$, $\mathbb{C}$). The family satisfies

$$c_v(\mu_i)^{-1}c_v(\chi) = c_v(\mu_i), \qquad v \notin V_{\mu_i},$$

since $\mu_i$ is $\chi$-self dual, and it determines $\mu_i$ uniquely, by the theorem of strong multiplicity one.

As an example, we shall describe the classification of $\chi$-self dual, unitary, cuspidal automorphic representations of GL(2).

*Example.* Suppose that $E$ is a quadratic extension of $F$, and that $\theta$ is an idèle class character of $E$. We assume that $\theta$ is not fixed by Gal($E/F$). Then there is a (unique) $\chi$-self dual, unitary, cuspidal automorphic representation $\mu = \mu(\theta)$ of GL(2) such that

$$c(\mu) = \{c_v(\mu) = \rho(c_v(\theta))\colon\ v \notin V_\theta\},$$

where $\theta$ is regarded as an automorphic representation of the group $K_E = \mathrm{Res}_{E/F}(\mathrm{GL}(1))$, and $\rho$ is the standard two dimensional representation of ${}^L K_E$. Conversely, suppose that $\mu$ is a $\chi$-self dual, unitary, cuspidal automorphic representation of GL(2). We write $\chi_\mu$ for the central character of $\mu$. It follows from the definitions that $\chi_\mu^2 = \chi^2$, or in other words, that the idèle class character $\eta_\mu = \chi_\mu\chi^{-1}$ of $F$ has order one or two. If $\eta_\mu \neq 1$, it is known that $\mu$ equals $\mu(\theta)$, for an idèle class character $\theta$ of the class field $E$ of $\eta_\mu$. In this case, we shall say that $\mu$ is of *orthogonal type*. If $\eta_\mu = 1$, $\mu$ is to be regarded as symplectic, for the obvious reason that GL(2) $\cong$ GSp(2). This is really the generic case, since if $\mu$ is *any* automorphic representation with central character $\chi_\mu$ equal to $\chi$, $\mu$ is automatically $\chi$-self dual.

For the group GL(4), some $\chi$-self dual, unitary, cuspidal automorphic representations can also be described in relatively simple terms. They are given by the following theorem of Ramakrishnan.

THEOREM: [Ra]. *Let $E$ be an extension of $F$ of degree at most two, and set*

$$H_E = \begin{cases} \mathrm{GL}(2) \times \mathrm{GL}(2), & \text{if } E = F, \\ \mathrm{Res}_{E/F}(\mathrm{GL}(2)), & \text{if } E \neq F. \end{cases}$$

*Let $\rho$ be the homomorphism from ${}^L H_E$ to the group* $\mathrm{GL}(4, \mathbb{C}) \cong \mathrm{GL}(M_2(\mathbb{C}))$ *defined by setting*

$$\rho(g_1, g_2)X = g_1 X^t g_2, \qquad g_1, g_2 \in \mathrm{GL}(2, \mathbb{C}),$$

*and*

$$\rho(\sigma)X = \begin{cases} X, & \text{if } \sigma_E = 1, \\ {}^t X, & \text{if } \sigma_E \neq 1, \end{cases}$$

*for any $X \in M_2(\mathbb{C})$, and for any $\sigma \in \Gamma_F$ with image $\sigma_E$ in $\Gamma_{E/F}$. Suppose that $\tau$ is a unitary, cuspidal automorphic representation of $H_E$ that is not a transfer from* GL(2) *(relative to the natural embedding of* $\mathrm{GL}(2, \mathbb{C})$ *into ${}^L H_E$), and whose central character is the pullback of $\chi$ (relative to the natural mapping from the center of $H_E$ to* GL(1)*). Then there is a unique $\chi$-self dual, unitary, cuspidal automorphic representation $\mu$ of* GL(4) *such that*

$$c(\mu) = \{c_v(\mu) = \rho(c_v(\tau)) \colon v \notin V_\tau\}.$$

We shall say that a $\chi$-self dual, unitary, cuspidal automorphic representation $\mu$ of GL(4) is of *orthogonal type* if it is given by the construction of Ramakrishnan's theorem. This is of course because $\mu$ is a transfer to GL(4) of a representation of the group GO(4).

Returning to our group $G \cong \mathrm{GSpin}(N+1)$, with $N$ even, we can try to construct objects $\psi$ for $G$ purely in terms of automorphic representations. We now define $\Psi_2(G, \chi)$ to be the set of formal (unordered) sums

$$\psi = \psi_1 \boxplus \cdots \boxplus \psi_r$$

of distinct, formal, $\chi$-self dual tensor products

$$\psi_i = \mu_i \boxtimes \nu_i, \qquad 1 \le i \le r,$$

of symplectic type. More precisely, $\nu_i$ is an irreducible representation of $\mathrm{SL}(2, \mathbb{C})$ of dimension $m_i$, and $\mu_i$ is a $\chi$-self dual, unitary, cuspidal automorphic representation of $\mathrm{GL}(m_i)$ that is of symplectic type if $n_i$ is odd and orthogonal type if $n_i$ is even, for integers $m_i$ and $n_i$ such that

$$N = N_1 + \cdots + N_r = m_1 n_1 + \cdots + m_r n_r.$$

To complete the definition, one would of course have to be able to say what it means for $\mu_i$ to be of symplectic or orthogonal type. In general, this must be done in terms of whether a certain automorphic $L$-function for $\mu_i$ (essentially the symmetric square or skew-symmetric square) has a pole at $s = 1$. The necessary consistency arguments for such a characterization are inductive, and require higher cases of the fundamental lemma, even when $N = 4$. However, if $m_i$ equals either 2 or 4, we can give an ad hoc characterization. In these cases, we have already defined what it means for $\mu_i$ to be of orthogonal type. We declare $\mu_i$ to be of symplectic type simply if it is not of orthogonal type. This expedient allows us to construct the family $\Psi_2(G, \chi)$ in the case that $N = 4$.

Suppose that $N$ is such that the set $\Psi_2(G, \chi)$ has been defined, and that

$$\psi = (\mu_1 \boxtimes \nu_1) \boxplus \cdots \boxplus (\mu_r \boxtimes \nu_r)$$

is an element in this set. For any $i$, $\mu_i$ comes with a family $c(\mu_i)$ of Frobenius–Hecke conjugacy classes in $\mathrm{GL}(m_i, \mathbb{C})$. The representation $\nu_i$ of $\mathrm{SL}(2, \mathbb{C})$ gives rise to its own family

$$c(\nu_i) = \left\{ c_v(\nu_i) = \nu_i \begin{pmatrix} |\varpi_v|^{\frac{1}{2}} & 0 \\ 0 & |\varpi_v|^{-\frac{1}{2}} \end{pmatrix} : \ v \notin V_{F,\infty} \right\}$$

of conjugacy classes in $\mathrm{GL}(n_i, \mathbb{C})$. The tensor product family

$$c(\psi_i) = \{ c_v(\psi_i) = c_v(\mu_i) \otimes c_v(\nu_i) \colon \ v \notin V_{\mu_i} \}$$

is then a family of semisimple conjugacy classes in the group $\mathrm{GL}(N_i, \mathbb{C}) = \mathrm{GL}(m_i n_i, \mathbb{C})$. Taking the direct sum over $i$, we obtain a family

$$c(\psi) = \bigoplus_{i=1}^{r} c(\psi_i) = \left\{ c_v(\psi) = \bigoplus_{i=1}^{r} c_v(\psi_i) \colon \ v \notin V_\psi \right\}$$

of semisimple conjugacy classes in $\mathrm{GL}(N, \mathbb{C})$. This is of course parallel to the family of conjugacy classes constructed with the earlier interpretation of $\psi$ as a representation of $L_F \times \mathrm{SL}(2, \mathbb{C})$. We also set

$$\mathcal{S}_\psi = (\mathbb{Z}/2\mathbb{Z})^{r-1},$$

as before. We can then define a character

$$\varepsilon_\psi : \mathcal{S}_\psi \to \{\pm 1\}$$

in terms of symplectic root numbers by copying the prescription in [A1, §8].

The local Langlands conjecture has now been proved for the general linear groups $\mathrm{GL}(m_i)$ [HT], [He]. We can therefore identify any local component

$$\psi_v = (\mu_{1,v} \boxtimes \nu_1) \boxplus \cdots \boxplus (\mu_{r,v} \boxtimes \nu_r)$$

of an element $\psi \in \Psi_2(G, \chi)$ with an $N$-dimensional representation of the group $L_{F_v} \times \mathrm{SL}(2, \mathbb{C})$. In the process of proving the classification theorem stated below, one shows that if $\mu_i$ is either symplectic or orthogonal (in the sense alluded to above), the same holds for the local components $\mu_{i,v}$ (as representations of $L_{F_v}$). It follows that $\psi_v$ can be identified with a homomorphism from $L_{F_v} \times \mathrm{SL}(2,\mathbb{C})$ into $\widehat{G}$. In particular, we can define the groups $S_{\psi_v}$ and $\mathcal{S}_{\psi_v}$ as before, in terms of the centralizer of the image of $\psi_v$. Moreover, there is a canonical homomorphism $s \to s_v$ from $\mathcal{S}_\psi$ to $\mathcal{S}_{\psi_v}$. The groups $\mathcal{S}_\psi$ and $\mathcal{S}_{\psi_v}$ are always abelian, and in fact are products of groups $\mathbb{Z}/2\mathbb{Z}$.

**5.** We shall now specialize to the case that $N = 4$. Thus, $\widehat{G}$ is isomorphic to $\mathrm{GSp}(4,\mathbb{C})$, and

$$G \cong \mathrm{GSpin}(5) \cong \mathrm{GSp}(4).$$

As we have noted, the set $\Psi_2(G,\chi)$ can be defined explicitly in this case in terms of certain cuspidal automorphic representations of general linear groups.

The object of this article has been to announce the following classification theorem for automorphic representations of $G$. The theorem is contingent upon cases of the fundamental lemma that are in principle within reach, and I should also admit, the general results in [A3] that have still to be written up in detail.

Classification Theorem. (i) *There exist canonical local packets*

$$\Pi_{\psi_v}, \qquad \psi \in \Psi_2(G,\chi), \qquad v \in V_F,$$

*of (possibly reducible) representations of the groups $G(F_v)$, together with injections*

$$\pi_v \to \xi_{\pi_v}, \qquad \pi_v \in \Pi_{\psi_v},$$

*from these packets to the associated finite groups $\widehat{\mathcal{S}}_{\psi_v}$ of characters on $\mathcal{S}_{\psi_v}$.*

(ii) *The automorphic discrete spectrum attached to $\chi$ has an explicit decomposition*

$$L^2_{\mathrm{disc}}(G(F)\backslash G(\mathbb{A}),\chi) = \bigoplus_{\psi\in\Psi_2(G,\chi)} \ \bigoplus_{\{\pi\in\prod_\psi:\ \xi_\pi=\varepsilon_\psi\}} \pi$$

*in terms of the global packets*

$$\Pi_\psi = \{\pi = \bigotimes_v \pi_v :\ \pi_v \in \Pi_{\psi_v},\ \xi_{\pi_v} = 1 \textit{ for almost all } v\}$$

*of (possibly reducible) representations of $G(\mathbb{A})$, and corresponding characters*

$$\xi_\pi : s \to \prod_v \xi_{\pi_v}(s_v), \qquad s \in \mathcal{S}_\psi, \qquad \pi \in \Pi_\psi,$$

*on the groups $\mathcal{S}_\psi$.*

(iii) *The global packets*

$$\Pi_\psi, \qquad \psi \in \Pi_2(G,\chi),$$

*are disjoint, in the sense that no irreducible representation of* $G(\mathbb{A})$ *is a constituent of representations in two distinct packets. Moreover, if* $\psi$ *belongs to the subset* $\Psi_{ss,2}(G, \chi)$ *of elements in* $\Psi_2(G, \chi)$ *that are trivial on* SL(2, $\mathbb{C}$)*, the packet* $\Pi_\psi$ *contains only irreducible representations. Thus, for any* $\psi \in \Psi_{ss,2}(G, \chi)$*, any representation* $\pi \in \Pi_\psi$ *occurs in* $L^2_{\text{disc}}(G(F)\backslash G(\mathbb{A}), \chi)$ *with multiplicity* 1 *or* 0.

*Remarks.* 1. The local packets $\Pi_{\psi_v}$ in part (i) are defined by the endoscopic transfer of characters. More precisely, the characters of representations in $\Pi_{\psi_v}$ are defined in terms of Langlands–Shelstad (and Kottwitz–Shelstad) transfer mappings of functions, and the groups $\mathcal{S}_{\psi_v}$. I do not know whether the representations in $\Pi_{\psi_v}$ are generally irreducible. However, in the case that $\psi$ is unramified at $v$, one can at least show that the preimage of the trivial character in $\widehat{\mathcal{S}}_{\psi_v}$ under the mapping $\pi_v \to \xi_{\pi_v}$ is irreducible.

2. If $\psi$ belongs to the complement of $\Psi_{ss,2}(G, \chi)$ in $\Psi_2(G, \chi)$, the representations in the packet $\Pi_\psi$ are all nontempered. On the other hand, if $\psi$ belongs to $\Psi_{ss,2}(G, \chi)$, the generalized Ramanujan conjecture (applied to the groups GL(2) and GL(4)) implies that the representations in the packet $\Pi_\psi$ are tempered. Thus, the multiplicity assertion at the end of the theorem pertains to what ought to be the tempered constituents of $L^2_{\text{disc}}(G(F)\backslash G(\mathbb{A}), \chi)$. If $\psi$ is a more general element in $\Psi_2(G, \chi)$, and if the direct sum of the representations in each local packet $\Pi_{\psi_v}$ is multiplicity free, the irreducible constituents of the representations in $\Pi_\psi$ also occur with multiplicity 1 or 0.

The multiplicity formula of the theorem is a quantitative description of the decomposition of the discrete spectrum. The general structure of the parameters $\psi$ also provides useful qualitative information about the spectrum. We shall conclude with a list of the six general families of automorphic representations that occur in the discrete spectrum. In each case, we shall describe the relevant parameters $\psi$, the corresponding families of Frobenius–Hecke conjugacy classes, the groups $\mathcal{S}_\psi$, and the sign characters $\varepsilon_\psi$ on $\mathcal{S}_\psi$. (The characters $\varepsilon_\psi$ are in fact trivial for all but one of the six families.) We shall write $\nu(n)$ for the irreducible representation of SL(2, $\mathbb{C}$) of dimension $n$. Observe that the Frobenius–Hecke conjugacy classes

$$c(\nu(n)) = \left\{ c_v(\nu(n)) = \begin{pmatrix} |\varpi_v|^{\frac{n-1}{2}} & & & 0 \\ & |\varpi_v|^{\frac{n-3}{2}} & & \\ & & \ddots & \\ 0 & & & |\varpi_v|^{-\frac{n-1}{2}} \end{pmatrix} \right\}$$

of $\nu(n)$ have positive real eigenvalues. This is in contrast to the case of a unitary, cuspidal automorphic representation $\mu$ of GL($m$), which according to the generalized

Ramanujan conjecture, has Frobenius–Hecke conjugacy classes

$$c(\mu) = \left\{ c_v(\mu) = \begin{pmatrix} c_v^1(\mu) & & 0 \\ & \ddots & \\ 0 & & c_v^m(\mu) \end{pmatrix} \right\}$$

whose eigenvalues lie on the unit circle.

We list the six families according to how they behave with respect to stability (for the multiplicities of representations $\pi \in \Pi_\psi$) and the implicit Jordan decomposition (for elements $\psi \in \Psi_2(G, \chi)$). I have also taken the liberty of assigning proper names to some of the families, which I hope give fair reflection of their history.

(a) Stable, semisimple (general type)

$$\psi = \psi_1 = \mu \boxtimes 1,$$

where $\mu$ is a $\chi$-self dual, unitary cuspidal automorphic representation of GL(4) that is not of orthogonal type,

$$c(\psi) = c(\mu) = \left\{ \begin{pmatrix} c_v^1(\mu) & & 0 \\ & \ddots & \\ 0 & & c_v^4(\mu) \end{pmatrix} \right\},$$

$$\mathcal{S}_\psi = 1,$$
$$\varepsilon_\psi = 1.$$

(b) Unstable, semisimple (Yoshida type [Y])

$$\psi = \psi_1 \boxplus \psi_2 = (\mu_1 \boxtimes 1) \boxplus (\mu_2 \boxtimes 1),$$

where $\mu_1$ and $\mu_2$ are *distinct*, unitary, cuspidal automorphic representations of GL(2) whose central characters satisfy $\chi_{\mu_1} = \chi_{\mu_2} = \chi$,

$$c(\psi) = c(\mu_1) \oplus c(\mu_2) = \left\{ \begin{pmatrix} c_v^1(\mu_1) & & & 0 \\ & c_v^1(\mu_2) & & \\ & & c_v^2(\mu_2) & \\ 0 & & & c_v^2(\mu_1) \end{pmatrix} \right\},$$

$$\mathcal{S}_\psi = \mathbb{Z}/2\mathbb{Z},$$
$$\varepsilon_\psi = 1.$$

(c) Stable, mixed (Soudry type [So])

$$\psi = \psi_1 = \mu \boxtimes \nu(2),$$

where $\mu = \mu(\theta)$ is a unitary, cuspidal automorphic representation of GL(2) of orthogonal type with $\chi_\mu^2 = \chi$,

$$\begin{aligned} c(\psi) &= c(\mu) \otimes c(\nu(2)) \\ &= \left\{ \begin{pmatrix} c_v^1(\mu)|\varpi_v|^{\frac{1}{2}} & & & 0 \\ & c_v^2(\mu)|\varpi_v|^{\frac{1}{2}} & & \\ & & c_v^1(\mu)|\varpi_v|^{-\frac{1}{2}} & \\ 0 & & & c_v^2(\mu)|\varpi_v|^{-\frac{1}{2}} \end{pmatrix} \right\}, \\ \mathcal{S}_\psi &= 1, \\ \varepsilon_\psi &= 1. \end{aligned}$$

(d) Unstable, mixed (Saito, Kurokawa type [Ku])

$$\psi = \psi_1 \boxplus \psi_2 = (\lambda \boxtimes \nu(2)) \boxplus (\mu \boxtimes 1),$$

where $\lambda$ is an idèle class character of $F$ and $\mu$ is a unitary, cuspidal automorphic representation of GL(2), with $\lambda^2 = \chi_\mu = \chi$,

$$\begin{aligned} c(\psi) &= (c(\lambda) \otimes c(\nu(2)) \oplus (c(\mu)) \\ &= \left\{ \begin{pmatrix} c_v(\lambda)|\varpi_v|^{\frac{1}{2}} & & & 0 \\ & c_v^1(\mu) & & \\ & & c_v^2(\mu) & \\ 0 & & & c_v(\lambda)|\varpi_v|^{-\frac{1}{2}} \end{pmatrix} \right\}, \\ \mathcal{S}_\psi &= \mathbb{Z}/2\mathbb{Z}, \\ \varepsilon_\psi &= \begin{cases} 1, & \text{if } \varepsilon\left(\frac{1}{2}, \mu \otimes \lambda^{-1}\right) = 1, \\ \text{sgn}, & \text{if } \varepsilon\left(\frac{1}{2}, \mu \otimes \lambda^{-1}\right) = -1, \end{cases} \end{aligned}$$

where sgn is the nontrivial character on $\mathbb{Z}/2\mathbb{Z}$.

(e) Unstable, almost unipotent (Howe, Piatetski-Shapiro type [HP])

$$\psi = \psi_1 \boxplus \psi_2 = (\lambda_1 \boxtimes \nu(2)) \boxplus (\lambda_2 \boxtimes \nu(2)),$$

where $\lambda_1$ and $\lambda_2$ are distinct idèle class characters of $F$ with $\lambda_1^2 = \lambda_2^2 = \chi$,

$$\begin{aligned} c(\psi) &= (c(\lambda_1) \otimes c(\nu(2))) \oplus (c(\lambda_2) \otimes c(\nu(2))) \\ &= \left\{ \begin{pmatrix} c_v(\lambda_1)|\varpi_v|^{\frac{1}{2}} & & & 0 \\ & c_v(\lambda_2)|\varpi_v|^{\frac{1}{2}} & & \\ & & c_v(\lambda_2)|\varpi_v|^{-\frac{1}{2}} & \\ 0 & & & c_v(\lambda_1)|\varpi_v|^{-\frac{1}{2}} \end{pmatrix} \right\}, \\ \mathcal{S}_\psi &= \mathbb{Z}/2\mathbb{Z}, \\ \varepsilon_\psi &= 1. \end{aligned}$$

(f) Stable, almost unipotent (one dimensional type)

$$\psi = \psi_1 = \lambda \boxtimes \nu(4),$$

where $\lambda$ is an idèle class character of $F$ with $\lambda^4 = \chi$,

$$\begin{aligned} c(\psi) &= c(\lambda) \otimes c(\nu(4)) \\ &= \left\{ \begin{pmatrix} c_v(\lambda)|\varpi_v|^{\frac{3}{2}} & & & 0 \\ & c_v(\lambda)|\varpi_v|^{\frac{1}{2}} & & \\ & & c_v(\lambda)|\varpi_v|^{-\frac{1}{2}} & \\ 0 & & & c_v(\lambda)|\varpi_v|^{-\frac{3}{2}} \end{pmatrix} \right\}, \\ \mathcal{S}_\psi &= 1, \\ \varepsilon_\psi &= 1. \end{aligned}$$

---

## REFERENCES


[A1] J. Arthur, Unipotent automorphic representations, *Astérisque* **171–172** (1989), 13–71.

[A2] ________, A stable trace formula III. Proof of the main theorems, *Ann. of Math.* (to appear).

[A3] ________, *Automorphic Representations of Classical Groups,* in preparation.

[AG] J. Arthur and S. Gelbart, Lectures on automorphic L-functions. I, L- functions and Arithmetic, *London Math. Soc. Lecture Note Series,* vol. 153, Cambridge Univ. Press, London, 1991, pp. 2–21.

[B] A. Borel, Automorphic L-functions, in Automorphic Forms, Representations and L-functions. I, *Proc. Sympos. Pure Math., Amer. Math. Soc.* **33** (1979), 27–61.

[BWW] J. Ballmann, R. Weissauer and U. Weselmann, Remarks on the fundamental lemma for stable twisted endoscopy of classical groups, preprint.

[CKPS] J. Cogdell, H. Kim, I. Piatetski-Shapiro and F. Shahidi, On lifting from classical groups to $\mathrm{GL}_N$, *Inst. Hautes Études Sci. Publ. Math.* **93** (2001), 5–30.

[F] Y. Flicker, Matching of Orbital Integrals on GL(4) and GSp(2), *Memoirs of Amer. Math. Soc.,* 1999.

[GRS] D. Ginzburg, S. Rallis and D. Soudry, Generic automorhic forms on $\mathrm{SO}(2n+1)$: Functorial lift to $\mathrm{GL}(2n)$, endoscopy and base change, *Intl. Math. Res. Notices* **14** (2001), 729–764.

[Ha] T. Hales, The fundamental lemma for Sp(4), *Proc. Amer. Math. Soc.* **125** (1997), 301–308.

[HT] M. Harris and R. Taylor, On the geometry and cohomology of some simple Shimura varieties, *Ann. of Math. Stud.,* vol. 151, Princeton Univ. Press, 2001.

[He] G. Henniart, Une preuve simple des conjectures de Langlands de $\mathrm{GL}(n)$ sur un corps $p$-adique, *Invent. Math.* 139 (2000), 439–455.

[HP] R. Howe and I. Piatetski-Shapiro, A counterexample to the "generalized Ramanujan conjecture" for (quasi-)split groups, Automorphic Forms, Representations, and L-functions, *Proc. Sympos. Pure Math. Amer. Math. Soc.* **33** (1979), 315–322.

[Ko] R. Kottwitz, Stable trace formula: cuspidal tempered terms, *Duke Math. J.* **51** (1984), 611–650.

[Ku] N. Kurokawa, Examples of eigenvalues of Hecke operators on Siegel cusp forms of degree two, *Invent. Math.* **49** (1978), 149–165.

[Ra] D. Ramakrishnan, Modularity of solvable Artin representations of GO(4)-type, *Int. Math. Res. Not.* **1** (2002), 1–54.

[Ro] B. Roberts, Global L-packets for GSp(2) and theta lifts, *Doc. Math.* **6** (2001), 247–314.

[Sc] M. Schröder, *Zählen der Punkte* mod $p$ *einer Schimuravarietät zu GSp4 durch die* $L^2$*-Spurformel von Arthur: Die Kohomologie der Zentralisatoren halbeinfacher Elemente und Orbitalintegrale* *auf halbeifachen Elementen zu gewissen Heckeoperatoren,* Inauguraldissertation, Mannheim, 1993.

[So] D. Soudry, *Automorphic forms on* GSp(4), in *Festschrift in honor of I. I. Piatetski-Shapiro on the occasion of his sixtieth birthday,* part II, 1989, 291–303.

[W] J.-L. Waldspurger, Homogénéité de certaines distributions sur les groupes p-adiques, *Inst. Hautes Études Sci. Publ. Math.* **81** (1995), 22–72.

[Y] H. Yoshida, Siegel's modular forms and the arithmetic of quadratic forms, *Invent. Math.* **60** (1980), 193–248.

CHAPTER 5

# ELLIPTIC CURVES, HILBERT MODULAR FORMS, AND THE HODGE CONJECTURE

By Don Blasius

---

## 1. Introduction.

**1.1.** Recall the following special case of a foundational result of Shimura ([S2, Theorems 7.15 and 7.16]):

Theorem. *Let $f$ be a holomorphic newform of weight 2 with rational Fourier coefficients $\{a_n(f) | n \geq 1\}$. There exists an elliptic curve $E$ defined over $\mathbf{Q}$ such that, for all but finitely many of the primes $p$ at which $E$ has good reduction $E_p$, the formula*

$$a_p(f) = 1 - N_p(E) + p$$

*holds. Here $N_p(E)$ denotes the number of points of $E_p$ over the field with p elements.*

**1.2.** The first result of this type is due to Eichler ([E]) who treated the case where $f = f_{11}$ is the unique weight 2 newform for $\Gamma_0(11)$ and $E$ is the compactified modular curve for this group. Later, in several works, Shimura showed that the Hasse-Weil zeta functions of special models (often called *canonical models*) of modular and quaternionic curves are, at almost all finite places $v$, products of the $v$-Euler factors attached to a basis of the Hecke eigenforms of the given level. These results give at once computations of the zeta functions of the Jacobians of these curves since $H_l^1(C) = H_l^1(Jac(C))$ for a smooth projective curve. However, it was only in late 1960's that the correspondence between individual forms and geometry came to be emphasized. In particular, in his proof of the above Theorem, Shimura identified $L(f, s)$ as the zeta function of a one dimensional factor $E_f$ of the Jacobian variety. He also treated the case where the $a_n(f)$ are not rational; then $E_f$ is replaced by a higher dimensional factor (or, alternatively, quotient) $A_f$ of the Jacobian ([S3] and [S4]).

By Tate's conjecture ([F]), the above correspondence $f \rightarrow E_f$ determines $E_f$ up to an isogeny defined over $\mathbf{Q}$. Further, by works of Igusa, Langlands, Deligne, Carayol, and others, a completed result is known: the conductor $N_E$ of E coincides with the conductor $N_f$ of f, the above formula holds for all p such that $(N_f, p) = 1$, and corresponding but more complicated statements (the local Langlands correspondence) hold at the primes $p$ which divide $N_f$.

---

Manuscript received October 22, 2002.

**1.3.** Our goal here is to give a *conditional* generalisation of Shimura's result to totally real fields $F$, i.e. to Hilbert modular forms. Thus, we replace $f$ by a cuspidal automorphic representation $\pi = \pi_\infty \otimes \pi_f$ of the adele group $GL_2(A_F)$. The weight 2 condition generalises to the requirements that (i) $\pi_\infty$ belong to the lowest discrete series as a representation of $GL_2(F \otimes \mathbf{R})$, and (ii) $\pi$ have central character $\omega_\pi$ equal to the inverse of the usual idelic norm. In the classical language of holomorphic forms on (disjoint unions of) products of upper-half planes, condition (i) asserts that on each product the form be of diagonal weight $(2, \ldots, 2)$. For each finite place $v$ of $F$ at which $\pi$ is unramified, we have the Hecke eigenvalue $a_v(\pi)$. We assume that the $a_v(\pi)$ belong to $\mathbf{Q}$ for all such v, and that the Hecke operators are so normalized that, for almost all $v$, the polynomial $T^2 - a_v(\pi)T + N_v$ has its zeros at numbers of size $N_v^{1/2}$, where $N_v$ denotes the number of elements in the residue field of $F$ at $v$. Here is the now standard conjectural generalisation of Shimura's result.

*Existence Conjecture* 1.4. For $\pi$ as above, there exists an elliptic curve $E$ defined over $F$ such that for all but finitely many of the finite places $v$ of $F$ at which $E$ has good reduction $E_v$,

$$a_v(\pi) = 1 - N_v(E) + N_v$$

holds. Here $N_v(E)$ is the number of points of $E_v$ over the residue field at $v$. It should be noted that if the Existence Conjecture is proven, then the appropriate statements of the Langlands correspondence hold at all places. (This follows from [T1], [T2], and the Cebotarev theorem.)

**1.5.** Central to our argument is an unproved hypothesis of Deligne: the *Absolute Hodge Conjecture* ([D4]). This conjecture can be stated in several ways. For us, its categorical formulation is most directly useful. To recall it, let $M_{\mathbf{C}}$ denote the tensor category of motives for absolute Hodge cycles defined over $\mathbf{C}$ (cf. [D4]). By Hodge theory, the usual topological cohomology functor on varieties, which attaches to each projective smooth complex variety $X$ its total cohomology ring $H_B^*(X, \mathbf{Q})$, takes values in the tensor category of polarisable rational Hodge structures. It extends to the category $M_{\mathbf{C}}$ and we denote this extended functor by $\omega_B$. The rational Hodge structure attached to a motive $M$, defined over a subfield $L$ of $\mathbf{C}$, is $M_B = \omega_B(M \otimes \mathbf{C})$, where $M \otimes \mathbf{C}$ is the base change of $M$ to $\mathbf{C}$.
In this language the Absolute Hodge Conjecture asserts simply:

*The functor $\omega_B$ is fully faithful.*

In fact we shall use precisely the assertion:

*If M and N are motives for absolute Hodge cycles defined over* $\mathbf{C}$*, and $M_B$ is isomorphic to $N_B$ as Hodge structure, then M is isomorphic to N.*

Of course, the Absolute Hodge Conjecture (AHC) is trivially a consequence of the usual Hodge Conjecture. The AHC was proved for all abelian varieties over subfields of **C** (in particular, for all products of curves over **C**) by Deligne, and as such it has been of great utility in the theory of Shimura varieties. We need to use it in this paper for a product of a Picard modular surface (e.g. an arithmetic quotient of the unit ball) and an abelian variety. In such a case the conjecture is unknown.

**1.6.** The main result here is the following

THEOREM 1. *Suppose that the Absolute Hodge Conjecture is true. Then the Existence Conjecture is true.*

**1.7.** For background it is essential to recall the known cases of the Existence Conjecture.

**1.7.1.** It is an easy consequence of work of Hida ([H]) and Faltings ([F]) in the cases covered by the following hypotheses **QC**:

**QC1.** [F:Q] is odd, or
**QC2.** $\pi$ has a finite place at which $\pi_v$ belongs to the discrete series.

In Section 2 below this case is deduced from Hida's work.

**1.7.2.** The conjecture is also known for all forms $\pi$ of CM type. Here $\pi$ is defined to be of CM type if there exists a non-trivial idele class character $\epsilon$ (necessarily of order 2) of $F$ such that the representation $\pi \otimes \epsilon$ is isomorphic to $\pi$. See 2.2 below for some comments.

**1.7.3.** However, to our knowledge, the existence conjecture is not known for the non-CM-type everywhere unramified representations $\pi$ in the case where $F$ is a real quadratic extension of **Q**, except in the case where $\pi$ is the quadratic twist of a base change from **Q**. One may regard these forms as the test case for any construction.

**1.8.** In recent years, many automorphic correspondences, more sophisticated than the Jacquet-Langlands correspondence, have been proven. It is natural to ask whether, at least in principle, there should exist some other such functorialities to a group of Hermitian symmetric type from which the sought $E$'s can be directly constructed. However, some new ideas will be needed. Indeed, we have the following folklore result:

THEOREM 2. *Every simple abelian variety whose $H^1$ Hodge structure occurs in $H^1$ of a Shimura variety is isogenous to a base change of a factor of the Jacobian of a quaternionic Shimura curve.*

The proof of this easy negative result is sketched, if not proven, below in 2.3. Further, it is not hard to see (proof omitted!) that there do exist elliptic curves over $F$, e.g. for $F$ real quadratic, which are not isogenous over $\mathbf{C}$ to any such factor. Hence, it seemed reasonable to investigate whether any known principles (e.g. the Hodge Conjecture) could provide the additional abelian varieties needed for the Existence Conjecture.

**1.9.** Here is a brief outline of the proof of Theorem 1. There are 4 main steps:

**1.9.1.** We use a sequence of functorialities to find an orthogonal rank 3 motive $M$ in the second cohomology of a Picard modular surface ([LR]) which has the same L-function as the symmetric square of a base change of $\pi$.

**1.9.2.** From the weight 2 Hodge structure $M_B$ of $M$, we construct a rank 2 rational Hodge structure $R$ of type (1,0) (0,1) whose symmetric square is isomorphic to $M_B$. We let $A = A_{\mathbf{C}}$ be an elliptic curve over $\mathbf{C}$ whose $H^1_B(A)$ is isomorphic as Hodge structure to $R$.

**1.9.3.** Using the Absolute Hodge Conjecture, we descend $A$ to a curve, also denoted $A$, defined over a number field $L$ which contains $F$.

**1.9.4.** Using the existence ([T1, BR2]) of a two dimensional $l$-adic representation attached to $\pi$, we find a $D$ of dimension 1 inside the Weil restriction of scalars of $A$ from $L$ to $F$; by construction, $D$ has the correct $l$-adic representations.

**1.10.** Since the result is in any case conditional, and for simplicity, we have usually treated only the case of rational Hecke eigenvalues where the sought variety really is an elliptic curve. However, the proof extends, with some changes, to the general case of forms with $\pi_\infty$ belonging to the lowest discrete series. At the suggestion of the referee, we have indicated significant changes needed for the case of general Hecke field $T_\pi$, which is defined as the number field generated by the $a_v(\pi)$. In this case it is more convenient to state the conjecture cohomologically:

*For $\pi$ of weight $(2, \ldots, 2)$, having Hecke field $T_\pi$, there exists an abelian variety $A_\pi$, defined over $F$, such that $End(A_\pi) = T_\pi$, and such that for all but finitely many of the finite places $v$ of $F$ at which $A_\pi$ has good reduction,*

$$L_v\left(H_l^1(A_\pi), T_\pi, s\right) = L_v(\pi, s).$$

Here $L_v(H_l^1(A_\pi), T_\pi, s)$ is the L-function denoted $\zeta(s; A_\pi/F, T_\pi)$ in ([S2, Section 7.6]), and is denoted by $L_v(H_l^1(B_\pi), \sigma_l, s)$ below in 2.1, for the case $\sigma_l = id$.

**1.11.** Finally, we note that in several lectures on this topic the proof was given with alternative first steps using quaternionic surfaces. The proof given here is a little simpler, but the earlier construction is still useful. In particular, it enables the unconditional proof of the Ramanujan conjecture at all places for holomorphic Hilbert modular forms which are discrete series at infinity ([B]).

*Notations* 1.12. Throughout the chapter, $F$ denotes a totally real field, $K_0$ is a fixed quadratic imaginary extension of $\mathbf{Q}$ and $K = FK_0$. The letter $L$ will denote a number field whose definition depends on context. The symbol $\pi$ denotes a fixed cuspidal automorphic representation, as above, of $GL(2, A_F)$, i.e. which is (i) of lowest discrete series type at infinity, and (ii) has central character equal to the inverse of the norm.

*Acknowledgments.* This work was largely done during a visit to SFB 478 in Münster, Germany; this hospitality was greatly appreciated.

## 2. Known cases of the existence conjecture and negative background.

**2.1. The case QC.** Hida ([H, Theorem 4.12]) generalised most of Shimura's result (1.1), to the curves defined by quaternion algebras over totally real fields. Thus, to use Hida's work, one first invokes the Jacquet-Langlands correspondence ([JL]) to find an automorphic representation $\pi_Q$ with the same L-function as $\pi$ but on the adele group associated to the muliplicative group of a suitable quaternion algebra $Q$ over $F$. Here suitable means that the canonical models of the associated arithmetic quotients are curves defined over $F$. The hypotheses **QC** above are necessary and sufficient for such a pair $(Q, \pi_Q)$ to exist.

Let $T = T_\pi$ be the Hecke field of $\pi$. It is a number field which is either totally real or a totally imaginary quadratic extension of a totally real field. By Hida ([H, Prop. 4.8]) there exists an abelian variety $B_\pi$ and a $T_\pi$-subalgebra $T$ of $End(B_\pi)$, which is isomorphic to a direct sum of number fields, such that $H^1_l(B_\pi)$ is a free rank 2 $T \otimes \mathbf{Q}_l$-module. Further, for almost all finite places $v$ of $F$, all $l$ prime to $v$, and all morphisms $\sigma_l : T \to \overline{\mathbf{Q}}_l$,

$$L_v\left(H^1_l(B_\pi), \sigma_l, s\right) = L_v(\pi, \sigma_l, s)^d$$

where $d = [T : T_\pi]$, the $L$-factor on the left hand side is that of the $\sigma_l$-eigensubspace of $H^1_l(B_\pi) \otimes \overline{\mathbf{Q}}_l$, and that on the right hand side is the L-factor with coefficients in $\sigma_l(T_\pi)$ defined by applying $\sigma_l$ to the coefficients of $L_v(\pi, s)$. (The equality makes sense if one recalls the usual convention that $\overline{\mathbf{Q}}$ is identified once and for all with subfields of both $\mathbf{C}$ and $\overline{\mathbf{Q}}_l$ via fixed embeddings.)

By ([BR1]), each $\sigma_l$-eigensubspace of $H^1_l(B_\pi) \otimes \overline{\mathbf{Q}}_l$, is absolutely irreducible. Hence, the commutant of the image of Galois in $End_{\mathbf{Q}_l}(H^1_l(B_\pi) \otimes \overline{\mathbf{Q}}_l)$ is isomorphic to the matrix algebra $M_d(T_\pi \otimes \overline{\mathbf{Q}}_l)$. By the Tate conjecture ([F]), this means that $End(B_\pi)$ is a simple algebra with center $T_\pi$ and $T$ is a maximal commutative semisimple subalgebra. By Albert's classification, $End(B_\pi)$ is isomorphic either to (i) $M_e(D)$ where $D$ is a quaternion algebra with center $T_\pi$, or (ii) $M_d(T_\pi)$.

This follows because we know in general that the rank over $T$ of the topological cohomology $H^1_B(B_\pi)$ is a multiple of k, if $[D:T_\pi]=k^2$. To exclude the former case if $T_\pi=\mathbf{Q}$ is easy: $D$ is split at all finite places and so, since there is only 1 infinite place, it must be split everywhere. However, if $T_\pi \neq \mathbf{Q}$, we need to use a standard argument which relies upon the fact that $F$ has a real place. Let $B_0$ be a simple factor of $B_\pi$. Then $End(B_0)=D$. Note that complex conjugation defines a continous involution of the set of complex points $B_0(\mathbf{C})$ of $B_0$. Denote the associated involution on the rank 1 $D$ module $H^1_B(B_0)$ by $Fr_\infty$. Evidently, $Fr_\infty$ has order 2, has eigenvalues 1 and -1 with equal multiplicity (since it interchanges the holomorphic and antiholomorphic parts of the Hodge splitting) and commutes with $T_\pi$. To conclude, note that the commutant of $D$ in $End(B_0)$ is isomorphic to the opposite quaternion algebra $D^{op}$ to $D$, and that $Fr_\infty$ is a non-scalar element of this algebra. Since $Fr_\infty$ is non-scalar, the algebra $\mathbf{Q}[X]/(X^2-1)$ embeds in $D^{op}$. But this algebra is not a field, and hence this case cannot occur.

Thus, $End(B_\pi)=M_d(T_\pi)$. If $A_\pi$ denotes any simple factor, we have $End(A_\pi)=T_\pi$ and finally

$$L_v\left(H^1_l(A_\pi),\sigma_l,s\right)=L_v(\pi,\sigma_l,s)$$

for all $\sigma_l$, almost all $v$, and all $l$ which are prime to $v$. Hence:

THEOREM 3. *The Existence Conjecture holds for all $\pi$ which satisfy the hypotheses* **QC**.

*Remark.* It would be interesting to show that we can take $d=1$ without invoking the Tate conjecture.

**2.2. The CM case.** This follows from the work of Casselman ([S1]), the Tate conjecture for abelian varieties of CM type, and the fact that the holomorphic cusp forms of CM type (of weight 2) are exactly those associated by theta series (or automorphic induction) to algebraic Hecke characters of totally imaginary quadratic extensions of F and having a CM type for their infinity type. In more detail, if $\pi$ is of CM type, then there exists a totally imaginary quadratic extension $J$ of $F$ and a Hecke character $\rho$ of $J$ such that $L(\pi,s)=L(\rho,s)$, where the equality is one of formal Euler products over places of $F$. Since the Hecke eigenvalues of $\pi$ are rational, the field generated by the values of $\rho$ on the finite ideles of $J$ is a quadratic imaginary extension $T$ of $\mathbf{Q}$. By Casselman's theorem ([S1], Theorem 6), there exists an elliptic curve $E$ defined over J, having complex multiplication by $\mathcal{O}_T$ over $J$, such that $L(\rho,s)L(\overline{\rho},s)$ is the zeta function of $H^1(E)$. Let $RE$ denote the restriction scalars (in the sense of Weil) from $J$ to $F$ of $E$. Let for each finite place $\lambda$ of $K$, $\rho_\lambda$ be the $\lambda$-adic representation of $Gal(\overline{\mathbf{Q}}/\mathbf{Q})$ attached to $\rho$ by Weil. Then the induced representation $Ind^J_F(\rho_\lambda)$ has a $\mathbf{Q}_l$-rational character and hence, since $F$ has a real place, can be defined over $\mathbf{Q}_l$. Note the L-function of the compatible system of all $Ind^J_F(\rho_\lambda)$ is $L(\pi,s)$. Since $H^1_l(E)$ is a free rank 2 $K\otimes\mathbf{Q}_l$-module, and the Galois action commutes with this structure, the Galois action on $H_l^1(E)$ is the direct sum of two copies of the $\mathbf{Q}_l$ model of $Ind_F^J(\rho_\lambda)$. Hence the commutant of the image of Galois is $M_2(\mathbf{Q}_l)$. Since this holds for all $l$, and using the Tate conjecture, we conclude that $End(RE)$ is $M_2(\mathbf{Q})$ and $RE$ is isogenous to a square $E_0 \times E_0$. Evidently, we have $L(\pi, s) = L(H^1(E_0), s)$.

**2.3.** We sketch the proof of Theorem 2. Let $S$ be any Shimura variety in the sense of Deligne's axioms ([D1]). The condition that $S$ have $H_B^1(S, \mathbf{Q}) \neq 0$ is very restrictive. Indeed, unless $dim(S) = 1$, we have

$$H_B^1(S, \mathbf{Q}) = IH_B^1(S, \mathbf{Q}) = H_2^1(S, \mathbf{Q})$$

where $IH^*$ denotes the usual intersection cohomology and $H_2^1$ denotes the $L^2$ cohomology. Now using the Künneth formula, we can assume that $S$ is defined by an algebraic group $G$ over $\mathbf{Q}$ whose semisimple part $G_{ss}$ is almost simple. In a standard way (using the Matsushima formula, the Künneth formula for continuous cohomology, the Vogan-Zuckermann classification, and the strong approximation theorem), we see that $G_{ss}(\mathbf{R})$ can have at most one non-compact factor $G_1$ which must itself be of real rank 1. By the classification of groups of Hermitian symmetric type, the only possibilities for $G_1$ are that it be isogenous to $SU(n, 1)$ for $n \geq 1$. Suppose $n > 1$. Then by ([MR] and [BR3], Prop. 7.2) $H_B^1(S, \mathbf{Q})$ is of CM type and so is the associated Picard variety. On the other hand, if $n = 1$, then $S$ is a curve, and it is not hard to see, from the definition of reflex fields and the computation of the zeta function of these curves, that every factor of the Jacobian of $S$ is isogenous to the base change of a factor of the Jacobian of a quaternionic Shimura curve. This completes our sketch of the proof.

## 3. Some functorialities.

**3.1.** We assume throughout the paper that $[F : \mathbf{Q}] > 1$. Further, from now until the last Section of the paper we insist that

(i) $K$ is unramified quadratic over $F$,

(ii) $\pi$ is unramified at every finite place of $F$,

(iii) the central character $\omega_\pi$ of $\pi$ is $|*|_F^{-1}$, the inverse of the idelic norm.

These conditions will impose no restriction on our final result. Indeed, the first condition may be achieved, starting from any quadratic extension $K$ of $F$, by a cyclic totally real base change from $F$ to an extension $F'$, suitably ramified at the places of $F$ where $K$ is ramified, so that $K' = KF'$ is unramified over $F'$. For (ii), if $\pi_v$ belongs to the discrete series at any finite place $v$, then the condtion **QC2** is satisfied, and there is nothing to prove. On the other hand, if $\pi_v$ belongs to the principal series at each finite place, then there is always an abelian totally real base change which kills all ramification, as may easily be seen by using the fact that the Galois representations attached to $\pi$ by Taylor ([T1, T2]) satisfy the local Langlands correspondence. Finally, since $\omega_\pi$ differs from $| * |_F^{-1}$ by a totally even character $\psi$ of finite order, a base change to the field $F_\psi$ associated to $\psi$ by class-field theory establishes (iii). Of course, for the main case of this paper, it is part of our initial assumption. Note in any case that it ensures, since $\pi$ is non-CM, that $T_\pi$ is totally real.

**3.2.** Jacquet and Gelbart ([GJ]) have proven a correspondence $Sym^2$ from non-CM cuspidal automorphic representations of $GL(2, A_F)$ to cuspidal automorphic representations of $GL(3, A_F)$. The underlying local correspondence is elementary to describe, at least at the finite places, since $\pi$ is unramified. For such a place $v$, recall that the Hecke polynomial at $v$ of $\pi$ is

$$H_v(\pi)(T) = T^2 - a_v(\pi)T + N_v,$$

which we factor as $H_v(\pi)(T) = (T - \alpha_v(\pi))(T - \beta_v(\pi))$. Similarly, each cuspidal automorphic representation $\Pi$ of $GL(3, A_F)$ has Hecke polynomials

$$H_v(\Pi)(T) = (T - r_v)(T - s_v)(T - t_v)$$

for $v$ which are unramified for $\Pi$. Define

$$H_v^2(\pi)(T) = (T - \alpha^2)(T - N_v)(T - \beta_v^2).$$

Then by ([GJ]) there exists a unique cuspidal automorphic representation $Sym^2(\pi)$ of $GL(3, A_F)$ such that

$$H_v(Sym^2(\pi)) = H_v^2(\pi)$$

for all finite $v$.

An analogous result holds at the infinite places. Here the groups are $GL(2, \mathbf{R})$ and $GL(3, \mathbf{R})$ whose irreducible admissible representations are classified by conjugacy classes of semisimple homomorphisms $\sigma_v : W_{\mathbf{R}} \to GL(k, \mathbf{C})$, $(k = 2, 3)$, where $W_{\mathbf{R}}$ is the Weil group of $\mathbf{R}$. Of course, given $\sigma_v : W_{\mathbf{R}} \to GL(2, \mathbf{C})$, there is a naturally defined class $Sym^2(\sigma_v) : W_{\mathbf{R}} \to GL(3, \mathbf{C})$. (One definition is $Sym^2 = (\sigma_v) \otimes (\sigma_v)/det(\sigma_v)$.) Then the correspondence at infinite $v$ is $\sigma_v(Sym^2(\pi)) = Sym^2(\sigma_v)$.

**3.3.** There is a base change correspondence ([AC, Theorem 4.2])

$$BC_F^K \{\text{cusp forms on } \mathrm{GL}_3(\mathbf{A}_\mathrm{F})\} \longmapsto \{\text{cusp forms on } \mathrm{GL}_3(\mathbf{A}_\mathrm{K})\}$$

which takes cusp forms to cusp forms. For $\pi_3$ a cuspidal automorphic representation of $GL(3, A_K)$, $BC_F^K(\pi_3)$ is characterized by the equality, for all but finitely many $v$,

$$L_w\big(BC_F^K(\pi_3), s\big) = L_v(\pi_3, s)L_v(\pi_3 \otimes \epsilon_{K/F}, s),$$

where $\epsilon_{K/F}$ is the idele class character of $A_F^*$ associated by class-field theory to $K/F$. An analogous result holds at all places. At the infinite places, the groups are $GL(3, \mathbf{R})$ and $GL(3, \mathbf{C})$ whose irreducible admissible representations are classified by semisimple homomorphisms $\sigma : W_{\mathbf{R}} \to GL(3, \mathbf{C})$ and $\sigma : W_{\mathbf{C}} = \mathbf{C}^* \to GL(3, \mathbf{C})$, respectively. Then

$$\sigma\big(\big(BC_F^K(\pi_3)\big)_w\big) = \sigma(\pi_3)_v | W_{\mathbf{C}}$$

if $w$ lies over $v$.

**3.4.** Let $V$ be a vector space of dimension 3 over $K$ and let $H : V \times V \to K$ be a non-degenerate Hermitian form relative to $K/F$. Let $G = U(H)$ denote the unitary group of $H$ as an algebraic group over $F$. Assume that $H$ is chosen so that $G$ is quasi-split. Let $G^* = GU^*(H)$ be the group of rational similitudes of $H$. Then the base change of $G$ to $K$ (as algebraic group) is isomorphic to $GL(3, K)$ as algebraic group over $K$. In [R2] Rogawski established base change correspondences, also denoted $BC_F^K$, between automorphic forms on $G(A_F)$ and on $GL(3, A_K)$. Let $\eta$ be the algebraic automorphism of $G$ defined by $\eta(g) =^t \overline{g}^{-1}$ for all $g$ in $GL(3, K)$ and extend $\eta$ to $GL(3, A_K)$ in the natural way. Then a cusp form $\Pi_3$ on $GL(3, A_K)$ is in the image of base change $BC_F^K$ from a global L-packet of automorphic representations of $G$ iff $\Pi_3 \circ \eta$ is isomorphic to $\Pi_3$. Let $\Pi_3 = BC_F^K(Sym^2(\Pi)) \otimes |det|$. Then evidently $\Pi_3 \circ \eta \cong \Pi_3$ since this is so almost everywhere locally (since the form is a base change, the conjugation can be ignored; since the map $g \to^t g^{-1}$ takes a local unitary $\Pi_w$ to its contragredient, it suffices to note that the local components of $BC_F^K(Sym^2(\pi)) \otimes |det|$ are self-dual). By [R2] there exists an L-packet $\Pi(G)$ of $G(A_F)$ such that $BC_F^K(\Pi(G)) = BC_F^K(Sym^2(\Pi)) \otimes |det|$. Further, at each infinite place $v$, the members of the local L-packet $\Pi(G)_v$ belong to the discrete series; they are exactly the 3 discrete series representations $\pi(G)_v$ such that

$$dim(H^2(Lie(G), k_\infty, \pi(G)_v)) = 1.$$

Denoting any of these $\Pi(G)_v$ by $\Pi(G)_\infty$, put

$$\Pi(G)_\infty = \{\pi^+, \pi^-, \pi^0\},$$

where the members are respectively holomorphic, antiholomorphic, and neither, for the usual choice of complex structure on the symmetric space (=unit ball) attached to $G$. Since $BC_F^K(Sym^2(\Pi))$ is cuspidal, $\Pi(G)$ is stable and its structure is easy to describe. If $\pi(G) \in \Pi(G)$, then $\pi(G) = \pi_\infty(G) \otimes \pi_f(G)$. The $\pi_f(G)$ is independent of the choice of $\pi(G) \in \Pi(G)$, and the $\pi_\infty(G)$ is any one of $3^g$ representations of $G(F \otimes \mathbf{R})$ which arise as external tensor products of the elements of the local L-packets $\Pi(G)_v$.

**3.5.** Now let $G_1$ be the inner form of $G$ which is

(i) isomorphic to $G$ at all finite places,

(ii) isomorphic to $U(2,1)$, e.g. quasi-split, at the archimedean place $v_1$ of $F$ defined by the given embedding of $F$ into $\mathbf{R}$,

(iii) isomorphic to $U(3)$, e.g. compact, at the other archimedean places of $F$.

Then $G_1$ is an anisotropic group, also defined by a Hermitian form $H_1$; let $G_1^*$ be the associated group of rational similitudes. In [R2], Rogawski proved a Jacquet-Langlands type of correspondence between L-packets on $G$ and $G_1$. Since $\Pi(G)$ is stable cuspidal and $\Pi(G)_v$ is discrete series at each infinite place, there is a unique L-packet, all of whose members are automorphic, $\Pi(G_1)$ on $G_1$ such that

(i) $\Pi(G_1)_f = \Pi(G)_f$,

(ii) $\Pi(G_1)_{\infty_1} = \Pi(G)_\infty$,

(iii) $\Pi(G_1)_v$ consists solely of the trivial representation for all archimedean $v$ other than $\infty_1$.

Further, (i) the central character of each member of $\Pi(G_1)$ is trivial and (ii) the multiplicity of any $\pi(G_1)$ in the discrete automorphic spectrum of $G_1$ is one. (These results are not stated explicitly in Chapter 14 of [R2] but they follow easily from Theorem 14.6.1 (comparison of traces) and Theorem 13.3.3, and in any case are well-known.)

**3.6.** Each automorphic representation $\pi(G_1)$ in $\Pi(G_1)$ extends uniquely to an automorphic representation $\pi(G_1^*)$ of $G_1^*$ with trivial central character. Thus, we obtain on $G_1^*$ a set $\Pi(G_1^*)$ of 3 automorphic representations with isomorphic finite parts $\pi(G_1^*)_f$ and whose infinite parts are identified, via projection of $G_1^*(\mathbf{R})$ onto the factor corresponding to $F_{\infty_1}$ as the members of $\Pi(G)_\infty = \{\pi^+, \pi^-, \pi^0\}$. For more details see [BR1, R1].

## 4. Picard modular surfaces.

**4.1.** The group $G_1^*$ defines a compact Shimura variety $Sh$ whose field of definition is easily seen to be $K$. Let $U$ be an open compact subgroup of $(G_1^*)_f$. Then $Sh$ is the projective limit over such $U$ of projective varieties $Sh_U$, each of which is defined over $K$ and consists of a finite disjoint union of projective algebraic surfaces. (It is customary to refer to any of the $Sh_K$ as a Picard modular surface. See [D1, LR] for background.)

Let $U(1)$ be an open compact which is a product of hyperspecial maximal compact subgroups at each finite place, and let $U$ be a normal subgroup of $U(1)$, sufficiently small so that $Sh_U$ is non-singular. For a $(G_1^*)_f$ module $V$, let $V^{U(1)}$

denote the subspace of $U(1)$ invariants. It is a module for the Hecke algebra $\mathbf{H}_{U(1)}$ of compactly supported bi-$U(1)$ invariant functions on $(G_1^*)_f$. Note that $V^{U(1)}$ is a module for $\mathbf{H}_{U(1)}$.

The degree 2 cohomology $H_B^2(Sh, \overline{\mathbf{Q}})$ of $Sh$ decomposes as a direct sum of isotypic $\pi(G_1^*)_f$ modules. Now let $V = H_B^2(Sh, \overline{\mathbf{Q}})(\pi(G_1^*)_f)$ denote the $\pi(G_1^*)_f$-isotypic component. Then, as usual, $V^{U(1)}$ is identified, using the Matsushima formula, with a 3 dimensional subspace of $H_B^2(Sh_U, \overline{\mathbf{Q}})$; it is an isotypic component for a representation $(\pi(G_1^*)_f)^{U(1)}$ of $\mathbf{H}_{U(1)}$.

**4.2.** **Q**-structure. Note that $V$ has a natural $\overline{\mathbf{Q}}$ structure coming from $H_B^2(Sh, \overline{\mathbf{Q}})$.

LEMMA. *The subspace $V^{U(1)}$ of $V$ is defined over* **Q**.

*Proof.* Let $\tau$ be an automorphism of **C** and let $V^{U(1)\tau}$ be the conjugate of $V^{U(1)}$ inside $H_B^2(Sh, \overline{\mathbf{Q}}) = H_B^2(Sh, \mathbf{Q}) \otimes \overline{\mathbf{Q}}$. Then $V^{U(1)\tau}$ is the $((\pi(G_1^*)_f)^\tau)^{U(1)}$ isotypic subspace of $H_B^2(Sh, \overline{\mathbf{Q}})$, so it is enough to show that

$$\left(\pi\left(G_1^*\right)_f\right)^\tau = \pi\left(G_1^*\right)_f.$$

Since these representations are unramified, and since, by the discussion of [R2, 12.2], unramified local L-packets for $G_1$ consist of single elements, it is enough to check, for each finite place $v$, the equality of Langlands classes

$$\sigma_v\left(\left(\pi\left(G_1^*\right)_f\right)^\tau\right) = \sigma_v\left(\pi\left(G_1^*\right)_f\right).$$

However, it is not hard to check that

$$\sigma_v\left(\left(\pi\left(G_1^*\right)_f\right)^\tau\right) = \sigma_v\left(\left(\pi\left(G_1^*\right)_f\right)\right)^\tau.$$

By the discussion of [R1, 4.1-2], these classes are in Galois equivariant bijection with associated L-factors. In our case, as is easily seen, the factor is given by

$$L_w\left(BC_F^K(Sym^2(\pi)), s\right),$$

where $w$ is any extension of $v$. Since $\pi$ has rational Hecke eigenvalues $L(BC_F^K(Sym^2(\Pi)), s)$ is an Euler product over reciprocals of Dirichlet polynomials with rational coefficients. Hence

$$\sigma_v\left(\left(\pi\left(G_1^*\right)_f\right)\right)^\tau = \sigma_v\left(\left(\pi\left(G_1^*\right)_f\right)\right.$$

and the result follows. □

**4.2.1.** The case $T_\pi \neq \mathbf{Q}$. Here, instead of the above Lemma, one shows that the smallest subspace $W_B$ of $H_B^2(Sh, \overline{\mathbf{Q}})$ whose scalar extension to $\overline{\mathbf{Q}}$ contains $V^{U(1)}$ is a 3-dimensional $T_\pi$ vector space, where $T_\pi$ is identified with the quotient of $\mathbf{H}_{U(1)}$ acting on $W_B$. The argument requires nothing new.

**4.3. Hodge structure.** Let $M_B$ denote the $\mathbf{Q}$ vector space provided by Lemma 4.2, such that

$$M_B \otimes \overline{\mathbf{Q}} = V^{U(1)}.$$

By the stability of the L-packet, each tensor product $\sigma^* \otimes \pi_f$ with $\sigma^* \in \Pi(G)_{\infty 1}$ is automorphic. The bigraded Matsushima formula ([BW]) shows that the Hodge decomposition of $M_B$ has the form

$$M_B \otimes \mathbf{C} = M^{(2,0)} \bigoplus M^{(1,1)} \bigoplus M^{(0,2)}$$

where the factors, each of dimension 1, are the contributions of $\sigma^+$, $\sigma^0$, and $\sigma^-$ to the cohomology, respectively.

**4.4. Motive.** For each rational prime $l$, $M_l = M_B \otimes \mathbf{Q}_l$ is identified with a summand of the $l$-adic etale cohomology $H^2_B(Sh \times \overline{\mathbf{Q}}, \mathbf{Q}_l)$. Since the action of the Hecke algebra on $H^2_B(Sh, \mathbf{Q})$ is semisimple, there exists an element $e$ which acts as an idempotent on $H^2_B(Sh, \mathbf{Q})$ and whose image is $M_B$. Interpreting, as usual, the action of Hecke operators via algebraic correspondences (c.f. [DM, BR2]), we see that the pair

$$M = \left(H^2_B(Sh, \mathbf{Q}), e\right)$$

defines a Grothendieck motive with associated $\infty$-tuple of realizations

$$M_r = (M_B, M_{DR}; M_l, l\ prime).$$

Here $M_B$ is regarded as a rational Hodge structure, $M_{DR}$, a graded $K$-vector space, is the De Rham cohomology of $M$, and each $M_l$ is a $Gal(\overline{\mathbf{Q}}/\mathbf{Q})$-module. Each module is the image of the action of $e$ in the corresponding cohomology theory of $Sh$. Since the classes of algebraic cycles are absolute Hodge cycles, every Grothendieck motive is also a motive for absolute Hodge cycles. Henceforth, by abuse of language, we regard $M_r$ as such motive, and all motives will be motives for absolute Hodge cycles.

**4.5. Polarization.** There is ([DM]) a well-known Tate twist operation which for any motive M and any $n \in \mathbf{Z}$ defines a new motive $M(n)$; we recall the properties of this operation only as needed. Further, any motive is polarisable. In our case, this means that there is a non-degenerate symmetric morphism of Hodge structures

$$\Psi_M : M_B(1) \otimes M_B(1) \to \mathbf{Q}$$

whose associated bilinear form has signature $(1, 2)$. Here $\mathbf{Q}$ has the "trivial" Hodge structure of type $(0, 0)$, and the Hodge decomposition of $M_B(1)$ coincides with that of $M_B$, but with each pair $(p, q)$ of the bigrading replaced by $(p - 1, q - 1)$. The form induces, for each $l$, a $Gal(\overline{\mathbf{Q}}/K)$-equivariant map

$$\Psi_l : M_l(1) \otimes M_l(1) \to \mathbf{Q}_l$$

where $\mathbf{Q}_l$ is the trivial Galois module. A similar remark applies to $M_{DR}$, but we will have no need of it. Likewise, an explicit realisation of the polarisation is given by the restriction to M of the cup-product on $H^2(Sh)$, but we don't need this fact.

**4.5.1.** If $T_\pi \neq \mathbf{Q}$, then the above construction provides us with a motive $M$ of rank 3 over $T_\pi$ such that $T_\pi$ acts on each component of the Hodge decomposition by the regular representation. The polarization $\Psi_M$ has the form $\Psi_M = Tr_{K/F}(\Psi_0)$ with a $T_\pi$-linear symmetric bilinear form $\Psi_0$ taking values in $T_\pi \otimes$. The $T_\pi \otimes \mathbf{R}$-valued form $\Psi_M \otimes \mathbf{R}$ decomposes as a sum, indexed by the embeddings of $T_\pi$ into $\mathbf{R}$ of forms of signature $(1, 2)$. In particular, the orthogonal group of this form is quasi-split at each infinite place.

**4.6. Construction of an elliptic curve over C.** For a Hodge structure $H$, let $Sym^2(H)$ denote its symmetric square.

PROPOSITION. *There exists, up to isomorphism, a unique two dimensional rational Hodge structure $H_B$ having Hodge types $(1, 0)$ and $(0, 1)$ such that $Sym^2(H_B)$ is isomorphic to $M_B$.*

*4.7. Proof.* 4.7.1. We follow [D2], Sections 3 and 4. Let $C^+ = C^+(M_B(1))$ be the even Clifford algebra of the quadratic module $M_B(1)$. A Hodge structure on a rational vector space $X$ is given by a morphism of real algebraic groups $h : \mathbf{S} \to Aut(X \otimes \mathbf{R})$ where $\mathbf{S} = R_{\mathbf{C}/\mathbf{R}} G_m$ is the Weil restriction of scalars from $\mathbf{C}$ to $\mathbf{R}$ of the multiplicative group. Let $h = h_{M_B(1)}$ denote the morphism so defined for $M$. There is a canonical morphism with kernel $G_m$ of the algebraic group $(C^+)^*$ into the group $SO(M_B(1), \Psi)$. Of course, $im(h)$ lies in $SO(M_B(1), \Psi)(\mathbf{R})$. The morphism $h$ lifts uniquely to a morphism $h^+ : \mathbf{S} \to (C^+)^* \otimes \mathbf{R}$ such that the associated Hodge structure on $C^+$ has types $(1, 0)$ and $(0, 1)$ only. Note that in our case $C^+$ is a quaternion algebra with center $\mathbf{Q}$. □

**4.7.2.** We must show that $C^+$ is a split algebra. If so, then denoting by $W$ an irreducible 2 dimensional module for $C^+$, we know by [D2], 3.4 that $End(W)/center$ is isomorphic to $\Lambda^2(M_B(1))$, which is itself isomorphic to $M_B(1)$. (Note that the center of $End(W)$ is a rational sub-Hodge structure of type $(0, 0)$.) Hence $Sym^2(W) = (W \otimes W)/\Lambda^2(W)$ is isomorphic to $M_B$.

**4.7.3.** To show that $C^+$ is split, we use the surface $Sh$ and R. Taylor's l-adic representations $\rho_l^T$ of $Gal(\overline{\mathbf{Q}}/F)$.

LEMMA 4.7.4. *The representations $\rho_l^T$ and $Sym^2(\rho_l^T)$ are irreducible and remain irreducible when restricted to any finite subgroup of $Gal(\overline{\mathbf{Q}}/F)$.*

*Proof.* Evidently, if the result holds for $Sym^2(\rho_l^T)$, then it holds for $\rho_l^T$. The former case will follow from [BR1], Theorem 2.2.1(b), once we exclude the case (iib) of that Theorem, i.e. that $Sym^2(\rho_l^T)$ is potentially abelian. To see this, observe that, since $Sym^2(\rho_l^T)$ occurs as the Galois action on $M_l$ for a motive $M$ whose $M_B$ has 3 Hodge types (2, 0), (1, 1), and (0, 2), the Hodge-Tate theory shows that the semisimple part of the Zariski-closure of the image of Galois, over any finite extension $L$ of $K$, contains a non-trivial torus over $\mathbf{C}_l$. Considering now $\rho_l^T$, this means that if $\rho_l^T$ is potentially abelian, the connected component of the Zariski closure over $\mathbf{C}_l$ is a non-central torus $S$ of $GL(2)$. Hence the image of $\rho_l^T$ must lie in the normalizer $N$ of $S$ inside $GL(2)$. Such an $N$ is either abelian or has an abelian subgroup of index 2, and so the image of Galois is either abelian or is non-abelian but has an abelian subgroup of index 2. The first case contradicts ([BR1, Prop. 2.3.1], [T2]), whereas the second means that $\pi$ is of CM type. □

For each $l$, let $\eta_l = Sym^2(\rho_l^T|_K)(1)$. Each $\eta_l$ acts on $\mathbf{Q}_l^3$ which is isomorphic to the module $M_l(1)$ as Galois module. Let $G_l$ denote the Zariski closure of the image of $\eta_l$ in $End(\mathbf{Q}_l^3)$.

LEMMA 4.7.5. *$G_l$ is a quasi-split special orthogonal group.*

*Proof.* The image of $Sym^2 \otimes (det^{-1}) : GL(2) \to GL(3)$ is a quasi-split special orthogonal group $SO(3)_{qs}$. Hence we need only check that the image of $\eta_l$, automatically contained in this group, has Zariski closure equal to it. But $(\eta_l)|_L$ is irreducible for all finite extensions $L$ of $K$. Hence $G_l$ is a connected irreducible algebraic subgroup of $SO(3)_{qs}$. Fortunately, the only such subgroup is $SO(3)_{qs}$ itself. □

**4.7.6. Completion of proof that $C^+$ is split.** Now we know that the Zariski closure of the Galois action on each $M_l(1)$ is the quasi-split orthogonal group. On the other hand, this irreducible action preserves the form $\Psi_l$. Since an irreducible orthogonal representation can preserve at most one quadratic form (up to homothety), this shows that the special orthogonal group of $\Psi_l$ is $SO(3)_{qs}$ for all $l$. Hence for all primes $l$, the algebra $C^+ \otimes \mathbf{Q}_l$ is split. Hence it must be split at infinity as well, and so is a matrix algebra.

**4.7.7.** If $T_\pi \neq \mathbf{Q}$, we must construct a rational Hodge structure on the underlying rational vector space of a 2 dimensional $T_\pi$ vector space $H_B$ so that (i) only Hodge types $(1, 0)$ and $(0, 1)$ occur, and (ii) $T_\pi$ acts via endomorphisms of the Hodge structure. The argument of 4.7.1 provides such a $T_\pi$-linear Hodge structure $h^+$ on $C^+$, which is now quaternion algebra over $T_\pi$. Similarly, the argument through 4.7.5, using the $T_\pi$-linear $Sym^2$, extends without difficulty to show that $C^+$ is split at all finite places. Finally, since $SO(\Psi_M)$ is quasi-split at each infinite place, so is $(C^+)^*$. So $(C^+)^*$ is everywhere locally, and hence globally, split.

**4.8.** The Hodge structure $H_B$ defines a unique isogeny class of elliptic curves over **C** such that for any member $A$ of this class, $H^1_B(A)$ is isomorphic as Hodge structure to $H_B$. We now choose and fix one such $A$.

**4.8.1.** If $T_\pi \neq \mathbf{Q}$, $H_B$ defines an isogeny class of abelian varieties over **C**, such that for any member $A = A_\pi$ of this class, $H^1_B(A)$ is isomorphic as Hodge structure to $H_B$. Indeed, the $T_\pi$ action renders $H_B$ automatically polarisable, which is all that needed to be checked.

## 5. Descent of $A$ to $\overline{\mathbf{Q}}$.

**5.1.** Let $\tau$ be an automorphism of **C** over $\overline{\mathbf{Q}}$. Let $\tau(A)$ be the conjugate of $A$ by $\tau$.

THEOREM 4. *Suppose the Absolute Hodge Conjecture holds. Then $\tau(A)$ is isogenous to $A$.*

PROPOSITION 5.2. *Let $E_1$ and $E_2$ be elliptic curves over* **C** *and suppose that the Hodge structures $Sym^2(H^1_B(E_1))$ and $Sym^2(H^1_B(E_2))$ are isomorphic. Then $E_1$ is isogenous to $E_2$.*

*5.3 Proof.* $E_1$ is isogenous to $E_2$ if $H^1_B(E_1)$ is isomorphic to $H^1_B(E_2)$ as rational Hodge structure. As in 4.7.1, let $h_1$ and $h_2$ be the Hodge structure morphisms defined for $H^1_B(E_1)$ and $H^1_B(E_2)$; these actions are each separately equivalent over **R** to the tautological action of $\mathbf{C}^* = \mathbf{S}(R)$ on **C**, for a suitable choice of isomorphism of **C** with $\mathbf{R}^2$. Let $V_j = H^1_B(E_j)$ $(j \in \{1, 2\})$, and let

$$h = (h_1, h_2) : \mathbf{S} \to GL(V_1, \mathbf{R}) \times GL(V_2, \mathbf{R})$$

be the product morphism. Let $H$ be the smallest algebraic subgroup of $GL(V_1, \mathbf{Q}) \times GL(V_2, \mathbf{Q})$ which contains $im(h)$ over **R**. Then $H$ is a reductive algebraic group; it is the Mumford-Tate group of $E_1 \times E_2$ and the isomorphism classes of its algebraic representations over **Q** are in natural bijection with the isomorphism classes of rational Hodge structures contained in all tensor powers of all sums of $V_1$, $V_2$, and their duals. The projection of $H$ to a factor is the Mumford-Tate group of the corresponding curve. □

Put $W = (V_1 \otimes V_2) \otimes (V_1 \otimes V_2)$. Then of course $W \cong (V_1 \otimes V_1) \otimes (V_2 \otimes V_2)$. Since the action of $H$ on $V_j \otimes V_j$ factors through the projection onto $GL(V_j)$, we see that the action of H on $V_j \otimes V_j$ decomposes as a sum of the 1-dimensional representation $det \circ pr_j$ and the 3-dimensional representation $Sym^2(V_j) \circ pr_j$. Further, the representations $det(pr_1)$ and $det(pr_2)$ are isomorphic since the corresponding Hodge structures are 1-dimensional, of type (1, 1), and there is up to isomorphism only 1 such Hodge structure. Since, by assumption, $Sym^2(V_1)$ is isomorphic to $Sym^2(V_2)$, and $Sym^2(V_1)^* \cong Sym^2(V_1) \otimes (det(pr_1))^{-2}$, we see that

$Sym^2(V_1) \otimes Sym^2(V_2)$ decomposes as the sum of a 1-dimensional $(det(pr_1))^2$, and an 8-dimensional representation. Hence $W$ contains at least two copies of the representation $(det(pr_1))^2$. Now, $V_1$ and $V_2$ are non-isomorphic as Hodge structures if they are non-isomorphic as $H$-modules if $V_1 \otimes V_2$ contains no 1-dimensional summand isomorphic to $det(pr_1)$. This follows since $(V_1)^* \cong V_1 \otimes (det(pr_1)^{-1})$, and the morphisms of Hodge structure are exactly the vectors in $V_1^* \otimes V_2$ on which $H$ acts via the trivial representation. Suppose now $V_1 \otimes V_2 \cong A \oplus B$ with $A$ and $B$ 2-dimensional sub-Hodge structures. Evidently, one of the factors is purely of type $(1, 1)$, and so the associated $H$ representation is isomorphic to $det(pr_1) \oplus det(pr_1)$. Hence $V_1$ is isomorphic to $V_2$. (Of course, in this case we know more: both have complex multiplication, since if $\alpha$ and $\beta$ are two non-homothetic elements of $Hom(E_1, E_2)$, $(\hat{\beta}) \circ \alpha$ is a non-scalar endomorphism of $E_1$.) Thus, we must only exclude the possibility that $V_1 \otimes V_2$ is an irreducible Hodge structure, i.e. corresponds to an irreducible representation of $H$. But in this case $V_1 \otimes V_2 \otimes V_1 \otimes V_2$ contains exactly one copy of $(det(pr_1))^2$, by Schur's Lemma. Since we have seen above that the multiplicity of $det(pr_1)$ is at least 2, this case cannot occur.

*5.4. Remark.* This Proposition is easily adapted to the general case, provided one works always in the category of Hodge structures which are $T_\pi$-modules, working $T_\pi$-linearly.

**5.5. Proof of Theorem 4.** By construction, the Hodge structure $Sym^2(H^1(A))$ is isomorphic to that of $M_B(\pi_f)$. Hence, by the Absolute Hodge Conjecture, there is an isomorphism in the category of motives over $\mathbf{C}$:

$$\phi : Sym^2(H^1(A)) \to M_B(\pi_f) \otimes \mathbf{C}.$$

Let $\tau$ be an element of $Aut(\mathbf{C}/\overline{\mathbf{Q}})$. Conjugating by $\tau$, we get an isomorphism $\phi^\tau : (Sym^2(H^1(A)))^\tau \to M_B(\pi_f)^\tau$. Since $M_B(\pi_f)$ is defined over $\overline{\mathbf{Q}}$, $M_B(\pi_f)^\tau = M_B(\pi_f)$. Further, $(Sym^2(H^1(A)))^\tau \cong Sym^2(H^1(A^\tau))$. Hence $Sym^2(H^1(A^\tau)) \cong Sym^2(H^1(A))$. In particular, the Hodge structures $Sym^2(H_B^1(A^\tau))$ and $Sym^2(H_B^1(A))$ are isomorphic. By the Proposition 5.2, $A^\tau$ is isogenous to $A$.

COROLLARY 5.6. *Suppose the Absolute Hodge Conjecture holds. Then $A$ admits a model over a finite extension $L$ of* $\mathbf{Q}$.

*Proof.* The complex isogeny class of $A$ contains only countably many complex isomorphism classes of elliptic curves. For an elliptic curve $B$ let $j(B)$ be its $j$-invariant. Then $B_1$ is isomorphic over $\mathbf{C}$ to $B_2$ iff $j(B_1) = j(B_2)$. Furthermore, $j(\tau(B)) = \tau(j(B))$ for all $B$. Hence, considering all automorphisms $\tau$ of $\mathbf{C}$ over $\overline{\mathbf{Q}}$, the set $\{\tau(j(A))\}$ is countable. Let on the other hand $z$ be any non-algebraic complex number. Then it is well-known that the set $\{\tau(z)\}$ is uncountable. Hence, $j(A)$ must be in $\overline{\mathbf{Q}}$, i.e. $\{\tau(j(A))\}$ is finite. Let $L = Q(j(A))$. Since any elliptic curve $B$ admits a Weierstrass model over $Q(j(B))$, we are done. □

*Remark* 5.7. In the general case $T_\pi \neq \mathbf{Q}$, we can, changing $A$ within its isogeny class as needed, give $A$ a principal polarisation $P_A$. Then the set of all $j$-invariants is replaced by the quasi-projective variety $M_d$, defined over $\mathbf{Q}$, which parametrises all isomorphism classes $\mathcal{B}$ of pairs $(B, P_B)$ where $B$ has dimension $d$, and $P_B$ is a principal polarisation of $B$. Then $\mathcal{A} = (A, P_A)$ defines a point $\nu(\mathcal{A})$ of $M_d$. The variety $M_d$ is defined over $\mathbf{Q}$. Further, if $\tau$ is an automorphism of $\mathbf{C}$ over $\overline{\mathbf{Q}}$, $\tau(\mathcal{A}) = (\tau(\mathrm{A}), \tau(P_{\mathrm{A}}))$ is also a principally polarised abelian variety and $\tau(\nu(\mathcal{A})) = \nu(\tau(\mathcal{A}))$. Again, the set of all isomorphism classes of pairs $\mathcal{B} = (\mathcal{B}, P_{\mathcal{B}})$ where $B$ is isogenous to $A$ is countable. Since $\tau(A)$ *is* isogenous to $A$, this means that the set of all $\tau(\nu(\mathcal{A}))$ is countable. Hence $\nu(\mathcal{A})$ has algebraic coordinates. (This is standard. Choosing a suitable hyperplane $H$, defined over $\mathbf{Q}$ which does not contain $\nu((\mathcal{A}))$, the countable set $\tau(\nu(\mathcal{A}))$ is contained in an affine variety defined over $\mathbf{Q}$. But if $V$ is an affine variety defined over $\mathbf{Q}$, and $v$ is a point of $V$ with complex coordinates such that $\tau(v)$ is countable, then the coordinates of $v$ are all algebraic.) Now let $L_0$ be the number field generated by the coordinates of $\nu(\mathcal{A})$; it is called the field of moduli of $\mathcal{A}$. But it is known that every polarised smooth projective variety admits a model over a finite algebraic extension of its field of moduli. So $(A, P_A)$, and hence $A$ alone, is definable over a finite extension $L$ of $L_0$. This completes our sketch of the general case.

## 6. Comparison of $H_l^1(A)$ and $\rho_l^T(\pi)$.

**6.1.** Fix a prime $l$. The two dimensional $\mathbf{Q}_l$-vector space $H_l^1(A)$ is a $Gal(\overline{\mathbf{Q}}/L)$-module. Recall that $V_l^T(\pi)$ denotes the 2-dimensional $Gal(\overline{\mathbf{Q}}/F)$-module attached to $\pi$ ([T1, BR2]); the Galois action has been denoted $\rho_l^T(\pi_f)$.

PROPOSITION 6.2. *Suppose the Absolute Hodge Conjecture holds. Then there exists a finite extension $L_1$ of $L$, containing $K$, such that $V_l^T(\pi)|_{L_1}$ is isomorphic as $Gal(\overline{\mathbf{Q}}/L_1)$-module to $H_l^1(A)|_{L_1}$.*

*6.3. Proof.* The motive $Sym^2(H^1(A))$ is defined over $L$, and by construction, there is an isomorphism $\iota_B$ of Hodge structures between $M_B$ and $Sym^2(H^1(A))_B$. Regarding $\iota_B$ as an element of $M_B^* \otimes Sym^2(H^1(A))_B$, it is a rational class of type $(0, 0)$. By Deligne's theorem, $Gal(\overline{\mathbf{Q}}/L)$ acts continuously, via a finite quotient group, on the $\mathbf{Q}$-vector subspace of all rational classes of type $(0, 0)$. Let $L_1$ be a finite extension of $L$, containing $K$, such that $Gal(\overline{\mathbf{Q}}/L)$ acts trivially on $\iota_{\mathrm{B}}$. Now, for a rational prime $l$, $\iota_{\mathrm{B}}$ defines also an isomorphism

$$\iota_l : (M \otimes_K L_1)_l \to \left(Sym^2(H^1(A)) \otimes_L L_1\right)_l$$

which is $Gal(\overline{\mathbf{Q}}/L_1)$-equivariant. □

**6.4.** Since the restriction to $K$ of

$$Sym^2\left(\rho^T(\pi)_l\right) : Gal(\overline{\mathbf{Q}}/F) \to Aut\left(V_l^T\right)$$

is isomorphic to $M_l$, we now know that, over $L_1$, $Sym^2(\rho^T(\pi)_l)|L_1$ is isomorphic as a Galois module to $Sym^2(H^1(A))_l$. Since the dual of $Sym^2(\rho^T(\pi)_l)|L_1$ is isomorphic to $(Sym^2(\rho^T(\pi)_l)|L_1)(2)$, this means that

$$\left(Sym^2\left(\rho^T(\pi)_l\right)|L_1\right) \otimes \left(Sym^2\left(H^1(A)\right)_l\right)$$

contains $\chi_l^{-2}$, where $\chi_l$ is the $l$-adic cyclotomic character. Let $\mathbf{Q}_l(-1)$ be $\mathbf{Q}_l$ with Galois action given by $\chi_l$.

**6.5.** We now proceed as in the conclusion of the proof of the preceding proposition. Let $V_1 = H_l^1(A)$ and let $V_2 = V_l^T|L_1$. Then

$$V_1 \otimes V_1 \otimes V_2 \otimes V_2$$

is isomorphic to

$$\left(Sym^2(V_1) \oplus \mathbf{Q}_l(-1)\right) \otimes \left(Sym^2(V_2) \oplus \mathbf{Q}_l(-1)\right)$$

which, from the above, contains the square of the inverse of the l-adic cyclotomic character $\chi_l^{-2}$ with multiplicity two. Hence, putting $W = V_1 \otimes V_2$, we see that $W \otimes W$ contains $\chi^{-2}$ with multiplicity two. If $W$ is absolutely (i.e. $\overline{\mathbf{Q}}_l$) irreducible as a Galois module, then $W \otimes W$ contains $\chi_l^{-2}$ with multiplicity one. So $W$ is reducible. If $W$ decomposes as $X \oplus Y$, with $X$ and $Y$ irreducible and two dimensional, then consider the exterior square

$$\Lambda = \Lambda^2(X \oplus Y).$$

This decomposes as

$$det(X) \oplus X \otimes Y \oplus det(Y).$$

On the other hand

$$\Lambda = \Lambda^2(V_1 \otimes V_2)$$

which decomposes as

$$\Lambda = (Sym^2(V_1) \otimes det(V_2)) \oplus (det(V_1) \otimes Sym^2(V_2)).$$

The first calculation shows that $\Lambda$ has at least two 1-dimensional summands, and, since $Sym^2(V_1) \cong Sym^2(V_2)$ is irreducible, the second shows that $\Lambda$ has no 1-dimensional summands, this case is impossible. Hence $W = U \oplus Z$ with a 1-dimensional summand $Z$.

**6.6.** To finish the argument, denote the Galois action on $Z$ by $\psi$. Then $V_1^* \otimes (V_2 \otimes \psi^{-1})(-1)$ contains the trivial representation. Since $V_1$ is irreducible, this means that $V_1$ is isomorphic to $(V_2 \otimes \psi^{-1})(-1)$. Since $det(V_1)$ and $det(V_2)$ both have the Galois action given by $\chi_l^{-1}$, we conclude by taking determinants, that $(\psi\chi_l)^2 = 1$. Set $\mu = \psi\chi_l$. Enlarge $L_1$ by a finite extension $L_2$ such that $\mu|L_2$ is trivial. Then over $L_2$, $V_2 \cong V_1$. Relabeling $L_2$ as $L_1$, we are done.

## 7. Completion of the Construction.

**7.1. Preliminary.** We now remove the conditions, in force since Section 3, that $\pi$ be unramified over $F$, and $FK_0 = K$ be unramified over $F$, and we reinterpret the preceding constructions as commencing from the base change of $\pi$ to the solvable totally real extension $F_1$ of $F$, over which $BC_F^{F_1}(\pi)$ is unramified, and such that $F_1 K_0$ is unramified. Further, we henceforth let $L$ (not $L_1$) denote any number field which is a field of definition of $A$ and which satisfies the conclusion of the previous proposition.

**7.2.** Finally, let $B = R_{L/F}(A)$. Then the $Gal(\overline{\mathbf{Q}}/F)$-module $H_l^1(B)$ is isomorphic to the induced module $Ind_{L/F}(H_l^1(A))$. However, $H_l^1(A)$ is isomorphic to the restriction $Res_{L/F}(V_l^T)$ of the 2-dimensional $Gal(\overline{\mathbf{Q}}/F)$-module $V_l^T = V_l^T(\pi)$. Hence $H_l^1(B)$ is isomorphic to $V_l^T \otimes \Pi_{L/F}$ where $\Pi_{L/F}$ is the permutation representation defined by the action of $Gal(\overline{\mathbf{Q}}/F)$ on the set of $F$-linear embeddings of $L$ into $\overline{\mathbf{Q}}$.

**7.3.** Since the trivial representation occurs in $\Pi_{L/F}$, we see that $V_l^T$ occurs in $H_l^1(B)$. Let $\tau$ be any non-trivial irreducible constituent of $\Pi_{L/F} \otimes \overline{\mathbf{Q}}_l$. Then we claim that $\rho_l^T$ is not a constituent of $\rho_l^T \otimes \tau$. To see this, just note that the multiplicity in question is the dimension of $((V_l^T)^* \otimes V_l^T \otimes \tau)^{Gal}$ where $Gal = Gal(\overline{\mathbf{Q}}/F)$. Since $(V_l^T)^* \otimes V_l^T = 1 \oplus Ad(V_l^T)$, and $1 \otimes \tau$ is irreducible, it is enough to check that $Ad(V_l^T) \otimes \tau$ contains no Galois invariants. But if $J$ denotes the kernel of $\tau$, then the restriction of $Ad(V_l^T) \otimes \tau$ to $J$ is isomorphic to the direct sum of 3 copies of (the restriction to $J$ of) $Ad(V_l^T)$. Since we have seen that $Ad(V_l^T) = Sym^2(V_l^T) \otimes \omega_\pi^{-1}$, and $Sym^2(V_l^T)$ remains irreducible on restriction to $J$, there are no invariants. Hence the multiplicity of $V_l^T$ in $H_l^1(B)$ is 1.

**7.4.** Let $D$ be the smallest abelian subvariety of $B$ such that $H_l^1(D)$ contains the unique submodule of $H_l^1(B)$ which is isomorphic to $V_l^T$. We will show that $D$ is an elliptic curve. Evidently $D$ is simple and so $End(D)$ is a division algebra. In fact, $End(D)$ is a field. To see this let $Z$ be the center of $End(D)$ and let $dim_Z(End(D)) = n^2$. Then over $Z \otimes \overline{\mathbf{Q}}_l$, $H_l^1(D) \otimes \overline{\mathbf{Q}}_l$ is a free module over the the matrix algebra $M_n(Z \otimes \overline{\mathbf{Q}}_l)$. Hence each irreducible Galois submodule of $H_l^1(D) \otimes \overline{\mathbf{Q}}_l$ must occur at least n times. Since $V_l^T$ occurs once, this means $n = 1$, i.e. $End(D) = Z$.

**7.5.** Now we must show that $Z = \mathbf{Q}$. To see this, note that $H_l^1(D)$ is a free $Z \otimes \mathbf{Q}_l$-module. Put $Z \otimes \mathbf{Q}_l = Z_1 \oplus \ldots \oplus Z_t$, with local fields $Z_j$. Let $e_j$ denote the idempotent of $Z \otimes \mathbf{Q}_l$ which has image the factor $Z_j$. Choose the indexing so that the $Z_1$ module $e_1(H_l^1(D))$ contains the $\mathbf{Q}_l$-submodule $W$ isomorphic to $V_l^T$. Since the Galois action on $V_l^T$ is irreducible, there is an embedding of

$V_l^T \otimes Z_1$ into $e_1(H_l^1(D))$. But if $[Z_1 : \mathbf{Q}_l] > 1$, the commutant of this image would be non-abelian. Since $Z$ is a field, this means $Z_1 = \mathbf{Q}_l$. Since l is arbitrary, the Cebotarev theorem ([CF], Exercise 6.2) forces $Z = \mathbf{Q}$. Thus the commutant of the image of Galois in $End(H^1(D)) \otimes \overline{\mathbf{Q}_l}$ is $\overline{\mathbf{Q}_l}$. This means that $H_l^1(D)$ is isomorphic to $V_l^T$, and so $D$ is in fact the sought elliptic curve. This completes the construction.

**7.6.** If $T_\pi \neq \mathbf{Q}$, the arguments of Sections 6 and 7 proceed essentially unchanged, albeit $T_\pi$-linearly, using the free rank 2 $T_\pi \otimes \mathbf{Q}_l$-adic representations $V_l^T$ of Taylor.

---

## REFERENCES


[AC] J. Arthur and L. Clozel, *Simple algebras, base change, and the advanced theory of the trace formula.* Annals of Mathematics Studies, 120, Princeton University Press, Princeton, NJ, 1989.

[B] D. Blasius, Hilbert modular forms and the Ramanujan Conjecture, preprint, June, 2003.

[BR1] D. Blasius and J. Rogawski, *Tate classes and arithmetic quotients of the two-ball.* The zeta functions of Picard modular surfaces, 421–444, Univ. Montréal, Montréal, QC, 1992.

[BR2] D. Blasius and J. Rogawski, Motives for Hilbert modular forms. *Invent. Math.* **114** (1993), no. 1, 55–87.

[BR3] D. Blasius and J. Rogawski, *Zeta functions of Shimura varieties,* Motives (Seattle, WA, 1991), Proc. Sympos. Pure Math., vol. 55, Part 2, Amer. Math. Soc., Providence, RI, 1994, pp. 525–571.

[BW] A. Borel and N. Wallach, *Continuous Cohomology, Discrete Subgroups, and Representations of Reductive Lie Groups*, Annals of Mathematics Studies 94, Princeton University Press, Princeton, N.J., 1980.

[CF] J.W.S. Cassels and A. Fröhlich, *Algebraic Number Theory,* Thompson Book Company, Washington, D.C., 1967.

[D1] P. Deligne, *Travaux de Shimura. Séminaire Bourbaki 1970–71, Exposé 389, Lecture Notes in Math.* Vol. 244. Springer-Verlag, Berlin-New York, 1971.

[D2] P. Deligne, La conjecture de Weil pour les surfaces K3. *Invent. Math.* **15** (1972), 206–226.

[D3] P. Deligne, Théorie de Hodge. II, *Publ. Math. IHES* **40,** (1972), 5–57.

[D4] P. Deligne, *Hodge cycles on abelian varieties, Hodge cycles, motives, and Shimura varieties. Lecture Notes in Math.* Vol. 900. Springer-Verlag, Berlin-New York, 1982.

[DM] P. Deligne and J.S. Milne, *Tannakian Categories.* Hodge cycles, motives, and Shimura varieties. Lecture Notes in Mathematics, 900. Springer-Verlag, Berlin-New York, 1982.

[DMOS] P. Deligne, J. Milne, A. Ogus and K. Shih, *Hodge cycles, motives, and Shimura varieties.* Lecture Notes in Mathematics, 900. Springer-Verlag, Berlin-New York, 1982.

[E] Martin Eichler, Quaternäre quadratische Formen und die Riemannsche Vermütung für die Kongruenzzetafunktion. *Arch. Math.* **5,** (1954), 355–366.

[F] G. Faltings, Endlichkeitssätze für abelsche Varietäten über Zahlkörpern. *Invent. Math.* **73** (1983), no. 3, 349–366.

[GJ] S. Gelbart and H. Jacquet, A relation between automorphic representations of GL(2) and GL(3). *Ann. Sci. École Norm.* Sup. (4) **11** (1978), no. 4, 471–542.

[H] H. Hida, On abelian varieties with complex multiplication as factors of the Jacobians of Shimura curves. *Amer. J. Math.* **103** (1981), no. 4, 726–776.

[JL] H. Jacquet and R. P. Langlands, *Automorphic forms on* GL(2). Lecture Notes in Mathematics, vol. 114. Springer-Verlag, Berlin-New York, 1970.

[L] R.P. Langlands, *Base change for* GL(2). Annals of Mathematics Studies, 96, Princeton University Press, Princeton, N.J., 1980.

[LR] R.P. Langlands and D. Ramakrishnan, *The zeta functions of Picard modular surfaces.* Université de Montréal, Centre de Recherches Mathématiques, Montréal, QC, 1992.

[MR] R. Murty and D. Ramakrishnan, *The Albanese of unitary Shimura varieties.* The zeta functions of Picard modular surfaces. Université de Montréal, Centre de Recherches Mathématiques, Montréal, QC, 1992.

[O] T. Oda, *Periods of Hilbert modular surfaces.* Progress in Mathematics, 19. Birkhä user, Boston, Mass., 1982.

[R1] J. Rogawski, *Analytic expression for the number of points mod p.* The zeta functions of Picard modular surfaces, 65–109, Univ. Montréal, Montréal, QC, 1992.

[R2] J. Rogawski, *Automorphic representations of unitary groups in three variables.* Annals of Mathematics Studies, 123. Princeton University Press, Princeton, NJ, 1990.

[S1] G. Shimura, *On the zeta-function of an abelian variety with complex multiplication.* Ann. of Math. (2) 94 (1971), 504–533.

[S2] G. Shimura, *Introduction to the arithmetic theory of automorphic functions.* Kanô Memorial Lectures, No. 1. Publications of the Mathematical Society of Japan, No. 11. Iwanami Shoten, Publishers, Tokyo; Princeton University Press, Princeton, N.J., 1971.

[S3] Goro Shimura, On elliptic curves with complex multiplication as factors of the Jacobians of modular function fields. *Nagoya Math. J.* **43** (1971), 199–208.

[S4] G. Shimura, On the factors of the jacobian variety of a modular function field. *J. Math. Soc. Japan* **25** (1973), 523–544.

[T1] R. Taylor, On Galois representations associated to Hilbert modular forms. *Invent. Math.* **98** no. 2, (1989), 265–280.

[T2] R. Taylor, *On Galois representations associated to Hilbert modular forms. II.* Elliptic curves, modular forms, & Fermat's last theorem (Hong Kong, 1993), 185–191, Ser. Number Theory, I, Internat. Press, Cambridge, MA, 1995.

## CHAPTER 6

# ON THE GLOBAL GROSS-PRASAD CONJECTURE FOR YOSHIDA LIFTINGS

By Siegfried Böcherer, Masaaki Furusawa, and Rainer Schulze-Pillot

---

*Prof. J. Shalika on his 60th birthday*

**Introduction.** In two articles in the Canadian Journal [16, 17], B. Gross and D. Prasad proclaimed a global conjecture concerning the decomposition of an automorphic representation of an adelic special orthogonal group $G_1$ upon restriction to an embedded orthogonal group $G_2$ of a quadratic space in smaller dimension and also its local counterpart. In the local situation, one can summarize the conjecture by saying that the occurrence of $\pi_2$ in the restriction of $\pi_1$ depends on the $\epsilon$-factor attached to the representation $\pi_1 \otimes \pi_2$; in the global situation, assuming the existence of the local nontrivial invariant functional at all places and its nonvanishing on the spherical vector at almost all unramified places, one considers a specific linear functional given by a period integral. This period integral is then conjectured to give a nontrivial functional if and only if the central critical value of the $L$-function attached to $\pi_1 \otimes \pi_2$ is nonzero. In particular in the case when $G_1$ is the group of an $n$-dimensional nondegenerate quadratic space $V$ and $G_2$ is the group of an $(n-1)$-dimensional subspace $W$ of $V$, they showed that in low dimensions $(n \leq 4)$ known results can be interpreted as evidence for this conjecture, using the well known isomorphisms for orthogonal groups in low dimensions.

The case $n = 5$ has been treated in the local situation by Prasad [27]; it can also be reinterpreted using these isomorphisms: The split special orthogonal group in dimension 5 is isomorphic to the projective symplectic similitude group $\mathrm{PGSp}_2$, and the spin group of the 4-dimensional split orthogonal group is $\mathrm{SL}_2 \times \mathrm{SL}_2$. Prasad then showed that for forms on $\mathrm{PGSp}_2$ that are lifts from the orthogonal group of a 4-dimensional space, the situation can be understood in terms of the seesaw dual reductive pair (in Kudla's sense)

$$\begin{array}{ccc} GSp_2 & & GO(4) \times GO(4). \\ | & \times & | \\ G(SL_2 \times SL_2) & & GO(4) \end{array}$$

Part of this work was done during a stay of all three authors at the IAS, Princeton, supported by von Neumann Fund, Bell Fund and Bankers Trust Fund. Furusawa was also supported by Sumitomo Foundation. We thank the IAS for its hospitality.

S. Böcherer and R. Schulze-Pillot thank D. Prasad and the Harish Chandra Research Institute, Allahabad, India for their hospitality. Schulze-Pillot's visit to HCRI was also supported by DFG.

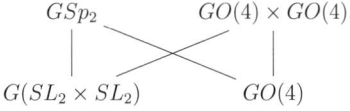

In classical terms, the analogous global question leads to the problem to determine those pairs of cuspidal elliptic modular forms (eigenforms of almost all Hecke operators) that can occur as summands if one decomposes the restriction of a cuspidal Siegel modular form $F$ of degree 2 (that is an eigenform of almost all Hecke operators) to the diagonally embedded product of two upper half planes into a sum of (products of) eigenforms of almost all Hecke operators (for a discussion of the problems that arise in translating the representation theoretic statement into a classical statement see below and Remark 2.13). One can also rephrase this as the problem of calculating the period integral

$$\int_{(\Gamma\backslash\mathbf{H})\times(\Gamma\backslash\mathbf{H})} F\left(\begin{pmatrix} z_1 & 0 \\ 0 & z_2 \end{pmatrix}\right)\overline{f_1(z_1)f_2(z_2)}d^*z_1d^*z_2$$

for two elliptic Hecke eigenforms $f_1, f_2$. The $L$-function that should occur then according to the conjecture of Gross and Prasad is the degree 16 $L$-function associated to the tensor product of the 4-dimensional representation of the $L$-group Spin(4) of SO(5) with the two 2-dimensional representation associated to two copies of SO(3) (due to the decomposition of SO(4) or rather its covering group mentioned above). We denote this $L$-function as $L(Spin(F), f_1, f_2, s)$.

In this reformulation it is natural to go beyond the original question of nonvanishing and to try and get an explicit formula connecting the $L$-value in question with the period integral. In general, it seems rather difficult to calculate the period integral as above, since little is known about the restriction of Siegel modular forms to the diagonally embedded product of two upper half planes. One should also point out that an integral representation for the degree 16 $L$-function in question is not known yet. However, for theta series of quadratic forms the restriction to the diagonal of a degree two theta series becomes simply the product of the degree one theta series in the variables $z_1, z_2$, which allows one to get a calculation started. Thus here we only consider Siegel modular forms (of trivial character) that arise as linear combinations of theta series of quaternary quadratic forms. Such Siegel modular forms, if they are eigenforms, are called Yoshida liftings, attached to a pair of elliptic cusp forms or, equivalently, to a pair of automorphic forms on the multiplicative group of an adelic quaternion algebra. These liftings have been investigated in [32, 4, 8], and the connection between the trilinear forms on the spaces of automorphic forms on the multiplicative group of an adelic quaternion algebra and the triple product $L$-function has been investigated in [20, 15, 7]. If one combines these results and applies them to the present situation, it turns out that the period integral in question can indeed be explicitly calculated in terms of the central critical value of the $L$-function mentioned; in the case of a Yoshida lifting attached to the pair $h_1, h_2$ of elliptic cusp forms, this $L$-function is seen to split into the product $L(h_1, f_1, f_2, s)L(h_2, f_1, f_2, s)$, so that the central critical value becomes the product of the central critical values of these two triple product $L$-functions. We prove a formula that expresses the square of the period integral explicitly as the product of these two central critical values, multiplied by an explicitly known non-zero factor. We reformulate the obtained identity in a way which makes sense as well for an arbitrary Siegel modular form $F$ in terms of the original $L(Spin(F), f_1, f_2, s)$, in place of $L(h_1, f_1, f_2, s)L(h_2, f_1, f_2, s)$, hoping that such an identity indeed holds for any Siegel modular form $F$. At present we cannot prove it except for the case when $F$ is a Siegel or suitable Klingen Eisenstein series of level 1. In the case when $F$ is the Saito-Kurokawa lifting of an elliptic Hecke eigenform $h$, one sees easily that the period integral is zero unless one has $f_1 = f_2 = f$; in this case the period integral can be transformed into the Petersson inner product of the restriction of the first Fourier Jacobi coefficient of $F$ to the upper half plane with $f$ and then leads us to a conjectural identity for the square of this Petersson inner product with the central critical value of $L(h, f, f; s)$.

Our calculation leaves the question open whether it can happen that the period integral vanishes for the classical modular forms considered but is non-zero for other functions in the same adelic representation space. Viewed locally, this amounts to the question whether an invariant nontrivial linear functional on the local representation space is necessarily non-zero at the given vector. At the infinite place we can exhibit such a vector (depending on the weights given) by applying a suitable differential operator to the Siegel modular form considered. At the finite places not dividing the level, it comes down to the question whether (for an unramified representation) a nontrivial invariant linear functional is necessarily nonzero at the spherical (or class 1) vector invariant under the maximal compact subgroup. This is generally expected, at least for generic representations. We intend to come back to this question in future work.

We also investigate the situation where the pair $f_1, f_2$ and the product $\mathbf{H} \times \mathbf{H}$ are replaced by a Hilbert modular form and the modular embedding of a Hilbert modular surface; in terms of the Gross-Prasad conjecture this amounts to replacing the split orthogonal group of a 4-dimensional space from above by a nonsplit (but quasisplit) orthogonal group that is split at infinity. It turns out that one gets an analogous result; we prove this only in the simplest case when all modular forms involved have weight 2, the class number of the quadratic field involved is 1 and the order in a quaternion algebra belonging to the situation is a maximal order. The proof for the general case should be possible in an analogous manner.

**1. Yoshida liftings and their restriction to the diagonal.** For generalities on Siegel modular forms we refer to [Fre1]. For a symplectic matrix $M = \left(\begin{smallmatrix} A & B \\ C & D \end{smallmatrix}\right) \in GSp_n(\mathbf{R})$ (with $n \times n$-blocks $A, B, C, D$) we denote by $(M, Z) \mapsto M < Z >= (AZ + B)(CZ + D)^{-1}$ the usual action of the group $G^+Sp(n, \mathbf{R})$ of proper symplectic similitudes on Siegel's upper half space $\mathbf{H}_n$.

We shall mainly be concerned with Siegel modular forms for congruence subgroups of type

$$\Gamma_0^{(n)}(N) = \left\{ \begin{pmatrix} A & B \\ C & D \end{pmatrix} \in Sp(n, \mathbf{Z}) \mid C \equiv 0 \bmod N \right\}.$$

The space of Siegel modular forms (and cusp forms respectively) of degree $n$ and weight $k$ for $\Gamma_0^{(n)}(N)$ will be denoted by $M_n^k(N)$ $(S_n^k(N))$, for a vector valued modular form transforming according to the representation $\rho$ the weight $k$ above should be replaced by $\rho$. By $<,>$ we denote the Petersson scalar product.

We recall from [4, 8, 5] some notations concerning the Yoshida liftings whose restrictions we are going to study in this article. For details we refer to the cited articles. We consider a definite quaternion algebra $D$ over $\mathbf{Q}$ and an Eichler order $R$ of square free level $N$ in it and decompose $N$ as $N = N_1N_2$ where $N_1$ is the product of the primes that are ramified in $D$. On $D$ we have the involution $x \mapsto \overline{x}$, the (reduced) trace $\mathrm{tr}(x) = x + \overline{x}$ and the (reduced) norm $n(x) = x\overline{x}$.

The group of proper similitudes of the quadratic form $q(x) = n(x)$ on $D$ is isomorphic to $(D^\times \times D^\times)/Z(D^\times)$ (as algebraic group) via

$$(x_1, x_2) \mapsto \sigma_{x_1,x_2} \text{ with } \sigma_{x_1,x_2}(y) = x_1 y x_2^{-1},$$

the special orthogonal group is then the image of

$$\{(x_1, x_2) \in D^\times \times D^\times \mid n(x_1) = n(x_2)\}.$$

We denote by $H$ the orthogonal group of $(D, n)$ and by $H^+$ the special orthogonal group.

For $\nu \in \mathbf{N}$ let $U_\nu^{(0)}$ be the space of homogeneous harmonic polynomials of degree $\nu$ on $\mathbf{R}^3$ and view $P \in U_\nu^{(0)}$ as a polynomial on

$$D_\infty^{(0)} = \{x \in D_\infty \mid \mathrm{tr}(x) = 0\}$$

by putting

$$P\left(\sum_{i=1}^{3} x_i e_i\right) = P(x_1, x_2, x_3)$$

for an orthonormal basis $\{e_i\}$ of $D_\infty^{(0)}$ with respect to the norm form $n$. The space $U_\nu^{(0)}$ is known to have a basis of rational polynomials (i.e., polynomials that take rational values on vectors in $D^{(0)} = D_\infty^{(0)} \cap D$).

The group $D_\infty^\times/\mathbf{R}^\times$ acts on $U_\nu^{(0)}$ through the representation $\tau_\nu$ (of highest weight $(\nu)$) given by

$$(\tau_\nu(y))(P)(x) = P(y^{-1}xy).$$

Changing the orthonormal basis above amounts to replacing $P$ by $(\tau_\nu(y))(P)$ for some $y \in D_\infty^\times$.

By $\langle\langle\ ,\ \rangle\rangle_0$ we denote the suitably normalized invariant scalar product in the representation space $U_\nu^{(0)}$.

For $\nu_1 \geq \nu_2$ the $H^+(\mathbf{R})$-space

$$U_{\nu_1}^{(0)} \otimes U_{\nu_2}^{(0)}$$

(irreducible of highest weight $(\nu_1+\nu_2, \nu_1-\nu_2)$) is isomorphic to the $H^+(\mathbf{R})$-space $U_{\nu_1,\nu_2}$ of $\mathbf{C}[X_1, X_2]$-valued harmonic forms on $D_\infty^2$ transforming according to the representation of $GL_2(\mathbf{R})$ of highest weight $(\nu_1+\nu_2, \nu_1-\nu_2)$.

An intertwining map $\Psi$ has been given explicitly in [5, Section 3]; for $\mathbf{x} = (x_1, x_2) \in D_\infty^2$ the polynomial $\Psi(Q)(\mathbf{x}) \in \mathbf{C}[X_1, X_2]$ is homogeneous of degree $2\nu_2$. We write now for $Q \in U_{\nu_1}^{(0)} \otimes U_{\nu_2}^{(0)}$

$$\Psi(Q)(\mathbf{x}) = \sum_{\alpha_1+\alpha_2=2\nu_2} c_{\alpha_1\alpha_2}(\mathbf{x}, Q) X_1^{\alpha_1} X_2^{\alpha_2}. \tag{1.1}$$

The map $\mathbf{x} \mapsto c_{\alpha_1\alpha_2}(\mathbf{x}, Q)$ is (for fixed $Q$) a polynomial in $x_1, x_2$ that is harmonic of degree $\alpha_1' = \alpha_1 + \nu_1 - \nu_2$ in $x_1$ and harmonic of degree $\alpha_2' = \alpha_2 + \nu_1 - \nu_2$ in $x_2$, and for $h \in H^+(\mathbf{R})$ we have

$$c_{\alpha_1\alpha_2}(h\mathbf{x}, Q) = c_{\alpha_1\alpha_2}(\mathbf{x}, h^{-1}Q).$$

The irreducibility of the space $U_{\nu_1}^{(0)} \otimes U_{\nu_2}^{(0)}$ implies that this map is nonzero for some $Q$. We denote by $U_\alpha$ the space of harmonic polynomials of degree $\alpha$ on $D_\infty$ with invariant scalar product $\langle\langle\ ,\ \rangle\rangle$. If $\alpha$ is even, the $H(\mathbf{R})$-spaces $U_\alpha$ and $U_{\alpha/2}^{(0)} \otimes U_{\alpha/2}^{(0)}$ are isomorphic and will be identified.

The map

$$(Q, R_1, R_2) \mapsto \langle\langle c_{\alpha_1\alpha_2}(\cdot, Q), R_1 \otimes R_2 \rangle\rangle \tag{1.2}$$

for $Q \in U_{\nu_1}^{(0)} \otimes U_{\nu_2}^{(0)}$, $R_1 \in U_{\alpha_1'}$, $R_2 \in U_{\alpha_2'}$ defines then a nontrivial invariant trilinear form for the triple of $H(\mathbf{R})$-spaces $((U_{\nu_1}^{(0)} \otimes U_{\nu_2}^{(0)}), U_{\alpha_1'}, U_{\alpha_2'})$.

LEMMA 1.1. *Let integers $\nu_1 \geq \nu_2$ and $\beta_1, \beta_2$ be given for which*

$$\beta_1' = \beta_1 + \nu_1 - \nu_2, \beta_2' = \beta_2 + \nu_1 - \nu_2$$

*are even. Then there exists a nontrivial $H(\mathbf{R})$-invariant trilinear form $T$ on the space $U_{\nu_1,\nu_2} \otimes U_{\beta_1'} \otimes U_{\beta_2'}$ if and only if there exist integers $\alpha_1, \alpha_2, \gamma$ such that $\beta_i = \alpha_i + \gamma$ and $\alpha_1 + \alpha_2 = 2\nu_2$ holds. This form is unique up to scalar multiples and can be decomposed as*

$$T = T_1^{(0)} \otimes T_2^{(0)}$$

*with (up to scalars) unique nontrivial invariant trilinear forms*

$$T_i^{(0)} = T_{i,\beta_1',\beta_2'}^{(0)}$$

*on*

$$U_{\nu_i}^{(0)} \otimes U_{\beta_1'/2}^{(0)} \otimes U_{\beta_2'/2}^{(0)}.$$

*In particular, for $\gamma = 0$ and $T$ fixed, the trilinear form given in (1.2) is proportional to $T$ (with a nonzero factor $\tilde{c}(\nu_1, \nu_2, \alpha_1, \alpha_2)$).*

*Proof.* Decomposing

$$U_{\beta_1'} = U^{(0)}_{\beta_1'/2} \otimes U^{(0)}_{\beta_1'/2}, U_{\beta_2'} = U^{(0)}_{\beta_2'/2} \otimes U^{(0)}_{\beta_2'/2}$$

as a $D_\infty^\times/\mathbf{R}^\times \times D_\infty^\times/\mathbf{R}^\times$-space one sees that $T$ as asserted exists if and only if there are nontrivial invariant trilinear forms

$$T_i^{(0)} = T^{(0)}_{i,\beta_1',\beta_2'}$$

on

$$U^{(0)}_{\nu_i} \otimes U^{(0)}_{\beta_1'/2} \otimes U^{(0)}_{\beta_2'/2}$$

for $i = 1, 2$. In this case $T$ decomposes as

$$T = T_1^{(0)} \otimes T_2^{(0)}.$$

□

The $T_i^{(0)}$ are known to exist if and only if the triples

$$(\beta_1'/2, \beta_2'/2, \nu_1), (\beta_1'/2, \beta_2'/2, \nu_2)$$

are balanced, i.e. the numbers in either triple are the lengths of the sides of a triangle (and then the form is unique up to scalars); they are unique up to scalar multiplication. It is then easily checked that the numerical condition given above is equivalent to the existence of nonnegative integers $\alpha_1, \alpha_2, \gamma$ satisfying $\beta_i = \alpha_i + \gamma$ and $\alpha_1 + \alpha_2 = 2\nu_2$. Consider now the Gegenbauer polynomial $G^{(\alpha)}(x, x') =$ obtained from

$$G_1^{(\alpha)}(t) = 2^\alpha \sum_{j=0}^{[\frac{\alpha}{2}]} (-1)^j \frac{1}{j!(\alpha - 2j)!} \frac{(\alpha - j)!}{2^{2j}} t^{\alpha - 2j}$$

by

$$\tilde{G}^{(\alpha)}(x, x') = 2^\alpha (n(x)n(x'))^{\alpha/2} G_1^{(\alpha)} \left( \frac{\mathrm{tr}(x\overline{x'})}{2\sqrt{n(x)n(x')}} \right)$$

and normalize the scalar product on $U_\alpha$ such that $G^{(\alpha)}$ is a reproducing kernel, i.e.

$$\langle\langle G^{(\alpha)}(x, x'), Q(x) \rangle\rangle_\alpha = Q(x')$$

for all $Q \in U_\alpha$. Then for $\alpha_1, \alpha_2, \alpha_1', \alpha_2'$ as above and some fixed $Q \in U_{\nu_1,\nu_2}$ the map

$$(x_1, x_2) \mapsto T(Q, G^{(\alpha_1')}(x_1, \cdot), G^{(\alpha_2')}(x_2, \cdot))$$

defines a polynomial $R_Q(x_1, x_2)$ in $x_1, x_2$ that is harmonic of degree $\alpha_1' = \alpha_1 + \nu_1 - \nu_2$ in $x_1$ and harmonic of degree $\alpha_2' = \alpha_2 + \nu_1 - \nu_2$ in $x_2$, and for $h \in H^+(\mathbf{R})$ we have $R_Q(h\mathbf{x}) = R_{h^{-1}Q}(\mathbf{x})$.

As above we can therefore conclude that one has

$$c_{\alpha_1\alpha_2}(\mathbf{x}, Q) = \tilde{c}(\nu_1, \nu_2, \alpha_1, \alpha_2)T(Q, G^{(\alpha_1')}(x_1, \cdot)), G^{(\alpha_2')}(x_2, \cdot), \tag{1.3}$$

where the factor of proportionality $\tilde{c}(\nu_1, \nu_2, \alpha_1, \alpha_2)$ is not zero.
We denote by

$$\mathcal{A}(D_{\mathbf{A}}^{\times}, R_{\mathbf{A}}^{\times}, \nu)$$

the space of functions $\varphi : D_{\mathbf{A}}^{\times} \to U_{\nu}^{(0)}$ satisfying $\varphi(\gamma x u) = \tau_{\nu}(u_{\infty}^{-1})\varphi(x)$ for $\gamma \in D_{\mathbf{Q}}^{\times}$ and $u = u_{\infty}u_f \in R_{\mathbf{A}}^{\times}$, where

$$R_{\mathbf{A}}^{\times} = D_{\infty}^{\times} \times \prod_p R_p^{\times}$$

is the adelic group of units of $R$. These functions are determined by their values on the representatives $y_i$ of a double coset decomposition

$$D_{\mathbf{A}}^{\times} = \cup_{i=1}^{r} D^{\times} y_i R_{\mathbf{A}}^{\times}$$

(where we choose the $y_i$ to satisfy $y_{i,\infty} = 1$ and $n(y_i) = 1$).

The natural inner product on the space $\mathcal{A}(D_{\mathbf{A}}^{\times}, R_{\mathbf{A}}^{\times}, \nu)$ is given by

$$\langle \varphi, \psi \rangle = \sum_{i=1}^{r} \frac{\langle\langle \varphi(y_i), \psi(y_i) \rangle\rangle_0}{e_i},$$

where $e_i = \mid (y_i R y_i^{-1})^{\times} \mid$ is the number of units of the order $R_i = y_i R y_i^{-1}$ of $D$.

On the space $\mathcal{A}(D_{\mathbf{A}}^{\times}, R_{\mathbf{A}}^{\times}, \nu)$ we have for $p \nmid N$ (hermitian) Hecke operators $\tilde{T}(p)$ (given explicitly by the $\mathrm{End}(U_{\nu}^{(0)})$-valued Brandt matrices $(B_{ij}(p))$) and for $p \mid N$ involutions $\tilde{w}_p$ commuting with the Hecke operators and with each other.

For $i = 1, 2$ and $\nu_1 \geq \nu_2$ with $\nu_1 - \nu_2$ even we consider now functions $\varphi_i$ in $\mathcal{A}(D_{\mathbf{A}}^{\times}, R_{\mathbf{A}}^{\times}, \nu_i)$.

The Yoshida lifting (of degree 2) of the pair $(\varphi_1, \varphi_2)$ is then given as

(1.4) $Y^{(2)}(\varphi_1, \varphi_2)(Z)(X_1, X_2) =$

$$= \sum_{i,j=1}^{r} \frac{1}{e_i e_j} \sum_{(x_1,x_2)\in(y_i R y_j^{-1})^2} \Psi(\varphi_1(y_i) \otimes \varphi_2(y_j))(x_1, x_2)(X_1, X_2) \times$$

$$\times \exp\left(2\pi i \mathrm{tr}\left(\begin{pmatrix} n(x_1) & tr(\overline{x_1}x_2) \\ tr(\overline{x_1}x_2) & n(x_2) \end{pmatrix} Z\right)\right).$$

This is a vector valued holomorphic Siegel modular form for the group $\Gamma_0^{(2)}(N)$ with trivial character and with respect to the representation $\sigma_{2\nu_2} \otimes \det^{\nu_1-\nu_2+2}$ (where $\sigma_{2\nu_2}$ denotes the $2\nu_2$-th symmetric power representation of $GL_2$).

If we consider the restriction of such a modular form to the diagonal $\left(\begin{smallmatrix} z_1 & 0 \\ 0 & z_2 \end{smallmatrix}\right)$, the coefficient of $X_1^{\alpha_1} X_2^{\alpha_2}$ becomes a function $F^{(\alpha_1,\alpha_2)}(z_1, z_2)$ which is in both variables a scalar valued modular form for the group $\Gamma_0(N)$ with trivial character of weight

$$\alpha_1 + \nu_1 - \nu_2 + 2 \text{ in } z_1, \quad \alpha_2 + \nu_1 - \nu_2 + 2 \text{ in } z_2.$$

In particular the weights in the variables $z_1, z_2$ add up to $2\nu_1 + 4$ for each pair $(\alpha_1, \alpha_2)$ with $\alpha_1 + \alpha_2 = 2\nu_2$ and the coefficient of $X_1^{\alpha_1} X_2^{\alpha_2}$ vanishes unless

$$\alpha_1' = \alpha_1 + \nu_1 - \nu_2, \quad \alpha_2' = \alpha_2 + \nu_1 - \nu_2$$

are even.

If $f_1, f_2$ are elliptic modular forms of weights $k_1, k_2$ we define then

$$\left\langle F\left(\left(\begin{smallmatrix} z_1 & 0 \\ 0 & z_2 \end{smallmatrix}\right)\right), f_1(z_1) f_2(z_2) \right\rangle_{k_1,k_2}$$

to be the double Petersson product

$$\left\langle \langle F^{(\alpha_1,\alpha_2)}(z_1, z_2), f_1(z_1)\rangle_{k_1}, f_2(z_2)\right\rangle_{k_2}$$

if

$$k_1 = \alpha_1 + \nu_1 - \nu_2 + 2, \quad k_2 = \alpha_2 + \nu_1 - \nu_2 + 2$$

for some $\alpha_1, \alpha_2$ with $\alpha_1 + \alpha_2 = 2\nu_2$ and to be zero otherwise (this definition coincides with the Petersson product of the corresponding automorphic forms on the groups $Sp_2(\mathbf{A})$ or $Sp_2(\mathbf{R})$ (restricted to the naturally embedded $SL_2(\mathbf{A}) \times SL_2(\mathbf{A})$) and $SL_2(\mathbf{A})$ (or the respective real groups)).

We will mainly consider Yoshida liftings for pairs of forms that are eigenforms of all Hecke operators and of all the involutions. It is then easy to see that $Y^{(2)}(\varphi_1, \varphi_2)(Z)$ is identically zero unless $\varphi_1, \varphi_2$ have the same eigenvalue under the involution $\tilde{w}_p$ for all $p \mid N$. The precise conditions under which the lifting is nonzero have been stated in [8].

We will finally need some facts about the correspondence studied e.g. in [10, 22, 29, 24] between modular forms for $\Gamma_0(N)$ (with trivial character) and automorphic forms on the adelic quaternion algebra $D_{\mathbf{A}}^\times$.

We consider the essential part

$$\mathcal{A}_{\text{ess}}\left(D_{\mathbf{A}}^\times, R_{\mathbf{A}}^\times, \nu\right)$$

consisting of functions $\varphi$ that are orthogonal to all $\psi \in \mathcal{A}(D_{\mathbf{A}}^\times, (R'_{\mathbf{A}})^\times, \nu)$ for orders $R'$ strictly containing $R$; this space is invariant under the $\tilde{T}(p)$ for $p \nmid N$ and the $\tilde{w}_p$ for $p \mid N$ and hence has a basis of common eigenfunctions of all the $\tilde{T}(p)$ for $p \nmid N$ and all the involutions $\tilde{w}_p$ for $p \mid N$. Being the components of eigenvectors of a rational matrix with real eigenvalues the values of these eigenfunctions are real, (i.e., polynomials with real coefficients in the vector valued case) when suitably normalized.

Moreover the eigenfunctions are in one to one correspondence with the newforms in the space

$$S^{2+2\nu}(N)$$

of elliptic cusp forms of weight $2+2\nu$ for the group $\Gamma_0(N)$ that are eigenfunctions of all Hecke operators (if $\tau$ is the trivial representation and $R$ is a maximal order one has to restrict here to functions orthogonal to the constant function 1 on the quaternion side in order to obtain cusp forms on the modular forms side). This correspondence (Eichler's correspondence) preserves Hecke eigenvalues for $p \nmid N$, and if $\varphi$ corresponds to $f \in S^{2+2\nu}(N)$ then the eigenvalue of $f$ under the Atkin-Lehner involution $w_p$ is equal to that of $\varphi$ under $\tilde{w}_p$ if $D$ splits at $p$ and equal to minus that of $\varphi$ under $\tilde{w}_p$ if $D_p$ is a skew field. The correspondence can be explicitly described by associating to $\varphi$ the modular form

$$h(z) = \sum_{i,j=1}^{r} \frac{1}{e_i e_j} \sum_{x \in (y_i R y_j^{-1})} (\varphi(y_i) \otimes \varphi(y_j))(x) \exp(2\pi i n(x) z)$$

(where as above $\varphi(y_i) \otimes \varphi(y_j)$ denotes the harmonic polynomial in $U_{2\nu}$ obtained by identifying $U_\nu^{(0)} \otimes U_\nu^{(0)}$ with $U_{2\nu}$).

An extension of Eichler's correspondence to forms $\varphi$ as above that are not essential but eigenfunctions of all the involutions $\tilde{w}_p$ has been given in [21, 7].

**2. Computation of periods.** Our goal is the computation of the periods

$$\langle Y^{(2)}(\varphi_1, \varphi_2)\left(\begin{pmatrix} z_1 & 0 \\ 0 & z_2 \end{pmatrix}\right), f_1(z_1) f_2(z_2)\rangle_{k_1,k_2}$$

defined above for elliptic modular forms $f_1, f_2$ for the group $\Gamma_0(N)$. For this we study first how the vanishing of this period integral depends on the eigenvalues of the functions involved under the Atkin-Lehner involutions or their quaternionic and Siegel modular forms counterparts.

For a Siegel modular form $F$ for the group $\Gamma_0^{(2)}(N)$ we let the Atkin-Lehner involutions with respect to the variables $z_1, z_2$ act on the restriction of $F$ to the diagonal matrices $\left(\begin{smallmatrix} z_1 & 0 \\ 0 & z_2 \end{smallmatrix}\right)$ and denote by $F(\left(\begin{smallmatrix} z_1 & 0 \\ 0 & z_2 \end{smallmatrix}\right)) \mid \widetilde{W}_p$ the result of this action.

LEMMA 2.1. *Let $N$ be squarefree and let $F$ be a vector valued Siegel modular form of degree 2 for the representation $\rho$ of $GL_2(\mathbf{C})$ of highest weight $(\lambda_1, \lambda_2)$ in the space $\mathbf{C}[X_1, X_2]_{\lambda_1-\lambda_2}$ of homogeneous polynomials of degree $\lambda_1 - \lambda_2$ in $X_1, X_2$ with respect to $\Gamma_0^{(2)}(N)$ and assume for $p \mid N$ that the restriction of $F$ to the diagonal is an eigenform of $\widetilde{W}_p$ with eigenvalue $\tilde{\epsilon}_p^{(0)}$. Let $f_1, f_2$ be elliptic cusp forms of weights $k_1, k_2$ for $\Gamma_0(N)$ that are eigenforms of the Atkin-Lehner involution $w_p$ with eigenvalues $\epsilon_p^{(1)}, \epsilon_p^{(2)}$. Then the period integral*

$$\left\langle F\left(\begin{pmatrix} z_1 & 0 \\ 0 & z_2 \end{pmatrix}\right), f_1(z_1) f_2(z_2)\right\rangle_{k_1,k_2} \tag{2.1}$$

*is zero unless one has $\tilde{\epsilon}_p^{(0)} \epsilon_p^{(1)} \epsilon_p^{(2)} = 1$.*

*Proof.* Applying the Atkin-Lehner involution $w_p$ to both variables $z_1, z_2$ one sees that this is obvious. □

We can view the condition of Lemma 2.1 as a (necessary) local condition for the nonvanishing of the period integral at the finite primes dividing the level, with a similar role being played at the infinite primes by the condition that modular forms of the weights $k_1, k_2$ of $f_1, f_2$ appear in the decomposition of the restriction of the vector valued modular form $F$ to the diagonal (or of a suitable form in the representation space of $F$, see below).

LEMMA 2.2. *Let $f_1, f_2, h_1, h_2$ be modular forms for $\Gamma_0(N)$ that are eigenfunctions of all Atkin-Lehner involutions for the $p \mid N$ with $f_1, f_2$ cuspidal. Let $h_1, h_2$ have the same eigenvalue $\epsilon'_p$ for all the $w_p$ for the $p \mid N$ and denote by $\epsilon_p^{(1)}, \epsilon_p^{(2)}$ the Atkin-Lehner eigenvalues at $p \mid N$ of $f_1, f_2$.*

*For a factorization $N = N_1N_2$ where $N_1$ has an odd number of prime factors let $D_{N_1}$ be the quaternion algebra over $\mathbf{Q}$ that is ramified precisely at $\infty$ and the primes $p \mid N_1$ and $R_{N_1}$ an Eichler order of level $N$ in $D_{N_1}$.*

*Let $\varphi_1^{(N_1)}, \varphi_2^{(N_1)}$ be the forms in $\mathcal{A}((D_{N_1})_{\mathbf{A}}^\times, (R_{N_1})_{\mathbf{A}}^\times, \tau_i)(i = 1, 2)$ corresponding to $h_1, h_2$ under Eichler's correspondence.*

*Then the period integral*

$$\left\langle Y^{(2)}(\varphi_1^{(N_1)}, \varphi_2^{(N_1)}) \left( \begin{pmatrix} z_1 & 0 \\ 0 & z_2 \end{pmatrix} \right), f_1(z_1)f_2(z_2) \right\rangle_{k_1,k_2} \tag{2.2}$$

*is zero unless $\epsilon'_p\epsilon_p^{(1)}\epsilon_p^{(2)} = -1$ holds for precisely those $p$ that divide $N_1$; in particular it is always zero unless $\prod_{p\mid N} \epsilon'_p\epsilon_p^{(1)}\epsilon_p^{(2)} = -1$ holds.*

*Proof.* For each factorization of $N$ as above we denote by $\widetilde{\epsilon_p}(N_1)$ the eigenvalue under $\tilde{w}_p$ of $\varphi_1^{(N_1)}, \varphi_2^{(N_1)}$, we have $\widetilde{\epsilon_p}(N_1) = -\epsilon'_p$ for the $p$ dividing $N_1$ and $\widetilde{\epsilon_p}(N_1) = \epsilon'_p$ for the $p$ dividing $N_2$. Hence the product $\widetilde{\epsilon_p}(N_1)\epsilon_p^{(1)}\epsilon_p^{(2)}$ is 1 for all $p$ dividing $N$ if $N_1$ is the product of the primes $p \mid N$ such that $\epsilon'_p\epsilon_p^{(1)}\epsilon_p^{(2)} = -1$ and is $-1$ for at least one $p \mid N$ otherwise; in particular a decomposition for which $\widetilde{\epsilon_p}(N_1)\epsilon_p^{(1)}\epsilon_p^{(2)} = 1$ for all $p \mid N$ holds and $N_1$ has an odd number of prime factors exists if and only if we have $\prod_{p\mid N} \widetilde{\epsilon_p}(N_1)\epsilon_p^{(1)}\epsilon_p^{(2)} = -1$.

The $\widetilde{W}_p$-eigenvalue of the restriction of $Y^{(2)}(\varphi_1^{(N_1)}, \varphi_2^{(N_1)})$ to the diagonal is $\widetilde{\epsilon_p}(N_1)$ by the result of Lemma 9.1 of [4] on the eigenvalue of $Y^{(2)}(\varphi_1^{(N_1)}, \varphi_2^{(N_1)})$ under the analogue for Siegel modular forms of the Atkin-Lehner involution. The assertion then follows from the previous lemma. □

For simplicity we will in the sequel assume that $h_1, h_2, f_1, f_2$ are all newforms of (squarefree) level $N$; essentially the same results can be obtained for more general quadruples of forms of squarefree level using the methods of [7].

LEMMA 2.3. *Let $N \neq 1$ be squarefree, $D, R$ as described in Section 1, let $f_1, f_2$ be normalized newforms of weights $k_1, k_2$ for the group $\Gamma_0(N)$. Let $\varphi_1, \varphi_2 \in \mathcal{A}(D_{\mathbf{A}}^\times, R_{\mathbf{A}}^\times, \tau_i)$ be as above. Assume that the (even) weights $k_1, k_2$ of $f_1, f_2$ can be written as $k_i = \alpha_i + \nu_1 - \nu_2 + 2 = \alpha_i' + 2$ with nonnegative integers $\alpha_i$ satisfying $\alpha_1 + \alpha_2 = 2\nu_2$ and denote for $i = 1, 2$ by $\psi_i$ the $U^{(0)}_{\alpha_i'/2}$-valued form in $\mathcal{A}(D_{\mathbf{A}}^\times, R_{\mathbf{A}}^\times, \tau_{\alpha_i'})$ corresponding to $f_i$ under Eichler's correspondence.*

*Then the period integral*

$$\left\langle Y^{(2)}(\varphi_1, \varphi_2)\left(\begin{pmatrix} z_1 & 0 \\ 0 & z_2 \end{pmatrix}\right), f_1(z_1) f_2(z_2)\right\rangle_{k_1, k_2} \tag{2.3}$$

*has the (real) value*

$$c\langle f_1, f_1\rangle\langle f_2, f_2\rangle \left(\sum_{j=1}^{r} T^{(0)}_{\nu_1, \alpha_1', \alpha_2'}(\varphi_1(y_i) \otimes \psi_1(y_i) \otimes \psi_2(y_i))\right) \times \left(\sum_{j=1}^{r} T^{(0)}_{\nu_2, \alpha_1', \alpha_2'}(\varphi_2(y_i) \otimes \psi_1(y_i) \otimes \psi_2(y_i))\right), \tag{2.4}$$

*with a nonzero constant $c$ depending only on $\nu_1, \nu_2, k_1, k_2$.*

*Proof.* The coefficient of $X_1^{\alpha_1} X_2^{\alpha_2}$ of the $(i, j)$-term in $F(z_1, z_2) = Y^{(2)}(\varphi_1, \varphi_2)$ $\left(\begin{smallmatrix} z_1 & 0 \\ 0 & z_2 \end{smallmatrix}\right)$ is (with $Q_{ij} := \varphi_1(y_i) \otimes \varphi_2(y_j))$ equal to

$$\tilde{c}(\nu_1, \nu_2, \alpha_1, \alpha_2) \sum_{(x_1, x_2) \in I_{ij}} T\left(Q_{ij}, G^{(\alpha_1')}(x_1, \cdot), G^{(\alpha_2')}(x_2, \cdot)\right) \times \exp(2\pi i n(x_1) z_1) \exp(2\pi i n(x_2) z_2) \tag{2.5}$$

by (1.3). We write

$$\Theta_{ij}^{(\alpha')}(z)(x') = \sum_{x \in I_{ij}} G^{(\alpha')}(x, x') \exp(2\pi i n(x) z) \tag{2.6}$$

for the $U_{\alpha'}$-valued theta series attached to $I_{ij}$ and the Gegenbauer polynomial $G^{(\alpha)}(x, x')$ and rewrite (2.5) as

$$\tilde{c}(\nu_1, \nu_2, \alpha_1, \alpha_2) T(Q_{ij}, \Theta_{ij}^{(\alpha_1')}(z_1), \Theta_{ij}^{(\alpha_2')}(z_2)). \tag{2.7}$$

The $ij$-term of the period integral (2.3) becomes then

$$\tilde{c}(\nu_1, \nu_2, \alpha_1, \alpha_2) T(Q_{ij}, \langle\Theta_{ij}^{(\alpha_1')}(z_1), f_1(z_1)\rangle, \langle\Theta_{ij}^{(\alpha_2')}(z_2), f_2(z_2)\rangle, \tag{2.8}$$

which by (3.13) of [6] and the factorization

$$T = T_{\nu_1, \nu_2, \alpha_1', \alpha_2'} = T^{(0)}_{\nu_1, \alpha_1', \alpha_2'} \otimes T^{(0)}_{\nu_2, \alpha_1', \alpha_2'} = T_1^{(0)} \otimes T_2^{(0)} \tag{2.9}$$

is equal to

$$\tilde{c}(\nu_1,\nu_2,\alpha_1,\alpha_2)\langle f_1, f_1\rangle\langle f_2, f_2\rangle T_1^{(0)}(\varphi_1(y_i),\psi_1(y_i),\psi_2(y_i)) \times T_2^{(0)}(\varphi_2(y_j),\psi_1(y_j),\psi_2(y_j)). \tag{2.10}$$

Summation over $i, j$ proves the assertion. The value computed is real since the values of the $\varphi_i, \psi_i$ are so and since $T_0$ is known to be real. □

THEOREM 2.4. *Let $h_1, h_2, f_1, f_2, \psi_1, \psi_2$ be as in Lemma 2.2, Lemma 2.3 with Atkin-Lehner eigenvalues $\epsilon'_p$ for $h_1, h_2$ and $\epsilon_p^{(1)}, \epsilon_p^{(2)}$ for $f_1, f_2$; assume $\prod_{p|N}\epsilon'_p\epsilon_p^{(1)}\epsilon_p^{(2)} = -1$. Let $D$ be the quaternion algebra over $\mathbf{Q}$ which is ramified precisely at the primes $p \mid N$ for which $\epsilon'_p\epsilon_p^{(1)}\epsilon_p^{(2)} = -1$ holds and $R$ an Eichler order of level $N$ in $D$, let $\varphi_1, \varphi_2$ be the forms in $\mathcal{A}(D_{\mathbf{A}}^\times, R_{\mathbf{A}}^\times, \tau_{1,2})$ corresponding to $h_1, h_2$ under Eichler's correspondence.*

*Then the square of the period integral*

$$\left\langle Y^{(2)}(\varphi_1,\varphi_2)\left(\begin{pmatrix} z_1 & 0 \\ 0 & z_2\end{pmatrix}\right), f_1(z_1)f_2(z_2)\right\rangle_{k_1,k_2} \tag{2.11}$$

*is equal to*

$$\frac{c}{\langle h_1,h_1\rangle\langle h_2,h_2\rangle} L\left(h_1, f_1, f_2; \frac{1}{2}\right) L\left(h_2, f_1, f_2; \frac{1}{2}\right), \tag{2.12}$$

*where $c$ is an explicitly computable nonzero number depending only on $\nu_1, \nu_2, k_1, k_2, N$ and the triple product $L$-function $L(h, g, f; s)$ is normalized to have its functional equation under $s \mapsto 1-s$.*

*In particular the period integral is nonzero if and only if the central critical value of $L(h_1, f_1, f_2; s)L(h_2, f_1, f_2; s)$ is nonzero.*

*Proof.* The choice of the decomposition $N = N_1N_2$ made above implies that we can use Theorem 5.7 of [7] to express the right hand side of (2.4) by the product of central critical values of the triple product $L$-functions associated to $(h_1, f_1, f_2), (h_2, f_1, f_2)$. The Petersson norms of $f_1, f_2$ appearing in Theorem 5.7 of [7] cancel against those appearing in the proof of Lemma 2.3. □

*Remark* 2.5. (a) If the product $\prod_{p|N}\epsilon'_p\epsilon_p^{(1)}\epsilon_p^{(2)}$ is $+1$ we know from [7] that the sign in the functional equation of the triple product $L$-functions $L(h_1, f_1, f_2; s)$, $L(h_2, f_1, f_2; s)$ is $-1$ and hence the central critical values are zero; from Lemma 2.2 we know that for any Yoshida lifting $F$ associated to $h_1, h_2$ as in Lemma 2.2 the Petersson product of the restriction of $F$ to the diagonal and $f_1(z_1), f_2(z_2)$ is zero as well.

(b) It should be noticed that given $h_1, h_2$ there are $2^{\omega(N)-1}$ possible choices of the quaternion algebra with respect to which one considers the Yoshida lifting associated to $h_1, h_2$. All these Yoshida liftings are different, but have the same Satake parameters for all $p \nmid N$. Given $f_1, f_2$ with $\prod_{p|N} \epsilon'_p \epsilon_p^{(1)} \epsilon_p^{(2)} = -1$ there is then precisely one choice of quaternion algebra that leads to a nontrivial result for the period integral, all the others give automatically zero by Lemma 2.2. The choice of this quaternion algebra should be seen as variation of the Vogan $L$-packet of the $p$-adic component of the adelic representation generated by the Siegel modular form for the $p \mid N$ in such a way that the resulting $L$-packet satisfies the local Gross-Prasad condition for the split 5-dimensional and 4-dimensional orthogonal groups.

We want to rephrase the result of Theorem 2.4 in order to replace the factor of comparison $\langle h_1, h_1\rangle\langle h_2, h_2\rangle$ occurring by a factor depending only on $F = Y^{(2)}(\varphi_1, \varphi_2)$ instead of $h_1, h_2$. Concerning the symmetric square $L$-function of $F$ occurring in the following corollary we remind the reader that we view $F$ as an automorphic form on the adelic orthogonal group of the 5-dimensional quadratic space $V$ of discriminant 1 over $\mathbf{Q}$ that contains a 2-dimensional totally isotropic subspace.

COROLLARY 2.6. *Under the assumptions of Theorem 2.4 and the additional assumption that $h_1, h_2$ are not proportional, the value of (2.12) is equal to:*

$$\frac{c\langle F, F\rangle}{L^{(N)}(F, Sym^2, 1)} L\left(h_1, f_1, f_2; \frac{1}{2}\right) L\left(h_2, f_1, f_2; \frac{1}{2}\right), \tag{2.13}$$

*where again $c$ is an explicitly computable nonzero constant depending only on the levels and weights involved and $L^{(N)}(F, Sym^2, s)$ is the $N$-free part of the $L$ function of $F$ with respect to the symmetric square of the 4-dimensional representation of the $L$-group of the group $SO(V)$ ($V$ as above).*

*Proof.* Since $h_1, h_2$ are not proportional, the Siegel modular form $F$ is cuspidal and the Petersson product $\langle F, F\rangle$ is well defined. From [4, Proposition 10.2] we recall that $\langle F, F\rangle$ is (up to a nonzero constant) equal to the the residue at $s = 1$ of the $N$-free part $D_F^{(N)}(s)$ of the degree 5 $L$-function associated to $F$ (normalizing the $\varphi_i$ to $\langle\varphi_i, \varphi_i\rangle = 1$; the formulas given in [4] generalize easily to the situation where the $\varphi_i$ take values in harmonic polynomials). It is also well known that $\langle h_i, h_i\rangle$ is equal (up to a nonzero constant depending only on weights and levels) to $D_{h_i}^{(N)}(1)$ where $D_{h_i}^{(N)}(s)$ is the symmetric square $L$-function associated to $h_i$. Comparing the parameters of the $L$-functions $L^{(N)}(F, \mathrm{Sym}^2, s)$ and $D_F^{(N)}(s)D_{h_1}^{(N)}(s)D_{h_2}^{(N)}(s)$ we see that the value of $L^{(N)}(F, \mathrm{Sym}^2, s)$ at $s = 1$ is equal to the residue at $s = 1$ of $D_F^{(N)}(s)D_{h_1}^{(N)}(s)D_{h_2}^{(N)}(s)$, which gives the assertion. □

*Remark.* We can as well view $L^{(N)}(F, \mathrm{Sym}^2, s)$ as the exterior square of the degree 5 $L$-function associated to $F$.

Let us discuss now two degenerate cases:

COROLLARY 2.7. *(a) Under the assumptions of Theorem 2.4 replace* $\varphi_2$ *by the constant function* $(\Sigma_i \frac{1}{e_i})^{-1}$ *(and hence* $h_2(z)$ *by the Eisenstein series* $E(z) = (\Sigma_{i,j} \frac{1}{e_i e_j})^{-1} \Sigma_{i,j} \frac{1}{e_i e_j} \Theta_{ij}^{(0)}(z)$*). Then the period integral*

$$\left\langle Y^{(2)}(\varphi_1, \varphi_2)\left(\begin{pmatrix} z_1 & 0 \\ 0 & z_2 \end{pmatrix}\right), f_1(z_1) f_2(z_2) \right\rangle_{k_1, k_2} \tag{2.14}$$

*is zero unless* $f_1 = f_2 =: f$*, in which case its square is equal to the value at* $s = 1$ *of*

$$\frac{c\langle F, F\rangle}{L^{(N)}(F, Sym^2, s)} L\left(h_1, f, f; s - \frac{1}{2}\right) L\left(E, f, f; s - \frac{1}{2}\right), \tag{2.15}$$

*where again* $c$ *is an explicitly computable nonzero constant depending only on the levels and weights involved and* $L(E, f_1, f_2; s)$ *is defined in the same way as the triple product* $L$*-function for a triple of cusp forms, setting the* $p$*-parameters of* $E$ *equal to* $p^{1/2}, p^{-1/2}$ *for* $p \nmid N$*.*

*(b) Under the assumptions of Theorem 2.4 let* $h = h_1 = h_2$*. Then the period integral (2.11) is equal to*

$$\frac{c}{\langle h, h\rangle} L\left(h, f_1, f_2; \frac{1}{2}\right), \tag{2.16}$$

*where* $c$ *is a nonzero constant depending only on the levels and weights involved.*

*Proof.* Both assertions are obtained in the same way as Theorem 2.4 and Corollary 2.6; notice that in case a) (with $\omega(N)$ denoting the number of prime factors of $N$) both $L(E, f, f; s - \frac{1}{2})$ and $L^{(N)}(F, \mathrm{Sym}^2, s)$ are of order $\omega(N) - 1$ at $s = 1$. □

*Remark.* (a) The form of our result given in Corollary 2.6 could in principle be true for any Siegel modular form $F$ instead of a Yoshida lifting if one replaces $L(h_1, f_1, f_2; \frac{1}{2}) L(h_2, f_1, f_2; \frac{1}{2})$ by the value $L(Spin(F), f_1, f_2, \frac{1}{2})$ of the spin $L$-function mentioned in the introduction. There is, however, not much known about the analytic properties of $L^{(N)}(F, \mathrm{Sym}^2, s)$, in particular this $L$- function might have a zero or a pole at $s = 1$.

(b) In the degenerate case of Corollary 2.7 a) the Yoshida lifting $F$ is the Saito-Kurokawa lifting associated to $h_1$. The result of that case could also be true in the case that $h_1, f_1, f_2$ are of level 1 and $F$ is the Saito-Kurokawa lifting of $h_1$, but we cannot prove this at present (except for the vanishing of the period integral in the case that $f_1 \neq f_2$, which is easily proved).

We notice that in that last case (as well as in the related case of a Yoshida lifting of Saito-Kurokawa type) the period integral is seen to be equal to the Petersson product $\langle \phi_1(\tau, 0), f(\tau)\rangle$, where $\phi_1(\tau, z)$ is the first Fourier Jacobi coefficient of $F$.

In the case of Corollary 2.7 b) the Yoshida lift $F$ can be viewed as an Eisenstein series of Klingen type associated to $h$, in particular its image under Siegel's $\Phi$-operator is equal to $h$ (more precisely, the Klingen Eisenstein series in question is a sum of Yoshida liftings associated to various quaternion algebras of level dividing $N$ (see [6]), where the other contributions yield a vanishing period integral). Notice that this Eisenstein series is vector valued if the weight $k$ of $h$ is $> 2$; the usual scalar valued Klingen Eisenstein series leads (in the case of level 1) to a similar formula with $L(h, f_1, f_2; \frac{k-1}{2})$ instead of $L(h, f_1, f_2; \frac{1}{2})$.

We can obtain a result similar to that of Theorem 2.4 for more general weights $k_1, k_2$ of the modular forms $f_1, f_2$. For this, remember that according to Lemma 1.1 the value given in (2.4) for the period integral in question also makes sense if one replaces throughout $\alpha_1', \alpha_2'$ by $\beta_1' = \alpha_1' + \gamma, \beta_2' = \alpha_2' + \gamma$ for some fixed $\gamma > 0$; the forms $f_1, f_2$ then having weights $k_i = \alpha_i' + 2 + \gamma$ for $i = 1, 2$. As noticed above, our period integral becomes 0 in this situation. We can, however, modify the function $Y^{(2)}(\varphi_1, \varphi_2)(Z)$ by a differential operator $\tilde{\mathcal{D}}^{\gamma}_{2,\alpha_1,\alpha_2}$ in such a way that $\tilde{\mathcal{D}}^{\gamma}_{2,\alpha_1,\alpha_2} Y^{(2)}(\varphi_1, \varphi_2)$ is a function on $\mathbf{H} \times \mathbf{H}$ that is a modular form of weights $k_1, k_2$ of $z_1, z_2$ as described above and yields a value for the period integral of the same form as the one given in form (2.4).

More precisely, we have:

PROPOSITION 2.8. *For nonnegative integers $k$, $r$ and $l$ with $k \geq 2$ and any partition $l = a + b$, there exists a (nonzero) holomorphic differential operator $\mathcal{D}^r_{k,a,b}$ (polynomial in $X_1^2 \frac{\partial}{\partial z_1}, X_1 X_2 \frac{\partial}{\partial z_{12}}, X_2^2 \frac{\partial}{\partial z_2}$, evaluated in $z_{12} = 0$) mapping $\mathbf{C}[X_1, X_2]_l$-valued functions on $\mathbf{H}_2$ to $\mathbf{C} \cdot X_1^{a+r} X_2^{b+r}$-valued functions on $\mathbf{H} \times \mathbf{H}$ and satisfying*

$$\mathcal{D}^r_{k,a,b}\big(F \mid_{k,l} M_1^{\uparrow} M_2^{\downarrow}\big) = \big(\mathcal{D}^r_{k,a,b} F\big) \mid^{z_1}_{k+a+r} M_1 \mid^{z_2}_{k+b+r} M_2$$

*for all $M_1, M_2 \in SL(2, \mathbf{R})$; here the upper indices $z_1$ and $z_2$ at the slash-operator indicate the variable, with respect to which one has to apply the elements of $SL(2, \mathbf{R})$ and $\uparrow\downarrow$ denote the standard embedding of $SL(2) \times SL(2)$ into $Sp(2)$ given by*

$$\begin{pmatrix} a & b \\ c & d \end{pmatrix}^{\uparrow} \times \begin{pmatrix} A & B \\ C & D \end{pmatrix}^{\downarrow} = \begin{pmatrix} a & 0 & b & 0 \\ 0 & A & 0 & B \\ c & 0 & d & 0 \\ 0 & C & 0 & D \end{pmatrix}.$$

*Of course one can consider $\mathcal{D}^r_{k,a,b} F$ as a $\mathbf{C}$-valued function.*

*Remark* 2.9. One may indeed show, that there exists (up to multiplication by a constant) precisely one such nontrivial holomorphic differential operator.

Corollary 2.10. *The differential operator $\mathcal{D}^r_{k,a,b}$ defined above gives rise to a map*

$$\tilde{\mathcal{D}}^r_{k,a,b} : M^2_{k,l}\big(\Gamma^2_0(N)\big) \longrightarrow M^1_{k+a+r}\big(\Gamma^1_0(N)\big) \otimes M^1_{k+b+r}\big(\Gamma^1_0(N)\big)$$

*of spaces of modular forms. (It is easy to see that for $r > 0$ this map actually goes into spaces of cusp foms).*

Corollary 2.11. *Denoting by $Sym_2(\mathbf{C})$ the space of complex symmetric matrices of size 2 we define a polynomial function*

$$Q : Sym_2(\mathbf{C}) \longrightarrow \mathbf{C}[X_1, X_2]_{2r}$$

*by*

$$\mathcal{D}^r_{k,a,b} e^{tr(TZ)} = Q(T) e^{t_1 z_1 + t_2 z_2}$$

*where*

$$Z = \begin{pmatrix} z_1 & z_3 \\ z_3 & z_2 \end{pmatrix} \in \mathbf{H}_2.$$

*Furthermore we assume (with $k = \frac{m}{2} + \nu$) that $P : \mathbf{C}^{(m,2)} \longrightarrow \mathbf{C}[X_1, X_2]_l$ is a polynomial function satisfying*

*a) $P$ is pluriharmonic*

*b) $P((X_1, X_2)A) = \rho_{\nu,l}(A) P(X_1, X_2)$ for all $A \in GL(2, \mathbf{C})$.*

*Then, for $\mathbf{Y}_1, \mathbf{Y_2} \in \mathbf{C}^m$*

$$(\mathbf{Y}_1, \mathbf{Y}_2) \longmapsto \left\{ P(\mathbf{Y}_1, \mathbf{Y}_2) \cdot Q\left(\begin{pmatrix} \mathbf{Y}_1^t \mathbf{Y}_1 & \mathbf{Y}_1^t \mathbf{Y}_2 \\ \mathbf{Y}_2^t \mathbf{Y}_1 & \mathbf{Y}_2^t \mathbf{Y}_2 \end{pmatrix}\right) \right\}_{a+r+\nu, b+r+\nu}$$

*defines an element of $H_{a+r+\nu}(m) \otimes H_{b+r+\nu}(m)$, where $H_\mu(m)$ is the space of harmonic polynomials in $m$ variables (for the standard quadratic form), homogeneous of degree $\mu$ and for any $R \in \mathbf{C}[X_1, X_2]_{l+2r}$ we denote by $\{R\}_{\alpha,\beta}$ the coefficient of $X_1^\alpha X_2^\beta$ in $R$, $\alpha + \beta = l + 2r$.*

The proof of Corollary 2.11 is a vector-valued variant of similar statements in [2] and [9, p. 200], using the proposition above and the characterization of harmonic polynomials by the Gauß-transform; we leave the details of proof to the reader.

*Proof (of Proposition 2.8).* We start from a Maaß-type differential operator $\delta_{k+l}$ which maps $\mathbf{C}[X_1, X_2]_l$-valued functions on $\mathbf{H}_2$ to $\mathbf{C}[X_1, X_2]_{l+2}$-valued ones and satisfies

$$(\delta_{k+l} F) \mid_{k,l+2} M = \delta_{k+l} \left(F \mid_{k,l} M\right)$$

for all $M \in Sp(2, \mathbf{R})$.

It is well known, how such operators arise from elements of the universal enveloping algebra of the complexified Lie algebra of $Sp(2, \mathbf{R})$, see e.g. [19] . In our case (we refer to [3] for details) we can describe these operators quite explicitly in terms of the simple operators

$$DF := \left(\frac{1}{2\pi i}\frac{\partial}{\partial Z}\right)[\mathbf{X}]$$

and

$$NF := \left(-\frac{1}{4\pi}(ImZ)^{-1}F\right)[\mathbf{X}].$$

Here $\mathbf{X}$ stands for the column vector $\binom{X_1}{X_2}$. Then we define

$$\delta_k F = kNF + DF.$$

It is remarkable (and already incorporated in our notation!) that $\delta_{k+l}$ depends only on $k+l$.
The iteration

$$\delta_{k+l}^r := \delta_{k+l+2r-2} \circ \cdots \circ \delta_{k+l+2} \circ \delta_{k+l}$$

can also be described explicitly by

$$\delta_{k+l}^r = \sum_{i=0}^{r} \frac{\Gamma(k+l+r)}{\Gamma(k+l+r-i)}\binom{r}{i} N^i D^{r-i}.$$

For a function $F : \mathbf{H}_2 \longrightarrow \mathbf{C}[X_1, X_2]_l$ and a decomposition $l = a+b$ we put

$$\nabla_{k+l}^r(a,b)F =: X_1^{a+r}X_2^{b+r} - \text{coefficient of } \left(\delta_{2,k+l}^r F\right)_{|\mathbf{H}\times\mathbf{H}}.$$

Then $\nabla$ has already the transfomation properties required in the proposition, i.e.

$$\nabla_{k+l}^r(a,b)\big(F\mid_{k,l} M_1^{\uparrow}M_2^{\downarrow}\big) = \big(\nabla_{k+l}^r(a,b)F\big)\mid_{k+a+r}^{z_1} M_1 \mid_{k+b+r}^{z_2} M_2$$

for $M_1,\ M_2 \in SL(2,\mathbf{R})$.
Moreover, if $F$ is in addition a holomorphic function on $\mathbf{H}_2$, then $\nabla_{k+l}^r(a,b)F$ is a nearly holomorphic function in the sense of Shimura (with respect to both variables $z_1$ and $z_2$), as polynomials in $\frac{1}{y_1}$ and $\frac{1}{y_2}$ they are of degree $\leq r$. Shimura's structure theorem on nearly holomorphic functions [30] says that all nearly holomorphic functions on $\mathbf{H}$ can be obtained from holomorphic functions by applying Maaß-type operators

$$\delta_k := \frac{k}{2iy} + \frac{\partial}{\partial z}$$

and their iterates. This however is only true if the weight (i.e. $k+a+r$ or $k+r+b$) is bigger than $2r$, which is not necessarily true in our situation. We therefore use a weaker version of Shimura's theorem (see [31, Theorem 3.3]), valid under the assumption "$w > 1+r$", where $w$ is the weight at hand and $r$ is the degree of the nearly holomorphic function: Every such function $f$ on $\mathbf{H}$ of degree $\leq r$ has an expression

$$f = f_{hol} + L_w(\tilde{f})$$

where $f_{hol}$ is holomorphic and $\tilde{f}$ is again nearly holomorphic of degree $\leq r$; in this expression

$$L_w := \delta_{w-2}\left(y^2\frac{\partial}{\partial\bar{z}}\right) = \frac{w}{2i}y\frac{\partial}{\partial\bar{z}} + y^2\frac{\partial^2}{\partial z\partial\bar{z}}$$

is a "Laplacian" of weight $w$ commuting with the $|_w$-action of $SL(2,\mathbf{R})$. We also point out that $f_{hol}$ is uniquely determined by $f$ (in particular, $f = f_{hol}$, if $f$ is holomorphic) and we have $(f\mid_w M)_{hol} = (f_{hol})\mid_w M$ for all $M \in SL(2,\mathbf{R})$.

If we apply this statement to $\nabla^r_{k+l}(a,b)F$, considered as function of $z_1$ and $z_2$, we get an expression of type

$$\nabla^r_{k+l}(a,b)F = f + L^{z_1}_{k+a+r}g_1 + L^{z_2}_{k+r+b}g_2 + L^{z_1}_{k+a+r}L^{z_2}_{k+r+b}h$$

where $f, g_1, g_2, h$ are nearly holomorphic functions on $\mathbf{H}\times\mathbf{H}$, $f$ being holomorphic in both variables, $g_1$ holomorphic in $z_2$, $g_2$ holomorphic in $z_1$. Note that (due to our assumption $k \geq 2$ ) Shimura's theorem is applicable here. An inspection of Shimura's proof (which is quite elementary for our case) shows that $f$ is indeed of the form $f = \mathbf{D}F$, where $\mathbf{D}$ is a polynomial $p$ in $\frac{\partial}{\partial z_1}, \frac{\partial}{\partial z_{12}}, \frac{\partial}{\partial z_2}$, evaluated in $z_{12} = 0$. This polynomial does not depend on $F$ at all and it has the required transformation properties.

It remains however to show that $\mathbf{D}$ is not zero:

For this purpose, we consider the special function

$$\mathbf{z}^r_{12} : \begin{cases} \mathbf{H}_2 \longrightarrow \mathbf{C}[X_1, X_2]_l \\ Z \longmapsto z^r_{12}X_1^aX_2^b \end{cases}.$$

It is easy to see that $\nabla^r_{k+l}(a,b)(\mathbf{z}^r_{12})$ is then equal to the constant function $r!$, therefore

$$\nabla^r_{k+l}(a,b)\big(\mathbf{z}^r_{12}\big) = \mathbf{D}\big(\mathbf{z}^r_{12}\big) = r!,$$

in particular, $\mathbf{D}$ is nonzero and we may put

$$\mathcal{D}^r_{k,a,b} = p\left(X_1^2\frac{\partial}{\partial z_1}, X_1X_2\frac{\partial}{\partial z_{12}}, X_2^2\frac{\partial}{\partial z_2}\right),$$

evaluated at $z_{12} = 0$. □

We can then prove in the same way as above:

COROLLARY 2.12. *The assertions of Lemma 2.3 and Theorem 2.4 remain true if $f_1, f_2$ have weights $k_i = \alpha_i' + 2 + \gamma (i = 1, 2)$ with some $\gamma > 0$, if one replaces*

$$Y^{(2)}(\varphi_1,\varphi_2)\left(\begin{pmatrix} z_1 & 0 \\ 0 & z_2 \end{pmatrix}\right)$$

*by*

$$\tilde{\mathcal{D}}^{\gamma}_{2,\alpha_1,\alpha_2} Y^{(2)}(\varphi_1, \varphi_2)(z_1, z_2).$$

*Remark* 2.13. Application of the differential operator to $Y^{(2)}((\varphi_1, \varphi_2)(Z))$ before restriction to the diagonal does not change the $Sp_2(\mathbf{R})$-representation space of that function, i.e., we have found a different function in the same representation space whose period integral assumes the value that is predicted by the conjecture of Gross and Prasad. More precisely, (using remark 2.9 and some additional considerations) one can show that the vanishing of this predicted value is already sufficient for the vanishing of the period integral for all triples $F'$, $f_1'$, $f_2'$ of functions in the Harish-Chandra modules generated by the original functions $F$, $f_1$, $f_2$. To obtain a similar statement for the local representations at the finite places not dividing the level one would have to show that a nonvanishing invariant linear functional on the tensor product of the representations is not zero on the product of the spherical (or class 1) vectors invariant under the maximal compact subgroup. This is expected to be true as well; we plan to come back to these problems in future work.

**3. Restriction to an embedded Hilbert modular surface.** To avoid technical difficulties we deal here only with the simplest case: The quaternion algebra $D$ is ramified at all primes $p$ dividing the level $N$ and we have $\nu_1 = \nu_2 = 0$, i.e., the Yoshida lifting is a scalar valued Siegel modular form of weight 2 and the order $R$ we are considering is a maximal order. We put $F = \mathbf{Q}(\sqrt{N})$ and assume that $N$ is such that the class number of $F$ is 1. We denote by $\Delta$ the discriminant of $F$, by $a \mapsto a^{\sigma}$ its nontrivial automorphism and consider the basis $1, w$ with $w = \frac{\Delta+\sqrt{\Delta}}{2}$ of the ring $\mathfrak{o}_F$ of $F$. Denoting by $C$ the matrix

$$C := \begin{pmatrix} 1 & 1 \\ w & \bar{w} \end{pmatrix}$$

we have the usual modular embedding

(3.1) $$\iota : (z_1, z_2) \mapsto C \begin{pmatrix} z_1 & 0 \\ 0 & z_2 \end{pmatrix} {}^tC$$

of $\mathbf{H} \times \mathbf{H}$ into the Siegel upper half plane $\mathbf{H}_2$ and

(3.2) $$\tilde{\iota} : \begin{pmatrix} a & b \\ c & d \end{pmatrix} \mapsto \begin{pmatrix} C & 0 \\ 0 & {}^tC^{-1} \end{pmatrix} \begin{pmatrix} a & 0 & b & 0 \\ 0 & a^{\sigma} & 0 & b^{\sigma} \\ c & 0 & d & 0 \\ 0 & c^{\sigma} & 0 & d^{\sigma} \end{pmatrix} \begin{pmatrix} C^{-1} & 0 \\ 0 & {}^tC \end{pmatrix}$$

from $SL_2(F)$ into $Sp_2(\mathbf{Q})$.

We have then for $\gamma \in SL_2(F)$ :

(3.3) $$\tilde{\iota}(\gamma)(\iota((z_1, z_2))) = \tilde{\iota}(\gamma((z_1, z_2)))$$

with the usual actions of the groups $SL_2(F)$ on $\mathbf{H} \times \mathbf{H}$ and of $Sp_2(\mathbf{Q})$ on the Siegel upper half plane.
We put now

$$J = \begin{pmatrix} 1 & 0 & 0 & 0 \\ 0 & 0 & 0 & 1 \\ 0 & 0 & 1 & 0 \\ 0 & 1 & 0 & 0 \end{pmatrix}$$

and consider for a Siegel modular form $f$ of weight $k$ for the group $\Gamma \subseteq Sp_2(\mathbf{Q})$ the function

$$\tilde{f}(\tau_1, \tau_2) := f|_k J(\iota(\tau_1, \tau_2)). \tag{3.4}$$

Writing $\iota_0 = J \circ \iota$, $\widetilde{\iota_0}(\gamma) := J\iota(\gamma)J^{-1}$ we see that $\tilde{f}$ is a Hilbert modular form for the group $\widetilde{\iota_0}^{-1}(\Gamma)$.

By calculating $\widetilde{\iota_0}(\gamma)$ explicitly for $\gamma \in SL_2(F)$ one checks that $\widetilde{\iota_0}(\gamma)$ is in $\Gamma_0^{(2)}(N)$ if and only if $\gamma$ is in $SL_2(\mathfrak{o}_F \oplus \mathfrak{d})$, the group of matrices $\left(\begin{smallmatrix} a & b \\ c & d \end{smallmatrix}\right)$ with $a, d \in \mathfrak{o}_F, c \in \mathfrak{d}, b \in \mathfrak{d}^{-1}$, where $\mathfrak{d}$ is the different of $F$.

If $L$ is a $\mathbf{Z}$-lattice of (even) rank $m = 2k$ with quadratic form $q$ and associated bilinear form $B(x, y) := q(x + y) - q(x) - q(y)$ satisfying $q(L) \subseteq \mathbf{Z}$ and $Nq(L^\#)\mathbf{Z} = \mathbf{Z}$ ($N$ is the level of $(L, q)$) then it is shown in [4] that

(3.5)

$$\vartheta^{(2)}(L, q, Z)|_k J = c_1 \sum_{x_1 \in L, x_2 \in L^\#} \exp\left(2\pi i \mathrm{tr}\left(\begin{pmatrix} q(x_1) & B(x_1, x_2)/2 \\ B(x_1, x_2)/2 & q(x_2) \end{pmatrix} Z\right)\right),$$

where $c_1$ is a nonzero constant depending only on the genus of $(L, q)$ and where $\vartheta^{(2)}(L, q, Z)$ is the usual theta series of degree 2 of $(L, q)$. One checks therefore that, writing $K = \{x_1 + x_2 w \in L \otimes F \mid x_1 \in L, x_2 \in L^\#\}$, we have

$$\vartheta^{(2)}(L, q, \widetilde{\iota_0}(z_1, z_2)) = \vartheta(K, q, (z_1, z_2)), \tag{3.6}$$

where we denote by $\vartheta(K, q, (z_1, z_2)) = \sum_{y \in K} \exp(2\pi i(q(y)z_1 + q(y)^\sigma z_2))$ the theta series of the $\mathfrak{o}_F$-lattice $K$ with the extended form $q$ on it.

It is again easily checked that $L^\# \subseteq N^{-1}L$ implies that $K$ is an integral unimodular $\mathfrak{o}_F$-lattice, and it is well known that then the theta series $\vartheta(K, q, (z_1, z_2))$ is a modular form of weight $k$ for the group $SL_2(\mathfrak{o}_F \oplus \mathfrak{d})$.

LEMMA 3.1. *Let $\tilde{D}$ be a quaternion algebra over $F$ ramified at both infinite primes and let $\tilde{R}$ be a maximal order in $\tilde{D}$. Let*

$$\mathcal{A}(\tilde{D}_{\mathbf{A}}^\times, \tilde{R}_{\mathbf{A}}^\times, 0) =: \mathcal{A}(\tilde{D}_{\mathbf{A}}^\times, \tilde{R}_{\mathbf{A}}^\times)$$

*be defined in the same way as in Section 1 for D and let $\mathcal{A}(\tilde{D}_{\mathbf{A}}^\times, \tilde{R}_{\mathbf{A}}^\times)$ be equipped with the natural action of Hecke operators $T(\mathfrak{p})$ for the $\mathfrak{p}$ not dividing $N$ described* *by Brandt matrices as explained in* [11]. *Then by associating to a Hecke eigenform* $\psi \in \mathcal{A}(\tilde{D}_{\mathbf{A}}^{\times}, \tilde{R}_{\mathbf{A}}^{\times})$ *the Hilbert modular form*

$$f(z_1, z_2) = \int_{(\tilde{D}^{\times}\backslash\tilde{D}_{\mathbf{A}}^{\times})\times(\tilde{D}^{\times}\backslash\tilde{D}_{\mathbf{A}}^{\times})} \psi(y)\psi(y')\vartheta\big(y'\tilde{R}y^{-1}, (z_1, z_2)\big)dy\,dy'$$

*one gets a bijective correspondence between the* $\psi$ *as above and the Hecke eigenforms of weight 2 and trivial character for the group* $\Gamma_0(\mathfrak{n}, \mathfrak{d})$ *of precise level* $\mathfrak{n}$ *giving an explicit realization of the correspondence of Shimizu und Jacquet/Langlands* [29, 24]. *Here* $\mathfrak{n}$ *denotes the product of the prime ideals ramified in* $\tilde{D}$ *and* $\Gamma_0(\mathfrak{n}, \mathfrak{d})$ *is the subgroup of* $SL_2(\mathfrak{o}_F \oplus \mathfrak{d})$ *whose lower left entries are in* $\mathfrak{n}\mathfrak{d}$.

*The function* $\rho(y, y')$ *on* $\tilde{D}_{\mathbf{A}}^{\times} \times \tilde{D}_{\mathbf{A}}^{\times}$ *given by setting* $\rho(y, y')$ *equal to the Petersson product of* $f$ *with* $\vartheta(y'\tilde{R}y^{-1}, (z_1, z_2))$ *is proportional to* $(y, y') \mapsto \psi(y)\psi(y')$.

*Proof.* The first part of this Lemma is due to Shimizu [29] (taking into account that by [11] the group $\Gamma_0(\mathfrak{n}, \mathfrak{d})$ is the correct transformation group for the theta series in question). Let $\hat{\rho}$ be the function on the adelic orthogonal group of $\tilde{D}$ induced by $\rho$ and let $\hat{\psi}$ be the function on the adelic orthogonal group of $\tilde{D}$ induced by $(y, y') \mapsto \psi(y)\psi(y')$. The function $\hat{\psi}$ generates an irreducible representation space of $\tilde{D}_{\mathbf{A}}^{\times}$ whose theta lifting to $SL_2(F_{\mathbf{A}})$ is generated by $f$, and $\hat{\rho}$ is a vector in the theta lifting of this latter representation of $SL_2(F_{\mathbf{A}})$, which by [25] coincides with the original representation space generated by $\hat{\psi}$. Since both $\hat{\rho}, \hat{\psi}$ are invariant under the same maximal compact subgroup of $\tilde{D}_{\mathbf{A}}^{\times}$, the uniqueness of such a vector implies that they must coincide up to proportionality. That $\hat{\rho}$ is not zero follows from the obvious fact that $f$ by its construction can not be orthogonal to all the theta series. □

LEMMA 3.2. *With the above notations let* $\varphi_1, \varphi_2$ *in* $\mathcal{A}(D_{\mathbf{A}}^{\times}, R_{\mathbf{A}}^{\times}, 0)$ *be Hecke eigenforms with the same eigenvalue under the involutions* $\widetilde{w_p}$ *for the* $p \mid N$ *with associated newforms* $h_1, h_2$ *of weight 2 and level* $N$. *Let* $f$ *be a Hilbert modular form of weight 2 for the group* $SL_2(\mathfrak{o}_F \oplus \mathfrak{d})$ *that corresponds in the way described in Lemma 3.1 to the function* $\psi \in \mathcal{A}(\tilde{D}_{\mathbf{A}}^{\times}, \tilde{R}_{\mathbf{A}}^{\times})$ *for* $\tilde{D} = D \otimes F$ *and* $\tilde{R}$ *being the maximal order in* $\tilde{R}$ *containing* $R$. *Then the value of the period integral*

$$\int_{SL_2(\mathfrak{o}_F \oplus \mathfrak{d})\backslash H\times H} \big(Y^{(2)}(\varphi_1, \varphi_2)\big) \mid_2 J(\iota((z_1, z_2)))f((z_1, z_2))dz_1dz_2 \tag{3.7}$$

*is equal to*

$$c_2\langle f, f\rangle\Big(\sum_i \varphi_1(y_i)\psi(y_i)\Big)\Big(\sum_i \varphi_2(y_i)\psi(y_i)\Big), \tag{3.8}$$

*where we identify* $y_i$ *with* $y_i \otimes 1 \in \tilde{D}$ *and where* $c_2$ *is some constant depending only on* $N$.

In order to interpret the value obtained in (3.8) in the same way as in Section 2 as the central critical value of an $L$-function, we review briefly the integral representation of the $L$-function that one obtains when one replaces in a triple $(h, f_1, f_2)$ of elliptic cusp forms the pair $(f_1, f_2)$ by one Hilbert cusp form $f$ for a real quadratic field.

For the moment, both the Hilbert cusp form $f$ and the elliptic cusp form $h$ can be of arbitrary even weight $k$. Now we consider the Siegel type Eisenstein series of weight $k$, defined on $\mathbf{H}_3$ by

$$E_3^k(W, s) = \sum_{\gamma=\begin{pmatrix} * & * \\ C & D\end{pmatrix} \in \Gamma_0^3(N)_\infty \backslash \Gamma_0^3(N)} det(CW + D)^{-k} det(\Im(\gamma < W >)^s)$$

Here and in the sequel we denote by $G_\infty$ the subgroup of $G$ defined by "$C = 0$," where $G$ is any group of symplectic matrices.

We restrict this Eisenstein series to $W = \begin{pmatrix} \tau & 0 \\ 0 & Z\end{pmatrix}$ with $\tau \in \mathbf{H}$, $Z \in \mathbf{H_2}$ and furthermore we consider then the modular embedding with respect to $Z$.

In this way we get a function $E(\tau, z_1, z_2, s)$, which behaves like a modular form for $\tau$ and like a Hilbert modular form for $(z_1, z_2)$ of weight $k$. We want to compute the twofold integral

$$I(f, h, s) := \int_{SL_2(\mathfrak{o}_F \oplus \mathfrak{d}) \backslash \mathbf{H}^2} \int_{\Gamma_0(N) \backslash \mathbf{H}} \overline{h(\tau) f(z_1, z_2)} E(\tau, z_1, z_2, s) d\tau^* dz_1^* dz_2^*$$

where $dz^* = y^{k-2} dx dy$ for $z = x + iy \in \mathbf{H}$.

This can be done in several ways: One can relate this integral to similar ones in [26] or in [14] (both these works are in an adelic setting) or one can try to do it along classical lines as in [13, 28, 9]. We sketch the latter approach here (for class number one, $h$ being a normalized newform of level $N$).

The inner integration over $\tau$ (which can be done with $Z \in \mathbf{H}_2$ instead of the embedded $(z_1, z_2)$) is the same as in the papers mentioned above, producing an $L$-factor $L_2(h, 2s + 2k - 2)$ (with $L_2(,)$ denoting the symmetric square $L$-function) times a Klingen-type Eisenstein series $E_{2,1}(h, s)$, which is defined as follows: We denote by $C_{2,1}$ the maximal parabolic subgroup of $Sp(2)$ for which the last line is of the form $(0, 0, 0, *)$ and we put $C_{2,1}(N) = C_{2,1}(\mathbf{Q}) \cap \Gamma_0^2(N)$. Furthermore we define a function $h_s(Z)$ on $\mathbf{H}_2$ by

$$h_s(Z) = h(z_1) \left( \frac{det(Y)}{y_1} \right)^s,$$

where $z_1 = x_1 + iy_1$ denotes the entry in the upper left corner of $Z = X + iY \in \mathbf{H}_2$. Then we put

$$E_{2,1}(h, s)(Z) = \sum_{\gamma \in C_{2,1} \backslash \Gamma_0^2(N)} h_s(Z) \mid_k \gamma$$

To do the second integration, one needs information on certain cosets: This is the only new ingredient entering the picture:

LEMMA 3.3. *A complete set of representatives for* $C_{2,1}(N)\backslash\Gamma_0^2(N)$ *is given by*

$$\{d(M)J^{-1}\tilde{\iota}_0(\gamma)\}$$

*with* $\gamma$ *running over* $SL_2(\mathfrak{o}_F\oplus\mathfrak{d})_\infty\backslash SL_2(\mathfrak{o}_F\oplus\mathfrak{d})$, *and* $M=\left(\begin{smallmatrix} * & * \\ u & v\end{smallmatrix}\right)$ *running over those elements of* $SL(2,\mathbf{Z})_\infty\backslash SL(2,\mathbf{Z})$ *with* $v\equiv 0(N)$, *where* $d$ *denotes the standard embedding of* $GL(2)$ *in* $Sp(2)$ *given by* $d(M)=\left(\begin{smallmatrix} (M^{-1})^t & 0 \\ 0 & M\end{smallmatrix}\right)$.

This lemma is related to the double coset decomposition

$$C_{2,1}(N)\backslash\Gamma_0^2(N)/\tilde{\iota}_0(SL_2(\mathfrak{o}_F\oplus\mathfrak{d}))$$

and somewhat analogous to the coset decomposition in [28, p. 692]; we omit the proof.

We may now do the usual unfolding to get

$$\int\limits_{SL_2(\mathfrak{o}_F\oplus\mathfrak{d})\backslash\mathbf{H}^2}\overline{h(z_1,z_2)}\widetilde{E_{2,1}(h,*,s)}(z_1,z_2)\,dz_1^*dz_2^* \tag{3.9}$$

$$=\int\limits_{SL_2(\mathfrak{o}_F\oplus\mathfrak{d})_\infty\backslash\mathbf{H}^2}\sum_{M=\left(\begin{smallmatrix} * & * \\ v & u\end{smallmatrix}\right)} h\left((v,u)C\begin{pmatrix} z_1 & 0 \\ 0 & z_2\end{pmatrix}C^t\begin{pmatrix} v \\ u\end{pmatrix}\right)$$

$$\times\overline{f(z_1,z_2)}\left(\frac{Dy_1y_4}{(v+u\omega)^2y_1+(v+u\bar{\omega})^2y_2}\right)^s dz_1^*dz_2^*.$$

Using the Fourier expansions of $f$ and $F$,

$$h(z)=\sum_{n=1}^{\infty}a(n)e^{2\pi inz},\qquad f(z_1,z_2)=\sum_{\nu\in\mathbf{0}_F,\nu\gg 0}A(\nu)e^{2\pi i tr(\nu\cdot z)}$$

one can (after some standard calculations) write the integral above as

$$\gamma(s)\sum_n\sum_{\sim\backslash(u,v)}a(n)\overline{A(n(v+u\omega)^2)}n^{-s-2k+2}N(v+u\omega)^{-2s-2k+2}$$

where we use the following equivalence relation: two pairs $(u,v)$ and $(u',v')$ are called equivalent iff $v+u\omega$ and $v'+u'\omega$ are equal up to a unit from $\mathfrak{o}_F$ as a factor.

Assume now in addition that $h$ is a normalized eigenfunction of all Hecke operators; then we define the $L$-function $L(h\otimes f,s)$ as an Euler product over all primes $p$ with Euler factors (at least for $p$ coprime to $N$)

$$L_p(h\otimes f,s):=L_p^{Asai}\left(f,\alpha_p p^{-s}\right)L_p^{Asai}\left(f,\beta_p p^{-s}\right)$$

where we use the Euler factors $L_p^{Asai}(f, s)$ of the Asai-L-function attached to $f$ (see [1]) and $\alpha_p$ and $\beta_p$ are the Satake-p-parameters attached to the eigenform $h$ (normalized to have absolute values $p^{\frac{k-1}{2}}$). We will write later $L(h, f; s)$ to denote the shift of this $L$-function that is normalized to have functional equation under $s \mapsto 1 - s$.

By standard calculation, we see that the integral above is, after multiplication by $L_2(h, 2s + 2k - 2)$, equal to the $L$-function $L(h \otimes f, s + 2k - 2)$ (up to elementary factors; the condition $v \equiv 0(N)$ also creates some extra contribution for $p$-Euler factors with $p \mid N$). This calculation of course requires some formal calculations similar to those given e.g. in [13].

*Remark* 3.4. If the class number $H$ of $F$ is different from one, then the orbit structure is more complicated. One gets $H$ different sets of representatives of the type described in the lemma above (each one twisted by a matrix in $SL_2(F)$ mapping a cusp into $\infty$). After unfolding, one gets then a Dirichlet series also involving Fourier coefficients of $f$ at all the $H$ different cusps. If we assume that $f$ is the first component (i.e. the one corresponding to the principal ideal class) in a tuple of $H$ Hilbert modular forms such that the corresponding adelic modular form is an eigenform of all Hecke operators, then it is possible (but quite unpleasant) to transfer that Dirichlet series into the Euler product in question.

Now we return to the case of weight 2. We can compute the integral $I(f, h, s)$ at $s = 0$ not only by unfolding as above but also by using the Siegel-Weil formula for the Eisenstein series in the integrand. Then one gets in the same way as in [7] that the square of the right hand side of (3.8) is (up to an explicit constant) the product of the central critical values of the $L$-functions attached to the pairs $h_1$, $f$ and $h_2$, $f$ as above:

THEOREM 3.5. *Let $\varphi_1, \varphi_2, h_1, h_2, f, \psi$ be as in Lemma 3.2. Then the square of the period integral*

$$\int_{SL_2(\mathfrak{o}_F \oplus \mathfrak{d}) \backslash \mathbf{H} \times \mathbf{H}} \left(Y^{(2)}(\varphi_1, \varphi_2)\right) \mid_2 J(\iota((z_1, z_2))) f((z_1, z_2))\, dz_1 dz_2 \tag{3.10}$$

*is equal to*

$$\frac{c_3}{\langle h_1, h_1 \rangle \langle h_2, h_2 \rangle} L(h_1, f; 1/2) L(h_2, f; 1/2), \tag{3.11}$$

*where $c_3$ is an explicitly computable nonzero number depending only on $N$ and the product L-function $L(h, f; s)$ is normalized to have its functional equation under $s \mapsto 1 - s$.*

*In particular the period integral is nonzero if and only if the central critical value of $L(h_1, f; s) L(h_2, f, s)$ is nonzero.*

SIEGFRIED BÖCHERER
KUNZENHOF 4B
79117 FREIBURG
GERMANY
*E-mail:* boech@siegel.math.uni-mannheim.de

MASAAKI FURUSAWA
DEPARTMENT OF MATHEMATICS
GRADUATE SCHOOL OF SCIENCE
OSAKA CITY UNIVERSITY
SUGIMOTO 3–3–138, SUMIYOSHI-KU
OSAKA 558-8585, JAPAN
*E-mail:* furusawa@sci.osaka-cu.ac.jp

RAINER SCHULZE-PILLOT
FACHRICHTUNG 6.1 MATHEMATIK
UNIVERSITÄT DES SAARLANDES (GEB. 27)
POSTFACH 151150
66041 SAARBRÜCKEN
GERMANY
*E-mail:* schulzep@math.uni-sb.de

## REFERENCES

[1] T. Asai, On certain Dirichlet Series Associated with Hilbert Modular Forms and Rankin's Method, *Math. Ann.* **226** (1977), 81–94.

[2] S. Böcherer, Über die Fourier-Jacobi-Entwicklung Siegelscher Eisensteinreihen. II, *Math. Z.* **189** (1985), 81–100.

[3] S. Böcherer, T. Satoh, and T. Yamazaki, On the pullback of a holomorphic differential operator and its application to vector-valued Eisenstein series. *Comm. Math. Univ. S. Pauli* **41** (1992), 1–22.

[4] S. Böcherer and R. Schulze-Pillot, Siegel modular forms and theta series attached to quaternion algebras. *Nagoya Math. J.* **121** (1991), 35–96.

[5] S. Böcherer and R. Schulze-Pillot, Mellin transforms of vector valued theta series attached to quaternion algebras, *Math. Nachr.* **169** (1994), 31–57.

[6] S. Böcherer and R. Schulze-Pillot, Vector valued theta series and Waldspurger's theorem, *Abh. Math. Sem.* Hamburg **64** (1994), 211–233.

[7] S. Böcherer and R. Schulze-Pillot, On the central critical value of the triple product L-function. In: Number Theory 1993–94, 1–46. Cambridge University Press, 1996.

[8] S. Böcherer and R. Schulze-Pillot, Siegel modular forms and theta series attached to quaternion algebras II. *Nagoya Math. J.* **147** (1997), 71–106.

[9] S. Böcherer and R. Schulze-Pillot, Squares of automorphic forms on quaternion algebras and central critical values of $L$-functions of modular forms. *J. Number Theory* **76,** (1999) 194–205.

[10] M. Eichler, The basis problem for modular forms and the traces of the Hecke operators, p. 76–151 in Modular functions of one variable $I$, Lecture Notes Math. 320, Berlin-Heidelberg-New York, 1973.

[11] M. Eichler, On theta functions of real algebraic number fields. *Acta Arith.* **33** (1977), no. 3, 269–292.
[12] E. Freitag, Siegelsche Modulfunktionen, Berlin-Heidelberg-New York, 1983.
[13] P. Garrett, Decomposition of Eisenstein series: Rankin triple products. Annals of Math. **125,** (1987) 209–235.
[14] P. Garrett, Integral representations of certain $L$-functions attached to one, two, and three modular forms. University of Minnesota Technical report 86–131 (1986).
[15] B. H. Gross and S. S. Kudla, Heights and the central critical values of triple product $L$-functions. *Compositio Math.* **81** (1992), no. 2, 143–209.
[16] B. H. Gross and D. Prasad, On the decomposition of a representation of $SO_n$ when restricted to $SO_{n-1}$. *Canad. J. Math.* **44** (1992), no. 5, 974–1002.
[17] B. H. Gross and D. Prasad, On irreducible representations of $SO_{2n+1} \times SO_{2m}$. *Canad. J. Math.* **46** (1994), no. 5, 930–950.
[18] W. Hammond, The modular groups of Hilbert and Siegel, *Am. J. Math.* **88** (1966), 497–516.
[19] M. Harris, Special values of zeta functions attached to modular forms. *Ann. Scient. Ec. Norm. Sup.* **14,** (1981) 77–120.
[20] M. Harris and S. Kudla, The central critical value of a triple product $L$-function, *Annals of Math.* **133** (1991), 605–672.
[21] K. Hashimoto, On Brandt matrices of Eichler orders, *Mem. School Sci. Engrg. Waseda Univ.* No. **59** (1995), 143–165 (1996).
[22] H. Hijikata and H. Saito, On the representability of modular forms by theta series, p. 13–21 in Number Theory, Algebraic Geometry and Commutative Algebra, in honor of Y. Akizuki, Tokyo, 1973.
[23] T. Ibukiyama, On differential operators on automorphic forms and invariant pluriharmonic polynomials. *Comm. Math. Univ. S. Pauli* **48,** (1999) 103–118.
[24] H. Jacquet and R. Langlands, Automorphic forms on $GL(2)$, Lect. Notes in Math. 114, Berlin-Heidelberg-New York, 1970.
[25] C. Moeglin, Quelques propriétes de base des series theta, *J. of Lie Theory* **7** (1997), 231–238.
[26] I. Piatetski-Shapiro and S. Rallis, Rankin triple $L$-functions. *Compos. Math.* **64,** (1987) 31–115.
[27] D. Prasad, Some applications of seesaw duality to branching laws, *Math. Annalen* **304,** (1996) 1–20.
[28] T. Satoh, Some remarks on triple $L$-functions. *Math. Ann.* **276,** (1987) 687–698.
[29] H. Shimizu, Theta series and automorphic forms on $GL_2$. *J. of the Math. Soc. of Japan* **24** (1972), 638–683.
[30] G. Shimura, The special values of the zeta functions associated with cusp forms, *Comm. pure appl. Math.* **29,** (1976) 783–804.
[31] G. Shimura, Differential operators, holomorphic projection, and singular forms, *Duke Math. J.* **76,** (1994) 141–173.
[32] H. Yoshida, Siegel's modular forms and the arithmetic of quadratic forms. *Invent. Math.* **60** (1980), no. 3, 193–248.

CHAPTER 7

# SUMS OF TWISTED $GL(3)$ AUTOMORPHIC L-FUNCTIONS

By DANIEL BUMP, SOLOMON FRIEDBERG, and JEFFREY HOFFSTEIN

*This paper is dedicated to Professor Joseph Shalika*

*Abstract.* Let $F$ be a number field and $\pi$ be an automorphic representation on $GL_r(A_F)$. In this paper we consider weighted sums of quadratic twists of the $L$-function for $\pi$, $\sum_d L(s, \pi, \chi_d)\, a(s, \pi, d)\, Nd^{-w}$, where $\chi_d$ is a quadratic character roughly attached to $F(\sqrt{d})/F$. We analyze the properties which the weights $a(s, \pi, d)$ must satisfy if this is to satisfy a (certain, non-abelian) group of functional equations in $(s, w)$, and show that there is a unique family of weight functions with this property for $r \leq 3$. We describe these weights in detail when $r = 3$. As an application we give, for cuspidal $\pi$ on $GL_3(A_{\mathbb{Q}})$, a new proof of the holomorphicity of the symmetric square $L$-function of $\pi$. We also prove that if $\pi'$ is a cuspidal automorphic representation of $GL(2)$ over $\mathbb{Q}$ then infinitely many quadratic twists of the adjoint square L-function of $\pi'$ are nonvanishing at the center of the critical strip.

**0. Introduction.** This paper concerns double Dirichlet series which may be expressed as weighted sums of quadratic twists of $L$-functions. In [BFH], a class of double Dirichlet series was proposed as follows. Let $\pi$ be an automorphic representation of $GL(r, A)$, where $A$ is the adele ring of the global field $F$, let $\chi_\pi$ be the central character of $\pi$, and let $\omega$ be an idèle class character of $A^\times/F^\times$. Roughly speaking, and restricting ourselves to the *quadratic* case, the Dirichlet series of interest is:

$$Z_0(s, w) = \sum_d L(s, \pi, \chi_d)\, \omega((d))\, Nd^{-w} \tag{0.1}$$

where $d$ runs through classes of $F^\times$ modulo squares, $Nd$ is the absolute norm, $\chi_d$ is the quadratic character attached to $F(\sqrt{d})$, and $L(s, \pi, \chi_d)$ denotes the twisted $L$-function of $\pi$. As is explained in [BFH], the expectation is that this function of two variables will satisfy two functional equations generating a finite group if $r \leq 3$ and an infinite group if $r \geq 4$.

Unfortunately (0.1) is only an approximation to the actual Dirichlet series which we want. To explain why, let us sketch a method of studying such a series. First, each $L$-series being summed has a functional equation. Taking into account the power of the conductor of $\chi_d$ which occurs in the epsilon-factor for $\pi \otimes \chi_d$, one sees that there should be a functional equation for $Z_0(s, w)$ under the

1991 *Mathematics Subject Classification.* Primary 11F66, Secondary 11F70, 11M41, 11N75.

*Key words ad phrases.* Automorphic representation, double Dirichlet series, quadratic twist, twisted $L$-function, mean value.

Research supported in part by NSF grants DMS-9970841 (Bump), DMS-9970118 (Friedberg), and DMS-0088921 (Hoffstein).

transformation $(s, w) \to (1 - s, w + r(s - 1/2))$. Second, writing $L(s, \pi, \chi_d) = \sum c(m)\, \chi_d(m)\, Nm^{-s}$, where the sum is over a set of representatives for the nonzero integral ideals of $F$, the series $Z_0(s, w)$ may be written

$$Z_0(s, w) = \sum_d \sum_m c(m)\, \chi_d(m)\, \omega((d))\, Nm^{-s}\, Nd^{-w}.$$

Now if $m$ is an integer, then quadratic reciprocity says roughly that $\chi_d(m)$ is equal to $\chi_m(d)$. Thus after an interchange of summation, the inner sum is roughly of the form $\sum_d \chi_m(d)\, \omega((d))\, Nd^{-w}$, and this is roughly a $GL_1$ $L$-function in $w$. This suggests that the series should have a second functional equation under the transformation $(s, w) \to (s + w - 1/2, 1 - w)$! If this can be made rigorous, then for $r \leq 3$ it is not hard to check that these two functional equations allow one to (meromorphically) continue $Z_0(s, w)$ to a tube domain in $\mathbb{C}^2$ whose convex hull is all of $\mathbb{C}^2$. Putting in a finite set of factors to cancel possible poles, one has a holomorphic function on this tube domain, and hence by the continuation theorem for functions of several complex variables which are holomorphic on a tube domain [Ho, Theorem 2.5.10], a function which continues to $\mathbb{C}^2$.

Now we can see the issues to be addressed. Indeed, for the interchange to yield a $GL_1$ sum, one must sum over *all* $d$ giving nonzero integral ideals of $F$, rather than over a set of representatives for $F^\times/(F^\times)^2$. This is problematic for several reasons. First, it is natural to attach quadratic characters only to elements of $F^\times$, not to ideals. In essence, issues of units and class number arise. In addition quadratic reciprocity must be formulated precisely. Third, if one is to sum over all $d$, this raises the question of how to get the desired functional equations. Even in the class-number-one case, the conductor of $\chi_d$ involves only the square-free-part of $d$. So if one is to sum over non-square-free $d$, one must include with the $L$-functions of (0.1) a second "correction factor" $a(s, \pi, d)$ which also has a functional equation under $s \to 1 - s$, this second functional equation giving the corresponding power of the square-part of $d$. In a similar way there must be such factors in $w$, after the interchange. Do such factors, behaving properly with respect to the functional equations and allowing the interchange of summation, exist? Are they unique?

The first two obstructions above, involving class number and quadratic reciprocity, have been resolved by Fisher and Friedberg [FF1, FF2]. We will describe their results briefly below. Then in this paper we will study these last questions by describing and analyzing the (essentially local) relations which must be satisfied by the correction factors for this sketch of proof to go through. Using this analysis, we will show that the last two questions may be answered in the affirmative for $r \leq 3$—the correction factors do exist, and they are unique.

A different approach to the existence of the correction factors has recently been given by Fisher and Friedberg [FF1, FF2]; this approach does not consider the issue of their uniqueness. Correction factors also arise in the work of Siegel [Si] ($r = 1$) and Bump, Friedberg and Hoffstein [BFH] and Friedberg and Hoffstein [FH] ($r = 2$).

Let us now explain how to resolve the issues from algebraic number theory as in the work of Fisher and Friedberg. Let $\mathfrak{O}$ be the ring of integers of $F$, and let $S$ be a finite set of places containing all even places and all archimedean places, and such that the ring of $S$-integers $\mathfrak{O}_S$ has class number 1. (In the application to double Dirichlet series, we will suppose that $\pi$ and $\omega$ are unramified outside $S$.) If $T$ is any set of places, let $I(T)$ denote the ideals of $\mathfrak{O}$ prime to $T$. Let $\mathtt{Div}(F)$ be the free abelian group generated by the places of $F$, and let $C = \sum_{v \in S} n_v v \in \mathtt{Div}(F)$, with $n_v = \mathrm{ord}_v(4)$ if $v$ is even and $n_v = 1$ otherwise. Let $H_C$ be the (narrow) ray class group modulo $C$, and let $\mathcal{E}_0 \subseteq \mathcal{I}(S)$ be a set of representatives for a basis for $H_C \otimes \mathbb{Z}/2\mathbb{Z}$ as a vector space over $\mathbb{F}_2$. For each $E_0 \in \mathcal{E}_0$ choose $m_{E_0} \in F^\times$ such that $E_0\mathfrak{O}_S = m_{E_0}\mathfrak{O}_S$. Let $\mathcal{E} \subseteq \mathcal{I}(S)$ be the full set of representatives for $H_C \otimes \mathbb{Z}/2\mathbb{Z}$ obtained by taking all possible products of distinct elements of $\mathcal{E}_0$. If $E = \prod_{E_0 \in \mathcal{E}_0} E_0^{n_{E_0}}$ (each $n_{E_0} = 0$ or 1), then let $m_E = \prod_{E_0 \in \mathcal{E}_0} m_{E_0}^{n_{E_0}}$, so that $E\mathfrak{O}_S = m_E\mathfrak{O}_S$. Also, let $(\frac{a}{*})$ be the power residue symbol (Jacobi symbol) attached to the extension $F(\sqrt{a})$ of $F$. Then one has

PROPOSITION 0.1 (FISHER-FRIEDBERG). *Let $I, I_1 \in \mathcal{I}(S)$ be coprime. Write $I = (m)EG^2$ with $E \in \mathcal{E}$, $m \equiv 1 \bmod C$ and $G \in I(S)$, $(G, I_1) = 1$. Then the quadratic power residue symbol $(\frac{mm_E}{I_1})$ is defined. If $I = (m')E'G'^2$ is another such decomposition, then $E' = E$ and $(\frac{m'm_E}{I_1}) = (\frac{mm_E}{I_1})$.*

In view of this Proposition, let us define (following Fisher and Friedberg) the quadratic symbol $(\frac{I}{I_1})$ by $(\frac{I}{I_1}) = (\frac{mm_E}{I_1})$ and the quadratic character $\chi_I$ by $\chi_I(I_1) = (\frac{I}{I_1})$. This depends on the choices above, but we suppress this from the notation. Let $S_I$ denote the support of the conductor of $\chi_I$. One may check that if $I = I'(I'')^2$, then $\chi_I(I_1) = \chi_{I'}(I_1)$ whenever both are defined. This allows one to extend $\chi_I$ to a character of all ideals of $\mathcal{I}(S \cup S_I)$.

PROPOSITION 0.2 (FISHER-FRIEDBERG). *Reciprocity – Let $I, I_1 \in \mathcal{I}(S)$ be disjoint, and $\alpha(I, I_1) = \chi_I(I_1)\chi_{I_1}(I)$. Then $\alpha(I, I_1)$ depends only on the images of $I$ and $I_1$ in $H_C \otimes \mathbb{Z}/2\mathbb{Z}$.*

Using these results, let us reformulate our problem as follows. Let $\pi$ be a cuspidal automorphic representation of $GL_r(A)$ unramified outside $S$, and let $\omega$ be an idèle class character also unramified outside $S$, where $S$ is as above. Regard $\omega$ as a character of $\mathcal{I}(S)$. For each $d \in I(S)$, let $L(s, \pi, \chi_d)$ denote the *partial* $L$-function for $\pi \otimes \chi_d$ with the places in $S \cup S_d$ removed. Then we will study the sum

$$Z(s, w; \pi, \omega) = \sum_d L(s, \pi, \chi_d)\, a(s, \pi, d)\, \omega(d)\, Nd^{-w}, \tag{0.2}$$

the sum over $d \in I(S)$, where $a(s, \pi, d)$ is a correction factor to be determined.

One may first ask how this object depends on the choices in the definition of the quadratic characters $\chi_I$ above, indeed whether or not this series is natural. As shown in Fisher-Friedberg, if one allows $\omega$ to vary over twists by characters of the finite group $H_C \otimes \mathbb{Z}/2\mathbb{Z}$ and similarly considers twists of $\pi$ by such characters, then the resulting series span a finite dimensional vector space which is independent of all choices. Thus the problem is natural; moreover, one may establish the two functional equations for different bases of this vector space.

For $x \in H_C \otimes \mathbb{Z}/2\mathbb{Z}$, let $\delta_x$ be the function of $\mathcal{I}(S)$ which projects to the characteristic function of $x$. For the first functional equation, a basis for the vector space above is given by

$$Z(s, w; \pi, \delta_x \omega) = \sum_{[d] \in x} L(s, \pi, \chi_d)\, a(s, \pi, d)\, \omega(d)\, Nd^{-w}, \tag{0.3}$$

where the sum is over $d \in \mathcal{I}(S)$ projecting to $x$. It suffices to establish a functional equation for each such series. By the theory of $GL_r$ $L$-functions, each $L$-function in (0.3) satisfies a functional equation of the form

$$L(s, \pi, \chi_d) = \epsilon(s, \pi \otimes \chi_d)\, L(1 - s, \tilde{\pi}, \chi_d) \prod_{v \in S} \frac{L(1 - s, \tilde{\pi}_v, \chi_{d,v})}{L(s, \pi_v, \chi_{d,v})}. \tag{0.4}$$

Let $E \in \mathcal{E}$ represent $x$. Since $d$ is in the class of $x$, we have $d = (f)EG^2$ where $f \in F^\times$ satisfies $f \equiv 1 \bmod C$ and where $G \in \mathcal{I}(S)$. Since $f \equiv 1 \bmod C$, $f$ is a square in the local ring $\mathfrak{O}_v$ for all $v \in S$. It follows that the local quadratic characters $\chi_{d,v} = \chi_{fm_E,v} = \chi_{E,v}$ for all $v \in S$. Write $d = d_0 d_1^2$ with $d_0, d_1 \in \mathcal{I}(S)$, $d_0$ square-free. Recall that each epsilon-factor in (0.4) is of the form $AB^{s-1/2}$ for some $A, B$. To obtain a functional equation for (0.3), we are concerned with the dependence of these quantities on $d$. It can be shown that the central value $A$ for the twisted epsilon-factor in (0.4) is $\chi_\pi(d_0)$ times a quantity depending only on $E$. This may be established since outside $S \cup S_d$ the epsilon-factor is a product of $GL_1$ epsilon-factors, and these may be computed by comparing Gauss sums. The remaining part of each epsilon factor, $B^{s-1/2}$, may also be studied by reducing to $GL_1$. One finds that the dependence of $B$ on $d$ is given by $Nd_0^{-r(s-1/2)}$. Hence each $L$-function in (0.3) transforms under $s \to 1 - s$ by $\chi_\pi(d_0) Nd_0^{-r(s-1/2)}$ times a factor which may be pulled out of the sum. To achieve a functional equation for this sum, we must find a correction factor transforming by a suitable multiple of $d_1$. That is, we seek correction factors $a$ satisfying the condition

$$a(s, \pi, d) = (Nd_1)^{-2r(s-1/2)}\, \chi_\pi(d_1^2)\, a(1 - s, \tilde{\pi}, d). \tag{0.5}$$

The correction factor is to be given by a finite Dirichlet polynomial, i.e. a polynomial in $Np^{-s}$, and the primes $p$ must divide $d_1$. Equation (0.5) gives the first condition on these polynomials.

For the second functional equation, it is sufficient to study $Z(s, w; \pi\delta_x, \omega\delta_y)$ with $x, y \in H_C \otimes \mathbb{Z}/2\mathbb{Z}$ since (by the orthogonality relations) such series span the vector space of double Dirichlet series. Let $L(s, \pi, \chi_d) = \sum_m c(m)\, \chi_d(m)\, Nm^{-s}$ where the sum is now over $m \in \mathcal{I}(S)$. Then after interchange of summation, one arrives at

$$(0.6)\qquad Z(s,w;\pi\delta_x,\omega\delta_y)=\sum_{[m]\in x}\sum_{[d]\in y} c(m)\,\chi_d(m)\,\omega(d)\,a(s,\pi,d)\,Nm^{-s}\,Nd^{-w}.$$

Since $[m]\in x$, $[d]\in y$, we have $\chi_d(m)=\alpha(y,x)\,\chi_m(d)$ by Proposition 0.2. Thus the inner sum gives a sum of the $GL_1$ $L$-functions $L(w,\omega\rho\chi_m)$ (the characters $\rho$ summing to give the characteristic function $\delta_y$), provided the correction factors behave appropriately. More precisely, we seek to write (0.6) in the form

$$(0.7)\qquad \sum L(w,\omega\rho\chi_m)\,b(w,\omega\rho,\pi,m)\,m^{-s},$$

for certain (different) correction factors $b(w,\omega\rho,\pi,m)$. (Here we have absorbed the coefficients $c(m)$ into the correction factors $b(w,\omega\rho,\pi,m)$.) Once again, since $\chi_m$ depends only on the square-free part of $m$, for this to be a sum of terms with a functional equation we must impose conditions on the $b(w,\omega\rho,\pi,m)$. If $m=m_0m_1^2$ with $m_0,m_1\in\mathcal{I}(S)$, $m_0$ square-free, then we require

$$(0.8)\qquad b(w,\omega\rho,\pi,m)=(Nm_1)^{1-2w}\,\omega(m_1^2)\,b\left(1-w,\omega^{-1}\rho,\pi,m\right).$$

(Note that $\rho^2=1$.) The correction factor $b(w,\omega\rho,\pi,m)$ is also to be given by a finite Dirichlet polynomial, this time in the primes dividing $m_1$. Equation (0.8) gives a condition on these polynomials.

In Section 1 we combine the two conditions (0.5), (0.8) and the relation among the $a$'s, $b$'s, $\pi$, and $\omega$ necessary for the interchange of summation to work, and analyze the resulting situation. We derive relations among the coefficients of the polynomials which give the correction factors, and show that these lead to a (very complicated) system of recursion relations upon the coefficients. This allows us to establish the uniqueness of the correction factor for $r\le 3$. Then in Section 2 we present a solution to the relations when $r=3$.

In Section 3, we illustrate the utility of these double Dirichlet series by giving an application to the estimation of sums of twisted $GL_3$ $L$-functions. In this section we restrict to $F=\mathbb{Q}$ and $\chi_\pi^2=1$ for convenience. We first give details of the continuation of $Z(s,w;\pi,\omega)$ and describe the precise pole. As (0.7) suggests, when $\omega^2=1$ there is a pole at $w=1$ whose residue essentially gives the contribution from the terms where $m$ is a square. This residue is (up to bad prime factors) precisely $\zeta(6s-1)\,L(2s,\pi,\vee^2)$, where the last factor is the symmetric square $L$-function for $\pi$!

The analytic properties of the symmetric square L-function have already been studied by the Langlands-Shahidi method (Langlands [L] and Shahidi [Sh]) and by the Rankin-Selberg method (see Patterson and Piatetski-Shapiro [PP] and Bump and Ginzburg [BG]). The present method seems fundamentally different from either of these. We will prove:

THEOREM 0.3. *(i) Let $f$ be an automorphic cuspidal representation of $GL(3)$ over $\mathbb{Q}$ whose central character has square 1. Let $M$ be a finite set of primes including 2, $\infty$, and the primes dividing the level of $f$. Let $L_M(s,f,\vee^2)$ denote the* *symmetric square L-series with the Euler factors corresponding to primes dividing M removed. Then* $\zeta(3s-1)\, L_M(s, f, \vee^2)$ *is an analytic function of s for all s, with the possible exception of* $s = 1$*, where a pole of order 1 can occur, and at* $s = 2/3$ *where a pole comes from the zeta factor.*

*(ii) Restrict to the case when* $L_M(s, f, \vee^2)$ *has a pole of order 1 at* $s = 1$*. Let d run through a sequence of discriminants such that d falls into a fixed quadratic residue class mod p for every prime dividing M* (mod8 *if* $p = 2$*). Then there exist infinitely many d in this sequence such that* $L(1/2, f, \chi_d) \neq 0$*, where the L-series referred to is the twist of the L-series of f by the character* $\chi_d$.

From this we deduce:

THEOREM 0.4. *Let* $\pi'$ *be a cuspidal automorphic representation of* $GL(2)$ *over* $\mathbb{Q}$*. Then there exist infinitely many quadratic characters* $\chi_d$ *as in Theorem 0.3 (ii) such that* $L(1/2, \pi', Ad^2 \otimes \chi_d) \neq 0$.

Indeed, we may apply Theorem 0.3 (ii) to the adjoint (Gelbart-Jacquet [GJ]) lift $\pi$ of $\pi'$ to $GL(3)$. The symmetric square L-function of this is the Riemann zeta function times a twist of the symmetric fourth power L-function of $\pi'$. Now the (twisted) symmetric fourth power L-function is automorphic on $GL(5)$ by Kim and Shahidi [KS], hence does not vanish at $s = 1$ by Jacquet and Shalika [JS1]. Therefore the symmetric square of $\pi$ has a pole there and Theorem 0.3 is applicable.

Actually Theorem 0.3 (ii) is not more general than this special case. Indeed Ginzburg, Rallis and Soudry [GRS] have shown that if the symmetric or exterior square L-function of an automorphic form on $GL(r)$ has a pole, then the automorphic form is a classical lift. When $r = 3$, this means that in order for the symmetric square L-function to have a pole, the automorphic cuspidal representation $\pi$ of $GL(3)$ must be a quadratic twist of the adjoint (Gelbart-Jacquet [GJ]) lift of an automorphic representation $\pi'$ of $GL(2)$.

In Theorem 3.8 we establish a somewhat more refined mean value result from which Theorem 0.3 is derived. Another work which utilizes the results of this paper to give estimates for sums of $L$-functions is Diaconu, Goldfeld and Hoffstein [DGH].

*Acknowledgments.* The authors wish to warmly thank Adrian Diaconu, Benji Fisher and Dorian Goldfeld for helpful conversations. We also wish to thank the Institute for Advanced Study and the Banker's Trust Company Foundation for support.

**1. The interchange of summation.** Let $\pi$ be an automorphic representation of $GL_r(A)$ and $\omega$ be an idèle class character, both unramified outside $S$. Recall that if $d = d_0 d_1^2$ with $d_0$ square free, then $\chi_d = \chi_{d_0}$. As above we write the partial $L$-function $L(s, \pi, \chi_d) = \sum_m c(m)\, \chi_{d_0}(m)\, Nm^{-s}$, where the sum is over $m \in \mathcal{I}(S)$

and we define $\chi_{d_0}(m) = 0$ if $d_0$ and $m$ are not coprime. Suppose that the correction factor $a(s, \pi, d)$ is given by a Dirichlet polynomial $P_{d_0,d_1}(s)$ in the primes dividing $d_1$. Then

$$Z(s, w; \pi, \omega) = \sum_{\substack{d=d_0d_1^2 \\ d_0 \text{ squarefree}}} \sum_{m} c(m)\, \chi_{d_0}(m)\, Nm^{-s}\, P_{d_0,d_1}(s)\, \omega(d)\, Nd^{-w}. \tag{1.1}$$

As in Section 0, we seek polynomials $P$ and $Q$, the latter giving the correction factors $b(w, \omega, \pi, m)$, such that

(1.2)
$$Z(s, w; \pi, \omega) = \sum_{x,y} \alpha(x, y) \sum_{\substack{m=m_0m_1^2\in[x] \\ m_0 \text{ squarefree}}} \sum_{d\in[y]} \omega(d)\, \chi_{m_0}(d)\, Nd^{-w}\, Q_{m_0,m_1}(w)\, Nm^{-s}.$$

Here the first sum is over $x, y \in H_C \otimes \mathbb{Z}/2\mathbb{Z}$. We shall call the equality of (1.1) and (1.2) the *basic identity*. Denote the above quantity $Z(s, w)$ for conciseness.

The degrees of the polynomials $P$, $Q$ (if they exist) are determined by the relations (0.5), (0.8) respectively. We introduce the notation

$$P_{d_0,d_1}(s) = \prod_{p^\alpha \| d_1} \left(1 + a^{(\alpha)}_{d_0,p}\, Np^{-s} + a^{(\alpha)}_{d_0,p^2}\, Np^{-2r\alpha s} + \cdots + a^{(\alpha)}_{d_0,p^{2r\alpha}} Np^{-2r\alpha s}\right),$$

$$Q_{m_0,m_1}(w) = \prod_{p^\alpha \| m_1} \Big(b^{(\alpha)}_{m_0,1} + \omega(p)\, b^{(\alpha)}_{m_0,p}\, Np^{-w} + \omega(p^2)\, b^{(\alpha)}_{m_0,p^2}\, Np^{-2w} + \cdots$$
$$+ \omega(p^{2\alpha})\, b^{(\alpha)}_{m_0,p^{2\alpha}} Np^{-2\alpha w}\Big).$$

The coefficients of $P$ here depend on $\pi$ but are independent of $\omega$ by hypothesis, while those of $Q$ depend on $\pi$, $\omega$. (As we shall soon see, with this normalization of $Q$ the coefficients $b^{(\alpha)}_{m_0,p^j}$ turn out to be independent of $\omega$.) We normalize the coefficients of $P$ to begin with 1 as shown.

In this section we investigate the relations that the coefficients $a^{(\alpha)}_{d_0,p^j}$ and $b^{(\alpha)}_{m_0,p^j}$ must satisfy if they are to give the basic identity.

PROPOSITION 1.1. *Let $t_0$, $u_0$, $p \in \mathcal{I}(S)$ be squarefree with $p$ prime and $t_0$, $u_0$, $p$ pairwise coprime. Then the coefficients $a^{(\alpha)}_{d_0,d_1}$ and $b^{(\alpha)}_{m_0,m_1}$ must satisfy the following relations for all $e, f \geq 0$:*

$$c(t_0) \sum_{k=0}^{2e+1} c(p^k)\, \chi_{u_0}(p)^{k+1}\, a^{(f)}_{u_0,p^{2e+1-k}} = b^{(e)}_{pt_0,p^{2f}}. \tag{1.3}$$

$$c(t_0)\, \chi_p(t_0)\, a^{(f)}_{u_0p,p^{2e+1}} = \chi_{u_0}(p)\, b^{(e)}_{t_0p,p^{2f+1}}. \tag{1.4}$$

$$c(t_0) \left[\sum_{k=0}^{2e} c(p^k)\, \chi_{u_0}(p)^k\, a^{(f)}_{u_0,p^{2e-k}} - a^{(f-1)}_{u_0p,p^{2e}}\right] = b^{(e)}_{t_0,p^{2f}}. \tag{1.5}$$

*(Here* $a^{(-1)}_{u_0 p, p^{2e}} \equiv 0$ *by definition.)*

$$c(t_0)\, \chi_p(t_0) \left[ a^{(f)}_{u_0 p, p^{2e}} - \sum_{k=0}^{2e} c(p^k)\, \chi_{u_0}(p)^k\, a^{(f)}_{u_0, p^{2e-k}} \right] = b^{(e)}_{t_0, p^{2f+1}}. \tag{1.6}$$

*In particular, for* $\delta = 0, 1$

$$b^{(e)}_{t_0 p^\delta, p^f} = c(t_0)\, \chi_p(t_0)^f\, b^{(e)}_{p^\delta, p^f}$$

$$b^{(e)}_{t_0 p^\delta, 1} = c(t_0 p^{\delta+2e}),$$

*and for all* $m_0 \in \mathcal{I}(S)$ *the coefficients* $b^{(e)}_{m_0, p^f}$ *are independent of* $\omega$.

Note that the relations (1.3)–(1.6) are also consistent with the equation

$$a^{(f)}_{u_0 p^\delta, p^e} = \chi_{u_0}(p)^e\, a^{(f)}_{p^\delta, p^e} \qquad \text{for } \delta = 0, 1.$$

*Proof.* We equate coefficients in the basic identity. Note that

$$\alpha(ab, xy) = \alpha(a, x)\alpha(a, y)\alpha(b, x)\alpha(b, y). \tag{1.7}$$

Fix a prime $p$.

First, consider the coefficient in $Z(s, w)$ of

$$Np^{-(2e+1)s-2fw}\, Nt_0^{-s}\, Nu_0^{-w} \qquad (e, f \geq 0).$$

In (1.1), this arises from summands with $(d_0, d_1) = (u_0, p^f)$. One must sum the contributions for $m = t_0 p^k$ with $0 \leq k \leq 2e+1$. In (1.2), this term arises when $(m_0, m_1) = (pt_0, p^e)$. Thus in (1.2), we must have $(d, p) = 1$ (since otherwise $\chi_{pt_0}(d) = 0$) so in fact $d = u_0$. Comparing terms we obtain

$$c(t_0)\, \omega\left(u_0 p^{2f}\right)\, \chi_{u_0}(t_0) \sum_{k=0}^{2e+1} c(p^k)\, \chi_{u_0}(p)^k\, a^{(f)}_{u_0, p^{2e+1-k}}$$

$$= \alpha(pt_0, u_0)\, \omega\left(u_0 p^{2f}\right)\, \chi_{pt_0}(u_0)\, b^{(e)}_{pt_0, p^{2f}}.$$

Making use of Proposition 0.2, we have $\alpha(pt_0, u_0)\, \chi_{pt_0}(u_0) = \chi_{u_0}(pt_0)$. Canceling the factor $\omega(u_0 p^{2f})\, \chi_{u_0}(t_0)$ from both sides, one obtains equation (1.3).

Second, consider the coefficient of

$$Np^{(2e+1)s-(2f+1)w}\, Nt_0^{-s}\, Nu_0^{-w} \qquad (e, f \geq 0).$$

In (1.1), this arises from summands with $(d_0, d_1) = (u_0 p, p^f)$. In (1.2), it arises when $(m_0, m_1) = (t_0 p, p^e)$. This time, we must have $(m, p) = 1$ and hence $m = t_0$ in (1.1), and $(d, p) = 1$ so $d = u_0$ in (1.2). The factor $\chi_{u_0 p}(t_0)$ appears in (1.1) and the factor $\alpha(t_0 p, u_0)\chi_{t_0 p}(u_0) = \chi_{u_0}(t_0 p)$ in (1.2). Canceling $\omega(u_0 p^{2f+1})\, \chi_{u_0}(t_0)$, the relation obtained is simply (1.4). Later we will see that both sides of (1.4) are 0 for $r \leq 3$.

Third, for the coefficients corresponding to an even power of $Np^{-s}$, it is best to multiply by a zeta factor in $w$ first. Thus we consider the coefficient of

$$Np^{-2es-2fw}\, Nt_0^{-s}\, Nu_0^{-w} \qquad (e, f \geq 0)$$

in

$$\left(1 - \chi_p(t_0)\,\omega(p)\,Np^{-w}\right)\, Z(s, w).$$

Substituting the expression for $Z(s, w)$ from (1.1), we get contributions to this coefficient when $(d_0, d_1) = (u_0, p^f)$ *and* when $(d_0, d_1) = (u_0 p, p^{f-1})$. For this second term, one must have $(m, p) = 1$ so in that case $m = t_0$. On the other hand, substituting the expression from (1.2), one gets a contribution from $(m_0, m_1) = (t_0, p^e)$, and then the sum telescopes! (Here one makes use of (1.7).) Canceling $\omega(u_0 p^{2f})\,\chi_{u_0}(t_0)$, we arrive at the relation (1.5).

Fourth, consider the coefficient of

$$Np^{-2es-(2f+1)w}\, Nt_0^{-s}\, Nu_0^{-w} \qquad (e, f \geq 0)$$

in $(1 - \chi_p(t_0)\,\omega(p)\,Np^{-w})Z(s, w)$. Once again, substituting in the two sums for $Z(s, w)$, the contributions in (1.1) arise from $(d_0, d_1) = (u_0 p, p^f)$ and $(d_0, d_1) = (u_0, p^f)$. In the first of these one necessarily has $m = t_0$. The contributions in (1.2) arise from $(m_0, m_1) = (t_0, p^e)$. Once again, this second sum telescopes. After canceling $\omega(u_0 p^{2f+1})\,\chi_{u_0}(t_0)$, one obtains the relation (1.6). □

The relations of Proposition 1.1 completely determine the $b$'s from the $a$'s. Surprisingly, they do more: they determine the $a$'s as well when $r \leq 3$! Before explaining this, let us record the relations on the coefficients of $P$, $Q$ implied by (0.5) and (0.8). These state that:

$$a^{(\alpha)}_{d_0, p^i} = \chi_\pi(p)^{2\alpha}\, Np^{i-r\alpha}\, \tilde{a}^{(\alpha)}_{d_0, p^{2r\alpha-i}}, \qquad 0 \leq i \leq 2r\alpha \tag{1.8}$$

$$b^{(\alpha)}_{m_0, p^j} = Np^{j-\alpha}\, b^{(\alpha)}_{m_0, p^{2\alpha-j}}, \qquad 0 \leq j \leq 2\alpha. \tag{1.9}$$

Here and below we let $\tilde{a}$ denote the family of coefficients associated with the contragredient representation $\tilde{\pi}$ of $\pi$. We also let $\tilde{c}$ be defined by $L(s, \tilde{\pi}, \chi_d) = \sum \tilde{c}(m)\,\chi_d(m)\,Nm^{-s}$. (In the proof of (S2) below we assume that $\tilde{a}^{(\alpha)}_{d_0, p^i} = 0$ implies $a^{(\alpha)}_{d_0, p^i} = 0$. This is consistent with the formal equality of (1.1) and (1.2) for all $\pi$ which we are seeking.) Since $\tilde{a}^{(\alpha)}_{d_0, 1} = 1$ and $b^{(\alpha)}_{m_0, 1} = c(m_0 p^{2\alpha})$, we deduce that

$$a^{(\alpha)}_{d_0, p^{2r\alpha}} = \chi_\pi(p)^{2\alpha}\, Np^{r\alpha}, \qquad b^{(\alpha)}_{n_0, p^{2\alpha}} = Np^{\alpha}\, c\left(m_0 p^{2\alpha}\right). \tag{1.10}$$

We turn now to the relations which follow from Proposition 1.1.

THEOREM 1.2. *Let $u_0, p \in \mathcal{I}(S)$ be squarefree with $p$ prime and $(p, u_0) = 1$. Then the coefficients $a^{(\alpha)}_{d_0, p^j}$ must satisfy the following relations, which completely determine them if $r \leq 3$.*

*First, the coefficients are stable in the sense that:*

(S1) $a^{(f)}_{u_0,p^e} = a^{(f')}_{u_0,p^e}$ *for* $f, f' \geq \left\lfloor \frac{e+1}{2} \right\rfloor$.

(S2) $a^{(f)}_{u_0 p, p^{2e+1}} = 0$ *for* $f > 0$ *and* $e \geq 0$, *if* $r \leq 3$.

(S3) $a^{(f)}_{u_0 p, p^{2e}} = a^{(f+1)}_{u_0 p, p^{2e}}$ *for* $f \geq e$.

*Second, the coefficients satisfy the relations:*

(R1) $a^{(f)}_{u_0,p^{2e+1}} = -\sum_{k=1}^{2e+1} c(p^k)\,\chi_{u_0}(p)^k\, a^{(f)}_{u_0,p^{2e+1-k}}$ *for* $f > e$.

(R2) $a^{(e)}_{u_0,p^{2e+1}} = \chi_{u_0}(p)\, N p^e\, c(p^{2e+1}) - \sum_{k=1}^{2e+1} c(p^k)\,\chi_{u_0}(p)^k\, a^{(e)}_{u_0,p^{2e+1-k}}$ *for* $e \geq 1$.

(R3)

$$a^{(f)}_{u_0,p^{2e+1}} = N p^{2f-e} \sum_{k=0}^{2e+1} c(p^k)\,\chi_{u_0}(p)^k\, a^{(e-f)}_{u_0,p^{2e+1-k}} - \sum_{k=1}^{2e+1} c(p^k)\,\chi_{u_0}(p)^k\, a^{(f)}_{u_0,p^{2e+1-k}}$$

*for* $f \leq e < 2f$.

(R4) $a^{(f)}_{u_0,p^{2e}} = a^{(f-1)}_{u_0 p,p^{2e}} - \sum_{k=1}^{2e} c(p^k)\,\chi_{u_0}(p)^k\, a^{(f)}_{u_0,p^{2e-k}}$ *for* $f > e$.

(R5) $a^{(e)}_{u_0,p^{2e}} = N p^e\, c(p^{2e}) + a^{(e-1)}_{u_0 p,p^{2e}} - \sum_{k=1}^{2e} c(p^k)\,\chi_{u_0}(p)^k\, a^{(e)}_{u_0,p^{2e-k}}$.

(R6)

$$a^{(f)}_{u_0,p^{2e}} = a^{(f-1)}_{u_0 p,p^{2e}} - \sum_{k=1}^{2e} c(p^k)\,\chi_{u_0}(p)^k\, a^{(f)}_{u_0,p^{2e-k}} + N p^{2f-e}\left[\sum_{k=0}^{2e} c(p^k)\,\chi_{u_0}(p)^k\, a^{(e-f)}_{u_0,p^{2e-k}} - a^{(e-f-1)}_{u_0 p,p^{2e}}\right]$$

*for* $f \leq e < 2f$.

(R7) $a^{(f)}_{u_0 p,p^{2e}} = \sum_{k=0}^{2e} c(p^k)\,\chi_{u_0}(p)^k\, a^{(f)}_{u_0,p^{2e-k}}$ *for* $f \geq e$.

(R8)

$$a^{(f)}_{u_0 p,p^{2e}} = \sum_{k=0}^{2e} c(p^k)\,\chi_{u_0}(p)^k\, a^{(f)}_{u_0,p^{2e-k}} + N p^{2f+1-e}\left[a^{(e-f-1)}_{u_0 p,p^{2e}} - \sum_{k=0}^{2e} c(p^k)\,\chi_{u_0}(p)^k\, a^{(e-f-1)}_{u_0,p^{2e-k}}\right]$$

*for* $f < e < 2f + 1$.

*Proof.* In this proof, we use the relations of Proposition 1.1 with $t_0 = 1$ and the functional equation (1.9) extensively. The coefficients with even and odd indices will be treated separately, since they satisfy separate relations per Proposition 1.1.

First let us obtain stability and also some of the $a$'s. Observe that $b^{(e)}_{m_0,p^{2f}} = 0$ for $f > e$ (see equation (0.8) or (1.9)). Applying this to (1.5) we see that (since

$c(1) \neq 0$)

$$a^{(f-1)}_{u_0 p, p^{2e}} = \sum_{k=0}^{2e} c(p^k)\, \chi_{u_0}(p)^k\, a^{(f)}_{u_0, p^{2e-k}}$$

for $f > e$. Similarly, $b^{(e)}_{1,p^{2f+1}} = 0$ for $f \geq e$, so (1.6) gives

$$a^{(f)}_{u_0 p, p^{2e}} = \sum_{k=0}^{2e} c(p^k)\, \chi_{u_0}(p)^k\, a^{(f)}_{u_0, p^{2e-k}}.$$

Thus we conclude that relation (R7) holds and also that stability (S3) holds. We also deduce that the sum

$$T(e, f) = \sum_{k=0}^{2e} c(p^k)\, \chi_{u_0}(p)^k\, a^{(f)}_{u_0, p^{2e-k}}$$

satisfies $T(e, f) = T(e, f')$ for $f, f' \geq e$.

Next we turn to relation (1.3). Once again the right side vanishes for $f > e$. For example, when $e = 0$ we obtain $a^{(f)}_{u_0,p} = -\chi_{u_0}(p)\, c(p)$ for $f > 0$. We also deduce that

$$T_o(e, f) = \sum_{k=0}^{2e+1} c(p^k)\, \chi_{u_0}(p)^k\, a^{(f)}_{u_0, p^{2e+1-k}}$$

satisfies $T_o(e, f) = T_o(e, f')$ for $f, f' > e$.

We say that the coefficients $a^{(f)}_{d_0, p^k}$ are stable if they are independent of $f$ for $f$ sufficiently large (depending on $k$). To establish this, alternating between $T_o$ and $T$, we obtain the stability of the $a$'s up to the top one in each sum. Since the entire expression is stable, the top term must be too. For example, since $T(1, f) = T(1, f')$ for $f, f' \geq 1$, we conclude that $a^{(f)}_{u_0, p^2}$ is stable for $f \geq 1$. Then since $T_o(1, f) = T_o(1, f')$ for $f, f' > 1$ we conclude that $a^{(f)}_{u_0, p^3}$ is stable for $f \geq 2$. Continuing in this way, we obtain stability relation (S1).

Next, relation (1.4) implies that

$$a^{(f)}_{u_0 p, p^{2e+1}} = 0 \qquad f \geq e, \tag{1.11}$$

$$b^{(e)}_{p, p^{2f+1}} = 0 \qquad e \geq rf. \tag{1.12}$$

These quantities are then 0 for $r \leq 3$. (This may be true in general but we do not check it here.) Indeed, if $r = 1$ this is immediate from the above, as $b^{(e)}_{p,p^{2f+1}} = 0$ for $e \geq f$ and for $f \geq e$. For $r \geq 2$, since $b^{(e)}_{t_0 p, p} = 0$ for $e \geq 0$, the functional equation (1.9) in the $b$'s implies that $b^{(e)}_{t_0 p, p^{2e-1}} = 0$ for $e \geq 0$. By relation (1.4), $a^{(f)}_{u_0 p, p^{2e+1}} = 0$ for $f = e - 1$. Since $a^{(f)}_{u_0 p, p^{2e+1}} = 0$ for $e \leq f + 1$, the functional equation (1.8) in the $a$'s gives

$$a^{(f)}_{u_0 p, p^{2e+1}} = 0 \qquad \text{for} \quad \frac{2rf - (2e+1) - 1}{2} \leq f + 1.$$

This last condition may be rewritten as $f(r-1) \le e+2$. Thus by relation (1.4) once again,

$$b^{(e)}_{t_0 p, p^{2f+1}} = 0 \qquad \text{for} \quad 2f+1 \le 2\frac{e+2}{r-1} + 1.$$

For $r = 2, 3$ this holds for $2f+1 \le e$. By the functional equation (1.9) it follows that all $b^{(e)}_{t_0 p, p^{2f+1}} = 0$, hence by (1.4), all $a^{(f)}_{u_0 p, p^{2e+1}} = 0$. This establishes stability condition (S2).

We now deduce the relations for the remaining relations (R) on the $a$'s.

First, relation (R1) is obtained from relation (1.3) since the right hand side is 0 when $f > e$. Similarly, relation (R2) is obtained from relation (1.3) with $e = f$; using $b^{(e)}_{p,p^{2e}} = Np^e\, c(p^{2e+1})$ (see (1.10)). As for (R3), we note that (1.9) gives

$$b^{(e)}_{p,p^{2f}} = Np^{2f-e}\, b^{(e)}_{p,p^{2e-2f}}.$$

Thus (1.3) gives

$$\sum_{k=0}^{2e+1} c(p^k)\, \chi_{u_0}(p)^k\, a^{(f)}_{u_0, p^{2e+1-k}} = Np^{2f-e} \sum_{k=0}^{2e+1} c(p^k)\, \chi_{u_0}(p)^k\, a^{(e-f)}_{u_0, p^{2e+1-k}}.$$

Moving the left hand summands with $k \ge 1$ to the right hand side, relation (R3) follows. Note that the relation only determines $a^{(e)}_{u_0, p^{2e+1}}$ if $0 \le e - f < f$ or $f \le e < 2f$. (If $e - f < 0$ the right hand side vanishes and we get relation (R1).)

Similarly we obtain the relations for $a^{(f)}_{u_0, p^{2e}}$ from relation (1.5) above. If $f > e$ the right side of (1.5) vanishes. Solving for the $a^{(f)}_{u_0, p^{2e}}$ gives precisely (R4). If $e = f$, the right side of (1.5) equals $Np^e\, c(p^{2e})$ and proceeding as above we get (R5). Finally to get (R6), we use the functional equation

$$b^{(e)}_{1,p^{2f}} = Np^{2f-e}\, b^{(e)}_{1,p^{2(e-f)}}.$$

Thus (1.5) gives precisely (R6) when one uses this, equating the expressions in the $a$'s corresponding to the above equation. Note that one gets information only when $e - f < f$ and the indexing with $e - f \le 0$ is already captured in (R4), (R5) above.

Lastly, one determines the coefficients $a^{(f)}_{u_0 p, p^{2e}}$ from equation (1.6). If $f \ge e$ the right side of (1.6) equals zero, which immediately gives (R7), as already noted above. To get (R8), one may use the functional equation

$$b^{(e)}_{1,p^{2f+1}} = Np^{2f-e+1}\, b^{(e)}_{1,p^{2(e-f-1)+1}}.$$

Substituting the expressions which, by (1.6), correspond to each side of this equation and solving, one arrives at (R8). One needs $0 \le e - f - 1 < f$ to extract information from this relation.

This concludes the proof of Theorem 1.2. □

Note that none of these arguments above uses a specific value for the rank $r$, except for our treatment of the vanishing of $a^{(f)}_{u_0 p, p^{2e+1}}$ and $b^{(e)}_{t_0 p, p^{2f+1}}$. However the relations given do not determine all other coefficients when $r \ge 4$. In contrast, in the case $r \leq 3$, the relations in Theorem 1.2 overdetermine the coefficients. It is not hard to check that if there is a set of solutions to the relations given in Proposition 1.1 and Theorem 1.2, then the basic identity does indeed hold.

## 2. The coefficients for $r = 3$.

THEOREM 2.1. *The combinatorial problem of Section 1 has a unique solution when $r = 3$.*

Uniqueness has already been noted. To prove existence, we describe the coefficients explicitly.

Fix a prime $p$ and squarefree $u_0 \in \mathcal{I}(S)$ prime to $p$. Let $\alpha$, $\beta$ and $\gamma$ be the Langlands parameters of $\pi_p$. Thus $c(p^k)$ is the $k$-th complete symmetric polynomial in $\alpha$, $\beta$ and $\gamma$. More generally, let $c(p^{k_1}, p^{k_2})$ be the Schur polynomial

$$c(p^{k_1}, p^{k_2}) = \frac{\begin{vmatrix} \alpha^{k_1+k_2+2} & \beta^{k_1+k_2+2} & \gamma^{k_1+k_2+2} \\ \alpha^{k_2+1} & \beta^{k_2+1} & \gamma^{k_2+1} \\ 1 & 1 & 1 \end{vmatrix}}{\begin{vmatrix} \alpha^2 & \beta^2 & \gamma^2 \\ \alpha & \beta & \gamma \\ 1 & 1 & 1 \end{vmatrix}}.$$

Then $c(p^k) = c(p^k, 1)$, $\chi_\pi(p) = \alpha\beta\gamma$ and $\tilde{c}(p^k) = \chi_\pi(p)^{-k} c(1, p^k)$. We will also use the notation

$$[k_1, k_2] = \chi_{u_0}(p^{k_1+2k_2}) c(p^{k_1}, p^{k_2}).$$

We sometimes omit the comma if it is obvious, so $[10] = [1, 0]$. Each coefficient $[k_1, k_2]$ is the character of an irreducible representation of $GL(3, \mathbb{C})$ evaluated at the semisimple conjugacy class with eigenvalues $\alpha' = \chi_{u_0}(p)\alpha$, $\beta' = \chi_{u_0}(p)\beta$, $\gamma' = \chi_{u_0}(p)\gamma$. We define $[n, m] = 0$ if either $n$ or $m$ is negative.

If $\lambda = 1$ or $p$, we want to compute $a^{(\theta)}_{\lambda u_0, p^k}$, where $0 \leq k \leq 6\theta$. The formulas are simpler when $\lambda = p$. The general formula in this case is:

$$a^{(\theta)}_{pu_0, p^{2k}} = \sum_{r \geq 0} N p^{k+r} \left( \sum_{\substack{r_1, r_2 \geq 0 \\ r_1 + 2r_2 \leq k - 3r \\ 2r_1 + r_2 \leq 3\theta - k - 3r \\ r_1 - r_2 \equiv k (\mathrm{mod} 3)}} \chi_\pi(p)^{(2/3)(k - r_1 - 2r_2)} [2r_1, 2r_2] \right). \tag{2.1}$$

Though the sum over $r$ in (2.1) is infinite, almost all terms are zero since the inner summation conditions are empty for $r$ sufficiently large. The odd coefficients $a^{(\theta)}_{pu_0, p^{2k+1}}$ vanish as noted in Section 2. From formula (2.1) it is not difficult to check the functional equation (1.8). One may verify the other relations such as (R1–R6) in Theorem 1.2 using these formulas and Pieri's formula in the form

$$[t, 0][r_1, r_2] = \sum_{\substack{m, n \geq 0 \\ n \leq r_1 \\ 0 \leq t - m - n \leq r_2}} [r_1 + m - n, r_2 + 2n + m - t]$$

and its special case $[r_1, r_2] = [r_1, 0][0, r_2] - [r_1 - 1, 0][0, r_2 - 1]$.

The formula for $a^{(\theta)}_{u_0, p^{2k}}$ is given by:

$$
\begin{aligned}
a^{(\theta)}_{u_0,p^{2k}} = &\sum_{r=1}^{\infty} Np^{k+r-1} \sum_{\substack{r_1+2r_2 \le 2k-6r+6 \\ 2r_1+r_2 \le 6\theta-2k-6r+6 \\ r_1 \equiv r_2 \equiv 0 \pmod 2 \\ r_1-r_2 \equiv 2k \pmod 3}} \chi_\pi(p)^{(1/3)(2k-r_1-2r_2)}[r_1, r_2] \\
&+ \sum_{r=0}^{\infty} Np^{k+r-1} \sum_{\substack{r_1+2r_2 \le 2k-6r \\ 2r_1+r_2 \le 6\theta-2k-6r-3 \\ r_1 \equiv 0, r_2 \equiv 1 \pmod 2 \\ r_1-r_2 \equiv 2k \pmod 3}} \chi_\pi(p)^{(1/3)(2k-r_1-2r_2)}[r_1, r_2] \\
&+ \sum_{r=0}^{\infty} Np^{k+r-1} \sum_{\substack{r_1+2r_2 \le 2k-6r-3 \\ 2r_1+r_2 \le 6\theta-2k-6r \\ r_1 \equiv 1, r_2 \equiv 0 \pmod 2 \\ r_1-r_2 \equiv 2k \pmod 3}} \chi_\pi(p)^{(1/3)(2k-r_1-2r_2)}[r_1, r_2] \\
&+ \sum_{r=0}^{\infty} Np^{k+r-1} \sum_{\substack{r_1+2r_2 \le 2k-6r-3 \\ 2r_1+r_2 \le 6\theta-2k-6r-3 \\ r_1 \equiv r_2 \equiv 1 \pmod 2 \\ r_1-r_2 \equiv 2k \pmod 3}} \chi_\pi(p)^{(1/3)(2k-r_1-2r_2)}[r_1, r_2].
\end{aligned}
\tag{2.2}
$$

Once again, though the sums in (2.2) are infinite, almost all terms are zero since each of the four inner summation conditions is empty for $r$ sufficiently large. The formula for $a^{(\theta)}_{u_0, p^{2k+1}}$ is given by:

$$
\begin{aligned}
a^{(\theta)}_{u_0,p^{2k+1}} = -\Bigg( &\sum_{r=0}^{\infty} Np^{k+r-1} \sum_{\substack{r_1+2r_2 \le 2k-6r-2 \\ 2r_1+r_2 \le 6\theta-2k-6r-4 \\ r_1 \equiv r_2 \equiv 0 \pmod 2 \\ r_1-r_2 \equiv 2k+1 \pmod 3}} \chi_\pi(p)^{(1/3)(2k+1-r_1-2r_2)}[r_1, r_2] \\
&+ \sum_{r=1}^{\infty} Np^{k+r-1} \sum_{\substack{r_1+2r_2 \le 2k-6r+4 \\ 2r_1+r_2 \le 6\theta-2k-6r+5 \\ r_1 \equiv 0, r_2 \equiv 1 \pmod 2 \\ r_1-r_2 \equiv 2k+1 \pmod 3}} \chi_\pi(p)^{(1/3)(2k+1-r_1-2r_2)}[r_1, r_2] \\
&+ \sum_{r=1}^{\infty} Np^{k+r-1} \sum_{\substack{r_1+2r_2 \le 2k-6r+7 \\ 2r_1+r_2 \le 6\theta-2k-6r+2 \\ r_1 \equiv 1, r_2 \equiv 0 \pmod 2 \\ r_1-r_2 \equiv 2k+1 \pmod 3}} \chi_\pi(p)^{(1/3)(2k+1-r_1-2r_2)}[r_1, r_2] \\
&+ \sum_{r=1}^{\infty} Np^{k+r-1} \sum_{\substack{r_1+2r_2 \le 2k-6r+7 \\ 2r_1+r_2 \le 6\theta-2k-6r+5 \\ r_1 \equiv r_2 \equiv 1 \pmod 2 \\ r_1-r_2 \equiv 2k+1 \pmod 3}} \chi_\pi(p)^{(1/3)(2k+1-r_1-2r_2)}[r_1, r_2] \Bigg).
\end{aligned}
\tag{2.3}
$$

The formulas for the $b$'s are simpler than those for the $a$'s. We have, for $k$ even,

$$\begin{aligned}
b^{(\theta)}_{p,p^k} = &\sum_{j=-k/2}^{\min(k/2,2\theta-3k/2)} \sum_{\substack{0\le r_2\le\min\left(\frac{k}{2}-|j|,\theta-\frac{k}{2}-|k+j-\theta|\right)\\ r_2\equiv j-k/2 \pmod 2}} Np^{k+j}\\
&\times \tfrac{1}{2}\min\left(\frac{k}{2}-|j|-r_2+2,\theta-\frac{k}{2}-|k+j-\theta|-r_2+2,\right.\\
&\left.-\frac{3k}{2}-3j+2\theta+2-r_2\right)\\
&\times \chi_\pi(p)^{(k/2)+j-r_2}\left[2\theta-\frac{3k}{2}-3j-r_2+1,2r_2\right],
\end{aligned} \tag{2.4}$$

while $b^{(\theta)}_{p,p^k}=0$ if $k$ is odd. Also if $k$ is even:

$$\begin{aligned}
b^{(\theta)}_{1,p^k} = &\sum_{j=-k/2}^{\min(k/2,2\theta-3k/2)} \sum_{\substack{0\le r_2\le\min\left(\frac{k}{2}-|j|,\theta-\frac{k}{2}-|k+j-\theta|\right)\\ r_2\equiv j-k/2 \pmod 2}} Np^{k+j}\\
&\times \tfrac{1}{2}\min\left(\frac{k}{2}-|j|-r_2+2,\theta-\frac{k}{2}-|k+j-\theta|-r_2+2,\right.\\
&\left.-\frac{3k}{2}-3j+2\theta+2-r_2\right)\\
&\times \chi_\pi(p)^{(k/2)+j-r_2}\left[2\theta-\frac{3k}{2}-3j-r_2,2r_2\right]
\end{aligned} \tag{2.5}$$

If $k$ is odd:

$$\begin{aligned}
b^{(\theta)}_{1,p^k} = -&\sum_{j=-\frac{k-1}{2}}^{\min\left(\frac{k-1}{2},2\theta-\frac{3k+1}{2}\right)} \sum_{\substack{0\le r_2\le\min\left(\frac{k-1}{2}-|j|,\theta-\frac{k+1}{2}-|k+j-\theta|\right)\\ r_2\equiv j-(k-1)/2 \pmod 2}} Np^{k-1+j}\\
&\times \tfrac{1}{2}\min\left(\frac{k-1}{2}-|j|-r_2+2,\theta-\frac{k+1}{2}-|k+j-\theta|-r_2\right.\\
&\left.+2,2\theta-\frac{3(k-1)}{2}-3j-r_2\right)\\
&\times \chi_\pi(p)^{\left(\frac{k-1}{2}\right)+j-r_2}\left[2\theta-\frac{3(k-1)}{2}-3j-r_2,2r_2\right].
\end{aligned} \tag{2.6}$$

Next we offer some remarks about the stable $a$-coefficients. Let $F(k)$ be the $k$-th symmetric power of the symmetric square character $[2,0]$. This is a homogeneous polynomial of degree $2k$ in $\alpha'$, $\beta'$ and $\gamma'$ given by the formula

$$F(k) = \sum_{\substack{r_1+2r_2 \le k \\ r_1+2r_2 \equiv k \pmod 3}} [2r_1, 2r_2].$$

Define $F(k) = 0$ if $k < 0$. Using (S1)–(S3) the stable coefficients may be defined by $a^{\text{stab}}_{\lambda, p^k} = a^{(\theta)}_{\lambda, p^k}$ whenever $\theta \ge k/2$. Then one may check that:

$$\begin{aligned} a^{\text{stab}}_{u_0 p, p^{2k}} &= Np^k\, F(k) + Np^{k+1}\, \chi_\pi(p)^2\, F(k-3) \\ &\quad + Np^{k+2}\, \chi_\pi(p)^4\, F(k-6) + \cdots \\ a^{\text{stab}}_{u_0, p^{2k}} &= a^{\text{stab}}_{u_0 p, p^{2k}} + [0, 1] a^{\text{stab}}_{u_0 p, p^{2k-2}}, \\ a^{\text{stab}}_{u_0, p^{2k+1}} &= -\big([1, 0] a^{\text{stab}}_{u_0, p^{2k}} + \chi_\pi(p)\, a^{\text{stab}}_{u_0 p, p^{2k-2}}\big). \end{aligned} \tag{2.7}$$

Introducing a topic to be developed further in the next section, we observe now that the symmetric square L-function appears as a residue of the double Dirichlet series, and one can see its functional equation this way. Indeed, the group of functional equations is generated by

$$(s, w) \mapsto \left(s + w - \tfrac{1}{2}, 1 - w\right), \qquad (s, w) \mapsto \left(1 - s, w + 3s - \tfrac{3}{2}\right).$$

One element of this dihedral group of order twelve is:

$$(s, w) \mapsto \left(\tfrac{3}{2} - s - w, w\right),$$

so when taking the residue at $w = 1$, we obtain a Dirichlet series having a functional equation under $s \to \frac{1}{2} - s$. Now

$$\sum_{t=0}^{2\theta} Np^{-t} b^{(\theta)}_{1, p^t} = F(\theta) + Np\, F(\theta - 3) + Np^2 F(\theta - 6) + \dots. \tag{2.8}$$

The significance of (2.8) is that this Dirichlet series is

$$L(2s, f, \vee^2)\, \zeta(6s - 1).$$

This follows from (2.7) since

$$Np^\theta \sum_{t=0}^{2\theta} Np^{-t} b^{(\theta)}_{1, p^t} = a^{\text{stab}}_{p, p^{2\theta}}.$$

We have also investigated the local matters reported on here for $r = 4$. Globally and locally this is a fundamentally different situation, for in that case the group of functional equations is infinite. We find that there the local combinatorial problem, discussed above for $r = 3$, does have a solution which is almost but not quite unique. In the Dirichlet polynomial $P$ the middle coefficient $a^{(\alpha)}_{1, p^{4\alpha}}$ is not determined and can be chosen freely—there will still be a solution to the combinatorial problem of Section 1. The residue will be the symmetric square times a Dirichlet series with a natural boundary. Nevertheless computer computations suggest that there might be a choice which makes this Dirichlet series as simple as possible. Let $[n_1, n_2, n_3]$ denote the L-group character with highest weight $\sum n_i \varpi_i$ in terms of the fundamental dominant weights $\varpi_i$. Assuming for simplicity that the central character $\chi_\pi$ is trivial, this preferred choice would seem to be:

$$a^{(1)}_{1,p^4} = [0,2,0]Np^2 + [1,0,1]Np + 1,$$

$$\begin{aligned} a^{(2)}_{1,p^8} &= Np^4([2,0,2] + [0,4,0] + 1) + Np^3([0,1,2] + [1,0,1] + [1,2,1] \\ &\quad + [2,1,0]) + Np^2[0,2,0]. \end{aligned}$$

If we make these choices then the residue is a series

$$\sum_\alpha Z[\alpha]\, Np^{-2\alpha s},$$

where in terms of the symmetric square $[2,0,0]$ we have:

$$\begin{aligned} Z[1] &= [2, 0.0], \\ Z[2] &= \vee^2([2,0,0]), \\ Z[3] &= \vee^3([2,0,0]) + [0,0,2]\, Np, \\ Z[4] &= \vee^4([2,0,0]) + [2,0,2]\, Np + Np^2, \\ Z[5] &= \vee^5([2,0,0]) + ([4,0,2] + [0,2,2] - [0,1,0])\, Np + 2[2,0.0]\, Np^2. \end{aligned}$$

For different choices of coefficients these formulae would be considerably more complicated.

**3. An application to automorphic forms on $GL(3)$.** In this section our objective will be to illustrate the technique of multiple Dirichlet series in the particular case of automorphic $L$-series on $GL(3)$ defined over $\mathbb{Q}$. We change slightly the notation from Sections 0–2 in order to state our results in a way which closely resembles the classical $GL(2)$ theory.

Let $f$ be an automorphic cusp form on $GL(3)$ over $\mathbb{Q}$ of level $N$ corresponding to the automorphic representation $\pi$. All $L$-series in this section will have the archimedean factor removed (we shall deal with this factor explicitly and separately below). Assume that $f$ is an eigenfunction of all the Hecke operators. Let $S$ be a finite set of primes including 2 and the primes dividing $N$ and let $M = \prod_{p \in S} p$. As remarked before, the $L$-series associated to $f$ has the form

$$L(s, f) = \sum_1^\infty \frac{c(m)}{m^s}$$

for certain generalized Fourier coefficients $c(m)$, and decomposes into the Euler product

$$L(s, f) = \prod_{p} \left(1 - \alpha_p p^{-s}\right)^{-1} \left(1 - \beta_p p^{-s}\right)^{-1} \left(1 - \gamma_p p^{-s}\right)^{-1}, \tag{3.1}$$

the product being over all primes $p$ of $\mathbb{Q}$.

It is well known that $L(s, f)$ possesses a functional equation:

$$N^{s/2} G_f(s) L(s, f) = \epsilon_f N^{(1-s)/2} G_f(1-s) L(1-s, \tilde{f}), \tag{3.2}$$

where $\tilde{f}$ denotes the contragredient of $f$ and $G_f$ is a product of gamma factors depending on $f$.

If $f$ is twisted by a character $\chi$ of conductor $D$ the associated $L$-series becomes

$$\begin{aligned} &L(s, f, \chi) \\ &= \textstyle\prod_p \left(1 - \chi(p)\alpha_p p^{-s}\right)^{-1} \left(1 - \chi(p)\beta_p p^{-s}\right)^{-1} \left(1 - \chi(p)\gamma_p p^{-s}\right)^{-1}, \end{aligned} \tag{3.3}$$

and the functional equation is given by

$$\begin{aligned} &(D^3 N)^{s/2} G_{f,\chi}(s) L(s, f, \chi) \\ &\quad = \epsilon_f \tau(\chi)^3 \psi_N(D) (D^3 N)^{(1-s)/2} G_{f,\chi}(1-s) L(1-s, \tilde{f}, \overline{\chi}). \end{aligned} \tag{3.4}$$

Here the gamma factors have an additive shift that depends on the sign of $\chi(-1)$, $\psi_N$ is the Dirichlet character associated to the idèle class character $\chi_\pi$, and $\tau(\chi)$ is the Gauss sum associated to $\chi$ normalized to have absolute value equal to 1. This is simply (0.4) restated in classical language. In this section we will also assume that $\psi_N$ is quadratic or trivial of conductor dividing $N$ (this assumption simplifies our work but is not essential for it).

In what follows we will not need a very explicit description of $G_{f,\chi}$ but we will find the following upper bound convenient. For $\sigma_1 > \sigma_2$ and $t$ real it follows from Stirling's formula that for large $|t|$, independent of $\chi$,

$$|G_{f,\chi}(\sigma_1 + it)/G_{f,\chi}(\sigma_2 - it)| << (|t| + 1)^{3(\sigma_1 - \sigma_2)/2}. \tag{3.5}$$

Here the implied constant depends only on the eigenvalues of $f$.

When all finite primes are included in the product (3.1) the functional equation (3.2) has its optimal form. However, it is often convenient to omit factors corresponding to "bad" primes, for example 2 and those dividing the level $N$. For $M$, $S$ as above we denote the $L$-series with Euler factors corresponding to primes dividing $M$ removed as follows:

$$
\begin{aligned}
L_M(s, f, \chi) & \\
&= \prod_{p \notin S} \left(1 - \chi(p)\alpha_p p^{-s}\right)^{-1} \left(1 - \chi(p)\beta_p p^{-s}\right)^{-1} \left(1 - \chi(p)\gamma_p p^{-s}\right)^{-1} \\
&= L(s, f, \chi) \prod_{p \in S} \left(1 - \chi(p)\alpha_p p^{-s}\right)\left(1 - \chi(p)\beta_p p^{-s}\right)\left(1 - \chi(p)\gamma_p p^{-s}\right).
\end{aligned}
\tag{3.6}
$$

A particularly interesting class of twists are the quadratic characters described by $\chi_d(*) = (\frac{d}{*})$, where $d$ is a squarefree nonzero integer and $\chi_d$ is the real character associated to the quadratic field $\mathbb{Q}(\sqrt{d})$. We will assume in the following that the character $\omega$ that we twist by is quadratic.

When twisted by $\chi_d$, the $L$-series $L(s, f, \chi_d)$ will have a functional equation of the form (3.4) when $\chi_d$ is a primitive character. This corresponds to the case where $d$ is the squarefree part of a fundamental discriminant. In the preceding sections, it was noted that when $d$ is *not* squarefree, it is possible to complete $L(s, f, \chi_d)$ by multiplying by a certain Dirichlet polynomial in such a way that the resulting product has a functional equation of precisely the same form (3.4), with $D$ replaced by $|d|$ or $|4d|$.

Our object will be to obtain the analytic continuation in $s$, $w$, and an estimate for the growth in vertical strips $w = \sigma + it$, for fixed $\sigma$ and $s$, of the following double Dirichlet series. Let $l_1, l_2 > 0, l_1, l_2 | M$ and $a_1, a_2 \in \{1, -1\}$ and let $\chi_{a_1 l_1}, \chi_{a_2 l_2}$ be the quadratic characters corresponding to $a_1 l_1, a_2 l_2$ as defined above. For $l_1, l_2 > 0, l_1, l_2 | M$ and $a_1, a_2 \in \{1, -1\}$ define

$$
Z_M(s, w; \chi_{a_2 l_2}, \chi_{a_1 l_1}) = \sum_{(d,M)=1} \frac{L_M(s, f, \chi_{d_0}\chi_{a_1 l_1})\chi_{a_2 l_2}(d_0) P_{a_1 l_1 d_0, d_1}(s)}{d^w},
\tag{3.7}
$$

where we sum over $d > 0$ and use the decomposition $d = d_0 d_1^2$, with $d_0 > 0$ squarefree. In the notation of Section 1, $Z_M(s, w; \chi_{a_2 l_2}, \chi_{a_1 l_1}) = Z(s, w; \pi, \omega)$. Here $\pi$ corresponds to $f \otimes \chi_{a_1 l_1}$ and $\chi_{a_2 l_2}$ replaces $\omega$.

Proceeding as in [DGH], the following proposition will provide a useful way of collecting the properties of $Z_M(s, w; \chi_{a_2 l_2}, \chi_{a_1 l_1})$. For a positive integer $M$, define

$$
\mathrm{Div}(M) = \left\{ a \cdot l \;\middle|\; a = \pm 1,\ 1 \le l,\ l | M \right\},
$$

which has cardinality $2d(M) = 2\sum_{d|M} 1$. Let $\vec{\mathbf{Z}}_{M,f}(s, w; \chi_{a_2 l_2}, \chi_{\mathrm{Div}(M)})$ denote the $2d(M)$ by 1 column vector whose $j^{th}$ entry is $Z_M(s, w; \chi_{a_2 l_2}, \chi^{(j)})$, where $\chi^{(j)}(j = 1, 2, \ldots, 2d(M))$ ranges over the characters $\chi_{a_1 l_1}$ with $a_1 = \pm 1,\ 1 \le l_1,\ l_1 | M$. Then, we will prove

PROPOSITION 3.1. *There exists a $2d(M)$ by $2d(M)$ matrix $\Phi^{(a_2l_2)}(w)$ such that for any fixed $w$, $w \neq 1$, and for any $s$ with sufficiently large real part (depending on $w$)*

$$\prod_{p|(M/l_2)} \left(1 - p^{-2+2w}\right) \cdot \vec{\mathbf{Z}}_{M,f}(s, w; \chi_{a_2l_2}, \chi_{\mathrm{Div}(M)})$$
$$= \Phi^{(a_2l_2)}(w)\vec{\mathbf{Z}}_{M,f}(s + w - 1/2, 1 - w; \chi_{a_2l_2}, \chi_{\mathrm{Div}(M)}).$$

*The entries of $\Phi^{(a_2l_2)}(w)$, denoted by $\Phi^{(a_2l_2)}_{i,j}(w)$, are meromorphic functions in $\mathbb{C}$. Also, the only possible poles of the $\Phi^{(a_2l_2)}_{i,j}(w)$ are canceled by trivial zeros of the $L$-series appearing in the numerator of $\vec{\mathbf{Z}}_{M,f}(s + w - 1/2, 1 - w; \chi_{a_2l_2}, \chi_{\mathrm{Div}(M)})$.*

*Proof.* By the basic identity (1.1), (1.2), writing $n = n_0 n_1^2$, with squarefree $n_0 > 0$, we have

(3.8)
$$Z_M(s, w; \chi_{a_2l_2}, \chi_{a_1l_1}) = \sum_{(n,M)=1} \frac{L_M(w, \tilde{\chi}_{n_0}\chi_{a_2l_2})\chi_{a_1l_1}(n_0)c(n_0n_1^2)\, Q^{(a_2l_2)}_{n_0,n_1}(w)}{n^s}.$$

Here $\tilde{\chi}_{n_0}$ denotes the quadratic character $(*/n_0)$. Now

$$L_M(w, \tilde{\chi}_{n_0}\chi_{a_2l_2}) = L(w, \tilde{\chi}_{n_0}\chi_{a_2l_2}) \cdot \prod_{p|M} \left(1 - \tilde{\chi}_{n_0}\chi_{a_2l_2}(p)p^{-w}\right), \tag{3.9}$$

where $L(w, \tilde{\chi}_{n_0}\chi_{a_2l_2})$ satisfies the functional equation

$$\begin{aligned} &G_\epsilon(w)(n_0l_2D_{a_2l_2})^{w/2}L(w, \tilde{\chi}_{n_0}\chi_{a_2l_2}) \\ &\quad = G_\epsilon(1-w)(n_0l_2D_{a_2l_2})^{(1-w)/2}L(1-w, \tilde{\chi}_{n_0}\chi_{a_2l_2}). \end{aligned} \tag{3.10}$$

Here $\epsilon = \tilde{\chi}_{n_0}\chi_{a_2l_2}(-1)$,

$$G_\epsilon(w) = \begin{cases} \pi^{-w/2}\Gamma(w/2) & \text{if} \quad \epsilon = 1 \\ \pi^{-(w+1)/2}\Gamma((w+1)/2) & \text{if} \quad \epsilon = -1, \end{cases} \tag{3.11}$$

and

$$D_{a_2l_2} = \begin{cases} 1 & \text{if} \quad a_2l_2 \equiv 1 \pmod 4 \\ 4 & \text{otherwise.} \end{cases} \tag{3.12}$$

Combining this with the functional equation for $Q$ implied by (0.8), we obtain

$$Z_M(s, w; \chi_{a_2l_2}, \chi_{a_1l_1})$$
$$= \sum_{a_3=1,-1} \sum_{(n,M)=1,\, n\equiv a_3\,(4)} \frac{G_{\epsilon(a_3a_2l_2)}(1-w)(l_2D_{a_2l_2})^{1/2-w}}{G_{\epsilon(a_3a_2l_2)}(w)n^{s+w-1/2}}$$
$$\times \chi_{a_1l_1}(n_0)L_M(1-w, \tilde{\chi}_{n_0}\chi_{a_2l_2})c(n_0n_1^2)Q^{(a_2l_2)}_{n_0,n_1}(1-w) \cdot \prod_{p|(M/l_2)} \left(1 - \tilde{\chi}_{n_0}\chi_{a_2l_2}(p)p^{-w}\right)$$
$$\times \prod_{p|(M/l_2)} \left(1 - \tilde{\chi}_{n_0}\chi_{a_2l_2}(p)p^{-1+w}\right)^{-1}.$$

Here $\epsilon(a)$ denotes the sign of $a$. Note that we are leaving out terms in the product where $p|l_2$ as the character vanishes here.

Multiplying by $\prod_{p|(M/l_2)}(1-p^{-2+2w})$, reorganizing, and proceeding exactly as in the proof of Proposition 4.2 in [DGH] we obtain (in the case of $a_2l_2 \equiv 1 \pmod 4$)

$$\prod_{p|(M/l_2)} \left(1 - p^{-2+2w}\right) \cdot Z_M(s, w; \chi_{a_2l_2}, \chi_{a_1l_1})$$
$$= \frac{1}{2} \cdot l_2^{1/2-w} \cdot \sum_{l_3,l_4|(M/l_2)} \mu(l_3)\chi_{a_2l_2}(l_3l_4)l_3^{-w}l_4^{-1+w} \sum_{a_3=1,-1} \frac{G_{\epsilon(a_3a_2l_2)}(1-w)}{G_{\epsilon(a_3a_2l_2)}(w)}$$
$$\times \big(Z_M(s+w-1/2, 1-w; \chi_{a_2l_2}, \chi_{a_1l_1l_3l_4}) + a_3Z_M(s+w-1/2, 1-w; \chi_{a_2l_2}, \chi_{-a_1l_1l_3l_4})\big). \tag{3.13}$$

In the case $a_2l_2 \equiv -1,\ 2 \pmod 4$, just the behavior at the finite place 2 changes. The observation concerning the poles of $\Phi^{(a_2l_2)}_{i,j}(w)$ follows from the fact that the only source of possible poles on the right hand side of (3.13) are the gamma factors $G_{\epsilon(a_3a_2l_2)}(1-w)$. These, however, are canceled by trivial zeros of $L$ series.

This completes the proof of Proposition 3.1. □

The function $Z_M(s, w; \chi_{a_2l_2}, \chi_{a_1l_1})$ defined in (3.7) also possesses a functional equation as $s \to 1-s$. Let $d(M)$ be as before, and let $\vec{\mathbf{Z}}_{M,f}(s, w; \chi_{\mathrm{Div}(M)}, \chi_{a_1l_1})$ denote the $2d(M)$ by 1 column vector whose $j^{th}$ entry is $Z_M(s, w; \chi^{(j)}, \chi_{a_1l_1})$, where $\chi^{(j)}$ $(j = 1, 2, \ldots, 2d(M))$ ranges over the characters $\chi_{a_2l_2}$ with $a_2 = \pm 1,\ 1 \le l_2,\ l_2|M$.

Then we have the following.

PROPOSITION 3.2. *There exists a $2d(M)$ by $2d(M)$ matrix $\Psi^{(a_1 l_1)}(s)$ such that for any fixed $s$, $s \neq 1$, and for any $w$ with sufficiently large real part (depending on $s$)*

$$\vec{\mathbf{Z}}_{M,f}(s, w; \chi_{\mathrm{Div}(M)}, \chi_{a_1 l_1}) \cdot \prod_{p|(M/l_1)} \left(1 - \tilde{\alpha}_p^2 p^{-2+2s}\right)\left(1 - \tilde{\beta}_p^2 p^{-2+2s}\right)\left(1 - \tilde{\gamma}_p^2 p^{-2+2s}\right)$$
$$= \Psi^{(a_1 l_1)}(s) \vec{\mathbf{Z}}_{M,\tilde{f}}(1 - s, w + 3s - 3/2; \chi_{\mathrm{Div}(M)}, \chi_{a_1 l_1}).$$

*The entries of $\Psi^{(a_1 l_1)}(s)$, denoted by $\Psi_{i,j}^{(a_1 l_1)}(s)$, are meromorphic functions in $\mathbb{C}$. The only possible poles of $\Psi_{i,j}^{(a_1 l_1)}(s)$ are canceled by trivial zeros of the L-series occurring in the numerator of $\vec{\mathbf{Z}}_{M,\tilde{f}}(1 - s, w + 3s - 3/2; \chi_{\mathrm{Div}(M)}, \chi_{a_1 l_1})$.*

*Proof.* First, write

$$(3.14)\quad L_M(s, f, \chi_{d_0}\chi_{a_1 l_1}) = L(s, f, \chi_{d_0}\chi_{a_1 l_1})$$
$$\cdot \prod_{p|(M/l_1)} \left(1 - \alpha_p \chi_{d_0}\chi_{a_1 l_1}(p) p^{-s}\right)\left(1 - \beta_p \chi_{d_0}\chi_{a_1 l_1}(p) p^{-s}\right)\left(1 - \gamma_p \chi_{d_0}\chi_{a_1 l_1}(p) p^{-s}\right)$$
$$= L(s, f, \chi_{d_0}\chi_{a_1 l_1})$$
$$\cdot \sum_{l_\alpha|(M/l_1)} \mu(l_\alpha)\chi_{a_1 d_0 l_1}(l_\alpha) l_\alpha^{-s} \prod_{p|l_\alpha} \alpha_p$$
$$\cdot \sum_{l_\beta|(M/l_1)} \mu(l_\beta)\chi_{a_1 d_0 l_1}(l_\beta) l_\beta^{-s} \prod_{p|l_\beta} \beta_p \sum_{l_\gamma|(M/l_1)} \mu(l_\gamma)\chi_{a_1 d_0 l_1}(l_\gamma) l_\gamma^{-s} \prod_{p|l_\gamma} \gamma_p.$$

By (3.4)

$$(3.15)\quad \begin{aligned} &L(s, f, \chi_{d_0}\chi_{a_1 l_1}) \\ &= \epsilon_f \psi_N(a_1 d_0 l_1)((d_0 l_1 D_{a_1 d_0 l_1})^3 N)^{(1/2-s)} \tfrac{G_{f,\epsilon}(1-s)}{G_{f,\epsilon}(s)} L(1 - s, \tilde{f}, \chi_{d_0}\chi_{a_1 l_1}), \end{aligned}$$

where $G_{f,\epsilon}$ is a product of gamma factors depending only on $f$ and on the sign of $a_1 d_0 l_1$, and $D_{a_1 d_0 l_1}$ is given by (3.12).

On the other side of the functional equation,

$$L(1 - s, \tilde{f}, \chi_{a_1 d_0 l_1}) = L_M(1 - s, \tilde{f}, \chi_{a_1 d_0 l_1}) \prod_{p|(M/l_1)}$$
$$\left(1 - \tilde{\alpha}_p \chi_{a_1 d_0 l_1}(p) p^{-1+s}\right)^{-1}\left(1 - \tilde{\beta}_p \chi_{a_1 d_0 l_1}(p) p^{-1+s}\right)^{-1}\left(1 - \tilde{\gamma}_p \chi_{a_1 d_0 l_1}(p) p^{-1+s}\right)^{-1},$$

so

$$L(1-s,\tilde{f},\chi_{a_1d_0l_1})\prod_{p|(M/l_1)}\left(1-\tilde{\alpha}_p^2p^{-2+2s}\right)\left(1-\tilde{\beta}_p^2p^{-2+2s}\right)\left(1-\tilde{\gamma}_p^2p^{-2+2s}\right)$$

$$= L_M(1-s,\tilde{f},\chi_{a_1d_0l_1})$$

$$\cdot\prod_{p|(M/l_1)}\left(1+\tilde{\alpha}_p\chi_{a_1d_0l_1}(p)p^{-1+s}\right)\left(1+\tilde{\beta}_p\chi_{a_1d_0l_1}(p)p^{-1+s}\right)\left(1+\tilde{\gamma}_p\chi_{a_1d_0l_1}(p)p^{-1+s}\right)$$

$$= L_M(1-s,\tilde{f},\chi_{a_1d_0l_1})\sum_{l_{\tilde{\alpha}}|(M/l_1)}\chi_{a_1d_0l_1}(l_{\tilde{\alpha}})l_{\tilde{\alpha}}^{-1+s}\prod_{p|l_{\tilde{\alpha}}}\tilde{\alpha}_p$$

$$\cdot\sum_{l_{\tilde{\beta}}|(M/l_1)}\chi_{a_1d_0l_1}(l_{\tilde{\beta}})l_{\tilde{\beta}}^{-1+s}\prod_{p|l_{\tilde{\beta}}}\tilde{\beta}_p\sum_{l_{\tilde{\gamma}}|(M/l_1)}\chi_{a_1d_0l_1}(l_{\tilde{\gamma}})l_{\tilde{\gamma}}^{-1+s}\prod_{p|l_{\tilde{\gamma}}}\tilde{\gamma}_p.$$

Combining the above with (3.14) and (3.15) we obtain

(3.16)

$$Z_M(s,w;\chi_{a_2l_2},\chi_{a_1l_1})\prod_{p|(M/l_1)}\left(1-\tilde{\alpha}_p^2p^{-2+2s}\right)\left(1-\tilde{\beta}_p^2p^{-2+2s}\right)\left(1-\tilde{\gamma}_p^2p^{-2+2s}\right)$$

$$=\sum_{(d,M)=1}\epsilon_f\psi_N(a_1d_0l_1)((l_1D_{a_1d_0l_1})^3N)^{(1/2-s)}\frac{G_{f,\epsilon}(1-s)}{G_{f,\epsilon}(s)}\cdot\frac{L_M(1-s,\tilde{f},\chi_{d_0a_1l_1})}{d^{w+3s-3/2}}$$

$$\cdot P_{d_0,d_1}^{(a_1l_1)}(1-s)\chi_{a_2l_2}(d_0)\sum_{l_\alpha|(M/l_1)}\mu(l_\alpha)\chi_{a_1d_0l_1}(l_\alpha)l_\alpha^{-s}\prod_{p|l_\alpha}\alpha_p$$

$$\cdot\sum_{l_\beta|(M/l_1)}\mu(l_\beta)\chi_{a_1d_0l_1}(l_\beta)l_\beta^{-s}\prod_{p|l_\beta}\beta_p\sum_{l_\gamma|(M/l_1)}\mu(l_\gamma)\chi_{a_1d_0l_1}(l_\gamma)l_\gamma^{-s}\prod_{p|l_\gamma}\gamma_p$$

$$\cdot\sum_{l_{\tilde{\alpha}}|(M/l_1)}\chi_{a_1d_0l_1}(l_{\tilde{\alpha}})l_{\tilde{\alpha}}^{-1+s}\prod_{p|l_{\tilde{\alpha}}}\tilde{\alpha}_p$$

$$\cdot\sum_{l_{\tilde{\beta}}|(M/l_1)}\chi_{a_1d_0l_1}(l_{\tilde{\beta}})l_{\tilde{\beta}}^{-1+s}\prod_{p|l_{\tilde{\beta}}}\tilde{\beta}_p\sum_{l_{\tilde{\gamma}}|(M/l_1)}\chi_{a_1d_0l_1}(l_{\tilde{\gamma}})l_{\tilde{\gamma}}^{-1+s}\prod_{p|l_{\tilde{\gamma}}}\tilde{\gamma}_p.$$

This decomposes into a linear combination of the functions $\widetilde{Z}_M(1-s,w+3s-3/2;\chi^*,\chi_{a_1l_1})$ where the character $\chi^*$ takes one of the two forms

$$\psi_N\chi_{l_\alpha l_\beta l_\gamma l_{\tilde{\alpha}}l_{\tilde{\beta}}l_{\tilde{\gamma}}}\chi_{a_2l_2},\qquad \psi_N\chi_{-1}\chi_{l_\alpha l_\beta l_\gamma l_{\tilde{\alpha}}l_{\tilde{\beta}}l_{\tilde{\gamma}}}\chi_{a_2l_2}.$$

Here we have denoted by $\widetilde{Z}_M(s,w;\chi_{\alpha_2l_2},\chi_{\alpha_1l_1})$, the double Dirichlet series formed with $\tilde{f}$ rather than $f$. The result then follows as in the previous proposition and as in [DGH] and we omit the precise, rather complicated, expression for $\Psi^{(a_1l_1)}(s)$. This completes the proof of Proposition 3.2. □

We now continue the analytic continuation of $Z_M(s,w;\chi_{\alpha_2l_2},\chi_{\alpha_1l_1})$. Our approach will follow a parallel route to that of [DGH], differing only in some details caused by the fact that the Ramanujan conjecture is not known for automorphic forms on $GL(3)$. We refer to [DGH] for background and definitions that will be used from the theory of several complex variables, and for some details of arguments that we will omit below.

We will repeatedly apply the functional equations given in Propositions 3.1, 3.2. To do this conveniently, we define two involutions on $\mathbb{C} \times \mathbb{C}$:

$$\alpha\colon (s, w) \to (1 - s, w + 3s - 3/2) \quad \text{and} \quad \beta\colon (s, w) \to (s + w - 1/2, 1 - w).$$

Then $\alpha$, $\beta$ generate $D_{12}$, the dihedral group of order 12, and $\alpha^2 = \beta^2 = 1$, $(\alpha\beta)^6 = (\beta\alpha)^6 = 1$. Note that $\alpha\beta \neq \beta\alpha$.

We will find it useful in the following to define three regions $R_1$, $R_2$, $R_3$ as follows: Write $s$, $w$ as $s = \sigma + it$, $w = \nu + i\gamma$.

The tube region $R_1$ is defined to be the set of all points $(s, w)$ such that $(\sigma, \nu)$ lie strictly above the polygon determined by $(-2/5, 37/10)$, $(3/2, 0)$, and the rays $\nu = -3\sigma + 5/2$ for $\sigma \leq -2/5$ and $\nu = -\sigma + 3/2$ for $\sigma \geq 3/2$.

The tube region $R_2$ is defined to be the set of all points $(s, w)$ such that $(\sigma, \nu)$ lie strictly above the polygon determined by $(-1/2, 3)$ and $(3/2, 0)$ and the rays $\nu = -2\sigma + 2$ for $\sigma \leq -1/2$, and $\nu = -\sigma + 3/2$ for $\sigma \geq 3/2$.

The tube region $R_3$ is defined to be the set of all points $(s, w)$ such that $(\sigma, \nu)$ lie strictly above the line $\nu = -2\sigma + 2$.

These regions are related by the involutions $\alpha$, $\beta$ as described in the following proposition. The proof, a simple exercise, is omitted.

PROPOSITION 3.3. *The regions $R_1$ and $\alpha(R_1)$ have a non-empty intersection, and the convex hull of $R_1 \cup \alpha(R_1)$ equals $R_2$. Similarly, $R_2$ and $\beta(R_2)$ have a non-empty intersection and the convex hull of $R_2 \cup \beta(R_2)$ equals $R_3$. Finally, $R_3$ and $\alpha(R_3)$ have a non-empty intersection and the convex hull of $R_3 \cup \alpha(R_3)$ equals $\mathbb{C}^2$.*

We will begin by demonstrating

PROPOSITION 3.4. *Let $R_1$ be the tube region defined above. The functions*

$$(w - 1)Z_M(s, w; \chi_{a_2 l_2}, \chi_{a_1 l_1}) \quad \textit{and} \quad (w - 1)\widetilde{Z}_M(s, w; \chi_{a_2 l_2}, \chi_{a_1 l_1})$$

*are analytic in $R_1$.*

*Proof.* Consider first the expression for $Z_M(s, w; \chi_{a_2 l_2}, \chi_{a_1 l_1})$ given in (3.7). We first need to determine when $L_M(s, f, \chi_{d_0} \chi_{a_1 l_1})$ converges absolutely. A well known upper bound for the size of the coefficients $c(n)$ of $f$ (and $\tilde{c}(n)$ of $\tilde{f}$ ) is $c(n) \ll |n|^{2/5+\epsilon}$. This follows from the properties of the Rankin-Selberg convolution of $f$ with $\tilde{f}$ (in particular from the presence of nine gamma factors in the functional equation). As a consequence, $L_M(s, f, \chi_{d_0} \chi_{a_1 l_1})$ converges absolutely for $\sigma > 7/5$ and is bounded above by a constant independent of $d_0$. See Jacquet, Piatetski-Shapiro and Shalika [JPS1] and Jacquet and Shalika [JS2] for the Rankin-Selberg method for $GL(n)$. Over a number field a weaker estimate could instead be obtained from Jacquet, Piatetski-Shapiro and Shalika [JPS2].

Now suppose that the sum were restricted only to squarefree $d = d_0$. It is clear then that $Z_M(s, w; \chi_{a_2l_2}, \chi_{a_1l_1})$ would converge absolutely in the region $\sigma > 7/5, \nu > 1$. Reflecting $s$ to $1 - s$ (using (3.4)) and noting that the missing Euler factors do not affect convergence, one sees that there would be convergence in the region $\sigma < -2/5, \nu > -3\sigma + 5/2$ and consequently, by the Phragmen-Lindelöf principle, there would be convergence in the region lying between these two areas and above the line connecting $(-2/5, 37/10)$ and $(7/5, 1)$. It follows from (2.1–2.3) that we have the bound

$$P_{a_1l_1d_0,d_1}(s) \ll 1$$

for $\sigma > 7/5$. In fact, more precisely, $P_{a_1l_1d_0,d_1}(s) \ll 1$ will hold whenever $\sigma > 1/2 + \eta$, where $c(m) \ll_\epsilon m^{\eta+\epsilon}$ is an upper bound for the coefficients of the $GL(3)$ form $f$. Thus, for example, if $f$ is the adjoint square lift of a form on $GL(2)$ then $\eta = 2/5$ will produce the estimate above.

Because of this, and the functional equation (0.5) applied to $P_{a,l,d_0,d}(s)$, precisely the same estimates apply as we sum over all $d$. Consequently, $Z_M(s, w; \chi_{a_2l_2}, \chi_{a_1l_1})$ converges above the given lines. As $f$ is cuspidal $L_M(s, f, \chi_{d_0}\chi_{a_1l_1})$ has no poles in this region.

Noting that the expression converges when $\nu > 1, \sigma > 7/5$, we now change the order of summation to the form given in (3.8). The location of the pole of the Rankin-Selberg convolution of $f$ with itself implies the absolute convergence of $L_M(s, f)$ for $\sigma > 1$. Consequently, applying the usual convexity estimates for $L(w, \chi_{n_0})$, the corresponding estimate

$$Q^{(a_2l_2)}_{n_0,n_1}(w) \ll c(n_0n_1^2)$$

(which follows from (2.4–2.6)) and functional equations for $c(n_0n_1^2)Q^{(a_2l_2)}_{n_0,n_1}(w)$, we see $Z_M(s, w; \chi_{a_2l_2}, \chi_{a_1l_1})$ converges for $\sigma > 1$ when $\nu > 1$, for $\sigma > (-1/2)\nu + 3/2$ when $0 \le \nu \le 1$, and for $\sigma > -\nu + 3/2$ when $\nu < 0$. The factor $w - 1$ cancels the pole at $w = 1$. The regions described above overlap and thus by the convexity principle for several complex variables (see Proposition 4.6 of [DGH]) $Z_M(s, w; \chi_{a_2l_2}, \chi_{a_1l_1})P(s, w)$ has an analytic continuation to the convex closure of the regions, which is $R_1$ described above. An identical argument applies when $f$ is replaced by $\tilde{f}$.

This completes the proof of Proposition 3.4. □

Our plan is now to apply the involutions $\alpha$, $\beta$, $\alpha$ in that order to $R_1$, and use Propositions 3.1 and 3.2 to extend the analytic continuation to $\mathbb{C}^2$. To aid in this, it will be useful to introduce some additional notation to make the content of these propositions a bit clearer and easier to apply. Let

$$A(s,w) \equiv A_M(s,w) = \prod_{p|M} \left(1-\tilde{\alpha}_p^2 p^{-2+2s}\right)\left(1-\tilde{\beta}_p^2 p^{-2+2s}\right)\left(1-\tilde{\gamma}_p^2 p^{-2+2s}\right)$$

and

$$B(s,w) \equiv B_M(s,w) = \prod_{p|M} (1-p^{-2+2w}), \tag{3.17}$$

and let

$$\tilde{\Psi}^{(a_1 l_1)}(s,w) = \Psi^{(a_1 l_1)}(s) \prod_{p|l_1} \left(1-\tilde{\alpha}_p^2 p^{-2+2s}\right)\left(1-\tilde{\beta}_p^2 p^{-2+2s}\right)\left(1-\tilde{\gamma}_p^2 p^{-2+2s}\right),$$

$$\tilde{\Phi}^{(a_2 l_2)}(s,w) = \Phi^{(a_2 l_2)}(w) \prod_{p|l_2} (1-p^{-2+2w}).$$

The following is a reformulation of the content we require now from Propositions 3.1 and 3.2. For $(s,w)$ such that both sides are contained in a connected region of analytic continuation for $(w-1)Z_M(s,w;\chi_{a_2 l_2},\chi_{a_1 l_1})$

$$A(s,w)\vec{\mathbf{Z}}_{M,f}(s,w;\chi_{\mathrm{Div}(M)},\chi_{a_1 l_1}) = \tilde{\Psi}^{(a_1 l_1)}(s,w)\vec{\mathbf{Z}}_{M,f}(\alpha(s,w);\chi_{\mathrm{Div}(M)},\chi_{a_1 l_1}) \tag{3.18}$$

and

$$B(s,w)\vec{\mathbf{Z}}_{M,f}(s,w;\chi_{a_2 l_2},\chi_{\mathrm{Div}(M)}) = \tilde{\Phi}^{(a_2 l_2)}(s,w)\vec{\mathbf{Z}}_{M,f}(\beta(s,w);\chi_{a_2 l_2},\chi_{\mathrm{Div}(M)}). \tag{3.19}$$

Next we prove the analytic continuation of $Z_M(s,w;\chi_{a_2 l_2},\chi_{a_1 l_1})$.

PROPOSITION 3.5. *Let*

$$\mathcal{P}(s,w) = w(w-1)(3s+w-5/2)(3s+2w-3)(3s+w-3/2).$$

*Then the following product has an analytic continuation to an entire function in* $\mathbb{C}^2$:

$$\begin{aligned} &Z_M^*(s,w;\chi_{a_2 l_2},\chi_{a_1 l_1}) := \\ &A(s,w)A(\alpha(s,w))A(\beta(s,w))A(\beta\alpha(s,w))B(s,w)B(\alpha(s,w))\mathcal{P}(s,w) \\ &\times Z_M(s,w;\chi_{a_2 l_2},\chi_{a_1 l_1}). \end{aligned}$$

*Proof.* Let $P(s,w) = w-1$. In Proposition 3.4 we established the analytic continuation of $Z_M(s,w;\chi_{a_2 l_2},\chi_{a_1 l_1})P(s,w)$ and $\widetilde{Z}_M(s,w;\chi_{a_2 l_2},\chi_{a_1 l_1})P(s,w)$ in $R_1$. As $\alpha^2 = 1$, $\tilde{\tilde{f}} = f$ and $\tilde{\Psi}^{(a_1 l_1)}(s,w)$ is meromorphic in $\mathbb{C}^2$, it follows that

$$\tilde{\Psi}^{(a_1 l_1)}(s,w)\vec{\mathbf{Z}}_{M,f}(\alpha(s,w);\chi_{\mathrm{Div}(M)},\chi_{a_1 l_1})P(\alpha(s,w))$$

is a meromorphic function in $\alpha(R_1)$. As any poles of $\tilde{\Psi}^{(a_1l_1)}(s,w)$ arise from gamma factors and are canceled by trivial zeros of the $L$-series appearing in

$$\vec{\mathbf{Z}}_{M,\tilde{f}}(\alpha(s,w);\chi_{\mathrm{Div}(M)},\chi_{a_1l_1}),$$

we can conclude from Proposition 3.4 and (3.18) that

$$A(s,w)P(s,w)P(\alpha(s,w))\vec{\mathbf{Z}}_{M,f}(s,w;\chi_{\mathrm{Div}(M)},\chi_{a_1l_1})$$

is analytic in $R_1 \cup \alpha(R_1)$, $R_1$ and $\alpha(R_1)$ having a substantial intersection (containing $\Re(s)$, $\Re(w) > 1$). Thus by the convexity principle this function is analytic in $R_2$, the convex hull of the union.

A similar argument applies to $\vec{\mathbf{Z}}_{M,\tilde{f}}(s,w;\chi_{\mathrm{Div}(M)},\chi_{a_1l_1})$. The rest of the argument proceeds exactly as in the proof of Proposition 4.10 of [DGH]. The only change is a slight difference in the definition of the region $R_1$, which does not affect the argument. This completes the proof of Proposition 3.5. □

We will now use the analytic continuation and functional equations (3.18), (3.19) for the function $\vec{\mathbf{Z}}_{M,f}(s,w;\chi_{\mathrm{Div}(M)},\chi_{a_1l_1})$ to locate poles and obtain an estimate for the growth of $Z_M(1/2,w;\chi_{a_2l_2},\chi_{a_1l_1})$ in a vertical strip. Before doing this, however, we need some additional notation.

Let $\vec{\mathbf{Z}}_{M,f}(s,w)$ denote the $4d(M)^2$–dimensional column vector consisting of the concatenation of the $2d(M)$ column vectors $\vec{\mathbf{Z}}_{M,f}(s,w;\chi_{a_2l_2},\chi_{\mathrm{Div}(M)})$ for $a_2 \in \{1,-1\}$ and all $l_2|M$. Then by Propositions 3.1 and 3.2, combined with (3.18), (3.19), there exist $4d(M)^2$ by $4d(M)^2$ matrices $\Phi_M(s,w)$, $\Psi_M(s,w)$ such that

$$A_M(s,w)\vec{\mathbf{Z}}_{M,f}(s,w) = \Psi_M(s,w)\vec{\mathbf{Z}}_{M,\tilde{f}}(\alpha(s,w)) \tag{3.20}$$

and

$$B_M(s,w)\vec{\mathbf{Z}}_{M,f}(s,w) = \Phi_M(s,w)\vec{\mathbf{Z}}_{M,f}(\beta(s,w)). \tag{3.21}$$

Here $A_M(s,w)$, $B_M(s,w)$ are given by (3.17). The matrices $\Phi_M(s,w)$, $\Psi_M(s,w)$ are constructed from blocks of $\tilde{\Phi}^{(a_2l_2)}(s,w)$ and $\tilde{\Psi}^{(a_1l_1)}(s,w)$ on the diagonal.

We now remark that the function $Z_M^*(1/2,w;\chi_{a_2l_2},\chi_{a_1l_1})$, defined in Proposition 3.5, is of finite order. This straightforward statement about a one–variable problem requires the theory of several complex variables for the proof. As the proof in this context is virtually identical to that given in detail in Proposition 4.11 of [DGH] we will omit it.

We now show:

Proposition 3.6. *Let $w = v + it$. For $\epsilon > 0$, $-3/4 - \epsilon \le v$, and any $a_1, a_2 \in \{1,-1\}$, $l_1, l_2|M$, the function $Z_M(1/2,w;\chi_{a_2l_2},\chi_{a_1l_1})$ is an analytic function of $w$, except for poles at $w = \frac{3}{4}, 1$. If $(l_1,\ l_2) = 1$ or $2$ and $|t| > 1$, then it satisfies the* *upper bounds*

$$Z_M\left(\frac{1}{2}, \nu + it; \chi_{a_2 l_2}, \chi_{a_1 l_1}\right) \ll_\epsilon (NM^3)^{1/4+\epsilon},$$

*for* $7/4 + \epsilon < \nu$, *and*

$$Z_M\left(\frac{1}{2}, \nu + it; \chi_{a_2 l_2}, \chi_{a_1 l_1}\right) \ll_\epsilon (NM^3)^{1/4+\epsilon} M^{a(7/4-\nu)+\nu_1(\epsilon)} |t|^{b(7/4-\nu)+\nu_2(\epsilon)}$$

*for* $-3/4 - \epsilon \leq \nu \leq 7/4 + \epsilon$. *The constants* $a, b$ *are positive and explicitly computable. Also the functions* $\nu_1(\epsilon)$, $\nu_2(\epsilon)$ *are explicitly computable functions satisfying*

$$\lim_{\epsilon \to 0} \nu_1(\epsilon) = \lim_{\epsilon \to 0} \nu_2(\epsilon) = 0.$$

*Proof.* The result stated here is rather crude, but suffices for our purposes. It would be an easy matter to refine these estimates but as we are not aiming for an optimal unweighted mean value estimate this is all that we require.

The first bound in the region $7/4 + \epsilon < \nu$ follows immediately from the standard convexity estimate $L(1/2, f, \chi_D) \ll (ND^3)^{1/4+\epsilon}$, where $N$ is the level of $f$ and $D$ is the conductor of $\chi_D$. The bound for $-3/4 - \epsilon \leq \nu \leq 7/4 + \epsilon$ is more difficult to obtain. The approach is to first obtain a bound for $Z_M(\frac{1}{2}, -3/4 - \epsilon + it; \chi_{a_2 l_2}, \chi_{a_1 l_1})$. We then apply a convexity argument (using the finite order property mentioned above) to complete the proof for $-3/4 - \epsilon \leq \nu \leq 7/4 + \epsilon$. The argument is identical to that presented in the proof of Proposition 4.12 of [DGH] and so will be omitted. We will remark, however, that the only possible poles of $Z_M(1/2, w; \chi_{a_2 l_2}, \chi_{a_1 l_1})$ are at locations corresponding to zeros of the factors $A$, $B$ and $\mathcal{P}$ appearing on the right hand side of the product given in Proposition 3.5. All can be eliminated as possibilities except for $w = 1$ and $w = 3/4$. This completes the proof of Proposition 3.6. □

It now remains to calculate the order of the pole of $Z_M(1/2, w; \chi_{a_2 l_2}, \chi_{a_1 l_1})$ and compute the leading coefficient in the Laurent expansion at $w = 1$.

To do this we first compute the residue of $Z_M(s, w; \chi_{a_2 l_2}, \chi_{a_1 l_1})$ at $w = 1$ for $s$ in a neighborhood of $1/2$. Our result is summarized in

PROPOSITION 3.7. *For* $s$ *in a neighborhood of* $1/2$ *but* $s \neq 1/2$, *the function* $Z_M(s, w; \chi_{a_2 l_2}, \chi_{a_1 l_1})$ *is analytic at* $w = 1$ *whenever* $a_2 l_2 \neq 1$. *When* $a_2 l_2 = 1$, $Z_M(s, w; 1, \chi_{a_1 l_1})$ *has a pole of order 1 at* $w = 1$ *with residue*

$$R_M(s; \chi_{a_1 l_1}) = \lim_{w \to 1} (w - 1) Z_M(s, w; 1, \chi_{a_1 l_1})$$

*given by:*

$$R_M(s; \chi_{a_1 l_1}) = \prod_{p|M} (1 - p^{-1}) L_M(2s, f, \vee^2) \zeta_M(6s - 1).$$

*Here* $L_M(2s, f, \vee^2)$, $\zeta_M(6s-1)$ *denote the symmetric square L-function of* $f$ *at* $2s$ *and the zeta function at* $6s-1$*, each with Euler factors corresponding to primes dividing* $M$ *removed.*

*Proof.* Taking real parts of $w, s$ sufficiently large to guarantee absolute convergence we apply the basic identity translated into the equality of equations (3.7) and (3.8). It follows that for the real part of $s$ sufficiently large

$$\lim_{w\to 1}(w-1)Z_M(s,w;\chi_{a_2l_2},\chi_{a_1l_1})$$

$$= \sum_{(n,M)=1,n>0} \frac{\lim_{w\to 1}(w-1)L_M(w,\tilde{\chi}_{n_0}\chi_{a_2l_2})\chi_{a_1l_1}(n_0)c\big(n_0n_1^2\big)\, Q^{(a_2l_2)}_{n_0,n_1}(w)}{n^s}.$$

Now $\lim_{w\to 1}(w-1)L_M(w,\tilde{\chi}_{n_0}\chi_{a_2l_2})$ vanishes unless $\chi_{n_0}\chi_{a_2l_2}$ is the trivial character. This can only happen if $n_0 = 1$ and $a_2l_2 = 1$. Thus the residue vanishes unless $a_2l_2 = 1$. If $a_2l_2 = 1$ then all the terms on the right hand side of the above will vanish unless $n_0 = 1$, in which case $L_M(w,\tilde{\chi}_{n_0}\chi_{a_2l_2}) = \zeta_M(w)$ and the residue is $\prod_{p|M}(1-p^{-1})$.

After taking the residue, the right hand side of the above reduces to

$$\prod_{p|M}(1-p^{-1}) \sum_{(n_1,M)=1,n_1>0} \frac{c\left(n_1^2\right) Q^{(1)}_{1,n_1}(1)}{n_1^{2s}}.$$

Referring to (2.7) we see that each $p$-part sums to the $p$-part of

$$L_M(2s, f, \vee^2)\,\zeta_M(6s-1)$$

as claimed. □

Recall now that the product

$$A(s,w)A(\alpha(s,w))A(\beta(s,w))A(\beta\alpha(s,w))B(s,w)B(\alpha(s,w))\mathcal{P}(s,w)$$
$$\times Z_M(s,w;\chi_{a_2l_2},\chi_{a_1l_1})$$

was shown in Proposition 3.5 to be an entire function of $s, w$. Specializing to $w = 1$ and removing the $w-1$ factor, the remaining factors of $\mathcal{P}(s,1)$ are $(3s-3/2)(3s-1)(3s-1/2)$. Thus the only possible poles of the residue $R_M(s;\chi_{a_1l_1})$ are at zeros of $A(s,1)A(\alpha(s,1))A(\beta(s,1))A(\beta\alpha(s,1))B(s,1)B(\alpha(s,1))$ or at $s = 1/2, 1/3, 1/6$. The possible poles corresponding to zeros of Euler factors of $A$, $B$ are eliminated as in Proposition 3.6. The factor $3s-1$ cancels the simple pole of $\zeta_M(6s-1)$ at $s = 1/3$. The possible pole at $s = 1/6$ is easily eliminated and we are left with the possibility of a simple pole at $s = 1/2$. This will occur if and only if $L_M(2s, f, \vee^2)$ has a simple pole at $s = 1/2$. We have thus completed the proof of the first part of Theorem 0.3.

We now proceed with the proof of a mean value result from which the second half of Theorem 0.3 will follow. We will work with the case that $L_M(2s, f, \vee^2)$ has a simple pole at $s = 1/2$ and denote the nonzero residue as $R_M$, so

$$R_M = \lim_{s\to 1/2}(s - 1/2)L_M(2s, f, \vee^2). \tag{3.22}$$

The polar line $w = 1$ is reflected into the lines $w = -3s + 5/2$ and $2w = -3s + 3$ by the functional equations $\alpha$, $\beta$. The third of these gives a single polar line passing through $(1/2, 3/4)$ while $w = 1$, $w = -3s + 5/2$ give two polar lines intersecting at $(1/2, 1)$. The behavior of $Z_M(s, w; 1, \chi_{a_1l_1})$ near $(1/2, 1)$ is given by a sum of contributions from the two polar lines:

(3.23)

$$\begin{aligned} &Z_M(s, w; 1, \chi_{a_1l_1})\\ &= \frac{A_0}{(w-1)(s-1/2)} + \frac{A_1(s)}{w-1} + \frac{A_0'}{(w+3s-5/2)(s-1/2)} + \frac{A_1'(s)}{w+3s-5/2}\\ &\quad + H(s, w), \end{aligned}$$

where

$$A_0 = \lim_{s\to 1/2}\lim_{w\to 1}(s-1/2)(w-1)Z_M(s, w; 1, \chi_{a_1l_1}).$$

$$A_0' = \lim_{s\to 1/2}\lim_{w\to 5/2-3s}(s-1/2)(w+3s-5/2)Z_M(s, w; 1, \chi_{a_1l_1}),$$

and $H(s, w)$ is an analytic function of $s, w$ near $(1/2, 1)$. Let $s \to 1/2$ in (3.23) while $w \neq 1$. We know from the left hand side of (3.23) that this must remain analytic. However, summing the two leading fractions, we see that this can only happen if $A_0' = -A_0$, in which case (3.23) reduces, at $s = 1/2$, to

$$Z_M(1/2, w; 1, \chi_{a_1l_1}) = \frac{3A_0}{(w-1)^2} + \frac{B_1}{w-1} + I(w), \tag{3.24}$$

where $B_1$ is computable and $I(w)$ is an analytic function of $w$ near $w = 1$.

We will now state the mean value theorem and do the remaining bit of work necessary to complete the proof.

THEOREM 3.8. *Let $l_1, l_2 > 0$, $l_1, l_2 | M$ and $a_1, a_2 \in \{1, -1\}$ and let the L-series $L_M(1/2, f, \chi_{d_0}\chi_{a_1l_1})$ and correction factors $P_{a_1l_1d_0,d_1}(1/2)$ be as defined above. Then*

$$\sum_{d>0,(d,M)=1} L_M(1/2, f, \chi_{d_0}\chi_{a_1l_1})\chi_{a_2l_2}(d_0)P_{a_1l_1d_0,d_1}(1/2)e^{-d/x}$$

$$= \delta_{1,a_2l_2}\prod_{p|M}(1-p^{-1})R_M\zeta_M(2)(x\log x) + \delta_{1,a_2l_2}C_Mx + C_M'x^{3/4} + C_M'' + O(x^{-3/4}).$$

*Here* $R_M = \lim_{s\to 1/2}(s-1/2)L_M(2s, f, \vee^2)$, $C_M, C'_M, C''_M$ *are computable constants and* $\delta_{1,a_2l_2} = 1$ *if* $a_2l_2 = 1$ *and is 0 otherwise.*

*Proof.* Applying the integral transform

$$\frac{1}{2\pi i}\int_{2-i\infty}^{2+i\infty}\Gamma(w)x^w\,dw = e^{-1/x},$$

valid for $x > 0$, we obtain first

$$\frac{1}{2\pi i}\int_{2-i\infty}^{2+i\infty} Z_M(1/2, w; \chi_{a_2l_2}, \chi_{a_1l_1})\Gamma(w)x^w\,dw$$
$$= \sum_{d>0,(d,M)=1} L_M(1/2, f, \chi_{d_0}\chi_{a_1l_1})\chi_{a_2l_2}(d_0)P_{a_1l_1d_0,d_1}(1/2)e^{-d/x}.$$

Moving the line of integration to $\Re(w) = -3/4-\epsilon$, for $\epsilon > 0$, we pick up from the pole at $w = 1$ a polynomial of the form $x(A_0\log x + C_M)$, where the constants $A_0, C_M$ are computable and

$$A_0 = \lim_{s\to 1/2}(s-1/2)R_M(s;\chi_{a_1l_1}) = \prod_{p|M}(1-p^{-1})R_M\zeta_M(2). \tag{3.25}$$

Passing the pole at w = 3/4 we pick up an additional contribution of $C'_M x^{3/4}$ and at $w = 0$ we get a contribution of $C''_M$. The integral at $\Re(w) = -3/4-\epsilon$ converges absolutely by the upper bound estimate of Proposition 3.6 (as $\Gamma(w)$ decays exponentially), and contributes an error on the order of $x^{-3/4-\epsilon}$. This completes the proof of Theorem 3.8.

The remainder of Theorem 0.3 follows after taking a linear combination of different $\chi_{a_1l_1}, \chi_{a_2l_2}$ to sieve out $d_0$ with particular quadratic residue restrictions. As the leading term in the mean value estimate is independent of $a_1l_1$ and vanishes if $a_2l_2 \neq 1$, for any class of $d$ satisfying the restrictions of Theorem 0.3 there will be a nonzero leading term in a sum over this restricted class.

It follows from the explicit description (2.1–2.3) that $P_{a_1l_1d_0,d_1}(1/2) \ll_{d_0,\epsilon} d_1^{2\eta+\epsilon}$ where, as mentioned previously, $c(m) \ll_\epsilon m^{\eta+\epsilon}$ is any valid upper bound for the coefficients of the $GL(3)$ form $f$. As a consequence, the correction factors $P_{a_1l_1d_0,d_1}(1/2)$ attached to a single $d_0$ can only add a contribution on the order of $x^{1/2+\eta+\epsilon}$. As the bound $\eta = 2/5$ is true, and $1/2+2/5 < 1$, there must be nonvanishing for infinitely many distinct $d_0$. □

Let us remark that since the symmetric square $L$-function has a pole, the sum of the quadratic twists is on the order of $x\log x$, as we have seen in Theorem 3.8, with no conditions at all on the signs of the functional equations. By contrast, if there were no pole, then the corresponding sums would be on the order of $x$, and could conceivably lack a main term if two colliding main terms had opposite signs.

Thus to obtain a nonvanishing theorem the precise analysis in that case is more delicate, and we do not carry it out here.

---

## REFERENCES


[BFH] D. Bump, S. Friedberg and J. Hoffstein, On some applications of automorphic forms to number theory, *Bulletin of the A.M.S.* **33** (1996), 157–175.

[BG] D. Bump and D. Ginzburg, Symmetric square L-functions on GL(r), *Ann. of Math.* **136** (1992), 137–205.

[DGH] A. Diaconu, D. Goldfeld and J. Hoffstein, Multiple Dirichlet series and moments of zeta and L-functions (to appear in *Compositio Math.*).

[FF1] B. Fisher and S. Friedberg, Double Dirichlet series over function fields (to appear in *Compositio Math.*).

[FF2] ———, Sums of twisted GL(2) L-functions over function fields, *Duke Math. J.* **117** (2003), 543–570.

[FH] S. Friedberg and J. Hoffstein, Nonvanishing theorems for automorphic L-functions on GL(2), *Ann. of Math.* **142** (1995), 385–423.

[GJ] S. Gelbart and H. Jacquet, A relation between automorphic representations of GL(2) and GL(3), *Ann. Sci. École Norm. Sup.* (4) **11** (1978), 471–542.

[GRS] D. Ginzburg, S. Rallis and D. Soudry, On explicit lifts of cusp forms from $GL_m$ to classical groups, *Ann. of Math.* (2) **150** (1999), 807–866.

[Ho] L. Hörmander, *An introduction to complex analysis in several variables* (Third edition), North-Holland Publishing Co., Amsterdam, 1990.

[JPS1] H. Jacquet, I. Piatetskii-Shapiro and J. Shalika, Automorphic forms on GL(3), I and II. *Ann. of Math.* **109** (1979), 169–212, 213–258.

[JPS2] ———, Rankin-Selberg convolutions, *Amer. J. Math.* **105** (1983), 367–464.

[JS1] H. Jacquet and J. Shalika, A non-vanishing theorem for zeta functions of $GL_n$, *Invent. Math.* **38** (1976), 1–16.

[JS2] ———, *Rankin-Selberg convolutions: Archimedean theory,* in Festschrift in honor of I. I. Piatetski-Shapiro on the occasion of his sixtieth birthday, Part I (Ramat Aviv, 1989) (1990), 125–207.

[KS] H. Kim and F. Shahidi, Cuspidality of symmetric powers with applications, *Duke Math. J.* **112** (2002), 177–197.

[L] R. Langlands, *Euler Products,* Yale Mathematical Monographs, 1971.

[PP] S. Patterson and I. Piatetski-Shapiro. The symmetric-square L-function attached to a cuspidal automorphic representation of $GL_3$, *Math. Ann.* **283** (1989), 551–572.

[Sh] F. Shahidi, The functional equation satisfied by certain L-functions, *Compositio Math.* **37** (1978), 171–207.

[Si] C. L. Siegel, Die Funktionalgleichungen einiger Dirichletscher Reihen, *Math. Zeitschrift* **63** (1956), 363–373.

CHAPTER 8

# HARMONIC ANALYSIS OF THE SCHWARTZ SPACE OF $\Gamma\backslash SL_2(\mathbb{R})$

By BILL CASSELMAN

---

*To Joe Shalika with fond memories of a fruitful if brief collaboration*

This is the fourth in a series of papers (the earlier ones are [Casselman:1989], [Casselman:1993] and [Casselman:1999]) intended to present eventually a new way of proving, among other things, the well known results of Chapter 7 of [Langlands:1977] on the completeness of the spectrum arising from cusp forms and Eisenstein series. That Langlands' results have lasted for nearly 40 years without major improvements is testimony to their depth, but—to (mis)quote Peter Sarnak, who had some recent work of Joseph Bernstein's in mind—it is time to reconsider the theory.

In the first of this series of papers I attempted to pursue with some force an idea apparently due originally to Godement—that from an analyst's point of view the theory of automorphic forms is essentially the study of the Schwartz space of $\Gamma\backslash G$ and its dual. In the next two, I looked at subgroups of $SL_2(\mathbb{R})$ in this perspective. In one of these two I attempted to explain in terms of tempered distributions certain features of the theory—integral formulas such as that for the volume of $\Gamma\backslash G$ and the Maass-Selberg formula—which might have seemed up to then coincidental. In the other I found a new derivation of the Plancherel measure in the case of rank one groups.

This is presumably the last of the series in which I try to explain how a few new ideas, tailored principally to the case of higher rank, may be applied in the simplest case, that of $SL_2(\mathbb{R})$. In this paper, the principal result will be a theorem of Paley-Wiener type for the Schwartz space $\mathcal{S}(\Gamma\backslash SL_2(\mathbb{R}))$, from which the completeness theorem (due originally in this case, I imagine, to Selberg) follows easily.

Paley-Wiener theorems of this sort have been proven before. The earliest result that I am aware of can be found in the remarkable paper [Ehrenpreis-Mautner:1959]. Ehrenpreis and Mautner defined the Schwartz space of $SL_2(\mathbb{Z})\backslash\mathcal{H}$ (where $\mathcal{H}$ is the upper half-plane) and characterized functions in it by their integrals against cusp forms and Eisenstein series. Their formulation and their proof both depended strongly on properties of the Riemann $\zeta$-function, and it was not at all apparent how to generalize their results to other than congruence subgroups. In fact, their dependence on properties of $\zeta(s)$ disguised the essentially simple nature of the problem. I was able to find a generalization of their result in [Casselman:1984], in the course of trying to understand the relationship between Paley-Wiener theorems and cohomology. In spite of the title of that paper, the arguments there are valid for an arbitrary arithmetic subgroup acting on $\mathcal{H}$, and indeed only a few slight modifications would be required in order to deal with arbitrary arithmetic groups of rational rank one.

In this paper, I will prove a slightly stronger result than that in [Casselman: 1984], but by methods which I have developed in the meantime to apply also to groups of higher rank. The point is not so much to prove the new result itself, which could have been done by the methods of [Casselman:1984], but to explain how the new methods work in a simple case.

What is new here? In the Paley-Wiener theorem I envisage in general, a crucial role is played by a square-integrability condition on the critical line. In the earlier work, I followed Langlands' argument in shifting contours *in towards* the critical line to deal with square-integrability before I moved contours *out from* the critical line in order to derive estimates on the growth of certain functions near a cusp. This duplication of effort was annoying. Since then, in [Casselman:1999], I have been able to obtain with no contour movement a Plancherel theorem which implies the square-integrability condition directly. Both here and earlier I move contours in order to evaluate a certain constant term. The most difficult point in this is to take the first nearly infinitesimal step off the critical line. In the earlier paper I used a very special calculation (the Maass-Selberg formula) to do that. That argument, although surprisingly elementary (depending only on the integrability of $1/\sqrt{x}$ near $x = 0$) will unfortunately not work in all situations which arise for groups of higher rank. In this paper I replace that argument by a very general one, one closely related to a more or less standard one in the theory of Laplace transforms of distributions.

Roughly speaking, the arguments of this series differ from Langlands' in that whereas he moved contours to evaluate the inner product of two Eisenstein series of functions of compact support, I move them to evaluate a constant term. There are many virtues to this new technique, but most of them will appear clearly only for groups of higher rank. One virtue, likely to be appreciated by those familiar with Langlands' work, is that in the new arguments each Eisenstein series residue actually contributes to the spectrum, whereas in Langlands' argument (illustrated by his well known example of $G_2$) there occurs a certain complicating cancellation of residues which makes it difficult to understand their significance. I would like to think that the new arguments will eventually make it possible to calculate residues of Eisenstein series by computer, something not easy to see how to do at the moment.

## 0. Introduction.

Let

$$G = \mathrm{SL}_2(\mathbb{R})$$

$$P = \text{the subgroup of upper triangular matrices in } G$$

$$N = \text{the subgroup of unipotent matrices in } P$$

$A =$ the group of positive diagonal matrices in $G$, which may be identified with the multiplicative group $\mathbb{R}^{\text{pos}}$

$K =$ the maximal compact subgroup of rotation matrices

$$\begin{bmatrix} \cos\theta & -\sin\theta \\ \sin\theta & \cos\theta \end{bmatrix}.$$

I assume $\Gamma$ to be a discrete subgroup of $\mathrm{SL}_2(\mathbb{R})$ of "arithmetic type." In addition, a few extra conditions will be put on $\Gamma$ in order to simplify the argument without significant loss of generality. The precise assumptions we make on $\Gamma$ are:

- The group $\Gamma$ has a single cusp at $\infty$;
- the intersection $\Gamma \cap P$ consists of all matrices of the form

$$\begin{bmatrix} \pm 1 & n \\ 0 & \pm 1 \end{bmatrix}$$

where $n$ varies over all of $\mathbb{Z}$.

The effect of these assumptions is to allow a reasonable simplification in notation, without losing track of the most important ideas. Of course there is at least one group satisfying these conditions, namely $\mathrm{SL}_2(\mathbb{Z})$.

Let $\mathcal{H}$ be the upper half-plane $\{z \in \mathbb{C} \mid \Im(z) > 0\}$. The group $G$ acts on it on the left:

$$g = \begin{bmatrix} a & b \\ c & d \end{bmatrix} : z \longmapsto \frac{az + b}{cz + d}.$$

This action preserves the non-Euclidean metric $(dx^2 + dy^2)/y^2$ and the non-Euclidean measure $dx\,dy/y^2$ on $\mathcal{H}$.

The group $\Gamma \cap P$ stabilizes each domain $\mathcal{H}_T = \{y \geq T\}$, and for $T$ large enough the projection from $\Gamma \cap P\backslash\mathcal{H}_T$ to $\Gamma\backslash\mathcal{H}$ is injective. Its image is a neighborhood of the cusp $\infty$. That $\Gamma$ has a single cusp means that the complement of this image is compact.

The isotropy subgroup of $i$ in $\mathcal{H}$ is the subgroup $K$, and the quotient $G/K$ may therefore be identified with $\mathcal{H}$. Let $G_T$ be the inverse image of $\mathcal{H}_T$, which consists of those $g$ in $G$ with an Iwasawa factorization

$$g = \begin{bmatrix} 1 & x \\ 0 & 1 \end{bmatrix} \begin{bmatrix} a & 0 \\ 0 & a^{-1} \end{bmatrix} k$$

where $a^2 > T$ and $k$ lies in $K$. For large $T$ the quotient $\Gamma \cap P\backslash G_T$ embeds into $\Gamma\backslash G$ with compact complement.

In the rest of this paper, $T$ will be assumed to be large enough so that this embedding occurs.

The area of $\Gamma \cap P\backslash\mathcal{H}_T$ can be calculated explicitly, and it is finite. Because the complement is compact, the area of $\Gamma\backslash\mathcal{H}$ and the volume of $\Gamma\backslash G$ are both finite as well. Define $\delta$ to be the function pulled back from the $y$-coordinate on $\mathcal{H}$. In terms of the Iwasawa factorization $G = NAK$, $\delta(nak) = \delta(a) = |\det \mathrm{Ad}_{\mathfrak{n}}(a)|$, where

$$\delta(a) = x^2 \text{ if } a = \begin{bmatrix} x & 0 \\ 0 & x^{-1} \end{bmatrix}.$$

The character $\delta$ is also the modulus character of $P$.

The length of the non-Euclidean circle around $i$ and passing through $iy$ is $(y - y^{-1})/2$. This means that in terms of the Cartan factorization $G = KAK$ we have an integral formula

$$\int_G f(g)\,dg = \int_K \int_A \int_K f(k_1 a k_2) \left( \frac{\delta(a) - \delta^{-1}(a)}{2} \right) dk_1\, da\, dk_2$$

with a suitable measure assigned to $K$. On $G$ we define the norm

$$\|k_1 a k_2\| = \max |\delta(a)|, |\delta(a)|^{-1}.$$

This is the same as

$$\sup_{\|v\|=1} \|gv\| \qquad (v \in \mathbb{R}^2)$$

and satisfies the inequality

$$\|gh\| \le \|g\|\,\|h\|.$$

The function $\|g\|^{-(1+\epsilon)}$ is then integrable on $G$ for $\epsilon > 0$.

A function $f$ on $\Gamma\backslash G$ is said to be of *moderate growth* at $\infty$ if $f = O(\delta^m)$ on the regions $G_T$ for some integer $m > 0$, and *rapidly decreasing* at $\infty$ if it is $O(\delta^{-m})$ for all $m$. The *Schwartz space* $\mathcal{S}(\Gamma\backslash G)$ is that of all smooth right-$K$-finite functions $f$ on $\Gamma\backslash G$ with all $R_X f$ $(X \in U(\mathfrak{g}))$ rapidly decreasing at $\infty$. Because of the condition of $K$-finiteness, any function in $\mathcal{S}(\Gamma\backslash G)$ may be expressed as a finite sum of components transforming on the right by a character $\chi$ of $K$:

$$\begin{aligned} \mathcal{S}(\Gamma\backslash G) &= \bigoplus \mathcal{S}(\Gamma\backslash G)_\chi \\ \mathcal{S}(\Gamma\backslash G)_\chi &= \{ f \in \mathcal{S}(\Gamma\backslash G) \mid f(gk) = \chi(k) f(g) \text{ for all } k \in K, g \in G \}. \end{aligned}$$

If $\chi = 1$, we are looking at functions on $\Gamma\backslash\mathcal{H}$.

The problem that this paper deals with is how to characterize the functions in the Schwartz space by their integrals against various automorphic forms, and especially Eisenstein series. There are technical reasons why the Paley-Wiener theorem for $\mathcal{S}(\Gamma\backslash\mathcal{H})$ is simpler than the one for $\mathcal{S}(\Gamma\backslash G)$, and for that reason I will discuss the first case in detail, then go back and deal with the extra complications needed to deal with $\mathcal{S}(\Gamma\backslash G)$. But in order to give an idea of what's going on, I'll explain in the next section a few of the simplest possible Paley-Wiener theorems.

**The simplest Paley-Wiener theorems.** It will be useful to keep in mind a few elementary theorems of the kind we are looking for.

(1) Define the Schwartz space $\mathcal{S}$ of the group $A$, identified here with the multiplicative group of positive real numbers $\mathbb{R}^{\text{pos}}$, to be made up of those smooth functions $f(x)$ on $A$ satisfying the condition that it and all its derivatives vanish rapidly at 0 and $\infty$ in the sense that

$$\left|f^{(n)}(x)\right| = O(x^m)$$

for all non-negative integers $n$ and all integers $m$, whether positive or negative. Then for all $s$ in $\mathbb{C}$ we can define the Fourier transform

$$F(s) = \widehat{f}(s) = \int_0^\infty f(x)x^{-s}\,\frac{dx}{x}.$$

It turns out to be holomorphic in all of $\mathbb{C}$, and since the Fourier transform of the multiplicative derivative $x\,df/dx$ is $s\widehat{f}$ it satisfies the growth condition

$$F(\sigma + it) = O\left(\frac{1}{1+|t|^m}\right)$$

for all $m > 0$, uniformly in vertical bands of finite width. Conversely, if $F(s)$ is any entire function satisfying these growth conditions, then for any real $\sigma$

$$f(x) = \frac{1}{2\pi i}\int_{\sigma - i\infty}^{\sigma + i\infty} F(s)x^s\,dx$$

will be a function in $\mathcal{S}(A)$, independent of $\sigma$, whose Fourier transform is $F$. The proof depends on a clearly justifiable shift of contour of integration.

(2) A second result will turn out to look even more similar to that for arithmetic quotients. Define $L_{\mathcal{S}}^{2,\infty}(A)$ to be the space of all smooth functions $f$ on $(0, \infty)$ such that (a) $f$ and all its derivatives vanish of infinite order at 0; (b) $f$ and all its multiplicative derivatives are square-integrable on $A$. Condition (a) implies that the Fourier transform is defined and holomorphic in the region $\Re(s) > 0$. On the other hand, condition (b) implies that the Fourier transform of $f$ on the line $\Re(s) = 0$ exists as a square-integrable function. The relationship between the two definitions of $F(s)$ on $\Re(s) = 0$ and $\Re(s) > 0$ is that uniformly on bounded horizontal strips the function $F(\sigma + it)$ approaches the function $F(it)$ in the $L^2$-norm. In these circumstances we have the following result:

THEOREM. *If $f(x)$ lies in $L_{\mathcal{S}}^{2,\infty}$ then its Fourier transform $F(s)$ satisfies the following conditions:*

- *$F(s)$ is holomorphic in the half plane $\Re(s) > 0$;*
- *it satisfies the condition*

$$F(\sigma + it) = O\left(\frac{1}{\sqrt{\sigma}(1+|t|)^n}\right)$$

*for $\sigma > 0$ and all $n > 0$, uniformly on horizontally bounded vertical strips;*

• *the restriction of $F$ to $\Re(s) = 0$ is square-integrable, and the weak limit of the distributions*

$$F_\sigma(it) = F(\sigma + it)$$

*as $\sigma \to 0$.*

*Conversely, if $F(s)$ is a function satisfying these conditions then it is the Fourier transform of the function*

$$f(x) = \frac{1}{2\pi i} \int_{\sigma - i\infty}^{\sigma + i\infty} F(s) x^s \, ds$$

*which does not depend on the choice of $\sigma > 0$, and which lies in $L_S^{2,\infty}$.*

The natural proof of this relies on results from the last section of this paper, and much of the argument duplicates what I shall say about the analogous (and more difficult) result for the upper half plane. I leave it as an exercise.

**Quotients of the upper half-plane.** For the group $\mathbb{R}^{\mathrm{pos}}$, the results stated in the previous section are just some of many analogous results, most notably one characterizing functions of compact support by their Fourier transforms. But for quotients of symmetric spaces by arithmetic subgroups I do not know whether a result for functions of compact support is possible even in principle. A Paley-Wiener theorem for functions of rapid decrease may be the only natural one to consider.

In this section I'll explain the Paley-Wiener theorem for $\mathcal{S}(\Gamma\backslash\mathcal{H})$. The definition of the space $\mathcal{S}(\Gamma\backslash\mathcal{H})$ involves lifting a function $f$ on $\Gamma\backslash\mathcal{H}$ to a function $F$ on $\Gamma\backslash G$ and then considering the right derivatives $R_X F$. But the functions in $\mathcal{S}(\Gamma\backslash\mathcal{H})$ may be more concretely identified with those smooth functions on $\Gamma\backslash\mathcal{H}$ satisfying the condition that

$$\Delta^n f = O(y^{-m})$$

for all positive integers $n$ and $m$, where $\Delta$ is the non-Euclidean Laplace operator

$$\Delta = y^2 \left( \frac{\partial^2}{\partial x^2} + \frac{\partial^2}{\partial y^2} \right).$$

This definition of the Schwartz space is the one used by Ehrenpreis and Mautner, and in my 1984 paper (Proposition 2.3) I showed that this notion is equivalent to the one given earlier. That equivalence will not play a role here except in so far as it ties the result of Ehrenpreis and Mautner to mine.

Any smooth function $f(z)$ on $\Gamma\backslash\mathcal{H}$ may be expanded in a Fourier series

$$f(x + iy) = \sum_{-\infty}^{\infty} f_n(y) e^{2\pi i n x}.$$

If $f$ is any smooth function on $\Gamma\backslash\mathcal{H}$ which is of *uniform moderate growth* in the sense that for some fixed $m$

$$\Delta^n f = O(y^m)$$

for all $n > 0$, then all coefficient functions $f_n(y)$ for $n \neq 0$ vanish rapidly as $y \to \infty$, and more generally the difference between $f(y)$ and $f_0(y)$ also vanishes rapidly. In other words, the asymptotic behavior of $f(y)$ as $y \to \infty$ is controlled by the *constant term* $f_0(y)$. Furthermore, as we shall see later, the Schwartz space decomposes into a sum of two large pieces—the *cuspidal* component, that of functions whose constant terms vanish identically, and the *Eisenstein* component orthogonal to the cuspidal one. The cuspidal component is a discrete sum of eigenspaces of the Laplace operator, and is of no particular interest in this discussion.

The spectrum of $\Delta$ is continuous on the Eisenstein component. The functions which for $\Gamma\backslash\mathcal{H}$ play the role of the characters $x^s$ on $\mathbb{R}^{\mathrm{pos}}$ are the *Eisenstein series*. For every $s$ with $\Re(s) > 1$ the series

$$E_s(z) = \sum_{\Gamma\cap P\backslash\Gamma} y(\gamma(z))^s$$

converges to an eigenfunction of $\Delta$ on $\Gamma\backslash\mathcal{H}$, with eigenvalue

$$\Delta(s) = s(s-1) = (s-1/2)^2 - 1/4.$$

When $\Gamma = \mathrm{SL}_2(\mathbb{Z})$, for example, this series was first defined by Maass, and can be expressed more explicitly as

$$E_s(z) = \sum_{c>0,\ \gcd(c,d)=1} \frac{y^s}{|cz+d|^{2s}}.$$

For all $\Gamma$, the function $E_s$ continues meromorphically in $s$ to all of $\mathbb{C}$. In the right-hand half-plane $\Re(s) \geq 1/2$ there is always a simple pole at $s = 1$, and there may be a few more simple poles on $(1/2, 1)$. The constant term of $E_s$ is of the form

$$y^s + c(s)y^{1-s}$$

where $c(s)$ is a meromorphic function on $\mathbb{C}$. For $\Gamma = \mathrm{SL}_2(\mathbb{Z})$

$$c(s) = \frac{\xi(2s-1)}{\xi(2s)}, \qquad \xi(s) = \pi^{-s/2}\Gamma(s/2)\zeta(s).$$

In this case the behaviour of $E_s$ for $\Re(s) < 1/2$ is therefore related to the Riemann hypothesis, and ought to be considered, whenever possible, as buried inside an impenetrable box. The function $E_s$ satisfies the functional equation

$$E_s = c(s)E_{1-s}$$

so that $s$ and $1-s$ contribute essentially the same automorphic forms to $\Gamma\backslash\mathcal{H}$. From this equation for $E_s$ it follows that $c(s)$ satisfies the functional equation

$$c(s)c(1-s) = 1.$$

In the region $\Re(s) > 1/2$, $s \notin (1/2, 1]$ the Eisenstein series can be constructed by a simple argument relying only on the self-adjointness of the operator $\Delta$ on $\Gamma\backslash\mathcal{H}$. The rough idea is this:

Let $\chi(y)$ be a function on $(0, \infty)$ which is identically 1 for large $y$, and non-vanishing only for large $y$. The product $\chi(y)y^s$ may be identified with a function $Y_s$ on $\Gamma\backslash\mathcal{H}$. Choose $s$ such that $\Re(s) > 1/2$, $s \notin (1/2, 1]$, and let $\lambda = s(s-1)$. Then $X_s = (\Delta - \lambda)Y_s$ will have compact support on $\Gamma\backslash\mathcal{H}$, since $\Delta y^s = \lambda y^s$. For $\lambda \notin (-\infty, 0]$ (the spectrum of $\Delta$) let

$$F_s(z) = -(\Delta - \lambda)^{-1} X_s.$$

Then

$$E_s(z) = F_s(z) + Y_s(z).$$

In other words, for $s$ in this region the function $E_s$ is uniquely determined by the conditions that (a) $\Delta E_s = \lambda E_s$ and (b) $E_s - y^s$ is square-integrable near $\infty$. (This is explained in more detail in [Colin de Verdière:1981].) The theory of self-adjoint operators also guarantees that

$$\begin{aligned}
\|\Delta - \lambda\|^2 &= \|\Delta - \Re(\lambda)\|^2 + |\Im(\lambda)|^2 \\
\|\Delta - \lambda\| &\geq |\Im(\lambda)| \\
&= |2\sigma t| \quad (s = 1/2 + \sigma + it) \\
\|\Delta - \lambda\|^{-1} &\leq |2\sigma t|^{-1},
\end{aligned}$$

which implies that $\|F_s\| = O(|2\sigma t|^{-1})$.

For $T$ large enough we can define the *truncation* of an automorphic form $F(z)$ in the region $y \geq T$. On the quotient $\Gamma\backslash\mathcal{H}_T$ the truncation $\Lambda^T F$ is the difference between $F$ and its constant term. Because the asymptotic behavior of $F$ is controlled by its constant term, this is always square-integrable. For Eisenstein series there exists the explicit *Maass-Selberg formula* for the inner product of two truncations. For generic $s$ and $t$ it asserts that

$$\left\langle \Lambda^T E_s, \Lambda^T E_t \right\rangle = \frac{T^{s+t-1} - c(s)c(t)T^{1-s-t}}{s+t-1} - \frac{c(s)T^{1-s+t} - c(t)T^{1-t+s}}{s-t}.$$

Formally, the expression on the right is

$$\int_0^T (y^s + c(s)y^{1-s})(y^t + c(t)y^{1-t})\, y^{-2}\, dy.$$

This apparent accident is explained in [Casselman:1993]. When $s = 1/2 + \sigma + i\tau$ and $t = \bar{s}$ it becomes

$$\|\Lambda^T E_s\|^2 = \frac{T^{2\sigma} - |c(s)|^2 T^{-2\sigma}}{2\sigma} - \frac{c(s)T^{-2i\tau} - c(t)T^{2i\tau}}{2i\tau}.$$

This formula makes more precise the idea that the behavior of $E_s$ is determined by that of $c(s)$, and *vice versa*. That this must always be positive, for example, implies that $|c(s)|$ must be bounded at $\pm i\infty$ in the region $\sigma > 0$, and that the poles of $E_s$ and $c(s)$ have to be simple in that region. (See [Langlands:1966], or the proof of Proposition 3.7 in [Casselman:1984] for more detail.)

The *Fourier-Eisenstein transform* of $f$ in $\mathcal{S}(\Gamma\backslash\mathcal{H})$ is

$$F(s) = \widehat{f}(s) = \int_{\Gamma\backslash\mathcal{H}} f(z) E_{1-s}(z) \frac{dx\,dy}{y^2}.$$

It follows immediately from properties of $E_s$ that

(PW1) *The function $F(s)$ satisfies the functional equation*

$$F(1-s) = c(s)F(s).$$

(PW2) *The function $F(s)$ is meromorphic everywhere in $\mathbb{C}$, holomorphic in the half-plane $\Re(s) \le 1/2$ except for possible simple poles in $[0, 1/2)$ corresponding to those of $E_{1-s}$.*

There are also a few other significant and more subtle properties of $F(s)$.

(PW3) *The function $F(s)$ is square-integrable on $\Re(s) = 1/2$.*

(PW4) *In any region*

$$\sigma_0 < \Re(s) < 1/2, \qquad |\Im(s)| > \tau$$

*we have for all $m > 0$*

$$|F(s)| = O\left(\frac{1}{|1/2-\sigma|\,|t|^m}\right) \quad (s = \sigma + it).$$

The first follows from the following result, a Plancherel formula for the critical line, which is far more basic:

*For $\Phi(s)$ a function of compact support on the line $\Re(s) = 1/2$, the integral*

$$E_\Phi(z) = \frac{1}{2\pi i}\int_{1/2-i\infty}^{1/2+i\infty} \Phi(s)E_s(z)\,ds$$

*defines a square-integrable function on $\Gamma\backslash\mathcal{H}$ with*

$$\frac{1}{2}\,\|E_\Phi\|^2 = \frac{1}{2\pi i}\int_{1/2-i\infty}^{1/2+i\infty} |\Phi(s)|^2\,ds.$$

This is well known. The usual proof (as in [Langlands:1966]) relies on contour movement, but in [Casselman:1999] it is proven directly. At any rate, given this, we can verify property (PW3). First of all it implies that $E_\Phi$ can be defined as an

$L^2$ limit for arbitrary functions in $L^2(1/2 + i\mathbb{R})$. Second, for $f$ in $\mathcal{S}(\Gamma\backslash\mathcal{H})$ we can calculate that

$$\begin{aligned} \|f\| \, \|E_\Phi\| &\geq \langle f, E_\Phi \rangle \\ &= \frac{1}{2\pi i} \int_{1/2-i\infty}^{1/2+i\infty} \Phi(s) \, \langle f, E_s \rangle \, ds \\ &= \frac{1}{2\pi i} \int_{1/2-i\infty}^{1/2+i\infty} \Phi(s) \widehat{f}(1-s) \, ds \end{aligned}$$

for any square-integrable $\Phi$, which implies that $\widehat{f}$ itself is square-integrable.

As for property (PW4), it can be proven either from the spectral inequality mentioned above, or from the Maass-Selberg formula. This property is used in moving contours of integration; the second proof, which asserts a more precise result than the other, allows an elementary argument in doing this (see [Casselman:1984]), but in higher rank becomes invalid. The first is therefore preferable. In both proofs, we begin by writing

$$\langle f, E_s \rangle = \langle f, \Lambda^T E_s \rangle + \langle f, C^T E_s \rangle$$

and arguing separately for each term. For the first term, use the spectral construction of $E_s$ described earlier. An estimate for the second term follows easily from an argument about the multiplicative group.

THEOREM. *Let $F(s)$ be any function on $\mathbb{C}$ such that all $\Delta^n(s)F(s)$ satisfy conditions (PW1)–(PW4), and for each pole $s$ in $[0, 1/2)$ let $F^\#(s)$ be the residue of $F$ there. Then*

$$f(z) = -\sum F^\#(s) E_s + \frac{1}{2} \frac{1}{2\pi i} \int_{1/2-i\infty}^{1/2+i\infty} F(s) E_s(z) \, ds$$

*lies in $\mathcal{S}(\Gamma\backslash\mathcal{H})$ and has Fourier-Eisenstein transform $F$.*

Note that because $c(s)c(1-s) = 1$, if $c$ has a pole at $1-s$ then $c(s) = 0$, $E_s$ is well defined, and its constant term is exactly $y^s$. The integral is to be interpreted as the limit of finite integrals

$$\frac{1}{2\pi i} \int_{1/2-iT}^{1/2+iT} F(s) E_s \, ds$$

which exists as a square-integrable function on $\Gamma\backslash\mathcal{H}$ by the Plancherel formula explained above. In fact, it lies in the space $A_{\mathrm{umg}}(\Gamma\backslash\mathcal{H})$. This is proven directly in [Casselman:1984], but follows easily from extremely general reasoning about $L^{2,\infty}(\Gamma\backslash G)$ (Theorem 1.16 and Proposition 1.17 of [Casselman:1989]). This argument is recalled in a simplified form later in this paper.

What this means is that in order to determine whether $f(x)$ lies in $\mathcal{S}(\Gamma\backslash\mathcal{H})$ we can look at its constant term.

The constant term of the integral is

$$\frac{1}{2\pi i}\int_{1/2-i\infty}^{1/2+i\infty}\left[\frac{F(s)y^s + c(s)F(s)y^{1-s}}{2}\right] ds$$

(suitably interpreted as a limit) which is equal to

$$\frac{1}{2\pi i}\int_{1/2-i\infty}^{1/2+i\infty}\left[\frac{F(s)y^s + c(1-s)F(1-s)y^s}{2}\right] ds = \frac{1}{2\pi i}\int_{1/2-i\infty}^{1/2+i\infty} F(s)y^s\, ds$$

by (PW2). The most difficult step in the whole proof is to justify replacing the integral

$$\frac{1}{2\pi i}\int_{1/2-i\infty}^{1/2+i\infty} F(s)y^s\, ds$$

by the integral

$$\frac{1}{2\pi i}\int_{\sigma-i\infty}^{\sigma+i\infty} F(s)y^s\, ds$$

for some number $\sigma$ very close to $1/2$. This can be done by the results in the final section of this paper. Once this step has been taken, the growth conditions on $F(s)$ in vertical bands allow us to move arbitrarily far to the left, picking up residues as we go. Recall that the constant term of $E_s$ is $y^s$ at a pole of $F(s)$. These residues cancel out with the residues in the formula for $f(z)$. Therefore the constant term of $f(z)$ is equal to

$$\frac{1}{2\pi i}\int_{\sigma-i\infty}^{\sigma+i\infty} F(s)y^s\, ds$$

for arbitrary $\sigma \ll 0$, which implies that it vanishes rapidly as $y \to \infty$. A classical result from the theory of the Laplace transform finishes off the Proposition.

**The constant term.** In this section, I begin consideration of $\Gamma\backslash G$ instead of $\Gamma\backslash\mathcal{H}$. Some points are simpler, and in fact some of the claims for $\Gamma\backslash\mathcal{H}$ are best examined in the current context. The principal complication is that notation is more cumbersome.

For any reasonable function $f$ on $\Gamma\backslash G$ define its *constant term* to be the function on $N(\Gamma\cap P)\backslash G$ defined by the formula

$$f_P(g) = \int_{\Gamma\cap N\backslash N} f(xg)\, dx.$$

If $f$ lies in $\mathcal{S}(\Gamma\backslash G)$ then $f_P$ will be bounded on all of $N(\Gamma\cap P)\backslash G$ and in addition satisfy an inequality

$$R_X f(g) = O(\delta(g)^{-m})$$

on $G_T$, for all $X$ in $U(g)$ and $m > 0$.

Define $A_{\mathrm{umg}}(\Gamma\backslash G)$ to be the space of all functions of *uniform* moderate growth on $\Gamma\backslash G$—those smooth functions $F$ for which there exists a single $m > 0$ with

$$|R_X F(g)| = O(\delta(g)^m)$$

on $G_T$, for all $X$ in $U(\mathfrak{g})$.

If $F$ lies in $A_{\mathrm{umg}}(\Gamma\backslash G)$ and $f$ lies in $\mathcal{S}(\Gamma\backslash G)$ then the product $Ff$ will lie in $\mathcal{S}(\Gamma\backslash G)$, and hence may be integrated. The two spaces are therefore in duality. It is shown in [Casselman:1989] that the space $A_{\mathrm{umg}}(\Gamma\backslash G)$ may be identified with the Gårding subspace of the dual of $\mathcal{S}(\Gamma\backslash G)$, the space of *tempered distributions* on $\Gamma\backslash G$.

For large $T$, the *truncation* $\Lambda^T F$ of a continuous function $F$ on $\Gamma\backslash G$ at $T$ is what you get from $F$ by chopping away its constant term on $G_T$. More precisely, if $\Phi$ is any function on $N(\Gamma\cap P)\backslash G$ define $C^T\Phi$ to be the product of $\Phi$ and the characteristic function of $G_T$, and then for $F$ on $\Gamma\backslash G$ set

$$C^T F(g) = \sum_{\Gamma\cap P\backslash\Gamma} C^T F_P(\gamma g)$$

$$\Lambda^T F = F - C^T F.$$

The sum $F = \Lambda^T F + C^T F$ is orthogonal.

One of the basic results in analysis on $\Gamma\backslash G$ is that

*if $F$ lies in $A_{\mathrm{umg}}(\Gamma\backslash G)$ then $\Lambda^T F$ is rapidly decreasing at $\infty$.*

**Analysis on $\mathbf{N(\Gamma\cap P)\backslash G}$.** The space $N(\Gamma\cap P)\backslash G$ plays the same role for $\Gamma\backslash G$ that $A \cong N(\Gamma\cap P)\backslash G/K$ plays for $\Gamma\backslash\mathcal{H}$. And analysis on $N(\Gamma\cap P)\backslash G$ still looks much like analysis on the multiplicative group $\mathbb{R}^{\mathrm{pos}}$. One can be phrased literally in terms of the other since we can look at irreducible $K$-eigenspaces, and $N\backslash G/K \cong A/\{\pm 1\} \cong \mathbb{R}^{\mathrm{pos}}$.

For each $s$ in $\mathbb{C}$ define the space

$$I_s = \{f \in C^\infty(G) \mid f \text{ is } K\text{-finite, } f(pg) = \delta^s(p)f(g) \text{ for all } p \in P, g \in G\}.$$

Right derivation makes this into the *principal series* representation of $(\mathfrak{g}, K)$ parametrized by the character $p \mapsto \delta^s(p)$. It has a basis made up of functions $f_{n,s}$ where

$$f_{n,s}(pk) = \delta^s(p)\varepsilon^n(k)$$

where

$$\varepsilon\colon \begin{bmatrix} c & -s \\ s & c \end{bmatrix} \longmapsto c + is.$$

If

$$\kappa = \begin{bmatrix} 0 & -1 \\ 1 & 0 \end{bmatrix}$$

$$X_+ = (1/2)\begin{bmatrix} 1 & -i \\ -i & -1 \end{bmatrix}$$

$$X_- = (1/2)\begin{bmatrix} 1 & i \\ i & -1 \end{bmatrix}$$

then on $I_s$

$$R_\kappa f_{n,s} = ni f_{n,s}$$

$$R_{X_+} f_{n,s} = (s + n/2) f_{n+2,s}$$

$$R_{X_-} f_{n,s} = (s - n/2) f_{n-2,s}.$$

These are generators of $U(\mathfrak{g})$, and therefore every element of $U(\mathfrak{g})$ acts on $I_s$ by a polynomial function of $s$.

The representation of $(\mathfrak{g}, K)$ on $I_s$ is irreducible for almost all $s$, and the Casimir operator $\mathfrak{C}$ acts as the scalar $\Delta(s) = s(s-1)$ on it. Elements of $I_0$ may be identified with functions on $\mathbb{P}^1(\mathbb{R})$, those of $I_1$ with smooth 1-densities on $\mathbb{P}^1(\mathbb{R})$. With a suitable choice of measures, the integral formula

$$\int_{N(\Gamma\cap P)\backslash G} f(x)\,dx = \int_{P\backslash G} \bar{f}(x)\,dx$$

is valid, where

$$\bar{f}(x) = \int_A \delta_P^{-1}(a) f(ax)\,da$$

lies in $I_1$.

The product of an element of $I_s$ and one in $I_{1-s}$ lies in $I_1$, and may then be integrated. The space $I_{1-s}$ is therefore the contragredient of $I_s$. If $\Re(s) = 1/2$ so that $s = 1/2 + it$, then $1 - s = 1/2 - it = \bar{s}$; the representation of $(\mathfrak{g}, K)$ on $I_s$ is therefore unitary.

Let

$$\mathcal{I} = \text{ the space of } K\text{-finite functions on } K \cap P\backslash K\,.$$

Since $G = PK$, restriction to $K$ is a $K$-covariant isomorphism of $I_s$ with $\mathcal{I}$. Thus as vector spaces and as representations of $K$, all the $I_s$ may be identified with each other. It therefore makes sense to say that they form a holomorphic family, or that the representation of $\mathfrak{g}$ varies holomorphically with $s$. Restriction to $K$ can be used to define a norm on the $I_s$. For $f$ in $I_s$ with the decomposition $f = \sum f_\chi$ into

$K$-components, define

$$\|f\|^2 = \int_{K\cap P\backslash K} |f(k)|^2\, dk = \sum_\chi \|f_\chi\|^2.$$

For $\Re(s) = 1/2$, $\|f\|$ is the same as the norm induced by the identity of $I_{1-s}$ with the contragredient of $I_s$, the $G$-invariant Hilbert space norm on $I_s$.

Fourier analysis decomposes functions on $N(\Gamma \cap P)\backslash G$ into its components in the spaces $I_s$. As with classical analysis, there are several variants.

**A Paley-Wiener theorem.** Suppose $\varphi$ to be a smooth $K$-finite function on $N(\Gamma \cap P)\backslash G$ which is rapidly decreasing at infinity on $N(\Gamma \cap P)\backslash G$ in both directions, in the sense that for any integer $m$ whatsoever (positive or negative) $R_X\varphi = O(\delta^m)$ for all $X$ in $U(\mathfrak{g})$. Then for any $s$ in $\mathbb{C}$ we can define an element $\widehat{\varphi}_s$ in $I_s$ by the condition

$$\langle \widehat{\varphi}_s, \psi \rangle = \int_{N(\Gamma\cap P)\backslash G} \varphi(x)\psi(x)\, dx$$

for each $\psi$ in $I_{1-s}$. More explicitly, we can write the integral as

$$\int_{N(\Gamma\cap P)\backslash G} \varphi(x)\psi(x)\, dx = \int_{P\backslash G}\int_A \delta_P^{-1}(a)\varphi(ax)\psi(ax)\, da\, dx$$

so that

$$\widehat{\varphi}_s(g) = \int_0^\infty \delta(a)^{-s}\varphi(ag)\, da.$$

The function $\widehat{\varphi}_s$ will determine a section of $I$ over all of $\mathbb{C}$, rapidly decreasing at $\pm i\infty$. We can recover $\varphi$ from the functions $\widehat{\varphi}_s$ by the formula

$$\varphi(g) = \frac{1}{2\pi i}\int_{\Re(s)=\sigma} \widehat{\varphi}_s(g)\, ds$$

for any real number $\sigma$.

If $\varphi$ and $\psi$ are two such functions on $N(\Gamma \cap P)\backslash G$ then their inner product can be calculated from their Fourier transforms by the formula

$$\int_{N(\Gamma\cap P)\backslash G} \varphi(g)\psi(g)\, dg = \frac{1}{2\pi i}\int_{1/2-i\infty}^{1/2+i\infty} \left\langle \widehat{\varphi}_s, \widehat{\psi}_{1-s} \right\rangle ds.$$

The map taking $\varphi$ to $\widehat{\varphi}$ is an isomorphism of $\mathcal{S}(N(\Gamma \cap P)\backslash G)$ with that of all holomorphic sections $\Phi_s$ of $I_s$ over all of $\mathbb{C}$ satisfying the condition that for all $m > 0$ we have

$$\|\Phi_{\sigma+it}\| = O\left(\frac{1}{1+|t|^m}\right)$$

uniformly on horizontally bounded vertical strips.

*The Laplace transform.* Suppose $\varphi$ to be a smooth function on $N(\Gamma \cap P)\backslash G$, such that each right derivative $R_X\varphi$ is bounded overall and rapidly decreasing at $\infty$ (but not necessarily at 0). These conditions are satisfied, for example, by the constant terms of functions in $\mathcal{S}(\Gamma\backslash G)$. Then for any $s$ in $\mathbb{C}$ with $\Re(s) < 1/2$ the integral

$$\widehat{\varphi}_s(g) = \int_0^\infty \delta(a)^{-s}\varphi(ag)\,da$$

converges and defines a function in $I_s$. In other words, we now have a holomorphic section of $I_s$ over the region $\Re(s) < 1/2$, which can reasonably be called the *Laplace transform* of $\varphi$. When only one $K$-component is involved, this amounts to the usual Laplace transform on the multiplicative group $\mathbb{R}^{\text{pos}}$. Standard arguments from the theory of the Laplace transform on the multiplicative group of positive reals then imply that the function $\varphi$ can be recovered from $\widehat{\varphi}$:

$$\varphi(g) = \frac{1}{2\pi i}\int_{\sigma-i\infty}^{\sigma+i\infty} \widehat{\varphi}_s(g)\,ds$$

for all $\sigma < 1/2$. The integral over each line makes sense because under the assumptions on $\varphi$ the magnitude of $\widehat{\varphi}(s)$ decreases rapidly at $\pm i\infty$. In particular, if $\widehat{\varphi}(s)$ vanishes identically, then $\varphi = 0$. This is a consequence of our assumption that $\Gamma \cap P$ contains $\pm 1$—without this assumption we would have to take into account characters of $A$ not necessarily trivial on $\pm 1$.

In particular:

*If $f$ lies in $\mathcal{S}(\Gamma\backslash G)$ and the Laplace transform of $f_P$ vanishes, then so does $f_P$.*

*Square-integrable functions.* The map taking $f$ in from $\mathcal{S}(N(\Gamma \cap P)\backslash G)$ to $\widehat{f}$ extends to an isomorphism of $L^2(N(\Gamma \cap P)\backslash G)$ (square-integrable half-densities) with the space $L^2(1/2 + i\mathbb{R}, \mathcal{I})$ of all square-integrable functions $\Phi$ on $1/2 + i\mathbb{R}$ with values in $\mathcal{I}$, i.e. those such that

$$\frac{1}{2\pi i}\int_{1/2-i\infty}^{1/2+i\infty} \|\Phi(s)\|^2\,ds < \infty.$$

**Eisenstein series.** Suppose $\Phi$ to be an element of $I_s$ with $\Re(s) > 1$. Then the *Eisenstein series*

$$E(\Phi) = \sum_{\Gamma\cap P\backslash\Gamma} \Phi(\gamma g)$$

will converge absolutely to a function of uniform moderate growth—in fact, an automorphic form—on $\Gamma\backslash G$.

Let $\iota_s$ be the identification of $\mathcal{I}$ with $I_s$, extending $\varphi$ on $K \cap P\backslash K$ to $\varphi_s = \iota_s\varphi$ on $N(\Gamma \cap P)\backslash G$ where

$$\varphi_s(pk) = \delta^s(p)\varphi(k).$$

Then the composite

$$E_s(\varphi) = E(\varphi_s)$$

will vary holomorphically for $s$ with $\Re(s) > 1$.

The map

$$E_s\colon\ \mathcal{I} \to A_{\mathrm{umg}}(\Gamma\backslash G)$$

continues meromorphically to all of $\mathbb{C}$, defining where it is holomorphic a $(\mathfrak{g}, K)$-covariant map from $I_s$ to $\mathcal{A}(\Gamma\backslash G)$. It is holomorphic in the region $\Re(s) \geq 1/2$ except for a simple pole at $s = 1$ and possibly a few more simple poles on the line segment $(1/2, 1)$.

The constant term of $E(\varphi_s)$ is for generic $s$ a sum

$$\varphi_s + \tau(\varphi_s)$$

where $\tau$ is a covariant $(\mathfrak{g}, K)$ map from $I_s$ to $I_{1-s}$. Let $\tau_s$ be the composite

$$\tau_s\colon\ \varphi \to \varphi_s \to \tau(\varphi_s)|K.$$

It is meromorphic in $s$. For $\Re(s) > 1/2$ and $s \notin [1/2, 1]$, the Eisenstein series $E(\varphi_s)$ is determined uniquely by the conditions that (1) near $\infty$ it is the sum of $\varphi_s$ and something square-integrable; (2) it is an eigenfunction of the Casimir operator in $U(\mathfrak{g})$. As a result of uniqueness, the Eisenstein series satisfies a *functional equation*

$$E_s(\varphi) = E_{1-s}(\tau_s\varphi).$$

In any event, the operator $\tau_s$ satisfies the condition $\tau_s\tau_{1-s} = 1$, and is a unitary operator when $\Re(s) = 1/2$. When $\varphi \equiv 1$ and $\Gamma = \mathrm{SL}_2(\mathbb{Z})$, as I have already mentioned, $\tau_s(\varphi)$ is related to the Riemann $\zeta$ function. In this case, the functional equation for the Eisenstein series is implied by—but does not imply—that for $\xi(s)$. Poles of $E_s$ in the region $\Re(s) < 1/2$ will in this case arise from zeroes of $\zeta(s)$.

It is not important in this context to know exactly what happens to the left of the critical line $\Re(s) = 1/2$. This is just as well, because this is uncharted—and perhaps unchartable—territory.

The truncation $\Lambda^T E$ of any Eisenstein series $E$ will be square-integrable. There is a relatively simple formula, called the *Maass-Selberg formula,* for the inner product of two of these. For generic values of $s$ and $t$ we have a formal rule

$$\left\langle \Lambda^T \Phi_s, \Lambda^T \Psi_t \right\rangle = -\int_{N(\Gamma\cap P)\backslash G_T} \langle \varphi_s, \psi_t\rangle \, dx$$

where $\Phi_s$ lies in the image of $E_s$, etc., and $\varphi_s$ is its constant term. The integral is defined by analytic continuation and, if necessary, l'Hôpital's rule. If we take $\Psi$ to be the conjugate of $\Phi$, we get a formula for $\|\Lambda^T\Phi_s\|$.

If $\varphi$ is $K$-invariant, there is always a pole of $E(\varphi_s)$ at $s = 1$, and its residue is a constant function whose value is related to the volume of $\Gamma\backslash G$. For other $K$-eigenfunctions there will be no poles at $s = 1$. These phenomena occur because the trivial representation of $(\mathfrak{g}, K)$ is a quotient of $I_1$ and embeds into $I_0$. Since $I_s$ is irreducible for $1/2 < s < 1$, poles in $(1/2, 1)$ will occur simultaneously for all $K$-components of $I_s$.

**The cuspidal decomposition.** A function $F$ on $\Gamma\backslash G$ is said to be *cuspidal* if its constant term vanishes identically. If $F$ lies in $A_{\mathrm{umg}}(\Gamma\backslash G)$ and it is cuspidal then it will lie in $\mathcal{S}(\Gamma\backslash G)$. Define $\mathcal{S}_{\mathrm{cusp}}$ to be the subspace of cuspidal functions in $\mathcal{S}(\Gamma\backslash G)$.

If $\varphi$ lies in $\mathcal{S}(N(\Gamma\cap P)\backslash G)$ then the Eisenstein series

$$E_\varphi(g) = \sum_{\Gamma\cap P\backslash\Gamma} \varphi(\gamma g)$$

will converge to a function in $\mathcal{S}(\Gamma\backslash G)$, and the map from $\mathcal{S}(N(\Gamma\cap P)\backslash G)$ to $\mathcal{S}(\Gamma\backslash G)$ is continuous. Define $\mathcal{S}_{\mathrm{Eis}}$ to be the closure in $\mathcal{S}(\Gamma\backslash G)$ of the image of $\mathcal{S}(N(\Gamma\cap P)\backslash G)$.

PROPOSITION. *The Schwartz space $\mathcal{S}(\Gamma\backslash G)$ is the direct sum of its two subspaces $\mathcal{S}_{\mathrm{cusp}}$ and $\mathcal{S}_{\mathrm{Eis}}$.*

As a preliminary:

LEMMA. *The space $L^{2,\infty}(\Gamma\backslash G)$ is contained in $A_{\mathrm{umg}}$.*

I recall that the space $L^{2,\infty}$ is that of all functions $\Phi$ on $\Gamma\backslash G$ such that the distributional derivatives $R_X F$ $(X \in U(\mathfrak{g}))$ are all square-integrable. It is to be shown that every $\Phi$ in $L^{2,\infty}$ is a smooth function on $\Gamma\backslash G$ and that for some single $m > 0$ independent of $\Phi$ we have

$$R_X\Phi(g) = O(\delta^m(g))$$

on $G_T$, for all $X \in U(\mathfrak{g})$.

A much more general result is proven in [Casselman:1989] (Proposition 1.16 and remarks afterwards), but circumstances here allow a simpler argument.

*Proof.* According to the Decomposition Theorem (see §1.2 of [Cartier:1974]) we can express the Dirac $\delta$ at 1 as

$$\delta_1 = \sum \xi_i * f_i$$

where $\xi_i$ are in $U(\mathfrak{g})$, the $f_i$ in $C_c^k(G)$, and $k$ is arbitrarily high. As a consequence, every $\Phi$ in $L^{2,\infty}$ can be expressed as a sum of vectors $R_f F$, where $F$ lies in $L^{2,\infty}$ and $f$ in $C_c^k(G)$. Furthermore, if

$$\Phi = \sum R_{f_i} F_i$$

then

$$R_X \Phi = \sum R_{Xf_i} F_i.$$

It therefore suffices to prove that for some $m > 0$, all $f$ in $C_c(G)$, and all $F$ in $L^{2,\infty}$ the convolution $R_f F$ is continuous on $\Gamma\backslash G$ and satisfies

$$R_f F(g) = O(\delta^m(g))$$

on $G_T$. On a fundamental domain of $\Gamma$, the function $\delta(g)$ and the norm $\|g\|$ are asymptotically equivalent, hence it is sufficient to verify

$$R_f F(g) = O(\|g\|^m).$$

Formally we can write

$$\begin{aligned} R_f F(g) &= \int_G F(gx) f(x)\, dx \\ &= \int_{\Gamma\backslash G} F(y) \sum_\Gamma f(g^{-1}\gamma y)\, dy \\ &= \langle F, \Theta_{L_g f}\rangle \\ &\le \|F\|\, \|\Theta_{L_g f}\| \end{aligned}$$

where $\Theta$ is the map taking $f$ in $C_c(G)$ to

$$\Theta_f(y) = \sum_\Gamma f(\gamma y).$$

There are only a finite number of non-zero terms in this series, which therefore converges to a continuous function of compact support on $\Gamma\backslash G$, so the formal calculation at least makes sense.

Since

$$\|\Theta_f\| \le \mathrm{vol}(\Gamma\backslash G)^{1/2} \sup_{\Gamma\backslash G} \left|\Theta_f(x)\right|$$

we must find a bound on the values of $\Theta_f$, and then see how the bound for $\Theta_{L_g f}$ changes with $g$.

Choose a compact open subgroup $U$ such that

$$\Gamma \cap U^{-1} \cdot U = \{1\}$$

and let

$$\|U\| = \max_{u\in U} \|u\|.$$

Then for $u$ in $U$, $\gamma$ in $\Gamma$, $x$ in $G$

$$\begin{aligned} \|u\gamma x\| &\le \|u\|\,\|\gamma x\| \\ &\le \|U\|\,\|\gamma x\| \\ \frac{1}{\|\gamma x\|} &\le \frac{\|U\|}{\|u\gamma x\|}. \end{aligned}$$

and

$$\sum_\Gamma \frac{1}{\|\gamma x\|^{1+\epsilon}} \le \frac{\|U\|^{1+\epsilon}}{\text{meas}(U)} \int_{U\gamma x} \frac{1}{\|y\|^{1+\epsilon}}\,dy.$$

If $C_{f,\epsilon} = \max \|x\|^{1+\epsilon}\,|f(x)|$ then

$$\begin{aligned} \left|\Theta_f(x)\right| &\le \sum_\Gamma \left|f(\gamma x)\right| \\ &\le \sum_\Gamma \frac{C_{f,\epsilon}}{\|\gamma x\|^{1+\epsilon}} \\ &= C_{f,\epsilon} \sum_\Gamma \frac{1}{\|\gamma x\|^{1+\epsilon}} \\ &\le C_{f,\epsilon} \frac{\|U\|^{1+\epsilon}}{\text{meas}(U)} \sum_\Gamma \int_{U\gamma x} \frac{1}{\|y\|^{1+\epsilon}}\,dy \\ &\le C_{f,\epsilon} \frac{\|U\|^{1+\epsilon}}{\text{meas}(U)} \int_G \frac{1}{\|y\|^{1+\epsilon}}\,dy \end{aligned}$$

(since the $U\gamma x$ are disjoint) and

$$\left|\Theta_{L_g f}(x)\right| \le K \|g\|^{1+\epsilon} C_{f,\epsilon}$$

for a constant $K > 0$ depending only on $\epsilon$. Everything we want to know follows from this. □

*Proof of the Proposition.* Let $L^2_{\text{cusp}}$ be the subspace of functions in $L^2(\Gamma\backslash G)$ whose constant terms vanish, and $L^2_{\text{Eis}}$ its orthogonal complement. Any $f$ in $\mathcal{S}$ can be expressed as a sum of two corresponding components

$$f = f_{\text{cusp}} + f_{\text{Eis}}$$

where *a priori* each component is known only to lie in $L^2$. But the first component lies in $L^{2,\infty} \subseteq A_{\text{umg}}$ and has constant term equal to 0, so lies itself in $\mathcal{S}$. Therefore the second does, too. This proves that

$$\mathcal{S} = \mathcal{S}_{\text{cusp}} \oplus \left(\mathcal{S} \cap L^2_{\text{Eis}}\right).$$

It remains to be shown that the second component here is the closure of the functions $E(\varphi)$ with $\varphi$ in $\mathcal{S}(N(\Gamma \cap P)\backslash G)$.

For this, because of the Hahn-Banach theorem, it suffices to show that if $\Phi$ is a tempered distribution which is equal to 0 on both $\mathcal{S}_{\text{cusp}}$ and all the $E(\varphi)$, then it is 0. On the one hand, the constant term of $\Phi$ vanishes, and therefore so does that of every $R_f\Phi$, which since it lies in $A_{\text{umg}}$ must also lie in $\mathcal{S}_{\text{cusp}}$. But on the other hand, the orthogonal complement of $\mathcal{S}_{\text{cusp}}$ is $G$-stable, so all these $R_f\Phi$ also lie in this complement. But since they themselves are cuspidal, they must vanish, too. However, $\Phi$ is the weak limit of $R_f\Phi$ if $f$ converges weakly to the Dirac distribution $\delta_1$. Therefore $\Phi$ itself vanishes. □

An analogous result for groups of arbitrary rank, essentially a reformulation of a result due to Langlands, is proven in [Casselman:1989].

**Definition of the Fourier-Eisenstein transform.** Suppose $f$ to be in $\mathcal{S}(\Gamma\backslash G)$. For $s \in \mathbb{C}$ where the Eisenstein series map $E_{1-s}$ is holomorphic, define its Fourier-Eisenstein transform $\widehat{f}(s)$ to be the unique element of $I_s$ such that

$$\langle f, E_{1-s}(\varphi)\rangle = \langle \widehat{f}(s), \iota_s\varphi\rangle$$

for every $\varphi$ in $\mathcal{I}$. The section $F = \widehat{f}$ of $\mathcal{I}$ is meromorphic in $s$ and has poles where $E_{1-s}$ does. It clearly satisfies this condition:

(PW1) $F(1-s) = \tau_s F(s)$.

The next step is to investigate more carefully the singularities of $F(s)$. They will only occur at the poles of $E_{1-s}$. In the region $\Re(s) < 1/2$, which is all we will care about, they are simple. What can we say about its residues in that region?

(PW2) *The function $F(s)$ has simple poles on $[0, 1/2)$ where $E_{1-s}$ does, and the residue $F^{\#}(s)$ at such a pole lies in the image of the residue of $\tau_{1-s}$.*

*Proof.* For $\Re(s) < 0$ we have a simple rearrangement of a converging series that shows

$$\langle f, E_{1-s}(\varphi)\rangle_{\Gamma\backslash G} = \langle f_P, \varphi_{1-s}\rangle_{N(\Gamma\cap P)\backslash G}$$

so that $\widehat{f} = 0$ if $f_P = 0$. The kernel of this transform is therefore precisely the subspace $\mathcal{S}_{\text{cusp}}$ of "The cuspidal decomposition", and the transform is completely determined by its restriction to $\mathcal{S}_{\text{Eis}}$. The space $\mathcal{S}_{\text{Eis}}$ is the closure of the image of the functions $E_f$ for $f$ in $\mathcal{S}(\Gamma\backslash G)$. Any particular $K$-constituent in $I_s$ is finite-dimensional, so the image of all of $\mathcal{S}(\Gamma\backslash G)$ in $I_s$ under the Fourier-Eisenstein transform is the same as the image of the functions $E_f$ for $f$ in $\mathcal{S}_{N(\Gamma\cap P)\backslash G}$.

If $f$ lies in $\mathcal{S}_{N(\Gamma\cap P)\backslash G}$ we can express it as

$$\frac{1}{2\pi i}\int_{\sigma-i\infty}^{\sigma+i\infty} \widehat{f}(s)\,ds$$

for any real $\sigma$. If we choose $\sigma > 1$ this gives us

$$E_f = \frac{1}{2\pi i} \int_{\sigma - i\infty}^{\sigma + i\infty} E(\widehat{f}(s))\, ds$$

and then

$$\begin{aligned}
[E_f]_P &= \frac{1}{2\pi i} \int_{\sigma - i\infty}^{\sigma + i\infty} E(\widehat{f}(s))_P\, ds \\
&= \frac{1}{2\pi i} \int_{\sigma - i\infty}^{\sigma + i\infty} \widehat{f}(s) + \tau_s \widehat{f}(s)\, ds \\
&= \frac{1}{2\pi i} \int_{\sigma - i\infty}^{\sigma + i\infty} \widehat{f}(s)\, ds + \frac{1}{2\pi i} \int_{\sigma - i\infty}^{\sigma + i\infty} \tau_s \widehat{f}(s)\, ds \\
&= \frac{1}{2\pi i} \int_{1-\sigma - i\infty}^{1-\sigma + i\infty} \widehat{f}(s) + \tau_{1-s} \widehat{f}(1-s)\, ds.
\end{aligned}$$

In the last step we move the contour of one integral and make a substitution of $1 - s$ for $s$ in the other. This implies that *the Fourier-Eisenstein transform of $E_f$* is $\widehat{f}(s) + \tau_{1-s}\widehat{f}(1-s)$. If we take residues of this expression at a pole, we obtain (PW2). □

Keep in mind that since $\tau_s \tau_{1-s} = 1$, on this image $E_s$ is well defined and $\tau_s = 0$. Hence the constant term of $E(F_s)$ will just be $F_s$ itself.

**The Plancherel theorem.** Suppose $\varphi_s$ to be a smooth function of compact support on the critical line $\Re(s) = 1/2$ with values in $\mathcal{I}$. Define the Eisenstein series $E_\varphi$ to be

$$E_\varphi = \frac{1}{2\pi i} \int_{1/2 - i\infty}^{1/2 + i\infty} E(\varphi_s)\, ds.$$

It will be a smooth function on $\Gamma\backslash G$. The *Plancherel Formula* for $\Gamma\backslash G$ asserts that it will be in $L^2(\Gamma\backslash G$ and that its $L^2$-norm will be given by the equation

$$\frac{1}{2} \|E_\varphi\|^2 = \frac{1}{2\pi i} \int_{1/2 - i\infty}^{1/2 + i\infty} \|\varphi_s\|^2\, ds.$$

As a consequence, the map $\varphi \mapsto E_\varphi$ extends to one from $L^2(1/2 + i\mathbb{R})$ to $L^2(\Gamma\backslash G)$. The principal consequence of the Plancherel theorem for our purposes is this:

(PW3) *For $f$ in $\mathcal{S}(\Gamma\backslash\mathcal{H})$ the function $\widehat{f}(s)$ is square-integrable on $1/2 + i\mathbb{R}$ in the sense that*

$$\frac{1}{2\pi i} \int_{1/2 - i\infty}^{1/2 + i\infty} \|\widehat{f}(s)\|^2\, ds < \infty.$$

*Proof.* For $\varphi$ of compact support

$$\begin{aligned}\langle f, E_\varphi\rangle &= \frac{1}{2\pi i}\int_{\Re(s)=1/2} \langle f, E(\varphi_s)\rangle\, ds \\ &= \frac{1}{2\pi i}\int_{\Re(s)=1/2} \langle \widehat{f}_{1-s}, \varphi_{1-s}\rangle\, ds \\ &\le \|f\|\,\|E_\varphi\| \\ &= \frac{1}{2}\,\|f\|\,\|\varphi\|\end{aligned}$$

so $\widehat{f}(s)$ extends to a continuous functional on $L^2(1/2 + i\mathbb{R})$, and must itself lie in $L^2(1/2 + i\mathbb{R})$ by Radon-Nikodym. □

**Spectral considerations.** The Casimir operator is self-adjoint on any one $K$-component of $L^2(\Gamma\backslash\mathcal{H})$. A standard argument about self-adjoint operators implies that

$$\|\mathfrak{C} - \lambda\|^{-1} \le |\Im(\lambda)|^{-1}$$

and here

$$\Im(s(s-1)) = 2\sigma t, \qquad s = 1/2 + \sigma + it$$

$$\|\mathfrak{C} - s(s-1)\|^{-1} \le \frac{1}{2|\sigma t|}.$$

The construction of Eisenstein series in, for example, [Colin de Verdière:1981] shows then that

$$\|\Lambda^T E(\varphi_s)\| = O\left(\frac{1}{2\sigma|t|}\right).$$

Since we can write

$$E(\varphi_s) = \Lambda^T E(\varphi_s) + C^T(\varphi_s)$$

we have

$$\langle f, E(\varphi_{1-s})\rangle = \big\langle f, \Lambda^T E(\varphi_{1-s})\big\rangle + \big\langle f, C^T E(\varphi_{1-s})\big\rangle$$
$$|\,\langle f, E(\varphi_{1-s})\rangle\,| \le \|f\|\,\big\|\Lambda^T E(\varphi_{1-s})\big\| + \big|\big\langle f, C^T E(\varphi_{1-s})\big\rangle\big|.$$

The second term involves an easy calculation on $\mathbb{R}^{\mathrm{pos}}$, and since the same reasoning applies to all $\mathfrak{C}^n f$ we deduce

(PW4) *In any subregion of* $\Re(s) < 1/2$, $|\Re(s)| > \tau$ *bounded to the left*

$$\|F(s)\| = O\left(\frac{1}{\sigma|t|^m}\right)$$

*for all* $m > 0$, *where* $s = 1/2 - \sigma + it$.

**The Paley-Wiener theorem.** If $f$ lies in $\mathcal{S}$ then so does every $\mathfrak{C}^m f$. Define $PW(\Gamma\backslash G)$ to be the space of all meromorphic functions $F(s)$ with values in $\mathcal{I}$ such that every $\Phi(s) = \Delta(s)^m F(s)$ satisfies (PW1)–(PW4). These translate to the following conditions on $F(s)$ itself:

- $F(1-s) = \tau_s F(s)$
- $F(s)$ has only simple poles on $[0, 1/2)$ in the region $\Re(s) \le 0$, located among the poles of $E_{1-s}$. The residue $F^\#(s)$ at $s$ lies in the image of $\tau_{1-s}$.
- The restriction of any $s^m F(s)$ to $(1/2 + i\mathbb{R})$ is square-integrable.
- In any region $s = 1/2 - \sigma + it$ with $\sigma$ bounded, $t$ bounded away from 0, we have

$$\|F(s)\| = O\left(\frac{1}{\sigma |t|^m}\right).$$

for all $m > 0$.

For $F$ in $PW(\Gamma\backslash G)$, let $F^\#(s)$ be its residue at any $s$ in $[0, 1/2)$. Define

$$\mathcal{E}(F) = -\sum E_s(F^\#(s)) + \frac{1}{2}\frac{1}{2\pi i}\int_{1/2-i\infty}^{1/2+i\infty} E_s(F(s))\, ds.$$

THEOREM. *(1) The map $\mathcal{E}$ has image in $\mathcal{S}(\Gamma\backslash G)$. (2) If $F = \widehat{f}$ then $\mathcal{E}(F)$ has the same constant term as $f$.*

*Proof.* It comes to showing that the constant term of $\mathcal{E}(F)$ is

$$\frac{1}{2\pi i}\int_{\sigma-i\infty}^{\sigma+i\infty} F(s)\, ds$$

for $\sigma \ll 0$. The crucial point, as before, is that we are allowed to move contours by the results of the last section. □

**Cusp forms.** We now have a map from $\mathcal{S}(\Gamma\backslash G)$ to a space of meromorphic sections of $\mathcal{I}$ satisfying certain conditions, with an inverse map back from the space of such sections to $\mathcal{S}(\Gamma\backslash G)$. The kernel of this map is precisely the subspace of functions in $\mathcal{S}(\Gamma\backslash G)$ whose constant term vanishes identically. This is the subspace of *cusp forms*. We therefore have an explicit version of the direct sum decomposition

$$\mathcal{S}(\Gamma\backslash G) = \mathcal{S}_{\text{cusp}} \oplus \mathcal{S}_{\text{Eis}}.$$

The space of cusp forms is itself a direct sum of irreducible $G$-representations, each with finite multiplicity. If $\pi$ is one of these components, then the map $f \mapsto \langle f, v\rangle$ ($v \in V_\pi$) induces a map from $\mathcal{S}(\Gamma\backslash G)$ to the dual of a cuspidal representation

$\pi$. The cuspidal component of $\mathcal{S}(\Gamma\backslash G)$ is a kind of Schwartz discrete sum of irreducible unitary representations of $G$. In order to say more we must know about the asymptotic distribution of cusp forms. But that is another story.

**A calculus exercise.** In the next section we shall need this result:

LEMMA. *Suppose $f(x)$ to be a function in $C^{r+1}(0, \rho]$, such that for some $\kappa_{r+1}$*

$$\left|f^{(r+1)}(x)\right| \leq \frac{\kappa_{r+1}}{x^r}$$

*for all $0 < x \leq \rho$. Then*

$$f_0 = \lim_{x\to 0} f(x)$$

*exists, and*

$$|f_0| \leq A\,\kappa_{r+1} + \sum_{0\leq k\leq r} \frac{\rho^k}{k!}|f^{(k)}(\rho)|$$

*for some positive coefficient A independent of $f$.*

In effect, the function $f(x)$ extends to a continuous function on all of $[0, \rho]$.

As an illustration of the Lemma, let $\ell(x) = x\log x - x$. We have on the one hand

$$\begin{aligned}
\ell(x) &= x\log x - x\\
\ell'(x) &= \log x\\
\ell''(x) &= \frac{1}{x}\\
\ell'''(t) &= -\frac{1}{x^2}\\
\ell^{(p)}(x) &= (-1)^p\frac{(p-2)!}{x^{p-1}}\\
\ell^{(r+1)}(x) &= (-1)^{r+1}\frac{(r-1)!}{x^r}\,,
\end{aligned}$$

and on the other $\lim_{x\to 0}\ell(x) = 0$. The function $\ell(x)$ will play a role in the proof of the Lemma.

*Proof.* It is an exercise in elementary calculus. The cases $r = 0$, $r \geq 1$ are treated differently. Begin by recalling the elementary criterion of Cauchy: *If $f(x)$ is continuous in $(0, \rho]$ then $\lim_{x\to 0} f(x)$ exists if and only if for every $\varepsilon > 0$ we can find $\delta > 0$ such that $|f(y) - f(z)| < \varepsilon$ whenever $0 < y, z < \delta$.* □

(1) *The case* $r = 0$. By assumption, $f$ is $C^1$ on $(0, \rho]$ and $f'$ is bounded by $\kappa_1$ on that interval. For any $y, z$ in $(0, \rho]$.

$$f(z) - f(y) = \int_y^z f'(x)\,dx, \qquad |f(z) - f(y)| \le \kappa_1 |z - y|.$$

Therefore Cauchy's criterion is satisfied, and the limit $f_0 = \lim_{x\to 0} f(x)$ exists. Furthermore

$$\begin{aligned} f_0 &= -f(\rho) + \int_0^\rho f'(x)\,dx \\ |f_0| &\le |f(\rho)| + \kappa_1 \rho. \end{aligned}$$

(2) *The case* $r > 0$. For any $y$ in $(0, \rho]$ we can write

$$\begin{aligned} f(\rho) - f(y) &= \int_y^\rho f'(x_1)\,dx_1 \\ f(y) &= -\int_y^\rho f'(x_1)\,dx_1 + f(\rho). \end{aligned}$$

We extend this by repeating the same process with $f'(x_1)$ etc. to get

$$\begin{aligned} f'(x_1) &= -\int_{x_1}^\rho f''(x_2)\,dx_2 + f'(\rho) \\ f(y) &= -\int_y^\rho f'(x_1)\,dx_1 + f(\rho) \\ &= -\int_y^\rho \left(-\int_{x_1}^\rho f''(x_2)\,dx_2 + f'(\rho)\right) dx_1 + f(\rho) \\ &= \int_y^\rho \int_{x_1}^\rho f''(x_2)\,dx_2\,dx_1 + (y-\rho) f'(\rho) + f(\rho) \\ &= -\int_y^\rho \int_{x_1}^\rho \int_{x_2}^\rho f'''(x_3)\,dx_3\,dx_2\,dx_1 + \frac{(y-\rho)^2}{2} f''(\rho) \\ &\quad + (y-\rho) f'(\rho) + f(\rho) \\ &= \quad \dots \\ &= (-1)^p \int_y^\rho \cdots \int_{x_{p-1}}^\rho f^{(p)}(x_p)\,dx_p \dots dx_1 \\ &\quad + \frac{(y-\rho)^{p-1}}{(p-1)!} f^{(p-1)}(\rho) + \frac{(y-\rho)^{p-2}}{(p-2)!} f^{(p-2)}(\rho) + \dots + f(\rho). \end{aligned}$$

This is the familiar calculation leading to Taylor series at $\rho$. If we apply this also to $z$ in $(0, \rho]$ and set $p = r + 1$ we get by subtraction

$$\begin{aligned} f(y) - f(z) = (-1)^{r+1} \int_y^z \cdots \int_{x_r}^{\rho} f^{(r+1)}(x_{r+1})\, dx_{r+1} \ldots dx_1 \\ + [(y-\rho)^r - (z-\rho)^r] \frac{f^{(r)}(\rho)}{r!} \\ + [(y-\rho)^{r-1} - (z-\rho)^{r-1}] \frac{f^{(r-1)}(\rho)}{(r-1)!} \\ + \cdots + [y - z]\, f'(\rho). \end{aligned}$$

In order to apply Cauchy's criterion, we must show how to bound

$$\left| \int_y^z \cdots \int_{x_r}^{\rho} f^{(r+1)}(x_{r+1})\, dx_{r+1} \ldots dx_1 \right| \le \int_y^z \cdots \int_{x_r}^{\rho} \frac{\kappa_{r+1}}{x_{r+1}^r}\, dx_{r+1} \ldots dx_1 .$$

We do not have to do a new calculation to find an explicit formula for the iterated integral

$$K_{y,z,r} = \int_y^z \cdots \int_{x_r}^{\rho} \frac{1}{x_{r+1}^r}\, dx_{r+1} \ldots dx_1 .$$

If we set $f = \ell$ above we get

$$\begin{aligned} \ell(y) &= y \log y - y \\ &= \int_y^{\rho} \cdots \int_{x_r}^{\rho} \frac{(r-1)!}{x_{r+1}^r}\, dx_{r+1} \ldots dx_1 \\ &\quad + \frac{(y-\rho)^r}{r!} \ell^{(r)}(\rho) + \frac{(y-\rho)^{r-1}}{(r-1)!} \ell^{(r-1)}(\rho) + \cdots + \ell(\rho)\,, \end{aligned}$$

so that

$$\begin{aligned} K_{y,\rho,r} &= \int_y^{\rho} \cdots \int_{x_r}^{\rho} \frac{(r-1)!}{x_{r+1}^r}\, dx_{r+1} \ldots dx_1 \\ &= \ell(y) - \frac{(y-\rho)^r}{r!} \ell^{(r)}(\rho) - \frac{(y-\rho)^{r-1}}{(r-1)!} \ell^{(r-1)}(\rho) - \cdots - \ell(\rho). \\ K_{y,z,r} &= K_{z,\rho,r} - K_{y,\rho,r} \end{aligned}$$

Since $\ell(x)$ is continuous on $[0, \rho]$ we may now apply Cauchy's criterion in the other direction to see that the limit $f_0$ exists. Furthermore, the bound on $f^{(r+1)}$ together with the equation for $f(y) - f(z)$ enable us to to see that

$$|f_0| \le \kappa_{r+1} \, |K_{0,\rho,r+1}| + \sum_0^r \frac{\rho^k}{k!} \left| f^{(k)}(\rho) \right| .$$

This concludes the proof of the Lemma.

**Moving contours.** In the proof of Paley-Wiener theorems for the Schwartz space of arithmetic quotients, it is necessary to allow a change of contour of integration which is not obviously justifiable. This is a consequence of the following very general result. In this paper I require only the special case $n = 1$, but it is only slightly more difficult to deal with the general case, which will be needed for Paley-Wiener theorems for groups of higher rank.

For the next result, for $\varepsilon > 0$ let

$$\sum_{\varepsilon} = \{s \in \mathbb{C}^n \mid 0 < \Re(s_i) < \varepsilon\}$$

and

$$\overline{\sum}_{\varepsilon} = \{s \in \mathbb{C}^n \mid 0 \leq \Re(s_i) < \varepsilon\}.$$

THEOREM. *Suppose $\Phi(s)$ to be holomorphic in $\Sigma_\varepsilon$. Suppose in addition that for some positive integers $m$ and $r \geq 0$ it satisfies an inequality*

$$\Phi(\sigma + it) = O\left(\frac{1 + \|t\|^m}{\prod \sigma_i^r}\right).$$

*Thus for a fixed $s$ in $\Sigma_\varepsilon$ the function $t \mapsto \Phi(s + it)$ is of moderate growth and therefore defines by integration a tempered distribution $\Phi_s$. For every $s$ in the region $\overline{\Sigma}_\varepsilon$ the weak limit*

$$\Phi_s = \lim_{x \in \Sigma_\varepsilon, x \to s} \Phi_x$$

*exists as a tempered distribution. If the tempered distribution $\varphi_0$ is the inverse Fourier transform of $\Phi_0$, then for every $s$ in $\overline{\Sigma}_\varepsilon$ the product distribution $\varphi_s = e^{-\langle s, \bullet\rangle}\varphi_0$ is tempered and has Fourier transform $\Phi_s$.*

*Proof.* It is a straightforward modification of that of a similar result to be found on p. 25 in volume II of the series on methods of mathematical physics by by Mike Reed and Barry Simon (which also contains an implicit version of the Lemma in the previous section).

Suppose for the moment that $s = 0$, and choose $\lambda$ a real point in $\Sigma_\varepsilon$. Suppose $\Psi(t)$ to be a function in the Schwartz space $\mathcal{S}(\mathbb{R}^n)$. For each $x$ in $(0, 1]$ let

$$f_\lambda(x) = \int_{\mathbb{R}^n} \Phi(x\lambda + it)\Psi(t)\, dt,$$

i.e., integration against $\Psi$ on the space $\Re(s) = x\lambda$. Then

$$\begin{aligned}
f'_\lambda(x) &= \int_{\mathbb{R}^n} \frac{d}{dx}\Phi(x\lambda + it)\Psi(t)\, dt \\
&= \int_{\mathbb{R}^n} \sum \lambda_k \left[\frac{\partial\, \Phi}{\partial s_k}\right](x\lambda + it)\Psi(t)\, dt \\
&= \int_{\mathbb{R}^n} \sum \lambda_k \frac{1}{i}\frac{\partial}{\partial t_k}\big[t \mapsto \Phi(x\lambda + it)\big]\Psi(t)\, dt \\
&= i\int_{\mathbb{R}^n} \Phi(x\lambda + it) \sum \lambda_k \frac{\partial\, \Psi(t)}{\partial t_k}\, dt \quad \text{(integration by parts)} \\
&= \int_{\mathbb{R}^n} \Phi(x\lambda + it) D_\lambda \Psi(t)\, dt
\end{aligned}$$

□

where

$$D_\lambda = i \sum \lambda_k \frac{\partial}{\partial t_k}.$$

Therefore for all $p$

$$f_\lambda^{(p)}(x) = \int_{\mathbb{R}^n} \Phi(x\lambda + it) D_\lambda^p \Psi(t)\, dt.$$

The assumptions on $\Phi$ and $\Psi$ ensure that for all large integers $k$ and suitable $C_{m+k}$

$$\left|\Phi(x\lambda + it) D_\lambda^p \Psi(t)\right| \le C \frac{1 + \|t\|^m}{x^{nr} \prod \lambda_k^r} \frac{C_{m+k}}{1 + \|t\|^{m+k}}$$

$$\left|f_\lambda^{(p)}(x)\right| \le \frac{1}{x^{nr}} \frac{C C_{m+k}}{\prod \lambda_k^r} \int_{\mathbb{R}^n} \frac{1 + \|t\|^m}{1 + \|t\|^{m+k}}\, dt.$$

The Lemma can therefore be applied to $f_\lambda(x)$ to see that $f_\lambda(0)$ exists and depends continuously on the norms of $\Psi$, therefore defining in limit the tempered distribution

$$\langle \Phi_{0,\lambda}, \Psi \rangle = \lim_{x \to 0} \langle \Phi_{x\lambda}, \Psi \rangle$$

where

$$\langle \Phi_\sigma, \Psi \rangle = \int_{\mathbb{R}^n} \Phi(\sigma + it) \Psi(t)\, dt.$$

Define $\varphi_{0,\lambda}$ to be the inverse Fourier transform of $\Phi_{0,\lambda}$, a tempered distribution on $\mathbb{R}^n$. It remains to be shown that the product $\varphi_\sigma$ of $e^{-\sigma x}$ and $\varphi_{0,\lambda}$ is also tempered for $\sigma$ in $\Sigma_\varepsilon$, and that $\Phi_\sigma$ is the Fourier transform of $\varphi_\sigma$. This will prove among other things that $\Phi_{0,\lambda}$ doesn't actually depend on the choice of $\lambda$.

Choose a function $\psi$ in $C_c^\infty(\mathbb{R}^n)$. Its Fourier transform

$$\Psi(s) = \int_{\mathbb{R}^n} \psi(x) e^{-\langle s, x \rangle}\, dx$$

will be entire, satisfying inequalities

$$|\Psi(s)| = O\left(\frac{1}{1 + \|\Im(s)\|^m}\right)$$

for every $m > 0$, uniformly on vertical strips $\|\Re(s)\| < C$.

Then for every $\sigma$ in $\mathbb{R}^n$ with $\sigma_i > 0$ the product $e^{-\langle \sigma, x \rangle} \psi(x)$ will also be of compact support with Fourier transform

$$\Psi_\sigma(s) = \int_{-\infty}^{\infty} e^{-\langle \sigma, x \rangle} e^{-\langle s, x \rangle} f(x)\, dx = \Psi(\sigma + s).$$

Recall that if $\varphi$ is a tempered distribution on $\mathbb{R}^n$ and and $\psi$ in $\mathcal{S}(\mathbb{R}^n)$ with Fourier transforms $\Phi$ and $\Psi$ then (expressing it formally)

$$\langle \varphi, \psi \rangle = \left(\frac{1}{2\pi i}\right)^n \int_{(i\mathbb{R})^n} \Phi(s) \Psi(-s)\, ds.$$

Thus

$$\langle \varphi_\sigma, \psi(x) \rangle = \left\langle \varphi_0, e^{-\langle \sigma, x \rangle} \psi(x) \right\rangle$$

$$= \lim_{x \to 0} \left( \frac{1}{2\pi i} \right)^n \int_{(i\mathbb{R})^n} \Phi(x\lambda + it) \Psi(\sigma - it)\, dt.$$

We change of contour of integration from $\sigma + (i\mathbb{R})^n$ to $(i\mathbb{R})^n$, which is permissible by our assumptions. The calculation continues

$$\langle \varphi_\sigma, \psi(x) \rangle = \lim_{x \to 0} \left( \frac{1}{2\pi i} \right)^n \int_{(i\mathbb{R})^n} \Phi(x\lambda + \sigma + iu) \Psi(-iu)\, du$$

$$= \left( \frac{1}{2\pi i} \right)^n \int_{(i\mathbb{R})^n} \Phi(\sigma + iu) \Psi(-iu)\, du.$$

This result implies that the limit of $\Phi_s$ as $s$ approaches 0 does not depend on the way in which the limit is taken, since $\Phi_0 = e^{\langle s, x \rangle} \Phi_s$ for all $s$ in $\Sigma_\varepsilon$.

Dealing with an arbitrary $s$ in $\overline{\Sigma}_\varepsilon$ is straightforward, since $e^{-\langle s, \bullet \rangle} \varphi_0$ is clearly tempered.

This concludes the proof of the Theorem.

COROLLARY. Suppose $\Phi(s)$ to be holomorphic in the region $\Sigma_\varepsilon$, having as continuous limit as $\Re(s) \to 0$ a function in $L^2((i\mathbb{R})^n)$. Assume that for some integer $r > 0$ it satisfies an inequality

$$|\Phi(\sigma + it)| \le \frac{C_m}{(1 + \|t\|^m) \prod \sigma_i^r}$$

*for all $m > 0$ in the region $\Sigma_\varepsilon$. Then*

$$\lim_{T \to \infty} \left( \frac{1}{2\pi i} \right)^n \int_{\|s\| \le T} \Phi(s) e^{\langle s, x \rangle}\, ds = \left( \frac{1}{2\pi i} \right)^n \int_{\Re(s) = \sigma} \Phi(s) e^{\langle s, x \rangle}\, ds$$

for any $\sigma$ in $\Sigma_\varepsilon$.

*The limit here is to be the limit in the $L^2$ norm of the functions*

$$\varphi_T(x) = \left( \frac{1}{2\pi i} \right)^n \int_{\|s\| \le T} \Phi(s) e^{\langle s, x \rangle}\, ds.$$

Formally, this is just a change of contours, but a direct argument allowing this does not seem possible. Instead, apply the Theorem to the function $\Phi(s)$, using the hypotheses to compute its inverse Fourier transform in two ways.

To apply the results of this section to the principal results of this paper, a change from additive to multiplicative coordinates is necessary. Thus $e^{\langle s, x \rangle}$ is replaced by $x^s = \prod x_k^{s_k}$.

Bill Casselman
Mathematics Deparment
University of British Columbia
Vancouver, Canada
*E-mail:* cass@math.ubc.ca

---

## REFERENCES

[1] P. Cartier, 'Vecteurs différentiables dans les représentations unitaires des groupes de Lie', exposé 454 of Séminaire Bourbaki 1974/75.

[2] W. Casselman, *Automorphic forms and a Hodge theory for congruence subgroups of $SL_2(\mathbb{Z})$*, Lecture Notes in Mathematics, **1041**, Springer-Verlag, 1984, 103–140.

[3] ———, Introduction to the Schwartz space of $\Gamma\backslash G$, Can. J. Math. **Vol. 40** (1989), 285–320.

[4] ———, Extended automorphic forms on the upper half plane, Math. Ann. **296** (1993), 755–762.

[5] ———, On the Plancherel measure for the continuous spectrum of the modular group, in *Proceedings of Symposia in Pure Mathematics* **66** (1999), A. M. S., 19–25.

[6] Y. Colin de Verdière, Une nouvelle démonstration du prolongement méromorphe des séries d'Eisenstein, *C. R. Acad. Sci. Paris* **293** (1981), 361–363.

[7] L. Ehrenpreis and I. Mautner, The Fourier transform on semi-simple Lie groups III, Transactions A. M. S. **90** (1959), 431–484. Section 9 is concerned with the Schwartz space of $\mathrm{SL}_2(\mathbb{Z})\backslash\mathcal{H}$ and a Paley-Wiener theorem characterizing its elements by their Fourier-Eisenstein transforms.

[8] R. P. Langlands, Eisenstein series, in Algebraic groups and discontinuous subgroups, Proceedings of Symposia in Pure Mathematics **9** (1966), A. M. S., 235–252.

[9] ———, On the functional equations satisfied by Eisenstein series, Lecture Notes in Mathematics **514**, Springer-Verlag, 1977. This amounts to a re-edition of mimeographed notes first distributed in 1965, on which Langlands' talks at the 1965 Boulder conference were based.

[10] M. Reed and B. Simon, Fourier analysis, self-adjointness, volume II of the series Methods of mathematical physics, Academic Press, 1972.

CHAPTER 9

# EQUIDISTRIBUTION DES POINTS DE HECKE

By LAURENT CLOZEL and EMMANUEL ULLMO

---

**1. Introduction.** André et Oort ont formulé un analogue de la conjecture de Manin-Mumford, démontrée par Raynaud [32] [33], pour les points à multiplication complexe de l'espace des modules $\mathbf{A}_{g,1}$ des variétés abéliennes principalement polarisées. Dans les deux cas ces conjectures s'énoncent sous la forme: Une composante irréductible de l'adhérence de Zariski d'un ensemble de points spéciaux est une sous-variété spéciale.

La nouvelle preuve de la conjecture de Manin-Mumford, via la conjecture de Bogomolov [39] [41] et l'équidistribution des petits points [37], suggère que l'on peut essayer d'attaquer la conjecture de André et Oort via des théorèmes d'équidistribution ayant une signification modulaire. Dans cette optique, on s'attend à une réponse positive à la question suivante:

QUESTION 1.1. *Soit $x_n$ une suite "générique" de points à multiplication complexe sur $\mathbf{A}_{g,1}(\overline{\mathbb{Q}})$. Cela signifie que pour toute sous-variété $Y \subset \mathbf{A}_{g,1}$ avec $Y \neq \mathbf{A}_{g,1}$, l'ensemble $\{n \in \mathbb{N} \mid x_n \in Y(\overline{\mathbb{Q}})\}$ est fini. Pour $x \in \mathbf{A}_{g,1}(\overline{\mathbb{Q}})$, on note $O(x)$ l'orbite sous Galois de $x$. Pour $x \in \mathbf{A}_{g,1}(\mathbb{C})$, on note $\delta_x$ la mesure de Dirac au point $x$. Est-il vrai que la suite de mesures*

$$\mu_n = \frac{1}{Card(O(x_n))} \sum_{y \in O(X_n)} \delta_y$$

*converge faiblement vers la mesure $Sp(2g)(\mathbb{R})$-invariante sur $\mathbf{A}_{g,1}(\mathbb{C})$?*

Dans le cas $g = 1$ et pour les points ayant multiplication complexe par l'anneau des entiers $O_K$ d'un corps de nombres quadratique imaginaire, une réponse affirmative à cette question est donnée par un théorème de Duke [16].

Dans ce texte nous démontrons des énoncés d'équidistribution de ce type pour les suites de mesures associées à des correspondance de Hecke. Les résultats que nous avons en vue concernent les groupes $G_1 = GL_n$ et $G_2 = GSp(2g)$. On rappelle que $Sp(2g)$ désigne le groupe laissant invariant la forme symplectique sur un espace de dimension $2g$ de matrice

$$J = \begin{pmatrix} 0 & -1_g \\ 1_g & 0 \end{pmatrix}$$

et que $GSp(2g)$ désigne le groupe de similitude associé. Soit $\Gamma_i = G_i(\mathbb{Z})$.

---

Manuscript received October 17, 2002.

On considère pour tout entier $N$ positif et tout $r \in [1, \dots, n]$ la double classe

$$T_{r,N} = \Gamma_1 \mathrm{Diag}(N, \dots, N, 1 \dots, 1)\Gamma_1 \tag{1}$$

où $\mathrm{Diag}(N, \dots, N, 1 \dots, 1)$ désigne la matrice diagonale dont les $r$ premiers termes sont égaux à $N$ et les $(n-r)$ suivants sont égaux à 1. Pour $G_2$, on considère la double classe

$$T_N = \Gamma_2 \mathrm{Diag}(N, \dots, N, 1 \dots, 1)\Gamma_2 \tag{2}$$

où $\mathrm{Diag}(N, \dots, N, 1, \dots, 1)$ désigne la matrice diagonale dont les $g$ premiers termes sont égaux à $N$ et les $g$ suivants sont égaux à 1. On note $|T_{r,N}|$ et $|T_N|$ les degrés de $T_{r,N}$ et $T_N$.

On note $Z_i$ le centre de $G_i$ et $L^2(\Gamma_i \backslash G_i(\mathbb{R}), \mathbf{1})$ l'espace des fonctions $Z_i$-invariantes de carré intégrable modulo le centre sur $\Gamma_i \backslash G_i(\mathbb{R})$ pour la mesure $G_i(\mathbb{R})$-invariante normalisée. On peut alors voir $T_{r,N}$ et $T_N$ comme des opérateurs sur $L^2(\Gamma_i \backslash G_i(\mathbb{R}), \mathbf{1})$. On note alors $\overline{T}_{r,N} = \frac{T_{r,n}}{|T_{r,n}|}$ l'opérateur normalisé et on définit de même $\overline{T}_N$.

On peut aussi voir $T_{r,N}$ comme une correspondance sur $\Gamma_1 Z_1(\mathbb{R}) \backslash G_1(\mathbb{R})$. Pour tout $x \in G_1(\mathbb{R})$, on note $T_{r,N}.x$ l'ensemble de points correspondant de $\Gamma_1 \backslash G_1(\mathbb{R})$. Par abus de notation, on note de la même manière l'image de ces points dans $\Gamma_1 Z_1(\mathbb{R}) \backslash G_1(\mathbb{R})$. La même notation s'applique à $T_N$.

Soit $G$ un groupe réductif connexe défini et déployé sur $\mathbb{Q}$, $\Gamma = G(\mathbb{Z})$ et soit $Z$ son centre. Soit $\mu_n$ une suite de mesure de masse 1 sur $\Gamma Z(\mathbb{R}) \backslash G(\mathbb{R})$. On normalise la mesure de Haar $d\mu(g)$ sur $G$ de sorte que

$$\int_{Z(\mathbb{R})\Gamma\backslash G(\mathbb{R})} d\mu(g) = 1.$$

Soit $C_0(\Gamma Z(\mathbb{R}) \backslash G(\mathbb{R}))$ l'espace des fonctions continues sur $\Gamma Z(\mathbb{R}) \backslash G(\mathbb{R})$ qui tendent vers 0 à l'infini. On dit que $\mu_n$ converge faiblement vers $d\mu(g)$ si pour tout $f \in C_0(\Gamma Z(\mathbb{R}) \backslash G(\mathbb{R}))$, on a

$$\lim_{n\to\infty} \mu_n(f) = \int_{Z(\mathbb{R})\Gamma\backslash G(\mathbb{R})} f(g) d\mu(g).$$

Si $E$ est un ensemble de points de $\Gamma Z(\mathbb{R}) \backslash G(\mathbb{R})$ de cardinal $n_E$, on note $\mu_E$ la mesure $\frac{1}{n_E} \Sigma_{y \in E} \delta_y$, ou $\delta_x$ désigne la mesure de Dirac de support $x$. Soit $E_n$ une suite d'ensembles finis de points de $\Gamma Z(\mathbb{R}) \backslash G(\mathbb{R})$. On dit que les $E_n$ sont équidistribués pour la mesure de Haar si la suite $\mu_{E_n}$ converge faiblement vers $d\mu(g)$.

On montre dans ce texte les deux théorèmes suivants:

THÉORÈME 1.2. *Soient $n \geq 3$, $G = G_1 = GL_n$ et $r \in [1, \dots, n-1]$.*

*(a) Pour tout $f \in L^2(\Gamma_1 \backslash G_1(\mathbb{R}), \mathbf{1})$ et tout nombre premier $p$, on a*

$$\|\overline{T}_{r,p} f - \int_{Z_1(\mathbb{R})\Gamma_1\backslash G_1(\mathbb{R})} f(g) d\mu(g)\| \leq n!\ p^{-\frac{min(r,n-r)}{2}} \|f\|. \tag{3}$$

*En particulier, quand $N$ tend vers l'infini, parmi les entiers sans facteurs carrés $\overline{T}_{r,N} f$ tend, pour la convergence $L^2$, vers la fonction constante égale à $\int_{Z_1(\mathbb{R})\Gamma_1\backslash G_1(\mathbb{R})} f(g)d\mu(g)$.*

*(b) Pour tout $x \in \Gamma_1\backslash G_1(\mathbb{R})$ la suite des $T_{r,N}.x$ est équidistribuée pour la mesure de Haar.*

Pour décrire les résultats analogues pour $G_2 = GSp(2g)$, on introduit la "constante de Ramanujan" $\theta$ relative au groupe $SL_2$. La définition précise sera donné dans la partie 6 de ce texte. Disons seulement qu'avec nos normalisations, on a $\theta = 0$ si la conjecture de Ramanujan pour les composantes locales des représentations automorphes cuspidales de $SL_2(F)$, pour un corps de nombres arbitraire $F$ est vérifiée. L'estimation triviale de Hecke donne $\theta = 1$, le théorème de Gelbart-Jacquet [19] donne $0 \le \theta \le \frac{1}{2}$ et la meilleure estimée valable pour tout corps de nombres (Shahidi [35]) donne $0 \le \theta \le \frac{2}{5}$. Notons que Luo, Rudnick et Sarnak [26] [27] ont récemment obtenu la même borne par une autre méthode qui a l'avantage de donner aussi des résultats de ce type aux places archimédiennes. Une meilleure estimée, valable uniquement pour $F = \mathbb{Q}$ est donné dans [4]; elle sera utilisé dans la partie concernant la courbe modulaire $X(1)$.

THÉORÈME 1.3. *Soit $G = G_2 = GSp(2g)$.*

*(a) Pour tout $f \in L^2(\Gamma_2\backslash G_2(\mathbb{R}), \mathbf{1})$ et tout nombre premier $p$ on a*

$$\|\overline{T}_p f - \int_{Z_2(\mathbb{R})\Gamma_1\backslash G_2(\mathbb{R})} f(g)d\mu(g)\| \le 2^g\, p^{-\frac{g(1-\theta)}{2}} \|f\|. \tag{4}$$

*En particulier, quand $N$ tend vers l'infini parmi les entiers sans facteurs carrés, $\overline{T}_{r,N} f \to \int_{Z_2(\mathbb{R})\Gamma_2\backslash G_2(\mathbb{R})} f(g)d\mu(g)$ au sens $L^2$.*

*(b) Pour tout $x \in \Gamma_2\backslash G_2(\mathbb{R})$ la suite des $T_N.x$ est équidistribuée pour la mesure de Haar.*

Notons que les estimations pour la norme $L^2$ dans les deux théorèmes précédents sont essentiellement optimales: En formant des séries d'Eisenstein à partir de la représentation triviale du sous-groupe de Lévi $GL_{n-1} \times GL_1$ de $GL_n$, on fabrique des fonctions dans $L^2(\Gamma\backslash GL_n(\mathbb{R}), \mathbf{1})$, dans l'orthogonale des fonctions constantes, pour lesquelles les bornes obtenues dans le théorème (1.2) sont optimales en ce qui concerne la puissance de $p$. De même en formant des séries d'Eisenstein à partir de la représentation triviale du sous-groupe de Lévi $GL_1 \times GSp(2g-2)$ on voit que si la conjecture de Ramanujan pour $SL_2$ est vraie (donc $\theta = 0$) les estimées du théorème (1.3) sont aussi essentiellement optimales.

Décrivons le plan de ce texte et des démonstrations de ces théorèmes.

Dans la deuxième partie nous traitons le cas de la courbe modulaire $Y(1) = \Gamma\backslash\mathbb{H}$ quotient du demi-plan de Poincaré par $\Gamma = SL_2(\mathbb{Z})$. Cette partie est indépendante du reste du texte mais elle permet de comprendre la srtucture de la preuve des théorèmes (1.2) et (1.3) dans ce cadre élémentaire. Le résultat découle du fait que les fonctions intevenant dans la décomposition spectrale sont propres pour les opérateurs de Hecke et que l'on dispose des bornes [4] pour les valeurs propres de ces opérateurs. Nous donnons deux applications de ces résultats: Dans la section 2.3 nous étendons les résultats de Duke à des discriminants non fondamentaux. Ceci donne donc une réponse complète à la question 1 quand $g = 1$. Dans la section 2.4, on utilise les résultats précédents pour montrer qu'une application propre de $Y(1) = SL(2, \mathbb{Z})\backslash\mathbb{H}$ dans $Y(1)$ qui commute en tant que correspondance à trois opérateurs de Hecke est soit l'identité soit l'application obtenue par passage au quotient de l'application $z \to -\bar{z}$ du demi-plan de Poincaré $\mathbb{H}$.

Dans la troisième partie, on décrit la décomposition spectrale de l'espace $L^2(Z(\mathbb{R})\Gamma\backslash G(\mathbb{R}))$ en fonction des vecteurs sphériques des représentation automorphes de $G(\mathbf{A})$. On explique aussi comment les opérateurs de Hecke agissent sur les vecteurs sphériques de ces représentations via la transformée de Sataké. On verra que la transformée de Sataké s'exprime sous une forme très simple en fonctions des paramètres des représentations (non ramifiées) intervenant dans la décomposition spectrale pour les opérateurs de Hecke des théorèmes 1.2 et 1.3. Ceci explique dans quelles autres situations notre méthode est susceptible de s'adapter.

Dans la quatrième partie nous démontrons la partie $L^2$ du théorème 1.2 pour $GL_n$. La difficulté nouvelle par rapport au cas classique provient de la présence du spectre résiduel. Le théorème de Moeglin et Waldspurger [30] permet de décrire tous les paramètres des représentations automorphes intervenant dans la décomposition spectrale pour $GL_n$ à partir des paramètres de représentations automorphes cuspidales de $GL_m$ pour $m \leq n$. On utilise alors les approximations de la conjecture de Ramanujan dues à Jacquet et Shalika pour $n \geq 3$ [22] et à Gelbart et Jacquet [19] pour $n = 2$. Le résultat combinatoire à la base de la démonstration du théorème est donné dans la proposition (4.2). Notons qu'une meilleure approximation de la conjecture de Ramanujan pour $GL_m$ comme celles obtenues dans [26] [27] ne changerait pas nos estimations.

Pour $GSp(2g)$, on ne dispose pas d'une description aussi agréable des paramètres des représentations intervenant dans la décomposition spectrale. Nous donnons dans la cinquième partie deux démonstrations de l'analogue en $p$ du principe de restriction de Burger, Li et Sarnak [8] [9]: Pour tout groupe semi-simple, simplement connexe $G$ défini sur $\mathbb{Q}$, on définit le spectre automorphe $\hat{G}_p^{aut}$. C'est un sous-ensemble du dual unitaire $\hat{G}_p$ de $G_p = G(\mathbb{Q}_p)$. On montre alors

THÉORÈME 1.4. *Soit $H$ un sous-groupe semi-simple de $G$. Soit $\pi \in \hat{G}_p^{aut}$ et $\pi' \in \hat{H}_p$. Si $\pi'$ est faiblement contenue dans $\pi|_{H_p}$, alors $\pi' \in \hat{H}_p^{aut}$.*

Nous utilisons cet énoncé dans la sixième partie avec $G = Sp(2g)$ et $H = SL_2^g$ pour controler la croissance des fonctions sphériques associées aux représentations non ramifiés intervenant dans la décomposition spectrale de $G$ grâce aux formules explicites pour les fonctions sphériques associés aux représentations automorphes sphériques sur $SL_2$. On interprète ensuite ces résultats comme des bornes sur le spectre de l'opérateur $T_p$. On termine cette partie en faisant le lien avec les conjectures d'Arthur qui permettent aussi de prévoir le spectre de $T_p$.

Dans la septième partie, nous étendons les techniques précédentes à des groupes possédant un sous-système de racines fortement orthogonal de rang maximal. On explicite le cas de $G = SO(2g+1)$.

Dans la dernière partie on complète la démonstration des théorèmes 1.2 et 1.3. Quand les fonctions test sont suffisament régulières nous estimons la vitesse de convergence dans les énoncés des théorèmes 1.2 b et 1.3 b.

Les problèmes considérés dans cet article, ainsi que les méthodes que nous utilisons ont été introduits pour la première fois de façon systématique par Burger, Li et Sarnak [8], [9]. En particulier la section 5 ne fait qu'étendre aux places finies, ainsi qu'ils l'avaient eux même envisagé, un résultat fondamental de Burger et Sarnak [9], alors que les sections 6 et 7 adaptent leur méthode de restriction à un groupe maximal du groupe ambiant.

Nous avons appris après avoir complété ce travail qu'une partie des résultats concernant $GL_n$ et l'opérateur $T_{1,p}$ ont été annoncés avec une esquisse de preuve par Sarnak [34] dans son rapport au congrès international de Kyoto. Des précisions pour le cas $n = 2$ et $n = 3$ sont données par Chiu dans [13]. Comme notre méthode pour $GL_n$ est plus précise (grâce à notre utilisation du théorème de Moeglin et Waldspurger), concerne plus d'opérateurs de Hecke, et vu l'absence d'une preuve détaillée des résultats de [34], il nous a semblé utile de rédiger nos résultats sur $GL_n$.

Nous tenons à remercier P. Gille, P. Michel, P. Sarnak, S.Kudla et S. Zhang pour d'utiles conversations relatives à ce travail. Après la rédaction de ce travail est apparu le papier de Oh [31] qui permet d'améliorer les résultats de notre § 6. Pour ceci nous renvoyons le lecteur à: L. Clozel, H. Oh, E. Ullmo, Equidistribution des points de Hecke, Inventiones Math **114** (2001), 327–351. Comme les méthodes introduites dans cet article ont depuis été utilisées—et l'article cité—dans de nombreux papiers ultérieurs, nous avons néanmoins conservé la présente rédaction.

*Acknowledgments.* L'un des auteurs (L.C.) se souvient avec gratitude de l'hospitalité de Joseph Shalika à l'université de Johns Hopkins. Tous deux sont heureux de lui dédier ce travail en témoignage d'admiration.

**2. Le cas classique.** Soient $\mathbb{H}$ le demi plan de Poincaré, $\Gamma = SL(2,\mathbb{Z})$. On note $Y(1)$ la courbe modulaire $Y(1) = \Gamma\backslash\mathbb{H}$ et $X(1)$ la courbe propre obtenue en rajoutant la pointe $\infty$ à $Y(1)$. On note $d\mu_0 = \frac{3}{\pi}\frac{dx\,dy}{y^2}$ la mesure de Poincaré et $D_0 = y^2(\frac{\partial^2}{\partial x^2} + \frac{\partial^2}{\partial y^2})$ le laplacien associé. On note $L^2(\Gamma\backslash\mathbb{H}, d\mu_0)$ l'espace des fonctions $\Gamma$–invariantes de carré intégrable pour la mesure de Poincaré. On note

$C_c(\Gamma\backslash\mathbb{H})$ (resp $C_0(\Gamma\backslash\mathbb{H})$) l'espace des fonctions continues à support compact (resp des fonction continues tendant vers 0 en l'infini). On notera aussi $D(\Gamma\backslash\mathbb{H})$ l'espace de fonctions $\Gamma$–invariantes bornées de classe $C^\infty$ telles que $D_0 f$ est aussi bornée et de classe $C^\infty$.

On dispose sur $X(1)$ des correspondances de Hecke $T_n$ définies pour tout entier $n$ par

$$T_n.z = \sum_{ad=n} \sum_{0\leq b<d} \frac{az+b}{d}. \tag{5}$$

La correspondance $T_n$ est de degré $\sigma_1(n)$ où pour tout nombre complexe $s$ on a $\sigma_s(n) = \sum_{d/n} d^s$. Pour toute fonction $f$ sur $X(1)$ on définit $T_n f$ par la formule

$$T_n f(z) = \sum_{y\in T_n.z} f(y). \tag{6}$$

THÉORÈME 2.1. *(a) Pour toute fonction $f$ dans $L^2(\Gamma\backslash\mathbb{H}, d\mu_0)$ et tout $\epsilon > 0$, il existe une constante $C_\epsilon$, ne dépendant que de $\epsilon$, telle que:*

$$\left\| \frac{1}{\sigma_1(n)} T_n f - \int_{\Gamma\backslash\mathbb{H}} f(\zeta) d\mu_0(\zeta) \right\| \leq C_\epsilon n^{-\frac{1}{2}+\frac{5}{28}+\epsilon} \| f \|. \tag{7}$$

*En particulier on a une convergence au sens $L^2$ de $T_n f$ vers $\int_{\Gamma\backslash\mathbb{H}} f(\zeta) d\mu_0(\zeta)$ quand $n$ tend vers l'infini.*

*(b) Pour toute fonction $f \in D(\Gamma\backslash\mathbb{H})$, pour tout $z \in \Gamma\backslash\mathbb{H}$ et tout $\epsilon > 0$, il existe une constante $C_{\epsilon,z,f}$ telle que*

$$\left| \frac{1}{\sigma_1(n)} T_n f(z) - \int_{\Gamma\backslash\mathbb{H}} f(\zeta) d\mu_0(\zeta) \right| \leq C_{\epsilon,z,f} n^{-\frac{1}{2}+\frac{5}{28}+\epsilon}. \tag{8}$$

*Quand $z$ varie dans un compact, la constante $C_{\epsilon,z,f}$ peut être rendue indépendante de $z$.*

*(c) Pour tout $f \in C_0(\Gamma\backslash\mathbb{H})$ et tout $z \in \Gamma\backslash\mathbb{H}$, on a*

$$\lim_{n\to+\infty} \frac{1}{\sigma_1(n)} T_n f(z) = \int_{\Gamma\backslash\mathbb{H}} f(\zeta) d\mu_0(\zeta). \tag{9}$$

*Cette convergence est uniforme sur les compacts.*

**2.1. Décomposition spectrale de $L^2(\Gamma\backslash\mathbb{H}, d\mu_0)$.** On rappelle que $L^2(\Gamma\backslash\mathbb{H}, d\mu_0)$ désigne l'espace des fonctions $\Gamma$–invariantes de carré intégrable pour la mesure de Poincaré. Cet espace se décompose via l'opérateur $\mathrm{D}_0$ sous la forme

$$L^2(\Gamma\backslash\mathbb{H}, d\mu_0) = \oplus_{n\geq 0} \mathbb{C}[\varphi_n] \oplus \mathcal{E}$$

où $\varphi_n$ est une famille orthonormée de fonctions propres de $\mathrm{D}_0$ de valeurs propres associées $-\lambda_n$ et $\mathcal{E}$ est la partie relative au spectre continu. La famille non bornée $0 = \lambda_0 < \lambda_1 \leq \lambda_2 \ldots$ forme le spectre discret de l'opérateur $\mathrm{D}_0$. On note $s_n$ et $r_n$

les nombres complexes vérifiant $\lambda_n = s_n(1 - s_n) = 1/4 + r_n^2$ ($r_n$ est réel d'après Roelcke). Il est possible de choisir les $\varphi_n$ propres pour tout les opérateurs de Hecke. Nous supposerons ce choix fait dans la suite.

La partie relative au spectre continu est donnée par l'isomométrie suivante:

$$\begin{aligned} E : L^2(\mathbb{R}_+) &\to \mathcal{E} \\ h &\to \frac{1}{\sqrt{2\pi}} \int_0^{+\infty} h(t) E_\infty\left(z, \frac{1}{2} + it\right) dt \end{aligned} \tag{10}$$

où $L^2(\mathbb{R}_+)$ désigne les fonctions de $\mathbb{R}_+$ de carré intégrable pour la mesure de Lebesgue et $E_\infty(z, s)$ est la série d'Eisenstein en la pointe $\infty$, donnée par la formule

$$E_\infty(z, s) = \frac{1}{2} \sum_{(m,n)=1} \frac{1}{|mz + n|^{2s}}.$$

Soit $\alpha \in L^2(\Gamma\backslash\mathbb{H}, d\mu_0)$ de décomposition spectrale

$$\alpha(z) = \sum_{n \geq 0} A_n \varphi_n(z) + \int_0^{+\infty} h(t) E_\infty\left(z, \frac{1}{2} + it\right) dt \tag{11}$$

alors

$$A_n = (\alpha, \varphi_n) = \int_X \alpha(z) \overline{\varphi}_n(z)\, d\mu_0(z)$$

$$h(t) = \frac{1}{2\pi} \int_X \alpha(z) E_\infty\left(z, \frac{1}{2} - it\right) d\mu_0(z)$$

(au moins si $\alpha$ est à décroissancce assez rapide, par exemple si $\alpha \in D(\Gamma\backslash\mathbb{H})$) et sa norme $L^2$ est donnée par

$$\|\alpha\|^2 = \sum_n |A_n|^2 + 2\pi \int_0^{+\infty} |h(t)|^2\, dt. \tag{12}$$

Si on suppose de plus que la fonction $\alpha(z)$ appartient à $D(\Gamma\backslash\mathbb{H})$, alors le développement (11) est absolument convergent et uniformément convergent sur les compacts [21] théorèmes (4.7) et (7.3).

**2.2. Preuve du théorème dans le cas classique.** Pour démontrer le théorème 2.1, on commence par traiter le cas où on a $f = \varphi_n$ pour un $n \in \mathbb{N}$ et le cas où $f$ est dans la partie continue du spectre de $D_0$. On finit en utilisant la décomposition spectrale (11).

Pour $f = \varphi_0$, c'est à dire pour les fonctions constantes le théorème 2.1 est trivial car $T_n$ est une correspondance de degré $\sigma_1(n)$.

LEMME 2.2. *Pour tout $k \geq 1$, et tout $z \in Y(1)$, on a*

$$\lim_{n \to \infty} \frac{T_n \varphi_k(z)}{\sigma_1(n)} = 0 = \int_{\Gamma\backslash\mathbb{H}} \varphi_k d\mu_0. \tag{13}$$

*Preuve.* On rappelle que $\varphi_n$ est une fonction propre de tout les opérateurs de Hecke. On définit $\alpha_k(n)$ par

$$T_n.\varphi_k = \alpha_k(n)\varphi_k.$$

La conjecture de Ramanujan-Petersson dans ce cadre prévoirait l'estimation:

$$|\alpha_k(n)| \leq d(n)n^{1/2},$$

où $d(n) = \sigma_0(n)$ est le nombre de diviseurs de $n$. La meilleure borne connue est [4]:

$$|\alpha_k(n)| \leq d(n)n^{1/2+\frac{5}{28}}. \tag{14}$$

On en déduit que l'on a pour tout $\epsilon > 0$ et $n$ assez grand

$$\left|\frac{T_n\varphi_k(z)}{\sigma_1(n)}\right| = \left|\frac{\alpha_k(n)}{\sigma_1(n)}f(z)\right| \leq n^{-1/2+\frac{5}{28}+\epsilon}|f(z)|. \tag{15}$$

Ceci termine la preuve du Lemme 2.2.

Lemme 2.3. *Soit $f$ une fonction dans $\mathcal{E} \cap D(\Gamma\backslash\mathbb{H})$. On a:*

$$\lim_{n\to\infty} \frac{T_n f(z)}{\sigma_1(n)} = 0 = \int_{\Gamma\backslash\mathbb{H}} f d\mu_0. \tag{16}$$

*Preuve.* On peut alors trouver une fonction $h(t)$ dans $L^2(\mathbb{R}_+)$ telle que:

$$f(z) = \int_0^\infty h(t)E_\infty\left(z, \frac{1}{2} + it\right)dt$$

et cette dernière intégrale est absolument convergente. On rappelle que la série d'Eisenstein $E_\infty(z, s)$ est propre pour tous les opérateurs de Hecke et que l'on a la relation:

$$T_n E_\infty(z, s) = n^s\sigma_{1-2s}(n)E_\infty(z, s).$$

On en déduit alors la relation

$$T_n f(z) = n^{1/2}\int_0^\infty n^{it}\sigma_{-2it}(n)h(t)E_\infty\left(z, \frac{1}{2} + it\right)dt. \tag{17}$$

On en déduit alors que pour tout $\epsilon > 0$ et tout $n$ assez grand, on a l'inégalité:

$$\left|\frac{T_n f(z)}{\sigma_1(n)}\right| \leq n^{-1/2+\epsilon}\int_0^\infty |h(t)E_\infty(z, s)|dt. \tag{18}$$

Ceci termine la preuve du lemme 2.3.

*Preuve du Théorème 2.1.* Soit $f$ une fonction dans $L^2(\Gamma\backslash\mathbb{H}, d\mu_0))$. On écrit sa décomposition spectrale sous la forme

$$f(z) = \sum_{k\geq 0} A_k\varphi_k(z) + \int_0^{+\infty} h(t)E_\infty\left(z, \frac{1}{2} + it\right) dt. \tag{19}$$

Si on pose

$$J_n = \left\| \frac{T_n f}{\sigma_1(n)} - \int_{\Gamma\backslash\mathbb{H}} f(\zeta)d\mu_0(\zeta) \right\|,$$

on obtient alors

$$J_n = \left\| \sum_{k\geq 1} \frac{A_k\alpha_k(n)\phi_k}{\sigma_1(n)} + \frac{n^{\frac{1}{2}}}{\sigma_1(n)} \int_0^\infty n^{it}\sigma_{-2it}(n)h(t)E_\infty(z,s)\,dt \right\|.$$

D'après l'expression de la norme $L^2$ donnée à l'équation (12), on a:

$$J_n^2 = \frac{1}{\sigma_1(n)^2} \sum_{k\geq 1} |A_k|^2|\alpha_k(n)|^2 + 2\pi \frac{n}{\sigma_1(n)^2} \int_0^\infty |h(t)|^2|\sigma_{-2it}(n)|^2 dt.$$

On termine la preuve de la première partie du théorème (2.1) en utilisant l'estimation sur les valeurs propres $\alpha_k(n)$ donnée dans l'équation (14).

On suppose maintenant que $f \in D(\Gamma\backslash\mathbb{H})$ et on fixe $z \in \Gamma\backslash\mathbb{H}$. En utilisant les équations (15) et (18) on obtient pour tout $\epsilon > 0$ et tout $n$ assez grand la majoration:

$$\left| \frac{T_n f(z)}{\sigma_1(n)} - A_0 \right| \leq n^{-1/2+5/28+\epsilon} \left( \sum_{k\geq 1} |A_k\varphi_k(z)| + \int_0^\infty |h(t)E_\infty(z,s)|dt \right).$$

Ceci termine la preuve de la deuxième partie du théorème (2.1) car $A_0 = \int_{\Gamma\backslash\mathbb{H}} f(z)d\mu_0(z)$ et l'expression dans le membre de droite de la dernière inégalité a un sens car $f \in D(\Gamma\backslash\mathbb{H})$ assure que sa décomposition spectrale converge absolument. L'assertion sur la constante $C_{\epsilon,z,f}$ résulte du fait que pour $f \in D(\Gamma\backslash\mathbb{H})$, la décomposition spectrale converge uniformément sur les compacts.

On suppose maintenant que $f \in C_0(\Gamma\backslash\mathbb{H})$. Soient $z \in \Gamma\backslash\mathbb{H}$ et $\epsilon > 0$. On peut alors trouver $\phi \in D(\Gamma\backslash\mathbb{H})$ telle que

$$\sup_{x\in\Gamma\backslash\mathbb{H}} |f(x) - \phi(x)| \leq \epsilon$$

D'après la deuxième partie du théorème 2.1, pour tout $n$ assez grand, on a

$$\left| \frac{T_n\phi(z)}{\sigma_1(n)} - \int_{\Gamma\backslash\mathbb{H}} \phi\, d\mu_0 \right| \leq \epsilon.$$

On pose $I_n = |\frac{T_n f(z)}{\sigma_1(n)} - \int_{\Gamma\backslash\mathbb{H}} f d\mu_0|$. On déduit de ce qui précède que

$$I_n \leq \left|\frac{T_n f(z)}{\sigma_1(n)} - \frac{T_n\phi(z)}{\sigma_1(n)}\right| + \left|\frac{T_n\phi(z)}{\sigma_1(n)} - \int_{\Gamma\backslash\mathbb{H}} \phi d\mu_0\right| + \left|\int_{\Gamma\backslash\mathbb{H}} (\phi - f) d\mu_0\right| \leq 3\epsilon.$$

Ceci termine la preuve de la troisième partie du théorème 2.1.

**2.3. Equidistribution des points CM.** Dans cette partie, nous allons étendre les résultats de Duke [16] sur l'équidistribution des points CM sur $Y = \Gamma\backslash\mathbb{H}$ pour donner une réponse à la question (1) dans le cas $g = 1$. Le travail de Duke permet de traiter les discriminants fondamentaux. Les techniques précédentes permettent d'aller dans la direction opposée: On fixe le corps de multiplication complexe et on fait varier l'ordre dans l'anneau d'entiers.

Soient $d$ un entier sans facteurs carrés, $K_d = \mathbb{Q}(\sqrt{-d})$, $O_{K_d}$ son anneau d'entiers et $h_d$ l'ordre du groupe de Picard de $O_{K_d}$. Les ordres de $O_{K_d}$ sont de la forme $O_{K_{d,f}} = \mathbb{Z} + f O_{K_d}$ pour un unique conducteur $f \geq 1$. On note $h_{d,f}$ le cardinal du groupe de Picard de $O_{K_{d,f}}$.

Soit $\Lambda_{d,f}$ l'ensemble des points ayant multiplication complexe par $O_{K_{d,f}}$; c'est un ensemble de cardinal $h_{d,f}$.

THÉORÈME 2.4. *Pour toute fonction $\phi \in C_0(Y)$, on a*

$$\frac{1}{h_{d,f}} \sum_{y\in\Lambda_{d,f}} \phi(y) \longrightarrow \int_{\Gamma\backslash\mathbb{H}} \phi(x) d\mu_0(x), \tag{20}$$

*quand $df \to \infty$.*

Comme l'ensemble $C_c^\infty(Y)$ des fonctions $C^\infty$ sur $Y$, à support compact, sont denses dans $C_0(Y)$, il suffit de prouver le théorème pour $\phi \in C_c^\infty(Y)$. En changeant $\phi$ par $\phi - \int_{\Gamma\backslash\mathbb{H}} \phi(x) d\mu_0(x)$ on se ramène au cas où $\int_{\Gamma\backslash\mathbb{H}} \phi(x) d\mu_0(x) = 0$. Soit

$$\phi(z) = \sum_{n\geq 1} A_n \phi_n(z) + \int_0^\infty h(t) E_\infty\left(z, \frac{1}{2} + it\right) dt \tag{21}$$

sa décomposition spectrale. On sait alors que pour tout $A$, on a $A_n = O(\frac{1}{n^A})$ et $h(t) = O(\frac{1}{t^A+1})$. Le théorème est alors conséquence de la proposition suivante:

PROPOSITION 2.5. *Il existe $A > 0$ tel que pour tout $\epsilon > 0$, il existe $C_\epsilon$ tel que:*

$$\left|\frac{1}{h_{d,f}} \sum_{y\in\Lambda_{d,f}} \phi_n(y)\right| \leq \frac{C_\epsilon n^A}{d^{\frac{1}{28}-\epsilon} f^{\frac{1}{2}-\frac{5}{28}-\epsilon}} \tag{22}$$

*et*

$$\left|\frac{1}{h_{d,f}} \sum_{y\in\Lambda_{d,f}} E_\infty\left(y, \frac{1}{2} + it\right)\right| \leq \frac{C_\epsilon (t^A + 1)}{d^{\frac{1}{28}-\epsilon} f^{\frac{1}{2}-\epsilon}}. \tag{23}$$

Quand $f = 1$, la proposition précédente est due à Duke. Pour tout $d$ sans facteurs carrés et tout $n \in \mathbb{N}$, on note $R_d(n)$ le nombre d'idéaux entiers de norme $n$ dans $O_{K_d}$. Pour tout couple $(d, f)$, on note $w_{d,f}$ le cardinal de $O^*_{K_{d,f}}/\mathbb{Z}^*$. Nous aurons besoin du lemme suivant dû à Zhang [42] (prop 3.2.1).

LEMME 2.6. *On a l'égalité entre diviseurs sur $Y$ suivante:*

$$T_f \frac{\Lambda_{d,1}}{w_{d,1}} = \sum_{c/f} R_d\left(\frac{f}{c}\right) \frac{\Lambda_{d,c}}{w_{d,c}}. \tag{24}$$

On va traiter le cas où $w_{d,c} = 1$ pour tout $c$ et on laisse au lecteur scrupuleux le soin d'écrire les modifications nécéssaires. Soit $\epsilon_d$ le caractère quadratique associé à l'extension $\mathbb{Q}(\sqrt{-d})/\mathbb{Q}$. On a alors la relation:

$$R_d(n) = \sum_{k/n} \epsilon_d(k) \tag{25}$$

où $\epsilon_d$ désigne le caractère quadratique de l'extension $\mathbb{Q}(\sqrt{-d})$ de $\mathbb{Q}$.

Faisons un court rappel sur les fonctions arithmétiques. Notre référence est [1]. Soit $f(n)$ et $g(n)$ deux fonctions arithmétiques, on définit le produit de convolution $f * g(n)$ par la formule

$$f * g(n) = \sum_{d/n} f(d) g\left(\frac{n}{d}\right).$$

Ce produit est alors associatif, commutatif et on dispose d'un élément neutre $I(n)$ qui est la fonction telle que $I(1) = 1$ et $I(n) = 0$ si $n \neq 1$. Si $f(1) \neq 0$, il existe une unique fonction arithmétique $f^{-1}$ telle que $f * f^{-1} = I$. Si $\mu(n)$ désigne la fonction de Moebius et $u(n)$ la fonction constante $u(n) = 1$, on a la relation $\mu * u = I$.

La relation (25) s'écrit alors dans ce langage $R_d(n) = \epsilon_d * u(n)$. La formule explicite pour le calcul de l'inverse nous donne alors la relation

$$R_d^{-1}(m) = \mu * (\mu \epsilon_d). \tag{26}$$

On déduit de cela que pour tout $\epsilon > 0$, on a $|R_d^{-1}(m)| = O(m^\epsilon)$.

Soit $\phi$ une fonction sur $Y = \Gamma \backslash \mathbb{H}$. On rappelle que si $E$ est un ensemble fini de points de $Y$, on définit $\phi(E)$ par la formule

$$\phi(E) = \sum_{y \in E} \phi(y).$$

On suppose que $\phi$ est propre pour tout les opérateurs de Hecke. On écrit alors $T_c \phi = \lambda_c(\phi)\phi$. On définit alors les deux fonctions arithmétiques

$$h(m) = T_m \phi(\Lambda_{d,1}) \tag{27}$$

et

$$g(m) = \phi(\Lambda_{d,m}). \tag{28}$$

On a donc la relation $h = R_d * g$ donc $g = R_d^{-1} * h$. Un simple calcul prouve que l'on a:

$$\sum_{y \in \Lambda_{d,f}} \phi(y) = \sum_{y \in \Lambda_{d,1}} \sum_{c/f} R_d^{-1}\left(\frac{f}{c}\right) \lambda_c(\phi)\phi(y). \tag{29}$$

Si on prend $\phi = 1$ dans cette équation,

$$h_{d,f} = \mathrm{card}(\Lambda_{d,f}) \geq h_d f. \tag{30}$$

Quand $\phi = \phi_n$ est une forme de Maass ou une série d'Eisenstein, Duke montre ([16] équations (6-5) et (6.6)) l'existence d'un $A > 0$ tel

$$\left| \frac{1}{h_d} \sum_{y \in \Lambda_{d,1}} \phi_n(y) \right| \leq n^A O(|d|^{-\frac{1}{28}+\epsilon})$$

et

$$\left| \frac{1}{h_d} \sum_{y \in \Lambda_{d,1}} E_\infty\left(z, \frac{1}{2} + it\right) \right| \leq t^A O(|d|^{-\frac{1}{28}+\epsilon}).$$

On finit alors la preuve de la proposition en utilisant les estimations données pour les valeurs propres de Hecke des formes de Maass et des séries d'Eisenstein dans les équations (14) et (17) et l'estimée $R_d^{-1}(c) = O(c^\varepsilon)$ pour tout $\varepsilon > 0$.

**2.4. Application: Un théorème de rigidité.** Nous allons déduire du théorème 2.1 le résultat de rigidité suivant. Soient $\Gamma = SL(2, \mathbb{Z})$ et $Y = \Gamma \backslash \mathbb{H}$. Soit

$$h : Y \longrightarrow Y$$

une application continue. Soit $n$ un entier, on dit que $h$ commute à l'opérateur $T_n$ si pour tout $z \in Y$, l'ensemble $T_n(h(z))$ coïncide (avec ses multiplicités) avec $h(T_n(z))$. Soit $T_{-1}$ l'application de $Y$ dans $Y$ obtenue par passage au quotient de l'application $(z \to -\overline{z})$ de $\mathbb{H}$. On vérifie que $T_{-1}$ commute avec tout les opérateurs de Hecke. Le but de cette partie est de montrer le résultat suivant:

THÉORÈME 2.7. *Soient $(p_1, p_2, p_3)$ trois nombres premiers distincts. Soit $h : Y \to Y$ une application continue et propre commutant avec $T_{p_i}$ ($i \in \{1, 2, 3\}$). Alors $h$ est l'identité ou $h = T_{-1}$.*

LEMME 2.8. *Soit $p$ un nombre premier. Soit $h : Y \to Y$ une application propre continue commutant à un opérateur de Hecke $T_p$, alors $h$ préserve la mesure invariante sur $Y$.*

Comme $h$ est propre, $h$ preserve $C_c(Y)$. Le lemme se déduit alors de la convergence faible de la suite de mesure $\mu_n$ associée à l'ensemble $E_n = (T_p)^n.z_0$ vers la mesure invariante $d\mu_0$ qui se déduit du théorème 2.1. (Voir § 8 dans le cas le plus général.)

On rappelle que l'on a une décomposition $L^2(Y, d\mu_0) = L^2_{disc} \oplus \mathcal{E}$ où $L^2_{disc}$ désigne la partie discrète et $\mathcal{E}$ désigne la partie continue du spectre. Comme $h$ commute avec un opérateur $T_p$ qui est borné et autoadjoint il conserve son spectre discret et son spectre continu. On en déduit alors que:

LEMME 2.9. *L'application*

$$H : L^2(Y, d\mu_0) \to L^2(Y, d\mu_0)$$

*définie par* $H(f)(z) = f(h(z))$ *préserve* $L^2_{disc}$ *et* $\mathcal{E}$.

LEMME 2.10. *Soit* $h : Y \to Y$ *une application vérifiant les hypothèses du théorème 2.4. Pour tout* $z \in Y$ *et tout* $t \in \mathbb{R}$, *on a*

$$E_\infty\left(h(z), \frac{1}{2} + it\right) = c(t)E_\infty\left(z, \frac{1}{2} + it\right) \tag{31}$$

*pour une fonction* $c(t)$ *continue sur* $\mathbb{R}$ *telle que* $|c(t)| = 1$.

Soit $E$ l'isométrie entre $L^2(\mathbb{R}^+)$ et $\mathcal{E}$ donnée dans (10). Fixons $T > 0$. Pour tout $\alpha \in L^2[0, T]$, on définit $K_T(\alpha) = E^{-1}(H(E(\alpha)))$. On a donc la relation

$$\int_0^T \alpha(t)E_\infty\left(h(z), \frac{1}{2} + it\right) dt = \int_0^\infty K_T(\alpha)(t))E_\infty\left(z, \frac{1}{2} + it\right) dt. \tag{32}$$

Soit $p$ un des trois nombres premiers $(p_1, p_2, p_3)$. On pose

$$\eta_p(t) = p^{it} + p^{-it}.$$

On a donc

$$T_p E_\infty\left(z, \frac{1}{2} + it\right) = p^{\frac{1}{2}}\eta_p(t)E_\infty\left(z, \frac{1}{2} + it\right).$$

Comme $h$ commute à $T_p$, on a

$$K_T(\alpha\eta_p)(t) = \eta_p(t)K_T(\alpha)(t). \tag{33}$$

On commence par montrer:

LEMME 2.11. *Soit* $\mathcal{A}$ *l'algèbre des fonctions continues complexes sur* $[0, T]$ *engendrée par les fonctions constantes et* $(\eta_{p_1}, \eta_{p_2}, \eta_{p_3})$. *Alors* $\mathcal{A}$ *est dense dans l'ensemble* $C([0, T])$ *des fonctions continues complexes sur* $[0, T]$.

*Preuve.* Puisque $\overline{\eta_p}(t) = \eta_p(t)$, il suffit par le théorème de Stone-Weierstrass de voir que $\mathcal{A}$ sépare les points. Or si $\eta_p(t) = \eta_p(t')$, il existe $k \in \mathbb{Z}$ tel que $t = \pm t' + \frac{2k\pi}{\ln p}$.

Si une telle égalité est vérifiée pour trois nombres premiers distincts, on obtient $t = t'$. Ceci termine la preuve du lemme (2.11).

Posons $K^T(\alpha) = K_T(\alpha)|_{[0,T]}$. D'après l'équation (33), $K^T$ commute à la multiplication par les fonctions continues sur $[0, T]$. On en déduit que $K^T$ est la multiplication par la fonction $c_T(t) = K^T(1)$. Si $T' > T$, la restriction à $[0, T]$ de $c_{T'}$ coïncide avec $c_T$. On en déduit que $c_T(t)$ est indépendant de $T$. On note alors $c(t)$ la fonction sur $\mathbb{R}^+$ qui vaut $c_T(t)$ pout tout $T > t$. Soient maintenant $\alpha \in L^2(\mathbb{R}^+)$ et $\chi_T$ la fonction caractéristique de $[0, T]$. En appliquant ce qui précède à la fonction $\alpha\chi_T$ et en faisant tendre $T$ vers l'infini, on trouve l'égalité entre fonctions de $\mathcal{E}$ suivante:

$$\int_0^\infty \alpha(t) E_\infty\left(h(z), \frac{1}{2} + it\right) dt = \int_0^\infty \alpha(t) c(t) E_\infty\left(z, \frac{1}{2} + it\right) dt. \tag{34}$$

Pour des choix convenables de $\alpha(t)$, à $z$ fixé, on a des convergences absolues dans les intégrales précédentes. On en déduit que

$$E_\infty\left(h(z), \frac{1}{2} + it\right) = c(t) E_\infty\left(z, \frac{1}{2} + it\right). \tag{35}$$

Comme les séries d'Eisenstein sont analytiques en $t$ ([21] théorème 6-11) et non identiquement nulles en $z$ si $t \neq 0$ ([21] proposition 6-12), la fonction $c(t)$ est analytique pour tout $t \neq 0$. Comme $h$ préserve la mesure invariante, on a presque partout $|c(t)| = 1$. Par continuité de $c(t)$, on obtient bien $|c(t)| = 1$. Ceci termine la preuve du lemme (2.10).

Lemme 2.12. *On a* $c(t) = 1$.

*Preuve.* On note, pour tout $s \in \mathbb{C}$, $\theta(s) = \pi^{-s}\Gamma(s)\zeta(2s)$ et $\phi(s) = \frac{\theta(s)}{\theta(1-s)}$. L'équation fonctionelle des séries d'Eisenstein ([21] chapitre 3) donne:

$$\overline{E_\infty\left(z, \frac{1}{2} + it\right)} = E_\infty\left(z, \frac{1}{2} - it\right) = \phi\left(\frac{1}{2} - it\right) E_\infty\left(z, \frac{1}{2} + it\right). \tag{36}$$

En écrivant cette égalité pour $h(z)$ et en utilisant la définition de $c(t)$, on trouve

$$\overline{c(t) E_\infty\left(z, \frac{1}{2} + it\right)} = \phi\left(\frac{1}{2} - it\right) c(t) E_\infty\left(z, \frac{1}{2} + it\right). \tag{37}$$

Comme pour tout $t \neq 0$, $E_\infty(z, \frac{1}{2} + it)$ n'est pas identiquement nulle et que la fonction $\phi(\frac{1}{2} + it)$ ne s'annule pas (car $\theta(s)$ ne s'annule pas pour $\mathrm{Re}(s) = \frac{1}{2}$) les équations (36) et (37) impliquent que pour $t \neq 0$, $c(t) = \overline{c(t)}$. Comme $|c(t)| = 1$ et $c(t)$ est continue, on en déduit que $c = \pm 1$. Par prolongement analytique, on a pour tout $s$ où $E_\infty(z, s)$ est holomorphe

$$E_\infty(h(z), s) = \pm E_\infty(z, s).$$

Pour $s$ réel et assez grand la série d'Eisenstein est positive donc $c(t) = 1$. On a donc

$$E_\infty(h(z), s) = E_\infty(z, s). \tag{38}$$

On va déduire la démonstration du théorème de ce qui précède et du comportement de séries d'Eisenstein pour $s$ réel tendant vers $+\infty$. En utilisant l'expression explicite de la série d'Eisenstein pour $\mathrm{Re}(s) > 1$, $z = x + iy$:

$$E_\infty(z, s) = \frac{1}{2} \sum_{(c,d)\in\mathbb{Z}\ |(c,d)=1} \frac{y^s}{|cz+d|^{2s}}, \tag{39}$$

on montre facilement le lemme suivant

Lemme 2.13. *Soit $D = \{z \in \mathbb{H}$ tels que $|z| > 1$ , $-\frac{1}{2} < Re(z) < \frac{1}{2}\}$ le domaine usuel pour $\Gamma$. Soit $z = x + iy \in D$. Pour $s$ réel, $s \to +\infty$, on a le développement asymptotique*

$$E_\infty(z, s) = y^s(1 + |z+1|^{-2s} + |z-1|^{-2s} + |z|^{-2s} + O(\rho^s)), \tag{40}$$

*avec $\rho < Inf(|z+1|^{-2}, |z-1|^{-2}) < 1$.*

On déduit de cela (et du fait que pour $z \in D$, on a $|z \pm 1| > 1$) que si $(z = x + iy, z' = x' + iy') \in D$ et $E_\infty(z, s) = E_\infty(z', s)$, alors $y = y'$ et

$$\{|z+1|, |z-1|\} = \{|z'+1|, |z'-1|\}.$$

Il en résulte que $z' = z$ ou $z' = -\bar{z}$. Nous pouvons maintenant compléter la preuve du théorème.

Soit $\Delta \subset X$ l'image de $D$, c'est un ouvert de mesure totale dans $X$. Soit $\Delta' = h^{-1}(\Delta)$. Comme $h$ préserve la mesure, c'est encore un ouvert de mesure totale dans $X$. Soit $D'$ l'image inverse de $\Delta'$ dans $D$, c'est un ouvert de mesure totale dans $D$ et on peut considérer $h$ comme une application continue de $D'$ dans $D$. Il résulte du lemme 2.4 et de l'équation (38) que pour $z \in D'$, on a $h(z) = z$ ou $h(z) = -\bar{z}$. On en déduit par continuité de $h$ que pour tout $z \in D$ on a $h(z) = z$ ou $h(z) = -\bar{z}$. Il en résulte que $h$ peut être vu comme une application continue de $D$ dans $D$. Soit

$$D_- = \{z \in D; \mathrm{Re}(z) < 0\} \text{ et } D_+ = \{z \in D; \mathrm{Re}(z) > 0\}.$$

Par connexité, on a $h(D_-) = D_-$ ou $h(D_-) = D_+$. Comme $h$ préserve la mesure, on en déduit enfin que $h$ est donné par l'identité ou par $z \to -\bar{z}$ sur $D_- \cup D_+$ et donc partout.

## 3. Préliminaire.

**3.1 Interprétation adélique.** Soit $G$ un groupe réductif déployé défini sur $\mathbb{Z}$ et $\Gamma = G(\mathbb{Z})$. On a en vue $G = G_1 = GL_n$ ou $G = G_2 = GSp(2g)$. Soit $\mathbf{A}$ l'anneau des adèles de $\mathbb{Q}$ et $\mathbf{A_f}$ celui des adèles finies. On note $G_f = G(\mathbf{A_f})$. Pour un nombre premier $p$, on note $\mathbb{Q}_p$ le corps des nombres $p$-adiques et $\mathbb{Z}_p$ son anneau d'entiers.

Soit $K_p = G(\mathbb{Z}_p)$, $K_f = \Pi_p K_p$. On se donne de plus un compact maximal $K_\infty$ de $G(\mathbb{R})$ et on pose $K = K_\infty \times K_f$. Soit $G^+(\mathbb{R})$ la composante connexe de l'élément neutre de $G(\mathbb{R})$, on suppose que l'on a une décomposition

$$G(\mathbf{A}) = G(\mathbb{Q})G^+(\mathbb{R})K_f.$$

C'est le cas pour $G = G_1$ et $G = G_2$. On a alors:

$$\Gamma = G(\mathbb{Q}) \cap (G(\mathbb{R}) \times K_f)$$

et $\Gamma$ est un sous-groupe discret de $G(\mathbb{R})$. On a une bijection entre l'espace des fonctions sur $G(\mathbb{Q})\backslash G(\mathbf{A})/K_f$ et l'espace des fonctions sur $\Gamma\backslash G(\mathbb{R})$ donnée par

$$\alpha \longrightarrow (x \to \alpha(x \times 1_f)), \tag{41}$$

où $1_f$ désigne l'adèle finie dont toutes les composantes valent 1. On note $f \to \phi_f$ la bijection réciproque.

Soit $H_p = H(G_p, K_p)$, l'algèbre de Hecke locale des fonctions sur $G_p = G(\mathbb{Q}_p)$ bi-invariantes sous $K_p$ et de support contenu dans une union finie de doubles classes de la forme $K_p g_p K_p$, $g_p \in G_p$. On pose

$$H_f(G_f, K_f) = \bigotimes_p H(G_p, K_p).$$

On dispose d'une action par convolution de $H_f(G_f, K_f)$ sur l'espace des fonctions sur $G(\mathbb{Q})\backslash G(\mathbf{A})/K_f$.

Soient $r \in [1, n-1]$, $ch_{r,p} \in H(GL_n(\mathbb{Q}_p), K_p)$ la fonction caractéristique de la double classe

$$K_p \mathrm{Diag}(p, \ldots, p, 1, \ldots, 1) K_p$$

et $T'_{r,p}$ l'opérateur associé sur les fonctions sur $GL_n(\mathbb{Q})\backslash GL_n(\mathbf{A})/K_f$. On a alors la relation

$$T'_{r,p}\phi_f = \phi_{T_{r,p}\cdot f} \tag{42}$$

De même soit $ch_p \in H(GSp(2g)(\mathbb{Q}_p), K_p)$, la fonction caractéristique de la double classe

$$K_p \mathrm{Diag}(p, \ldots, p, 1, \ldots, 1) K_p$$

et $T'_p$ l'opérateur associé, on a

$$T'_p\phi_f = \phi_{T_p\cdot f}. \tag{43}$$

**3.2. Décomposition spectrale de $L^2(\Gamma\backslash G(\mathbb{R}), 1)$.** Soit $G$ un groupe algébrique réductif défini et déployé sur $\mathbb{Q}$. Le but de cette partie est d'expliquer les informations sur la décomposition spectrale de l'espace $L^2(\Gamma\backslash G(\mathbb{R}), \mathbf{1})$ dont nous aurons besoin dans la suite de ce texte. Les références pour cette partie sont [3], [29]. On garde les notations de la partie précédente. On fixe un parabolique minimal $P_0 = B$ de composant de Lévi $M_0$.

Soit $P$ un parabolique standard (donc défini sur $\mathbb{Q}$ et contenant $P_0$). On note $N_P$ le radical unipotent de $P$ et $M_P$ l'unique composant de Lévi de $P$ contenant $M_0$ et $T_P$ le tore maximal (déployé) qui est central dans $M_P$. Quand il n'y a pas d'ambiguïté sur $P$, on se permet de supprimer les indices "$P$" dans les notations précédentes. On rappelle que l'on sait définir une relation d'équivalence, la relation d'*association,* pour les paraboliques standard [3].

On note $X^*(M)$ le groupe des caractères rationnels de $M$,

$$\mathfrak{a}_M^* = X^*(M) \otimes_{\mathbb{Z}} \mathbb{C}$$

et $X_*(M) = \mathrm{Hom}_{\mathbb{Z}}(X^*(M), \mathbb{Z})$. Pour tout $\chi \in X^*(M)$, on définit un homomorphisme continu $|\chi|$ de $M(\mathbf{A})$ dans $\mathbb{C}^*$ en associant à tout $m = (m_l) \in M(\mathbf{A})$ le nombre complexe $|\chi|(m) = \Pi_l |m_l^{\chi_l}|_l$ et on note

$$M^1 = \cap_{\chi \in X^*(M)} \mathrm{Ker}|\chi|.$$

On note $X_M$ le groupe des homomorphimes continus de $M(\mathbf{A})$ dans $\mathbb{C}^*$ qui sont triviaux sur $M^1$ et $X_M^G$ le sous-groupe de $X_M$ formé des homomorphismes qui sont de plus triviaux sur le centre $Z(\mathbf{A})$ ($Z$ désignant le centre de $G$). On peut alors identifier $X_M^G$ à un sous-espace vectoriel de $\mathfrak{a}_M^*$ [29]. On note $\mathrm{Im}(X_M^G)$ la partie imaginaire de $X_M^G$.

On rappelle que $L^2(\Gamma\backslash G(\mathbb{R}), \mathbf{1})$ désigne l'espace des fonctions sur $\Gamma\backslash G(\mathbb{R})$ qui sont triviales sur le centre $Z(\mathbb{R})$ de $G(\mathbb{R})$ et qui sont de carré intégrable modulo le centre. On a alors la décomposition de Langlands qui réalise $L^2(\Gamma\backslash G(\mathbb{R}), \mathbf{1})$ comme un sous-module de

$$\bigoplus_{[P]} \bigoplus_{\tau} \int_{\mathrm{Im}(X_{M_P}^G)} \mathrm{Ind}_{P(\mathbf{A})}^{G(\mathbf{A})} (\tau \otimes s \otimes 1)^{K_f} d\mu_\tau(s) \tag{44}$$

où $[P]$ parcourt les classes d'association de paraboliques, $\tau$ décrit les représentations automorphes irréductibles intervenant dans la partie discrète de la représentation régulière de $M(\mathbf{A})$ sur $L^2(T(\mathbb{R})^0 M(\mathbb{Q})\backslash M(\mathbf{A}))$. Dans la somme directe (44), $L^2(\Gamma\backslash G(\mathbb{R}), \mathbf{1})$ est l'espace des fonctions vérifiant les équations fonctionnelles données par les opérateurs d'entrelacement [29].

La représentation

$$\mathrm{Ind}_{M(\mathbf{A})N(\mathbf{A})}^{G(\mathbf{A})} (\tau \otimes s \otimes 1)$$

est l'induite unitaire de la représentation de $P(\mathbf{A}) = M(\mathbf{A})N(\mathbf{A})$ qui est l'image inverse de la représentation $\tau \otimes s$. Cet espace s'envoie dans l'espace $L^2(G(\mathbb{Q})\backslash G(\mathbf{A}), 1)$ via la théorie des séries d'Eisenstein et on prend les $K_f$-invariants.

**3.3. Action des opérateurs de Hecke.** Pour comprendre l'action des opérateurs de Hecke sur les fonctions intervenant dans la décomposition spectrale (44), il faut décrire les représentations qui interviennent dans cette décomposition ainsi que l'action de $H(G_p, K_p)$ sur les vecteurs $K_f$-invariants de ces représentations.

Soit $\pi$ une telle représentation, on a une décomposition sous la forme d'un produit tensoriel restreint $\pi = \hat{\otimes}\pi_l$ [18] et la composante en $p$ est une représentation unitaire non ramifiée de $G_p = G(\mathbb{Q}_p)$. On sait alors que $\pi_p$ est unitairement induite à partir d'un caractère non ramifié $\chi$ de $T_0(\mathbb{Q}_p)$.

Soit $N(T_0)$ le normalisateur de $T_0$ et $W \simeq N(T_0)/T_0$ le groupe de Weyl de $G$. Le caractère $\chi$ n'est défini qu'à l'action de $W$ près. On note $X^*(T_0)$, le groupe des caractères rationnels de $T_0$ et $X_*(T_0)$ celui des cocaractères. On note $< \, , \, >$ l'accouplement de dualité sur $X_*(T_0) \times X^*(T_0)$. Soit ${}^LG$ le $L$-groupe associé à $G_p$ (voir [5]). C'est un groupe réductif connexe sur $\mathbb{C}$ dont le système de racines est dual de celui de $G$. On fixe un tore maximal $\hat{T}_0$ dans un sous-groupe de Borel de ${}^LG$. On a alors un isomorphisme

$$X^*(\hat{T}_0) \simeq X_*(T_0).$$

Soit $(t_1, \ldots, t_n)$ les coordonnées sur $\hat{T}_0$. La transformée de Sataké induit un isomorphisme de l'algèbre de Hecke locale $H(G_p, K_p) \otimes \mathbb{C}$ avec l'algèbre $\mathbb{C}[X^*(\hat{T}_0)]^W$ des polynômes en les $t_i^{\epsilon_i}$, ($\epsilon_i = +1$ ou $-1$), qui sont invariants par le groupe de Weyl. On note $Sf$ la transformée de Sataké de $f \in H(G_p, K_p)$.

On sait par ailleurs qu'il existe une bijection entre les caractères non ramifiés de $T_0(\mathbb{Q}_p)$ modulo l'action de $W$ et les classes de conjugaison d'éléments semi-simples dans ${}^LG(\mathbb{C})$. On note $t_\chi = (\alpha_1, \ldots, \alpha_n)$ un représentant de la classe de conjugaison d'éléments semi-simples associée à $\chi$. On sait que l'espace des vecteurs $K_p$-invariants de $\pi$ est propre pour l'action des $f \in H(G_p, K_p)$ de vecteur propre $Sf(\alpha_1, \ldots, \alpha_n)$.

Le calcul de la transformée de Sataké peut se révéler très complexe; pour les opérateurs de Hecke utilisés dans ce texte elle prend une forme particulièrement simple. Soit $\phi \subset X^*(T_0)$ l'ensemble des racines de $G_p$, $\phi^+$ le sous-ensemble des racines positives et $\rho \in X^*(T_0)$ la demi somme des racines positives. On note $P^+ \subset X_*(T_0)$ la chambre de Weyl positive définie par

$$P^+ = \{\lambda \in X_*(T_0) \mid <\lambda, \alpha> \geq 0 \text{ pour tout } \alpha \in \phi^+\}.$$

On a alors une décomposition de la forme

$$G_p = \cup_{\lambda \in P^+} K_p \lambda(p) K_p$$

et les fonctions caractéristiques $ch_\lambda$ de $K_p\lambda(p)K_p$ pour $\lambda \in P^+$ forment une base de $H(G_p, K_p)$. Par ailleurs les $\lambda \in P^+ \subset X^*(\hat{T}_0)$ paramètrent les représentations irréductibles de dimensions finies $V_\lambda$ de ${}^LG$; $\lambda$ est le plus haut poids de $V_\lambda$. Si $\lambda$ est un poids minuscule de ${}^LG$ la transformée de Sataké de $ch_\lambda$ prend la forme suivante [24]:

$$S\,ch_\lambda = p^{<\lambda,\rho>}\mathrm{Tr}(V_\lambda). \tag{45}$$

On vérifie que l'on est dans cette situation quand $G = GL_n$ ou $G = GSp(2g)$ avec les opérateurs de Hecke considéré dans ce texte.

Quand $G = GL_n$, on peut prendre $t_\chi = (t_1, \ldots, t_n)$ avec

$$t_i = \chi(\mathrm{Diag}(1, \ldots, 1, p, 1, \ldots, 1)). \tag{46}$$

La représentation associé à $ch_{r,p}$ est la puissance exterieure $r$-ième de la représentation standard de ${}^L G = GL_n(\mathbb{C})$ et on trouve

$$S\,ch_{r,p}(\alpha_1, \ldots, \alpha_n) = p^{\frac{r(n-r)}{2}} \sum_{1\le i_1<i_2<\cdots<i_r\le n} \prod_j \alpha_{i_j}. \tag{47}$$

On est aussi dans cette situation pour $G = GSp(2g)$ et $ch_p$. Le résultat est donné dans le lemme 6.4. Pour le groupe $G = SO(2g+1)$, le résultat est donné dans le lemme (7.2).

## 4. Le cas de $GL_N$.

**4.1. Le théorème de Moeglin-Waldspurger.** Soit $n \geq 2$ un entier et $G = GL_n$. Les notations générales des parties précédentes s'appliquent. Les sous-groupes de Lévi standard de $G$ sont de la forme

$$M = GL_{N_1} \times \cdots \times GL_{N_r}$$

pour des entiers $(N_1, \ldots, N_r)$ tels que $n = \sum N_i$. On note $P$ le parabolique standard de composant de Lévi $M$ formé de matrices triangulaires supérieures et $N$ son radical unipotent. Le but de cette partie est de décrire, grâce au théorème de Moeglin et Waldspurger [30] les paramètres des représentations non ramifiées qui interviennent dans la décomposition spectrale (44). Ces paramètres ne sont définis qu'a l'action du groupe symétrique $S_n$ près.

Soit $\pi = \hat{\otimes}_l \pi_l$ une représentation automorphe cuspidale de $GL_n(\mathbf{A})$. Soit $p$ une place telle que $\pi_p$ soit non ramifiée. Soit $\chi$ un caractère de $T_0(\mathbb{Q}_p)$ associé et $t_\chi = (t_1, \ldots, t_n)$ (donnée par (46)). La conjecture de Ramanujan-Petersson prévoit que pour tout $j \in [1, \ldots, n]$, on a $|t_j| = 1$. Une approximation de cette conjecture est donné par un théorème de Jacquet et Shalika [22]:

$$p^{-\frac{1}{2}} < |t_i| < p^{\frac{1}{2}}. \tag{48}$$

Si $n = 2$, on a une meilleure estimée [19]:

$$p^{-\frac{1}{4}} < |t_i| < p^{\frac{1}{4}}. \tag{49}$$

Ces estimations sont en fait valable sur tout corps de nombres et de meilleures estimations sont données dans [26] [27] [35], mais ces améliorations ne nous seront pas utiles dans la suite.

Soit $a$ et $b$ deux entiers et $N = ab$. On pose

$$M' = GL_a \times \cdots \times GL_a \subset GL_N$$

et $P'$ le parabolique associé comme précédemment. Soit $\sigma = \hat{\otimes}\sigma_l$ une représentation automorphe cuspidale unitaire de $Gl_a(\mathbf{A})$. Pour tout $s \in \mathbb{C}$, on note $\sigma[s]$ la représentation telle que pour tout $g \in GL_a(\mathbf{A})$ on ait

$$\sigma[s](g) = |\det|^s \sigma(g).$$

Soit

$$\theta = \sigma\left[\frac{b-1}{2}\right] \otimes \sigma\left[\frac{b-3}{2}\right] \otimes \cdots \otimes \sigma\left[\frac{1-b}{2}\right],$$

(il s'agit de produits tensoriels externes) c'est une représentation de $M'(\mathbf{A})$. Par un théorème de Langlands, on sait que la représentation de $GL_N(\mathbf{A})$:

$$Ind_{P'(\mathbf{A})}^{GL_N(\mathbf{A})} \theta \otimes \mathbf{1}$$

admet un unique quotient irréductible $J(\sigma)$. Jacquet et Shalika [23] montrent que cette représentation intervient dans le spectre discret de $GL_N(\mathbf{A})$. Le théorème de Moeglin-Waldspurger [30] nous assure que quand on fait varier $a$ parmi les diviseurs de $N$ et $\sigma$ parmi les représentations automorphes cuspidales on obtient tout le spectre discret de $GL_N(\mathbf{A})$.

En une place $p$, où $\sigma$ est non ramifiée, on note $(t_1, \ldots, t_a)$ les paramètres de $\sigma_p$. La représentation $J(\sigma)_p$ est alors non ramifiée et a pour paramètres:

$$(50)\quad t_{\sigma,a,b} = \left(t_1 p^{\frac{b-1}{2}}, \ldots, t_a p^{\frac{b-1}{2}}, t_1 p^{\frac{b-3}{2}}, \ldots, t_a p^{\frac{b-3}{2}}, \ldots, t_1 p^{\frac{1-b}{2}}, \ldots, t_a p^{\frac{1-b}{2}}\right).$$

On revient maintenant à la situation intiale où $n = N_1 + N_2 + \cdots + N_r$. On suppose que pour tout $i \in [1, \ldots, r]$ on a des entiers, $a_i$ et $b_i$ tels que $N_i = a_i b_i$. On se donne des représentations cuspidales automorphes $\sigma_i = \hat{\otimes}\sigma_{i,l}$ de $GL_{a_i}(\mathbf{A})$ et on fixe une place $p$ pour laquelle $\sigma_{i,p}$ est non ramifiée. On note $(t_{i,1}, \ldots, t_{i,a_i})$ les paramètres de $\sigma_{i,p}$. Soit

$$\tau = J(\sigma_1) \otimes \cdots \otimes J(\sigma_r)$$

la représentation discrète de $M(\mathbf{A})$ obtenue par la construction précédente. Pour tout $s \in \mathrm{Im} X_M^G$ telle que la représentation

$$\pi = \mathrm{Ind}_{P(\mathbf{A})}^{G(\mathbf{A})} \tau \otimes s \otimes \mathbf{1} = \hat{\otimes}\pi_l$$

soit irréductible et non ramifiée en $p$, les paramètres de $\pi_p$ sont de la forme

$$t_\pi = (t'_{\sigma_1,a_1,b_1}, \ldots, t'_{\sigma_i,a_i,b_i})$$

avec

$$(51)\qquad t'_{\sigma_i,a_i,b_i} = \left(t'_{i,1} p^{\frac{b_i-1}{2}}, \ldots, t'_{i,a_i} p^{\frac{b_i-1}{2}}, \ldots, t'_{i,1} p^{\frac{1-b_i}{2}}, \ldots, t'_{i,a_i} p^{\frac{1-b_i}{2}}\right)$$

où les $t'_{i,j}$ sont des nombres complexes tels que $|t'_{i,j}| = |t_{i,j}|$.

On définit dans la section suivante la notion de matrices de Hecke admissibles. La définition est faite de telle sorte que, au vu de la discussion précédente, des estimées de Jacquet-Shalika et du fait que les représentations automorphes cuspidales ont un caractère central unitaire on a montré le lemme:

LEMME 4.1. *Les paramètres locaux en $p$ d'une représentation $\pi$ intervenant dans la décomposition spectrale (44) sont les coefficients d'une matrice de Hecke admissible.*

**4.2. Matrices de Hecke admissibles.** Soit $n$ un entier positif, $n = \sum_{i=1}^{r} a_i b_i$ pour des entiers $a_i$ et $b_i$ positifs. On suppose que si $i \leq j$ alors $b_i \geq b_j$. Pour tout $i \in [1, \ldots, r]$, on se donne des nombres complexes $(t_{i,1}, \ldots, t_{i,a_i})$ avec les propriétés suivantes:

$$\text{Pour tout couple } (i, j) \in [1, \ldots, r] \times [1, \ldots, a_i] \text{ on a } p^{\frac{-1}{2}} < |t_{i,j}| < p^{\frac{1}{2}}. \tag{52}$$

$$\text{Pour tout } i \in [1, \ldots, r] \text{ on a } \prod_{j=1}^{a_i} |t_{i,j}| = 1. \tag{53}$$

$$\text{Si } a_i = 2 \text{ on a } |t_{i,1}| < p^{\frac{1}{4}} \text{ et } |t_{i,2}| < p^{\frac{1}{4}}. \tag{54}$$

On note $D_i$ la matrice diagonale de $GL_{a_i b_i}$:

$$\text{Diag}\left(t_{i,1} p^{\frac{b_i-1}{2}}, \ldots, t_{i,a_i} p^{\frac{b_i-1}{2}}, t_{i,1} p^{\frac{b_i-3}{2}}, \ldots, t_{i,a_i} p^{\frac{b_i-3}{2}}, \ldots, t_{i,1} p^{\frac{1-b_i}{2}}, \ldots, t_{i,a_i} p^{\frac{1-b_i}{2}}\right).$$

On note $D = \text{Diag}(D_1, \ldots, D_r)$. On dit qu'une matrice de ce type est une matrice de Hecke admissible. Les $t_{i,j}$ sont appelés paramètres cuspidaux de $D$. Une matrice de Hecke est dite de type Ramanujan si tous ses paramètres cuspidaux sont de module 1. Pour toute matrice de Hecke admissible $D = \text{Diag}(x_1, \ldots, x_n)$, on note $|D| = \text{Diag}(|x_1|, \ldots, |x_n|)$ et $D_R$ la matrice obtenue à partir de $D$ en remplaçant tous les paramètres cuspidaux par 1. On constate que $|D|$ est une matrice de Hecke admissible et que $D_R$ est une matrice de Hecke de type Ramanujan. On note $D_{triv}$ la matrice $D_{triv} = \text{Diag}(p^{\frac{n-1}{2}}, p^{\frac{n-3}{2}}, \ldots, p^{\frac{1-n}{2}})$.

On se donne de plus un polynôme homogène symétrique

$$P_{r_1,\ldots,r_n}(x_1, \ldots, x_n) \tag{55}$$

obtenu par symétrisation à partir du monôme $\prod_{i=1}^{n} x_i^{r_i}$. On suppose que si $i \leq j$ alors $r_i \geq r_j$. Si $D = \text{Diag}(a_1, \ldots, a_n) \in GL_n$, on note

$$P_{r_1,\ldots,r_n}(D) = P_{r_1,\ldots,r_n}(a_1, \ldots, a_n).$$

Le but de cette partie est de montrer la proposition suivante.

Proposition 4.2. *Pour tout $n \geq 3$ et pour toute matrice de Hecke admissible $D$ telle que $|D| \neq D_{triv}$, on a:*

$$\Gamma(D) = \frac{|P_{r_1,\ldots,r_n}(D)|}{|P_{r_1,\ldots,r_n}(D_{triv})|} \leq n!p^{-\sum_{i=1}^{[\frac{n+1}{2}]} \frac{r_i - r_{n+1-i}}{2}} \tag{56}$$

*où $[x]$ désigne la partie entière de $x$ et $\Gamma(D)$ est défini par cette égalité.*

*Remarque* 4.3. L'exposant en $p$ est optimal, comme on le voit en calculant $P_{r_1,\ldots,r_n}(D_0)$ où $D_0$ désigne la matrice de Hecke admissible de type Ramanujan $D_0 = \mathrm{Diag}(p^{\frac{(n-1)-1}{2}}, \ldots, p^{\frac{1-(n-1)}{2}}, 1)$.

Soit $D = \mathrm{Diag}(\alpha_1, \ldots, \alpha_n)$ une matrice de Hecke admissible. On choisit une permutation $\sigma$ du groupe symétrique $S_n$ telle que

$$\beta_1 = |\alpha_{\sigma(1)}| \geq \beta_2 = |\alpha_{\sigma(2)}| \geq \cdots \geq \beta_n = |\alpha_{\sigma(n)}|.$$

On note $t_i = p^{\theta_i}$ les paramètres cuspidaux de $\beta_i$. On note

$$\mathrm{Diag}(\alpha'_i, \ldots, \alpha'_n) = D_R$$

et on choisit une permutation $\sigma'$ de $S_n$ telle que

$$\beta'_1 = \alpha'_{\sigma'(1)} \geq \cdots \geq \beta'_n = \alpha'_{\sigma'(n)}.$$

On écrit $\beta'_i = p^{\gamma_i}$, on remarque que l'on a la relation $\gamma_{n+1-i} = -\gamma_i$. On en déduit le

Lemme 4.4. *Pour toute matrice de Hecke admissible $D$ on a:*

$$\Gamma(D) \leq n!\ p^{\sum_{i=1}^{n} r_i\theta_i + \sum_{i=1}^{[\frac{n+1}{2}]} (r_i - r_{n+1-i})\left(\gamma_i - \frac{n-(2i-1)}{2}\right)}. \tag{57}$$

Soient $a$ et $b$ deux entiers naturels positifs. On pose $\delta(1, b) = 0$, $\delta(2, b) = \frac{b}{4}$ et pour tout $a > 2$

$$\delta(a, b) = \min\left(\frac{(a-1)b}{2}, \frac{ab}{4}\right).$$

On commence par montrer

Lemme 4.5. *On a l'inégalité:*

$$\sum_{i=1}^{n} r_i\theta_i \leq \sum_{i=1}^{r} \delta(a_i, b_i)(r_1 - r_n). \tag{58}$$

*Preuve.* On fixe $i \in [1,\ldots,r]$ et on note, pour $j \in [1,\ldots,a_i]$, $\theta'_j$ le nombre réel tel que $|t_{i,j}| = p^{\theta'_j}$. On peut supposer (et on supposera) que l'on a les relations

$$\frac{1}{2} \geq \theta'_1 \geq \theta'_2 \geq \cdots \geq \theta'_{a_i} \geq -\frac{1}{2}.$$

Dans la somme $\Lambda = \sum_{i=1}^{n} r_i\theta_i$, chaque $\theta'_j$ intervient $b_i$ fois. On note $\sigma_{(i,j,k)}$ l'entier de $[1,\ldots,n]$ qui est l'indice pour lequel $\theta'_j$ intervient pour la $k$-ième fois dans la somme $\Lambda$. On a alors, pour tout $j \in [1,\ldots,a_i]$ et tout $1 \leq k \leq k' \leq b_i$, l'inégalité $\sigma_{(i,j,k)} \leq \sigma_{(i,j,k')}$. Par ailleurs, on peut supposer que tout $1 \leq j \leq j' \leq a_i$ et tout $k \in [1,\ldots,b_i]$ l'inégalité: $\sigma_{(i,j,k)} \leq \sigma_{(i,j',k)}$.

En utilisant (53), on trouve que

$$\Lambda_i = \sum_{j=1}^{a_i}\sum_{k=1}^{b_i} r_{\sigma_{(i,j,k)}}\theta'_j = \sum_{j=1}^{a_i-1}\sum_{k=1}^{b_i}\left(r_{\sigma_{(i,j,k)}} - r_{\sigma_{(i,a_i,k)}}\right)\theta'_j.$$

Ce qui donne en utilisant (52):

$$\Lambda_i \leq \frac{(a_i-1)b_i}{2}(r_1 - r_n) \tag{59}$$

et si $a_i = 2$ on obtient en utilisant (54):

$$\Lambda_i \leq \frac{b_i}{4}(r_1 - r_n). \tag{60}$$

Par ailleurs, si $a_i > 2$ et $a'_i$ désigne le nombre de $\theta'_j$ qui sont positifs, on a:

$$\sum_{i=1}^{a'_i}\theta'_i = -\sum_{i=a'_i+1}^{a_i}\theta'_i \leq \min\left(\frac{a'_i}{2}, \frac{a_i - a'_i}{2}\right) \leq \frac{a_i}{4}.$$

On en déduit alors que

$$\Lambda_i \leq b_i(r_1 - r_n)\left(\sum_{i=1}^{a'_i}\theta'_i\right) \leq \frac{a_ib_i}{4}(r_1 - r_n). \tag{61}$$

Ceci termine la preuve du lemme 4.5 quand on a remarqué que $\Lambda = \sum_{i=1}^{r}\Lambda_i$.

Lemme 4.6. *Pour tout $n \geq 4$ et pour toute matrice de Hecke admissible $D$ de $GL_n$ telle que $|D| \neq D_{triv}$, on a*

$$\sum_{i=1}^{n} r_i\theta_i + (r_1 - r_n)\left(\gamma_1 - \frac{n-1}{2}\right) \leq -\frac{r_1 - r_n}{2}. \tag{62}$$

*Preuve.* D'après la définition des matrices de Hecke admissible et la convention $b_1 \geq b_2 \geq \cdots \geq b_r$, on a $\gamma_1 = \frac{b_1 - 1}{2}$. D'après le lemme (4.5), il suffit de montrer que

$$\lambda = \frac{b_1 - n}{2} + \sum_{i=1}^{r} \delta(a_i, b_i) \le -\frac{1}{2}.$$

On a

$$\lambda \le \frac{b_1 - n}{2} \sum_{i=1}^{r} \frac{(a_i - 1)b_i}{2} = -\sum_{i=2}^{r} \frac{b_i}{2}.$$

Ceci prouve le lemme si $r \ge 2$. Si $r = 1$, on a $n = a_1 b_1$ et $a_1 \neq 1$ (sinon $|D| = D_{triv}$). Si $a_1 = 2$, on a $b_1 \ge 2$ et $\lambda = -\frac{b_1}{4} \le -\frac{1}{2}$; ce qui montre le lemme dans ce cas. Si $a_1 > 2$, on a

$$\lambda \le \frac{b_1(2 - a_1)}{4} \le -\frac{1}{2}$$

dès que $n \neq 3$. Ceci termine la preuve du lemme (4.6).

La preuve de la proposition (4.2) pour $n \ge 5$ est alors conséquence du lemme (4.2) et du lemme suivant:

LEMME 4.7. *Pour tout $n \ge 5$ et pour toute matrice de Hecke admissible $D$ de $GL_n$ telle que $|D| \neq D_{triv}$, on a*

$$\sum_{i=2}^{[\frac{n+1}{2}]} (r_i - r_{n+1-i}) \left( \gamma_i - \frac{n - (2i - 1)}{2} \right) \le - \sum_{i=2}^{[\frac{n+1}{2}]} \frac{r_i - r_{n+1-i}}{2}. \tag{63}$$

*Preuve.* On note $\epsilon = \frac{1}{2}$ si $n$ impair et $\epsilon_n = 0$ si $n$ pair. Pour tout $j \in [1, \ldots, [\frac{n+1}{2}]]$, on pose $\delta_j = \gamma_{[\frac{n+1}{2}]+1-j}$. On a alors en posant

$$\mu = \sum_{i=2}^{[\frac{n+1}{2}]} (r_i - r_{n+1-i}) \left( \gamma_i - \frac{n - (2i - 1)}{2} \right) \tag{64}$$

la relation

$$\mu = \sum_{j=1}^{[\frac{n-1}{2}]} (r_{[\frac{n+1}{2}]+1-j} - r_{[\frac{n}{2}]+j}) \left( \delta_j + \frac{1}{2} + \epsilon_n - j \right). \tag{65}$$

On note pour tout $i \in [1, \ldots, r]$

$$A_i = \left\{ \frac{b_i - 1}{2}, \frac{b_i - 3}{2}, \ldots, \frac{1}{2} \right\} \text{ si } b_i \text{ paire}$$

et

$$A_i = \left\{ \frac{b_i - 1}{2}, \frac{b_i - 3}{2}, \ldots, 0 \right\} \text{ si } b_i \text{ impaire.}$$

Pour tout $\alpha \in \frac{1}{2}\mathbb{N}$ différent de 0, on note

$$\lambda_\alpha = \sum_{i|\alpha\in A_i} a_i$$

et

$$\lambda_0 = \left[\frac{1 + \sum_{i|0\in A_i} a_i}{2}\right].$$

La suite des $\delta_j$ est alors la suite croissante formée d'éléments $\alpha \in \frac{1}{2}\mathbb{N}, 0 \leq \alpha \leq \frac{b_1-1}{2}$, où chaque $\alpha$ apparaît $\lambda_\alpha$ fois. On a pour tout $j$ l'inégalité $\delta_{j+1} - \delta_j \leq 1$.

On voit alors qu'il suffit de prouver que pour tout $j \geq 2$, on a

$$\delta_j + 1 - j \leq -\frac{1}{2}$$

et que $\delta_1 - \frac{1}{2} \leq -\frac{1}{2}$ car si $n$ est impair alors $r_{[\frac{n+1}{2}]} - r_{[\frac{n}{2}]} = 0$. Le lemme (4.7) s'obtient facilement avec cette remarque quand $\lambda_0 \geq 2$ ou $\lambda_0 = 1$ et $n$ pair.

Si $\lambda_0 = 1$ et $n$ impair alors $\lambda_{\frac{1}{2}} \neq 0$ sinon on a $|D| = D_{triv}$. On a donc $\delta_1 = 0$ et pour tout $j \geq 2$, $\delta_j \leq \frac{1}{2} + j - 2$. On a donc pour tout $j \geq 2$ l'inégalité $\delta_j + 1 - j \leq -\frac{1}{2}$. Ceci démontre le lemme dans ce cas car $n$ étant impair il n'y a rien à vérifier pour $j = 1$.

Si $\lambda_0 = 0$, tous les $b_i$ sont pairs et donc $n$ est lui aussi pair. On a $\delta_1 = \frac{1}{2}$ et

$$\lambda_{\frac{1}{2}} = \sum_{i=1}^{r} a_i \geq 2$$

car dans le cas contraire on a $r = 1$ et $a_1 = 1$ et on retrouve $|D| = D_{triv}$. On en déduit que $\delta_2 = \frac{1}{2}$ et que pour tout $j \geq 2$, on a $\delta_j \leq \frac{1}{2} + j - 2$. On en déduit alors que pour tout $j \geq 2$, on a $\delta_j + \frac{1}{2} - j \leq -1$. Ceci termine la preuve du lemme et de la proposition car comme $n \geq 5$ il y a au moins 3 termes dans la somme $\mu$ et on a l'inégalité $r_1 \geq r_2 \geq \cdots \geq r_n$.

Il reste à montrer la proposition (4.2) pour $n = 3$ et $n = 4$.

Traitons le cas $n = 4$. On reprend les notations précédentes. La preuve précédente prouve la proposition (4.2) si $\gamma_2 = 0$ (utiliser les lemmes (4.4), (4.5) et (4.6). On peut donc supposer $b_1 \geq 2$. Si $b_1 = 4$ alors $|D| = D_{triv}$. Si $b_1 = 3$ alors $|D| = D_0$ et on a vu à la remarque (4.3) que l'estimée de la proposition (4.2) est optimale. Il reste à regarder les cas $b_1 = a_1 = 2$ et $b_1 = b_2 = 2$ (donc $a_1 = a_2 = 1$). Dans ces deux cas, il existe $\theta \in [0, \frac{1}{4}]$ tel que

$$|D| = \mathrm{Diag}(p^{\frac{1}{2}+\theta}, p^{\frac{1}{2}-\theta}, p^{\frac{-1}{2}+\theta}, p^{\frac{-1}{2}-\theta}).$$

On a alors

$$P_{r_1,r_2,r_3,r_4}(D) \leq 4!\, p^{(\frac{1}{2}+\theta)(r_1-r_4)+(\frac{1}{2}-\theta)(r_2-r_3)}.$$

Il suffit alors de prouver l'inégalité:

$$(r_1 - r_4)\left(-\frac{1}{2} + \theta\right) + (r_2 - r_3)\left(\frac{1}{2} - \theta\right) \leq 0$$

qui est une conséquence de l'inégalité $r_1 \geq r_2 \geq r_3 \geq r_4$.

Dans le cas $n = 3$, on regarde suivant les valeurs des $(a_i, b_i)$. Le cas le plus difficile est $a_1 = 3$, $b_1 = 1$, on laisse les autres cas au lecteur. On a alors

$$D = \mathrm{Diag}(p^{\theta_1}, p^{\theta_2}, p^{\theta_3})$$

avec $\theta_1 + \theta_2 + \theta_3 = 0$ et $\frac{1}{2} \geq \theta_1 \geq \theta_2 \geq \theta_3 \geq -\frac{1}{2}$. On doit alors prouver que

$$I = (r_1 - r_3)\left(\theta_1 - \frac{1}{2}\right) + (r_2 - r_3)\theta_2 \leq 0.$$

Ceci est clair si $\theta_2 \leq 0$. Si $\theta_2 \geq 0$, alors comme $r_1 \geq r_2$, on trouve

$$I \leq (r_1 - r_3)\left(-\frac{1}{2} + \theta_1 + \theta_2\right) = (r_1 - r_3)\left(-\frac{1}{2} - \theta_3\right) \leq 0.$$

Ceci termine la preuve de la proposition (4.2).

**4.3. Preuve du théorème 1.2.** Le but de cette partie est de prouver la première partie du théorème 1.2. Soit donc $n \geq 3$ un entier. On écrit la décomposition orthogonale:

$$L^2(\Gamma\backslash GL_n(\mathbb{R}), \mathbf{1}) = \mathbb{C}.\mathbf{1} \oplus L$$

où $\mathbb{C}.\mathbf{1}$ désigne l'espace des fonctions constantes et $L$ désigne l'orthogonal des constantes. Soit $\pi = \hat{\otimes}\pi_l$ une représentation automorphe de $GL_n(\mathbf{A})$ intervenant dans la décomposition spectrale (44). La composante en $p$, $\pi_p$ est non ramifiée et est unitairement induite à partir d'un caractère non ramifié $\chi$ de $T_0(\mathbb{Q}_p)$. On note $t_\chi = (t_1, t_2, , \ldots, t_n)$ les paramètres de $\pi_p$. On sait d'après le lemme (4.1) que la matrice diagonale $D_\chi$ ayant $(t_1, \ldots, t_n)$ comme coefficients est une matrice de Hecke admissible au sens de la partie précédente.

On pose

$$P_r(X_1, X_2, \ldots, X_n) = \sum_{1 \leq i_1 < i_2 < \cdots < i_r \leq n} X_{i_1} X_{i_2} \ldots X_{i_n}.$$

Avec les notations de l'équation (55), on a

$$P_r(X_1, X_2, \ldots, X_n) = P_{1,\ldots,1,0,\ldots,0}(X_1, X_2, \ldots, X_n).$$

En utilisant (41) et (42), on voit que l'action de $\overline{T}_p$ sur la partie relative à $\pi$ dans la décomposition spectrale (44) est donnée par l'action de $\frac{ch_{r,p}}{\deg(ch_{r,p})}$ sur la composante en $p$ d'un vecteur $K_f$-invariant de $\pi$. D'après l'équation (47) cette action est donnée par la multiplication par

$$\lambda_{\pi,p} = \frac{P_r(t_1, \ldots, t_n)}{P_r(D_{triv})}.$$

Si $\pi$ n'est pas la représentation triviale, on a $|D_\chi| \neq D_{triv}$. D'après la proposition (4.2) on a

$$|\lambda_{r,p}| \leq n!\,p^{-\frac{\min_{(r,n-r)}}{2}}.$$

La première partie du Théorème (1.2) découle de la décomposition spectrale (44) et de cette inégalité.

**5. Le principe de restriction de Burger, Li et Sarnak aux places finies.** Soit $G$ un groupe semi-simple connexe et simplement connexe défini sur $\mathbb{Q}$ et $H$ un sous-groupe semi-simple connexe de $G$. Le but de ce paragraphe est d'étendre aux places finies un résultat de Burger et Sarnak ([9]: voir aussi [8]) formulé pour la place archimédienne.

Soit donc $p$ un nombre premier, et $G_p = G(\mathbb{Q}_p)$. Posons $G_\infty = G(\mathbb{R})$. Soit $\widehat{G}_p$ le dual unitaire de $G_p$. Soit $\Gamma \subset G_\infty \times G_p$, un sous-groupe de congruence: donc

$$\Gamma = G(\mathbb{Q}) \cap K \tag{66}$$

où $K$ est un sous-groupe compact ouvert de $G(\mathbb{A}_f^p)$; $\Gamma$ est naturellement plongé dans $G_\infty \times G_p$.

Le groupe $G_\infty \times G_p$ opère à droite sur l'espace de formes automorphes $L^2(\Gamma\backslash G_\infty \times G_p)$.

Par restriction, on obtient une représentation de $G_p$. Son support (Dixmier [14], chapitre 18) est un sous-ensemble fermé de $\widehat{G}_p$. Quand $\Gamma$ varie parmi les sous-groupes de congruence, on obtient ainsi une famille de fermés de $\widehat{G}_p$; la clôture de leur réunion est par définition le **spectre automorphe** de $G_p$, noté $\widehat{G}_p^{\text{aut}}$.

Rappelons que si $\pi$ est une représentation unitaire d'un groupe localement compact, et $\rho$ une représentation irréductible, $\rho$ est **faiblement contenue** dans $\pi$ ([14]: 18.1.3) si et seulement si $\rho$ appartient au support de $\pi$.

Les notations et les définitions qui précèdent s'appliquent au groupe $H$.

THÉORÈME 5.1. *Soit $\pi \in \widehat{G}_p^{\text{aut}}$ et soit $\pi' \in \widehat{H}_p$. Supposons $\pi'$ faiblement contenue dans $\pi|_{H_p}$. Alors $\pi' \in \widehat{H}_p^{\text{aut}}$.*

Si l'on remplace $p$ par la valuation archimédienne, ceci est le Théorème 1.1 (a) de Burger et Sarnak. Notons le Corollaire suivant: (Théorème 1.1 (b) de [9]):

COROLLAIRE 5.2. *Si $\pi, \rho \in \widehat{G}_p^{\text{aut}}$ et $\pi' \in \widehat{G}_p$ est faiblement contenue dans $\pi \otimes \rho$, $\pi' \in \widehat{G}_p^{\text{aut}}$.*

Il suffit en effet de remarquer que le produit tensoriel extérieur $\pi \boxtimes \rho \in (G_p \times G_p)^\wedge$ est dans le spectre automorphe. Le groupe $G_p$ étant de type 1, $(G_p \times G_p)^\wedge = \widehat{G}_p \times \widehat{G}_p$ [14], avec la topologie produit. Pour tout couple $(\Gamma, \Delta)$ de sous-groupes de congruence de $G_\infty \times G_p$, la représentation associée de $G_p \times G_p$ est un produit tensoriel (extérieur) et son support est le produit des supports. De tels couples étant cofinaux dans la famille des sous-groupes de congruence de $(G_\infty \times G_p)^2$, on en déduit que $(G_p \times G_p)^{\wedge,\text{aut}} = \widehat{G}_p^{\text{aut}} \times \widehat{G}_p^{\text{aut}}$. Le Corollaire résulte du Théorème à l'aide du plongement diagonal.

Nous déduirons en fait le théorème 5 d'une version "$S$-arithmétique" du résultat de Burger-Sarnak. Soit $S$ un ensemble fini de places de $\mathbb{Q}$ contenant $\infty$, $G_S = \Pi_{v\in S} G(\mathbb{Q}_v)$. On définit de façon analogue $\widehat{G}_S^{\text{aut}}$, à l'aide de sous-groupes $S$-arithmétiques. Même notation pour $H$.

THÉORÈME 5.3. *Soient* $\pi \in \widehat{G}_S^{\text{aut}}$ *et* $\pi' \in \widehat{H}_S$. *Si* $\pi'$ *est faiblement contenue dans* $\pi|_{H_S}$, $\pi' \in \widehat{H}_S^{\text{aut}}$.

**5.1. Réduction au cas $S$-arithmétique.** Ce paragraphe est consacré à démontrer que le Théorème 5.3 implique le Théorème 5.1. On prend $S = \{p, \infty\}$. La démonstration va reposer sur la décomposition de $L^2(\Gamma\backslash G_\infty \times G_p)$ donnée par les séries d'Eisenstein, rappelée dans la section 3.2.

Si le sous-groupe de $S$-congruences $\Gamma$ est fixé, on a:

$$L^2(\Gamma\backslash G_\infty \times G_p) \subset \bigoplus_M \bigoplus_\tau \int_{i\mathfrak{a}_M^*} \text{Ind}(\tau, s)^K d\mu_\tau(s) \tag{67}$$

où $M$ parcourt l'ensemble des sous-groupes de Levi de $G$ à association près, $\tau$ parcourt l'ensemble des représentations de $M(\mathbb{Q})\backslash M(\mathbb{A})$ – modulo torsion par les caractères non ramifiés – dans le spectre discret, $i\mathfrak{a}_M^*$ paramètre les caractères unitaires non ramifiés de $M(\mathbb{A})$, et $\text{Ind}(\tau, s) = \text{Ind}_{M(\mathbb{A})N(\mathbb{A})}^{G(\mathbb{A})}(\tau \otimes s)$ est envoyé dans l'espace des formes automorphes par l'opérateur d'Eisenstein. (Il faut tenir compte des isométries induites par les opérateurs d'entrelacement, mais ceci n'aura pas d'influence sur les arguments suivants.) Le groupe $K \subset G(\mathbb{A}_f^p)$ est associé à $\Gamma$ (cf. 66) et l'on prend les $K$-invariants dans $\text{Ind}(\tau, s)$: sous $G_\infty \times G_p$, $\text{Ind}(\tau, s)^K$ est donc une somme finie de représentations de la forme $\text{Ind}(\tau_\infty \otimes \tau_p, s)$ (cf. 44) dans le § 3.2).

D'après (67), une représentation $\pi_p$ de $G_p$ est alors dans le support de $L^2(\Gamma\backslash G_\infty \times G_p)$ si, et seulement si, c'est une limite pour la topologie de Fell [40] de sous-représentations de $\text{Ind}_{M(\mathbb{Q}_p)N(\mathbb{Q}_p)}^{G(\mathbb{Q}_p)}(\tau_p \otimes s)$. On a un résultat analogue pour une représentation $\pi_\infty \otimes \pi_p$ de $G_\infty \times G_p$.

Pour $s$ générique, $\text{Ind}(\tau_\infty \otimes \tau_p, s)$ est irréductible. Soit alors $\pi_p = \text{Ind}(\tau_p, s)$ et supposons que l'induite $\pi_\infty \otimes \pi_p := \text{Ind}(\tau_\infty \otimes \tau_p, s)$ est irréductible.

Soit $\pi_p' \in \widehat{H}_p$ faiblement contenue dans $\pi_p$, et choisissons $\pi_\infty' \in \widehat{H}_\infty$ faiblement contenue dans $\pi_\infty$. La représentation $\pi_\infty \otimes \pi_p$ de $H_\infty \otimes H_p$ étant un produit tensoriel, et les deux groupes étant de type 1, sa décomposition spectrale est le produit des deux décompositions de $\pi_\infty$, $\pi_p$. En particulier $\pi_\infty' \otimes \pi_p'$ est faiblement contenue dans $\pi_\infty \otimes \pi_p|_{H_{\infty,p}}$. D'après le Théorème 5.3, $\pi_\infty' \otimes \pi_p'$ appartient au spectre automorphe de $H_\infty \times H_p$. Ceci implique a fortiori que $\pi'_p$ appartient au spectre automorphe de $H_p$.

Considérons maintenant le cas général; soit $\pi_p \in \widehat{G}_p^{\text{aut}}$. Alors d'une part $\pi_p$ est limite (selon les groupes de $(\infty, p)$-congruence $\Gamma$) de représentations $\pi_p^\Gamma$ apparaissant dans l'espace des formes automorphes $L^2$; par ailleurs pour $\Gamma$ fixé, $\pi_p^\Gamma$ est limite de représentations apparaissant dans (67); on peut de plus supposer que pour celles-ci l'induite $\text{Ind}(\tau_n \otimes \tau_p, s)$ est irréductible. Par conséquent le Théorème 5.1 sera démontré (modulo le Théorème 5.3) si l'on prouve:

LEMME 5.4. *Soit $A$ un ensemble filtrant d'indices, et soit $\pi = \lim_{\alpha \in A} \pi^\alpha$ dans $\widehat{G}_p$. Supposons que pour tout $\alpha$ le support de $\pi^\alpha|_{H_p}$ est contenu dans $\widehat{H}_p^{\text{aut}}$. Alors le support de $\pi|_{H_p}$ est contenu dans $\widehat{H}_p^{\text{aut}}$.*

*Démonstration.* Le lemme est plus général: $H \subset G$ est un sous-groupe fermé d'un groupe localement compact, $C \subset \widehat{H}$ est un fermé, et $\text{Supp}(\pi^\alpha|_H) \subset C$; on en déduit que $\text{Supp}(\pi|_H) \subset C$. Par définition ceci veut dire que tout coefficient diagonal $f_\pi$ de $\pi$, restreint à $H$ est limite de fonctions $\varphi_\beta$, les $\varphi_\beta$ étant sommes finies (positives) de coefficients diagonaux de représentations de $C$:

$$f_\pi(h) = \lim_\beta \varphi_\beta(h) \tag{68}$$

la convergence étant uniforme sur tout compact de $H$. Notons que l'on pourrait prendre des **suites** d'indices, les topologies sur tous les groupes considérés ici étant séparables (mais non métrisables!).

Mais $\pi^\alpha \to \pi$ dans $\widehat{G}$, ce qui veut dire que

$$f_\pi(g) = \lim_\alpha f_\alpha(g) \tag{69}$$

$f_\alpha$ étant une somme finie de coefficients diagonaux de $\pi_\alpha$, la convergence étant uniforme sur tout compact de $G$. Enfin, puisque le support de $\pi^\alpha$ est contenu dans $C$, on a pour tout $\alpha$:

$$f_\alpha(h) = \lim_\beta \varphi_\beta^\alpha(h), \tag{70}$$

les $\varphi_\beta^\alpha$ étant des sommes de coefficients de $C$. Alors (68) résulte de (69) et (70). Ceci termine la démonstration.

Nous donnerons deux démonstrations du Théorème 5. La première imite celle de Burger et Sarnak [9]; elle semble nécessiter l'hypothèse que $G(\mathbb{R})$ est non compact. La seconde ne fait pas cette hypothèse, mais, dans le cas où $G$ est obtenu par restriction des scalaires à partir d'une forme de $SL(2, k)$ où $k$ est un corps de nombres, nécessite une approximation de la conjecture de Ramanujan pour $G$.

**5.2. Première démonstration.** Dans cette section, nous supposons donc que $G$ est **quasi-simple** sur $\mathbb{Q}$ et que $G(\mathbb{R})$ est non compact. Pour simplifier les notations, nous démontrons le Théorème 5.3 dans le cas où $S = \{\infty, p\}$ ce qui suffira pour les besoins du reste de l'article.

Nous fixons un sous-groupe de congruence $\Gamma \subset G_\infty \times G_p$ (cf. (5.1)). Si $\delta \in G(\mathbb{Q})$, la double classe $\Gamma\delta\Gamma = T_\delta$ opère de la façon habituelle sur $\mathcal{L}_\Gamma := L^2(\Gamma\backslash G)$; pour abréger nous écrivons $G = G_\infty \times G_p$ si cela ne prête pas à confusion. Si

$$\Gamma\delta\Gamma = \coprod_i \Gamma\delta_i, \tag{71}$$

alors

$$(f|T_\delta)(g) = \sum f(\delta_i g) \quad (g \in G). \tag{72}$$

Soit $\deg(T_\delta)$ le degré de $T_\delta$, égal au cardinal de $\Gamma\backslash\Gamma\delta\Gamma$. Nous appellerons **opérateur de Hecke** positif une combinaison linéaire finie à coefficients entiers $\geq 0$ de $T_\delta$; son degré est alors défini par additivité.

Lemme 5.5. *[9] $T_\delta$ est un opérateur borné dans $\mathcal{L}_\Gamma$, de norme*

$$\|T_\delta\| = \deg(T_\delta).$$

*Démonstration.* Soit $\Delta = \delta\Gamma\delta^{-1} \cap \Gamma$, $\Delta' = \Gamma \cap \delta^{-1}\Gamma\delta$. Alors $T = T_\delta$ est obtenu par composition à partir du diagramme

$$\begin{array}{ccc} \Delta\backslash G & \xrightarrow[{[\delta]}]{} & \Delta'\backslash G \\ {\scriptstyle\pi}\downarrow & & \downarrow{\scriptstyle\rho} \\ \Gamma\backslash G & \longrightarrow & \Gamma\backslash G \end{array} \tag{73}$$

où $[\delta]$ est l'application $f(g) \mapsto f(\delta g)$ et $\pi$, $\rho$ sont les projections naturelles. On a alors:

$$Tf = \rho_*\delta_*\pi^* f \quad (f \in \mathcal{L}_\Gamma) \tag{74}$$

$$(Tf, h)_\Gamma = (\rho_*\delta_*\pi^* f, h)_\Gamma = (\delta_*\pi^* f, \rho^* h)_{\Delta'} \tag{75}$$

d'où

$$\begin{aligned} |(Tf, h)|_\Gamma &\leq \|\delta_*\pi^* f\|_{\Delta'}\|\rho^* h\|_{\Delta'} \\ &\leq \|\pi^* f\|_\Delta \|\rho^* h\|_{\Delta'} \\ &\leq [\Gamma : \Delta]\|f\|_\Gamma \|h\|_\Gamma. \end{aligned} \tag{76}$$

(On a utilisé le fait que $[\delta]$ induit une isométrie, que $\|\pi^* f\|_\Delta^2 = [\Gamma : \Delta]\|f\|_\Gamma^2$, et que $[\Gamma : \Delta] = [\Gamma : \Delta']$). Le Lemme résulte de ce calcul, et du fait que $T$ opère sur les constantes par $\deg(T)$.

Le Théorème 5.3 va résulter de la proposition suivante.

Si $f \in \mathcal{L}_\Gamma$, soit $f^0$ la projection de $f$ sur l'espace des fonctions invariantes par $G$:

$$f^0(g) = \int_{\Gamma\backslash G} f(x)dx \quad \forall g \in G. \tag{77}$$

(Rappelons que $\Gamma\backslash G$ est de volume fini. On normalise la mesure de façon que $\mathrm{vol}(\Gamma\backslash G) = 1$.) Si $T$ est un opérateur de Hecke, soit $\widetilde{T} = \deg(T)^{-1}T$ l'opérateur **normalisé** associé. Enfin, si $X$ est un espace localement compact, soit $C_0(X)$ l'espace des fonctions continues sur $X$ tendant vers 0 à l'infini.

PROPOSITION 5.6. *Il existe un opérateur de Hecke auto-adjoint $T : \mathcal{L}_\Gamma \to \mathcal{L}_\Gamma$ tel que $\widetilde{T}^m f \to f^0$ pour la topologie faible sur $\mathcal{L}_\Gamma$ si $f \in \mathcal{L}_\Gamma$.*

Nous renvoyons à Burger et Sarnak [9] pour la démonstration du Théorème 5.3 à partir de la Proposition 5.6. Nous démontrons maintenant la Proposition, sous l'hypothèse que $G$ est quasi-simple et $G(\mathbb{R})$ non compact.

Soit $F$ un ensemble fini de nombres premiers tel que $G$ admette un plongement dans $GL(N)$ défini sur l'anneau $\mathbb{Z}_{(F)}$ des $F$-entiers, et soit $q \notin F$ tel que $G(\mathbb{Q}_q)$ ne soit pas compact. D'après le théorème d'approximation forte, $G(\mathbb{Q})$ est dense dans $G(\mathbb{A}^q)$. Soit $\mathbb{A}_F = \Pi_{\ell\in F}\mathbb{Q}_\ell$, $S = F \cup \{p, q\}$ et $\mathcal{O}^S = \Pi_{\ell\notin S}\mathbb{Z}_\ell$. Alors $G(\mathbb{Q}) \cap G(\mathcal{O}^S) = G(\mathbb{Z}_{(S)})$, et ce groupe est dense dans $G(\mathbb{R}) \times G(\mathbb{Q}_p) \times G(\mathbb{A}_F)$; en particulier il est dense dans $G(\mathbb{R}) \times G(\mathbb{Q}_p)$.

D'après Borel et Serre [7], $G(\mathbb{Z}_{(S)})$ est un groupe de type fini. Soit $\{\varepsilon_1, \ldots, \varepsilon_r\}$ un ensemble fini de générateurs, et posons

$$T = \sum_{i=1}^{r} \left(T_{\varepsilon_i} + T_{\varepsilon_i^{-1}}\right), \tag{78}$$

opérant sur $\mathcal{L}_\Gamma$. La norme $L^2$ de $T$ est égale à son degré

$$k = \sum_i \left\|T_{\varepsilon_i}\right\| + \left\|T_{\varepsilon_i^{-1}}\right\|.$$

On a alors pour $f \in \mathcal{L}_\Gamma$:

$$T^m f(g) = \sum_{j=1}^{k^m} f\left(\delta_j^{(m)} g\right). \tag{79}$$

LEMME 5.7. *Pour tout m, on peut écrire*

$$T^m f(g) = \sum_{\substack{i_1,\ldots,i_m \\ x_1,\ldots,x_m}} f\left(\varepsilon_{i_1}^{x_1} \cdots \varepsilon_{i_m}^{x_m} g\right) + \sum_{j'} f\left(\delta_{j'}^{(m)} g\right) \tag{80}$$

*où $i = (i_1, \ldots, i_m)$ parcourt les applications de $[1, m]$ vers $[1, r]$ et $x = (x_1, \ldots, x_m)$ parcourt $\{\pm 1\}^m$.*

En effet $Tf(g) = \sum_{i=1}^{r}\{f(\varepsilon_i g) + f(\varepsilon_i^{-1} g)\} + \sum_{j'} f(\delta_{j'}, g)$, d'où le résultat par itération. Une assertion plus forte figure dans [9] (voir après (2.7) de leur article) mais n'a pas l'air correcte sauf si $\Gamma \subset G(\mathbb{Z}_{(S)})$ (avec leur notation).

On vérifie que $T_{\varepsilon_i^{-1}}$ est l'adjoint de $T_{\varepsilon_i}$; l'opérateur $\widetilde{T}$ est donc autoadjoint. Son spectre est contenu dans $[-1, +1]$ d'après le Lemme 5.5. Il suffit alors (voir [9]) de montrer que $(-1)$ n'est pas une valeur propre de $\widetilde{T}$ et que 1 n'apparaît qu'avec multiplicité 1, correspondant à l'espace des fonctions constantes. Soit donc

$$\mathcal{L}_1^{\perp} = \left\{ f \in \mathcal{L}_\Gamma : \widetilde{T} f = f, \int_{\Gamma \backslash G} f \, dg = 0 \right\} \tag{81}$$

$$\mathcal{L}_{-1} = \{ f \in \mathcal{L}_\Gamma : \widetilde{T} f = -f \}. \tag{82}$$

Noter que $\widetilde{T}$ commute à l'action de $G$ par translations à droite. En convolant avec des fonctions lisses à support compact sur $G$, on en déduit que $\mathcal{L}_1^{\perp} \cap C(\Gamma\backslash G)$ et $\mathcal{L}_{-1} \cap C(\Gamma\backslash G)$ sont denses dans $\mathcal{L}_1^{\perp}$ et $\mathcal{L}_{-1}$.

Si $\mathcal{L}_1^{\perp}$ ou $\mathcal{L}_{-1}$ est non-nul, il contient donc une fonction continue que l'on peut supposer à valeurs réelles puisque $T$ est un opérateur réel. Noter que si $f \in \mathcal{L}_{-1}$ alors $\int_{\Gamma\backslash G} f(g) dg = 0$ car, 1 désignant la fonction constante unité, on a $(f, 1) = -(\widetilde{T} f, 1) = -(f, \widetilde{T} 1) = -(f, 1)$. Par ailleurs, si $\widetilde{T} f = \pm f$, on a alors

$$(f, f) = |(\widetilde{T} f, f)| \leq (|\widetilde{T} f|, |f|) \leq (\widetilde{T}|f|, |f|), \tag{83}$$

la dernière inégalité résultant immédiatement de la définition de $\widetilde{T}$. Mais $(\widetilde{T}|f|, |f|) \leq \|f\|^2$; on en déduit que $(\widetilde{T}|f|, |f|) = (f, f) = (|f|, |f|)$. Puisque $\widetilde{T}$ est de norme 1, ceci implique que $\widetilde{T}|f| = |f|$.

Par conséquent la démonstration se ramène au.

LEMME 5.8. *Si $f$ est une fonction continue réelle, intégrable et d'intégrale nulle sur $\Gamma\backslash G$ et $\widetilde{T}|f| = |f|$ alors $f = 0$.*

C'est le seul endroit où la démonstration requiert un argument nouveau. Nous aurons besoin du théorème des valeurs intermédiaires pour des fonctions continues sur $\Gamma\backslash G$!

LEMME 5.9. *Si $f$ est une fonction continue à valeurs réelles prenant des valeurs positives et négatives sur $\Gamma\backslash G$, $f$ prend la valeur 0.*

*Démonstration.* Supposons l'inverse, et soit

$$s(g) = s(g_\infty, g_p) (\text{avec } g = (g_\infty, g_p) \in G(\mathbb{R}) \times G(\mathbb{Q}_p))$$

le signe de $f$. Pour $g_p$ fixé, $s(g_\infty, g_p)$ est une fonction constante de $g_\infty$. On a donc $s(g_\infty, g_p) = s(g_p) \in \pm 1$. Mais $s$ est alors une fonction sur $G_p$ invariante par $\Gamma$.

Puisque $G_\infty$ n'est pas compact, $\Gamma$ est dense dans $G_p$, toujours d'après le théorème d'approximation forte. Donc $s$ est constante, contrairement à l'hypothèse.

Nous terminons la démonstration à l'aide de l'argument de Burger-Sarnak [9]. Soit $f$ comme dans le Lemme 5.8. D'après le Lemme 5.9 il existe $g \in \Gamma\backslash G$ tel que $|f(g)| = 0$. On a alors $T^m|f|(g) = 0$ pour tout $m$. D'après le Lemme 5.9, ceci implique que $f(\varepsilon_{i_1}^{x_1} \cdots \varepsilon_{i_m}^{x_m} g) = 0$ pour tout monôme en les $\varepsilon_i^{\pm 1}$. Puisque les $\varepsilon_i$ engendrent $G(\mathbb{Z}_{(S)})$ qui est dense dans $G$, $f = 0$.

**5.3. Deuxième démonstration.** Nous donnons maintenant une nouvelle démonstration de la Proposition 5.6 et donc des théorèmes 5.1 et 5.3. Notons que si $G_p$ (donc $H_p$) est compact le Théorème 5.1 est trivial car **toute** représentation de $H_p$ est automorphe; ceci résulte par exemple de la formule des traces de Selberg, de façon très simple puisque $H$ est alors anisotrope (e.g., [12]). De même, le Théorème 5.3 est trivial si $G(\mathbb{A}_S)$ est compact. Soit $\Gamma \subset G(\mathbb{A}_S)$ un sous-groupe de $S$-congruences, et définissons comme auparavant $\mathcal{L}_\Gamma = L^2(\Gamma\backslash G)$; soit $\mathcal{L}_\Gamma^\perp$ l'orthogonal de l'espace des fonctions constantes. Nous pouvons donc supposer que $G(\mathbb{A}_S)$ n'est **pas** compact; $G$ est supposé quasi-simple sur $\mathbb{Q}$. Alors le Théorème 5.3 résulte du résultat suivant, plus fort que la Proposition 5.6.

THÉORÈME 5.10. *Il existe un opérateur de Hecke positif $T$ (cf. avant le Lemme 5.5) et auto-adjoint tel que $\widetilde{T} f = f$ pour $f$ constante sur $\Gamma\backslash G$ et $\|\widetilde{T}|_{\mathcal{L}_\Gamma^\perp}\| < 1$.*

La norme est la norme forte d'opérateur:

$$\|\widetilde{T} f\| \leq \|\widetilde{T}|_{\mathcal{L}_\Gamma^\perp}\| \, \|f\|, \quad f \in \mathcal{L}_\Gamma^\perp. \tag{84}$$

La démonstration est la suivante. Pour simplifier nous supposons toujours que $S = \{\infty, p\}$. Par approximation forte $G(\mathbb{Q})$ est dense dans $G(\mathbb{A}_f^p)$. Si $\Gamma$ est défini par (1) on en déduit que l'algèbre de Hecke classique définie par $(G(\mathbb{Q}), \Gamma)$ (cf. Shimura [36]) coïncide avec l'algèbre des fonctions bi-$K$-invariantes sur $G(\mathbb{A}_f^p)$, munie du produit de convolution. On normalise la mesure de Haar par $\mathrm{vol}(K) = 1$.

Fixons alors une place $q$ telle que $G$ soit déployé sur $\mathbb{Q}_q$, que $G$ soit défini sur $\mathbb{Z}_{(q)}$, et que $K = K^q K_q$, $K_q \subset G(\mathbb{Q}_q)$ étant le sous-groupe hyperspécial $G(\mathbb{Z}_q)$. L'opérateur de Hecke cherché va appartenir à la composante locale $\mathcal{H}(G_q, K_q)$ de $\mathcal{H}(G(\mathbb{A}_f^p), K)$.

Puisque $G$ est quasi-simple, $G$ est obtenu par restriction des scalaires (pour une extension finie $F$ de $\mathbb{Q}$) d'un groupe absolument quasi-simple $G^0/F$. Sous nos hypothèses, $q$ est une place décomposée dans $F$ et

$$G(\mathbb{Q}_q) = \prod_{v|q} G^0(F_v),$$

chaque facteur étant isomorphe à $G^d(\mathbb{Q}_q)$ ou $G^d$ est le groupe déployé simplement connexe de même système de racines que $G^0$. Nous allons en fait travailler dans l'algèbre de Hecke $\mathcal{H}(G^0(F_v), K_v) = \mathcal{H}(G^d(\mathbb{Q}_q), K_q)$, avec des notations évidentes. Si $G^d \neq SL(2)$, ce groupe est de rang $\geq 2$. Il suffit alors de démontrer:

LEMME 5.11. *Si* **G** *est un groupe déployé de rang* $\geq 2$ *sur* $\mathbb{Q}_q$ *et*

$$K = \mathbf{G}(\mathbb{Z}_q) \subset G = \mathbf{G}(\mathbb{Q}_q),$$

*il existe* $\varphi \in \mathcal{H}(G, K)$ *positive, à coefficients entiers et auto-adjointe* ($\varphi(g) = \varphi(g^{-1})$) *et* $K < 1$ *tels que*

$$\|\pi(\varphi)\| \leq K \ \deg(\varphi) \tag{85}$$

*si* $\pi \in \widehat{G}$ *est différente de la représentation triviale.*

*Démonstration.* Soit $\widehat{G}_{nr}$ l'ensemble des représentations non-ramifiées de $G$. Alors $\widehat{G}_{nr}$ est un sous-ensemble quasi-compact de $\widehat{G}$. Si $T \subset G$ est un tore maximal déployé et $W$ le groupe de Weyl associé, $\widehat{G}_{nr}$ s'identifie à un sous-espace compact $C$ de $\widehat{T}/W$, sa topologie étant la topologie quotient. En particulier $\widehat{G}_{nr}$ est séparé (pour tous ces faits voir Tadič [38]). Par hypothèse la représentation triviale $\mathbb{C}$ (associée à la "demi-somme des racines" $\rho \in \widehat{T}$) est isolée dans $C$. Soit $C^0 = C - \{\rho\}$. Soient enfin $\mathcal{C}$ l'algèbre de Hecke des fonctions localement constantes à support compact sur $G$, et $\mathcal{A} \subset \mathcal{C}$ le sous-ensemble formé des fonctions positives d'intégrale 1. (On normalise la mesure de Haar sur $G$ par $\mathrm{vol}(K) = 1$). Soit $\mathcal{H} = \mathcal{H}(G, K)$.

Puisque le rang de $G$ est $\geq 2$, la représentation triviale est isolée dans $\widehat{G}$ ([28], Ch. III) et donc a fortiori dans $\widehat{G}_{nr}$. D'après ([28], Prop. III.1.3) il existe, pour tout $\pi \in C^0$, une fonction $\varphi \in \mathcal{A}$ telle que $\|\pi(\varphi)\| < 1$ – la norme étant toujours la norme d'opérateur. Si $1_K$ est la fonction caractéristique de $K$, on a alors $\|\pi(1_K * \varphi * 1_K)\| < 1$ avec $\psi = 1_K * \varphi * 1_K \in \mathcal{H}(G, K)$.

Noter que $\pi(\psi)$ est simplement donnée par l'action de $\psi$ sur le vecteur $K$-invariant de $\pi$; en particulier $\|\pi(\psi)\|$ est une fonction continue de $\pi \in C^0$. Par compacité on en déduit un recouvrement fini de $C^0$ par des ouverts $U_i$, et des fonctions $\psi_i$ telles que $\|\pi(\psi_i)\| \leq K_i < 1$ si $\pi \in U_i$. On a $\psi_i \in \mathcal{A}$ et donc $\|\pi(\psi_i)\| \leq 1$ pour tout $\pi$ unitaire. Si on pose

$$\varphi = \frac{1}{2N} \sum_{i=1}^{N} \psi_i,$$

alors $\varphi(g) + \varphi(g^{-1})$ vérifie:

$$\begin{aligned} &\deg \varphi = 1 \\ &\|\pi(\varphi)\| \leq K, \ K = \tfrac{\sup(K_i)+(N-1)}{N}, \end{aligned} \tag{86}$$

pour tout $\pi \in C^0$. De plus $\pi(\varphi) = 0$ si $\pi \in \widehat{G}$ est ramifiée. Le Lemme s'en déduit en approchant $\varphi$ par des fonctions à valeurs rationnelles.

Si $G$ est une forme de $SL(2)/F$, le Lemme 5.11 reste vrai si on considère $\pi \in \widehat{G}_v^{\mathrm{aut}}$, puisque l'on dispose d'une approximation de la conjecture de Ramanujan.

Plus précisément, la clôture dans $(\widehat{G}_v)_{nr}$ de $(\widehat{G}_v)^{\text{aut}}$ est disjointe de la représentation triviale.

Complétons alors la démonstration du Théorème 5.10. Ecrivons

$$\mathcal{L}_\Gamma = L^2(\Gamma\backslash G) = \mathbb{C} \oplus \mathcal{L}_\Gamma^\perp \tag{87}$$

où $\mathbb{C}$ désigne l'espace des constantes; $\mathcal{L}_\Gamma$ est une représentation de $G$. Soit $v$ la place au-dessus de $q$ choisie plus haut, et écrivons $K = K_v K^v$ ($K_v \subset G^0(F_v)$) et soit $\Gamma_v = G(\mathbb{Q}) \cap K^v$. Alors

$$\mathcal{L}_{\Gamma_v} = L^2(G(\mathbb{Q})\backslash G^0(\mathbb{A}_F)/K^v) \tag{88}$$

(notations évidentes) est une représentation de $G \times G^0(F_v)$, et $\mathcal{L}_\Gamma$ est l'espace des $K_v$-invariants dans $\mathcal{L}_{\Gamma_v}$. Par approximation forte, $\mathcal{L}_\Gamma^\perp = (\mathcal{L}_{\Gamma_v}^\perp)^{K_v}$ où $\mathcal{L}_{\Gamma_v}^\perp$ est l'orthogonal de l'espace des constantes.

La théorie des séries d'Eisenstein donne une décomposition

$$\mathcal{L}_{\Gamma_v} = \mathbb{C} \oplus \int_{\widehat{G}^0(F_v)} m(\pi_v)\pi_v \, d\mu(\pi_v) \tag{89}$$

que nous n'expliciterons pas, mais où l'intégrale porte sur l'espace des représentations automorphes non triviales dans $\widehat{G}^0(F_v)$. L'opérateur $T$ associé à la fonction $\varphi$ du Lemme 5.11 opère alors sur $\mathcal{L}_{\Gamma_v}^{K_v}$ décomposé selon (89) par

$$T = \deg(T) \oplus \int_{\widehat{G}^0_{nr}(F_v)} m(\pi_v)\pi_v(\varphi) d\mu(\pi_v); \tag{90}$$

$\pi_v(\varphi)$ est un scalaire de norme $\leq K \deg T$. D'où le Théorème 5.10.

*Remarque.* La démonstration est essentiellement différente de celle de Burger-Sarnak puisque l'opérateur $T$ est de nature locale. Noter en revanche que $T$ est loin d'être explicite (voir la démonstration de Margulis pour la Proposition III.1.3 citée ci-dessus). Si $G$ est **adjoint** plutôt que simplement connexe, les fonctions de Hecke associées aux poids minuscules du groupe dual $\widehat{G}$—s'il y en a—fournissent des opérateurs **explicites** ayant les propriétés du Lemme 5.11 (voir les sections 6 7). Il serait intéressant d'obtenir de telles fonctions **explicites** pour un groupe arbitraire.

**5.4. Extension au cas non simplement connexe.** Pour une application ultérieure, montrons enfin que l'on peut se dispenser de l'hypothèse que $G$ est simplement connexe dans les Théorèmes 5.1 et 5.3. Utilisons la démonstration donnée dans la section 5.3. Soient $S \ni \infty$ un ensemble de places et $\Gamma \subset G(\mathbb{A}_S)$ un sous-groupe de congruence. Il s'agit de produire un opérateur de Hecke autoadjoint ayant la propriété de la Proposition 5.6. Soit $G_{sc}$ le revêtement (fini) simplement connexe de $G$.

Si $K^S \subset G(\mathbb{A}^S)$ est compact-ouvert, et $\Gamma = G(\mathbb{Q}) \cap K^S$, on a maintenant

$$G(\mathbb{Q})\backslash G(\mathbb{A})/K^S = \coprod_i \Gamma_i \backslash G_S \tag{91}$$

$$L^2(G(\mathbb{Q}\backslash G(\mathbb{A})/K^S) = \bigoplus_i \mathcal{L}_{\Gamma_i} \tag{92}$$

(unions et sommes finies, $\Gamma$ étant l'un des $\Gamma_i$). Les opérateurs de Hecke globaux vont a priori permuter les $\mathcal{L}_{\Gamma_i}$.

Il résulte de (91) qu'il existe un ensemble fini $T \supset S$ tel que

$$G(\mathbb{A}) = G(\mathbb{Q})G_T K^T \tag{93}$$

(avec $K^S = K_T^S K^T$, les notations étant évidentes) et donc un sous-groupe de congruence $\Delta \subset G_T$ tel que

$$\Delta\backslash G_T = G(\mathbb{Q})\backslash G(\mathbb{A})/KT. \tag{94}$$

Choisissons une place $v \notin T$ ayant les propriétés de la section 5.3. Nous ferons agir sur $\mathcal{L}_\Delta$ l'algèbre de Hecke $\mathcal{H}(G_v, K_v)$.

A priori cette algèbre n'est pas toute obtenue à l'aide d'opérateurs de Hecke "classiques" donnés par des doubles classes $\Delta\gamma\Delta$, $\gamma \in G(\mathbb{Q})$, puisque l'approximation forte fait défaut. Mais soit $\Delta_{sc} \subset G_{sc}(\mathbb{Q})$ l'image inverse de $\Delta$: c'est un sous-groupe de congruence de $G_{sc}(\mathbb{Q})$ puisque les fibres de $G_{sc}(\mathbb{A}) \to G(\mathbb{A})$ sont compactes. On vérifie alors que l'algèbre de Hecke classique $\mathcal{H}(G_{sc}(\mathbb{Q}), \Delta_{sc})$ est munie d'un homomorphisme naturel vers $\mathcal{H}(G(\mathbb{Q}), \Delta)$.

En particulier l'algèbre de Hecke $\mathcal{H}_v^{sc} = \mathcal{H}(G_{sc}(k_v), \widetilde{K}_v)$ (où $\widetilde{K}_v$ est l'image inverse de $K_v$), contenue dans $\mathcal{H}(G_{sc}(\mathbb{Q}), \Delta_{sc})$, s'envoie injectivement sur une sous-algèbre de $\mathcal{H}(G(\mathbb{Q}), \Delta)$ et aussi de $\mathcal{H}(G(k_v), K_v)$. On peut alors utiliser de tels opérateurs de Hecke à la place $v$: ils ont les propriétés utilisées dans la section 5.3.

Il reste à vérifier qu'on peut construire de tels opérateurs ayant les propriétés du Lemme 5.11. Puisque la fonction $\varphi$ provient de $\mathcal{H}(G_{sc}(k_v), \widetilde{K}_v)$, elle n'aura les propriétés cherchées que pour $\pi \in \widehat{G}_v$ telle que $\pi|G_{sc}$ soit différente de la représentation triviale. Ceci revient à dire que $\pi$ n'est pas un caractère non ramifié de $G(k_v)$.

Revenons alors à la démonstration de Burger-Sarnak [9]. Ils considèrent une fonction $f \in C_0(\Gamma\backslash G(\mathbb{R}))$ – ici $C_0(\Delta\backslash G(\mathbb{A}_T))$ – et montrent, modulo le Lemme 5.11, que les coefficients matriciels diagonaux

$$\psi(h) = \int_{\Delta\backslash G(\mathbb{A}_T)} f(g)\overline{f}(gh)dg \quad (L \in H(\mathbb{A}_s)) \tag{95}$$

sont limites de coefficients diagonaux de représentations dans $\widehat{H}^{\mathrm{aut}}(\mathbb{A}_s)$. Si $f$ appartient à l'espace $C_{00}(\Delta\backslash G(\mathbb{A}_T))$ des fonctions orthogonales à tous les caractères non ramifiés, abéliens de $G(k_v)$, il en est de même de $\psi$ considérée comme fonction de $h \in H(\mathbb{A}_S\varphi)$, et le résultat est impliqué par la démonstration de [9].

En général, la décomposition spectrale pour $\mathcal{L}_\Delta = L^2(\Delta\backslash G(\mathbb{A}_T))$ montre que $\mathcal{L}_\Delta$ est somme directe (hilbertienne) de $\mathcal{L}_\Delta \subset C_{00}$ et d'une somme finie de caractères abéliens de $G(\mathbb{A}_T)$. (Utiliser l'approximation forte pour $G_{sc}$). Il est clair que cette décomposition reste vraie dans $C(\Delta\backslash G(\mathbb{A}_T))$. Puisque la restriction à $H(\mathbb{A}_T)$ d'un caractère abélien est trivialement automorphe, ceci termine la démonstration dans ce cas: le support de $\mathcal{L}_\Delta|_{H(\mathbb{A}_T)}$ est contenu dans $(\widehat{H}_T)^{\mathrm{aut}}$.

Revenant alors à notre situation originale, remarquons que

$$L^2(G(\mathbb{Q})\backslash G(\mathbb{A})/K^s) = \mathcal{L}_\Delta^{K_T^s}.$$

On en déduit que cet espace est supporté dans le dual automorphe de $G_S$, comme dans la réduction du Théorème 5.1 au Théorème 5.3.

## 6. Composantes locales des représentations automorphes de *Sp*(2*g*).

**6.1. Introduction.** Dans ce chapitre nous allons utiliser les résultats du chapitre 5, ainsi qu'un argument, encore dû à Burger et Sarnak, relatif aux fonctions sphériques, pour exhiber des bornes non triviales sur les composantes locales (non ramifiées) des représentations automorphes de $Sp(2g)$ sur un corps de nombres $k$. Nous interprétons ensuite ce résultat comme une borne sur le spectre de l'opérateur standard $T_{\mathfrak{p}}$ opérant dans les formes automorphes sur $GSp(2g)$, $\mathfrak{p}$ étant une place de $k$.

Les bornes que nous obtiendrons dépendent de l'approximation connue (voir l'Introduction, ainsi que la section 3) de la conjecture de Ramanujan pour $SL(2)$. Dans la section 6.4, nous expliquons comment les conjectures d'Arthur [2, 3] permettent de prévoir effectivement le spectre de $T_{\mathfrak{p}}$. Nous montrons que notre estimée est optimale, au vu des conjectures d'Arthur, si la conjecture de Ramanujan est connue pour $SL(2)$, et qu'en général elle donne une approximation des estimées vraies proportionnelle à l'approximation connue de la conjecture de Ramanujan (toujours pour $SL(2)$).

**6.2. Composantes locales des représentations automorphes de *Sp*(2*g*).** Nous notons $Sp(2g)$ le groupe laissant invariante la forme symplectique sur un espace de dimension $2g$ et de matrice

$$\begin{pmatrix} 0 & -1_g \\ 1_g & 0 \end{pmatrix} \tag{96}$$

et $GSp(2g)$ le groupe de similitudes associé.

Soit $G$ le groupe $Sp(2g)$; $G$ contient un tore déployé $T$ donné par:

$$T = \left\{ \begin{pmatrix} x_1 & & & & & \\ & \ddots & & & & \\ & & x_g & & & \\ & & & x_1^{-1} & & \\ & & & & \ddots & \\ & & & & & x_g^{-1} \end{pmatrix} \right\} \cong \mathbb{G}_m^g \tag{97}$$

ainsi qu'un sous-groupe $H \cong SL(2)^g$, chaque facteur de $H$ étant plongé dans $G$ par

$$\begin{pmatrix} a & b \\ c & d \end{pmatrix} \mapsto \begin{pmatrix} 1 & & & & & & & & \\ & \ddots & & & & & & & \\ & & 1 & & & & & & \\ & & & a & & & b & & \\ & & & & \ddots & & & & \\ & & & & & 1 & & & \\ & & & & & & 1 & & \\ & & & & & & & \ddots & \\ & & & c & & & d & & \\ & & & & & & & & \ddots \\ & & & & & & & & & 1 \end{pmatrix}, \quad \begin{pmatrix} a & b \\ c & d \end{pmatrix} \in SL(2). \tag{98}$$

Noter que $T$ est encore un tore déployé maximal de $H$.

Soit maintenant $F$ un corps $p$-adique. Nous pouvons considérer $G, H, T$ comme des groupes définis sur $F$; $G(\mathcal{O}_F) = K$ est un sous-groupe hyperspécial de $G(F)$, que nous noterons souvent $G$. Une représentation non ramifiée de $G$ est alors unitairement induite à partir d'un caractère non ramifié $\chi$ de $T = T(F)$; $\chi$ est alors uniquement déterminé par

$$t_i = \chi \begin{pmatrix} 1 & & & & & & & & \\ & \ddots & & & & & & & \\ & & \varpi & & & & & & \\ & & & \ddots & & & & & \\ & & & & 1 & & & & \\ & & & & & \ddots & & & \\ & & & & & & \varpi^{-1} & & \\ & & & & & & & \ddots & \\ & & & & & & & & 1 \end{pmatrix} \tag{99}$$

où $\varpi \in F$ est une uniformisante; soit $t_\chi = (t_1, \ldots, t_g) \in (\mathbb{C}^\times)^g$. Nous écrivons parfois $\widehat{T}$ pour $(\mathbb{C}^\times)^g$ qui est ainsi identifié naturellement au tore dual [5] de $T$. Noter que le groupe de Weyl $W$ de $(G, T)$ est $\{\pm 1\}^g \rtimes \mathfrak{S}_g$, $(\pm 1)^g$ opérant sur les composantes par $x_i \mapsto x_i^{\pm 1}$. Le groupe de Weyl de $(H, T)$ est $(\pm 1)^g$.

Une base de racines simples pour $(G, T)$ est donnée par

$$\Delta_G = \left\{x_1 x_2^{-1}, \ldots, x_{g-1} x_g^{-1}, x_g^2\right\}.$$

Une base $\Delta_H$ pour $(H, T)$ est donné par $\{x_1^2, \ldots, x_g^2\}$. Les demi-sommes de racines positives associées sont

(100) $$\rho_G = x_1^g x_2^{g-1} \cdots x_g$$

(101) $$\rho_H = x_1 \cdots x_g.$$

On pose $\delta_G = |\rho_G|$, $\rho_H = |\rho_H|$.

Puisque les réseaux de caractères et de cocaractères de $T$ et de $\widehat{T}$ sont en dualité, les coracines de $(G, T)$ définissent naturellement des caractères de $\widehat{T}$. Notons $\alpha_1, \ldots \alpha_g$ les racines de $\Delta_G$. Alors

(102) $$\Delta^\vee = \left\{\alpha_1^\vee = t_1 t_2^{-1}, \alpha_2^\vee = t_2 t_3^{-1}, \ldots, \alpha_g^\vee = t_g\right\}.$$

Notons enfin $T_+$ le domaine fondamental dans $T$ pour l'action de $W$ (i.e., $T = \cup_{w \in W} w T_+$; les $w T_+$ ne sont bien sûr pas disjoints, ni même d'intersection négligeable pour un corps $p$-adique):

(103) $$T_+ = \{(x_1, \ldots x_g) = |x_1| \le |x_2| \le \cdots \le |x_g| \le 1\}.$$

On a alors

(104) $$\mathrm{val}(x_1) \ge \mathrm{val}(x_2) \ge \cdots \ge \mathrm{val}(x_y) \ge 0,$$

val( ) désignant la valuation normalisée de $F$.

Rappelons une partie de la théorie des fonctions sphériques, selon Harish-Chandra (cf. Cartier [10] et Casselman [11]). Soient $\chi$ un caractère non ramifié de $T = T(F)$, $\pi_\chi$ la représentation associée et $\varphi_\chi$ la fonction sphérique attachée à $\pi_\chi$. Soit $t = t_\chi \in \widehat{T}$ paramétrant $\chi$; $t$ n'est défini que modulo l'action de $W$.

Il existe alors, modulo $W$, un choix de $t$ vérifiant

(105) $$|t_1| \ge \cdots \ge |t_g| \ge 1.$$

Le caractère $\chi$ associé vérifie alors

(106) $$\begin{gathered} |\chi(x)| \ge 1 \quad (x \in T_+) \\ |\chi(x)| \ge |\chi(wx)| \quad (x \in T_+, w \in W). \end{gathered}$$

Il peut exister plusieurs $t$ vérifiant (105), mais les valeurs absolues $|t_i|$ sont alors uniquement définies. Soient $\{\chi_1, \ldots, \chi_r\}$ les caractères de $T$ associés – donc $r \le |W|$, et les $|\chi_i|$ ont la même valeur absolue. Alors $\varphi_\chi$ admet un terme principal au sens suivant: si $\varphi, \psi$ sont deux fonctions sur $T_+$ écrivons $\varphi \sim \psi$ si $\varphi, \psi$ sont équivalentes pour $x \in T_+$ et $|x^{\alpha_i}| \to 0$ pour toute racine simple $\alpha_i$. Par ailleurs, une fonction $L(x)$ sur $T_+$ est logarithmique si elle est invariante par $T(\mathcal{O}_F)$ et s'étend en un polynôme sur le réseau $T(F)/T(\mathcal{O}_F)$. On sait alors qu'il existe des fonctions logarithmiques $L_i$ sur $T_+$ telles que

$$\varphi_\chi(x) \sim \sum_{i=1}^{r} \chi_i(x)\delta_G(x)L_i(x). \tag{107}$$

Noter que $\delta_G(x) \leq 1$ sur $T_+$ et que $\delta_G(\chi) \to 0$ si $|x^{\alpha_i}| \to 0$. Pour la démonstration, voir Casselman ([11], Théorème 4.3.3). (Casselman ne donne pas explicitement le développement (107), mais celui-ci se déduit de son résultat et du cas, trivial, des représentations de dimension finie du tore $T$).

Nous supposons maintenant donné un corps de nombres – noté pour l'instant $k$ – et une place finie $v$ de $k$ telle que $F \cong k_v$. Supposons que $\pi$ appartienne au spectre **automorphe** $\widehat{G}^{\text{aut}} = \widehat{G}(k_v)^{\text{aut}}$, relativement au groupe global $G/k$. Alors la décomposition spectrale de $\pi|_H$ est une intégrale sur $\widehat{H}^{\text{aut}}$. Notons que si $e \in \mathcal{H}$ (où $\mathcal{H}$ désigne l'espace de $\pi$) est le vecteur $K$-invariant, la décomposition qui en résulte pour $e$ ne porte que sur les représentations non-ramifiées de $H$. Notons, comme dans le paragraphe 5.3, $\widehat{H}_{nr}^{\text{aut}}$ le spectre automorphe non ramifié pour $H$. D'après les travaux de Gelbart-Jacquet [19], on a alors $\widehat{H}_{nr}^{\text{aut}} = (\widehat{SL(2,F)}_{nr}^{\text{aut}})^g$, avec

$$\widehat{SL(2,F)}_{nr}^{\text{aut}} \subset \{\mathbb{C}\} \cup \{\tau(z) : |z| = 1 \text{ ou } z \in [q^{-1/2}, q^{1/2}]\}. \tag{108}$$

On a noté $q$ le cardinal du corps résiduel pour $F$, et $\tau(z)$ la représentation de $SL(2)$ induite à partir de

$$\begin{pmatrix} x & 0 \\ 0 & x^{-1} \end{pmatrix} \mapsto z^{\text{val}(x)}, \quad x \in F^\times. \tag{109}$$

Si la décomposition spectrale de $\pi|_H$ contenait, avec une mesure non nulle, un facteur relatif à la représentation triviale, $\mathcal{H}$ contiendrait un vecteur invariant par l'un des facteurs $SL(2, F)$. C'est impossible d'après un théorème de Howe et Moore [20]. Si $C^0$ est le produit, dans $\widehat{H} = \Pi_{i=1}^g \widehat{SL(2,F)}$, des duaux automorphes des facteurs $SL(2, F)$ dont on retranche la représentation triviale, on a donc pour $\pi$ non triviale, $m(\tau)$ désignant une multiplicité:

$$\pi|_H = \int_{C^0} m(\tau)\tau d\mu(\tau) \oplus (\text{partie ramifiée}). \tag{110}$$

Soit $e$ un vecteur sphérique unitaire dans $\mathcal{H}$. Selon (110), on a donc, presque partout, $e(\tau) \in \mathcal{H}_\tau^{m(\tau)}$ et

$$(e, e) = \int_{C^0} (e(\tau), e(\tau))d\mu(\tau). \tag{111}$$

Si $h \in H$, $h \cdot e(\tau)$ reste dans le sous-espace irréductible de $\mathcal{H}_\tau^{m(\tau)}$, isomorphe à $\mathcal{H}_\tau$, engendré par $e(\tau)$. Pour $h \in H$ on a donc en particulier

$$\varphi_\pi(h) = (he, e) = \int_{C^0} \varphi_\tau(h)(e(\tau), e(\tau))d\mu(\tau) \tag{112}$$

où $\varphi_\tau$ est la fonction sphérique associée à $\tau$.

Notons maintenant $T_+^H$ le domaine fondamental pour $H$:

$$T_+^H = \{(x_1, \dots x_g) : |x_i| \leq 1\}. \tag{113}$$

Noter que $T_+ \subset T_+^H$. (Il est éclairant de dessiner l'inclusion correspondante dans $X_*(T) \otimes \mathbb{R}$ pour $g = 2$). Nous aurons besoin de contrôler la croissance des fonctions $F_\tau$ intervenant dans (112) sur $T_+^H$. Rappelons l'expression des fonctions sphériques sur $SL(2)$. Si $z \in \mathbb{C}^\times$ et si $\tau(z)$ est la représentation associée de $SL(2, F)$ (cf. (108)), on a, $\psi_z$ étant la fonction sphérique:

$$\psi_z(x) = \frac{1}{1+q^{-1}} \quad q^{-v}\left(z^v \frac{1-q^{-1}z^{-1}}{1-z^{-1}} + z^{-v}\frac{1-q^{-1}z}{1-z}\right) \tag{114}$$

où $x \in F^\times$ est associé à $\left(\begin{smallmatrix} x & 0 \\ 0 & x^{-1}\end{smallmatrix}\right)$ et $v = \mathrm{val}(x)$. Pour $z = 1$ cette expression doit être interprétée par prolongement analytique.

Introduisons la "constante de Ramanujan" (sic) relative à $k$, i.e., le plus petit $\theta$ tel qu'on ait $q^{-\theta} \leq |z| \leq q^\theta$ si $v$ est une place de $k$ et si $\tau(z) = \tau_v(z)$ est la composante locale d'une représentation cuspidale non ramifiée de $SL(2, \mathbb{A}_k)$. On a donc:

$$0 \leq \theta \leq \frac{1}{2}(\text{Gelbart-Jacquet}), \tag{115}$$

$\theta = 1$ correspondant à l'estimée triviale "de Hecke."

LEMME 6.1. *Pour $|x| \leq 1$ on a*

$$|\psi_z(x)| \leq C\ |x|^{1-\theta}L(x) \tag{116}$$

*uniformément en $z \in \mathbb{C}^\times$ tel que $|z| = 1$ ou $q^{-\theta} \leq |z| \leq q^\theta$, $C$ étant une constante et $L$ une fonction logarithmique sur $F^\times$ indépendantes de $z$.*

Noter que ceci est l'assertion (107) pour $SL(2)$, l'uniformité en sus. On le démontre de la façon suivante. Si $|z| = 1$ on sait que $|\psi_z(x)| \leq \psi_1(x)$ et qu'une telle estimée est satisfaite, avec $\theta = 0$ (Harish-Chandra). On peut donc supposer que $z \in ]1, q^\theta]$. Posons $z = q^\alpha$ avec $\alpha \in ]0, \theta]$. On peut supposer $v > 0$. En développant la fraction rationnelle (114), on obtient l'autre expression de $\psi_z(x)$:

$$(1+q^{-1})\psi_z(x) = q^{-v}z^v\{(1+z^{-1}+\cdots+z^{-2v}) - \tag{117}$$

$$-q^{-1}z^{-1}(1+z^{-1}+\cdots+z^{-2v+2})\}. \tag{118}$$

Pour $z = q^{\alpha}$ on en déduit:

$$(1 + q^{-1})\psi_z(x) \leq q^{(\alpha-1)v}(2v + 1) \tag{119}$$

d'où une majoration du type (116).

Pour $x = (x_1, \ldots x_g) \in T_+^H$ et $\tau = \tau(z_1, \ldots z_g)$ figurant dans (112) on a donc:

$$|\varphi_\tau(x)| \leq C\ |x|^{1-\theta} L(x) \tag{120}$$

en posant $|x| = |x_1 \cdots x_g|$, $C$ étant une constante et $L$ une fonction logarithmique. On en déduit:

$$|\varphi_\pi(x)| \leq C\ |x|^{1-\theta} L(x) \tag{121}$$

avec une nouvelle constante.

Comparons maintenant ceci avec l'expression (107). Puisque (121) et (107) sont des sommes finies d'exponentielles – logarithmes, l'inégalité (121) donne alors le:

Lemme 6.2. *Pour chacun des caractéres dominants $\chi_i$ figurant dans (107), et pour tout $x \in T_+$, on a*

$$|\chi_i(x)\delta_G(x)| \leq |x|^{1-\theta} = |\delta_H(x)|^{1-\theta}. \tag{122}$$

Revenons à nos expressions explicites: $T_+$ est défini par

$$\mathrm{val}(x_1) \geq \mathrm{val}(x_2) \geq \cdots \geq \mathrm{val}(x_g) \geq 0,$$

et $|\chi_i(x)|\delta_G(x)\delta_H(x)^{\theta-1}$ est donné en fonction de $v_i = \mathrm{val}(x_i)$ par

$$(v_1, \ldots, v_g) \mapsto \left(|t_1|q^{-(g-1)-\theta}\right)^{v_1} \left(|t_2|q^{-(g-2)-\theta}\right)^{v_2} \cdots \left(|t_g|q^{-\theta}\right)^{v_g}. \tag{123}$$

Alors (122) est équivalent à

$$\begin{aligned} |t_1|q^{-(g-1)-\theta} &\leq 1 \\ |t_1t_2|q^{-(2g-3)-2\theta} &\leq 1 \\ &\vdots \\ |t_1t_2\cdots t_g|q^{-\frac{g(g-1)}{2}-g\theta} &\leq 1. \end{aligned} \tag{124}$$

Rappelons par ailleurs que

$$|t_1| \geq |t_2| \geq \cdots |t_g| \geq 1.$$

On a donc enfin démontré le résultat suivant.

Théorème 6.3. *Soient $k$ un corps de nombres, $F$ un complété $p$-adique de $k$, et $\pi$ une représentation non triviale, non ramifiée de $G(F) = Sp(2g, F)$ qui est composante locale d'une représentation de $G(\mathbb{A}_k)$ apparaissant dans $L^2(G(k)\backslash G(\mathbb{A}_k))$.*

*Alors* $t_\pi = (t_1, \ldots, t_g) \in (\mathbb{C}^\times)^g$, *normalisé par*

$$|t_1| \geq |t_2| \geq \cdots \geq |t_g| \geq 1,$$

*vérifie*

$$\begin{aligned} |t_1| &\leq q^{g-1+\theta} \\ |t_1 t_2| &\leq q^{2g-3+2\theta} \\ &\vdots \\ |t_1 t_2 \cdots t_g| &\leq q^{\frac{g(g-1)}{2}+g\theta} = q^{\frac{g(g+1)}{2}-g(1-\theta)} \end{aligned} \tag{125}$$

*où* $\theta$ *est la constante de Ramanujan pour* $SL(2, k)$.

Noter que l'on peut bien sûr déduire de ceci des estimées pour les $|t_i|$ mais celles-ci ne semblent pas intéressantes. Comparons en revanche les majorations obtenues à la matrice $t_{\mathbb{C}}$ associée à la représentation triviale. D'après (100) on a

$$t_{\mathbb{C}} = (q^g, q^{g-1}, \ldots q). \tag{126}$$

Donc $\theta = 1$ (estimée triviale pour $SL(2)$) correspond à l'estimée évidente disant que pour une représentation unitaire $|t_\pi|$ est dans l'enveloppe convexe de l'orbite sous $W$ de $|t_{\mathbb{C}}|$. L'amélioration par rapport à ceci est linéaire en $\theta$. Pour $\theta = 0$, nous verrons dans la section 6.4 que l'on obtient les bornes prévues par les conjectures d'Arthur.

**6.3. Valeurs propres de l'opérateur standard $T_{\mathbf{p}}$.** Nous notons maintenant $G$ le groupe de **similitudes** symplectiques $GSp(2g)$. Nous utiliserons $G_0, \ldots,$ $T_0, \ldots$ pour les données attachées à $Sp(2g)$.

Le centre $Z$ de $G$ s'identifie à $GL(1)$ et son groupe dérivé à $G_0$. Soit $T \cong \mathbb{G}_m^{g+1}$ le tore maximal déployé de $G$ donné par

$$T = \left\{ \begin{pmatrix} x_1 & & & & & \\ & \ddots & & & & \\ & & x_g & & & \\ & & & y_1 & & \\ & & & & \ddots & \\ & & & & & y_g \end{pmatrix} : x_1 y_1 = \cdots x_g y_g = z \right\} \tag{127}$$

$z$ est alors le rapport de similitude. Le groupe de Weyl $W = W(G, T)$ est le groupe engendré par les permutations des coordonnées (à la fois en $x$ et en $y$) et par les transpositions $(x_i, y_i) \mapsto (y_i, x_i)$.

Une description non homogène de $T$ est obtenue en posant:

$$T = \left\{ \begin{pmatrix} x_1 & & & & & \\ & \ddots & & & & \\ & & x_g & & & \\ & & & zx_1^{-1} & & \\ & & & & \ddots & \\ & & & & & zy_g^{-1} \end{pmatrix} \right\} \cong \mathbb{G}_m^{g+1}. \tag{128}$$

On rappelle que $F$ désigne un corps $p$-adique. Une représentation non-ramifiée $\pi$ de $GSp(2g, F)$ est déterminée par l'orbite sous $W$ d'un caractère non ramifié $\chi$ de $T(F)$.

Une fonction $\varphi$ dans l'algèbre de Hecke de $G(F)$ par rapport à $G(\mathcal{O}_F)$ est déterminée par sa transformée de Satake $\widehat{\varphi}$, une fonction sur $\widehat{T} \cong (\mathbb{C}^\times)^{g+1}$ invariante par $W$. Alors $\varphi$ opère dans $\pi(\chi)$ par $\widehat{\varphi}(t_\chi)$ où $t_\chi$ est défini comme en 6.2.

Selon la paramétrisation non homogène (128), on a alors

$$\widehat{\varphi} \in \mathbb{C}\left[t_1, t_1^{-1}, \ldots, t_g, t_g^{-1}, s, s^{-1}\right]^W.$$

Le groupe $W$ opère en permutant les indices, et par le groupe d'ordre $2^g$, engendré par les symétries données à permutation près par

$$(t_1, \ldots t_g, s) \mapsto \left(t_1^{-1}, t_2, \ldots t_g, t_1 s\right). \tag{129}$$

Notons $\Phi_v$ la fonction caractéristique de la double classe

$$K \begin{pmatrix} \varpi & & & & & \\ & \ddots & & & & \\ & & \varpi & & & \\ & & & 1 & & \\ & & & & \ddots & \\ & & & & & 1 \end{pmatrix} K \tag{130}$$

dans $G = G(F)$, avec $K = G(\mathcal{O}_F)$.

LEMME 6.4. *On a*

$$\widehat{\Phi}_v(t_1, \ldots t_g, s) = q^{\frac{1}{4}g(g+1)} \sum_I \Big(\prod_{i \in I} t_i\Big) s = q^{\frac{1}{4}g(g+1)} s \prod_{i=1}^{g} (1 + t_i) \tag{131}$$

*où $I$ parcourt les sous-ensembles de* $\{1, \ldots g\}$.

Nous renvoyons à Duke, Howe et Li pour une démonstration explicite ([17], Lemme 2.3); ceci est un lemme standard dans la théorie des variétés de Shimura, voir [24].

Considérons en particulier le cas où $t \in \widehat{T}$ est associé à la représentation unité de $G$.

Soit $\chi$ le caractère de $T(F)$ associé à $t$. Alors $\chi$ est trivial sur le centre $Z$, donc

$$\chi(x, \dots x, x^2) = 1 \ (x \in F^\times), \tag{132}$$

donc

$$t_1 \cdots t_g s^2 = 1. \tag{133}$$

Par ailleurs $\chi(x_1, \dots x_g, 1)$ est donné par (100). En normalisant $t$ selon (105) on a donc:

$$(t_1, \dots t_g) = (q^g, \dots q) \tag{134}$$

soit $s = q^{-\frac{g(g+1)}{4}}$ et enfin:

(135)
$$\widehat{\Phi}(t_1, \dots, t_g, s) = \deg(\Phi_v) = \sum_{I \subset \{1,\dots g\}} \prod_{i \in I} q^i = (1+q)(1+q^2)\cdots(1+q^g).$$

Considérons maintenant la décomposition spectrale de $L^2(G(k)\backslash G(\mathbb{A}_k), \varepsilon)$ où $\varepsilon$ est un caractère unitaire de $Z(k)\backslash \mathbb{Z}(\mathbb{A}_k) \cong k^\times \backslash \mathbb{A}_k^\times$. Notons $\mathcal{L}_\varepsilon$ cet espace, et $\mathcal{L}_\varepsilon^\perp$ le sous-espace orthogonal de l'espace des caractères abéliens (caractères $\omega$ de $k^\times \backslash \mathbb{A}_k^\times$, considérés comme des caractères de $G(\mathbb{A}_k)$ via le rapport de similitude; on doit donc avoir alors $\omega^2 = \varepsilon$).

Proposition 6.5. *Soit $\pi$ une représentation unitaire de $G(\mathbb{A}_k)$ intervenant dans la décomposition spectrale de $\mathcal{L}_\varepsilon^\perp$, et non ramifiée en $v$. Alors, si*

$$\pi(\Phi_v)e_v = \lambda_v \, e_v, \tag{136}$$

*$e_v$ étant le vecteur non ramifié de $\pi_v$, on a*

$$|\lambda_v| \leq 2^g q^{\frac{g(g+1)}{2} - \frac{g}{2}(1-\theta)}, \tag{137}$$

*$\theta$ étant la constante de Ramanujan pour $SL(2, \mathbb{A}_k)$.*

*Démonstration.* D'après l'argument donné pour la représentation triviale, on a $|t_1 \cdots t_g \, s^2| = 1$. Par ailleurs on peut supposer que $(t_1, \dots t_g) \in (\mathbb{C}^\times)^g$ vérifie

$$|t_1| \geq \cdots \geq |t_g| \geq 1. \tag{25}$$

Par conséquent la valeur absolue de la somme dans (131) est majorée par $2^g |t_1 \cdots t_g \, s|$. Or $|s| = |t_1 \cdots t_g|^{-1/2}$ d'où enfin $|t_1 \cdots t_g \, s| = |t_1 \cdots t_g|^{1/2}$. D'après le Lemme 6.4 et le Théorème 6.3, on en déduit

$$|\lambda_v| \leq 2^g q^{\frac{g(g+1)}{2} - \frac{g}{2}(1-\theta)}. \tag{138}$$

A titre de vérification, noter que $\theta = 1$ conduit à $|\lambda_v| \leq 2^g q^{\frac{g(g+1)}{2}}$ – estimée grossière du degré de $\Phi_v$ – et que, pour $g = 1$, $\theta = 0$ donne

$$|\lambda_v| \leq 2q^{\frac{1}{2}}, \tag{139}$$

c'est-à-dire la conjecture de Ramanujan.

Corollaire 6.6. *Si la conjecture de Ramanujan* ($\theta = 0$) *est vraie pour k, on a*

$$|\lambda_v| \leq 2^g\, q^{\frac{g^2}{2}}. \tag{140}$$

**6.4. Interprétation au vu des conjectures d'Arthur.** Dans ce paragraphe, nous expliquons comment interpréter la Proposition 6.5 et le Corollaire 6.6 au vu des conjectures générales d'Arthur, et des propriétés (connues et conjecturées) de la cohomologie des variétés de modules associées à $GSp(2g)$. Nous supposerons pour simplifier que $k = \mathbb{Q}$.

Le groupe dual [5] de $G = GSp(2g)$ est un groupe réductif complexe $\widehat{G}$, ayant $\widehat{T}$ comme tore maximal, et ayant comme système de racines $R(\widehat{G}, \widehat{T})$ les coracines de $(G, T)$ – vues comme sous-groupes à un paramètre de $T$, donc comme caractères de $\widehat{T}$. On vérifie que c'est le groupe $GSpin(2g+1)$ des similitudes spinorielles en rang $g$; le groupe dérivé est donc le revêtement spinoriel de $SO(2g+1)$.

D'après Arthur [2], [3] il devrait exister un groupe conjectural $L_{\mathbb{Q}}$ (dont les représentations irréductibles de degré $n$ paramétrent les représentations cuspidales de $GL(n, \mathbb{Q})$) ayant la propriété suivante. Toute représentation automorphe $\pi$ de $GSp(2g, \mathbb{A}_{\mathbb{Q}})$ devrait être associée à une représentation continue

$$\psi : L_{\mathbb{Q}} \times SL(2, \mathbb{C}) \to \widehat{G}. \tag{141}$$

Si $F = \mathbb{Q}_p$ est un complété $p$-adique de $\mathbb{Q}$, le groupe de Weil $W_{\mathbb{Q}_p}$ doit être muni d'un homomorphisme $\iota_p : W_{\mathbb{Q}_p} \to L_{\mathbb{Q}}$.

Si $\pi_p$ est non ramifiée, $\psi \circ \iota_p$ devrait être non ramifiée. Par ailleurs, on a un homomorphisme naturel $j : W_{\mathbb{Q}_p} \to SL(2, \mathbb{C})$ donné par

$$w \mapsto \begin{pmatrix} |w|^{1/2} & 0 \\ 0 & |w|^{-1/2} \end{pmatrix} \tag{142}$$

où $|w|$ est la norme sur $\mathbb{Q}_p^\times$, composée avec l'isomorphisme du corps de classes local. Soit

$$\psi(\mathrm{Frob}_p) = \psi(\iota_p(\mathrm{Frob}_p), j(\mathrm{Frob}_p)) \in \widehat{G} \tag{143}$$

où $\mathrm{Frob}_p \in W_p$ est un élément de Frobenius; ceci est bien défini conjecturalement si $\pi_p$ est non ramifiée. Alors la classe de conjugaison associée à $\psi(\mathrm{Frob}_p)$ dans $\widehat{G}$ devrait être semi-simple et correspondre, par la paramétrisation de Satake, à $\pi_p$. Enfin, si $\pi$ est unitaire, l'image de $L_{\mathbb{Q}}$ doit être d'adhérence compacte. Par conséquent, les valeurs absolues des valeurs propres de $\psi(\mathrm{Frob}_p)$ sont uniquement déterminées par $\psi|_{SL(2,\mathbb{C})}$.

Les valeurs absolues des matrices de Hecke

$$t(\pi_p) \in \widehat{T} \subset \widehat{G}$$

d'une représentation de $G$ correspondent donc, par composition avec le plongement:

$$\mathrm{Frob}_p \mapsto \begin{pmatrix} p^{1/2} & 0 \\ 0 & p^{-1/2} \end{pmatrix}, \tag{144}$$

aux représentations $SL(2) \to \widehat{G}$, donc aux homomorphismes

$$SL(2) \to SO(2g+1),$$

i.e., aux représentations orthogonales de degré $(2g+1)$ de $SL(2)$. Une telle représentation est une représentation de $SL(2)$ dont tous les constituants irréductibles de multiplicité impaire sont orthogonaux, i.e., de degré impair.

Puisque on considère des représentations de $GSp(2g)$ de caractère central unitaire, la taille des matrices de Hecke est déterminée par leur restriction à $Sp(2g)$. Soit alors

$$G_0 = Sp(2g), \quad T_0 = \left\{ \begin{pmatrix} x_1 & & & & & \\ & \ddots & & & & \\ & & x_g & & & \\ & & & x_1^{-1} & & \\ & & & & \ddots & \\ & & & & & x_g^{-1} \end{pmatrix} \right\} \tag{145}$$

et $\widehat{T}_0 \subset \widehat{G}_0 = SO(2g+1)$. On a alors naturellement

$$T_0 = \left\{ \begin{pmatrix} t_1 & & & & & & \\ & \ddots & & & & & \\ & & t_g & & & & \\ & & & 1 & & & \\ & & & & t_1^{-1} & & \\ & & & & & \ddots & \\ & & & & & & t_g^{-1} \end{pmatrix} : t_i \in \mathbb{C}^\times \right\}, \tag{146}$$

les coracines de $(G_0, T_0)$ s'identifiant aux racines $\{t_i t_j^{-1} : i \neq j\}$ et $\{t_i^{\pm 1}\}$ de $(\widehat{G}_0, \widehat{T}_0)$—cf. (102).

Soit alors $r : SL(2) \to SO(2g+1)$ une représentation. On peut supposer que $r$ envoie le tore diagonal de $SL(2)$ vers $\widehat{T}_0$.

Pour $h = \begin{pmatrix} x & 0 \\ 0 & x^{-1} \end{pmatrix} \in SL(2)$, on peut supposer à conjugaison près dans $\widehat{G}$ que

$$r(h) = \begin{pmatrix} x^{m_1} & & & & & & \\ & \ddots & & & & & \\ & & x^{m_g} & & & & \\ & & & 1 & & & \\ & & & & x^{-m_1} & & \\ & & & & & \ddots & \\ & & & & & & x^{-m_g} \end{pmatrix} \tag{147}$$

avec $m_1 \geq \cdots \geq m_g$.

Alors la représentation $\psi = 1 \otimes r$ de $L_{\mathbb{Q}} \times SL(2)$ donne pour matrice de Hecke

$$t_p = \Psi(\mathrm{Frob}_p) = \begin{pmatrix} p^{m_1/2} & & & & & \\ & p^{m_2/2} & & & & \\ & & \ddots & & & \\ & & & 1 & & \\ & & & & \ddots & \\ & & & & & p^{-m_g/2} \end{pmatrix}. \tag{148}$$

Si $r$ est la représentation irréductible (donc orthogonale) de degré $(2g + 1)$, on obtient donc:

$$t_p = \begin{pmatrix} p^g & & & & & & & \\ & p^{g-1} & & & & & & \\ & & \ddots & & & & & \\ & & & p & & & & \\ & & & & 1 & & & \\ & & & & & p^{-g} & & \\ & & & & & & \ddots & \\ & & & & & & & p^{-1} \end{pmatrix}, \tag{149}$$

c'est-à-dire la matrice de Hecke de la représentation triviale.

LEMME 6.7. *Pour $r$ orthogonale mais non irréductible, on a*

$$m_1 + \cdots + m_g \leq g(g - 1). \tag{150}$$

*Démonstration.* Si $\rho$ est une représentation irréductible de $SL(2)$ de plus haut poids $h \mapsto x^m$ $(m \geq 0)$, la somme des poids $\geq 0$ de $\rho$ est $\frac{m(m+2)}{4}$ si $m$ est pair et $\frac{(m+1)^2}{4}$ si $m$ est impair. En paramétrant les composantes irréductibles de $r$ par leur degré $d = m + 1$, on a donc

$$M = m_1 + \cdots + m_g = \sum_{d \text{ impair}} \frac{d^2 - 1}{4} + \sum_{d \text{ pair}} \frac{d^2}{4} \tag{151}$$

$$\sum d = 2g + 1, \tag{152}$$

les degrés $d$ pouvant bien sûr intervenir avec des multiplicités.

L'inégalité (150) s'écrit alors

$$\sum_{d \equiv 1} (d^2 - 1) + \sum_{d \equiv 0} d^2 \leq 4g^2 - 4g \tag{153}$$

où les congruences sont modulo 2.

Soit $e$ l'un des degrés $d$ apparaissant dans (153). Alors

$$\sum d^2 \leq e^2 + (2g + 1 - e)^2$$

donc (153) sera vérifiée si

$$e^2 + (2g + 1 - e)^2 \leq 4g^2 - 4g, \tag{154}$$

soit

$$2e^2 - 2(2g + 1)e + 8g + 1 \leq 0. \tag{155}$$

Cette équation est symétrique si on remplace $e$ par $2g + 1 - e$. Le polynôme quadratique est égal en $e = 3$ à $-4g + 13 < 0$ si $g > 3$, ce que nous supposons pour l'instant.

Si la partition $2g + 1 = \sum d$ viole (153), on doit donc avoir, pour tout $d$, $d = 1$, 2 ou $d = 2g, 2g - 1$. (On a supposé $r$ réductible, donc $d = 2g + 1$ est exclu). Les seules partitions de $2g + 1$ à considérer sont donc $(1, 1, 2g - 1)$, $(2, 2g - 1)$ et $(1, 2g)$. (Les partitions de la forme $1^a 2^b$ sont aisément exclues) La seconde et la troisième sont exclues car la représentation de degré 2 ou $2g$, qui apparaîtrait avec multiplicité 1, est symplectique. Il reste la première; dans ce cas, revenant à (153), on voit qu'on a bien

$$\sum (d^2 - 1) = -3 + 2 + (2g - 1)^2 = (2g - 1)^2 - 1 = 4g^2 - 4g. \tag{156}$$

Si $g = 2$, en se rappelant que les représentations de degré pair doivent apparaître avec multiplicité paire, on voit que les seules partitions à considérer modulo les arguments précédents sont $5 = 2 + 2 + 1, 5 = 3 + 1 + 1, 5 = 1 + 1 + \cdots + 1$, qui vérifient (153), avec égalité pour les deux premières. Pour $g = 3$, on doit considérer les partitions $3^2 \cdot 1$, $3 \cdot 2^2$, $3 \cdot 1^4$, $2^2 \cdot 1^3$ et $1^7$. Toutes vérifient l'inégalité stricte dans (153).

Notons la conséquence du calcul, dont nous ne ferons pas usage ici:

LEMME 6.8. *Si $g \geq 3$, la seule représentation $r$ donnant l'égalité dans (150) est celle de degrés* $(1, 1, 2g - 1)$.

*Si $g = 2$, les représentations de degrés* $(2, 2, 1)$ *et* $(3, 1, 1)$ *conviennent.*

Nous pouvons maintenant interpréter le Corollaire 3.5 à l'aide des conjectures d'Arthur. Si $\pi$ est une représentation de $Sp(2g, \mathbb{A}_{\mathbb{Q}})$ non triviale apparaissant dans l'espace des formes de carré intégrable, sa matrice de Hecke $t_p$ en une place non ramifiée doit avoir des valeurs absolues données par (148), pour une représentation $r$ de $SL(2)$ vérifiant (150). On a donc

$$|t_1 \cdots t_g| \leq p^{\frac{g(g-1)}{2}} = p^{\frac{g(g+1)}{2} - g}. \tag{157}$$

C'est l'estimée du Théorème 6.3 (pour $\theta = 0$) et on en déduit le Corollaire 6.6.

Noter que les conjectures d'Arthur impliquent donc que l'estimée associée à $\theta = 0$ est optimale pour $Sp(2g)$, puisque les paramètres associés à la partition considérée sont présents. On obtiendrait des formes automorphes correspondant à ces matrices de Hecke en formant des séries d'Eisenstein à partir de la représentation triviale du sous-groupe de Levi $Sp(2(g-1)) \times GL(1)$ de $Sp(2g)$.

Terminons en indiquant une autre interprétation, géométrique, de l'estimée du Corollaire 6.6. Soient $N$ un entier $\geq 1$, et $X_N = \mathcal{A}_g(N)$ l'espace de modules des variétés abéliennes principalement polarisées munies d'une structure de niveau $N$. Alors la fonction $\Phi_p$ du 6.3 opère sur la cohomologie $L^2$, $H^\bullet_{(2)}(X_N, \mathbb{C})$; celle-ci, d'après la conjecture de Zucker, est isomorphe à la cohomologie d'intersection $IH^\bullet(\overline{X}_N, \mathbb{C})$, $\overline{X}_N$ étant la compactification de Baily-Borel.

Les variétés $X_N$, $\overline{X}_N$ sont définies sur $\mathbb{Q}$ et les valeurs propres $\alpha_p$ de $\Phi_p$ devraient être liées à celle d'un opérateur de Frobenius en $p$, opérant dans $IH^\bullet(\overline{X}_N \otimes \overline{\mathbb{Q}}, \mathbb{Q}_\ell)$. La cohomologie d'intersection vérifie la dualité de Poincaré, et, d'après un théorème d'annulation bien connu [6], est nulle en degré $\leq g$ – en omettant les classes de Chern associées à la représentation triviale de $Sp(2g, \mathbb{A}_{\mathbb{Q}})$. En dehors de la partie triviale de la cohomologie, les propriétés de pureté des valeurs propres de Frobenius doivent alors impliquer $\alpha_p \leq p^{\frac{g(g+1)}{2} - \frac{g}{2}} = p^{\frac{g^2}{2}}$ puisque la dimension de $X_N$ est $\frac{g(g+1)}{2}$. Le facteur $2^g$ du Corollaire 6.6 correspond, comme pour $GL(2)$, au degré maximal des représentations galoisiennes de $\mathrm{Gal}(\overline{\mathbb{Q}}/\mathbb{Q})$ apparaissant dans la cohomologie d'intersection, ici égal à $2^g$. On renvoie à Kottwitz [24] pour une description précise (mais conjecturale) de la cohomologie d'intersection de $\mathcal{A}_g(N)$ en termes de formes automorphes. Noter qu'il résulte de Li [25] que $IH^g(\overline{X}_N, \mathbb{C})$ est effectivement non nul, pour $N$ convenable (et contient des classes différentes des classes de Chern). Ceci explique de nouveau pourquoi le Corollaire 6.6 est optimal.

**6.5. Preuve du théorème 1.3.** La preuve du théorème 1.3 à partir de ce qui précède est identique à la preuve du théorème 1.2. On écrit la décomposition orthogonale:

$$L^2(\Gamma_2 \backslash G_2(\mathbb{R}), \mathbf{1}) = \mathbb{C}\mathbf{1} \oplus L,$$

où $\mathbb{C}\mathbf{1}$ désigne l'espace des fonctions constantes et $L$ son orthogonal.

Soit $\pi = \hat{\otimes}\pi_p$ une représentation automorphe irréductible de $GSp(2g)(\mathbb{A})$ intervenant dans la décomposition spectrale (44). La composante $\pi_p$ de $\pi$ en $p$ est non ramifiée. Soit $\chi$ un caractère associé et

$$t_\chi = (t_1, \ldots, t_g, s)$$

les paramètres de $\pi_p$. On a défini $ch_p$ avant l'équation (43). D'après (43), l'action de $\overline{T}_p$ sur la partie relative à $\pi$ de la décomposition spectrale (44) est donnée par l'action de $\frac{ch_p}{\deg(ch_p)}$ sur la composante en $p$ d'un vecteur $K_f$-invariant de $\pi$. D'après la section 3.3, le lemme 6.4 et l'équation (135), cette action est donnée par multiplication par

$$\overline{\lambda}_p = \frac{\widehat{ch}_p(t_1, \ldots t_g, s)}{\prod_{i=1}^{g}(1 + p^i)}.$$

D'après la proposition 6.5, si $\pi$ n'est pas la représentation triviale, on a

$$|\widehat{ch}_p(t_1, \ldots, t_g, s)| \leq 2^g p^{\frac{g(g+1)}{2} - \frac{g(1-\theta)}{2}}.$$

On déduit la première partie du théorème 1.3 en utilisant la décomposition spectrale (44) et cette inégalité.

## 7. Généralisation. Le cas des groupes orthogonaux.

**7.1. Le cas des groupes fortement orthogonaux.** L'élément crucial dans les majorations obtenues dans la partie 6 est le fait que Sp($g$) contient un produit de $g$ facteurs $SL(2)$. Cet argument s'applique à d'autres groupes; il est remarquable qu'en particulier (pour des groupes déployés) cela est le cas quand $G$ est associé à des variétés de Shimura. Nous expliquons l'argument dans le cas général, puis nous dérivons des majorations explicites dans le cas de $SO(2g + 1)$ où, pour simplifier, nous supposerons $g$ pair.

Soit donc $G$ un groupe simple déployé de rang $g$ sur un corps de nombres $k$, et supposons qu'il existe un homomorphisme de noyau fini $\varphi : SL(2)^g \to G$, d'image $H$. Soit $T$ un tore maximal déployé de $H$, donc de $G$. Les racines de $T$ associées aux sous-groupes unipotents de $H$ forment alors un système de racines (de type $A_1^g$) fortement orthogonales. (Rappelons que ceci veut dire que, pour deux d'entre elles, soit $\alpha$, $\beta$, $\pm\alpha \pm \beta$ n'est pas une racine).

En particulier, le groupe de Weyl $W(G, T)$ doit contenir l'élément $(-1)$, et l'on sait en fait que ceci est équivalent à l'existence de $g$ racines fortement orthogonales (e.g., Warner [40] p. 93). Si $k$ est totalement réel, ceci revient à supposer que $G(k \otimes_{\mathbb{Q}} \mathbb{R})$ a une série discrète: en particulier ceci est vrai pour les groupes déployés (sur des corps totalement réels) dont l'espace symétrique est hermitien.

Avec les notations ci-dessus, notons $R(G, T)$ et $R(H, T)$ les systèmes de racines de $G$ et $H$, contenus dans le groupe $X^*$ des caractères de $T$. On choisit une base $\Delta_G$ de $R(G, T)$ (et donc un choix de racines positives); il existe une unique base

$\Delta_H$ de $R(H,T)$ formé de racines positives. Si $X_*$ est le réseau des cocaractères de $T$, dual de $X^*$, on écrira

$$\mathfrak{t} = X_* \otimes_{\mathbb{Z}} \mathbb{R}, \ \widehat{\mathfrak{t}} = X^* \otimes_{\mathbb{Z}} \mathbb{R}. \tag{158}$$

$$\widehat{T} = \mathrm{Hom}(X_*, \mathbb{C}^\times). \tag{159}$$

Si $v$ est une place non archimédienne de $k$ et $F = k_v$ on écrira simplement $T$ pour $T(F)$. On note $q$ le cardinal du corps résiduel. Soit

$$T^+ = \{x \in T \mid |x^\alpha| \leq 1 \quad \forall \alpha \in \Delta_G\} \tag{160}$$

$$T_H^+ = \{x \in T \mid |x^\alpha| \leq 1 \quad \forall \alpha \in \Delta_H\} \tag{161}$$

de sorte que $T^+ \subset T_H^+$. Par ailleurs, soit $\Delta_G^\vee \subset X_*(T) = X^*(\widehat{T})$ l'ensemble des racines duales, et

$$\widehat{T}_+ = \{t \in \widehat{T} : |t^\beta| \geq 1 \quad \forall \beta \in \Delta_G^\vee\}. \tag{162}$$

On identifie $X_*(T)$ à $T(F)/T(\mathcal{O})$ en envoyant $u \in X_*(T)$ sur $u(\varpi)$, où $\varpi$ est une uniformisante. Ainsi $\widehat{T}$ s'identifie au groupe des caractères non ramifiés de $T$. On vérifie alors que l'image inverse de $T^+$ dans $X_*(T)$ est l'intersection de celui-ci avec la chambre de Weyl

$$C^+ \subset \mathfrak{t}, \ C^+ = \{X \in \mathfrak{t} : (X, \alpha) \geq 0 \quad \forall \alpha \in \Delta_G\}. \tag{163}$$

Notons $Re$ l'application logarithme de $\widehat{T}$ vers $\widehat{\mathfrak{t}}$ donnée par

$$|\chi(u)| = q^{(Re\,\chi, u)} \quad (u \in X_*(T), \chi \in \widehat{T}). \tag{164}$$

Alors

$$\widehat{T}_+ = \{t \in \widehat{T} : Re t \in \widehat{C}^+\} \tag{165}$$

où

$$\widehat{C}^+ = \{Y \in \widehat{\mathfrak{t}} : (Y, \beta) \geq 0 \quad \forall \beta \in \Delta_G^\vee\}. \tag{166}$$

Enfin soit $\delta_G$, $\delta_H$ les caractères à valeurs positives de $T$ donnés par la racine carrée du module du produit des racines positives relatives à $G$, $H$. Noter que leurs valeurs sont $\leq 1$ sur $T^+$. Un calcul facile montre que

$$Re\,\delta_G = -\rho_G, \ Re\,\delta_H = -\rho_H \tag{167}$$

où $\rho_G, \rho_H \in \widehat{\mathfrak{t}}$ sont les demi-sommes de racines positives de $G$, $H$.

Soit alors $\chi \in \widehat{T}$ un caractère non ramifié, et $\varphi_\chi$ la fonction sphérique associée. L'expression (108) de la fonction sphérique donne ici, si $\chi \in \widehat{T}_+$:

$$\varphi_\chi(x) \sim \sum_{i=1}^{r} \chi_i(x)\delta_G(x)L_i(x) \quad (x \in T^+) \tag{168}$$

où les $\chi_i$ sont les conjugués de $\chi$ par $W(G, T)$ contenus dans $\widehat{T}_+$ et les $L_i$ sont des fonctions polynomiales sur $T(F)/T(\mathcal{O}) = X_*(T)$.

Supposons que la représentation non ramifiée $\pi_\chi$ associée à $\chi$ est automorphe (i.e., $\in \widehat{G}(F)_{\text{aut}}$ au sens de la section 5) et n'est pas un caractère abélien. Si $\theta$ est la constante de Ramanujan pour $k$, il résulte alors des résultats de la section 5 et des arguments donnés autour du Lemme 6.2, que

$$|\varphi_\chi(x)| \leq C\delta_H(x)^{1-\theta}L(x) \quad \left(x \in T_H^+\right) \tag{169}$$

pour une fonction polynomiale $L$. D'après (168) et (169):

$$|\chi(x)|\delta_G(x)\delta_H(x)^{\theta-1} \leq 1 \quad (x \in T^+). \tag{170}$$

Vu la description de $T^+$ donnée avant (163), on en déduit d'après (164) et (167):

$$(Re\,\chi - \rho_G + (1-\theta)\rho_H, X) \leq 0 \quad (X \in C^+), \tag{171}$$

c'est-à-dire:

$$Re\,\chi = \rho_G + (\theta - 1)\rho_H - \sum_{\alpha \in \Delta_G} \lambda_\alpha \alpha, \ \lambda_\alpha \geq 0. \tag{172}$$

On a donc démontré le résultat suivant. Notons $\leq$ l'ordre sur $\widehat{\mathfrak{t}}$ associé aux racines positives de $G$.

THÉORÈME 7.1. *Soient $G$ un groupe simple et déployé sur $k$ et $H \subset G$ un sous-groupe isogène à $SL(2)^g$ où $g$ est le rang de $G$; soit $F = k_v$ une complétion non archimédienne de $k$.*

*Soit $\chi \in \widehat{T}$ un caractère associé à une représentation automorphe non abélienne et non ramifiée de $G(F)$, et supposons que $Re\,\chi \in \widehat{C}^+$. Alors $Re\,\chi \leq \rho_G + (\theta - 1)\rho_H$.*

En particulier, on voit que sous la conjecture de Ramanujan ($\theta = 0$) on devrait avoir

$$Re\,\chi \leq \rho_G - \rho_H = \rho_G - \frac{1}{2}\sum_\beta \beta \tag{173}$$

pour tout système $\{\beta\}$ de $g$ racines positives fortement orthogonales.

Dans le paragraphe qui suit, nous expliciterons la relation (173) dans le cas des groupes orthogonaux. On verra dans ce cas que (pour $T \subset G$, $\Delta_G$ fixés) il existe un système $\Phi$ de racines fortement orthogonales telle que la relation (173) déduite de $\Phi$ implique toutes les relations analogues. Autrement dit, si $\Phi'$ est un autre système, on a, avec des notations évidentes:

$$\rho_\Phi \geq \rho_{\Phi'}. \tag{174}$$

Hee Oh [31] a vérifié cette propriété pour tous les systèmes de racines.

**7.2. Application au groupe $G = SO(2g+1)$.** Considérons maintenant le cas où $G = SO(2g+1)$; nous supposerons pour simplifier $g = 2h$ pair. On suppose $G$ associé à la forme quadratique sur $k^{2g+1}$ de matrice

$$Q = \begin{pmatrix} & & 1_g \\ & 1 & \\ 1_g & & \end{pmatrix}. \tag{175}$$

Soit $T = \left\{ \begin{pmatrix} x_1 & & & & & & \\ & \ddots & & & & & \\ & & x_g & & & & \\ & & & 1 & & & \\ & & & & x_1^{-1} & & \\ & & & & & \ddots & \\ & & & & & & x_g^{-1} \end{pmatrix} \right\} \cong \mathbb{G}_m^g \subset G.$

Les calculs relatifs aux racines sont duaux de ceux du § 6.

On a un système de racines simples

$$\Delta_G = \left\{ x_1 x_2^{-1}, \ldots, x_{g-1} x_g^{-1}, x_g \right\}. \tag{176}$$

Notant comme ci-dessus $T = T(F)$, on a donc

$$T_+ = \{(x_1, \ldots, x_g) : |x_1| \leq |x_2| \leq \cdots \leq |x_g| \leq 1\}. \tag{177}$$

On a $\widehat{T} = (\mathbb{C}^\times)^g$, un caractère $\chi \in \widehat{T}$ étant associé à $(t_1, \ldots, t_g)$ avec

$$t_i = \chi \begin{pmatrix} 1 & & & & & & & & \\ & 1 & & & & & & & \\ & & \varpi & & & & & & \\ & & & \ddots & & & & & \\ & & & & 1 & & & & \\ & & & & & 1 & & & \\ & & & & & & \varpi^{-1} & & \\ & & & & & & & \ddots & \\ & & & & & & & & 1 \end{pmatrix}. \tag{178}$$

Les coracines ont une base donnée par

$$\Delta_G^\vee = \left\{ t_1 t_2^{-1}, \ldots, t_{g-1} t_g^{-1}, t_g^2 \right\}, \tag{179}$$

donc $\widehat{T}_+ = \{t \in \widehat{T} : |t_1| \geq \cdots \geq |t_g| \geq 1\}$.

Notons enfin que les racines positives associés à la base $\Delta_G$ sont données par

$$R^+(G, T) = \left\{ x_i x_j^{-1}, x_i x_j \right\}_{i<j} \cup \{x_i\}_{1 \leq i \leq g}. \tag{180}$$

Un système $\Phi$ de racines positives fortement orthogonales est, comme on le vérifie aisément, de la forme suivante. Soit $\{1, \dots, g\} = \coprod_{k=1}^{h} \{i_k, j_k\}$ une partition de $\{1, \dots, g\}$ en ensembles à deux éléments; on suppose $i_k < j_k$. Alors:

$$\Phi = \cup_{k=1}^{h} \left\{x_{i_k} x_{j_k}^{-1}, x_{i_k} x_{j_k}\right\} \tag{181}$$

est fortement orthogonal. On a alors:

$$\rho_G^2(x) = x_1^{2g-1} x_2^{2g-3} \cdots x_{g-1}^3 x_g \tag{182}$$

$$\rho_H^2(x) = \prod_{k=1}^{h} x_{i_k}^2, \tag{183}$$

et si $\rho_G$ et $\rho_H$ sont vus comme des éléments de $\frac{1}{2} X^*(T) \subset \widehat{\mathfrak{t}}$ (cf. après (167)), où $X^*(T)$ est identifié à $\mathbb{Z}^g$:

$$\rho_G = \left(\frac{2g-1}{2}, \frac{2g-3}{2}, \dots, \frac{2g+1-21}{2}, \dots, \frac{1}{2}\right). \tag{184}$$

$$\rho_H = (0, \dots 1, 0, \dots 1, \dots 0, \dots) \tag{185}$$

les coefficients 1 apparaissant pour $i = i_k$.

Si $\chi \in \widehat{T}_+$ et $Y = Re\, \chi \in \widehat{t} = \mathbb{R}^g$, $Y$ vérifie

$$Y_1 \geq Y_2 \geq \cdots \geq Y_g \geq 0, \tag{186}$$

et par définition

$$Y_i = \frac{\ell n |t_i|}{\ell n q}. \tag{187}$$

Enfin, $Y \in \widehat{\mathfrak{t}}$ est positif (pour l'ordre associé aux racines) si

$$\begin{aligned} Y_1 & \geq 0 \\ Y_1 + Y_2 & \geq 0 \\ Y_1 + \cdots + Y_g & \geq 0. \end{aligned} \tag{188}$$

Noter que $\rho_\Phi = \rho_H$, où $H$ est associé à $\Phi$, ne dépend que des indices $i_k$ dans (181). Soit $\Phi$ un système fortement orthogonal tel que $\{i_1, \dots i_k\} = \{1, \dots h\}$. On a donc

$$\rho_\Phi = (1, \dots 1, 0, \dots 0). \tag{189}$$

Si $\Phi'$ est un autre système fortement orthogonal et $\rho_{\Phi'} = (Y_1, \dots, Y_g)$, on a, pour tout $i \leq h$, $Y_1 + \cdots + Y_i \leq i$, et pour tout $i > h$, $Y_1 + \cdots + Y_i \leq h$. Ceci montre d'après (188) que $\rho_\Phi \geq \rho_{\Phi'}$, comme on l'a remarqué après le Théorème 4.1.

L'inégalité

$$Re\, \chi \le \rho_G + (\theta - 1)\rho_H,$$

où $H$ est associé à $\Phi'$ arbitraire, est donc impliquée par l'unique inégalité associée à $\Phi$ comme ci-dessus. En explicitant (188), on obtient donc les majorations suivantes pour $Y_i = \frac{\ell n(t_i)}{\ell n q}$:

$$\begin{aligned} Y_1 &\le \tfrac{1}{2}(2g-1) + \theta - 1 \\ Y_1 + Y_2 &\le \tfrac{1}{2}(2g - 1 + 2g - 3) + 2(\theta - 1) \\ &\vdots \\ Y_1 + \cdots + Y_h &\le \tfrac{1}{2}(2g - 1 + 2g - 3 + \cdots + (2g + 1 - 2h)) + h(\theta - 1) \\ Y_1 + \cdots + Y_{h+1} &\le \tfrac{1}{2}(2g - 1 + \cdots + (2g + 1 - 2(h+1)) + h(\theta - 1) \\ &\vdots \\ Y_1 + \cdots + Y_g &\le \tfrac{1}{2}((2g-1) + (2g-3) + \cdots + 1) + h(\theta - 1). \end{aligned} \tag{190}$$

Donnons une application de cette estimée à la trace d'un opérateur de Hecke naturel.

Soit $\widehat{G} = Sp(2g)$ le groupe dual de $G$, de sorte que $\widehat{G} \supset \widehat{T}$. Les poids dominants des représentations irréductibles de dimension finie de $\widehat{G}$ correspondent alors aux éléments (dominants) de $X_*(T)$. En particulier, la représentation standard de degré $2g$ de $\widehat{G}$, dont le poids maximal est minuscule, correspond à un cocaractère $\mu \in X_*(T)$. Soit $a = \mu(\varpi) \in T(F)$ et soit $\Phi_v$ la fonction caractéristique de $K_v a K_v$, où $K_v = G(\mathcal{O}_v)$. On a alors:

Lemme 7.2. *(Langlands, Kottwitz). Si $\varphi \mapsto \widehat{\varphi}$ ($\varphi \in \mathcal{H}(G_v, K_v), \widehat{\varphi} \in \mathbb{C}[\widehat{T}]^W$) désigne la transformée de Satake,*

$$\widehat{\Phi}_v(t_1, \ldots t_g) = q^{\frac{2g-1}{2}} \sum_{i=1}^{g} \left(t_i + t_i^{-1}\right). \tag{191}$$

Pour la démonstration, voir Kottwitz [24] Théorème. 2.1.3.

Des inégalités (190) on déduit alors la majoration grossière (On laisse au lecteur le soin de dériver des majorations plus fines de (190) et d'inégalités de convexité):

Corollaire 7.3. *Si $\pi_v$ est une représentation non ramifiée et non abélienne de $G(F)$ apparaissant dans l'espace des formes automorphes,*

$$\widehat{\Phi}_v(\pi_v) = tr(\pi_v(\Phi_v)) \tag{192}$$

*vérifie:*

$$|\widehat{\Phi}_v(\pi_v)| \le 2g\, q^{2g-1+\theta-1}. \tag{193}$$

*En particulier, sous l'hypothèse de Ramanujan pour $SL(2)$:*

$$|\widehat{\Phi}_v(\pi_v)| \le 2g\, q^{2g-2}. \tag{194}$$

On notera que si $\pi_v$ est la représentation triviale, la matrice de Hecke $t \in \widehat{T}$ est

$$t_{\mathbb{C}} = \begin{pmatrix} q^{\frac{2g-1}{2}} & & & & & & \\ & q^{\frac{2g-3}{2}} & & & & & \\ & & \ddots & & & & \\ & & & q & & & \\ & & & & q^{-1} & & \\ & & & & & \ddots & \\ & & & & & & q^{\frac{1-2g}{2}} \end{pmatrix} \tag{195}$$

(on a considéré $\widehat{T}$, de la façon naturelle, comme plongé dans $\widehat{G}$) et donc:

$$\widehat{\Phi}_v(\mathbb{C}) = q^{2g-1} + q^{2g-2} + \cdots + 1 \asymp q^{2g-1}. \tag{196}$$

L'amélioration donnée par le Corollaire est donc en $q^{\theta-1}$, $q^{-1/2}$ inconditionnellement (Gelbart-Jacquet) et $q^{-1}$ sous l'hypothèse de Ramanujan.

On notera que cette majoration est de nouveau essentiellement optimale (en supposant $\theta = 0$) au vu des conjectures d'Arthur. On renvoie au chapître 6.4 pour le cas de $Sp(2g)$. Pour $G = SO(2g+1)$, $\widehat{G} = Sp(2g)$, les valeurs absolues des valeurs propres des matrices de Hecke de représentations automorphes (au sens de la décomposition de $L^2$) sont données par la recette donnée à la section 6.4 appliquée à des représentations $r$: $SL(2) \to \widehat{G}$. Si $r$ est réductible et symplectique, de degré $(2g)$, son poids maximal $m = \mathrm{Max}(m_i)$ (notation (147) est $m = 2g - 3$. Si $\pi_v$ vérifie l'hypothèse du Corollaire 7.3, les valeurs absolues des valeurs propres de sa matrice de Hecke $t_{\pi_v} \in \widehat{G}$ sont alors majorées par $q^{\frac{2g-3}{2}}$, d'où (194) d'après le Lemme 7.2. On voit en particulier que (194) donne l'ordre de grandeur correct.

Comme dans le cas de $Sp(2g)$, nous ne savons pas si l'on dispose d'assez d'informations sur le dual unitaire pour dériver (190) des résultats locaux. Noter cependant que (190) est, dans le cas de $SO(2g+1)$, plus faible que les conjectures d'Arthur. Par exemple (190) donne, sous l'hypothèse de Ramanujan:

$$|t_1 \cdots t_g| \leq q^{\frac{g}{2}-\frac{1}{2}} q^{\frac{1}{2}\{(2g-3)+(2g-5)+\cdots+1\}} \tag{197}$$

alors que les conjectures d'Arthur impliquent

$$|t_1 \cdots t_g| \leq q^{\frac{1}{2}\{(2g-3)+(2g-5)+\cdots+1\}}. \tag{198}$$

**8. Equidistribution des points de Hecke.** Dans cette section nous expliquons comment on obtient la deuxième partie des théorèmes 1.2 et 1.3 à partir des résultats pour la norme $L^2$.

Soit $G$ un groupe réductif défini sur $\mathbb{Q}$. On note $\Gamma = G(\mathbb{Z})$ et $Z$ le centre de $G$. On note $C_c = C_c(\Gamma Z(\mathbb{R})\backslash G(\mathbb{R}))$ l'ensemble des fonctions continues à support compact dans $\Gamma Z(\mathbb{R})\backslash G(\mathbb{R})$.

Soit $E$ un sous ensemble fini, invariant à droite par $\Gamma$, de $\Gamma\backslash G(\mathbb{R})$. Soit $|E|$ son cardinal. On dispose alors d'un opérateur $T_E$ sur $L^2(\Gamma Z(\mathbb{R})\backslash G(\mathbb{R}))$ définie par

$$T_E f(z) = \frac{1}{|E|}\sum_{m\in E} f(mz).$$

On note encore $T_E$ l'opérateur sur $C_c$ obtenue par restriction. Si $h \in G(\mathbb{Q})$ et $E = \Gamma h \Gamma = \cup_{i=1}^{r} \Gamma h_i$, on note $T_h = T_E$.

PROPOSITION 8.1. *Soit $E_n$ une suite de sous ensembles finis, invariants à droite par $\Gamma$, de $\Gamma\backslash G(\mathbb{R})$. On suppose que pour toute fonction*

$$f \in L^2(\Gamma Z(\mathbb{R})\backslash G(\mathbb{R}))$$

*on a*

$$\lim_{n\to\infty} \|T_{E_n} f - \int_{\Gamma Z(\mathbb{R})\backslash G(\mathbb{R})} f(x) d\mu(x)\| = 0.$$

*Pour toute fonction $f \in C_c$ et tout $z_0 \in \Gamma Z(\mathbb{R})\backslash G(\mathbb{R})$ on a la convergence simple*

$$\lim_{n\to\infty} T_{E_n} f(z_0) = \int_{\Gamma Z(\mathbb{R})\backslash G(\mathbb{R})} f(x) d\mu(x).$$

On note $< \, , \, >$ le produit scalaire sur $L^2(\Gamma Z(\mathbb{R})\backslash G(\mathbb{R}))$. Par l'inégalité de Cauchy-Schwartz on montre que pour des fonctions $f$ et $g$ dans

$$L^2(\Gamma Z(\mathbb{R})\backslash G(\mathbb{R}))$$

on a:

$$\lim_{n\to\infty} < T_{E_n} f, g > = < f, 1 > < 1, g > . \tag{199}$$

Soient $f \in C_c$ et $\epsilon > 0$. On identifie $f$ à une fonction $\Gamma$-invariante sur $Z(\mathbb{R})\backslash G(\mathbb{R})$; $f$ est alors uniformément continue pour la structure uniforme invariante à droite. Par uniforme continuité de $f$, il existe un voisinage $U_\epsilon(z_0)$ de $z_0$ telle que pour tout $z \in U_\epsilon(z_0)$ et tout $g \in G$ on ait

$$||f(gz) - f(gz_0)| \leq \epsilon. \tag{200}$$

Soit $\psi \in C_c$ une fonction positive dont le support est contenue dans $U_\epsilon(z_0)$ telle que

$$\int_{\Gamma Z(\mathbb{R})\backslash G(\mathbb{R})} \psi(x) d\mu(x) = 1.$$

D'après l'équation (199), pour tout $n$ assez grand, on a

$$| < T_{E_n} f, \psi > - \int_{\Gamma Z(\mathbb{R})\backslash G(\mathbb{R})} f(x) d\mu(x)| \leq \epsilon.$$

d'après l'équation (200), on a

$$| < T_{E_n} f, \psi > - T_{E_n} f(z_0)| \leq \epsilon.$$

On déduit des deux dernières inégalités que l'on a bien

$$\lim_{n\to\infty} T_{E_n} f(z_0) = \int_{\Gamma Z(\mathbb{R})\backslash G(\mathbb{R})} f(x) d\mu(x).$$

**8.1. Estimées ponctuelles.** Les Théorèmes 1.2 et 1.3 (b) ne donnent pas la vitesse de convergence de $\overline{T}_N x$ vers la mesure de Haar. Dans cette section nous montrons comment estimer la vitesse de convergence quand $\overline{T}_N x$ est évalué sur une fonction suffisamment différentiable.

Dans la section 2.2 (voir Théorème 2.1 (b)) une telle estimée était déduite d'évaluations uniformes des valeurs des séries d'Eisenstein. Nous ne savons pas si de telles estimées sont accessibles en général. Nous remplaçons donc cet argument par l'usage d'une paramétrix pour un opérateur elliptique d'ordre 2 sur $G(\mathbb{R})$. Un tel argument semble avoir été utilisé pour la première fois par Duflo et Labesse [15] p. 199.

Gardons les notations du début de la section 8. On notera simplement $T$ un opérateur de Hecke $T_E$, normalisé. On suppose que $\Gamma \subset G(\mathbb{Q})$ est arithmétique.

Notons $d$ la dimension de $Z(\mathbb{R})\backslash G(\mathbb{R}) = PG(\mathbb{R})$, que l'on suppose connexe. Pour une fonction $L^2$ $f$ sur $Z(\mathbb{R})\Gamma\backslash G(\mathbb{R})$, soit $f^0$ sa projection sur l'espace des constantes.

Soient $\mathfrak{g}$ l'algèbre de Lie de $PG(\mathbb{R})$ et $\mathcal{Z}$ le centre de son algèbre enveloppante $U(\mathfrak{g})$. Soit $\mathfrak{g} = \mathfrak{k} \oplus \mathfrak{p}$ une décomposition de Cartan de $\mathfrak{g}$. Soient $Z_\alpha$, $Y_\beta$ des vecteurs de $\mathfrak{k}$ et $\mathfrak{p}$ tel que $B(Z_\alpha, Z_\alpha) = -1$ et $B(Y_\beta, Y_\beta) = +1$ et autrement orthogonaux. Alors $C_G = -\Sigma Z_\alpha^2 + \Sigma Y_\beta^2$ est un élément de $\mathcal{Z}$ et $C_K = -\Sigma Z_\alpha^2$ appartient à $\mathcal{Z}_K$. Notons $D = C_G - 2C_K$: $D$ est un opérateur différentiel elliptique dans $U(\mathfrak{g})$.

Soit $m$ un entier $\geq \frac{d+1}{2}$. D'après la théorie de la paramétrix [15] il existe des fonctions $\psi \in \mathcal{D}(PG(\mathbb{R}))$, $\varphi \in \mathcal{D}^{2m-d-1}(PG(\mathbb{R}))$-$\mathcal{D}^i$ désignant l'espace des fonctions $i$ fois différentiables à support compact et $\mathcal{D}$ l'espace des fonctions $C^\infty$ à support compact – telles que l'on ait sur $PG(\mathbb{R})$:

$$D^m * \varphi = \delta + \psi \tag{201}$$

$\delta$ étant la mesure de Dirac à l'origine.

Notons $\mathcal{D}(m)$ l'espace des fonctions sur $Z(\mathbb{R})\Gamma\backslash G(\mathbb{R})$ telles que $f$ et $f * D^m$ soient de carré intégrable.

PROPOSITION 8.2. *Soit $\omega \subset Z(\mathbb{R})\Gamma\backslash G(\mathbb{R})$ un sous-ensemble compact. Il existe alors des constantes $C_1(\omega)$ et $C_2(\omega)$ telles que, pour tout $f \in \mathcal{D}(m)$ et $x \in \omega$, et tout opérateur de Hecke $T$:*

$$|Tf(x) - f^0(x)| \leq C_1(\omega)\|Tf - f^0\|_2 + C_2(\omega) \cdot \|T(f * D^m) - f^0 * D^m\|_2. \tag{202}$$

Corollaire 8.3. *Si* $(T_n)$ *est une suite d'opérateurs de Hecke normalisés tels que* $\|T_n f - f^0\|_2 \le \varepsilon(n)\|f\|_2$ $(f \in L^2)$ *alors pour tout* $f \in \mathcal{D}(m)$, $x \in Z(\mathbb{R})\Gamma\backslash G(\mathbb{R})$

$$|T_n f(x) - f^0(x)| \le C\,\varepsilon(n) \tag{203}$$

*la constante C (dépendant de f) étant uniforme pour x dans un compact.*

Ceci permet d'évaluer les restes dans les résultats d'équidistribution des Théorèmes 1.2 et 1.3.

Corollaire 8.4. *(Notation du Théorème 1.2). Si* $f \in \mathcal{D}(\frac{n^2}{2})$ *alors*

$$|\overline{T}_{r,p} f(x) - f^0(x)| \le C\,p^{-\frac{\min(r,n-r)}{2}} \tag{204}$$

*pour tout r, p, la constante C étant uniforme pour x dans un compact.*

Corollaire 8.5. *(Notations du Thm 1.3). Si* $f \in \mathcal{D}(g^2 + \frac{g+1}{2})$,

$$|\overline{T}_p f(x) - f^0(x)| \le C\,p^{-\frac{g(1-\theta)}{2}} \tag{205}$$

*uniformément sur tout compact.*

*Démonstration de la Proposition 8.2.* D'après (201) on peut écrire pour toute fonction $f$ sur $X = Z(\mathbb{R})\Gamma\backslash G(\mathbb{R})$ et même pour toute distribution (les distributions apparaissant dans (201) étant à support compact):

$$f = (f * D^m) * \varphi - f * \psi. \tag{206}$$

L'opérateur $T$, opérant à gauche, commute avec les convolutions, et on a donc

$$Tf = T(f * D^m) * \varphi - (Tf) * \psi. \tag{207}$$

Considérons par exemple le second terme de (207). La fonction $Tf * \psi$ est convolée d'une fonction $L^2$ et d'une fonction à support compact, et il en est de même pour $f^0 * \psi$. On en déduit qu'elle est continue et que

$$|Tf * \psi(x) - f^0 * \psi(x)| \le C_1(\omega) \cdot \|Tf - f^0\|_2 \tag{208}$$

pour $x$ dans un compact $\omega$. De façon analogue, $\varphi \in \mathcal{D}^{2m-d-1}$ étant continue, on a

(209)
$$|T(f * D^m) * \varphi(x) - (f * D^m)^0 * \varphi(x)| \le C_2(\omega)\|T(f * D^m) - (f * D^m)^0\|_2.$$

Noter que le même argument appliqué à la décomposition (206) montre que $f$ est continue (Bien sûr, ceci n'est que le lemme de Sobolev dans ce contexte.) D'après (208) et (209),

$$|Tf(x) + f^0 * \psi(x) - (f * D^m)^0 * \varphi(x)| \tag{210}$$

admet la majoration indiquée dans la Proposition 2. Il reste à remarquer que la projection sur les constantes commute avec la convolution, de sorte que

$$f^0 * \psi - (f * D^m)^0 * \varphi = (f * \psi - f * D^m * \varphi)^0 = -f^0. \tag{211}$$

---

## RÉFÉRENCES

---


[1] T. M. Apostol, Analytic Number Theory, Springer Verlag, New York, 1976.

[2] J. Arthur, Unipotent Automorphic Representations: Conjectures. Orbites unipotentes et représentations II: Groupes p-adiques et réels, Astérisque **171–172** (1989), 13–71.

[3] J. Arthur, Eisenstein Series and the Trace Formula. I, *Proceedings of Symposia in Pure Mathematics* **33** (1979), 253–274.

[4] D. Bump, W. Duke, J. Hoffstein, and H. Iwaniec, An Estimate for the Hecke Eigenvalues of Maass Forms, *International Mathematics Research Notices* **4** (1992), 75–81.

[5] A. Borel, Automorphic L–Functions. *Proceedings of Symposia in Pure Mathematics II*, **33** (1979), 27–61.

[6] A. Borel and N. Wallach. Continuous Cohomology, Discrete Subgroups and Representation of Reductive Groups. *Ann. Math. Studies* **94**, (1980), 5.

[7] A. Borel and J.-P Serre, Cohomologie d'immeubles et de Groupes S-arithmétiques, *Topology* **15,** 211–232 (1976).

[8] M. Burger, J. S. Li, and P. Sarnak, Ramanujan Duals and Automorphic Spectrum, *Bull. of the A.M.S.* **26** (1992), 253–257.

[9] M. Burger and P. Sarnak, Ramanujan Duals. II, *Invent. Math.* **106** (1991), 1–11.

[10] P. Cartier, Representations of $p$-Adic Groups: A Survey *Proc. Sympos. Pure Math.* **33** (1979), 111–1555.

[11] W. Casselman, Introduction to the theory of admissible representations of $p$-adic reductive groups, manuscrit non publié.

[12] L. Clozel, On Limit Multiplicities of Discrete Series Representations in Spaces of Automorphic Forms, *Inv. Math.* **83** (1986), 265–284.

[13] P. Chiu, Covering with Hecke Points, *Journ. Numb. Theory* **53** (1995), 25–44.

[14] J. Dixmier, Les $C^*$-algèbres et Leurs Représentations, Gauthier-Villars, Paris, 1969.

[15] M. Duflo and J.-P. Labesse, Sur la Formule Des Traces de Selberg, *Ann. Sc. de l'Ecole Normale Suprieure* **4** (1971), 193–284.

[16] W. Duke, Hyperbolic Distribution Problems and Half-Integral Weight Maass Forms, *Invent. math.* **92** (1988), 73–90.

[17] W. Duke, R. Howe, and J.-S. Li, Estimating Hecke Eigenvalues of Siegel Modular Forms. *Duke Math. Journ.* **67** (1992), 219–240.

[18] M. Flath, Decomposition of Representations into Tensor Products. I, *Proceedings of Symposia in pure Mathematics* **33** (1979), 179–183.

[19] S. Gelbart and H. Jacquet, A Relation Between Automorphic Representations of $GL_2$ and $GL_3$, *Ann. Sci. Ecole Norm. Sup.* **11** (1978), 471–552.

[20] R. Howe, and C. Moore, Asymptotic Properties of Unitary Representations, *J. Func. Anal.* **32** (1979), 72–96.

[21] H. Iwaniec, Introduction to the Spectral Theory of Automorphic Forms, *Publ. Revista Matemática Iberoamericana.*

[22] H. Jacquet and J. A. Shalika, On Euler Products and the Classification of Automorphic Representations. *I, II, Amer, J. Math.* **103** (1981), 499–557, 777–815.

[23] H. Jacquet and J. A. Shalika, Sur le Spectre Résiduel du Groupe Linéaire. C.R *Acad. Sci. Paris,* **293** (1981), 541–543.

[24] R. Kottwitz, Shimura Varieties and Twisted Orbital Integrals, *Math. Ann.* **269** (1984), 287–300.

[25] J.-S. Li, Non-Vanishing Theorems for the Cohomology of Certain Arithmetic Quotients, *J. Reine und Angewandte Math.* **428** (1992), 111–217.

[26] W. Luo, Z. Rudnick, and P. Sarnak, On Selberg's Eigenvalue Conjecture, *Geom. Funct. Anal.* **5** (1995), 387–401.

[27] W. Luo, Z. Rudnick, and P. Sarnak, On the Generalized Ramanujan Conjecture for $GL_n$, à paraître dans *Proceedings of Symposia in Pure Mathematics,* vol. 66, II, Amer. Math. Soc., Providence, RI, 1999, pp. 301–310.

[28] G. A. Margulis, Discrete Subgroups of Semi-Simple Lie Groups, Springer-Verlag, Berlin, (1989).

[29] C. Moeglin and J.-L. Waldspurger, *Décomposition Spectrale et Séries d'Eisenstein.* Basel, Boston, Birkauser (1994).

[30] C. Moeglin and J.-L. Waldspurger, Le Spectre Résiduel de $GL_n$, *Ann. Sci. Ecole Norm. Sup.* **22** (1989), 605–694.

[31] H. Oh, Uniform Pointwise Bounds for Matrix Coefficients of Unitary Representations and Applications to Kazhdan Constants, preprint. 1999.

[32] M. Raynaud, Courbes Sur Une Variété Abélienne et Points de Torsion, *Invent. Math.* **71** (1983), 207–223.

[33] M. Raynaud, Sous-Variétés D'une Variété Abélienne et Points de Torsion, Arithmetic and Geometry: Papers Dedicated to I. R. Shafarevich on the Occasion of His Sixtieth Birthday, vol. **1**, (J. Coates and S. Helgason), (eds.), Birkhäuser (1983).

[34] P. Sarnak, Diophantine Problems and Linear Groups, *Proceedings of the International Congress of Mathematicians*, Vol. **1,** (Kyoto, 1990), Springer-Verlag, Tokyo, 1991, pp. 459–471.

[35] F. Shahidi, On the Ramanujan Conjecture and Finiteness of Poles for Certain $L$-Functions, *Ann. of Maths.* **127** (1988), 547–584.

[36] G. Shimura, Introduction to the Arithmetic Theory of Automorphic Functions, Iwanami Shoten, Publishers and Princeton University Press, 1971.

[37] L. Szpiro, E. Ullmo, S. Zhang, Equirépartition des petits points, *Invent. Math.* **127** (1997), 337–347.

[38] M. Tadic, Geometry of Dual Spaces of Reductive Groups (Non-Archimedean Case), *J. d'Analyse Math.* **51** (1988), 139–181.

[39] E. Ullmo, Positivité et Dicrétion Des Points Algébriques Des Courbes, *Ann. of Maths* **147** (1998), 167–179.

[40] G. Warner, Harmonic Analysis on Semi-Simple Lie Groups I, II. Springer-Verlag, Berlin, 1972.

[41] S. Zhang, Equidistribution of Small Points on Abelian Varieties, *Ann. of Maths* **147** (1998), 159–165.

[42] ———, Heights of Heegner Points on Shimura Curve, preprint, 1999.

CHAPTER 10

# REMARKS ON RANKIN-SELBERG CONVOLUTIONS

By James W. Cogdell and Ilya I. Piatetski-Shapiro

*Dedicated to Joe Shalika*

In this paper we would like to present two types of results on the theory of Rankin-Selberg convolution $L$-functions for $GL_n \times GL_m$. Both families of results are based on the foundational work of Shalika with Jacquet and the second author of this paper [10, 11, 12, 13] on the analysis of these $L$-functions via the theory of integral representations.

In the first section we present results on the local archimedean Rankin-Selberg convolutions. This section was written in response to a question of D. Ramakrishnan as to whether the local $L$-function as defined by Jacquet and Shalika in [13] was indeed the "correct" factor in the sense that it is precisely the standard archimedean Euler factor which is determined by the poles of the family of local integrals using either $K$-finite data or smooth data (i.e., without passing to the Casselman-Wallach completion). In Section 1 we answer this affirmatively as a consequence of showing that the ratio of the local integral divided by the $L$-function is continuous in the appropriate topology, uniformly on compact subsets of $\mathbb{C}$. As a consequence we establish a non-vanishing result for this ratio which is necessary for the completion of the global theory of Rankin-Selberg convolutions.

In the second section we complete global theory of Rankin-Selberg convolutions from the point of view of integral representations. This section was motivated by the comment of Jacquet that, although known to the experts, this completion had never appeared in print. Most of the necessary results can be found in the paper [12] by Jacquet and Shalika, though not always explicitly stated. One missing ingredient was the non-vanishing result for the archimedean Rankin-Selberg integrals alluded to above. With this in hand, in Section 2 we combine the global results of [10, 12] with the local results of [11, 13] and Section 1 of this paper to give a proof of the fact that the global $L$-functions $L(s, \pi \times \pi')$ are *nice*, in the sense that they have meromorphic continuation, are bounded in vertical strips, and satisfy a global functional equation, within the context of integral representations. Actually, we are only able to establish the boundedness in vertical strips within the method for $m = n$ and $m = n - 1$. Outside of these cases we must rely on the results of Gelbart and Shahidi [6]. In addition we establish the location of poles for these $L$-functions, giving the proof of Jacquet, Piatetski-Shapiro, and Shalika

JWC was partially supported by the NSA. IIPS was partially supported by the NSF.

of these results alluded to in the appendix of [14]. If one combines these results with the strong multiplicity one results of [12] and the converse theorems [2, 3] we can consider the basic global theory of Rankin-Selberg convolutions via integral representations to now be essentially complete, with the exception of the cases of boundedness in vertical strips alluded to above.

*Acknowledgments.* We would like to thank D. Ramakrishnan for asking us the question which led to the first part of this paper. We would like to thank H. Jacquet for pointing out the need for the second part of this paper and for providing us with the sketch of his proof with Shalika of Theorem 1.3.

**1. Archimedean Rankin–Selberg convolutions.** This section complements the material in the paper of Jacquet and Shalika [13] and is meant to show that indeed the results there are enough for most applications. Unless otherwise noted, the notation is as in [13].

**1.1. An extension of Dixmier–Malliavin.** Let $E$ be a Fréchet space, $G$ a real Lie group, $\mathfrak{g}$ its complexified Lie algebra, and $\pi$ a continuous representation of $G$ on $E$. Let $\{p_j\}$ be a set of seminorms on $E$ defining the topology on $E$.

Let $E^\infty$ be the smooth vectors of $E$. Let $U(\mathfrak{g})$ be the universal enveloping algebra of $\mathfrak{g}$ and let $\{u_i\}$ be a basis of $U(\mathfrak{g})$. The topology on $E^\infty$ is defined by the seminorms $q_{i,j}(\xi) = p_j(\pi(u_i)\xi)$ for $\xi \in E^\infty$. With this topology, $E^\infty$ is again a Fréchet space [1]. For convenience, reindex the family $\{q_{i,j}\}$ by a single index $\{q_i\}$.

Let $\xi_k \to \xi_0$ be a convergent sequence in $E^\infty$. The purpose of this section is to prove the following extension of Theorem 3.3 of [4]. Our proof is a variation of that in [4] which we follow.

PROPOSITION 1.1. *There exists a finite set of functions $f_j \in C_c^\infty(G)$ and a collection of vectors $\xi_{k,j} \in E^\infty$ such that $\xi_k = \sum \pi(f_j)\xi_{k,j}$ for all $k \geq 0$ and such that for each $j$, $\xi_{k,j}$ converge to $\xi_{0,j}$ in $E^\infty$.*

*Proof.* Since $E^\infty$ is linear, it suffices to consider the case $\xi_0 = 0$.

Let $\{X_1, \ldots, X_m\}$ be a basis of $\mathfrak{g}$ with the property that under the map

$$(t_1, \ldots, t_m) \mapsto e^{t_1 X_1} \ldots e^{t_m X_m}$$

from $\mathbb{R}^m$ to $G$ the open set $(-1, 1)^m$ is mapped diffeomorphically onto an open set $\Omega$ of $G$.

LEMMA 1.1. *For each choice of seminorm $q_i$ and non-negative integer $n$ the set of real numbers $\{q_i(\pi(X_1)^{2n}\xi_k)\}$ is bounded.*

*Proof.* Since $\pi(X_1)^{2n}$ acts continuously and the seminorm $q_i$ is continuous, the sequence $q_i(\pi(X_1)^{2n}\xi_k)$ converges to $q_i(\pi(X_1)^{2n}0) = 0$. Hence the sequence of real numbers $q_i(\pi(X_1)^{2n}\xi_k)$ is bounded. □

Let $M_{n,i}$ be an upper bound for $\{q_i(\pi(X_1)^{2n}\xi_k)\}$.

LEMMA 1.2. *There exist positive real numbers $\beta_n$ such that the sum $\sum_n \beta_n M_{n,i}$ is convergent for all $i$.*

*Proof.* For each $i$ there are positive numbers $\beta_{n,i}$ such that $\sum_n \beta_{n,i} M_{n,i}$ converges. Let $\beta_n^{(k)} = \min_{1\le i\le k} \beta_{n,i}$ and set $\beta_n = \beta_n^{(n)}$. Then

$$\sum_n \beta_n M_{n,i} = \sum_{n\le i} \beta_n M_{n,i} + \sum_{n>i} \beta_n M_{n,i}.$$

For $n > i$, $\beta_n = \min_{1\le j\le n} \beta_{n,j} \le \beta_{n,i}$. So

$$\sum_{n>i} \beta_n M_{n,i} \le \sum_{n>i} \beta_{n,i} M_{n,i} < \infty.$$ □

Now let $\epsilon \in (0, \frac{1}{2}]$. Then by Lemma 2.5 and Remark 2.6 of [4] there is a sequence of positive numbers $\alpha_n$ and functions $g(t)$ and $h(t)$ in $C_c^\infty(\mathbb{R})$, supported in $(-\epsilon, \epsilon)$ such that

$$\sum_n \alpha_n M_{n,i} < \infty \text{ for all } i$$

and

$$\sum_{n=0}^{p} (-1)^n \alpha_n \delta_0^{(2n)} * g \to \delta_0 + h$$

in the space $\mathcal{E}'(\mathbb{R})$ of compactly supported distributions on $\mathbb{R}$. $\delta_0$ is the Dirac measure supported at the origin of $\mathbb{R}$.

The measures $g(t)\,dt$ and $h(t)\,dt$ induce measures $\mu_1$ and $\nu_1$ on $G$ under the map $\mathbb{R} \to G$ given by $t \mapsto e^{tX_1}$. Then

$$\mu_1 * \sum_{n=0}^{p} (-1)^n \alpha_n X_1^{2n} = \sum_{n=0}^{p} (-1)^n \alpha_n X_1^{2n} * \mu_1 \to \delta_e + \nu_1$$

in the space $\mathcal{E}'(G)$ of compactly supported distributions on $G$ and

$$\pi(\mu_1) \sum_{n=0}^{p} (-1)^n \alpha_n \pi(X_1)^{2n} \xi_k \to \xi_k + \pi(\nu_1)\xi_k$$

in the weak topology on $E$.

However, by our choice of $\alpha_n$, $\sum_{n=0}^{\infty} q_i(\alpha_n \pi(X_1)^{2n}\xi_k) < \infty$ for each seminorm $q_i$. Therefore $\sum_{n=0}^{p}(-1)^n \alpha_n \pi(X_1)^{2n}\xi_k$ converges to a vector $\eta_k$ in $E^\infty$. Therefore we have

$$\xi_k = \pi(\mu_1)\eta_k - \pi(\nu_1)\xi_k$$

for each $\xi_k$.

LEMMA 1.3. *The sequence $\eta_k$ converges to* 0 *in $E^\infty$.*

*Proof.* By continuity of the seminorms,

$$q_i(\eta_k) \leq \lim_{p\to\infty} \sum_{n=0}^{p} \alpha_n q_i\big(\pi(X_1)^{2n}\xi_k\big) \leq \sum_{n=0}^{\infty} \alpha_n q_i\big(\pi(X_1)^{2n}\xi_k\big).$$

Since the sum $\sum \alpha_n q_i(\pi(X_1)^{2n}\xi_k)$ is absolutely convergent, we can interchange limit and summation to obtain

$$\lim_{k\to\infty} \sum_{n=0}^{\infty} \alpha_n q_i(\pi(X_1)^{2n}\xi_k) = 0.$$

Therefore $\lim_{k\to\infty} q_i(\eta_k) = 0$ for all $q_i$. Hence $\eta_k \to 0$ in $E^\infty$. □

Now apply the same process for $X_2$ through $X_m$. In this way we obtain a finite collection of measures $\{\mu_{i,j}\}$, where each $\mu_{i,j}$ is the image of a measure $g_{i,j}(t_i)\,dt_i$ under the map $t_i \mapsto e^{t_i X_i}$ as above, and sequences $\xi_{k,j}$ such that

$$\xi_k = \sum_j \pi(\mu_{1,j} * \cdots * \mu_{m,j})\xi_{j,k}$$

for each $k$ with $\lim_{k\to\infty} \xi_{j,k} = 0$ for each $j$.

The measure $\mu_{1,j} * \cdots * \mu_{m,j}$ on $G$ is then the image of the measure on $\mathbb{R}^m$ given by $g_{1,j}(t_1)\cdots g_{m,j}(t_m)\,dt_1\cdots dt_m$. If $g_j(t_1,\cdots,t_m) = g_{1,j}(t_1)\cdots g_{m,j}(t_m)$ then $g_j$ is smooth with compact support in $(-\epsilon,\epsilon)^m$. Hence by our choice of basis on $\mathfrak{g}$ the image of the measure $g_j(t)\,dt$ on $\mathbb{R}^m$ will be of the form $f_j(g)\,dg$ on $G$ with $f_j \in C_c^\infty(G)$.

Hence we now have a finite collection of $f_j \in C_c^\infty(G)$ and $\xi_{k,j} \in E^\infty$ such that

$$\xi_k = \sum_j \pi(f_j)\xi_{k,j}$$

with the sequence $\xi_{k,j}$ now converging to 0 in $E^\infty$ for each $j$.

This completes the proof of the proposition. □

**1.2. Continuity of the archimedean local integral.** Let $F$ be either $\mathbb{R}$ or $\mathbb{C}$. Let $\psi$ be a non-trivial additive character of $F$. Let $GL_r = GL_r(F)$. Let $(\pi, V)$ be a finitely generated admissible smooth representation of moderate growth of $GL_n$, as in [1, 13]. Let $V_o$ denote the space of $K_n$–finite vectors, i.e., the underlying Harish-Chandra module. Similarly, let $(\sigma, E)$ be a finitely generated admissible smooth representation of moderate growth of $GL_m$, and $E_o$ its underlying Harish-Chandra module. Note that both $V$ and $E$ are Fréchet spaces and equal to their spaces of smooth vectors.

We further assume that $\pi$ and $\sigma$ are of Whittaker type as in [13], with continuous Whittaker functionals $\lambda_\pi$ with respect to $\psi$ and $\lambda_\sigma$ with respect to $\psi^{-1}$.

We will let $(\pi \otimes \sigma, V \otimes E)$ denote the algebraic tensor product of $(\pi, V)$ and $(\sigma, E)$. We let $(\pi \hat{\otimes} \sigma, V \hat{\otimes} E)$ denote the (projective) topological tensor product. Then $(\pi \hat{\otimes} \sigma, V \hat{\otimes} E)$ is the again an admissible smooth representation of moderate growth of $GL_n \times GL_m$ and is in fact the Casselman-Wallach completion of the algebraic tensor product [1, 13]. (Note: This notation is slightly different from that of [13] where they use $\otimes$ for the topological tensor product.)

The linear functional $\mu = \lambda_\pi \otimes \lambda_\sigma$ is a continuous Whittaker functional on $V \otimes E$ and extends to a Whittaker functional on $V \hat{\otimes} E$ [13]. For each $v \in V \hat{\otimes} E$ let

$$W_v(g, g') = \mu(\pi(g) \hat{\otimes} \sigma(g') v)$$

and let $\mathcal{W}(\pi \hat{\otimes} \sigma, \psi)$ be the space spanned by all such functions. Then $\mathcal{W}(\pi \hat{\otimes} \sigma, \psi) \supset \mathcal{W}(\pi, \psi) \otimes \mathcal{W}(\sigma, \psi^{-1})$.

As in [13], define for $W \in \mathcal{W}(\pi \hat{\otimes} \sigma, \psi)$ and $\Phi \in \mathcal{S}(F^n)$

$$\Psi(s; W, \Phi) = \int_{N_n \backslash GL_n} W(g, g) \Phi(e_n g) \, |\det(g)|^s \, dg \qquad \text{if } n = m$$

$$\Psi(s; W) = \int_{N_m \backslash GL_m} W\left(\begin{pmatrix} g & \\ & I_{n-m} \end{pmatrix}, g\right) |\det(g)|^{s-(n-m)/2} \, dg \qquad \text{if } n > m$$

$$\Psi(s; W, j) = \int_{N_m \backslash GL_m} \int_X W\left(\begin{pmatrix} g & & \\ x & I_j & \\ & & I_{k+1} \end{pmatrix}, g\right) |\det(g)|^{s-(n-m)/2} dx \, dg \quad \text{if } n > m$$

where $j + k = n - m - 1$. These are all absolutely convergent for $Re(s) \gg 0$.

Define $\tilde{W}(g, g') = W(w_n g^\iota, w_m g'^\iota)$, where $w_r$ is the long Weyl element $\begin{pmatrix} & & 1 \\ & \cdot^{\cdot^{\cdot}} & \\ 1 & & \end{pmatrix}$ and $\iota$ is the outer automorphism of $GL_r$, namely $g \mapsto g^\iota = {}^t g^{-1}$. Then $\tilde{W}$ is in the Whittaker model of $V^\iota \hat{\otimes} E^\iota = (V \hat{\otimes} E)^\iota$. Then we have the functional equation:

$$\Psi(1 - s; \tilde{W}, \hat{\Phi}) = \omega_\sigma(-1)^{n-1} \gamma(s, \pi \times \sigma, \psi) \Psi(s; W, \Phi) \qquad \text{if } n = m$$

$$\Psi(1 - s; \rho(w_{n,m}) \tilde{W}, j) = \omega_\sigma(-1)^{n-1} \gamma(s, \pi \times \sigma, \psi) \Psi(s; W, k) \qquad \text{if } n > m$$

where $j + k = n - m - 1$ and

$$\gamma(s, \pi \times \sigma, \psi) = \frac{\varepsilon(s, \pi \times \sigma, \psi) L(1 - s, \pi^{\iota} \times \sigma^{\iota})}{L(s, \pi \times \sigma)}.$$

Note that here $L(s, \pi \times \sigma)$ is as in [13], i.e., it is the factor attached to the pair $(\pi, \sigma)$ by the (arithmetic) Langlands classification.

The purpose of this section is to prove the following result.

THEOREM 1.1. *Let $\Lambda_s$, respectively $\Lambda_{s,\Phi}$, be the linear functional on $V \hat{\otimes} E$ defined by*

$$\Lambda_s(v) = \frac{\Psi(s; W_v)}{L(s, \pi \times \sigma)} \qquad \text{if } n > m$$

$$\Lambda_{s,\Phi}(v) = \frac{\Psi(s; W_v, \Phi)}{L(s, \pi \times \sigma)} \qquad \text{if } n = m$$

*for $v \in V \hat{\otimes} E$. Then $\Lambda_s$, respectively $\Lambda_{s,\Phi}$, is continuous on $V \hat{\otimes} E$, uniformly for $s$ in a compact set.*

Note that we claim the continuity for all $s$, not just for those $s$ for which the local integral is absolutely convergent.

We begin by recalling the following result of [13].

LEMMA 1.4. *Let $f \in C_c^{\infty}(GL_n \times GL_m)$. Then there exists a seminorm $\beta$ on $V \hat{\otimes} E$ and a gauge $\xi$ on $GL_n \times GL_m$ depending only on $f$ such that*

$$|\rho(f) W_v(g, g')| \leq \beta(v) \xi(g, g')$$

*for all $v \in V \hat{\otimes} E$.*

*Proof.* The proof is word for word the same as the proof of Proposition 2.1 in [13]. □

We will prove the theorem in the case $n > m$. The proof in the case $n = m$ is the same, with the obvious modifications.

PROPOSITION 1.2. *For $s$ in the half plane of absolute convergence, the functional $v \mapsto \Psi(s; W_v)$ is continuous on $V \hat{\otimes} E$, uniformly for $s$ in a compact set.*

*Proof.* Since the functional is evidently linear, it is enough to show that the sequence $\Psi(s; W_{v_k})$ converges to 0 whenever $v_k \to 0$ in $V \hat{\otimes} E$, uniformly for $s$ in a compact set.

By Proposition 1.1, there exists a finite collection of functions $f_j \in C_c^\infty(GL_n \times GL_m)$ and sequences $v_{k,j}$ in $V\hat{\otimes}E$ such that $v_k = \sum_j \pi\hat{\otimes}\sigma(f_j)v_{k,j}$ for each $k$ and $v_{k,j} \to 0$ for each $j$.

Then we have

$$W_{v_k}(g, g') = \sum_j \rho(f_j) W_{v_{k,j}}(g, g')$$

so that by Lemma 1.4

$$|W_{v_k}(g, g')| \leq \sum_j \beta_j(v_{k,j})\xi_j(g, g')$$

for seminorms $\beta_j$ and gauges $\xi_j$ depending only on $f_j$. Then

$$\begin{aligned} |\Psi(s; W_{v_k})| &= \left| \int W_{v_k}\left(\begin{pmatrix} g & \\ & I_{n-m} \end{pmatrix}, g\right) \right| \det(g)|^{s-(n-m)/2}\, dg| \\ &\leq \int \left| W_{v_k}\left(\begin{pmatrix} g & \\ & I_{n-m} \end{pmatrix}, g\right) \right| |\det(g)|^{Re(s)-(n-m)/2}\, dg \\ &\leq \sum_j \beta_j(v_{k,j}) \int \xi_j\left(\begin{pmatrix} g & \\ & I_{n-m} \end{pmatrix}, g\right) |\det(g)|^{Re(s)-(n-m)/2}\, dg. \end{aligned}$$

In this last expression, each integral involving a gauge $\xi_j$ is absolutely convergent for $Re(s) \gg 0$, uniformly for $s$ in compact sets. Since the seminorms $\beta_j$ are continuous on $V\hat{\otimes}E$ and since each sequence $v_{k,j} \to 0$ as $k \to \infty$ we have $|\Psi(s; W_{v_k})|$ converges to 0 as $k \to \infty$ uniformly for $s$ in a compact set. □

COROLLARY. *For $s$ in the realm of absolute convergence of the local integrals, the functional $\Lambda_s(v) = \Psi(s; W_v)/L(s, \pi \times \sigma)$ is continuous on $V\hat{\otimes}E$, uniformly for $s$ in a compact set.*

Repeating the proof we also obtain the following.

COROLLARY. *For $s$ in the realm of absolute convergence of the local integrals, the functional $\Lambda_{s,j}(v) = \Psi(s; W_v, j)/L(s, \pi \times \sigma)$ is continuous on $V\hat{\otimes}E$, uniformly for $s$ in a compact set.*

From this we obtain:

COROLLARY. *The functional $\tilde{\Lambda}_{s,j}(v) = \Psi(1-s; \rho(w_{n,m})\tilde{W}_v, j)/L(1-s, \pi^\iota \times \sigma^\iota)$ is continuous on $V\hat{\otimes}E$, uniformly for $s$ in a compact set, in the domain $Re(s) \ll 0$.*

We are now ready to prove the theorem.

*Proof.* [Proof of Theorem 1.1] By the first Corollary, we have that the functional $\Lambda_s(v)$ is continuous in a domain $Re(s) > B'$, uniformly for $s$ in a compact set. If we let $\Lambda'_s(v) = \Lambda_s(v)e^{s^2}$ then $\Lambda'_s$ will also be continuous in this domain with the same uniformity.

Let $B > B'$. Then on the line $Re(s) = B$ we have a uniform estimate $|\Lambda'_s(v)| \leq c_B\Psi(B; |W_v|)$. To see this, write

$$|\Lambda'_s(v)| = |\Psi(s; W_v)| \left| \frac{e^{s^2}}{L(s, \pi \times \sigma)} \right|.$$

On the line $Re(s) = B$, the function $e^{(B+it)^2} L(B + it, \pi \times \sigma)^{-1}$ is rapidly decreasing as $|t| \to \infty$. Hence there is a constant $c_B$ so that $|e^{(B+it)^2} L(B + it, \pi \times \sigma)^{-1}| \leq c_B$. On the other hand, it is elementary that $|\Psi(s; W)| \leq \Psi(B; |W|)$. This gives the estimate.

By the functional equation, we have

$$\begin{aligned}\Lambda'_s(v) &= \frac{\Psi(s; W)e^{s^2}}{L(s, \pi \times \sigma)} \\ &= \omega_\sigma(-1)^{n-1}\varepsilon(s, \pi \times \sigma, \psi)^{-1} \frac{\Psi(1 - s; \rho(w_{n,m})\tilde{W}, n - m - 1)e^{s^2}}{L(1 - s, \pi^\iota \times \sigma^\iota)} \\ &= \tilde{\Lambda}'_{s,n-m-1}(v).\end{aligned}$$

It follows from the third corollary that $\Lambda_{s,n-m-1}$ is continuous in a halfplane $Re(s) < A'$, hence so are $\Lambda'_s(v)$ and $\Lambda_s(v) = \Lambda'_s(v)e^{-s^2}$, with uniformity on compact subsets of $Re(s) < A'$.

Arguing as above, if $A < A'$ we have a uniform bound on the line $Re(s) = A$ of the form $|\Lambda'_s(v)| = |\tilde{\Lambda}'_{s,n-m-1}(v)| \leq c_A\Psi(1 - A; |\rho(w_{n,m})\tilde{W}|, n - m - 1)$.

Consider now the behavior of $\Lambda'_s(v)$ in the strip $A \leq Re(s) \leq B$. The function $\Lambda'_s(v)$, as a function of $s$, grows sufficiently slowly that we may apply Phragmen–Lindelöf to the strip $A \leq Re(s) \leq B$ and we obtain the estimate

$$|\Lambda'_s(v)| \leq \max(c_B\Psi(B; |W_v|), c_A\Psi(1 - A; |\rho(w_{n,m})\tilde{W}|, n - m - 1))$$

in this strip. Now suppose that $v_k$ is a sequence converging to 0 in $V\hat{\otimes}E$. Then the proof of Proposition 1.2 shows that both the contributions $\Psi(B; |W_{v_k}|)$ and $\Psi(1 - A; |\rho(w_{n,m})\tilde{W}_{v_k}|, n - m - 1)$ go to 0 as $k \to \infty$. Hence $\Lambda'_s(v_k)$ converges to 0 in the strip, uniformly for all $s$. Hence $\Lambda'_s$ is continuous for $s$ in the strip, and uniformly so. Then $\Lambda_s(v) = \Lambda'_s(v)e^{-s^2}$ will be continuous on this strip, uniformly for $s$ in a compact set.

This completes the proof of the theorem. □

**1.3. Applications.** In this section we would like to present our applications to the analytic properties of the local Rankin-Selberg convolutions, which in turn are needed for the completion of the global theory of Rankin-Selberg convolutions in the following section.

We keep the notation of Section 1.2. Recall that $V_o$ and $E_o$ are the underlying Harish-Chandra modules of $V$ and $E$. Let $\mathcal{W}_o(\pi, \psi)$ be the subspace of $\mathcal{W}(\pi, \psi)$ spanned by the Whittaker functions associated to vectors in $V_o$, and similarly for $\mathcal{W}_o(\sigma, \psi^{-1})$.

THEOREM 1.2. (i) *For each* $W \in \mathcal{W}_o(\pi, \psi)$ *and* $W' \in \mathcal{W}_o(\sigma, \psi^{-1})$ *the ratio*

$$e(s; W, W') = \frac{\Psi(s; W, W')}{L(s, \pi \times \sigma)}$$

*is an entire function of* $s$.

(ii) *For every* $s_0 \in \mathbb{C}$ *there is a choice of* $W_0 \in \mathcal{W}_o(\pi, \psi)$ *and* $W_0' \in \mathcal{W}_o(\sigma, \psi^{-1})$ *such that* $e(s_0; W_0, W_0') \neq 0$.

*Proof.* We have that $V_o$ is dense in $V$ and $E_o$ is dense in $E$. The Casselman-Wallach completion of the Harish-Chandra module $V_o \otimes E_o$ is $V \hat{\otimes} E$. Hence $V_o \otimes E_o$ is dense in $V \hat{\otimes} E$.

Statement (i) now follows from Theorem 11.1 of [13].

Statement (ii) follows from Theorem 11.1 of [13] and Theorem 2.1 above. By Theorem 11.1 of [13] we know that $L(s, \pi \times \sigma)$ is obtained by $\Psi(s; W)$ for some $W = W_v$ with $v \in V \hat{\otimes} E$. For this $v$, $\Lambda_s(v) = \Psi(s; W_v)/L(s, \pi \times \sigma) = 1$. Since $V_o \otimes E_o$ is dense, there will be a vector $\tilde{v} \in V_o \otimes E_o$ for which $\Lambda_{s_0}(\tilde{v})$ is close to 1 and in particular is non-zero. Writing $\tilde{v}$ as a sum of decomposable tensors, we find a vector $v_0 \otimes v_0'$ such that $\Lambda_{s_0}(v_0 \otimes v_0') \neq 0$. But

$$\Lambda_{s_0}(v_0 \otimes v_0') = \frac{\Psi(s_0; W_{v_0}, W'_{v_0'})}{L(s_0, \pi \times \sigma)} = e(s_0; W_{v_0}, W'_{v_0'}).$$

Hence (ii). □

The same proof yields the following corollary.

COROLLARY. (i) *For each pair* $W \in \mathcal{W}(\pi, \psi)$ *and* $W' \in \mathcal{W}(\sigma, \psi^{-1})$ *the ratio*

$$e(s; W, W') = \frac{\Psi(s; W, W')}{L(s, \pi \times \sigma)}$$

*is an entire function of* $s$.

(ii) *For every* $s_0 \in \mathbb{C}$ *there is a choice of* $W_0 \in \mathcal{W}(\pi, \psi)$ *and* $W_0' \in \mathcal{W}(\sigma, \psi^{-1})$ *such that* $e(s_0; W_0, W_0') \neq 0$.

These results show that the $L$-function $L(s, \pi \times \sigma)$ as defined in [13] not only cancels all poles of the local integrals, but also dividing by it introduces no extraneous zeros. Hence this is the minimal standard Euler factor which cancels all poles in the local integrals, even for the $K$-finite vectors, as in the non-archimedean case [11].

The continuity of the local integrals also plays a role in proving the following result of Stade [16, 17] and Jacquet and Shalika (unpublished).

THEOREM 1.3. *In the cases $m = n$ and $m = n - 1$ there exist a finite collection of $K$–finite functions $W_i \in \mathcal{W}_o(\pi, \psi)$, $W_i' \in \mathcal{W}_o(\sigma, \psi^{-1})$, and $\Phi_i \in \mathcal{S}(F^n)$ if necessary such that*

$$L(s, \pi \times \sigma) = \sum \Psi(s; W_i, W_i') \quad or \quad L(s, \pi \times \sigma) = \sum \Psi(s; W_i, W_i', \Phi_i).$$

In the case where both $\pi$ and $\sigma$ are unramified, Stade shows that one obtains the $L$-function exactly with the $K$–invariant Whittaker functions (and Schwartz function if necessary). Our results are not needed in this case.

In the general case, Jacquet has provided us with a sketch of his argument with Shalika. First one proves that the integrals involving $K$–finite functions are equal to the product of a polynomial and the $L$-factor. It suffices to prove this for principal series, since the other representations embed into principal series. For principal series one proceeds by an induction argument on $n$, however one must prove the $m = n$ and $m = n - 1$ cases simultaneously. The (essentially formal) arguments needed are to be found in the published papers of Jacquet and Shalika. The polynomials in question then form an ideal and the point now is to show this ideal is the full polynomial ring. This is then implied by Theorem 1.2 (ii) above.

**2. Global Rankin–Selberg convolutions.** It was recently pointed out to us by Jacquet that the global theory of Rankin–Selberg convolutions via integral representations has never appeared in print. We would like to take this opportunity to at least partially correct this situation. All of the necessary global foundational material can be found in [10] and [12] and the necessarily local results are in [11] and [13] with the addition of the material in Section 1 above.

Let $k$ be a global field, $\mathbb{A}$ its ring of adeles, and fix a non-trivial continuous additive character $\psi = \otimes\psi_v$ of $\mathbb{A}$ trivial on $k$.

Let $(\pi, V_\pi)$ be a unitary cuspidal representation of $GL_n(\mathbb{A})$ and $(\pi', V_{\pi'})$ a unitary cuspidal representation of $GL_m(\mathbb{A})$. Since they are irreducible we have restricted tensor product decompositions $\pi \simeq \otimes'\pi_v$ and $\pi' \simeq \otimes'\pi_v'$ with $(\pi_v, V_{\pi_v})$ and $(\pi_v', V_{\pi_v'})$ irreducible admissible smooth generic unitary representations of $GL_n(k_v)$ and $GL_m(k_v)$ [5, 7, 8]. Let $\omega = \otimes'\omega_v$ and $\omega' = \otimes'\omega_v'$ be their central characters. These are both continuous characters of $k^\times\backslash\mathbb{A}^\times$.

**2.1. Global Eulerian integrals for $GL_n \times GL_m$.** Let us first assume that $m < n$. Then the results we need can be found in Part II of [12]. Let $\varphi \in V_\pi$ and $\varphi' \in V_{\pi'}$ be two cusp forms. The integral representations in this situation are of

Hecke type and essentially involve the integration of these cusp forms against a factor of $|\det|^s$, that is, a type of generalized Mellin transform.

In $GL_n$, let $P_n$ denote the mirabolic subgroup, that is, the stabilizer of the row vector $(0, \ldots, 0, 1)$. Let $N_n$ be the subgroup of upper triangular unipotent matrices, that is, the unipotent radical of the standard Borel subgroup. In the usual way, the additive character $\psi$ defines a non-degenerate character of $N_n$ through its abelianization. Let $Y_{n,m}$ be the unipotent radical of the standard parabolic subgroup attached to the partition $(m + 1, 1, \ldots, 1)$. Then $\psi$ defines a character of $Y_{n,m}(\mathbb{A})$ trivial on $Y_{n,m}(k)$ since $Y_{n,m} \subset N_n$. The group $Y_{n,m}$ is normalized by $GL_{m+1} \subset GL_n$ and the mirabolic subgroup $P_{m+1} \subset GL_{m+1}$ is the stabilizer in $GL_{m+1}$ of the character $\psi$.

Definition. *If $\varphi(g)$ is a cusp form on $GL_n(\mathbb{A})$ define the projection operator $\mathbb{P}^n_m$ from cusp forms on $GL_n(\mathbb{A})$ to cuspidal functions on $P_{m+1}(\mathbb{A})$ by*

$$\mathbb{P}^n_m\varphi(p) = |\det(p)|^{-\left(\frac{n-m-1}{2}\right)} \int_{Y_{n,m}(k)\backslash Y_{n,m}(\mathbb{A})} \varphi\left(y\begin{pmatrix} p & \\ & I_{n-m-1}\end{pmatrix}\right) \psi^{-1}(y)\, dy$$

*for $p \in P_{m+1}(\mathbb{A})$.*

This function $\mathbb{P}^n_m\varphi$ is essentially the same as the function denoted $V_{\varphi,m}$ in Part II of [12]. As the integration is over a compact domain, the integral is absolutely convergent. We first analyze the behavior on $P_{m+1}(\mathbb{A})$. From Section 3.1 of Part II of [12] we find the proofs of the following Lemmas.

Lemma 2.1. *The function $\mathbb{P}^n_m\varphi(p)$ is a cuspidal function on $P_{m+1}(\mathbb{A})$.*

Lemma 2.2. *Let $\varphi$ be a cusp form on $GL_n(\mathbb{A})$. Then for $h \in GL_m(\mathbb{A})$, $\mathbb{P}^n_m\varphi\begin{pmatrix} h & \\ & 1\end{pmatrix}$ has the Fourier expansion*

$$\mathbb{P}^n_m\varphi\begin{pmatrix} h & \\ & 1\end{pmatrix} = |\det(h)|^{-\left(\frac{n-m-1}{2}\right)} \sum_{\gamma\in N_m(k)\backslash GL_m(k)} W_\varphi\left(\begin{pmatrix} \gamma & 0\\ 0 & I_{n-m}\end{pmatrix}\begin{pmatrix} h & \\ & I_{n-m}\end{pmatrix}\right)$$

*with convergence absolute and uniform on compact subsets.*

We now have the prerequisites for writing down a family of Eulerian integrals for cusp forms $\varphi$ on $GL_n$ twisted by automorphic forms on $GL_m$ for $m < n$. Let $\varphi \in V_\pi$ be a cusp form on $GL_n(\mathbb{A})$ and $\varphi' \in V_{\pi'}$ a cusp form on $GL_m(\mathbb{A})$. (Actually, we could take $\varphi'$ to be an arbitrary automorphic form on $GL_m(\mathbb{A})$.) Consider the integrals

$$I(s;\varphi,\varphi') = \int_{GL_m(k)\backslash GL_m(\mathbb{A})} \mathbb{P}^n_m\varphi\begin{pmatrix} h & 0\\ 0 & 1\end{pmatrix} \varphi'(h)|\det(h)|^{s-1/2}\, dh.$$

The integral $I(s;\varphi,\varphi')$ is absolutely convergent for all values of the complex parameter $s$, uniformly in compact subsets, since the cusp forms are rapidly decreasing. Hence it is entire and bounded in any vertical strip.

Let us now investigate the Eulerian properties of these integrals. We first replace $\mathbb{P}^n_m\varphi$ by its Fourier expansion.

$$\begin{aligned} I(s;\varphi,\varphi') &= \int_{GL_m(k)\backslash GL_m(\mathbb{A})} \mathbb{P}^n_m\varphi\begin{pmatrix} h & 0 \\ 0 & I_{n-m}\end{pmatrix}\varphi'(h)|\det(h)|^{s-1/2}\,dh \\ &= \int_{GL_m(k)\backslash GL_m(\mathbb{A})} \sum_{\gamma\in N_m(k)\backslash GL_m(k)} W_\varphi\left(\begin{pmatrix} \gamma & 0 \\ 0 & I_{n-m}\end{pmatrix}\begin{pmatrix} h & 0 \\ 0 & I_{n-m}\end{pmatrix}\right) \\ &\quad \times \varphi'(h)|\det(h)|^{s-(n-m)/2}\,dh. \end{aligned}$$

Since $\varphi'(h)$ is automorphic on $GL_m(\mathbb{A})$ and $|\det(\gamma)| = 1$ for $\gamma \in GL_m(k)$ we may interchange the order of summation and integration for $Re(s) \gg 0$ and then recombine to obtain

$$I(s;\varphi,\varphi') = \int_{N_m(k)\backslash GL_m(\mathbb{A})} W_\varphi\begin{pmatrix} h & 0 \\ 0 & I_{n-m}\end{pmatrix}\varphi'(h)|\det(h)|^{s-(n-m)/2}\,dh.$$

This integral is absolutely convergent for $Re(s) \gg 0$ by the gauge estimates of [10, Section 13] and this justifies the interchange.

Let us now integrate first over $N_m(k)\backslash N_m(\mathbb{A})$. Recall that for $n \in N_m(\mathbb{A}) \subset N_n(\mathbb{A})$ we have $W_\varphi(ng) = \psi(n)W_\varphi(g)$. Hence we have

$$\begin{aligned} I(s;\varphi,\varphi') &= \int_{N_m(\mathbb{A})\backslash GL_m(\mathbb{A})}\int_{N_m(k)\backslash N_m(\mathbb{A})} W_\varphi\left(\begin{pmatrix} n & 0 \\ 0 & I_{n-m}\end{pmatrix}\begin{pmatrix} h & 0 \\ 0 & I_{n-m}\end{pmatrix}\right) \\ &\quad \times\varphi'(nh)\,dn\,|\det(h)|^{s-(n-m)/2}\,dh \\ &= \int_{N_m(\mathbb{A})\backslash GL_m(\mathbb{A})} W_\varphi\begin{pmatrix} h & 0 \\ 0 & I_{n-m}\end{pmatrix} \\ &\quad \times \int_{N_m(k)\backslash N_m(\mathbb{A})} \psi(n)\varphi'(nh)\,dn\,|\det(h)|^{s-(n-m)/2}\,dh \\ &= \int_{N_m(\mathbb{A})\backslash GL_m(\mathbb{A})} W_\varphi\begin{pmatrix} h & 0 \\ 0 & I_{n-m}\end{pmatrix} W'_{\varphi'}(h)\,|\det(h)|^{s-(n-m)/2}\,dh \\ &= \Psi(s;W_\varphi,W'_{\varphi'}) \end{aligned}$$

where $W'_{\varphi'}(h)$ is the $\psi^{-1}$-Whittaker function on $GL_m(\mathbb{A})$ associated to $\varphi'$, i.e.,

$$W'_{\varphi'}(h) = \int_{N_m(k)\backslash N_m(\mathbb{A})} \varphi'(nh)\psi(n)\,dn,$$

and we retain absolute convergence for $Re(s) \gg 0$.

From this point, the fact that the integrals are Eulerian is a consequence of the uniqueness of the Whittaker model for $GL_n$ [9, 15]. Take $\varphi$ a smooth cusp form in a cuspidal representation $\pi$ of $GL_n(\mathbb{A})$. Assume in addition that $\varphi$ is factorizable, i.e., in the decomposition $\pi = \otimes' \pi_v$ of $\pi$ into a restricted tensor product of local representations, $\varphi = \otimes \varphi_v$ is a pure tensor. Then there is a choice of local Whittaker models so that $W_\varphi(g) = \prod W_{\varphi_v}(g_v)$. Similarly for decomposable $\varphi'$ we have the factorization $W'_{\varphi'}(h) = \prod W'_{\varphi'_v}(h_v)$.

If we substitute these factorizations into our integral expression, then since the domain of integration factors $N_m(\mathbb{A})\backslash GL_m(\mathbb{A}) = \prod N_m(k_v)\backslash GL_m(k_v)$ we see that our integral factors into a product of local integrals

$$\Psi(s; W_\varphi, W'_{\varphi'}) = \prod_v \int_{N_m(k_v)\backslash GL_m(k_v)} W_{\varphi_v}\begin{pmatrix} h_v & 0 \\ 0 & I_{n-m} \end{pmatrix} W'_{\varphi'_v}(h_v)\, |\det(h_v)|_v^{s-(n-m)/2}\, dh_v.$$

If we denote the local integrals by

$$\Psi_v(s; W_{\varphi_v}, W'_{\varphi'_v}) = \int_{N_m(k_v)\backslash GL_m(k_v)} W_{\varphi_v}\begin{pmatrix} h_v & 0 \\ 0 & I_{n-m} \end{pmatrix} W'_{\varphi'_v}(h_v)\, |\det(h_v)|_v^{s-(n-m)/2}\, dh_v,$$

which converges for $Re(s) \gg 0$ by the gauge estimate of [10, Proposition 2.3.6], we see that we now have a family of Eulerian integrals.

Now let us return to the question of a functional equation. The functional equation is essentially a consequence of the existence of the outer automorphism $g \mapsto \iota(g) = g^\iota = {}^t g^{-1}$ of $GL_n$. If we define the action of this automorphism on automorphic forms by setting $\widetilde{\varphi}(g) = \varphi(g^\iota) = \varphi(w_n g^\iota)$ and let $\widetilde{\mathbb{P}}^n_m = \iota \circ \mathbb{P}^n_m \circ \iota$ then our integrals naturally satisfy the functional equation

$$I(s; \varphi, \varphi') = \widetilde{I}(1 - s; \widetilde{\varphi}, \widetilde{\varphi}')$$

where

$$\widetilde{I}(s; \varphi, \varphi') = \int_{GL_m(k)\backslash GL_m(\mathbb{A})} \widetilde{\mathbb{P}}^n_m \varphi \begin{pmatrix} h & \\ & 1 \end{pmatrix} \varphi'(h) |\det(h)|^{s-1/2}\, dh.$$

We have established the following result.

THEOREM 2.1. *Let $\varphi \in V_\pi$ be a cusp form on $GL_n(\mathbb{A})$ and $\varphi' \in V_{\pi'}$ a cusp form on $GL_m(\mathbb{A})$ with $m < n$. Then the family of integrals $I(s; \varphi, \varphi')$ define entire functions of $s$, bounded in vertical strips, and satisfy the functional equation*

$$I(s; \varphi, \varphi') = \widetilde{I}(1 - s; \widetilde{\varphi}, \widetilde{\varphi}').$$

*Moreover the integrals are Eulerian and if $\varphi$ and $\varphi'$ are factorizable, we have*

$$I(s; \varphi, \varphi') = \prod_v \Psi_v(s; W_{\varphi_v}, W'_{\varphi'_v})$$

*with convergence absolute and uniform for $Re(s) \gg 0$.*

The integrals occurring in the right hand side of our functional equation are again Eulerian. One can unfold the definitions to find first that

$$\widetilde{I}(1-s;\widetilde{\varphi},\widetilde{\varphi}') = \widetilde{\Psi}(1-s;\rho(w_{n,m})\widetilde{W}_\varphi,\widetilde{W}'_{\varphi'})$$

where the unfolded global integral is

$$\widetilde{\Psi}(s;W,W') = \int\int W\begin{pmatrix} h & & \\ x & I_{n-m-1} & \\ & & 1\end{pmatrix} dx\, W'(h)\,|\det(h)|^{s-(n-m)/2}\, dh$$

with the $h$ integral over $N_m(\mathbb{A})\backslash GL_m(\mathbb{A})$ and the $x$ integral over $M_{n-m-1,m}(\mathbb{A})$, the space of $(n-m-1)\times m$ matrices, $\rho$ denoting right translation, and $w_{n,m}$ the Weyl element $w_{n,m} = \begin{pmatrix} I_m & \\ & w_{n-m}\end{pmatrix}$ with $w_{n-m} = \begin{pmatrix} & & 1 \\ & \cdot^{\cdot^{\cdot}} & \\ 1 & & \end{pmatrix}$ the standard long Weyl element in $GL_{n-m}$. Also, for $W\in\mathcal{W}(\pi,\psi)$ we set $\widetilde{W}(g) = W(w_n g^\iota)\in\mathcal{W}(\widetilde{\pi},\psi^{-1})$. The extra unipotent integration is the remnant of $\widetilde{\mathbb{P}}^n_m$. As before, $\widetilde{\Psi}(s;W,W')$ is absolutely convergent for $Re(s)\gg 0$. For $\varphi$ and $\varphi'$ factorizable as before, these integrals $\widetilde{\Psi}(s;W_\varphi,W'_{\varphi'})$ will factor as well. Hence we have

$$\widetilde{\Psi}(s;W_\varphi,W'_{\varphi'}) = \prod_v \widetilde{\Psi}_v(s;W_{\varphi_v},W'_{\varphi'_v})$$

where

$$\widetilde{\Psi}_v(s;W_v,W'_v) = \int\int W_v\begin{pmatrix} h_v & & \\ x_v & I_{n-m-1} & \\ & & 1\end{pmatrix} dx_v\, W'_v(h_v)|\det(h_v)|^{s-(n-m)/2}\, dh_v$$

where now with the $h_v$ integral is over $N_m(k_v)\backslash GL_m(k_v)$ and the $x_v$ integral is over the matrix space $M_{n-m-1,m}(k_v)$. Thus, coming back to our functional equation, we find that the right hand side is Eulerian and factors as

$$\widetilde{I}(1-s;\widetilde{\varphi},\widetilde{\varphi}') = \widetilde{\Psi}(1-s;\rho(w_{n,m})\widetilde{W}_\varphi,\widetilde{W}'_{\varphi'}) = \prod_v \widetilde{\Psi}_v(1-s;\rho(w_{n,m})\widetilde{W}_{\varphi_v},\widetilde{W}'_{\varphi'_v}).$$

Now consider the case of $m=n$. Then the results we need can essentially be found in Part I of [12]. Let $(\pi,V_\pi)$ and $(\pi',V_{\pi'})$ be two unitary cuspidal representations of $GL_n(\mathbb{A})$. Let $\varphi\in V_\pi$ and $\varphi'\in V_{\pi'}$ be two cusp forms. The integral representation in this situation is an honest Rankin–Selberg integral and will involve the integration of the cusp forms $\varphi$ and $\varphi'$ against a particular type of Eisenstein series on $GL_n(\mathbb{A})$.

To construct the Eisenstein series as in Part I of [12] we observe that $P_n\backslash GL_n \simeq k^n - \{0\}$. If we let $\mathcal{S}(\mathbb{A}^n)$ denote the Schwartz–Bruhat functions on $\mathbb{A}^n$, then each $\Phi\in\mathcal{S}$ defines a smooth function on $GL_n(\mathbb{A})$, left invariant by $P_n(\mathbb{A})$, by $g\mapsto \Phi((0,\ldots,0,1)g) = \Phi(e_n g)$. Let $\eta$ be a unitary idele class character. (For our application $\eta$ will be determined by the central characters of $\pi$ and $\pi'$.) Consider the function

$$F(g, \Phi; s, \eta) = |\det(g)|^s \int_{\mathbb{A}^\times} \Phi(ae_n g)|a|^{ns}\eta(a)\, d^\times a.$$

If we let $P_n' = Z_n P_n$ be the parabolic of $GL_n$ associated to the partition $(n-1, 1)$ then one checks that for $p' = \begin{pmatrix} h & y \\ 0 & d \end{pmatrix} \in P_n'(\mathbb{A})$ with $h \in GL_{n-1}(\mathbb{A})$ and $d \in \mathbb{A}^\times$ we have,

$$\begin{aligned} F(p'g, \Phi; s, \eta) &= |\det(h)|^s |d|^{-(n-1)s}\eta(d)^{-1} F(g, \Phi; s, \eta) \\ &= \delta_{P_n'}^s(p')\eta^{-1}(d) F(g, \Phi; s, \eta), \end{aligned}$$

with the integral absolutely convergent for $Re(s) > 1/n$, so that if we extend $\eta$ to a character of $P_n'$ by $\eta(p') = \eta(d)$ in the above notation we have that $F(g, \Phi; s, \eta)$ is a smooth section of the normalized induced representation $Ind_{P_n'(\mathbb{A})}^{GL_n(\mathbb{A})}(\delta_{P_n'}^{s-1/2}\eta^{-1})$. Since the inducing character $\delta_{P_n'}^{s-1/2}\eta^{-1}$ of $P_n'(\mathbb{A})$ is invariant under $P_n'(k)$ we may form Eisenstein series from this family of sections by

$$E(g, \Phi; s, \eta) = \sum_{\gamma \in P_n'(k)\backslash GL_n(k)} F(\gamma g, \Phi; s, \eta).$$

If we replace $F$ in this sum by its definition we can rewrite this Eisenstein series as

$$\begin{aligned} E(g, \Phi; s, \eta) &= |\det(g)|^s \int_{k^\times\backslash\mathbb{A}^\times} \sum_{\xi \in k^n - \{0\}} \Phi(a\xi g)|a|^{ns}\eta(a)\, d^\times a \\ &= |\det(g)|^s \int_{k^\times\backslash\mathbb{A}^\times} \Theta_\Phi'(a, g)|a|^{ns}\eta(a)\, d^\times a \end{aligned}$$

and this first expression is convergent absolutely for $Re(s) > 1$ [12].

The second expression essentially gives the Eisenstein series as the Mellin transform of the Theta series

$$\Theta_\Phi(a, g) = \sum_{\xi \in k^n} \Phi(a\xi g),$$

where in the above we have written

$$\Theta_\Phi'(a, g) = \sum_{\xi \in k^n - \{0\}} \Phi(a\xi g) = \Theta_\Phi(a, g) - \Phi(0).$$

This allows us to obtain the analytic properties of the Eisenstein series from the Poisson summation formula for $\Theta_\Phi$, namely

$$\begin{aligned} \Theta_\Phi(a, g) &= \sum_{\xi \in k^n} \Phi(a\xi g) = \sum_{\xi \in k^n} \Phi_{a,g}(\xi) \\ &= \sum_{\xi \in k^n} \widehat{\Phi_{a,g}}(\xi) = \sum_{\xi \in k^n} |a|^{-n}|\det(g)|^{-1}\hat{\Phi}(a^{-1}\xi {}^t g^{-1}) \\ &= |a|^{-n}|\det(g)|^{-1}\Theta_{\hat{\Phi}}(a^{-1}, {}^t g^{-1}) \end{aligned}$$

where the Fourier transform $\hat{\Phi}$ on $\mathcal{S}(\mathbb{A}^n)$ is defined by

$$\hat{\Phi}(x) = \int_{\mathbb{A}^\times} \Phi(y)\psi(y^t x)\, dy.$$

This allows us to write the Eisenstein series as

$$E(g, \Phi, s, \eta) = |\det(g)|^s \int_{|a|\geq 1} \Theta'_\Phi(a, g)|a|^{ns}\eta(a)\, d^\times a$$
$$+ |\det(g)|^{s-1} \int_{|a|\geq 1} \Theta'_{\hat{\Phi}}(a, {}^t g^{-1})|a|^{n(1-s)}\eta^{-1}(a)\, d^\times a + \delta(s)$$

where

$$\delta(s) = \begin{cases} 0 & \text{if } \eta \text{ is ramified} \\ -c\Phi(0)\frac{|\det(g)|^s}{s+i\sigma} + c\hat{\Phi}(0)\frac{|\det(g)|^{s-1}}{s-1+i\sigma} & \text{if } \eta(a) = |a|^{in\sigma} \text{ with } \sigma \in \mathbb{R} \end{cases}$$

with $c$ a non-zero constant. From this we derive easily the basic properties of our Eisenstein series [12, Part I, Section 4].

PROPOSITION 2.1. *The Eisenstein series $E(g, \Phi; s, \eta)$ has a meromorphic continuation to all of $\mathbb{C}$ with at most simple poles at $s = -i\sigma, 1 - i\sigma$ when $\eta$ is unramified of the form $\eta(a) = |a|^{in\sigma}$. As a function of $g$ it is smooth of moderate growth and as a function of $s$ it is bounded in vertical strips (away from the possible poles), uniformly for $g$ in compact sets. Moreover, we have the functional equation*

$$E(g, \Phi; s, \eta) = E(g^\iota, \hat{\Phi}; 1 - s, \eta^{-1})$$

*where $g^\iota = {}^t g^{-1}$.*

Note that under the center the Eisenstein series transforms by the central character $\eta^{-1}$.

Now let us return to our Eulerian integrals. Let $\pi$ and $\pi'$ be our irreducible cuspidal representations. Let their central characters be $\omega$ and $\omega'$. Set $\eta = \omega\omega'$. Then for each pair of cusp forms $\varphi \in V_\pi$ and $\varphi' \in V_{\pi'}$ and each Schwartz-Bruhat function $\Phi \in \mathcal{S}(\mathbb{A}^n)$ set

$$I(s; \varphi, \varphi', \Phi) = \int_{Z_n(\mathbb{A})\, GL_n(k)\backslash\, GL_n(\mathbb{A})} \varphi(g)\varphi'(g)E(g, \Phi; s, \eta)\, dg.$$

Since the two cusp forms are rapidly decreasing on $Z_n(\mathbb{A})\, GL_n(k)\backslash\, GL_n(\mathbb{A})$ and the Eisenstein is only of moderate growth, we see that the integral converges absolutely for all $s$ away from the poles of the Eisenstein series and is hence meromorphic. It will be bounded in vertical strips away from the poles and satisfies the functional equation

$$I(s; \varphi, \varphi', \Phi) = I(1 - s; \widetilde{\varphi}, \widetilde{\varphi}', \hat{\Phi}),$$

coming from the functional equation of the Eisenstein series, where we still have $\widetilde{\varphi}(g) = \varphi(g^{\iota}) = \varphi(w_n g^{\iota}) \in V_{\widetilde{\pi}}$ and similarly for $\widetilde{\varphi}'$.

These integrals will be entire unless we have $\eta(a) = \omega(a)\omega'(a) = |a|^{in\sigma}$ is unramified. In that case, the residue at $s = -i\sigma$ will be

$$\operatorname*{Res}_{s=-i\sigma} I(s; \varphi, \varphi', \Phi) = -c\Phi(0) \int_{Z_n(\mathbb{A})\, GL_n(\mathbb{A}) \backslash GL_n(\mathbb{A})} \varphi(g)\varphi'(g)|\det(g)|^{-i\sigma}\, dg$$

and at $s = 1 - i\sigma$ we can write the residue as

$$\operatorname*{Res}_{s=1-i\sigma} I(s; \varphi, \varphi', \Phi) = c\hat{\Phi}(0) \int_{Z_n(\mathbb{A})\, GL_n(k) \backslash GL_n(\mathbb{A})} \widetilde{\varphi}(g)\widetilde{\varphi}'(g)|\det(g)|^{i\sigma}\, dg.$$

Therefore these residues define $GL_n(\mathbb{A})$ invariant pairings between $\pi$ and $\pi' \otimes |\det|^{-i\sigma}$ or equivalently between $\widetilde{\pi}$ and $\widetilde{\pi}' \otimes |\det|^{i\sigma}$. Hence a residue can be non-zero only if $\pi \simeq \widetilde{\pi}' \otimes |\det|^{i\sigma}$ and in this case we can find $\varphi$, $\varphi'$, and $\Phi$ such that indeed the residue does not vanish.

We have yet to check that our integrals are Eulerian. To this end we take the integral, replace the Eisenstein series by its definition, and unfold:

$$\begin{aligned} I(s; \varphi, \varphi', \Phi) &= \int_{Z_n(\mathbb{A})\, GL_n(k) \backslash GL_n(\mathbb{A})} \varphi(g)\varphi'(g)E(g, \Phi; s, \eta)\, dg \\ &= \int_{Z_n(\mathbb{A})\, P_n'(k) \backslash GL_n(\mathbb{A})} \varphi(g)\varphi'(g)F(g, \Phi; s, \eta)\, dg \\ &= \int_{Z_n(\mathbb{A})\, P_n(k) \backslash GL_n(\mathbb{A})} \varphi(g)\varphi'(g)|\det(g)|^{s} \int_{\mathbb{A}^{\times}} \Phi(ae_n g)|a|^{ns}\eta(a)\, da\, dg \\ &= \int_{P_n(k) \backslash GL_n(\mathbb{A})} \varphi(g)\varphi'(g)\Phi(e_n g)|\det(g)|^{s}\, dg. \end{aligned}$$

We next replace $\varphi$ by its Fourier expansion in the form

$$\varphi(g) = \sum_{\gamma \in N_n(k) \backslash P_n(k)} W_{\varphi}(\gamma g)$$

and unfold to find

$$\begin{aligned} I(s; \varphi, \varphi', \Phi) &= \int_{N_n(k) \backslash GL_n(\mathbb{A})} W_{\varphi}(g)\varphi'(g)\Phi(e_n g)|\det(g)|^{s}\, dg \\ &= \int_{N_n(\mathbb{A}) \backslash GL_n(\mathbb{A})} W_{\varphi}(g) \int_{N_n(k) \backslash N_n(\mathbb{A})} \varphi'(ng)\psi(n)\, dn\, \Phi(e_n g)|\det(g)|^{s}\, dg \\ &= \int_{N_n(\mathbb{A}) \backslash GL_n(\mathbb{A})} W_{\varphi}(g)W'_{\varphi'}(g)\Phi(e_n g)|\det(g)|^{s}\, dg \\ &= \Psi(s; W_{\varphi}, W'_{\varphi'}, \Phi). \end{aligned}$$

This expression converges for $Re(s) \gg 0$ by the gauge estimates as before.

To continue, we assume that $\varphi$, $\varphi'$ and $\Phi$ are decomposable tensors under the isomorphisms $\pi \simeq \otimes' \pi_v$, $\pi' \simeq \otimes' \pi'_v$, and $\mathcal{S}(\mathbb{A}^n) \simeq \otimes' \mathcal{S}(k_v^n)$ so that we have $W_\varphi(g) = \prod_v W_{\varphi_v}(g_v)$, $W'_{\varphi'}(g) = \prod_v W'_{\varphi'_v}(g_v)$ and $\Phi(g) = \prod_v \Phi_v(g_v)$. Then, since the domain of integration also naturally factors we can decompose this last integral into an Euler product and now write

$$\Psi(s; W_\varphi, W'_{\varphi'}, \Phi) = \prod_v \Psi_v(s; W_{\varphi_v}, W'_{\varphi'_v}, \Phi_v),$$

where

$$\Psi_v(s; W_{\varphi_v}, W'_{\varphi'_v}, \Phi_v) = \int_{N_n(k_v)\backslash GL_n(k_v)} W_{\varphi_v}(g_v) W'_{\varphi'_v}(g_v) \Phi_v(e_n g_v) |\det(g_v)|^s \, dg_v,$$

still with convergence for $Re(s) \gg 0$ by the local gauge estimates. We have now established the following result.

THEOREM 2.2. *Let $\varphi \in V_\pi$ and $\varphi' \in V_{\pi'}$ cusp forms on $GL_n(\mathbb{A})$ and let $\Phi \in \mathcal{S}(\mathbb{A}^n)$. Then the family of integrals $I(s; \varphi, \varphi', \Phi)$ define meromorphic functions of $s$, bounded in vertical strips away from the poles. The only possible poles are simple and occur iff $\pi \simeq \widetilde{\pi}' \otimes |\det|^{i\sigma}$ with $\sigma$ real and are then at $s = -i\sigma$ and $s = 1 - i\sigma$ with residues as above. They satisfy the functional equation*

$$I(s; \varphi, \varphi', \Phi) = I(1 - s; \widetilde{\varphi}, \widetilde{\varphi}', \hat{\Phi}).$$

*Moreover, for $\varphi$, $\varphi'$, and $\Phi$ factorizable we have that the integrals are Eulerian and we have*

$$I(s; \varphi, \varphi', \Phi) = \prod_v \Psi_v(s; W_{\varphi_v}, W'_{\varphi'_v}, \Phi_v)$$

*with convergence absolute and uniform for $Re(s) \gg 0$.*

We remark in passing that the right hand side of the functional equation also unfolds as

$$\begin{aligned} I(1 - s; \widetilde{\varphi}, \widetilde{\varphi}', \hat{\Phi}) &= \int_{N_n(\mathbb{A})\backslash GL_n(\mathbb{A})} \widetilde{W}_\varphi(g) \widetilde{W}'_{\varphi'}(g) \hat{\Phi}(e_n g) |\det(g)|^{1-s} \, dg \\ &= \prod_v \Psi_v(1 - s; \widetilde{W}_{\varphi_v}, \widetilde{W}'_{\varphi'_v}, \hat{\Phi}) \end{aligned}$$

with convergence for $Re(s) \ll 0$.

**2.2. The Global $L$-function.** Let $S$ be the finite set of places of $k$, containing the archimedean places $S_\infty$, such that for all $v \notin S$ we have that $\pi_v$, $\pi'_v$, and $\psi_v$ are unramified.

For each place $v$ of $k$ local factors $L(s, \pi_v \times \pi'_v)$ and $\varepsilon(s, \pi_v \times \pi'_v, \psi_v)$ have been defined through the local theory of Rankin-Selberg convolutions in [11] for non-archimedean $v$ and in [13] for archimedean $v$. Then we can at least formally define

$$L(s,\pi\times\pi')=\prod_v L(s,\pi_v\times\pi'_v)\quad\text{and}\quad \varepsilon(s,\pi\times\pi')=\prod_v \varepsilon(s,\pi_v\times\pi'_v,\psi_v).$$

We need to discuss convergence of these products. Let us first consider the convergence of $L(s,\pi\times\pi')$. For those $v\notin S$, so $\pi_v,\pi'_v$, and $\psi_v$ are unramified, Jacquet and Shalika have explicitly computed the local factor in [12, Part I, Section 2; Part II, Section 1]. They show

$$L(s,\pi_v\times\pi'_v)=\det(I-q_v^{-s}A_{\pi_v}\otimes A_{\pi'_v})^{-1}$$

where $A_{\pi_v}$ and $A_{\pi'_v}$ are the associated Satake parameters, and that the eigenvalues of $A_{\pi_v}$ and $A_{\pi'_v}$ are all of absolute value less than $q_v^{1/2}$ [12, Part I, Corollary 2.5]. Thus, as in [12, Theorem 5.3], the partial (or incomplete) $L$-function

$$L^S(s,\pi\times\pi')=\prod_{v\notin S}L(s,\pi_v\times\pi'_v)=\prod_{v\notin S}\det(I-q_v^{-s}A_{\pi_v}\otimes A_{\pi'_v})^{-1}$$

is absolutely convergent for $Re(s)\gg 0$. Thus the same is true for $L(s,\pi\times\pi')$.

*Remark.* The local calculation alluded to above is actually the computation of the local integral with the unramified Whittaker functions. For $v\notin S$, in the Whittaker models there will be unique normalized $K=GL(\mathfrak{o}_v)$–fixed Whittaker functions, $W_v^\circ\in\mathcal{W}(\pi_v,\psi_v)$ and $W_v'^\circ\in\mathcal{W}(\pi'_v,\psi_v^{-1})$, normalized by $W_v^\circ(e)=W_v'^\circ(e)=1$. When $n=m$ let $\Phi=\Phi_v^\circ$ be the characteristic function of the lattice $\mathfrak{o}_v{}^n\subset k_v^n$. What Jacquet and Shalika show is that

$$\det(I-q_v^{-s}A_{\pi_v}\otimes A_{\pi'_v})^{-1}=\begin{cases}\Psi(s;W_v^\circ,W_v'^\circ) & m<n\\ \Psi(s;W_v^\circ,W_v'^\circ,\Phi_v^\circ) & m=n\end{cases}$$

and hence $\det(I-q_v^{-s}A_{\pi_v}\otimes A_{\pi'_v})$ divides $L(s,\pi_v\times\pi'_v)^{-1}$. To see that this actually calculates the $L$-function, one needs to combine this calculation with Proposition 9.4 of [11].

For the $\varepsilon$–factor, it follows from the local calculation cited above and the local functional equation [11, Theorem 2.7 (iii)] that $\varepsilon(s,\pi_v\times\pi'_v,\psi_v)\equiv 1$ for $v\notin S$ so that the product is in fact a finite product and there is no problem with convergence. The fact that $\varepsilon(s,\pi\times\pi')$ is independent of $\psi$ can either be checked by analyzing how the local $\varepsilon$–factors vary as you vary $\psi$, as is done in [2, Lemma 2.1], or it will follow from the global functional equation presented below.

**2.3. The basic analytic properties.** Our first goal is to show that these $L$-functions have nice analytic properties.

THEOREM 2.3. *The global L–functions $L(s,\pi\times\pi')$ are nice in the sense that*
(1) *$L(s,\pi\times\pi')$ has a meromorphic continuation to all of $\mathbb{C}$,*

(2) *the extended function is bounded in vertical strips (away from its poles),*
(3) *they satisfy the functional equation*

$$L(s,\pi\times\pi')=\varepsilon(s,\pi\times\pi')L(1-s,\widetilde{\pi}\times\widetilde{\pi}').$$

To do so, we relate the $L$-functions to the global integrals.

Let us begin with continuation. In the case $m<n$ for every $\varphi\in V_\pi$ and $\varphi'\in V_{\pi'}$ we know the integral $I(s;\varphi,\varphi')$ converges absolutely for all $s$. From the unfolding in Section 2.1 and the local calculation mentioned above we know that for $Re(s)\gg 0$ and for appropriate choices of $\varphi$ and $\varphi'$ we have

$$\begin{aligned}I(s;\varphi,\varphi')&=\prod_v \Psi_v(s;W_{\varphi_v},W_{\varphi'_v})\\&=\left(\prod_{v\in S}\Psi_v(s;W_{\varphi_v},W_{\varphi'_v})\right)L^S(s,\pi\times\pi')\\&=\left(\prod_{v\in S}\frac{\Psi_v(s;W_{\varphi_v},W_{\varphi'_v})}{L(s,\pi_v\times\pi'_v)}\right)L(s,\pi\times\pi')\\&=\left(\prod_{v\in S}e_v(s;W_{\varphi_v},W_{\varphi'_v})\right)L(s,\pi\times\pi').\end{aligned}$$

We know that each $e_v(s;W_v,W'_v)$ is entire. For non-archimedean $v$ this follows from [11, Theorem 2.3] and for archimedean $v$ this follows from Theorem 1.2 above and its corollary. Hence $L(s,\pi\times\pi')$ has a meromorphic continuation. If $m=n$ then for appropriate $\varphi\in V_\pi$, $\varphi'\in V_{\pi'}$, and $\Phi\in\mathcal{S}(\mathbb{A}^n)$ we again have

$$I(s;\varphi,\varphi',\Phi)=\left(\prod_{v\in S}e_v(s;W_{\varphi_v},W'_{\varphi'_v},\Phi_v)\right)L(s,\pi\times\pi').$$

Once again, since each $e_v(s;W_v,W'_v,\Phi_v)$ is entire, $L(s,\pi\times\pi')$ has a meromorphic continuation.

Let us next turn to the functional equation. This will follow from the functional equation for the global integrals given above and the local functional equations [11, Theorem 2.7 (iii)] and [13, Theorem 5.1 (ii)]. We will consider only the case where $m<n$ since the other case is entirely analogous. The functional equation for the global integrals is simply

$$I(s;\varphi,\varphi')=\tilde{I}(1-s;\widetilde{\varphi},\widetilde{\varphi}').$$

Once again we have for appropriate $\varphi$ and $\varphi'$

$$I(s;\varphi,\varphi')=\left(\prod_{v\in S}e_v(s;W_{\varphi_v},W'_{\varphi'_v})\right)L(s,\pi\times\pi')$$

while on the other side

$$\tilde{I}(1-s;\widetilde{\varphi},\widetilde{\varphi}')=\left(\prod_{v\in S}\tilde{e}_v(1-s;\rho(w_{n,m})\widetilde{W}_{\varphi_v},\widetilde{W}'_{\varphi'_v})\right)L(1-s,\widetilde{\pi}\times\widetilde{\pi}').$$

However, by the local functional equations, for each $v\in S$ we have

$$\begin{aligned}\tilde{e}_v(1-s;\rho(w_{n,m})\widetilde{W}_v,\widetilde{W}'_v)&=\frac{\widetilde{\Psi}(1-s;\rho(w_{n,m})\widetilde{W}_v,\widetilde{W}'_v)}{L(1-s,\widetilde{\pi}\times\widetilde{\pi}')}\\&=\omega'_v(-1)^{n-1}\varepsilon(s,\pi_v\times\pi'_v,\psi_v)\frac{\Psi(s;W_v,W'_v)}{L(s,\pi\times\pi')}\\&=\omega'_v(-1)^{n-1}\varepsilon(s,\pi_v\times\pi'_v,\psi_v)e_v(s,W_v,W'_v).\end{aligned}$$

Combining these, we have

$$L(s,\pi\times\pi')=\left(\prod_{v\in S}\omega'_v(-1)^{n-1}\varepsilon(s,\pi_v\times\pi'_v,\psi_v)\right)L(1-s,\widetilde{\pi}\times\widetilde{\pi}').$$

Now, for $v\notin S$ we know that $\pi'_v$ is unramified, so $\omega'_v(-1)=1$, and also that $\varepsilon(s,\pi_v\times\pi'_v,\psi_v)\equiv 1$. Therefore

$$\begin{aligned}\prod_{v\in S}\omega'_v(-1)^{n-1}\varepsilon(s,\pi_v\times\pi'_v,\psi_v)&=\prod_{v}\omega'_v(-1)^{n-1}\varepsilon(s,\pi_v\times\pi'_v,\psi_v)\\&=\omega'(-1)^{n-1}\varepsilon(s,\pi\times\pi')\\&=\varepsilon(s,\pi\times\pi')\end{aligned}$$

and we indeed have

$$L(s,\pi\times\pi')=\varepsilon(s,\pi\times\pi')L(1-s,\widetilde{\pi}\times\widetilde{\pi}').$$

Note that this implies that $\varepsilon(s,\pi\times\pi')$ is independent of $\psi$ as well.

Let us now turn to the boundedness in vertical strips. For the global integrals $I(s;\varphi,\varphi')$ or $I(s;\varphi,\varphi,\Phi)$ this simply follows from the absolute convergence. For the $L$-function itself, the paradigm is the following. For every finite place $v\in S$, by the definition of the local $L$-function as the generator of the fractional ideal spanned by the local integrals [11, Theorem 2.7 (ii)] we know that there is a choice of finite collections $W_{v,i}$, $W'_{v,i}$, and if necessary $\Phi_{v,i}$ such that

$$\begin{aligned}L(s,\pi_v\times\pi'_v)&=\sum\Psi(s;W_{v,i},W'_{v'i})\quad\text{or}\\L(s,\pi_v\times\pi'_v)&=\sum\Psi(s;W_{v,i},W'_{v'i},\Phi_{v,i}).\end{aligned}$$

If $m=n-1$ or $m=n$ then by the results of Stade [16, 17] or the unpublished work of Jacquet and Shalika presented in Theorem 1.3 above we know that we have similar statements for $v\in S_\infty$. Hence if $m=n-1$ or $m=n$ there are finite global choices $\varphi_i$, $\varphi'_i$, and if necessary $\Phi_i$ such that

$$L(s,\pi\times\pi')=\sum I(s;\varphi_i,\varphi'_i)\quad\text{or}\quad L(s,\pi\times\pi')=\sum I(s;\varphi_i,\varphi'_i,\Phi_i).$$

Then the boundedness in vertical strips for the $L$-functions follows from that of the global integrals.

However, if $m < n-1$ then all we know at those $v \in S_\infty$ is that there is a function $W_v \in \mathcal{W}(\pi_v \hat{\otimes} \pi'_v, \psi_v) = \mathcal{W}(\pi_v, \psi_v) \hat{\otimes} \mathcal{W}(\pi'_v, \psi_v^{-1})$ or a finite collection of such functions $W_{v,i}$ and of $\Phi_{v,i}$ such that

$$L(s, \pi_v \times \pi'_v) = I(s; W_v) \quad \text{or} \quad L(s, \pi_v \times \pi'_v) = \sum I(s; W_{v,i}, \Phi_{v,i}).$$

To make the above paradigm work for $m < n-1$ one possibility would be to rework the theory of global Eulerian integrals for cusp forms in $V_\pi \hat{\otimes} V_{\pi'}$. This is naturally the space of smooth vectors in an irreducible unitary cuspidal representation of $GL_n(\mathbb{A}) \times GL_m(\mathbb{A})$. So we would need to extend the global theory of integrals parallel to Jacquet and Shalika's extension of the local integrals in the archimedean theory. There seems to be no obstruction to carrying this out, and then we would obtain boundedness in vertical strips for $L(s, \pi \times \pi')$ in general within the context of integral representations. However, if one approaches these $L$-functions by the method of constant terms and Fourier coefficients of Eisenstein series, then Gelbart and Shahidi have shown a wide class of automorphic $L$-functions, including ours, to be bounded in vertical strips [6]. Thus the boundedness in vertical strips is true, even if we must go "outside the method" for this fact at this point.

**2.4. Poles of $L$-functions.** Let us determine where the global $L$-functions can have poles. The poles of the $L$-functions will be related to the poles of the global integrals. Recall from Section 2.2 that in the case of $m < n$ we have that the global integrals $I(s; \varphi, \varphi')$ are entire and that when $m = n$ then $I(s; \varphi, \varphi', \Phi)$ can have at most simple poles and they occur at $s = -i\sigma$ and $s = 1 - i\sigma$ for $\sigma$ real when $\pi \simeq \tilde{\pi}' \otimes |\det|^{i\sigma}$. As we have noted above, the global integrals and global $L$-functions are related, for appropriate $\varphi$, $\varphi'$, and $\Phi$, by

$$I(s; \varphi, \varphi') = \left( \prod_{v \in S} e_v(s; W_{\varphi_v}, W'_{\varphi'_v}) \right) L(s, \pi \times \pi')$$

or

$$I(s; \varphi, \varphi', \Phi) = \left( \prod_{v \in S} e_v(s; W_{\varphi_v}, W'_{\varphi'_v}, \Phi_v) \right) L(s, \pi \times \pi').$$

On the other hand, for any $s_0 \in \mathbb{C}$ and any $v$ there is a choice of local $W_v$, $W'_v$, and $\Phi_v$ such that the local factors $e_v(s_0; W_v, W'_v) \neq 0$ or $e_v(s_0; W_v, W'_v, \Phi_v) \neq 0$. For archimedean $v$ this is Theorem 1.2 (ii) and its corollary. For non-archimedean $v$ this follows from the definition of the $L$-function as the generator of the fractional ideal spanned by the local integrals. As noted above this implies that there are finite collections $W_{v,i}$, $W'_{v,i}$, and $\Phi_{v,i}$ if necessary such that

$$L(s, \pi_v \times \pi'_v) = \sum \Psi(s; W_{v,i}, W'_{v'i}) \quad \text{or}$$
$$L(s, \pi_v \times \pi'_v) = \sum \Psi(s; W_{v,i}, W'_{v'i}, \Phi_{v,i})$$

which is equivalent to

$$1 = \sum e(s; W_{v,i}, W'_{v'i}) \quad \text{or} \quad 1 = \sum e(s; W_{v,i}, W'_{v'i}, \Phi_{v,i}).$$

Hence for any choice of $s_0 \in \mathbb{C}$ one of the $e(s_0; W_{v,i}, W'_{v'i})$ or $e(s_0; W_{v,i}, W'_{v'i}, \Phi_{v,i})$ must be non-vanishing. So as we vary $\varphi$, $\varphi'$ and $\Phi$ at the places $v \in S$ we see that division by these factors can introduce no extraneous poles in $L(s, \pi \times \pi')$, that is, in keeping with the local characterization of the $L$-factor in terms of poles of local integrals, globally the poles of $L(s, \pi \times \pi')$ are precisely the poles of the family of global integrals $\{I(s; \varphi, \varphi')\}$ or $\{I(s; \varphi, \varphi', \Phi)\}$. Hence from Theorems 2.1 and 2.2 we have:

THEOREM 2.4. *If $m < n$ then $L(s, \pi \times \pi')$ is entire. If $m = n$, then $L(s, \pi \times \pi')$ has at most simple poles and they occur iff $\pi \simeq \widetilde{\pi}' \otimes |\det|^{i\sigma}$ with $\sigma$ real and are then at $s = -i\sigma$ and $s = 1 - i\sigma$.*

If we apply this with $\pi' = \widetilde{\pi}$ we obtain the following corollary.

COROLLARY. *$L(s, \pi \times \widetilde{\pi})$ has simple poles at $s = 0$ and $s = 1$.*

Since a general, not necessarily unitary, cuspidal representation $\pi$ is always of the form $\pi = \pi^u \otimes |\det|^r$ with $\pi^u$ unitary cuspidal, these results extend in a straightforward way to all cuspidal representations. In particular, this gives the proof of Jacquet, Piatetski-Shapiro, and Shalika of these results which was alluded to in the appendix of [14], where these results were proven using the technique of Eisenstein series.

DEPARTMENT OF MATHEMATICS, OKLAHOMA STATE UNIVERSITY, STILLWATER, OK 74078
*E-mail:* cogdell@math.okstate.edu

DEPARTMENT OF MATHEMATICS, YALE UNIVERSITY, NEW HAVEN, CT 06520
*E-mail:* ilya@math.yale.edu

---

REFERENCES

[1] W. CASSELMAN, Canonical extensions of Harish-Chandra modules to representations of $G$, *Can. J. Math.* **XLI** (1989), 385–438.

[2] J. W. COGDELL AND I. I. PIATETSKI-SHAPIRO, Converse Theorems for $GL_n$, *Publ. Math. IHES* **79** (1994), 157–214.

[3] ———, Converse Theorems for $GL_n$, II *J. reine angew. Math.* **507** (1999), 165–188.

[4] J. DIXMIER AND P. MALLIAVIN, Factorisations de fonctions et de vecteurs indéfiniment différentiables, *Bull. Sc. Math.*, 2$^e$ série **102** (1978), 305–330.

[5] D. FLATH, Decomposition of representations into tensor products, *Proc. Sympos. Pure Math.* **33**, part 1, (1979), 179–183.

[6] S. GELBART AND F. SHAHIDI, Boundedness of automorphic $L$-functions in vertical strips, *J. Amer. Math. Soc.* **14** (2001), 79–107.

[7] I. M. GELFAND, M. I. GRAEV, AND I. I. PIATETSKI-SHAPIRO, Representations of adele groups, *Dokl. Akad. Nauk SSSR* **156** (1964), 487–490; *Soviet Math. Doklady* **5** (1964), 657–661.

[8] ———, Representation Theory and Automorphic Functions, Academic Press, San Diego, 1990.

[9] I. M. GELFAND AND D. A. KAZHDAN, Representations of $GL(n, K)$ where $K$ is a local field, in *Lie Groups and Their Representations*, edited by I. M. Gelfand. John Wiley & Sons, New York–Toronto, 1971, 95–118.

[10] H. JACQUET, I. I. PIATETSKI-SHAPIRO, AND J. SHALIKA, Automorphic forms on $GL(3)$, I & II, *Ann. Math.* **109** (1979), 169–258.

[11] ———, Rankin-Selberg Convolutions, *Am. J. Math.* **105** (1983), 367–464.

[12] H. JACQUET AND J. SHALIKA, On Euler products and the classification of automorphic representations, *Amer. J. Math.* I: **103** (1981), 499–588; II: **103** (1981), 777–815.

[13] ———, Rankin-Selberg Convolutions: Archimedean Theory, in "Festschrift in Honor of I.I. Piatetski-Shapiro, Part I," Weizmann Science Press, Jerusalem, 1990, 125–207.

[14] C. MOEGLIN AND J-L. WALDSPURGER, Le spectre résiduel de $GL(n)$, *Ann. scient. Éc. Norm. Sup.*, 4ᵉ série **22** (1989), 605–674.

[15] J. SHALIKA, The multiplicity one theorem for $GL(n)$, *Ann. Math.* **100** (1974), 171–193.

[16] E. STADE, Mellin transforms of $GL(n, \mathbb{R})$ Whittaker functions, *Amer. J. Math.* **123** (2001), 121–161.

[17] ———, Archimedean $L$-factors on $GL(n) \times GL(n)$ and generalized Barnes integrals, *Israel J. Math.* **127** (2002), 201–219.

CHAPTER 11

# ON SOME GEOMETRIC CONSTRUCTIONS RELATED TO THETA CHARACTERISTICS

By BENEDICT H. GROSS and JOE HARRIS

---

*To J. Shalika, on his 60th birthday*

**0. Introduction.** The theory of quadratic forms over the field of 2 elements has many mathematical applications, from finite group theory to algebraic topology. Here we pursue a connection discovered by Mumford, relating theta characteristics on an algebraic curve to quadratic forms on the vector space of 2-torsion points in its Jacobian.

We develop the algebraic and combinatorial aspects of quadratic forms in the first three sections, then review some of the theory of theta characteristics in section 4. The last three sections use this theory to investigate some classical geometric constructions on curves of genus 2 and 3.

Some of the material in sections 2 and 3 appears in the 19th century literature (cf. for example [C1], [C2], and [W]), and has been abstracted in expository articles (cf. [Sa]). Similarly, versions of the geometric constructions in sections 5–7 have appeared in several excellent modern expositions (cf. [G–H] and [D–O]).

*Acknowledgments.* We wish to thank Igor Dolgachev, who guided us to much of the existing literature.

**1. Quadratic forms.** Throughout this paper, $k = \mathbb{Z}/2\mathbb{Z}$ is the field with 2 elements. Let $V$ be a vector space of dimension $2g$ over $k$. We fix a nondegenerate, strictly alternating from $\langle , \rangle : V \otimes V \to k$. Thus $\langle v, v \rangle = 0$ for all $v \in V$, and the map $v \mapsto f_v(u) = \langle u, v \rangle$ gives an isomorphism from $V$ to its dual space $\mathrm{Hom}(V, k)$.

The symplectic space $(V, \langle , \rangle)$ is uniquely determined up to isomorphism by its dimension $2g$. Let $Sp(V)$ be the group of all $k$-linear isomorphisms $T : V \to V$ which satisfy $\langle Tv, Tu \rangle = \langle v, u \rangle$ for all $v, u \in V$. The group $Sp(V)$ is generated by the transvections:

$$T_u(v) = v + \langle v, u \rangle u \tag{1.1}$$

where $u \neq 0$ in $V$, and these form a single conjugacy class of involutions in $Sp(V)$. The finite group $Sp(V)$ has order $2^{g^2}(2^{2g} - 1)(2^{2g-2} - 1) \cdots (2^2 - 1)$ ([A]).

A subspace $X \subset V$ is isotropic if $\langle x, x' \rangle = 0$ for all $x, x' \in X$. The maximal isotropic spaces all have dimension $g$, and may be completed to an isotropic decomposition of $V$:

$$V = X \oplus Y \tag{1.2}$$

with $X$ and $Y$ isotropic of dimension $g$. The pairing $\langle , \rangle$ on $V$ puts the subspaces $X$ and $Y$ in duality. The isotropic decompositions (1.2) of $V$ are all conjugate under $Sp(V)$, and the stability subgroup of a fixed decomposition is isomorphic to $GL(X)$ ([A]). If $\langle e_1, \dots, e_g \rangle$ is a basis for $X$ and $\langle f_1, \dots, f_g \rangle$ is the dual basis of $Y$, the vectors $\langle e_1, \dots, e_g; f_1, \dots, f_g \rangle$ give a symplectic basis for $V$.

We say a function $q : V \to k$ is a quadratic form on $V$ (relative to the fixed sympletic form $\langle , \rangle$) provided that

$$q(v + u) + q(v) + q(u) = \langle v, u \rangle \tag{1.3}$$

for all $v, u \in V$. If $V = X \oplus Y$ is any isotropic decomposition, the function

$$q_0(x + y) = \langle x, y \rangle \tag{1.4}$$

defines a quadratic form on $V$. In terms of a symplectic basis:

$$q_0 \left( \sum_{i=1}^{g} \alpha_i e_i + \sum_{i=1}^{g} \beta_i f_i \right) = \sum_{i=1}^{g} \alpha_i \beta_i . \tag{1.5}$$

Let $QV$ denote the set of all quadratic forms on $V$, relative to $\langle , \rangle$. Then $QV$ is a principal homogeneous space for $V$: if $q \in QV$ and $v \in V$ we define the quadratic form $q + v$ by

$$q + v(u) = q(u) + \langle v, u \rangle . \tag{1.6}$$

Similarly, if $q$ and $q'$ are two elements of $QV$ there is a unique vector $v = q + q'$ such that

$$\langle v, u \rangle = q(u) + q'(u). \tag{1.7}$$

This gives the disjoint union $W = V \cup QV$ the structure of a $k$-vector space of dimension $2g + 1$, which contains $V$ as a subspace of codimension 1.

The group $Sp(V)$ acts on the set $QV$ by the formula $q \mapsto Tq$, where

$$Tq(Tv) = q(v). \tag{1.8}$$

This gives a linear action of $Sp(V)$ on $W$, and we have an exact sequence of $Sp(V)$-modules

$$0 \to V \to W \to k \to 0. \tag{1.9}$$

We define the Arf invariant $a : QV \to k$ as follows. Let $\langle e_1, \dots, e_g; f_1, \dots, f_g \rangle$ be a symplectic basis of $V$. For $q \in QV$, let

$$a(q) = \sum q(e_i) q(f_i). \tag{1.10}$$

*A priori*, this depends on the symplectic basis chosen, but we have the following:

PROPOSITION 1.11. *The Arf invariant $a(q)$ does not depend on the symplectic basis chosen to define it. If $q$ is defined by an isotropic decomposition, as in (1.4), then $a(q) = 0$.*

*We have the formulas*

$$a(Tq) = a(q) \qquad T \in Sp(V)$$

$$a(q+v) = a(q) + q(v) \qquad v \in V.$$

*The group $Sp(V)$ has 2 orbits on $QV$, the $2^{g-1}(2^g+1)$ even forms $q$ with $a(q) = 0$, and the $2^{g-1}(2^g-1)$ odd forms $q$ with $a(q) = 1$.*

*Proof.* Since $Sp(V)$ acts transitively on the collection of symplectic bases, and is generated by the transvections $T_u$, it suffices to check that

$$\sum q(e_i)q(f_i) = \sum q(T_u e_i)q(T_u f_i).$$

This is an amusing exercise, which uses the identity $\alpha^2 = \alpha$ in $k$.

If $q = q_0$ is defined using an isotropic decomposition (1.4), then $q = 0$ on the subspaces $X$ and $Y$. Hence $a(q) = 0$.

The formula $a(Tq) = a(q)$ follows from the independence of basis. To prove that $a(q+v) = a(q) + q(v)$, extend $v = e_1$ to a symplectic basis of $V$ and use (1.10) to calculate $a(q+v)$.

One shows, by induction on $g$, that the form $q_0$ defined by (1.4)–(1.5) has $2^{g-1}(2^g+1)$ zeroes on $V$. Since $a(q_0) = 0$, we have $a(q_0+v) = q_0(v)$. Hence there are $2^{g-1}(2^g+1)$ forms $q = q_0 + v$ with $a(q) = 0$, and $2^{g-1}(2^g-1)$ forms $q$ with $a(q) = 1$.

Now fix $q$, and consider the action of the involutions $T_u \in Sp(V)$. If $q(u) = 1$, we find that $T_u q = q$. If $q(u) = 0$, then $T_u q = q + u$. It follows that the group $Sp(V)$ acts transitively on the set of forms with either Arf invariant. □

COROLLARY 1.12. *For $q \in QV$, the following conditions are all equivalent*

1. *$a(q) = 0$.*
2. *$q$ has $2^{g-1}(2^g+1)$ zeroes on $V$.*
3. *There is an isotropic decomposition $V = X \oplus Y$ such that $q = 0$ on the subspaces $X$ and $Y$.*

COROLLARY 1.13. *The stabilizer $O(V,q) \subset Sp(V)$ of a form $q \in QV$ has order*

$$2^{g^2-g+1}(2^{2g-2}-1)(2^{2g-4}-1)\cdots(2^2-1)(2^g-1) \qquad \text{if } a(q) = 0$$
$$2^{g^2-g+1}(2^{2g-2}-1)(2^{2g-4}-1)\cdots(2^2-1)(2^g+1) \qquad \text{if } a(q) = 1.$$

*The transvection $T_u$ lies in $O(V,q)$ if $q(u) = 1$.*

COROLLARY 1.14. *If $g \geq 2$, the sequence (1.9) of $Sp(V)$-modules $0 \to V \to W \to k \to 0$ is **not** split.*

**2. Aronhold sets.** Recall that $W = V \cup QV$ is a $k$-vector space of dimension $2g+1$ over $k$. Let $S = \{q_1, q_2, \dots, q_{2g+1}\}$ be a set of linearly independent vectors, all of which lie in the coset $QV$. Relative to this basis, any vector $w \in W$ has a unique expression $w = \Sigma \alpha_i q_i$ with $\alpha_i = 0, 1$ in $\mathbb{Z}$. We define $\#w = \Sigma \alpha_i$, so $0 \leq \#w \leq 2g+1$. If $w$ lies in the coset $QV$, then $\#w$ is odd.

We say $S$ is an Aronhold set provided that the Arf invariant of any element $q = \Sigma \alpha_i q_i$ in $QV$ depends only on the residue class of the odd integer $\#q$ (mod 4). If $S$ is an Aronhold set, we must have $a(q_1) = a(q_2) = \cdots = a(q_{2g+1})$, as these are the forms $q$ with $\#q = 1$. Also, there is a unique form $q_S = \Sigma q_i$ with $\#q_S = 2g+1$, and this form must satisfy $a(q_S) \equiv a(q_i) + g \pmod 2$.

PROPOSITION 2.1. *There exist Aronhold sets* $S = \{q_1, \dots, q_{2g+1}\}$ *with*

$$a(q_i) = \begin{cases} 0 & g \equiv 0, 1 \pmod 4 \\ 1 & g \equiv 2, 3 \pmod 4 \end{cases}.$$

*The group $Sp(V)$ acts transitively on the collection of Aronhold sets in $W$, and the stabilizer of $S$ is the full symmetric group*

$$\mathrm{Sym}(S) \hookrightarrow O(V, q_S) \hookrightarrow Sp(V).$$

*Proof.* Define the vector space $N$ of dimension $2g+1$ over $k$, with basis $S = \{n_1, n_2, \dots, n_{2g+1}\}$. Let $M$ be the subspace $\{\Sigma \alpha_i n_i : \Sigma \alpha_2 \equiv 0\}$ of codimension 1. The bilinear form on $N$

$$\left\langle \sum \alpha_i n_i, \sum \beta_i n_i \right\rangle = \sum \alpha_i \beta_i$$

is strictly alternating and nondegenerate on $M$.

For $n = \Sigma \alpha_i n_i$ in $N$, with $\alpha_i = 0, 1$ in $\mathbb{Z}$, define $\#n = \Sigma \alpha_i$. Put

$$a(n_i) = \begin{cases} 0 & g \equiv 0, 1 \pmod 4 \\ 1 & g \equiv 2, 3 \pmod 4 \end{cases}.$$

For $n \in N - M$, $\#n$ is odd and we define

$$a(n) \equiv a(n_i) + \left( \frac{\#n - 1}{2} \right).$$

Clearly $a(n)$ depends only on $\#n$ (mod 4). We now show that, like the Arf invariant, $a(n)$ takes the value zero $2^{g-1}(2^g+1)$ times on $N - M$, and the value one on the remaining $2^{g-1}(2^g-1)$ elements. □

LEMMA 2.2.

$$\sum_{k \equiv 1 \pmod 4} \binom{2g+1}{k} = \begin{cases} 2^{g-1}(2^g+1) & g \equiv 0, 1 \pmod 4 \\ 2^{g-1}(2^g-1) & g \equiv 2, 3 \pmod 4 \end{cases}.$$

*Proof.* We have

$$\sum_{k \text{odd}} \binom{2g+1}{k} = 2^{2g}.$$

If $i^2 = -1$ in $\mathbb{C}$, we also have

$$(1+i)^{2g+1} = (2i)^g(1+i) = 2^g\left(i^g + i^{g+1}\right) = \sum \binom{2g+1}{k} i^k.$$

Taking the coefficient of $i$ gives the identity

$$\sum_{k \equiv 1 \pmod 4} \binom{2g+1}{k} - \sum_{k \equiv 3 \pmod 4} \binom{2g+1}{k} = \begin{cases} 2^g & g \equiv 0, 1 \pmod 4 \\ -2^g & g \equiv 2, 3 \pmod 4 \end{cases}.$$

Adding this to the first identity in the proof gives the desired formula.

For $n \in N - M$, we define the function $f_n : M \to k$ by the formula $f_n(m) = a(n+m) + a(n)$. □

LEMMA 2.3. *The function $f_n$ is a quadratic form on $M$ associated to the symplectic form $\langle , \rangle$. The Arf invariant of $f_n$ is $a(n)$.*

*Proof.* We must show that

$$f_n(m_1 + m_2) + f_n(m_1) + f_n(m_2) = \langle m_1, m_2 \rangle.$$

We first check this for the special case when $n = n_S = \sum n_i$ has $\#n = 2g + 1$. In this case, we write $f$ for the function $f_n$ and observe that we have the simple formula

$$f(m) \equiv \frac{\#m}{2} \pmod 2.$$

Since $\#(m_1 + m_2) = \#m_1 + \#m_2 - 2\#(m_1 \cap m_2)$ with $\#(m_1 \cap m_2) = \sum \alpha_i(m_1) \alpha_i(m_2) \equiv \langle m_1, m_2 \rangle$, we have

$$f(m_1 + m_2) + f(m_1) + f(m_2) = \langle m_1, m_2 \rangle$$

as desired.

In general, $n = n_S + m'$ and $f_n(m) = f(m + m') + f(m')$. We must show that the four-term sum

$$f(m_1 + m_2 + m') + f(m_1 + m') + f(m_2 + m') + f(m')$$

is equal to $\langle m_1, m_2 \rangle$. By the above identity for $f$, this sum is equal to

$$f(m_1 + m_2) + \langle m', m_1 + m_2 \rangle + f(m_1) + \langle m', m_1 \rangle + f(m_2) + \langle m', m_2 \rangle.$$

Since $\langle m', m_1 + m_2\rangle = \langle m', m_1\rangle + \langle m', m_2\rangle$, the sum is equal to $f(m_1+m_2) + f(m_1) + f(m_2) = \langle m_1, m_2\rangle$ as desired.

The Arf invariant of $f_n$ clearly depends only on $\#n \pmod 4$, so it suffices to check that it is correct when $n = n_S$ has $\#n = 2g+1$. An argument similar to Lemma 2.2 shows that the function $f$ has $2^{g-1}(2^g+1)$ zeroes on $M$ when $g \equiv 3$, 4 (mod 4) and $2^{g-1}(2^g-1)$ zeroes on $M$ when $g \equiv 1, 2 \pmod 4$. Hence the Arf invariant of the form $f$ is equal to $a(n_S) \equiv a(n_i) + g \pmod 2$.

We now complete the proof of Proposition 2.1. Since $M$ and $V$ are both nondegenerate symplectic spaces of dimension $2g$, there is a linear isomorphism $T : M \to V$ which satisfies $\langle Tm_1, Tm_2\rangle = \langle m_1, m_2\rangle$ for $m_1, m_2 \in M$. Via $T$, we may identify the elements $n$ of $N - M$ with quadratic forms $q = T(n)$ on $V : q(Tm) = f_n(m)$. The fact that $q$ is a quadratic form, with Arf invariant $a(n) \equiv a(n_i) + (\frac{\#n-1}{2})$, follows from Lemma 2.3.

The induced map $T : N \to W$ is a linear isomorphism, and the images $q_i = T(n_i)$ give an Aronhold set $S$ in $W$. The transitivity of $Sp(V)$ on Aronhold sets, and the stability subgroup follow similarly. This completes the proof of Proposition 2.1. □

We give some examples of Aronhold sets $S$ for small values of $g$. When $g = 1$, there are 3 even forms $q$ on $V$. The group $Sp(V)$ is isomorphic to $S_3$ via its permutation representation on the set $S = \{q_1, q_2, q_3\}$ of even forms. This is the unique Aronhold set in $W$, and $q_S = q_1 + q_2 + q_3$ is the unique odd form on $V$.

When $g = 2$ there are 6 odd forms $q$ on $V$, and the group $Sp(V)$ is isomorphic to $S_6$ via its permutation representation on the set of odd forms. The Aronhold sets $S = \{q_1, q_2, \ldots, q_5\}$ are the 5-subsets of the set of odd forms, and $q_S$ is the unique odd form not in $S$. The group $O(V, q_S)$ is isomorphic to the symmetric group $\mathrm{Sym}(S) = S_5$.

When $g = 3$ there are 28 odd forms $q$ on $V$. An Aronhold set $S = \{q_1, q_2, \ldots, q_7\}$ consists of 7 odd forms which give a basis for $W = V \cup QV$, such that

$$q_i + q_j + q_k \quad \text{is even} \quad i \neq j \neq k$$

$$q_i + q_j + q_k + q_\ell + q_m \quad \text{is odd} \quad i \neq j \neq k \neq \ell \neq m$$

$$q_S = \sum_{i=1}^{7} q_i \quad \text{is even.}$$

The following simpler criterion is often useful.

Proposition 2.4. *Let $S = \{q_1, q_2, \ldots, q_7\}$ be a set of seven distinct odd forms on $V$, such that $q_i + q_j + q_k$ is even for all $i \neq j \neq k$. Then $S$ is an Aronhold set in $W$.*

*Proof.* The hypothesis implies that the 21 vectors $v_{ij} = q_i + q_j$ are nonzero and distinct in $V$. For if $v_{ij} = v_{k,\ell}$, then $q_i + q_j + q_k + q_\ell = 0$ in $W$, and the form $q_i + q_j + q_k = q_\ell$ would be odd.

If $k \neq i, j$ then $q_k(v_{ij}) = a(q_k) + a(q_i + q_j + q_k) = 1$. On the other hand, $q_i(v_{ij}) = q_j(v_{ij}) = a(q_i) + a(q_j) = 0$. Hence $q_S = \sum_{i=1}^{7} q_i$ takes the value 1 on the 21 vectors $v_{ij}$, and is not equal to any of the $q_i$. It follows that the vectors $\{q_i\}$ are linearly independent in $W$, so give a basis.

The forms $q_i + q_j + q_k$ with $i \neq j \neq k$ are hence all distinct, and give $\binom{7}{3} = 35$ of the 36 even forms on $V$. It therefore suffices to prove that $q_S$ is even.

Of the 21 forms $q_i + q_j + q_k + q_\ell + q_m$ with $i \neq j \neq k \neq \ell \neq m$, at least 20 must be odd. Assume $r = q_1 + q_2 + q_3 + q_4 + q_5$ is odd. Since $q_S = r + v_{67}$, $a(q_S) = a(r) + q_S(v_{67}) = 1 + 1 = 0$ as desired. □

COROLLARY 2.5. *The 21 vectors $v_{ij} = q_i + q_j$ in $V$ determine the Aronhold set $S = \{q_i\}$ in $W$.*

*Proof.* Indeed, there is a *unique* even form $q$ $(= q_S)$ which takes the value 1 on the $v_{ij}$. Let $\{v_1, \ldots, v_7\}$ be the remaining vectors where $q(v_i) = 1$, and let $q_i = q + v_i$. These are the 7 forms of $S$.

The group $S_7 = \mathrm{Sym}(S)$ has index eight in the group $O(V, q_S)$, so for each even form $q$ there are precisely 8 Aronhold sets $S$ which satisfy $qS = q$. (This gives $288 = 36.8$ Aronhold sets $S$ in all.) The action of $O(V, q_S)$ on these 8 sets $S$ gives an isomorphism $O(V, q_S) \simeq S_8$. Finally, each odd form $q_1$ appears in precisely two Aronhold sets $S$ with $q_S = q$, so in 72 Aronhold sets in all.

If $\langle e_1, e_2, e_3; f_1, f_2, f_3 \rangle$ is a symplectic basis for $V$ such that the even form $q$ is given by

$$q\left(\sum \alpha_i e_i + \sum \beta_i f_i\right) = \sum \alpha_i \beta_i$$

then an explicit Aronhold set $S$ with $q_S = q$ is given by

$$q_1 = q + e_1 + f_1 + f_2$$
$$q_2 = q + e_2 + f_2$$
$$q_3 = q + e_3 + f_1 + f_2 + f_3$$
$$q_4 = q + e_1 + e_2 + f_1 + f_3$$
$$q_5 = q + e_1 + e_3 + f_2 + f_3$$
$$q_6 = q + e_2 + e_3 + f_3$$
$$q_7 = q + e_1 + e_2 + e_3 + f_1.$$

□

*Note.* We have chosen the name Aronhold set for $S$ in view of Aronhold's work on the 28 bitangents to a smooth plane quartic (the notion of an Aronhold set appears in Coble's book [C2, §26] as a "normal fundamental set"). The relationship between this work and the theory of quadratic forms on $V$ when $g = 3$ will be explained in §4.

We define an *Aronhold basis* of $W$ to be an *ordered* Aronhold set $\langle q_1, q_2, \dots, q_{2g+1}\rangle$ in $QV$. The group $Sp(V)$ acts simply-transitively on Aronhold bases of $W$, and on symplectic bases $\langle e_1, \dots, e_g; f_1, \dots, f_g\rangle$ of $V$. We may identify these principal homogeneous spaces for $Sp(V)$ by associating to each Aronhold basis the symplectic basis:

$$\begin{aligned} e_1 &= q_1 + q_2 & f_1 &= q_1 + q_{2g+1} \\ e_2 &= q_3 + q_4 & f_2 &= q_1 + q_2 + q_3 + q_{2g+1} \\ &\ \vdots & &\ \vdots \\ e_g &= q_{2g-1} + q_{2g} & f_g &= q_1 + q_2 + \cdots + q_{2g-1} + q_{2g+1} \end{aligned}$$

**3. A bipartite graph.** Many of the combinatorial questions involving quadratic forms on $V$ can be studied by a consideration of a certain bipartite graph $\Gamma$. The two sets of vertices of $\Gamma$ correspond respectively to the elements of $V$ and $QV$. The two vertices $v$ and $q$ are joined by an edge if $a(q + v) = 1$.

The group $Sp(V) \propto W$ acts as an automorphism of $\Gamma$. The subgroup $W$ permutes the vertices simply-transitively, and $Sp(V)$ preserves the vertex $v = 0$. The subgroup $O(V, q)$ of $Sp(V)$ preserves the two vertices $v = 0$ and $q$.

Let $q$ be a fixed *even* quadratic form, and let $\sigma = \sigma_q$ be the involution of $\Gamma$ given by translation by $q$. Since $w$ is not connected to $w + q$, $\sigma$ fixes no edge of $\Gamma$, and the quotient graph $\Delta = \Gamma/\langle\sigma\rangle$ has no loops.

The combinatorial graph $\Delta$ has $2^{2g}$ vertices, indexed by $v \in V$. The vertices $v$ and $u$ are connected by an edge if $a(q + v + u) = 1$. Therefore, $\Delta$ is regular with valency

$$2^{g-1}(2^g - 1) = \#\{v : q(v) = 1\}.$$

The group $O(V, q) \propto V$ acts as automorphism of the quotient graph $\Delta$. The subgroup $V$ permutes the vertices simply-transitively, and $O(V, q)$ preserves the vertex $v = 0$. It acts on the complete subgraph $\Delta(0)$, which is, by definition, the induced graph on the star of $v = 0$ in $\Delta$.

We give some examples of $\Delta$ and $\Delta(0)$ for small $g$. When $g = 1$, $\Delta$ is the graph

where $v$ is the unique vector in $V$ with $q(v) = 1$. In this case, $O(V, q)$ is the group $S_2$ of order 2. In this case, $O(V, q)$ is the group $S_2$ of order 2.

When $g = 2$, $\Delta$ has 16 vertices and 48 edges. There is a unique partition $\{v_1, v_2, v_3\}$, $\{v_4, v_5, v_6\}$ of the six vectors $v$ with $q(v) = 1$ into two 3-element subsets, such that each element in the first 3-subset is orthogonal to each element in the second 3-subset. Indeed, if $\langle e_1, e_2; f_1, f_2\rangle$ is a symplectic basis for $V$ and $q(\alpha_1 e_1 + \alpha_2 e_2 + \beta_1 f_1 + \beta_2 f_2) = \alpha_1\beta_1 + \alpha_2\beta_2$, we have the partition

$$\begin{cases} v_1 = e_1 + f_1 \\ v_2 = e_2 + f_2 + e_1 \\ v_3 = e_2 + f_2 + f_1 \end{cases}$$

$$\begin{cases} v_4 = e_2 + f_2 \\ v_5 = e_1 + f_1 + e_2 \\ v_6 = e_1 + f_1 + f_2. \end{cases}$$

The subgraph $\Delta(0)$ is given by

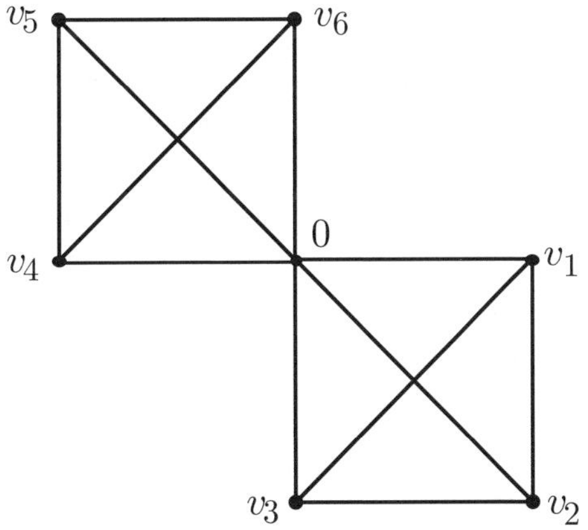


and the group $O(V, q)$ is isomorphic to the subgroup $(S_3 \times S_3) \cdot 2$ of $S_6 = Sp(V)$.

When $g = 3$ the graph $\Delta$ has $64 = 2^6$ vertices and $896 = 2^7 \cdot 7$ edges. There are 28 vectors $v$ with $q(v) = 1$. Let $S = \{q_1, q_2, \dots, q_7\}$ be an Aronhold set with $q = q_S = \Sigma q_i$, and write $q_i = q + v_i$. Since $q(v_i) = a(q_i) = 1$, the seven vertices $v_i$ lie in $\Delta(0)$. Since $q(v_i + v_j) = a(\Sigma_{k\neq i,j} q_k) = 1$, the vertices $v_i$ and $v_j$ are connected by an edge in $\Delta(0)$. Hence $\Delta(0)$ contains the complete graph $\Delta_S$ on the eight vertices $\{0, v_1, v_0, \dots, v_7\}$:

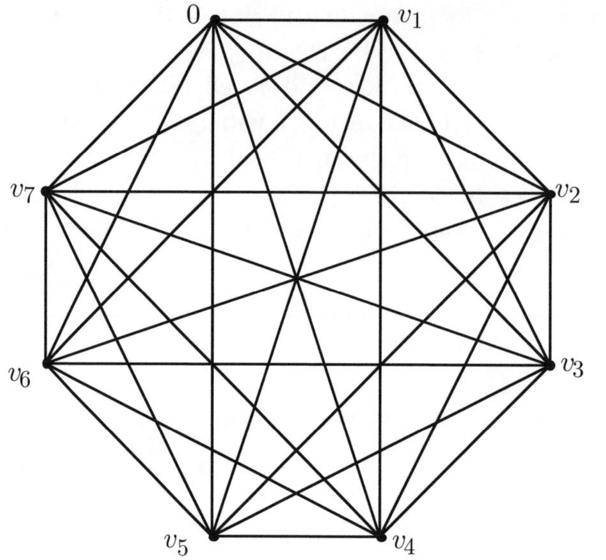


There are 8 choices for $S$ with $q_S = q$. If $S' \neq S$, the complete graphs $\Delta_S$ and $\Delta_{S'}$ meet along a single edge with vertex $v = 0$ in $\Delta(0)$. Of the 28 vertices $v \neq 0$ in $\Delta(0)$, each is connected to 12 vertices $w \neq 0$. The group $O(V, q) = S_8$ acts on $\Delta(0)$; the subgroup $S_7$ stabilizing $\Delta_S$ permutes the 7 vertices $\{v_1, v_2, \ldots, v_7\}$, as well as the 7 remaining complete graphs $\Delta_{S'}$.

**4. Theta characteristics.** Following Mumford [M], we recall how the theory of theta characteristics is linked to the theory of quadratic forms over $\mathbb{Z}/2\mathbb{Z}$.

Let $C$ be a complete, nonsingular algebraic curve of genus $g$, defined over an algebraically closed field of characteristic $\neq 2$. Let $\mathrm{Pic}(C)$ be the group of divisor classes on $C$, and $\mathrm{Pic}^n(C)$ the divisor classes of degree $n$. If $d \in \mathrm{Pic}(C)$, we write $\mathcal{L}(d)$ for the corresponding line bundle on $C$, and $h^0(d)$ for the dimension of the space of sections $H^0(C, \mathcal{L}(d))$.

Let

$$V = \{v \in \mathrm{Pic}^0(C) : 2v = 0\} \tag{4.1}$$

be the classes killed by 2. This is a vector space of dimension $2g$ over $k = \mathbb{Z}/2\mathbb{Z}$. It has a nondegenerate symplectic form defined by the Weil pairing. If $d$ and $e$ are divisors with disjoint support in the classes of $v$ and $u$, and $2d = \mathrm{div}(f)$, $2e = \mathrm{div}(g)$, then

$$(-1)^{\langle v,u\rangle} = f(e)/g(d).$$

Let $\kappa$ be the canonical class in $\mathrm{Pic}^{2g-2}(C)$. The set of theta characteristics

$$QV = \{q \in \mathrm{Pic}^{g-1}(C) : 2q = \kappa\} \tag{4.2}$$

is a principal homogeneous space for $V$ which can be identified with the set of quadratic forms on $V$. The form $q$ is given by the formula

(4.3) $$q(v) \equiv h^0(q+v) + h^0(q) \pmod 2.$$

The Arf invariant of $q$ is given by

(4.4) $$a(q) \equiv h^0(q) \pmod 2.$$

The vector space $W = V \cup QV$ appears as a subgroup of order $2^{2g+1}$ in $\mathrm{Pic}(C)/\mathbb{Z}\kappa$, when $g \neq 1$.

We consider some examples in low genus. When $g = 0$, there is a single (even) characteristic $q$. In this case $C \simeq \mathbb{P}^1$, $\mathcal{L}(q) \simeq \mathcal{O}(-1)$, and $h^0(q) = 0$.

When $g = 1$, $C$ is an elliptic curve and $\kappa = 0$. In this case, $QV = V$. There are three even characteristics $q = v \neq 0$ with $h^0(q) = 0$, and one odd characteristic with $h^0(q) = 1$.

When $g = 2$, the canonical series $|\kappa|$ gives a two-fold covering $C \to \mathbb{P}^1$. Let $\tau$ be the hyperelliptic involution of $C$ and let $\{p_1, p_2, \ldots, p_6\}$ be the fixed points of $\tau$. The six odd characteristics $q$ correspond to the line bundles $\mathcal{L}(p_i)$ and satisfy $h^0(q) = 1$. The 10 even characteristics $q$ correspond to the line bundles $\mathcal{L}(p_i + p_j - p_k)$, where $\{p_i, p_j, p_k\}$ is a 3-subset of the 6 fixed points, well-defined up to complementation. They satisfy $h^0(q) = 0$.

When $g = 3$, the canonical series $|\kappa|$ gives a morphism $\pi : C \to \mathbb{P}^2$. If $C$ is hyperelliptic, this map is 2-to-1 onto a smooth conic $D \subset \mathbb{P}^2$. In this case, there is a distinguished even theta characteristic $q_C$ with $h^0(q_C) = 2$; we have $\mathcal{L}(q_C) = \pi^*\mathcal{O}_D(1)$. Let $\tau$ be the hyperelliptic involution of $C$, and let $\{p_1, p_2, \ldots, p_8\}$ be its 8 fixed points. The 28 odd theta characteristics $q$ correspond to the line bundles $\mathcal{L}(p_i + p_j)$, and satisfy $h^0(q) = 1$.

If $C$ is not hyperelliptic, the canonical series embeds $C$ as a smooth quartic in $\mathbb{P}^2$. The 28 odd theta characteristics $q$ correspond to the 28 bitangent lines to this quartic; the corresponding line bundles are $\mathcal{L}(p + r)$ where $p$ and $r$ are the two points of double tangency.

Classically, an Aronhold set $S$ consisted of 7 bitangents $\{(p_1, r_1), \ldots, (p_7, r_7)\}$ on $C \subseteq \mathbb{P}^2$, with the property that no 6 points of the form $\{p_i, r_i, p_j, r_j, p_k, r_k\}$ $i \neq j \neq k$ were on the intersection of $C$ with a conic. Equivalently, $S$ is a collection of 7 odd theta characteristics $\{q_1, \ldots, q_7\}$ such that $q_i + q_j + q_k - \kappa$ is an even characteristic, for $i \neq j \neq k$. The remaining 21 odd characteristics have the form $q_i + q_j + q_k + q_\ell + q_m - 2_\kappa$, so the remaining 21 bitangents to $C$ can be obtained as the residual intersection of $C$ with a cubic passing through 10 points of the form $\{p_i, r_i, p_j, r_j, p_k, r_k, p_\ell, r_\ell, p_m, r_m\}$.

The Aronhold set $S = \{q_1, \ldots, q_7\}$ determines an even characteristic $q_s = q_1 + q_2 + \cdots + q_7 - 3_\kappa$. Since $C$ is not hyperelliptic, $h^0(q_S) = 0$. The linear series $|\kappa + q_S|$ embeds $C$ as a nonsingular sextic curve in $\mathbb{P}^3$, and Hesse showed that the 28 lines $\overline{pr}$ in $\mathbb{P}^3$ given by the odd theta characteristics meet in 8 distinct points $Y = \{y_1, y_2, \ldots, y_8\}$. The scheme $Y$ is the base locus of a net of quadrics $\{Q_\lambda\}_{\lambda \in \mathbb{P}^2}$

in $\mathbb{P}^3$, whose discriminant locus defines the smooth quartic curve $C$ in $\mathbb{P}^2$:

$$C = \{\lambda \in \mathbb{P}^2 : \det Q_\lambda = 0\}.$$

The lines $\overline{pr}$ in $\mathbb{P}^3$ give the complete graph on the 8 points of $Y = \cap_\lambda Q_\lambda$, and the Aronhold set $S$ is a complete fan from one of the vertices. We will recover and extend these results in §6.

**5. Curves of genus 2.** Let $C$ be a nonsingular, complete curve of genus 2 defined over an algebraically closed field of characteristic $\neq 2$. Let $A = \mathrm{Pic}^0(C)$ be the Jacobian of $C$, and let $\kappa$ be the canonical class in $\mathrm{Pic}^2(C)$. We have the symplectic space $V = \{v \in A : 2v = 0\}$ of dimension 4 over $\mathbb{Z}/2\mathbb{Z}$, and identify $QV$ with the set of 16 theta characteristics $\{q \in \mathrm{Pic}^1(C) : 2q = \kappa\}$ as in §4.

In this section, we will show how $C$ gives rise to a $K3$-surface $X$, together with a very ample line bundle $\mathcal{L}$ of degree 8 on $X$. We will also see how to associate to each even theta characteristic $q$ on $C$ an Enriques involution $\sigma = \sigma_q$ of $X$ fixing $\mathcal{L}$, and will study a configuration of 16 rational curves on the quotient Enriques surface $Y = X/\langle\sigma_q\rangle$, whose dual graph is the combinatorial graph $\Delta$ described in §3. The subgraph $\Delta(0)$ will be used to produce an elliptic fibration $f : Y \to \mathbb{P}^1$ with two double fibres, each of Kodaira type $I_3$.

Our treatment follows [GH, Ch. 6]. Let $\tau$ be the hyperelliptic involution of $C$, which induces the automorphism $-1$ on $A$. Besides the 16 points of $V$ on $A$, which are fixed by $-1$, we have the 16 curves $C_q$ of genus 2 corresponding to classes in $QV$.

$$C_q = \{(p) - q : p \in C\}. \tag{5.1}$$

The involution $-1$ of $A$ fixes $C_q$, and induces its hyperelliptic involution. If $q \neq q'$, then $C_q$ and $C_{q'}$ meet in 2 points of $V$, and the point $v \in V$ lies on the curve $C_q$ if $a(q + v) = 1$.

The class of the divisor $E = 2 \cdot (C_q)$ in $\mathrm{Pic}(A)$ is independent of the choice of $q \in QV$; this is the linear series usually denoted $|2\Theta|$. The line bundle $\mathcal{L}(E)$ is ample and satisfies $\mathcal{L}(E)^2 = 8$, $h^0(\mathcal{L}(E)) = 4$. The sections of $\mathcal{L}(E)$ are all fixed by $-1$ and give a projective embedding of the Kummer surface $K = A/\langle -1\rangle$.

$$K \hookrightarrow \mathbb{P}^3 \tag{5.2}$$

where the image is a hypersurface of degree 4. This quartic has 16 ordinary double points, at the images of elements in $V$. The curves $C_q$ on $A$ give 16 rational curves $C_q/\langle\tau_q\rangle$ on $K$, which map to conics in the 16 reducible hyperplane sections.

Let $X$ be the blow-up of $K$ at the 16 double points $V$. This is abstractly a $K3$ surface, with 32 obvious rational curves

$$\begin{array}{c} \text{the 16 proper transforms } D_q \text{ of the curves } C_q/\langle\tau_q\rangle \\ \text{the 16 exceptional divisors } D_V. \end{array} \tag{5.3}$$

These all satisfy $D^2 = -2$, and the dual graph of the configuration is the bipartite graph $\Gamma$:

$$D_v \cdot D_q = \begin{cases} 1 & \text{if } a(q+v) = 1 \\ 0 & \text{if } a(q+v) = 0. \end{cases} \tag{5.4}$$

Next, we show that the finite group $W = V \cup QV \cong (\mathbb{Z}/2\mathbb{Z})^5$ acts as automorphisms of $X$ and permutes the 32 curves $D_w$ simply-transitively. The involutions $\sigma_v$ associated to points $v \neq 0$ in $V$ are induced by the translations $a \mapsto a + v$ on $A$. They have 8 fixed points on $X$—the images of the 16 points $a$ of order 4 on $A$ which satisfy $2a = v$.

The involutions $\sigma_q$ associated to forms $q \in QV$ are more subtle to define, as they act only on $X$ (not on $A$ or $K$). We specify $\sigma_q$ by insisting that it permutes the curves $D_w$ in the obvious manner: $\sigma_q(D_w) = D_{w+q}$. This gives an involution of the subgroup of $NS(X)$ spanned by the curves $D_w$, which is free of rank 17, and extends uniquely to an involution of the Hodge structure on $H^2(X)$, acting trivially on the image of $H^2(A)$. By the Torelli theorem for K3 surfaces, $\sigma_q$ arises from a unique automorphism of $X$.

Let $B$ be the blow-up of $A$ at the 16 points $v \in V$. Then $-1$ lifts to an involution of $B$, and $B/\langle -1 \rangle = X$. The branch divisor of the 2-fold cover $B \to X$ is equal to $\Sigma_v(D_v)$, so this class is divisible by 2 in $\mathrm{Pic}(X) = NS(X)$. Since $\mathrm{Pic}(X)$ is torsion-free, the class $\frac{1}{2}\Sigma(D_v)$ is well-defined in $\mathrm{Pic}(X)$.

For $q \in QV$, define the class

$$H_q = 4(D_q) + 2 \sum_{a(q+v)=1} (D_v) - \frac{1}{2} \sum_v (D_v) \tag{5.5}$$

in $\mathrm{Pic}(X)$. Then

$$H_q \cdot D_w = 1 \qquad w \in W = V \cup QV. \tag{5.6}$$

Hence $H = H_q$ is independent of the choice of $q$, and fixed under the action of $W$ on $\mathrm{Pic}(X)$. The associated line bundle $\mathcal{L} = \mathcal{L}(H)$ satisfies

$$\mathcal{L} \cdot \mathcal{L} = 8, \qquad h^0(\mathcal{L}) = 6. \tag{5.7}$$

This is the very ample class in $\mathrm{Pic}(X)$ determined by $C$.

The sections of $\mathcal{L}$ give a projective embedding

$$X \hookrightarrow \mathbb{P}^5 \tag{5.8}$$

where the image has degree 8. The 32 rational curves $D_w$ are mapped to lines in $\mathbb{P}^5$, which lie on the 80 reducible hyperplane sections (each of which contains 4 lines $D_v$ and 4 lines $D_q$). [GH, Ch. 6].

To understand the projective representation of $W$ on $\mathbb{P}^5 = \mathbb{P}(H^0(X, \mathcal{L}))$, we introduce a central extension $U$ of $W$ which acts linearly on $H^0(X, \mathcal{L})$. Let $U$ be a 6-dimensional vector space over $k = \mathbb{Z}/2\mathbb{Z}$, with basis $\langle u_1, u_2, \ldots, u_6 \rangle$, and let

$\langle q_1, q_2, \dots, q_6\rangle$ be the odd quadratic forms on $V$. We have an exact sequence of $Sp(V) = S_6$–modules

$$\begin{aligned} 0 \to k \to U \to W \to 0 \\ u_i \mapsto q_i \\ 1 \mapsto \sum u_i. \end{aligned} \tag{5.9}$$

The projective representation of $W$ lifts uniquely to a linear representation of $U$ on $H^0(X, \mathcal{L})$, such that the space of sections decomposes as the direct sum of the six lines $L_i = \{s \in H^0(X, \mathcal{L}) : u_i(s) = -s\}$. Using this decomposition, we obtain eigencoordinates $\langle x_1, x_2, \dots, x_6\rangle$ on $\mathbb{P}^5$ such that $X$ appears as the intersection of 3-diagonal quadrics [GH, 768–769]:

$$\sum \alpha_i x_i^2 = \sum \beta_i x_i^2 = \sum \gamma_i x_i^2 = 0.$$

If $q = q_i$ is odd, $\sigma_q$ fixes a hyperplane section of $X$, which is a canonical curve of genus 5. If $q$ is even, $\sigma_q$ is given by the action of $u_i \cdot u_j \cdot u_k$, where $\{q_i, q_j, q_k\}$ is the 3-subset of the odd forms well determined by $q$ up to complementation. In this case, $\sigma_q$ is fixed-point free on $X$.

We summarize the results obtained so far.

PROPOSITION 5.10. *Associated to a curve $C$ of genus 2 is a $K3$ surface $X$, together with a very ample line bundle $\mathcal{L}$ on $X$ which satisfies $\mathcal{L} \cdot \mathcal{L} = 8$.*

*The surface $X$ contains 32 rational curves $D_w$ corresponding to elements $w \in W$. The dual graph of the configuration of these curves is the bipartite graph $\Gamma$. The group $W$ acts as automorphisms of $X$, fixes the class $\mathcal{L}$ in* Pic$(X)$*, and permutes the curves $D_w$ simply-transitively.*

*The sections of $\mathcal{L}$ give a projective embedding $X \hookrightarrow \mathbb{P}^5$ in which $W$ acts as the group $\langle \pm 1\rangle^6/\langle -1, -1, \dots, -1\rangle$ of sign changes on the 6 coordinates. The surface $X$ is the complete intersection of 3 diagonal quadrics, and the curves $D_w$ are mapped to lines in $\mathbb{P}^5$.*

*If $q$ is an even theta characteristic on $C$, the involution $\sigma_q$ involves 3 sign changes on $\mathbb{P}^5$ and is fixed-point free on $X$.*

Now let $q$ be a fixed even characteristic on $C$, and write $\sigma = \sigma_q$ for the corresponding Enriques involution of $X$. The quotient $Y = X/\langle\sigma\rangle$ is an Enriques surface, with 16 rational curves $D_v$ (the image of either $D_v$ or $D_{v+q}$) indexed by $v \in V$. The group $V$ acts as automorphisms of $Y$ and permutes the curves $D_v$ simply-transitively. The dual graph of the configuration of these curves is the graph $\Delta = \Gamma/\langle\sigma\rangle$ studied in §3.

$$\begin{cases} D_v^2 = 2 \\ D_v \cdot D_u = \begin{cases} 1 & \text{if} \quad a(q+v+u) = 1 \\ 0 & \text{if} \quad a(q+v+u) = 0. \end{cases} \end{cases}$$

Let $T = \{q_1, q_2, q_3\}$ and $T' = \{q_4, q_5, q_6\}$ be the two 3-subsets of odd characteristics determined by $q$. We have $q_1 + q_2 + q_3 = q_4 + q_5 + q_6 = q$ in $W$. The embedding $X \hookrightarrow \mathbb{P}^5 = \mathbb{P}(V)$ gives rise to two coverings:

$$\pi_T : Y \to \mathbb{P}^2 = \mathbb{P}(V_{123})$$
$$\pi_T{}' : Y \to \mathbb{P}^2 = \mathbb{P}(V_{456}).$$

Each has degree 4, and the line bundle $\pi_T^* \mathcal{O}(1) \otimes \pi_{T'}^* \mathcal{O}(1)^{-1}$ is isomorphic to the canonical bundle $\Omega_Y^2$ of $Y$.

Write $q_i = q + v_i$, and consider the two divisors

$$E = D_{v_1} + D_{v_2} + D_{v_3}$$
$$E' = D_{v_4} + D_{v_5} + D_{v_6}$$

on $Y$. By our picture of $\Delta(0)$ in §3, these give disjoint configurations of $-2$ curves on $Y$:

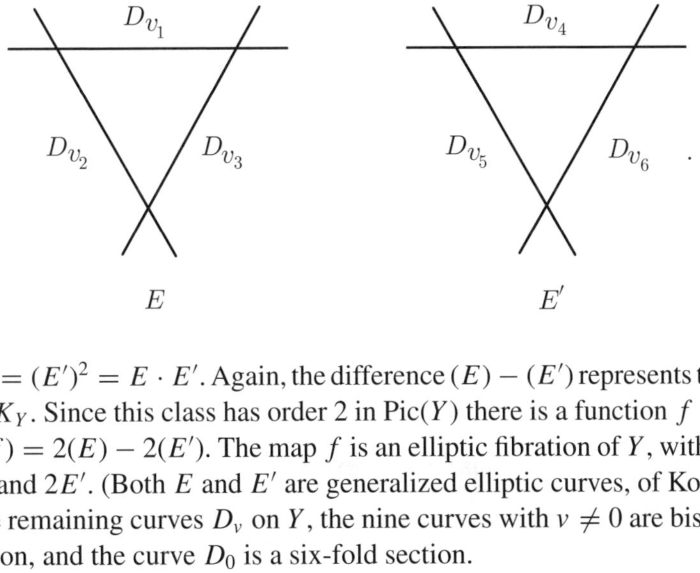


Hence $E^2 = (E')^2 = E \cdot E'$. Again, the difference $(E) - (E')$ represents the canonical class $K_Y$. Since this class has order 2 in Pic$(Y)$ there is a function $f : Y \to \mathbb{P}^1$ with div$(f) = 2(E) - 2(E')$. The map $f$ is an elliptic fibration of $Y$, with 2 double fibres $2E$ and $2E'$. (Both $E$ and $E'$ are generalized elliptic curves, of Kodaira type $I_3$.) Of the remaining curves $D_v$ on $Y$, the nine curves with $v \neq 0$ are bisections of this fibration, and the curve $D_0$ is a six-fold section.

**6. Curves of genus 3.** In this section $C$ is a nonsingular, complete curve of genus 3 over an algebraically closed field of characteristic $\neq 2$, which is not hyperelliptic. Let $q$ be an even theta characteristic on $C$, and let $K = 2q$ be the canonical class. We study the sesquicanonical embedding of $C$ into $\mathbb{P}^3$ given by the very ample linear series $|3q| = |K + q|$ of degree 6.

To understand the geometry of the image $B$ of $C$ in $\mathbb{P}^3$, which is a sextic space curve, we have to arrive at it from another point of view, that of symmetric determinantal representations of the plane quartic.

To start with, suppose we are given a *net of quadrics* in $\mathbb{P}^3$, which we define to be an inclusion

$$\varphi : V \hookrightarrow \mathrm{Sym}^2 W^*$$

of a three-dimensional vector space $V$ into the space of symmetric bilinear forms on a four-dimensional vector space $W$, up to the action of $\mathrm{Aut}(V)$ and $\mathrm{Aut}(W)$ on the left and right. We can immediately associate to such an object the locus

$$C = \{[v] \in \mathbb{P}V : \varphi(v) \text{ is singular}\} \subset \mathbb{P}V \cong \mathbb{P}^2.$$

Given that this is the zero locus of the determinant of a symmetric $4 \times 4$ matrix of linear forms on $\mathbb{P}V$, we expect it to be a quartic curve; we will call the net *typical* if it is a smooth plane quartic.

It is natural to ask whether every plane quartic curve $C$ may be realized in this fashion, and if in turn $C$ determines the net. The answers to these two questions are, respectively, "yes" and "not quite." In fact, we need only a little extra structure beyond the specification of $C$ to specify the net; and once we have said what that is, we will readily prove that we have a one-to-one correspondence between nets and curves with this structure.

Now, suppose our net is typical. It follows then that the rank of $\varphi(v)$ is never less than 3: the variety in $\mathrm{Sym}^2 W^*$ of quadrics of rank 3 or less is singular along the locus of quadrics of rank strictly less than 3. From this we see that we can further associate to our net a line bundle on $C$: the line bundle $M$ whose fiber at any point $p = [v]$ on $C$ is the kernel of $\varphi(v)$, viewed as a map from $W$ to $W^*$. In a similar way, we can associate another line bundle $N$ on $C$, whose fiber at $p$ is the cokernel of $\varphi(v)$.

To be more precise, our net $\varphi$ may be viewed as giving a morphism of locally free sheaves on $\mathbb{P}V$:

$$\psi : \mathcal{O}_{\mathbb{P}V} \otimes W \to \mathcal{O}_{\mathbb{P}V}(1) \otimes W^*.$$

The cokernel of this morphism of sheaves on $\mathbb{P}V$ is supported on $C$; in fact, it is just the line bundle $N$. We thus have an exact sequence of sheaves on $\mathbb{P}^2$:

$$0 \to \mathcal{O}_{\mathbb{P}V} \otimes W \to \mathcal{O}_{\mathbb{P}V}(1) \otimes W^* \to N \to 0. \tag{6.1}$$

When we restrict to $C$—that is, tensor with $\mathcal{O}_C$—this does not remain exact; rather we get a four-term exact sequence

$$0 \to M \to \mathcal{O}_C \otimes W \to \mathcal{O}_C(1) \otimes W^* \to N \to 0. \tag{6.2}$$

From this we see in particular that

$$c_1(N) - c_1(M) = c_1(\mathcal{O}_C(1) \otimes W^*),$$

i.e.,

$$N \otimes M^* \cong \mathcal{O}_C(4).$$

At the same time, the symmetry of the map $\tilde{\varphi}$ gives another relation between $N$ and $M$. Dualizing the four-term sequence of locally-free sheaves on $C$, we get a sequence

$$0 \to N^* \to \mathcal{O}_C(-1) \otimes W \to \mathcal{O}_C \otimes W^* \to M^* \to 0,$$

where the map in the middle is simply the transpose of $\tilde{\varphi}$. But this is simply the map $\tilde{\varphi}$ again, tensored with (the identity map on) $\mathcal{O}_C(-1)$; in other words, tensoring this with $\mathcal{O}_C(1)$ we have the same sequence as before, and thus deduce that

$$M^*(1) \cong N \quad \text{and (equivalently)} \quad N^*(1) \cong M.$$

Combining this with our earlier relation, we see that

$$N \otimes N \cong N \otimes M^*(1) \cong \mathcal{O}_C(5)$$

and correspondingly

$$M^* \otimes M^* \cong \mathcal{O}_C(3),$$

in other words, the line bundle $L = M^*$ is sesquicanonical, and can be written as

$$L \cong K_C \otimes \Theta$$

for some theta-characteristic $\Theta$ on $C$. Moreover, tensoring the exact sequence (1) above with $\mathcal{O}_{\mathbb{P}V}(2)$ we arrive at

$$0 \to \mathcal{O}_{\mathbb{P}V}(-2) \otimes W \to \mathcal{O}_{\mathbb{P}V}(-1) \otimes W^* \to N(-2) \to 0,$$

and we may deduce, given that the left-hand term has no cohomology whatsoever, that

$$h^0(C, \Theta) = h^0(C, N(-2)) = 0,$$

i.e., that $\Theta$ is an even theta-characteristic. We have thus described a map

$$\left\{ \begin{matrix} \text{nets of quadrics in} \\ \mathbb{P}^3 \end{matrix} \right\} \xrightarrow{\alpha} \left\{ \begin{matrix} \text{smooth nonhyperelliptic curves } C \\ \text{of genus 3 with even theta-characteristics } \Theta \end{matrix} \right\}.$$

LEMMA 6.3. *The map $\alpha$ is a bijection.*

*Proof.* We have to exhibit an inverse, that is, associate to a pair $(C, \Theta)$ a net of quadrics.

We will give a relatively concrete approach. We start with an abstract curve $C$ and an even theta-characteristic $\Theta$ on $C$; by way of notation, we will denote by $C$ again its canonical image in $\mathbb{P}V \cong \mathbb{P}^2$, and by $B$ the image of $C$ in $\mathbb{P}W \cong \mathbb{P}^3$ under the map $\varphi = \varphi_L$ associated to the line bundle $L = K_C \otimes \Theta$ on $C$.

Choose a basis $\sigma_1, \ldots, \sigma_4$ for $H^0(C, L)$. For every pair $i$, $j$ the product $\sigma_i \sigma_j$ will be a section of

$$2 \cdot L = 2(K_C + \Theta) = 3 \cdot K_C = \mathcal{O}_C(3),$$

and so will be the restriction to $C$ of a unique cubic polynomial $F_{ij}(X)$ on $\mathbb{P}^2$ (viewed as a section of $\mathcal{O}_{\mathbb{P}^2}(3)$).

Now, consider the (symmetric) $4 \times 4$ matrix

$$\Phi = \begin{pmatrix} F_{1,1} & F_{1,2} & F_{1,3} & F_{1,4} \\ F_{2,1} & F_{2,2} & F_{2,3} & F_{2,4} \\ F_{3,1} & F_{3,2} & F_{3,3} & F_{3,4} \\ F_{4,1} & F_{4,2} & F_{4,3} & F_{4,4} \end{pmatrix}.$$

On $C$, this matrix has rank one—that is, every $2 \times 2$ minor vanishes on $C$; every $3 \times 3$ minor vanishes to order 2 on $C$ and the determinant of $\Phi$ vanishes to order 3. In fact, since the determinant of $\Phi$ is a homogeneous polynomial of degree 12 it follows from the last that it must be (up to scalars) simply the cube of the quartic polynomial $G(X)$ defining $C$.

Now let $\Psi_0$ be the matrix of cofactors of $\Phi$. By what we have just said, every entry of $\Psi$ is divisible by $G(X)^2$; set

$$\Psi = \frac{\Psi_0}{G(X)^2}.$$

$\Psi$ is then a symmetric matrix of linear forms on $\mathbb{P}^2$. Now, since

$$\Phi \cdot \Psi_0 = \det(\Phi) \cdot I,$$

we have

$$\det(\Phi) \cdot \det(\Psi_0) = G(X)^3 \cdot \det(\Psi_0) = G(X)^{12},$$

that is, the determinant of $\Psi_0$ is $G(X)^9$ and hence

$$\det(\Psi) = G(X).$$

We thus arrive at a symmetric $4 \times 4$ matrix $\Psi$ of linear forms on $\mathbb{P}^2$—that is, a net of quadrics in $\mathbb{P}^3$—whose discriminant curve is $C$. To complete the proof of the Lemma, then it remains to see that the theta characteristic on $C$ associated to this net is indeed $\Theta$.

To do this, we go back initially to the matrix $\Phi$. The restriction of this matrix to $C$ we may view as a map of vector bundles

$$\omega : \mathcal{O}_C(-3)^{\oplus 4} \to \mathcal{O}_C{}^{\oplus 4}$$

having rank 1 everywhere. Moreover, at each point $p \in C$ the image of this map is simply the one-dimensional subspace of $\mathbb{C}^4$ corresponding to the point $p \in B$, thus the image of $\varphi$ is the line bundle $L^{-1}$.

In the same way, we may view the matrix $\Psi$ of cofactors (after dividing out by $G(X)^2$) as a vector bundle map

$$\psi : \mathcal{O}_C{}^{\oplus 4} \to \mathcal{O}_C(1)^{\oplus 4},$$

having rank 3 everywhere on $C$; and the kernel of this map, at each point $p \in C$, will be simply the subspace of $\mathbb{C}^4$ spanned by any of the rows of $\Phi$, that is to say, again the one-dimensional subspace corresponding to the point $p$ on $B$. Thus

$$\mathrm{Ker}(\psi) = L^{-1} = -K_C - \Theta,$$

and so the theta characteristic associated to our net is indeed $\Theta$.

Now suppose we are given a smooth non-hyperelliptic curve $C$ of genus 3 and an even theta characteristic $\Theta$ on $C$. Let $V = H^0(C, K_C)^*$ and $W = H^0(C, K_C \otimes \Theta)^*$, and let $\varphi : V \to \mathrm{Sym}^2 W^*$ be the associated net of quadrics. By way of notation, we will denote by $C$ again its canonical image in $\mathbb{P}V \cong \mathbb{P}^2$, and by $B$ its image in $\mathbb{P}W \cong \mathbb{P}^3$.

The association to the pair $(C, \Theta)$ of the net of quadrics does two things. First of all, it allows us to realize the curve $B \subset \mathbb{P}W$ directly: $B$ is simply the locus of singular points of the singular quadrics $Q_P$ in the net; as we will see, this description will be instrumental in describing the geometry of $B$. Secondly, we see that the ambient space $\mathbb{P}W = \mathbb{P}^3$ of the curve $B$ carries additional structure: in particular, it contains 8 distinguished points, the base points of the net. □

LEMMA 6.4. *A net $\{Q_P\}$ of quadrics in $\mathbb{P}^3$ is typical if and only if the base locus of the net is zero-dimensional, reduced, and in linear general position; that is, it consists of 8 distinct points, no four coplanar.*

*Proof.* The proof is based on one simple observation: in the space $\mathbb{P}^9$ of all quadrics in $\mathbb{P}^3$, the locus $\Sigma$ of singular quadrics is smooth exactly along the open subset of quadrics of rank 3, that is, cones $Q$ over smooth plane conics; and the projective tangent space $\mathbb{T}_Q(\Sigma) \subset \mathbb{P}^9$ to $\Sigma$ at such a point $Q$ is simply the hyperplane in $\mathbb{P}^9$ of quadrics containing the vertex $Q_{\mathrm{sing}}$ of $Q$. It follows that a net of quadrics is typical if and only if it satisfies the two conditions

i. it contains no quadrics of rank 2 or less; and
ii. no singular point of any quadric of the net is a base point of the net.

Now, suppose first that the tangent space to the intersection $\Gamma = \cap Q_P$ at a point $r$ is positive-dimensional; let $v \in T_r(\mathbb{P}^3)$ be a tangent vector to $\Gamma$ at this point. Then not only do all the quadrics $Q_p$ contain the point $r$, the partial derivatives of their defining equations in the direction $v$ all vanish; it follows that at least one of them, say $Q_0$, is singular at $r$. But by our initial remark, this means the net is not typical.

Similar, suppose that the base locus of the net contains four coplanar points $p_1, \ldots, p_4 \in H$. If three of them are colinear the base locus of the pencil will be positive-dimensional, and hence the net cannot be typical by the above; so we may assume this is not the case. It follows that there are only two conics in $H$ containing $p_1, \ldots, p_4$, that is, the restriction map from our net of quadrics in $\mathbb{P}^3$ to $H$ must have a kernel. Thus at least one of the quadrics in our net must contain $H$, and hence have rank at most 2; so our net cannot be typical.

The reverse implication likewise follows immediately from our remark. Given any quadric $Q_0$ in our net, we can write the base locus of the net as a complete intersection $\Gamma = Q_0 \cap Q_1 \cap Q_2$; if $\Gamma$ consists of 8 distinct points it follows that $Q_0$ must be smooth at each of them. Similarly, if $\Gamma$ does not contain a planar subscheme of degree 4, no union of two planes can contain it, so no quadric in the net can have rank less than 4, and our net must be typical. This completes the proof of the lemma. □

We have thus seen that to a sesquicanonical curve $B \subset \mathbb{P}^3$ of genus 3 we may associate a configuration of 8 points in $\mathbb{P}^3$, in linear general position, and (by the initial remark in the proof of the lemma) disjoint from $B$. What is the relationship between these 8 points and the curve? It turns out to be a beautiful one. Briefly, it is this: the 28 lines joining the 8 points pairwise—which a priori need not meet the curve $B \subset \mathbb{P}^3$ at all—all turn out to be bisecants to the curve $B$; and *the pairs of points of incidence of these lines with B are exactly the odd theta-characteristics of B*.

It will take us the next few pages to establish these facts. To start with, let $p$ and $q$ be two of the 8 points, and consider first the line $L = \overline{pq}$ containing them. Since $L$ contains two base points of the net, the net cuts out on $L$ a fixed divisor; thus the kernel of the restriction map of our net to the line must be two-dimensional; or in other words, the net contains a pencil of quadrics containing $L$. Let $M \subset \mathbb{P}V$ be the line corresponding to this pencil. Now, what does a pencil of quadrics in $\mathbb{P}^3$ containing a line $L$ look like? The answer is that if no element of the pencil has rank 2 or less it has as base; locus the union of the line $L$ with a twisted cubic curve $T$ meeting $L$ twice or tangent to it once. To see this, observe first that the base locus $\Psi$ of the pencil must be one-dimensional, if no element of the pencil is reducible; in particular, the pencil will consist of all the quadrics containing $\Psi$. Now let $T$ be the curve residual to $L$ in this intersection. Let $Q$ be a general quadric of the net, so that $L \subset Q$ is a curve of type (1,0) on $Q$ and $T$ is correspondingly a curve of type (1,2). Either $T$ is irreducible, in which case it is a twisted cubic curve meeting $L$ in a total of $(T \cdot L)_Q = 2$ points counting multiplicity; or it is reducible, in which case it contains a planar curve of degree 2 and we see that the pencil contains a reducible member.

From this picture it is easy to see what the singular elements of the pencil must be. For one thing, all the lines of a quadric cone pass through its vertex; so if the union $T \cup L$ lies on a quadric cone, of course the vertex $r$ of that cone must lie on $L$. At the same time, projection of $T$ from $r$ must be a conic, so $r$ must lie on $T$ as well; thus $r$ must be one of the points of intersection of $T$ with $L$. One thing that follows from this is that the line $L$ must have been a chord to $B$, meeting $B$ in two points $a$ and $b$.

Next, since the pencil has only one or two singular elements rather than the expected four, the line $M \subset \mathbb{P}^2$ must be in special position with respect to the curve $C \subset \mathbb{P}^2$ of singular elements of the net: specifically, it can meet $C$ in at most two points. In fact, we can see directly that it must be a bitangent (a hyperflex is considered a bitangent). We can do this in two ways: by the same sort of argument as given above, we can argue that since the base locus of the pencil is singular at the vertex of each singular element of the pencil, the curve $C$ must be tangent to $M$ everywhere they meet, that is, the intersection $M \cap C$ is everywhere nonreduced. Alternatively, we can observe that since we have a marked ruling on each quadric of the pencil (namely, the one containing $L$) the discriminant of the pencil can vanish only to even order.

The conclusion in any case is this: the line $L = \overline{pq}$ is a chord to $B$; the line $M$ is a bitangent line to $C$; and the two points of intersection of $M$ with $C$ are the points at which the line $L$ meets $B$. Moreover, since there are 28 lines joining the eight base points of the net pairwise, the converse is also true: for any odd theta-characteristic $\Xi = \mathcal{O}_C(a + b)$, the line $\overline{ab}$ will contain two of the base points of the pencil.

We thus have a correspondence between the odd theta characteristics on $C$ and the pairs of base points of the net. In particular, labelling the eight base points of the net $p_1, \ldots, p_8$ is equivalent to labeling of the 28 odd theta characteristics on $C$.

Let us take a moment out and consider in some more detail this correspondence between the odd theta characteristics on $C$ and the pairs of base points of the net, that is, edges of the complete octagon with vertices $p_1, \ldots, p_8$. For this purpose, we will for $1 \leq i, j \leq 8$ identify the theta characteristic cut by the line $p_i p_j$ by $E_{i,j}$. Moreover, so as not to confuse the notions of equality in $\mathrm{Pic}(C)$ and equality in the group $W$ in $\mathrm{Pic}(C)/\mathbb{Z}K_C$ associated to the curve $C$ in the preceding section, we will denote by $q_0$ and $q_{i,j}$ the elements of $W$ corresponding to $\Theta$ and $E_{i,j}$.

We may make one preliminary observation: since, as a net of quadrics varies among all typical nets, the monodromy action on the base points is the full symmetric group on 8 letters. It follows that for any subset of these edges, the sum of the corresponding theta characteristics will be zero or nonzero (if it is an even sum) and odd or even (it is an odd sum) depending only on the configuration of the corresponding edges. We will try to say in some cases which it is.

Now, our first observation is simple: for any triple of base points of the net, the sum of the corresponding divisors $E_{i,j}$, $E_{i,k}$ and $E_{j,k}$ is a hyperplane section of the curve $B$, so that in $\mathrm{Pic}(C)$ we have

$$E_{i,j} + E_{i,k} + E_{j,k} = K_C + \Theta$$

and correspondingly in the group $W$ we have

$$(*_{i,j,k}) \qquad q_{i,j} + q_{i,k} + q_{j,k} = q_0.$$

Next, since the theta characteristics $q_{i,j}$ are all distinct, the pairwise sums $q_{i,j} + q_{k,\ell}$ are all nonzero. The distinction is this: if two theta characteristics correspond to incident lines $p_i p_j$ and $p_j p_k$, then since $E_{i,j} + E_{j,k} + \Theta = K_C + E_{i,k}$, we have

$$q_0(q_{i,j} + q_{j,k}) = 1,$$

i.e., the sum of the theta characteristics corresponding to two incident lines is not a zero of the quadratic form $q_0$ on $V$. Conversely, the sum of the theta characteristics corresponding to two disjoint lines is not a zero of the quadratic form $q_0 : q_0(q_{i,j} + q_{k,\ell}) = 1$ would mean that the divisor

$$E_{i,j} + E_{k,\ell} + \Theta - K_C = K_C + \Theta - E_{i,j} - E_{k,\ell}$$

was effective; but since the complete linear series $|K_C + \Theta|$ is cut on $B$ by planes, this is not the case. (In fact, by monodromy considerations, the 216 pairwise sums of theta characteristics corresponding to incident lines must be evenly distributed among the 28 vectors $\nu \in V$ such that $q_0(\nu) = 1$, that is, each must occur 6 times; and the 210 pairwise sums of theta characteristics corresponding to skew lines must include all 35 zeroes of the form $q_0$ other than 0, each occurring 8 times.)

Now, fix a single base point $p_i$, say $p_8$, and for each $i$, $j$ consider the vector $\nu_{i,j} = q_{i,8} + q_{j,8}$. We claim that the pairwise sums of the vectors $\nu_{i,j}$ are all distinct, that is, the four-fold sum $q_{i,8} + q_{j,8} + q_{k,8} + q_{\ell,8} \neq 0$ for any $i, j, k, \ell$ distinct. To see this, simply observe that by the equalities $(*_{i,j,8})$ and $(*_{k,\ell,8})$ above, $\nu_{i,j} = q_{i,8} + q_{j,8} = q_0 + q_{i,j}$ and likewise $\nu_{k,\ell} = q_0 + q_{k,\ell}$; thus

$$\nu_{i,j} + \nu_{k,\ell} = q_{i,8} + q_{j,8} + q_{k,8} + q_{\ell,8} = q_{i,j} + q_{k,\ell} \neq 0.$$

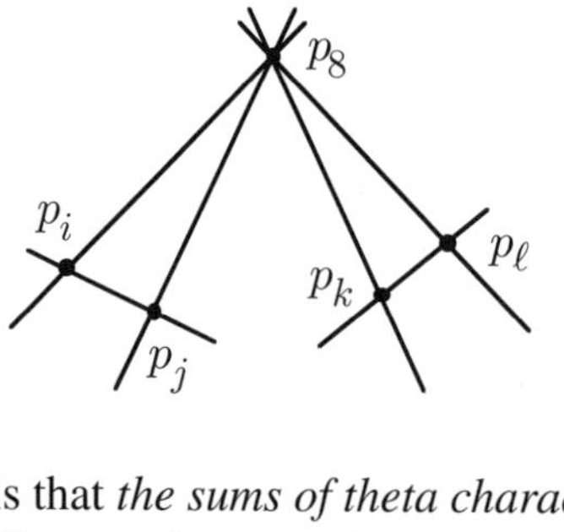


It also follows from this that *the sums of theta characteristic corresponding to three concurrent lines are all even;* given any distinct $i, j, k$ between 1 and 7, we can write $q_{i,8} + q_{j,8} + q_{k,8} = q_{i,8} + q_{j,k} + q_0$, so that the parity of $q_{i,8} + q_{j,8} + q_{k,8}$ is $q_0(q_{i,8} + q_{j,k}) = 0$.

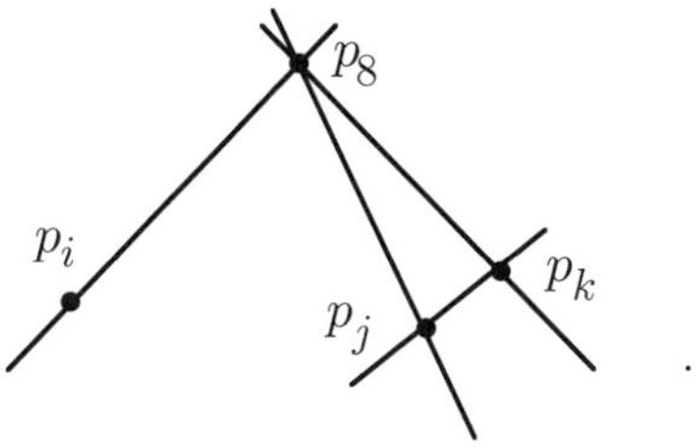


In the same way, we can determine the parity of the sum of any three theta characteristics in terms of the configuration of the corresponding lines: if we have theta characteristics $q_{i,j}$ and $q_{j,k}$ corresponding to two incident lines and one $q_{\ell,m}$ corresponding to a line skew to both, we may add the relation $(*_{i,j,k})$ to deduce that

$$q_{i,j} + q_{j,k} + q_{\ell,m} = q_{j,k} + q_{\ell,m} + q_0$$

is even. Similarly, if we have three theta characteristics $q_{i,j}$, $q_{j,k}$, and $q_{k,\ell}$ corresponding to lines forming a chain, we may add the relations $(*_{i,j,k})$ and $(*_{i,k,\ell})$ to deduce that

$$q_{i,j} + q_{j,k} + q_{k,\ell} = q_{i,\ell}$$

is odd. The last case, that of three theta characteristics corresponding to mutually skew lines, yields an even theta characteristic, as will be seen in a moment.

To conclude, we consider further sums of theta characteristics corresponding to concurrent lines. First, fix one base point $p_8$ as before, and consider the sum of all seven theta characteristics $E_{i,8}$ To describe this, note that projection from the point $p_8$ maps $B \subset \mathbb{P}^3$ to a plane curve $\bar{B}$. Being of degree 6, the curve $\bar{B}$ has arithmetic genus $\binom{5}{2} = 10$; since it has geometric genus 3, we would expect $\bar{B}$ to have $10 - 3 = 7$ nodes. In fact we can see them all: they are exactly the images of the lines $\overline{pp_i}$. By adjunction, the canonical series on $B$ will be cut out by the series of plane cubics passing through the nodes of $\bar{B}$; in other words, we have a linear equivalence

$$K_B = \mathcal{O}_B(3)(-E_{1,8} - \cdots - E_{7,8}).$$

But $\mathcal{O}_B(1) = K_B \otimes \Theta$, and this amounts to saying that

$$E_{1,8} + \cdots + E_{7,8} = 3K_C + 3\Theta - K_C = 3K_C + \Theta;$$

in other words,

$$q_{1,8} + \cdots + q_{7,8} = q_0.$$

Next, for any $i_1, \ldots, i_5 \subset (1, \ldots, 7)$, consider the corresponding subset $E_{i_1,8}, \ldots, E_{i5,8}$ of five of the seven theta characteristics $E_i$. Since any five of the lines $pp_i$ lie on a quadric (the cone over the conic containing the images of the $p_i$ under projection from $p$), we see that $h^0(\mathcal{O}_B(2)(-E_{i_1} - \cdots - E_{i_5})) > 0$; this translates into the assertion that

$$\begin{aligned} 0 &\leq 2(K_C + \Theta) - E_{i_1,8} - \cdots - E_{i_5,8} \\ &= E_{i_1,8} + \cdots + E_{i_5,8} - 2K_C \end{aligned},$$

that is, the sum $E_{i_18} + \cdots + E_{i_8} - 2K_C$ *is an odd theta characteristic* $\Omega$. In fact, we can say which one it is: if $i, j \in \{1, \cdots, 7\}$ are the two indices not included in the subset $\{i_1, \cdots, i_5\}$, we can add the relation $(*_{i,j,8})$ to the relation $q_{1,8} + \cdots + q_{7,8} = q_0$ above to conclude that

$$q_{i_1,8} + \cdots + q_{i_5,8} = q_{i,j}.$$

Note that the assertions verified above imply (somewhat redundantly, in fact) that the theta characteristics corresponding to the seven lines through one of the base points of the net *form an Aronhold set with associated even theta characteristic* $\Theta$. That is, the seven lines through a base point give an Aronhold set $q_1, \ldots, q_7$ and $q_S = \Sigma_{i=1}^7 q_i = 3K$ is the class of $\Theta$. We may thus refine the correspondence $\alpha$ above further, to arrive at bijections:

$$\left\{\begin{matrix}\text{typical nets of quadrics}\\ \text{in } \mathbb{P}^3 \text{ with choice of}\\ \text{one base point}\end{matrix}\right\} \overset{\alpha'}{\longleftrightarrow} \left\{\begin{matrix}\text{smooth nonhyperelliptic}\\ \text{curves } C \text{ of genus 3}\\ \text{with Aronhold set}\end{matrix}\right\}$$

and

$$\left\{\begin{matrix}\text{typical nets of quadrics}\\ \text{in } \mathbb{P}^3 \text{ with labeled}\\ \text{base points}\end{matrix}\right\} \overset{\alpha''}{\longleftrightarrow} \left\{\begin{matrix}\text{smooth nonhyperelliptic}\\ \text{curves } C \text{ of genus 3}\\ \text{with full level 2 structure}\end{matrix}\right\}.$$

**7. Del Pezzo surfaces.** Now that we have established this correspondence, let us take it one step further and consider how a general net of quadrics in $\mathbb{P}^3$ may be specified. Of course, a net is specified by its base points, but it is not the case conversely that any 8 points in $\mathbb{P}^3$ form the base of a net of quadrics. What is true, however, is that seven points in linear general position do impose independent conditions on quadrics, so that there will be exactly a net of quadrics containing them. We accordingly make the

*Definition.* We will say that a collection $\Phi = \{p_1, \ldots, p_7\}$ of seven points in $\mathbb{P}^3$ is typical if they are in linear general position, and the net of quadrics containing them is typical.

Note that there are two ways that a configuration of seven points in linear general position may fail to be typical in this sense. First of all, the base locus $\Gamma$ of the net of quadrics containing them may be positive-dimensional. Now, if this is the case, the degree of $\Gamma$ can be at most 3; so it must contain either a line, a plane conic, or a twisted cubic curve. In fact, it cannot contain a conic, since the restriction map from our net to the plane containing the conic would have to have a kernel, i.e., the net would have to include reducible quadrics and seven points in linear general position cannot lie on the union of two planes. Similarly, $\Gamma$ cannot contain a line, by the same argument applied to the plane spanned by that line and any of the points $p_i$ not lying on it. Thus, $\Gamma$ can be positive-dimensional only if it contains a twisted cubic curve $X$, in which case the net is simply the net of quadrics containing $X$ and $\Gamma = X$; in fact this will occur if and only if the points $p_1, \ldots, p_7$ lie on a twisted cubic curve.

If indeed the base locus of the net determined by $\Phi$ is zero-dimensional, the points $p_1, \ldots, p_7$ may still fail to be typical if that base locus is nonreduced, that is, contains one of the points $p_i$ multiply. By what we have said, this will in turn be the case only if some member of the net is singular at the point $p_i$; which is in turn equivalent to saying that the projection of the remaining 6 points $\{p_j : j \neq i\}$ from $p_i$ lie on a conic. Since, if all seven points lie on a twisted cubic, the projection from any one of the points of the remaining six lie on the conic that is the image under projection of the twisted cubic, we have the

LEMMA 7.1. *A collection $\{p_1, \ldots, p_7\}$ of seven points in $\mathbb{P}^3$ in linear general position is typical if and only if the projection of any six from the remaining point does not lie on a conic.*

Now that we have characterized typical collections of 7 points in $\mathbb{P}^3$, we can extend our correspondence. Tautologously, an unordered collection $\Phi$ of 7 points in $\mathbb{P}^3$ determines a typical net, and of course determines also one base point of that net, namely the one not in $\Phi$. Similarly, an ordered collection $\Psi$ of 7 points in $\mathbb{P}^3$ determines a typical net together with an ordering of the base points of that net. We thus have two further correspondences, expressed in the

THEOREM 7.2. *We have bijections*

$$\left\{\begin{matrix}\text{smooth nonhyperelliptic}\\ \text{curves } C \text{ of genus 3}\\ \text{with Aronhold set}\end{matrix}\right\} \overset{\beta'}{\longleftrightarrow} \left\{\begin{matrix}\text{unordered collections of}\\ \text{7 typical points in } \mathbb{P}^3,\\ \text{modulo } PGL_4\end{matrix}\right\}$$

and

$$\left\{\begin{matrix}\text{smooth nonhyperelliptic}\\ \text{curves } C \text{ of genus 3}\\ \text{with full level 2 structure}\end{matrix}\right\} \overset{\beta''}{\longleftrightarrow} \left\{\begin{matrix}\text{ordered collections of 7}\\ \text{typical points in } \mathbb{P}^3,\\ \text{modulo } PGL_4\end{matrix}\right\}.$$

Moreover, in terms of the structure of coarse moduli space on all four sets, these two maps are isomorphisms of varieties.

Note in particular that the moduli space of ordered collections of 7 typical points in $\mathbb{P}^3$ is rational: there is a unique element of $PGL_4$ sending the first 5 of the seven points to the standard points [1,0,0,0], [0,1,0,0], [0,0,1,0], [0,0,0,1], and [1,1,1,1] so that this moduli space is in fact isomorphic to an open subset of $\mathbb{P}^3 \times \mathbb{P}^3$. We thus have the

COROLLARY 7.3. *The moduli space $\mathcal{M}_3[2]$ of curves of genus 3 with full level 2 structure is rational. In particular, the locus of curves C of genus 3 defined over $\mathbb{Q}$, all of whose line bundles of order 2 and all of whose theta characteristics are defined over $\mathbb{Q}$, is Zariski dense in the moduli space of curves of genus 3.*

For the next step, suppose we are given a typical collection $\Psi = \{p_1, \ldots, p_7\}$ of 7 points in $\mathbb{P}^3$, either ordered or not. We can associate to this the base locus $\Gamma$ of the net of quadrics they determine, which is a collection of 8 points $p_1, \ldots, p_8$, any seven of which are typical. Now, consider in turn the projection of the original collection $\Phi$ from the eight point $p_8$ to the plane. This will be a configuration $\Psi$ of 7 points in the plane, no three of which will be collinear and no six of which will lie on a conic, since any subset of seven of the eight points $p_1, \ldots, p_8$ is typical. We thus make yet another

*Definition.* We will say that a collection $\Psi = \{q_1, \ldots, q_7\} \subset \mathbb{P}^2$ of seven points in the plane is typical if no three are collinear and no six lie on a conic and we observe that we have a natural map $\gamma$ from the space $\mathcal{C}_{3,7}$ of typical 7-tuples of points in $\mathbb{P}^3$ modulo $PGL_4$ to the space $\mathcal{C}_{2,7}$ of typical 7-tuples of points in $\mathbb{P}^2$ modulo $PGL_3$, defined simply by sending a typical 7-tuple $p_1, \ldots, p_7$ to the projection of $p_1, \ldots, p_7$ from the eighth point of intersection of the quadrics containing them.

Now, it may seem at first glance that, in projecting seven points $p_1, \ldots, p_7 \in \mathbb{P}^3$ to $\mathbb{P}^2$ from the eighth base point of the net they determine we are necessarily losing information. Remarkably (to us, anyway) this is not the case; in fact, we claim that *$\gamma$ is an isomorphism.*

To prove this, we will exhibit an explicit inverse map $\gamma' : \mathcal{C}_{2,7} \to \mathcal{C}_{3,7}$. This goes as follows: suppose we are given a typical configuration $\Psi = \{q_1, \ldots, q_7\} \subset \mathbb{P}^2$ of seven points in the plane. Such a collection of points imposes independent conditions on cubic curves in the plane, so that it lies exactly on a net of cubics; and moreover this net has no other base points beyond the $q_i$. Now, choose a general pencil $\mathcal{D} = \{C_\lambda\}$ of cubics in this net; it will have nine base points, consisting of $q_1, \ldots, q_7$ and two further points $r$ and $s$. Let $\varphi : \mathbb{P}^2 \to \mathbb{P}^3$ be the rational map (regular on $\mathbb{P}^2 - \{r, s\}$) given by the web of conics through $r$ and $s$, and let $p_i = \varphi(q_i) \in \mathbb{P}^3$, $i = 1, \ldots, 7$. We will then set

$$\gamma'(\Psi) = \Gamma = \{p_1, \ldots, p_7\} \subset \mathbb{P}^3.$$

We have then the

LEMMA 7.4. *$\gamma'$ is a well-defined map from $\mathcal{C}_{2,7}$ to $\mathcal{C}_{3,7}$ (that is, up to $PGL_4$ the configuration $\{p_1, \ldots, p_7\} \subset \mathbb{P}^3$ does not depend on the choice of pencil), and $\gamma$ and $\gamma'$ are inverse isomorphisms.*

*Proof.* It will turn out to be simpler to establish a more refined bijection. We have seen in the definition of $\gamma'$ that the data of a typical configuration $\Psi \subset \mathbb{P}^2$ of seven points in the plane, together with a choice of $\mathcal{D}$ of pencil of cubics in the net $|\mathcal{I}_\Psi(3)|$ containing $\Psi$, determines a configuration $\Gamma \subset \mathbb{P}^3$ of seven points in space together with a quadric in the net $|\mathcal{I}_\Gamma(2)|$ containing $\Gamma$, namely, the image $Q = \varphi(\mathbb{P}^2)$. □

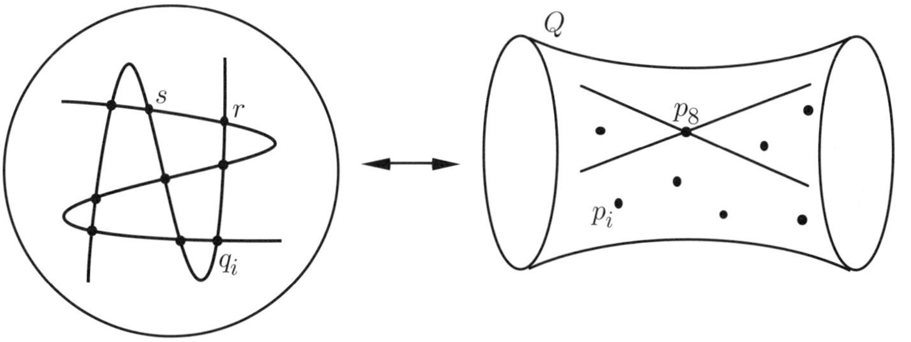


Note also that $\varphi$ blows up the points $r$ and $s$ and collapses the line $L = \overline{rs}$ joining them to a point, which we will call $p_8$; the rulings of the quadric $Q$ correspond to the pencils of lines through $r$ and $s$. At the same time, $\varphi$ carries cubics through $r$ and $s$—and in particular the cubics in our pencil—into quartic curves in $\mathbb{P}^3$; more specifically, since $C_\lambda$ meets the general line through $r$ or $s$ in two other points, this will be a curve of type (2,2) on $Q$. Also since every cubic $C_\lambda$ in the pencil meets $L$ at one point beyond $r$ and $s$, their images $E_\lambda \subset Q$ will all pass through the point $p_8$. The curves $E_\lambda$ thus form a pencil on $Q$ with base points $p_1, \ldots, p_8$; and this pencil is the restriction to $Q$ of a net of quadrics in $P^3$ with these base points. In particular, $p_1, \ldots, p_8$ form the base of a net of quadrics in $\mathbb{P}^3$, so that $\gamma(\gamma'(\Psi)) = \pi_{p_8}(\Gamma) = \Psi$.

Conversely, suppose we are given a typical configuration $\Gamma = \{p_1, \ldots, p_7\} \subset \mathbb{P}^3$ together with a general quadric $Q$ in the net $|\mathcal{I}_\Psi(3)|$; let $p_8 \in Q$ be the eighth base point of the net and $M$ and $N \subset Q$ the two lines of $Q$ passing through the point $p_8$. Now, the net of quadrics through $\Gamma$ cuts on $Q$ a pencil $\varepsilon = \{E_\lambda\}$ of curves of type (2,2) on $Q$ (in particular, quartic curves of arithmetic genus 1), whose base locus consists of $p_1, \ldots, p_8$. When we project from $p_8$, the curves $E_\lambda$ are mapped isomorphically into a pencil $\mathcal{D} = \{C_\lambda\}$ of plane cubic curves, whose base points will include the images $q_i = \pi(p_i)$, $i = 1, \ldots, 7$. The other two base points of $\mathcal{D}$ will be the images $r$ and $s \in \mathbb{P}^2$ of the two lines $L$ and $M$: since each curve $E_\lambda$

meets $M$ and $N$ in one point other than $p_8$, their images $C_\lambda$ will all contain the points $r$ and $s$.

In sum, we see that we have a bijection

$$\left\{\begin{array}{c}(\Gamma, Q) : \Gamma \text{ is a typical} \\ \text{configuration of 7} \\ \text{points in } \mathbb{P}^3, \text{ and } Q \supset \Gamma \\ \text{a quadric, modulo } PGL_4\end{array}\right\} \overset{\tilde{\gamma}}{\longleftrightarrow} \left\{\begin{array}{c}\Psi, \mathcal{D} : \Psi \text{ is a typical} \\ \text{configuration of 7} \\ \text{points in } \mathbb{P}^2; \mathcal{D} \subset |\mathcal{I}_\Psi(3)| \\ \text{a pencil, modulo } PGL_3\end{array}\right\}$$

that induces the bijections

$$\left\{\begin{array}{c}\text{ordered collections of 7} \\ \text{typical points in } \mathbb{P}^3, \\ \text{modulo } PGL_4\end{array}\right\} \underset{\gamma'}{\overset{\gamma}{\rightleftarrows}} \left\{\begin{array}{c}\text{ordered collections of} \\ \text{7 typical points in } \mathbb{P}^2, \\ \text{modulo } PGL_3\end{array}\right\}.$$

We should mention in passing that $\gamma$ and $\gamma'$ represent a very special case of the *Gale transform;* see for example [E-P].

Now that we have arrived at configuration of seven points in the plane, what can we do with those? One answer from classical algebraic geometry is immediate: we can blow them up to obtain a quadric *del Pezzo* surface. This is what we will do next, after a short interlude to discuss del Pezzo surfaces in general.

For our present purposes, it will make sense to define a *del Pezzo* surface to be simply one whose anticanonical bundle is ample. It is then a classical result that any such surface $S$ is either $\mathbb{P}^1 \times \mathbb{P}^1$ or the blow-up of $\mathbb{P}^2$ at $m \leq 8$ points $p_i$, no three collinear and no six on a conic. In the latter case, we will denote by $\ell$ the pullback to $S$, via the blow-up map, of the class of a line in $\mathbb{P}^2$. Note that this is not intrinsic to the abstract surface $S$, but will (as we will see) depend on the representation $\pi : S \to \mathbb{P}^2$ of $S$ as a blow-up of $\mathbb{P}^2$. We will denote by $E_i$ the exceptional divisor lying over the point $p_i$, and its class by $e_i$; so that, for example, the anticanonical class is given by

$$H = -K_S \sim 3\ell - \sum e_i.$$

Given that the self-intersection of $\ell$ on $S$ is 1, as it is on $\mathbb{P}^2$, the self-intersection of $e_i$ is $-1$, and $(\ell \cdot e_i) = 0$ for each $i$, we see that the self-intersection of $-K_S$ is $9 - m$; we will call this quantity the *degree* of $S$ and denote it $d$. In that vein, we will call a *line* of $S$ any curve with intersection number 1 with $H$. We can easily list the lines on $S$: they are

(1) the exceptional divisor $E_i$;

(2) the proper transforms of the lines joining two of the points $p_i$;

(3) the proper transforms of the conics containing five of the points $p_i$;

(4) the proper transforms of the cubics double at one of the points $p_i$ and containing six others;

(5) the proper transforms of the quartics double at three of the points $p_i$ and containing five others; and

(6) the proper transforms of the quintics double at six of the points $p_i$ and containing two others.

Note that the last two are possible only when $m = 8$, the last three only when $m \geq 7$, and so on.

It is not hard to see directly that the points $p_i$ impose independent conditions on cubic curves in the plane, so that the dimension of the anticanonical series will be exactly $9 - m = d$. Now, in case $d \geq 3$, this series will in fact be very ample, giving an embedding of $S$ in $\mathbb{P}^d$ (the images of these maps are what is often referred to as del Pezzo surfaces, i.e., the definition of del Pezzo may in some sources require the anticanonical series to be very ample rather than merely ample).

The principal case of interest to us at present, however, is the case of $m = 7$, that is, del Pezzos of degree 2. In this case, the anticanonical series gives a regular map $\varphi : S \to \mathbb{P}^2$, expressing $S$ as a double cover of the plane. It is not hard to see what the branch divisor $C \subset \mathbb{P}^2$ must be, in any of several ways. For one thing, the inverse image $E$ of a general line $L \subset \mathbb{P}^2$ will be a smooth curve of genus

$$\frac{(-K_S \cdot (-K_S + K_S))}{2} + 1 = 1;$$

inasmuch as the map $\varphi$ expresses $E$ as a double cover of a line branched at its points of intersection with $C$, we conclude that $C$ must be a quartic curve. $C$ moreover must be smooth since $S$ is. Alternatively, we can apply the Riemann-Hurwitz formula: if $R \subset S$ is the ramification divisor, the canonical line bundle of $S$ is given by

$$\begin{aligned} K_S &= \varphi^*(K_{\mathbb{P}^2})(R) \\ &\sim 3 \cdot K_S + R, \end{aligned}$$

we conclude that $R \sim -2K_S$, and in particular that the degree of the image $C = \varphi(R)$ is

$$(R \cdot -K_S) = 2(K_S \cdot K_S) = 4.$$

Thus, a del Pezzo of degree 2 is a double cover of $\mathbb{P}^2$ branched along a smooth quartic. Conversely, if $\varphi : S \to \mathbb{P}^2$ is a double cover branched along a quartic, the canonical bundle of $S$ is given by Riemann-Hurwitz as

$$K_S = \varphi^*(\mathcal{O}_{\mathbb{P}^2}(-1));$$

and since the map $\varphi$ is finite we conclude that $-K_S$ is indeed ample, so $S$ is a del Pezzo surface.

Note that the expression $\varphi : S \to \mathbb{P}^2$ of $S$ as a double cover defines a regular involution $\psi$ on $S$ exchanging the sheets of this map; $\psi$ may also be viewed as a birational involution of the plane $\mathbb{P}^2$ of which $S$ is the blow-up. (Here and elsewhere there is some potential confusion between the two copies of $\mathbb{P}^2$ floating around, inasmuch as $\varphi$ defines a rational map $\mathbb{P}^2 \to \mathbb{P}^2$ of degree 2.) This involution is known as the *Geyser involution,* and has the following alternative description. Given a general point $q \in \mathbb{P}^2$, the cubic curves passing through the eight points $p_1, \ldots, p_7$ and $q$ will form a pencil, and this pencil will have one other base point $r \in \mathbb{P}^2$; the involution is the one sending $q$ to $r$.

Next, we would like to discuss the lines on $S$. We already know them in terms of the description of $S$ as a blow-up of $\mathbb{P}^2$; they are

(1) the 7 exceptional divisors $E_1$;

(2) the 21 proper transforms of the lines joining two of the points $p_i$;

(3) the 21 proper transforms of the conics containing five of the points $p_i$;

(4) the 7 proper transforms of the cubics double at one of the points $p_i$ and containing six others.

We would now like to describe their images under the double cover $\varphi$. This is in fact easy: since by definition a line of $S$ is a curve on $S$ mapped by $\varphi$ one-to-one onto a line in $\mathbb{P}^2$, the inverse images in $S$ of the image in $\mathbb{P}^2$ of any line on $S$ must consist of exactly two lines on $S$. In particular, the 56 lines of $S$ map onto exactly 28 lines in $\mathbb{P}^2$. Moreover, inasmuch as the inverse images of these 28 lines in $\mathbb{P}^2$ are reducible, they cannot have any points of odd intersection multiplicity with the branch divisor $C$ of $\varphi$; thus the 28 images of the lines of $S$ must be exactly the 28 bitangent lines to the curve $C$.

Note finally that labeling all 56 lines of $S$ (equivalently, choosing a set of generators for $\mathrm{Pic}(S)$) amounts to labeling the 28 bitangents to $C$, so that a quadric del Pezzo surface together with a set of generators for its Picard group gives us the data of curve of genus 3 with full level two structure. If we do not identify the elements of $\mathrm{Pic}(S)$, but only the expression $\Pi : S \to \mathbb{P}^2$ of $S$ as a blow-up of $\mathbb{P}^2$ at 7 points (that is, equivalently, specify only the unordered set $\{E_1, \ldots, E_7\}$ of exceptional divisors), we arrive at a curve $C$ of genus 3, together with the specification of an Aronhold set. And finally, if we specify only the abstract surface $S$, and not its representation as a blow-up of the plane, we find the curve $C$, but no further level 2 structure on it.

We have now come, as promised, all the way around. To recap the journey:

We start with a *smooth, nonhyperelliptic curve of genus* 3, *with full level* 2 *structure*. This structure, as we have seen in preceding sections, is equivalent to specifying an *ordered Aronhold set* $\Theta_1, \ldots, \Theta_7$ *of odd theta characteristics on* $C$.

We may then associate to this data a typical net $\varphi$ of quadrics in $\mathbb{P}^3$ with base points labeled $p_1, \ldots, p_8$. The discriminant curve of this net—that is, the set of singular quadrics in the net—will be $C \subset \mathbb{P}^2$; and the locus in $\mathbb{P}^3$ of vertices of singular quadrics in the net will be the image $B \subset \mathbb{P}^3$ of the curve $C$, embedded by the linear series $K_C + \Theta$ where $\Theta$ is the even theta characteristic associated to our Aronhold set (in other words, $\Theta = \Sigma\Theta_i$, in the group $W$). The lines joining the points $p_i$ pairwise will cut on $B$ the 28 odd theta characteristics, with our Aronhold set $\Theta_1, \ldots, \Theta_7$ cut by the seven lines $p_i p_8$ through the point $p_8$.

Now, a typical net $\varphi$ of quadrics is determined by any seven of its base points; thus we may associate to $\varphi$ simply the typical configuration $p_1, \ldots, p_7$ of seven ordered points in $\mathbb{P}^3$.

Next, we may project the configuration $p_1, \ldots, p_7$ from $p_8$ to obtain a typical configuration of seven points $q_1, \ldots, q_7$ in the plane. Again, it is far from clear at first that this is equivalent data, but it is, as may be seen either from the identification of $\{q_1, \ldots, q_7\} \subset \mathbb{P}^2$ with the Gale transform of $\{p_1, \ldots, p_7\} \subset \mathbb{P}^3$ or the explicit (if complicated) inverse given by the cycle of associations here.

Now, a typical configuration $\Gamma = \{q_1, \ldots, q_7\} \subset \mathbb{P}^2$ of seven ordered points in the plane determines a del Pezzo surface $S$, together with a standard basis for the Picard group of $S$, namely, we take $S = \mathrm{Bl}_\Gamma(\mathbb{P}^2)$ to be the blow-up of the plane at $q_1, \ldots, q_7$ and the generators of $\mathrm{Pic}(S)$ the pullback of the class of a line in $\mathbb{P}^2$ and classes of the seven exceptional divisors $E_i$.

Finally, a del Pezzo surface determines a smooth plane quartic, namely the branch divisor $C \subset \mathbb{P}^2$ of the anticanonical map $\varphi_{-K} : S \to \mathbb{P}^2$. Moreover, the seven generators in a normalized set correspond to seven exceptional divisors, whose images in $\mathbb{P}^2$ will be bitangent lines to $C$ forming an Aronhold set. Thus we arrive once more at a smooth nonhyperelliptic curve $C$ with an ordered Aronhold set, that is to say, full level 2 structure. We may represent the various stages in this cycle in a diagram:

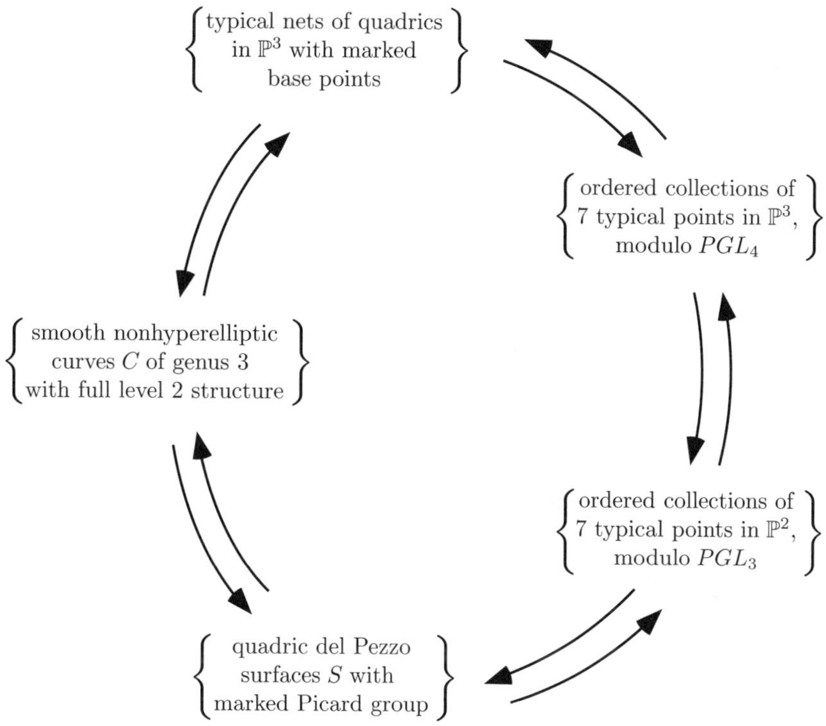

Alternatively, we may go through an analogous cycle of objects and associations starting with the specification on $C$ only of an unordered Aronhold set. In this case we get a typical net of quadrics in $\mathbb{P}^3$ with one distinguished base point, but no ordering of the remaining seven; this gives us in turn an unordered collection of 7 typical points in $\mathbb{P}^3$, then an unordered collection of 7 typical points in $\mathbb{P}^2$, then a quadric del Pezzo surface with a choice of regular birational map $\pi : S \to \mathbb{P}^2$ (equivalently, a divisor class $L$ with $L^2 = 1$ and $L \cdot K_S = -3$). Finally, the branch divisor of the anticanonical map of $S$ is a plane quartic curve, on which the seven exceptional divisors of the map $\pi$ cut an unordered Aronhold set of odd theta characteristics. In other words, we have the analogous diagram

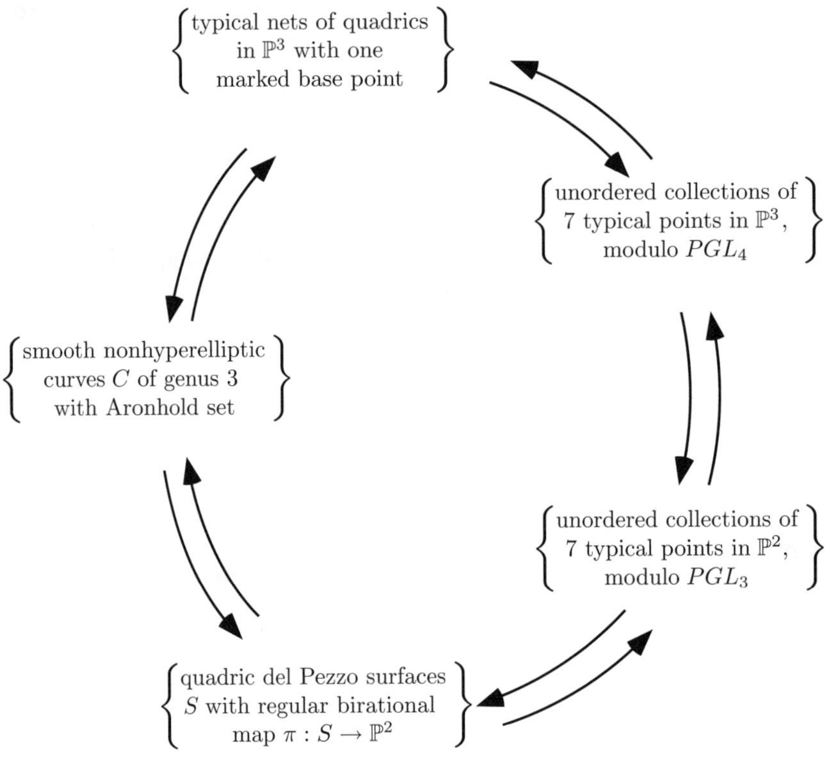


Lastly, if we specify only the curve $C$ and an even theta characteristic $\Theta$ on $C$, we get a typical net of quadrics in $\mathbb{P}^3$, but no distinguished base point; and we cannot complete the cycle in any way.

There are a number of questions, we can ask about this correspondence. For one, all the various moduli spaces referred to in the diagrams above have known compactifications—some, several. To what extent do these associations extend to regular isomorphisms between these moduli spaces? To mention one particularly interesting case, we could enlarge the set of smooth nonhyperelliptic curves of genus 3 to all smooth curves of genus three, and try to extend the definition of the maps $\alpha$, $\beta$ and $\gamma$ to this locus. What happens to the net of quadrics when the curve becomes hyperelliptic? What happens to its eight base points?

Similarly, if we consider some singular curves of genus 3—for example, to take the simplest cases, nodal plane quartics—what happens to the nets, to their base points, and to the del Pezzo surfaces?

Finally, it is known that (some) plane curves with theta characteristics, as well a nets of quadrics are related to the theory of vector bundles on $\mathbb{P}^2$. In the particular case relevant to the paper one takes (stable) rank 2 bundles with $c_1 = 0$ and $c_2 = 4$. Then the quartic curves which arise in this way can be characterized as *Luroth* quartics; see again [B].

*E-mail:* gross@math.harvard.edu

*E-mail:* harris@math.harvard.edu

---

## REFERENCES

[A] E. Artin, Geometric Algebra, Interscience 1957

[B] W. Barth, Moduli of vector bundles on the projective plane, *Invent. Math.* **42** (1977), 63–91

[C-vL] P. Cameron and J. H. van Lint, *Designs, Graphs, Codes and their Links,* Cambridge 1991

[C1] A. Coble, An application of finite geometry to the characteristic theory of the odd and even theta functions, *Trans. A.M.S.* **7** (1906), 241–276

[C2] A. Coble, *Algebraic geometry and theta functions*, A.M.S. Colloquium Publications 1929

[D-O] I. Dolgachev and D. Ortland, Point sets in projective spaces and theta functions, *Astérisque* **165** (1988)

[E-P] D. Eisenbud and S. Popescu, The projective geometry of the Gale transform, *J. Algebra* **230** (2000), no. 1, 127–173

[G-H] P. Griffiths and J. Harris, *Principles of Algebraic Geometry,* Wiley-Interscience 1978

[MH] J. Milnor and D. Husemoller, *Symmetric Bilinear Forms,* Springer-Verlag 1973

[M] D. Mumford, *Theta characteristics on an algebraic curve,* Ann. Ec. Norm. Sup. 1971 p. 181–192

[Sa] N. Saavedra-Rivano, *Finite geometries in the theory of theta characteristics,* L'Ens. Math. **22** (1976), p. 191–218

[S] J.-P. Serre, *Linear Representations of Finite Groups,* Springer-Verlag 1977

[W] H. Weber, *Lehrbuch der Algebra* Band 2, Braunschweig, 1896

CHAPTER 12

# CAN $p$-ADIC INTEGRALS BE COMPUTED?

By Thomas C. Hales

*Abstract.* This article gives an introduction to arithmetic motivic integration in the context of $p$-adic integrals that arise in representation theory. A special case of the fundamental lemma is interpreted as an identity of Chow motives.

**1. Introduction.** This article raises a question in its title, and the short answer to the question is that it still has not been answered. However, tools have now been developed to answer questions such as this, and this article gives an introduction to some of these tools.

This article will concentrate on a particular family of integrals that arise in connection with the representation theory of reductive groups. These are orbital integrals. The clear expectation is that these integrals can be computed, for reasons that will be explained below.

This article will also touch on the fundamental lemma, which is a conjectural identity that holds between certain orbital integrals. This article will include a statement of the fundamental lemma in a special case.

The first sections may seem misplaced because they describe some methods that are not in current use in representation theory, but by the end of the article, their relevance will be established.

The central question in my research for some time is the question of how to use a computer to prove theorems, particularly theorems in geometry. I hope to show that there is some interesting geometry that arises in connection with $p$-adic integration, and that computers can enhance our understanding of that geometry.

In Sections 2, 4, and 5, three major threads will be introduced: Tarski's decision procedure for the real numbers, $p$-adic integration, and motives. The other sections will tie these threads together in the context of the fundamental lemma and $p$-adic orbital integrals.

*Acknowledgments.* I would like to thank Carol Olczak for her assistance in preparing this manuscript.

Manuscript received May 21, 2002

This article is based on a lecture at IAS, April 6, 2001 http://www.math.ias.edu/amf/

Version 5/19/02.

Research supported in part by the NSF

**2. Tarski's decision procedure.** Around 1930, Tarski proved a decision procedure for sentences in the elementary theory of real closed fields [2, 3, 29].

Tarski's result can be formulated precisely in terms of a first-order language. The language is built from the fifteen symbols.

$$\begin{array}{cccc} 0 & 1 & + & * \\ ( & ) & = & < \\ \forall & \exists & x & ' \\ \wedge & \vee & \neg & \end{array}$$

We will not go into the details of the syntax of the language [12, 14]. Each $x$ is followed by zero or more primes, and primes only occur after $x$ or another prime. We abbreviate $x$ followed by $n$ primes to $x_n$. The language contains variables $x_n$, and the constants 0, 1. The variables and constants can be added and multiplied (symbols $+$ and $*$). Polynomial expressions can be compared with the predicates $=$ and $<$. The quantifiers ($\forall$ and $\exists$) should be understood as ranging over the real numbers (or a complete ordered field). For example, the assertion that a quadratic polynomial has a root can be written in this formal language as

$$\neg(x' = 0) \wedge \exists x(x' * x * x + x'' * x + x''' = 0) \tag{1}$$

The formal language quickly becomes cumbersome, and we allow ourselves certain informal shorthand conventions, for example, writing Formula 1 as

$$a \neq 0 \wedge \exists x(ax^2 + bx + c = 0),$$

whenever a translation back into a formal statement of the language is clear.

Many things are noticeably absent from this little first-order language. There is no way to express particular real numbers in this language such as $\pi = 3.14159\ldots$, $e = 2.71828\ldots$, $\ln(2)$. There is no notion of set. There are no quantifiers that range over subsets of the real numbers (for example, there are no quantifiers over the integers). There are no transcendental functions such as the cosine function. There is no calculus or integration (except for formal derivatives of polynomials and the like).

Tarski's result can be expressed as an algorithm for the elimination of quantifiers in this first-order language. It takes a formula in this language and manipulates it by an entirely mechanical procedure into an equivalent form that contains no quantifiers ($\exists\ \forall$). The formula that this procedure gives as output is equivalent to the input in the sense that the same $n$-tuples of real numbers satisfy the two formulas.

For example, if we apply Tarski's procedure to Formula 1, it returns something equivalent to the quantifier-free formula

$$\begin{aligned} &\neg(x' = 0) \\ &\wedge(x'' * x'' - (1+1+1+1) * x' * x'' > 0 \\ &\vee x'' * x'' - (1+1+1+1) * x' * x' = 0)) \end{aligned}$$

or less formally,

$$a \neq 0 \wedge (b^2 - 4ac \geq 0).$$

In other words, Tarski's procedure determines that a quadratic equation has a real root if and only if the discriminant is non-negative. In a similar way, the truth value of all sentences in this language can be decided: the truth value of an equivalent sentence without quantifiers is trivially determined.

Here is a more difficult example, drawn from [3, page 7]. When is a quartic polynomial semi-definite? Tarski's algorithm takes the (formal translation of)

$$\forall x(x^4 + px^2 + qx + r \geq 0)$$

and returns a formula equivalent to

$$\begin{aligned}
&(256r^3 - 128p^2r^2 \quad + 144pq^2r \\
&\qquad\qquad\qquad\qquad + 16p^4r - 27q^4 - 4p^3q^2 \geq 0 \\
&\wedge \\
&\quad 8pr - 9q^2 - 2p^3 \quad \leq 0) \\
&\vee \\
&\quad (27q^2 + 8p^3 \geq 0 \wedge 8pr - 9q^2 - 2p^3 \geq 0) \\
&\wedge \\
&\quad r \geq 0.
\end{aligned}$$

Tarski's original algorithm is very slow, but in 1975 George Collins found a vastly improved method of quantifier elimination. Further improvements are mentioned in the survey article [3].

The methods have improved to the point that the algorithms are of practical importance. For instance, in robotics, quantifier elimination can be used to determine whether two moving objects will collide [3]. Mathematica 4.0 implements an experimental package in quantifier elimination [28]. There are highly nontrivial problems in discrete geometry that can be expressed in this little first-order language (for example, the dodecahedral conjecture [21]). The strategy is to squeeze nontrivial assertions into this little language, and then let the general algorithms prove the results.

**3. Pas's language.** This article is concerned, however, with $p$-adic quantifier elimination and not with Tarski's quantifier elimination over the reals. The first early results on quantifier elimination can be found in articles by Ax-Kochen and Ershov ([1] and [13]). The approach that we follow grows out of the article *Decision procedures for real and p-adic fields* by Paul J. Cohen in 1969 (see [4]). Cohen's work on $p$-adic quantifier elimination was refined and extended by various people (Denef [6], Macintyre [27], and Pas [30]). We will describe $p$-adic quantifier elimination as it is developed by Pas.

Pas defines a first-order language for complete Henselian rings that is analogous to Tarski's first-order language for the theory of complete ordered fields. It contains the following tokens

$$\begin{array}{cccccc} 0 & 1 & + & * & & \\ ( & ) & = & < & & \\ \forall & \exists & x & m & \xi & ' \\ \wedge & \vee & \neg & & & \\ \text{ord} & \text{ac} & & & & . \end{array}$$

The language consists of syntactically well-formed formulas in this language. There are three sorts of variables $x, x', x''$ (which we abbreviate to $x_0, x_1$, etc.), $m, m', m''$ (which we abbreviate to $m_i$) and $\xi$, $\xi'$, $\xi''$, etc. (which we abbreviate to $\xi_i$).

In the interpretations of this language, there are three algebraic structures: a valued field (such as a $p$-adic field), a value group (the target of the valuation, which will typically be the additive group of the integers), and a residue field. The variables $x_i$ are of the valued-field sort, the variables $m_i$ are of the additive group sort, and the variables $\xi_i$ are of the residue field sort. Correspondingly, there are three sorts of quantification depending on the sort of variable the quantified is attached to. The constant 0 comes in three sorts: ($0_x$, $0_m$, and $0_\xi$). These are interpreted as the zero element in the valued field, the additive value group, and the residue field, respectively. The addition symbol $+$ is overloaded in that it is interpreted as addition in the valued field, addition in the value group, or addition in the residue field, according to its arguments. (The syntax requires the arguments to $+$ to be of the same sort.)

The function name *ord* is interpreted as the valuation on the field. If the model is a $p$-adic field, *ord* is interpreted as the normalized valuation on the field. The function name *ac* is interpreted as an angular component function. On the units in the ring of integers, the interpretation is the mapping from the units to its nonzero residue in the residue field. On general nonzero elements, it is interpreted as the function that scales its argument by a power of a uniformizer to make it a unit and then takes its image in the residue field. (Although a uniformizer is used to construct the interpretation of the function *ac*, the uniformizer itself does not appear in Pas's language.) Expressions involving "$<$" are restricted to the additive group sort.

One of the design requirements of this language is that it be small enough for there to be a quantifier elimination procedure. By results of Gödel, this would not be possible if the language were to encompass the full arithmetic theory of the integers [16]. For this reason, the language is restricted to the additive theory of the value group. That is, integer products such as $m * m'$ are prohibited in the language. Integer expressions may be compared through equality and inequality ($=$ and $<$). According to a result proved by Presburger in 1929, a decision procedure exists for the additive theory of the integers ([31]).

Just as in the case of the first-order theory of the reals, much is missing from the language. For instance, there is no uniformizer in the language, so we cannot express $p$-adic expansions of numbers in the valued field. As in the case of the reals, there is no notation that would allow us to express sets in this language. It is impossible to express field extensions directly (only indirectly through polynomials defining the roots, for instance). Most of Galois theory and local class field theory will be inexpressible.

However, this language is small enough for there to be a procedure of quantifier elimination. In 1989, Pas, building on earlier results, proved that the quantifiers of the valued field sort can be eliminated, in the sense that an algorithm exists to produce an equivalent formula without quantifiers of the valued-field sort. Pas's language gives quantifier elimination of quantifiers of the valued field sort. To eliminate all quantifiers, Pas's result must be combined with Presburger's quantifier elimination on the additive theory of the integers, and with the theory of Galois stratification for quantifiers of the residue field sort. (Equivalence here means in the sense that for any complete henselian ring with a residue field of characteristic zero, the two formulas have the same set of solutions. Although Pas's procedure requires the residue field to have characteristic zero, Pas, Denef, and Loeser are able to apply these results to $p$-adic fields with residue fields of positive characteristic. This involves the use of ultrafilters and ultraproducts. Finitely many primes are discarded in the process.)

One of the main applications of Pas's language and its quantifier elimination procedure has been to the theory of $p$-adic integration. For example, Pas's original article contains results about the Igusa local zeta function, which is a $p$-adic integral ([30]).

**4. $p$-adic integration.** Let $F$ be a $p$-adic field of characteristic zero. Let $\mathfrak{g}$ be a reductive Lie algebra defined over $F$, $X$ a regular semisimple element of $\mathfrak{g}(F)$. Let $f$ be a function of compact support on $\mathfrak{g}(F)$. We consider the stable orbit $O^{st}(X)$ of $X$ (meaning the $F$-points of the orbit of $X$ over an algebraic closure). We pick an invariant measure $\mu$ on the orbit. The integral of $f$ over $O^{st}(X)$ is called an orbital integral. A fundamental problem is to compute

$$\int_{O^{st}(X)} f \, d\mu.$$

These integrals arise repeatedly in the representation theory of $p$-adic groups, in places such as the trace formula. The conjectural fundamental lemma (it will be discussed in Section 7.4) is an identity of orbital integrals. The fact that the fundamental lemma has resisted all efforts to prove it is closely related to the difficulty of computing orbital integrals.

**4.1. An example in $so(5)$.** An example will illustrate the nature of these integrals. Let $\mathbb{F}_q$ be the residue field. Assume that its characteristic is not 2. Let $\mathfrak{g} = so(5)$. Assume that $X$ has that property that the valuation of $\alpha(X)$ is independent of the root $\alpha$. Assume that

$$|\alpha(X)| = q^{-r/2},$$

for an odd integer $r$. Viewing $X$ as a linear transformation on a 5-dimensional vector space, the roots of the characteristic polynomial of $X$ are

$$0, \pm t_1, \pm t_2.$$

Let $R_X$ be the quadratic polynomial in $k[\lambda]$ with roots the reduction mod a uniformizer $\varpi_F$ of

$$t_i^2/\varpi_F^r.$$

We have an elliptic curve $E_X$ over the finite field $k$ given by

$$y^2 = R_X(\lambda^2).$$

There are test functions $f$ so that (for appropriate normalizations of measures) we have

$$\int_{O^{st}(X)} f\, d\mu = A(q) + B(q)|E_X(k)|, \tag{2}$$

for some rational functions of $q$: $A$ and $B \neq 0$. (See [19].) The rational functions $A$ and $B$ depend on $f$. This special case gives an indication of what orbital integrals can give.

What does it mean to calculate the orbital integral? The naive and completely unsatisfactory answer is that a calculation of an orbital integral is to take a particular $p$-adic field, a particular element $X$ and to program a computer to find the complex number expressed on the right-hand side of Equation 2. A satisfactory answer to what it should mean to calculate the orbital integral is to find the rational functions $A$ and $B$, and to give the elliptic curve $E_X$. In other words, what is really needed is a symbolic computation that gets at the underlying variety (in this case an elliptic curve). This is the sense in which I intend the question asked in the title "Can $p$-adic integrals be computed?"

If we examine this example more closely, we might ask what features of the problem made this calculation possible? The first obvious feature is that as we vary the parameter $X$, the elliptic curves do not change erratically; rather, they vary within a nice family of elliptic curves over the finite field.

The second noteworthy feature is that as we move from local field to local field, we obtain (in some sense) the "same family" of elliptic curves in each case. It is this consistency as we go from one local field to another that makes it reasonable to hope that a computer algorithm might be found to compute the orbital integrals for all local fields. We can view this as a single elliptic curve $E$ that is defined over $\mathbb{Q}(a, b)$:

$$y^2 = x^4 + ax^2 + b,$$

or as a family parameterized by $a$ and $b$. All elliptic curves $E_X$ for all $p$-adic orbital integrals for all local fields come as various specializations of this family.

To carry this example farther, we might look at some of the identities that are predicted by Langlands's principle of functoriality. One such identity (that is needed for applications of the trace formula) predicts an equality of orbital integrals between $so(5)$ and $sp(4)$. It has the form

$$\int_{O^{st}(X),so(5)} f\,d\mu = \int_{O^{st}(Y),sp(4)} f'\,d\mu'.$$

The data for $sp(4)$ is similar to that data for $so(5)$. The elements $X$ and $Y$ are related through their characteristic polynomials $P_X$ (resp. $P_Y$):

$$P_X(\lambda) = \lambda P_Y(\lambda).$$

(Note that 0 is always a root of $P_X$.) (The function $f'$ has to be related to $f$ in a suitable way.) When these integrals are computed we find elliptic curves for $sp(4)$ as well, and the identity of orbital integrals holds if and only if we have an identity of the following form

$$A(q) + B(q)|E_X(k)| = A(q) + B(q)|E'_Y(k)|.$$

It turns out that the elliptic curves $E_X$ and $E'_Y$ are not isomorphic (they have different $j$-invariants), but they can be proved to have the same number of points by producing an isogeny between $E_X$ and $E'_Y$.

This isogeny can be expressed as a single isogeny between two elliptic curves $E$ and $E'$ over $\mathbb{Q}(a, b)$. The conclusion of this discussion is that this particular identity of orbital integrals holds for all $p$-adic fields because of an identity of two Chow motives over $\mathbb{Q}(a, b)$: that is, $E$ is isogenous to $E'$.

What this suggests is a general hope that there are global objects attached to $p$-adic integrals. The global object should be something like a Chow motive. Identities of $p$-adic integrals should be consequences of identities of Chow motives.

**5. Motives.** We are finally in the position to give a precise definition of what it means to compute a $p$-adic integral. We state it as a thesis:

THESIS 5.1. *The computation of a p-adic integral is an effective algorithm to obtain the underlying virtual Chow motive.*

This thesis is incoherent unless virtual Chow motives are associated with general families of $p$-adic integrals. That this should be so was articulated by Loeser in Strasbourg in 1999 [25].

PRINCIPLE 5.2. *(Denef-Loeser Principle) All "natural" p-adic integrals are motivic.*

Without committing Denef and Loeser to any particular definition of "natural," as representation theorists, we like to think that the important integrals that arise in representation theory are natural. We are thus led to an investigation of motivic underpinnings of $p$-adic integrals.

**5.1. An example of motivic integration.** Motivic integration was introduced by Kontsevich in a lecture in Orsay in 1995 [22]. The properties of motivic integration have been developed in a series of fundamental articles by Denef and Loeser [7], [8], [9], [11]. In fact, my entire article is nothing but an application of the beautiful circle of ideas that they develop.

To describe the theory in a few words, I will describe motivic integration by analogy with $p$-adic integration. Consider the following elementary $p$-adic integral:

$$\int_{\mathbb{F}_q[[t]]} |x|^m\, dx = \sum_{\ell=0}^{\infty} |\varpi^\ell|^{m+1} \int_{|u|=1} \frac{du}{|u|}$$

$$= (1 + q^{-(m+1)} + q^{-2(m+1)} \cdots)(1 - q^{-1}).$$

The answer is independent of the field, so we are tempted to write for any field (say a field of characteristic zero):

$$\int_{k[[t]]} |x|^m\, dx = \sum_{\ell=0}^{\infty} |\varpi^\ell|^{m+1} \int_{|u|=1} \frac{du}{u}$$

$$= (1 + q^{-(m+1)} + q^{-2(m+1)} \cdots)(1 - q^{-1}).$$

The only difficulty is in the interpretation of $q$. Kontsevich supplies the answer with motivic integration: it is a symbol. More specifically, in the case of finite fields it is a symbol attached to the affine line $\mathbb{A}^1$, and for general fields we can continue to view it as a symbol attached to the affine line. To mark the change of context from $p$-adic fields to more general fields, we replace the symbol $q$ with $\mathbb{L}$ (to suggest the Lefschetz motive).

**5.2. Rings of virtual motives.** Section 5.1 describes a simple example of motivic integration. You integrate much in the same way as with $p$-adic integration, but whenever in $p$-adic integration it becomes necessary to count points on a variety, with motivic integration you introduce a new symbol for that variety and move on. Although a single symbol ($q$ or $\mathbb{L}$) suffices for the one example that was shown, motivic integration in general will require a host of new symbols. Integration should be linear, so relations must be introduced among the symbols to make motivic integration linear. In that example, $q$ (or $\mathbb{L}$) occurs as a denominator, and this will require us to invert $\mathbb{L}$ in the ring we construct. In that example, the answer is a limit (that is, an infinite sum), and this will require us to complete the ring we construct.

**6. Rings of motives.** Let $k$ be a field of characteristic zero. Let $\mathrm{Sch}_k$ be the category of varieties over $k$. Let $K_0(\mathrm{Sch}_k)$ be the Grothendieck ring of varieties over $k$. It is the commutative ring generated by symbols

$$[S]$$

for each variety $S$ over $k$. The relations are

$$[S \times S'] = [S][S']$$

and if $S'$ is closed in $S$, then

$$[S] = [S \setminus S'] + [S'].$$

We let $\mathbb{L} = [\mathbb{A}^1]$. Let $K_0(\mathrm{Sch}_k)_{loc}$ be the ring obtained by inverting $\mathbb{L}$.

We let $\mathrm{Mot}_{\mathrm{k},\bar{\mathbb{Q}}}$ be the category of Chow motives over $k$ with coefficients in $\bar{\mathbb{Q}}$, an algebraic closure of $\mathbb{Q}$. (This category is described in detail in [32].) The objects in this category are triples

$$(S, p, n)$$

where $S$ is a variety over $k$, $p$ is a projection operator over $\bar{\mathbb{Q}}$, and $n$ is an integer. This category is an additive category, but it is not abelian [32].

Let $K_0(\mathrm{Mot}_{\mathrm{k},\bar{\mathbb{Q}}})$ be the Grothendieck group of the additive category $\mathrm{Mot}_{\mathrm{k},\bar{\mathbb{Q}}}$. The set of generators of this group is the set of objects in $\mathrm{Mot}_{\mathrm{k},\bar{\mathbb{Q}}}$. By a fundamental result of Gillet and Soulet [15] and Guillén and Navarro Aznar [17], there is a homomorphism of rings

$$K_0(\mathrm{Sch}_k) \to K_0(\mathrm{Mot}_{\mathrm{k},\bar{\mathbb{Q}}})$$

that takes the symbol $[S]$ of a smooth projective variety $S$ to the generator associated with $(S, \mathrm{id}, 0)$, where id is the identity projection operator. The image of $\mathbb{L}$ under this homomorphism is invertible. Thus, the homomorphism extends to $K_0(\mathrm{Sch}_k)_{loc}$. Let $K_0^v(\mathrm{Mot}_{\mathrm{k},\bar{\mathbb{Q}}})_{loc}$ be the image of this homomorphism.

There is a filtration $F^m K_0^v(\mathrm{Mot}_{\mathrm{k},\bar{\mathbb{Q}}})_{loc}$ on this group given by $S/\mathbb{L}^i \in F^m$ iff $\dim S - i \leq -m$. We let $\hat{K}_0^v(\mathrm{Mot}_{\mathrm{k},\bar{\mathbb{Q}}})_{loc}$ be the completion with respect to this filtration. This is the ring in which motivic integrals take their values (or sometimes its tensor product with $\mathbb{Q}$).

**6.1. Arithmetic motivic integration.** In 1999, Denef and Loeser developed an arithmetic theory of motivic integration in [10]. (This theory is distinct from a geometric theory of motivic integration that was developed earlier.) In their article, Denef and Loeser describe the three threads introduced in Sections 2, 4, and 5, and show how they relate to one another. In their article, they make two fundamental discoveries:

(1) Motives can be attached to formulas in Pas's language.

(2) The trace of Frobenius on the motive equals the $p$-adic integral over the $p$-adic set defined by the formula.

The process is represented schematically in Figure 1. A formula in Pas's language can be interpreted two ways, leading to two different integrals. First the formula can be interpreted over a $p$-adic field. The $p$-adic set of points that satisfy the formula has a volume. The formula can also be interpreted over a henselian field

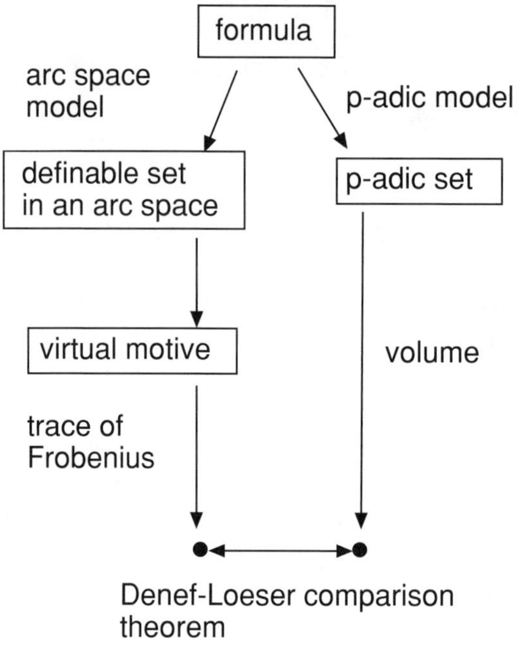


*Figure 12.1 The Denef-Loeser comparison theorem*

(such as $\mathbb{C}((t))$). The set of points that satisfy the formula has a motivic volume (an element of the ring $\hat{K}_0^v(\text{Mot}_{k,\bar{\mathbb{Q}}})_{loc}$).

The comparison theorem of Denef and Loeser asserts that the trace of Frobenius against this virtual motive is equal to the $p$-adic volume of the $p$-adic set. In particular, if a $p$-adic set has the special form given by the set of points satisfying a formula in Pas's language, then a motive can be attached to it. It is this comparison theorem that will permit us to show that interesting $p$-adic integrals have a motivic interpretation.

**7. The fundamental lemma.** All the sections until now have been an extended introduction to provide context for the results I am about to describe. Roughly speaking, I have found that orbital integrals can be placed into the framework of Denef and Loeser.

**7.1. Strips.** Let $F$ be a $p$-adic field of characteristic 0. Let $k$ be the residue field of $F$. Fix parameters $n$, $k$, and $r$ satisfying the following conditions:

- $n$ is a positive integer.
- $k$ is an integer $k \leq n$.
- $r$ is a rational number. Write it as $r = \ell/h$, with $\ell$ and $h$ relatively prime.

Let $\mathfrak{g} = so(2n+1)$. There are endoscopic Lie algebras

$$\mathfrak{h} = so(2k+1) \times so(2n-2k+1).$$

That is, we take a product of two orthogonal Lie algebras, whose ranks add up to that of $\mathfrak{g}$. (Endoscopy was originally defined in terms of groups, but it has become common practice to follow the practice of Waldspurger and to pass to the Lie algebras.)

We define a subset of $\mathfrak{g}$ that I will call a strip. It depends on the parameter $r$. Define strip($r$) to be the set of all $X \in \mathfrak{g}$ such that $|\alpha(X)| = q^{-r}$ for all roots $\alpha$. These elements are called elements *equal valuation*.

Write the characteristic polynomial $P_X(\lambda)$ as

$$P_X(\lambda) = \lambda P_X^0(\lambda).$$

Let $\bar{\mathbb{F}}_q$ be an algebraic closure of $\mathbb{F}_q$. Let $R_X$ be the separable polynomial in $\mathbb{F}_q[\lambda]$ with roots in $\bar{\mathbb{F}}_q$ given by the reduction of the elements

$$t_i^h/\omega_F^\ell,$$

where $t_i$ are the nonzero roots of $P_X$ (that is, the roots of $P_X^0$). The elements $t_i^h$ have been multiplied by an appropriate power of the uniformizer, so that they become units. As a result, the roots of $R_X$ are nonzero.

**7.2. Aside on equal valuation.** In this article, we do not justify our restriction to this special kind of element. Without going into the details, it seems to me that the study of orbital integrals can be divided into two quite different parts. The part discussed in this article is that of elements of equal valuation. It seems that geometric methods such as motivic integration are very important for this part.

For elements of nonequal valuation, it seems that a quite different set of methods will be relevant. Here issues such as homogeneity (generalizing the results of Waldspurger [34] and DeBacker [5]) and descent for orbital integrals [24] should be relevant.

This is currently merely speculation, but it is the reason that I am restricting to elements of equal valuation. It seems that different groups could study these two different kinds of orbital integrals with little interaction and few shared methods.

**7.3. A hope**

CONJECTURE 7.1. *If $X$ and $X'$ are elements in* strip($r$) *such that $R_X = R_{X'}$, then their orbital integrals are equal.*

We represent strip($r$) schematically as a long rectangular strip (a union of semisimple orbits). Around the conjugacy class of $X$ we can draw a tube (a thickened neighborhood) of all the elements in the strip with the same reduced characteristic polynomial $R_X$. (Figure 2.) The function $X \mapsto R_X$ thus partitions strip($r$) into tubes.

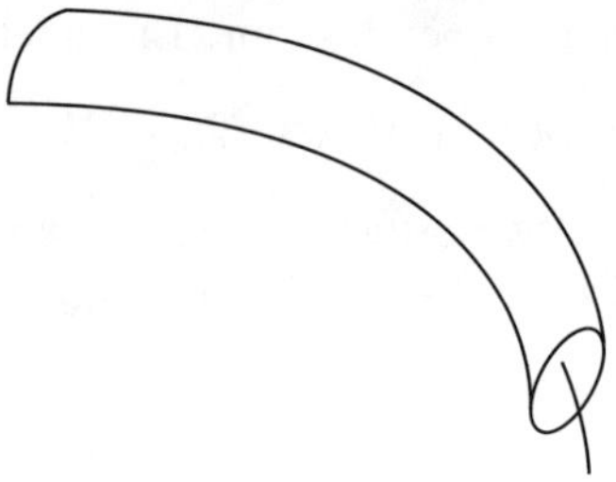

*Figure 12.2* A tube

**7.4. A $p$-adic fundamental lemma.** Langlands and Shelstad define a transfer factor $\Delta(X, Y, Z)$ on $so(2n+1) \times so(2k+1) \times so(2(n-k)+1)$. On the strip$(r)$ it has the form

$$q^c \operatorname{sign}(X, Y, Z)$$

for some constant $c = c(n, k, r)$ and some function sign taking values in $\{0, 1, -1\}$. Let $O_F$ be the ring of integers of the $p$-adic field $F$. They conjecture that for appropriate normalizations of measures [20], we have the following special case of the fundamental lemma:

CONJECTURE 7.2. *(Langlands-Shelstad) For all $Y$ and $Z$ regular semisimple such that there exists a regular semisimple $X$ in $\mathfrak{g}$ such that $P_X^0 = P_Y^0 P_Z^0$, we have*

$$q^c \Sigma_X \int_{O(X)\cap\mathfrak{g}(O_F)} \operatorname{sign}(X, Y, Z) = \int_{O^{st}(Y)\times O^{st}(Z)\cap\mathfrak{h}(O_F)} 1.$$

The sum runs over representatives of all regular semisimple conjugacy classes. The function sign $(X, Y, Z)$ is zero unless $P_X^0 = P_Y^0 P_Z^0$. It is enough to restrict the sum to such representatives.

PROPOSITION 7.3. *If $F$ is a field of sufficiently large residual characteristic, then the sign of the transfer factor in $so(2n+1)$,*

$$\operatorname{sign}^{-1}(x), \ \textit{for } x \in \{0, 1, -1\},$$

*is given by a formula in the language of rings.*

The proof will be given in a separate article. The surprising thing about this calculation is that the full strength of Pas's language is not required. That is, the transfer factor is expressed without the functions ord and ac, and without quantifiers over the additive group and residue field. This means that we can define a transfer factor for any field.

The formula for $\mathrm{sign}^{-1}\{-1, 1\}$ does not require quantifiers. It is the set of elements $(X, Y, Z)$ with

$$\lambda P_X = P_Y P_Z$$

for which $X$ is regular, which is expressed as the nonvanishing of the resultant

$$\mathrm{resultant}\,(P_X, P'_X) \neq 0.$$

The starting point for the proof of the proposition is Waldspurger's simplified formula for the transfer factors on the Lie algebra of classical groups [33].

From this proposition, the Denef-Loeser construction gives us a $+1$-motive in $\hat{K}_0^v(\mathrm{Mot}_{\mathrm{k},\bar{\mathbb{Q}}})_{loc}$. It also gives a $-1$-motive in the same set. In this sense, we can affirm that the Langlands-Shelstad transfer factor is a motive.

**7.5. Orbital integrals.** A serious difficulty that we encounter in the study of orbital integrals is that individual orbits of semisimple elements are not given by a formula in Pas's language. In fact, the characteristic polynomial

$$P_X \in F[\lambda]$$

has $p$-adic coefficients, which cannot be expressed in the language. (Without a uniformizer in the language, we cannot express the $p$-adic expansion of the coefficients.)

Our only hope is to use the fact that orbital integrals are locally constant. We place each orbit into a larger tube, where the tube is large enough to be defined by a formula in Pas's language. Each tube is defined by the set of elements with a given reduced characteristic polynomial with coefficients in the residue field (see Section 7.1):

$$R_X \in \mathbb{F}_q[\lambda].$$

To patch these together into a global object, we let $k$ be a finite extension of $\mathbb{Q}$, with ring of integers $O_k$. Each polynomial $R_X$ is a specialization of a polynomial

$$\dot{R}_X \in S[\lambda],$$

where $S$ is the coordinate ring over $O_k$ of the set of regular orbits

$$Z = Z_r = \mathfrak{a}/\mathrm{Ad}\,A$$

on an appropriate Lie algebra $\mathfrak{a}$ and group $A$, defined over $O_k$. We can then use $\dot{R}_X$ to define a formula in

$$\mathcal{L}_{pas}(S).$$

(This denotes the Pas's language extended by a symbolic constant for each element of $S$, as described in [14, Section 6.3].)

The formula for the set of elements in the tube with transfer factor equal to $+1$ gives, by the construction of Denef and Loeser, a Chow motive

$$\Theta^{G,+}_{n,k,r}.$$

The negative part of the tube gives a second Chow motive

$$\Theta^{G,-}_{n,k,r}.$$

The tube on the endoscopic groups gives a third Chow motive

$$\Theta^{H,st}_{n,k,r}.$$

Recall that the $p$-adic transfer factor has the form

$$\pm q^{c}$$

for some constant $c = c(n, k, r)$. We are thus able to formulate a motivic fundamental lemma:

CONJECTURE 7.4. *Given n, k, and r, we have*

$$\mathbb{L}^{c}\left(\Theta^{G,+}_{n,k,r} - \Theta^{G,-}_{n,k,r}\right) = \Theta^{H,st}_{n,k,r}$$

*in*

$$\hat{K}^{v}_{0}(M_{\mathbb{Q}(Z_r),\bar{\mathbb{Q}}})_{loc,\mathbb{Q}}.$$

This single identity of Chow motives governs the fundamental lemma over the entire strip$(r)$ at almost all places.

*Remark* 7.5. The Denef-Loeser comparison theorem relates the trace of Frobenius on these motives to the traditional fundamental lemma. The Denef-Loeser comparison theorem in its current form is not quite strong enough to deduce the fundamental lemma from its motivic form. However, I hope that these are relatively minor obstacles that future research should be able to surmount.

First of all, we need the Denef-Loeser comparison theorem for the finitely generated extension $\mathbb{Q}(Z_r)/\mathbb{Q}$. Denef and Loeser give two comparison theorems, one for $p$-adic integration on local fields of positive characteristic, and another on local fields in characteristic zero. The comparison theorem in characteristic zero assumes that the field is a finite extension of $\mathbb{Q}$ and hence it cannot be applied to $\mathbb{Q}(Z_r)$. (Denef and Loeser, via private communication have informed me that they can relax this restriction for $p$-adic fields of characteristic zero that are unramified over $\mathbb{Q}$. This is expected to be sufficient for applications to the fundamental lemma.)

The second restriction of the Denef-Loeser comparison theorem is that if $R$ is a normal domain with field of fractions $\mathbb{Q}(Z_r)$, then the comparison theorem holds at all closed points $x$ of Spec $R_f$ for some non-explicit element $f$ of $R$. It is possible that $f$ blocks a comparison of some elements of the $p$-adic field.

*Remark* 7.6. One of the most interesting aspects of this calculation is that it shows that there are two local-global pathways. The local-global pathway connecting automorphic representation theory with the representation theory of local fields is a well-established part of the Langlands program. It generalizes the pathway between global and local class field theory. The Denef-Loeser apparatus gives a genuinely new pathway between global and local objects. (To see that it is a different pathway, observe that a global regular semisimple element in a reductive group

$$\gamma \in G(\mathbb{Q})$$

lies in an unramified Cartan subgroup at almost every place. However, the Denef-Loeser construction quite often globalizes local data that is everywhere ramified.)

Thus, we have global problems in automorphic representation theory that are localized by the first pathway, and then globalized again by the second pathway to obtain conjectural identities of Chow motives.

**8. Open problems.** We conclude this article with four problems that are raised by the study of $p$-adic integrals from the vantage point of motivic integration.

PROBLEM 8.1. *Give effective algorithms to find the Chow motives*

$$\Theta^{*,*}_{n,k,r}.$$

By solving this problem, we succeed in computing $p$-adic orbital integrals in the sense proposed in this article. The quantifier elimination procedures (Pas's algorithm [30], Presburger's algorithm [12], and Galois stratification [14]) are entirely algorithmic. Thus, I hope that this first problem can be settled.

PROBLEM 8.2. *Prove the hope of Conjecture 7.1: if $R_X = R_{X'}$, then the orbital integrals of $O^{st}(X)$ and $O^{st}(X')$ are equal.*

In unpublished work, Clifton Cunningham has made progress toward a solution of this second problem.

PROBLEM 8.3. *Extend the results to degenerate elements $X$ that do not lie in any strip($\mathfrak{r}$). In particular, find finitely many motives over finitely generated extensions of $\mathbb{Q}$ that govern the fundamental lemma for all $X \in \mathfrak{g}$ over almost all completions of any number field.*

This seems to me to be a difficult problem. As an earlier section states (see 7.2), the methods involved here seem to be methods of generalized homogeneity laws in the spirit of Waldspurger and DeBacker ([34] and [5]).

PROBLEM 8.4. *Prove the motivic fundamental lemma.*

If the first three problems can be solved, then we have an algorithm to compute the Chow motives that govern the fundamental lemma for a given group. However, more is needed. First there is the problem of equality: given two Chow motives, is there an algorithm to determine if they are equal?

Second, there is the problem of induction. Even if we have an algorithm to check the fundamental lemma for one group, how do we give a proof for all reductive groups at once? Here it seems to me that we need to develop a deeper understanding of the motives that arise in connection with the fundamental lemma.

**9. Conclusion.** The Denef-Loeser apparatus of arithmetic motivic integration seems to mesh well with certain $p$-adic integrals that arise in representation theory.

We should investigate how far motives permeate representation theory of $p$-adic groups. If we believe with Denef and Loeser that all natural $p$-adic integrals are motivic, then the influence of the motivic point of view will be far-reaching. One can speculate that many of the basic objects of representation theory (such as Harish-Chandra characters) have a motivic nature.

The hope is that the motivic interpretation will allow us to calculate $p$-adic integrals that have resisted all efforts until now.

## REFERENCES


[1] J. Ax and S. Kochen, Diophantine problems over local fields, I, II, *Amer. J. Math* **87** (1965), 605–648; III, *Ann. Math.* **83** (1966), 437–456.

[2] S. Basu, R. Pollack, and M.-F. Roy, On the combinatorial and algebraic complexity of Quantifier Elimination, *Proceedings of the Foundations of Computer Science,* 1994, pp. 632–641.

[3] B. F. Caviness and J.R. Johnson (eds.), *Quantifier Elimination and Cylindrical Algebraic Decomposition,* Springer, 1998.

[4] P. J. Cohen, Decision procedures for real and $p$-adic fields, *Comm. Pure Appl. Math.* **22** (1969), 131–151.

[5] S. DeBacker, Homogeneity results for invariant distributions of a reductive $p$-adic group, preprint.

[6] J. Denef, $p$-adic semi-algebraic sets and cell decomposition, *J. Reine Angew. Math.* **369** (1986), 165–166.

[7] J. Denef and F. Loeser, Motivic Igusa zeta functions, *Journal of Algebraic Geometry* **7** (1998) 505–537.

[8] ———, Germs of arcs on singular algebraic varieties and motivic integration, *Inventiones Mathematicae* **135** (1999) 201–232.

[9] ———, Motivic exponential integrals and a motivic Thom-Sebastiani Theorem, *Duke Mathematical Journal,* **99** (1999) 285–309.

[10] ———, Definable sets, motives, and $p$-adic integrals, [math.AG/9910107], *Journal of the Amer. Math. Soc.* **14** (2001) 429–469.

[11] ———, Motivic integration, quotient singularities and the McKay correspondence, to appear in *Compositio Math.*

[12] H. Enderton, *A Mathematical Introduction to Logic,* Academic Press, 1972.

[13] Ju. L. Eršov, On the elementary theory of maximal normed fields, *Soviet Math. Dokl.* **6** (1965), 1390–1393.

[14] M. D. Fried and M. Jarden, *Field Arithmetic,* Springer, 1986.

[15] H. Gillet and C. Soulé, Descent, motives, and $K$-theory, *J. reine angew. Math.* **478** (1996), 127–176.

[16] K. Gödel, *On Formally Undecidable Propositions Of Principia Mathematica And Related Systems,* Dover reprint edition, 1972.

[17] F. Guillén and V. Navarro Aznar, Un critère d'extension d'un foncteur défini sur les schémas lisses, preprint.

[18] T. Hales, A simple definition of transfer factors, *Contemporary Math.* **145** (1993), 109–134.

[19] ———, Harmonic analysis and hyperelliptic curves, in *Representation Theory and Analysis on Homogeneous Spaces,* ed. by S. Gindikin, et al., *Comtemp. Math,* AMS, 1994.

[20] ———, Can $p$-adic integrals be computed, slides and video of a lecture at IAS, 6 Apr 2001, http://www.math.ias.edu/amf/.

[21] ———, Some algorithms arising in the proof of the Kepler conjecture, to appear.

[22] M. Kontsevich, lecture at Orsay, 7 Dec 1995.

[23] R. P. Langlands and D. Shelstad, On the definition of transfer factors, *Math. Ann.* **278** (1987), 219–271.

[24] ———, Descent for Transfer Factors, in *The Grothendieck Festscrift,* vol II, ed. by R. Cartier et al., 1991.

[25] F. Loeser, lecture at Strasbourg, 1999.

[26] E. Looijenga, Motivic Measures, arXiv.math.AG/0006220 v2 21 Oct 2000.

[27] Macintyre, On definable subsets of $p$-adic fields, *J. Symb. Logic* **41** (1976), 605–610.

[28] Quantifier Elimination—from MathWorld, http://mathworld.wolfram.com/QuantifierElimination.html

[29] B. Mishra, *Computational Real Algebraic Geometry, Handbook of Discrete and Computational Geometry,* CRC Press, 1997.

[30] J. Pas, Uniform $p$-adic cell decomposition and local zeta functions, *J. Reine Angew. Math.* **399** (1989), 137–172.

[31] M. Presburger, Über die Vollständigkeit eines gewissen Systems der Arithmetik ganzer Zahlen, in *welchem die Addition als einzige Operation hervortritt, Comptes-rendus du I Congrès des Mathématiciens des Pays Slaves,* Warsaw, **395** (1929), 92–101.

[32] A. Scholl, Classical motives, in *Motives,* U. Jannsen, S. Kleiman, J.-P. Serre Ed., *Proc. Symp. Pure Math.,* Vol 55 Part 1 (1994), 163–187.

[33] J.-L. Waldspurger, Intégrales orbitales nilpotentes et endoscopie pour les groupes classiques non ramifiés, *Astérisque,* vol. 269, SMF, 2001.

[34] ———, Quelques résultats de finitude concernant les distributions invariantes sur les algèbres de Lie $p$-adiques, preprint.

CHAPTER 13

# OCCULT PERIOD INVARIANTS AND CRITICAL VALUES OF THE DEGREE FOUR *L*-FUNCTION OF *GSp*(4)

By MICHAEL HARRIS

---

*To Joe Shalika*

**0. Introduction.** Let $G$ be the similitude group of a four-dimensional symplectic space over $\mathbb{Q}$:

$$G = \{g \in GL(4) \mid {}^t g J g = \lambda(g) J\} \tag{0.1}$$

where $J = \begin{pmatrix} 0 & I_2 \\ -I_2 & 0 \end{pmatrix}$ is the standard alternating form of dimension 4 and the homomorphism $\lambda : G \to \mathbb{G}_m$ is defined by (0.1). Let $\pi$ be a cuspidal automorphic representation of $G$, $\pi = \pi_\infty \otimes \bigotimes_p \pi_p$, where $p$ runs over rational primes. Let $S$ be the set of finite primes for which $\pi_p$ is not a spherical representation, together with the archimedean prime. The Langlands dual group of $G$ can be identified with $G$ itself, hence it makes sense to speak of the Langlands Euler product

$$L^S(s,\pi) = \prod_{p \notin S} L(s, \pi_p, r), \tag{0.2}$$

where $r : G \to GL(4)$ is the tautological representation. When $\pi_\infty$ is in the holomorphic discrete series and $S$ is empty, an integral representation for $L(s,\pi)$ was discovered by Andrianov [An], with a functional equation (at least) when all $\pi$ are unramified. An adelic version of Andrianov's construction, valid in principle for all $\pi$, and over any number field, was discovered by Piatetski-Shapiro a few years later, but was only published in 1997 [PS]. The article [PS] defines local Euler factors at ramified primes as well, and obtains local and global functional equations as in Tate's thesis. More generally, if $\mu$ is a Hecke character of $\mathbf{A}^\times/\mathbb{Q}^\times$, then one can define the twisted (partial) $L$-function $L^S(s,\pi,\mu) = \prod_{p\notin S} L(s,\pi_p,\mu_p,r)$. The Andrianov-Piatetski-Shapiro method gives integral representations and functional equations for these $L$-functions as well.

The constructions of [An] and [PS] are based on the Fourier expansion of forms in $\pi$. Let $P \subset G$ be the Siegel parabolic

$$P = \left\{ \begin{pmatrix} A & B \\ 0 & D \end{pmatrix} \right\} \subset G,$$

$U \subset P$ the unipotent radical, isomorphic to the additive group $S_2$ of symmetric $2 \times 2$ matrices. Let $\psi : \mathbf{A}/\mathbb{Q} \to \mathbb{C}^\times$ be a non-trivial additive character, $\beta \in S_2(\mathbb{Q})$, $\psi_\beta : U(\mathbf{A})/U(\mathbb{Q}) \to \mathbb{C}^\times$ the character $u \to \psi(tr(\beta u))$, where $u \in U(\mathbf{A})$ is viewed as an element of $S_2(\mathbf{A})$ as above and $tr$ is the usual trace. For $f \in \pi$, we can write

$$f(ug) = \sum_{\beta} f_\beta(g)\psi_\beta(u), \; u \in U(\mathbf{A}), \tag{0.3}$$

where $f_\beta$ is a smooth function on $G(\mathbf{A})$. We say $\beta$ is in the *support* of $\pi$ if $f_\beta \neq 0$ for some $f \in \pi$.

Write $P = MU$, with $M$ the Levi component, isomorphic to $GL(2) \times \mathbb{G}_m$. We assume $\det(\beta) \neq 0$, and let $D = D_\beta \subset M$ denote the identity component of the stabilizer of the linear form $u \to tr(\beta u)$ under the adjoint action. Then there is a unique quadratic semi-simple algebra $K = K_\beta$ over $\mathbb{Q}$ such that $D = R_{K/\mathbb{Q}}\mathbb{G}_m$. (For all this, as well as what follows, see [PS].) Let $N_\beta = \{n \in U \mid tr(\beta n) = 0\}$, and define $R = DU \subset P \subset G$. Let $H$ be the subgroup of $R_{K/\mathbb{Q}}GL(2)$ defined by the following cartesian diagram:

$$\begin{array}{ccc} H & \longrightarrow & R_{K/\mathbb{Q}}GL(2) \\ \downarrow & & \det \downarrow \\ \mathbb{G}_{m,\mathbb{Q}} & \longrightarrow & R_{K/\mathbb{Q}}\mathbb{G}_{m,K}. \end{array}$$

There is an embedding $H \to G$ such that $H \cap R = DN$ [PS, Prop. 2.1]. We let $\lambda_H$ denote the composition of the similitude character $\lambda$ with this embedding.

One chooses a $\beta$ in the support of $\pi$ and a Hecke character $\nu$ of $D(\mathbf{A})/D(\mathbb{Q})$, and constructs a standard Eisenstein series $E = E^\Phi(h; \mu, \nu, s)$ with $h \in H_\beta(\mathbb{Q})\backslash H_\beta(\mathbf{A})$, meromorphic in $s$, and depending on additional data $\Phi$ to be specified below. Let $Z_G$ denote the center of $G$. Then the family of integrals

$$Z(f, \Phi, \mu, \nu, s) = \int_{Z_G(\mathbf{A})H(\mathbb{Q})\backslash H(\mathbf{A})} f(h)E^\Phi(h; \mu, \nu, s)\, dh \tag{0.4}$$

has an Euler product whose local factors are almost everywhere given by $L(s, \pi_p, r)$:

$$Z(f, \Phi, \mu, \nu, s) = a(\pi, \beta, \nu) \prod_{w \in S} Z_w(f, \Phi, \mu, \nu, s) \prod_{p \notin S} L(s, \pi_p, \mu, r), \tag{0.5}$$

assuming of course $f$ and $\Phi$ to be factorizable data. Here $a(\pi, \beta, \nu)$ is a coefficient to be explained below. Note that the $L$-function on the right-hand side of (0.5) does not depend on the choice of $\nu$. Starting in §1, $\beta$ will be assumed isotropic, so $D$ will be a split torus and $\nu$ can be written as a pair of Hecke characters of $\mathbf{A}^\times/\mathbb{Q}^\times$.

The group $G$ is attached to a three-dimensional Shimura variety $Sh$, isomorphic to the Siegel modular variety of genus 2, with a canonical model over $\mathbb{Q}$ (see §2 for the formula). Holomorphic automorphic forms on $G$ can be regarded as sections of automorphic vector bundles on $Sh$, and possess a rational structure over $\mathbb{Q}$. In the setting of [An], in which $\pi$ is holomorphic discrete series, $\beta$ is necessarily a positive-definite (or negative-definite) symmetric matrix, $K$ is an imaginary quadratic field, and $H(\mathbb{R})$ is basically the same as $GL(2, \mathbb{C})$, so no Shimura variety is attached to $H$. Thus the formula (0.5) admits no clear interpretation in terms of algebraic geometry at integral points $s$. This is not surprising, since Deligne's conjecture expresses the critical values of $L(s, \pi)$ in terms of the determinant of a $2 \times 2$ matrix of periods, involving forms from distinct elements of the $L$-packet (conjecturally) attached to $\pi$. In particular, the critical values should involve periods of functions in $\pi' = \pi_\infty^{nh} \otimes \bigotimes_p \pi_p'$ where $\pi_p' = \pi_p$ for almost all $p$ but $\pi_\infty^{nh}$ belongs to the non-holomorphic discrete series associated to $\pi_\infty$. Anyway, no one knows how to construct periods of general cohomological automorphic forms on $G$.

On the other hand, suppose $\beta$ is of signature $(1, 1)$, so that $\pi_\infty$ is in the non-holomorphic discrete series. Then $K$ is either a real quadratic field or $K = \mathbb{Q} \oplus \mathbb{Q}$, $H$ corresponds to a Shimura subvariety $Sh_H \subset Sh$, and for special values of $s$ and appropriate choices of the auxiliary data the Eisenstein series $E^\Phi(h; \mu, \nu, s)$ are *nearly holomorphic* automorphic forms, in Shimura's sense [S2]. The modest goal of this note is to show that, for certain choices of $f$, the global zeta integral in (0.4), when $s = m$ is a critical value in Deligne's sense, can be interpreted as a cup product in coherent cohomology on $Sh_H$. This then expresses the special values of the corresponding $L$-function in terms of intrinsic coherent cohomological invariants of $\pi$, the occult period invariants of the title, related to the coefficients $a(\pi, \beta, \nu)$ for varying $\nu$. These invariants are doubly occult: in the first place, because they cannot be defined merely by reference to the abstract representation $\pi$ but depend on its realization in coherent cohomology; in the second place, because they are not (yet) known to be non-trivial in any specific case.

The invariants $a(\pi, \beta, \nu)$ are obtained by comparing a rational structure on the space $\pi$ defined in terms of coherent cohomology with one defined in terms of the Bessel model of $\pi$ attached to the pair $(\beta, \nu)$. More precisely, in the absence of a natural choice of archimedean local data, the invariant is the product $a(\pi, \beta, \nu) Z_\infty(f, \Phi, \mu, \nu, m)$. Coherent cohomological invariants of this type already appeared in [H2] in connection with Rankin-Selberg $L$-functions for Hilbert modular forms. Similar invariants, making use of the rational structure on topological cohomology of non-hermitian locally symmetric spaces, were related by Hida [Hida] to Rankin-Selberg $L$-functions for $GL(2)$ over totally imaginary number fields; Grenié [G] has recently obtained a partial generalization of Hida's work to $GL(n)$ for $n > 2$. The most intriguing discovery in the present paper is that odd and even critical values are related to $a(\pi, \beta, \nu)$ for different choices of $\nu$, an observation consistent with Deligne's conjecture.

The ideas in this note date back to 1988. The appearance of [PS] has made their publication more reasonable, and in view of the conjectures of Furusawa and Shalika [FS] on the products of two central special values of $L(s, \pi)$, publication may actually be of some use. The reader should nevertheless bear in mind that nothing in this paper should be considered definitive. In particular, the heuristic arguments presented here are vacuous unless one knows, first, that the archimedean zeta integrals in (0.5) do not vanish for a cohomological choice of data; and more crucially, that the global invariant (Bessel coefficient) $a(\pi, \beta, \nu)$ does not vanish for arithmetically interesting characters $\nu$. Local non-vanishing should not be too hard to establish, but I haven't tried to do so. On the other hand, I have no idea how to prove non-vanishing of the global invariant.

*Acknowledgment.* I began writing this article in the summer of 2001, during a visit to the Hong Kong University of Science and Technology; I thank Jian-Shu Li and the HKUST for their generous hospitality. I also thank David Soudry for his very pertinent answers to a number of questions about Piatetski-Shapiro's article and its relation to other integral representations of automorphic $L$-functions, Masaaki Furusawa for comments regarding his conjectures with Shalika, and Daniel Bertrand for the reference for the discussion in (2.7). I thank the referee for a careful reading. Finally, it is indeed a pleasure to dedicate this paper to Joe Shalika, in whose Princeton algebra course I first encountered the notion of a group representation, and who continues to inspire and surprise.

**1. Motives for *GSp*(4).** Notation is as in the introduction. We make the following departure from convention. Let $K_\infty = Z_G(\mathbb{R})U(2)$ be the standard maximal compact (mod center) *connected* subgroup of $G(\mathbb{R})$, the stabilizer of the point $i \cdot I_2$ in the Siegel upper half space of genus 2. For us, it is more convenient to let a cuspidal automorphic representation of $G$ be an irreducible $(\mathfrak{g}, K_\infty) \times G(\mathbf{A}_f)$-submodule of the cusp forms on $G(\mathbb{Q})\backslash G(\mathbf{A})$. In the cases of interest to us, it generally takes two automorphic representations of this kind to make one of the usual kind, simply because $G(\mathbb{R})$ is disconnected.

*Henceforward we assume $\beta$ to be isotropic over* $\mathbb{Q}$, and write $H$ for $H_\beta$. (There is also a theory for general indefinite $\beta$, but it seems to give less complete results.) Then up to isomorphism,

$$H = \{(g_1, g_2) \in GL(2) \times GL(2)\ |det(g_1) = det(g_2)\}.$$

Thus, letting $Sh(GL(2))$ denote the standard Shimura variety attached to $GL(2)$ (the tower of modular curves of all levels), there is a natural embedding

$$Sh_H \hookrightarrow Sh(GL(2)) \times Sh(GL(2))$$

rational over $\mathbb{Q}$. Let $pr_1$ and $pr_2$ denote the two projections of $Sh_H$ to the Shimura variety $Sh(GL(2))$ attached to $GL(2)$ (the tower of modular curves), corresponding to the composition of the $H \to GL(2) \times GL(2)$ with projection on the first and second factors respectively. Automorphic vector bundles on $Sh(GL(2))$ are denoted $F_{k,d}$ for pairs of integers $k \equiv d \pmod 2$. Given a triple $(k, \ell, -c)$ in $\mathbb{Z}$, with $k + \ell \equiv -c \pmod 2$, $F_{(k,\ell,-c)}$ on $Sh_H$ is the pullback via $(pr_1, pr_2)$ of the external tensor product $F_{k,d_1} \otimes F_{\ell,d_2}$ for any pair of integers $(d_1, d_2)$ such that $d_1 \equiv k \pmod 2$, $d_2 \equiv \ell \pmod 2$, and $d_1 + d_2 = -c$.

To any irreducible finite-dimensional representation $(\rho, V_\rho)$ of $G$ we can associate a local system $\tilde{V}_\rho$ in $\mathbb{Q}$-vector spaces over the Shimura variety $Sh$. Note that

$\rho$ can be realized over $\mathbb{Q}$. We write

$$_K Sh(\mathbb{C}) = G(\mathbb{Q})\backslash G(\mathbf{A})/K_\infty K,$$

$$Sh(\mathbb{C}) = \varprojlim_K {}_K Sh(\mathbb{C}),$$

where $K_\infty = Z_G(\mathbb{R})U(2)$ is as above and $K$ runs over open compact subgroups of $G(\mathbf{A}_f)$. Then

$$\tilde{V}_{\rho,\ell}(\mathbb{C}) = \varprojlim_K G(\mathbb{Q})\backslash G(\mathbf{A}) \times V_\rho(\mathbb{Q})/K_\infty K, \tag{1.1}$$

where $G(\mathbb{Q})$ acts diagonally on $G(\mathbf{A}) \times V_\rho(\mathbb{Q})$. There are compatible right actions of $G(\mathbf{A}_f)$ on *Sh* and $\tilde{V}_\rho$, and hence on the cohomology of the former with coefficients in the latter. The middle dimensional interior cohomology

$$H_!^3(Sh, \tilde{V}_\rho) = Im\left[H_c^3(Sh, \tilde{V}_\rho) \otimes \overline{\mathbb{Q}} \to H^3(Sh, \tilde{V}_\rho) \otimes \overline{\mathbb{Q}}\right]$$

decomposes as the direct sum of irreducible $\overline{\mathbb{Q}}[G(\mathbf{A}_f)]$-modules with finite multiplicities. One expects that, for the "general" $\overline{\mathbb{Q}}[G(\mathbf{A}_f)]$-module $\pi_f$, the space

$$M(\pi_f) = H_!^3(Sh, \tilde{V}_\rho)[\pi_f] = Hom_{G(\mathbf{A}_f)}\left(\pi_f, H_!^3(Sh, \tilde{V}_\rho)\right)$$

is four-dimensional, and moreover one has

$$H_!^3(Sh, \tilde{V}_\rho)[\pi_f] = Hom_{G(\mathbf{A}_f)}\left(\pi_f, H_{cusp}^3(Sh, \tilde{V}_\rho)\right)$$

where $H_{cusp}^3$ is the image in $H_!^3$ of the cuspidal cohomology. This is still not completely known, although significant partial results have been proved by Taylor, Laumon, and Weissauer. We assume $\pi = \pi_\infty \otimes \pi_f$ to have this property, which we call *stability* (at infinity); it also presupposes a multiplicity one property about which relatively little is known.

Tensoring over $\overline{\mathbb{Q}}$ with $\mathbb{C}$, the interior cohomology acquires a Hodge decomposition

$$H_!^3(Sh, \tilde{V}_\rho) \otimes_{\overline{\mathbb{Q}}} \mathbb{C} = \bigoplus_{i=0}^{3} H_!^i\left(Sh, E_\rho^{3-i}\right) \otimes_{\mathbb{Q}} \mathbb{C},$$

hence $M(\pi_f)$ has the analogous decomposition

$$M(\pi_f) = \bigoplus_{i=0}^{3} H_!^i\left(Sh, E_\rho^{3-i}\right) \otimes_{\mathbb{Q}} \mathbb{C}[\pi_f]. \tag{1.2}$$

Here $E_\rho^j$, $j = 0, 1, 2, 3$, is a locally free coherent sheaf (automorphic vector bundle) over *Sh*, defined over $\mathbb{Q}$, which we can describe explicitly in terms of the highest weights of $\rho$. We choose a maximal compact (mod center) torus $T \subset K_\infty$ and a positive root system as in [HK], to which the reader is referred for details of the following construction. Suppose $(\rho, \tilde{V}_\rho)$ has highest weight $(a, b, c)$, with $a \geq b \geq 0$ in $\mathbb{Z}$ and $c$ an integer congruent to $a + b$ modulo 2. An automorphic vector bundle is associated to an irreducible algebraic representation of $K_\infty$, hence to a triple of integers $(a', b', c')$ with $a' \geq b'$ and the same parity condition. Then $E^j_\rho$ is associated to the triple $\Lambda^j_\rho$, where

$$(1.3)\ \Lambda^0_\rho = (a, b, c); \quad \Lambda^1_\rho = (a, -b-2, c); \quad \Lambda^2_\rho = (b-1, -a-3, c);$$
$$\Lambda^3_\rho = (-b-3, -a-3, c)$$

[HK, Table 2.2.1].

For future reference, we note that the Hodge numbers corresponding to the Hodge decomposition (1.2) are given by

$$(1.4) \qquad (a+b+3+\delta, \delta), \quad (a+2+\delta, b+1+\delta),$$
$$(b+1+\delta, a+2+\delta), \quad (\delta, a+b+3+\delta)$$

where $\delta = \frac{c-a-b}{2}$ and the weight is $w = 3+c$. It will be most convenient to fix $c = a+b$; then $\delta = 0$ and the weight is $a+b+3$.

The notation $H^i_!$ is slightly abusive. Indeed, to define $H^i_!(Sh, E^j_\rho)$, one has first to replace ${}_KSh$ at finite level $K$ by a smooth projective toroidal compactification ${}_KSh_\Sigma$ such that ${}_KSh_\Sigma - {}_KSh$ is a divisor with normal crossings, and to replace $E^j_\rho$ by a pair of canonically defined extensions, $E^{j,sub}_\rho \subset E^{j,can}_\rho$ to vector bundles over $Sh_\Sigma$. (Here and below, the subscript $\Sigma$ designates an unspecified datum used to define a toroidal compactification, as in [H1].) Then [H1, §2]

$$H^i_!(Sh, E^j_\rho) = \varinjlim_K Im\big[H^i({}_KSh_\Sigma, E^{j,sub}_\rho) \to H^i({}_KSh_\Sigma, E^{j,can}_\rho)\big]$$

is independent of the choices of $\Sigma$ (hence the direct limit makes sense) and is an admissible $G(\mathbf{A}_f)$-module, with a canonical $\mathbb{Q}$-rational structure. Thus any irreducible admissible $G(\mathbf{A}_f)$-module $\pi_f$ that occurs in $H^i_!(Sh, E^j_\rho)$ can be realized over $\overline{\mathbb{Q}}$, and we can define $H^i_!(Sh, E^{3-i}_\rho) \otimes_{\mathbb{Q}} \overline{\mathbb{Q}}[\pi_f]$ as well as $H^i_!(Sh, E^{3-i}_\rho) \otimes_{\mathbb{Q}} \mathbb{C}[\pi_f]$. A strengthening of Hodge theory due to various people in various forms (Zucker, Faltings, ... ) then yields the decomposition (1.2). Stability at infinity comes down to

Hypothesis (1.5). *For $i = 0, 1, 2, 3$,*

$$H^i_!(Sh, E^{3-i}_\rho) \otimes_{\mathbb{Q}} \overline{\mathbb{Q}}[\pi_f] = H^i_{cusp}(Sh, E^{3-i}_\rho) \otimes_{\mathbb{Q}} \overline{\mathbb{Q}}[\pi_f]$$

*is of dimension one.*

Here $H^i_{cusp}$ is the image of the cusp forms in $H^i_!$ (cf.[H1]). Equality of the two spaces in (1.5) can be taken as part of the hypothesis; in any case, it will be automatic in the applications (see the next paragraph). In Hypothesis (1.5) we can replace $\overline{\mathbb{Q}}$ by $\mathbb{C}$; the two versions are equivalent. Actually, for most of what we have in mind it suffices to assume the different Hodge components have the same dimension. In any case, there is a discrete series $L$-packet $(\pi^j_\rho, j = 0, 1, 2, 3)$ of

$(\mathfrak{g}, K_\infty)$-modules and an isomorphism

$$(1.6) \qquad H^i_{cusp}\big(Sh, E^{3-i}_\rho\big) \otimes_{\mathbb{Q}} \mathbb{C}[\pi_f] = Hom_{(\mathfrak{g},K_\infty)\times G(\mathbf{A}_f)}\big(\pi^i_\rho \otimes \pi_f, \mathcal{A}_{cusp}(G)\big)$$

where $\mathcal{A}_{cusp}(G)$ is the space of cusp forms on $G(\mathbb{Q})\backslash G(\mathbf{A})$ [BHR]. For future reference, we let $H^i_{cusp}(Sh, E^{3-i}_\rho) \otimes_{\mathbb{Q}} \overline{\mathbb{Q}}(\pi_f)$ denote the image of any non-zero $G(\mathbf{A}_f)$-morphism $\pi_f \to H^i_{cusp}(Sh, E^{3-i}_\rho) \otimes_{\mathbb{Q}} \overline{\mathbb{Q}}$; by Hypothesis (1.5) such a morphism, an element of $H^i_{cusp}(Sh, E^{3-i}_\rho) \otimes_{\mathbb{Q}} \overline{\mathbb{Q}}[\pi_f]$ (note the difference in notation!), is unique up to (algebraic) scalar multiples.

We now return to $\pi$ as in the introduction and assume $\pi = \pi^2_\rho \otimes \pi_f$, so $\pi$ contributes to coherent cohomology in degree 2, i.e., to $H^2_{cusp}(Sh, E^1_\rho) \otimes_{\mathbb{Q}} \overline{\mathbb{Q}}$. We assume $a > b > 0$ (strict inequality), so that $H^2_{cusp}(Sh, E^1_\rho) = H^2_!(Sh, E^1_\rho)$ (cf. [MT], 2.1, Proposition 1). Moreover, we have a natural inclusion [H1, §3]

$$(1.7) \qquad H^2_{cusp}\big(Sh, E^1_\rho\big) \subset \varinjlim_K H^2\big({}_K Sh_\Sigma, E^{1,sub}_\rho\big).$$

The space on the right-hand side is denoted $\tilde{H}^2(Sh, E^{1,sub}_\rho)$, as in [H1, HK].

Let $Sh_H$ be as in the introduction. We may assume $T$ to be a maximal torus in $H = H_\beta$; then $T(\mathbb{R})$ is a maximal connected compact (mod center) subgroup of $H(\mathbb{R})$, and automorphic vector bundles on $Sh_H$ correspond to weights of the torus $T$. Specifically, we let $T = H \cap K_\infty$, the stabilizer of the point $(i, i)$ in the product of two upper half-planes; we occasionally also write $K_{H,\infty} = T(\mathbb{R})$. Thus any triple of integers $\Lambda^\# = (r, s, c)$, with $c \equiv r + s \pmod 2$, defines an automorphic vector bundle $F_{\Lambda^\#} = F_{(r,s,c)}$ on $Sh_H$. As in (1.6), we can define toroidal compactifications $Sh_{H,\Sigma_H}$ of $Sh_H$, and canonical and subcanonical extensions $F^{can}_{\Lambda^\#}$ and $F^{sub}_{\Lambda^\#}$ over $Sh_{H,\Sigma_H}$. For $i \in \mathbb{Z}$, we define $\Lambda^\#(i) = (a - i, i - b - 2, c)$ (for $\rho = (a, b, c)$ as above). Let $\iota : Sh_H \to Sh$ be the embedding. For $0 \le i \le a + b + 2$ there are homomorphisms [HK, (2.6.3)]

$$\iota^*\big((E^1_\rho)^{sub}\big) \to F^{sub}_{\Lambda^\#(i)}$$

giving rise to homomorphisms

$$(1.8) \qquad \psi_i : \tilde{H}^2_{cusp}\big(Sh, (E^1_\rho)^{sub}\big) \to \tilde{H}^2\big(Sh_H, F^{sub}_{\Lambda^\#(i)}\big).$$

It follows from (1.7) that one can actually lift $\psi_i$ to a homomorphism

$$(1.9) \qquad H^2_{cusp}\big(Sh, E^1_\rho\big) \to \tilde{H}^2\big(Sh_H, F^{sub}_{\Lambda^\#(i)}\big) \xrightarrow{\sim} \big[\tilde{H}^0\big(Sh_H, F^{can}_{(i-a-2,b-i,-c)}\big)\big]^*,$$

where the isomorphism is given by Serre duality ([H1], Corollary 2.3; see §8 of [H1] for the shift by 2). The right-hand side is isomorphic to the space of (pairs of) classical holomorphic modular forms on $Sh_H$, including Eisenstein series, of weight $(a + 2 - i, i - b)$. Unless $b + 1 \le i \le a + 1$ the right-hand side is therefore uninteresting for our purposes.

Let $f \in \pi$ belong to the lowest $K_\infty$-type subspace of $\pi_\infty$, which has highest weight $(a+3, -b-1, c)$ [HK, Table 2.2.1]. We assume $f$ is a weight vector for $T$ with character $(a+2-i, i-b, c)$, hence upon restriction to $H$ can pair non-trivially with a section $g \in \tilde{H}^0(Sh_H, F^{can}_{(i-a-2,b-i,-c)})$, to yield a complex number $< f, g >$. If both $f$ and $g$ are rational over $\overline{\mathbb{Q}}$, then so is $< f, g >$. (Cohomology classes in $H^2_{cusp}(Sh, E^1_\rho)$ are defined by vector-valued automorphic forms via a normalized Dolbeault isomorphism, as in [H1]. By $f$ being rational we mean $f$ is the weight component of character $(a+2-i, i-b, c)$ of a rational element of $H^2_{cusp}(Sh, E^1_\rho)$). We want to identify the zeta integral of (0.4) with such a pairing. Unfortunately, the Eisenstein series we need is in general not a holomorphic section of $F^{can}_{(i-a-2,b-i,-c)}$, so a priori the pairing cannot be interpreted in terms of coherent cohomology.

**(1.10) Maass operators.** Our main results are based on the algebraic interpretation of the Maass operators due to Katz, and developed in the present language in [H2] and elsewhere. In this section $k$ and $\ell$ are positive integers. Let $\Omega^1_i$, $i = 1, 2$ denote the pullback via $pr_i$ to $Sh_H$ of the cotangent bundle of $Sh(GL(2))$, and let $jet^{r_1,r_2}(F_{(-k,-\ell,-c)})$ denote the pullback via $(pr_1, pr_2)$ of $jet^{r_1} F_{-k,d_1} \otimes jet^{r_2} F_{-\ell,d_2}$. Let

$$j^{r_1,r_2} : F^{can}_{(-k,-\ell,-c)} \to jet^{r_1,r_2}\left(F^{can}_{(-k,-\ell,-c)}\right)$$

denote the canonical differential operator of order $r_1 + r_2$, and let

$$(1.10.1) \qquad \underline{Split}(r_1, r_2) : jet^{r_1,r_2}\left(F^{can,\infty}_{(-k,-\ell,-c)}\right) \longrightarrow$$

$$\left[Sym^{r_1}\left(\Omega^1_1\right) \otimes Sym^{r_2}\left(\Omega^1_2\right) \otimes F_{(-k,-\ell,-c)}\right]^{can,\infty} \xrightarrow{\sim} F^{can,\infty}_{(-k-2r_1,-\ell-2r_2,-c)}$$

be the canonical splitting of the Hodge filtration in the category of $H(\mathbf{A}_f)$-equivariant $C^\infty$ vector bundles (cf. [H2, 2.5]). Then $\delta_{r_1,r_2} = \underline{Split}(r_1, r_2) \circ j^{r_1,r_2}$ corresponds to the classical Maass operator in two variables. Explicitly, let $E_k$ and $E_\ell$ be holomorphic modular forms on $GL(2, \mathbb{Q})\backslash GL(2, \mathbf{A})$ of weights $k$ and $\ell$, respectively, corresponding (via the trivializations of [H2]) to sections $f_k \in H^0(Sh(GL(2)), F_{-k,d_1})$ and $f_\ell \in H^0(Sh(GL(2)), F_{-\ell,d_2})$, respectively. Then $\delta_{r_1,r_2}(pr_1, pr_2)^*(f_k \otimes f_\ell)$ is the $C^\infty$ section of $F^{can,\infty}_{(-k-2r_1,-\ell-2r_2,-c)}$ corresponding to the restriction to $H$ of

$$D^{r_1}_k E_k \otimes D^{r_2}_\ell E_\ell$$

where for any $j$, $d_j$ is the first order differential operator on the upper half-plane

$$d_j = \frac{1}{2\pi i}\left(\frac{d}{dz} + \frac{j}{2iy}\right)$$

and $D^{r_1}_k = d_{k+2r_1-2} \circ \ldots d_{k+2} \circ d_k$, $D^{r_2}_\ell = d_{\ell+2r_2-2} \circ \ldots d_{\ell+2} \circ d_\ell$.

The map (1.10.1) is a splitting in the $C^\infty$ category of a short exact sequence of automorphic vector bundles:

$$(1.10.2)\quad 0 \to \left[Sym^{r_1}(\Omega_1^1) \otimes Sym^{r_2}(\Omega_2^1) \otimes F_{(-k,-\ell,-c)}\right]^{can} \to jet^{r_1,r_2}(F_{(-k,-\ell,-c)})^{can} \to jet^{r_1-1,r_2-1}(F_{(-k,-\ell,-c)})^{can} \to 0$$

where the final term is defined by analogy with that in the middle.

PROPOSITION 1.10.3. *Let dh denote a Haar measure on* $Z_G(\mathbf{A})H(\mathbb{Q})\backslash H(\mathbf{A})$. *There is a constant* $c(dh) \in \mathbb{R}^\times$ *with the following property. Let* $\pi \subset \mathcal{A}_0(G)$ *be as in the introduction. Suppose* $f \in \pi$ *defines a* $\overline{\mathbb{Q}}$*-rational cohomology class (also denoted* $f$*) in* $H^2_{cusp}(Sh, E^1_\rho)$. *Let i be an integer,* $b+1 \leq i \leq a+1$, *and suppose*

$$a+2-i = k+2r_1, \qquad i-b = \ell+2r_2$$

*with positive (resp. non-negative) integers* $k, \ell$ *(resp.* $r_1, r_2$*). Let* $E_k$, $E_\ell$ *be holomorphic modular forms on* $GL(2,\mathbb{Q})\backslash GL(2,\mathbf{A})$, *as above, and suppose the corresponding* $f_k$ *and* $f_\ell$ *are* $\overline{\mathbb{Q}}$*-rational. Then*

$$c(dh)\cdot \int_{Z_G(\mathbf{A})H(\mathbb{Q})\backslash H(\mathbf{A})} f(h) D_k^{r_1} E_k \otimes D_\ell^{r_2} E_\ell(h)dh$$

*lies in* $\overline{\mathbb{Q}}$.

*Proof.* With $k, \ell, r_j$, and $i$ as in the statement of the proposition, the first non-zero term on the left of (1.10.2) can be identified with $F^{can}_{(-k-2r_1,-\ell-2r_2,-c)} = F^{can}_{(-2-a+i,i-b)}$. Tensoring with $F^{sub}_{\Lambda^\#(i)} = F^{sub}_{(a-i,i-b-2,c)}$ we obtain an exact sequence

$$0 \to F^{sub}_{(2,2)} \to \left[jet^{r_1,r_2}(F_{(-k,-\ell,-c)}) \otimes F_{(a-i,i-b-2,c)}\right]^{sub} \to \left[jet^{r_1-1,r_2-1}(F_{(-k,-\ell,-c)}) \otimes F_{(a-i,i-b-2,c)}\right]^{sub} \to 0,$$

which we rewrite

$$0 \to F^{sub}_{(2,2)} \to (J^{r_1,r_2})^{sub} \to (J^{r_1-1,r_2-1})^{sub} \to 0.$$

Now $F^{sub}_{(2,2)} \xrightarrow{\sim} \Omega^1_{Sh_H}$ is the dualizing sheaf. Taking the long exact sequence of cohomology, we have

$$(1.10.4)\quad \ldots \to \tilde{H}^1\left(Sh_H, (J^{r_1-1,r_2-1})^{sub}\right) \to \tilde{H}^2\left(Sh_H, \Omega^1_{Sh_H}\right) \to \tilde{H}^2\left(Sh_H, (J^{r_1,r_2})^{sub}\right) \to \tilde{H}^2\left(Sh_H, (J^{r_1-1,r_2-1})^{sub}\right) \to 0$$

which by Serre duality (cf. [H1], Corollary 2.3 for the duality between $^{sub}$ and $^{can}$) yields

$$0 \to \tilde{H}^0\left(Sh_H, \Omega^1_{Sh_H} \otimes (J^{r_1-1,r_2-1})^{sub,*}\right) \to \tilde{H}^0\left(Sh_H, \Omega^1_{Sh_H} \otimes (J^{r_1,r_2})^{sub,*}\right) \to \tilde{H}^0\left(Sh_H, \mathcal{O}^{can}_{Sh_H}\right) \to \tilde{H}^1\left(Sh_H, \Omega^1_{Sh_H} \otimes (J^{r_1-1,r_2-1})^{sub,*}\right) \to \cdots.$$

Here we use the superscript * to denote duality. Now $\Omega^1_{Sh_H} \otimes (J^{r_1-1,r_2-1})^{sub,*}$ has a finite filtration whose associated graded object is a sum of line bundles of the form

$$\Omega^1_{Sh_H} \otimes F^{sub,*}_{(i+a-2+2e_j,b-i+2f_j)} = F^{can}_{(2-a-i-2e_j,i-b-2f_j)}$$

where $0 \leq e_j \leq r_1 - 1$, $0 \leq f_j \leq r_2 - 1$. By our choice of $r_1$ and $r_2$, each term is of the form $F_{\alpha,\beta}$ with $\alpha, \beta \geq 2$. It is known (cf. [H1, §8]) that $H^1(Sh_H, F^{can}_{\alpha,\beta}) = 0$ for $\alpha, \beta \geq 2$, hence (1.10.4) becomes

$$(1.10.5) \qquad 0 \to \tilde{H}^0\big(Sh_H, \mathcal{O}^{can}_{Sh_H}\big)^* \to \tilde{H}^2\big(Sh_H, (J^{r_1,r_2})^{sub}\big)$$
$$\to \tilde{H}^0\big(Sh_H, \Omega^1_{Sh_H} \otimes (J^{r_1-1,r_2-1})^{sub,*}\big) \to 0.$$

The term $\tilde{H}^0(Sh_H, \mathcal{O}^{can}_{Sh_H})^*$ is a sum of one-dimensional representations of $G(\mathbf{A}_f)$. On the other hand, by filtering $\Omega^1_{Sh_H} \otimes (J^{r_1-1,r_2-1})^{sub,*}$ as before, one sees that the $H(\mathbf{A}_f)$-representation on $\tilde{H}^0(Sh_H, \Omega^1_{Sh_H} \otimes (J^{r_1-1,r_2-1})^{sub,*})$ has a filtration by representations corresponding to holomorphic modular forms of positive weight, hence its Jordan-Hölder series contains no one-dimensional constituents. (One can make the filtration finite by restricting to the subrepresentation generated by vectors of fixed level $K \subset H(\mathbf{A}_f)$.) It follows that the natural map from the middle term of (1.10.5) to its $H(\mathbf{A}_f)$-coinvariants factors through a non-trivial, $\overline{\mathbb{Q}}$-rational map from the middle term

$$(1.10.6) \qquad I_{r_1,r_2} : \tilde{H}^2\big(Sh_H, (J^{r_1,r_2})^{sub}\big) \to \big[\tilde{H}^0\big(Sh_H, \mathcal{O}^{can}_{Sh_H}\big)^*\big]_{H(\mathbf{A}_f)}$$
$$= \big[\tilde{H}^0\big(Sh_H, \mathcal{O}^{can}_{Sh_H}\big)^*\big]^{H(\mathbf{A}_f)},$$

where the right-hand side is a one-dimensional space generated by the constant function 1. In particular, if $\phi$ is a rapidly decreasing Dolbeault cocycle representing a class $[\phi] \in \tilde{H}^2(Sh_H, (J^{r_1,r_2})^{sub})$ [H1], then $\phi \mapsto I_{r_1,r_2}[\phi]$ factors through projection on $K_{H,\infty} = T(\mathbb{R})$-invariants.

Putting together all these maps, we obtain a $\mathbb{Q}$-rational, $H(\mathbf{A}_f)$-invariant pairing

$$H^2_{cusp}\big(Sh, (E^1_\rho)^{sub}\big) \otimes \tilde{H}^0\big(Sh_H, F_{(-k,-\ell,-c)}\big)$$
$$\overset{\iota^*\otimes j^{r_1,r_2}}{\to} \tilde{H}^2\big(Sh_H, \iota^*\big((E^1_\rho)^{sub}\big)\big) \otimes \tilde{H}^0\big(Sh_H, jet^{r_1,r_2}(F_{(-k,-\ell,-c)})^{can}\big)$$
$$\overset{\psi_i}{\to} \tilde{H}^2\big(Sh_H, F^{sub}_{\Lambda^\#(i)}\big) \otimes \tilde{H}^0\big(Sh_H, jet^{r_1,r_2}(F_{(-k,-\ell,-c)})^{can}\big)$$
$$\overset{\cup}{\to} \tilde{H}^2\big(Sh_H, (J^{r_1,r_2})^{sub}\big)$$
$$\overset{I_{r_1,r_2}}{\to} \big[\tilde{H}^0\big(Sh_H, \mathcal{O}^{can}_{Sh_H}\big)^*\big]^{G(\mathbf{A}_f)} \overset{\sim}{\longrightarrow} \mathbb{C}.$$

Applying this composition to the rapidly decreasing Dolbeault cocycle represented by $f \otimes E_{k,\ell}$, we find that

$$I_{r_1,r_2}(\psi_i(f) \otimes j^{r_1,r_2}(E_{k,\ell})) \in \overline{\mathbb{Q}}.$$

But we have seen that $I_{r_1,r_2}$ factors through projection on the $T$-invariants. Thus $I_{r_1,r_2}$ factors through

$$\begin{aligned}\psi_i(f) \otimes j^{r_1,r_2}(E_{k,\ell}) &\mapsto (1 \otimes \underline{Split}(r_1, r_2))(\psi_i(f) \otimes j^{r_1,r_2}(E_{k,\ell})) \\ &= \psi_i(f) \otimes \delta^{r_1,r_2} E_{k,\ell}.\end{aligned}$$

Finally, the Serre duality pairing is expressed in terms of integration of Dolbeault cocycles with growth conditions [H1, Proposition 3.8]. The Proposition now follows from the definitions. □

**2. Interlude on Deligne's conjecture.** For motives, their $L$-functions, and their Deligne period invariants, we refer to [D]. Let $w = a + b + 3 = c + 3$ and let $\mu$ be a Hecke character of finite order. Let $\pi$ be as above, with central character $\xi_\pi = \xi_{0,\pi} \cdot |\bullet|^{-c}$, where $|\bullet|$ is the idèle norm; then $\xi_{0,\pi}$ is a Hecke character of finite order. We postulate the existence of a motive $M(\pi_f)$ with coefficients in some number field $E(\pi_f)$ of rank four, unramified outside $S$, of weight $w$, such that

$$L^S(s, M(\pi_f)) = L^S\left(s - \frac{3}{2}, \pi\right) \tag{2.1}$$

as Euler products away from $S$. The Hodge numbers are given by (1.4), with $\delta = 0$. If we are satisfied to work with motives for absolute Hodge cycles, as in [D], then the existence of $M(\pi_f)$ as indicated is roughly equivalent to Hypothesis 1.5, once one has overcome scruples regarding cohomology with support, given results of Taylor, Laumon, and Weissauer on the cohomology of the genus two Siegel modular variety. The functional equation of [PS], relating (the completed $L$-function) $L(s, \pi, \mu)$, to $L(1 - s, \hat{\pi}, \mu^{-1}) = L(c + 1 - s, \pi, \mu^{-1} \cdot \xi_{0,\pi}^{-1})$, becomes an equation relating $L(s, M(\pi_f), \mu)$ (one completes using the local factors of [PS]) to $L(w + 1 - s, M'(\pi_f), \mu^{-1})$, where

$$M'(\pi_f) = \hat{M}(\pi_f)(-w) = M \otimes M(\xi_{0,\pi}^{-1}) \tag{2.2}$$

where $\hat{*}$ designates duality and $(w)$ denotes Tate twist. Indeed, there is a non-degenerate bilinear pairing

$$M(\pi_f) \otimes M(\pi_f) \to M(\xi_{0,\pi})(-w). \tag{2.3}$$

in any realization, where $M(\xi_{0,\pi})$ is the rank one motive attached to the Dirichlet character $\xi_{0,\pi}$ (note that the values of $\xi_{0,\pi}$ are contained in the coefficient field $E(\pi_f)$). For instance, in the $\ell$-adic realizations, it suffices by Chebotarev's density theorem to verify this locally for all $p \notin S$, and this follows from (2.1) and the characteristic fact that

$$\hat{\pi}_f \xrightarrow{\sim} \pi_f \otimes \lambda \circ \xi_\pi^{-1} \tag{2.4}$$

which reflects the fact that $GSp(4)$ is its own Langlands dual group.

From (2.3) we derive the isomorphism (2.2). More generally, we let $M(\pi_f, \mu)$ be the motive whose $L$-function is $L(s - \frac{w}{2}, \pi, \mu)$ (one can obtain the $\ell$-adic realization by twisting by $\mu$ composed with the similitude character, since the Galois representation takes values in $GSp(4)$); then

$$\hat{M}(\pi_f, \mu) \xrightarrow{\sim} M(w) \otimes M((\mu \cdot \xi_{0,\pi})^{-1}).$$

Let $m$ be any integer. By standard calculations (as in [D] (5.1.8)) one verifies the following relations for the Deligne periods:

$$c^{+}(M(\pi_f, \mu)(m)) = (2\pi i)^{2m} g(\mu)^2 c^{\pm}(M(\pi_f)) \tag{2.5}$$

where $g(\mu)$ is a Gauss sum and $\pm = (-1)^{m+e(\mu_\infty)}$ where $e(\mu_\infty) = 0$ if $\mu_\infty$ is trivial and $= 1$ otherwise.

By (1.4) and standard hypotheses (e.g., [D,5.2]) the archimedean Euler factors in the functional equation for $L(s, M(\pi_f), \mu)$ are given, independently of $\mu$, by $\Gamma_{\mathbb{C}}(s)\Gamma_{\mathbb{C}}(s - b - 1)$, with $\Gamma_{\mathbb{C}}(s) = 2 \cdot (2\pi)^{-s}\Gamma(s)$ for the Euler Gamma function. The *critical values* of $L(s, M(\pi_f), \mu)$, in Deligne's sense, are then the integers $m \in [b + 2, a + 2]$. The right half of the critical set, accessible by combining the geometric considerations of §1 with the calculations in terms of Bessel models, is then the set of integers in $[\frac{a+b}{2} + 2, a + 2]$. Note that the central value $m = \frac{a+b}{2} + 2$ is critical if and only if $c = a + b$ is even.

By (2.5), Deligne's conjecture for the special values of $L(s, M(\pi_f, \mu))$ can be stated uniformly in terms of the Deligne periods $c^{\pm}(M(\pi_f))$ and elementary factors. Let $E(\pi_f, \mu)$ be the field generated by $E(\pi_f)$ and the values of $\mu$. We consider $L(s, M(\pi_f, \mu))$ as a function with values in $\mathbb{C} \otimes E(\pi_f, \mu)$, as in [D]. Then we have

2.6 (Deligne's Conjecture). *For* $m \in \left[\frac{a+b}{2} + 2, a + 2\right] \cap \mathbb{Z}$,

$$L(m, M(\pi_f, \mu))/(2\pi i)^{2m} g(\mu)^2 c^{\pm}(M(\pi_f)) \in E(\pi_f, \mu),$$

*with* $\pm = (-1)^{m+e(\mu_\infty)}$.

*(2.7) Remark.* When $\mu$ is fixed, Deligne's conjecture thus relates the odd and even critical values to distinct, presumably transcendental, invariants. When the motive does not have additional symmetries one expects the periods $c^{+}(M(\pi_f))$ and $c^{-}(M(\pi_f))$ to be algebraically independent. For example, suppose $M(\pi_f)$ is of the form $Sym^3(M)$, where $M$ is the motive attached to an elliptic modular form of weight $k > 2$. (Combining the proof by Kim and Shahidi of the existence of the symmetric cube lift from $GL(2)$ to $GL(4)$ with any of a number of methods (e.g. [GRS], or earlier unpublished results of Jacquet, Piatetski-Shapiro, and Shalika) for associating generic representations of classical groups to self-dual forms on $GL(4)$, one can construct at least part of the motive $Sym^3(M)$ on $GSp(4)$. See [KS, §9].) Deligne's calculations in [D, Prop. 7.7] identify $c^{+}(Sym^3(M))$ (resp. $c^{-}(Sym^3(M))$) with $(2\pi i)^{-1}c^{+}(M)^3c^{-}(M)$ (resp. $(2\pi i)^{-1}c^{-}(M)^3c^{+}(M)$), up to rational factors. On the

other hand, it follows from a generalization of a theorem of Th. Schneider, due to Bertrand (and subsequently vastly generalized by Wüstholz), that when $M$ is the motive attached to a modular form of weight 2, then $c^+(M)/c^-(M)$ is transcendental (cf. [B], Corollary 1, p. 35). In this case it follows easily that $c^+(Sym^3(M))$ and $c^-(Sym^3(M))$ are also algebraically independent over $\overline{\mathbb{Q}}$.

**3. Bessel models and zeta integrals.** Notation is as in the Introduction. Recall that we are assuming $\beta$ isotropic over $\mathbb{Q}$, so that $K = \mathbb{Q} \oplus \mathbb{Q}$. If $h = (g_1, g_2) \in H(F)$, for some field $F$, then $\lambda_H(h) = det(g_1) = det(g_2)$. Let $\mu$ and $\nu = (\nu_1, \nu_2)$ be Hecke characters of $\mathbf{A}^\times/\mathbb{Q}^\times$ and $(\mathbf{A}^\times/\mathbb{Q}^\times)^2$, respectively. Let $V$ denote the free $K$-module $K^2$, $\mathcal{S}(V_\mathbf{A})$ the space of Schwartz-Bruhat functions on the adeles of $V$. To $\Phi \in \mathcal{S}(V_\mathbf{A})$ one can assign an Eisenstein series $E^\Phi(h; \mu, \nu, s)$ as a function of $(h, s) \in H(\mathbb{Q})\backslash H(\mathbf{A}) \times \mathbb{C}$, meromorphic in $s$ but holomorphic for $Re(s) >> 0$; the normalizations are given in [PS, p. 270]. With this notation, the zeta integral $Z(f, \Phi, \mu, \nu, s)$ of (0.4) is defined.

We define an adelic character $\alpha_{\nu,\beta}$ of $R = DU$ by

$$\alpha_{\nu,\beta}(du) = \nu(d)\psi_\beta(u),\ d \in D(\mathbf{A}),\ u \in U(\mathbf{A}).$$

With this definition, and for $f \in \pi$, let

$$W_f(g) = W_f^{\beta,\nu}(g) = \int_{Z_G(\mathbf{A})R(\mathbb{Q})\backslash R(\mathbf{A})} f(rg)\alpha_{\nu,\beta}^{-1}(r)dr. \tag{3.1}$$

As in [PS], we refrain from normalizing measures, only requiring that measures on groups over non-archimedean local fields take algebraic values. The formulas to follow only hold for consistent choices of measures. The interesting question is to normalize the archimedean measure in an arithmetically meaningful way, in connection with hypothesis (3.2.2) below. Since we do not calculate the archimedean zeta integral explicitly, we do not address this question.

The map $f \mapsto W_f$ is a $G(\mathbf{A})$-equivariant homomorphism from $\pi$ to the space of functions $W$ on $G(\mathbf{A})$ satisfying

$$W(rg) = \alpha_{\nu,\beta}(r)W(g). \tag{3.2}$$

If this map is non-zero, it is called a $(\beta, \nu)$-Bessel model ([PS] refers to it as a generalized Whittaker model, but the terminology "Bessel model" appears in other articles of Piatetski-Shapiro and seems to be more widely used).

Let $\xi_\nu$ denote the restriction of $\nu$ to the ideles of $\mathbb{Q}$, embedded (diagonally) in the ideles of $K$. Let $\xi_\pi$ denote the central character of $\pi$, also a Hecke character of $\mathbb{Q}$. If $\pi$ has a global $(\beta, \nu)$-Bessel model, then necessarily $\xi_\nu = \xi_\pi$. Moreover, each local component $\pi_w$ has a local $(\beta, \nu_w)$-Bessel model, i.e., a map $\ell_{\beta,\nu_w}$ to the space of functions on $G(\mathbb{Q}_w)$ satisfying the analogue of (3.2). One knows (cf. Theorem 3.1 of [PS] and the references given there) that local Bessel models are unique. Thus if $f = \otimes f_w \in \pi = \bigotimes \pi_w$ is a factorizable function, we can factor $W_f = \otimes_w W_{f,w} = \otimes_w W_{f,w}^{\beta,\nu}$, and this gives rise to the Euler product factorization of the zeta integral (0.4). For details see [PS, §5], and the discussion below.

***(3.3) Hypotheses.***

(3.3.1) *$\beta$ is isotropic over $\mathbb{Q}$.*

(3.3.2) *$\mu$ is a character of finite order.*

(3.3.3) *$\pi$ has a $(\beta, \nu)$-Bessel model.*

These hypotheses are not all of the same nature. Hypotheses (3.3.1), already introduced in §1, and (3.3.2) carry no commitment, whereas (3.3.3) is an existence hypothesis.

The $\mathbb{Q}$-isotropic non-degenerate symmetric matrices $\beta$ form a single conjugacy class under the adjoint action of the rational points of the Levi component of $P$. If $\pi$ is a theta lift from maximally isotropic $O(4)$ or $O(6)$ then $\beta$ is in the support of $\pi$ for any $\beta$ in this conjugacy class [R, (I) §3]. Such a $\pi$ thus has a $(\beta, \nu)$-Bessel model for some $\nu$, but not necessarily the ones we introduce below.

**(3.4) Arithmetic Eisenstein series.** Let $B \subset GL(2)$ be the standard Borel subgroup, and choose an Iwasawa decomposition $GL(2, \mathbf{A}_f) = B(\mathbf{A}_f)K_f$, with $K_f = \prod_p GL(2, \mathbb{Z}_p)$. We choose a pair of integers $(k, \gamma)$ with $k \equiv \gamma \pmod 2$, $k > 0$, and a Dirichlet character $\bar{\mu}$, viewed as an adelic Hecke character of finite order, satisfying

(3.4.1) $$\bar{\mu}_\infty(-1) = (-1)^k.$$

Define the character

(3.4.2) $$\chi_{k,\gamma,\bar{\mu}} : B(\mathbf{A}) \to \mathbb{C}^\times;\ \chi_{k,\gamma,\bar{\mu}}\begin{pmatrix} a & b \\ 0 & d \end{pmatrix} = |ad|^{\frac{k+\gamma}{2}}|d|^{-k}\bar{\mu}(d),$$

where $|\bullet|$ is the idèle norm. Let $I_{k,\gamma,\bar{\mu}}$ be the induced representation $Ind_{B(\mathbf{A})}^{GL(2,\mathbf{A})}\chi_{k,\gamma,\bar{\mu}}$ where here and below we work with *non-normalized* induction.

Write $K_{GL(2),\infty}$ for the stabilizer $Z_{GL(2)} \cdot SO(2)$ of the point $i$ in the upper half-plane. Let $\phi \in I_{k,\gamma,\bar{\mu}}$ and write $\phi = \phi_\infty \otimes \phi_f$, and always assume $\phi_\infty$ to be $K_{GL(2),\infty}$-finite. By the Iwasawa decomposition for $GL(2, \mathbb{R})$, $\phi_\infty$ is determined by its restriction to $K_{GL(2),\infty}$, and can be written as a finite sum

$$\phi_\infty(t) = \sum_\kappa a_\kappa \kappa(t),$$

where $a_\kappa \in \mathbb{C}$ and $\kappa$ runs through characters of $T(\mathbb{R})$ whose restriction to the center $K_{GL(2),\infty} \cap B(\mathbb{R})$ of $GL(2, \mathbb{R})$ coincides with $\chi_{k,\gamma,\bar{\mu}}$. Say $\phi$ is *pure* (of type $\kappa$) if $\phi_\infty$ is isotypic for character $\kappa$ (i.e., $a_{\kappa'} = 0$ for $\kappa' \neq \kappa$). For each $\kappa$ as above, the unique pure $\phi_\infty$ of type $\kappa$ with $\phi_\infty(1) = 1$ is called a *canonical automorphy factor*, and is denoted $\phi_\kappa$. For exactly one $\kappa$ (namely $\kappa = k$, in an appropriate normalization; cf. [H2, §3]) $\phi_\kappa$ is a holomorphic automorphy factor; if $\phi$ is pure for this $\kappa$ we call $\phi$ holomorphic.

Suppose $k > 2$. Then for any $K_{GL(2),\infty}$-finite function $\phi \in I_{k,\gamma,\bar{\mu}}$ we can define an Eisenstein series $E_{k,\gamma,\bar{\mu}}(\phi)$ on $GL(2,\mathbb{Q})\backslash GL(2,\mathbf{A})$ by the absolutely convergent formula

$$E_{k,\gamma,\bar{\mu}}(\phi, g) = \sum_{\alpha \in B(\mathbb{Q})\backslash GL(2,\mathbb{Q})} \phi(\alpha g).$$

When $\phi_\infty$ is a holomorphic automorphy factor then $E_{k,\gamma,\bar{\mu}}(\phi)$ is a holomorphic Eisenstein series (of classical weight $k$); we denote it $E_{k,\gamma,\bar{\mu}}(\phi_f)$ to stress that $\phi_\infty$ is fixed. When $k = 1$, one can define a holomorphic Eisenstein series $E_{k,\gamma,\bar{\mu}}(\phi)$ by analytic continuation, and when $k = 2$ one can still define $E_{k,\gamma,\bar{\mu}}(\phi)$ by analytic continuation, provided $\bar{\mu}$ is a non-trivial character, which we will henceforth assume for simplicity.

We say $\phi$ is *arithmetic* if $\phi_\infty$ is pure of type $\kappa$ with $a_\kappa \in \overline{\mathbb{Q}}$, and if $\phi_f$ takes values in $\overline{\mathbb{Q}}$; we then say $E_{k,\gamma,\bar{\mu}}(\phi)$ is arithmetic (though not necessarily holomorphic). The following Lemma is a special case of the results used in §3 of [H2].

LEMMA 3.4.3. *Suppose $\phi$ is arithmetic and holomorphic. Then, with normalizations as in [H2], the normalized holomorphic Eisenstein series $(2\pi i)^{\frac{k+\gamma}{2}} E_{k,\gamma,\bar{\mu}}(\phi_f)$ corresponds, as in §1, to a $\overline{\mathbb{Q}}$-rational section of the line bundle $F_{-k,-\gamma}$ on $Sh(GL(2))$.*

Write $\underline{k}$ for the triple $(k, \gamma_1, \bar{\mu}_1)$. For two pairs $(\underline{k}_1, \phi_1)$ and $(\underline{k}_2, \phi_2)$, with $\underline{k}_1 = (k, \gamma_1, \bar{\mu}_1)$, $\underline{k}_2 = (\ell, \gamma_1, \bar{\mu}_1)$, we let $E_{\underline{k}_1,\underline{k}_2}(\phi_{1,f}, \phi_{2,f})$ denote the restriction of $E_{\underline{k}_1}(\phi_{1,f}) \otimes E_{\underline{k}_2}(\phi_{2,f})$ to $H(\mathbf{A}) \subset GL(2,\mathbf{A}) \times GL(2,\mathbf{A})$. For any pair $(r_1, r_2)$ of non-negative integers, we let

$$E^{(r_1,r_2)}_{\underline{k}_1,\underline{k}_2}(\phi_{1,f}, \phi_{2,f}) = D_k^{r_1} \otimes D_\ell^{r_2} E_{\underline{k}_1,\underline{k}_2}(\phi_{1,f}, \phi_{2,f}).$$

where $D_k^{r_1}$ and $D_\ell^{r_2}$ are the Maass operators on the first and second factors of $GL(2,\mathbf{A}) \times GL(2,\mathbf{A})$, respectively, normalized as in §1. Then $E^{(r_1,r_2)}_{\underline{k}_1,\underline{k}_2}(\phi_{1,f}, \phi_{2,f})$ is a (special value of a) real analytic Eisenstein series on $H$, which is *nearly holomorphic* in Shimura's sense—in other words, is contained in a representation generated by holomorphic representation of classical weight $(k + 2r_1, \ell + 2r_2)$ in the two variables. In any case, $E^{(r_1,r_2)}_{\underline{k}_1,\underline{k}_2}(\phi_{1,f}, \phi_{2,f})$ belongs to a representation of $H(\mathbf{A})$ isomorphic to the restriction of $Ind_{B(\mathbf{A})}^{GL(2,\mathbf{A})} \chi_{\underline{k}_1} \otimes Ind_{B(\mathbf{A})}^{GL(2,\mathbf{A})} \chi_{\underline{k}_2}$, independent of the choice of $(r_1, r_2)$.

One can define arithmeticity for Eisenstein series on $H(\mathbf{A})$ corresponding to $K_{H,\infty}$-finite functions in (the restriction to $H(\mathbf{A})$ of) $I_{k,\gamma_1,\bar{\mu}_1} \otimes I_{\ell,\gamma_2,\bar{\mu}_2}$; holomorphy has already been defined.

COROLLARY 3.4.4. *Suppose $\phi_{1,f}$ and $\phi_{2,f}$ take values in $\overline{\mathbb{Q}}$. Then $(2\pi i)^{r_1+r_2} E^{(r_1,r_2)}_{\underline{k}_1,\underline{k}_2}(\phi_{1,f}, \phi_{2,f})$ is arithmetic.*

This follows from standard formulas for the action of Maass operators on canonical automorphy factors (cf. [S1]); the power of $2\pi i$ comes from our normalization of these operators.

Let $B'$ denote the upper triangular subgroup of $H$. The integral representation (0.4) of $L(s, M(\pi_f), \mu)$, taking into account the shift (2.1), uses an Eisenstein series $E(h, \Phi, \mu, \nu, s)$ induced from the character

$$\chi_{s,\mu,\nu}\left(\begin{pmatrix} a_1 & b_1 \\ 0 & d_1 \end{pmatrix}, \begin{pmatrix} a_2 & b_2 \\ 0 & d_2 \end{pmatrix}\right) = \mu(a_1/d_2) \cdot |a_1/d_2|^{s-1} \cdot \nu_1^{-1}(d_1)\nu_2^{-1}(d_2) \tag{3.4.5}$$

of $B'(\mathbf{A})$, where we have the relation $a_1 d_1 = a_2 d_2$. Here $\mu$, $\nu_1$, and $\nu_2$ are Hecke characters of $\mathbb{Q}^\times \backslash \mathbf{A}^\times$, satisfying the single relation

$$\nu_1 \cdot \nu_2 = \xi_\pi = \xi_{\pi,0} \cdot |\bullet|^{-c}. \tag{3.4.6}$$

We assume $\mu$ to be of finite order, as above, and write $\nu_i = \nu_{i,0} \cdot |\nu_i|$, $i = 1, 2$, where $\nu_{i,0}$ is of finite order and $|\nu_i(t)| = |t|^{\alpha_i}$ is a power of the norm; then (3.4.6) implies $\alpha_1 + \alpha_2 = -c$. The argument $\Phi$ belongs to the space $\mathcal{S}(V_{\mathbf{A}})$ of Schwartz-Bruhat functions on $V_{\mathbf{A}}$, where $V = (\mathbb{Q}^2)^2$ corresponding to the realization of $H$ as a subgroup of $GL(2, \mathbb{Q}^2)$. In what follows, we let $(x_i, y_i)$, $i = 1, 2$, denote the standard rational coordinates on $V$, so that the identity subgroup of the stabilizer $T$ of the quadratic form $Q(v) = x_1^2 + y_1^2 + x_2^2 + y_2^2$, with $v = ((x_1, y_1), (x_2, y_2))$, is the stabilizer of the point $(i, i)$ in the product of two upper half-planes. The Schwartz-Bruhat function $\Phi_\infty$ defines a $K_{H,\infty}$-finite Eisenstein series $E(h, \Phi, \mu, \nu, s)$, for our choice of $K_{H,\infty}$, provided it is of the form $P(v)e^{-\pi Q(v)}$, where $P(v)$ is a polynomial and $e^{-\pi Q(v)}$ is the standard Gaussian. More precisely, let $I_K$ denote $\mathbf{A}^\times \times \mathbf{A}^\times$, as in [PS], and define

$$f^\Phi(h; \mu, \nu, s) = \mu(\det h)|\det h|^{s-1} \int_{I_K} \Phi((0, t)h)|t_1 t_2|^{s-1} \mu(t_1 t_2)\nu_1(t_1)\nu_2(t_2)\, d^\times t \tag{3.4.7}$$

as in [PS, §5], where $t = (t_1, t_2) \in \mathbf{A}^\times \times \mathbf{A}^\times$ and $|\bullet|$ is the idèle norm; we have incorporated the shift (2.1). It is then clear that $\Phi_\infty = e^{-\pi Q(v)}$ gives rise to a vector fixed by the maximal compact subgroup $K_{H,\infty}^c$ of $K_{H,\infty}$. More generally, $K_{H,\infty}$ acts linearly on the space of polynomials on $V(\mathbb{R})$, and if $P(v)$ is isotypic for $K_{H,\infty}$ then $\Phi_\infty = P(v)e^{-\pi Q(v)}$ yields a vector isotypic of the same type (for $K_{H,\infty}^c$). We let $\mathcal{E}(\chi_{s,\mu,\nu})$ denote the space of Eisenstein series of the form $E(h, \Phi, \mu, \nu, s)$.

We only consider Schwartz-Bruhat functions $\Phi(v_1, v_2)$ on $V(\mathbf{A}) = \mathbf{A}^2 \oplus \mathbf{A}^2$ that factor as $\Phi((v_1, v_2)) = \Phi_1(v_1)\Phi_2(v_2)$, and such that each $\Phi_i$ factors as $\prod_w \Phi_{i,w}$ over the places of $\mathbb{Q}$. It follows from (3.4.7) that

$$f^\Phi(1; \mu, \nu, s) = \int_{I_K} \Phi((0, t))|t_1 t_2|^{s-1} \mu(t_1 t_2)\nu(t)\, d^\times t$$

factors as a product of local Tate integrals

$$\prod_{i=1}^{2}\prod_{w}\int_{\mathbb{Q}_w^\times}\Phi_{i,w}(0,t_{i,w})|t_{i,w}|^{s-1}\mu(t_{i,w})\nu_{i,w}(t_{i,w})\,d^\times t_{i,w}$$

(3.4.8)

$$=\prod_{i=1}^{2}\prod_{w\in S}Z_w(\Phi_{i,w},s,\mu_{i,w},\nu_{i,w})\cdot L^S(s-1,\mu\cdot\nu_i),$$

where $S$ is a finite set of bad primes, including the archimedean primes, $Z_w(\Phi_{i,w}, s, \mu_{i,w}, \nu_{i,w})$ is just the local factor on the first line of (3.4.8), and $L^S(s-1, \mu\cdot\nu_i)$ is the partial Dirichlet $L$-series.

**(3.4.9).** Set $\bar{\mu}_i = (\mu\cdot\nu_{i,0})^{-1}, i = 1, 2$. For any automorphic representation $\sigma$ of $H(\mathbf{A})$ and any Hecke character $\xi$ of $\mathbf{A}^\times$, we write $\sigma\otimes\xi$ for $\sigma\otimes\xi\circ\lambda$. We fix positive integers $k$ and $\ell$ as above. Comparing (3.4.2) and (3.4.5), we obtain

LEMMA 3.4.9.1. *Let* $s = m$. *For any* $(r_1, r_2)$, $E^{(r_1,r_2)}_{\underline{k}_1,\underline{k}_2}(\phi_{1,f}, \phi_{2,f})$ *belongs to* $\mathcal{E}(\chi_{m,\mu,\nu})\otimes\mu^{-1}$ *provided*

(3.4.9.2) $$m-1=k-\alpha_1=\ell-\alpha_2=\frac{k+\ell+\gamma_1+\gamma_2}{2}.$$

*Moreover, any holomorphic vector in* $\mathcal{E}(\chi_{s,\mu,\nu})\otimes\mu^{-1}$ *is of the form* $E_{\underline{k}_1,\underline{k}_2}(\phi_{1,f}, \phi_{2,f})$ *for some choice of* $\phi_1, \phi_2$.

The first part is a trivial computation, whereas the second part follows from the fact that the holomorphic subspace of the archimedean component of $Ind_{B'(\mathbf{A})}^{H(\mathbf{A})}\chi_{m,\mu,\nu}\otimes\mu^{-1}$ is of dimension one. Given $s$ and $\nu$, the weight $(k, \ell)$ of the holomorphic vector is determined by (3.4.9.2), as is the sum $\gamma_1+\gamma_2$; the individual $\gamma_i$ are only visible on $GL(2)\times GL(2)$, and not on the subgroup $H$.

More precisely, it follows from (3.4.6) and (3.4.9.2) that

(3.4.9.3) $$\gamma_1+\gamma_2=-(\alpha_1+\alpha_2)=c;\quad \alpha_1=\frac{k-\ell-c}{2};\quad \alpha_2=\frac{\ell-k-c}{2}.$$

This and the congruences

(3.4.9.4) $$\gamma_1\equiv k \pmod 2,\quad \gamma_1\equiv\ell \pmod 2)$$

are the only restrictions on our choices. In order to obtain nearly holomorphic Eisenstein series, we also need to suppose (3.4.1), which is equivalent to

(3.4.9.5) $$\bar{\mu}_{i,\infty}(-1)=(-1)^{\gamma_i}.$$

As the reader will verify, (3.4.9.5) is compatible with (3.4.6).

Say the (factorizable) Schwartz-Bruhat function

$$\Phi((v_1,v_2))=\prod_{i=1}^{2}\prod_{w}\Phi_{i,w}(v_{i,w})$$

is *arithmetic* if $\Phi_{i,\infty}$ is of the form $P_i(v_i)e^{-\pi Q_i(v)}$ with $P_i$ a homogeneous polynomial of fixed degree $h_i$ with $\overline{\mathbb{Q}}$ coefficients, and if $\Phi_{i,w}$ for finite primes $w$ takes values in $\overline{\mathbb{Q}}$. It follows easily from (3.4.8), (3.4.9.5), and the classical formulas for special (critical) values of Dirichlet $L$-functions that, if $\Phi$ is arithmetic, then there is a constant $c^1_\infty$, depending only on $\mu_\infty$, $\nu_{i,\infty}$, the $h_i$, and $m$, such that the Eisenstein series $E(h, \Phi, \mu, \nu, m)$ is arithmetic, in the sense introduced above (3.4.3). Indeed, $c^1_\infty$ can be taken to be an integral power of $2\pi i$ (more precisely, a product of two integral powers of $2\pi i$, one coming from the factors $L^S(m-1, \mu\nu_i)$, the other coming from the archimedean zeta integrals), which can easily be determined explicitly. It's pointless to be more precise, though, since our final result will involve an archimedean zeta integral about which nothing is known.

LEMMA 3.4.9.6. *Suppose the Schwartz-Bruhat function $\Phi$ is arithmetic, and suppose the Eisenstein series $E(h, \Phi, \mu, \nu, m)$ and $E^{(r_1,r_2)}_{\underline{k}_1,\underline{k}_2}(\phi_{1,f}, \phi_{2,f})$ for some pair of non-negative integers $r_1$, $r_2$ and (any) finite data $\phi_{i,f}$ are of the same $K^c_{H,\infty}$-type. Then under (3.4.9.2), there is a constant $c^2_\infty(m)$, depending only on $\mu_\infty$, $\nu_{i,\infty}$, the $h_i$, and $m$, such that $c^2_\infty(m)E(h, \Phi, \mu, \nu, m)$ corresponds to the image under $D^{r_1}_k \otimes D^{r_2}_\ell$ of a $\overline{\mathbb{Q}}$-rational section of the line bundle $F_{(-k,-\ell,-c)}$ on $Sh_H$.*

*Proof.* An easy consequence of Lemma 3.4.9.1 and Corollary 3.4.4; the twist by $\mu^{-1}$ in (3.4.9.1) has no effect on the rationality over $\overline{\mathbb{Q}}$. □

***(3.5) The main theorem.*** *(3.5.0) Hypotheses, recalled.* As above, $a > b > 0$ is a pair of positive integers, $E^1_\rho$ the automorphic vector bundle on $Sh$ associated to the $K_\infty$-type with highest weight $\Lambda^1_\rho = (a, -b-2, a+b)$, and $\pi$ is a cuspidal automorphic representation of $G$ such that $H^2_!(Sh, E^1_\rho)[\pi_f]$ is of dimension one. Let $\mu$ be a Hecke character of $\mathbf{A}^\times/\mathbb{Q}^\times$ of finite order, and $\nu = (\nu_1, \nu_2)$ a pair of Hecke characters of $\mathbf{A}^\times/\mathbb{Q}^\times$, satisfying the relation (3.4.6). Let $\beta$ be a $\mathbb{Q}$-isotropic symmetric $2\times 2$-matrix with non-zero determinant; we assume $\pi$ admits a $(\beta, \nu)$-Bessel model. Various choices of $\nu$ will be made in the following discussion, depending on the special value in question and the sign of $\mu$. Finally, the integers $k, \ell, \gamma_1, \gamma_2$, are chosen subject to the restrictions (3.4.9.3) and (3.4.9.4).

To motivate the main theorem, we first work it out in the special cases

(3.5.1(A)) $\qquad \mu(-1) = 1, \quad a - b \equiv 0 \pmod 4;$

(3.5.1(B)) $\qquad \mu(-1) = 1, \quad a - b \equiv 2 \pmod 4.$

In both cases $c = a + b$ is even, so that $\xi_{0,\pi} = \nu_{1,0}\cdot\nu_{2,0}$. Odd and even critical values are treated by two separate calculations: (i) $k \equiv \ell \equiv 1 \pmod 2$ and (ii) $k \equiv \ell \equiv 2 \pmod 2$. In case A(i) or B(ii), we choose $i = \frac{a+b}{2} + 1$, so that $a + 2 - i = i - b$, i.e. the middle of the range considered in Proposition 1.10.3, and choose $k = \ell$ and $r_1 = r_2$ subject to the hypotheses of that proposition.

(3.5.1 A(i)/B(ii)) $$1 \le k = \ell \le \frac{a-b}{2} + 1, \quad 2r_1 = 2r_2 = \frac{a-b}{2} + 1 - k.$$

In case A(ii), we choose $i = \frac{a+b}{2}$, take $k = \ell$ and $r_2 = r_1 - 1$, so that

$$(3.5.1\ \mathrm{A(ii)/B(i)}) \qquad 2 \leq k = \ell \leq \frac{a-b}{2}, \quad 2r_1 = 2r_2 - 2 = \frac{a-b}{2} - k.$$

In case A(i)/B(ii) (resp. A(ii)/B(i)) we choose $\gamma_1$ and $\gamma_2$ odd (resp. even) subject to (3.4.9.4), and we *fix* $\nu_1$ and $\nu_2$ satisfying (3.4.9.2) and (3.4.9.5); in any case the weights $\alpha_i$ are determined by (3.4.9.3). We let $\nu^{odd}$ (resp. $\nu^{even}$) denote the fixed pair $(\nu_1, \nu_2)$ in case A(i)/B(ii) (resp. A(ii)/B(i)). We define $m$ by (3.4.9.2); then corresponding to (3.5.1) we have

$$(3.5.2\ \mathrm{A(i)/B(ii)}) \qquad m = \frac{a+b}{2} + k + 1 = \frac{a+b}{2} + 2, \frac{a+b}{2} + 4, \ldots a+2;$$

$$(3.5.2\ \mathrm{A(ii)/B(i)}) \qquad m = \frac{a+b}{2} + k + 1 = \frac{a+b}{2} + 3, \frac{a+b}{2} + 5, \ldots a+1.$$

The union of these two sets is precisely the right half of the critical set for $L(s, M(\pi_f, \mu))$, as determined in §2.

Assume $\Phi$ and $f \in \pi$ satisfy the hypotheses of Lemma 3.4.9.6 and Proposition 1.10.3, respectively, with the choices of $k, \ell, \ldots$ as above. In particular, $f$ is identified with a class in $H^2_{cusp}(Sh, E^1_\rho) \otimes_{\mathbb{Q}} \overline{\mathbb{Q}}(\pi_f)$. Assume $f = \otimes_w f_w$ is factorizable, with $f_w$ in the (abstract) representation $\pi_w$, as in the discussion preceding (3.3), and write $W_f(g) = \prod_w W^{\beta,\nu}_{f_w}(g_w)$, where $W_f$ is the generalized "Bessel function" of type $(\beta, \nu)$ defined by (3.1). The function $W_{f_w}(g_w)$ can be defined as $\ell_{\beta,\nu}(\pi_w(g_w)f_w)$, where $\ell_{\beta,\nu,w}$ is a (fixed) Bessel functional on $\pi_w$, as defined above. Recall that $f$ is of $K_\infty$-type $(a+3, -b-1, c)$. This is the lowest $K_\infty$-type $\tau_{\pi_\infty}$ in $\pi_\infty$, hence is of multiplicity one. Moreover, we have assumed $f$ to be a weight vector for $T(\mathbb{R})$ with character $(a+2-i, i-b, c)$, with the $i$ just specified. The corresponding weight subspace $\tau_{\pi_\infty}(a+2-i, i-b, c)$ is of dimension one. We arbitrarily choose a non-zero vector $f_{\pi_\infty}(a+2-i, i-b, c) \in \tau_{\pi_\infty}(a+2-i, i-b, c)$, and let $W^{(a+2-i,i-b,c)}_{\pi_\infty}$ denote its image under the (also arbitrarily chosen) non-zero $(\beta, \nu_\infty)$-Bessel functional $\ell_{\beta,\nu,\infty}$.

Now it follows as in [BHR] that the representations $\pi_p$, for $p$ finite, can all be realized over $\overline{\mathbb{Q}}$. Moreover, the local Hecke characters $\nu_p$ take algebraic values. It follows that the local Bessel functionals $\ell_{\beta,\nu,p}$ can be chosen to take $\overline{\mathbb{Q}}$-rational vectors in $\pi_w$ to functions $W_{f_p}$ in $C^\infty(G(\mathbb{Q}_p), \overline{\mathbb{Q}})$. Recall that we have defined $H^2_{cusp}(Sh, E^1_\rho) \otimes_{\mathbb{Q}} \overline{\mathbb{Q}}(\pi_f) \subset H^2_{cusp}(Sh, E^2_\rho) \otimes_{\mathbb{Q}} \overline{\mathbb{Q}}$ following (1.6), and that it is isomorphic to $\pi_f$. It then follows easily from the unicity of the Bessel model that:

PROPOSITION 3.5.2. *There exists a constant $a(\pi, \beta, \nu) \in \mathbb{C}^\times$, well-defined up to $\overline{\mathbb{Q}}^\times$-multiples, such that the global Bessel functional $f \mapsto W_f$ takes $H^2_{cusp}(Sh, E^1_\rho) \otimes_{\mathbb{Q}} \overline{\mathbb{Q}}(\pi_f) \xrightarrow{\sim} \pi_f$ to the space of $(\beta, \nu)$-Bessel functions on $G(\mathbf{A})$ of the form*

$$a(\pi, \beta, \nu) \cdot W^{(a+2-i,i-b,c)}_{\pi_\infty} \otimes W_{\pi_{fin}}$$

*where $W_{\pi_{fin}}(g_f) \in \overline{\mathbb{Q}}$ for all $g_f \in G(\mathbf{A}_f)$.*

The constant $a(\pi, \beta, \nu)$ is the occult period invariant of the title. Under Hypothesis (1.5) it depends only on $\pi_f$. Using the argument that follows, one can show that this remains true even without Hypothesis (1.5), provided there are non-vanishing special values in the critical range.

We can now explain the Euler factorization in (0.5). For $p \notin S$ we have arranged that $\Phi_p$ as well as $W_{f_p}$ are standard unramified data and the local zeta integral is just the local Euler factor $L_p(s, M(\pi_f, \mu)) = L(s - \frac{3}{2}, \pi_p, \mu, r)$. For $p \in S$ finite we have

(3.5.3)

$$Z_p(f, \Phi, \mu, \nu, s) = \int_{N_p \backslash H_p} W_{f_p}(h_p)\Phi_p((0, 1)h_p)\mu(\det h_p)|\det h_p|^{s-1} dh_p.$$

Here $H_p = H(\mathbb{Q}_p)$, $N_p = N_\beta(\mathbb{Q}_p)$, with $N_\beta$ as in the introduction, and $dh_p$ is a rational-valued Haar measure. The integral converges absolutely for $Re(s)$ sufficiently large, and extends analytically to a rational function of $s$, still denoted $Z_p(f, \Phi, \mu, \nu, s)$.

LEMMA 3.5.4. *Suppose $\Phi_p$ is arithmetic and $W_{f_p}$ takes algebraic values. Then for any integer $m$, $Z_p(f, \Phi, \mu, \nu, m) \in \overline{\mathbb{Q}}$. Moreover, for an appropriate choice of arithmetic data $\Phi$ and $W_{f_p}$ we can arrange that $Z_p(f, \Phi, \mu, \nu, m) \in \overline{\mathbb{Q}}^\times$.*

*Proof.* The first assertion is proved by in [H2, Lemma 3.4.2]. The non-vanishing of the local zeta integral at $m$ for some (not necessarily arithmetic) choice of data is implicit in Proposition 3.2 of [PS], and is proved by standard arguments. Since the arithmetic data define $\overline{\mathbb{Q}}$-structures on the Schwartz-Bruhat and Bessel spaces, and since the zeta integral is bilinear as a function of $W_{f_p}$ and $\Phi_p$, the second assertion then follows from the first. □

Finally, we let

$$(3.5.5)\quad Z_\infty(f, \Phi, \mu, \nu, s) = \int_{N_\beta(\mathbb{R})\backslash H(\mathbb{R})} W_{\pi_\infty}^{(a+2-i, i-b, c)}(h_\infty)\Phi_\infty((0, 1)h_\infty) \mu(\det h_\infty)|\det h_\infty|^{s-1} dh_\infty.$$

With these choices, the Euler product in the form (0.5) follows from Proposition 3.5.2.

Combining (1.10.3), (3.4.9.6), (3.5.2), (3.5.4), and (0.5), we obtain our main theorem:

THEOREM 3.5.5. *Let $m$ be in the right-hand half of the critical set of the $L$-function $L(s, M(\pi_f, \mu))$. Suppose $\mu(-1) = 1$ and $c = a + b$ is even. There is a constant $c_\infty^3(m) \in \mathbb{C}^\times$, well defined up to $\overline{\mathbb{Q}}^\times$-multiples, with the following property.*

*In case A(i)/B(ii), for m in the list (3.5.2) A(i)/B(ii), we have*

$$\left(c_\infty^3(m)\right)^{-1} a(\pi, \beta, \nu^{odd}) Z_\infty(f, \Phi, \mu, \nu^{odd}, m) L(m, M(\pi_f, \mu)) \in \overline{\mathbb{Q}}.$$

*In case A(ii)/B(i), for m in the list (3.5.2) A(ii)/B(i), we have*

$$\left(c_\infty^3(m)\right)^{-1} a(\pi, \beta, \nu^{even}) Z_\infty(f, \Phi, \mu, \nu^{odd}, m) L(m, M(\pi_f, \mu)) \in \overline{\mathbb{Q}}.$$

We have incorporated the constant $c(dh)$ of Proposition 1.10.3 into our new constant $c_\infty^3(m)$.

*Remarks.* Note that this theorem is roughly compatible with Deligne's conjecture, in that, up to the "elementary factor" $(c_\infty^3(m))^{-1} Z_\infty(f, \Phi, \mu, \nu^*, m)$, the special value is determined by the parity of $m$. Here and below $*$ denotes "odd" or "even". Of course this theorem is vacuous if (3.3.3) fails for all $\nu$ satisfying (3.4.9.4) and (3.4.9.5). Even if the appropriate Bessel model exists, the theorem is still vacuous if our normalized archimedean zeta factor $Z_\infty(f, \Phi, \mu, \nu^*, m)$ vanishes. It should not be too difficult to determine at least whether the non-holomorphic discrete series $\pi_\infty$ has non-vanishing $(\beta, \nu_\infty)$ Bessel models, and then the calculation of the archimedean zeta factor should not be too taxing. We suspect the theorem is not vacuous, because of the formal fit with Deligne's conjecture, but we have not carried out the necessary archimedean calculations, and we have nothing to say about the global hypothesis (3.3.3).

The product $a(\pi, \beta, \nu^*) Z_\infty(f, \Phi, \mu, \nu^*, m)$ does not depend on the choice of the vector $W_{\pi_\infty}^{(a+2-i, i-b, c)}$ (of given $K_\infty$ and $T$-type) in the $(\beta, \nu_\infty)$-subspace, but one expects there is a natural choice for which $Z_\infty(f, \Phi, \mu, \nu^*, m)$ is an algebraic multiple of some power of $\pi$ for the indicated values of $m$. Then $a(\pi, \beta, \nu^*)$ should be directly related to $c^\pm(M(\pi_f))$, where the relation of $*$ to $\pm$ depends on the parity of $\frac{a+b}{2}$.

**(3.5.6) The remaining cases.** We now assume $c = a + b$ odd, so $w = c + 3$ is even. As above, we distinguish two cases:

(3.5.6.1(C)) $\qquad \mu(-1) = 1, \quad a - b \equiv 1 \pmod 4;$

(3.5.6.1(D)) $\qquad \mu(-1) = 1, \quad a - b \equiv 3 \pmod 4.$

It is then natural to choose $i = \frac{a+b+3}{2}$, so that $a + 2 - i = i - b - 1$, one of two points closest to the middle of the range of Proposition 1.10.3. Then $k$ and $\ell$ necessarily have opposite parity: $k$ is odd in case (C) and even in case (D). Again there are two calculations, according as (i) $k = \ell - 1$ or (ii) $k = \ell + 1$. We have

(3.5.6.1C(i)) $\quad (k, \ell) = (1, 2), (3, 4), \ldots, \left(\frac{a-b+1}{2}, \frac{a-b+3}{2}\right),$

$$2r_1 = 2r_2 = \frac{a-b+1}{2} - k;$$

(3.5.6.1D(i)) $(k, \ell) = (2, 3), (4, 5), \ldots, \left(\frac{a-b+1}{2}, \frac{a-b+3}{2}\right)$,

$$2r_1 = 2r_2 = \frac{a-b+1}{2} + 1 - k;$$

(3.5.6.1C(ii)) $(k, \ell) = (3, 2), (5, 4), \ldots, \left(\frac{a-b+1}{2}, \frac{a-b-1}{2}\right)$,

$$2r_1 = 2r_2 - 2 = \frac{a-b+1}{2} - k;$$

(3.5.6.1D(ii)) $(k, \ell) = (2, 1), (4, 3), \ldots, \left(\frac{a-b+1}{2}, \frac{a-b-1}{2}\right)$,

$$2r_1 = 2r_2 - 2 = \frac{a-b+1}{2} + 1 - k.$$

In case C(i)/D(i) (resp. C(ii)/D(ii)) we have $(\alpha_1, \alpha_2) = (\frac{-1-c}{2}, \frac{1-c}{2})$ (resp. $(\alpha_1, \alpha_2) = (\frac{1-c}{2}, \frac{-1-c}{2})$). We fix $\nu$ consistent with these values of $\alpha_i$ and satisfying (3.4.9.5), as before, and denote them $\nu^{(i)}$ and $\nu^{(ii)}$, respectively. Defining $m$ by (3.4.9.2), we obtain

(3.5.6.2C(i)) $$m = k + 1 - \alpha_1 = \frac{a+b+3}{2} + 1, \frac{a+b+3}{2} + 4, \ldots, a+2;$$

(3.5.6.2D(i)) $$m = \frac{a+b+3}{2} + 2, \frac{a+b+3}{2} + 5, \ldots, a+2;$$

(3.5.6.2C(ii)) $$m = \frac{a+b+3}{2} + 2, \ldots, a+1;$$

(3.5.6.2D(ii)) $$m = \frac{a+b+3}{2} + 1, \ldots, a+1;$$

The analogue of Proposition 3.5.2 remains true, under the hypotheses (3.3), and we conclude

THEOREM 3.5.7. *Let $m$ be in the right-hand half of the critical set of the $L$-function $L(s, M(\pi_f, \mu))$. Suppose $\mu(-1) = 1$ and $c = a + b$ is odd. In case C(i)/D(i), for $m$ in the corresponding lists (3.5.6.2), we have*

$$\left(c_\infty^3(m)\right)^{-1} a(\pi, \beta, \nu^{(i)}) Z_\infty(f, \Phi, \mu, \nu^{(i)}, m) L(m, M(\pi_f, \mu)) \in \overline{\mathbb{Q}}.$$

*In case C(ii)/D(ii), for $m$ in the corresponding lists (3.5.6.2), we have*

$$\left(c_\infty^3(m)\right)^{-1} a(\pi, \beta, \nu^{(ii)}) Z_\infty(f, \Phi, \mu, \nu^{(ii)}, m) L(m, M(\pi_f, \mu)) \in \overline{\mathbb{Q}}.$$

**(3.6) Period relations and the case $\mu(-1) = -1$.** We note first that Theorems 3.5.5 and 3.5.6.3 make no reference to the finite part of the character $\nu$. In other words, assuming there are non-vanishing special values, we can write $a(\pi, \beta, \nu) = a(\pi, \beta, \nu_\infty)$ for the $\nu_\infty$ in question, which in turn is determined by the signs of $\bar{\mu}_i/\mu$ and the pair $(\alpha_1, \alpha_2)$. Presumably there are refinements, in which $a(\pi, \beta, \nu)$ is determined up to $\mathbb{Q}$ rather than $\overline{\mathbb{Q}}$, which would be sensitive to the full character $\nu$. Finally, for the critical values corresponding to Eisenstein series of weight 2 we have made the assumption that $\bar{\mu}$ is nontrivial; this places an implicit restriction on the choice of $\nu$.

We have chosen to assume $|k - \ell| \leq 2$, with $i$ close to the center of the available range, in order to cover the largest possible number of special values. This choice determines $\nu_\infty$ via (3.4.9.2) and (3.4.9.5). However, we can repeat the above argument, using other values of $i$, or using other sequences of pairs $(k, \ell)$ satisfying the appropriate congruence conditions. For example, in case A(ii)/B(i) we can take $\ell = k + 2$. This leads to a different value for $\nu_\infty$, hence to a relation between the corresponding $a(\pi, \beta, \nu)$ (assuming they do not vanish). We have no interpretation to propose for this phenomenon.

The characters $\nu_\infty$ used in cases A(i)/B(ii) and in A(ii)/B(i) differ only in the signs. In both cases $k = \ell$ and $\alpha_1 = \alpha_2 = -\frac{c}{2}$; however $\nu_{i,\infty}(-1) = -1$ in cases (A(i)/B(ii)), whereas $\nu_{i,\infty}(-1) = 1$ in cases A(ii)/B(i). Now suppose $\mu(-1) = -1$. Then (3.4.9.5) requires that the signs change; i.e. that $\nu_{i,\infty}(-1) = 1$ (resp. $= -1$) in cases A(i)/B(ii) (resp. in cases A(ii)/B(i)). We leave it to the reader to verify, using (2.5), that this is completely consistent with what is predicted by Deligne's conjecture, and to check the cases (C) and (D).

UFR DE MATHÉMATIQUE, UNIVERSITÉ PARIS 7, 2 PL. JUSSIEU
75251 PARIS CEDEX 05 FRANCE
INSTITUT DE MATHÉMATIQUES DE JUSSIEU, U.M.R. 7586 DU CNRS. MEMBRE, INSTITUT UNIVERSITAIRE DE FRANCE

## REFERENCES

[An] A. N. Andrianov, Zeta functions and the Siegel modular forms, *Proc. Summer School Bolyai-Janos Math. Soc.*, Budapest (1970), Halsted, N.Y. (1975), 9–20.

[B] D. Bertrand, Endomorphismes de groupes algébriques; applications arithmétiques, *Progress in Math.* **31** (1983), 1–45.

[BHR] D. Blasius, M. Harris, D. Ramakrishnan, Coherent cohomology, limits of discrete series, and Galois conjugation, *Duke. Math. J.* **73** (1994), 647–685.

[D] P. Deligne, Valeurs de fonctions $L$ et périodes d'intégrales, *Proc. Symp. Pure Math.* **33** (2) (1979), 313–346.

[FS] M. Furusawa and J. Shalika, The fundamental lemma for the Bessel and Novodvorsky subgroups of $GSp(4)$. I, *C.R.A.S. Paris* **328** (1999), 105–110; II, *C.R.A.S. Paris* **331** (2000), 593–598.

[GRS] D. Ginzburg, S. Rallis, and D. Soudry, On explicit lifts of cusp forms from $GL_m$ to classical groups, *Ann. of Math.* **150** (1999), 807–866.

[G] L. Grenié, Valeurs speciales de fonctions $L$ de $GL(r) \times GL(r)$, Ph.D. thesis, Université Paris, 2000.

[H1] M. Harris, Automorphic forms of $\bar{\partial}$-cohomology type as coherent cohomology classes, *J. Diff. Geom.* **32** (1990), 1–63.

[H2] ———, Period invariants of Hilbert modular forms. I, *Lect. Notes Math.* **1447** (1990), 155–202.

[HK] M. Harris and S. S. Kudla, Arithmetic automorphic forms for the nonholomorphic discrete series of $GSp(2)$, *Duke Math. J.* (1992), 59–121.

[Hida] H. Hida, On the critical values of $L$-functions of $GL(2)$ and $GL(2) \times GL(2)$, *Duke Math. J.* **74** (1994), 431–529.

[KS] H. H. Kim and F. Shahidi, Functorial products for $GL_2 \times GL_3$ and the symmetric cube for $GL_2$, *Ann. of Math.* **155** (2002), 837–893.

[MT] A. Mokrane and J. Tilouine, Cohomology of Siegel varieties with $p$-adic integral coefficients and applications, *Asterisque* **280** (2002).

[PS] I. I. Piatetski-Shapiro, $L$-functions for $GSp_4$, *Pacific. J. Math.* Olga Taussky-Todd memorial issue (1997), 259–275.

[R] S. Rallis, On the Howe duality conjecture, *Compositio Math.* **51** (1984), 333–394.

[S1] G. Shimura, The special values of the zeta functions associated with cusp forms, *Comm. Pure Appl. Math.* **29** (1976), 783–804.

[S2] ———, On a class of nearly holomorphic automorphic forms, *Ann. of Math.* **123** (1986), 347–406.

CHAPTER 14

# ON A CONJECTURE OF JACQUET

By Michael Harris and Stephen S. Kudla

*For Joe Shalika, with our admiration and appreciation*

**0. Introduction.** Let $k$ be a number field and let $\pi_i$, $i = 1, 2, 3$, be cuspidal automorphic representations of $GL_2(\mathbb{A})$ such that the product of their central characters is trivial. Jacquet then conjectured that the central value $L(\frac{1}{2}, \pi_1 \otimes \pi_2 \otimes \pi_3)$ of the triple product L–function is nonzero if and only if there exists a quaternion algebra $B$ over $k$ and automorphic forms $f_i^B \in \pi_i^B$ such that the integral

$$(0.1) \qquad I\big(f_1^B, f_2^B, f_3^B\big) = \int_{Z(\mathbb{A})B^\times(k)\backslash B^\times(\mathbb{A})} f_1^B(b) f_2^B(b) f_3^B(b)\, d^\times b \neq 0,$$

where $\pi_i^B$ is the representation of $B^\times(\mathbb{A})$ corresponding to $\pi_i$ via the Jacquet-Langlands correspondence.

In a previous paper [4], we proved this conjecture in the special case where $k = \mathbb{Q}$ and the $\pi_i$'s correspond to a triple of holomorphic newforms. Our method was based on a combination of the Garrett, Piatetski-Shapiro, Rallis integral representation of the triple product L-function with the extended Siegel-Weil formula and the seesaw identity. The restriction to holomorphic newforms over $\mathbb{Q}$ arose from (i) the need to invoke the Ramanujan Conjecture to control the poles of some bad local factors and (ii) the use of a version of the Siegel-Weil formula for similitudes. In this note, we show that, thanks to the recent improvement on the Ramanujan bound due to Kim-Shahidi [11], together with a slight variation in the setup of (ii), our method yields Jacquet's conjecture in general.

Since the exposition in [4] was specialized from the start to the case of interest for certain arithmetic applications, we will briefly sketch the method in general in the first few sections. We then prove the facts required about the extended Siegel-Weil formula.

Several authors have considered interpretations of the vanishing of the central value $L(\frac{1}{2}, \pi_1 \otimes \pi_2 \otimes \pi_3)$. Here we mention only the work of Dihua Jiang, [9], who gave an intriguing relation with a period of an Eisenstein series on $G_2$ and the recent Princeton thesis of Thomas Watson, [26], who applies these central values to problems in "quantum chaos."

*Acknowledgments.* The authors would like to thank the IHP in Paris where this project was realized during the special program on "geometric aspects of automorphic forms" in June of 2000. We also thank the referee for helpful comments, and in particular for reminding us that the completion of the proof of Proposition 5.3 made implicit use of recent results of Loke [16].

**1. The integral representation of the triple product L-function.** Let $G = GSp_6$ be the group of similitudes of the standard 6-dimensional symplectic vector space over $k$, and let $P = MN$ be the Siegel parabolic subgroup of $G$. For $a \in GL_3$, $b \in \mathrm{Sym}_3$ and $\nu$ a scalar, let

$$(1.1)\quad m(a) = \begin{pmatrix} a & \\ & {}^t a^{-1} \end{pmatrix}, \quad n(b) = \begin{pmatrix} 1 & b \\ & 1 \end{pmatrix}, \quad \text{and} \quad d(\nu) = \begin{pmatrix} 1 & \\ & \nu \end{pmatrix} \in G.$$

Let $K_G = K_{G,\infty} \cdot K_{G,f}$ be the standard maximal compact subgroup of $G(\mathbb{A})$. For $s \in \mathbb{C}$, let $\lambda_s$ be the character of $P(\mathbb{A})$ defined by

$$(1.2)\qquad \lambda_s(d(\nu)n(b)m(a)) = |\nu|^{-3s}\,|\det(a)|^{2s}.$$

Let $I(s) = I_P^G(\lambda_s)$ be the normalized induced representation of $G(\mathbb{A})$, consisting of all smooth $K_G$-finite functions $\Phi_s$ on $G(\mathbb{A})$ such that

$$(1.3)\qquad \Phi_s(d(\nu)n(b)m(a)g, s) = |\nu|^{-3s-3}\,|\det(a)|^{2s+2}\,\Phi_s(g).$$

The Eisenstein series associated to a section $\Phi_s \in I(s)$ is defined for $\mathrm{Re}(s) > 2$ by

$$(1.4)\qquad E(g, s, \Phi_s) = \sum_{\gamma \in P(k)\backslash G(k)} \Phi_s(\gamma g),$$

and the normalized Eisenstein series is

$$(1.5)\qquad E^*(g, s, \Phi_s) = b_G(s) \cdot E(g, s, \Phi_s),$$

where $b_G(s) = \zeta_k(2s+2)\,\zeta_k(4s+2)$, as in [17]. Note that the central character of $E(g, s, \Phi_s)$ is trivial. These functions have meromorphic analytic continuations to the whole $s$-plane and have no poles on the unitary axis $\mathrm{Re}(s) = 0$. In particular, the map

$$(1.6)\qquad E^*(0) : I(0) \longrightarrow \mathcal{A}(G), \qquad \Phi_0 \mapsto (g \mapsto E^*(g, 0, \Phi_s))$$

gives a $(\mathfrak{g}_\infty, K_{G,\infty}) \times G(\mathbb{A}_f)$–intertwining map from the induced representation $I(0)$ at $s = 0$ to the space of automorphic forms on $G$ with trivial central character.

Let

$$(1.7)\qquad \begin{aligned} \mathbf{G} &= (GL_2 \times GL_2 \times GL_2)_0 \\ &= \{(g_1, g_2, g_3) \in (GL_2)^3 \mid \det(g_1) = \det(g_2) = \det(g_3)\}. \end{aligned}$$

This group embeds diagonally in $G = GSp_6$. For automorphic forms $f_i \in \pi_i, i = 1, 2, 3$, let $F = f_1 \otimes f_2 \otimes f_3$ be the corresponding function on $\mathbf{G}(\mathbb{A})$. The global zeta integral [17] is given by

$$(1.8)\qquad Z(s, F, \Phi_s) = \int_{Z_G(\mathbb{A})\mathbf{G}(k)\backslash \mathbf{G}(\mathbb{A})} E^*(\mathbf{g}, s, \Phi_s)\, F(\mathbf{g})\, d\mathbf{g}.$$

Suppose that the automorphic forms $f_i \in \pi_i$ have factorizable Whittaker functions $W_i^\psi = \otimes_v W_{i,v}^\psi$ and that the section $\Phi_s$ is factorizable. Let $S$ be a finite set of places of $k$, including all archimedean places, such that, for $v \notin S$,

(i) the fixed additive character $\psi$ of $\mathbb{A}/k$ has conductor $\mathcal{O}_{k,v}$ at $v$.

(ii) $\pi_{i,v}$ is unramified, $f_i$ is fixed under $K_v = GL_2(\mathcal{O}_{k,v})$, and $W_{i,v}^\psi(e) = 1$.

(iii) $\Phi_{s,v}$ is right invariant under $G(\mathcal{O}_{k,v}) = K_{G,v}$ and $\Phi_{s,v}(e) = 1$.

Then

$$Z(s, F, \Phi_s) = L^S\left(s + \frac{1}{2}, \pi_1 \otimes \pi_2 \otimes \pi_3\right) \cdot \prod_{v \in S} Z_v(s, W_v^\psi, \Phi_{s,v}), \tag{1.9}$$

for local zeta integrals $Z_v(s, W_v^\psi, \Phi_{s,v})$, where $W_v^\psi = W_{1,v}^\psi \otimes W_{2,v}^\psi \otimes W_{3,v}^\psi$. Here

$$Z(s, W_v^\psi, \Phi_{s,v}) = \int_{Z_G(k_v)\mathbf{M}(k_v)\backslash \mathbf{G}(k_v)} \Phi_{s,v}(\delta\, g)\, W_v^\psi(g)\, dg, \tag{1.10}$$

where $\delta \in G(k)$ is a representative for the open orbit of $\mathbf{G}$ in $P\backslash G$, cf. for example [2], and

$$\mathbf{M} = \left\{ \left( \begin{pmatrix} 1 & x_1 \\ & 1 \end{pmatrix}, \begin{pmatrix} 1 & x_2 \\ & 1 \end{pmatrix}, \begin{pmatrix} 1 & x_3 \\ & 1 \end{pmatrix} \right) \in \mathbf{G} \mid x_1 + x_2 + x_3 = 0 \right\}. \tag{1.11}$$

Here $L^S(s, \pi_1 \otimes \pi_2 \otimes \pi_3)$ is the triple product L-functions with the factors for $v \in S$ omitted.

**2. Local zeta integrals.** In this section, we record some consequences of recent results of Kim and Shahidi [11] on the Ramanujan estimate for the $\pi_i$'s. We begin by recalling relevant aspects of the local theory of the triple product, as recently completed by Ikeda and Ramakrishnan. In the following proposition by "local Euler factor" at a finite place $v$ of $k$ we mean a function of the form $P(q_v^{-s})^{-1}$, where $P$ is a polynomial, $P(0) = 1$, and $q_v$ is the order of the residue field; at an archimedean field we mean a finite product of Tate's local Euler factors for $GL(1)$.

PROPOSITION 2.1. *Let $v$ be a place of $k$ and let $\pi_{i,v}$, $i = 1, 2, 3$, be a triple of admissible irreducible representations of $GL(2, k_v)$ that arise as local components at $v$ of cuspidal automorphic representations $\pi_i$.*

(i) *There exists a local Euler factor $L(s, \pi_{1,v} \otimes \pi_{2,v} \otimes \pi_{3,v})$ such that, for any local data $(W_v^\psi, \Phi_{s,v})$, the quotient*

$$\tilde{Z}_v\big(s, W_v^\psi, \Phi_{s,v}\big) = Z_v\big(s, W_v^\psi, \Phi_{s,v}\big) \cdot L\left(s + \frac{1}{2}, \pi_{1,v} \otimes \pi_{2,v} \otimes \pi_{3,v}\right)^{-1}$$

*is entire as a function of $s$.*

(ii) *Let* $\sigma_{i,v}$, $i = 1, 2, 3$, *be the representations of the Weil-Deligne group of* $\mathbb{k}_v$ *associated to* $\pi_{i,v}$ *by the local Langlands correspondence. Then*

$$L(s, \pi_{1,v} \otimes \pi_{2,v} \otimes \pi_{3,v}) = L(s, \sigma_{1,v} \otimes \sigma_{2,v} \otimes \sigma_{3,v}).$$

(iii) *For any finite place* $v$, *there is a local section* $\Phi_{s,v}$ *and a Whittaker function* $W_v^{\psi} = W_{1,v}^{\psi} \otimes W_{2,v}^{\psi} \otimes W_{3,v}^{\psi}$, *such that*

$$Z(s, \Phi_{s,v}, W_v^{\psi}) \equiv 1.$$

(iv) *For any archimedean place* $v$, *there exists a finite collection of Whittaker functions* $W_v^{\psi,j}$ *and of sections* $\Phi_{s,v}^j$, *holomorphic in a neighborhood of* $s = 0$ *such that*

$$\sum_j Z(0, \Phi_{s,v}^j, W_v^{\psi,j}) = 1.$$

*Proof.* For $v$ nonarchimedean, assertion (i) is proved in §3, Appendix 3, of [17]; see [7], p. 227 for a concise statement. For $v$ real or complex, (i) and (ii) were proved in several steps by Ikeda, of which the crucial one is [8], Theorem 1.10. Assertion (ii) in general is due to Ramakrishnan, [21], Theorem 4.4.1. For the moment, the hypothesis that the $\pi_{i,v}$ embed in global cuspidal representations seems to be necessary.

Assertions (iii) and (iv) are contained in Proposition 3.3 of [17]. □

PROPOSITION 2.2. (i) *For any triple* $\pi_i$ *of cusp forms for* $GL_2$ *over* $\mathbb{k}$, *and for any place* $v$, *the local Langlands L-factor* $L(s, \pi_{1,v} \otimes \pi_{2,v} \otimes \pi_{3,v})$ *is holomorphic at* $s = \frac{1}{2}$.

(ii) *For any place* $v$, *for any triple of Whittaker functions* $W_{i,v}^{\psi}$ *in the Whittaker spaces of* $\pi_{i,v}$, *and for any section* $\Phi_{s,v} \in I_v(s)$, *holomorphic in a neighborhood of* $s = 0$, *the local zeta integral* $Z(s, W_v^{\psi}, P_{s,v})$ *is holomorphic in a neighborhood of* $s = 0$.

*Proof.* This follows from the results of Kim and Shahidi. We sketch the simple argument, quoting the proof of Proposition 3.3.2 of [21]. Let $\sigma_{i,v}$ correspond to $\pi_{i,v}$ as in the previous proposition. For the present purposes we can assume each $\pi_i$ to be unitary. Indeed, this can be arranged by twisting $\pi_i$ by a (unique) character of the form $|\cdot|^{a_i}$, where $|\cdot|$ is the idèle norm and $a_i \in \mathbb{C}$. Since the product of the central characters of $\pi_i$ is trivial, we have $a_1 + a_2 + a_3 = 0$, so the triple product $L$-factor is left unaffected.

To each $\pi_{i,v}$, necessarily generic and now assumed unitary, we can assign an index $\lambda_{i,v}$ which measures the failure of $\pi_{i,v}$ to be tempered; we have $\lambda_{i,v} = t$ if $\pi_{i,v}$ is a complementary series attached to $(\mu|\cdot|^t, \mu|\cdot|^{-t})$ with $t > 0$ and $\mu$ unitary, $\lambda_{i,v} = 0$ otherwise. Then, according to [21], (3.3.10),

$$L(s, \pi_{1,v} \otimes \pi_{2,v} \otimes \pi_{3,v}) \tag{2.1}$$

is holomophic for $\mathrm{Re}\,(s) > \lambda(\pi_{1,v}) + \lambda(\pi_{2,v}) + \lambda(\pi_{3,v})$.

Now (i) follows from (2.1) and the Kim-Shahidi estimate $\lambda(\pi_{i,v}) < \frac{5}{34}$ for all $i$ and all $v$ [11], whereas (ii) follows from (i), and Proposition 2.1 (i) and (ii). □

By (1.8), (1.9), and the holomorphy of $E^*(g, s, \Phi)$ on the unitary axis, the expression

$$L^S\left(s + \frac{1}{2}, \pi_1 \otimes \pi_2 \otimes \pi_3\right) \cdot \prod_{v \in S} Z_v(s, F, \Phi_{s,v}) = \int_{Z_G(\mathbb{A})\mathbf{G}(k)\backslash\mathbf{G}(\mathbb{A})} E^*(\mathbf{g}, s, \Phi_s)\, F(\mathbf{g})\, d\mathbf{g} \tag{2.2}$$

is holomorphic at $s = 0$ for all choices of data $F$ and $\Phi_s$. By varying the data for places in $S$ and applying (iii) and (iv) of Proposition 2.1, it follows that the partial Euler product $L^S(s + \frac{1}{2}, \pi_1 \otimes \pi_2 \otimes \pi_3)$ is holomorphic at $s = 0$. By (i) of Proposition 2.2, the Euler product $L(s + \frac{1}{2}, \pi_1 \otimes \pi_2 \otimes \pi_3)$ over all finite places is holomorphic at $s = 0$, and we obtain the identity

$$L\left(\frac{1}{2}, \pi_1 \otimes \pi_2 \otimes \pi_3\right) \cdot \prod_{v \in S} Z_v^*\left(0, W_v^\psi, \Phi_{s,v}\right) = \int_{Z_G(\mathbb{A})\mathbf{G}(k)\backslash\mathbf{G}(\mathbb{A})} E^*(\mathbf{g}, 0, \Phi_s)\, F(\mathbf{g})\, d\mathbf{g} \tag{2.3}$$

where

$$Z_v^*\left(s, W_v^\psi, \Phi_{s,v}\right) = \begin{cases} \tilde{Z}_v\left(s, W_v^\psi, \Phi_{s,v}\right) & \text{if } v \in S_f, \\ Z_v\left(s, W_v^\psi, \Phi_{s,v}\right) & \text{if } v \in S_\infty. \end{cases} \tag{2.4}$$

COROLLARY 2.3. *$L(\frac{1}{2}, \pi_1 \otimes \pi_2 \otimes \pi_3) = 0$ if and only if*

$$\int_{Z_G(\mathbb{A})\mathbf{G}(k)\backslash\mathbf{G}(\mathbb{A})} E^*(\mathbf{g}, 0, \Phi_s)\, F(\mathbf{g})\, d\mathbf{g} = 0,$$

*for all choices of $F \in \Pi = \pi_1 \otimes \pi_2 \otimes \pi_3$ and $\Phi_s \in I(s)$.*

Of course, relation (2.3) gives a formula for $L(\frac{1}{2}, \pi_1 \otimes \pi_2 \otimes \pi_3)$ for a suitable choice of $F$ and $\Phi_s$.

**3. The Weil representation for similitudes.** The material of this section is a slight variation on that of section 5 of [3]. We consider only the case of the dual pair $(GO(V), GSp_6)$ where the space $V$ has square discriminant.

Let $B$ be a quaternion algebra over $k$ (including the possibility $B = M_2(k)$), and let $V = B$ be a 4-dimensional quadratic space over $k$ where the quadratic form is given by $Q(x) = \alpha\, \nu_B(x)$, where $\nu_B$ is the reduced norm on $B$ and $\alpha \in k^\times$. Note that the isomorphism class of $V$ is determined by $B$ and the sign of $\alpha$ at the set $\Sigma_\infty(B)$ of real archimedean places of $k$ at which $B$ is division. Let $H = GO(V)$ and let $H_1 = O(V)$ be the kernel of the scale map $\nu : H \to \mathbb{G}_m$. Let $G = GSp_6$, and let $G_1 = Sp_6$ be the kernel of the scale map $\nu : G \to \mathbb{G}_m$. We want to extend the standard Weil representation $\omega = \omega_\psi$ of $H_1(\mathbb{A}) \times G_1(\mathbb{A})$ on the Schwartz space $S(V(\mathbb{A})^3)$. First, there is a natural action of $H(\mathbb{A})$ on $S(V(\mathbb{A})^3)$ given by

$$L(h)\varphi(x) = |\nu(h)|^{-3}\,\varphi(h^{-1}x). \tag{3.1}$$

For $g_1 \in G_1(\mathbb{A})$ one has

$$L(h)\omega(g_1)L(h)^{-1} = \omega(d(\nu)g_1 d(\nu)^{-1}), \tag{3.2}$$

where $\nu = \nu(h)$, and $d(\nu)$ is as in section 1. Therefore, one obtains a representation of the semidirect product $H(\mathbb{A}) \ltimes G_1(\mathbb{A})$ on $S(V(\mathbb{A})^3)$. Let

$$R = \{(h, g) \in H \times G \mid \nu(h) = \nu(g)\}. \tag{3.3}$$

Then there is an isomorphism

$$R \longrightarrow H \ltimes G_1, \qquad (h, g) \mapsto (h, d(\nu(g))^{-1}g) = (h, g_1), \tag{3.4}$$

(this defines a map $g \mapsto g_1$) and a representation of $R(\mathbb{A})$ on $S(V(\mathbb{A})^3)$ given by

$$\omega(h, g)\varphi(x) = (L(h)\omega(g_1)\varphi)(x) = |\nu(h)|^{-3}(\omega(g_1)\varphi)(h^{-1}x). \tag{3.5}$$

The theta distribution $\Theta$ on $S(V(\mathbb{A})^3)$ is invariant under $R(k)$, since, for $(h, g) \in R(k)$,

$$\begin{aligned} \Theta(\omega(h, g)\varphi) &= \sum_{x \in V(k)^3} |\nu(h)|^{-3}(\omega(g_1)\varphi)(h^{-1}x) \\ &= \sum_{x \in V(k)^3} (\omega(g_1)\varphi)(x) \\ &= \Theta(\omega(g_1)\varphi) = \Theta(\varphi), \end{aligned} \tag{3.6}$$

since $g_1 \in G_1(k)$. The theta kernel, defined for $(h, g) \in R(\mathbb{A})$ by

$$\theta(h, g; \varphi) = \sum_{x \in V(k)^3} \omega(h, g)\varphi(x), \tag{3.7}$$

is thus left $R(k)$ invariant.

*Remark* 3.1. Aside from a shift in notation, the convention here is essentially the same as in section 5 of [3] and section 3 of [4], *except that* we take pairs $(h, g)$ here versus $(g, h)$ there. Compare (3.5) above with (5.1.5) of [3]. It turns out that this seemingly slight shift in convention will be crucial for the extension of the Siegel-Weil formula to similitudes, as we will see below.

Note that the set of archimedean places $\Sigma_\infty(B)$ introduced above is the set of all real archimedean places of $k$ at which $V$ is definite (positive or negative). Then,

$$\begin{aligned} G(\mathbb{A})^+ &:= \{g \in G(\mathbb{A}) \mid \nu(g) \in \nu(H(\mathbb{A}))\} \\ &= \{g \in G(\mathbb{A}) \mid \nu(g)_v > 0,\ \forall v \in \Sigma_\infty(B)\}. \end{aligned} \tag{3.8}$$

For $g \in G(\mathbb{A})^+$, and $\varphi \in S(V(\mathbb{A})^3)$, and for $V$ anisotropic over $k$, i.e., for $B \neq M_2(k)$, the theta integral is defined by

$$I(g, \varphi) = \int_{H_1(k)\backslash H_1(\mathbb{A})} \theta(h_1 h, g; \varphi)\, dh_1, \tag{3.9}$$

where $h \in H(\mathbb{A})$ with $\nu(h) = \nu(g)$. It does not depend on the choice of $h$.

In the case $B = M_2(k)$, the theta integral must be defined by regularization. If $k$ has a real place, the procedure outlined on p. 621 of [4], [15], using a certain differential operator to kill support, can be applied. An analogous procedure using an element of the Bernstein center can be applied at a nonarchimedean place, [25]. We omit the details.

LEMMA 3.2. (i) *(Eichler's norm Theorem) If $\alpha \in \nu(H(\mathbb{A})) \cap k^\times$, then there exists an element $h \in H(k)$ with $\nu(h) = \alpha$.*

(ii) *The theta integral is left invariant under $G(\mathbb{A})^+ \cap G(k)$.*

(iii) *The theta integral has trivial central character, i.e., for $z \in Z_G(\mathbb{A}) \subset G(\mathbb{A})^+$, $I(zg, \varphi) = I(g, \varphi)$.*

*Proof.* (i) is a standard characterization of $\nu(H(k))$ in the present case. To check (ii), given $\gamma \in G(\mathbb{A})^+ \cap G(k)$, choose $\gamma' \in H(k)$ with $\nu(\gamma') = \nu(\gamma)$. Then

$$\begin{aligned} I(\gamma g, \varphi) &= \int_{H_1(k)\backslash H_1(\mathbb{A})} \theta(h_1 \gamma' h, \gamma g; \varphi)\, dh_1 \\ &= \int_{H_1(k)\backslash H_1(\mathbb{A})} \theta(\gamma' h_1 h, \gamma g; \varphi)\, dh_1 \\ &= I(g, \varphi), \end{aligned} \tag{3.10}$$

via the left invariance of the theta kernel under $(\gamma', \gamma) \in R(k)$. Here, in the next to last step, we have conjugated the domain of integration $H_1(k)\backslash H_1(\mathbb{A})$ by the element $\gamma' \in H(k)$.

Finally, the proof of (iii) is just like that of Lemma 5.1.9 (ii) in [3]. □

Since $G(\mathbb{A}) = G(k)G(\mathbb{A})^+$, it follows that $I(g, \varphi)$ has a unique extension to a left $G(k)$-invariant function on $G(\mathbb{A})$. Moreover, for any $g_0 \in G(\mathbb{A})^+$, we have

$$I(gg_0, \varphi) = I(g, \omega(h_0, g_0)\varphi), \tag{3.11}$$

where $h_0 \in H(\mathbb{A})$ with $\nu(h_0) = \nu(g_0)$. In particular, if $h_1 \in H_1(\mathbb{A})$, then

$$I(g, \omega(h_1)\varphi) = I(g, \varphi). \tag{3.12}$$

**4. The Siegel-Weil formula for $(GO(V), GSp_6)$.** First we recall the Siegel-Weil formula for $(O(V), Sp_6)$. The results of [15] on the regularized Siegel-Weil formula were formulated over a totally real number field, since, at a number of points, we needed facts about degenerate principal series, intertwining operators, etc. which had not been checked for complex places. The proof in the case of the central value of the Siegel-Eisenstein series is simpler than the general case, and the additional facts needed at complex places are easy to check. In the rest of this section, we will state the results for an arbitrary number field $\Bbbk$. A sketch of the proof of Theorem 4.1 below for such a field $\Bbbk$ will be given in the Appendix below.

Let $I_1(s) = I_{P_1}^{G_1}(\lambda_s)$ be the global induced representation of $G_1(\mathbb{A}) = Sp_6(\mathbb{A})$ induced from the restriction of the character $\lambda_s$ of $P(\mathbb{A})$ to $P_1(\mathbb{A}) = P(\mathbb{A}) \cap G_1(\mathbb{A})$.

For a global quadratic space $V$ of dimension 4 over $\Bbbk$ associated to a quaternion algebra $B$, as in the previous section, there is a $(\mathfrak{g}_{1,\infty}, K_{G_{1,\infty}}) \times G_1(\mathbb{A}_f)$-equivariant map

$$S(V(\mathbb{A})^3) \longrightarrow I_1(0), \qquad \varphi \mapsto [\varphi], \tag{4.1}$$

where

$$[\varphi](g_1) = (\omega(g_1)\varphi)(0). \tag{4.2}$$

The image, $\Pi_1(V)$, is an irreducible summand of the unitarizable induced representation $I_1(0)$. By the results of Rallis [19], Kudla-Rallis [14], [13], and the appendix,

$$\Pi_1(V) \simeq S(V(\mathbb{A})^3)_{O(V)(\mathbb{A})}, \tag{4.3}$$

the space of $H_1(\mathbb{A}) = O(V)(\mathbb{A})$-coinvariants. One then has a decomposition

$$I_1(0) = (\oplus_V \Pi_1(V)) \oplus (\oplus_{\mathcal{V}} \Pi_1(\mathcal{V})), \tag{4.4}$$

into irreducible representations of $G_1(\mathbb{A})$, as $V$ runs over the isomorphism classes of such spaces and as $\mathcal{V}$ runs over the incoherent collections, obtained by switching one local component of a $\Pi_1(V)$, cf. [12].

The Siegel-Weil formula of [4], asserts the following in the present case.

THEOREM 4.1. (i) *The* $(\mathfrak{g}_{1,\infty}, K_{G_{1,\infty}}) \times G_1(\mathbb{A}_f)$*-intertwining map*

$$E_1(0) : I_1(0) \longrightarrow \mathcal{A}(G_1), \qquad \Phi_0 \mapsto (g_1 \mapsto E(g_1, 0, \Phi_s))$$

*has kernel* $\oplus_{\mathcal{V}} \Pi_1(\mathcal{V})$.

(ii) *For a section* $\Phi_s \in I_1(s)$ *with* $\Phi_0 = [\varphi]$ *for some* $\varphi \in S(V(\mathbb{A})^3)$,

$$E(g_1, 0, \Phi_s) = 2\, I(g_1, \varphi), \tag{SW}$$

*for the theta integral as defined in §3.*

As explained in the previous section, the theta integral can be extended to an automorphic form on $G(\mathbb{A})$. We will see presently that it coincides with an Eisenstein series on $G(\mathbb{A})$.

Restriction of functions from $G(\mathbb{A})$ to $G_1(\mathbb{A})$ yields an isomorphism $I(s) \overset{\sim}{\rightarrow} I_1(s)$, which is intertwining for the right action of $G_1(\mathbb{A})$. Here $I(s)$ is the induced representation of $G(\mathbb{A})$ defined in section 1. The inverse map is given by $\Phi_s \mapsto \Phi_s^{\sim}$ where

$$\Phi_s^{\sim}(g) = |\nu(g)|^{-3s-3}\Phi_s(g_1), \tag{4.5}$$

for $g_1 = d(\nu(g))^{-1}g$, as above. The decomposition (4.4) into $G_1(\mathbb{A})$-irreducibles yields a decomposition

$$I(0) = (\oplus_B \Pi(B)) \oplus (\oplus_{\mathcal{B}} \Pi(\mathcal{B})), \tag{4.6}$$

into irreducible representations of $G(\mathbb{A})$, where, for a global quaternion algebra $B$ over $k$,

$$\Pi(B) = \oplus_V \Pi(V) \tag{4.7}$$

where $V$ runs over the non-isomorphic spaces associated to $B$ (i.e., different multiples of the norm form) and $\Pi(V)$ denotes the image of $\Pi_1(V)$ under the inverse of the restriction isomorphism. Note that there are $2^{|\Sigma_\infty(B)|}$ such $V$'s. In effect, at a real archimedean place $v$, the local induced representation $I(0)_v$ has a decomposition into irreducible $(\mathfrak{g}_{1,v}, K_{G_{1,v}})$-modules

$$I(0)_v = \Pi(4,0)_v \oplus \Pi(2,2)_v \oplus \Pi(0,4)_v \tag{4.8}$$

according to signatures. The space $\Pi(2,2)_v$ is actually stable under $(\mathfrak{g}_v, K_{G,v})$, as is the sum $\Pi(4,0)_v \oplus \Pi(0,4)_v$, and

$$\Pi(B)_v = \begin{cases} \Pi(2,2)_v & \text{if } B_v \simeq M_2(\mathbb{R}). \\ \Pi(4,0)_v \oplus \Pi(0,4)_v & \text{if } B_v \text{ is division.} \end{cases} \tag{4.9}$$

The summands $\Pi(\mathcal{B})$ are defined similarly.

THEOREM 4.2. (i) *The $(\mathfrak{g}_\infty, K_{G,\infty}) \times G(\mathbb{A}_f)$-intertwining map*

$$E(0) : I(0) \longrightarrow \mathcal{A}(G), \qquad \Phi_s \mapsto (g \mapsto E(g, 0, \Phi_s))$$

*has kernel $\oplus_{\mathcal{B}} \Pi(\mathcal{B})$.*

(ii) *For a section $\Phi_s \in I(s)$ with $\Phi_0 \in \Pi(V)$ so that $\Phi_0 = [\varphi]^{\sim}$ for some $\varphi \in S(V(\mathbb{A})^3)$,*

$$E(g, 0, \Phi_s) = 2\, I(g, \varphi), \tag{GSW}$$

*for the theta integral as defined in §3.*

*Proof.* For $g_0 \in G(\mathbb{A}_f)$, we have

$$E(gg_0, s, \Phi_s) = E(g, s, r_s(g_0)\Phi_s), \tag{4.10}$$

where $r_s$ denotes the action in the induced representation $I(s)$ by right translation. Taking the value at $s = 0$, we obtain

$$E(gg_0, 0, \Phi_s) = E(g, 0, r_s(g_0)\Phi_s). \tag{4.11}$$

Note that this value depends only on $\Phi_0$ and $r_0(g_0)\Phi_0$.

LEMMA 4.3. *For $\varphi \in S(V(\mathbb{A})^3)$, let $[\varphi] \in I_1(0)$ be defined by (4.2) and let $[\varphi]^\sim$ be the corresponding function in $I(0)$ under the inverse of the restriction isomorphism.*

(i) *For $g \in G(\mathbb{A})^+$,*

$$[\varphi]^\sim(g) = (\omega(h, g)\varphi)(0),$$

*where $h \in GO(V)(\mathbb{A})$ with $\nu(h) = \nu(g)$.*

(ii) *For $g_0 \in G(\mathbb{A}_f)$,*

$$r_0(g_0)[\varphi]^\sim = [\omega(h_0, g_0)\varphi]^\sim,$$

*where $h_0 \in GO(V)(\mathbb{A}_f)$ with $\nu(h_0) = \nu(g_0)$.*

*Proof.* For (i), we have

$$\begin{aligned} [\varphi]^\sim(g) &= [\varphi]^\sim(d(\nu)g_1) \\ &= |\nu|^{-3}[\varphi](g_1) \\ &= |\nu|^{-3}(\omega(g_1)\varphi)(0) \\ &= (L(h)\omega(g_1)\varphi)(0) \\ &= (\omega(h, g)\varphi)(0). \end{aligned} \tag{4.12}$$

For (ii),

$$\begin{aligned} (r_0(g_0)[\varphi]^\sim)(g) &= |\nu|^{-3}[\varphi]^\sim(g_1g_0) \\ &= |\nu|^{-3}(\omega(h_0, g_1g_0)\varphi)(0) \\ &= |\nu|^{-3}(\omega(g_1)\omega(h_0, g_0)\varphi)(0) \\ &= |\nu|^{-3}[\omega(h_0, g_0)\varphi](g_1) \\ &= [\omega(h_0, g_0)\varphi]^\sim(g). \end{aligned} \tag{4.13}$$

Thus, if $\Phi_0 = [\varphi]^\sim$, then $r_0(g_0)\Phi_0 = [\omega(h_0, g_0)\varphi]^\sim$. Since $G(\mathbb{A}) = G(k)Z_G(\mathbb{A})G_1(\mathbb{A})G(\mathbb{A}_f)$, we have, by (4.11),

$$\begin{aligned} E(g, 0, \Phi_s) &= E(\gamma z g_1 g_0, 0, \Phi_s) \\ &= E(g_1, 0, r_s(g_0)\,\Phi_s), \end{aligned} \tag{4.14}$$

the value at $g_1$ of the Siegel-Eisenstein series attached to $\omega(h_0, g_0)\varphi \in S(V(\mathbb{A})^3)$. On the other hand, by (3.11),

$$I(g, \varphi) = I(\gamma z g_1 g_0, \varphi) \tag{4.15}$$
$$= I(g_1, \omega(h_0, g_0)\,\varphi).$$

Thus (GSW) follows from (SW). □

**5. Proof of Jacquet's conjecture.** Applying the Siegel-Weil formula for similitudes to the basic identity (2.3), we obtain

$$L\left(\frac{1}{2}, \pi_1 \otimes \pi_2 \otimes \pi_3\right) \cdot Z^*(F, \Phi)$$
$$= \int_{Z_{\mathbf{G}}(\mathbb{A})\mathbf{G}(k)\backslash\mathbf{G}(\mathbb{A})} E^*(\mathbf{g}, 0, \Phi_s)\, F(\mathbf{g})\, d\mathbf{g} \tag{5.1}$$
$$= 2\zeta_k(2)^2 \sum_V \int_{Z_{\mathbf{G}}(\mathbb{A})\mathbf{G}(k)\backslash\mathbf{G}(\mathbb{A})} I(\mathbf{g}, \varphi^V)\, F(\mathbf{g})\, d\mathbf{g}$$

where

$$Z^*(F, \Phi) = \prod_{v \in S} Z_v^*\big(0, W_v^{\psi}, \Phi_{s,v}\big), \tag{5.2}$$

and where $\varphi^V \in S(V(\mathbb{A})^3)$, and, in fact, only a finite set of $V$'s occurs in the sum. More precisely, in the decomposition (4.6),

$$\Phi_0 = \sum_V [\varphi^V]^{\sim} + \text{ terms in the } \Pi(\mathcal{B})\text{'s } \in I(0), \tag{5.3}$$

where the quaternion algebras $B$ associated to $V$'s are split outside the set $S$, due to condition (iii) in the definition of $S$ in section 1. We thus have the following reformulation of Corollary 2.3, generalizing Proposition 5.6 of [4]:

COROLLARY 5.1. $L(\frac{1}{2}, \pi_1 \otimes \pi_2 \otimes \pi_3) = 0$ *if and only if*

$$\int_{Z_{\mathbf{G}}(\mathbb{A})\mathbf{G}(k)\backslash\mathbf{G}(\mathbb{A})} I(\mathbf{g}, \varphi^V)\, F(\mathbf{g})\, d\mathbf{g}$$

*vanishes for all choices of* $F \in \Pi = \pi_1 \otimes \pi_2 \otimes \pi_3$, *all choices of quadratic spaces* $V$ *attached to quaternion algebras* $B$ *over* $k$, *and all choices of* $\varphi^V \in S(V(\mathbb{A})^3)$.

Now consider the integral in the last line of (5.1) for a fixed $\varphi = \varphi^V$. To apply the seesaw identity, we set

$$H = GO(V)$$
$$\mathbf{H} = \{(h_1, h_2, h_3) \in H^3 \mid \nu(h_1) = \nu(h_2) = \nu(h_3)\},$$

$$
\begin{aligned}
\mathbf{R} &= \{(\mathbf{h}, \mathbf{g}) \in \mathbf{H} \times \mathbf{G} \mid \nu(\mathbf{h}) = \nu(\mathbf{g})\} \\
R_0 &= \{(h, \mathbf{g}) \in H \times \mathbf{G} \mid \nu(h) = \nu(\mathbf{g})\},
\end{aligned} \tag{5.4}
$$

and hence have the seesaw pair:

$$
\begin{array}{cccccc}
I(\cdot, \varphi; F) & (GO(V)^3)_0 = \mathbf{H} & & G = GSp_6 & I(\cdot, \varphi) \\
& \uparrow & \searrow \; \swarrow & \uparrow & \\
1\!\!1 & GO(V) = H & & \mathbf{G} = \left(GL_2^3\right)_0 & F.
\end{array} \tag{5.5}
$$

There are representations of both $R(\mathbb{A})$ and $\mathbf{R}(\mathbb{A})$ on $S(V(\mathbb{A})^3)$, and the restriction of these representations to the common subgroup $R_0(\mathbb{A})$ coincide.

For $F$ a cuspidal automorphic form on $\mathbf{G}(\mathbb{A})$ and for $\mathbf{h} \in \mathbf{H}(\mathbb{A})$, let

$$
I(\mathbf{h}, \varphi; F) = \int_{\mathbf{G}_1(k)\backslash \mathbf{G}_1(\mathbb{A})} \theta(\mathbf{h}, \mathbf{g}_1\mathbf{g}; \varphi)\, F(\mathbf{g}_1\mathbf{g})\, d\mathbf{g}_1, \tag{5.6}
$$

where $\mathbf{g} \in \mathbf{G}(\mathbb{A})$ with $\nu(\mathbf{g}) = \nu(\mathbf{h})$.

LEMMA 5.2. (Seesaw identity)

$$
\int_{Z_G(\mathbb{A})\mathbf{G}(k)\backslash \mathbf{G}(\mathbb{A})} I(\mathbf{g}, \varphi)\, F(\mathbf{g})\, d\mathbf{g} = \int_{Z_H(\mathbb{A})H(k)\backslash H(\mathbb{A})} I(h, \varphi; F)\, dh.
$$

*Proof.* Note that $Z_G(\mathbb{A})\mathbf{G}(k)\backslash \mathbf{G}(\mathbb{A}) \simeq Z_G(\mathbb{A})\mathbf{G}(k)^+\backslash \mathbf{G}(\mathbb{A})^+$, and that

$$
Z_G(\mathbb{A})\mathbf{G}(k)^+\mathbf{G}_1(\mathbb{A})\backslash \mathbf{G}(\mathbb{A})^+ \simeq Z_H(\mathbb{A})H(k)\backslash H(\mathbb{A}) \simeq \mathbb{A}^{\times,2}k^{\times,+}\backslash \mathbb{A}^{\times,+} =: C, \tag{5.7}
$$

is compact, where $\mathbb{A}^{\times,+} = \nu(H(\mathbb{A}))$ and $k^{\times,+} = \nu(H(k))$. Fixing a Haar measure $dc$ giving $C$ volume 1, we have

$$
\begin{aligned}
&\int_{Z_G(\mathbb{A})\mathbf{G}(k)\backslash \mathbf{G}(\mathbb{A})} I(\mathbf{g}, \varphi)\, F(\mathbf{g})\, d\mathbf{g} \\
&= \int_C \int_{\mathbf{G}_1(k)\backslash \mathbf{G}_1(\mathbb{A})} \int_{H_1(k)\backslash H_1(\mathbb{A})} \theta(h_1 h(c), \mathbf{g}_1\mathbf{g}(c); \varphi)\, F(\mathbf{g}_1\mathbf{g}(c))\, dh_1\, d\mathbf{g}_1\, dc \qquad (5.8) \\
&= \int_{Z_H(\mathbb{A})H(k)\backslash H(\mathbb{A})} I(h, \varphi; F)\, dh,
\end{aligned}
$$

generalizing the proof of Proposition 7.1.4 of [3]. □

To apply the seesaw identity to the restriction to $\mathbf{G}(\mathbb{A})$ of a function $F \in \Pi = \pi_1 \otimes \pi_2 \otimes \pi_3$, we recall the description, from sections 7 and 8 of [4], of the corresponding space of functions $\Theta(\Pi)$ on $\mathbf{H}(\mathbb{A})$ spanned by the $I(\mathbf{h}, \varphi; F)$'s for $F \in \Pi$ and $\varphi \in S(V(\mathbb{A})^3)$. Note that one obtains the same space by fixing a nonzero $F$ and only varying $\varphi$ [5].

The action of $B^\times \times B^\times$ on $V = B$, $\rho(b_1, b_2)x = b_1 x b_2^{-1}$ determines an extension

$$(5.9) \qquad 1 \longrightarrow \mathbb{G}_m \longrightarrow \left(B^\times \times B^\times\right) \rtimes \langle \mathbf{t} \rangle \longrightarrow H = GO(V) \longrightarrow 1$$

where the involution $\mathbf{t}$ acts on $V$ by $\rho(\mathbf{t})(x) = x^\iota$ and on $B^\times \times B^\times$ by $(b_1, b_2) \mapsto (b_2^\iota, b_1^\iota)^{-1}$. Write

$$(5.10) \qquad \tilde{H} = \left(B^\times \times B^\times\right) \rtimes \langle \mathbf{t} \rangle \qquad \text{and} \qquad \tilde{H}^0 = B^\times \times B^\times,$$

and let $\tilde{\mathbf{H}}$ and $\tilde{\mathbf{H}}^0$ be the analogous groups for $\mathbf{H} = (GO(V)^3)_0$. Thus, we have the diagram

$$(5.11) \qquad \begin{array}{ccccc} & & \tilde{\Theta}(\Pi) & & \Theta(\Pi) \\ \tilde{\mathbf{H}}^0 & \hookrightarrow & \tilde{\mathbf{H}} & \longrightarrow & \mathbf{H} \\ \uparrow & & \uparrow & & \uparrow \\ \tilde{H}^0 & \hookrightarrow & \tilde{H} & \longrightarrow & H \end{array} .$$

For an irreducible cuspidal automorphic representation $\pi$ of $GL_2(\mathbb{A})$, let $\pi^B$ be the associated automorphic representation of $B^\times(\mathbb{A})$ under the Jacquet-Langlands correspondence. We take $\pi^B$ to be zero if $\pi$ does not correspond to a representation of $B^\times(\mathbb{A})$. Similarly, let $\Pi^B = \pi_1^B \otimes \pi_2^B \otimes \pi_3^B$ be the corresponding representation of $B^\times(\mathbb{A})^3$, or zero if some factor does not exist. Note that the central character of $\Pi^B$ is trivial, and so, $(\Pi^B)^\vee \simeq \Pi^B$, where $(\Pi^B)^\vee$ is the contragradient of $\Pi^B$. Thus we can view the space of functions $\Pi^B$ on $B^\times(\mathbb{A})^3$ as the automorphic realization of both $\Pi^B$ and its contragredient.

The following result is proved in [4], sections 7 and 8, based on the work of Shimizu and Prasad.

Proposition 5.3. (i) $\Theta(\Pi)$ *is either zero or a cuspidal automorphic representation of* $\mathbf{H}(\mathbb{A})$ *and is nonzero if and only if* $\Pi^B$ *is nonzero.*

(ii) *As spaces of functions on* $\tilde{\mathbf{H}}^0(\mathbb{A})$,

$$\tilde{\Theta}(\Pi)\big|_{\tilde{\mathbf{H}}^0(\mathbb{A})} = \left(\Pi^B \otimes (\Pi^B)^\vee\right)\Big|_{\tilde{\mathbf{H}}^0(\mathbb{A})}.$$

For fixed $F \in \Pi$ and $\varphi \in S(V(\mathbb{A})^3)$, we let $\tilde{I}(\cdot, \varphi; F)$ denote the pullback of $I(\cdot, \varphi; F)$ to $\tilde{\mathbf{H}}(\mathbb{A})$, and, via (ii) of Proposition 5.3, we write the restriction of this function to $\tilde{\mathbf{H}}^0(A)$ as

$$(5.12) \qquad \tilde{I}((\mathbf{b}_1, \mathbf{b}_2), \varphi; F) = \sum_r I^{1,r}(\mathbf{b}_1, \varphi; F)\, I^{2,r}(\mathbf{b}_2, \varphi; F)$$

for functions $I^{i,r}(\cdot, \varphi; F) \in \Pi^B$ and $\mathbf{b}_i \in B^\times(\mathbb{A})^3$. The seesaw then gives

$$\begin{aligned}
&\int_{Z_G(\mathbb{A})\mathbf{G}(k)\backslash\mathbf{G}(\mathbb{A})} I(\mathbf{g}, \varphi)\, F(\mathbf{g})\, d\mathbf{g} \\
&= \int_{Z_H(\mathbb{A})H(k)\backslash H(\mathbb{A})} I(h, \varphi; F)\, dh \\
(5.13)\quad &= \int_{Z_{\tilde{H}^0}(\mathbb{A})\tilde{H}(k)\backslash\tilde{H}(\mathbb{A})} \tilde{I}(h, \varphi; F)\, dh \\
&= \int_{Z_{\tilde{H}^0}(\mathbb{A})\tilde{H}^0(k)\backslash\tilde{H}^0(\mathbb{A})} \tilde{I}(h, \varphi; F)\, dh \\
&= \sum_r \int_{\mathbb{A}^\times B^\times(k)\backslash B^\times(\mathbb{A})} I^{1,r}(b_1, \varphi; F)\, db_1 \cdot \int_{\mathbb{A}^\times B^\times(k)\backslash B^\times(\mathbb{A})} I^{2,r}(b_2, \varphi; F)\, db_2.
\end{aligned}$$

The fact that the integral over $Z_{\tilde{H}^0}(\mathbb{A})\tilde{H}(k)\backslash\tilde{H}(\mathbb{A})$ in the third line can be replaced by the integral over $Z_{\tilde{H}^0}(\mathbb{A})\tilde{H}^0(k)\backslash\tilde{H}^0(\mathbb{A})$ is (7.3.2), p. 632 of [4]. Its proof in section 8.6, p. 636 of [4] depends on Prasad's uniqueness theorem [18] for invariant trilinear forms. This theorem was recently completed by H. Y. Loke [16], who treated general triples of admissible irreducible representations of $GL(2,\mathbb{R})$ and $GL(2,\mathbb{C})$. This is the only place in [4] where Prasad's uniqueness theorem is used, although it was an important motivation for the article as well as for Jacquet's conjecture. Thus the calculation in (5.13) is valid for all number fields and for all triples of cuspidal automorphic representations.

Finally, we observe that the integrals in the last line of (5.13) are finite linear combinations of the quantities $I(f_1^B, f_2^B, f_3^B)$ of (0.1). By (ii) of Proposition 5.3, every such quantity can be obtained as an integral $\int_{\mathbb{A}^\times B^\times(k)\backslash B^\times(\mathbb{A})} I^{1,r}(b_1, \varphi; F)\, db_1$ for some $\varphi$, $F$ and $r$.

Jacquet's conjecture now follows upon combining this observation with Corollary 5.1 (compare the proof of Theorem 7.4 in [4]).

*Remark* 5.4. In fact, by Prasad's uniqueness theorem, if the root number

$$\epsilon\left(\frac{1}{2}, \pi_1 \otimes \pi_2 \otimes \pi_3\right) = 1, \tag{5.14}$$

then there is a unique $B$ for which $\Pi^B \neq 0$ and for which the space of global invariant trilinear forms on $\Pi^B$ has dimension 1. The *automorphic* trilinear form is given by integration over $\mathbb{A}^\times B^\times(k)\backslash B^\times(\mathbb{A})$ is then nonzero if and only if $L(\frac{1}{2}, \pi_1 \otimes \pi_2 \otimes \pi_3) \neq 0$. Choose $f_i^B \in \pi_i^B$, $i = 1, 2, 3$, such that

$$I\big(f_1^B, f_2^B, f_3^B\big) \neq 0. \tag{5.15}$$

For any nonzero $F \in \Pi$, we can choose $\varphi \in S(V(\mathbb{A}))^3$ such that

$$\tilde{I}((\mathbf{b}, \mathbf{b}'), \varphi; F) = f_1^B(b_1) f_2^B(b_2) f_3^B(b_3)\, f_1^B(b_1') f_2^B(b_2') f_3^B(b_3'), \tag{5.16}$$

where $\mathbf{b} = (b_1, b_2, b_3)$ and $\mathbf{b}' = (b_1', b_2', b_3')$. We then obtain

$$L\left(\frac{1}{2}, \pi_1 \otimes \pi_2 \otimes \pi_3\right) \cdot Z^*(F, \Phi) = 2\zeta_k(2)^2\, I\big(f_1^B, f_2^B, f_3^B\big)^2. \tag{5.17}$$

where, $\Phi$ is determined by $\varphi$, and $Z^*(F, \Phi) \neq 0$. Of course, this identity is only useful when one has sufficient information about the function $\varphi$ and the product of local zeta integrals $Z^*(F, \Phi)$. This was a main concern in [4].

On the other hand, when the root number $\epsilon(\frac{1}{2}, \pi_1 \otimes \pi_2 \otimes \pi_3) = -1$, then there is no $\Pi^B$ which supports an invariant trilinear form, and the central value of the triple product L-function vanishes due to the sign in the functional equation.

**Appendix: The Siegel-Weil formula for general $k$.** In this appendix, we will sketch the proof of Theorem 4.1 for an arbitrary number field $k$, indicating the additional facts which are needed when $k$ has complex places.

First, suppose that $v$ is a complex place of $k$ and consider the local degenerate principal series representation $I_{1,v}(0)$ of $G_{1,v} = Sp_3(\mathbb{C})$ and the Weil representation of $G_{1,v}$ on $S(V_v^3)$, where $V_v \simeq M_2(\mathbb{C})$ with $Q(x) = \det(x)$.

LEMMA A.1. (i) *$I_{1,v}(0)$ is an irreducible unitarizable representation of $G_{1,v}$.*

(ii) *(Coinvariants) The map $S(V_v^3) \to I_{1,v}(0)$, $\varphi \mapsto [\varphi]$, analogous to (4.1) induces an isomorphism*

$$S\big(V_v^3\big)_{H_{1,v}} \xrightarrow{\sim} I_{1,v}(0).$$

*Here $H_1 = O(V)$.*

*Remark.* Statement (i) is in Sahi's paper, [23], Theorem 3A. The proof of (ii) was explained to us by Chen-bo Zhu (he also directed us to [23]. We wish to thank him for his help on these points) [28], and is based on the method of [27].

In the case $B = M_2(k)$, the theta integral must be defined by regularization, cf. the remarks before Lemma 3.2 above.
We write

$$I_{\text{reg}}(g_1, \varphi) = \begin{cases} I(g_1, \varphi) & \text{if } V \text{ is anisotropic,} \\ B_{-1}(g_1, \varphi) & \text{if } V \text{ is isotropic,} \end{cases} \tag{A.1}$$

where $B_{-1}$ is as in (5.5.24) of [15], except that we normalize the auxillary Eisenstein-series $E(h, s)$ to have residue 1 at $s_0'$. The key facts which we need are the following:

LEMMA A.2. (i) *The map $I_{\text{reg}} : S(V(\mathbb{A})^3) \to \mathcal{A}(G_1)$ factors through the space of coinvariants $S(V(\mathbb{A})^3)_{H_1(\mathbb{A})} = \Pi(V)$.*

(ii) *For all* $\beta \in \mathrm{Sym}_3(\mathbb{k})$,

$$I_{\mathrm{reg},\beta}(g_1,\varphi) = \frac{1}{2} \cdot \int_{H_1(\mathbb{A})} \omega(g_1)\varphi(h^{-1}x)\,dh,$$

*where* $x \in V(\mathbb{k})^3$ *with* $Q(x) = \beta$.

The second statement here is Corollary 6.11 of [15]; it asserts that the nonsingular Fourier coefficients behave as though no regularization were involved.

Next we have the analogue of Lemma 4.2, p. 111 of [20]; the main point of the proof is the local uniqueness, and the "submersive set" argument for the archimedean places carries over for a complex place.

LEMMA A.3. *For* $\beta \in \mathrm{Sym}_3(\mathbb{k})$ *with* $\det(\beta) \neq 0$, *let* $\mathcal{T}_\beta$ *be the space of distributions* $T \in S(V(\mathbb{A})^3)'$ *such that*

(i) $T$ *is* $H(\mathbb{A})$*–invariant.*

(ii) *For all* $b \in \mathrm{Sym}_3(\mathbb{A}_f)$,

$$T(\omega(n(b)\varphi)) = \psi_\beta(b)\,T(\varphi),$$

*where* $\psi_\beta(b) = \psi(\mathrm{tr}(\beta b))$.

(iii) *For an archimedean place* $v$ *of* $\mathbb{k}$ *and for all* $X \in \mathfrak{n} = \mathrm{Lie}(N)$,

$$T(\omega(X)\varphi) = d\psi_\beta(X) \cdot T(\varphi).$$

*Then* $\mathcal{T}_\beta$ *has dimension at most* 1 *and is spanned by the orbital integral*

$$T(\varphi) = \int_{H_1(\mathbb{A})} \varphi(h^{-1}x)\,dh,$$

*where* $x \in V(\mathbb{k})^3$ *with* $Q(x) = \beta$. *In particular,* $\mathcal{T}_\beta = 0$ *if and only if there is no such* $x$.

*Sketch of the Proof of Theorem 4.1.* First consider a global space $V$ associated to a quaternion algebra $B$. We have two intertwining maps

$$\text{(A.2)} \qquad E_1(0) : \Pi(V) \longrightarrow \mathcal{A}(G_1) \qquad \text{and} \qquad I_{\mathrm{reg}} : \Pi(V) \longrightarrow \mathcal{A}(G_1)$$

from the irreducible representation $\Pi(V) \simeq S(V(\mathbb{A})^3)_{H(\mathbb{A})}$ of $G_1(\mathbb{A})$ to the space of automorphic forms. For a nonsingular $\beta \in \mathrm{Sym}_3(\mathbb{k})$, the distributions obtained by taking the $\beta$th Fourier coefficient of the composition of the projection $S(V(\mathbb{A})^3) \to \Pi(V)$ with each of the embeddings in (A.2) satisfy the conditions of Lemma A.3 and hence are proportional. In particular, the $\beta$-th Fourier of the Eisenstein series vanishes unless $\beta$ is represented by $V$. By the argument of pp. 111–115 of [20], the constant of proportionality is independent of $\beta$ and so there is a constant $c$ such that $E_1(g, 0, [\varphi]) - c \cdot I_{\mathrm{reg}}(g, \varphi)$ has vanishing nonsingular Fourier coefficients. But then the argument at the top of p. 28 of [15], cf. also, [20], implies that this difference is identically zero.

In the case of a component $\Pi(\mathcal{V}) \subset I_1(0)$, the nonsingular Fourier coefficients of $E(g, 0, \Phi)$ vanish by the argument on p. 28 of [15], so, again by "nonsingularity" the map $E_1(0)$ must vanish on $\Pi(\mathcal{V})$.

---

## REFERENCES


[1] P. Garrett, Decomposition of Eisenstein series: Rankin triple products, *Annals of Math.* **125** (1987), 209–235.

[2] B. Gross and S. Kudla, Heights and the central critical values of triple product L-functions, *Compositio Math.* **81** (1982), 143–209.

[3] M. Harris and S. Kudla, Arithmetic automorphic forms for the nonholomorphic discrete series of GSp(2), *Duke Math. J.* **66** (1992), 59–121.

[4] ———, The central critical value of a triple product L-function, *Annals of Math.* **133** (1991), 605–672.

[5] R. Howe and I. I. Piatetski–Shapiro, Some examples of automorphic forms on $Sp_4$, *Duke Math. J.* **50** (1983), 55–106.

[6] T. Ikeda, On the functional equations of triple L-functions, *J. Math. Kyoto Univ.* **29** (1989), 175–219.

[7] ———, On the location of poles of the triple L-functions, *Compositio Math.* **83** (1992), 187–237.

[8] ———, On the gamma factor of the triple L-function I, *Duke Math. J.* **97** (1999), 301–318.

[9] Dihua Jiang, Nonvanishing of the central critical value of the triple product L-functions, *Internat. Math. Res. Notices* **2** (1998), 73–84.

[10] H. H. Kim and F. Shahidi, Holomorphy of Rankin triple L-functions; special values and root numbers for symmetric cube L-functions, *Israel J. Math.* **120** (2000), 449–466.

[11] ———, Functorial products for $GL_2 \times GL_3$ and functorial symmetric cube for $GL_2$, *C. R. Acad. Sci. Paris Sér. I Math.* **331** (2000), 599–604.

[12] S. Kudla, Central derivatives of Eisenstein series and height pairings, *Ann. of Math.* **146** (1997,) 545–646.

[13] S. Kudla and S. Rallis, Degenerate principal series and invariant distributions, *Israel J. Math.* **69** (1990), 25–45.

[14] ———, Ramified degenerate principal series, *Israel J. Math.* **78** (1992). 209–256.

[15] ———, A regularized Siegel-Weil formula: The first term identity, *Annals of Math.* **140** (1994), 1–80.

[16] H. Y. Loke, Trilinear forms of $gl_2$, *Pacific J. Math.* **197** (2001), 119–144.

[17] I. I. Piatetski-Shapiro and S. Rallis, Rankin triple L-functions, *Compositio Math.* **64** (1987), 31–115.

[18] D. Prasad, Trilinear forms for representations of *GL*(2) and local (epsilon)-factors, *Compositio Math.* **75** (1990), 1–46.

[19] S. Rallis, On the Howe duality conjecture, *Compositio Math.* **51** (1984), 333–399.

[20] ———, L-functions and the oscillator representation, *Lecture Notes in Math.* **1245,** Springer–Verlag, New York, 1987.

[21] D. Ramakrishnan, Modularity of the Rankin-Selberg *L*-series, and multiplicity one for *SL*(2), *Annals of Math.* **152** (2000), 45–111.

[22] B. Roberts, The theta correspondence for similitudes, *Israel J. Math.* **94** (1996), 285–317.

[23] S. Sahi, Jordan algebras and degenerate principal series, *Crelle's Jour.* **462** (1995), 1–18.

[24] H. Shimizu, Theta series and automorphic forms on $GL_2$, *Jour. Math. Soc. Japan* **24** (1972), 638–683.

[25] V. Tan, A regularized Siegel-Weil formula on U(2,2) and U(3), *Duke Math. J.* **94** (1998), 341–378.

[26] T. Watson, Rankin triple products and quantum chaos, *Thesis, Princeton Univ.* (2002).

[27] Chen-bo Zhu, Invariant distributions of classical groups, *Duke Math. Jour.* **65** (1992).

[28] ———, Private Communication.

CHAPTER 15

# INTEGRAL REPRESENTATION OF WHITTAKER FUNCTIONS

By HERVÉ JACQUET

---

*To Joseph Shalika*

**1. Introduction.** Recently, the converse theorem has been used to prove spectacular results in the theory of automorphic representations for the group $GL(n)$ (see [KS], [CKPSS] for instance). The converse theorem ([CPS1] & [CPS2]) is based in part on a careful analysis of the properties of the Rankin-Selberg integrals at infinity ([JS]). The simplest example of such an integral takes the form

$$\Psi(s, W, W') = \int_{N_n\backslash G_n} W\left[\begin{pmatrix} g & 0 \\ 0 & 1 \end{pmatrix}\right] \mid \det g \mid^{s-1/2} W'(g)dg.$$

Here $W$ is in the Whittaker model $\mathcal{W}(\pi, \psi)$ of a unitary generic representation $\pi$ of $GL(n, F)$ and $W'$ in $\mathcal{W}(\pi', \overline{\psi})$ where $\pi'$ is a unitary generic representation of $GL(n-1, F)$ (see below for unexplained notations). One of the difficulties of the theory is that the representations $\pi$ and $\pi'$ need not be tempered. Thus one is led to consider holomorphic fiber bundles of representations $(\pi_u)$ and $(\pi'_{u'})$, for instance, non-unitary principal series. Correspondingly, the functions $W = W_u$ and $W' = W'_{u'}$ depend also on $u$ and $u'$. They are associated with sections of the fiber-bundle of representations at hand. Rather than standard sections (with a constant restriction to the maximal compact subgroup), we consider convolutions of standard sections with smooth functions of compact support. It is difficult to prove that the integrals $\Psi(s, W_u, W'_{u'})$ are meromorphic functions of $(s, u, u')$. The elaborate technics of [JS] were designed to go around this difficulty. In particular, there, the analytic properties of the integral as functions of $s$, as well as their functional equations, were found to be equivalent to a family of identities (depending on $(u, u')$) which were then established by analytic continuation with respect to the parameters $(u, u')$. In the present note, we first find integral representations for $W_u$ and $W'_{u'}$ which converge for **all values** of the parameters $(u, u')$. In particular, it is easy to obtain estimates for $W_u$ and $W'_{u'}$ which are uniform in $(u, u')$. Then, using these integral representations, we show that the integrals at hand are meromorphic functions of $(s, u, u')$. This being established, one can use the methods of [JS] to prove the functional equations. We do not repeat this step here because it is now much easier: one first proves that the integrals are meromorphic functions

---

Research supported in part by NSF grant DMS 9619766.

of $(s, u, u')$ and then one proves the functional equation. In contrast, in [JS], we add to prove simultaneously the analytic continuation and the functional equation. Moreover, it was difficult to obtain estimates for the functions. Having the functional equation at our disposal, we obtain the more precise result that the integral is a holomorphic function of $(s, u, u')$ times the appropriate $\Gamma$ factor. We emphasize that the $\Gamma$ factor is itself a meromorphic function of $(s, u, u')$. Thus the proofs here are much simpler than in [JS].

Another advantage of the present approach is that we obtain directly the properties of the integrals for smooth vectors. In [JS], we first established the properties of the integrals for $K_n$-finite vectors and then used the automatic continuity theorem (Casselman and Wallach, see [W] 11.4) to extend the results to smooth vectors. Needless to say, we use extensively (and most of the time implicitly) the existence of a canonical topological model for representations of $GL(n)$ ([W], Chapter 11).

In addition, we have now at our disposal the very complete results on the Whittaker integrals contained in the remarkable book of N. Wallach ([W], Chapter 15). We use them extensively.

Since the publication of [JS], there have been several papers containing related results ([D], [S], [St]).

In this note we will not discuss the more subtle question of proving that the appropriate $\Gamma$ factor can be obtained in terms of the integral. In [JS], it is established that the $\Gamma$ factor, in other words, the factor $L(s, \pi \times \pi')$, is equal to an integral of the form

$$\int W\left[\begin{pmatrix} g & 0 \\ 0 & 1 \end{pmatrix}, g\right] \mid \det g \mid^{s-1/2} dg,$$

where $W$ is a function on $GL(n, F) \times GL(n-1, F)$ which belongs to the Whittaker model of $\pi \otimes \pi'$. In other words, the function $W$ corresponds to a smooth vector for the representation $\pi \otimes \pi'$ which needs not be of the form $\sum_i v_i \otimes v_i'$. One could use this to prove directly that the global $L$-function $L(s, \pi \times \pi')$ is entire and bounded in vertical strips and similarly for the other Rankin-Selberg integrals. Of course, this is no longer needed as direct proofs from the theory of Eisenstein series are now available ([GS], see also [RS]). Nonetheless, the following question is still of interest, namely, to show that

$$L(s, \pi \times \pi') = \sum_i \Psi(s, W_i, W_i'),$$

where the functions $W_i$ and $W_i'$ are respectively $K_n$ and $K_{n-1}$ finite. We will show this is the case in another paper. A more subtle question is to identify precisely the functions $W_i$ and $W_i'$ (see [St1] & [St2] for a special case). The integrals attached to the pair of integers $(n, n)$ have analogous properties. However, for the Rankin-Selberg integrals attached to pairs $(n, m)$ with $m < n-1$ the $L$-factor cannot be obtained in terms of $K$-finite vectors but only in terms of smooth vectors in the tensor product representation ([JS]).

*Acknowledgments.* Finally, it is a pleasure to dedicate this note to Joseph Shalika as a memento of a long, fructuous, and most enjoyable collaboration. The proof given here is similar to, but different of, the original unpublished proof of the results of [JS]. Moreover, a suggestion of Piatetski-Shapiro is used in a somewhat different form. Thus I must thank both of my former collaborators without being able to pinpoint precisely their contribution.

The paper is arranged as follows. In section 2, we review results on Whittaker linear forms for the principal series. In section 3, we present the main ideas of our construction in the case of the group $GL(2)$. Section 4 contains auxiliary but crucial results. In section 5, we give an integral representation for our sections which leads to the integral representation of Whittaker functions in section 6. In section 7 we give the main properties of the Whittaker functions. In sections 8 to 10, we prove the main result on Rankin-Selberg integrals for the pairs $(n, n)$. Finally, in section 11, we give a few indications on how to treat the case of the Rankin-Selberg integrals for the other cases.

**2. Whittaker linear form for principal series.** Let $F$ be the field of real or complex numbers. We denote by $| z |_F$ or simply $\alpha_F(z)$ the module of $z \in F$. Thus $\alpha_F(z) = z\overline{z}$ if $F$ is complex. We will denote by $\psi$ a non-trivial additive character of $F$ and by $dx$ the corresponding self-dual Haar measure on $F$. We will set

$$d^\times x = \frac{dx}{| x |_F} L(1, 1_F).$$

We denote by $\mathbb{U}_1$ the subgroup of $z \in \mathbb{C}$ such that $z\overline{z} = 1$.

We will denote by $G_n$ the group $GL(n, F)$, by $A_n$ the subgroup of diagonal matrices, by $B_n$ the group of upper triangular matrices, by $N_n$ the group of upper triangular matrices with unit diagonal. We write $\overline{B}_n$ and $\overline{N}_n$ for the corresponding groups of lower triangular matrices. All these groups are regarded as algebraic groups over $F$. We denote by $K_n$ the standard maximal compact subgroup of $G_n$. We often write $G_n$ for $G_n(F)$ and so on for the other groups. We define a character $\theta_n : N_n \to \mathbb{U}_1$ by

$$\theta_n(v) = \psi\left(\sum_{1 \le i \le n-1} v_{i,i+1}\right).$$

When there is no confusion, we often drop the index $n$ from the notations.

We shall say that a character of module 1 of $F^\times$ is normalized, if its restriction to $\mathbb{R}_+^\times$ is trivial. Hence if $F = \mathbb{R}$ every normalized character is either trivial or of the form $t \mapsto t \mid t \mid^{-1}$. If $F = \mathbb{C}$ then every normalized character has the form

$$z \mapsto \left(\frac{z}{(z\overline{z})^{\frac{1}{2}}}\right)^n \text{ or } z \mapsto \left(\frac{\overline{z}}{(z\overline{z})^{\frac{1}{2}}}\right)^n$$

with $n \in \mathbb{N}$. Let $\mu = (\mu_1, \mu_2, \dots, \mu_n)$ be a $n$-tuple of normalized characters of $F^\times$. Given $u= (u_1, u_2, \dots, u_n)$ a $n$-tuple of complex numbers, we consider the representation of $G = GL(n, F)$

(1) $$\Xi_u = I(\mu_1\alpha^{u_1}, \mu_2\alpha^{u_2}, \ldots, \mu_n\alpha^{u_n})$$

induced by the quasi-characters $\mu_i\alpha^{u_i}$ and the group $\overline{B}_n$. A function $f$ in the space of the representation is a smooth function $f$ on $G_n(F)$ with complex values such that

$$f(\overline{v}ag) = \mu(a) \prod_{1\leq\leq i\leq n-1} \mid a_i \mid^{u_i - \frac{n-i}{2}} f(g),$$

for all $v \in \overline{N}_n, a = \mathrm{diag}(a_1, a_2, \ldots, a_n) \in A_n$ and $g \in G_n$; we have written $\mu(a) = \prod_i \mu_i(a_i)$.

We will use Arthur's standard notations. Therefore let

$$\mathfrak{a}^* = \mathrm{Hom}_F(A_n, F^\times) \otimes_{\mathbb{Z}} \mathbb{R} \simeq \mathbb{R}^n$$

be the real vector space generated by the algebraic characters of $A_n$. Let $\mathfrak{a}$ be the dual vector space. Let $\rho \in \mathfrak{a}^*$ be the half sum of the roots positives for $B_n$ (i.e., the roots in $N_n$). We also denote by $\alpha_i$ the simple roots. We have a map $H : G_n \to \mathfrak{a}$ defined by $e^{\langle H(a), u\rangle} = \prod \mid a_i \mid^{u_i}$ for $a \in A_n$, and $H(vak) = H(a)$, for $v \in N_n$, $k \in K_n$, $a \in A_n$. We also define $H' : G_n \to \mathfrak{a}$ by $H'(\overline{v}ak) = H(a)$, for $v \in \overline{N}_n$, $k \in K_n$, $a \in A_n$. Thus a function $f$ in the space of the representation is a smooth function such that

$$f(g) = f(\overline{v}ak) = \mu(a) f(k) e^{\langle H'(g), u-\rho\rangle >}.$$

Such a function is determined by its restriction to $K_n$. Let therefore $V(\mu)$ be the space of smooth functions $f : K_n \to \mathbb{C}$ such that

$$f(ak) = \prod \mu_i(a_i) f(k)$$

for all $a \in A \cap K$. If $f \in V(\mu)$ then for every $u$ the function $f_u$ defined by

$$f_u(\overline{v}ak) = f(k) e^{\langle H(a), u-\rho\rangle}$$

is in the induced representation. Such a section of the fiber bundle of the representations is called a **standard section.**

Suppose that $f_u$ is a standard section. Let $\phi$ be a smooth function of compact support on $G(F)$. Then the function $f_{\phi,u}$ defined by

(2) $$f_{\phi,u}(g) := \int_{G(F)} f_u(gx)\phi(x)dx$$

is, for every $u$, an element of the corresponding induced representation. More precisely, its value on $g \in K_n$ is given by

$$\int_{G(F)} f_u(x)\phi(g^{-1}x)dx = \int_{N\times K\times A^+} f(k)\phi(g^{-1}\overline{n}ak) e^{\langle H(a), u-\rho\rangle} d\overline{n}dadk.$$

We often say that such a section is a **convolution section.** For $f \in V(\mu)$, we define the **Whittaker integral**

$$\mathcal{W}_u(f) := \int f_u(v)\overline{\theta}(v)dv. \tag{3}$$

It converges when $\Re(u_{i+1} - u_i) > 0$ for $1 \leq i \leq n-1$ and extends to an entire function of $u$ ([W], Chapter 11). We claim the same is true of the integral

$$\int f_{\phi,u}(n)\overline{\theta}(n)dn. \tag{4}$$

Indeed, we appeal to a simple lemma, which will be constantly used in this paper.

LEMMA 1. *Let $V$ be a locally convex complete topological vector space. Let $\Omega$ be an open subset of $\mathbb{C}^n$. Suppose that $A : \Omega \to V$ is a continuous holomorphic map. Suppose that we are given a map $\lambda : \Omega \times V \to \mathbb{C}$, which is continuous and such that, for every $v \in V$, the map $s \mapsto \lambda(s, v)$ is holomorphic, and for every $s$, the map $v \mapsto A(s, v)$ is linear. Then the map $s \mapsto \lambda(s, A(s))$ is holomorphic.*

*Proof.* Indeed the map $(s_1, s_2) \mapsto \lambda(s_1, A(s_1))$ from $\Omega \times \Omega$ to $\mathbb{C}$ is continuous and separately holomorphic. Hence it is holomorphic. Thus its restriction to the diagonal is holomorphic. □

We will write $\mathcal{W}_u(f_u)$ and $\mathcal{W}_u(f_{\phi,u})$ for the above integrals. Recall ([W], Chapter 11) that for a given $u$, the integral (3) defines a **non-zero** linear form on the space $V(\mu)$. Moreover, within a constant factor, it is the only continuous linear form $\mathcal{W}$ such that, for all $v \in N_n$,

$$\mathcal{W}(\Xi_u(v)f) = \theta(v)\mathcal{W}(f).$$

We denote by $\mathcal{W}(\Xi_u, \psi)$ the space spanned by the functions

$$g \mapsto \mathcal{W}_u(\Xi_u(g)f_u).$$

At least when $\Xi_u$ is irreducible, this is the **Whittaker model** of $\Xi_u$.

**3. The case of $GL(2)$.** For an introduction, we review the case of $GL(2)$ which was considered in previous papers (for instance [GJR]). Let $f_u(\bullet)$ be a section, for instance, a standard section. Thus

$$f_u\left[\begin{pmatrix} a_1 & 0 \\ x & a_2 \end{pmatrix} k\right] = \mu_1(a_1) \mid a_1 \mid^{u_1 - 1/2} \mu_2(a_2) \mid a_2 \mid^{u_2 + 1/2} f_u(k).$$

If $\phi$ is a smooth function of compact support then the convolution of $f_u$ and $\phi$ is a new section $f_{\phi,u}$ defined by

$$f_{\phi,u}(g) = \int f_u(gx)\phi(x)dx = \int f(x)\phi(g^{-1}x)dx = \int f_u(x^{-1})\check{\phi}(xg)dx$$

where we set $\check{\phi}(g) := \phi(g^{-1})$. To compute this integral we set

$$x = k \begin{pmatrix} 1 & 0 \\ y & 1 \end{pmatrix} \begin{pmatrix} a_1 & 0 \\ 0 & a_2 \end{pmatrix}.$$

Then

$$dx = dk dy d^\times a_1 d^\times a_2$$

and the integral becomes

$$\int f_u(k^{-1}) \check{\phi} \left[ k \begin{pmatrix} 1 & 0 \\ y & 1 \end{pmatrix} \begin{pmatrix} a_1 & 0 \\ 0 & a_2 \end{pmatrix} g \right]$$
$$\mu_1(a_1)^{-1} \mid a_1 \mid^{-u_1+1/2} \mu_2(a_2) \mid a_2 \mid^{-u_2-1/2} dk dy d^\times a_1 d^\times a_2.$$

Let us set

$$g_u(g) = \int f(k^{-1}) \check{\phi} \left[ k \begin{pmatrix} 1 & 0 \\ y & 1 \end{pmatrix} \begin{pmatrix} 1 & 0 \\ 0 & a_2 \end{pmatrix} g \right]$$
$$\mu_2 \left( \det \left[ \begin{pmatrix} 1 & 0 \\ y & 1 \end{pmatrix} \begin{pmatrix} 1 & 0 \\ 0 & a_2 \end{pmatrix} g \right] \right) \left| \det \left[ \begin{pmatrix} 1 & 0 \\ y & 1 \end{pmatrix} \begin{pmatrix} 1 & 0 \\ 0 & a_2 \end{pmatrix} g \right] \right|^{-u_2-1/2}$$
$$dk dy d^\times a_2.$$

This function of $g$ is invariant on the left under the subgroup $P$ of matrices of the form

$$\begin{pmatrix} 1 & 0 \\ * & * \end{pmatrix}.$$

Thus there is a function $\Phi_u[\bullet]$ on $F^2$ such that

$$\Phi_u[(1, 0)g] = g_u(g).$$

In particular, the function

$$(x, y) \mapsto \Phi_u[(x, y)]$$

has support in a fixed compact set of $F^2 - 0$. Then

$$f_{\phi,u}(g) = \int g_u \left[ g \begin{pmatrix} t & 0 \\ 0 & 1 \end{pmatrix} \right] \mu_2 \mu_1^{-1}(t) \mid t \mid^{u_2-u_1+1} d^\times t \, \mu_2(\det g) \mid \det g \mid^{u_2+1/2}$$
$$= \int \Phi_u[(t, 0)g] \mid t \mid^{u_2-u_1+1} d^\times t \, \mu_2(\det g) \mid \det g \mid^{u_2+1/2}.$$

We now compute

$$\mathcal{W}_u(f_{\phi,u}) = \int f_{\phi,u} \left[ \begin{pmatrix} 1 & x \\ 0 & 1 \end{pmatrix} \right] \psi(-x) dx$$
$$= \int \int \Phi_u [(t, tx)] \mu_2 \mu_1^{-1}(t) \mid t \mid^{u_2-u_1+1} d^\times t \psi(-x) dx.$$

This double integral converges absolutely for $\Re u_2 > \Re u_1$. If we change $x$ to $xt^{-1}$ we get:

$$\int \mathcal{F}_1(\Phi_u)[(t, t^{-1})]\mu_2 \cdot \mu_1^{-1}(t) \mid t \mid^{u_2-u_1} d^\times t$$

where we denote by a $\mathcal{F}_1$ (or simply $\hat{\Phi}$) the partial Fourier transform with respect to the second variable:

$$\mathcal{F}_1(\Phi)(x, y) = \int \Phi(x, z)\psi(-zx)dz.$$

It is then clear that this new integral converges for all values of $u$. For the case of $GL(2)$, it is the integral representation alluded to in the introduction.

To obtain a more general formula, we introduce two representations $l_2$ and $\hat{l}_2$ of $GL(2, F)$ on $\mathcal{S}(F^2)$ by

$$l_2(g)\Phi(X) = \Phi(Xg)\,,\ \hat{l}_2(g)\mathcal{F}_1(\Phi) = \mathcal{F}_1(l_2(g)\Phi).$$

Then $W_u(g) := \mathcal{W}_u(\Xi_u(g)f_{\phi,u})$ is given by

$$W_u(g) = \int \hat{l}_2(g)\mathcal{F}_1(\Phi_u)[(t, t^{-1})]\mu_2\mu_1^{-1}(t) \mid t \mid^{u_2-u_1} d^\times t$$
$$\mu_2(\det g) \mid \det g \mid^{u_2+\frac{1}{2}} .$$

For instance, if $\alpha = \mathrm{diag}(\alpha_1, \alpha_2)$, then

$$W_u(\alpha) = \int \mathcal{F}_1(\Phi_u)\big[\big(t\alpha_1, t^{-1}\alpha_2^{-1}\big)\big]\mu_2\mu_1^{-1}(t) \mid t \mid^{u_2-u_1} d^\times t$$
$$\mu_2(\alpha_1\alpha_2) \mid \alpha_1\alpha_2 \mid^{u_2} \mid \alpha_1 \mid^{1/2} \mid \alpha_2 \mid^{-1/2} .$$

**4. A reduction step.** We will need an elementary but crucial result which in the case of $GL(2)$ describes the space of function represented by the integrals $\mathcal{W}_u(\Xi_u(g)f_{\phi,u})$. We will set $\mathbb{Z}_F = \mathbb{Z}$ if $F$ is real and $\mathbb{Z}_F = \mathbb{Z}/2$ if $F$ is complex.

PROPOSITION 1. *Let $\Omega$ be an open, connected, relatively compact set of $\mathbb{C}$, the closure of which is contained in $\mathbb{C} - \mathbb{Z}_F$. Let $\mathbb{X}$ be an auxiliary real vector space of finite dimension. Given $\Phi \in \mathcal{S}(F^2 \oplus \mathbb{X})$ and $u \in \Omega$ there are $\Phi_{1,u}$ and $\Phi_{2,u}$ in $\mathcal{S}(F \oplus \mathbb{X})$ such that, for all $u \in \Omega$,*

$$w_{\Phi,u}(t_1, t_2 : X) := \int \Phi(t_1 t, t_2 t^{-1} : X) \mid t \mid^u \mu(t) d^\times t$$

*is equal to*

$$\Phi_{1,u}(t_1t_2 : X) \mid t_1 \mid^{-u} \mu^{-1}(t_1) + \Phi_{2,u}(t_1t_2 : X) \mid t_2 \mid^u \mu(t_2).$$

*One can choose the functions $\Phi_{i,u}(t : X)$ in such a way that the maps $u \mapsto \Phi_{i,u}(\bullet : \bullet)$ are holomorphic maps from $\Omega$ to $\mathcal{S}(F \oplus \mathbb{X})$. Furthermore, if $\Phi$ remains in a bounded set, then the functions $\Phi_{i,u}$ remain in a bounded set. Finally, if $\Phi = \Phi_s$ depends* *holomorphically on $s \in \Omega'$, $\Omega'$ open in $\mathbb{C}^n$, and remains in a bounded set for all values of s, one can choose the functions $\Phi_*$ to depend holomorphically on $(u, s)$ and to remain in a bounded set.*

*Proof.* We begin the proof with a variant of the Borel lemma, as expounded in [H], Theorem 1.2.6. □

LEMMA 2. *Let $f_j$ be a sequence of holomorphic functions on an open, connected, set $\Omega$ of $\mathbb{C}^n$. Assume each $f_j$ is bounded. For any $\epsilon > 0$ there is a smooth function $f(u,t)$ on $\Omega \times ]-\epsilon, \epsilon[$ with the following properties. The function is holomorphic in u and, for each $j \geq 0$,*

$$\frac{\partial f^j}{\partial t^j}(u, 0) = f_j(u).$$

*Finally the support of $f$ has a compact projection on the second factor.*

*Proof of the Lemma.* We choose a smooth function $g$ of compact support contained in $]-\epsilon, \epsilon[$, with $g(t) = 1$ for $t$ sufficiently close to 0. Next we choose $0 < \epsilon_j < 1$ and set

$$g_j(u,t) = g\left(\frac{t}{\epsilon_j}\right)\frac{t^j}{j!} f_j(u).$$

We have then, for $\alpha < j$,

$$\left|\frac{\partial^\alpha g_j(u,t)}{\partial t^\alpha}\right| \leq C_{\alpha,j} \sup_\Omega \mid f_j \mid \epsilon_j^{j-\alpha} \leq C_j \sup_\Omega \mid f_j \mid \epsilon_j$$

where $C_j = \sup_{\alpha<j} C_{\alpha,j}$ depends only $j$ (and our choice of $g$). If

$$\epsilon_j \leq \frac{1}{C_j \sup_\Omega \mid f_j \mid 2^j}$$

then, for all $\alpha < j$,

$$\left|\frac{\partial t^\alpha g_j(u,t)}{\partial t^\alpha}\right| \leq 2^{-j}.$$

The function

$$f(u,t) = \sum_j g_j(u,t)$$

has the required property. □

In the same way, one proves the following lemma.

LEMMA 3. *Let $f_{j,k}$ be a double sequence of holomorphic functions on an open set $\Omega$ of $\mathbb{C}^n$. Assume each $f_{j,k}$ is bounded. For any open disc $D$ of center 0 in $\mathbb{C}$,* *there is a smooth function $f(u,t)$ on $\Omega \times D$ with the following properties. The function is holomorphic in u and, for each $j \geq 0, k \geq 0$,*

$$\frac{\partial f^{j+k}}{\partial t^j \partial \bar{t}^k}(u,0) = f_{j,k}(u).$$

*Finally the support of $f$ has a compact projection on the second factor.*

Another variant of the Borel lemma is as follows.

LEMMA 4. *Let $f_j$ be a sequence of Schwartz functions on some vector space $\mathbb{U}$. For any $\epsilon > 0$ there is a Schwartz function $f$ on $\mathbb{U} \oplus \mathbb{R}$, supported on $\mathbb{U} \times [-\epsilon, \epsilon]$ such that, for every $j \geq 0$,*

$$\frac{\partial f^j}{\partial t^j}(u,0) = f_j(u).$$

Indeed, we may view a Schwartz function on $\mathbb{U}$ as a smooth function on the sphere $S$ in $\mathbb{U} \oplus \mathbb{R}$ which, in addition, vanishes as well as all its derivatives at some point $P_0 \in S$. We construct the required function $f$ as before as a sum of a series

$$f(s,t) = \sum_j g\left(\frac{t}{\epsilon_j}\right)\frac{t^j}{j!} f_j(s)$$

which can be differentiated term wise. In particular $f$ and all its derivatives in the $s$ variables vanish at any point of the form $(P_0, t)$.

We go back to the proof of the proposition. To be definite we assume that $F$ is complex. The real case is somewhat simpler. We recall that an integral

$$\int \Phi(t) \mid t \mid^s d^\times t$$

converges absolutely for $\Re s > 0$ and extends meromorphically with a simple pole at $s = 0$ and residue $\Phi(0)$. Let $\xi$ be a normalized character. To find the poles and residues of an integral of the form

$$\int \Phi(t) \mid t \mid^s \xi(t) d^\times t$$

in the half plane $\Re s > -M - 1$, we choose a smooth function of compact support $\phi_0$ on $F$ equal to 1 near 0. We write

$$\begin{aligned}\Phi(t) &= \phi_0(t)\Phi(t) + (1 - \phi_1)\Phi(t)\\ &= \phi_0(t) \sum_{n_1+n_2 \leq M} \frac{t^{n_1}\bar{t}^{n_2}}{n_1! n_2!} \frac{\partial^{n_1+n_2}\Phi}{\partial t^{n_1} \partial \bar{t}^{n_2}}(0) + r(t).\end{aligned}$$

Then the integral is the sum of

$$\int r(t) \mid t \mid^s \xi(t) d^\times t$$

which has no pole in the half plane and

$$\sum_{n_1+n_2 \leq M} \frac{1}{n_1!n_2!} \frac{\partial^{n_1+n_2}\Phi}{\partial t^{n_1} \partial \bar{t}^{n_2}}(0) \int \phi_0(t) \mid t \mid^s \xi(t) t^{n_1} \bar{t}^{n_2} d^\times t.$$

Each term contributes (at most) one simple pole at any point $s$ such that there are integers $n_1 \geq 0, n_2 \geq 0$ with

$$\mid t \mid^s \xi(t) t^{n_1} \bar{t}^{n_2} \equiv 1.$$

The residue is then

$$\frac{1}{n_1!n_2!} \frac{\partial^{n_1+n_2}\Phi}{\partial t^{n_1} \partial \bar{t}^{n_2}}(0).$$

For instance, if $\xi(z) = (z/\sqrt{z\bar{z}})r$ with $r \geq 0$, there is a pole at any point $-\frac{r}{2} - n$ with $n \geq 0$ integer. The residue is given by the above formula with $n_1 = n$ and $n_2 = n + r$.

Coming back to our integral, we will first prove the Proposition when there is no auxiliary space $\mathbb{X}$ and no dependence on some complex parameter $s$. We remark that, after a change of variables, we can reduce ourselves to the case where $t_2 = 1$, in other words, study the function $w_u(a) := w_{\Phi,u}(a, 1)$. We then have to show that

$$w_u(a) = \Phi_{1,u}(a) \mid a \mid^{-u} \mu^{-1}(a) + \Phi_{2,u}(a).$$

After a change of variables, we find

$$\int w_u(a) \mid a \mid^s \xi(a) d^\times a$$
$$= \int\int \Phi(t_2, t_1) \mid t_2 \mid^s \xi(t_2) d^\times t_2 \mid t_1 \mid^{s-u} \xi\mu^{-1}(t_1) d^\times t_1.$$

Consider for one moment the following function of two variables:

$$A(s_2, s_1) := \int\int \Phi(t_2, t_1) \mid t_2 \mid^{s_2} \xi(t_2) d^\times t_2 \mid t_1 \mid^{s_1} \xi\mu^{-1}(t_1) d^\times t_1.$$

It is a meromorphic function of $(s_2, s_1)$ with hyperplane singularities. There are singularities on some of the hyperplanes

$$s_2 + z_2 = 0,\ z_2 \in \mathbb{Z}_F, z_2 \geq 0$$

and on some of the hyperplanes

$$s_1 + z_1 = 0,\ z_1 \in \mathbb{Z}_F, z_1 \geq 0.$$

Now consider the function $A(s, s-u)$. In $\mathbb{C}^2$, a singular hyperplane

$$s + z_2 = 0$$

and a singular hyperplane

$$s - u + z_1 = 0$$

intersect only at points where $u \in \mathbb{Z}_F$. In particular, if $u$ is not in $\mathbb{Z}_F$, then $A(s, s-u)$, viewed as a function of $s$, has only simple poles. At a pole where $s$ is such that

$$| t_2 |^s \xi(t_2) t_2^{n_1} \overline{t}_2^{n_2} \equiv 1$$

the residue is equal to

$$\frac{1}{n_1! n_2!} \int \frac{\partial^{n_1+n_2}\Phi}{\partial t_2^{n_1} \partial \overline{t}_2^{n_2}}(0, t_1) \mid t_1 \mid^{-u} t_1^{-n_1} \overline{t}_1^{-n_2} \mu^{-1}(t_1) d^\times t_1.$$

Note that the integral defines a holomorphic function of $u$ in the complement of $\mathbb{Z}_F$.

Likewise, at a pole where $s$ is such that

$$| t_1 |^{s-u} \xi\mu^{-1}(t_1) t_1^{m_1} \overline{t}_1^{m_2} \equiv 1$$

the residue is equal to

$$\frac{1}{m_1! m_2!} \int \frac{\partial^{m_1+m_2}\Phi}{\partial t_1^{m_1} \partial \overline{t}_1^{m_2}}(t_2, 0) \mid t_2 \mid^{u} \mu(t_2) t_2^{-m_1} \overline{t}_2^{-m_2} d^\times t_2.$$

Applying the variant of Borel lemma given above, we can find smooth functions of compact support $\Phi_{2,u}(a)$ and $\Phi_{1,u}(a)$, depending holomorphically on $u$, such that, for all $u \in \Omega$,

$$\frac{\partial^{n_1+n_2}\Phi_{2,u}}{\partial a^{n_1} \partial \overline{a}^{n_2}}(0) = \int \frac{\partial^{n_1+n_2}\Phi}{\partial t_2^{n_1} \partial \overline{t}_2^{n_2}}(0, t_1) \mid t_1 \mid^{-u} t_1^{-n_1} \overline{t}_1^{-n_2} \mu^{-1}(t_1) d^\times t_1 \tag{5}$$

and

$$\frac{\partial^{m_1+m_2}\Phi_{1,u}}{\partial a^{m_1} \partial \overline{a}^{m_2}}(0) = \int \frac{\partial^{m_1+m_2}\Phi}{\partial t_1^{m_1} \partial \overline{t}_1^{m_2}}(t_2, 0) \mid t_2 \mid^{u} \mu(t_2) t_2^{-m_1} \overline{t}_2^{-m_2} d^\times t_2. \tag{6}$$

The difference between

$$\int w_u(a) \mid a \mid^s \xi(a) d^\times a$$

and

$$B_u := \int \Phi_{2,u}(a) \mid a \mid^s \xi(a) d^\times a + \int \Phi_{1,u}(a) \mid a \mid^{s-u} \xi\mu^{-1}(a) d^\times a \tag{7}$$

is then, by construction, an entire function of $s$. It is rapidly decreasing in a vertical strip and also rapidly decreasing with respect to $\xi$. It follows there is a smooth function $\Psi_u(a)$ on $F$, rapidly decreasing at infinity, with zero derivatives at 0 such that the above difference is

$$\int \Psi_u(a) \mid a \mid^s \xi(a) d^\times a$$

for all $\xi$. Furthermore, the function depends holomorphically on $u$. We obtain our claim with $\Phi_{2,u}$ replaced by $\Phi_{2,u} + \Psi_u$.

If there is an auxiliary vector space $X$ and dependence on a complex parameter $s$ we use the variants of the Borel lemma to choose the functions $\Phi_{i,u}(x : X : s)$ in such a way that

$$\begin{aligned}
&\frac{\partial^{n_1+n_2}\Phi_{2,u}}{\partial a^{\,n_1}\partial \overline{a}^{\,n_2}}(0 : X : s) \\
&\quad = \int \frac{\partial^{n_1+n_2}\Phi}{\partial t_2^{n_1}\partial \overline{t}_2^{n_2}}[(0, t_1) : X : s] \mid t_1 \mid^{-u} t_1^{-n_1}\overline{t}_1^{-n_2}\mu^{-1}(t_1) d^\times t_1
\end{aligned}$$

and

$$\begin{aligned}
&\frac{\partial^{m_1+m_2}\Phi_{1,u}}{\partial a^{\,m_1}\partial \overline{a}^{\,m_2}}(0 : X : s) \\
&\quad = \int \frac{\partial^{m_1+m_2}\Phi}{\partial t_1^{m_1}\partial \overline{t}_1^{m_2}}[(t_2, 0) : X : s] \mid t_2 \mid^{u} \mu(t_2) t_2^{-m_1}\overline{t}_2^{-m_2} d^\times t_2. \qquad \square
\end{aligned}$$

In fact, we will need the following supplement to the previous proposition.

Proposition 2. *Let $0 < a$ be real a number not in $\mathbb{Z}_F$ and $\mathfrak{S}$ the strip $\{u \mid -a < \Re u < a\}$. Let $P(u) = P(-u)$ be the polynomial $\prod(u - u_j)$, where the product is over all $u_j \in \mathfrak{S} \cap \mathbb{Z}_F$. Given $\Phi \in \mathcal{S}(F^2 \oplus \mathbb{X})$ set as before*

$$w_{\Phi,u}(t_1, t_2 : X) := \int \Phi(t_1 t, t_2 t^{-1}) \mid t \mid^u \mu(t) d^\times t.$$

*For each $u \in \mathfrak{S}$, there are $\Phi_{1,u}$ and $\Phi_{2,u}$ in $\mathcal{S}(F \oplus \mathbb{X})$ such that, for $u \in \mathfrak{S}$,*

$$\begin{aligned}
&P(u) w_{\Phi,u}(t_1, t_2 : X) \\
&\quad = \Phi_{1,u}(t_1 t_2 : X) \mid t_1 \mid^{-u} \mu^{-1}(t_1) + \Phi_{2,u}(t_1 t_2 : X) \mid t_2 \mid^{u} \mu(t_2).
\end{aligned}$$

*One can choose the functions in such a way that the maps $u \mapsto \Phi_{i,u}$ from $\mathfrak{S}$ to $\mathcal{S}(F \oplus \mathbb{X})$ are holomorphic and, furthermore, the functions $\Phi_{i,u}$ remain in a bounded set if $\Phi$ does. Finally, if $\Phi$ depends holomorphically on $s \in \Omega'$, $\Omega'$ open in $\mathbb{C}^m$, and remains in a bounded set for all values of $s$, one can choose the functions $\Phi_*$ to depend holomorphically on $(u, s)$ and to remain in a bounded set.*

*Proof.* The proof is similar to the proof of the previous proposition. The only difference is that we choose the functions $\Phi_{i,u}$ such that they satisfy, instead of (5) and (6) the following relations, for all $u \in \mathfrak{S}$,

$$\frac{\partial^{n_1+n_2}\Phi_{2,u}}{\partial a^{n_1}\partial \overline{a}^{n_2}}(0) = P(u)\int \frac{\partial^{n_1+n_2}\Phi}{\partial t_2^{n_1}\partial \overline{t}_2^{n_2}}(0, t_1) \mid t_1 \mid^{-u} t_1^{-n_1}\overline{t}_1^{-n_2}\mu^{-1}(t_1)d^\times t_1$$

and

$$\frac{\partial^{m_1+m_2}\Phi_{1,u}}{\partial a^{m_1}\partial \overline{a}^{m_2}}(0) = P(u)\int \frac{\partial^{m_1+m_2}\Phi}{\partial t_1^{m_1}\partial \overline{t}_1^{m_2}}(t_2, 0) \mid t_2 \mid^{u} \mu(t_2)t_2^{-m_1}\overline{t}_2^{-m_2}d^\times t_2.$$

Again, the right-hand sides integrals are holomorphic functions of $s$ in $\mathfrak{S}$. To finish the proof as before, we need only check that the difference between

$$A_u(s) := P(u)\int w_u(a) \mid a \mid^s \xi(a)d^\times a$$

and $B_u(s)$ defined in (7) is, for each $u \in \mathfrak{S}$, a holomorphic function of $u$. For $u \notin \mathbb{Z}_F$, this follows directly as before from the constructions. Consider now a $u_0 \in \mathfrak{S} \cap \mathbb{Z}_F$. Then $A_{u_0}(s) \equiv 0$. Thus we need to check that $B_{u_0}(s)$ is a holomorphic function of $s$. We write $P(u) = P_1(u)(u - u_0)$. Consider then a potential pole $s_0$ of the first term in $B_u$. This means that, for suitable integers $n_1, n_2 \geq 0$,

$$\mid a \mid^{s_0} \xi(a)a^{n_1}\overline{a}^{n_2} \equiv 1. \tag{8}$$

The residue of the first term is then $P_1(u_0)$ times

$$\frac{u-u_0}{n_1!n_2!}\int \frac{\partial^{n_1+n_2}\Phi}{\partial t_2^{n_1}\partial \overline{t}_2^{n_2}}(0, t_1) \mid t_1 \mid^{-u} t_1^{-n_1}\overline{t}_1^{-n_2}\mu^{-1}(t_1)d^\times t_1\Bigg|_{u=u_0}.$$

This is zero unless $u_0$ is a singularity of the integral, that is, for suitable integers $m_1 \geq 0, m_2 \geq 0$,

$$\mid t_1 \mid^{-u_0} t_1^{-n_1}\overline{t}_1^{-n_2}\mu^{-1}(t_1)t_1^{m_1}\overline{t}_2^{m_2} \equiv 1. \tag{9}$$

The residue is then

$$-\frac{P_1(u_0)}{n_1!n_2!m_1!m_2!}\frac{\partial^{n_1+n_2+m_1+m_2}\Phi}{\partial t_2^{n_1}\partial \overline{t}_2^{n_2}\partial t_1^{m_1}\partial \overline{t}_1^{m_2}}(0, 0).$$

On the other hand, we have from (8) and (9)

$$\mid a \mid^{s_0-u_0} \xi\mu^{-1}(a)a^{m_1}\overline{a}^{m_2} \equiv 1.$$

Thus $s_0$ is a potential pole of the second term with residue

$$P_1(u_0)\frac{(u-u_0)}{m_1!m_2!}\int \frac{\partial^{m_1+m_2}\Phi}{\partial t_1^{m_1}\partial \overline{t}_1^{m_2}}(t_2, 0) \mid t_2 \mid^{u} \mu(t_2)t_2^{-m_1}\overline{t}_2^{-m_2}d^\times t_2\Bigg|_{u=u_0}.$$

From (8) and (9) we get

$$\mid t_2 \mid^{u_0} \mu(t_2)t_2^{-m_1}\overline{t}_2^{-m_2}t_2^{n_1}\overline{t}_2^{n_2} \equiv 1.$$

We see that $u_0$ is a pole of this integral and so the residue of the second term at $s_0$ is

$$\frac{P_1(u_0)}{n_1!n_2!m_1!m_2!}\frac{\partial^{n_1+n_2+m_1+m_2}\Phi}{\partial t_2^{n_1}\partial\overline{t}_2^{n_2}\partial t_1^{m_1}\partial\overline{t}_1^{m_2}}(0,0).$$

We see then that the sum of the two terms has no pole at $s_0$. The same analysis applies to the second term. Our assertion follows. □

We will need a coarse majorization of our functions.

PROPOSITION 3. *For $\Phi \in \mathcal{S}(F^2 \oplus \mathbb{X})$ set*

$$w_{\Phi,u}(t_1,t_2:X) := \int \Phi(t_1 t, t_2 t^{-1}:X) \mid t \mid^u \mu(t) d^\times t.$$

*Suppose that $\Phi$ is in a bounded set and $\Re u$ in a compact set. Then, there is $M > 0$ such that, for every $N > 0$, there is $\phi \geq 0$ in $\mathcal{S}(\mathbb{X})$ such that*

$$|w_{\Phi,u}(t_1,t_2:X)| \leq \mid t_2 \mid^{\Re u} \phi(X)\frac{\mid t_1t_2 \mid^{-M}}{(1+\mid t_1t_2 \mid)^N}$$

*If $\Phi$ is in a bounded set and u purely imaginary, there is, for every $N > 0$, a function $\phi \geq 0$ in $\mathcal{S}(\mathbb{X})$ such that*

$$|w_{\Phi,u}(t_1,t_2:X)| \leq \phi(X)\frac{A+B\mid \log \mid t_1t_2 \mid\mid}{(1+\mid t_1t_2 \mid)^N}.$$

*Proof.* Say $F$ is complex. We may as well study

$$w_{\Phi,u}(a:X) := \int \Phi(at, t^{-1}:X) \mid t \mid^u \mu(t) d^\times t.$$

We may bound $\Phi$ in absolute value by a product $\Phi_0\Psi(X)$ with fixed non-negative Schwartz functions and replace $u$ by its real part. Then the integral is bounded, for all $N_1$, $N_2$, by a constant times $\Psi(X)$ times

$$\int \frac{\mid t \mid^u}{(1+\mid at \mid)^{N_1}(1+\mid t^{-1} \mid)^{N_2}} d^\times t \leq \mid a \mid^{-N_1} \int \frac{\mid t \mid^{N_2-N_1+u}}{(1+\mid t \mid)^{N_2}} d^\times t.$$

The first assertion follows then by considering separately the case $\mid a \mid \leq 1$ and the case $\mid a \mid \geq 1$.

For the second assertion we may assume $u = 0$. We need only consider the estimate for $\mid a \mid \leq 1$. We write

$$\int \frac{1}{(1+\mid at \mid)^{N_1}\left(1+\mid t^{-1} \mid\right)^{N_2}} d^\times t$$

as the sum of

$$\int_{|t|\leq 1} \frac{\mid t \mid^{N_2}}{(1+\mid at \mid)^{N_1}(1+\mid t \mid)^{N_2}} d^\times t$$

and

$$\int_{|t|\geq 1} \frac{|\, t \,|^{N_2}}{(1+ |\, at \,|)^{N_1} (1+ |\, t \,|)^{N_2}} d^\times t.$$

The first integral is bounded independently of $a$. The second is bounded by

$$\int_{|t|\geq 1} \frac{1}{(1+ |\, at \,|)^{N_1}} d^\times t = \int_{|t|\geq |a|} \frac{1}{(1+ |\, t \,|)^{N_1}} d^\times t$$

which is in turn bounded by a polynomial of degree 1 in $\log |\, a \,|$. □

**5. Integral representation of sections.** Our goal in this section is to obtain an integral representation of sections of the form $f_{\phi,u}$. Assume that $\phi = \phi_1 * \phi_2 * \cdots * \phi_{n-1}$ with $\phi_i \in \mathcal{D}(G_n)$. According to [DM], every element of $\mathcal{D}(G_n)$ is a sum of such products, so there is no loss of generality. Let $f_u$ be a standard section. We consider the section $f_{\phi,u}$ defined by

$$f_{\phi,u}(g) = \int f_u(gx)\phi(x)dx \tag{10}$$

or, more explicitly,

$$f_{\phi,u}(g) = \int f_u[gx_1x_2 \cdots x_{n-1}]\phi_1(x_1)\phi_2(x_2) \cdots \phi_{n-1}(x_{n-1})dx_1dx_2 \cdots dx_n.$$

We set

$$\mathbb{V}_n := M(1 \times 2, F) \times M(2 \times 3, F) \times \cdots \times M(n-1 \times n, F). \tag{11}$$

We denote by $R_i$ the set of matrices of rank $i$ in $M(i \times i+1, F)$.

PROPOSITION 4. *There is a function*

$$\Phi_u\, [X_1 : X_2 : \cdots : X_{n-1}]$$

*on $\mathbb{C}^n \times \mathbb{V}_n$ with the following properties. It is a smooth function, holomorphic in $u$. The projection of its support on the $i-th$ factor is contained in a fixed compact subset of $R_i$. Finally, for every $u$,*

$$f_{\phi,u}(g) = |\det g\,|^{u_n + \frac{n-1}{2}}\, \mu_n(\det g)$$
$$\times \int \Phi_u \left[(g_1, 0)g_2^{-1} : (g_2, 0)g_3^{-1} : \cdots : (g_{n-2}, 0)g_{n-1}^{-1} : (g_{n-1}, 0)g\right]$$
$$\prod_{i=1}^{i=n-1} |\det g_i\,|^{u_{i+1}-u_i+1}\, \mu_{i+1}\mu_i^{-1}(\det g_i)dg_i\, ;$$

*in this integral, each variable $g_i$ is integrated over $GL(i, F)$.*

Note that the integral is convergent for all $u$ under the restricted assumption on the support of $\Phi$. This would not be true for a Schwartz function or even a function of compact support.

In what follows, by abuse of language, we will suppress the characters $\mu_i$ from the notations. The reader will easily re-establish them by replacing $\mid \det g \mid^{u_n}$ by $\mu_n(\det g) \mid g \mid^{u_n}$ and so on. It will be more convenient to prove the result for somewhat more general sections. Namely, we consider functions of the form

$$f_u[g, g_1, g_2, \dots, g_{n-1}]$$

which are smooth functions on $\mathbb{C}^n \times GL(n)^n$ and holomorphic in $u$; we assume that, for fixed $u$ and fixed $(g_i)$, the function

$$g \mapsto f_u[g, g_1, g_2, \dots, g_{n-1}]$$

belongs to the space of $\Xi_u$. The projection of the support of $f$ on the $i+1$-th factor is contained in a fixed compact set of $GL(n)$. Then

$$g \mapsto \int f_u[gx_1x_2\cdots x_{n-1}, x_1, x_2, \dots, x_{n-1}]dx_1dx_2\cdots dx_{n-1} \tag{12}$$

is the type of section for which we prove our integral representation, by induction on $n$.

Thus we assume our assertion established for sections of this type and the integer $n-1$. In fact, for the purpose of carrying out our induction, we should consider more generally functions of the form

$$f_{u,s}[g, g_1, g_2, \dots, g_{n-1}, h]$$

where the $s$ is an auxiliary complex parameter and the last variable $h$ is in some auxiliary manifold $S$; the function depends holomorphically on $s$ and the projection of its support on $S$ is contained in a fixed compact set. Then the function $\Phi_u$ of the Proposition would also depend on $s$ and $h$ with obvious properties. We simply ignore this complication.

Consider a section of the type (12). It is convenient to introduce

$$f_u^1[g, x_1, x_2, \dots, x_{n-1}; s] = f_u\left[g, x_1^{-1}, x_2, \dots, x_{n-1}\right]$$

and then the integral defining (12) takes the form

$$\int f_u^1\left[x_1x_2\cdots x_{n-1}, x_1^{-1}g, x_2, \dots, x_{n-1}\right]dx_1dx_2\cdots dx_{n-1}.$$

Next, we use the Iwasawa decomposition of $x_1$, in the form

$$x_1 = p_1k_1,$$

$$p_1 = \begin{pmatrix} g_{n-1} & 0 \\ X & t \end{pmatrix} = \begin{pmatrix} g_{n-1} & 0 \\ 0 & 1 \end{pmatrix}\begin{pmatrix} 1_{n-1} & 0 \\ X & t \end{pmatrix} \tag{13}$$

with $k_1 \in K_n$, $g_{n-1} \in GL(n-1)$, $t \in F^\times$. After a change of variables, the integral becomes

$$\int f_u^1 \left[ \begin{pmatrix} g_{n-1} & 0 \\ 0 & 1 \end{pmatrix} x_2 \cdots x_{n-1}, k_1^{-1} p_1^{-1} g, k_1^{-1} x_2, \ldots, x_{n-1} \right]$$
$$\mid t \mid^{u_n - \frac{n-1}{2}} dk_1 dg_{n-1} d^\times t dX dx_2 \cdots dx_{n-1}.$$

If we set

$$f_u^2(g, x_1, x_2, \ldots, x_{n-1}) = \int f_u^1(g, k_1^{-1} x_1, k_1^{-1} x_2, \ldots, x_{n-1}) dk_1$$

then, after integrating over $k_1$, we obtain

$$\int f_u^2 \left[ \begin{pmatrix} g_{n-1} & 0 \\ 0 & 1 \end{pmatrix} x_2 \cdots x_{n-1}, p_1^{-1} g, x_2, \ldots, x_{n-1} \right]$$
$$\mid t \mid^{u_n - \frac{n-1}{2}} dg_{n-1} d^\times t dX dx_2 \cdots dx_{n-1}.$$

To continue we set

$$x_i = \begin{pmatrix} m_i & 0 \\ 0 & 1 \end{pmatrix} p_i,\ m_i \in GL(n-1, F),\ p_i \in GL(n-1, F)\backslash GL(n, F).$$

Then

$$dx_i = dm_i d_r p_i$$

where $d_r p$ is an invariant measure on $GL(n-1, F)\backslash GL(n, F)$. After a change of variables, we get

$$\int f_u^2 \left[ \begin{pmatrix} g_{n-1} m_2 m_3 \cdots m_{n-1} & 0 \\ 0 & 1 \end{pmatrix} p_{n-1},\ p_1^{-1} g, \right.$$
$$\left. \begin{pmatrix} m_2 & 0 \\ 0 & 1 \end{pmatrix} p_2, p_2^{-1} \begin{pmatrix} m_3 & 0 \\ 0 & 1 \end{pmatrix} p_3, \ldots, p_{n-2}^{-1} \begin{pmatrix} m_{n-1} & 0 \\ 0 & 1 \end{pmatrix} p_{n-1} \right]$$
$$\mid t \mid^{u_n - \frac{n-1}{2}} dg_{n-1} d^\times t dX \bigotimes_{i=2}^{n-1} dm_i \bigotimes_{i=2}^{n-1} d_r p_i.$$

We introduce a new function $f_u^3$ on

$$\mathbb{C}^n \times GL(n-1) \times GL(n) \times GL(n-1)^{n-2}$$

which is defined by

$$\mid \det g_{n-1} \mid^{-1/2} \prod_{i=2}^{n-1} \mid \det m_i \mid^{1/2} f_u^3 [g_{n-1}, g, m_2, m_3, \ldots, m_{n-1}] :$$
$$= \int f_u^2 \left[ \begin{pmatrix} g_{n-1} & 0 \\ 0 & 1 \end{pmatrix} p_{n-1},\ g, \right.$$
$$\left. \begin{pmatrix} m_2 & 0 \\ 0 & 1 \end{pmatrix} p_2, p_2^{-1} \begin{pmatrix} m_3 & 0 \\ 0 & 1 \end{pmatrix} p_3, \ldots, p_{n-2}^{-1} \begin{pmatrix} m_{n-1} & 0 \\ 0 & 1 \end{pmatrix} p_{n-1} \right] \bigotimes_{i=2}^{i=n-1} d_r p_i.$$

It is clear that $f_u^3$ is a smooth function, holomorphic in $u$, and the projection of its support on each linear factor except the first one is contained in a fixed compact set. Moreover, the function

$$g_{n-1} \mapsto f_u^3\,[g_{n-1}, g, m_2, m_3, \ldots, m_{n-1}]$$

belongs to the space of the representation of $GL(n-1)$ determined by $(\mu_1, \mu_2, \ldots, \mu_{n-1})$ and $(u_1, u_2, \ldots u_{n-1})$.

Finally, we set

$$f_u^4(g_{n-1}, g)$$
$$= \int f_u^3(g_{n-1}m_2m_3\cdots m_{n-1}, g, m_2, m_3, \ldots, m_{n-1})dm_2dm_3\cdots dm_{n-1}.$$

The section (12) can be represented by the integral

$$\int f_u^4\left[g_{n-1}, p_1^{-1}g\right] \mid \det g_{n-1}\mid^{-1/2} \mid t\mid^{u_n-\frac{n-1}{2}} dg_{n-1}d^\times t dX. \tag{14}$$

We apply the induction hypothesis to the function $g_{n-1} \mapsto f_u^4(g_{n-1}, g)$. There is a function $\Phi_u^1$ on

$$\mathbb{C}^n \times \mathbb{V}_{n-1} \times GL(n)$$

such that

$$\begin{aligned} f_u^4(g_{n-1}, g) =& \mid \det g_{n-1}\mid^{u_{n-1}+\frac{n-2}{2}} \\ &\times \int \Phi_u^1\left[(g_1, 0)g_2^{-1} : (g_2, 0)g_3^{-1} : \cdots : (g_{n-3}, 0)g_{n-2}^{-1} : (g_{n-2}, 0)g_{n-1} : g\right] \\ &\prod_{i=1}^{i=n-2} \mid \det g_i\mid^{u_{i+1}-u_i+1} dg_i\,; \end{aligned} \tag{15}$$

the projection of its support on each factor of $\mathbb{V}_{n-1}$ or the factor $GL(n)$ is contained in a fixed compact set; more precisely, for each factor of $\mathbb{V}_{n-1}$, a compact set of $R_i$.

Finally, we can change variables in the integral (14) where $p_1$ is given by (13) to obtain

$$\int f_u^4\left[g_{n-1}^{-1}, \begin{pmatrix} 1_{n-1} & 0 \\ X & t \end{pmatrix}\begin{pmatrix} g_{n-1} & 0 \\ 0 & 1 \end{pmatrix} g\right]$$
$$\mid \det g_{n-1}\mid^{1/2} \mid t\mid^{-u_n-\frac{n-1}{2}} dg_{n-1}d^\times t dX.$$

The above integral can be written as

$$\begin{aligned} &\int f_u^4\left[g_{n-1}^{-1}, p\begin{pmatrix} g_{n-1} & 0 \\ 0 & 1 \end{pmatrix} g\right]\left|\det p\begin{pmatrix} g_{n-1} & 0 \\ 0 & 1 \end{pmatrix} g\right|^{-u_n-\frac{n-1}{2}} \\ &dg_{n-1}d_r p \mid \det g_{n-1}\mid^{u_n+\frac{n}{2}} \mid \det g\mid^{u_n+\frac{n-1}{2}} \end{aligned} \tag{16}$$

where the integral in $p$ is over the group $P$ of matrices of the form

$$\begin{pmatrix} 1_{n-1} & 0 \\ * & * \end{pmatrix}$$

and $d_r p$ is a right invariant measure. In terms of $\Phi_u^1$ this can be written as

$$\begin{aligned} &| \det g \,|^{u_n+\frac{n-1}{2}} \\ &\times \int \Phi_u^1 \Big[(g_1,0)g_2^{-1} : (g_2,0)g_3^{-1} : \cdots : (g_{n-3},0)g_{n-2}^{-1} : (g_{n-2},0)g_{n-1}^{-1} \\ &\quad : p\begin{pmatrix} g_{n-1} & 0 \\ 0 & 1 \end{pmatrix} g\Big] \left|\det p\begin{pmatrix} g_{n-1} & 0 \\ 0 & 1 \end{pmatrix} g\right|^{-u_n-\frac{n-1}{2}} \\ &\quad \prod_{i=1}^{i=n-1} | \det g_i \,|^{u_{i+1}-u_i+1}\, dg_i\, d_r p. \end{aligned}$$

There is a smooth function $\Phi_u$ on $\mathbb{C}^n \times \mathbb{V}_n$ such that

$$\begin{aligned} &\Phi_u(X_1, X_2, \ldots, X_{n-1}, (1_{n-1}, 0)g) \\ &\quad = \int \Phi_u^1(X_1, X_2, \ldots, X_{n-1}, pg)\, |\det pg|^{-u_n-\frac{n-1}{2}}\, d_r p. \end{aligned} \tag{17}$$

As before, the projection of its support on the $i-th$ linear factor of $\mathbb{C}^n \times \mathbb{V}_n$ is contained in a fixed compact set of $R_i$. If we combine (14) and the formula just before it we arrive at our conclusion.

We will need estimates on the function $\Phi_u$ of the previous proposition.

PROPOSITION 5. *Suppose that $\phi$ is the convolution product of $(n-1)$ functions, each of which is in a bounded set $\mathcal{B}$ of $\mathcal{D}(G_n)$ and the standard section $f_u$ (or rather its restriction to $K_n$) is in a bounded set of $\mathcal{V}(\mu)$. Fix a multi strip $\mathfrak{S} = \{u \mid -A < \Re u_i < A, 1 \le i \le n\}$ in $\mathbb{C}^n$. Then, there is a bounded set $\mathcal{C}$ of $\mathcal{S}(\mathbb{V}^n)$ such that $\Phi_u \in \mathcal{C}$, for all $u$ in $\mathfrak{S}$. Moreover, the map*

$$(f_u, \phi_1, \phi_2, \ldots, \phi_{n-1}) \mapsto \Phi_u$$

*is continuous.*

*Proof.* We prove the first assertion. The second assertion has a similar proof. Our construction shows that the functions $\Phi_u$ have a support contained in a fixed compact set of $\mathbb{V}_n$. Thus we have only to show that their derivatives are bounded uniformly when $u$ is in $\mathfrak{S}$. This follows from the following lemma. □

LEMMA 5. *Let $\mathfrak{S}$ be a strip in $\mathbb{C}$ and $\mathcal{B}$ a bounded set of $\mathcal{D}(G_n)$. For $\phi \in \mathcal{D}(G_n)$, $u \in \mathbb{C}$, set*

$$g_u(g) = | \det g \,|^u\, \mu(\det g)\phi(g)$$

*and*

$$g_u^1 = \int_P g_u(pg)d_r p.$$

*Let $R_{n-1}$ be the open set of matrices of maximal rank in $M((n-1)\times n, F)$. Define a function $\Phi_u \in \mathcal{S}(M((n-1)\times n, F))$ with compact support contained in $R_{n-1}$ by*

$$\Phi_u[(1_{n-1}, 0)g] = g_u^1(g).$$

*If $u$ is in a strip $\mathfrak{S}$ and $\phi$ in a bounded set $\mathcal{B}$, the function $\Phi_u$ remains in a bounded set of $\mathcal{S}(M((n-1)\times n, F))$.*

*Proof of the Lemma.* We have to show the derivatives of $\Phi_u$ are uniformly bounded. The group $SL(n, F)$ is transitive on $R_{n-1}$ thus the above relation for $g \in SL(n, F)$ already determines $\Phi_u$. If $X$ is in the enveloping algebra of $SL(n, F)$ (viewed as a real Lie group) it is easy to see that for $u \in \mathfrak{S}$ and $\phi \in \mathcal{B}$, the function $\rho(X)g_u$ is uniformly bounded. The same assertion is thus true for the function $g_u^1$. Since $R_{n-1}$ is isomorphic to $P \cap SL(n, F)\backslash SL(n, F)$ as a manifold, it will suffice to show that for every differential operator $\xi$ on $P \cap SL(n, F)\backslash SL(n, F)$ the function $\xi g_u^1$ is uniformly bounded for $g \in SL(n, F)$ and $u$ in the given strip. This is true for an operator of the form $\xi = \rho(X)$ where $X$ is in the enveloping algebra of $SL(n, F)$. Thus, in turn, it will suffice to show that every differential operator on $P \cap SL(n, F)\backslash SL(n, F)$ can be written as a linear combination of operators of the form $\rho(X)$ with smooth coefficients. For instance, one may use the fact that the map $(g, k) \mapsto gk$ from $GL(n-1) \times K \cap SL(n)$ to $GL(n, F)$ passes to the quotients and defines an isomorphism of a quotient of $GL(n-1) \times K \cap SL(n)$ with $P \cap SL(n, F)\backslash SL(n, F)$. □

**6. Integral representation of Whittaker functions.** In this section, our goal is to obtain an absolutely convergent integral formula for $\mathcal{W}_u(f_{\phi,u})$. To that end, we introduce more notations. The groups $GL(n-1, F)$ and $GL(n, F)$ operate on the space of Schwartz functions on $M((n-1)\times n, F)$ as follows:

$$l_n(g_n).\Psi.r_{n-1}(g_{n-1})[X] = \Psi[g_{n-1}Xg_n].$$

As the notation indicates, $l_n$ is a left action and $r_{n-1}$ a right action. We may identify the space of $(n-1)\times n$ matrices ($n-1$ rows, $n$ columns) to the direct sum of the space $\mathcal{B}_l(n-1, F)$ of lower triangular matrices of size $(n-1)\times(n-1)$ and the space $\mathcal{B}_u(n-1, F)$ of upper triangular matrices of size $(n-1)\times(n-1)$. For instance, for $n = 3$, the matrix

$$\begin{pmatrix} x_{1,1} & x_{1,2} & x_{1,3} \\ x_{2,1} & x_{2,2} & x_{2,3} \end{pmatrix}$$

corresponds to the pair of matrices

$$\left(\begin{pmatrix} x_{1,1} & 0 \\ x_{2,1} & x_{2,2} \end{pmatrix}, \begin{pmatrix} x_{1,2} & x_{1,3} \\ 0 & x_{2,3} \end{pmatrix}\right).$$

Accordingly, if $\Phi$ is a Schwartz function on $M((n-1) \times n, F)$ we define its partial Fourier transform $\mathcal{F}_{n-1}\Phi$ or simply $\hat{\Phi}$ by

$$\hat{\Phi}(b_1, b_2) = \int_{\mathcal{B}_u(n-1,F)} \Phi(b_1 \oplus b')\psi(-\mathrm{tr}(b_2 b'))db'.$$

It is thus a function on $\mathcal{B}_l(n-1, F) \oplus \mathcal{B}_l(n-1, F)$. We can use this partial Fourier transform to define two new representations $\hat{r}_{n-1}, \hat{l}_n$ by

$$\mathcal{F}_{n-1}\left(l_n(g_n).\Psi.r_{n-1}(g_{n-1})\right) = \hat{l}_n(g_n).\hat{\Psi}.\hat{r}_{n-1}(g_{n-1}),$$

on the space of Schwartz functions on $\mathcal{B}_l(n-1, F) \oplus \mathcal{B}_l(n-1, F)$.

The following formula will be very useful. Let $\alpha = \mathrm{diag}(\alpha_1, \alpha_2, \ldots, \alpha_n)$ be a diagonal matrix of size $n$. Set

$$\alpha^b = \mathrm{diag}(\alpha_1, \alpha_2, \ldots, \alpha_{n-1}),\ \alpha^e = \mathrm{diag}(\alpha_2, \alpha_3, \ldots, \alpha_n). \tag{18}$$

We will denote by $\alpha^{-b}$ and $\alpha^{-e}$ the inverses of the matrices $\alpha^b, \alpha^e$. Then:

$$l_n(\alpha)\Phi(b_1 \oplus b') = \Phi(b_1\alpha^b \oplus b'\alpha^e) \tag{19}$$

$$\hat{l}_n(\alpha)\hat{\Phi}(b_1, b_2) = \hat{\Phi}(b_1\alpha^b, \alpha^{-e}b_2) \mid \alpha_2 \mid^{-1} \mid \alpha_3 \mid^{-2} \cdots \mid \alpha_n \mid^{-(n-1)}. \tag{20}$$

We can define analogous representations of the appropriate linear groups on the space of Schwartz functions on

$$\mathbb{V}_n := M(1 \times 2, F) \oplus M(2 \times 3, F) \oplus \cdots \oplus M(n-1 \times n, F).$$

They are denoted by $r_1, r_2, \ldots, r_{n-1}, l_2, l_3, \ldots, l_n$. For instance, if $n = 3$, then

$$l_2(g_2)l_3(g_3)\Phi\left[(x_1, y_1) : \begin{pmatrix} x_{1,1} & x_{1,2} & x_{1,3} \\ x_{2,1} & x_{2,2} & x_{2,3} \end{pmatrix}\right] r_2(g_2')$$

$$= \Phi\left[(x_1, y_1)g_2 : g_2' \begin{pmatrix} x_{1,1} & x_{1,2} & x_{1,3} \\ x_{2,1} & x_{2,2} & x_{2,3} \end{pmatrix} g_3\right].$$

We can also define the partial Fourier transform $\mathcal{F}(\Phi) = \mathcal{F}_1\mathcal{F}_2 \cdots \mathcal{F}_{n-1}(\Phi)$ of a function $\Phi \in \mathcal{S}(\mathbb{V}^n)$. This partial Fourier transform is then a function on the direct sum:

(21) $\mathbb{U}_n :=$

$$\mathcal{B}_l(1, F) \oplus \mathcal{B}_l(1, F) \oplus \mathcal{B}_l(2, F) \oplus \mathcal{B}_l(2, F) \oplus \cdots \oplus \mathcal{B}_l(n-1, F) \oplus \mathcal{B}_l(n-1, F).$$

We have also representations $\hat{r}_1, \hat{r}_2, \dots, \hat{r}_{n-1}, \hat{l}_2, \dots, \hat{l}_n$ on the space $\mathcal{S}(\mathbb{U}^n)$. For $\Phi \in \mathcal{S}(\mathbb{V}^n)$ we define $\mathcal{F}^A(\Phi) \in \mathcal{S}(\mathbb{U}_n)$ by

$$\mathcal{F}^A(\Phi) = \int \hat{l}_2(k_2)\hat{l}_3(k_3)\cdots\hat{l}_{n-1}(k_{n-1})\mathcal{F}(\Phi)\hat{r}_2\left(k_2^{-1}\right)\hat{r}_3\left(k_3^{-1}\right)\cdots\hat{r}_{n-1}\left(k_{n-1}^{-1}\right) \prod_{i=2}^{i=n-1} \mu_i^{-1}\mu_{i-1}(\det k_i)dk_i \tag{22}$$

the integral being over the product $K_2 \times K_3 \times \cdots \times K_{n-1}$.

THEOREM 1. *Suppose that $f_{\phi,u}$ is the section represented by the integral of Proposition 4. Set*

$$\Psi_u = \mathcal{F}^A(\Phi_u).$$

*Then*

$$\mathcal{W}_u(f_{\phi,u}) = \int \Psi_u\left[a_1 a_2^{-b}, a_1^{-1}a_2^e : a_2 a_3^{-b}, a_2^{-1}a_3^e : \cdots : a_{n-2}a_{n-1}^{-b}, a_{n-2}^{-1}a_{n-1}^e : a_{n-1}, a_{n-1}^{-1}\right] \prod_{i=1}^{i=n-1} \mu_{i+1}\mu_i^{-1}(\det a_i) \mid \det a_i \mid^{u_{i+1}-u_i} da_i.$$

*Here $a_i$ is integrated over $A_i$.*

Before we embark on the proof of the theorem we write down a more general formula, which follows from the theorem.

$$\mathcal{W}_u(\Xi_u(g)f_{\phi,u}) = \int \hat{l}_n(g).\Psi_u\left[a_1 a_2^{-b}, a_1^{-1}a_2^e : a_2 a_3^{-b}, a_2^{-1}a_3^e : \cdots : a_{n-2}a_{n-1}^{-b}, a_{n-2}^{-1}a_{n-1}^e : a_{n-1}, a_{n-1}^{-1}\right] \prod_{i=1}^{i=n-1} \mu_{i+1}\mu_i^{-1}(\det a_i) \mid \det a_i \mid^{u_{i+1}-u_i} da_i \times \mid \det g \mid^{u_n+\frac{n-1}{2}} \mu_n(\det g). \tag{23}$$

In particular, for a diagonal matrix $\alpha$,

$$\mathcal{W}_u(\Xi_u(\alpha)f_{\phi,u}) = \int \Psi_u\left[a_1 a_2^{-b}, a_1^{-1}a_2^e : a_2 a_3^{-b}, a_2^{-1}a_3^e : \cdots : a_{n-2}a_{n-1}^{-b}, a_{n-2}^{-1}a_{n-1}^e : a_{n-1}\alpha^b, a_{n-1}^{-1}\alpha^{-e}\right] \tag{24}$$

$$\prod_{i=1}^{i=n-1} \mu_{i+1}\mu_i^{-1}(\det a_i) \mid \det a_i \mid^{u_{i+1}-u_i} da_i$$
$$\times \mid \det \alpha \mid^{u_n} \mu_n(\det \alpha) e^{\langle H(\alpha),\rho\rangle}.$$

*Proof.* As before, we suppress the characters $\mu_i$ from the notations. Our task can be summarized as follows.

LEMMA 6. *Suppose $\Phi$ is a smooth function on $\mathbb{V}_n$ with compact support contained in $\prod R_i$. Define*

$$f(g) = \int \Phi\left[(g_1,0)g_2^{-1} : (g_2,0)g_3^{-1} : \cdots : (g_{n-2},0)g_{n-1}^{-1} : (g_{n-1},0)g\right]$$
$$\prod_{i=1}^{i=n-1} \mid \det g_i \mid^{u_{i+1}-u_i+1} \mu_{i+1}\mu_i^{-1}(\det g_i)dg_i.$$

*Then, for $\Re(u_{i+1}-u_i) > 0$ for all $i$,*

$$\int_{N_n} f(vg)\overline{\theta}_n(v)dv$$
$$= \int \hat{l}_n(g)\Psi\left[a_1a_2^{-b}, a_1^{-1}a_2^e : a_2a_3^{-b}, a_2^{-1}a_3^e :\right.$$
$$\left.\cdots : a_{n-2}a_{n-1}^{-b}, a_{n-2}^{-1}a_{n-1}^e : a_{n-1}, a_{n-1}^{-1}\right]$$
$$\prod_{i=1}^{i=n-1} \mu_{i+1}\mu_i^{-1}(\det a_i) \mid \det a_i \mid^{u_{i+1}-u_i} da_i$$

*where $\Psi = \mathcal{F}^A(\Phi)$.*

We prove the lemma by induction on $n$. The case $n=2$ was treated in section 3. We may assume $n>2$ and the lemma true for $n-1$. Since $\mathcal{F}^A(l_n(g)\Phi) = \hat{l}_n(g)\mathcal{F}^A(\Phi)$, it suffices to prove the formula for $g=1$. We first compute

$$w(g) := \int_{N_{n-1}} f\left[\begin{pmatrix} v & 0 \\ 0 & 1 \end{pmatrix} g\right]\overline{\theta}_{n-1}(v)du. \tag{25}$$

After a change of variables we find $w(g)$ is equal to

$$\iint \Phi\left[(g_1,0)g_2^{-1} : (g_2,0)g_3^{-1} : \cdots : (g_{n-2},0)vg_{n-1}^{-1} : (g_{n-1},0)g\right]$$
$$\prod_{i=1}^{i=n-1} \mid \det g_i \mid^{u_{i+1}-u_i+1} dg_i\overline{\theta}_{n-1}(v)dv.$$

We are led to introduce

$$\Omega(g_{n-1}:X):=\int \Phi\left[(g_1,0)g_2^{-1}:(g_2,0)g_3^{-1}:\cdots:(g_{n-2},0)g_{n-1}:X\right] \prod_{i=1}^{i=n-2} |\det g_i|^{u_{i+1}-u_i+1}\,dg_i \tag{26}$$

and

$$\Xi(g_{n-1}:X):=\int \Omega_u(vg_{n-1}:X)\overline{\theta}_{n-1}(v)dv. \tag{27}$$

Then

$$w(g)=\int \Xi(g_{n-1}^{-1},(g_{n-1},0)g)\,|\det g_{n-1}|^{u_n-u_{n-1}+1}\,dg_{n-1}.$$

We can apply the induction hypothesis to the function $g_{n-1}\mapsto\Omega(g_{n-1}:X)$. Before we do, we remark that the representation $\hat{l}_{n-1}$ of $GL(n-1)$ on $\mathcal{S}(\mathbb{U}_{n-1})$ and the representation $r_{n-1}$ of $GL(n-1)$ on $\mathcal{S}(M(n-1)\times n,F)$ give corresponding representations on the space of Schwartz functions on $\mathbb{U}_{n-1}\oplus M((n-1)\times n,F)$. By the induction hypothesis, there is a Schwartz function $\Psi^1$ on that direct sum such that

$$\Xi(g_{n-1}:X) = \int \hat{l}_{n-1}(g_{n-1}).\Psi^1\left[a_1a_2^{-b},a_1^{-1}a_2^{e}:a_2a_3^{-b},a_2^{-1}a_3^{e}:\cdots:a_{n-2},a_{n-2}^{-1}:X\right] \prod_{i=1}^{i=n-2}|\det a_i|^{u_{i+1}-u_i}\,da_i.$$

It follows that

$$w(g)=\int \hat{l}_{n-1}(g_{n-1}^{-1}).\Psi^1\left[a_1a_2^{-b},a_1^{-1}a_2^{e}:a_2a_3^{-b},a_2^{-1}a_3^{e}:\cdots:a_{n-2},a_{n-2}^{-1}:(g_{n-1},0)g\right] \prod_{i=1}^{i=n-2}|\det a_i|^{u_{i+1}-u_i}\,da_i\,|\det g_{n-1}|^{u_n-u_{n-1}+1}\,dg_{n-1}. \tag{28}$$

At this point, we use the Iwasawa decomposition of $GL(n-1)$ to write $g_{n-1}=k_{n-1}b_{n-1}$ with $k_{n-1}\in K_{n-1}$ and $b\in B_{n-1}$. Then $dg_{n-1}=dk_{n-1}d_rb_{n-1}$, where $d_rb_{n-1}$ is a right invariant measure on $B_{n-1}$. Recalling our notational convention, we replace $\Psi^1$ by the Schwartz function $\Psi^2$ defined by

$$\Psi^2=\int_{K_{n-1}}\hat{l}_{n-1}(k_{n-1}^{-1}).\Psi_u^1.r_{n-1}(k_{n-1})\mu_n\mu_{n-1}^{-1}(\det k_{n-1})dk_{n-1}$$

or, somewhat more explicitly,

$$\Psi^2(\bullet,X)=\int_{K_{n-1}}\hat{l}_{n-1}(k_{n-1}^{-1}).\Psi^1(\bullet,k_{n-1}X)\mu_n\mu_{n-1}^{-1}(\det k_{n-1})dk_{n-1}.$$

We keep in mind the relation

$$\Xi(vg_{n-1}, X) = \Xi(g_{n-1}, X)\theta_{n-1}(v).$$

It is equivalent to the relation

$$\int \hat{l}_{n-1}(vg_{n-1}).\Psi^2\left[a_1 a_2^{-b}, a_1^{-1}a_2^e : a_2 a_3^{-b}, a_2^{-1}a_3^e : \cdots : a_{n-2}, a_{n-2}^{-1} : X\right]$$

$$\prod_{i=1}^{i=n-2} \mid \det a_i \mid^{u_{i+1}-u_i} da_i$$

$$= \theta_{n-1}(v)$$

$$\times \int \hat{l}_{n-1}(g_{n-1}).\Psi^2\left[a_1 a_2^{-b}, a_1^{-1}a_2^e : a_2 a_3^{-b}, a_2^{-1}a_3^e : \cdots : a_{n-2}, a_{n-2}^{-1} : X\right]$$

$$\prod_{i=1}^{i=n-2} \mid \det a_i \mid^{u_{i+1}-u_i} da_i .$$

Formula (28) for $w(g)$ becomes

(29)

$$\int \hat{l}_{n-1}\left(b_{n-1}^{-1}\right).\Psi^2\left[a_1 a_2^{-b}, a_1^{-1}a_2^e : a_2 a_3^{-b}, a_2^{-1}a_3^e : \cdots : a_{n-2}, a_{n-2}^{-1} : (b_{n-1}, 0)g\right]$$

$$\prod_{i=1}^{i=n-2} \mid \det a_i \mid^{u_{i+1}-u_i} da_i$$

$$\mid \det b_{n-1} \mid^{u_n - u_{n-1}+1} d_r b_{n-1} .$$

Now we set

$$b_{n-1} = \begin{pmatrix} a_{1,1} & x_{1,2} & x_{1,3} & \cdots & x_{1,n-1} \\ 0 & a_{2,2} & x_{2,3} & \cdots & x_{2,n-1} \\ \cdots & \cdots & \cdots & \cdots & \cdots \\ 0 & \cdots & \cdots & a_{n-2,n-2} & x_{n-2,n-1} \\ 0 & \cdots & \cdots & 0 & a_{n-1,n-1} \end{pmatrix},$$

$$a_{n-1} = \operatorname{diag}\left(a_{1,1}, a_{2,2}, \ldots, a_{n-1,n-1}\right).$$

Then

$$d_r b_{n-1} = da_{n-1} \bigotimes dx_{i,j} \mid a_{2,2} \mid^{-1} \mid a_{3,3} \mid^{-2} \cdots \mid a_{n-1,n-1} \mid^{-(n-2)} .$$

We now use formula (29) to compute

$$\int f(v)\overline{\theta}_n(v)dv = \int w\left[\begin{pmatrix} 1_{n-1} & U \\ 0 & 1 \end{pmatrix}\right] \psi(-\epsilon_{n-1}U)dU$$

where $U$ is integrated over the space of $(n-1)$-columns and

$$\epsilon_{n-1} = (0, 0, \ldots, 0, 1).$$

We find

$$\int \hat{l}_{n-1}\big(b_{n-1}^{-1}\big).\Psi_u^2\left[a_1a_2^{-b}, a_1^{-1}a_2^e : a_2a_3^{-b}, a_2^{-1}a_3^e : \cdots : a_{n-2}, a_{n-2}^{-1} : (b_{n-1}, b_{n-1}U)\right]$$

$$\prod_{i=1}^{i=n-2} \mid \det a_i \mid^{u_{i+1}-u_i} da_i$$

$$\mid \det b_{n-1} \mid^{u_n-u_{n-1}+1} d_rb_{n-1}\psi(-\epsilon_{n-1}U)dU.$$

We change $U$ to $b_{n-1}^{-1}U$. We get

$$\int \hat{l}_{n-1}\big(b_{n-1}^{-1}\big).\Psi_u^2\left[a_1a_2^{-b}, a_1^{-1}a_2^e : a_2a_3^{-b}, a_2^{-1}a_3^e : \cdots : a_{n-2}, a_{n-2}^{-1} : (b_{n-1}, U)g\right]$$

$$\prod_{i=1}^{i=n-2} \mid \det a_i \mid^{u_{i+1}-u_i} da_i$$

$$\mid \det b_{n-1} \mid^{u_n-u_{n-1}} d_rb_{n-1}\psi\big(-\epsilon_{n-1}b_{n-1}^{-1}U\big)dU$$

or, more explicitly, denoting by $U_i$, $1 \le i \le n-1$, the entries of $U$,

$$\int \hat{l}_{n-1}\big(a_{n-1}^{-1}\big).\Psi^2\left[a_1a_2^{-b}, a_1^{-1}a_2^e : a_2a_3^{-b}, a_2^{-1}a_3^e : \cdots : a_{n-2}, a_{n-2}^{-1} : (b_{n-1}, U)\right]$$

$$\prod_{i=1}^{i=n-2} \mid \det a_i \mid^{u_{i+1}-u_i} da_i$$

$$\psi\left(-a_{1,1}^{-1}x_{1,2} - a_{2,2}^{-1}x_{2,3}\cdots - a_{n-2,n-2}^{-1}x_{n-2,n-1} - a_{n-1,n-1}^{-1}U_{n-1}\right)$$

$$\mid \det b_{n-1} \mid^{u_n-u_{n-1}} d_rb_{n-1}dU.$$

By (20), we may bring $\hat{l}_{n-1}(a_{n-1}^{-1})$ "inside" to get

$$\hat{l}_{n-1}\big(a_{n-1}^{-1}\big).\Psi^2\big[\bullet : a_{n-2}, a_{n-2}^{-1} : \bullet\big]$$
$$= \Psi_2\big[\bullet : a_{n-2}a_{n-1}^{-b}, a_{n-2}^{-1}a_{n-1}^e : \bullet\big] \mid a_{2,2} \mid\mid a_{3,3} \mid^2 \cdots \mid a_{n-1,n-1} \mid^{n-2}.$$

Thus we find for our integral

$$\int \Psi_u^2\left[a_1a_2^{-b}, a_1^{-1}a_2^e : a_2a_3^{-b}, a_2^{-1}a_3^e : \cdots : a_{n-2}a_{n-1}^{-b}, a_{n-2}^{-1}a_{n-1}^e : b_{n-1}, U)\right]$$

$$\prod_{i=1}^{i=n-1} \mid \det a_i \mid^{u_{i+1}-u_i} da_i$$

$$\psi\left(-a_{1,1}^{-1}x_{1,2} - a_{2,2}^{-1}x_{2,3}\cdots - a_{n-2,n-2}^{-1}x_{n-2,n-1} - a_{n-1,n-1}^{-1}U_{n-1}\right)$$

$$\bigotimes dx_{i,j} \bigotimes dU_i.$$

Finally, we set $\Psi = \mathcal{F}_{n-1}(\Psi^2)$. Then the above integral can be written as

$$\int \Psi\left[a_1 a_2^{-b}, a_1^{-1} a_2^e : a_2 a_3^{-b}, a_2^{-1} a_3^e : \cdots : a_{n-2} a_{n-1}^{-b}, a_{n-2}^{-1} a_{n-1}^e : a_{n-1}, a_{n-1}^{-1})\right]$$

$$\prod_{i=1}^{i=n-1} \mid \det a_i \mid^{u_{i+1}-u_i} da_i .$$

To finish the proof of the lemma we remark that

$$\begin{aligned}
\Psi &= \mathcal{F}_{n-1}(\Psi^2) \\
&= \mathcal{F}_{n-1}\left(\int \hat{l}_{n-1}\left(k_{n-1}^{-1}\right)(\Psi^1) r_{n-1}(k_{n-1}) \mu_n \mu_{n-1}^{-1}(\det k_{n-1}) dk_{n-1}\right) \\
&= \int \mathcal{F}_{n-1}\left(\hat{l}_{n-1}\left(k_{n-1}^{-1}\right)(\Psi^1) r_{n-1}(k_{n-1})\right) \mu_n \mu_{n-1}^{-1}(\det k_{n-1}) dk_{n-1} \\
&= \int \hat{l}_{n-1}\left(k_{n-1}^{-1}\right)\left(\mathcal{F}_{n-1}(\Psi^1)\right) \hat{r}_{n-1}(k_{n-1}) \mu_n \mu_{n-1}^{-1}(\det k_{n-1}) dk_{n-1}
\end{aligned}$$

so that inductively we see that $\Psi = \mathcal{F}^A(\Phi)$ as claimed.

**7. Properties of the Whittaker functions.** In this section we will use the integral representation to obtain a very precise description of the behavior of a Whittaker function on the diagonal subgroup. Our starting point is an investigation of the type of integrals that represent Whittaker functions. We let $\mathbb{A}_n$ be the vector space of diagonal matrices with $n$-entries, a space we may also identify to $F^n$. We consider the direct sum

$$\mathbb{W}_n := \mathbb{A}_1 \oplus \mathbb{A}_1 \oplus \mathbb{A}_2 \oplus \mathbb{A}_2 \oplus \mathbb{A}_3 \oplus \mathbb{A}_3 \oplus \cdots \oplus \mathbb{A}_{n-1} \oplus \mathbb{A}_{n-1} \tag{30}$$

and a function $\Psi \in \mathcal{S}(\mathbb{W}_n \oplus \mathbb{X})$, where $\mathbb{X}$ is an auxiliary real vector space. We consider the function on $A_n \times \mathbb{X}$ defined by the following integral:

$$w_{\Psi,u}(m : X) :=\mid \det m \mid^{u_n} \mu_n(\det m) \tag{31}$$

$$\int_{A_1 \times A_2 \times \cdots \times A_{n-1}} \Psi\left[a_1 a_2^{-b}, a_1^{-1} a_2^e : a_2 a_3^{-b}, a_2^{-1} a_3^e : \cdots :\right.$$

$$\left. a_{n-2} a_{n-1}^{-b}, a_{n-2}^{-1} a_{n-1}^e : a_{n-1} m^b, a_{n-1}^{-1} m^{-e} : X\right]$$

$$\prod_{i=1}^{i=n-1} \mu_i \mu_{i-1}^{-1}(a_i) \mid \det a_i \mid^{u_i - u_{i-1}} da_i .$$

Let $\sigma_u$ be the $n$-dimensional representation of $F^\times$ (or the Weil group of $F$) defined by

$$\sigma_u(t) = \bigoplus \mu_i(t) \mid t \mid_F^{u_i} .$$

In particular, $\sigma_0 = \bigoplus \mu_i(t)$. For $1 \le j \le n$, the representation $\bigwedge^j \sigma_u$ decomposes into the sum of the characters

$$t \mapsto \mu_{i_1}\mu_{i_2}\cdots\mu_{i_j}(t) \mid t \mid^{u_{i_1}+u_{i_2}+\cdots+u_{i_j}}$$

with

$$1 \le i_1 < i_2 < \cdots < i_j \le n.$$

We consider the algebraic co-characters of $A_n$ defined by the fundamental co-weights:

$$\check{\varpi}_j(m) = (\overbrace{m, m, \ldots, m}^{j}, 1, \ldots 1).$$

Thus $\alpha_i(\check{\varpi}_j(m)) = m^{\delta_{i,j}}$. Below we consider over all complex characters $\xi$ of $A_n$ such that, for each $1 \le i \le n$, the character $\xi \circ \check{\varpi}_i$ is a component of $\bigwedge^i \sigma_u$. If we write $\xi(m) = \prod_{1 \le i \le n} \xi_i(m_i)$ where the $m_i$ are the entries of $m$, this amounts to say that every product $\xi_1\xi_2\cdots\xi_j$ is a component of $\bigwedge^j(\sigma_u)$. The characters $\xi$ depend on $u$ so we will often write them as $\xi_u$. Thus $\xi_{u,1}(t) = \mu_i(t) \mid t \mid^{u_i}$ for a suitable $i$ and

$$\xi_{u,1}\xi_{u,2}\cdots\xi_{u,n}(t) = \mu_1\mu_2\cdots\mu_n(t) \mid t \mid^{u_1+u_2+\cdots u_n}.$$

Proposition 6. *We fix a multi strip*

$$\mathfrak{S} = \{u \mid -A < \Re u_i < A, 1 \le i \le n\},\ A \notin \mathbb{Z}_F. \tag{32}$$

*With the above notations, there is a polynomial $P$, product of linear factors, with the following properties. For $u \in \mathfrak{S}$,*

$$P(u)w_{\Psi,u}(m : X) = \sum_{\xi_u} \Phi_{\xi_u}(m^b m^{-e} : X)\xi_u(m),\ \Phi_{\zeta u} \in \mathbb{S}(\mathbb{A}_{n-2} \oplus X),$$

*where the sum is over all the characters $\xi_u$ of $A_n$ of the above type. Suppose that $\Psi$ remains in a bounded set. Then one can choose the functions $\Phi_*$ in a bounded set. Suppose that $\Psi$ depends holomorphically on $s \in \Omega$, where $\Omega$ is open in $\mathbb{C}^m$, and remains in a bounded set for all values $s$. Then one can choose the functions $\Phi_*$ to depend holomorphically on $(u, s)$ and to remain in a bounded set.*

*Proof.* We prove the proposition by induction on $n$. The case $n = 2$ is Proposition 2 of section 4. Thus we assume $n > 2$ and the result true for $(n-1)$. Set $u' = (u_1, u_2, \ldots, u_{n-1})$ and let $\tau_u$ be the representation of degree $(n-1)$ of $F^\times$ associated with $u'$ and the characters $\mu_i$, $1 \le i \le n-1$.

As before, we suppress the characters $\mu_i$ from the notations. Consider the strip $\mathfrak{S}' = \{u' \mid -A < \Re u_i < A, 1 \le i \le n-1\}$. Let $Q'(u')$ be the polynomial whose existence is guaranteed by the induction hypothesis. Then the product of $Q(u')$ $\mid \det a_{n-1} \mid^{u_{n-1}}$ and the integral

$$\int \Psi\left[a_1 a_2^{-b}, a_1^{-1} a_2^e : a_2 a_3^{-b}, a_2^{-1} a_3^e : \cdots : a_{n-2} a_{n-1}^b, a_{n-2}^{-1} a_{n-1}^{-e} : Y : X\right]$$
$$\prod_{i=1}^{i=n-2} \mid \det a_i \mid^{u_{i+1}-u_i} da_i$$

is a sum of terms of the form

$$\Phi_{u'}(a_{n-1}^b a_{n-1}^{-e} : Y : X)\eta(a_{n-1}).$$

The components $\eta_i$ of $\eta$ have the property that each product $\eta_1\eta_2\cdots\eta_j$ is a component of $\bigwedge^j \tau_u$. Using this result we see that the product $Q(u')w_{\Psi,u}(m : X)$ is a sum of terms of the following form

$$\mid \det m \mid^{u_n}$$
$$\times \int \Phi_{u'}\left[a_{n-1}^{-b} a_{n-1}^e : a_{n-1} m^b, a_{n-1}^{-1} m^{-e} : X\right] \eta^{-1}(a_{n-1}) \mid \det a_{n-1} \mid^{u_n} da_{n-1}.$$

More explicitly, in terms of the entries $b_i$ of the matrix $a_{n-1}$ and the entries $m_i$ of $m$, this expression reads

$$\mid m_1 m_2 \cdots m_{n-1} m_n \mid^{u_n} \int \Phi_{u'}\left[b_1^{-1} b_2, b_2^{-1} b_3, \ldots, b_{n-2}^{-1} b_n :\right.$$
$$\left. b_1 m_1, b_1^{-1} m_2^{-1}, b_2 m_2, b_2^{-1} m_3^{-1}, \ldots, b_{n-1} m_{n-1}, b_{n-1}^{-1} m_n^{-1} : X\right]$$
$$\eta_1^{-1}(b_1)\eta_2^{-1}(b_2)\cdots\eta_{n-1}^{-1}(b_n) \mid b_1 b_2 \cdots b_{n-1} \mid^{u_n} \bigotimes d^\times b_i.$$

For convenience, we have changed the order of the variables. In general, we remark that if $\Phi(x, y, z, X)$ is a Schwartz function of $(x, y, z, X)$, then the function $\Phi(yz, y, z, X)$ is still a Schwartz function of $(y, z, X)$. Applying this simple remark repeatedly, we see the above expression has the form:

$$(33) \quad \int \Phi_{u'}^1\left[b_1 m_1, b_1^{-1} m_2^{-1}, b_2 m_2, b_2^{-1} m_3^{-1}, \ldots, b_{n-1} m_{n-1}, b_{n-1}^{-1} m_n^{-1} : X\right]$$
$$\eta_1^{-1}(b_1)\eta_2^{-1}(b_2)\cdots\eta_{n-1}^{-1}(b_n) \mid b_1 b_2 \cdots b_{n-1} \mid^{u_n}$$
$$\bigotimes d^\times b_i \mid m_1 m_2 \cdots m_{n-1} m_n \mid^{u_n}$$

where $\Phi_{u'}^1$ is a new Schwartz function. Next we apply Proposition 2 of section 4 repeatedly and we see that there is a polynomial $R(u)$ such that the product of the previous expression and $R(u)$ is a sum of terms of the following form

$$\Phi_u\left(m_1 m_2^{-1}, m_2 m_3^{-1}, \ldots, m_{n-1} m_n^{-1}\right) \xi(m),$$

where the character $\xi(m)$ is obtained as the product of $|\det m|^{u_n}$ and a character obtained from the following table

$$\left\{\begin{matrix}\eta_1(m_1)\mid m_1\mid^{-u_n}\\ \eta_1(m_2)\mid m_2\mid^{-u_n}\end{matrix}\right\}\left\{\begin{matrix}\eta_2(m_2)\mid m_2\mid^{-u_n}\\ \eta_2(m_3)\mid m_3\mid^{-u_n}\end{matrix}\right\}\cdots$$

$$\left\{\begin{matrix}\eta_{n-1}(m_{n-1})\mid m_{n-1}\mid^{-u_n}\\ \eta_{n-1}(m_n)\mid m_n\mid^{-u_n}\end{matrix}\right\}.$$

In each column we choose in any way a character and multiply our choices together. If we evaluate $\xi$ on an element of the form

$$\mathrm{diag}(\overbrace{m_0, m_0, \dots, m_0}^{j}, 1, 1\dots, 1)$$

we find $\eta_1\eta_2\cdots\eta_j(m_0)$ if $j < n$ and $\eta_1\eta_2\cdots\eta_{n-1}(m_0)\mid m_0\mid^{u_n}$ if $j = n$. Thus the character $\xi$ has the required properties. □

We can state our main theorem. Indeed, it follows at once from the previous proposition and the integral representation of the Whittaker functions.

THEOREM 2. *Suppose that $f_u$ is a standard section and $\phi$ the convolution of $(n-1)$ elements $\phi_i$ of $\mathcal{D}(G_n)$. Fix a multi strip $\mathfrak{S} = \{u \mid -A < \Re u_i < A, 1 \le i \le n\}$ with $A \notin \mathbb{Z}_F$. Then there is a polynomial $P(u)$, product of linear factors, such that in the multi strip*

$$P(u)\mathcal{W}_u(\Xi_u(mk)f_{\phi,u}) = \sum_{\xi_u}\Phi_{\xi_u,u}\left(m_1m_2^{-1}, m_2m_3^{-1}, \dots, m_{n-1}m_n^{-1}, k\right)\xi_u(m)e^{\langle H(m),\rho\rangle}.$$

*For each $\xi_u$, and each $u$ the function*

$$\Phi_{\xi_u,u}(x_1, x_2, \dots, x_{n-1}, k)$$

*is in $\mathcal{S}(F^{n-1}\times K)$ and $u \mapsto \Phi_{\xi_u,u}$ is a holomorphic function with values in $\mathcal{S}(F^{n-1}\times K)$. Its values are in a bounded set if $f_u$ and the functions $\phi_i$ are each in a bounded set.*

If $W_u(g) = \mathcal{W}_u(\rho(mk)f_{\phi,u})$ then we set

$$\widetilde{W}_{-u} := W_u\left(w_n{}^tg^{-1}\right). \tag{34}$$

Note that $\widetilde{W}_{-u}(ng) = \overline{\theta}(n)\widetilde{W}_{-u}(g)$. Moreover, for $m \in A_n$, $k \in K_n$,

$$\widetilde{W}_u(mk) = W_u(wm^{-1}wk'),\ k' = w\,{}^tk^{-1} \in K_n.$$

We remark that, for $1 \le i \le n$,

$$\bigwedge^{n-j}\sigma_u \otimes \det\sigma_u^{-1} = \bigwedge^{j}\widetilde{\sigma}_u$$

and

$$\check{\varpi}_i(wt^{-1}w) = \check{\varpi}_{n-i}(t)\check{\varpi}_n(t)^{-1}.$$

It follows that the function $\widetilde{W}_u(g)$ has the same properties as the function $W_u$, except that the representation $\sigma_u$ is replaced by the contragredient representation, or, what amounts to the same, the characters $\mu_i$ are replaced by their inverses and $u$ by $-u$. In particular, we have the following proposition, the proof of which is immediate.

PROPOSITION 7. *Notations being as in the theorem, the product*

$$P(u)\widetilde{W}_{-u}(mk)$$

*can be written as*

$$\sum_{\xi_u} \widetilde{\Phi}_{\xi_u,u}\left(m_1m_2^{-1}, m_2m_3^{-1}, \ldots, m_{n-1}m_n^{-1}, k\right) \xi_u(m)^{-1}e^{\langle H(m),\rho\rangle}$$

*where the sum is the same as before and the functions $\widetilde{\Phi}_{\xi_u,u}$ have the same properties as above.*

We shall need estimates which are uniform in $u$ for the Whittaker functions. We first consider integrals of the form (31).

PROPOSITION 8. *Suppose that $\Psi$ is in a bounded set and $u$ in a multi strip. There are integers $N_i$, $1 \leq i \leq n-1$, with the following property. For any integer $M > 0$ there is a majorization*

$$|w_{\Psi,u}(m : X)| \leq\, | m_n |^{\Re(\sum_{1\leq i\leq n} u_i)} \prod_{i=1}^{i=n-1} \frac{\left|m_im_{i+1}^{-1}\right|^{-N_i}}{\left(1+\left|m_im_{i+1}^{-1}\right|\right)^M}\Phi_0(X)$$

*where $\Phi_0 \geq 0$ is in $\mathcal{S}(\mathbb{X})$. Suppose that $u$ is purely imaginary. Then there is a polynomial $Q$ in $\log(| m_i/m_{i+1} |)$ such that, for every integer $M$, there is a majorization*

$$|w_{\Psi,u}(m : X)| \leq \frac{Q(\log | m_i/m_{i+1} |)}{\prod_{i=1}^{i=n-1}\left(1+\left|m_im_{i+1}^{-1}\right|\right)^M}\Phi_0(X).$$

*Proof.* We prove the first assertion by induction on $n$. For the case $n = 2$, we have, dropping the characters $\mu_i$ from the notation,

$$w_{\Psi,u}(m : X) = | m_1m_2 |^{u_2} \int \Psi\left(a_1m_1, a_1^{-1}m_2^{-1} : X\right) | a_1 |^{u_2-u_1} d^\times a_1$$

$$= | m_2 |^{u_1+u_2} | m_1m_2^{-1} |^{u_2} \int \Psi(a_1m_1m_2^{-1}, a_1^{-1} : X) | a_1 |^{u_2-u_1} d^\times a_1$$

and then our assertion follows from Proposition 3 of Section 4. Thus we may assume $n > 2$ and our assertion true for $n-1$. We may also assume that $\Psi$ is bounded by a function $\geq 0$ which is a product and so we may ignore the dependence on $X$

and assume $\Psi \geq 0$ and the $u_i$ are real. Then, for $a_{n-1} \in A_{n-1}$ with entries $b_i$, the product of $\mid \det a_{n-1} \mid^{u_{n-1}}$ and the integral

$$\int \Psi \left[a_1 a_2^{-b}, a_1^{-1} a_2^e : a_2 a_3^{-b}, a_2^{-1} a_3^e : \cdots : a_{n-2} a_{n-1}^b, a_{n-2}^{-1} a_{n-1}^{-e} : Y\right]$$

$$\prod_{i=1}^{i=n-2} \mid \det a_i \mid^{u_{i+1}-u_i} da_i$$

is bounded by

$$\mid b_{n-1} \mid^{\sum_{1\leq i\leq n-1} u_i} \times \prod_{i=1}^{i=n-2} \frac{\left|b_i b_{i+1}^{-1}\right|^{-N_i}}{\left(1+\left|b_i b_{i+1}^{-1}\right|\right)^M} \Phi_0(Y).$$

Thus $\mid w_{\Psi,u}(m) \mid$ is bounded by

$$\mid \det m \mid^{u_n} \times \int \mid \det a_{n-1} \mid^{u_n} \mid b_{n-1} \mid^{-\sum_{1\leq i\leq n-1} u_i} \prod_{i=1}^{i=n-2} \frac{\left|b_i b_{i+1}^{-1}\right|^{N_i}}{\left(1+\left|b_i^{-1} b_{i+1}\right|\right)^M}$$

$$\Phi_0\left(a_{n-1} m^b, a_{n-1}^{-1} m^{-e}\right) da_{n-1}.$$

In turn, this is majorized by

$$\mid \det m \mid^{u_n} \times \int \mid \det a_{n-1} \mid^{u_n} \mid b_{n-1} \mid^{-\sum_{1\leq i\leq n-1} u_i} \prod_{i=1}^{i=n-2} \left|b_i b_{i+1}^{-1}\right|^{N_i+M}$$

$$\Phi_0\left(a_{n-1} m^b, a_{n-1}^{-1} m^{-e}\right) da_{n-1}.$$

This has the form

$$\mid \det m \mid^{u_n} \times \int \prod_{i=1}^{i=n-1} \mid b_i \mid^{s_i} \Phi_0\left(a_{n-1} m^b, a_{n-1}^{-1} m^{-e}\right) da_{n-1}$$

where each $s_i$ belongs to some fixed interval and

$$\sum_{i=1}^{i=n-1} s_i = n u_n - \sum_{i=1}^{i=n} u_i.$$

Explicitly, the integral reads (after changing the order of the variables)

$$\mid \det m \mid^{u_n} \times \int \Phi_0\left[b_1 m_1, b_1^{-1} m_2^{-1}, b_2 m_2, b_2^{-1} m_3^{-1}, \ldots, b_{n-1} m_{n-1}, b_{n-1}^{-1} m_n^{-1}\right]$$

$$\prod_{i=1}^{i=n-1} \mid b_i \mid^{s_i} d^\times b_i.$$

Changing variables, we get:

$$| \det m |^{u_n} | m_2 |^{-s_1} | m_3 |^{-s_2} \cdots | m_n |^{-s_{n-1}}$$

$$\times \int \Phi_0 \left[b_1 m_1 m_2^{-1}, b_1^{-1}, b_2 m_2 m_3^{-1}, b_2^{-1}, \ldots, b_{n-1} m_{n-1} m_n^{-1}, b_{n-1}^{-1}\right]$$

$$\prod_{i=1}^{i=n-1} | b_i |^{s_i} d^\times b_i,$$

or, finally,

$$\prod_{1 \le i \le n-1} | m_i m_{i+1}^{-1} |^{t_i} | m_n |^{\sum_{1 \le i \le n} u_i}$$

$$\int \Phi_0 \left[b_1 m_1 m_2^{-1}, b_1^{-1}, b_2 m_2 m_3^{-1}, b_2^{-1}, \ldots, b_{n-1} m_{n-1} m_n^{-1}, b_{n-1}^{-1}\right]$$

$$\prod_{i=1}^{i=n-1} | b_i |^{s_i} d^\times b_i,$$

where each one of the exponents $t_i$ is in some fixed interval. Our assertion follows then from repeated use of Proposition 3 in Section 4. This concludes the proof of the first assertion. The proof of the second assertion is similar. □

Using once more the integral representation of the Whittaker functions we obtain at once the following estimates.

PROPOSITION 9. *Suppose that $u$ is in a multi strip, $f_u$ in a bounded set and the functions $\phi_i$ in a bounded set. Then there are $N_i > 0$, $1 \le i \le n-1$, such that, for all $M > 0$,*

$$|\mathcal{W}_u(\Xi_u(mk) f_{\phi,u})| \le c_M \mid m_n \mid^{\Re \Sigma u_i} e^{\langle H(m), \rho \rangle} \prod_{i=1}^{i=n-1} \frac{| m_i/m_{i+1} |^{-N_i}}{\left(1 + |m_i m_{i+1}^{-1}|\right)^M}.$$

*If $u$ is purely imaginary, then there is a polynomial $Q$ in $n-1$ variables such that*

$$|\mathcal{W}_u \Xi_u(mk) f_{\phi,u})| \le | \det m |^{\Re \Sigma u_i} e^{\langle H(m), \rho \rangle} \frac{Q(\log(| m_i/m_{i+1} |)}{\prod_{i=1}^{i=n-1} \left(1 + |m_i m_{i+1}^{-1}|\right)^M}.$$

**8. Rankin-Selberg integrals for $GL(n) \times GL(n)$.** In this section we consider two n-tuples of (normalized) characters

$$(\mu_1, \mu_2, \ldots, \mu_n), \; (\mu'_1, \mu'_2, \ldots, \mu'_n)$$

and two elements $u$ and $u'$ of $\mathbb{C}^n$. We denote by $\mathcal{W}'_{u'}$ the linear form on the space $\mathcal{V}(\mu')$ which is defined by analytic continuation of the integral

$$\mathcal{W}'_{u'}(f_{u'}) = \int_{N_n} f_{u'}(v) \theta_n(v) dv.$$

We consider two corresponding convolution sections $f_{\phi,u}$ and $f_{\phi',u'}$ and we set

(35) $$W_u(g) = \mathcal{W}_u(\Xi_u(g)f_{\phi,u}),\ W'_{u'}(g) = \mathcal{W}'_{u'}(\Xi_{u'}(g)f_{\phi',u'}).$$

We study the corresponding Rankin-Selberg integral. We set

(36) $$\epsilon_n = (0, 0, 0, \dots, 0, 1).$$

We let and $\Phi$ be an element of $\mathcal{S}(F^n)$. We define

(37) $$\Psi(s, W_u, W'_{u'}, \Phi) := \int_{N_n\backslash G_n} W_u(g)W'_{u'}(g')\Phi[(\epsilon_n)g] \mid \det g \mid^s dg.$$

PROPOSITION 10. *If $(u, u')$ are in a multi strip then there is $s_0$ such that the integral converges absolutely in the right half plane $\Re s > s_0$, uniformly on any vertical strip contained in the half plane. If $u$ and $u'$ are purely imaginary, then the integral converges in the half plane $\Re s > 0$, uniformly on any vertical strip contained in the half plane.*

*Proof.* We write $g = ak$ with $dg = e^{\langle H(a), -2\rho\rangle} dadk$. We write $a$ in terms of the co-weights:

$$a = \prod_{1\le i\le n} \check{\varpi}_i(a_i) = \operatorname{diag}(a_1a_2\cdots a_n, a_2\cdots a_n, \dots, a_n)$$

and use the majorization of $W_u(ak)$ and $W'_{u'}(ak)$ from the previous section. We find that

$$W_u(ak)W'_{u'}(ak)e^{\langle H(a), -2\rho\rangle}$$

is majorized by

$$c_M \prod_{1\le i\le n-1} \frac{\mid a_i \mid^{-M_i}}{(1+\mid a_i \mid)^N} \mid a_n \mid^{\Re\left(\sum u_i + \sum u'_i\right)}$$

where the integer $N$ is arbitrarily large. On the other hand, for all $N$, $\Phi[(\epsilon_n)ak]$ is majorized by

$$c'_N \frac{1}{(1+\mid a_n \mid)^N}.$$

The integral is thus majorized by

$$\int \frac{\mid a_1 \mid^{\Re s - M_1}}{(1+\mid a_1 \mid)^N} \frac{\mid a_2 \mid^{2\Re s - M_2}}{(1+\mid a_2 \mid)^N} \cdots$$

$$\frac{\mid a_{n-1} \mid^{(n-1)\Re s - M_{n-1}}}{(1+\mid a_{n-1} \mid)^N} \frac{\mid a_n \mid^{\sum \Re u_i + \sum \Re u'_i + n\Re s}}{(1+\mid a_n \mid)^N} \otimes d^\times a_i$$

and the first assertion follows. The proof of the second assertion is similar. We simply majorize $W_u(ak)W'_{u'}(ak)e^{\langle H(a), -2\rho\rangle}$ by a polynomial in the variables $\log \mid a_i \mid, 1 \le i \le n-1$. □

Our next result will be improved upon in the next theorem.

PROPOSITION 11. *Let* $\sigma_u$ *and* $\sigma'_{u'}$ *be the n-dimensional representations of* $F^\times$ *(or the Weil group) defined by*

$$\sigma_u = \oplus \mu_i \alpha_F^{u_i}, \ \sigma'_{u'} = \oplus \mu'_i \alpha_F^{u'_i}.$$

*Then* $\Psi(s, W_u, W'_{u'}, \Phi)$ *extends to a meromorphic function of* $(s, u, u')$ *which is the product of*

$$\prod_{j=1}^{j=n} L\left(js, \bigwedge^j \sigma_u \otimes \bigwedge^j \sigma'_{u'}\right)$$

*and an entire function of* $(s, u, u')$.

*Proof.* Choose a multi strip $\mathfrak{S}$ of the form (32) in $\mathbb{C}^{2n}$. For $(u, u') \in \mathfrak{S}$ the integral converges in some halfspace $\mathfrak{A} = \{s \mid \Re s > s_0\}$ and thus defines a holomorphic function of $(s, u, u')$ in $\mathfrak{A} \times \mathfrak{S}$. As before let us write $g = ak$ and $a$ in terms of the co-weights. There are polynomials $P$, $P'$, products of linear factors, such that

$$P(u)P(u')W_u(ak)W'_{u'}(ak)e^{\langle H(a), -2\rho\rangle}$$

is a sum of terms of the following form

$$\Phi_u(a_1, a_2, \ldots, a_{n-1}, k)\Phi'_{u'}(a_1, a_2, \ldots, a_{n-1}, k)$$
$$\xi_{1,u}\xi'_{1,u'}(a_1)\xi_{2,u}\xi'_{2,u'}(a_2)\cdots\xi_{n-1,u}\xi'_{n-1,u'}(a_{n-1})\xi_{n,u}\xi'_{n,u'}(a_n)$$

where each $\xi_{i,u}$ (resp. $\xi'_{i,u'}$) is a component of $\bigwedge^i \sigma_u$ (resp. $\bigwedge^i \sigma'_{u'}$). On the other hand

$$\mid \det ak \mid^s = \mid a_1 \mid^s \mid a_2 \mid^{2s} \cdots \mid a_n \mid^{ns}$$

and

$$\Phi(\epsilon_n ak) = \phi(a_n, k)$$

where $\phi$ is a Schwartz function in one variable depending on $k$. The contribution of the term at hand to the total integral has thus the form

$$\int \left( \int \Phi_u(a_1, a_2, \ldots, a_{n-1}, k)\Phi'_u(a_1, a_2, \ldots, a_{n-1}, k)\phi(a_n, k)dk \right)$$
$$\prod_{i=1}^{n} \xi_{i,u}\xi'_{i,u'}(a_i) \mid a_i \mid^{is} d^\times a_i.$$

After integrating over $K_n$ we obtain a multivariate Tate integral (where in addition the Schwartz function depends holomorphically on $(u, u')$). Thus

$$P(u)P(u')\Psi(s, W_u, W'_{u'}, \Phi)$$

is a sum of terms of the form

$$h(s, u, u') \prod_{j=1}^{n} L(js, \xi_j \xi'_j)$$

where $h$ is holomorphic in $\mathbb{C} \times \mathfrak{S}$. This already shows that $\Psi$ extends to a meromorphic function on $\mathbb{C} \times \mathfrak{S}$. The only possible singularities are the singularities of the $L$-factors and the zeroes of $P(u)P'(u')$. Thus they are hyperplanes of the form:

$$js + u_{i_1} + u_{i_2} + \cdots + u_{i_j} + u'_{i'_1} + u'_{i'_2} + \cdots + u'_{i'_j} = z_0,$$

$$l(u) = z_1, \; l(u') = z_2$$

with $z_* \in \mathbb{Z}_F$. Consider a hyperplane of the form $l(u) = z_1$ which intersects $\mathbb{C} \times \mathfrak{S}$. Then it also intersects $\mathfrak{A} \times \mathfrak{S}$. However the function $\Psi(s, W_u, W'_{u'}, \Phi)$ is holomorphic in this region because the integral is convergent, thus this is not actually a singular hyperplane. Likewise for a hyperplane $l(u') = z_2$. We conclude that the only singularities of $\Psi(s, W_u, W'_{u'}, \Phi)$ are those of the $L$-factors and we are done. □

PROPOSITION 12. *Let $u_0$ and $u'_0$ be fixed elements of $\mathbb{C}^n$. Then the meromorphic function*

$$\Psi(s, W_{u_0}, W'_{u'_0}, \Phi)$$

*is bounded at infinity in vertical strips.*

*Proof.* Fix a multi strip $\mathfrak{S}$ of the form (32) in $\mathbb{C}^{2n+1}$. Only finitely many singular hyperplanes of $\Psi(s, W_u, W'_{u'})$ intersect $\mathfrak{S}$. Let $Q(s, u, u')$ be the product of the corresponding (non-homogeneous) linear forms, with the appropriate multiplicity, repeated according to their multiplicity. The product of $Q(s, u, u')$ and any of the Tate integrals considered in the previous proposition is actually bounded in $\mathfrak{S}$. Thus the product

$$k(s, u, u') := Q(s, u, u')P(u)P'(u')\Psi(s, W_u, W'_{u'})$$

is holomorphic and bounded in $\mathfrak{S}$. Now we choose slightly smaller multi strip $\mathfrak{S}'$ and $\mathfrak{S}$" such that

$$\mathfrak{S}' \subset \text{Closure}(\mathfrak{S}') \subset \mathfrak{S}\text{"} \subset \text{Closure}(\mathfrak{S}\text{"}) \subset \mathfrak{S}.$$

By the Cauchy integral formula, $k$ and all its derivatives are bounded in $\mathfrak{S}$". On the other hand, by the previous proposition

$$k(s, u, u') = P(u)P'(u')h(s, u, u')$$

where $h$ is holomorphic in $\mathfrak{S}$. We claim that $h$ and all its derivatives are bounded in Closure($\mathfrak{S}'$) which will prove the proposition. We think of $\mathfrak{S}$" as the product $\mathfrak{a} \times \mathfrak{b}$ where

$$\mathfrak{a} = \{s \mid -a < \Re s < a\}$$

and

$$\mathfrak{b} = \{(u, u')|\forall i \ -a < \Re u_i < a, \ \forall j \ -a < \Re u'_j < a\}.$$

Clearly, it will be enough to show that for every $(u_0, u'_0) \in \mathfrak{b}$, there is a tubular neighborhood $\mathfrak{u}$ of $(u_0, u'_0)$ in $\mathfrak{b}$ such that $h$ and all its derivatives are bounded in $\mathfrak{a} \times \mathfrak{u}$. Indeed, a finite number of the sets $\mathfrak{a} \times \mathfrak{u}$ cover Closure($\mathfrak{S}'$). Thus it will suffice to apply repeatedly the following lemma. □

LEMMA 7. *Let $\mathfrak{a} = \{s \in \mathbb{C}| - a < \Re s < a\}$ and let $\mathfrak{u}$ be a tubular open set in $\mathbb{C}^n$. Suppose that $k$ is a holomorphic function on $\mathfrak{a} \times \mathfrak{u}$ uniformly bounded as well as its derivatives. Suppose that*

$$h(s, u) = (l(u) - z)k(s, u)$$

*where $k$ is holomorphic and $l(u) = \sum_{i=1}^{n} u_i v_i$ with $(v_i)$ non-zero in $\mathbb{R}^n$ and $z$ is real. Given $u_0 \in \mathfrak{u}$, there is a tubular neighborhood $\mathfrak{u}_0$ of $u_0$ in $\mathfrak{u}$ such that $h$ and all its derivatives are bounded in $\mathfrak{a} \times \mathfrak{u}_0$.*

*Proof of the Lemma.* Perhaps we should recall that a tubular neighborhood of $u_0$ in $\mathbb{C}^n$ is a set of the form $\{u|\Re u \in \Omega\}$ where $\Omega$ is a neighborhood of $\Re u_0$ in $\mathbb{R}^n$. There is no harm in assuming $u_0$ real. Our assertion is trivial if $l(u_0) - z \neq 0$. Thus we may assume $l(u_0) = z$. After a real change of coordinates in the variables $u$, we may assume $u_0 = 0$, $z = 0$, and

$$k(s, z_1, z_2, \dots, z_n) = z_1 h(s, z_1, z_2, \dots, z_n)$$

and $\mathfrak{u}$ contains the set

$$\mathfrak{u}_0 = \{z|\forall i \ -b < \Re z_i < b\}.$$

Then

$$h(s, z_1, z_2, \dots, z_n) = \int_0^1 \frac{\partial}{\partial z_1} k(s, tz_1, z_2, \dots, z_n) dt$$

and this formula shows that $h$ and all its derivatives are bounded on $\mathfrak{u}_0$. □

If we fix $u_0$ and $u'_0$ then the functions $\Psi(s, W_{u_0}, W'_{u_0}, \Phi)$ are holomorphic multiples of $L(s, \sigma_{u_0} \times \sigma_{u'_0})$. One can improve the previous result as follows. Let $\mathcal{H}$ be the space of holomorphic multiples $h(s)$ of $L(s, \sigma_{u_0} \times \sigma_{u'_0})$ such that, for any strip $-a < \Re s < a$, and any polynomial $P(s)$ such that $P(s)L(s, \sigma_{u_0} \times \sigma_{u'_0})$ has no pole on $-a \leq \Re s \leq a$, the product $P(s)h(s)$ is bounded in the open strip. Then $\Psi(W_{u_0}, W'_{u_0}, \Phi)$ is in $\mathcal{H}$. In fact, we can prove even more. Suppose that $f$ and $f'$ are elements of $V(\mu)$ and $V(\mu')$, each in some bonded set $\mathcal{B}$ and $\mathcal{B}'$ respectively. Let

$$W(g) = \mathcal{W}_{u_0}(\Xi_{u_0}(g)f), \ W'(g) = \mathcal{W}'_{u_0}(\Xi_{u_0}(g)f').$$

Suppose that $\Phi$ is in some bounded set $\mathcal{C}$ of $\mathcal{S}(F^n)$. Then we claim

$$\Psi(s, W, W', \Phi)$$

is in a bounded set of $\mathcal{H}$. Since the spaces $V(\mu)$, $V(\mu')$ and $\mathcal{S}(F^n)$ are bornological spaces, this proves that the trilinear map

$$(f, f', \Phi) \mapsto \Psi(\bullet, W, W', \Phi)$$

from $V(\mu) \times V(\mu') \times \mathcal{S}(F^n)$ to $\mathcal{H}$ is continuous. To verify our claim we use the Dixmier Malliavin Lemma to write

$$f = \sum_i \Xi_{u_0}(\phi_i) f_i$$

with $\phi_i$ in a bounded set of $\mathcal{D}(G_n)$ and $f_i$ in a bounded set of $V(\mu)$. We consider the standard sections $f_{i,u}$ attached to the $f_i$, the convolution sections $f_{i,\phi_i,u}$, and the corresponding Whittaker functions $W_{i,u}$. Likewise we write

$$f' = \sum_j \Xi_{u'_0}(\phi'_j) f'_j$$

and define $W'_{j,u'}$. Then

$$\Psi(s, W, W') = \sum_{i,j} \Psi(s, W_{i,u_0}, W'_{j,u'_0}, \Phi).$$

Our construction shows that each term is in a bounded set of $\mathcal{H}$.

We now are ready to state our main theorem.

THEOREM 3. *The ratio*

$$\frac{\Psi(s, W_u, W'_{u'}, \Phi)}{L(s, \sigma_u \otimes \sigma'_{u'})} \tag{38}$$

*is a holomorphic function of* $(s, u, u')$ *in* $\mathbb{C}^{2n+1}$.

*Proof.* We recall the functional equation satisfied by the functions $\Psi$ ([JS]). We introduce, as before,

$$\widetilde{W}_u(g) := W_u(w\,{}^t g^{-1}),\ \widetilde{W}'_{u'}(g) := W'_{u'}(w\,{}^t g^{-1}).$$

The function $\widetilde{W}_u(g)$ has the same properties as the function $W_u$, except that the representation $\sigma_u$ is replaced by the contragredient representation $\widetilde{\sigma_u}$. The equation in question reads:

$$\frac{\Psi(\widetilde{W}_u, \widetilde{W}'_{u'}, \Phi)}{L(1-s, \widetilde{\sigma_u} \otimes \widetilde{\sigma'_{u'}})} = \epsilon(s, \sigma_u \otimes \sigma'_{u'}, \psi) \frac{\Psi(s, W_u, W'_{u'}; \Phi)}{L(s, \sigma_u \otimes \sigma'_u)}.$$

Fix $u$ and $u'$ purely imaginary. Then the integral defining $\Psi(s, W_u, W'_{u'}; \Phi)$ converges for $\Re s > 0$. Thus the right-hand side, which is a priori defined for $\Re s >> 0$

extends to a function holomorphic in the half plane $\Re s > 0$. Likewise, the left-hand size extends to a holomorphic function in the half plane $\Re s < 1$. We conclude that the right-hand side, for $u$, $u'$ fixed and imaginary, is actually an entire function of $s$.

Now by the previous proposition, the ratio (38) has the form

$$h(s, u, u') \prod_{i=2}^{i=n} L\left(s, \bigwedge^{i} \sigma_u \otimes \bigwedge^{i} \sigma'_{u'}\right) \tag{39}$$

where $h$ is holomorphic. The singularities of $L(s, \bigwedge^i \sigma_u \otimes \bigwedge^i \sigma'_{u'})$ are hyperplanes of the form:

$$is + u_{j_1} + u_{j_2} + \cdots + u_{j_i} + u'_{k_1} + u'_{k_2} + \cdots u'_{k_i} = -z$$

with $z \geq 0, z \in \mathbb{Z}_F, j_1 < j_2 < \cdots < j_i, k_1 < k_2 < \cdots < k_i$. Fix $u$ and $u'$ imaginary in such a way that the values of $s$ determined by these equations are all distinct. Then, for $\Re s >> 0$, the expression (39) is equal to the integral divided by $L(s, \sigma_u \otimes \sigma'_{u'})$. Thus for fixed imaginary $(u, u')$ it extends to an entire function of $s$. This forces $h(s, u, u') = 0$ for any $s$ satisfying one (and only one) of these equations. It follows that $h$ vanishes identically on any of these hyperplanes. Thus it is divisible by each one of the linear forms $is + u_{j_1} + u_{j_2} + \cdots + u_{j_i} + u'_{k_1} + u'_{k_2} + \cdots u'_{k_i} + z$. If these singularities have multiplicity this argument can be repeated. Our assertion follows. □

**9. Other series for the real case.** It remains to prove the analog of the previous theorem for representations which are (non-unitarily) induced by discrete series of $GL(2)$ and characters of $GL(1)$. Thus we assume from now on that $F$ is real.

We first recall the following result on the convergence of the integrals.

PROPOSITION 13. *Let $\pi$ and $\pi'$ be unitary irreducible tempered representations of $GL(n)$. For $W$ (resp. $W'$) in the Whittaker model of $\pi$ (resp. $\pi'$) and any $\Phi \in \mathcal{S}(F^n)$, the integral $\Psi(s, W, W', \Phi)$ converges absolutely for $\Re s > 0$, uniformly for $\Re s$ in a compact set.*

*Proof.* We think of the smooth vectors of $\pi$ as elements of the Whittaker model. Let $\lambda$ be the evaluation at $e$. Thus

$$W(g) = \lambda(\pi(g)W).$$

According to [W] (Lemma 15.2.3 and Theorem 15.2.5, page 375), the linear form $\lambda$ is tame for the pair $(B, A)$; thus there is $\Lambda_\pi \in \mathfrak{a}^*$ and a continuous semi-norm $q$ such that, for $a \in A^-$ and any smooth vector $v$,

$$\lambda(\pi(a)v) \leq e^{\langle \Lambda_\pi + \rho, H(a) \rangle} (1 + \| \log a \|)^d \, q(v).$$

Since $\pi$ is tempered, $\Lambda_\pi = \sum_{1\le i\le n-1} x_i\alpha_i$ with $x_i \ge 0$. We also recall that there is an integer $N$ such that $q(\pi(g)v) \le\| g \|^N q(v)$.

A similar result applies to $\pi'$ with a linear form $\Lambda_{\pi'}$, a semi-norm $q'$ and a number $N'$.

Now

$$\Psi(s, W, W', \Phi)$$
$$= \int_{A_{n-1}\times F^\times\times K_n} W\left[\begin{pmatrix} a & 0\\ 0 & 1\end{pmatrix}k\right] W'\left[\begin{pmatrix} a & 0\\ 0 & 1\end{pmatrix}k\right]\Phi[(0,0,\dots,0,a_n)k]$$
$$| a_n |^{ns} \omega\omega'(a_n)e^{\langle H(a),-2\rho\rangle} | \det a |^s \, da d^\times a_n dk,$$

where $\omega$ and $\omega'$ are the central characters. We write as before

$$a = \prod_{i=1}^{i=n-1} b_i,\ b_i = \check{\varpi}_i(a_i) = \operatorname{diag}(\overbrace{a_i, a_i, \dots, a_i}^{i}, 1, \dots, 1).$$

By the Dixmier-Malliavin Lemma we may assume that

$$W(g) = \int W_0(gx)\phi(x)dx$$

where $\phi$ is smooth of compact support. Then

$$W\left[\begin{pmatrix} a & 0\\ 0 & 1\end{pmatrix}k\right]$$
$$= \int W_0\left[\begin{pmatrix} a & 0\\ 0 & 1\end{pmatrix}n_1m_1k_1\right]\phi(n_1m_1k_1k^{-1})dn_1dm_1e^{H(\langle m_1\rangle,-2\rho\rangle}dk_1$$
$$= \int W_0\left[\begin{pmatrix} a & 0\\ 0 & 1\end{pmatrix}m_1k_1\right]\phi(n_1m_1k_1k^{-1})\theta(an_1a^{-1})dn_1dm_1e^{H(\langle m_1\rangle,-2\rho\rangle}dk_1.$$

After integrating over $n_1$ we see that this has the form:

$$\int W_0\left[\begin{pmatrix} a & 0\\ 0 & 1\end{pmatrix}m_1k_1\right]\phi_1(a_1, a_2, \dots a_{n-1}; m_1, k_1, k)dm_1dk_1,$$

where $\phi_1$ is a smooth function on $\mathbb{F}^{n-1}\times A_n\times K_n\times K_n$ whose support has a compact projection on $A_n$ and which is rapidly decreasing with respect to the first $n-1$ variables.

If $T$ is a subset (possibly empty) of the set $\{\alpha_1, \alpha_2, \dots, \alpha_{n-1}\}$ of simple roots (or, equivalently, a subset of $[1, n-1]$), we denote by $A(T)$ the set of diagonal matrices $a$ such that $| \alpha_i(a) |\le 1$ for $\alpha_i \in T$ and $| \alpha_i(a) |> 1$ for $\alpha_i \notin T$. Thus $A$ is the union of the set $A(T)$. It will suffice to prove that for each $T$ the integral

$$\int_{\begin{pmatrix} a & 0 \\ 0 & 1 \end{pmatrix} \in A(T)} W\left[\begin{pmatrix} a & 0 \\ 0 & 1 \end{pmatrix} k\right] W'\left[\begin{pmatrix} a & 0 \\ 0 & 1 \end{pmatrix} k\right] \Phi[(0, 0, \ldots, 0, a_n)k]$$

$$| a_n |^{ns} \omega\omega'(a_n) e^{\langle H(a), -2\rho\rangle} | \det a |^s \, da d^\times a_n dk$$

converges absolutely. We write

$$a = b_T b^T$$

where $b_T = \prod_{i \in T} b_i$. We have thus

$$\left| W_0 \left[\begin{pmatrix} a & 0 \\ 0 & 1 \end{pmatrix} m_1 k_1\right]\right| \leq e^{\langle H(b_T), \Lambda_\pi + \rho\rangle} (1 + \| \log b_T \|)^d q\left(\pi(b^T m_1 k_1) W_0\right)$$

or

$$\left| W_0 \left[\begin{pmatrix} a & 0 \\ 0 & 1 \end{pmatrix} m_1 k_1\right]\right| \leq e^{\langle H(b_T), \Lambda_\pi + \rho\rangle} (1 + \| \log b_T \|)^d \| b^T \|^N \| m_1 \|^N q(W_0).$$

Thus

$$\left| W \left[\begin{pmatrix} a & 0 \\ 0 & 1 \end{pmatrix} k\right]\right| \leq e^{\langle H(b_T), \Lambda_\pi + \rho\rangle} (1 + \| \log b_T \|)^d$$

$$\times \| b^T \|^N \int | \phi | (a_1, a_2, \cdots, a_{n-1}; m_1, k_1, k) \| m_1 \|^N dm_1 dk_1.$$

Thus after integrating over $m_1$ and $k_1$, we find that

$$\left| W \left[\begin{pmatrix} a & 0 \\ 0 & 1 \end{pmatrix} k\right]\right| \leq e^{\langle H(b_T), \Lambda_\pi + \rho\rangle} (1 + \| \log b_T \|)^d \| b^T \|^N \phi_0(a_1, a_2, \cdots, a_{n-1})$$

where $\phi_0 \geq 0$ is a Schwartz function. In turn $\phi_0$ is majorized by a Schwartz function $\phi_T(a_i)$ depending only on the $a_i$, $i \notin T$. There is a similar majorization for $W'$.

We see then that we have to check the convergence of the three following integrals, for $s > 0$:

$$\int e^{\langle H(b_T), \Lambda_\pi + \Lambda_{\pi'}} > (1 + \| \log b_T \|)^{d+d'} | \det b_T |^s \, db_T$$

$$\int \| b^T \|^{N+N'} | \det b^T |^s e^{\langle b^T, -2\rho\rangle} \phi_T(a_i) db^T$$

$$\int \phi_1(a_n) | a_n |^{ns} da_n.$$

The first one is in fact

$$\int_{|a_i| \leq 1, i \in T} \prod_{i \in T} | a_i |^{is + x_i + x_i'} (1 + \| (\log a_i) \|)^{d+d'} \bigotimes_{i \in T} d^\times a_i$$

thus converges for $\Re s > 0$ since $x_i \geq 0$, $x_i' \geq 0$. The convergence of the other integrals is trivial. □

We now consider a general induced representation. In a precise way, we consider unitary irreducible representations $\tau_i$, $1 \leq i \leq r$ of degree $r_i = 1, 2$ of the Weil group of $\mathbb{R}$; we assume that $\sum r_i = n$. To each $\tau_i$ is attached an irreducible unitary representation (quasi-square integrable if $r_i = 2$) $\pi_i$ of $GL(r_i, \mathbb{R})$. We choose a non-zero linear form $\mathcal{W}_i$ on the space of smooth vectors of $\pi_i$ such that

$$\mathcal{W}_i \left[ \pi_i \begin{pmatrix} 1 & x \\ 0 & 1 \end{pmatrix} v \right] = \psi_i(x) \mathcal{W}_i(v).$$

If the degree is 1 then $\mathcal{W}_i$ is the linear form taking the value 1 on $1 \in \mathbb{C}$.

If $v$ is a $r$-tuple of complex numbers we consider the representation

$$\Pi_v := I(\pi_1 \otimes \alpha^{v_1} \times \pi_2 \otimes \alpha^{v_2} \times \cdots \times \pi_r \otimes \alpha^{v_r})$$

induced from the lower parabolic subgroup of type $(r_1, r_2, \ldots, r_r)$. Let $\pi$ be the tensor product of the representations $\pi_i$ and $\mathcal{W}_\pi$ be the tensor product of the linear forms $\mathcal{W}_i$. We define a linear form $\mathcal{W}_v$ by the integral:

$$\mathcal{W}_v(f) = \int_{N_P} \mathcal{W}_\pi f(n) e^{\langle H'(n), v + \rho_P \rangle} \theta(n) dn$$

The integral converges when $\Re v_r > \Re v_{r-1} > \cdots \Re v_1$. It has analytic continuation to an entire function of $u$.

We set

$$\tau_0 := \bigoplus_{1 \leq i \leq r} \tau_i, \ \tau_v := \bigoplus_{1 \leq i \leq r} \tau_i \otimes \alpha^{v_i}.$$

If the degree of $\tau_i$ is 2 then the representation $\pi_i$ is in the discrete series of $GL(2, \mathbb{R})$ and is a subrepresentation of an induced representation of the form

$$I(\xi_i \otimes \alpha^{p_i}, \eta_i \otimes \alpha^{q_i})$$

where $\xi_i, \eta_i$ are normalized characters (of module 1), $p_i, q_i$ are real numbers with $q_i - p_i > 0$. Then we take

$$\mathcal{W}_i(f) = \int f \left[ \begin{pmatrix} 1 & x \\ 0 & 1 \end{pmatrix} \right] \psi_i(x) dx.$$

We may view $\Pi_v$ as a subrepresentation of an induced representation of the form

$$\Xi_u = I(\mu_1 \otimes \alpha^{u_1}, \mu_2 \otimes \alpha^{u_2}, \ldots, \mu_n \otimes \alpha^{u_n})$$

where the $\mu_i$ are characters of module 1 and the $u_i$ complex numbers obtained in the following way. If the degree of $\tau_i$ is one then we associate with $\tau_i$ a character $\mu_k = \pi_i$ and a complex number $u_k = v_i$. If the degree of $\tau_i$ is 2 then we associate to $\tau_i$ characters $\mu_k = \xi_i$, $\mu_{k+1} = \eta_i$, and complex numbers $u_k = p_i + v_i$, $u_{k+1} = q_i + v_i$. We remark that if the differences $\Re v_{i+1} - \Re v_i$ are positive and large enough then

$$\Re u_n > \Re u_{n-1} > \cdots > \Re u_1.$$

We write $u = T(v)$, $u_i = T_i(v)$. Thus $T$ is an affine linear transformation.

As before, for arbitrary $u$, the representations $\Xi_u$ operate on the same vector space $\mathcal{V}(\Xi_0)$ of scalar valued functions on $K$. Likewise, all the representations $\Pi_v$ operate on the same vector space $\mathcal{V}(\Pi_0)$ which is a subspace of $\mathcal{V}(\Xi_0)$. This being so, if $f$ is in $\mathcal{V}(\Pi_0)$ then we have the standard section $f_v$ of $\Pi_v$ and the standard section $f_{T(v)}$ of $\Pi_u$ and, in fact,

$$f_v = f_{T(v)}.$$

Moreover

$$\mathcal{W}_v(f_v) = \mathcal{W}_{T(u)}(f_{T(u)})$$

when $\Re v_{i+1} - \Re v_i >> 0$. This relation remains true for all values of $v$.

We now consider similar data for the group $GL(n')$. Suppose that

$$W_v(g) = \mathcal{W}_v(\Pi_v(g) f_{\phi,v})$$

where $\phi$ is the convolution of $(n-1)$ elements of $\mathcal{D}(G_n)$. We note that we may define, more generally,

$$W_u(g) = \mathcal{W}_u(\Xi_u(g) f_{\phi,u})$$

for $u \in \mathbb{C}^n$. Then

$$W_v(g) = W_{T(v)}(g).$$

Similarly suppose that

$$W'_{v'}(g) = \mathcal{W}'_{v'}(\Pi'_{u'} f'_{\phi',v'}).$$

THEOREM 4. *If* $\Re v$ *and* $\Re v'$ *are in compact sets, then there is* $s_0 \in \mathbb{R}$ *such that the integral*

$$\Psi(s, W_v, W'_{v'})$$

*converges for* $\Re s > s_0$*, uniformly for* $\Re s$ *in a compact set. If* $v$ *and* $v'$ *are purely imaginary then the integral converges for* $\Re s > 0$*. Finally, the integral has analytic continuation to a meromorphic function of* $(s, v, v')$ *which is of the form*

$$L(s, \tau_v \otimes \tau'_{v'}) \times H(s, v, v')$$

*where* $H$ *is a holomorphic function on* $\mathbb{C}^{r+r'+1}$.

We first remark that the vectors $(s, T(u), T'(u'))$ are in an affine subspace of $\mathbb{C}^{n+n'+1}$ defined by real equations. Moreover, that affine subspace is not contained in any singular hyperplane for the function

$$\Psi(s, W_u, W'_u).$$

We may therefore restrict this function to the affine subspace at hand. The result is clearly a meromorphic function of $(s, v, v')$ which is a holomorphic multiple of

$$L(s, \sigma_{T(v)} \otimes \sigma'_{T'(v')}).$$

In fact this product has the form

$$L(s, \sigma_{T(v)} \otimes \sigma'_{T'(v')}) = \prod_{1 \leq i \leq r, 1 \leq j \leq r'} P_{i,j}(s + v_i + v'_j) L(s, \tau_v \otimes \tau'_{v'}) \times$$

where each $P_{i,j}$ is a polynomial. Furthermore

$$\begin{aligned} &\epsilon(s, \sigma_{T(v)} \otimes \sigma'_{T'(v')}, \psi) \frac{L(s, \sigma_{T(v)} \otimes \sigma'_{T'(v')})}{L(1-s, (\sigma_{T(v)} \otimes \sigma'_{T'(v')})^{\sim})} \\ &= \epsilon(s, \tau_v \otimes \tau'_{v'}, \psi) \frac{L(s, \tau_v \otimes \tau'_{v'})}{L(1-s, \widetilde{\tau_v} \otimes \widetilde{\tau'_{v'}})}. \end{aligned}$$

It follows that the functional equation can be written

$$\frac{\Psi(1-s, \widetilde{W}_v, \widetilde{W'}_{v'} \hat{\Phi})}{L(1-s, \widetilde{\tau_v} \otimes \widetilde{\tau'_{v'}})} = \epsilon(s, \tau_v \otimes \tau'_{v'}, \psi) \frac{\Psi(s, W_v, W'_{v'} \Phi)}{L(s, \tau_v \otimes \tau'_{v'})}.$$

One then finishes the proof as in the previous section.

**10. Complements.** Consider now a standard section $f_{u,u'}$ of the tensor product representation $\Xi_u \otimes \Xi_{u'}$ (or $\Pi_v \otimes \Pi'_{v'}$) and a corresponding convolution section $f_{\phi,u,u'}$ where now $\phi$ is a smooth function of compact support on $G_n \times G_n$. We then consider the Whittaker integral

$$\mathcal{W}_{u,u'}(f_{u,u'}) = \int_{N_n \times N_n} f_{u,u'}(v, v') \theta(v) \overline{\theta}(v') dv dv'.$$

It is then easy to obtain directly an integral representation for the functions

$$W_{u,u'}(g, g') = \mathcal{W}_{u,u'}(\Xi_u \otimes \Xi_{u'}(g, g') f_{\phi,u,u'})$$

and obtain the analytic properties of the integral

$$\Psi(s, W_{u,u'}, \Phi) := \int_{N_n \backslash G_n} W_{u,u'}(g) \Phi(\epsilon_n g) \mid \det g \mid^s dg.$$

In the previous construction we pass from a standard section $f_u$ to a convolution section $f_{\phi,u}$ but in fact the construction is more general. Let us say that a section $f_u$ is **well behaved** if, for $u$ in any multi strip, the restriction of $f_u$ to $K_n$ is uniformly bounded as well as all its $K_n$-derivatives. At this point, we recall the space $\mathcal{S}(G_n)$ ([W] 7.1); it is the space of smooth functions $\Phi$ on $G_n$ such that for every $X$ and $Y$ in the enveloping algebra of $G_n$ and for every integer $N$,

$$\sup_{g \in G_n} \| g \|^N |\lambda(X)\rho(Y)\Phi(g)| < +\infty.$$

If $f_u$ is a standard section and $\Phi \in \mathcal{S}(G_n)$ then

$$f_{\Phi,u}(g) := \int f_u(gx)\Phi(x)dx$$

is a well behaved section. In addition, the group $G_n$ (and any Lie subgroup of $G_n$) operates by right and left translations on $\mathcal{S}(G_n)$ and the Dixmier-Malliavin Lemma applies to these representations. Thus if $\Phi$ is in $\mathcal{S}(G_n)$ then it can be written as a finite sum of convolution products

$$\Phi(g) = \sum_i \phi_i * \Phi_i,$$

with $\phi_i \in \mathcal{D}(G_n)$ and $\Phi_i \in \mathcal{S}(G_n)$. Suppose that $f_u$ is a standard section. Then the section $f_{\Phi,u}$ verifies

$$f_{\Phi,u}(g) = \sum_i \int f_{\Phi_i,u}(gx)\phi_i(x)dx.$$

It follows that our previous results apply to Whittaker functions defined by

$$W_u(g) = \mathcal{W}_u(\Xi_u(g)f_{\Phi,u})$$

with $f_u$ a standard section and $\Phi \in \mathcal{S}(G_n)$.

**11. Other Rankin-Selberg integrals.** In [JS] we associate to every pair of integers $(n, n')$ with $n' \leq n$ a family of Rankin-Selberg integrals with similar analytic properties. To be specific, let us consider only the integrals attached to principal series representations. Thus we consider a $n$-tuple of characters $(\mu_1, \mu_2, \ldots, \mu_n)$ and a $n'$-tuple $(\mu'_1, \mu'_2, \ldots, \mu'_{n'})$ and $u \in \mathbb{C}^n$, $u' \in \mathbb{C}^{n'}$. If $n' = n - 1$ the integrals can be treated as in the case $n = n'$. However, for $n' \leq n - 2$, an additional complication is that, in order to state the functions equation, one has to consider integrals $\Psi(s, W_u, W'_{u'}; j)$ which involve an auxiliary integration over the space $M(j \times n')$ of matrices with $j$ rows and $n'$ columns, for $0 \leq j \leq n - n' - 1$, namely:

(40)

$$\Psi(s, W, W'; j) = \int W\left[\begin{pmatrix} g & 0 & 0 \\ X & 1_j & 0 \\ 0 & 0 & 1_{n-n'-j} \end{pmatrix}\right] W'(g) \mid \det g \mid^{s-\frac{n-n'}{2}} dgdX.$$

The following lemma will allow us to reduce the study of the integrals $\Psi(s, W_u, W'_{u'}; j)$ to the study of the integrals $\Psi(s, W_u, W'_{u'}; 0)$.

LEMMA 8. *Suppose $n' \leq n - 2$ and $0 < j \leq n - n' - 1$. Let $f_u$ be a standard section. Let $\Phi \in \mathcal{S}(G_n)$, and $W_u$ the Whittaker function attached to the section $f_{\Phi,u}$. Then there is $\Phi_1 \in \mathcal{S}(G_n)$ such that the Whittaker function $W_{1,u}$ attached to the* *section $f_{\Phi_{1,u}}$ verifies*

$$\int W_u\left[\begin{pmatrix} g & 0 & 0 \\ X & 1_j & 0 \\ 0 & 0 & 1_{n-n'-j}\end{pmatrix}\right] dX = W_{1,u}\left[\begin{pmatrix} g & 0 & 0 \\ 0 & 1_j & 0 \\ 0 & 0 & 1_{n-n'-j}\end{pmatrix}\right]$$

*for all $g \in G_{n'}$.*

*Proof of the Lemma.* The proof is by descending induction on $j$. We show the induction step. By the Dixmier Malliavin Lemma we can write

$$\Phi(g) = \sum_i \int \Phi_i\left[\begin{pmatrix} 1_{n'} & 0 & Y & 0 \\ 0 & 1_j & 0 & 0 \\ 0 & 0 & 1 & 0 \\ 0 & 0 & 0 & 1_{n-n'-j-1}\end{pmatrix} g\right] \phi_i(-Y)dY,$$

with $\phi_i \in \mathcal{D}(F)$ and $\Phi_i \in \mathcal{S}(G_n)$. Then

$$W_u(g) = \sum_i \int W_{i,u}\left[g\begin{pmatrix} 1_{n'} & 0 & Y & 0 \\ 0 & 1_j & 0 & 0 \\ 0 & 0 & 1 & 0 \\ 0 & 0 & 0 & 1_{n-n'-j-1}\end{pmatrix}\right] \phi_i(Y)dY$$

where $W_{i,u}$ is the Whittaker function corresponding to the section $f_{\Phi_i,u}$. Then

$$\int W_u\left[\begin{pmatrix} g & 0 & 0 \\ 0 & 1_j & 0 \\ 0 & 0 & 1_{n-n'-j}\end{pmatrix}\begin{pmatrix} 1_{n'} & 0 & 0 & 0 \\ X_1 & 1_{j-1} & 0 & 0 \\ X_2 & 0 & 1 & 0 \\ 0 & 0 & 0 & 1_{n-n'-j}\end{pmatrix}\right] dX_2$$

$$= \sum_i \int W_{i,u}\left[\begin{pmatrix} g & 0 & 0 \\ 0 & 1_j & 0 \\ 0 & 0 & 1_{n-n'-j}\end{pmatrix}\begin{pmatrix} 1_{n'} & 0 & 0 & 0 \\ X_1 & 1_{j-1} & 0 & 0 \\ X_2 & 0 & 1 & 0 \\ 0 & 0 & 0 & 1_{n-n'-j}\end{pmatrix}\right]$$

$$dX_2\psi(X_2Y)\phi_i(Y)dY$$

$$= \sum_i \int W_{i,u}\left[\begin{pmatrix} g & 0 & 0 \\ 0 & 1_j & 0 \\ 0 & 0 & 1_{n-n'-j}\end{pmatrix}\begin{pmatrix} 1_{n'} & 0 & 0 & 0 \\ X_1 & 1_{j-1} & 0 & 0 \\ X_2 & 0 & 1 & 0 \\ 0 & 0 & 0 & 1_{n-n'-j}\end{pmatrix}\right]$$

$$dX_2\widehat{\phi_i}(-X_2)$$

$$= W_{0,u}\left[\begin{pmatrix} g & 0 & 0 \\ 0 & 1_j & 0 \\ 0 & 0 & 1_{n-n'-j}\end{pmatrix}\begin{pmatrix} 1_{n'} & 0 & 0 & 0 \\ X_1 & 1_{j-1} & 0 & 0 \\ 0 & 0 & 1 & 0 \\ 0 & 0 & 0 & 1_{n-n'-j}\end{pmatrix}\right]$$

where $W_{0,u}$ is the Whittaker function attached to the section $f_{\Phi_{0},u}$ and $\Phi_0$ is the element of $\mathcal{S}(G)$ defined by:

$$\Phi_0(g) = \sum_i \int \Phi_i \left[ \begin{pmatrix} 1_{n'} & 0 & 0 & 0 \\ 0 & 1_{j-1} & 0 & 0 \\ X_2 & 0 & 1 & 0 \\ 0 & 0 & 0 & 1_{n-n'-j} \end{pmatrix} g \right] dX_2 \widehat{\phi_i}(X_2).$$

The lemma follows. □

As for the integrals $\Psi(s, W_u, W'_{u'}; 0)$, they can be studied in the same way as before.

---

## REFERENCES

---


[CKPSS] J. W. Cogdell, H., Kim, I. I., Piatetski-Shapiro F. Shahidi, On lifting from classical groups to $GL_N$, Publ. Math. I.H.E.S. (to appear).

[CPS1] J. W. Cogdell, I. I. Piatetski-Shapiro, Converse theorems for $GL_n$. *Inst. Hautes tudes Sci. Publ. Math. No.* **79** (1994), 157–214.

[CPS2] J. W. Cogdell, I. I. Piatetski-Shapiro, Converse theorems for GL, II, J. reine angew. *Math.* **507** (1999), 165–188.

[D] A. Deitmar, Mellin transforms of Whittaker functions, preprint.

[DM] J. Dixmier P. Malliavin, Factorisations de fonctions et de vecteurs indéfiniment différentiables, *Bull. Sci. Math.* (2), **102** (1978), 307–330.

[GJR] S. Gelbart, H. Jacquet, J. Rogawski Generic representations for the unitary group in three variables, to appear in Israel J. Math.

[GS] S. Gelbart, F. Shahidi, Boundedness of automorphic $L$-functions in vertical strips. *J. Amer. Math. Soc.* **14** (2001), no. 1, 79–107 (electronic).

[H] L. Hörmander, The Analysis of Linear partial differential operators I, Springer-Verlag, 1983.

[JS] H. Jacquet, J. Shalika, Rankin Selberg convolutions: Archimedean theory. Festschrift in honor of I. I. Piatetski-Shapiro on the occasion of his sixtieth birthday, Part I (Ramat Aviv, 1989), 125–207, Israel Math. Conf. Proc., 2, Weizmann, Jerusalem, 1990.

[KS] H. Kim, F. Shahidi, Functorial products for $GL_2 \times GL_3$ and functorial symmetric cube for $\mathrm{GL}_2$, *C. R. Acad. Sci. Paris Sr. I Math.* **331** (2000), no. 8, 599–604.

[RS] Z. Rudnick, P. Sarnak, The pair correlation function of fractional parts of polynomials, *Comm. Math. Phys.* **194** (1998), no. 1, 61–70.

[S] D. Soudry, On the Archimedean theory of Rankin-Selberg convolutions for $SO_{2l+1} \times GL_n$, *Ann. Scient. Écoc. Norm. Sup.* **28** (1995), 161–224.

[St1] E. Stade, On explicit integral formulas for $\mathrm{GL}(n, \mathbb{R})$-Whittaker functions. [With an appendix by Daniel Bump, Solomon Friedberg and Jeffrey Hoffstein.] *Duke Math. J.* **60** (1990), no. 2, 313–362.

[St2] E. Stade, Mellin transforms of $\mathrm{GL}(n, \mathbb{R})$ Whittaker functions, *Amer. J. Math.* **123** (2001), no. 1, 121–161.

[W] N. R. Wallach, *Real Reductive Groups I & II,* Academic Press, Pure and Applied Mathematics Vol. 132 & 132 II.

CHAPTER 16

# A SPECTRAL IDENTITY FOR SKEW SYMMETRIC MATRICES

By Hervé Jacquet, Erez Lapid, and Stephen Rallis

---

*Dedicated to Joe Shalika on the occasion of his 60th birthday*

*Abstract.* We prove an identity between a spherical distribution of a residual Eisenstein series on $GL_{2n}$ with respect to the symplectic group and a weighted trace of a cuspidal representation on $GL_n$ with respect to a certain automorphic form.

**1. Introduction.** Let $G$ be a reductive group over a number field $F$ and let $H$ be the fixed point subgroup of an involutive automorphism of $G$ defined over $F$. For simplicity assume that $H$ is semisimple. Let $\mathbb{A}$ be the ring of adèles of $F$. If $\pi$ is a cuspidal representation of $G(\mathbb{A})$ one may consider the *period* linear form

$$\ell_H(\varphi) = \int_{H(F)\backslash H(\mathbb{A})} \varphi(h)\, dh$$

on the space of $\pi$. If $\ell_H \not\equiv 0$ then $\pi$ is said to be *distinguished* by $H$. These kinds of periods are interesting from both an arithmetic and an analytic point of view, and there is a great deal of literature about them. Many of them are related to special values of $L$-functions.

If $\pi$ is distinguished by $H$ we define the *spherical distribution* of $\pi$ with respect to $H$ by

$$\mathcal{B}^{\pi}_{\ell_H,\overline{\ell_H}}(f) = \sum_{\{\varphi\}} \ell_H(\pi(f)\varphi)\overline{\ell_H(\varphi)}$$

for any $f \in C_c^\infty(G(\mathbb{A}))$, where $\varphi$ runs over an orthonormal basis of $\pi$.

Distinguished representations (by one, or a family of period subgroups) are often characterized, or expected to be characterized, as functorial images, in the sense of Langlands, from a third group $G'$. As pointed out in [JLR93] the hypothetical group $G'$ should roughly speaking be characterized by the fact that its conjugacy classes are in one-to-one correspondence with the double coset space $H\backslash G/H$. (For the existence of $G'$ with this property, cf. [KR69].) In the case at hand $G = GL_{2n}$, $H$ is the symplectic group in $2n$ variables, $G' = GL_n$ and the map associates to a conjugacy class of $g \in G'$ the double coset $H\left(\begin{smallmatrix} g & 0 \\ 0 & 1_n \end{smallmatrix}\right)H$. If $\pi$ is a cuspidal automorphic representation of $G'(\mathbb{A})$ the corresponding distinguished

---

Manuscript received October 21, 2002.
First named author partially supported by NSF grant DMS 9988611.
Second named author partially supported by NSF grant DMS 0070611.
Third named author partially supported by NSF grant DMS 9970342.

automorphic representation $\Pi$ of $G(\mathbb{A})$ is not cuspidal but in fact is the residue of the Eisenstein series induced from $\pi\,|\det|^{\frac{1}{2}} \otimes \pi\,|\det|^{-\frac{1}{2}}$ (normalized induction) viewed as a representation of the Levi subgroup $M$ of the parabolic subgroup $P = M \cdot U$ of $G$ of type $(n, n)$.

The correspondence between conjugacy classes and double cosets suggests an identity of the form

$$(1)\quad \int_{H(F)\backslash H(\mathbb{A})} \int_{H(F)\backslash H(\mathbb{A})} K_f(h_1, h_2)\, dh_1\, dh_2 = \int_{G'(F)\backslash G'(\mathbb{A})^1} K_{f'}(x, x)\, dx$$

suitably regularized (with truncation, etc.) where $f$ and $f'$ are smooth compactly supported function on $G(\mathbb{A})$, $G'(\mathbb{A})$ respectively, with *"matching"* orbital integrals. Here, as usual, $K_f(x, y)$ (resp. $K_{f'}(x', y')$) is the automorphic kernel for $G$ (resp. $G'$) defined by

$$K_f(x, y) = \sum_{\gamma \in G(F)} f(x^{-1}\gamma y)$$

and $G'(\mathbb{A})^1 = \{g \in GL_n(\mathbb{A}) : |\det(g)| = 1\}$. Unfortunately, the identity (1) simply does not work. It turns out, for reasons which are quite mysterious to us, that instead, one needs to put a certain automorphic weight function in either side. More precisely, we expect to have (after suitable regularization)

$$(2)\quad \int_{H(F)\backslash H(\mathbb{A})} \int_{H(F)\backslash H(\mathbb{A})} K_f(h_1, h_2)\, dh_1\, dh_2 = \int_{G'(F)\backslash G'(\mathbb{A})^1} K_{f'}(x, x)\mathfrak{E}(x)\, dx$$

and

$$(3)\quad \int_{H(F)\backslash H(\mathbb{A})} \int_{H(F)\backslash H(\mathbb{A})} K_f(h_1, h_2)\Theta(h_2)\, dh_1\, dh_2 = \int_{G'(F)\backslash G'(\mathbb{A})^1} K_{f'}(x, x)\, dx$$

where $\mathfrak{E}$ is a certain degenerate Eisenstein series on $G'$, and $\Theta$ is a residual Eisenstein series on $H$ which is induced from the character $|\det|^{n-3/2}$ on the Siegel parabolic.

The first instance of such a comparison (and the only one up to now) was carried out in [JLR93] in a certain rank one situation.

Naively, it should follow from (2) that to every cuspidal representation $\pi$ of $G'$ corresponds an automorphic representation $\Pi$ of $G$ distinguished by $H$ and moreover we have an identity between the spherical distribution of $\Pi$ and the *weighted trace* of $\pi$, the latter defined by

$$\sum_{\varphi'} \int_{G'(F)\backslash G'(\mathbb{A})^1} \pi(f')\varphi'(x)\overline{\varphi'(x)}\mathfrak{E}(x)\, dx$$

where $\varphi'$ ranges over an orthonormal basis of $\pi$. Such an identity is the main result of this paper. The transfer map $f \mapsto f'$ is given explicitly in terms of the Harish-Chandra map. This reflects the elementary nature of the functoriality. The proof of this result is direct and does not use the comparison (2).

There are two main ingredients in the proof. The first is the fact that $\Pi$ is constructed explicitly from $\pi$ by means of residues of Eisenstein series and the inner product of two such forms is given by

$$(4)\qquad \int_{G(F)\backslash G(\mathbb{A})^1} E_{-1}(g,\varphi_1)\overline{E_{-1}(g,\varphi_2)}\,dg = \int_{\mathbf{K}}\int_{M(F)\backslash M(\mathbb{A})^1} M_{-1}\varphi_1(mk)\overline{\varphi_2(mk)}\,dm\,dk.$$

Here $E_{-1}(\bullet,\varphi)$ is the residual Eisenstein series induced from an automorphic form $\varphi \in \mathcal{A}(M(F)U(\mathbb{A})\backslash G(\mathbb{A}))_{\pi\otimes\pi}$ and $M_{-1}$ is the residue of the intertwining operator. See §2 for other unexplained notation.

The second ingredient is two expressions for the period of a residue of an Eisenstein series. The first one is

$$(5)\qquad \int_{H(F)\backslash H(\mathbb{A})} E_{-1}(h,\varphi)\,dh = \int_{\mathbf{K}_H}\int_{M_H(F)\backslash M_H(\mathbb{A})^1} \varphi(mk)\,dm\,dk$$

where $M_H \simeq GL_n$ is the Levi subgroup of the Siegel parabolic of $H$. This formula was proved in [JR92b]. Curiously enough, there is an alternative formula for the period, in terms of $M_{-1}$. This is our second main result:

$$\frac{\lambda_{-1}}{n}\cdot\int_{H\backslash H(\mathbb{A})} E_{-1}(h,\varphi)\,dh = \int_{K_H}\int_{M_H\backslash M_H(\mathbb{A})^1} M_{-1}\varphi(mk)\mathfrak{E}(m)\,dm\,dk$$

where $\lambda_{-1} = \mathrm{vol}(F^*\backslash\mathbb{I}_F^1)$. The proof of this identity is achieved using a regularization procedure. We compute

$$(6)\qquad \int_{H\backslash H(\mathbb{A})} E_{-1}(h,\varphi)\theta_{\Phi,\sigma}(h)\,dh$$

as a distribution of $\sigma$. Here $\theta_{\Phi,\sigma}(h)$ is a pseudo-Eisenstein series of $H$ which depends on a test function whose Fourier transform is $\sigma$, and which is induced from the identity representation on the "Heisenberg" parabolic $Q$ of $H$ (the stabilizer of a line). Unlike the more standard pseudo-Eisenstein series which are induced from cuspidal representations, these are not rapidly decreasing and care must be taken in dealing with them and with expressions like (6). We use the expansion

$$E_{-1}(h,\varphi) = \sum_{P_{(n,n-1,1)}\backslash P_{(2n-1,1)}} M_{-1}\varphi(\gamma g)$$

where $P_{(n_1,\ldots,n_k)}$ denotes the parabolic of type $(n_1,\ldots,n_k)$. After resolving certain convergence issues, we can utilize the standard Rankin-Selberg method to compute (6) as a sum of contributions from double cosets $P_{(n,n-1,1)}\backslash P_{(2n-1,1)}/Q$. This is a finite sum. Most of the terms vanish because of the cuspidality of the data. If $n$ is even, the term attached to the open double coset vanishes because it factors through a period of $\pi$ over $Sp_{n/2}$ and this is known to be zero. In all cases, the only contribution is from the trivial double coset, which is easy to compute.

The conjectural identity (3) suggests a spectral identity as well. However, we will not discuss it in this paper because it involves regularization of the divergent integral

$$\int_{H(F)\backslash H(\mathbb{A})} \varphi(h)\Theta(h)\, dh$$

where $\varphi \in \Pi$.

The formula (5) can be used to prove a simpler identity

$$\mathcal{B}^{\Pi}_{\ell_H,\overline{\mathcal{W}^{\psi}}}(f) = \mathcal{B}^{\pi}_{\mathcal{W}^{\psi'},\overline{\mathcal{W}^{\psi'}}}(f') \tag{7}$$

(with a different $f'$). Here $\psi'$ is a generic character of $N'$, $\mathcal{W}^{\psi'}$ is the $\psi'$-th Fourier coefficient along $N'$ viewed as a linear form on $\pi$, $\psi$ is the degenerate character on $N$ defined by

$$\psi\left(\begin{pmatrix} n_1 & 0 \\ u & n_2 \end{pmatrix}\right) = \psi'(n_1)\psi'(n_2) \tag{8}$$

and $\mathcal{W}^{\psi}$ is the $\psi$-th Fourier coefficient on $\Pi$. A trace formula approach for the above identity was discussed in [JR92a]. In other (more complicated) situations trace formulas of this type and problems derived from them are discussed extensively in the literature, cf. [Jac87], [Mao92], [Jac95], [Mao97], [Fli97], [Ngô99], [JN99], [JLR99], [LR00], [LR], [Lap]. For a survey see [Jac97], [Jac].

Formulas (2) and (3), as well as the corresponding spectral identities should have analogues in the general case. In order for our method to generalize to other cases, one needs an explicit construction of the functoriality involved, as well as explicit formulas for the scalar product and the period. There was a great deal of progress on explicit constructions of functoriality in recent years. It would be interesting if one can utilize this to get more sophisticated identities of distributions. However, currently there are situations where the trace formula is the only available means, and then, formulas like (2) and (3) seem to be indispensable to study distinguished representations.

## 2. Notation and preliminaries.

**2.1. Roots, weights and vector spaces.** Let $F$ be a number field and $\mathbb{A} = \mathbb{A}_F$ its ring of adèles. By our convention, we denote by the same letter a group over $F$ and its $F$-points. For any group $X$ over $F$ set $X(\mathbb{A})^1 = \cap \ker|\chi|$ where $\chi$ ranges over all rational characters of $X$ and let $\delta_X(\bullet)$ be the modulus function on $X(\mathbb{A})$. Throughout let $G$ be the group $GL_{2n}$ with $n > 1$. Let $\mathbb{V} = F^{2n}$ be the vector space of row vectors on which $G$ acts on the right. We will denote by $P_{(n_1,\dots,n_k)} = M_{(n_1,\dots,n_k)} \cdot U_{(n_1,\dots,n_k)}$ the standard parabolic subgroup of $G$ corresponding to the partition $2n = n_1 + \cdots + n_k$ with its standard Levi decomposition. Let $T_0$ be the diagonal subgroup of $G$ isomorphic to $(F^*)^{2n}$ and let $P_0 = T_0 \cdot U_0$ be the standard

Borel subgroup. The embedding

$$\mathbb{R} \hookrightarrow F \otimes_{\mathbb{Q}} \mathbb{R} \hookrightarrow \mathbb{A}_F \tag{9}$$

defined by $x \mapsto 1 \otimes x$, will be used to obtain a subgroup $A_0$ of $T_0(\mathbb{A})$ isomorphic to $(\mathbb{R}_+^*)^{2n}$.

We let

$$\Delta_0 = \{\alpha_1, \dots, \alpha_{2n-1}\}$$

be the set of simple roots of $G$, in the usual ordering. Let $\hat{\Delta} = \{\varpi_1, \dots, \varpi_{2n-1}\}$ be the set of fundamental weights. As usual, $\mathfrak{a}_0$ will be the real vector space generated by the co-characters of $T_0$, and $\mathfrak{a}_0^*$ its dual. We may think of $\mathfrak{a}_0$ as the set of diagonal matrices $\mathrm{diag}(a_1, \dots, a_{2n})$, or simply as $\mathbb{R}^{2n}$. For any standard parabolic $P = M \cdot U$ of $G$ we have the decomposition

$$\mathfrak{a}_0 = \mathfrak{a}_M \oplus \mathfrak{a}_0^M$$

and similarly for the dual spaces. We let $\Delta_0^M$ or $\Delta_0^P$ be the set of simple roots of $T_0$ in $U_0 \cap M$. Then $(\mathfrak{a}_0^M)^*$ is spanned by $\Delta_0^M$ and $\hat{\Delta}$ spans $(\mathfrak{a}_0^G)^*$. We let $\mathfrak{a}_M^G = \mathfrak{a}_M \cap \mathfrak{a}_0^G$. Set $\Delta_0^{(n_1,\dots,n_k)} = \Delta_0^{P_{(n_1,\dots,n_k)}}$. We denote by $H_0 : G(\mathbb{A}) \to \mathfrak{a}_0$ the standard height function of $G$. On $T_0(\mathbb{A})$ it is given by

$$e^{\langle \chi, H_0(t) \rangle} = \prod_v |\chi_v(t_v)|$$

for any rational character $\chi$ of $T_0$. It extends to $G(\mathbb{A})$ by the Iwasawa decomposition. Similarly, we have maps $H_M : G(\mathbb{A}) \to \mathfrak{a}_M$ for any Levi subgroup $M$. Let $A_M$ be the intersection of $A_0$ with the center of $M(\mathbb{A})$. Then $H_M$ defines an isomorphism between $A_M$ and $\mathfrak{a}_M$. Let $X \mapsto e^X$ be its inverse. Let $\rho_P \in \mathfrak{a}_M^*$ be such that $\delta_P(p) = e^{\langle 2\rho_P, H_M(p) \rangle}$ for all $p \in P(\mathbb{A})$. We let $\mathcal{S}$ be a Siegel set of $G$ of the form $\omega \times A_0(c_0) \times \mathbf{K}$ where $\omega$ is a certain compact subset of $P_0(\mathbb{A})^1$, $\mathbf{K}$ is the maximal compact of $G$ and

$$A_0(c_0) = \{a \in A_0 : \langle \alpha, H_0(a) \rangle > c_0 \text{ for all } \alpha \in \Delta_0\}.$$

Similarly, we have Siegel sets $\mathcal{S}^P = \omega \times A_0^P(c_0) \times \mathbf{K}$ where $A_0^P(c_0)$ is defined as $A_0(c_0)$ except that we impose the inequalities only for $\alpha \in \Delta_0^P$. We set $\mathcal{S}^{(n_1,\dots,n_k)} = \mathcal{S}^{P_{(n_1,\dots,n_k)}}$. We choose $\omega$ and $c_0$ appropriately so that $G(\mathbb{A}) = P \cdot \mathcal{S}^P$ for all $P$.

For any parabolic $P = M \cdot U$, an automorphic representation $\pi$ of $M(\mathbb{A})$ and a parameter $\lambda \in \mathfrak{a}_{M,\mathbb{C}}^*$ we let $\mathcal{A}(U(\mathbb{A})M\backslash G(\mathbb{A}))_{\pi,\lambda}$ be the space of smooth functions on $G(\mathbb{A})$, left invariant under $U(\mathbb{A}) \cdot M$ such that for any $g \in G(\mathbb{A})$ the function $m \mapsto e^{-\langle \lambda + \rho_P, H_P(m) \rangle}\varphi(mg)$ belongs to the space of $\pi$. We will always assume that $\pi$ is trivial on $A_M$. We also set $\mathcal{A}(U(\mathbb{A})M\backslash G(\mathbb{A}))_\pi = \mathcal{A}(U(\mathbb{A})M\backslash G(\mathbb{A}))_{\pi,0}$. The space $\mathcal{A}(U(\mathbb{A})M\backslash G(\mathbb{A}))_{\pi,\lambda}$ is isomorphic to (the smooth part of ) $\mathrm{Ind}_{P(\mathbb{A})}^{G(\mathbb{A})} \pi \otimes e^{\langle \lambda, H_M(\bullet) \rangle} = I(\pi, \lambda)$. This applies in particular to the identity representation which we will denote by $\mathbf{1}$.

**2.2. The setup.** Let $H$ be the group $Sp_n$ defined by the skew symmetric form $[\bullet, \bullet]$ corresponding to

$$\epsilon_{2n} = \begin{pmatrix} 0 & w_n \\ -w_n & 0 \end{pmatrix}$$

where

$$w_n = \begin{pmatrix} 0 & 0 & \dots & 0 & 1 \\ 0 & 0 & \dots & 1 & 0 \\ \dots & \dots & \dots & \dots & \dots \\ 0 & 1 & \dots & 0 & 0 \\ 1 & 0 & \dots & 0 & 0 \end{pmatrix}.$$

We view $H$ as a subgroup of $G$. The notation for $H$ will be similar to that of $G$, except that it will usually be appended by $H$. The torus $T_0 \cap H$ is a maximal split torus of $H$ and the vector space $\mathfrak{a}_0^H$ spanned by the co-characters of $T_0 \cap H$ is a subspace of $\mathfrak{a}_0$. The restriction of $H_0$ to $H(\mathbb{A})$ is the height function with respect to $H(\mathbb{A})$. We also get a surjection of $\mathfrak{a}_0^*$ onto $(\mathfrak{a}_0^H)^*$. Although this is not injective, we will often not distinguish between $\lambda \in \mathfrak{a}_0^*$ and its image in $(\mathfrak{a}_0^H)^*$. We will choose the Siegel set $\mathcal{S}$ so that the intersection $\mathcal{S}^H = \mathcal{S} \cap H(\mathbb{A})$ is a Siegel set for $H$ with $H(\mathbb{A}) = H \cdot \mathcal{S}^H$.

**2.3. Specific notation.** From now on we let $P = M \cdot U$ be the parabolic of $G$ of type $(n, n)$ and $P_H = P \cap H = M_H \cdot U_H$ the Siegel parabolic of $H$. Its Levi subgroup $M_H$ is identified with $GL_n$ by

$$m \mapsto \begin{pmatrix} m^* & \\ & m \end{pmatrix} \tag{10}$$

where $m^* = w_n^{-1}\, {}^t m^{-1} w_n$. The unipotent radical $U_H$ is given by

$$\left\{ u = \begin{pmatrix} 1 & X \\ 0 & 1 \end{pmatrix} \in U : w_n\, {}^t X w_n = X \right\}.$$

The weight $\varpi_n \in (\mathfrak{a}_M^G)^*$ corresponds to the character

$$\begin{pmatrix} m_1 & 0 \\ 0 & m_2 \end{pmatrix} \mapsto |\det(m_1)/\det(m_2)|^{1/2}.$$

Under the identification $\mathfrak{a}_0^H \hookrightarrow \mathfrak{a}_0$, $\mathfrak{a}_{M_H}$ is identified with $\mathfrak{a}_M^G$. Similarly for the dual spaces. We have

$$\rho_P = n\varpi_n,$$
$$\rho_{P_H} = \frac{n+1}{2}\varpi_n.$$

We will denote by $Q$ the intersection of $P_{(2n-1,1)}$ with $H$. It is a maximal parabolic subgroup of $H$ which is also given by $P_{(1,2n-2,1)} \cap H$. Its Levi decomposition is

$L \cdot V$ where $L = M_{(1,2n-1,1)} \cap H$ and $V = U_{(1,2n-2,1)} \cap H$. Also, $Q(\mathbb{A})^1 = H(\mathbb{A}) \cap P_{(2n-1,1)}(\mathbb{A})^1$ and the modulus function $\delta_Q$ is the restriction to $Q(\mathbb{A})$ of $\delta_{P_{(2n-1,1)}}$. It is given by

$$\begin{pmatrix} a & * & * \\ & * & * \\ & & a^{-1} \end{pmatrix} \mapsto |a|^{2n}.$$

Finally, $Q_1 = P_H \cap Q = L_1 V_1$ will be the parabolic of $H$ of co-rank 2 whose intersection $P_1$ with $M_H$ is a parabolic of type $(n-1, 1)$. We have $V_1 = (V_1 \cap M_H) \cdot U_H$ and $P_1 = L_1 \cdot (V_1 \cap M_H)$ is the Levi decomposition of $P_1$. Also, $Q_1 = P_1 \cdot U_H$.

The convention about Haar measures will be the following. On any discrete group we take the counting measure. For any unipotent group $N$ we take the Tamagawa measure so that $\mathrm{vol}(N\backslash N(\mathbb{A})) = 1$. On the maximal compact $\mathbf{K}$ we take the measure of total mass 1. We fix a Haar measure $dg$ on $G(\mathbb{A})$. The Haar measure on $M(\mathbb{A})$ will be determined by

$$\int_{G(\mathbb{A})} f(g)\,dg = \int_{\mathbf{K}} \int_{U(\mathbb{A})} \int_{M(\mathbb{A})} f(muk)\,dm\,du\,dk.$$

Writing $M = M_1 \times M_2$ with $M_1 \simeq M_2 \simeq GL_n$ we obtain a measure on $GL_n(\mathbb{A})$ by requiring that $dm = dm_1 \times dm_2$. We then get a measure on $M_H(\mathbb{A})$ by identifying it with $GL_n(\mathbb{A})$. In turn, this will define a Haar measure on $H(A)$ which is compatible with respect to the Iwasawa decomposition relative to $P_H$, where on the maximal compact $\mathbf{K}_H = \mathbf{K} \cap H(\mathbb{A})$ we take the measure of total mass 1. The measure on $\mathfrak{a}_M^G \simeq \mathfrak{a}_{M_H}$ will be the pull-back of $dx$ under $X \mapsto \langle \varpi_n, X \rangle$. This will define a Haar measure on $M_H(\mathbb{A})^1$ by the isomorphism $M_H(\mathbb{A})/M_H(\mathbb{A})^1 \simeq \mathfrak{a}_{M_H}$. On $G(\mathbb{A})^1$ we will take the measure so that the quotient measure on $G(\mathbb{A})/G(\mathbb{A})^1$ is the pull-back of $d^*t$ under $|\det \bullet|$. This will define a measure on $M(\mathbb{A}) \cap G(\mathbb{A})^1$ by the isomorphism $M(\mathbb{A})/(M(\mathbb{A}) \cap G(\mathbb{A})^1) \simeq G(\mathbb{A})/G(\mathbb{A})^1$. We also get a Haar measure on $M(\mathbb{A})^1$ using the identification $(M(\mathbb{A}) \cap G(\mathbb{A})^1)/M(\mathbb{A})^1 \simeq \mathfrak{a}_M^G$. On the idèles $\mathbb{I}_F$ we take the unnormalized Tamagawa measure. The measure on $\mathbb{I}_F^1$ will be taken so that the quotient measure on $\mathbb{I}_F/\mathbb{I}_F^1$ will be the pull-back of $d^*t$ under $|\bullet|_F$.

**2.4. Eisenstein Series.** We will consider various Eisenstein series. Let $\pi$ be a cuspidal automorphic representation of $G' = GL_n$, trivial on $\mathbb{R}_+^*$ and view $\pi \otimes \pi$ as a representation of $M(\mathbb{A})$. For any $\varphi \in \mathcal{A}(U(\mathbb{A})M\backslash G(\mathbb{A}))_{\pi\otimes\pi}$ we consider the Eisenstein series

$$E(g, \varphi, \lambda) = \sum_{\gamma \in P\backslash G} \varphi(\gamma g) e^{\langle \lambda, H_P(\gamma g)\rangle}$$

on $G$. The residue $E_{-1}(g, \varphi)$ of $E(g, \varphi, s\varpi_n)$ at $s = 1$ is a square integrable function on $G\backslash G(\mathbb{A})^1$. Its constant term is given by $M_{-1}\varphi$ where $M_{-1}$ is the residue of the intertwining operator at $\varpi_n$. Let $\Pi$ be the representation generated by these residues. It is irreducible and lies in the discrete spectrum of $L^2(G\backslash G(\mathbb{A})^1)$. The inner product formula (4) follows from spectral theory [MW95], or simply by taking residue in the inner product formula for truncated cuspidal Eisenstein series ([Art80], §4), and taking into account our conventions on measures.

We now recall the construction of normalized Eisenstein series on $GL_{\mathbb{W}}$ where $\mathbb{W}$ is an $m$-dimensional vector space over $F$. Let $0 \neq v_0 \in \mathbb{W}$ and $P_{v_0}$ be the parabolic subgroup of $GL_{\mathbb{W}}$ fixing the line $F \cdot v_0$ with unipotent radical $U_{v_0}$. Let $\Phi$ be a Schwartz-Bruhat function on $\mathbb{W}(\mathbb{A})$. For any $s \in \mathbb{C}$ consider the function

$$\phi_{\Phi,s}(g) = \phi_{\Phi,s}^{\mathbb{W},v_0}(g) = \int_{\mathbb{I}_F} \Phi(v_0 t g)|t|^{m(s+1)/2}\, d^*t \cdot |\det(g)|^{(s+1)/2}$$

on $GL_{\mathbb{W}}(\mathbb{A})$. We have

$$\phi_{\Phi,s}(pg) = e^{\langle (s+1)\rho_{P_{v_0}}, H_{P_{v_0}}(p)\rangle} \phi_{\Phi,s}(g) \quad p \in P_{v_0}(\mathbb{A}),\ g \in GL_{\mathbb{W}}(\mathbb{A}),$$

and thus we obtain a $GL_{\mathbb{W}}(\mathbb{A})^1$-equivariant map

$$\Phi \mapsto \phi_{\Phi,s}^{\mathbb{W}}$$

from the space $\mathcal{S}(\mathbb{W}(\mathbb{A}))$ of Schwartz-Bruhat functions on $\mathbb{W}(\mathbb{A})$ to $\mathcal{A}(U_{v_0}(\mathbb{A})P_{v_0}\backslash GL_{\mathbb{W}}(\mathbb{A}))_{\mathbf{1},s\rho_{P_{v_0}}}$. We set

$$\mathcal{E}_{\Phi}^{\mathbb{W}}(g,s) = \sum_{\gamma \in P_{v_0}\backslash GL_{\mathbb{W}}} \phi_{\Phi,s}^{\mathbb{W}}(\gamma g).$$

The series converges for $\operatorname{Re} s > 1$ and admits a meromorphic continuation. Whenever it is regular, it gives rise to an intertwining map

$$\mathcal{S}(\mathbb{W}(\mathbb{A})) \to \mathcal{A}(G\backslash G(\mathbb{A}))$$

which factors through $\operatorname{Ind}_{P_{v_0}(\mathbb{A})}^{G(\mathbb{A})} e^{\langle s\rho_{P_{v_0}}, H_{P_{v_0}}(\bullet)\rangle}$. We have

$$\mathcal{E}_{\Phi}^{\mathbb{W}}(g,s) = |\det(g)|^{(s+1)/2} \cdot \int_{F^*\backslash \mathbb{I}_F} \sum_{v \in \mathbb{W}-\{0\}} \Phi(vtg)|t|^{m(s+1)/2}\, d^*t$$

for $g \in GL_{\mathbb{W}}(\mathbb{A})$.

In particular, this applies to the group $G$ acting on $\mathbb{V} = F^{2n}$ with $v_0 = (0,\ldots,0,1)$ and $P_{v_0} = P_{(2n-1,1)}$. Since $G = P_{(2n-1,1)}H$ and $Q = P_{(2n-1,1)} \cap H$ we may write

$$\mathcal{E}_{\Phi}^{\mathbb{V}}(h,s) = \sum_{\gamma \in Q\backslash H} \phi_{\Phi,s}(\gamma h).$$

Thus, the restriction $\mathcal{E}_{\Phi}(\bullet, s)$ of $\mathcal{E}_{\Phi}^{\mathbb{V}}(\bullet, s)$ to $H(\mathbb{A})$ is an Eisenstein series and it gives an intertwining map

$$\mathcal{S}(\mathbb{V}(\mathbb{A})) \to \mathcal{A}(H\backslash H(\mathbb{A}))$$

which factors through $\operatorname{Ind}_{Q(\mathbb{A})}^{G(\mathbb{A})} e^{\langle s\rho_Q, H_Q(\bullet)\rangle}$.

We define

$$\Theta_\Phi(g) = \sum_{v \in \mathbb{V}} \Phi(vg)$$

and

$$\Theta_\Phi^*(g) = \Theta_\Phi(g) - \Phi(0).$$

Then

$$\mathcal{E}_\Phi(h, s) = \int_{F^* \backslash \mathbb{I}_F} \Theta_\Phi^*(th)|t|^{n(s+1)}\, d^*t$$

for $h \in H(\mathbb{A})$. By Poisson summation formula we have

$$\Theta_\Phi(th) = t^{-2n}\Theta_{\hat{\Phi}}(t^{-1}h)$$

where

$$\hat{\Phi}(x) = \int_{\mathbb{V}(\mathbb{A})} \Phi(y)\psi_0([x, y])\, dy$$

and $\psi_0$ is a fixed non-trivial character on $F\backslash\mathbb{A}_F$. Using the computation in Tate's thesis,

$$\begin{aligned}(11)\quad \mathcal{E}_\Phi(h, s) &= \int_1^\infty \int_{F^*\backslash\mathbb{I}_F^1} \Theta_\Phi^*(txh)\, d^*x\, t^{n(s+1)}\, d^*t - \lambda_{-1} \cdot \Phi(0)/(n(s+1)) \\ &\quad + \int_1^\infty \int_{F^*\backslash\mathbb{I}_F^1} \Theta_{\hat{\Phi}}^*(txh)\, d^*x\, t^{n(1-s)}\, d^*t + \lambda_{-1} \cdot \hat{\Phi}(0)/(n(s-1)) \\ &= \mathcal{E}_{\hat{\Phi}}(h, -s),\end{aligned}$$

where $\lambda_{-1} = \mathrm{vol}(F^*\backslash\mathbb{I}_F^1)$. The integrals appearing in (11) are entire functions of $s$ and rapidly decreasing on vertical strips.

Let $\mathbb{V}'$ be the subspace of $\mathbb{V}$ defined by the vanishing of the first $n$ coordinates and let $\Phi_0 \geq 0$ be the "standard" Schwartz-Bruhat function on $\mathbb{V}'(\mathbb{A})$. Then $M_H \simeq GL_{\mathbb{V}'}$ via (10) and we will denote the unramified Eisenstein series $\mathcal{E}_{\Phi_0}^{\mathbb{V}'}(m, s)$ on $M_H$ by $\mathfrak{E}(m, s)$.

We will also set for any $\Phi \in \mathcal{S}(\mathbb{V}(\mathbb{A}))$

$$(12)\qquad \mathbb{E}_\Phi(h, s) = \sum_{Q_1\backslash P_H} \phi_{\Phi,s'}^{\mathbb{V}}(\gamma h) = \sum_{P_1\backslash M_H} \phi_{\Phi,s'}^{\mathbb{V}}(\gamma h)$$

where $s' = (s-1)/2$. It is an automorphic form on $U_H(\mathbb{A})M_H\backslash H(\mathbb{A})$ which under left multiplication by $A_{M_H}$ transforms according to the character $e^{\langle (s'+1)\cdot\varpi_n, H_{M_H}(\bullet)\rangle}$. For any $h \in H(\mathbb{A})$ we have

$$\mathbb{E}_\Phi(mh, s) = \mathcal{E}_{\Phi_h}^{\mathbb{V}'}(m, s) \quad m \in M_H(\mathbb{A})^1$$

where $\Phi_h$ is the translate of $\Phi$ by $h$ (restricted to $\mathbb{V}'(\mathbb{A})$). As before, we get an intertwining map

$$\mathcal{S}(\mathbb{V}(\mathbb{A})) \to \mathcal{A}(U_H(\mathbb{A})M_H\backslash H(\mathbb{A}))$$

which factors through $\mathrm{Ind}_{Q(\mathbb{A})}^{H(\mathbb{A})} e^{\langle s'\rho_Q, H_Q(\bullet)\rangle}$.

LEMMA 1. *The exponents of* $\mathcal{E}(h,s)$ *along* $P_H$ *are* $(\pm s - (n-1)/2)\varpi_n$.

*Proof.* We first claim that $H = QP_H \cup Q\xi P_H$ where $\xi = \left(\begin{smallmatrix} 0 & -1_n \\ 1_n & 0 \end{smallmatrix}\right)$. Indeed, $Q\backslash H$ can be identified with the set of lines in $\mathbb{V}$ while $P_H$ is the stabilizer of $\mathbb{V}'$. Our statement amounts to saying that if $\mathbb{W}_1$, $\mathbb{W}_2$ are two lines not contained in $\mathbb{V}'$ then $\mathbb{W}_2 = \mathbb{W}_1 \cdot h$ where $h \in H$ stabilizes $\mathbb{V}'$. Since $(\mathbb{V}' \oplus \mathbb{W}_1, [\bullet,\bullet]) \simeq (\mathbb{V}' \oplus \mathbb{W}_2, [\bullet,\bullet])$ with an isomorphism taking $\mathbb{W}_1$ to $\mathbb{W}_2$ and $\mathbb{V}'$ to itself, this follows from Witt's Theorem.

It follows that

$$\mathcal{E}(h,s) = \sum_{P_1\backslash M_H} \phi_{\Phi,s}(\gamma h) + \sum_{\xi Q\xi^{-1}\cap P_H\backslash P_H} \phi_{\Phi,s}(\xi\gamma h)$$

and hence the constant term of $\mathcal{E}(h,s_1)$ along $P_H$ is the sum of

$$\sum_{P_1\backslash M_H} \phi_{\Phi,s}(\gamma h)$$

and

$$f(h) = \int_{U_H\backslash U_H(\mathbb{A})} \sum_{\gamma\in P_H\cap\xi^{-1}Q\xi\backslash P_H} \phi_{\Phi,s}(\xi\gamma u h)\,du.$$

The first summand is $\mathbb{E}_\Phi(\bullet, 2s+1)$. Its behavior under $A_{M_H}$ is given by $e^{\langle (s+1)\varpi_n, H_{M_H}(\bullet)\rangle}$ so the exponent along $P_H$ is $(s+1)\varpi_n - \rho_{P_H} = (s-(n-1)/2)\varpi_n$. To analyze the second summand, observe that $P_H \cap \xi^{-1}Q\xi = (M_H \cap \xi^{-1}Q\xi)(U_H \cap \xi^{-1}Q\xi)$. Thus,

$$\begin{aligned} f(h) &= \int_{U_H\backslash U_H(\mathbb{A})} \sum_{\gamma\in M_H\cap\xi^{-1}Q\xi\backslash M_H} \sum_{\delta\in U_H\cap\gamma^{-1}\xi^{-1}Q\xi\gamma\backslash U_H} \phi_{\Phi,s}(\xi\gamma\delta u h)\,du \\ &= \sum_{\gamma\in M_H\cap\xi^{-1}Q\xi\backslash M_H} \int_{U_H\cap\gamma^{-1}\xi^{-1}Q\xi\gamma\backslash U_H(\mathbb{A})} \phi_{\Phi,s}(\xi\gamma u h)\,du \\ &= \sum_{\gamma\in M_H\cap\xi^{-1}Q\xi\backslash M_H} \int_{U_H(\mathbb{A})\cap\xi^{-1}Q(\mathbb{A})\xi\backslash U_H(\mathbb{A})} \phi_{\Phi,s}(\xi u\gamma h)\,du. \end{aligned}$$

We have $U_H \cap \xi^{-1} Q \xi = \{\mathfrak{L}(x_1, \dots, x_n) = \begin{pmatrix} 1_n & \begin{matrix} x_n & 0 & 0 \\ \vdots & 0 & 0 \\ x_1 & \dots & x_n \end{matrix} \\ 0 & 1_n \end{pmatrix}\}$. Since $v_0 \xi \mathfrak{L}(x_1, \dots, x_n) = (0, \dots, 0, 1, x_1, \dots, x_n)$,

$$\int_{U_H(\mathbb{A}) \cap \xi^{-1} Q(\mathbb{A}) \xi \backslash U_H(\mathbb{A})} \phi_{\Phi,s}(\xi u h)\, du$$

$$= \int_{\mathbb{I}_F} \int_{(x_1, \dots, x_n) \in \mathbb{A}^n} \Phi(t(0, \dots, 0, 1, x_1, \dots, x_n)h)\, dx_1 \dots\, dx_n |t|^{n(s+1)}\, d^*t$$

$$= \int_{\mathbb{I}_F} (\Phi_h)^{\mathbb{V}/\mathbb{V}'}(tv_1)|t|^{ns}\, d^*t$$

where $v_1 = (0, \dots, 0, 1, *, \dots, *) \in \mathbb{V}/\mathbb{V}'$ and $\Phi^{\mathbb{V}/\mathbb{V}'} \in \mathcal{S}(\mathbb{V}/\mathbb{V}'(\mathbb{A}))$ is defined by

$$\Phi^{\mathbb{V}/\mathbb{V}'}(w) = \int_{\mathbb{V}'(\mathbb{A})} \Phi(w + v)\, dv.$$

Note that $(\Phi_{mh})^{\mathbb{V}/\mathbb{V}'}(\bullet) = |\det_{\mathbb{V}'} m|^{-1} \cdot (\Phi_h)^{\mathbb{V}/\mathbb{V}'}(\bullet m)$ for $m \in M_H(\mathbb{A})$. We may identify $M_H$ with $GL_{\mathbb{V}/\mathbb{V}'}$ and then the stabilizer of $Fv_1$ is $M_H \cap \xi^{-1} Q \xi$. We then have

$$f(mh) = \mathcal{E}^{\mathbb{V}/\mathbb{V}'}_{(\Phi_h)^{\mathbb{V}/\mathbb{V}'}}(m, s') |\det_{\mathbb{V}/\mathbb{V}'} m|^{1-s}, \quad m \in M_H(\mathbb{A})$$

with $s'$ such that $s = (s' + 1)/2$. Hence $f$ is an automorphic form on $U_H(\mathbb{A}) P_H \backslash H(\mathbb{A})$ whose behavior under $A_{M_H}$ is according to the character $e^{\langle (1-s)\varpi_n, H_{M_H}(\bullet) \rangle}$. Thus the exponent is $(1 - s)\varpi_n - \rho_{P_H}$ as required. □

## 3. The Symplectic Period.

Our first main theorem is the following:

THEOREM 1. *For any $\varphi \in \mathcal{A}(U(\mathbb{A})M \backslash G(\mathbb{A}))_{\pi \otimes \pi}$ and $\Phi \in \mathcal{S}(\mathbb{V}(\mathbb{A}))$ we have*

$$\frac{\lambda_{-1}}{n} \hat{\Phi}(0) \cdot \int_{H \backslash H(\mathbb{A})} E_{-1}(h, \varphi)\, dh$$

$$= \int_{K_H} \int_{M_H \backslash M_H(\mathbb{A})^1} M_{-1}\varphi(mk) \mathbb{E}_\Phi(mk, 3)\, dm\, dk.$$

*In particular, taking $\Phi$ to be the "standard" $\mathbf{K}$-invariant function on $\mathbb{V}(\mathbb{A})$,*

$$\frac{\lambda_{-1}}{n} \int_{H \backslash H(\mathbb{A})} E_{-1}(h, \varphi)\, dh = \int_{K_H} \int_{M_H \backslash M_H(\mathbb{A})^1} M_{-1}\varphi(mk) \mathfrak{E}(m, 3)\, dm\, dk.$$

Recall that the formula (5) proved in [JR92b] gives a different identity for the same expression!

The theorem will eventually be proved in the end of this paper. Let us first try to motivate and interpret the theorem. Let $\psi \in \mathcal{A}(U(\mathbb{A})M \backslash G(\mathbb{A}))_{\pi \otimes \pi, -\varpi_n}$. Consider the representation $\tau = \mathrm{Ind}_{Q(\mathbb{A})}^{H(\mathbb{A})} \delta_Q^{\frac{1}{2}}$ of $H(\mathbb{A})$ (normalized induction). The form

$$\iota : \Phi \in \mathcal{S}(\mathbb{V}(\mathbb{A})) \mapsto \int_{A_{M_H} U_H(\mathbb{A}) P_H \backslash H(\mathbb{A})} \psi(h)\, \mathbb{E}_\Phi(h, 3)\, dh$$

is well defined since under left multiplication by $a \in A_{M_H}$ the integrand behaves according to the character

$$e^{\langle(-1+n)\varpi_n + 2\varpi_n, H_{M_H}(a)\rangle} = e^{\langle(n+1)\varpi_n, H_{M_H}(a)\rangle} = \delta_{P_H}(a).$$

Clearly, $\iota$ factors through $\tau$ because $\Phi \mapsto \mathbb{E}_\Phi(h, 3)$ does. Thus we get a map

$$\Upsilon : \mathcal{A}(U(\mathbb{A})M\backslash G(\mathbb{A}))_{\pi\otimes\pi, -\varpi_n} \longrightarrow \tau^\vee$$

which is clearly $H(\mathbb{A})$-equivariant. On the other hand

$$\hat{\Phi}(0) = \int_{Q(\mathbb{A})\backslash H(\mathbb{A})} \phi_{\Phi,1}(h)\, dh$$

so that the form $\Phi \mapsto \hat{\Phi}(0)$ can be factored through $\tau$ as $\Delta \circ \phi_{\Phi,1}$ where $\Delta \in \tau^\vee$ is defined by integration over $Q(\mathbb{A})\backslash H(\mathbb{A})$ (or over $\mathbf{K}_H$). Up to a scalar, $\Delta$ is the unique $H(\mathbb{A})$-invariant vector in $\tau^\vee$.

We can now reformulate Theorem 1 as follows.

THEOREM 2. *The image of* $\Upsilon \circ M_{-1} : \mathcal{A}(U(\mathbb{A})M\backslash G(\mathbb{A}))_{\pi\otimes\pi,\varpi_n} \to \tau^\vee$ *is the identity subrepresentation. In fact,*

$$\Upsilon \circ M_{-1} = \frac{\lambda_{-1}}{n} \cdot (\ell_H \circ E_{-1}) \cdot \Delta$$

*where* $\ell_H$ *is the* $H(\mathbb{A})$*-invariant form given by integration over* $H\backslash H(\mathbb{A})$.

Theorem 1 together with formula (5) suggest a local analogue. Let $\pi$ be a unitary generic representation of $GL_n$ over a local field. We realize $\pi$ in its Whittaker model with respect to the character

$$\psi'\left(\begin{pmatrix} 1 & x_1 & \cdots & \cdots \\ & 1 & \ddots & \cdots \\ & & 1 & x_{n-1} \\ & & & 1 \end{pmatrix}\right) = \psi_0(x_1 + \cdots + x_{n-1}).$$

Define the linear form $l_{M_H}$ on $\pi \otimes \pi$ by

$$l_{M_H}(W_1 \otimes W_2) = \int_{N_{n-1}\backslash GL_{n-1}} W_1(g^*) W_2(g)\, dg$$

where $N_k$ is the unipotent radical of the standard Borel in $GL_k$. Then $l_{M_H}$ is $M_H$-invariant by [Ber84] and [GK75]. Thus the linear form

$$l_H(\varphi) = \int_{P_H\backslash H} l_{M_H}(\varphi(h))\, dh = \int_{K_H} l_{M_H}(\varphi(k))\, dk$$

on $I(\pi \otimes \pi, \varpi_n)$ is $H$-invariant. Define the linear form

$$l_{M_H,\mathfrak{E}}(W_1 \otimes W_2) = \int_{N_n\backslash GL_n} W_1(g^*)W_2(g)\Phi_0((0,\ldots,0,1)g)|\det(g)|^2\, dg.$$

The claim is that the linear form

$$\int_{P_H\backslash H} l_{M_H,\mathfrak{E}}[(M(\varpi_n)\varphi)(h)]\, dh = \int_{K_H} l_{M_H,\mathfrak{E}}[(M(\varpi_n)\varphi)(k)]\, dk$$

is $H$-invariant and proportional to $l_H$ where $M(\bullet)$ is the local intertwining operator. In fact, if $\pi$ is a local component of a cuspidal automorphic representation of $GL_n$ this follows from our global result.

**4. Identity of distributions.** Theorem 1 and the alternative formula can be used to give an identity of Bessel distributions. As mentioned in the introduction, this identity is the cuspidal part of the spectral side of a hypothetical identity

$$\int_{H\backslash H(\mathbb{A})}\int_{H\backslash H(\mathbb{A})} K_f(h_1,h_2)\, dh_1\, dh_2 = \int_{G'\backslash G'(\mathbb{A})^1} K_{f'}(x,x)\mathfrak{E}(x^*,3)\, dx$$

(suitably regularized to overcome convergence issues). We emphasize again that the main point of the paper is to avoid the trace formula approach.

**4.1. Bessel distributions and weighted traces.** Let $F$ be a non-Archimedean field and $(\pi, V)$ an admissible representation of $G(F)$. Let $V'$ be the dual of $V$ and let $(\pi^\vee, V^\vee)$ be the dual representation on the smooth part of $V'$. Then $(\pi^\vee, V^\vee)$ is admissible and for any $f \in C_c^\infty(G)$ and $l \in V'$ we have $l \circ \pi(f) \in V^\vee$. For any form $l$ of $V$ and $l'$ of $V^\vee$ we let

$$\mathcal{B}^\pi_{l,l'}(f) = l'\,[l \circ \pi(f)]$$

for $f \in C_c^\infty(G)$. It is called the Bessel distribution of $\pi$ with respect to $l, l'$.

Let $\mathrm{End}(V)$ be the smooth part (with respect to $G \times G$) of the space of linear maps from $V$ to $V$. Then $\mathrm{End}(V) \simeq (V \otimes V^\vee)^\vee \simeq V^\vee \otimes V$ as a representation of $G \times G$. We can view any bilinear form $\mathcal{L}$ on $V^\vee \times V$ as a linear form on $\mathrm{End}(V)$. The weighted trace of $\pi$ with respect to $\mathcal{L}$ is defined by

$$\mathcal{T}^\pi_{\mathcal{L}}(f) = \mathcal{L}(\pi(f))$$

for $f \in C_c^\infty(G)$. This makes sense since $\pi(f) \in \mathrm{End}(V)$. The usual trace is the weighted trace with respect to the standard pairing.

The relation between the weighted trace and Bessel distributions is the following. If $l$ and $l'$ are forms on $V$ and $V^\vee$ respectively then we may consider the linear form $l' \otimes l$ on $V^\vee \otimes V$ as a bilinear pairing on $V^\vee \times V$. Clearly

$$\mathcal{B}^{\pi}_{l,l'}(f) = \mathcal{T}^{\pi}_{l' \otimes l}(f). \tag{13}$$

On the other hand, consider the representation $\pi \otimes \pi^\vee$ on $V \otimes V^\vee$. As before any bilinear pairing $\mathcal{L}$ on $V^\vee \times V$ defines a linear form on $V^\vee \otimes V = (V \otimes V^\vee)^\vee$. Let $l$ be the standard pairing on $V \otimes V^\vee$. Then

$$\mathcal{B}^{\pi\otimes\pi^\vee}_{l,\mathcal{L}}(f_1 \otimes f_2) = \mathcal{T}^{\pi}_{\mathcal{L}}(f_2^\vee \star f_1) \tag{14}$$

where $f_2^\vee(g) = f_2(g^{-1})$. More generally, if $f$ is a function on $G \times G$ then

$$\mathcal{B}^{\pi\otimes\pi^\vee}_{l,\mathcal{L}}(f) = \mathcal{T}^{\pi}_{\mathcal{L}}(f') \tag{15}$$

where $f'$ is the function on $G$ defined by

$$f'(g') = \int_G f(xg', x)\, dx.$$

Let $\overline{V}$ be the vector space obtained by conjugating the scalar multiplication on $V$ (but keeping the set, the addition and the group action the same as $V$). The resulting representation of $G$ on $\overline{V}$ will be denoted by $\overline{\pi}$. We let $\sigma$ be the anti-linear isomorphism between $V$ and $\overline{V}$. As a mapping of sets $\sigma$ is the identity. We will also denote by $\sigma$ the map $\overline{V} \to V$.

Consider the case where $\pi$ is unitary and let $(\bullet, \bullet)$ be an invariant positive definite Hermitian form on $V$. The map

$$v \mapsto (\bullet, v)$$

defines an equivalence between $\overline{\pi}$ and $\pi^\vee$. Thus for linear forms $l, l'$ of $V$ we may consider $\mathcal{B}^{\pi}_{l,\overline{l'}}(f)$ where $\overline{l'}$ is the linear form on $\overline{V} \simeq V^\vee$ defined by $\overline{l'}(\sigma v) = \overline{l'(v)}$ for all $v \in V$. Then $\mathcal{B}^{\pi}_{l,\overline{l'}}(f) = \overline{l'(\sigma\,[l \circ \pi(f)])}$. We also have

$$\mathcal{B}^{\pi}_{l,\overline{l'}}(f) = \sum_{e_i} l(\pi(f)e_i)\overline{l'(e_i)}$$

where $e_i$ ranges over an orthonormal basis in $V$. Similarly,

$$\mathcal{T}^{\pi}_{\mathcal{L}}(f) = \sum_{e_i} \mathcal{L}(\sigma(e_i), \pi(f)e_i).$$

In the Archimedean case, let $\pi$ be a continuous representation on a Hilbert space $\mathfrak{H}$ (not necessarily unitary) and let $V$ be the space of smooth vectors in $\mathfrak{H}$ equipped with the Frechet topology determined by the semi-norms $v \mapsto \|\pi(X)v\|$ where $X \in \mathcal{U}(\mathfrak{g})$. Recall that $\pi(f)(\mathfrak{H}) \subset V$ for $f \in C_c^\infty(G)$. Let $\mathfrak{H}^\vee$ be the dual Hilbert space. Let $\pi^\vee$ be the dual representation on $\mathfrak{H}^\vee$ defined by $\pi^\vee(g)u(v) = u(\pi(g^{-1}v))$. Then $\pi^\vee$ is also continuous ([Wal88], §1.1.4) and $\mathfrak{H}^\vee \subset V'$ where $V'$ is the topological dual of $V$. We will still denote by $\pi^\vee$ the representation of $G$ on $V'$. Let $V^\vee$ be the space of smooth vectors in $\mathfrak{H}^\vee$. We claim that if $f \in C_c^\infty(G)$ then $\pi^\vee(f)V' \subset V^\vee$. Indeed, if $l \in V'$ then there exist $v_i \in \mathfrak{H}^\vee$, $X_i \in \mathcal{U}(\mathfrak{g})$ $i = 1, \ldots, m$ such that $l(v) = \sum_i v_i(\pi(X_i)v)$. Then $\pi^\vee(f)l = \sum \pi^\vee((X_i f^\vee)^\vee)v_i \in V^\vee$. The Bessel distribution $\mathcal{B}_{l,l'}(f)$ is defined as before for any $l \in V'$, $l' \in (V^\vee)'$ and $f \in C_c^\infty(G)$.

It is known (§11.6.7 in [Wal92]) that the representation $V$ is determined by the underlying $(\mathfrak{g}, K)$-module $V_K$ of $K$-finite vectors in $\mathfrak{H}$, provided that the latter is admissible and finitely generated. In that case, $V_K^\vee$, the space of $K$-finite vectors in $\mathfrak{H}^\vee$, is equal to the space of $K$-finite vectors in the algebraic dual of $V_K$, and is also finitely generated and admissible. In particular, $V^\vee$ is also determined by $V_K$.

If $\pi$ is unitary and (topologically) irreducible, then $V_K$ is admissible and irreducible. Let $\sigma : \mathfrak{H} \to \mathfrak{H}^\vee$ be the anti-linear isomorphism. Then $\sigma$ is an anti-linear intertwining operator which takes $V$ to $V^\vee$ and $V_K$ to $V_K^\vee$.

In the global case we start with a continuous representation $\pi$ on a Hilbert space $\mathfrak{H}$ and let $V = \cup V_{K_0}$ where $K_0$ range over all compact open subgroups of $G(\mathbb{A}_f)$ and

$$V_{K_0} = \{v \in \mathfrak{H} : v \text{ is fixed under } K_0 \text{ and } g_\infty \mapsto \pi(g_\infty)v \in C^\infty(G_\infty)\}.$$

Each $V_{K_0}$ is a Frechet space and we let $V' = \varprojlim V'_{K_0}$. Then as before, $\pi(f)\mathfrak{H} \subset V$ and $\pi^\vee(f)(V') \subset V^\vee$ for any $f \in C_c^\infty(G(\mathbb{A}))$. We can define the Bessel distribution as before. If $\pi$ is the restricted infinite tensor product of $\pi_v$ and (in the appropriate sense) $l = \otimes l_v$ and $l' = \otimes l'_v$ then

$$\mathcal{B}^\pi_{l,l'}(f) = \prod_v \mathcal{B}^{\pi_v}_{l_v,l'_v}(f_v)$$

for $f = \otimes f_v$. Similarly for weighted traces.

**4.2. The Comparison.** Let $\pi$ and $\Pi$ be as in §2.4 and view $\pi \otimes \pi$ as a representation of $M(\mathbb{A})$. We use the inner products on $G'\backslash G'(\mathbb{A})^1$, $M\backslash M(\mathbb{A})^1$, and $G\backslash G(\mathbb{A})^1$ respectively to identify $\overline{\pi}$, $\overline{\pi \otimes \pi}$ and $\overline{\Pi}$ with $\pi^\vee$, $(\pi \otimes \pi)^\vee$ and $\Pi^\vee$ respectively.

Let $\ell_H$ be the linear form on $\Pi$ defined by

$$\ell_H(\varphi) = \int_{H\backslash H(\mathbb{A})} \varphi(h)\, dh$$

and $\ell_{M_H}$ be the period over $M_H\backslash M_H(\mathbb{A})^1$ as a linear form on $\pi \otimes \pi$. Let also $\ell_{M_H,\mathfrak{E}}$ be the linear form on $\pi \otimes \pi$ defined by

$$\ell_{M_H,\mathfrak{E}}(\varphi) = \int_{M_H\backslash M_H(\mathbb{A})^1} \varphi(m)\mathfrak{E}(m, 3)\, dm$$

and let $\mathfrak{B}$ be the bilinear form on $\pi^\vee \times \pi \simeq \overline{\pi} \times \pi$ defined by

$$\mathfrak{B}(\phi, \phi') = \int_{G'\backslash G'(\mathbb{A})^1} \overline{\phi(x)}\phi'(x)\mathfrak{E}(x^*, 3)\, dx.$$

For any function $f'$ on $M(\mathbb{A})$ we define the function $\mathfrak{C}(f')$ on $G'(\mathbb{A})$ by

$$\mathfrak{C}(f')(g') = \int_{G'(\mathbb{A})} f'(xg', x^*)\, dx \quad g' \in G'(\mathbb{A}),$$

where we recall that $x^* = w_n {}^t x w_n$. We now state the relative trace – weighted trace identity between $\pi$ and $\Pi$.

THEOREM 3. *We have*

$$\frac{\lambda_{-1}}{n} \cdot \mathcal{B}^{\Pi}_{\ell_H, \bar{\ell}_H}(f) = \mathcal{B}^{\pi\otimes\pi}_{\ell_{M_H}, \bar{\ell}_{M_H, \mathfrak{C}}}(f'_{\mathbf{K}_H}) = \mathcal{T}^{\pi}_{\mathfrak{B}}(\mathfrak{C}(f'_{\mathbf{K}_H})) \tag{16}$$

*where $f'_{\mathbf{K}_H}$ is the function on $M(\mathbb{A})$ defined by*

$$f'_{\mathbf{K}_H}(m) = e^{\langle \varpi_n + \rho_P, H_M(m)\rangle} \cdot \int_{\mathbf{K}_H} \int_{\mathbf{K}_H} \int_{U(\mathbb{A})} f(k'muk)\, du\, dk'\, dk.$$

Note that $f'_{\mathbf{K}_H}$ is essentially the Harish-Chandra map from $G$ to $M$. We also have a "generic version" of this identity which is the spectral counterpart of a trace formula considered in [JR92a]. In fact, this will not require Theorem 1. Both versions use the formula (5). Recall that $\psi'$ is a generic character of $U'_0$ and $\psi$ is defined by (8). The $\psi$-th Fourier coefficient is denoted by $\mathcal{W}^{\psi}$ and similarly for $\psi'$.

THEOREM 4. *We have*

$$\mathcal{B}^{\Pi}_{\ell_H, \overline{\mathcal{W}^{\psi}}}(f) = \mathcal{B}^{\pi\otimes\pi}_{\ell_{M_H}, \overline{\mathcal{W}^{\psi'\otimes\psi'}}}(f') = \mathcal{B}^{\pi}_{\mathcal{W}^{\psi'}, \overline{\mathcal{W}^{\psi'}}}(\mathfrak{C}(f')) \tag{17}$$

*where $f'$ is the function on $M(\mathbb{A})$ defined by*

$$f'(m) = e^{\langle \varpi_n + \rho_P, H_M(m)\rangle} \cdot \int_{\mathbf{K}_H} \int_{U(\mathbb{A})} f(k'mu)\, du\, dk'.$$

The map $\varphi \mapsto E_{-1}(\bullet, \varphi)$ defines an intertwining map

$$E_{-1} : I(\pi \otimes \pi, \varpi_n) \to \Pi.$$

In principle, $E_{-1}$ is defined only for $K$-finite vectors. However, by a Theorem of Wallach [Wal92] any smooth vector can be written as $\pi(f)v$ with $v$ $K$-finite and $f \in S(G)$. Hence, $E_{-1}$ is defined for smooth $\varphi$ as well and moreover

(18) $E_{-1}$ viewed as a map between the smooth parts is onto.

We may identify $I(\pi \otimes \pi, \varpi_n)^\vee$ with $I((\pi \otimes \pi)^\vee, -\varpi_n)$ by integrating over $\mathbf{K}$. We get a dual map

$$E^\vee_{-1} : \overline{\Pi} \to I((\pi \otimes \pi)^\vee, -\varpi_n).$$

We also have an intertwining map

$$M_{-1} : I(\pi \otimes \pi, \varpi_n) \to I(\pi \otimes \pi, -\varpi_n).$$

The inner product formula (4) for residual Eisenstein series can be interpreted as the identity

$$E_{-1}^{\vee} \circ \sigma \circ E_{-1} = \sigma \circ M_{-1}$$

where on the left hand side $\sigma$ is the anti-linear isomorphism from $\Pi$ to $\overline{\Pi}$ and on the right hand side $\sigma$ denotes the anti-linear isomorphism from $I(\pi \otimes \pi, -\varpi_n)$ to $I(\overline{\pi \otimes \pi}, -\varpi_n)$.

Let us first prove the second equality in (16). Consider the isomorphism $\theta$ of $M$ defined by $\theta(x, y) = (x, y^*)$. We have

$$\mathcal{B}^{\pi\otimes\pi}_{\ell_{M_H}, \overline{\ell}_{M_H,\mathfrak{e}}}(f'_{\mathbf{K}_H}) = \mathcal{B}^{\pi\otimes\pi^*}_{\ell_{M_H}\circ\theta, \overline{\ell}_{M_H,\mathfrak{e}}\circ\theta}(\theta(f'_{\mathbf{K}_H}))$$

where $\pi^*$ is the (regular) representation on the space of functions $\varphi(\bullet^*)$ where $\varphi \in V_\pi$. The form $\ell_{M_H} \circ \theta$ is a non-degenerate invariant pairing on $\pi \times \pi^*$. Hence, we may use it to identify $\pi^*$ with $\pi^{\vee}$. The derived isomorphism $\pi^* \simeq \overline{\pi}$ is $\varphi \mapsto \overline{\varphi}$. The form $\overline{\ell}_{M_H,\mathfrak{e}} \circ \theta$ viewed as a form on $\overline{\pi} \otimes \pi$ becomes the bilinear map $\mathfrak{B}$. We can now invoke (15).

Similarly, the form $\overline{\mathcal{W}^{\psi'\otimes\psi'}} \circ \theta = \overline{\mathcal{W}^{\psi'}} \otimes (\overline{\mathcal{W}^{\psi'}})^*$ considered as a form on $\overline{\pi} \otimes \pi$ becomes $\overline{\mathcal{W}^{\psi'}} \otimes \mathcal{W}^{\psi'}$. Thus, by (15) and (13)

$$\begin{aligned}\mathcal{B}^{\pi\otimes\pi}_{\ell_{M_H}, \overline{\mathcal{W}^{\psi'\otimes\psi'}}}(f') &= \mathcal{B}^{\pi\otimes\pi^*}_{\ell_{M_H}\circ\theta, \overline{\mathcal{W}^{\psi'\otimes\psi'}}\circ\theta}(\theta(f')) \\ &= \mathcal{T}^{\pi}_{\overline{\mathcal{W}^{\psi'}}\otimes\mathcal{W}^{\psi'}}(\mathfrak{C}(f')) = \mathcal{B}^{\pi}_{\mathcal{W}^{\psi'}, \overline{\mathcal{W}^{\psi'}}}(\mathfrak{C}(f'))\end{aligned}$$

and the second equality of (17) follows.

We now turn to the first equalities in (16) and (17). Theorem 1 states the identity

$$\frac{\lambda_{-1}}{n} \cdot \ell_H \circ E_{-1} = \beta \circ M_{-1}$$

on $I(\pi \otimes \pi, \varpi_n)$ where

$$\beta(\varphi) = \int_{\mathbf{K}_H} \ell_{M_H,\mathfrak{e}}(\varphi(k))\, dk$$

where $\varphi$ takes values in $\pi \otimes \pi$. Note also that

$$\mathcal{W}^{\psi}(E_{-1}\varphi) = \mathcal{W}^{\psi'\otimes\psi'}\left[M_{-1}\varphi(e)\right].$$

The first equalities in (16) and in (17) will follow from the following more general statement.

THEOREM 5. *Suppose that $\gamma_1 \in \Pi'$ satisfies*

$$\gamma_1(E_{-1}v) = \beta_1((M_{-1}v)(e)) \quad v \in I(\pi \otimes \pi, \varpi_n)$$

*for some $\beta_1 \in (\pi \otimes \pi)'$. Then with the notation of Theorem 4*

$$\mathcal{B}^{\Pi}_{\ell_H, \overline{\gamma_1}}(f) = \mathcal{B}^{\pi\otimes\pi}_{\ell_{M_H}, \overline{\beta_1}}(f').$$

*Similarly, if* $\gamma_2 \in \Pi'$ *satisfies*

$$\gamma_2(E_{-1}v) = \beta_2(M_{-1}v)$$

*where* $\beta_2 \in I(\pi \otimes \pi, -\varpi_n)'$ *is given by*

$$\beta_2(\varphi) = \int_{\mathbf{K}_H} \alpha_2(\varphi(k))\, dk \tag{19}$$

*with* $\alpha_2 \in (\pi \otimes \pi)'$ *then*

$$\mathcal{B}^{\Pi}_{\ell_H,\overline{\gamma_2}}(f) = \mathcal{B}^{\pi\otimes\pi}_{\ell_{M_H},\overline{\alpha_2}}(f'_{\mathbf{K}_H}).$$

To prove the theorem we first need the following lemma which will be proved below.

LEMMA 2. *The linear form*

$$\varphi \mapsto \ell_H(\Pi(f)E_{-1}(\bullet, \varphi)) \tag{20}$$

*on* $I(\pi \otimes \pi, \varpi_n)$ *is given by* $\Psi \in I((\pi \otimes \pi)^\vee, -\varpi_n)$ *where*

$$\Psi(g) = \ell_{M_H} \circ \pi \otimes \pi((R_g f)') \tag{21}$$

*where* $R_g f(\bullet) = f(\bullet g)$.

Let us prove the second part of Theorem 5. The first part is very similar but easier. The lemma shows that the form $\psi = \ell_H \circ \Pi(f)$ on $\Pi$ satisfies $E^\vee_{-1}\psi = \Psi$. By (18) we can write $\sigma\psi = E_{-1}v$ with $v \in I(\pi \otimes \pi, \varpi_n)$. Then

$$\begin{aligned}\mathcal{B}^{\Pi}_{\ell_H,\gamma_2}(f) &= \overline{\gamma_2(\sigma\,[\ell_H \circ \Pi(f)])} = \overline{\gamma_2(\sigma\psi)} = \overline{\gamma_2(E_{-1}v)} = \overline{\beta_2(M_{-1}v)} \\ &= \overline{\beta_2(\sigma\sigma M_{-1}v)} = \overline{\beta_2(\sigma E^\vee_{-1} \circ \sigma \circ E_{-1}v)} = \overline{\beta_2(\sigma E^\vee_{-1}\psi)} = \overline{\beta_2(\sigma\Psi)}.\end{aligned}$$

By (19) this is equal to

$$\overline{\int_{\mathbf{K}_H} \alpha_2(\sigma(\Psi(k)))\, dk} = \int_{\mathbf{K}_H} \overline{\alpha_2}(\Psi(k))\, dk = \overline{\alpha_2}\left[\int_{\mathbf{K}_H} \Psi(k)\, dk\right].$$

Using (21) we obtain

$$\overline{\alpha_2}\left[\int_{\mathbf{K}_H} \ell_{M_H} \circ \pi \otimes \pi((R_k f)')\, dk\right] = \overline{\alpha_2}[\ell_{M_H} \circ \pi \otimes \pi(f'_{\mathbf{K}_H})]$$

which is equal to $\mathcal{B}^{\pi\otimes\pi}_{\ell_{M_H},\overline{\alpha_2}}(f'_{\mathbf{K}_H})$ as required.

*Proof of Lemma 2.* First, one easily checks that $\Psi$ lies in $I((\pi \otimes \pi)^\vee, -\varpi_n)$. Using (5) we write (20) as

$$\int_{\mathbf{K}_H} \int_{M_H \backslash M_H(\mathbb{A})^1} I(f, \pi \otimes \pi, \varpi_n)\varphi(lk')\, dl\, dk'.$$

Note that

$$I(f, \pi \otimes \pi, \varpi_n)\varphi(g) = \int_{G(\mathbb{A})} f(x)\varphi(gx)\,dx = \int_{G(\mathbb{A})} f(g^{-1}x)\varphi(x)\,dx,$$

so that we get

$$\int_{\mathbf{K}_H} \int_{M_H\backslash M_H(\mathbb{A})^1} \int_{G(\mathbb{A})} f(k'l^{-1}x)\varphi(x)\,dx\,dl\,dk'.$$

Next, we use the Iwasawa decomposition $x = muk$ with $u \in U(\mathbb{A})$, $m \in M(\mathbb{A})$ and $k \in \mathbf{K}$ to obtain

$$\begin{aligned}&\int_{\mathbf{K}_H} \int_{M_H\backslash M_H(\mathbb{A})^1} \int_{\mathbf{K}} \int_{U(\mathbb{A})} \int_{M(\mathbb{A})} f(k'l^{-1}muk)\varphi(muk)\,dm\,du\,dk\,dl\,dk' \\ &= \int_{\mathbf{K}_H} \int_{M_H\backslash M_H(\mathbb{A})^1} \int_{\mathbf{K}} \int_{U(\mathbb{A})} \int_{M(\mathbb{A})} f(k'muk)\varphi(lmuk)\,dm\,du\,dk\,dl\,dk' \\ &= \int_{\mathbf{K}_H} \int_{M_H\backslash M_H(\mathbb{A})^1} \int_{\mathbf{K}} \int_{U(\mathbb{A})} \int_{M(\mathbb{A})} f(k'muk)\varphi(lmk)\,dm\,du\,dk\,dl\,dk'.\end{aligned}$$

Viewing $\varphi$ as an element $I(\pi \otimes \pi, \varpi_n)$ we get

$$\begin{aligned}\int_{\mathbf{K}_H} \int_{\mathbf{K}} \int_{U(\mathbb{A})} \int_{M(\mathbb{A})} f(k'muk)\ell_{M_H}(\pi \otimes \pi(m)\varphi(k))e^{\langle \varpi_n + \rho_P, H_M(m)\rangle} \\ dm\,du\,dk\,dk' = \int_{\mathbf{K}} \ell_{M_H} \circ (\pi \otimes \pi)((R_k f)')(\varphi(k))\,dk = \Psi(\varphi)\end{aligned}$$

as required. □

**5. A Distributional formula.** We now return to the proof of Theorem 1. The main technical tool will be *pseudo-Eisenstein series* which we proceed to define in this context.

**5.1. Pseudo-Eisenstein series.** Let $\mathcal{P}(\mathbb{C})$ be the space of holomorphic functions of Paley-Wiener type. For each $\sigma \in \mathcal{P}(\mathbb{C})$ let

$$\mathcal{C}_\sigma(X) = \int_{i\mathbb{R}} \sigma(s)e^{(s+1)X}\,ds.$$

(The contour of integration can be shifted to any $\mathrm{Re}(s) = s_0$.) It is a compactly supported smooth function on $\mathbb{R}$. We recover $\sigma$ from $\mathcal{C}_\sigma$ by

$$\sigma(s) = \frac{1}{2\pi i} \int_{\mathbb{R}} \mathcal{C}_\sigma(X)e^{-(s+1)X}\,dX.$$

For any $\Phi \in \mathcal{S}(\mathbb{V}(\mathbb{A}))$ the function

$$\mathcal{F}_{\Phi,\sigma}(g) = \int_{\mathbb{I}_F} \Phi(v_0 t g)\mathcal{C}_\sigma(\log(|t|^n |\det g|^{\frac{1}{2}}))\,d^*t$$

is a function on $G(\mathbb{A})$ which is left invariant under $P_{(2n-1,1)}(\mathbb{A})^1$. Its restriction to $H(\mathbb{A})$ is left invariant under $Q(\mathbb{A})^1$. Let $\theta_{\Phi,\sigma}(\bullet)$ be the pseudo-Eisenstein series on $G(\mathbb{A})$ defined by

$$\theta_{\Phi,\sigma}(g) = \sum_{\gamma \in P_{(2n-1,1)} \backslash G} \mathcal{F}_{\Phi,\sigma}(\gamma g).$$

The sum is absolutely convergent. Indeed, we may write this as

$$\sum_{F^* \backslash (\mathbb{V} - \{0\})} \int_{\mathbb{I}_F} \Phi(vtg) \mathcal{C}_\sigma(\log(|t|^n |\det g|^{\frac{1}{2}}))\, d^*t$$

$$= \int_{F^* \backslash \mathbb{I}_F} \sum_{\mathbb{V} - \{0\}} \Phi(vtg) \mathcal{C}_\sigma(\log(|t|^n |\det g|^{\frac{1}{2}}))\, d^*t.$$

The convergence follows since $\Phi \in \mathcal{S}(\mathbb{V}(\mathbb{A}))$ and $\mathcal{C}_\sigma(\log(|\bullet|))$ is compactly supported on $F^* \backslash \mathbb{I}_F$. Clearly $\theta_{\Phi,\sigma}(\bullet)$ is a function on $G \backslash G(\mathbb{A})$. Roughly speaking, these functions approximate the constant function on $G(\mathbb{A})$ as $\sigma$ approaches the delta function at 1. As before, we may sum $\gamma$ over $Q \backslash H$ and the restriction of $\theta_{\Phi,\sigma}$ to $H(\mathbb{A})$ is a pseudo-Eisenstein series on $H \backslash H(\mathbb{A})$. We have

$$\theta_{\Phi,\sigma}(h) = \int_{\mathrm{Re}(s) = s_0} \sigma(s) \mathcal{E}_\Phi(h, s)\, ds$$

whenever $s_0 > 1$. These functions are not rapidly decreasing because we started with the identity representation on $L(\mathbb{A})$.

Let us analyze the distribution

$$\mathfrak{P}_\Phi(\sigma) = \int_{H \backslash H(\mathbb{A})} E_{-1}(h, \varphi) \theta_{\Phi,\sigma}(h)\, dh.$$

The auxiliary function $\Phi$ will be fixed and will often be suppressed from the notation. We will prove:

THEOREM 6. *The integral defining $\mathfrak{P}_\Phi(\sigma)$ is absolutely convergent and*

$$\mathfrak{P}_\Phi(\sigma) = 2\pi i \sigma(1) \cdot \int_{K_H} \int_{M_H \backslash M_H(\mathbb{A})^1} M_{-1}\varphi(mk)\, \mathbb{E}_\Phi(mk, 3)\, dm\, dk. \tag{22}$$

Recall that Theorem 1 relates the right hand side of (22) to the period of $E_{-1}$ over $H \backslash H(\mathbb{A})$. For the rest of the section we will show that Theorem 6 implies Theorem 1. We will prove Theorem 6 in §6 and §7.

We write $\mathfrak{P}(\sigma)$ as

$$\mathfrak{P}(\sigma) = \int_{H \backslash H(\mathbb{A})} E_{-1}(h, \varphi) \left( \int_{\mathrm{Re}(s) = s_0} \sigma(s) \mathcal{E}(h, s)\, ds \right) dh$$

for any $s_0 > 1$. The integral converges as an iterated integral. Fix $s_1$ real with $|s_1| < 1$. Using (11) we may shift the contour of integration to get

$$\int_{\mathrm{Re}(s)=s_0} \sigma(s)\mathcal{E}(h,s)\,ds = \frac{2\pi i\lambda_{-1}}{n}\sigma(1)\hat{\Phi}(0) + \int_{\mathrm{Re}(s)=s_1} \sigma(s)\mathcal{E}(h,s)\,ds.$$

We will need the following proposition which will be proved below.

PROPOSITION 1. *The integral*

$$\int_{H\backslash H(\mathbb{A})} E_{-1}(h,\varphi)\mathcal{E}(h,s)\,dh \tag{23}$$

*converges absolutely for any $s$ such that $|\mathrm{Re}(s)| < 1$. Moreover, there exists a constant $c$ such that*

$$\int_{H\backslash H(\mathbb{A})} |E_{-1}(h,\varphi)\mathcal{E}(h,s)|\,dh < c$$

*for all $s$ with $\mathrm{Re}(s) = s_1 < 1$.*

Since

$$\int_{H\backslash H(\mathbb{A})} E_{-1}(h,\varphi)\,dh$$

converges absolutely ([JR92b]), it will follow by changing the order of integration that

$$\mathfrak{P}_\Phi(\sigma) = \frac{2\pi i\lambda_{-1}}{n}\sigma(1)\cdot\hat{\Phi}(0)\cdot\int_{H\backslash H(\mathbb{A})} E_{-1}(h,\varphi)\,dh + \int_{\mathrm{Re}\ s=s_1} \sigma(s)\Psi_\Phi(s)\,ds$$

where

$$\Psi_\Phi(s) = \int_{H\backslash H(\mathbb{A})} E_{-1}(h,\varphi)\mathcal{E}_\Phi(h,s)\,dh.$$

We compare this with the expression given in Theorem 6 and invoke the following lemma to complete the proof of Theorem 1.

LEMMA 3. *Let $D$ be a linear form on $\mathcal{P}(\mathbb{C})$ which is given by*

$$D(f) = \alpha\cdot f(1) + \int_{Re\,s=0} f(s)g(s)\,ds$$

*where $g$ is a bounded function. Suppose that $D(f) = 0$ for all $f$. Then $\alpha = 0$ and $g(s) \equiv 0$.*

*Proof.* Since any polynomial is a multiplier of $\mathcal{P}(\mathbb{C})$ the linear form $D_1(f) = D((s-1)f)$ is well defined. It is given by integration against the slowly increasing function $g_1(s) = (s-1)g(s)$ on the unitary axis. The Fourier transform of $g_1(i\bullet)$, which is a tempered distribution, vanishes on all compactly supported smooth functions, and hence it is zero. Thus $g_1 \equiv 0$, which in turn implies that $g \equiv 0$, and $\alpha = 0$. □

*Remark.* It will also follow that

$$\int_{H\backslash H(\mathbb{A})} E_{-1}(h,\varphi)\mathcal{E}_\Phi(h,s)\,dh = 0$$

whenever the integral converges, namely for $|\mathrm{Re}(s)| < 1$.

We finally prove Proposition 1. We first prove that there exists $\Phi_1 \in \mathcal{S}(\mathbb{V}(\mathbb{A}))$ such that $|\Phi| \le \Phi_1$ and $|\hat{\Phi}| \le \hat{\Phi}_1$.

Indeed, by the Dixmier-Malliavin Theorem ([DM78]), we may assume that $\Phi = \Phi_2 \star \Phi_3$ with $\Phi_2, \Phi_3 \in \mathcal{S}(V(\mathbb{A}))$. (In fact, we may even assume that $\Phi_3 \in C_c^\infty(V(\mathbb{A}))$.) We can find $\Phi_4 \in \mathcal{S}(\mathbb{V}(\mathbb{A}))$ such that $|\Phi_2| \le \Phi_4$ and $|\Phi_3| \le \Phi_4^\vee$ where $\Phi_4^\vee(x) = \Phi_4(-x)$. Let $\Phi_5 = \Phi_4 \star \Phi_4^\vee$. Then $|\Phi| \le \Phi_5$ and $\hat{\Phi}_5 = |\hat{\Phi}_4|^2 \ge 0$. Similarly, there exists $\Phi_6 \in \mathcal{S}(\mathbb{V}(\mathbb{A}))$ such that $|\hat{\Phi}| \le \Phi_6$ and $\hat{\Phi}_6 \ge 0$. Let $\Phi_1 = \Phi_5 + \hat{\Phi}_6^\vee$. Then $|\Phi| \le \Phi_1$ and $|\hat{\Phi}| \le \Phi_6 \le \hat{\Phi}_1$ as required.

It follows from (11) that

$$|\mathcal{E}_\Phi(h,s)| \le |\mathcal{E}_{\Phi_1}(h,\mathrm{Re}(s))| + c$$

where the constant $c$ depends only on $\mathrm{Re}(s)$. Thus the second statement of the proposition would follow from the first.

We will use Lemma I.4.1 of [MW95] to bound the automorphic forms at hand by their cuspidal exponents. The only cuspidal exponent of $E_{-1}$ is $-\varpi_n$ along $P$. By that lemma there exists an integer $N_1 \ge 0$ and for any $\mu \in (\mathfrak{a}_0^M)^*$ a constant $c_1$ such that

$$|E_{-1}(g,\varphi)| \le c_1 \cdot e^{\langle -\varpi_n + \mu + \rho_P, H_0(g)\rangle}(1 + \|H_0(g)\|)^{N_1}$$

for all $g \in \mathcal{S}$ where $\|\bullet\|$ is a norm on $\mathfrak{a}_0$. This in particular applies to $g \in \mathcal{S}^H$ and $\mu \in (\mathfrak{a}_0^{M_H})^*$. Similarly, $\mathcal{E}(h,s_1)$ is bounded by

$$c_2 \cdot \sum e^{\langle \mu_i + \rho_0^H, H_0(h)\rangle}(1 + \|H_0(h)\|)^{N_2}$$

on $\mathcal{S}^H$ for some $c_2, N_2 > 0$, where $\mu_i$ are the exponents of $\mathcal{E}(h,s_1)$ along the Borel subgroup of $H$. Let $f(h)$ be the integrand of (23). By the integration formula on $H\backslash H(\mathbb{A})$ we have to bound

$$\int_{\mathbf{K}_H}\int_{A_0^H(c_0)}\int_{\omega_H} e^{-\langle 2\rho_0^H, H_0(a)\rangle}|f(tak)|\,dt\,da\,dk.$$

Using the bounds above, it remains to show that for every exponent $\nu$ of $\mathcal{E}(\bullet,s_1)$, $\nu - \rho_0^H - \varpi_n + \rho_P + \mu$ lies in the obtuse Weyl chamber of $(\mathfrak{a}_0^H)^*$ for some $\mu \in (\mathfrak{a}_0^{M_H})^*$. If $\lambda \in \mathfrak{a}_{M_H}^*$ we write $\lambda > 0$ if $\lambda$ is a positive multiple on $\varpi_n$. Projecting to $\mathfrak{a}_{M_H}^*$ it thus remains to show that

$$(-1+n)\varpi_n + \nu_{M_H} - \rho_{P_H} < 0$$

i.e., that

$$\nu_{M_H} < -\frac{n-3}{2}\varpi_n.$$

The possible $\nu_{M_H}$'s are the exponents of the constant term of $\mathcal{E}(h, s_1)$ along the Siegel parabolic. By Lemma 1 these are $(\pm s_1 - (n-1)/2)\varpi_n$. Hence,

$$\mu_M < (1 - (n-1)/2)\varpi_n = -\frac{n-3}{2}\varpi_n$$

as required.

**6. Convergence.** We now begin the proof of Theorem 6. We can rewrite $\mathfrak{P}(\sigma)$ as

$$\int_{Q\backslash H(\mathbb{A})} E_{-1}(h, \varphi)\mathcal{F}_\sigma(h)\, dh.$$

We utilize the expansion

$$E_{-1}(g, \varphi) = \sum_{P_{(n,n-1,1)}\backslash P_{(2n-1,1)}} M_{-1}\varphi(\gamma g) \quad g \in G(\mathbb{A})$$

where the series is absolutely convergent ([MW89], Lemma 1 in appendix). Note that $Q \subset P_{(2n-1,1)}$. Thus, $\mathfrak{P}(\sigma)$ is equal, at least formally, to

$$\sum_{\eta \in P_{(n,n-1,1)}\backslash P_{(2n-1,1)}/Q} \int_{Q_\eta\backslash H(\mathbb{A})} M_{-1}\varphi(\eta h)\mathcal{F}_\sigma(h)\, dh \tag{24}$$

where $Q_\eta = Q \cap \eta^{-1} P_{(n,n-1,1)}\eta$. To justify this we will prove the convergence of

$$\int_{Q\backslash H(\mathbb{A})} \left( \sum_{P_{(n,n-1,1)}\backslash P_{(2n-1,1)}} |M_{-1}\varphi(\gamma h)| \right) \cdot |\mathcal{F}_\sigma(h)|\, dh$$

which is the same as

$$\int_{H\backslash H(\mathbb{A})} \sum_{\delta \in Q\backslash H} \left( \sum_{\gamma \in P_{(n,n-1,1)}\backslash P_{(2n-1,1)}} |M_{-1}\varphi(\gamma\delta h)| \right) \cdot |\mathcal{F}_\sigma(\delta h)|\, dh$$

or also

$$\int_{H\backslash H(\mathbb{A})} \sum_{\delta \in P_{(2n-1,1)}\backslash G} \left( \sum_{\gamma \in P_{(n,n-1,1)}\backslash P_{(2n-1,1)}} |M_{-1}\varphi(\gamma\delta h)| \right) \cdot |\mathcal{F}_\sigma(\delta h)|\, dh \tag{25}$$

since $G = P_{(2n-1,1)}H$. Let $\Xi(\bullet)$ be the integrand in (25). It is a function on $H\backslash H(\mathbb{A})$.

We will use the following notation. Let $X$, $Y$, $Z$ be non-negative variables depending on unspecified parameters. We write $Z = \Gamma(X, Y)$ if for any $N > 0$ there exists $c, k > 0$, *depending only on* $N$, such that $Z \le c \max(1, X)^{-N} \max(1, Y)^k$.

We will prove the following estimates.

PROPOSITION 2. 1. *We have*

$$\sum_{\delta\in P_{(2n-1,1)}\backslash G} \Gamma(1, e^{-\langle\varpi_{2n-1}, H_0(\delta h)\rangle})|\mathcal{F}_{\Phi,\sigma}(\delta h)| \le c\cdot e^{\langle\varpi_n, H_0(h)\rangle} \tag{26}$$

*for some constant c and all $h\in\mathcal{S}^H$. In particular,*

$$\sum_{\delta\in P_{(2n-1,1)}\backslash G} |\mathcal{F}_{\Phi,\sigma}(\delta h)| \le c\cdot e^{\langle\varpi_n, H_0(h)\rangle}. \tag{27}$$

2. *We have*

$$\sum_{\gamma\in P_{(n,n-1,1)}\backslash P_{(2n-1,1)}-P_{(n,n-1,1)}} |M_{-1}\varphi(\gamma g)| = \Gamma(1, e^{-\langle\varpi_{2n-1}, H_0(g)\rangle}) \tag{28}$$

*for all $g\in\mathcal{S}^{(2n-1,1)}\cap G(\mathbb{A})^1$.*

3. *There exists a constant $c>0$ such that*

$$\Xi(h)\le c\cdot e^{\langle n\varpi_n, H_{M_H}(h)\rangle}$$

*for $h\in\mathcal{S}^H$.*

The convergence of (25) will follow from the last part of the proposition. Indeed, by the integration formula on the Siegel domain, it remains to bound

$$\int_{\mathbf{K}_H}\int_{A_0^H(c_0)}\int_{\omega_H} e^{-\langle 2\rho_0^H, H_0(a)\rangle} e^{\langle n\varpi_n, H_{M_H}(a)\rangle}\, dt\, da\, dk.$$

This will converge, provided that the exponent $\mu=-2\rho_0^H+n\varpi_n$ lies in the negative obtuse Weyl chamber of $\mathfrak{a}_0^H$. This is true since $\mu^{M_H}=-2\rho_0^{M_H}$ and $\mu_{M_H}=-2\frac{n+1}{2}\varpi_n+n\varpi_n=-\varpi_n$.

For the proof of Proposition 2 we will first prove a few auxiliary results.

LEMMA 4. *There exists a constant c such that*

$$\langle\varpi, H_0(\gamma g)\rangle\le\langle\varpi, H_0(g)\rangle+c$$

*for any $g\in\mathcal{S}^G$, $\gamma\in G$ and fundamental weight $\varpi\in\hat\Delta_0$.*

*Proof.* Write $g=uak$ an Iwasawa decomposition for $g$ and let $X=H_0(g)$. Let also $\gamma=u_1twu_2$ be a Bruhat decomposition for $\gamma$. Then

$$\begin{aligned}H_0(\gamma g)=H_0(wu_2ua)&=H_0(waw^{-1}wu')=wX+H_0(wu')\\&=H_0(g)+wX-X+H_0(wu')\end{aligned}$$

for a certain $u'\in U_0(\mathbb{A})$. It is standard that $H_0(wu')$ lies in a fixed translate of the negative obtuse Weyl chamber. The same is true for $wX-X$ since $X$ lies in a fixed translate of the positive (acute) Weyl chamber. □

We will use $\|\bullet\|$ to denote the norm function on $G(\mathbb{A})$ whose definition and basic properties can be found in [MW95], I.2.2.

LEMMA 5. *For $N >> 0$*

$$\sum_{\gamma \in G} \|\gamma g\|^{-N}$$

*is bounded uniformly in $g \in G(\mathbb{A})$.*

*Proof.* Choose a compact neighborhood $\Omega$ of 1 such that $\Omega\gamma \cap \Omega = \emptyset$ for all $1 \neq \gamma \in G$. Then there exists a constant $c$ such that

$$\sum_{\gamma \in G} \|\gamma g\|^{-N} \leq c \sum_{\gamma \in G} \|\omega \gamma g\|^{-N}$$

for all $g \in G(\mathbb{A})$ and $\omega \in \Omega$. Integrating this inequality over $\Omega$ and using the fact that $\cup_{\gamma \in G} \Omega\gamma$ is disjoint we get

$$\begin{aligned} \mathrm{vol}(\Omega) \cdot \sum_{\gamma \in G} \|\gamma g\|^{-N} &\leq c \sum_{\gamma \in G} \int_{\Omega} \|\omega \gamma g\|^{-N} \, d\omega \leq c \int_{G(\mathbb{A})} \|x g\|^{-N} \, dx \\ &= c \int_{G(\mathbb{A})} \|x\|^{-N} \, dx. \end{aligned}$$

The right hand side converges for $N >> 0$ since $\mathrm{vol}(\{x \in G(\mathbb{A}) : \|x\| \leq T\})$ is bounded by a power of $T$. □

The following is a reformulation of Proposition 6 of [JR92b].

LEMMA 6. *Fix a cusp form $\varphi$ on $GL_n$. Then for every fundamental weight $\varpi$ and $g \in GL_n(\mathbb{A})^1$ we have*

$$|\varphi(g)| = \Gamma(e^{\langle \varpi, H_0(g) \rangle}, 1).$$

*Proof of Proposition 2.* To prove the first part, we first recall that

$$e^{-\langle \varpi_{2n-1}, H_0(g) \rangle} = |v_0 g|$$

with

$$|v| = \prod_w |v|_w$$

where the product ranges over all places of $F$ and

$$|(x_1, \ldots, x_{2n})|_w = \begin{cases} \max_i(|x_i|_w) & w \text{ non-archimedean,} \\ \left(\sum_i |x_i|_w^2\right)^{\frac{1}{2}} & \text{otherwise.} \end{cases}$$

Thus, identifying $P_{(2n-1,1)}\backslash G$ with the set of lines in $\mathbb{V}$, the left hand side of (26) is majorized by (a constant multiple of)

$$\int_{F^*\backslash\mathbb{I}_F} \sum_{v\in\mathbb{V}-\{0\}} |\Phi(vth)| \max\left(1, |vh|\right)^k |\mathcal{C}_\sigma(\log(|t|^n))| \, d^*t$$

for some $k \geq 0$. Since $h \in \mathcal{S}^H$ it can be written as $h = ah_0$ with $a \in A_0^H(c_0)$ and $h_0$ in a certain compact set. Also, $t$ may be restricted to a compact set since $\mathcal{C}_\sigma$ has compact support. It therefore remains to show that there exists a constant $c$ such that

$$\sum_{v\in\mathbb{V}-\{0\}} |\Phi'(va)| \max(1, |va|)^k \leq c \cdot e^{\langle\varpi_n, H_0(a)\rangle}$$

for all $a \in A_0^H(c_0)$ and $\Phi'$ in the set $\{\Phi(\bullet x) : x \in C\}$ where $C$ is a fixed compact subset of $H(\mathbb{A})$. Since this is a bounded set of Schwartz-Bruhat functions, it is bounded by a fixed $0 \leq \Phi' \in \mathcal{S}(\mathbb{V}(\mathbb{A}))$, so it is enough to prove it for $\Phi'$. Thus, it remains to bound

$$\sum_{v=(x_1,\ldots,x_n,y_n,\ldots,y_1)\in F^{2n}} \Phi'(va) \max(1, |va|)^k.$$

Let $a = \operatorname{diag}(a_1, \ldots, a_n, a_n^{-1}, \ldots, a_1^{-1})$ with $a_i$'s positive real and bounded away from zero. We get

$$\sum_{(x_1,\ldots,x_n,y_n,\ldots,y_1)\in F^{2n}} \Phi'\left((x_1a_1, \ldots, x_na_n, y_na_n^{-1}, \ldots, y_1a_1^{-1})\right) \cdot \max\left(1, |(x_1a_1, \ldots, x_na_n, y_na_n^{-1}, \ldots, y_1a_1^{-1})|\right)^k. \tag{29}$$

We have to bound this by a constant multiple of $\prod_{i=1}^n |a_i|_F$. It follows from the Dixmier-Malliavin Theorem that $\hat{\Phi}$ is a sum of convolutions products of Schwartz functions, each with $2n$ factors. Thus, $\Phi$ is a sum of (ordinary) products, each with $2n$ factors. It follows immediately that there exist functions $\Phi_i^{(1)}, \Phi_i^{(2)} \in \mathcal{S}(\mathbb{A})$, $i = 1, \ldots, n$ such that

$$\Phi'(x_1, \ldots, x_n, y_n, \ldots, y_1) \leq \prod_{i=1}^n \Phi_i^{(1)}(x_i)\Phi_i^{(2)}(y_i).$$

Then (29) is bounded by the product over $i$ of

$$\sum_{x_i\in F} \Phi_i^{(1)}(x_ia_i) \max(1, |xa_i|)^k \times \sum_{y_i\in F} \Phi_i^{(2)}\left(y_ia_i^{-1}\right) \max(1, |ya_i^{-1}|)^k.$$

Since $|x_ia_i| = a_i \geq c_1$ if $x_i \neq 0$ this is bounded by

$$\left(\Phi_i^{(1)}(0) + \sum_{x_i\neq 0} \Phi_i^{(1)}(x_ia_i)a_i^k) \times (\Phi_i^{(2)}(0) + c_1^{-k} \sum_{y_i\neq 0} \Phi_2(y_ia_i^{-1})\right).$$

The first factor is bounded uniformly, while the second factor is bounded by a multiple of $|a_i|_F$.

Before proving the second part of Proposition 2 let us introduce the following notation. If $X$ and $Y$ are real variables, depending on unspecified parameters, we write $X = O(Y)$ if there exist constants $a$ and $b$ such that $X \leq a \max(0, Y) + b$. If $Y$ is non-negative, we write $X = \Omega(Y)$ if there exist constants $a > 0$, $b$ such that $X \geq aY + b$. The constants should be universal.

Let $\gamma \in P_{(2n-1,1)}$. Upon multiplying by an element in $P_{(n,n-1,1)}$ we can and shall assume that $\gamma g \in \mathcal{S}^{(n,n-1,1)}$. Let $X = H_0(\gamma g)$. Then

$$\langle \alpha, X \rangle = \Omega(0) \text{ for all } \alpha \in \Delta_0^{(n,n-1,1)}.$$

Note that $\{\alpha_n\} = \Delta_0^{(2n-1,1)} \setminus \Delta_0^{(n,n-1,1)}$. If $\langle \alpha_n, X \rangle >> 0$ then $\gamma g \in \mathcal{S}^{(2n-1,1)}$ and since $g \in \mathcal{S}^{(2n-1,1)}$, a standard result in reduction theory implies that $\gamma \in P_{(n,n-1,1)}$. (Cf. p. 941 of [Art78] lines 14–19. Note the following misprint on the end of line 17: $\mathcal{S}^{P_1}(T_0, \omega)$ should be replaced by $\mathcal{S}^{P}(T_0, \omega)$.) We conclude that if $\gamma \notin P_{(n,n-1,1)}$ then

$$\langle \alpha_n, X \rangle = O(0).$$

Let $M_1, M_2$ be the $GL_n$ blocks of $M$ so that $M = M_1 \times M_2$. We may write $X = X_1 + X_2 + X_M$ where $X_1 \in \mathfrak{a}_0^{M_1}$, $X_2 \in \mathfrak{a}_0^{M_2}$ and $X_M \in \mathfrak{a}_M$. Concretely, if

$$\begin{aligned} X &= (x_1, \ldots, x_n, y_1, \ldots, y_n) \text{ then} \\ X_1 &= (x_1 - \bar{x}, \ldots, x_n - \bar{x}, 0, \ldots, 0) \\ X_2 &= (0, \ldots, 0, y_1 - \bar{y}, \ldots, y_n - \bar{y}) \\ X_M &= (\bar{x}, \ldots, \bar{x}, \bar{y}, \ldots, \bar{y}) \end{aligned}$$

where $\bar{x} = \frac{1}{n} \sum_{i=1}^n x_i$, $\bar{y} = \frac{1}{n} \sum_{i=1}^n y_i$. We have

1. $\bar{x} + \bar{y} = 0$.
2. $x_i - x_{i+1} = \Omega(0)$ for all $1 \leq i < n$.
3. $y_i - y_{i+1} = \Omega(0)$ for all $1 \leq i < n - 1$.
4. $x_n - y_1 = O(0)$.

Recall the norm $\|\bullet\|$ on $\mathfrak{a}_0$ (not to be confused with the norm function on $G(\mathbb{A})$).

LEMMA 7. *There exists a fundamental weight $\varpi$ of $M_2$ (depending on $X$) such that*

$$\|X\| = O(\|X_1\|) + O(\langle \varpi, X_2 \rangle) + O(-\langle \varpi_{2n-1}, H_0(g) \rangle).$$

*Proof.* We divide into two cases

$y_1 \geq 0$: Let $i$ be the last index strictly less than $n$ so that $y_i \geq 0$. Then for $1 \leq j \leq i$ we have $y_j = \Omega(0)$ and hence $y_j = \Omega(|y_j|)$. Take $\varpi$ corresponding to the $i$-th root of $M_2$. Then

$$n\langle\varpi, X_2\rangle = (n-i)\cdot\sum_{1\le j\le i} y_j - i\cdot\sum_{i<j\le n} y_j = \Omega\left(\sum_{1\le j\le n}|y_j|\right) - n\cdot\max(0, y_n)$$

and hence

$$\|X\| = O\left(\|X_1\| + \sum_{1\le j\le n}|y_j|\right) = O(\|X_1\|) + O(\langle\varpi, X_2\rangle) + O(y_n).$$

$y_1 < 0$: In this case $y_j = O(0)$ for $1 \le j < n$ and we take $\varpi$ corresponding to the first root of $M_2$. We have

$$0 = \bar{x} + \bar{y} = x_1 + O(0) + O(y_n) = x_1 + O(y_n).$$

Using the relation (6) we get

$$\begin{aligned} n\langle\varpi, X_2\rangle &= n\cdot y_1 - \sum_{1\le j\le n} y_j = n\cdot x_n - O(0) + \sum_{1\le j\le n}|y_j| - O(y_n) \\ &= n\cdot x_1 - O(\|X_1\|) + \sum_{1\le j\le n}|y_j| - O(y_n) \\ &= -O(\|X_1\|) + \sum_{1\le j\le n}|y_j| - O(y_n) \end{aligned}$$

since $x_1 = -O(y_n)$. Hence,

$$\|X\| = O(\|X_1\| + \sum_{1\le j\le n}|y_j|) = O(\|X_1\|) + O(\langle\varpi, X_2\rangle) + O(y_n).$$

Finally note that $y_n = -\langle\varpi_{2n-1}, X\rangle = -\langle\varpi_{2n-1}, H_0(g)\rangle$. □

We return to the proof of the second part of Proposition 2. Let $\gamma g = ue^{X_M}m_1m_2k$ be the Iwasawa decomposition of $\gamma g$ with $u \in U(\mathbb{A})$, $m_1 \in M_1(\mathbb{A})^1$, $m_2 \in M_2(\mathbb{A})^1$ and $k \in K$. Then

$$|M_{-1}\varphi(\gamma g)| = |M_{-1}\varphi\left(ue^{X_M}m_1m_2k\right)| = e^{\langle\mu, X_M\rangle}|M_{-1}\varphi(m_1m_2k)|$$

where $\mu = -\varpi_n + n\varpi_n \in \mathfrak{a}_M^*$. Thinking of $M_{-1}\varphi$ as a tensor of cusp forms on $M_1$ and $M_2$ we may use the rapid decay in $M_1$, the fact that $m_1 \in \mathcal{S}^{M_1}$ and Lemma 6 applied to $M_2$ to majorize the above by

$$e^{\langle\mu, X_M\rangle}\cdot\Gamma(\|m_1\|, 1)\cdot\Gamma(e^{\langle\varpi, H_0^{M_2}(m_2)\rangle}, 1) = e^{\langle\mu, X_M\rangle}\cdot\Gamma(e^{\|X_1\|}, 1)\cdot\Gamma(e^{\langle\varpi, X_2\rangle}, 1).$$

This and Lemma 7 give

$$|M_{-1}\varphi(\gamma g)| = \Gamma(e^{\|X\|}, e^{-\langle\varpi_{2n-1}, H_0(g)\rangle}) = \Gamma(\|\gamma g\|, e^{-\langle\varpi_{2n-1}, H_0(g)\rangle})$$

because $\log(\|\gamma g\|) = O(\|X\|)$. It follows that the left hand side of (28) is bounded by

$$\sum_{\gamma \in G} \Gamma(\|\gamma g\|, e^{-\langle \varpi_{2n-1}, H_0(g)\rangle}).$$

It remains to invoke Lemma 5.

To prove the last part of the proposition fix $h \in \mathcal{S}^H \subset \mathcal{S}^G$ and let $\delta \in G$. Upon multiplying $\delta$ on the left by an element of $P_{(2n-1,1)}$ we may assume that $\delta h \in \mathcal{S}^{(2n-1,1)}$. Applying the second part to $g = \delta h$,

$$\sum_{\gamma \in P_{(n,n-1,1)} \backslash P_{(2n-1,1)}} |M_{-1}\varphi(\gamma\delta h)| = |M_{-1}\varphi(\delta h)| + \Gamma(1, e^{-\langle \varpi_{2n-1}, H_0(\delta h)\rangle}).$$

Also, since any cusp form is bounded, we have

$$|M_{-1}\varphi(\delta h)| \leq c_2 \cdot e^{\langle -\varpi_n + n\varpi_n, H_M(\delta h)\rangle} \leq c_3 \cdot e^{\langle -\varpi_n + n\varpi_n, H_M(h)\rangle}$$

for appropriate constants where the second inequality follows from Lemma 4. Thus

$$\Xi(h) \leq c_3 \cdot e^{\langle -\varpi_n + n\varpi_n, H_M(h)\rangle} \cdot \sum_{\delta \in P_{(2n-1,1)} \backslash G} |\mathcal{F}_{\Phi,\sigma}(\delta h)|$$
$$+ \sum_{\delta \in P_{(2n-1,1)} \backslash G} \Gamma(1, e^{-\langle \varpi_{2n-1}, H_0(\delta h)\rangle}) \cdot |\mathcal{F}_{\Phi,\sigma}(\delta h)|.$$

Using (28), the second sum is bounded by $e^{\langle \varpi_n, H_0(h)\rangle}$. Also, by (27) the first sum is bounded by $e^{\langle n\varpi_n, H_0(h)\rangle}$. It remains to note that

$$e^{\langle \varpi_n, H_0(h)\rangle} \leq c_4 \cdot e^{\langle n\varpi_n, H_0(h)\rangle}$$

for an appropriate constant.

**7. Double cosets.** The map $p \mapsto \mathbb{V}'p$ is a bijection between $P\backslash G$ and the set of $n$-dimensional vector subspaces $\mathbb{V}_0 \subset \mathbb{V}$. The $H$-orbit of $\mathbb{V}_0$ is determined by the rank $r$ of $[\bullet, \bullet]|_{\mathbb{V}_0}$ and thus the double cosets $P\backslash G/H$ are indexed by the even integers $0 \leq r \leq n$ ([JR92b]). Next, we analyze the double cosets $P_{(n,n-1,1)}\backslash P_{(2n-1,1)}/Q$. Since $P_{(n,n-1,1)} = P_{(2n-1,1)} \cap P_{(n,n)}$ the map $p \mapsto \mathbb{V}'p$ is a bijection between $P_{(n,n-1,1)}\backslash P_{(2n-1,1)}$ and the set $\mathcal{V}$ of $n$-dimensional vector subspaces $\mathbb{V}_0 \subset \mathbb{V}$ containing $v_0$. Recall that $Q$ is in fact the stabilizer in $H$ of $F \cdot v_0$. To every $\mathbb{V}_0 \in \mathcal{V}$ we associate the pair $(r, \epsilon)$ where $r = \text{rank}([\bullet, \bullet]|_{\mathbb{V}_0})$ and $\epsilon = \pm 1$ is a sign which is 1 if $v_0 \in \text{rad}([\bullet, \bullet]|_{\mathbb{V}_0})$ and $-1$ otherwise. Clearly, these are $Q$-invariant. Conversely, it follows from Witt's Theorem that the $Q$-orbits in $\mathcal{V}$ are determined by the pairs $(r, \epsilon)$. These $Q$-orbits, or what amounts to the same, the double cosets $P_{(n,n-1,1)}\backslash P_{(2n-1,1)}/Q$, are thus indexed by an even integer $0 \leq r \leq n$ and $\epsilon = \pm 1$. If $r = 0$, i.e. if $[\bullet, \bullet]|_{\mathbb{V}_0} = 0$ then necessarily $\epsilon = 1$, while if $r = n$ (with $n$ even) then necessarily $\epsilon = -1$. We will calculate the contribution of each double coset to (24).

**7.1. Case $r = 0$.** In this case we may take $\eta = 1$ and its contribution is

$$\int_{Q_1\backslash H(\mathbb{A})} M_{-1}\varphi(h)\mathcal{F}(h)\,dh.$$

Recalling the notation from §2.3, we compute this as

$$\begin{aligned}
&\int_{P_H(\mathbb{A})\backslash H(\mathbb{A})}\int_{Q_1\backslash P_H(\mathbb{A})} \delta_{P_H}(p)^{-1}M_{-1}\varphi(ph)\mathcal{F}_\sigma(ph)\,dh\\
&= \int_{\mathbf{K}_H}\int_{P_1\backslash M_H(\mathbb{A})}\int_{U_H\backslash U_H(\mathbb{A})} \delta_{P_H}(m)^{-1}M_{-1}\varphi(umk)\mathcal{F}_\sigma(umk)\,du\,dm\,dk\\
&= \int_{\mathbf{K}_H}\int_{P_1\backslash M_H(\mathbb{A})^1} M_{-1}\varphi(mk)\\
&\times\left(\int_{\mathfrak{a}_{M_H}} e^{\langle -2\rho_{P_H},X\rangle}e^{\langle -\varpi_n+\rho_P,X\rangle}\mathcal{F}_\sigma(e^X mk)\,dX\right)dm\,dk
\end{aligned}$$

since $\mathcal{F}_\sigma$ is invariant under $U_H(\mathbb{A})$. The inner integral is

$$\begin{aligned}
&\int_{\mathfrak{a}_{M_H}} e^{-2\langle\varpi_n,X\rangle}\mathcal{F}_\sigma(e^X mk)\,dX\\
&= \int_{\mathfrak{a}_{M_H}} e^{-2\langle\varpi_n,X\rangle}\int_{\mathbb{I}_F}\Phi(tv_0e^Xmk)\mathcal{C}_\sigma(\log|t|^n)\,d^*t\,dX\\
&= \int_{\mathbb{R}} e^{-2x}\int_{\mathbb{I}_F}\Phi(te^{-x/n}v_0mk)\mathcal{C}_\sigma(\log|t|^n)\,d^*t\,dx
\end{aligned}$$

using the variable $x = \langle\varpi_n, X\rangle$ and the embedding (9). After a change of variable we obtain

$$\begin{aligned}
&\int_{\mathbb{R}} e^{-2x}\int_{\mathbb{I}_F}\Phi(tv_0mk)\mathcal{C}_\sigma(\log|t|^n+x)\,d^*t\,dx\\
&= \int_{\mathbb{I}_F}\Phi(tv_0mk)\int_{\mathbb{R}} e^{-2x}\mathcal{C}_\sigma(\log|t|^n+x)\,dx\,d^*t\\
&= \int_{\mathbb{I}_F}\Phi(tv_0mk)\left(\int_{\mathbb{R}} e^{-2x}\mathcal{C}_\sigma(x)\,dx\right)|t|^{2n}\,d^*t\\
&= \int_{\mathbb{I}_F}\Phi(tv_0mk)\cdot 2\pi i\sigma(1)\cdot|t|^{2n}\,d^*t = 2\pi i\sigma(1)\cdot\phi^{\mathbb{V}}_{\Phi,1}(mk).
\end{aligned}$$

The contribution from $\eta = 1$ now becomes $2\pi i\sigma(1)$ times

$$\int_{K_H}\int_{M_H\backslash M_H(\mathbb{A})^1} M_{-1}\varphi(mk)\cdot \sum_{\gamma\in P_1\backslash M_H} \phi_{\Phi,1}(\gamma mk)\,dm\,dk$$
$$= \int_{K_H}\int_{M_H\backslash M_H(\mathbb{A})^1} M_{-1}\varphi(mk)\mathbb{E}_\Phi(mk,3)\,dm\,dk$$

by (12). This is the required contribution.

**7.2. $r = n$.** The computation will be similar to the one in [JR92b]. We take $\mathbb{V}_0$ to be

$$(x_1,\ldots,x_{n/2},0,\ldots,0,y_1,\ldots,y_{n/2}).$$

Then $\mathbb{V}_0 = \mathbb{V}'\eta$ where

$$\eta = \begin{pmatrix} & & 1_{n/2} & \\ & 1_{n/2} & & \\ 1_{n/2} & & & \\ & & & 1_{n/2}\end{pmatrix}.$$

Then

$$(30)\qquad \int_{Q_\eta\backslash H(\mathbb{A})} M_{-1}\varphi(\eta h)\mathcal{F}_\sigma(h)\,dh = \int_{\eta Q_\eta\eta^{-1}\backslash\eta H(\mathbb{A})\eta^{-1}} M_{-1}\varphi(h\eta)\mathcal{F}_\sigma(h\eta)\,dh.$$

The group $\eta H\eta^{-1}$ is the symplectic group attached to the skew symmetric matrix ${}^t\eta\epsilon_{2n}\eta = \left(\begin{smallmatrix}\epsilon_n & 0\\ 0 & \epsilon_n\end{smallmatrix}\right)$. Since $\eta\in P_{(2n-1,1)}$ we have $Q_\eta = H\cap\eta^{-1}P_{(n,n)}\eta\cap P_{(2n-1,1)}$. Thus,

$$\eta Q_\eta\eta^{-1} = \eta H\eta^{-1}\cap P_{(n,n)}\cap P_{(2n-1,1)}.$$

The group $\eta H\eta^{-1}\cap P_{(n,n)}$ is equal to

$$\left\{\begin{pmatrix} m_1 & 0\\ 0 & m_2\end{pmatrix} : m_1, m_2\in Sp_{n/2}\right\}.$$

Thus,

$$\eta Q_\eta\eta^{-1} = \left\{\begin{pmatrix} m_1 & 0\\ 0 & m_2\end{pmatrix} : m_1\in Sp_{n/2}, m_2\in Sp_{n/2}\cap P_{(n-1,1)}\right\}.$$

The integral (30) factors through

$$\int_{Sp_{n/2}\backslash Sp_{n/2}(\mathbb{A})} M_{-1}\varphi\left(\begin{pmatrix} m_1 & 0\\ 0 & m_2\end{pmatrix}h\eta\right)\mathcal{F}_\sigma\left(\begin{pmatrix} m_1 & 0\\ 0 & m_2\end{pmatrix}h\eta\right)dm_1$$

which is 0 because it is a symplectic period of a cusp form on $GL_n$ ([JR92b], Proposition 1).

**7.3. $0 < r < n$.** Let $\mathrm{proj}_M : P \to M$ be the canonical projection. Let $M_1$ be the subgroup

$$\left\{ \begin{pmatrix} A & 0 \\ 0 & 1_n \end{pmatrix} : A \in GL_n \right\}$$

of $M$ and let $T_1$ be the inverse image of $M_1$ under $\mathrm{proj}_M$, i.e. the group

$$\left\{ \begin{pmatrix} A & B \\ 0 & 1_n \end{pmatrix} : A \in GL_n,\ B \in M_{n\times n} \right\}.$$

$T_1$ is a normal subgroup of $P$ contained in $P_{(n,n-1,1)}$.

Consider first a representative $\delta$ of the double coset of $P\backslash G/H$ indexed by $r$. We may take $\delta$ to be the representative considered in ([JR92b],§4). The stabilizer $P_\delta = P \cap \delta H \delta^{-1}$ is given explicitly. In particular ([loc. cit.] p. 184, formula (22)), the unipotent radical $S$ of $P_\delta$ consists of matrices of the form

$$\begin{pmatrix} 1_{n-r} & X_1 & * & * \\ 0 & 1_r & * & * \\ 0 & 0 & 1_r & X_2 \\ 0 & 0 & 0 & 1_{n-r} \end{pmatrix}$$

where $X_1$ is an arbitrary $(n-r)\times r$ matrix and $X_2$ is an arbitrary $r \times (n-r)$ matrix. It follows that the intersection $N = S \cap T_1$ is a unipotent normal subgroup of $P_\delta$ whose image $R$ under $\mathrm{proj}_M$ is a unipotent radical of a proper parabolic subgroup of $M_1$ (of type $(n-r, r)$).

Let $\eta$ be a representative for one of the two double cosets in $P_{(n,n-1,1)}\backslash P_{(2n-1,1)}/Q$ corresponding to $(r, \pm 1)$. Then $\eta \in P\delta H$ and we write $\eta = p\delta h$ with $p \in P$ and $h \in H$. Let $N_1 = pNp^{-1}$.

PROPOSITION 3. 1. *$N_1$ is a normal unipotent subgroup of $P_{(n,n-1,1)} \cap \eta Q \eta^{-1}$.*
2. *$R_1 = \mathrm{proj}_M(N_1)$ is a unipotent radical of a proper parabolic of $M_1$.*

*Proof.* Since $N \subset T_1$ and $T_1$ is normal in $P$ we have

$$N_1 \subset T_1 \subset P_{(n,n-1,1)}.$$

Hence,

$$\eta^{-1} N_1 \eta \subset P_{(2n-1,1)}$$

since $\eta \in P_{(2n-1,1)}$. On the other hand

$$\eta^{-1} N_1 \eta = h^{-1}\delta^{-1} N \delta h \subset H.$$

All in all,

$$N_1 \subset P_{(n,n-1,1)} \cap \eta(H \cap P_{(2n-1,1)})\eta^{-1} = P_{(n,n-1,1)} \cap \eta Q \eta^{-1}.$$

The first part follows since

$$P_{(n,n-1,1)} \cap \eta Q \eta^{-1} \subset P \cap \eta H \eta^{-1} = p P_\delta p^{-1}.$$

The second part is true since

$$\mathrm{proj}_M(N_1) = \mathrm{proj}_M(pNp^{-1}) = \mathrm{proj}_M(p) R \,\mathrm{proj}_M(p^{-1}). \qquad \square$$

Let $B$ be the kernel of the map $\mathrm{proj}_M$ restricted to $N_1$ and let $\nu : R_1 \to B \backslash N_1$ be the inverse of the isomorphism defined by $\mathrm{proj}_M$. Finally let $N' = \eta^{-1} N_1 \eta$. It is a normal unipotent subgroup of $Q_\eta$. The contribution from $\eta$ is

$$\begin{aligned}
&\int_{Q_\eta \backslash H(\mathbb{A})} M_{-1}\varphi(\eta h)\mathcal{F}_\sigma(h)\,dh \\
&= \int_{Q_\eta(\mathbb{A}) \backslash H(\mathbb{A})} \int_{Q_\eta \backslash Q_\eta(\mathbb{A})} \delta_{Q_\eta}(q)^{-1} M_{-1}\varphi(\eta q h)\mathcal{F}_\sigma(qh)\,dq\,dh \\
&= \int_{Q_\eta(\mathbb{A}) \backslash H(\mathbb{A})} \int_{Q_\eta \cdot N'(\mathbb{A}) \backslash Q_\eta(\mathbb{A})} \delta_{Q_\eta}(q)^{-1} \\
&\quad \times \left( \int_{N' \backslash N'(\mathbb{A})} M_{-1}\varphi(\eta n' q h)\mathcal{F}_\sigma(n' q h)\,dn' \right) dq\,dh \\
&= \int_{Q_\eta(\mathbb{A}) \backslash H(\mathbb{A})} \int_{Q_\eta \cdot N'(\mathbb{A}) \backslash Q_\eta(\mathbb{A})} \delta_{Q_\eta}(q)^{-1} \\
&\quad \left( \int_{N' \backslash N'(\mathbb{A})} M_{-1}\varphi(\eta n' q h)\,dn' \right) \mathcal{F}_\sigma(qh)\,dq\,dh
\end{aligned}$$

because $\mathcal{F}_\sigma$ is left invariant under $Q(\mathbb{A})^1$. However,

$$\begin{aligned}
&\int_{N' \backslash N'(\mathbb{A})} M_{-1}\varphi(\eta n' h)\,dn' = \int_{N_1 \backslash N_1(\mathbb{A})} M_{-1}\varphi(n \eta h)\,dn \\
&= \int_{B(\mathbb{A}) \cdot N_1 \backslash N_1(\mathbb{A})} \int_{B \backslash B(\mathbb{A})} M_{-1}\varphi(b n \eta h)\,db\,dn = \int_{B(\mathbb{A}) \cdot N_1 \backslash N_1(\mathbb{A})} M_{-1}\varphi(n \eta h)\,dn \\
&= \int_{R_1 \backslash R_1(\mathbb{A})} M_{-1}\varphi(\nu(r) \eta h)\,dr = \int_{R_1 \backslash R_1(\mathbb{A})} M_{-1}\varphi(r \eta h)\,dr
\end{aligned}$$

since $\nu(r)$ and $r$ have the same projection onto $M$. The last term vanishes by cuspidality.

Department of Mathematics, Columbia University, New York NY 10027
*E-mail:* hj@math.columbia.edu

Einstein Institute of Mathematics, Hebrew University of Jerusalem, Jerusalem 91904, Israel
*E-mail:* lapid@cims.nyu.edu

Department of Mathematics, The Ohio State University, 231W 18th Ave, Columbus OH 43210
*E-mail:* haar@math.ohio-state.edu

## REFERENCES

[Art78] James G. Arthur, A trace formula for reductive groups. I. Terms associated to classes in $G(\mathbf{Q})$. *Duke Math. J.*, **45(4):**911–952, 1978.

[Art80] James Arthur, A trace formula for reductive groups. II. Applications of a truncation operator. *Compositio Math.*, **40(1):**87–121, 1980.

[Ber84] Joseph N. Bernstein, $P$-invariant distributions on $\mathrm{GL}(N)$ and the classification of unitary representations of $\mathrm{GL}(N)$ (non-Archimedean case). In *Lie group representations, II (College Park, Md., 1982/1983)*, Springer, Berlin, 1984, pp. 50–102.

[DM78] Jacques Dixmier and Paul Malliavin, Factorisations de fonctions et de vecteurs indéfiniment différentiables. *Bull. Sci. Math. (2)*, **102(4):**307–330, 1978.

[Fli97] Yuval Z. Flicker, Cyclic automorphic forms on a unitary group. *J. Math. Kyoto Univ.*, **37(3):**367–439, 1997.

[GK75] I. M. Gelfand and D. A. Kajdan, Representations of the group $\mathrm{GL}(n, K)$ where $K$ is a local field. In *Lie groups and their representations (Proc. Summer School, Bolyai János Math. Soc., Budapest, 1971)*, Halsted, New York, 1975, pp. 95–118.

[Jac87] Hervé Jacquet, Sur un résultat de Waldspurger. II. *Compositio Math.*, **63(3):**315–389, 1987.

[Jac95] ———, The continuous spectrum of the relative trace formula for GL(3) over a quadratic extension. *Israel J. Math.*, **89(1–3):**1–59, 1995.

[Jac97] ———, Automorphic spectrum of symmetric spaces. In *Representation theory and automorphic forms (Edinburgh, 1996)*, Amer. Math. Soc., Providence, RI, 1997, pp. 443–455.

[Jac] ———, A guide to relative trace formula, preprint.

[JLR93] Hervé Jacquet, King F. Lai, and Stephen Rallis, A trace formula for symmetric spaces. *Duke Math. J.*, **70(2):**305–372, 1993.

[JLR99] Hervé Jacquet, Erez Lapid, and Jonathan Rogawski, Periods of automorphic forms. *J. Amer. Math. Soc.*, **12(1):**173–240, 1999.

[JN99] Hervé Jacquet and Chen Nan, Positivity of quadratic base change $L$-functions. *Bulletin de la SMF*, **351(3):**1227–1255, 1999.

[JR92a] Hervé Jacquet and Stephen Rallis, Kloosterman integrals for skew symmetric matrices. *Pacific J. Math.*, **154(2):**265–283, 1992.

[JR92b] ———, Symplectic periods. *J. Reine Angew. Math.*, **423:**175–197, 1992.

[KR69] Bertram Kostant and Stephen Rallis, On orbits associated with symmetric spaces. *Bull. Amer. Math. Soc.*, **75:**879–883, 1969.

[Lap] Erez Lapid, On the spectral expansion of the relative trace formula, preprint.

[LR] Erez Lapid and Stephen Rallis, A spectral identity between symmetric places. *Israel J. Math.*, to appear.

[LR00] ———, Stabilization of periods of Eisenstein series and Bessel distributions on $GL(3)$ relative to U(3). *Doc. Math.*, **5:**317–350 (electronic), 2000.

[LR03] Erez Lapid and Jonathan Rogawski, Periods of Eisenstein series: the galois case. *Duke Math. J.*, **120** (2003), 153–226.

[Mao92] Zhengyu Mao, Relative Kloosterman integrals for the unitary group. *C. R. Acad. Sci. Paris Sér. I Math.*, **315(4)**:381–386, 1992.

[Mao97] ———, Airy sums, Kloosterman sums, and Salié sums. *J. Number Theory*, **65(2)**:316–320, 1997.

[MW89] C. Mœglin and J.-L. Waldspurger, Le spectre résiduel de GL($n$). *Ann. Sci. École Norm. Sup. (4)*, **22(4)**:605–674, 1989.

[MW95] ———, *Spectral decomposition and Eisenstein series*. Cambridge University Press, Cambridge, 1995. Une paraphrase de l'Écriture [A paraphrase of Scripture].

[Ngô99] Báo Châu Ngô, Faisceaux pervers, homomorphisme de changement de base et lemme fondamental de Jacquet et Ye. *Ann. Sci. École Norm. Sup. (4)*, **32(5)**:619–679, 1999.

[Wal88] Nolan R. Wallach, *Real reductive groups. I*. Academic Press Inc., Boston, MA, 1988.

[Wal92] ———, *Real reductive groups. II*. Academic Press Inc., Boston, MA, 1992.

**Note Added in Proof.** Using the same technique, additional trace identities of the type considered in this paper were recently proved by the second and third named authors [LR]. Moreover, the subtle convergence arguments of §6 can be simplified considerably.

**Acknowledgments.** Part of this work was done while the authors participated in the special semester "Formes Automorphes" held at the Institut Henri Poincaré in Paris during the first half of 2000. The authors would like to thank this Institute for its hospitality. The second-named author would also like to thank University of Köln for its support during his stay at IHP.

CHAPTER 17

# GENERIC REPRESENTATIONS AND LOCAL LANGLANDS RECIPROCITY LAW FOR $p$-ADIC $\mathrm{SO}_{2n+1}$

By DIHUA JIANG and DAVID SOUDRY

---

*To Professor J. Shalika with admiration*

**Introduction.** An irreducible admissible representation of a quasi-split reductive algebraic group $G(F)$, where $F$ is a $p$-adic field, is called *generic,* or more precisely, generic with respect to a given nondegenerate character of a maximal unipotent subgroup of $G(F)$, if it admits a non-zero Whittaker functional with respect to the given character. Such a functional is unique up to scalar multiples. This well known uniqueness property was proved by Gelfand and Kazhdan for $G = \mathrm{GL}(n)$ [G.K.] and by J. Shalika [Sh] for general $G$ (over any local field). As a result, a generic representation has a unique Whittaker model. In the last thirty years, the uniqueness of Whittaker models for irreducible generic representations of $G$ has played very important roles in both the local theory and the global theory of automorphic forms, and in particular, in the theory of automorphic L-functions. In this paper, we study generic representations of $p$-adic $\mathrm{SO}(2n+1)$ and the relation with the local Langlands reciprocity conjecture.

We state the local Langlands reciprocity conjecture for a $p$-adic split reductive group $G$ below, the general version of which can be found in [B]. Let $W_F$ be the Weil group associated to the the local field $F$. We take

$$W_F \times \mathrm{SL}_2(\mathbb{C})$$

as the Weil-Deligne group [A]. The relation between this version of the Weil-Deligne group and the version used in [B] is not difficult to figure out. One may find a relevant explicit discussion in [Kn]. It is known [Sp] that a split reductive group $G$ is uniquely determined, up to central isogeny, by the associated root datum. The Langlands dual group $G^\vee(\mathbb{C})$ of $G$ is a complex reductive group with root datum dual to that of $G$. For example, the Langlands dual group of $\mathrm{SO}(2n+1)$ is $\mathrm{Sp}_{2n}(\mathbb{C})$. Let $\Phi(G)$ be the set of conjugacy classes of admissible homomorphisms $\varphi$ from $W_F \times SL_2(\mathbb{C})$ to $G^\vee(\mathbb{C})$. For any $\varphi \in \Phi(G)$, we may decompose it into a direct sum of irreducible representations of $W_F \times \mathrm{SL}_2(\mathbb{C})$

$$\varphi = \oplus_i \phi_i \otimes S_{w_i+1}.$$

---

The first named author is partly supported by NSF grant 0098003, by the Sloan Research Fellowship and by McKnight Professorship at University of Minnesota. The second named author is partly supported by a grant from the Israel-USA Binational Sciences Foundation.

Then the admissibility of $\varphi$ means that the representations $\phi_i$ are continuous complex representations of $W_F$ with $\phi_i(W_F)$ consisting of semi-simple elements in $G^\vee(\mathbb{C})$ and $S_{w_i+1}$ is the irreducible algebraic complex representations of $\mathrm{SL}_2(\mathbb{C})$ of dimension $w_i + 1$. The elements in the set $\Phi(G)$ are called the *local Langlands parameters* for $G(F)$.

The local Langlands reciprocity conjecture asserts the existence of a parameterization of the set $\Pi(G)$ of all equivalence classes of irreducible admissible representations of $G(F)$ in terms of the set $\Phi(G)$ of the local Langlands parameters. More precisely, the conjecture can be stated as follows:

LOCAL LANGLANDS RECIPROCITY CONJECTURE. *For each local Langlands parameter $\varphi \in \Phi(G)$, there exists a finite subset $\Pi(\varphi)$, which is called the local $L$-packet attached to $\varphi$, of $\Pi(G)$ such that*

(1) *the local $L$-packet $\Pi(\varphi)$ is not empty,*

(2) *the reciprocity map taking $\varphi$ to $\Pi(\varphi)$ is one to one, and preserves the local factors, i.e.,*

$$L(\varphi, s) = L(\pi, s),$$

$$\epsilon(\varphi, s, \psi) = \epsilon(\pi, s, \psi),$$

*for $\pi \in \Pi(\varphi)$, where $\psi$ is a given nontrivial character of $F$, and*

(3) *the union of all local $L$-packets $\Pi(\varphi)$ gives a partition for $\Pi(G)$.*

For $G = \mathrm{GL}(n)$, this conjecture was proved by M. Harris and R. Taylor [H.T.] and by G. Henniart [H1] for the supercuspidal cases, while the reduction of the conjecture for $\mathrm{GL}(n)$ to supercuspidal cases was given by A. Zelevinsky [Z]. More discussions about this conjecture for $\mathrm{GL}(n)$ can be found in [K]. It is worthwhile to point out that when $G = \mathrm{GL}(n)$, each local $L$-packet $\Pi(\varphi)$ contains only one member. A main reason for this is that any irreducible supercuspidal representation of $\mathrm{GL}_n(F)$ is generic, i.e. has a non-zero Whittaker model. However, when $G$ is not $\mathrm{GL}(n)$, it is expected that each local $L$-packet $\Pi(\varphi)$ contains more than one member in general. It is clear that in order to completely understand the set $\Pi(G)$, one has to have explicit description for each local $L$-packet $\Pi(\varphi)$ in addtion to the local Langlands reciprocity conjecture. From Arthur's trace formula approach to study square integrable automorphic representations [A] or Shahidi's proof of Langlands conjecture on Plancherel measures [S1], it is important to know when a local $L$-packet has a generic member. Some further refined properties for local $L$-packets $\Pi(\varphi)$ are given in Borel's paper [B].

The objective of this paper is to study the local Langlands reciprocity conjecture for $G = \mathrm{SO}(2n+1)$ and the genericity of the local $L$-packets. Our main results can be stated as follows.

THEOREM A. *For each local Langlands parameter $\varphi \in \Phi(\mathrm{SO}_{2n+1})$, there exists an irreducible admissible representation $\sigma(\varphi)$ of $\mathrm{SO}_{2n+1}(F)$, such that*

(1) *$\sigma(\varphi)$ is uniquely determined by the parameter $\varphi$ and can be realized as the Langlands subquotient of an induced representation*

$$\delta(\Sigma_1) \times \cdots \times \delta(\Sigma_f) \rtimes \sigma^{(t)}$$

*where $\sigma^{(t)}$ is an irreducible generic tempered representation of $\mathrm{SO}_{2n^*+1}(F)$, and $\delta(\Sigma_i)$ is the essentially square integrable representation of $\mathrm{GL}_{n_i}(F)$ associated to the imbalanced segment $\Sigma_i$ for $i = 1, 2, \ldots, f$ $(n = n^* + \sum_{i=1}^{f} n_i)$; and*

(2) *the reciprocity map taking $\varphi$ to $\sigma(\varphi)$ is one-to-one and preserves relevant local factors*

$$L(\varphi \otimes \phi, s) = L(\sigma(\varphi) \times r(\phi), s) \tag{0.1}$$

$$\epsilon(\varphi \otimes \phi, s, \psi) = \epsilon(\sigma(\varphi) \times r(\phi), s, \psi), \tag{0.2}$$

*where $\phi$ is a local Langlands parameter for $\mathrm{GL}_l$, and $r$ is the reciprocity map for $GL_l$ (for any positive integer $l$) given in [H.T.] and [H1].*

Theorem A will be proved by establishing the explicit local Langlands functorial lift from SO($2n+1$) to GL($2n$) (Theorems 1.1, 2.1, 3.1, 4.1, and 5.1), by studying the fine structure of the local Langlands parameters (Proposition 6.1), and by using the local Langlands reciprocity conjecture for GL($n$) in [H.T.] and [H1].

We state here the explicit local Langlands functorial lift from irreducible tempered generic representations of SO($2n+1$) to GL($2n$) (Theorem 4.1), while the generalization of this theorem to irreducible generic representations of SO($2n+1$) is Theorem 5.1, which is more involved and not convenient to be stated here. Let $\tau$ be an irreducible unitary supercuspidal representation of $\mathrm{GL}_k(F)$ and $2m$ be an integer. We denote by $\Delta(\tau, m)$ the unique irreducible square-integrable subrepresentation of $GL_{k(2m+1)}(F)$ attached to the balanced segment $[\nu^{-m}\tau, \nu^m\tau]$ (see formula (0.11) for more details).

THEOREM B. *There exists a bijection $\ell$ between the set $\Pi^{(tg)}(\mathrm{SO}_{2n+1})$ of equivalence classes of irreducible tempered generic representations of $\mathrm{SO}_{2n+1}(F)$ and the set of equivalence classes of irreducible tempered representations of $\mathrm{GL}_{2n}(F)$ of the following form*

$$\Delta(\lambda_1, h_1) \times \Delta(\lambda_2, h_2) \times \cdots \times \Delta(\lambda_f, h_f) \tag{0.3}$$

*where $\lambda_1, \ldots, \lambda_f$ are irreducible unitary supercuspidal representations of $\mathrm{GL}_{k_{\lambda_i}}(F)$, respectively, and $2h_i$'s are non-negative integers, such that for $1 \leq i \leq f$*

(1) *if $\lambda_i \not\cong \widehat{\lambda_i}$, then $\Delta(\lambda_i, h_i)$ occurs in (0.3) as many times as $\Delta(\widehat{\lambda_i}, h_i)$ does,*

(2) *if the exterior square L-factor $L(\lambda_i, \Lambda^2, s)$ has a pole at $s = 0$, and $h_i \in \frac{1}{2} + \mathbb{Z}_{\geq 0}$, then $\Delta(\lambda_i, h_i)$ occurs an even number of times in (0.3),*

(3) *if the symmetric square L-factor $L(\lambda_i, \mathrm{sym}^2, s)$ has a pole at $s = 0$, and $h_i \in \mathbb{Z}_{\geq 0}$, then $\Delta(\lambda_i, h_i)$ occurs an even number of times in (0.3).*

*Moreover, we have*

$$L(\sigma \times \pi, s) = L(\ell(\sigma) \times \pi, s) \tag{0.4}$$

$$\epsilon(\sigma \times \pi, s, \psi) = \epsilon(\ell(\sigma) \times \pi, s, \psi) \tag{0.5}$$

*for all generic tempered representations* $\sigma$ *of* $\mathrm{SO}_{2n+1}(F)$ *and all irreducible generic representations* $\pi$ *of* $\mathrm{GL}_k(F)$ *with all* $k \in \mathbb{Z}_{>0}$.

We shall construct explicitly in §6 the representation $\sigma(\varphi)$ of $\mathrm{SO}_{2n+1}(F)$ in terms of the local Langlands parameter $\varphi$ and show that $\sigma(\varphi)$ is generic if and only if the induced representation

$$\delta(\Sigma_1) \times \cdots \times \delta(\Sigma_f) \rtimes \sigma^{(t)}$$

is irreducible. A characterization of the genericity of $\sigma(\varphi)$ will be given in the following theorems.

THEOREM C. *For each local Langlands parameter* $\varphi \in \Phi(SO_{2n+1})$, *the irreducible representation* $\sigma(\varphi)$ *of* $\mathrm{SO}_{2n+1}(F)$ *constructed in Theorem A enjoys the following properties.*

(1) *One can associate to* $\varphi$ *at most one irreducible generic representation of* $\mathrm{SO}_{2n+1}(F)$ *satisfying conditions (0.1) and (0.2) in Theorem A;*

(2) *If* $\varphi$ *is tempered, i.e.,* $\varphi(W_F)$ *is bounded in* $\mathrm{Sp}_{2n}(\mathbb{C})$, *then the representation* $\sigma(\varphi)$ *is generic; and*

(3) *The representation* $\sigma(\varphi)$ *is generic if and only if the local adjoint L-function*

$$L\big(\mathrm{Ad}_{\mathrm{Sp}_{2n}} \circ \varphi, s\big)$$

*is regular at* $s = 1$. *(This is the* SO(2$n$ + 1)*-case of a conjecture of Gross-Prasad [G.P.] and of Rallis [K].)*

*Remark.* Property (1) in Theorem C follows from the local converse theorem for generic representations of $\mathrm{SO}_{2n+1}(F)$, which was proved in [Jng.S.] and stated as Theorem 1.3 in this paper. In the language of local $L$-packets (in the sense of the local Langlands reciprocity conjecture), Property (1) means that for each given local Langlands parameter $\varphi \in \Phi(SO_{2n+1})$, the "associated" local $L$-packet $\Pi(\varphi)$ has at most one generic member. Property (2) can be interpreted as that if the parameter $\varphi \in \Phi(SO_{2n+1})$ is tempered, the "associated" local $L$-packet $\Pi(\varphi)$ has a generic member. This statement is a conjecture stated in [S1] and a basic assumption used in [A] for SO($2n + 1$) and will be proved in §4. For $G = \mathrm{U}(2, 1)$, the genericity of tempered $L$-packets was established by S. Gelbart, H. Jacquet, and J. Rogawski in [G.J.R.]. Finally, Property (3) can be understood as a criterion for the genericity of the "associated" local $L$-packet $\Pi(\varphi)$ in terms of the regularity of the adjoint L-functions at $s = 1$, which will be proved in §7.

As applications of our results to the global theory of automorphic forms, we prove the following theorems. Let $k$ be a number field and $\mathbb{A}$ be the ring of adeles associated to $k$.

THEOREM D. *If an irreducible cuspidal automorphic representation $\pi$ of $GL_{2n}(\mathbb{A})$ has the property that $L(\pi, \Lambda^2, s)$ has a pole at $s = 1$, then every local component of $\pi$ is symplectic, i.e., the local Langlands parameter attached to each local component of $\pi$ is symplectic.*

This result proves Conjecture III in [PR], which may be deduced at least heuristically from the global Langlands reciprocity conjecture or Arthur's conjectures. We shall prove it in §7. It is worth pointing out that this result is an important ingredient in the recent work of E. Lapid and S. Rallis [L.R.] proving the positivity of the central value of the standard L-functions attached to certain self-dual cuspidal automorphic representations of GL($2n$).

THEOREM E. *The weak lift from irreducible generic cuspidal automorphic representations $\sigma$ of $SO_{2n+1}(\mathbb{A})$ to automorphic representations $\Sigma$ of $GL_{2n}(\mathbb{A})$ (established in [C.K.PS.S]) is in fact Langlands functorial, i.e., is compatible with the local Langlands functorial lift at all local places.*

This theorem completes the Langlands functorial lift from irreducible generic cuspidal automorphic representations of $SO_{2n+1}$ to $GL_{2n}$. In [C.K.PS.S], J. Cogdell, H. Kim, I. Piatetski-Shapiro, and F. Shahidi proved that there exists a lift from irreducible generic cuspidal automorphic representations $\sigma$ of $SO_{2n+1}$ to $\pi$ of $GL_{2n}$ which is compatible with the local Langlands functorial lift at archimedean places, and at finite places where $\sigma_v$ is unramified, i.e., there exists a weak Langlands functorial lift for such $\sigma$. The image of this weak Langlands functorial lift was explicitly characterized by D. Ginzburg, S. Rallis, and D. Soudry in [G.R.S.]. We shall prove Theorem E in §7. Of course, it is still an open problem to establish the Langlands functorial lift from irreducible non-generic cuspidal automorphic representations of SO($2n+1$) to GL($2n$).

As we already mentioned above, one of the main ingredients in the proof of our main results here is to establish explicitly the local Langlands functorial lift from irreducible generic representations of $SO_{2n+1}(F)$ to $GL_{2n}(F)$ (Theorem 5.1). We recall from [Jng.S.] the explicit local Langlands functorial lift from irreducible generic supercuspidal representations of $SO_{2n+1}(F)$ to $GL_{2n}(F)$ in §1. In order to get Theorem 5.1, we have to go through the induction procedure step by step. Using Muic's classification of irreducible generic representations of $SO_{2n+1}(F)$ in [M1, 2], we carry out the explicit lift and verify the compatibility of relevant local factors as required in the local Langlands conjecture. This work will be done in Sections 2, 3, 4, and 5. Basically, §2 deals with generic square-integrable representations, §3 discusses elliptic tempered generic representations, §4 is for tempered generic representations (Theorem B), and §5 proves Theorem 5.1 which is for irreducible generic representations. In §6, we shall prove Theorem A, and Theorems C, D, and E will be completed in §7.

*Acknowledgments.* Finally, we would like to thank Professor Jian-Shu Li for his warm invitation to the Hong Kong University of Science and Technology in the summer, 2001. We thank the Department of Mathematics, at HKUST, for providing us with a very friendly and stimulating research atmosphere during our stay, where we finished essentially the first version of this paper. Finally, we thank the referee for his comments on the earlier version of this paper.

**Notations and some preliminaries.** Let $F$ be a non-archimedean local field of characteristic zero. $\mathrm{SO}_{2n+1}(F)$ denotes the group of $F$-points of the split group $\mathrm{SO}_{2n+1}$. We realize it as

$$\{g \in \mathrm{GL}_{2n+1}(F) \mid {}^t g J g = J\}$$

where $J = \begin{pmatrix} 0 & & 1 \\ & \cdot^{\cdot^{\cdot}} & \\ 1 & & 0 \end{pmatrix}$, a $(2n+1) \times (2n+1)$-matrix.

In this paper, all representations are assumed to be smooth. Given irreducible (smooth) representations $\tau_i$ of $\mathrm{GL}_{m_i}(F)$, where $i = 1, 2, \ldots, r$, we denote by

$$\tau_1 \times \tau_2 \times \cdots \times \tau_r \tag{0.6}$$

the representation of $\mathrm{GL}_n(F)$, $n = m_1 + m_2 + \cdots + m_r$ induced from the standard parabolic subgroup of type $(m_1, m_2, \ldots, m_r)$.

We follow the notation used in [Td] and [M1,2]. For an irreducible representation $\sigma$ of $\mathrm{SO}_{2m+1}(F)$, we denote by

$$\tau_1 \times \cdots \times \tau_r \rtimes \sigma, \tag{0.7}$$

the representation of $\mathrm{SO}_{2l+1}(F)$ $(l = m + \sum_{i=1}^{r} m_i)$ induced from the standard parabolic subgroup, whose Levi part is isomorphic to $\mathrm{GL}_{m_1}(F) \times \cdots \times \mathrm{GL}_{m_r}(F) \times \mathrm{SO}_{2m+1}(F)$, and from the representation $\tau_1 \otimes \cdots \otimes \tau_r \otimes \sigma$.

We fix a nontrivial character $\psi$ of $F$. A representation $\sigma$ of $\mathrm{SO}_{2n+1}(F)$ (in a space $V_\sigma$) has a Whittaker model with respect to $\psi$, if there is a nontrivial linear functional $\mathfrak{l}$ on $V_\sigma$, such that

$$\mathfrak{l}\left(\sigma\begin{pmatrix} z & x & y \\ & 1 & x' \\ & & z^* \end{pmatrix}(v)\right) = \psi(z_{12} + z_{23} + \cdots + z_{n-1,n} + x_n)\mathfrak{l}(v) \tag{0.8}$$

for all $v \in V_\sigma$, where $z$ is in the maximal upper unipotent subgroup of $\mathrm{GL}_n(F)$ so that $\begin{pmatrix} z & x & y \\ & 1 & x' \\ & & z^* \end{pmatrix}$ is an element in the standard maximal unipotent subgroup $U_n(F)$ of $\mathrm{SO}_{2n+1}(F)$. Since the diagonal subgroup of $\mathrm{SO}_{2n+1}(F)$ acts transitively on $U_n(F)$, we may replace the "standard nondegenerate character of $U_n(F)$ defined by $\psi$" in (0.8) by the following character

$$U_n(F) \ni \begin{pmatrix} z & x & y \\ & 1 & x' \\ & & z^* \end{pmatrix} \mapsto \psi(a_1 z_{12} + a_2 z_{23} + \cdots + a_{n-1} z_{n-1,n} + a_n x_n)$$

for given $a_1, \ldots, a_n \in F^*$. Thus, the notion of having a Whittaker model (for $\sigma$) does not depend on $\psi$ or on $a_1, \ldots, a_n \in F^*$. We simply say that $\sigma$ is generic if $\sigma$ admits a non-zero Whittaker functional as in (0.8).

Given a representation $\lambda$ of $\mathrm{GL}_k(F)$, we often denote $k = k_\lambda$.

We recall the classification of irreducible representations of $\mathrm{GL}_n(F)$ from [Z]. Let $\lambda$ be an irreducible, supercuspidal representation of $\mathrm{GL}_k(F)$. Consider a segment

$$\Sigma = [\lambda, \nu^r \lambda] = \{\lambda, \nu\lambda, \nu^2\lambda, \ldots, \nu^r\lambda\}, \tag{0.9}$$

where $r \in \mathbb{Z}_{\geq 0}$ (the set of all non-negative integers), and $\nu(\cdot) := |\det(\cdot)|_F$ ($|\cdot|_F$ denotes the absolute value character of $F^*$). The representation

$$\nu^r\lambda \times \cdots \times \nu\lambda \times \lambda$$

has a unique irreducible subrepresentation $\delta[\lambda, \nu^r\lambda]$, and it is essentially square-integrable. We say, as in p. 22 of [M2], that a segment $\Sigma$ is balanced if $\delta(\Sigma)$ is square-integrable. Balanced segments have the form

$$[\nu^{-m}\tau, \nu^m\tau], \tag{0.10}$$

where $\tau$ is an irreducible unitary supercuspidal representation of $\mathrm{GL}_{k_\tau}(F)$ and $2m \in \mathbb{Z}_{\geq 0}$. We denote, in this case,

$$\Delta(\tau, m) = \delta[\nu^{-m}\tau, \nu^m\tau] \tag{0.11}$$

for the unique irreducible square-integrable subrepresentation attached to $[\nu^{-m}\tau, \nu^m\tau]$. In general, one has, for a segment $[\nu^\alpha\tau, \nu^\beta\tau]$, $\tau$ as above, and $\alpha \in \mathbb{R}$, $\beta - \alpha \in \mathbb{Z}_{\geq 0}$,

$$\delta[\nu^\alpha\tau, \nu^\beta\tau] = \nu^{\frac{\alpha+\beta}{2}} \Delta\left(\tau, \frac{\beta - \alpha}{2}\right). \tag{0.12}$$

The dual of a segment $\Sigma = [\lambda, \nu^r\lambda]$ is the segment $\widehat{\Sigma} = [\nu^{-r}\widehat{\lambda}, \widehat{\lambda}]$. One has

$$\widehat{\delta(\Sigma)} \cong \delta(\widehat{\Sigma}). \tag{0.13}$$

Two segments $\Sigma, \Sigma'$ are said *linked* if $\Sigma \cup \Sigma'$ is a segment, $\Sigma \not\subset \Sigma'$ and $\Sigma' \not\subset \Sigma$. Every irreducible, generic representation of $\mathrm{GL}_m(F)$ has the form $\delta(\Sigma_1) \times \cdots \times \delta(\Sigma_r)$ where no two of $\{\Sigma_1, \ldots, \Sigma_r\}$ are linked. The set $\{\Sigma_1, \ldots, \Sigma_r\}$ is determined uniquely by the given irreducible, generic representation of $\mathrm{GL}_m(F)$.

Finally, the following calculations of local gamma and L-factors follow from [J.PS.S], [S1,3]. See also [M1]. Let $\Sigma_1, \ldots, \Sigma_r$ be a sequence of segments. Write $\delta(\Sigma_i) = \nu^{e_i}\delta_i$ as in (0.12), where $\delta_i = \Delta(\tau_i, m_i)$ is square-integrable. Assume that $e_1 \geq e_2 \geq \cdots \geq e_r$. Let $\sigma$ be the Langlands quotient of $\delta(\Sigma_1) \times \cdots \times \delta(\Sigma_r)$. Similarly let $\Sigma'_1, \ldots, \Sigma'_{r'}$ be another such sequence of segments; $\delta(\Sigma'_j) = \nu^{e'_j}\delta'_j$, $\delta'_j = \Delta(\tau'_j, m'_j)$, as above. Let $\sigma'$ be the Langlands quotient of $\delta(\Sigma'_1) \times \cdots \times \delta(\Sigma'_{r'})$. Then

$$\gamma(\sigma \times \sigma', s, \psi) = \prod_{\substack{1 \leq i \leq r \\ 1 \leq j \leq r'}} \gamma(\delta_i \times \delta'_j, s + e_i + e'_j, \psi) \tag{0.14}$$

$$L(\sigma \times \sigma', s) = \prod_{\substack{1 \leq i \leq r \\ 1 \leq j \leq r'}} L(\delta_i \times \delta'_j, s + e_i + e'_j) \tag{0.15}$$

$$\begin{aligned} &\gamma(\Delta(\tau_i, m_i) \times \Delta(\tau'_j, m'_j), s, \psi) \\ &\quad = \prod_{\substack{0 \leq k \leq 2m_i \\ 0 \leq k' \leq 2m'_j}} \gamma(\tau_i \times \tau'_j, s - m_i - m'_j + k + k', \psi) \end{aligned} \tag{0.16}$$

$$\begin{aligned} &L(\Delta(\tau_i, m_i) \times \Delta(\tau'_j, m'_j), s) \\ &\quad = \prod_{\ell=1}^{\min(2m_i+1, 2m'_j+1)} L(\tau_i \times \tau'_j, s + m_i + m'_j + 1 - \ell). \end{aligned} \tag{0.17}$$

**1. Supercuspidal generic representations.** The description of the local Langlands functorial lift from irreducible, supercuspidal, generic representations of $\mathrm{SO}_{2n+1}(F)$ to irreducible representations of $\mathrm{GL}_{2n}(F)$, is one of the main results of our previous work [Jng.S., Theorems 6.1, 6.4]. We summarize it here.

THEOREM 1.1 [JNG.S., THEOREM 6.1]. *There is a bijection $\ell$ between the set $\Pi^{(sg)}(\mathrm{SO}_{2n+1})$ of equivalence classes of irreducible, supercuspidal, generic representation of $\mathrm{SO}_{2n+1}(F)$, and the set of isomorphism classes of representations of $\mathrm{GL}_{2n}(F)$, which have the form*

$$\tau_1 \times \tau_2 \times \cdots \times \tau_r = \mathrm{Ind}_Q^{\mathrm{GL}_{2n}(F)}(\tau_1 \otimes \tau_2 \otimes \cdots \otimes \tau_r), \tag{1.1}$$

*where $Q$ is a standard parabolic subgroup of $\mathrm{GL}_{2n}(F)$ of type $(2n_1, \ldots, 2n_r)$ with $n = \sum_{i=1}^r n_i$, and for each $1 \leq i \leq r$, $\tau_i$ is an irreducible, supercuspidal representation of $\mathrm{GL}_{2n_i}(F)$, such that $L(\tau_i, \Lambda^2, s)$ has a pole at $s = 0$ and for $i \neq j$, $\tau_i \not\cong \tau_j$. The bijection $\ell$ preserves local $L$ and $\epsilon$ factors with $\mathrm{GL}$-twists, namely,*

$$L(\sigma \times \pi, s) = L(\ell(\sigma) \times \pi, s) \tag{1.2}$$

$$\epsilon(\sigma \times \pi, s, \psi) = \epsilon(\ell(\sigma) \times \pi, s, \psi) \tag{1.3}$$

*for any irreducible, supercuspidal, generic representation $\sigma$ of $\mathrm{SO}_{2n+1}(F)$ and any irreducible, generic representation $\pi$ of $\mathrm{GL}_k(F)$ ($k$ is any positive integer).*

*Remark* 1.1. 1. If $\tau$ is an irreducible supercuspidal representation of $GL_{2m+1}(F)$, then $L(\tau, \Lambda^2, s)$ is holomorphic. This is the reason that $\tau_i$'s in the theorem are representations of $GL_{2n_i}(F)$.

2. The local factors on the l.h.s. of (1.2), (1.3) are the ones defined by Shahidi for $SO_{2n+1} \times GL_k$. The local factors on the r.h.s. are for $GL_{2n} \times GL_k$. See [J.PS.S] and [S1]. See also [S2] for properties of local exterior square L-functions.

3. Note that the representation (1.1) is irreducible and tempered.

We characterized in [Jng.S.] the local Langlands parameters for the representations in the image of the local Langlands functorial lift from irreducible generic supercuspidal representations of $SO_{2n+1}(F)$ to $GL_{2n}(F)$, based on [H.T.] and [H1]. This provides the local Langlands parameters for irreducible generic supercuspidal representations of $SO_{2n+1}(F)$. In general, local Langlands parameters for $SO_{2n+1}(F)$ are defined as follows. General discussion of local Langlands parameters can be found in [A], [B], and [Kn]. Recall that the Langlands dual group of $SO_{2n+1}(F)$ is $Sp_{2n}(\mathbb{C})$.

*Definition 1.1 (Local Langlands Parameters for* $SO_{2n+1}(F)$*).* Let $W_F \times SL_2(\mathbb{C})$ be the Weil-Deligne group attached to the base field $F$. A local Langlands parameter $\varphi$ for $SO_{2n+1}(F)$ is a conjugacy class of admissible homomorphism from $W_F \times SL_2(\mathbb{C})$ to $Sp_{2n}(\mathbb{C})$. Here, if we decompose $\varphi$ into a direct sum of irreducible representations of $W_F \times SL_2(\mathbb{C})$

$$\varphi = \oplus_i \phi_i \otimes S_{w_i+1},$$

the admissibility of $\varphi$ means that $\phi_i$'s are continuous complex representations of $W_F$ with $\phi_i(W_F)$ consisting of semi-simple elements in $Sp_{2n}(\mathbb{C})$ and $S_{w_i+1}$ is the irreducible algebraic complex representation of $SL_2(\mathbb{C})$ of dimension $w_i + 1$. We denote by $\Phi(SO_{2n+1})$ the set of all local Langlands parameters for $SO_{2n+1}(F)$.

Let $\Phi^{(0)}(SO_{2n+1})$ be the subset of $\Phi(SO_{2n+1})$ consisting of all parameters of type

$$\varphi = \oplus_i \phi_i \otimes S_{w_i+1}$$

with the properties that (i) $w_i = 0$ for all $i$'s, (ii) $\phi_i \not\cong \phi_j$ if $i \neq j$, and (iii) for each $i$, $\phi_i$ is an irreducible element in $\Phi(SO_{2n_i+1})$ for some non-negative integer $n_i$. Then we have

THEOREM 1.2 [JNG.S., THEOREM 6.2]. *There is a bijection $y$ between $\Phi^{(0)}(SO_{2n+1})$ and the set $\Pi^{(sg)}(SO_{2n+1})$. The bijection $y$ preserves local factors as follows.*

$$L(\varphi \otimes \varphi', s) = L(y(\varphi) \times r(\varphi), s) \tag{1.4}$$

$$\epsilon(\varphi \otimes \varphi', s, \psi) = \epsilon(y(\varphi) \times r(\varphi'), s, \psi) \tag{1.5}$$

*for all irreducible, admissible, representations $\varphi'$ of $W_F$ of dimension $k$ with all $k \in \mathbb{Z}_{>0}$. Here $r(\varphi')$ is the supercuspidal representation of* $\mathrm{GL}_k(F)$*, corresponding to $\varphi'$ by the local Langlands reciprocity map for* $\mathrm{GL}(k)$ *as in [H.T.] and [H1].*

Note that for $\varphi$ as in Theorem 1.2, the composition of $\varphi$ with the embedding $\mathrm{Sp}_{2n}(\mathbb{C}) \subset \mathrm{GL}_{2n}(\mathbb{C})$ gives a degree $2n$ representation of $W_F$, which has a multiplicity one decomposition into irreducible representations

$$\phi_1 \oplus \cdots \oplus \phi_r \tag{1.6}$$

such that each summand $\phi_i$ is symplectic, say of degree $2n_i$ with $n_1 + \cdots + n_r = n$. It turns out that the corresponding supercuspidal representation $\tau_i = r(\phi_i)$ of $\mathrm{GL}_{2n_i}(F)$ is such that $L(\tau_i, \Lambda^2, s)$ has a pole at $s = 0$. We have

$$\begin{aligned} y(\varphi) &= \ell^{-1}\left(\mathrm{Ind}_Q^{\mathrm{GL}_{2n}(F)}(\tau_1 \otimes \cdots \otimes \tau_r)\right) \\ &= \ell^{-1}\left(\mathrm{Ind}_Q^{\mathrm{GL}_{2n}(F)}(r(\phi_1) \otimes \cdots \otimes r(\phi_r))\right) \\ &= \ell^{-1}(r(\phi_1 \oplus \cdots \oplus \phi_r)) \end{aligned} \tag{1.7}$$

where we keep denoting by $r$ the local Langlands reciprocity map for GL.

A key ingredient in the proof of the theorems above is the following local converse theorem [Jng.S.].

THEOREM 1.3. *Let $\sigma$ and $\sigma'$ be irreducible, generic representations of* $\mathrm{SO}_{2n+1}(F)$*. If for all irreducible, supercuspidal representations $\pi$ of* $\mathrm{GL}_k(F)$*, $1 \le k \le 2n-1$, we have*

$$\gamma(\sigma \times \pi, s, \psi) = \gamma(\sigma' \times \pi, s, \psi)$$

*then $\sigma \cong \sigma'$.*

We quote here some special cases of Shahidi's general theorem on the Multiplicativity of Twisted Gamma Factors [S3]. These special cases are also proved by Soudry in [Sd1,2] and will be needed in the rest of the paper.

THEOREM 1.4 [MULTIPLICATIVITY OF GAMMA FACTORS]. (1) *Suppose that an irreducible admissible generic representation $\sigma$ of* $\mathrm{SO}_{2n+1}(k)$ *is a subquotient of* $\mathrm{Ind}_{P_r}^{\mathrm{SO}_{2n+1}(k)}(\tau_r \otimes \sigma_{n-r})$*, the unitarily induced representation from a standard maximal parabolic subgroup $P_r$ of* $\mathrm{SO}_{2n+1}(k)$*, where $\tau_r$ is an admissible generic representation of* $\mathrm{GL}_r(k)$ *and $\sigma_{n-r}$ is an admissible generic representation of* $\mathrm{SO}_{2(n-r)+1}(k)$*. Then*

$$\gamma(\sigma \times \varrho, s, \psi) = \omega_\tau(-1)^n \gamma(\tau_r \times \varrho, s, \psi) \cdot \gamma(\sigma_{n-r} \times \varrho, s, \psi) \cdot \gamma(\tau_r^\vee \times \varrho, s, \psi),$$

*for any irreducible admissible generic representations* $\varrho$ *of* $\mathrm{GL}_l(k)$ *with* $l$ *being any positive integer, where* $\gamma(\tau_r \times \varrho, s, \psi)$ *and* $\gamma(\tau_r^\vee \times \varrho, s, \psi)$ *are the local gamma factors defined in [J.PS.S.].* ($\tau^\vee$ *is the contragredient representation of* $\tau$.)

(2) *Suppose that an irreducible admissible generic representation* $\varrho$ *of* $\mathrm{GL}_l(k)$ *is a subquotient of* $\mathrm{Ind}_{P_{r,l-r}(k)}^{\mathrm{GL}_l(k)}(\tau_r \otimes \tau_{l-r})$, *the unitarily induced representation from a standard maximal parabolic subgroup* $P_{r,l-r}$ *of* $\mathrm{GL}_l$, *where* $\tau_r$ *is an admissible generic representation of* $\mathrm{GL}_r(k)$ *and* $\tau_{l-r}$ *is an admissible generic representation of* $\mathrm{GL}_{l-r}(k)$. *Then*

$$\gamma(\sigma \times \varrho, s, \psi) = \gamma(\sigma \times \tau_r, s, \psi) \cdot \gamma(\sigma \times \tau_{l-r}, s, \psi),$$

*for any irreducible admissible generic representations* $\sigma$ *of* $\mathrm{SO}_{2n+1}(k)$.

In Sections 2, 3, 4, and 5, we will extend Theorems 1.1 and 1.2 to generic discrete series representations, to generic elliptic tempered representations, to generic tempered representations, and to generic representations, respectively.

**2. Discrete Series generic representations.** We first extend the local Langlands functorial lift $\ell$ from the set $\Pi^{(sg)}(\mathrm{SO}_{2n+1})$ of all equivalence classes of irreducible, supercuspidal, generic representations to the set $\Pi^{(dg)}(\mathrm{SO}_{2n+1})$ of all equivalence classes of irreducible, discrete series, generic representations of $\mathrm{SO}_{2n+1}(F)$. Then we write the local Langlands parameters for each member in $\Pi^{(dg)}(\mathrm{SO}_{2n+1})$. As we mentioned before, the main ingredients for this extension are Theorem 1.1 and the description given by Muic in [M1,2] of square-integrable generic representations of $\mathrm{SO}_{2n+1}(F)$.

THEOREM 2.1. *The bijection* $\ell$ *of Theorem 1.1 can be extended to a bijection (we still denote it by* $\ell$*) between the set* $\Pi^{(dg)}(\mathrm{SO}_{2n+1})$ *and the set of all equivalence classes of irreducible tempered representations of* $\mathrm{GL}_{2n}(F)$ *of the following form*

$$\Delta(\tau_1, m_1) \times \Delta(\tau_2, m_2) \times \cdots \times \Delta(\tau_r, m_r) \tag{2.1}$$

*where the balanced segments* $[\nu^{-m_i}\tau_i, \nu^{m_i}\tau_i]$ *are pairwise distinct, self-dual (i.e.,* $\tau_i \cong \widehat{\tau}_i$*) and satisfy the following properties that for each* $i$,

(1) *if* $L(\tau_i, \mathrm{sym}^2, s)$ *has a pole at* $s = 0$, *then* $m_i \in \frac{1}{2} + \mathbb{Z}_{\geq 0}$; *or*

(2) *if* $L(\tau_i, \Lambda^2, s)$ *has a pole at* $s = 0$, *then* $m_i \in \mathbb{Z}_{\geq 0}$.

*We have the following compatibility of local factors:*

$$L(\sigma \times \pi, s) = L(\ell(\sigma) \times \pi, s), \tag{2.2}$$

$$\epsilon(\sigma \times \pi, s, \psi) = \epsilon(\ell(\sigma) \times \pi, s, \psi) \tag{2.3}$$

*for any* $\sigma$ *in* $\Pi^{(dg)}(\mathrm{SO}_{2n+1})$ *and any irreducible generic representation* $\pi$ *of* $\mathrm{GL}_k(F)$ *with all* $k \in \mathbb{Z}_{>0}$.

*Remark* 2.1. An equivalent description of the image of the local Langlands functorial lift $\ell$ in Theorem 2.1 is the following. An irreducible representation $\rho$ of $\mathrm{GL}_{2n}(F)$ lies in the image of $\ell$ in Theorem 2.1 if and only if $\rho$ is tempered, and satisfies the following properties: for any irreducible, unitary, supercuspidal representation $\tau$ of $\mathrm{GL}_k(F)$ with $k = 1, 2, \ldots, 2n$,

(1) if $\tau \not\cong \widehat{\tau}$, then $L(\rho \times \tau, s)$ has no poles on the real line,

(2) if $\tau \cong \widehat{\tau}$ and $L(\rho \times \tau, s)$ is not holomorphic, then

(a) if $L(\tau, \mathrm{sym}^2, s)$ has a pole at $s = 0$, then $L(\rho \times \tau, s)$ has only simple poles, whose real parts lie inside $-\frac{1}{2} + \mathbb{Z}_{\leq 0}$ (where $\mathbb{Z}_{\leq 0}$ denotes the set of all non-positive integers);

(b) if $L(\tau, \Lambda^2, s)$ has a pole at $s = 0$, then $L(\rho \times \tau, s)$ has only simple poles, whose real parts be inside $\mathbb{Z}_{\leq 0}$.

Note that by [Z] any (irreducible) tempered representation $\rho$ of $\mathrm{GL}_{2n}(F)$ has the form

$$\Delta(\tau_1, m_1) \times \Delta(\tau_2, m_2) \times \cdots \times \Delta(\tau_r, m_r)$$

where $2m_i \in \mathbb{Z}_{\geq 0}$ and $\tau_i$ are irreducible, unitary, supercuspidal representations of some $\mathrm{GL}_{k_i}(F)$, respectively. The condition (1) in the last remark implies that the representations $\tau_i$ are self-dual, because of formula (0.17), and then conditions (a), (b) show that the description in the last remark is identical with that of the image of $\ell$ in Theorem 2.1.

*Proof of Theorem 2.1.* We start with a tempered representation $\rho$ of $\mathrm{GL}_{2n}(F)$ in the proposed image, and associate to it a discrete series generic representation $\sigma = \sigma_\rho$ (depending on $\rho$) of $\mathrm{SO}_{2n+1}(F)$, such that $\rho = \ell(\sigma)$ and (2.2) and (2.3) hold. It will be convenient to use the description of Remark 2.1 as the conditions characterizing the image.

The idea is to use the information about the poles on the real line of the local L-functions $L(\rho \times \tau, s)$, for all $\tau$ in the set $\Pi^{(ss)}(\mathrm{GL}_k)$ of equivalence classes of irreducible, self-dual, supercuspidal representations of $\mathrm{GL}_k(F)$ (with $k$ being any positive integer), to determine the structure of the tempered representation $\rho$. Denote by $P(\rho)$ the set of all such poles, i.e.,

$$P(\rho) := \{\tau \in \Pi^{(ss)}(\mathrm{GL}_k) \mid L(\rho \times \tau, s) \text{ has a pole in } \mathbb{R}; k \in \mathbb{Z}_{>0}\}.$$

It is known from the structure of tempered representations of GL that the set $P(\rho)$ is finite. For $\tau \in P(\rho)$, we list the real poles of $L(\rho \times \tau, s)$ as follows

$$-m_{d_\tau}(\tau) < \cdots < -m_2(\tau) < -m_1(\tau) \leq 0. \tag{2.4}$$

Let us put $d_\tau = 0$ if $L(\rho \times \tau, s)$ is holomorphic for $\tau$ irreducible, supercuspidal (self-dual or not). Consider the following subsets of $P(\rho)$

$$A(\rho) = \{\tau \in P(\rho) \mid L(\tau, \Lambda^2, s) \text{ has a pole at } s = 0, \text{ and } d_\tau \text{ is odd}\},$$

$$B(\rho) = \{\tau \in P(\rho) \mid L(\tau, \Lambda^2, s) \text{ has a pole at } s = 0, \text{ and } d_\tau \text{ is even}\},$$

$$C(\rho) = \{\tau \in P(\rho) \mid L(\tau, \mathrm{sym}^2, s) \text{ has a pole at } s = 0\}.$$

Note that since $\rho$ satisfies the assumptions of Remark 2.1, we have

$$P(\rho) = A(\rho) \cup B(\rho) \cup C(\rho).$$

In particular, for $\tau \in A(\rho) \cup B(\rho)$, we have $\{m_i(\tau)\}_{i=1}^{d_\tau} \subset \mathbb{Z}_{\geq 0}$, and for $\tau \in C(\rho)$, we have $\{m_i(\tau)\}_{i=1}^{d_\tau} \subset \frac{1}{2} + \mathbb{Z}_{\geq 0}$.

We will now construct a certain set of supercuspidal representations, which defines an irreducible supercuspidal representation of a certain Levi subgroup of $\mathrm{SO}_{2n+1}(F)$ and then we form the corresponding induced representation of $\mathrm{SO}_{2n+1}(F)$.

Since, for $\tau \in A(\rho)$, $L(\tau, \Lambda^2, s)$ has a pole at $s = 0$, $k_\tau$ is even. Write $k_\tau = 2n_\tau$. By Theorem 1.1, there exists a unique (up to equivalence) irreducible, supercuspidal, generic representation $\sigma_\rho^0$ of $\mathrm{SO}_{2n'+1}(F)$, $n' = \sum_{\tau \in A(\rho)} n_\tau$, such that

$$\ell\big(\sigma_\rho^{(0)}\big) = \mathop{\times}_{\tau \in A(\rho)} \tau$$

on $\mathrm{GL}_{2n'}(F)$. Consider the following three subsets of $A(\rho)$:

$$A_0(\rho) = \{\tau \in A(\rho) \mid d_\tau = 1 \text{ and } m_1(\tau) = 0\},$$

$$A_1(\rho) = \{\tau \in A(\rho) \mid d_\tau \geq 3 \text{ and } m_1(\tau) = 0\},$$

$$A_2(\rho) = \{\tau \in A(\rho) \mid m_1(\tau) \geq 1\}.$$

It is clear that they form a partition of $A(\rho)$. For $\tau \in A_1(\rho)$, we consider the following set of essentially square-integrable representations (of appropriate $GL_*(F)$)

$$\Delta_i(\tau) = \delta[\nu^{-m_{2i}(\tau)}\tau, \nu^{m_{2i+1}(\tau)}\tau], \quad i = 1, 2, \ldots, \frac{d_\tau - 1}{2}, \tag{2.5}$$

and for $\tau \in A_2(\rho)$, we consider

$$\Delta_0(\tau) = \delta[\nu\tau, \nu^{m_1(\tau)}\tau], \; \Delta_i(\tau) = \delta[\nu^{-m_{2i}(\tau)}\tau, \nu^{m_{2i+1}(\tau)}\tau], \quad i = 1, 2, \ldots, \frac{d_\tau - 1}{2}. \tag{2.6}$$

(Recall that $d_\tau$ is odd, and that $m_i(\tau) \in \mathbb{Z}_+$, for $\tau \in A(\rho)$.)

For $\tau \in B(\rho)$, define the following sequence of essentially square-integrable representations (of appropriate $\mathrm{GL}_*(F)$).

$$\Delta_i(\tau) = \delta[\nu^{-m_{2i-1}(\tau)}\tau, \nu^{m_{2i}(\tau)}\tau], \quad i = 1, 2, \ldots, \frac{d_\tau}{2}, \tag{2.7}$$

(note that here $d_\tau$ is even). Similarly, for $\tau \in C(\rho)$, if $d_\tau$ is odd, define the sequence

$$\Delta_0(\tau) = \delta\big[\nu^{1/2}\tau, \nu^{m_1(\tau)}\tau\big], \ \Delta_i(\tau) = \delta\big[\nu^{-m_{2i}(\tau)}\tau, \nu^{m_{2i+1}(\tau)}\tau\big], \quad i = 1, 2, \dots, \frac{d_\tau - 1}{2}, \tag{2.8}$$

and finally, for $\tau \in C(\rho)$, if $d_\tau$ is even, define

$$\Delta_i(\tau) = \delta[\nu^{-m_{2i-1}(\tau)}\tau, \nu^{m_{2i}(\tau)}\tau], \quad i = 1, 2, \dots, \frac{d_\tau}{2}. \tag{2.9}$$

Finally, denote

$$J_\tau = \begin{cases} \{1, 2, \dots, \frac{d_\tau - 1}{2}\}, & \text{in case (2.5)} \\ \{0, 1, 2, \dots, \frac{d_\tau - 1}{2}\}, & \text{in cases (2.6), (2.8)} \\ \{1, 2, \dots, \frac{d_\tau}{2}\}, & \text{in cases (2.7), (2.9).} \end{cases}$$

Now, let $\sigma_\rho$ be the unique generic constituent (actually, it is a subrepresentation) of

$$\Big( \times_{\tau \in P(\rho) \setminus A_0(\rho)} \ \times_{j \in J_\tau} \Delta_j(\tau) \Big) \rtimes \sigma_\rho^{(0)}. \tag{2.10}$$

Note that $\sigma_\rho$ is indeed on $SO_{2n+1}(F)$. By [Td] and [M2, Thm. 2.1, Prop. 2.1], $\sigma_\rho$ is square-integrable. We will review and recall the square integrability of $\sigma_\rho$ in detail after we show formula (2.2) and (2.3) hold for the pair $(\sigma_\rho, \rho)$.

We first show that

$$\gamma(\sigma_\rho \times \pi, s, \psi) = \gamma(\rho \times \pi, s, \psi) \tag{2.11}$$

for all irreducible, generic representations $\pi$ of $GL_k(F)$ with all $k \in \mathbb{Z}_{>0}$. By the multiplicativity property of gamma factors (Theorem 1.4), it is enough to show (2.11) for supercuspidal $\pi$. Since $\sigma_\rho$ is a constituent of (2.10), we have

$$\begin{aligned} \gamma(\sigma_\rho \times \pi, s, \psi) &= \Big[ \prod_{\tau \in P(\rho) \setminus A_0(\rho)} \prod_{j \in J_\tau} \gamma(\Delta_j(\tau) \times \pi, s, \psi) \gamma(\widehat{\Delta_j(\tau)} \times \pi, s, \psi) \Big] \\ &\quad \times \gamma\big(\sigma_\rho^{(0)} \times \pi, s, \psi\big) \\ &= \Big[ \prod_{\tau \in P(\rho) \setminus A_0(\rho)} \prod_{j \in J_\tau} \gamma(\Delta_j(\tau) \times \pi, s, \psi) \gamma(\widehat{\Delta_j(\tau)} \times \pi, s, \psi) \Big] \\ &\quad \times \prod_{\tau \in A(\rho)} \gamma(\tau \times \pi, s, \psi). \end{aligned} \tag{2.12}$$

Here we used Theorem 1.1 and the fact that $\ell(\sigma_\rho^{(0)}) = \times_{\tau \in A(\rho)} \tau$.

We split the product (2.12) into the following five terms

$$(\mathrm{I}): \prod_{\tau\in A_0(\rho)} \gamma(\tau\times\pi, s, \psi)$$

$$(\mathrm{II})_{\mathrm{i}}: \prod_{\tau\in A_i(\rho)} \gamma(\tau\times\pi, s, \psi) \prod_{j\in J_\tau} \gamma(\Delta_j(\tau)\times\pi, s, \psi)\gamma(\widehat{\Delta_j(\tau)}\times\pi, s, \psi), i = 1, 2$$

$$(\mathrm{III})_{\mathrm{B}}: \prod_{\tau\in B(\rho)} \prod_{j\in J_\tau} \gamma(\Delta_j(\tau)\times\pi, s, \psi)\gamma(\widehat{\Delta_j(\tau)}\times\pi, s, \psi)$$

$$(\mathrm{III})_{\mathrm{C}}: \prod_{\tau\in C(\rho)} \prod_{j\in J_\tau} \gamma(\Delta_j(\tau)\times\pi, s, \psi)\gamma(\widehat{\Delta_j(\tau)}\times\pi, s, \psi).$$

Now we consider $\gamma(\rho\times\pi, s, \psi)$. By our assumption on $\rho$, formula (2.4) and the multiplicativity of gamma factors, we have

$$\gamma(\rho\times\pi, s, \psi) = \prod_{i=1}^{r} \gamma(\Delta(\tau_i, m_i)\times\pi, s, \psi) \tag{2.13}$$

$$= \prod_{\tau\in P(\rho)} \prod_{i=1}^{d_\tau} \gamma(\Delta(\tau, m_i(\tau))\times\pi, s, \psi).$$

We have to show that the product in (2.13) consists exactly of the factors which appear in the five products above. Note that each term in the product (I) appears in (2.13) since for $\tau\in A_0(\rho)$, we have $d_\tau = 1$ and $m_1(\tau) = 0$.

Next, we have to consider a product formula for gamma factors related to the induced representation

$$\Delta_j(\tau)\times\widehat{\Delta_j(\tau)}.$$

The use of the $\times$ here may be confused with the Rankin product in the twisted gamma factor. To distinguish these, we use

$$[\Delta_j(\tau)\times\widehat{\Delta_j(\tau)}]$$

to indicate the induced representation in any formula here. Then we have

$$\gamma([\Delta_j(\tau)\times\widehat{\Delta_j(\tau)}]\times\pi, s, \psi) = \gamma(\Delta_j(\tau)\times\pi, s, \psi)\gamma(\widehat{\Delta_j(\tau)}\times\pi, s, \psi)$$

for $j\geq 1$. There are two cases. In the first case the representation $\Delta_j(\tau)$ appears in (2.5), (2.6), and (2.8). In this case we have

$$\Delta_j(\tau)\times\widehat{\Delta_j(\tau)} = \delta\big[\nu^{-m_{2j}(\tau)}\tau, \nu^{m_{2j+1}(\tau)}\tau\big]\times\delta\big[\nu^{-m_{2j+1}(\tau)}\tau, \nu^{m_{2j}(\tau)}\tau\big].$$

By [Z], the unique generic constituent of this representation is

$$\delta\big[\nu^{-m_{2j}(\tau)}\tau, \nu^{m_{2j}(\tau)}\tau\big]\times\delta\big[\nu^{-m_{2j+1}(\tau)}\tau, \nu^{m_{2j+1}(\tau)}\tau\big] \tag{2.14}$$

$$= \Delta(\tau, m_{2j}(\tau))\times\Delta(\tau, m_{2j+1}(\tau)).$$

By multiplicativity of gamma factors, we get

$$(2.15)\quad \gamma([\Delta_j(\tau) \times \widehat{\Delta_j(\tau)}] \times \pi, s, \psi) = \gamma(\Delta(\tau, m_{2j}(\tau)) \times \pi, s, \psi) \times \gamma(\Delta(\tau, m_{2j+1}(\tau)) \times \pi, s, \psi).$$

In the second case the representation $\Delta_j(\tau)$ appears in (2.7), (2.9). In this case we similarly get

$$(2.16)\quad \gamma([\Delta_j(\tau) \times \widehat{\Delta_j(\tau)}] \times \pi, s, \psi) = \gamma(\Delta(\tau, m_{2j-1}(\tau)) \times \pi, s, \psi) \times \gamma(\Delta(\tau, m_{2j}(\tau)) \times \pi, s, \psi).$$

We conclude from (2.15) that

$$(2.17)\quad \prod_{\tau \in A_1(\rho)} \gamma(\tau \times \pi, s, \psi) \prod_{j \in J_\tau} \gamma(\Delta_j(\tau) \times \pi, s, \psi)\gamma(\widehat{\Delta_j(\tau)} \times \pi, s, \psi)$$
$$= \prod_{\tau \in A_1(\rho)} \gamma(\tau \times \pi, s, \psi) \prod_{k=2}^{d_\tau} \gamma(\Delta(\tau, m_k(\tau)) \times \pi, s, \psi)$$
$$= \prod_{\tau \in A_1(\rho)} \prod_{k=1}^{d_\tau} \gamma(\Delta(\tau, m_k(\tau)) \times \pi, s, \psi).$$

This shows that the product of type $(\mathrm{II})_1$ appears in (2.13). Similarly, using (2.16), we find that the product of type $(\mathrm{III})_{\mathrm{B}}$ appears in (2.13), and also the following part of $(\mathrm{III})_{\mathrm{C}}$ appears in (2.13)

$$(2.18)\quad \prod_{\substack{\tau \in C(\rho) \\ d_\tau \text{ even}}} \prod_{j \in J_\tau} \gamma(\Delta_j(\tau) \times \pi, s, \psi)\gamma(\widehat{\Delta_j(\tau)} \times \pi, s, \psi)$$
$$= \prod_{\substack{\tau \in C(\rho) \\ d_\tau \text{ even}}} \prod_{i=1}^{d_\tau} \gamma(\Delta(\tau, m_i(\tau)) \times \pi, s, \psi).$$

The term $(\mathrm{II})_2$ is treated as in (2.17), except that we still have to consider that $j = 0$ and $\tau \in A_2(\rho)$ in (2.6). In this case, we have

$$\Delta_0(\tau) \times \widehat{\Delta_0(\tau)} \times \tau = \delta[\nu\tau, \nu^{m_1(\tau)}\tau] \times \delta[\nu^{-m_1(\tau)}\tau, \nu^{-1}\tau] \times \tau.$$

By [Z], the unique generic constituent of this representation is

$$\delta[\nu^{-m_1(\tau)}\tau, \nu^{m_1(\tau)}\tau] = \Delta(\tau, m_1(\tau)).$$

Hence we get

$$(2.19)\quad \gamma(\tau \times \pi, s, \psi)\gamma(\Delta_0(\tau) \times \pi, s, \psi)\gamma(\widehat{\Delta_0(\tau)} \times \pi, s, \psi) = \gamma(\Delta(\tau, m_1(\tau)) \times \pi, s, \psi).$$

Using (2.19) and (2.15) for $j \geq 1$, we get

$$(2.20) \quad \prod_{\tau \in A_2(\rho)} \gamma(\tau \times \pi, s, \psi) \prod_{j \in J_\tau} \gamma(\Delta_j(\tau) \times \pi, s, \psi)\gamma(\widehat{\Delta_j(\tau)} \times \pi, s, \psi)$$

$$= \prod_{\tau \in A_2(\rho)} \gamma(\Delta(\tau, m_1(\tau)) \times \pi, s, \psi) \prod_{i=2}^{d\tau} \gamma(\Delta(\tau, m_i(\tau)) \times \pi, s, \psi)$$

$$= \prod_{\tau \in A_2(\rho)} \prod_{i=1}^{d\tau} \gamma(\Delta(\tau, m_i(\tau) \times \pi, s, \psi).$$

Thus, the term $(\mathrm{II})_2$ appears in (2.13). Finally, the term of type $(\mathrm{III})_{\mathrm{C}}$, with $d_\tau$ odd is treated similarly to the last case and to (2.20). We only have to consider $j = 0$ and $\tau \in C(\rho)$ in (2.8). In this case we have

$$\Delta_0(\tau) \times \widehat{\Delta_0(\tau)} = \delta\big[\nu^{1/2}\tau, \nu^{m_1(\tau)}\tau\big] \times \delta\big[\nu^{-m_1(\tau)}\tau, \nu^{-1/2}\tau\big].$$

By [Z], the unique generic constituent of this representation is

$$\delta\big[\nu^{-m_1(\tau)}\tau, \nu^{m_1(\tau)}\tau\big] = \Delta(\tau, m_1(\tau)).$$

We get, as in (2.18) and (2.20)

$$(2.21) \quad \prod_{\substack{\tau \in C(\rho) \\ d_\tau \text{ odd}}} \prod_{j \in J_\tau} \gamma(\Delta_j(\tau) \times \pi, s, \psi)\gamma(\widehat{\Delta_j(\tau)} \times \pi, s, \psi)$$

$$= \prod_{\substack{\tau \in C(\rho) \\ d_\tau \text{ odd}}} \prod_{i=1}^{d_\tau} \gamma(\Delta(\tau, m_i(\tau)) \times \pi, s, \psi).$$

Multiplying (2.18) and (2.21), we see that $(\mathrm{III})_{\mathrm{C}}$ appears in (2.13). Since

$$P(\rho) = A(\rho) \cup B(\rho) \cup C(\rho),$$

the identity in (2.11) is now clear.

Note that the generic constituent $\sigma_\rho$ of (2.10) is uniquely determined by (2.11). This follows from Theorem 1.3. From (2.11), we also conclude that

$$(2.22) \quad L(\sigma_\rho \times \pi, s) = L(\rho \times \pi, s)$$

$$(2.23) \quad \epsilon(\sigma_\rho \times \pi, s, \psi) = \epsilon(\rho \times \pi, s, \psi),$$

that is, (2.2) and (2.3) hold for the pair $(\sigma_\rho, \rho)$. Indeed, we may assume that $\pi$ is also unitary. Rewrite (2.11) as

$$(2.24) \quad \epsilon(\sigma_\rho \times \pi, s, \psi)\frac{L(\sigma_\rho \times \widehat{\pi}, 1-s)}{L(\sigma_\rho \times \pi, s)} = \epsilon(\rho \times \pi, s, \psi)\frac{L(\rho \times \widehat{\pi}, 1-s)}{L(\rho \times \pi, s)}.$$

Note that $\rho$ is self-dual. By [C.S., p. 573], since $\sigma_\rho$ and $\pi$ are (in particular) tempered, $L(\sigma_\rho \times \pi, z)$ and $L(\sigma_\rho \times \widehat{\pi}, z)$ are holomorphic, for $Re(z) > 0$. This shows that, as polynomials in $q^{-s}$, $L(\sigma_\rho \times \pi, s)^{-1}$ and $L(\sigma_\rho \times \widehat{\pi}, 1-s)^{-1}$ are relatively prime.

Similarly, (since $\rho$ is tempered) $L(\rho \times \pi, s)^{-1}$ and $L(\rho \times \widehat{\pi}, 1-s)^{-1}$ are relatively prime. Since the $\epsilon$-factors are of the form $aq^{bs}$, we get (2.22) and (2.23).

We turn now to show that the generic constituent $\sigma_\rho$ of (2.10) is square-integrable. To this end we recall from [Td], [M2, Sec. 2] the structure of generic square-integrable representations of $\mathrm{SO}_{2n+1}(F)$.

Let $P'$ be a finite set of irreducible, supercuspidal, self-dual representations $\tau$ of $\mathrm{GL}_{k_\tau}(F)$. Assume that for each $\tau \in P'$, there is a sequence of segments

$$D_i(\tau) = \left[\nu^{-a_i(\tau)}\tau, \nu^{b_i(\tau)}\tau\right], \quad i = 1, 2, \ldots, e_\tau, \tag{2.25}$$

satisfying

$$2a_i(\tau) \in \mathbb{Z} \quad \text{and} \quad 2b_i(\tau) \in \mathbb{Z}_+, \tag{2.26}$$

and

(2.27)
$$a_1(\tau) < b_1(\tau) < a_2(\tau) < b_2(\tau) < a_3(\tau) < \cdots < b_{e_\tau - 1}(\tau) < a_{e_\tau}(\tau) < b_{e_\tau}(\tau).$$

Next, let $\sigma^{(0)}$ be an irreducible, supercuspidal, generic representation of $\mathrm{SO}_{2n'+1}(F)$. Assume that

(2.28) $(C1)$ if $L(\sigma^{(0)} \times \tau, s)$ has a pole at $s = 0$, then $-1 \le a_i(\tau) \in \mathbb{Z}\backslash\{0\}$, for $1 \le i \le e_\tau$;

(2.29) $(C0)$ if $L(\tau, \Lambda^2, s)$ has a pole at $s = 0$, but $L(\sigma^{(0)} \times \tau, s)$ is holomorphic at $s = 0$, then $a_i(\tau) \in \mathbb{Z}_{\ge 0}$, for $1 \le i \le e_\tau$;

(2.30) $(C\frac{1}{2})$ if $L(\tau, \mathrm{sym}^2, s)$ has a pole at $s = 0$, then $a_i(\tau) \in -\frac{1}{2} + \mathbb{Z}_{\ge 0}$, for $1 \le i \le e_\tau$.

Then the unique generic constituent of

$$\Big(\mathop{\times}_{\tau \in P'} \mathop{\times}_{i=1}^{e_\tau} \delta(D_i(\tau))\Big) \rtimes \sigma^{(0)} \tag{2.31}$$

is square-integrable [Td]. Assume that the representation (2.31) is on $\mathrm{SO}_{2n+1}(F)$. Then every discrete series generic representation of $\mathrm{SO}_{2n+1}(F)$ is obtained in this way for a unique set consisting of a finite set $P'$, segments $\{D_i(\tau) \mid 1 \le i \le e_\tau; \tau \in P'\}$ and a unique generic supercuspidal representation $\sigma^0$ [M2], satisfying conditions (2.26)–(2.30). □

*Remark* 2.2. If $L(\sigma^{(0)} \times \tau, s)$ has a pole at $s = 0$ (case C1), then $L(\tau, \Lambda^2, s)$ has a pole at $s = 0$. See [Jng.S., Lemma 3.1]. Thus (2.28) and (2.29) cover all possible cases, where $L(\tau, \Lambda^2, s)$ has a pole at $s = 0$.

It is now easy to see that the sequence of segments in (2.5)–(2.9), together with $\sigma_\rho^0$ satisfy (2.26)–(2.30) and hence $\sigma_\rho$ is square-integrable.

Conversely, we have to show that any irreducible square-integrable generic representation $\sigma$ of $\mathrm{SO}_{2n+1}(F)$ has a local Langlands functorial lift $\ell(\sigma)$, which is an irreducible admissible representation of $\mathrm{GL}_{2n}(F)$ satisfying the conditions stated in Theorem 2.1. We start with an irreducible square-integrable, generic representation $\sigma$ on $\mathrm{SO}_{2n+1}(F)$. We may assume that $\sigma$ is realized as the unique generic constituent of the induced representation defined by (2.31) for a unique set consisting of a finite set $P'$, segments $\{D_i(\tau) \mid 1 \le i \le e_\tau; \tau \in P'\}$ and a unique generic supercuspidal representation $\sigma^0$, satisfying conditions (2.26)–(2.30).

By Theorem 1.1, there exists a unique finite set $A$ of irreducible, supercuspidal, self-dual representations $\tau$ of $\mathrm{GL}_{k_\tau}(F)$ ($k_\tau$ must be even), such that $L(\tau, \Lambda^2, s)$ has a pole at $s = 0$, and

$$\ell(\sigma^{(0)}) = \bigtimes_{\tau \in A} \tau.$$

We expect the lift $\ell(\sigma)$ to be the generic constituent of

$$\Big(\bigtimes_{\tau \in P'} \bigtimes_{i=1}^{e_\tau} [\delta(D_i(\tau)) \times \delta(\widehat{D_i(\tau)})]\Big) \times \ell(\sigma^{(0)}). \tag{2.32}$$

Note that (2.32) is reducible in general, due to segment linkages. We want to write the generic constituent of (2.32) in form according to the classification theory in [Z]. We define the following sets

$$\begin{aligned} A_0 &= A \setminus P' \\ A_1 &= \{\tau \in A \cap P' \mid a_1(\tau) \in \mathbb{Z}_{>0}\} \\ A_2 &= \{\tau \in A \cap P' \mid a_1(\tau) = -1\}. \end{aligned} \tag{2.33}$$

It follows from Theorem 1.1 that $A$ is exactly the set of all irreducible supercuspidal representations $\tau$ of $\mathrm{GL}_{k_\tau}(F)$, such that $L(\sigma^{(0)} \times \tau, s)$ has a pole at $s = 0$, and hence, by (2.28), we have

$$A = A_0 \cup A_1 \cup A_2. \tag{2.34}$$

Next, we define

$$B = \{\tau \in P' \setminus A \mid L(\tau, \Lambda^2, s) \text{ has a pole at } s = 0\} \tag{2.35}$$

and

$$C = \{\tau \in P' \mid L(\tau, \mathrm{sym}^2, s) \text{ has a pole at } s = 0\}. \tag{2.36}$$

Clearly, we have

$$P' = A_1 \cup A_2 \cup B \cup C.$$

In order to figure out the unique generic constituent of (2.32), we have to consider the procedure of induction in stages. With this procedure, we can separate the linkages among the segments and determine the reducibility of each step.

First we consider

$$\times_{\tau\in A_1}\times_{i=1}^{e_\tau}[\delta(D_i(\tau))\times\delta(\widehat{D_i(\tau)})],$$

which is a "piece" of (2.32) and is called $A_1$-piece for convenience. For $\tau\in A_1$, since $a_1(\tau)\in\mathbb{Z}_{>0}$, we have from (2.27)

$$D_i(\tau)\cap\widehat{D_i(\tau)}=[\nu^{-a_i(\tau)}\tau,\ \nu^{b_i(\tau)}\tau]\cap[\nu^{-b_i(\tau)}\tau,\ \nu^{a_i(\tau)}\tau]=[\nu^{-a_i(\tau)}\tau,\ \nu^{a_i(\tau)}\tau] \tag{2.37}$$

and

$$D_i(\tau)\cup\widehat{D_i(\tau)}=[\nu^{-b_i(\tau)}\tau,\ \nu^{b_i(\tau)}\tau],$$

with $i=1,\ldots,e_\tau$. By [Z], the unique generic constituent of $\delta(D_i(\tau))\times\delta(\widehat{D_i(\tau)})$ is

$$\delta[\nu^{-a_i(\tau)}\tau,\ \nu^{a_i(\tau)}\tau]\times\delta[\nu^{-b_i(\tau)}\tau,\ \nu^{b_i(\tau)}\tau]=\Delta(\tau,a_i(\tau))\times\Delta(\tau,b_i(\tau)). \tag{2.38}$$

Then the $A_1$-piece has

$$\times_{\tau\in A_1}\Delta(\tau,0)\times\times_{i=1}^{e_\tau}[\Delta(\tau,a_i(\tau))\times\Delta(\tau,b_i(\tau))] \tag{2.39}$$

as the unique generic constituent. Put, for $\tau\in A_1$, $d_\tau=2e_\tau+1$ and

$$m_1(\tau)=0,m_{2i}(\tau)=a_i(\tau),m_{2i+1}(\tau)=b_i(\tau),\ i=1,2,\ldots,e_\tau. \tag{2.40}$$

Next, we consider the $A_2$-piece of (2.32)

$$\times_{\tau\in A_2}\times_{i=1}^{e_\tau}[\delta(D_i(\tau))\times\delta(\widehat{D_i(\tau)})].$$

For $\tau\in A_2$, we repeat (2.37) and (2.38) for $i=2,\ldots,e_\tau$. For $i=1$, since $a_1(\tau)=-1$, the set $D_1(\tau)\cap\widehat{D_1(\tau)}$ is empty. By [Z], the unique generic constituent of $\delta(D_1(\tau))\times\delta(\widehat{D_1(\tau)})\times\tau$ is $\Delta(\tau,b_1(\tau))$. Hence the $A_2$-piece has the following term as the unique generic constituent

$$\times_{\tau\in A_2}\Delta(\tau,b_1(\tau))\times\times_{i=2}^{e_\tau}[\Delta(\tau,a_i(\tau))\times\Delta(\tau,b_i(\tau))]. \tag{2.41}$$

Put, for $\tau\in A_2$, $d_\tau=2e_\tau-1$, and

$$m_1(\tau)=b_1(\tau),m_{2i-2}(\tau)=a_i(\tau),m_{2i-1}(\tau)=b_i(\tau),\ i=2,3,\ldots,e_\tau. \tag{2.42}$$

By repeating the "operation" in (2.37) and (2.38), we know that the $B$-piece

$$\bigtimes_{\tau\in B}\bigtimes_{i=1}^{e_\tau}[\delta(D_i(\tau))\times\delta(\widehat{D_i(\tau)})]$$

has the unique generic constituent

$$\bigtimes_{\tau\in B}\bigtimes_{i=1}^{e_\tau}[\Delta(\tau,a_i(\tau))\times\Delta(\tau,b_i(\tau))], \tag{2.43}$$

and for $\tau\in B$, we put $d_\tau=2e_\tau$ and

$$m_{2i-1}(\tau)=a_i(\tau), m_{2i}(\tau)=b_i(\tau),\ i=1,2,\ldots,e_\tau. \tag{2.44}$$

In the $C$-piece of (2.32), we can get the unique generic constituent in the same way, but we have two cases to be considered. First we know by repeating (2.37) and (2.38) that

$$\bigtimes_{\substack{a_1(\tau)\geq\frac{1}{2}\\ \tau\in C}}\bigtimes_{i=1}^{e_\tau}[\delta(D_i(\tau))\times\delta(\widehat{D_i(\tau)})]$$

has a unique generic constituent which is

$$\bigtimes_{\substack{a_1(\tau)\geq\frac{1}{2}\\ \tau\in C}}\bigtimes_{i=1}^{e_\tau}[\Delta(\tau,a_i(\tau))\times\Delta(\tau,b_i(\tau))]. \tag{2.45}$$

In this case, for such $\tau$, we put $d_\tau=2e_\tau$, and

$$m_{2i-1}(\tau)=a_i(\tau), m_{2i}(\tau)=b_i(\tau),\ i=1,2,\ldots,e_\tau. \tag{2.46}$$

Next we know by the same reason that

$$\bigtimes_{\substack{a_1(\tau)=-\frac{1}{2}\\ \tau\in C}}\bigtimes_{i=1}^{e_\tau}[\delta(D_i(\tau))\times\delta(\widehat{D_i(\tau)})]$$

has a unique generic constituent which is

$$\bigtimes_{\substack{a_1(\tau)=-\frac{1}{2}\\ \tau\in C}}\Delta(\tau,b_1(\tau))\times\bigtimes_{i=2}^{e_\tau}[\Delta(\tau,a_i(\tau))\times\Delta(\tau,b_i(\tau))]. \tag{2.47}$$

In this case, for such $\tau$, we put $d_\tau=2e_\tau-1$, and

$$m_1(\tau)=b_1(\tau), m_{2i-2}(\tau)=a_i(\tau), m_{2i-1}(\tau)=b_i(\tau),\ i=2,3,\ldots,e_\tau. \tag{2.48}$$

Finally, for $\tau \in A_0$, we put

$$d_\tau = 1 \qquad \text{and} \qquad m_1(\tau) = 0. \tag{2.49}$$

Note that $\{m_i(\tau)\}_{i=1}^{d_\tau}$ is always strictly increasing, i.e.,

$$\left[\nu^{-m_i(\tau)}\tau,\ \nu^{m_i(\tau)}\tau\right] \subsetneq \left[\nu^{-m_{i+1}(\tau)}\tau,\ \nu^{m_{i+1}(\tau)}\tau\right] \tag{2.50}$$

for $1 \le i \le d_\tau - 1$. It is clear that the unique generic constituent of (2.32) is

$$\ell(\sigma) = \mathop{\times}_{\tau\in P'\cup A_0} \mathop{\times}_{i=1}^{d_\tau} \Delta(\tau, m_i(\tau)). \tag{2.51}$$

This is the product of $\mathop{\times}_{\tau\in A_0} \tau$ and the products in (2.39), (2.41), (2.43), (2.45) and (2.47). From (2.51), it is clear that $\ell(\sigma)$ is of the form (2.1) and satisfies the requirements of Theorem 2.1. Looking at (2.40), (2.42), (2.44), (2.46), (2.48), (2.49) and comparing to (2.5)–(2.9), it is clear that, for $\rho = \ell(\sigma)$, given by (2.51), and $P(\rho) = P' \cup A_0$,

$$\{\Delta_j(\tau) \mid \tau \in P(\rho)\backslash A_0(\rho),\ j \in J_\tau\} = \{D_i(\tau) \mid \tau \in P',\quad 1 \le i \le e_\tau\}.$$

We can use the first part of the proof to conclude (2.2) and (2.3). This completes the proof of Theorem 2.1.

The following is the generalization of Theorem 1.2 to $\Pi^{(dg)}(\mathrm{SO}_{2n+1})$, the set of equivalence classes of irreducible discrete series generic representations of $\mathrm{SO}_{2n+1}(F)$, which provides the local Langlands parameters for the members in $\Pi^{(dg)}(\mathrm{SO}_{2n+1})$. Let $\Phi^{(d)}(\mathrm{SO}_{2n+1})$ be the subset of $\Phi(\mathrm{SO}_{2n+1})$ consisting of all the local Langlands parameters of type

$$\varphi = \oplus_i \phi_i \otimes S_{m_i+1}$$

where $\phi_i$'s are irreducible self-dual representations of $W_F$ of dimension $k_{\phi_i}$ and $S_{m_i+1}$'s are irreducible representations of $\mathrm{SL}_2(\mathbb{C})$ of dimension $m_i + 1$, satisfying the conditions that (a) the tensor products $\phi_i \otimes S_{m_i+1}$ are irreducible and symplectic, (b) $\phi_i \otimes S_{m_i+1}$ and $\phi_j \otimes S_{m_j+1}$ are not equivalent if $i \neq j$, and (c) the image $\varphi(W_F \times \mathrm{SL}_2(\mathbb{C}))$ is not contained in any proper Levi subgroup of $\mathrm{Sp}_{2n}(\mathbb{C})$. The local Langlands parameters in $\Phi^{(d)}(\mathrm{SO}_{2n+1})$ are called *discrete*.

THEOREM 2.2. *There is a bijection $y$ (which extends the one in Theorem 1.2) between the set $\Phi^{(d)}(\mathrm{SO}_{2n+1})$ of discrete parameters $\varphi$ and the set $\Pi^{(dg)}(\mathrm{SO}_{2n+1})$. The bijection $y$ preserves local factors as in (1.4), (1.5), i.e.,*

$$L(\varphi \otimes \varphi', s) = L(y(\varphi) \times r(\varphi')s) \tag{2.52}$$

$$\epsilon(\varphi \otimes \varphi', s, \psi) = \epsilon(y(\varphi) \times r(\varphi'), s, \psi) \tag{2.53}$$

*for all irreducible, admissible representations* $\varphi'$ *of* $W_F$ *of degree* $k$ *with* $k \in \mathbb{Z}_{>0}$, *where* $r$ *is the reciprocity map for* GL *as before.*

*Proof.* The proof is outlined in [M1, p. 714] assuming Conjectures 3.1 and 3.2 in [M1]. With Theorem 1.2 and Theorem 6.3 in [Jng.S.], we write down the proof in detail without assumption. Let $\varphi : W_F \times SL_2(\mathbb{C}) \longrightarrow Sp_{2n}(\mathbb{C})$ be a parameter in $\Phi^{(d)}(SO_{2n+1})$. We write it as a multiplicity one decomposition into irreducible summands

$$\varphi = \bigoplus_{i=1}^{r} \phi_i \otimes S_{2m_i+1}, \quad 2m_i \in \mathbb{Z}_{\geq 0}. \tag{2.54}$$

Since the tensor product $\phi_i \otimes S_{2m_j+1}$ is symplectic for $1 \leq i \leq r$, one knows that $\phi_i$ is self-dual, and it is symplectic (orthogonal, resp.) if and only if $S_{2m_i+1}$ is orthogonal (symplectic, resp.). Thus, for $1 \leq i \leq r$,

(2.55) if $L(\Lambda^2(\phi_i), s)$ has a pole at $s = 0$, then $m_i \in \mathbb{Z}_+$

(2.56) if $L(\mathrm{sym}^2(\phi_i), s)$ has a pole at $s = 0$, then $m_i \in \frac{1}{2} + \mathbb{Z}_+$.

Let $\tau_i = r(\phi_i)$ be the irreducible, self-dual, supercuspidal representation of $GL_{k_{\phi_i}}(F)$, corresponding to $\phi_i$. By Henniart's result [Jng.S., Thm. 6.3],

(2.57)

$$L(\Lambda^2(\phi_i), s) \text{ has a pole at } s = 0 \Longleftrightarrow L(\tau_i, \Lambda^2, s) \text{ has a pole at } s = 0$$

and

(2.58)

$$L(\mathrm{sym}^2(\phi_i), s) \text{ has a pole at } s = 0 \Longleftrightarrow L(\tau_i, \mathrm{sym}^2, s) \text{ has a pole at } s = 0.$$

We have

$$r(\phi_i \otimes S_{2m_i+1}) = \Delta(r(\phi_i), m_i) = \Delta(\tau_i, m_i); \quad r(\varphi) = \mathop{\times}_{i=1}^{r} \Delta(\tau_i, m_i).$$

Now $r(\varphi)$ has the form (2.1), and the conditions (2.55), (2.56), translated through (2.57), (2.58) are exactly the conditions (1), (2) of Theorem 2.1. Thus,

$$\sigma = \ell^{-1}(r(\varphi))$$

is a discrete series generic representation of $SO_{2n+1}(F)$, such that

$$L(\sigma \times r(\varphi'), s) = L(r(\varphi) \times r(\varphi'), s) = L(\varphi \otimes \varphi', s)$$
$$\epsilon(\sigma \times r(\varphi'), s, \psi) = \epsilon(r(\varphi) \times r(\varphi'), s, \psi) = \epsilon(\varphi \otimes \varphi', s, \psi)$$

for all irreducible admissible representations $\varphi'$ of $W_F$ of degree $k$ with $k \in \mathbb{Z}_{>0}$. Finally, we define $y(\varphi) = \sigma$.

Conversely, let $\sigma$ be a discrete series, generic representation of $SO_{2n+1}(F)$. Consider, by Theorem 2.1, the corresponding tempered representation $\ell(\sigma)$ of $GL_{2n}(F)$. We know that $\ell(\sigma)$ has the form (2.1) so that the requirements (1), (2) of Theorem 2.1 are satisfied. Let $\phi_i = r^{-1}(\tau_i)$ for $1 \leq i \leq r$, and define $\varphi$ by (2.54). By (2.57), (2.58), it is clear that $\varphi$ is as required by Theorem 2.2, and that $y(\varphi) = \sigma$. This proves Theorem 2.2. □

**3. Elliptic tempered generic representations.** An irreducible admissible representation $\sigma$ of $SO_{2n+1}(F)$ is called *elliptic* if the distribution character $\Theta_\sigma$ is not zero on the set of elliptic regular elements of $SO_{2n+1}(F)$. Elliptic representations of $p$-adic classical groups (in particular, of $SO_{2n+1}(F)$) have been extensively studied by Herb in [Hb]. For instance, all discrete series representations of $SO_{2n+1}(F)$ are elliptic. In the classification theory (a la Bernstein-Zelevinsky) of $p$-adic classical groups, elliptic representations play an important role as building blocks in the induction procedure from supercuspidal representations to general representations. This type of classification for generic representations of $Sp_{2n}(F)$ and $SO_{2n+1}(F)$ has been carried out in [M2]. In this section, we study the explicit local Langlands functorial lift from the set $\Pi^{(etg)}(SO_{2n+1})$ of equivalence classes of elliptic tempered generic representations of $SO_{2n+1}(F)$ to $GL_{2n}(F)$.

We recall the construction of elliptic tempered generic representations of $SO_{2n+1}(F)$ from [Hb] and [M2].

Let $\sigma^{(2)}$ be a discrete series generic representation of $SO_{2n''+1}(F)$. The image of the local Langlands functorial lift of $\sigma^{(2)}$ is given by Theorem 2.1, i.e.,

$$\ell(\sigma^{(2)}) = \rho^{(2)}$$

which is an irreducible tempered representation of $GL_{2n''}(F)$. Write

$$\rho^{(2)} = \bigtimes_{i=1}^{r} \Delta(\tau_i, m_i) = \bigtimes_{\tau \in P(\rho^{(2)})} \bigtimes_{j=1}^{d_\tau} \Delta(\tau, m_j(\tau)) \tag{3.1}$$

as in (2.1) and as in the proof of Theorem 2.1. Put

$$P'(\rho^{(2)}) = P(\rho^{(2)}) \setminus A_0(\rho^{(2)}).$$

Let $\beta_1, \ldots, \beta_c$ (with possible repetitions) be irreducible, self-dual, supercuspidal representations of $GL_{k_{\beta_1}}(F), \ldots, GL_{k_{\beta_c}}(F)$, respectively. Consider a sequence of pairwise inequivalent square-integrable representations

$$\{\Delta(\beta_i, e_i)\}_{i=1}^{c}, \quad 2e_i \in \mathbb{Z}_{\geq 0}$$

of $GL_{k_{\beta_i}(2e_i+1)}(F)$ $(i = 1, \ldots, c)$, satisfying the following properties: for $1 \leq i \leq c$,

$$\Delta(\beta_i, e_i) \notin \{\Delta(\tau, m_j(\tau)) \mid 1 \leq j \leq d_\tau, \quad \tau \in P'(\rho^{(2)})\} \tag{3.2}$$

and one of the following conditions holds

(3.3) $\quad * \ \Delta(\beta_i, e_i) \in A_2(\rho^{(2)})$, which implies that $e_i = 0$, or

(3.4) $\quad * \ L(\sigma^{(0)} \times \beta_i, s)$ has a pole at $s = 0$, and $e_i \geq 1$, or

(3.5) $\quad * \ L(\beta_i, \Lambda^2, s)$ has a pole at $s = 0$, $L(\sigma^{(0)} \times \beta_i, s)$ is holomorphic at $s = 0$, and $e_i \in \mathbb{Z}_{\geq 0}$, or

(3.6) $\quad * \ L(\beta_i, \mathrm{sym}^2, s)$ has a pole at $s = 0$, and $e_i \in \frac{1}{2} + \mathbb{Z}_{\geq 0}$.

Recall that $\sigma^{(0)}$ is the irreducible, supercuspidal, generic representation of $\mathrm{SO}_{2n'+1}(F)$, such that $\ell(\sigma^{(0)}) = \bigtimes_{\tau \in A(\rho^{(2)})} \tau$ as in (2.32).

Note that conditions (3.2), (3.3)–(3.6) may be replaced by conditions (3.2)′, (3.3)–(3.6), where condition (3.2)′ is given by

(3.2′)

$$\Delta(\beta_i, e_i) \notin \{\Delta(\tau, m_j(\tau)) \mid 1 \leq j \leq d_\tau, \tau \in P(\rho^{(2)})\} = \{\Delta(\tau_i, m_i) \mid 1 \leq i \leq r\}$$

The unique generic constituent $\sigma$ of

$$\Delta(\beta_1, e_1) \times \cdots \times \Delta(\beta_c, e_c) \rtimes \sigma^{(2)} \tag{3.7}$$

is a tempered elliptic representation of $\mathrm{SO}_{2n+1}(F)$. This is the way that all elliptic tempered generic representations of $\mathrm{SO}_{2n+1}(F)$ are obtained and the induction data $\{\Delta(\beta_i, e_i)\}_{i=1}^{c}$ and $\sigma^{(2)}$ satisfying conditions (3.2)–(3.6) are uniquely determined ($n = n'' + \sum_{i=1}^{c}(2e_i + 1)k_{\beta_i}$). Note also that if $e_i \in \frac{1}{2} + \mathbb{Z}_{\geq 0}$, then $\Delta(\beta_i, e_i)$ is a representation of $\mathrm{GL}_{2e_i'+2}(F)$, where $e_i = \frac{1}{2} + e_i'$ and $e_i' \in \mathbb{Z}_{\geq 0}$. This characterization of elliptic tempered generic representations of $\mathrm{SO}_{2n+1}(F)$ is essentially due to [Hb] and can be found in [M2, §3] for relevant discussion.

THEOREM 3.1. *The bijection $\ell$ of Theorem 2.1 can be extended to a bijection (which is still denoted by $\ell$) between the set $\Pi^{(etg)}(\mathrm{SO}_{2n+1})$ and the set of equivalence classes of tempered representations of $\mathrm{GL}_{2n}(F)$ of the following form*

$$\Delta(\lambda_1, h_1) \times \Delta(\lambda_2, h_2) \times \cdots \times \Delta(\lambda_f, h_f) \tag{3.8}$$

*where each representation $\Delta(\lambda_i, h_i)$ in (3.8) is self-dual, and appears in (3.8) either once or twice, and satisfies the following condition, for each $i$,*

(1) *if $L(\lambda_i, \mathrm{sym}^2, s)$ has a pole at $s = 0$, then $h_i \in \frac{1}{2} + \mathbb{Z}_{\geq 0}$*

(2) *if $L(\lambda_i, \Lambda^2, s)$ has a pole at $s = 0$, then $h_i \in \mathbb{Z}_{\geq 0}$.*

*Moreover, we have*

$$L(\sigma \times \pi, s) = L(\ell(\sigma) \times \pi, s), \tag{3.9}$$

$$\epsilon(\sigma \times \pi, s, \psi) = \epsilon(\ell(\sigma) \times \pi, s, \psi), \tag{3.10}$$

*for any* $\sigma$ *in* $\Pi^{(etg)}(\mathrm{SO}_{2n+1})$ *and any irreducible, generic representation* $\pi$ *of* $\mathrm{GL}_k(F)$ *with all* $k \in \mathbb{Z}_{>0}$.

The following remark is analogous to the one right after Theorem 2.1.

*Remark* 3.1. An equivalent description of the image of $\ell$ in Theorem 3.1 is the following. An irreducible representation $\rho$ of $\mathrm{GL}_{2n}(F)$ lies in the image of $\ell$ (in Theorem 3.1) if and only if $\rho$ is tempered and satisfies the following properties. Let $\tau$ be an irreducible, unitary, supercuspidal representation of $\mathrm{GL}_k(F)$, $k = 1, 2, \ldots, 2n$. Then

(1) if $\tau \ncong \widehat{\tau}$, then $L(\rho \times \tau, s)$ has no poles on the real line,

(2) if $\tau \cong \widehat{\tau}$ and $L(\rho \times \tau, s)$ is not holomorphic, then

(a) if $L(\tau, \mathrm{sym}^2, s)$ has a pole at $s = 0$, then $L(\rho \times \tau, s)$ has poles, of order at most two, whose real parts lie inside $-\frac{1}{2} + \mathbb{Z}_{\leq 0}$.

(b) if $L(\tau, \Lambda^2, s)$ has a pole at $s = 0$, then $L(\rho \times \tau, s)$ has poles, of order at most two, whose real parts lie inside $\mathbb{Z}_{\leq 0}$.

*Proof of Theorem 3.1.* Let $\rho$ be a representation of $GL_{2n}(F)$ of the form (3.8), satisfying the conditions of Theorem 3.1. We shall construct an irreducible elliptic tempered generic representation $\sigma_\rho$ of $\mathrm{SO}_{2n+1}(F)$ such that (3.9) and (3.10) hold for the pair $(\sigma_\rho, \rho)$. Let $\Delta(\tau_1, m_1), \ldots, \Delta(\tau_r, m_r)$ be the factors, which appear exactly once in (3.8), and let $\Delta(\beta_1, e_1), \ldots, \Delta(\beta_c, e_c)$ be the different factors, which appear twice in (3.8). Put

$$\rho^{(2)} = \Delta(\tau_1, m_1) \times \cdots \times \Delta(\tau_r, m_r). \tag{3.11}$$

It is a representation of $\mathrm{GL}_{2n''}(F)$ belonging to the image of the local Langlands functorial lift as given in Theorem 2.1. Hence there exists a unique (up to equivalence) irreducible discrete series generic representation $\sigma^{(2)}$ of $\mathrm{SO}_{2n''+1}(F)$, such that $\ell(\sigma^{(2)}) = \rho^{(2)}$. Let $\sigma_\rho$ be the unique generic constituent of

$$\Delta(\beta_1, e_1) \times \cdots \times \Delta(\beta_c, e_c) \rtimes \sigma^{(2)}.$$

In order to show that $\sigma_\rho$ is tempered and elliptic, we have to verify the conditions (3.2)′, (3.3)–(3.6).

If $\Delta(\beta_i, e_i) \in \{\Delta(\tau_i, m_i) \mid 1 \leq i \leq r\}$, then $\Delta(\beta_i, e_i)$ appears three times in (3.8), which is impossible. This verifies (3.2)′. Next, if $L(\beta_i, \mathrm{sym}^2, s)$ has a pole at $s = 0$, then by assumption, $e_i \in \frac{1}{2} + \mathbb{Z}_{\geq 0}$, which is (3.6).

Let $\sigma^{(0)}$ be the irreducible, generic, supercuspidal representation of $\mathrm{SO}_{2n'+1}(F)$ associated to the square-integrable representation $\sigma^{(2)}$ of $\mathrm{SO}_{2n''+1}(F)$ as in formula (2.31). In this case we can express the image of $\sigma^{(0)}$ under the lift map $\ell$ as $\ell(\sigma^{(0)}) = \bigtimes_{\tau \in A(\rho^{(2)})} \tau$ (Theorem 1.1). If $L(\beta_i, \Lambda^2, s)$ has a pole at $s = 0$, then, by assumption,

$e_i \in \mathbb{Z}_{\geq 0}$. If $L(\sigma^{(0)} \times \beta_i, s)$ is holomorphic, then this is (3.5), so assume that $L(\sigma^{(0)} \times \beta_i, s)$ has a pole at $s = 0$. This means that $\beta_i \in A(\rho^{(2)})$. If $e_i \geq 1$, then this is (3.4).

Assume that $e_i = 0$. Recall that $A(\rho^{(2)}) = A_0(\rho^{(2)}) \cup A_1(\rho^{(2)}) \cup A_2(\rho^{(2)})$. If $\beta_i = \Delta(\beta_i, 0) \in A_0(\rho^{(2)}) \cup A_1(\rho^{(2)})$, then $\Delta(\beta_i, 0)$ occurs three times, which is impossible by the assumption. Thus, in this case, we must have $\Delta(\beta_i, e_i) \in A_2(\rho^{(2)})$, which is (3.3). This proves that the given generic constituent $\sigma_\rho$ of (3.11) is elliptic and tempered. Furthermore, we have, for an irreducible, supercuspidal representation $\pi$ of $\mathrm{GL}_k(F)$, the compatibility of the twisted gamma factors:

$$\begin{aligned}
\gamma(\sigma_\rho \times \pi, s, \psi) &= \left[\prod_{i=1}^{c} \gamma(\Delta(\beta_i, e_i) \times \pi, s, \psi)^2\right] \gamma(\sigma^{(2)} \times \pi, s, \psi) \\
&= \left[\prod_{i=1}^{c} \gamma(\Delta(\beta_i, e_i) \times \pi, s, \psi)^2\right] \prod_{i=1}^{r} \gamma(\Delta(\tau_i, m_i) \times \pi, s, \psi) \\
&= \prod_{i=1}^{f} \gamma(\Delta(\lambda_i, h_i) \times \pi, s, \psi) \\
&= \gamma(\rho \times \pi, s, \psi).
\end{aligned}$$

We used the multiplicativity property of gamma factors (Theorem 1.4) and Theorem 2.1. Now we conclude (3.9) and (3.10) (with $\ell(\sigma_\rho) = \rho$), in the same way as we obtained (2.22) and (2.23), since $\sigma_\rho$ and $\rho$ are tempered.

Conversely, for any $\sigma$ in $\Pi^{(etg)}(\mathrm{SO}_{2n+1})$, we write $\sigma$ as in (3.1)–(3.7), and define

(3.12)
$$\ell(\sigma) = \Delta(\beta_1, e_1) \times \cdots \times \Delta(\beta_c, e_c) \times \ell(\sigma^{(2)}) \times \Delta(\beta_c, e_c) \times \cdots \times \Delta(\beta_1, e_1).$$

Put $\rho = \ell(\sigma)$. This is an irreducible, tempered representation of $\mathrm{GL}_{2n}(F)$. By (3.2)′, no factor $\Delta(\beta_i, e_i)$ appears in

$$\ell(\sigma^{(2)}) = \rho^{(2)} = \Delta(\tau_1, m_1) \times \cdots \times \Delta(\tau_r, m_r).$$

Thus, each $\Delta(\beta_i, e_i)$ appears in (3.12) twice, and each $\Delta(\tau_i, m_i)$ appears once. The conditions (1), (2) of Theorem 3.1 are satisfied by assumption. By the first part of this proof we get that $\sigma = \sigma_\rho$ and conditions (3.9) and (3.10) are satisfied. This proves Theorem 3.1. □

Now it is easy to write the local Langlands parameter for each member in $\Pi^{(etg)}(\mathrm{SO}_{2n+1})$. For any $\sigma$ in $\Pi^{(etg)}(\mathrm{SO}_{2n+1})$, the lift $\ell(\sigma)$ is given as in (3.12). The parameter for $\sigma$ is

$$\varphi = \left[\bigoplus_{i=1}^{c} \phi_i \otimes S_{2e_i+1}\right] \oplus \varphi_{\sigma^{(2)}} \oplus \left[\bigoplus_{i=1}^{c} \phi_i \otimes S_{2e_i+1}\right],$$

with the image $\varphi(W_F \times \mathrm{SL}_2(\mathbb{C}))$ in a proper Levi subgroup of $\mathrm{Sp}_{2n}(\mathbb{C})$, where $\varphi_{\sigma^{(2)}}$ is the parameter for the irreducible square-integrable generic representation $\sigma^{(2)}$ occurring in $\sigma$.

**4. Tempered generic representations.** From [M2, §4], we know that tempered generic representations of $\mathrm{SO}_{2n+1}(F)$ are either elliptic or induced from elliptic tempered generic representations of Levi subgroups. We continue with Muic's description of tempered generic representations of $\mathrm{SO}_{2n+1}(F)$ to get an explicit description of the image of the local Langlands functorial lift to $\mathrm{GL}_{2n}(F)$ from the set $\Pi^{(tg)}(\mathrm{SO}_{2n+1})$ of equivalence classes of irreducible tempered generic representations of $\mathrm{SO}_{2n+1}(F)$.

Let $\sigma^{(et)}$ be an elliptic, tempered, generic representation of $\mathrm{SO}_{2\widetilde{n}+1}(F)$. Put $\rho^{(et)} = \ell(\sigma^{(et)})$, which is a tempered representation of $\mathrm{GL}_{2\widetilde{n}}(F)$ as given in Theorem 3.1. We keep the notations of (3.1)–(3.7) and write $\sigma^{(et)}$ as the unique generic constituent of (3.7), and then $\rho^{(et)} = \ell(\sigma^{(et)})$ can be expressed as in (3.12).

Let $\eta_1, \dots, \eta_d$ (with possible repetitions) be irreducible unitary, supercuspidal representations of $\mathrm{GL}_{k_{\eta_1}}(F), \dots, \mathrm{GL}_{k_{\eta_d}}(F)$, respectively. We construct from the $\eta_i$'s a sequence of irreducible square-integrable representations $\{\Delta(\eta_i, p_i)\}_{i=1}^d$ of $\mathrm{GL}_{k_{\eta_i}(2p_i+1)}(F)$ with $2p_i \in \mathbb{Z}_{\geq 0}$ and $i = 1, \dots, d$, satisfying one of the following properties, for $1 \leq i \leq d$

(4.1) $\quad * \ \Delta(\eta_i, p_i) \in \{\Delta(\beta_j, e_j) \mid 1 \leq j \leq c\}$, or

(4.2) $\quad * \ \Delta(\eta_i, p_i) \in \{\Delta(\tau_j, m_j) \mid 1 \leq j \leq r\}$, or

(4.3) $\quad * \ \eta_i \not\cong \widehat{\eta_i}$, or

(4.4) $\quad * \ L(\eta_i, \Lambda^2, s)$ has a pole at $s = 0$, and $p_i \in \frac{1}{2} + \mathbb{Z}_{\geq 0}$, or

(4.5) $\quad * \ L(\eta_i, \mathrm{sym}^2, s)$ has a pole at $s = 0$, and $p_i \in \mathbb{Z}_{\geq 0}$.

Then the induced representation

$$\sigma = \Delta(\eta_1, p_1) \times \cdots \times \Delta(\eta_d, p_d) \rtimes \sigma^{(et)} \tag{4.6}$$

is an irreducible, tempered, generic representation of $\mathrm{SO}_{2n+1}(F)$. This is the way all tempered, generic representations of $\mathrm{SO}_{2n+1}(F)$ are obtained, and the induction data $\{\Delta(\eta_i, p_i)\}_{i=1}^d$ and $\sigma^{(et)}$ are uniquely determined ($n = \widetilde{n} + \sum_{i=1}^d (2p_i + 1)k_{\eta_i}$), up to replacements

$$\Delta(\eta_i, p_i) \leftrightarrow \Delta(\hat{\eta}_i, p_i)$$

in cases (4.3).

THEOREM 4.1. *The bijection $\ell$ of Theorem 3.1 extends to a bijection (which is still denoted by $\ell$) between the set $\Pi^{(tg)}(\mathrm{SO}_{2n+1})$ and the set of equivalence classes of tempered representations of $\mathrm{GL}_{2n}(F)$ of the following form*

$$\Delta(\lambda_1, h_1) \times \Delta(\lambda_2, h_2) \times \cdots \times \Delta(\lambda_f, h_f) \tag{4.7}$$

*where* $\lambda_1, \ldots, \lambda_f$ *are unitary, supercuspidal,* $2h_i \in \mathbb{Z}_{\geq 0}$*, such that for* $1 \leq i \leq f$

(1) *if* $\lambda_i \not\cong \widehat{\lambda}_i$*, then* $\Delta(\lambda_i, h_i)$ *occurs in (4.7) as many times as* $\Delta(\widehat{\lambda}_i, h_i)$ *does,*

(2) *if* $L(\lambda_i, \Lambda^2, s)$ *has a pole at* $s = 0$*, and* $h_i \in \frac{1}{2} + \mathbb{Z}_{\geq 0}$*, then* $\Delta(\lambda_i, h_i)$ *occurs an even number of times in (4.7),*

(3) *if* $L(\lambda_i, \mathrm{sym}^2, s)$ *has a pole at* $s = 0$*, and* $h_i \in \mathbb{Z}_{\geq 0}$*, then* $\Delta(\lambda_i, h_i)$ *occurs an even number of times in (4.7).*

*Moreover, we have*

$$L(\sigma \times \pi, s) = L(\ell(\sigma) \times \pi, s) \tag{4.8}$$

$$\epsilon(\sigma \times \pi, s, \psi) = \epsilon(\ell(\sigma) \times \pi, s, \psi) \tag{4.9}$$

*for all generic tempered representations* $\sigma$ *of* $\mathrm{SO}_{2n+1}(F)$ *and all irreducible generic representations* $\pi$ *of* $\mathrm{GL}_k(F)$ *with all* $k \in \mathbb{Z}_{>0}$.

*Remark* 4.1. As in the previous section, we can write an equivalent description in terms of poles of L-functions. An irreducible representation $\rho$ of $\mathrm{GL}_{2n}(F)$ lies in the image of the lift $\ell$ defined in Theorem 4.1 if and only if $\rho$ is tempered and satisfies the following properties:

Let $\tau$ be an irreducible, unitary, supercuspidal representation of $\mathrm{GL}_{k_\tau}(F)$, such that $L(\rho \times \tau, s)$ has poles. Because there is a real number $t$, such that $L(\rho \times \nu^{it}\tau, s)$ has real poles, we may assume that $L(\rho \times \tau, s)$ has real poles. Note that such a real pole lies in $\frac{1}{2}\mathbb{Z}_{\leq 0}$. Let $m \in \frac{1}{2}\mathbb{Z}_{\leq 0}$ be one of such poles of $L(\rho \times \tau, s)$. Then

(1) if $\tau \not\cong \widehat{\tau}$, then $s = m$ is also a pole of $L(\rho \times \widehat{\tau}, s)$, with the same multiplicity as that of $s = m$ for $L(\rho \times \tau, s)$;

(2) if $L(\tau, \Lambda^2, s)$ has a pole at $s = 0$, and $m \in -\frac{1}{2} + \mathbb{Z}_{\leq 0}$, then the order of the pole at $s = m$ of $L(\rho \times \tau, s)$ is even; and

(3) if $L(\tau, \mathrm{sym}^2, s)$ has a pole at $s = 0$, and $m \in \mathbb{Z}_{\leq 0}$, then the order of the pole at $s = m$ of $L(\rho \times \tau, s)$ is even.

*Proof of Theorem 4.1.* Let $\rho$ be a representation of $\mathrm{GL}_{2n}(F)$ of the form (4.7) satisfying conditions (1)–(3) of Theorem 4.1. We define the following sets $N$, $M$, and $R$ from the "factors" of the induced representation in (4.7):

* $N$ consists of $\Delta(\lambda_i, h_i)$'s with $1 \leq i \leq f$ such that $\widehat{\lambda}_i \not\cong \lambda_i$,

* $W$ consists of $\Delta(\lambda_i, h_i)$'s with $1 \leq i \leq f$ such that $L(\lambda_i, \Lambda^2, s)$ has a pole at $s = 0$ and $h_i \in \frac{1}{2} + \mathbb{Z}_{\geq 0}$, or $L(\lambda, \mathrm{sym}^2, s)$ has a pole at $s = 0$ and $h_i \in \mathbb{Z}_{\geq 0}$, and

* $R$ consists of $\Delta(\lambda_i, h_i)$'s with $1 \leq i \leq f$ such that $L(\lambda_i, \Lambda^2, s)$ has a pole at $s = 0$ and $h_i \in \mathbb{Z}_{\geq 0}$, or $L(\lambda_i, \mathrm{sym}^2, s)$ has a pole at $s = 0$ and $h_i \in \frac{1}{2} + \mathbb{Z}_{\geq 0}$.

These sets are taken with *multiplicities*. Denote by $\mu'_i$ the multiplicity of $\Delta(\lambda_i, h_i)$ in (4.7). For example, if $\Delta(\lambda_i, h_i)$ occurs in $N$, then it must be counted $\mu'_i$ times.

By assumption, for $\Delta(\lambda_i, h_i) \in W$, $\mu'_i = 2\mu_i$ is even. Let

$$\{\Delta(\lambda_{i_1}, h_{i_1}), \ldots, \Delta(\lambda_{i_u}, h_{i_u})\} \tag{4.10}$$

be the set of all different elements in $W$. Put

$$J_W(\rho) = \mathop{\times}_{j=1}^{u} \underbrace{\left(\Delta(\lambda_{i_j}, h_{i_j}) \times \cdots \times \Delta(\lambda_{i_j}, h_{i_j})\right)}_{\mu_{i_j} \text{ times}}. \tag{4.11}$$

Now, by assumption, if $\Delta(\lambda_i, h_i) \in N$, then $\widehat{\Delta(\lambda_i, h_i)} = \Delta(\widehat{\lambda}_i, h_i) \in N$. Write, in this case: $\widehat{\Delta(\lambda_i, h_i)} = \Delta(\lambda_{\hat{i}}, h_{\hat{i}})$, and by assumption the multiplicities of $\Delta(\lambda_i, h_i)$ and $\Delta(\lambda_{\hat{i}}, h_{\hat{i}})$ are equal, i.e., $\mu'_i = \mu'_{\hat{i}}$. Let

$$\{\Delta(\lambda_{z_1}, h_{z_1}), \widehat{\Delta(\lambda_{z_1}, h_{z_1})}, \ldots, \Delta(\lambda_{z_v}, h_{z_v}), \widehat{\Delta(\lambda_{z_v}, h_{z_v})}\} \tag{4.12}$$

be the set of all different elements in $N$. Put

$$J_N(\rho) = \mathop{\times}_{j=1}^{v} \underbrace{\left(\Delta(\lambda_{z_j}, h_{z_j}) \times \Delta(\lambda_{z_j}, h_{z_j}) \times \cdots \times \Delta(\lambda_{z_j}, h_{z_j})\right)}_{\mu'_{z_j} \text{ times}}. \tag{4.13}$$

Let

$$R_1 = \{\Delta(\lambda_i, h_i) \in R \mid \mu'_i = 2\mu_i + 1 \text{ is odd}\} \tag{4.14}$$

$$R_2 = \{\Delta(\lambda_i, h_i) \in R \mid \mu'_i = 2\mu_i \text{ is even}\}. \tag{4.15}$$

The sets are taken with multiplicities. Let

$$\{\Delta(\tau_1, m_1), \ldots, \Delta(\tau_r, m_r)\} := \{\Delta(\lambda_{x_1}, h_{x_1}), \ldots, \Delta(\lambda_{x_r}, h_{x_r})\} \tag{4.16}$$

be the set of all different elements of $R_1$, and let

$$\{\Delta(\beta_1, e_1), \ldots, \Delta(\beta_c, e_c)\} := \{\Delta(\lambda_{t_1}, h_{t_1}), \ldots, \Delta(\lambda_{t_c}, h_{t_c})\} \tag{4.17}$$

be the set of all different elements of $R_2$. We define

$$J_{R_1}(\rho) = \mathop{\times}_{j=1}^{r} \underbrace{\left(\Delta(\lambda_{x_j}, h_{x_j}) \times \cdots \times \Delta(\lambda_{x_j} h_{x_j})\right)}_{\mu_{x_j}} \tag{4.18}$$

$$J_{R_2}(\rho) = \mathop{\times}_{j=1}^{c} \underbrace{\left(\Delta(\lambda_{t_j}, h_{t_j}) \times \cdots \times \Delta(\lambda_{t_j} h_{t_j})\right)}_{\mu_{t_j}-1}. \tag{4.19}$$

For the remaining representations from $R_1$ and $R_2$, we do the following. By assumption, the induced representation

$$\rho^{(2)} = \Delta(\tau_1, m_1) \times \cdots \times \Delta(\tau_r, m_r) \tag{4.20}$$

is a representation of $GL_{2n''}(F)$ and satisfies the conditions of Theorem 2.1. Hence by Theorem 2.1, there is a unique, up to equivalence, irreducible discrete series generic representation $\sigma^{(2)}$ of $SO_{2n''+1}(F)$ such that $\ell(\sigma^{(2)}) = \rho^{(2)}$. Similarly, by assumption,

$$\rho^{(et)} = \Delta(\beta_1, e_1) \times \cdots \times \Delta(\beta_c, e_c) \times \rho^{(2)} \times \Delta(\beta_c, e_c) \times \cdots \times \Delta(\beta_1, e_1) \tag{4.21}$$

is a representation of $GL_{2\widetilde{n}}(F)$ and satisfies the conditions of Theorem 3.1. Hence, by Theorem 3.1, there is a unique (up to equivalence) irreducible generic tempered elliptic representation $\sigma^{(et)}$ of $SO_{2\widetilde{n}+1}(F)$, such that $\ell(\sigma^{(et)}) = \rho^{(et)}$. Note that $\sigma^{(et)}$ can be realized as the unique generic constituent of

$$\Delta(\beta_1, e_1) \times \cdots \times \Delta(\beta_c, e_c) \rtimes \sigma^{(2)} \tag{4.22}$$

as in (3.7). Finally we define

$$\sigma_\rho := J_N(\rho) \times J_W(\rho) \times J_{R_1}(\rho) \times J_{R_2}(\rho) \rtimes \sigma^{(et)}, \tag{4.23}$$

which is a representation of $SO_{2n+1}(F)$. Let $\Delta(\eta_1, p_1), \ldots, \Delta(\eta_d, p_d)$ be the list of all "factors", with repetitions, which appear in $J_N(\rho) \times J_W(\rho) \times J_{R_1}(\rho) \times J_{R_2}(\rho)$. Then we have

$$\sigma_\rho = \Delta(\eta_1, p_1) \times \cdots \times \Delta(\eta_d, p_d) \rtimes \sigma^{(et)}. \tag{4.24}$$

We now claim that the representation $\sigma_\rho$ in (4.24) satisfies conditions (4.1)–(4.5). Indeed, if, for $1 \le i \le d$, $\Delta(\eta_i, p_i)$ is a factor of $J_{R_2}(\rho)$, then (4.1) is satisfied, and if $\Delta(\eta_i, p_i)$ is a factor of $J_{R_1}(\rho)$, then (4.2) is satisfied. If $\Delta(\eta_i, p_i)$ is a factor of $J_N(\rho)$, then (4.3) is satisfied, and if $\Delta(\eta_i, p_i)$ is a factor of $J_W(\rho)$, then (4.4) or (4.5) are satisfied. Finally, $\sigma_\rho$ in (4.24) is of the form (4.6). By the multiplicativity of gamma factors (Theorem 1.4) and by Theorems 1.1, 2.1, and 3.1, we have, for an irreducible supercuspidal representation $\pi$ of $GL_k(F)$,

$$\begin{aligned}
\gamma(\sigma_\rho \times \pi, s, \psi) &= \left[\prod_{i=1}^{d} \gamma(\Delta(\eta_i, p_i) \times \pi, s, \psi) \cdot \gamma(\widehat{\Delta(\eta_i, p_i)} \times \pi, s, \psi)\right] \\
&\quad \times \gamma(\sigma^{(et)} \times \pi, s, \psi) \\
&= \left[\prod_{i=1}^{d} \gamma(\Delta(\eta_i, p_i) \times \pi, s, \psi)\gamma(\widehat{\Delta(\eta_i, p_i)} \times \pi, s, \psi)\right] \\
&\quad \times \gamma(\rho^{(et)} \times \pi, s, \psi) \\
&= \left[\prod_{i=1}^{d} \gamma(\Delta(\eta_i, p_i) \times \pi, s, \psi)\gamma(\widehat{\Delta(\eta_i, p_i)} \times \pi, s, \psi)\right]
\end{aligned}$$

$$\times \left[\prod_{i=1}^{c} \gamma(\Delta(\beta_i, e_i) \times \pi, s, \psi)^2\right] \prod_{i=1}^{r} \gamma(\Delta(\tau_i, m_i) \times \pi, s, \psi)$$

$$= \prod_{i=1}^{f} \gamma(\Delta(\lambda_i, h_i) \times \pi, s, \psi)$$

$$= \gamma(\rho \times \pi, s, \psi).$$

Finally we conclude, for $\ell(\sigma_\rho) = \rho$, (4.8) and (4.9) exactly as we concluded (2.22) and (2.23), since $\sigma_\rho$ and $\rho$ are tempered.

Conversely, let $\sigma$ be a tempered generic representation of $\mathrm{SO}_{2n+1}(F)$. Write $\sigma$ in the form (4.6), so that (4.1)–(4.5) are satisfied. Define

(4.25)

$$\ell(\sigma) = \Delta(\eta_1, p_1) \times \cdots \times \Delta(\eta_d, p_d) \times \ell(\sigma^{(et)}) \times \widehat{\Delta(\eta_d, p_d)} \times \cdots \times \widehat{\Delta(\eta_1, p_1)}.$$

Put $\rho = \ell(\sigma)$. This is clearly an irreducible tempered representation of $\mathrm{GL}_{2n}(F)$. It is now easy to verify that $\ell(\sigma)$ is of the form in (4.7), so that conditions (1)–(3) of Theorem 4.1 are satisfied, and by the first part of this proof, we get (4.8) and (4.9). This proves Theorem 4.1. □

In the following we extend Theorem 2.2 to the case of tempered generic representations of $\mathrm{SO}_{2n+1}(F)$. By Theorem 4.1, we know the local Langlands parameter for each member $\sigma$ in $\Pi^{(tg)}(\mathrm{SO}_{2n+1})$. More precisely, for each $\sigma$ in $\Pi^{(tg)}(\mathrm{SO}_{2n+1})$, we can express its lift $\ell(\sigma)$ as in (4.25) and hence the local Langlands parameter for $\sigma$ is

$$\varphi_{\sigma^{(et)}} \oplus \bigoplus_{i=1}^{d} [\varphi_{\eta_i} \otimes S_{2p_i+1} \oplus \hat{\varphi_{\eta_i}} \otimes S_{2p_i+1}].$$

Let $\Phi^{(t)}(\mathrm{SO}_{2n+1})$ be the subset of $\Phi(\mathrm{SO}_{2n+1})$ consisting of the local Langlands parameters $\varphi$ with the property that $\varphi(W_F)$ is bounded in $\mathrm{Sp}_{2n}(\mathbb{C})$. The parameters in $\Phi^{(t)}(\mathrm{SO}_{2n+1})$ are called *tempered.*

THEOREM 4.2. *The bijection of Theorem 2.2 extends to a bijection y (which is still denoted by y) between the set* $\Phi^{(t)}(\mathrm{SO}_{2n+1})$ *and the set* $\Pi^{(tg)}(\mathrm{SO}_{2n+1})$. *We have*

(4.26) $$L(\varphi \otimes \varphi', s) = L(y(\varphi) \times r(\varphi'), s)$$

(4.27) $$\epsilon(\varphi \otimes \varphi', s, \psi) = \epsilon(y(\varphi) \times r(\varphi'), s, \psi)$$

*for all irreducible admissible representations* $\varphi'$ *of* $W_F$ *of degree k with all* $k \in \mathbb{Z}_{>0}$. *Here r is the reciprocity map for* GL.

*Proof.* Let $\varphi$ be an admissible homomorphism of $W_F \times \mathrm{SL}_2(\mathbb{C})$ into $\mathrm{Sp}_{2n}(\mathbb{C})$, such that $\varphi(W_F)$ is bounded. Compose $\varphi$ with the embedding $\mathrm{Sp}_{2n}(\mathbb{C}) \hookrightarrow \mathrm{GL}_{2n}(\mathbb{C})$, and regard $\varphi$ as a $2n$-dimensional representation of $W_F \times \mathrm{SL}_2(\mathbb{C})$. Since the image $\varphi(W_F \times \mathrm{SL}_2(\mathbb{C}))$ preserves a nondegenerate skew-symmetric bilinear form, the representation $\varphi$ has a decomposition of following form:

$$\varphi = J'_N(\varphi) \oplus J'_W(\varphi) \oplus J'_{R_1}(\varphi) \oplus J'_{R_2}(\varphi) \oplus J_2(\varphi) \tag{4.28}$$

where each summand can be written explicitly as follows. The summand $J'_N(\varphi)$ is

$$J'_N(\varphi) = \bigoplus_{j=1}^{v} \mu'_{z_j}\left(\varphi_{z_j} \otimes S_{2h_{z_j}+1} \oplus \widehat{\varphi_{z_j}} \otimes S_{2h_{z_j}+1}\right) \tag{4.29}$$

with the properties that (i) $2h_{z_j} \in \mathbb{Z}_{\geq 0}$, (ii) $\mu'_{z_j} \in \mathbb{Z}_{>0}$ are the multiplicities, (iii) $\varphi_{z_j} \not\cong \widehat{\varphi_{z_j}}$, and (iv) $\varphi_{z_1}, \ldots, \varphi_{z_v}$ are pairwise non-equivalent irreducible bounded representations of $W_F$. The summand $J'_W(\varphi)$ has an expression

$$J'_W(\varphi) = \bigoplus_{j=1}^{u} 2\mu_{i_j}\left(\varphi_{i_j} \otimes S_{2h_{i_j}+1}\right) \tag{4.30}$$

with the properties that (i) $2h_{i_j} \in \mathbb{Z}_{\geq 0}$, (ii) $\mu_{i_j} \in \mathbb{Z}_{>0}$ are the half of the multiplicities, and (iii) $\varphi_{i_1}, \ldots, \varphi_{i_u}$ are pairwise non-equivalent irreducible, bounded, self-dual, representations of $W_F$, such that $\varphi_{i_j} \otimes S_{2h_{ij}+1}$'s are orthogonal. This means that for each $j$, either $\varphi_{i_j}$ is symplectic and $h_{i_j} \in \frac{1}{2} + \mathbb{Z}_{\geq 0}$, or $\varphi_{i_j}$ is orthogonal and $h_{i_j} \in \mathbb{Z}_{\geq 0}$. The summand $J'_{R_2}(\varphi)$ has form

$$J'_{R_2}(\varphi) = \bigoplus_{j=1}^{c} 2\mu_{t_j}\left(\varphi_{t_j} \otimes S_{2h_{t_j}+1}\right) \tag{4.31}$$

with the properties that (i) $2h_{t_j} \in \mathbb{Z}_{\geq 0}$, (ii) $\mu_{t_j} \in \mathbb{Z}_{>0}$ are the half of the multiplicities, and (iii) $\varphi_{t_1}, \ldots, \varphi_{t_c}$ are pairwise non-equivalent irreducible, bounded, self-dual, representation of $W_F$, such that $\varphi_{t_j} \otimes S_{2n_{t_j}+1}$'s are symplectic. This means that for each $j$, either $\varphi_{t_j}$ is symplectic and $h_{t_j} \in \mathbb{Z}_{\geq 0}$, or $\varphi_{t_j}$ is orthogonal and $h_{t_j} \in \frac{1}{2} + \mathbb{Z}_{\geq 0}$. Finally, the summands $J'_{R_1}(\varphi)$ and $J_2(\varphi)$ can be expressed as

$$J'_{R_1}(\varphi) \oplus J_2(\varphi) = \bigoplus_{j=1}^{r} (2\mu_{x_j} + 1)\left(\varphi_{x_j} \otimes S_{2h_{x_j}+1}\right) \tag{4.32}$$

$$J_2(\varphi) = \bigoplus_{j=1}^{r} \varphi_{x_j} \otimes S_{2h_{x_j}+1} \tag{4.33}$$

with the properties that (i) $2h_{x_j} \in \mathbb{Z}_{\geq 0}$, (ii) $\mu_{x_j} \in \mathbb{Z}_{\geq 0}$ ($2\mu_{x_j} + 1$ are the multiplicities), and (iii) $\varphi_{x_1}, \ldots, \varphi_{x_r}$ are pairwise non-equivalent irreducible, bounded, self-dual, representations of $W_F$, such that $\varphi_{x_j} \otimes S_{2h_{x_j}+1}$ is symplectic. Note that some of the sums in (4.29)–(4.31), (4.33) may be empty. Let $y(\varphi)$ be the unique irreducible generic constituent of the following induced representation of $\mathrm{SO}_{2n+1}(F)$

$$J_N(y(\varphi)) \times J_W(y(\varphi)) \times J_{R_1}(y(\varphi)) \times J^*_{R_2}(y(\varphi)) \rtimes y(J_2(\varphi)).$$

The induction data in the induced representation can be expressed as follows:

$$J_N(y(\varphi)) = \mathop{\times}_{j=1}^{v} \underbrace{\left(\Delta\left(r\left(\varphi_{z_j}\right), h_{z_j}\right) \times \cdots \times \Delta\left(r\left(\varphi_{z_j}\right), h_{z_j}\right)\right)}_{\mu'_{z_j}}$$

$$J_W(y(\varphi)) = \mathop{\times}_{j=1}^{u} \underbrace{\left(\Delta\left(r\left(\varphi_{i_j}\right), h_{i_j}\right) \times \cdots \times \Delta\left(r\left(\varphi_{i_j}\right), h_{i_j}\right)\right)}_{\mu_{i_j}}$$

$$J^*_{R_2}(y(\varphi)) = \mathop{\times}_{j=1}^{c} \underbrace{\left(\Delta\left(r\left(\varphi_{t_j}\right), h_{t_j}\right) \times \cdots \times \Delta\left(r\left(\varphi_{i_j}\right), h_{t_j}\right)\right)}_{\mu_{t_j}}$$

$$J_{R_1}(y(\varphi)) = \mathop{\times}_{j=1}^{r} \underbrace{\left(\Delta\left(r\left(\varphi_{t_j}\right), h_{t_j}\right) \times \cdots \times \Delta\left(r\left(\varphi_{x_j}\right), h_{x_j}\right)\right)}_{\mu_{x_j}}$$

and $y(J_2(\varphi))$ is the irreducible discrete series (or square-integrable) generic representation of $\mathrm{SO}_{2n''+1}(F)$ given by Theorem 2.2. It follows that $y(\varphi)$ is tempered and generic, and by the proof of Theorem 4.1, we have that $\ell(y(\varphi)) = r(\varphi)$ and hence

$$L(\varphi \otimes \varphi', s) = L(r(\varphi) \times r(\varphi'), s) = L(y(\varphi) \times r(\varphi'), s)$$
$$\epsilon(\varphi \otimes \varphi', s, \psi) = \epsilon(r(\varphi) \times r(\varphi'), s, \psi) = \epsilon(y(\varphi) \times r(\varphi'), s, \psi)$$

for any irreducible representation $\varphi'$ of $W_F$.

Conversely, let $\sigma$ be an irreducible tempered generic representation of $\mathrm{SO}_{2n+1}(F)$. Consider the tempered representation $\ell(\sigma)$ of $\mathrm{GL}_{2n}(F)$ given by Theorem 4.1. By the discussion right before the statement of Theorem 4.2 we have that $\varphi = r^{-1}(\ell(\sigma))$, which is a tempered local parameter in $\Phi^{(t)}(\mathrm{SO}_{2n+1})$. Thus $\ell(\sigma) = r(\varphi)$ and hence $y(\varphi) = \sigma$ (we keep using Theorem 1.3). This finishes the proof of Theorem 4.2. □

The following result is a direct consequence of our discussions above. It is stated as Part (2) of Theorem C in the Introduction.

THEOREM 4.3. *If a local Langlands parameter $\varphi$ in $\Phi(\mathrm{SO}_{2n+1})$ is tempered, i.e., the image $\varphi(W_F)$ is bounded in the Langlands dual group $\mathrm{Sp}_{2n}(\mathbb{C})$ of $\mathrm{SO}_{2n+1}$, then the representation $y(\varphi)$ constructed in Theorem 4.2 is an irreducible generic representation of $\mathrm{SO}_{2n+1}(F)$, satisfying conditions (4.26) and (4.27).*

We remark that since the representation $y(\varphi)$ constructed in Theorem 4.2 satisfies the conditions in the local Langlands reciprocity conjecture, if the conjectured local L-packet $\Pi(\varphi)$ were constructed, then Theorem 4.3 would mean that each tempered local L-packet contains a generic member. This last statement is a

$\mathrm{SO}_{2n+1}$-case of a general conjecture stated in [S1] and of a basic assumption used in [A]. It is clear that the explicit construction of local L-packets is beyond the methods used here. We refer to [M.W.] and [M.T.] for a recent progress related to this problem.

**5. Generic representations.** In this section, we complete the local Langlands functorial lift $\ell$ from the set $\Pi^{(g)}(\mathrm{SO}_{2n+1})$ of equivalence classes of irreducible generic representations of $\mathrm{SO}_{2n+1}(F)$ to $\mathrm{GL}_{2n}(F)$ and write the local Langlands parameter for each member in $\Pi^{(g)}(\mathrm{SO}_{2n+1})$. From Theorems 1.1, 2.1, 3.1, and 4.1, we see that the image of $\ell$ consists of irreducible generic representations of $\mathrm{GL}_{2n}(F)$. However, this is no longer true for general (non-tempered) members in $\Pi^{(g)}(\mathrm{SO}_{2n+1})$. Because of this new phenomenon, we include the discussion of Theorem 5.2 here, although it is not directly related to the main results of this paper.

By the classification theory in [Z], any irreducible representation of $\mathrm{GL}_{2n}(F)$ can be realized as a constituent of an induced representation

$$\delta(\Sigma_1) \times \cdots \times \delta(\Sigma_q)$$

of $\mathrm{GL}_{2n}(F)$. We shall study which irreducible constituent is in the image of the local Langlands lift $\ell$ from $\Pi^{(g)}(\mathrm{SO}_{2n+1})$. (This $\ell$ will be a natural extension of the map $\ell$ in Theorem 4.1.) By Theorem 4.1, we should consider only self-dual representations of $\mathrm{GL}_{2n}(F)$ of the following form

$$\delta(\Sigma_1) \times \cdots \times \delta(\Sigma_f) \times \rho^{(t)} \times \delta(\widehat{\Sigma}_f) \times \cdots \times \delta(\widehat{\Sigma}_1), \tag{5.1}$$

where $\rho^{(t)}$ is an irreducible, self-dual, tempered representation of $\mathrm{GL}_{2n^*}(F)$ and $\Sigma_i$'s are imbalanced segments (which are given more explicitly below).

Assume that $\rho^{(t)}$ is in the image of the local Langlands lift from $\mathrm{SO}_{2n^*+1}(F)$ as given in Theorem 4.1. This means that there is a unique irreducible generic tempered representation $\sigma^{(t)}$ of $\mathrm{SO}_{2n^*+1}(F)$ such that $\ell(\sigma^{(t)}) = \rho^{(t)}$. In the following discussion, we let $\rho^{(2)}$ be the lift of the unique irreducible square-integrable generic representation $\sigma^{(2)}$, which is related to $\sigma^{(t)}$ by the classification theory (see (3.7) and (4.6)), and let $\sigma^{(0)}$ be the irreducible generic supercuspidal representation occurring in $\sigma^{(2)}$ as in (2.31), whose lift is denoted by $\ell(\sigma^{(0)}) = \rho^{(0)}$. Then, by Theorems 1.1, 2.1, 3.1, and 4.1, the representation $\rho^{(t)}$ is completely determined, up to isomorphism, by the following three families of irreducible square-integrable representations of $\mathrm{GL}_*(F)$:

$$\{\Delta(\tau_j, m_j)\}_{j=1}^r, \quad \{\Delta(\beta_j, e_j)\}_{j=1}^c, \quad \{\Delta(\eta_j, p_j)\}_{j=1}^d. \tag{5.2}$$

We use the notations of the previous sections and write

$$\Sigma_1 = \left[\nu^{-q_1}\xi_1, \nu^{-q_1+w_1}\xi_1\right], \ \Sigma_2 = \left[\nu^{-q_2}\xi_2, \nu^{-q_2+w_2}\xi_2\right], \ldots, \tag{5.3}$$
$$\Sigma_f = \left[\nu^{-q_f}\xi_f, \nu^{-q_f+w_f}\xi_f\right],$$

where $\xi_1, \ldots, \xi_f$ are irreducible, unitary and supercuspidal (with possible repetitions), $q_i \in \mathbb{R}$, $w_i \in \mathbb{Z}_{\geq 0}$ and $q_i \neq \frac{w_i}{2}$.

*Definition 5.1.* Let $\{\Sigma_j\}_{j=1}^f$ and $\rho^{(t)}$ be given as above. The sequence $\{\Sigma_j\}_{j=1}^f$ is called an $\mathrm{SO}_{2n+1}$-*generic* sequence (of segments) with respect to $\rho^{(t)}$ if it satisfies the following conditions:

(1) the segment $\Sigma_i$ is not *linked* to either $\Sigma_j$ or $\widehat{\Sigma_j}$ for $1 \leq i \neq j \leq f$; and

(2) for $1 \leq i \leq f$, $\Sigma_i$ is not linked to any segment, which corresponds to a representation in any of the families

$$\{\Delta(\tau_j, m_j)\}_{j=1}^r, \quad \{\Delta(\beta_j, e_j)\}_{j=1}^c, \quad \{\Delta(\eta_j, p_j)\}_{j=1}^d,$$

which are determined by $\rho^{(t)}$ as in (5.2); and

(3) one of the following three conditions holds

(3a) $\xi_i \not\cong \widehat{\xi_i}$, *or*

(3b) $\Sigma_i$ *is linked to an element of* $A_2(\rho^{(2)})$, *or*

(3c) $(\xi_i, \sigma^{(0)})$ *is* $(C\alpha)$ $(\alpha = 0, \frac{1}{2}, 1)$, *but* $\pm\alpha \notin \{-q_i, -q_i+1, \ldots, -q_i+w_i\}$.

*Remark 5.1.* Condition (3c) means that

- if $L(\sigma^{(0)} \times \xi_i, s)$ has a pole at $s = 0$, then $\pm 1 \notin \{-q_i, -q_i+1, \ldots, -q_i+w_i\}$,
- if $L(\xi_i, \Lambda^2, s)$ has a pole at $s = 0$, but $L(\sigma^{(0)} \times \xi_i, s)$ has no pole at $s = 0$, then $0 \notin \{-q_i, -q_i+1, -q_i+w_i\}$,
- if $L(\xi_i, \mathrm{sym}^2, s)$ has a pole at $s = 0$, then $\pm\frac{1}{2} \notin \{-q_i, -q_i+1, \ldots, -q_i+w_i\}$.

From the data given above, if we put $\pi_i = \delta(\Sigma_i)$, $i = 1, \ldots, f$, then the representation $\sigma$ of $\mathrm{SO}_{2n+1}(F)$ defined by

$$\sigma := \pi_1 \times \pi_2 \times \cdots \times \pi_f \rtimes \sigma^{(t)} \tag{5.4}$$

is irreducible and generic. Moreover, all irreducible generic representations of $\mathrm{SO}_{2n+1}(F)$ are obtained in this way. The set $\{\pi_1, \ldots, \pi_f; \sigma^{(t)}\}$ is uniquely determined. This is the classification theorem proved and explained by Muic in §4 of [M2].

The natural candidate for the image of the local Langlands lift of $\sigma$ is the Langlands subquotient of the induced representation

$$\delta(\Sigma_1) \times \cdots \times \delta(\Sigma_f) \times \rho^{(t)} \times \delta(\widehat{\Sigma}_f) \times \cdots \times \delta(\widehat{\Sigma}_1)$$

of $\mathrm{GL}_{2n}(F)$. It is clear that this lift defined in this way preserves the twisted local $\gamma$-factors, L-factors and $\epsilon$-factors. After re-arranging the induction data, we may assume that the exponents of $\delta(\Sigma_1), \ldots, \delta(\Sigma_f)$ are positive and in non-increasing order, i.e., the Langlands induction data. Let us now reformulate the conditions of Definition 5.1 as follows (adding now Condition (L1)):

(L1) $\frac{w_1}{2} - q_1 \geq \frac{w_2}{2} - q_2 \geq \cdots \geq \frac{w_f}{2} - q_f > 0$,

(L2) The only possible linkages among the segments

$$\Sigma_1, \Sigma_2, \ldots, \Sigma_f, \widehat{\Sigma}_f, \ldots, \widehat{\Sigma}_2, \widehat{\Sigma}_1$$

may occur between $\Sigma_i$ and $\widehat{\Sigma}_i$ for some indices $i$ (this is Condition (1) in Definition 5.1),

(L3) The representations $\delta(\Sigma_i) \times \rho^{(t)}$ and $\delta(\widehat{\Sigma}_i) \times \rho^{(t)}$ are irreducible for all $1 \leq i \leq f$ (this is Condition (2) in Definition 5.1), and

(L4) Assume that $\xi_i$ is self-dual and $2q_i \in \mathbb{Z}$, such that if $L(\xi_i, \Lambda^2, s)$ has a pole at $s = 0$, then $q_i \in \mathbb{Z}$, and if $L(\xi_i, \mathrm{sym}^2, s)$ has a pole at $s = 0$, then $q_i \in \frac{1}{2} + \mathbb{Z}$. Then in this case, $\Sigma_i$ is not linked to $\widehat{\Sigma}_i$. Moreover, if $L(\rho^{(0)} \times \xi_i, s)$ has a pole at $s = 0$, then $-q_i \geq 2$, or $q_i = -1$ and $\xi_i \in A_2(\rho^{(2)})$. This covers Condition (3) in Definition 5.1, where $A_2(\rho^{(2)})$ is defined in the proof of Theorem 2.1.

We summarize what we got. For any given irreducible generic representation $\sigma$ of $\mathrm{SO}_{2n+1}(F)$, which is realized as in (5.4), we have the following induced representation of $\mathrm{GL}_{2n}(F)$ satisfying conditions (L1)–(L4),

$$\delta(\Sigma_1) \times \delta(\Sigma_2) \times \cdots \times \delta(\Sigma_f) \times \rho^{(t)} \times \delta(\widehat{\Sigma}_f) \times \cdots \times \delta(\widehat{\Sigma}_2) \times \delta(\widehat{\Sigma}_1), \tag{5.5}$$

which is completely determined by $\sigma$. It follows from conditions (L1)–(L4) and the classification theory in [Z] that the induced representation in (5.5) is irreducible if and only if there is no index $i \in \{1, \ldots, f\}$ such that $\Sigma_i$ is linked to $\widehat{\Sigma}_i$. In general, we take $\rho$ to be the Langlands quotient of (5.5). It follows that $\rho$ is the local Langlands lift of $\sigma$ and satisfies the compatibilities of local factors:

$$L(\sigma_\rho \times \pi, s) = L(\rho \times \pi, s) \tag{5.6}$$

$$\gamma(\sigma_\rho \times \pi, s, \psi) = \gamma(\rho \times \pi, s, \psi) \tag{5.7}$$

$$\epsilon(\sigma_\rho \times \pi, s, \psi) = \epsilon(\rho \times \pi, s, \psi) \tag{5.8}$$

for any irreducible supercuspidal representation $\pi$ of $\mathrm{GL}_*(F)$. (See [S1 p. 308], [S3], [J.PS.S. p. 458] for details on these local factors.) This extends the bijection $\ell$ in Theorem 4.1 to the set $\Pi^{(g)}(\mathrm{SO}_{2n+1})$. Hence we have:

THEOREM 5.1. *The bijection $\ell$ of Theorem 4.1 extends to a bijection (which is still denoted by $\ell$) between the set $\Pi^{(g)}(\mathrm{SO}_{2n+1})$ of equivalence classes of irreducible generic representations $\sigma$ of $\mathrm{SO}_{2n+1}(F)$ and the set of equivalence classes of irreducible self-dual representations $\rho = \ell(\sigma)$ of $\mathrm{GL}_{2n}(F)$, which are Langlands quotients of representations as in (5.5), satisfying conditions (L1)–(L4). This bijection preserves twisted local $\gamma$-, $L$-, and $\epsilon$-factors as in* (5.6)–(5.8). *In addition,*

*$\ell(\sigma)$ is generic if and only if there is no index $i \in \{1, \dots, f\}$ such that $\Sigma_i$ is linked to $\widehat{\Sigma}_i$ (notations as in (5.5)).*

The last statement in Theorem 5.1 is about the genericity of $\rho = \ell(\sigma)$. This is equivalent to the irreducibility of the induced representation in (5.5). The conditions characterizing this property depend on segments and L-functions. In general, by the classification theory in [Z], an irreducible self-dual generic representation $\rho$ of $\mathrm{GL}_{2n}(F)$ can be written as

$$\rho = \delta(\Sigma_1) \times \cdots \times \delta(\Sigma_q). \tag{5.9}$$

It is our interest here to determine, in terms of its segments, if $\rho$ lies in the image of the local Langlands lift $\ell$ from $\Pi^{(g)}(\mathrm{SO}_{2n+1})$.

For the sake of simplicity, we say that a segment $\Sigma$ *occurs in* $\rho$ if the associated representation $\delta(\Sigma)$ occurs in the induced data of $\rho$ in (5.9). Let $P_{bs}(\rho)$ be the set of all different irreducible, supercuspidal, self-dual representations $\lambda$ of $\mathrm{GL}_*(F)$ such that *balanced segments* of the form $[\nu^{-a}\lambda, \nu^a\lambda]$ occur in $\rho$ for some $a$ with $2a \in \mathbb{Z}_{\geq 0}$.

Let $P_W(\rho)$ be the subset of $P_{bs}(\rho)$, consisting of all $\lambda$'s with the property that if $L(\lambda, \Lambda^2, s)$ has a pole at $s = 0$, then there are segments $[\nu^{-a}\lambda, \nu^a\lambda]$, with $a \in \frac{1}{2} + \mathbb{Z}_{\geq 0}$, which occur in $\rho$, or if $L(\lambda, \mathrm{sym}^2, s)$ has a pole at $s = 0$, then there are segments $[\nu^{-a}\lambda, \nu^a\lambda]$, with $a \in \mathbb{Z}_{\geq 0}$, which occur in $\rho$. For any $\lambda \in P_W(\rho)$, we define a set

$$E_\lambda(\rho) \subset \frac{1}{2} \cdot \mathbb{Z}, \tag{5.10}$$

which satisfies the following conditions:

(1) if $L(\lambda, \Lambda^2, s)$ has a pole at $s = 0$, then $E_\lambda(\rho)$ consists of all different half-integers $a \in \frac{1}{2} + \mathbb{Z}_{\geq 0}$ such that the segment $[\nu^{-a}\lambda, \nu^a\lambda]$ occurs in $\rho$; and

(2) if $L(\lambda, \mathrm{sym}^2, s)$ has a pole at $s = 0$, then $E_\lambda(\rho)$ consists of all different integers $a \in \mathbb{Z}_{\geq 0}$ such that the segment $[\nu^{-a}\lambda, \nu^a\lambda]$ occurs in $\rho$.

For $a \in E_\lambda(\rho)$, we denote by $\mu_\lambda(a)$ the multiplicity of the factor $\Delta(\lambda, a)$ in $\rho$. Write $E_\lambda(\rho)$ in decreasing order

$$E_\lambda(\rho): \ a_\lambda(1) > a_\lambda(2) > \cdots > a_\lambda(n_\lambda) \tag{5.11}$$

and write the sequences of multiplicities

$$\mu(E_\lambda(\rho)): \ \{\mu_\lambda(a_\lambda(1)), \mu_\lambda(a_\lambda(2)), \dots, \mu_\lambda(a_\lambda(n_\lambda))\}. \tag{5.12}$$

Single out of $\mu(E_\lambda(\rho))$ the odd multiplicities as follows. Let

(5.13)
$$\mathcal{O}_{E_\lambda(\rho)} = \{1 \leq i \leq n_\lambda \mid \mu_\lambda(a_\lambda(i)) \ \text{ is odd}\} = \mathcal{O}^{(1)}_{E_\lambda(\rho)} \cup \mathcal{O}^{(2)}_{E_\lambda(\rho)} \cup \cdots \cup \mathcal{O}^{(l_\lambda)}_{E_\lambda(\rho)}$$

where $\mathcal{O}^{(j)}_{E_\lambda(\rho)}$ are the "connected components" of $\mathcal{O}_{E_\lambda(\rho)}$ in the sense that $\mathcal{O}^{(j)}_{E_\lambda(\rho)}$ consists of consecutive (i.e., differ by one) integers. We note that the largest integer of $\mathcal{O}^{(j)}_{E_\lambda(\rho)}$ is smaller than the smallest integer of $\mathcal{O}^{(j+1)}_{E_\lambda(\rho)}$ by at least two.

THEOREM 5.2. *(a) Let $\rho$ be an irreducible, self-dual, generic representation of* $\mathrm{GL}_{2n}(F)$. *Then $\rho$ is in the image of $\ell$ as in Theorem 5.1, if and only if $\mathcal{O}_{E_\lambda(\rho)}$ is empty for all $\lambda \in P_W(\rho)$.*

*(b) There is a bijection $\mathfrak{b}$ between the set of equivalence classes of irreducible generic representations of* $\mathrm{SO}_{2n+1}(F)$ *and the set of equivalence classes of irreducible, self-dual, generic representations $\rho$ of* $\mathrm{GL}_{2n}(F)$ *satisfying the properties that for $\lambda$ in $P_W(\rho)$,*

(1) $|\mathcal{O}^{(j)}_{E_\lambda(\rho)}|$ *is even, for* $1 \le j \le l_\lambda - 1$,

(2) *if* $L(\lambda, \mathrm{sym}^2, s)$ *has a pole at* $s = 0$, *then* $|\mathcal{O}^{(l_\lambda)}_{E_\lambda(\rho)}|$ *is even,*

(3) *if* $L(\lambda, \Lambda^2, s)$ *has a pole at* $s = 0$, *then* $|\mathcal{O}^{(l_\lambda)}_{E_\lambda(\rho)}|$ *may be even or odd; in case it is odd, write* $\mathcal{O}^{(l_\lambda)}_{E_\lambda(\rho)} = \{\alpha, \alpha+1, \alpha+2, \ldots, \alpha+2t_0\}$, *then* $\alpha + 2t_0 = n_\lambda$.

*This bijection preserves twisted local gamma factors:*

$$\gamma(\sigma \times \pi, s, \psi) = \gamma(\mathfrak{b}(\sigma) \times \pi, s, \psi)$$

*for all irreducible generic representations $\sigma$ (resp. $\pi$) of* $\mathrm{SO}_{2n+1}(F)$ *(resp.* $\mathrm{GL}_k(F)$*) with all* $k \in \mathbb{Z}_{>0}$.

We remark that Part (b) of Theorem 5.2 characterizes the set of irreducible self-dual generic representations of $\mathrm{GL}_{2n}(F)$, which share the same twisted local gamma factors with those in the image of the local Langlands lift $\ell$ in Theorem 5.1. However, they do not necessarily share the same twisted L-factors in general. This is one of the points which we have to keep in mind when thinking about the characterization of local Langlands functorial lifts.

*Proof of Part (b) of Theorem 5.2.* We first write the given irreducible self-dual generic representation $\rho$ of $\mathrm{GL}_{2n}(F)$ (in (5.9)) as follows

$$\rho \cong \delta(\Sigma_1) \times \cdots \times \delta(\Sigma_{f_1}) \times \rho_1 \times \delta(\widehat{\Sigma}_{f_1}) \times \cdots \times \delta(\widehat{\Sigma}_1) \tag{5.14}$$

where $\Sigma_1, \ldots, \Sigma_{f_1}$ are imbalanced segments, $\widehat{\Sigma}_1, \ldots, \widehat{\Sigma}_{f_1}$ are the respective duals, and $\rho_1$ is self-dual and is an induced representation associated to balanced segments. Since $\rho$ is irreducible and generic, there are no linkages among the segments, in particular among the imbalanced segments

$$\Sigma_1, \ldots, \Sigma_{f_1}; \widehat{\Sigma}_1, \ldots, \widehat{\Sigma}_{f_1}.$$

For $1 \le i \le f_1$, write $\Sigma_i$ and $\widehat{\Sigma}_i$ as

$$\Sigma_i = \left[\nu^{-q_i}\xi_i, \nu^{-q_i+w_i}\xi_i\right], \ \widehat{\Sigma}_i = \left[\nu^{-w_i+q_i}\widehat{\xi}_i, \nu^{(-w_i+q_i)+w_i}\widehat{\xi}_i\right], \tag{5.15}$$

where $\xi_i$ is an irreducible unitary supercuspidal representation of $\mathrm{GL}_*(F)$ and $q_i \in \mathbb{R}$, $w_i \in \mathbb{Z}_{\geq 0}$, $w_i \neq 2q_i$. By assumption on linkage,

$$\delta(\Sigma_1) \times \cdots \times \delta(\Sigma_{f_1}) \times \delta(\widehat{\Sigma}_{f_1}) \times \cdots \times \delta(\widehat{\Sigma}_1)$$

is irreducible. Note that

$$\nu^{q_i - \frac{w_i}{2}} \delta(\Sigma_i) = \Delta\left(\xi_i, \frac{w_i}{2}\right) \tag{5.16}$$

is square-integrable. We may assume that

$$\frac{w_1}{2} - q_1 \geq \frac{w_2}{2} - q_2 \geq \cdots \geq \frac{w_{f_1}}{2} - q_{f_1} > 0. \tag{5.17}$$

This implies that

$$\frac{w_1}{2} - q_1 \geq \cdots \geq \frac{w_{f_1}}{2} - q_{f_1} > 0 > q_{f_1} - \frac{w_{f_1}}{2} \geq \cdots \geq q_1 - \frac{w_1}{2}. \tag{5.18}$$

Note that $P_W(\rho) = P_W(\rho_1)$. We consider the structure of the segments occurring in $\rho_1$. Let $\lambda \in P_W(\rho_1)$. Consider $\mathcal{O}_{E_\lambda(\rho_1)} = \mathcal{O}_{E_\lambda(\rho)}$, as in (5.13). Let $1 \leq j \leq l_\lambda - 1$, and in case $|\mathcal{O}^{(l_\lambda)}_{E_\lambda(\rho)}|$ is even, we include $j = l_\lambda$, as well. Write

$$\mathcal{O}^{(j)}_{E_\lambda(\rho)} = \{\alpha_j, \alpha_j + 1, \alpha_j + 2, \ldots, \alpha_j + 2s_j - 1\}. \tag{5.19}$$

Thus, the self-dual representations

$$\Delta(\lambda, a_\lambda(\alpha_j)), \Delta(\lambda, a_\lambda(\alpha_j + 1)), \ldots, \Delta(\lambda, a_\lambda(\alpha_j + 2s_j - 1)) \tag{5.20}$$

occur in $\rho_1$, each one with odd multiplicity

$$\mu_\lambda(a_\lambda(\alpha_j)), \mu_\lambda(a_\lambda(\alpha_j + 1)), \ldots, \mu_\lambda(a_\lambda(\alpha_j + 2s_j - 1)),$$

respectively. Note that the segments, which correspond to the representations in (5.20) form a strictly decreasing sequence

$$\left[\nu^{-a_\lambda(\alpha_j)}\lambda, \nu^{a_\lambda(\alpha_j)}\lambda\right] \underset{\neq}{\supseteq} \cdots \underset{\neq}{\supseteq} \left[\nu^{-a_\lambda(\alpha_j+2s_j-1)}\lambda, \nu^{a_\lambda(\alpha_j+2s_j-1)}\lambda\right].$$

Following the same idea as in (2.37) and (2.38), we define then the following segments:

$$\Sigma^{(i)}_{\lambda,j} = \left[\nu^{-a_\lambda(\alpha_j+2i-1)}\lambda, \nu^{a_\lambda(\alpha_j+2i-2)}\lambda\right], \; i = 1, 2, \ldots, s_j. \tag{5.21}$$

In case $|\mathcal{O}^{(l_\lambda)}_{E_\lambda(\rho)}|$ is odd, (this implies that $L(\lambda, \Lambda^2, s)$ has a pole at $s = 0$), we have to deal with the case of $j = l_\lambda$. We write

$$\mathcal{O}^{(l_\lambda)}_{E_\lambda(\rho)} = \left\{\alpha_{l_\lambda}, \alpha_{l_\lambda} + 1, \ldots, \alpha_{l_\lambda} + 2t_{l_\lambda}\right\} \tag{5.22}$$

and define, for $i = 1, 2, \ldots, t_{l_\lambda}$,

(5.23)

$$\Sigma^{(i)}_{\lambda,l_\lambda} = \left[\nu^{-a_\lambda(\alpha_{l_\lambda}+2i-1)}\lambda, \nu^{a_\lambda(\alpha_{l_\lambda}+2i-2)}\lambda\right], \text{ and } \Sigma^{(t_{l_\lambda}+1)}_{\lambda,l_\lambda} = \left[\nu^{\frac{1}{2}}\lambda, \nu^{a_\lambda(\alpha_{l_\lambda}+2t_{l_\lambda})}\lambda\right].$$

Note that $\Sigma_{\lambda,j}^{(i)}$ is linked to $\widehat{\Sigma}_{\lambda,j}^{(i)}$, for all $i \geq 1$, but there are no other linkages among

$$\Sigma_{\lambda,j}^{(1)},\ \widehat{\Sigma}_{\lambda,j}^{(1)},\ \Sigma_{\lambda,j}^{(2)},\ \widehat{\Sigma}_{\lambda,j}^{(2)},\dots. \tag{5.24}$$

The reason for this is that when $i' > i$, we have

$$\Sigma_{\lambda,j}^{(i)} \supsetneq \Sigma_{\lambda,j}^{(i')} \cup \widehat{\Sigma}_{\lambda,j}^{(i')}.$$

For $\lambda \in P_W(\rho_1)$, rewrite (5.19) and (5.22) in a uniform way

$$\mathcal{O}_{E_\lambda(\rho)}^{(j)} = \{\alpha_j, \alpha_j + 1, \dots, \alpha_j + t_j\}$$

and put $r_j = [\frac{t_j+2}{2}]$. The segments in (5.21), (5.23) can now be indexed in a uniform way

$$\Sigma_{\lambda,j}^{(1)},\ \Sigma_{\lambda,j}^{(2)},\dots,\ \Sigma_{\lambda,j}^{(r_j)}. \tag{5.25}$$

In order to study the structure of $\rho_1$ with the local Langlands lift given in Theorem 4.1, we express $\rho_1$ as

$$\rho_1 \cong \left( \mathop{\times}_{\lambda \in P_W(\rho)} \mathop{\times}_{j=1}^{l_\lambda} \mathop{\times}_{i=0}^{t_j} \Delta(\lambda, a_\lambda(\alpha_j + i)) \right) \times \rho^{(t)} \tag{5.26}$$

where $\rho^{(t)}$ is tempered and self-dual.

We claim that $\rho^{(t)}$ lies in the image of the lift $\ell$ in Theorem 4.1. To prove this statement, we only need to verify conditions (2), (3) of Theorem 4.1, while condition (1) is clear from the self-duality of $\rho^{(t)}$. By the definition of $P_W(\rho_1)$, if a representation $\lambda$ in $P_{bs}(\rho_1)$ belongs to $P_W(\rho_1)$, then it satisfies the same conditions about the poles of L-functions as required in condition (2) or (3) in Theorem 4.1. It remains to show that for $\lambda$ in $P_W(\rho_1)$, the multiplicity of $\Delta(\lambda, a_\lambda(i))$ occurring in $\rho^{(t)}$ is even for $1 \leq i \leq n_\lambda$. Indeed, if $i \in \mathcal{O}_{E_\lambda(\rho)}$, then $i = \alpha_j + i'$, for some $1 \leq j \leq l_\lambda$, $0 \leq i' \leq t_j$, and since $\mu_\lambda(a_\lambda(i))$ is odd, then $\Delta(\lambda, a_\lambda(i))$ appears an even number of times (possibly zero) in $\rho^{(t)}$. If $i \notin \mathcal{O}_{E_\lambda(\rho)}$, then $\mu_\lambda(a_\lambda(i))$ is even and hence $\Delta(\lambda, a_\lambda(i))$ appears an even number of times in $\rho^{(t)}$. This justifies the above claim.

We assume that the above $\rho^{(t)}$ is an irreducible representation of $\mathrm{GL}_{2n^*}(F)$. Since it satisfies the conditions characterizing the image of the lift $\ell$ in Theorem 4.1, there is a unique, up to equivalence, irreducible generic tempered representation $\sigma^{(t)}$ of $\mathrm{SO}_{2n^*+1}(F)$, such that $\rho^{(t)} = \ell(\sigma^{(t)})$. From the given irreducible generic representation $\rho$ and the constructed representations $\rho_1$, $\rho^{(t)}$, and $\sigma^{(t)}$, we construct the following representation of $\mathrm{SO}_{2n+1}(F)$:

$$\sigma_\rho = \mathop{\times}_{i=1}^{f_1} \delta(\Sigma_i) \times \left( \mathop{\times}_{\lambda \in P_W(\rho)} \mathop{\times}_{j=1}^{l_\lambda} \mathop{\times}_{i=1}^{r_j} \delta(\Sigma_{\lambda,j}^{(i)}) \right) \rtimes \sigma^{(t)}. \tag{5.27}$$

We are going to show that this representation $\sigma_\rho$ is irreducible and generic, or equivalently that it has the form (5.4) and satisfies all conditions in Definition 5.1. Write $\sigma^{(t)}$ in the form (4.6) (with $n^*$ replacing $n$) so that (4.1)–(4.5) are satisfied. For the sake of the following discussion, we recall that $\rho^{(2)}$ is the lift of the unique irreducible square-integrable generic representation $\sigma^{(2)}$, which is related to $\sigma^{(t)}$ by the classification theory (see (3.7) and (4.6)), and $\sigma^{(0)}$ is the irreducible generic supercuspidal representation occurring in $\sigma^{(2)}$ as in (2.31), whose lift is denoted by $\ell(\sigma^{(0)}) = \rho^{(0)}$.

We already know that there is no linkage among the segments $\{\Sigma_i, \widehat{\Sigma}_j \mid 1 \le i, j \le f_1\}$ and there is no linkage among the segments

$$\{\Sigma^{(i)}_{\lambda,j}, \widehat{\Sigma}^{(i')}_{\lambda,j'} \mid \lambda \in P_W(\rho),\quad 1 \le j,\ j' \le l_\lambda,\ 1 \le i,\ i' \le r_j\}$$

except for the linkages between $\Sigma^{(i)}_{\lambda,j}$ and $\widehat{\Sigma}^{(i)}_{\lambda,j}$, for each $\lambda, j, i$ as above. (See (5.24).) There is no linkage between $\Sigma_i$ and $\Sigma^{(i')}_{\lambda,j}$ or $\widehat{\Sigma}^{(i')}_{\lambda,j}$. Indeed, this happens, if and only if $\Sigma_i$ is linked to one of the segments

$$\Sigma^{(i')}_{\lambda,j} \cup \widehat{\Sigma}^{(i')}_{\lambda,j} = [\nu^{-a_\lambda(\alpha_j+2i'-2)}\lambda,\ \nu^{a_\lambda(\alpha_j+2i'-2)}\lambda], \tag{5.28}$$

$$\Sigma^{(i')}_{\lambda,j} \cap \widehat{\Sigma}^{(i')}_{\lambda,j} = [\nu^{-a_\lambda(\alpha_j+2i'-1)}\lambda,\ \nu^{a_\lambda(\alpha_j+2i'-1)}\lambda]. \tag{5.29}$$

Recall that $\Sigma_i$ is not linked to $\widehat{\Sigma}_i$. Note that this is also valid in case $j = l_\lambda$ and $i' = t_{l_\lambda} + 1$, since (5.29) is empty in these cases. Since each of the last two segments occurs in $\rho$, together with $\Sigma_i$, this is impossible. All segments which occur in $\rho$ are not linked. Similarly, $\widehat{\Sigma}_i$ can not be linked to $\Sigma^{(i')}_{\lambda,j}$ or to $\widehat{\Sigma}^{(i')}_{\lambda,j}$. This verifies Condition (1) in Definition 5.1.

Note also, that the segments $\Sigma^{(i)}_{\lambda,j}$ verify Condition (3c) in Definition 5.1. Recall that $a_\lambda(\nu) \in \frac{1}{2} + \mathbb{Z}_{\ge 0}$, if $L(\lambda, \Lambda^2, s)$ has a pole at $s = 0$, and $a_\lambda(\nu) \in \mathbb{Z}_{\ge 0}$, if $L(\lambda, \mathrm{sym}^2, s)$ has a pole at $s = 0$. Let us show now that the segments $\Sigma^{(i)}_{\lambda,j}$ satisfy Condition (2) in Definition 5.1 (we use its notation here). Clearly, $\Sigma^{(i)}_{\lambda,j}$ can not be linked to any segment of the form $[\nu^{-e_t}\beta_t, \nu^{e_t}\beta_t]$ or $[\nu^{-m_t}\tau_t, \nu^{m_t}\tau_t]$ since the powers of $\nu$ just do not match. If $\Sigma^{(i)}_{\lambda,j}$ is linked to a segment of the form $[\nu^{-p_t}\eta_t, \nu^{p_t}\eta_t]$, then $\eta_t = \lambda \in P_W(\rho)$ and $p_t$ is of the form $a_\lambda(\nu)$, $1 \le \nu \le n_\lambda$. By construction, we have either $\Sigma^{(i)}_{\lambda,j} \supset [\nu^{-a_\lambda(\nu)}\lambda, \nu^{a_\lambda(\nu)}\lambda]$ or $\Sigma^{(i)}_{\lambda,j} \subset [\nu^{-a_\lambda(\nu)}\lambda, \nu^{a_\lambda(\nu)}\lambda]$ and hence there can not be a linkage. This verifies Condition (2) in Definition 5.1 for the segments $\Sigma^{(i)}_{\lambda,j}$. The segments $\Sigma_i$ verify Condition (2) in Definition 5.1 as well, since the segments

$$\{[\nu^{-p_j}\eta_j, \nu^{p_j}\eta_j]\}_{j=1}^{d},\quad \{[\nu^{-e_j}\beta_j, \nu^{e_j}\beta_j]\}_{j=1}^{c},\quad \{[\nu^{-m_j}\tau_j, \nu^{m_j}\tau_j]\}_{j=1}^{r}$$

are precisely the ones which occur in $\rho^{(t)}$. Since $\Sigma_i$ occurs, together with these segments in $\rho$ (by (5.15), (5.26)), we see that $\Sigma_i$ can not be linked to any of the segments above. It remains to show that the $\Sigma_i$ verify one of the three conditions (3a), (3b), (3c), in Definition 5.1. If (3a) is not satisfied for $\Sigma_i$ then $\xi_i$ is self-dual. In this case, if (3c) is not satisfied, we have to consider the following cases:

(i) If $L(\sigma^{(0)} \times \xi_i, s)$ has a pole at $s = 0$ (($C1$) case), and if 1, or $-1$ belongs to $E^{(i)} = \{-q_i, -q_i + 1, \ldots, -q_i + w_i\}$, then $q_i \in \mathbb{Z}$. If $-1 \in E^{(i)}$, it follows that $\Sigma_i$ is linked to $\widehat{\Sigma}_i$ since $-q_i + w_i = (-q_i + \frac{w_i}{2}) + \frac{w_i}{2} > \frac{w_i}{2}$ by assumption. But this contradicts Condition (1) in Definition 5.1, which we just verified. This implies that $1 \in E^{(i)}$. Using the last argument, one shows that $q_i = -1$. Thus, $\Sigma_i$ is linked to $[\xi_i]$. Now, by assumption,

$$\xi_i \in A(\rho^{(2)}) = A_0(\rho^{(2)}) \cup A_1(\rho^{(2)}) \cup A_2(\rho^{(2)}).$$

Since the elements of $A_0(\rho^{(2)}) \cup A_1(\rho^{(2)})$ occur in $\rho$ (even in $\rho^{(2)}$) together with $\Sigma_i$, we see that $\Sigma_i$ can not be linked to an element of $A_0(\rho^{(2)}) \cup A_1(\rho^{(2)})$. This forces $\xi_i \in A_2(\rho^{(2)})$, and hence $\Sigma_i$ verifies (3b).

(ii) If $L(\sigma^{(0)} \times \xi_i, s)$ has no pole at $s = 0$ while $L(\xi_i, \Lambda^2, s)$ still has a pole at $s = 0$, and $0 \in E^{(i)}$, then $q_i \in \mathbb{Z}$. Since $\Sigma_i$ is imbalanced, we conclude that $\Sigma_i$ is linked to $\widehat{\Sigma}_i$, which is impossible.

(iii) If $L(\xi_i, \mathrm{sym}^2, s)$ has a pole at $s = 0$, and one of $\frac{1}{2}, -\frac{1}{2}$ lies in $E^{(i)}$ (this implies that $q_i \in \frac{1}{2} + \mathbb{Z}$), then in both cases, we conclude, as before, that $\Sigma_i$ is linked to $\widehat{\Sigma}_i$, which is impossible.

This proves that $\sigma_\rho$ defined in (5.27) is an irreducible generic representation of $\mathrm{SO}_{2n+1}(F)$, as required.

We now verify that twisted local gamma factors are preserved. Using the multiplicativity property of gamma factors (Theorem 1.4), we have, for an irreducible supercuspidal representation $\pi$ of $\mathrm{GL}_k(F)$,

$$(5.30)\quad \gamma(\sigma_\rho \times \pi, s, \psi)$$
$$= \gamma(\sigma^{(t)} \times \pi, s, \psi) \cdot \prod_{i=1}^{f_1} \gamma(\delta(\Sigma_i) \times \pi, s, \psi)\gamma(\delta(\widehat{\Sigma}_i) \times \pi, s, \psi) \cdot$$
$$\cdot \prod_{\lambda \in P_W(\rho)} \prod_{j=1}^{l_\lambda} \prod_{i=1}^{r_j} \gamma\big(\delta\big(\Sigma_{\lambda,j}^{(i)}\big) \times \pi, s, \psi\big)\gamma\big(\delta\big(\widehat{\Sigma}_{\lambda,j}^{(i)}\big) \times \pi, s, \psi\big).$$

For any segment $\Sigma$, such that $\Sigma \cup \widehat{\Sigma}$ is a segment, we have

$$(5.31)\qquad \gamma(\delta(\Sigma) \times \pi, s, \psi)\gamma(\delta(\widehat{\Sigma}) \times \pi, s, \psi)$$
$$= \gamma(\delta(\Sigma \cup \widehat{\Sigma}) \times \pi, s, \psi)\gamma(\delta(\Sigma \cap \widehat{\Sigma}) \times \pi, s, \psi).$$

Using (5.28), (5.29), (5.31) and Theorem 4.1, we get in (5.30),

$$\gamma(\sigma_\rho \times \pi, s, \psi) = \gamma(\rho^{(t)} \times \pi, s, \psi) \cdot \prod_{i=1}^{f_1} \gamma(\delta(\Sigma_i) \times \pi, s, \psi)\gamma(\delta(\widehat{\Sigma}_i) \times \pi, s, \psi) \cdot$$
$$\cdot \prod_{\lambda \in P_W(\rho)} \prod_{j=1}^{l_\lambda} \prod_{i=1}^{r_j} \gamma(\Delta(\lambda, a_\lambda(\alpha_j + 2i - 2)) \times \pi, s, \psi)$$
$$\gamma(\Delta(\lambda, a_\lambda(\alpha_j + 2i - 1)) \times \pi, s, \psi).$$

By the structure of the representation $\rho_1$ in (5.26) and the structure of the representation $\rho$ in (5.14), and by the the multiplicativity property of gamma factors (Theorem 1.4), the above product of gamma factors equals

$$\prod_{i=1}^{f_1} \gamma(\delta(\Sigma_i) \times \pi, s, \psi)\gamma(\delta(\widehat{\Sigma}_i) \times \pi, s, \psi)\gamma(\rho_1 \times \pi, s, \psi) = \gamma(\rho \times \pi, s, \psi).$$

We proved that

$$\gamma(\sigma_\rho \times \pi, s, \psi) = \gamma(\rho \times \pi, s, \psi). \tag{5.32}$$

Hence we prove one direction in Part (b) of Theorem 5.2.

To complete the proof of Part (b) of Theorem 5.2, we start now with an irreducible generic representation $\sigma$ of $\mathrm{SO}_{2n+1}(F)$ and construct $\rho = \mathfrak{b}(\sigma)$ on $\mathrm{GL}_{2n}(F)$, as required. Write $\sigma$ in the form (5.4), so that all conditions in Definition 5.1 are satisfied.

Separate first from (5.4) the segments $\Sigma_i$ which are not linked to their duals. Reorder them, if necessary, and assume that these are $\Sigma_1, \dots, \Sigma_{f_1}$. Consider the remaining segments $\Sigma_{f_1+1}, \dots, \Sigma_f$. Of course, $\Sigma_i$ is linked to $\widehat{\Sigma}_i$, for $f_1 + 1 \le i \le f$. For such $i$, it is clear that neither (3a) nor (3b) can be satisfied, and hence (3c) is satisfied. This implies, for $f_1 + 1 \le i \le f$, that if $L(\xi_i, \Lambda^2, s)$ has a pole at $s = 0$, then one of the segments $\Sigma_i$ or $\widehat{\Sigma}_i$ has the form $[\nu^{-a}\xi_i, \nu^b\xi_i]$, where $b \in \frac{1}{2} + \mathbb{Z}_{\ge 0}$ and $a = -1/2$ or $a \in \frac{1}{2} + \mathbb{Z}_{\ge 0}$ and $a < b$; or if $L(\xi_i, \mathrm{sym}^2, s)$ has a pole at $s = 0$, then one of the segments $\Sigma_i$ or $\widehat{\Sigma}_i$ has the form $[\nu^{-a}\xi_i, \nu^b\xi_i]$ where $b \in \mathbb{N}$ and $0 \le a < b$. In either case, we may assume that $\Sigma_i$ has the indicated form, and rewrite, as in the beginning of this section, $\Sigma_i = [\nu^{-q_i}\xi_i, \nu^{-q_i+w_i}\xi_i]$. We have, for $f_1 + 1 \le i \le f$,

$$\Sigma_i \cup \widehat{\Sigma}_i = \left[\nu^{q_i-w_i}\xi_i, \nu^{-q_i+w_i}\xi_i\right] \tag{5.33}$$

and unless $L(\xi_i, \Lambda^2, s)$ has a pole at $s = 0$, and $q_i = -1/2$ (in which case $\Sigma_i \cap \widehat{\Sigma}_i$ is empty),

$$\Sigma_i \cap \widehat{\Sigma}_i = \left[\nu^{-q_i}\xi_i, \nu^{q_i}\xi_i\right]. \tag{5.34}$$

It is easy to see, by Condition (1) in Definition 5.1, that the segments of type (5.33) or (5.34) are not linked to any of the segments $\Sigma_j$, $\widehat{\Sigma}_j$, $1 \le j \le f_1$. Note also that in the induced representation

$$\mathop{\times}_{i=f_1+1}^{f} [\Delta(\xi_i, -q_i + w_i) \times \Delta(\xi_i, q_i)] \tag{5.35}$$

each segment occurs only once, otherwise, we get a contradiction to Condition (1) in Definition 5.1. If $q_i = -1/2$, $\Delta(\xi_i, q_i)$ is omitted from (5.35).

We define

$$(5.36)\qquad \rho = \mathop{\times}_{i=1}^{f_1} [\delta(\Sigma_i) \times \delta(\widehat{\Sigma}_i)] \times \mathop{\times}_{i=f_1+1}^{f} (\Delta(\xi_i, -q_i + w_i) \times \Delta(\xi_i, q_i)) \times \rho^{(t)}$$

(where $\rho^{(t)} = \ell(\sigma^{(t)})$). It is clear that $\rho$ is irreducible, self-dual and generic, and that the segments of type (5.33), (5.34) ($f_1 + 1 \leq i \leq f$) occur an odd number of times, in $\rho$ (again, such a segment occurs exactly once in (5.35) and an even number of times, possibly zero, in $\rho^{(t)}$ (Theorem 4.1)). It is clear that

$$(5.37)\qquad \{\lambda \in P_W(\rho) \mid \mathcal{O}_{E_\lambda(\rho)} \text{ is not empty}\} = \{\xi_i \mid f_1 + 1 \leq i \leq f\}.$$

It remains to show that $|\mathcal{O}^{(j)}_{E_{\xi_i}}|$ is even for $1 \leq j \leq l_{\xi_i} - 1$, and for $j = l_{\xi_i}$, in case $L(\xi_i, \mathrm{sym}^2, s)$ has a pole $s = 0$. The reason for this is that there can not exist a factor $\Delta(\xi_i, m)$ occurring in $\rho^{(t)}$, such that

$$\left[\nu^{-q_i}\xi_i, \nu^{q_i}\xi_i\right] \subsetneqq \left[\nu^{-m}\xi_i, \nu^{m}\xi_i\right] \subsetneqq \left[\nu^{q_i-w_i}\xi_i, \nu^{-q_i+w_i}\xi_i\right]$$

because this would imply that $\Sigma_i$ is linked to $[\nu^{-m}\xi_i, \nu^m\xi_i]$, contradicting Condition (2) in Definition 5.1. Similarly, there can not exist $j \neq i$, $f_1 + 1 \leq j \leq f$, such that either one of $[\nu^{q_j-w_j}\xi_j, \nu^{-q_j+w_j}\xi_j]$ or $[\nu^{-q_j}\xi_j, \nu^{q_j}\xi_j]$ lies properly between $[\nu^{-q_i}\xi_i, \nu^{q_i}\xi_i]$ and $[\nu^{q_i-w_i}\xi_i, \nu^{-q_i+w_i}\xi_i]$, since this would imply a linkage between $\Sigma_i$ and $\Sigma_j$, or $\Sigma_i$ and $\widehat{\Sigma}_j$, contradicting Condition (1) in Definition 5.1. Thus, $\Delta(\xi_i, q_i)$ is the immediate successor of $\Delta(\xi_i, -q_i + w_i)$ in the decreasing inclusion ordering of the corresponding segments. This means that $\Delta(\xi_i, -q_i + w_i)$ and $\Delta(\xi_i, q_i)$ correspond to consecutive elements in the appropriate connected component, say $\mathcal{O}^{(j)}_{E_{\xi_i}}$, of $\mathcal{O}_{E_{\xi_i}}$ (see (5.13)). In other words, the indices occur in $\mathcal{O}^{(j)}_{E_{\xi_i}}$ by pairs, i.e., $|\mathcal{O}^{(j)}_{E_{\xi_i}}|$ is even in all the cases under consideration.

The only extra case is when $q_i = 1/2$. In this case $L(\xi_i, \Lambda^2, s)$ has a pole at $s = 0$, and $[\nu^{-q_i}\xi_i, \nu^{q_i}\xi_i]$ is empty. This implies that the corresponding connected component is $\mathcal{O}^{(l_{\xi_i})}_{E_{\xi_i}}$ and $-q_i + w_i = n_{\xi_i}$ (see Condition (3) in Theorem 5.2). In this case $|\mathcal{O}^{(l_{\xi_i})}_{E_{\xi_i}}|$ is odd.

Therefore, we see that $\rho$ is of the form considered in Theorem 5.2. By the first part of the proof, it is clear that $\sigma \cong \sigma_\rho$, and by (5.32)

$$\gamma(\sigma \times \pi, s, \psi) = \gamma(\rho \times \pi, s, \psi)$$

for all irreducible, supercuspidal representations $\pi$ of $\mathrm{GL}_k(F)$ with all $k \in \mathbb{Z}_{>0}$. This completes the proof of Part (b) of Theorem 5.2. □

Before we start the proof of Part (a) of Theorem 5.2, we make the following remarks on the bijection $\mathfrak{b}$ in Part (b) of Theorem 5.2. Although we always have

$$\gamma(\sigma \times \pi, s, \psi) = \gamma(\mathfrak{b}(\sigma) \times \pi, s, \psi)$$

for all irreducible, supercuspidal representations $\pi$ of $\mathrm{GL}_k(F)$ with all $k \in \mathbb{Z}_{>0}$, the twisted L-functions may *not* be preserved by the bijection $\mathfrak{b}$, i.e., the following identity

$$L(\sigma \times \pi, s) = L(\mathfrak{b}(\sigma) \times \pi, s) \tag{5.38}$$

for any irreducible supercuspidal representation $\pi$ of $\mathrm{GL}_*(F)$, may *not* hold.

(1) Assume that there is no $\lambda \in P_W(\rho)$, such that $L(\lambda, \Lambda^2, s)$ has a pole at $s = 0$, and $|\mathcal{O}^{(l_\lambda)}_{E_\lambda(\rho)}|$ is odd. Then

$$L(\sigma_\rho \times \pi, s) = L(\rho \times \pi, s)$$

for all irreducible, supercuspidal representations $\pi$ of $\mathrm{GL}_k(F)$ with all $k \in \mathbb{Z}_{>0}$.

Indeed, the proof is the same as the one for the gamma factor. See (0.15), (0.17). Note that

$$\begin{aligned} L\big(\delta\big(\Sigma^{(i)}_{\lambda,j}\big) \times \pi, s\big) &= L(\lambda \times \pi, s + a_\lambda(\alpha_j + 2i - 2)) \\ &= L(\Delta(\lambda, a_\lambda(\alpha_j + 2i - 2)) \times \pi, s) \\ L\big(\delta\big(\widehat{\Sigma}^{(i)}_{\lambda,j}\big) \times \pi, s\big) &= L(\lambda \times \pi, s + a_\lambda(\alpha_j + 2i - 1)) \\ &= L(\Delta(\lambda, a_\lambda(\alpha_j + 2i - 1))\times, \pi, s) \end{aligned} \tag{5.39}$$

so that

$$\begin{aligned} &L\big(\delta\big(\Sigma^{(i)}_{\lambda,j}\big) \times \pi, s\big)L\big(\delta\big(\widehat{\Sigma}^{(i)}_{\lambda,j}\big) \times \pi, s\big) \\ &= L(\Delta(\lambda, a_\lambda(\alpha_j + 2i - 2)) \times \pi, s)L(\Delta(\lambda, a_\lambda(\alpha_j + 2i - 1)) \times \pi, s) \\ &= L([\Delta(\lambda, a_\lambda(\alpha_j + 2i - 2)) \times \Delta(\lambda, a_\lambda(\alpha_j + 2i - 1))] \times \pi, s). \end{aligned} \tag{5.40}$$

(2) The equality (5.40) breaks down in case $j = l_\lambda$, $i = r_j$ (i.e., $i = t_{l_\lambda} + 1$ in the notation of (5.23)). Here, we get

$$\begin{aligned} &L\big(\delta\big(\Sigma^{(t_{l_\lambda}+1)}_{\lambda,l_\lambda}\big) \times \pi, s\big)L\big(\delta\big(\widehat{\Sigma}^{(t_{l_\lambda}+1)}_{\lambda,l_\lambda}\big) \times \pi, s\big) \\ &= L(\lambda \times \pi, s + a_\lambda(\alpha_{l_\lambda} + 2t_{l_\lambda}))L\left(\lambda \times \pi, s - \frac{1}{2}\right) \\ &= L(\Delta(\lambda, a_\lambda(\alpha_{l_\lambda} + 2t_{l_\lambda})) \times \pi, s)L\left(\lambda \times \pi, s - \frac{1}{2}\right). \end{aligned}$$

In this case, $L(\Delta(\lambda, a_\lambda(\alpha_{l_\lambda} + 2t_{l_\lambda})) \times \pi, s)$ accounts for the factor $\Delta(\lambda, a_\lambda(\alpha_{l_\lambda} + 2t_{l_\lambda}))$, which appears in $\rho$, but $L(\lambda \times \pi, s - 1/2)$ does not account for a factor of $\rho$. Here we get

$$L(\sigma_\rho \times \pi, s) = L(\rho \times \pi, s) \prod_{\substack{\lambda \in P_W(\rho) \\ |\mathcal{O}^{(l_\lambda)}_{E_\lambda(\rho)}| \text{odd}}} L(\lambda \times \pi, s - 1/2). \tag{5.41}$$

(3) Even in the case of the first remark, there is a problem on the reciprocity map. Looking again at (5.39), the parameter of $\Delta(\lambda, a_\lambda(\alpha_j + 2i - 2))$ is $\varphi_\lambda \otimes S_{2a_\lambda(\alpha_j+2i-2)+1}$ where $\varphi_\lambda = r^{-1}(\lambda)$. The parameter $\varphi_\lambda \otimes S_{2a_\lambda(\alpha_j+2i-2)+1}$ is orthogonal. Since $\Delta(\lambda, a_\lambda(\alpha_j + 2i - 2))$ appears an odd number of times in $\rho$, the multiplicity of $\varphi_\lambda \otimes S_{2a_\lambda(\alpha_j+2i-2)+1}$ in the parameter of $\rho$ is odd, and hence, its isotypic component can not be symplectic. Thus, if for $\lambda \in P_W(\rho)$, $\mathcal{O}_{E_\lambda(\rho)}$ is not empty, $\sigma_\rho$ does not have the same local Langlands parameter as $\rho$ does. However, it has the same gamma factors as $\rho$ (see (5.32)) and even the same L-factors as $\rho$ (see (5.38)) in the case of the first remark.

*Proof of Part (a) of Theorem 5.2.* For a given irreducible generic representation $\sigma$ of $SO_{2n+1}(F)$, we have actually constructed two irreducible representations of $GL_{2n}(F)$ in Theorem 5.1 and Part (b) of Theorem 5.2, respectively. Let $\rho_{(1)} = \ell(\sigma)$ be the Langlands quotient constructed in Theorem 5.1, and $\rho_{(2)} = \mathfrak{b}(\sigma)$ be the irreducible generic representation of $GL_{2n}(F)$ as constructed in Part (b) of Theorem 5.2. We know from the proof of Theorem 5.1 and the proof of Part (b) of Theorem 5.2 that there exists an induced representation of $GL_{2n}(F)$ as in (5.5) constructed from $\sigma$, such that $\rho_{(2)}$ can be realized as its unique generic subrepresentation and $\rho_{(1)}$ as its unique Langlands quotient.

Let $\rho$ be an irreducible self-dual generic representation of $GL_{2n}(F)$, such that $\mathcal{O}_{E_\lambda(\rho)}$ is empty for all $\lambda \in P_W(\rho)$. We will show that there exists an irreducible generic representation $\sigma$ of $SO_{2n+1}(F)$, such that $\rho = \ell(\sigma)$. Now the construction of this $\sigma$ follows exactly from the proof of Part (b) of Theorem 5.2; this is in fact the most trivial case of the construction. We summarize the details. We first write $\rho$ in the form (5.14). Since $\mathcal{O}_{E_\lambda(\rho)}$ is empty for all $\lambda \in P_W(\rho)$, we know that $\rho_1 \cong \rho^{(t)}$ in (5.26). Next in (5.27),

$$\sigma_\rho = \left[\mathop{\times}_{i=1}^{f_1} \delta(\Sigma_i)\right] \rtimes \sigma^{(t)}.$$

Then $\rho = \mathfrak{b}(\sigma_\rho)$ is given by (5.36) with $f_1 = f$

$$\mathfrak{b}(\sigma_\rho) = \left[\mathop{\times}_{i=1}^{f} \delta(\Sigma_i)\right] \times \rho^{(t)} \times \left[\mathop{\times}_{i=1}^{f} \delta(\widehat{\Sigma}_i)\right].$$

On the other hand, by Theorem 5.1, $\ell(\sigma_\rho)$ is given by the Langlands quotient of the induced representation

$$\left[\mathop{\times}_{i=1}^{f} \delta(\Sigma_i)\right] \times \rho^{(t)} \times \left[\mathop{\times}_{i=1}^{f} \delta(\widehat{\Sigma}_i)\right]$$

which is equal to $\rho$.

Now assume that an irreducible self-dual generic representation $\rho$ lies in the image of the local Langlands functorial lift $\ell$, i.e., $\rho = \ell(\sigma)$ for some irreducible generic representation $\sigma$ of $SO_{2n+1}(F)$. Then by the proof of Theorem 5.1 and Part (b) of Theorem 5.2, the representation $\sigma$ determines uniquely an induced representation of $GL_{2n}(F)$ of the form (5.5) and $\rho = \ell(\sigma)$ is realized as its Langlands quotient and the representation $\mathfrak{b}(\sigma)$ is realized as its unique irreducible generic subrepresentation. Since $\rho = \ell(\sigma)$ is assumed to be generic, one knows that

$$\rho = \ell(\sigma) = \mathfrak{b}(\sigma) = \rho_{(1)} = \rho_{(2)}.$$

This implies that the induced representation of $GL_{2n}(F)$ of the form (5.5) determined by $\sigma$ is irreducible. We will now show that $\mathcal{O}_{E_\lambda(\rho)}$ is empty for all $\lambda \in P_W(\rho)$.

Assume that $\mathcal{O}_{E_\lambda(\rho)}$ is not empty for some $\lambda \in P_W(\rho)$. We write $\rho_{(2)}$ as in (5.36). As we remarked right after the proof of Part (b) of Theorem 5.2, the representations $\Delta(\lambda, a_\lambda(\alpha_j + i - 2))$ and $\Delta(\lambda, a_\lambda(\alpha_j + 2i - 1))$, which correspond to the consecutive points $\alpha_j + 2i - 2, \alpha_j + 2i - 1$ in $\mathcal{O}^{(j)}_{E_\lambda(\rho)}$, occur in $\rho_{(2)}$ (which is $\rho$ in Part (b) of Theorem 5.2) with odd multiplicities ($\mu_\lambda(a_\lambda(\alpha_j - 2i - 2)), \mu_\lambda(a_\lambda(\alpha_j + 2i - 1))$ respectively). We take one from the $\mu_\lambda(a_\lambda(\alpha_j - 2i - 2))$ ($\mu_\lambda(a_\lambda(\alpha_j + 2i - 1))$, resp.) pieces of $\Delta(\lambda, a_\lambda(\alpha_j + i - 2))$ ($\Delta(\lambda, a_\lambda(\alpha_j + 2i - 1))$, resp.), the rest occur with even multiplicities in $\rho^{(t)}_{(2)}$. As in (2.37) and (2.38), we can realize the irreducible induced representation

$$[\Delta(\lambda, a_\lambda(\alpha_j + 2i - 2)) \times \Delta(\lambda, a_\lambda(\alpha_j + 2i - 1))]$$

as the unique irreducible generic subrepresentation of the following induced representation

$$\delta\big(\big[\nu^{-a_\lambda(\alpha_j+2i-1)}\lambda,\ \nu^{a_\lambda(\alpha_j+2i-2)}\lambda\big]\big) \times \delta\big(\big[\nu^{-a_\lambda(\alpha_j+2i-2)}\lambda,\ \nu^{a_\lambda(\alpha_j+2i-1)}\lambda\big]\big), \tag{5.42}$$

for $i = 1, \ldots, s_j$ (see (5.21)). In case $j = l_\lambda$ and $|\mathcal{O}^{(l_\lambda)}_{E_\lambda(\rho)}|$ is odd, we realize the irreducible generic representation $\Delta(\lambda, a_\lambda(n_\lambda))$ as the unique irreducible generic subrepresentation of the following induced representation

$$\delta\big[\nu^{1/2}\lambda,\ \nu^{a_\lambda(n_\lambda)}\lambda\big] \times \delta\big[\nu^{-a_\lambda(n_\lambda)}\lambda,\ \nu^{-1/2}\lambda\big]. \tag{5.43}$$

Hence we realize the representation $\rho_{(2)}$ in (5.36) as the unique irreducible generic subrepresentation of the following induced representation

$$\delta(\Sigma_1) \times \delta(\Sigma_2) \times \cdots \times \delta(\Sigma_f) \times \rho^{(t)}_{(2)} \times \delta(\widehat{\Sigma}_f) \times \cdots \times \delta(\widehat{\Sigma}_2) \times \delta(\widehat{\Sigma}_1), \tag{5.44}$$

which is in Langlands induction data and satisfies Conditions (L1)–(L4) (as in (5.5)). By Theorem 5.1, the Langlands quotient of (5.44) equals $\ell(\sigma_{\rho_{(2)}})$. Note that $\sigma = \sigma_{\rho_{(1)}} = \sigma_{\rho_{(2)}}$. It follows that

$$\ell(\sigma) \neq \mathfrak{b}(\sigma).$$

This is a contradiction. This completes the proof of Part (a) of Theorem 5.2. □

We conclude this section by giving the Langlands parameter for an irreducible generic representation $\sigma$ of $SO_{2n+1}(F)$. Let $\sigma$ be an irreducible generic representation of $SO_{2n+1}(F)$ given in (5.4). Then the local Langlands functorial lift $\ell(\sigma)$ is realized as the unique Langlands quotient of (5.5). Hence the Langlands parameter for $\sigma$ is the admissible homomorphism from $W_F \times SL_2(\mathbb{C})$ to $Sp_{2n}(\mathbb{C})$ of following type

(5.45)

$$\varphi_\sigma = y^{-1}(\sigma^{(t)}) \oplus \bigoplus_{i=1}^{f} \left[ |\cdot|^{-q_i+\frac{w_i}{2}} r^{-1}(\xi_i) \otimes S_{w_i+1} \oplus |\cdot|^{q_i-\frac{w_i}{2}} r^{-1}(\widehat{\xi_i}) \otimes S_{w_i+1} \right],$$

where $y$ is the reciprocity map given in Theorem 4.2 for irreducible tempered generic representations in $\Pi^{(tg)}(SO_{2n+1})$ and $r$ is the reciprocity map for $GL_*(F)$ defined in [H.T.] and [H1], and as in Theorem 1.2 in Section 1. Note that $|\cdot|^s$ is the character of $W_F$ normalized as in [T] via the local class field theory.

**6. Representations attached to local langlands parameters.** In Sections 1–5, we have established explicitly the local Langlands functorial lift from irreducible generic representations of $SO_{2n+1}(F)$ to $GL_{2n}(F)$, and have written down the local Langlands parameter for each member in $\Pi^{(g)}(SO_{2n+1})$. Our current methods seem insufficient to establish the theory for general irreducible representations of $SO_{2n+1}(F)$. However, in this section, we shall construct for each local Langlands parameter $\varphi$ in $\Phi(SO_{2n+1})$ an irreducible representation of $SO_{2n+1}(F)$ as stated in Theorem A in the Introduction.

We first describe the structure of local Langlands parameters in $\Phi(SO_{2n+1})$.

PROPOSITION 6.1. *Let $\varphi : W_F \times SL_2(\mathbb{C}) \to Sp_{2n}(\mathbb{C})$ be a local Langlands parameter in $\Phi(SO_{2n+1})$. Then either $\varphi$ is a tempered parameter in $\Phi^{(t)}(SO_{2n+1})$, i.e., $\varphi(W_F)$ is bounded in $Sp_{2n}(\mathbb{C})$ or $\varphi$ can be decomposed as a direct sum*

(6.1) $$\varphi = \varphi^{(t)} \oplus \varphi^{(n)}$$

*where $\varphi^{(t)}$ is a tempered parameter in $\Phi^{(t)}(SO_{2n^*+1})$ ($n^* < n$) and $\varphi^{(n)}$ is a parameter in $\Phi(SO_{2(n-n^*)+1})$ satisfying the following conditions: there exist $f \in \mathbb{Z}_{>0}$, $w_1, \ldots, w_f \in \mathbb{Z}_{\geq 0}$, and $q_1, \ldots, q_f \in \mathbb{R}$, such that $q_i \neq \frac{w_i}{2}$, for $1 \leq i \leq f$, and*

(6.2) $$\varphi^{(n)} = \bigoplus_{i=1}^{f} \left[ |\cdot|^{\frac{w_i}{2}-q_i} \varphi_i \otimes S_{w_i+1} \oplus |\cdot|^{-\frac{w_i}{2}+q_i} \hat{\varphi}_i \otimes S_{w_i+1} \right]$$

*where for $1 \leq i \leq f$, $\varphi_i$ is an irreducible bounded representation of the Weil group $W_F$,*

$$\frac{w_i}{2} - q_i \geq \frac{w_{i+1}}{2} - q_{i+1} > 0$$

*for $i = 1, 2, \ldots, f-1$, and $|\cdot|^*$ is the character of $W_F$ normalized as in [T] via the local class field theory.*

*Proof.* Let $V = \mathbb{C}^{2n}$ be the nondegenerate symplectic space of dimension $2n$, equipped with the symplectic form $<,>$, which corresponds to the given parameter $\varphi$. Write

$$V = V_1 \oplus V_2$$

where $V_1$ is the direct sum of all irreducible subspaces, which are stable under $W_F \times \mathrm{SL}_2(\mathbb{C})$ and in which $\varphi(W_F)$ is bounded; and similarly, $V_2$ is the direct sum of all irreducible subspaces, which are stable under $W_F \times \mathrm{SL}_2(\mathbb{C})$ and in which $\varphi(W_F)$ is unbounded. We have to show that both subspaces $V_1$ and $V_2$ are nondegenerate with respect to the restriction of the nondegenerate symplectic form $<,>$.

Let $\mathrm{rad}(V_i)$ $(i = 1, 2)$ be the radical of $(V_i, <,> |_{V_i})$. Then $\mathrm{rad}(V_i)$ is stable under the action of $W_F \times \mathrm{SL}_2(\mathbb{C})$. We first prove $\mathrm{rad}(V_2)$ is zero. Let $v_2$ be any vector in $\mathrm{rad}(V_2)$, which lies in an irreducible summand, say of the form $\phi \otimes S_{w+1}$, where $\phi$ is an irreducible unbounded representation of $W_F$. We may write

$$\phi = |\cdot|^t \phi',$$

where $\phi'(W_F)$ is bounded and $0 \neq t \in \mathbb{R}$. Then for any $v_1 \in V_1$, we have

$$(6.3) \quad < v_2, \varphi(w^{-1})(v_1) >=< \varphi(w)(v_2), v_1 >= |w|^t < (\phi'(w) \otimes id)(v_2), v_1 >.$$

It is clear that $< v_2, \varphi(w^{-1})(v_1) >$ is bounded, but $|w|^t < (\phi'(w) \otimes id)(v_2), v_1 >$ is unbounded, unless

$$(6.4) \qquad < \varphi(w)(v_2), v_1 >= 0, \ \text{for all } v_1 \in V_1.$$

Since $v_1$ is arbitrary, we get

$$< v_2, v_1 >= 0, \ \text{for all } v_1 \in V_1.$$

Because $v_2$ is in $\mathrm{rad}(V_2)$, we have

$$< v_2, v >= 0, \ \text{for all } v \in V.$$

Since $V$ is nondegenerate, the vector $v_2$ must be zero. We proved that $V_2$ is nondegenerate since $\mathrm{rad}(V_2)$ is zero. The same proof works for showing that $V_1$ is nondegenerate, where this time we fix $v_1$ in $\mathrm{rad}(V_1)$ in (6.4) and let $v_2$ vary in an arbitrary irreducible summand of $V_2$.

Now denote by $\varphi^{(t)}$ the subrepresentation of $W_F \times \mathrm{SL}_2(\mathbb{C})$ on $V_1$ and by $\varphi^{(n)}$ the subrepresentation of $W_F \times \mathrm{SL}_2(\mathbb{C})$ on $V_2$. It is clear that $\varphi^{(t)}$ is in $\Phi^{(t)}(\mathrm{SO}_{2n^*+1})$. The representation $\varphi^{(n)}$ is a direct sum of irreducible representations of $W_F \times \mathrm{SL}_2(\mathbb{C})$ in each of which $\varphi^{(n)}(W_F)$ is unbounded. Since $\varphi^{(n)}$ is self-dual (symplectic, in particular), it is clear that it has the form as in (6.2). □

By using the structure of the local Langlands parameters given in Proposition 6.1, the special cases of local Langlands reciprocity law for $\mathrm{SO}_{2n+1}$ in Sections 1–5, and the local Langlands reciprocity law for GL in [H.T.] and in [H1], we obtain the following version of local Langlands reciprocity law for irreducible representations of $\mathrm{SO}_{2n+1}(F)$, which is a restatement of Theorem A in the Introduction.

THEOREM 6.1. *The local Langlands reciprocity map defined in Theorem 4.2 has a unique extension to a bijection $y$ between the set $\Phi(\mathrm{SO}_{2n+1})$ and the set of equivalence classes of irreducible admissible representations of $\mathrm{SO}_{2n+1}(F)$, which are Langlands quotients of induced representations*

$$\delta(\Sigma_1) \times \cdots \times \delta(\Sigma_f) \rtimes \sigma^{(t)} \tag{6.5}$$

*where $\sigma^{(t)}$ is an irreducible generic tempered representation of $\mathrm{SO}_{2n^*+1}(F)$, and*

$$\Sigma_1, \Sigma_2, \ldots, \Sigma_f$$

*are imbalanced segments, whose exponents are positive and in non-increasing order. This bijection $y$ preserves twisted local factors:*

$$L(\varphi \otimes \varphi', s) = L(y(\varphi) \times r(\varphi'), s) \tag{6.6}$$

$$\epsilon(\varphi \otimes \varphi', s, \psi) = \epsilon(y(\varphi) \times r(\varphi'), s, \psi) \tag{6.7}$$

*for all admissible homomorphisms $\varphi' : W_F \times \mathrm{SL}_2(\mathbb{C}) \to \mathrm{GL}_k(\mathbb{C})$ with all $k \in \mathbb{Z}_{>0}$, where $r$ is given in [H.T.] and in [H1].*

*Proof.* For a given $\varphi$ in $\Phi(\mathrm{SO}_{2n+1})$, we write, as in Proposition 6.1,

$$\varphi = \varphi^{(t)} \oplus \varphi^{(n)}.$$

By Theorem 4.2 and the local Langlands conjecture for $\mathrm{GL}_k(F)$ ([H.T.] and [H1]), we define

$$\sigma^{(t)} = y(\varphi^{(t)}) \tag{6.8}$$

$$\Sigma_i = [\nu^{-q_i} r(\varphi_i), \nu^{-q_i+w_i} r(\varphi_i)], \quad i = 1, 2, \cdots, f. \tag{6.9}$$

Then $\sigma^{(t)}$ is an irreducible, generic, tempered representation of $\mathrm{SO}_{2n^*+1}(F)$. Let now $y(\varphi)$ be the Langlands quotient of the representation in (6.5). It follows that the map $y$ preserves the twisted local factors as stated. It follows also from the construction, Theorem 4.2, [H.T.] and [H1] that the map $y$ is bijective. The uniqueness of such extension $y$ follows from the local converse theorem we proved in [Jng.S.] (Theorem 1.3). □

**7. Applications.** In this last section, we discuss applications of our results obtained in Sections 1–6 to both local theory and global theory of automorphic forms. In §7.1, we discuss the Gross-Prasad and Rallis conjecture for $\mathrm{SO}_{2n+1}(F)$ and in §7.2, we obtained interesting applications to automorphic representations.

**7.1. On a conjecture of Gross-Prasad and Rallis.** Let $G$ be a reductive group which is split over $F$ and $G^\vee(\mathbb{C})$ be its Langlands dual group. A local

Langlands parameter

$$\varphi : W_F \times \mathrm{SL}_2(\mathbb{C}) \to G^\vee(\mathbb{C}),$$

is called *generic* if there is an irreducible generic representation of $G(F)$ attached to $\varphi$ by the local Langlands reciprocity conjecture. In particular, if $G = \mathrm{GL}_k$, then $\varphi$ is generic if $r(\varphi)$ is generic, where the map $r$ is the local Langlands reciprocity map for $GL$ in [H.R.] and [H1]; and if $G = \mathrm{SO}_{2n+1}$, then $\varphi$ is generic if $y(\varphi)$ defined in Theorem 6.1 is generic.

In the following, we shall verify a conjecture of Gross-Prasad and of Rallis for $\mathrm{GL}_n(F)$ and for $\mathrm{SO}_{2n+1}(F)$ ([G.P., p. 977] and [K, p. 384]).

*Conjecture (Gross-Prasad; Rallis).* Let $G$ be a reductive group which is split over $F$ and $G^\vee(\mathbb{C})$ be its Langlands dual group. A local Langlands parameter

$$\varphi : W_F \times \mathrm{SL}_2(\mathbb{C}) \to G^\vee(\mathbb{C}),$$

is generic if and only if the associated local adjoint L-function $L(\mathrm{Ad}_{G^\vee} \circ \varphi, s)$ is regular at $s = 1$ (where $\mathrm{Ad}_{G^\vee}$ is the adjoint representation of $G^\vee(\mathbb{C})$ on the Lie algebra of $G^\vee(\mathbb{C})$.)

This conjecture is known for $\mathrm{GL}_n(F)$ ([K]). We prove it here first for the convenience of the reader.

PROPOSITION 7.1. *A local Langlands parameter*

$$\varphi : W_F \times \mathrm{SL}_2(\mathbb{C}) \to \mathrm{GL}_n(\mathbb{C})$$

*is generic, if and only if the associated adjoint L-function $L(\mathrm{Ad}_{\mathrm{GL}_n} \circ \varphi, s)$ is regular at $s = 1$.*

*Proof.* By the definition of the adjoint L-function in this case, one has

$$L(\mathrm{Ad}_{\mathrm{GL}_n} \circ \varphi, s) = L(r(\varphi) \times \widehat{r(\varphi)}, s).$$

Thus, for an irreducible representation $\pi$ of $\mathrm{GL}_n(F)$, it suffices to show that $L(\pi \times \hat{\pi}, s)$ is holomorphic at $s = 1$, if and only if $\pi$ is generic.

Write $\pi$ as the Langlands quotient of an induced representation

$$\delta(D_1) \times \cdots \times \delta(D_f)$$

where $D_i = [\nu^{-a_i}\xi_i, \nu^{b_i}\xi_i]$, $\xi_i$ is an irreducible unitary supercuspidal representation of $\mathrm{GL}_{n_i}(F)$ and

$$b_i - a_i \geq b_{i+1} - a_{i+1}, \quad \text{for} \quad i = 1, 2, \ldots, f-1.$$

We may assume that $a_i \in \mathbb{R}$ and $b_i \in -a_i + \mathbb{Z}_{\geq 0}$. By [Z], $\pi$ is generic if and only if there are no linkages among the segments $\{D_i\}_{i=1}^f$. We have

$$L(\pi \times \hat{\pi}, s) = \prod_{1 \leq i,j \leq f} L(\delta(D_i) \times \delta(\hat{D}_j), s).$$

Note that $L(\delta(D_i)\times\delta(\hat{D}_i),s)$ is holomorphic at $s=1$. Indeed, write

$$\delta(D_i)=\nu^{\frac{b_i-a_i}{2}}\Delta\left(\xi_i,\frac{a_i+b_i}{2}\right).$$

By (0.17), we have

$$\begin{aligned}L(\delta(D_i)\times\delta(\hat{D}_i),s)&=L\left(\Delta\left(\xi_i,\frac{a_i+b_i}{2}\right)\times\Delta\left(\hat{\xi}_i,\frac{a_i+b_i}{2}\right),s\right)\\&=\prod_{l=1}^{a_i+b_i+1}L(\xi_i\times\hat{\xi}_i,s+a_i+b_i+1-l),\end{aligned}$$

which is holomorphic and non-vanishing at $s=1$ (since $a_i+b_i+1-l\geq 0$).

It remains to show that for $i\neq j$, the segments $D_i$ and $D_j$ are linked if and only if

$$L(\delta(D_i)\times\delta(\hat{D}_j),s)L(\delta(D_j)\times\delta(\hat{D}_i),s) \tag{7.1}$$

has a pole at $s=1$. We may assume that $a_i+b_i\geq a_j+b_j$.

Again, by (0.17), we have

$$\begin{aligned}&L(\delta(D_i)\times\delta(\hat{D}_j),s)L(\delta(D_j)\times\delta(\hat{D}_i),s) \qquad (7.2)\\&=\prod_{l=1}^{a_j+b_j+1}L(\xi_i\times\hat{\xi}_j,s+b_i+a_j+1-l)L(\xi_j\times\hat{\xi}_i,s+a_i+b_j+1-l).\end{aligned}$$

Clearly, a necessary condition for the last product to have a pole at $s=1$, or for $D_i$ to be linked to $D_j$, is that

$$\xi_i\cong\xi_j.$$

Thus, we may assume that $\xi_i=\xi_j$. In this case, if the product in (7.2) has a pole at $s=1$, then there is at least one $1\leq l\leq a_j+b_j+1$ such that

$$l=2+a_i+b_j\quad\text{or}\quad l=2+a_j+b_i.$$

This implies that $a_i\in a_j+\mathbb{Z}$. If $l=2+a_j+b_i$, then

$$-a_j-1\leq b_i\leq b_j-1.$$

Since $a_i+b_i\geq a_j+b_j$, we conclude that $D_i$ and $D_j$ are linked in such a way that $b_i<b_j$. If $l=2+a_i+b_j$, then the same argument yields that $D_i$ and $D_j$ are linked in such a way that $b_i>b_j$.

Conversely, if $D_i$ and $D_j$ are linked, then we have either (1) $b_j>b_i$ and $a_j<a_i$ or (2) $b_j<b_i$ and $a_j>a_i$. In case (1), we have

$$a_j+b_i<a_j+b_j\leq a_i+b_i<a_i+b_j.$$

Then there is an integer $1\leq l\leq a_j+b_j+1$ such that $l=2+a_j+b_i$. For this value of $l$ the L-factor $L(\xi_i\times\hat{\xi}_j,s+b_i+a_j+1-l)$ has a pole at $s=1$. Since

$\frac{1}{L(\xi_j \times \hat{\xi}_i, s)}$ is a polynomial in $q^{-s}$, we know that the L-factor $L(\xi_j \times \hat{\xi}_i, s + a_i + b_j + 1 - l)$ does not vanish. Hence we know that the product in (7.2) has a pole at $s = 1$. The same argument works for case (2). This proves Proposition 6.2. □

Now we consider the conjecture for $\mathrm{SO}_{2n+1}(F)$. First the following characterization of the genericity of the local Langlands parameters is given by the classification of irreducible generic representations of $\mathrm{SO}_{2n+1}(F)$ in [M2].

PROPOSITION 7.2. *For any local Langlands parameter*

$$\varphi : W_F \times \mathrm{SL}_2(\mathbb{C}) \to \mathrm{Sp}_{2n}(\mathbb{C}),$$

*the representation $y(\varphi)$ defined in Theorem 6.1 is generic if and only if $\sigma^{(t)}$ and $\Sigma_i$ ($i = 1, 2, \ldots, f$) defined in (6.8) and (6.9) satisfy the conditions of Definition 5.1 (with $\rho^{(t)} = \ell(\sigma^{(t)})$).*

Now we prove the Gross-Prasad and Rallis conjecture for $\mathrm{SO}_{2n+1}(F)$, which is Part (3) of Theorem C in the Introduction.

THEOREM 7.1. *For any local Langlands parameter*

$$\varphi : W_F \times \mathrm{SL}_2(\mathbb{C}) \to \mathrm{Sp}_{2n}(\mathbb{C}),$$

*the representation $y(\varphi)$ defined in Theorem 6.1 is generic if and only if the associated adjoint L-function $L(\mathrm{Ad}_{\mathrm{Sp}_{2n}} \circ \varphi, s)$ is regular at $s = 1$.*

*Proof.* First we show that if $y(\varphi)$ is generic, then $L(\mathrm{Ad}_{\mathrm{Sp}_{2n}} \circ \varphi, s)$ is regular at $s = 1$. We use previous notations as well as those in Section 5.

Write $\varphi = \varphi^{(t)} \oplus \varphi^{(n)}$ as in (6.1) and from (6.2), we let

$$\theta = \bigoplus_{i=1}^{f} |\cdot|^{\frac{w_i}{2} - q_i} \varphi_i \otimes S_{w_i+1}, \tag{7.3}$$

so that $\varphi^{(n)} = \theta \oplus \hat{\theta}$. Then we have by a direct calculation that

$$\begin{aligned} L\big(\mathrm{Ad}_{\mathrm{Sp}_{2n}} \circ \varphi, s\big) = {} & L(\theta \otimes \hat{\theta}, s) \cdot L(\theta \otimes \varphi^{(t)}, s) \cdot L(\hat{\theta} \otimes \varphi^{(t)}, s) \cdot \\ & \cdot L\big(\mathrm{Ad}_{\mathrm{Sp}_{2n^*}} \circ \varphi^{(t)}, s\big) \cdot L(\mathrm{sym}^2 \circ \theta, s) \cdot L(\mathrm{sym}^2 \circ \hat{\theta}, s). \end{aligned} \tag{7.4}$$

We show that each factor in the above product is holomorphic at $s = 1$. By Theorems 4.1 and 4.2 and by [H.T.] and [H1], we have

$$\begin{aligned} L(\theta \otimes \hat{\theta}, s) &= L(r(\theta) \times \widehat{r(\theta)}, s), \\ L(\theta \otimes \varphi^{(t)}, s) &= L(r(\theta) \times \rho^{(t)}, s), \\ L(\hat{\theta} \otimes \varphi^{(t)}, s) &= L(\widehat{r(\theta)} \times \rho^{(t)}, s), \end{aligned} \tag{7.5}$$

where $\rho^{(t)} = \ell(\sigma^{(t)}) = \ell(y(\varphi^{(t)}))$ and $r(\theta) = \delta(\Sigma_1) \times \cdots \times \delta(\Sigma_f)$. By Conditions (1) and (2) of Definition 5.1, the representations $r(\theta)$ and $\pi = r(\theta) \otimes \rho^{(t)}$ are irreducible and generic. Hence by Proposition 7.1, we know that $L(\pi \times \hat{\pi}, s)$ is holomorphic at $s = 1$. Since $L(\pi \times \hat{\pi}, s)$ can be expressed as

$$(7.6) \quad L(\pi \times \hat{\pi}, s) = L(r(\theta) \times r(\hat{\theta}), s) \cdot L(r(\theta) \times \rho^{(t)}, s) \cdot L(r(\hat{\theta}) \times \rho^{(t)}, s) \cdot L(\rho^{(t)} \times \rho^{(t)}, s)$$

and the last L-factor in (7.6), $L(\rho^{(t)} \times \rho^{(t)}, s)$ does not vanish at $s = 1$, it follows that

$$L(\theta \otimes \hat{\theta}, s) \cdot L(\theta \otimes \varphi^{(t)}, s) \cdot L(\hat{\theta} \otimes \varphi^{(t)}, s)$$

is holomorphic at $s = 1$. Note that this product of three L-functions occurs in (7.4).

As in Theorem 6.1, we know that $\rho^{(t)} = \ell(\sigma^{(t)})$ is an irreducible tempered (generic) representation of $GL_{2n^*}(F)$. By Proposition 7.1, we have

$$L(\rho^{(t)} \times \widehat{\rho}^{(t)}, s) = L\big(\mathrm{Ad}_{\mathrm{GL}_{2n^*}} \circ \varphi^{(t)}, s\big)$$

is regular at $s = 1$. This implies the holomorphicity at $s = 1$ of $L(\mathrm{Ad}_{\mathrm{Sp}_{2n^*}} \circ \varphi^{(t)}, s)$, since $L(\mathrm{Ad}_{\mathrm{Sp}_{2n^*}} \circ \varphi^{(t)}, s)^{-1}$ divides $L(\mathrm{Ad}_{\mathrm{GL}_{2n^*}} \circ \varphi^{(t)}, s)^{-1}$ (as polynomials in $q^{-s}$).

It remains to show that both $L(\mathrm{sym}^2 \circ \theta, s)$ and $L(\mathrm{sym}^2 \circ \hat{\theta}, s)$ are holomorphic at $s = 1$. It is easy to show that $L(\mathrm{sym}^2 \circ \theta, s)$ is holomorphic at $s = 1$. Indeed, since $\theta$ has positive exponents (in Proposition 6.1), the L-function $L(\theta \otimes \theta, s)$ is holomorphic at $s = 1$. From the standard decomposition

$$L(\theta \otimes \theta, s) = L(\mathrm{sym}^2 \circ \theta, s) \cdot L(\Lambda^2 \circ \theta, s),$$

the L-factor $L(\mathrm{sym}^2 \circ \theta, s)$ must be holomorphic at $s = 1$, since $L(\Lambda^2 \circ \theta, s)$ does not vanish at $s = 1$.

Finally, let us show that $L(\mathrm{sym}^2 \circ \hat{\theta}, s)$ is regular at $s = 1$. Recall from (7.3) that

$$\theta = \bigoplus_{i=1}^{f} |\cdot|^{\frac{w_i}{2} - q_i} \varphi_i \otimes S_{w_i+1}.$$

We let $\theta_i := \varphi_i \otimes S_{w_i+1}$ in the following calculation. Then we have

$$(7.7) \quad L(\mathrm{sym}^2 \circ \hat{\theta}, s) = \prod_{i=1}^{f} L(\mathrm{sym}^2 \circ \hat{\theta}_i, s - w_i + 2q_i) \times \prod_{1 \le i < j \le f} L\left(\hat{\theta}_i \otimes \hat{\theta}_j, s - \frac{w_i + w_j}{2} + q_i + q_j\right).$$

For $1 \le i < j \le f$, we use the same argument as in the proof of Proposition 7.1 and obtain that

$$(7.8) \qquad L\left(\hat{\theta}_i \otimes \hat{\theta}_j, s - \frac{w_i + w_j}{2} + q_i + q_j\right) = L(\delta(\widehat{\Sigma}_i) \times \delta(\widehat{\Sigma}_j), s)$$

is holomorphic at $s = 1$ since $\Sigma_i$ is not linked to $\hat{\Sigma}_j$ by Condition (1) of Definition 5.1. We have to calculate then the L-factors $L(\mathrm{sym}^2 \circ \hat{\theta}_i, z)$, for $i = 1, 2, \ldots, f$ and $z = s - w_i + 2q_i$.

It is an easy calculation from linear algebra that

(7.9)
$$\mathrm{sym}^2 \circ (\hat{\varphi}_i \otimes S_{w_i+1}) = \big(\mathrm{sym}^2 \circ \hat{\varphi}_i\big) \otimes \big(\mathrm{sym}^2 \circ S_{w_i+1}\big) \oplus \big(\Lambda^2 \circ \hat{\varphi}_i\big) \otimes \big(\Lambda^2 \circ S_{w_i+1}\big).$$

Now, recall (see for example [F.H., sec. 11.2, 11.3]) the formulae:

$$\mathrm{sym}^2(\mathrm{sym}^m \mathbb{C}^2) = \bigoplus_{k=0}^{[\frac{m}{2}]} \mathrm{sym}^{2m-4k}\mathbb{C}^2 \tag{7.10}$$

$$\Lambda^2(\mathrm{sym}^m \mathbb{C}^2) = \bigoplus_{k=0}^{[\frac{m-1}{2}]} \mathrm{sym}^{2(m-1)-4k}\mathbb{C}^2 \tag{7.11}$$

and recall that $S_{w_i+1}$ is the irreducible representation $\mathrm{sym}^{w_i}$ of $\mathrm{SL}_2(\mathbb{C})$. We obtain the following formula

$$\mathrm{sym}^2(\hat{\varphi}_i \otimes S_{w_i+1}) = \left[\bigoplus_{k=0}^{[\frac{w_i}{2}]} \big(\mathrm{sym}^2 \circ \hat{\varphi}_i\big) \otimes S_{2w_i-4k+1}\right] \oplus \left[\bigoplus_{k'=0}^{[\frac{w_i-1}{2}]} \big(\Lambda^2 \circ \hat{\varphi}_i\big) \otimes S_{2(w_i-1)-4k+1}\right], \tag{7.12}$$

and an identity for L-factors

$$L\big(\mathrm{sym}^2 \circ \hat{\phi}_i, z\big) = \prod_{k=0}^{[\frac{w_i}{2}]} L\big(\mathrm{sym}^2 \circ \hat{\varphi}_i, z + w_i - 2k\big) \cdot \prod_{k=0}^{[\frac{w_i-1}{2}]} L\big(\Lambda^2 \circ \hat{\varphi}_i, z + w_i - 1 - 2k\big). \tag{7.13}$$

Replacing $z = s - w_i + 2q_i$ in the last identity of L-factors, we see that it is enough to show that all

$$L\big(\mathrm{sym}^2 \circ \hat{\varphi}_i, s + 2q_i - 2k\big) \text{ and } L\big(\Lambda^2 \circ \hat{\varphi}_i, s + 2q_i - 1 - 2k'\big)$$

are holomorphic at $s = 1$ for $k = 0, 1, \ldots, [\frac{w_i}{2}]$ and $k' = 0, 1, \ldots, [\frac{w_i-1}{2}]$.

Clearly, if $\varphi_i$ is not self-dual, then all these L-factors are holomorphic on the real line, and in particular at $s = 1$. Thus, we may assume that $\varphi_i$ is self-dual. Recall Henniart's result from Theorem 6.3 in [Jng.S.] that $L(\mathrm{sym}^2 \circ \varphi_i, z)$ has a pole at $z = 0$ if and only if $L(r(\varphi_i), \mathrm{sym}^2, z)$ has a pole at $z = 0$, and the same statement holds for $L(\Lambda^2 \circ \varphi_i, z)$ and $L(r(\varphi_i), \Lambda^2, z)$ at $z = 0$.

If $L(r(\varphi_i), \mathrm{sym}^2, s + 2q_i - 2k)$ has a pole at $s = 1$, for some $0 \leq k \leq [\frac{w_i}{2}]$, then $1 + 2q_i - 2k = 0$ since $r(\varphi_i)$ is an irreducible, self-dual, supercuspidal representation of $\mathrm{GL}_*(F)$. It follows that

$$(7.14) \qquad -q_i = \frac{1}{2} - k \in \frac{1}{2} + \mathbb{Z}_-.$$

Since $-q_i + w_i \geq -q_i + \frac{w_i}{2} > 0$, we conclude from (7.14) that $-q_i + w_i \in \frac{1}{2} + \mathbb{Z}_{\geq 0}$. Since $-q_i \leq \frac{1}{2}$ (by (7.14)), it follows that

$$\frac{1}{2} \in \{-q_i, -q_i + 1, \ldots, -q_i + w_i\}.$$

This contradicts Conditions (3b) and (3c) in Definition 5.1. Since Condition (3a) of Definition 5.1 is not valid in this case, we conclude that the L-factor $L(r(\varphi_i), \mathrm{sym}^2, s + 2q_i - 2k)$ is holomorphic at $s = 1$, for all $0 \leq k \leq [\frac{w_i}{2}]$.

If $L(r(\varphi_i), \Lambda^2, s + 2q_i - 1 - 2k)$ has a pole at $s = 1$ for some $0 \leq k \leq [\frac{w_i - 1}{2}]$, then

$$-q_i = -k \in \mathbb{Z}_-.$$

As before, $-q_i + w_i$ is a positive integer, and in particular,

$$-q_i + w_i \geq 1.$$

We conclude that

$$0, 1 \in \{-q_i, -q_i + 1, \ldots, -q_i + w_i\}.$$

This contradicts Condition (3c) of Definition 5.1. Since both Conditions (3a) and (3b) are not valid in this case, we conclude that the L-factor $L(r(\varphi_i), \Lambda^2, s + 2q_i - 1 - 2k)$ is holomorphic at $s = 1$ for all $0 \leq k \leq [\frac{w_i - 1}{2}]$. This completes the proof of the holomorphicity at $s = 1$ of the adjoint L-function $L(\mathrm{Ad}_{\mathrm{Sp}_{2n}} \circ \varphi, s)$ when $\varphi$ in $\Phi(\mathrm{SO}_{2n+1})$ is a generic parameter.

Conversely, we now prove that if a local Langlands parameter $\varphi$ in $\Phi(\mathrm{SO}_{2n+1})$ is not generic, then the adjoint L-function $L(\mathrm{Ad}_{\mathrm{Sp}_{2n}} \circ \varphi, s)$ must have a pole at $s = 1$.

Assume that $\varphi \in \Phi(\mathrm{SO}_{2n+1})$ is not generic. Then

$$\Sigma_1, \ldots, \Sigma_f, \text{ and } \rho^{(t)} = \ell(\sigma^{(t)})$$

do not satisfy the conditions in Definition 5.1. If Condition (1) is not satisfied, then there are $1 \leq i \neq j \leq f$ such that $\Sigma_i$ is linked to $\Sigma_j$ or $\widehat{\Sigma}_j$. By the proof of Proposition 7.1, the product

(7.15)
$$L(\delta(\Sigma_i) \times \delta(\widehat{\Sigma}_j), s)L(\delta(\Sigma_j) \times \delta(\widehat{\Sigma}_i), s)L(\delta(\Sigma_i) \times \delta(\Sigma_j), s)L(\delta(\widehat{\Sigma}_i) \times \delta(\widehat{\Sigma}_j), s)$$

has a pole at $s = 1$. This implies that $L(Ad \circ \varphi, s)$ has a pole at $s = 1$ since the product in (7.15) is a factor in $L(\mathrm{Ad}_{\mathrm{Sp}_{2n}} \circ \varphi, s)$.

If Condition (2) in Definition 5.1 is not satisfied, then the representation

$$r(\theta) \times \rho^{(t)}$$

is reducible and its Langlands quotient $\pi$ is non-generic. By Proposition 7.1, the product in (7.6) has a pole at $s = 1$. It is clear that the pole at $s = 1$ of the product in (7.6) must occur at the product of the first three factors, since the last one is holomorphic at $s = 1$. Since the product of the first three factors in (7.6) occurs in $L(\mathrm{Ad}_{\mathrm{Sp}_{2n}} \circ \varphi, s)$, we see that $L(\mathrm{Ad}_{\mathrm{Sp}_{2n}} \circ \varphi, s)$ has a pole at $s = 1$ in this case.

Finally, if there is an integer $1 \leq i \leq f$ such that Condition (3) of Definition 5.1 is not satisfied, then $\varphi_i$ is self-dual, $\Sigma_i$ is not linked to an element of $A_2(\rho^{(2)})$ and $r(\varphi_i)$ does not satisfy Condition (3c), where $\rho^{(2)} = \ell(\sigma^{(2)})$ and $\sigma^{(2)}$ is the irreducible discrete series generic representation occurring in $\sigma^{(t)}$. In the following, $\sigma^{(0)}$ is the irreducible supercuspidal generic representation occurring in $\sigma^{(2)}$.

Put $\xi_i = r(\varphi_i)$. Since $\xi_i$ is self-dual, $(\xi_i, \sigma^{(0)})$ is $(C\alpha)$ for $\alpha = 1, 0$, or $\frac{1}{2}$. We show in each case that

$$L\big(\mathrm{sym}^2 \circ \hat{\phi}_i, s - w_i + 2q_i\big)$$

has a pole at $s = 1$, which is equivalent to showing that the product

$$\prod_{k=0}^{[\frac{w_i}{2}]} L\big(\mathrm{sym}^2 \circ \hat{\varphi}_i, s + 2q_i - 2k\big) \cdot \prod_{k'=0}^{\left[\frac{w_i-1}{2}\right]} L\big(\Lambda^2 \circ \hat{\varphi}_i, s + 2q_i - 1 - 2k'\big) \tag{7.16}$$

has a pole at $s = 1$ (see (7.13)).

Assume that $(\xi_i, \sigma^{(0)})$ is $(C1)$ (in particular it is not $(C\frac{1}{2})$ or $(C0)$). Then at least one of $\pm 1$ is in

$$\{-q_i, -q_i + 1, \ldots, -q_i + w_i\}.$$

In particular, $q_i \in \mathbb{Z}$. Since $L(\xi_i, \Lambda^2, s)$ has a pole at $s = 0$, (7.16) will have a pole at $s = 1$ if there is a solution $0 \leq k' \leq [\frac{w_i-1}{2}]$ to the equation

$$1 + 2q_i - 2k' - 1 = 0,$$

that is, $k' = q_i$. We have to show that $0 \leq q_i \leq [\frac{w_i-1}{2}]$. If $q_i < 0$, then $q_i \leq -1$, and so $-q_i \geq 1$. This implies that $-q_i = 1$. Since $L(\sigma^{(0)} \times \xi_i, s)$ has a pole at $s = 1$, this means that $\Sigma_i$ is linked to an element of $A(\rho^{(2)})$. By the present assumption, we know that $\Sigma_i$ is linked to an element of $A_0(\rho^{(2)}) \cup A_1(\rho^{(2)})$. This implies, in particular, that $\Sigma_i$ is linked to a segment of $\rho^{(t)}$, which means that Condition (2) of Definition 5.1 is violated. But, in this case we already showed that $L(Ad \circ \varphi, s)$ has a pole at $s = 1$. Therefore, we may assume that $q_i \geq 0$. Since $q_i < \frac{w_i}{2}$, we get

$$0 \leq q_i \leq \left[\frac{w_i - 1}{2}\right], \quad q_i \in \mathbb{Z}.$$

Thus, the second product of (7.16) has a pole at $s = 1$, since its factor corresponding to $k' = q_i$ has a pole at $s = 1$.

Next, assume that $(\xi_i, \sigma^{(0)})$ is $(C0)$ (in particular it is not $(C\frac{1}{2})$ or $(C1)$). Then we must have

$$0 \in \{-q_i, -q_i + 1, \ldots, -q_i + w_i\}.$$

Hence $q_i \in \mathbb{Z}$. As in the last case, we have to show that

$$0 \leq q_i \leq \left[\frac{w_i - 1}{2}\right], \quad q_i \in \mathbb{Z}.$$

This is clear, since $-q_i \leq 0$ in this case (recall that $q_i < \frac{w_i}{2}$).

The last case is that $(\xi_i, \sigma^{(0)})$ is $(C\frac{1}{2})$ (in particular it is not $(C0)$ or $(C1)$). Then at least one of $\pm\frac{1}{2}$ lies in

$$\{-q_i, -q_i + 1, \ldots, -q_i + w_i\}.$$

In particular, $q_i \in \frac{1}{2} + \mathbb{Z}$ and $-q_i \leq \frac{1}{2}$. Since $L(\xi_i, \mathrm{sym}^2, s)$ has a pole at $s = 0$, (7.16) has a pole at $s = 1$ if there is a solution $0 \leq k \leq [\frac{w_i}{2}]$ to the equation

$$1 + 2q_i - 2k = 0,$$

which means that $k = q_i + \frac{1}{2}$. Thus, we have to show that

$$0 \leq q_i + \frac{1}{2} \leq \left[\frac{w_i}{2}\right], \quad q_i \in \frac{1}{2} + \mathbb{Z}.$$

This follows exactly from the assumption and the fact that $\frac{w_i}{2} - q_i > 0$. The proof of Theorem 7.1 is completed. □

**7.2. Applications to automorphic representations.** Let $k$ be a number field and $\mathbb{A}$ be the ring of adeles of $k$. Our first application concerns the (global) Langlands functorial lift from irreducible generic cuspidal automorphic representations of $\mathrm{SO}_{2n+1}(\mathbb{A})$ to $\mathrm{GL}_{2n}(\mathbb{A})$. Recall that the Langlands dual group of $\mathrm{SO}_{2n+1}$ is $\mathrm{Sp}_{2n}(\mathbb{C})$, and the functorial lift we are concerned with is attached to the natural embedding of $\mathrm{Sp}_{2n}(\mathbb{C})$ into $\mathrm{GL}_{2n}(\mathbb{C})$.

Let $\sigma$ be an irreducible generic cuspidal automorphic representation of $\mathrm{SO}_{2n+1}$ $(\mathbb{A})$. An irreducible automorphic representation $\pi$ of $\mathrm{GL}_{2n}(\mathbb{A})$ is called a *Langlands functorial lift* of $\sigma$ from $\mathrm{SO}_{2n+1}(\mathbb{A})$ if at all local places, the local component $\pi_v$ of $\pi$ is a local Langlands functorial lift of the local component $\sigma_v$, which is equivalent to the compatibility of local factors at all places. In general, the definition of Langlands functorial lift is given in terms of global L-packets [B]. A weak version of this Langlands functorial lift was established in [C.K.PS.S], which states that there is an irreducible automorphic representation $\pi$ of $\mathrm{GL}_{2n}(\mathbb{A})$ such that the local component $\pi_v$ of $\pi$ is a local Langlands functorial lift from the local component $\sigma_v$ of $\sigma$ at all archimedean places, and at almost all finite places where $\sigma_v$ are unramified. Such a weak version of a Langlands functorial lift is usually called a *weak lift*.

It follows from the strong multiplicity one theorem for GL of H. Jacquet and J. Shalika [J.S.] that the image of the weak lift of $\sigma$ is unique. The precise characterization of the structure of the image was given in [G.R.S]. In particular, the image of the weak lift is generic in the sense that it has a non-zero Whittaker-Fourier coefficient. It follows from [C.K.PS.S] that if $\pi$ is the weak lift of $\sigma$, then at each local place $v$ the twisted local gamma factors are preserved, i.e.,

$$\gamma(\sigma_v \times \tau, s, \psi) = \gamma(\pi_v \times \tau, s, \psi) \tag{7.17}$$

where $\tau$ is any irreducible supercuspidal representation of $\mathrm{GL}_l(k_v)$ for $l = 1, 2, \ldots,$ and $\psi$ is a given nontrivial additive character of $k_v$.

We remark that for representations of $p$-adic groups in general the preservation of the twisted local gamma factors as in (7.17) is generally not enough to assure that the two representations are related by the local Langlands functorial lift. See the discussion in the proof of Part (b) of Theorem 5.2 and the remarks afterwards. However, in the special case that $\sigma_v$ and $\pi_v$ are the local components of $\sigma$ and $\pi$, respectively, we can show that $\pi_v$ is the local Langlands lift of $\sigma_v$. This is done as follows by applying Theorem 5.1 in Section 5 and the argument in [C.K.PS.S] which prove the existence of the weak lift.

The idea is now very simple. By using the local Langlands functorial lift from $\mathrm{SO}_{2n+1}$ to $\mathrm{GL}_{2n}$ for irreducible representations at archimedean places and for irreducible unramified representations at finite places, an irreducible admissible representation $\pi'$ of $\mathrm{GL}_{2n}(\mathbb{A})$ was chosen in [C.K.PS.S] so that the associated L-functions with restricted twisting have the "nice" analytic properties as required for application of the Converse Theorem. Now Theorem 5.1 establishes the local Langlands functorial lift for irreducible generic representations of $\mathrm{SO}_{2n+1}(k_v)$ for *any* finite local place $v$. For each finite local place $v_0$ where $\sigma_{v_0}$ is ramified, we treat the local place $v_0$ in the same way as archimedean places are treated in [C.K.PS.S], i.e., we choose the local component $\pi'_{v_0}$ of $\pi'$ to be the one given by Theorem 5.1. Then we repeat the proof exactly as the one given in [C.K.PS.S] and conclude the existence of an irreducible automorphic representation $\pi$ which has the property that

$$\pi_v \cong \pi'_v$$

for all archimedean places and for almost all finite places where $\sigma_v$ are unramified, and also for the given local finite place $v_0$. Hence the weak lift so obtained is compatible with the local Langlands functorial lift at the finite place $v_0$. Since the image of the weak lift is uniquely determined by $\sigma$, we have thus proved that the weak lift given in [C.K.PS.S] is compatible with the local Langlands functorial lift at every local place. This is Theorem E in the Introduction.

THEOREM 7.2. *The weak (global) Langlands functorial lift from irreducible generic cuspidal automorphic representations of* $\mathrm{SO}_{2n+1}(\mathbb{A})$ *to* $\mathrm{GL}_{2n}(\mathbb{A})$ *given in*

*[C.K.PS.S] is compatible with the local Langlands functorial lift at every local place. In particular, at each finite place, the local Langlands lift is given explicitly by Theorem 5.1.*

As a consequence of Theorems 5.1, 5.2, and 7.2, we may give a certain description of the structure of local components of irreducible generic cuspidal automorphic representations of $\mathrm{SO}_{2n+1}(\mathbb{A})$ and of the local Langlands parameters of local components of irreducible self-dual cuspidal automorphic representations $\pi$ of $\mathrm{GL}_{2n}(\mathbb{A})$ whose exterior square L-function $L(\pi, \Lambda^2, s)$ has a pole at $s = 1$.

Let $\sigma$ be an irreducible generic cuspidal automorphic representation of $\mathrm{SO}_{2n+1}(\mathbb{A})$ and let $\sigma_v$ be the local component of $\sigma$ at a finite place $v$. Then $\sigma_v$ is generic, and hence $\sigma_v$ can be written as

$$\sigma_v = \delta_v(\Sigma_1) \times \cdots \times \delta_v(\Sigma_{f_v}) \rtimes \sigma_v^{(t)}, \tag{7.18}$$

so that $\Sigma_1, \ldots, \Sigma_{f_v}$ and $\sigma_v^{(t)}$ satisfy the Conditions of Definition 5.1 (with $\rho^{(t)} = \ell(\sigma^{(t)})$).

COROLLARY 7.1. *If $\sigma_v$ (as given in (7.18)) is the local component of an irreducible generic cuspidal automorphic representation $\sigma$ of $\mathrm{SO}_{2n+1}(\mathbb{A})$, then there is no $1 \le i \le f_v$ such that the segments $\Sigma_i$ and $\widehat{\Sigma}_i$ are linked.*

*Proof.* By Theorem 7.2, the local Langlands functorial lift $\ell(\sigma_v)$ constructed in Theorem 5.1 is a local component of an irreducible generic automorphic representation of $\mathrm{GL}_{2n}(\mathbb{A})$. In particular, $\ell(\sigma_v)$ is generic. Now the conclusion follows from Theorem 5.1 again. □

It is also interesting to characterize by the criterion in Part (a) of Theorem 5.2, the structure of the local components at finite places of an irreducible generic cuspidal automorphic representation $\sigma$ of $\mathrm{SO}_{2n+1}(\mathbb{A})$, but we will not make this explicit here.

Finally, we consider an irreducible, automorphic, self-dual, cuspidal representation $\pi$ of $\mathrm{GL}_{2n}(\mathbb{A})$ such that the exterior square L-function $L(\pi, \Lambda^2, s)$ has a pole at $s = 1$. By the backward lift from $\mathrm{GL}_{2n}$ to $\mathrm{SO}_{2n+1}$, established in [G.R.S], there exists an irreducible, automorphic, cuspidal, generic representation $\sigma$ of $\mathrm{SO}_{2n+1}(\mathbb{A})$, which has a weak lift to $\pi$. By Theorem 7.2, each local component $\pi_v$ of $\pi$ is the local Langlands functorial lift of $\sigma_v$. Therefore we have the following Theorem, which is Theorem D in the Introduction.

THEOREM 7.3. *Let $\pi$ be an irreducible, automorphic, self-dual, cuspidal representation $\pi$ of $\mathrm{GL}_{2n}(\mathbb{A})$, such that the exterior square L-function $L(\pi, \Lambda^2, s)$ has a pole at $s = 1$. Let $\varphi_v$ be the local Langlands parameter attached to the local component $\pi_v$ at $v$ of $\pi$ (by the local Langlands reciprocity law for $\mathrm{GL}_{2n}$ established in*

*[H.T.] and [H1]). Then for every local place v, $\varphi_v$ is symplectic, i.e., the parameter $\varphi_v$ is a local Langlands parameter for* $SO_{2n+1}$.

School of Mathematics, University of Minnesota, Minneapolis, MN 55455, U.S.A

School of Mathematical Sciences, Sackler Faculty of Exact Sciences, Tel Aviv University, Ramat-Aviv, 69978 Israel Tel Aviv 69978, Israel

---

## REFERENCES

[A] J. Arthur, Unipotent automorphic representations: conjectures, in Orbites et representations II, *Asterisque,* **171–172** (1989), 13–71.

[B] A. Borel, Automorphic $L$-functions, Automorphic forms, representations and $L$-functions *Proc. Sympos. Pure Math.,* **33,** Part 2, pp. 27–61, *Amer. Math. Soc., Providence, R.I.,* 1979.

[C.S.] W. Casselman and F. Shahidi, On the irreducibility of standard modules for generic representations, *Ann. Sci. Éc. Norm.* **31** (1998), 561–589.

[C.K.PS.S] J. Cogdell, H. Kim, I. Piatetski-Shapiro, and F. Shahidi, On lifting from classical groups to $\mathrm{GL}_N$, *Inst. Hautes Études Sci. Publ. Math. No.* **93** (2001), 5–30.

[F.H.] W. Fulton and J. Harris, Representation Theory, a first course, *Readings in Math., GTM,* Vol. **129,** Springer 1991.

[G.J.R.] S. Gelbart, H. Jacquet, and J. Rogawski, Generic representations for the unitary group in three variables, *Israel J. Math.* (to appear).

[G.K.] I. Gelfand and D. Kazhdan, Representations of $\mathrm{GL}_n(K)$, in Lie groups and their representations, Halsted, 1975.

[G.R.S] D. Ginzburg, S. Rallis, and D. Soudry, Generic automorphic forms on $\mathrm{SO}(2n+1)$; functorial lift, endoscopy and base change, *Internat. Math. Res. Notices* 2001, no. 14, 729–764.

[G.P.] B. Gross and D. Prasad, On the decomposition of a representation of $\mathrm{SO}_n$ when restricted to $\mathrm{SO}_{n-1}$, *Can. J. Math.* Vol. **44** (1992), 974–1002.

[H.T.] M. Harris and R. Taylor, On the geometry and cohomology of some simple Shimura varieties, *Ann. of Math. Studies,* Vol. **151,** Princeton University Press, 2001.

[H1] G. Henniart, Une preuve simple des conjectures de Langlands pour $\mathrm{GL}(n)$ sur un corps $p$-adique, *Inv. Math.* **139** (2000), no. 2, 439–455.

[Hb] R. Herb, Elliptic representations of $Sp(2n)$ and $SO(n)$, *Pacific J. Math.* **161** (1993), 347–358.

[J.PS.S.] H. Jacquet, I. Piatetski-Shapiro, and J. Shalika, Rankin-Selberg convolutions, *Amer. J. Math.* **105** (1983), 367–464.

[J.S.] H. Jacquet and J. Shalika, On Euler products and the classification of automorphic representations. I, II. *Amer. J. Math.* **103** (1981), no. 3, 499–558; no. 4, 777–815.

[Jng.S.] D. Jiang and D. Soudry, The local converse theorem for $\mathrm{SO}(2n+1)$ and applications, accepted by *Ann. of Math,* 2002.

[Kn] A. W. Knapp, Introduction to the Langlands program, *Representation theory and automorphic forms* (Edinburgh, 1996), 245–302, *Proc. Sympos. Pure Math.,* **61,** *Amer. Math. Soc., Providence, RI,* 1997.

[K] S. Kudla, The local Langlands correspondence: the non-archimedean case, *Motives. Proc. Symp. in Pure Math.,* Vol. **55,** Part 2, 1994, 365–397.

[L.R.] E. Lapid and S. Rallis, On the non-negativity of $L(1/2, \pi)$ for $SO(2n+1)$, submitted, (2001).

[M.W.] C. Moeglin and J.-P. Waldspurger, Paquets stables de representations temperees et de reduction unipotente pour $\mathrm{SO}(2n+1)$, preprint, 2001.

[M.T.] C. Moeglin and M. Tadic, Construction of discrete series for classical $p$-adic groups, preprint.

[M1] G. Muic, Some results on square integrable representations; irreducibility of standard representations, *IMRN(1998)* no. 14, 705–726.

[M2] ———, On generic irreducible representations of $\mathrm{Sp}(n, F)$ and $\mathrm{SO}(2n+1, F)$, *Glasnik Mat.* vol. **33** (53), (1998), 19–31.

[PR] D. Prasad and D. Ramakrishnan, On the global root numbers of $\mathrm{GL(n)} \times \mathrm{GL(m)}$, *Proc. Sympos. Pure Math.,* vol. **66** (2), Amer. Math. Soc., Providence, RI, 1999, pp. 311–330.

[S1] F. Shahidi, A proof of Langlands' conjecture on Plancherel measures; Complementary series for $p$-adic groups, *Annals of Math.* **132** (1990), 273–330.

[S2] ———, Twisted endoscopy and reducibility of induced representations for $p$-adic groups, *Duke Math. J.* **66** (1992), 1–41.

[S3] ———, On multiplicativity of local factors, in Festschrift in honor of I. Piatetski-Shapiro, Part 2, *Israel Math. Conf. Proc. 3,* vol. **3** (1990), 226–242.

[Sh] J. Shalika, The multiplicity one theorem for $\mathrm{GL}(n)$, *Ann. of Math.,* **100** (1974), 171–193.

[Sp] T. A. Springer, Linear algebraic groups, Second edition, *Progress in Mathematics,* **9.** Birkhäuser Boston, Inc., Boston, MA, 1998.

[Sd1] D. Soudry, Full multiplicativity of gamma factors for $\mathrm{SO}_{2l+1} \times \mathrm{GL}_n$, *Proceedings of the Conference on p-adic Aspects of the Theory of Automorphic Representations* (Jerusalem, 1998), *Israel J. Math.* **120** (2000), part B, 511–561.

[Sd2] ———, Rankin-Selberg convolutions for $\mathrm{SO}_{2l+1} \times \mathrm{GL}_n$: local theory, *Mem. Amer. Math. Soc.* **105** (1993), no. 500.

[Td] M. Tadic, On square integrable representations of classical p-adic groups, *Amer. J. Math.* **120** (1998), no. 1, 159–210.

[T] J. Tate, Number theoretic background, *Automorphic forms, representations and L-functions, Proc. Sympos. Pure Math.,* **33,** Part 2, pp. 3–26, *Amer. Math. Soc., Providence, R.I.,* 1979.

[Z] A. V. Zelevinsky, Induced representations of reductive $p$-adic groups, *On irreducible representations of* $\mathrm{GL}(n)$, *Ann. Sci. École. Norm. Sup.* **13** (1980), 165–210.

CHAPTER 18

# LARSEN'S ALTERNATIVE, MOMENTS, AND THE MONODROMY OF LEFSCHETZ PENCILS

By Nicholas M. Katz

---

*To Joe Shalika on his 60th birthday*

**Introduction.** We work over an algebraically closed field "$\mathbb{C}$" of characteristic zero. Let $V$ be a $\mathbb{C}$-vector space of dimension $N \geq 2$. We fix a (not necessarily connected) Zariski closed subgroup $G \subset GL(V)$ which is reductive (i.e., every finite-dimensional representation of $G$ is completely reducible). We are interested in criteria which guarantee that $G$ is one of the standard classical groups, i.e., that either $G$ is caught between $SL(V)$ and $GL(V)$, or that $G$ is one of $SO(V)$, $\mathrm{O}(V)$, or (if $\dim(V)$ is even) $\mathrm{Sp}(V)$.

Larsen's Alternative (cf. [Lar-Char] and [Lar-Normal]) is a marvelous criterion, in terms of having a sufficiently small "fourth moment," which guarantees that $G$ is either a standard classical group or is a finite group. We have already made use of this criterion in [Ka-LFM, page 113]. In that application, we were content with either alternative.

However, in many applications, especially to the determination of (Zariski closures of) geometric monodromy groups in explicitly given families, we want to be able to rule out the possibility that $G$ be finite. Failing this, we would at least like to have a better understanding of the cases in which $G$ can in fact be finite.

Part I of this paper represents very modest progress toward these two goals. Toward the first goal, we give criteria for ruling out the possibility that $G$ be finite. These criteria rely on the observation that if $G$ is finite and has a sufficiently small fourth moment, it must be primitive. This observation in turn allows us to bring to bear the classical results of Blichfeld and of Mitchell, and the more recent results of Wales and Zalesskii. Toward the second goal, we give examples of finite $G$ with a very low fourth moment.

In Part II, we apply the results proven in Part I to the monodromy of Lefschetz pencils. Start with a projective smooth variety $X$ of dimension $n+1 \geq 1$, and take the universal family of (or a sufficiently general Lefschetz pencil of) smooth hypersurface sections of degree $d$. By its monodromy group $G_d$, we mean the Zariski closure of the monodromy of the local system $\mathcal{F}_d$ on the space of all smooth, degree $d$, hypersurface sections, given by

$$H \mapsto H^n(X \cap H)/H^n(X).$$

Let us denote by $N_d$ the rank of this local system.

For $n$ odd, the monodromy group $G_d$ is the full symplectic group $\mathrm{Sp}(N_d)$, cf. [De-Weil II, 4.4.1 and 4.4.2[a]].

For $n = 0$, $X$ is a curve, $X \cap H$ is finite, $N_d + 1 = \mathrm{Card}\,((X \cap H)(\bar{k})) = d \times \deg(X)$, and the monodromy group $G_d$ is well known to be the full symmetric group $S_{N_d+1} := \mathrm{Aut}\,((X \cap H)(\bar{k}))$, cf. 2.4.4.

For $n \geq 2$ and even, the situation is more involved. Deligne proved [De-Weil II, 4.4.1, 4.4.2[s], and 4.4.9] that the monodromy group $G_d$ is either the full orthogonal group $O(N_d)$ or a finite reflection group, and that the only finite reflection groups that arise are the Weyl groups of root systems of type $A$, $D$, or $E$ in their standard representations. Deligne needed this more precise information for his pgcd theorem [De-Weil II, 4.5.1], where the $O(N_d)$ case was easy, but the finite case required case by case argument. Using the criteria developed in Part I, we show that the monodromy group $G_d$ is in fact the full orthogonal group $O(N_d)$ for all sufficiently large $d$ (more precisely, for all $d$ with $d \geq 3$ and $N_d > 8$, and also for all $d$ with $d \geq 7$ and $N_d > 2$, cf. 2.2.4, 2.2.15, and 2.3.6).

*Acknowledgments.* I would like to thank CheeWhye Chin for his assistance in using the computer program GAP [GAP] to compute moments of exceptional Weyl groups. I would also like to thank the referee, for suggesting Theorem 2.3.6.

## Part I: Group Theory

### 1.1. Review of Larsen's Alternative

1.1.1. Recall that $\mathbb{C}$ is an algebraically closed field of characteristic zero, $V$ is a $\mathbb{C}$-vector space of dimension $N \geq 2$, and $G$ is a Zariski closed, reductive subgroup of $GL(V)$.

1.1.2. For each pair $(a, b)$ of non-negative integers, we denote by $M_{a,b}(G, V)$ the dimension of the space of $G$-invariant vectors in $V^{\otimes a} \otimes (V^{\vee})^{\otimes b}$:

$$M_{a,b}(G, V) := \dim_{\mathbb{C}} (V^{\otimes a} \otimes (V^{\vee})^{\otimes b})^G. \tag{1.1.2.1}$$

We call $M_{a,b}(G, V)$ the $(a, b)$'th moment of $(G, V)$. For each even integer $2n \geq 2$, we denote by $M_{2n}(G, V)$ the $2n$'th absolute moment, defined by

$$M_{2n}(G, V) := M_{n,n}(G, V). \tag{1.1.2.2}$$

If $H$ is any subgroup of $G$, we have the a priori inequalities

$$M_{a,b}(G, V) \leq M_{a,b}(H, V) \tag{1.1.2.3}$$

for every $(a, b)$.

1.1.3. The reason for the terminology "moments" is this. If $\mathbb{C}$ is the field of complex numbers, and if $K \subset G(\mathbb{C})$ is a maximal compact subgroup of $G(\mathbb{C})^{an}$, then $K$ is Zariski dense in $G$ (Weyl's unitarian trick). If we denote by $dk$ the Haar measure on $K$ of total mass one, and by

$$\chi : G(\mathbb{C}) \to \mathbb{C}$$
$$\chi(g) := \mathrm{Trace}\,(g|V),$$

the character of $V$ as $G$-module, then we have the formulas

(1.1.3.1) $$M_{a,b}(G, V) = \int_K \chi(k)^a \overline{\chi}(k)^b dk,$$

(1.1.3.2) $$M_{2n}(G, V) = \int_K |\chi(k)|^{2n} dk.$$

Thus the terminology "moments" and "absolute moments".

1.1.4. The most computationally straightforward interpretation of the $2n$'th absolute moment $M_{2n}(G, V)$ is this. Decompose the $G$-module $V^{\otimes n}$ as a sum of irreducibles with multiplicities:

(1.1.4.1) $$V^{\otimes n} \cong \oplus_i m_i W_i.$$

Then by Schur's Lemma we have

(1.1.4.2) $$M_{2n}(G, V) = \Sigma_i (m_i)^2.$$

More precisely, given any decomposition of $V^{\otimes n}$ as a sum of (not necessarily irreducible) $G$-modules $V_i$ with (strictly positive integer) multiplicities $m_i$,

(1.1.4.3) $$V^{\otimes n} \cong \oplus_i m_i V_i,$$

we have the inequality

(1.1.4.4) $$M_{2n}(G, V) \geq \Sigma_i (m_i)^2,$$

with equality if and only if the $V_i$ are distinct irreducibles.

1.1.5. If $n$ is itself even, say $n = 2m$, there is another interpretation of $M_{4m}(G, V)$. Decompose the $G$-module $V^{\otimes m} \otimes (V^\vee)^{\otimes m} = \mathrm{End}\,(V^{\otimes m})$ as a sum of irreducibles with multiplicities:

(1.1.5.1) $$\mathrm{End}\,(V^{\otimes m}) \cong \oplus_i n_i W_i.$$

Then we have, again by Schur's Lemma,

(1.1.5.2) $$M_{4m}(G, V) = \Sigma_i (n_i)^2.$$

More precisely, given any decomposition of $\mathrm{End}\,(V^{\otimes m})$ as a sum of (not necessarily irreducible) $G$-modules $V_i$ with (strictly positive integer) multiplicities $n_i$,

(1.1.5.3) $$\mathrm{End}\,(V^{\otimes m}) \cong \oplus_i n_i V_i,$$

we have the inequality

(1.1.5.4) $$M_{2n}(G, V) \geq \Sigma_i (n_i)^2,$$

with equality if and only if the $V_i$ are distinct irreducibles.

THEOREM 1.1.6. *(Larsen's Alternative, cf. [Lar-Char], [Lar-Normal], [Ka-LFM, page 113]) Let $V$ be a $\mathbb{C}$-vector space of dimension $N \geq 2$, $G \subset GL(V)$ a (not necessarily connected) Zariski closed reductive subgroup of $GL(V)$.*

(1) *If $M_4(G, V) \leq 5$, then $V$ is $G$-irreducible.*

(2) *If $M_4(G, V) = 2$, then either $G \supset SL(V)$, or $G/(G \cap \text{scalars})$ is finite. If in addition $G \cap$ scalars is finite (e.g., if $G$ is semisimple), then either $G^0 = SL(V)$, or $G$ is finite.*

(3) *Suppose $<,>$ is a nondegenerate symmetric bilinear form on $V$, and suppose $G$ lies in the orthogonal group $O(V) := \mathrm{Aut}(V, <,>)$. If $M_4(G, V) = 3$, then either $G = O(V)$, or $G = SO(V)$, or $G$ is finite. If $\dim(V)$ is 2 or 4, then $G$ is not contained in $SO(V)$.*

(4) *Suppose $<,>$ is a nondegenerate alternating bilinear form on $V$ (such a form exists only if $\dim(V)$ is even), suppose $G$ lies in the symplectic group $\mathrm{Sp}(V) := \mathrm{Aut}(V, <,>)$, and suppose $\dim(V) > 2$. If $M_4(G, V) = 3$, then either $G = \mathrm{Sp}(V)$, or $G$ is finite.*

*Proof.* To prove 1), suppose that $V = V_1 \oplus V_2$ is the direct sum of two non-zero $G$-modules. Then we have a $G$-isomorphism

$$V^{\otimes 2} \cong (V_1)^{\otimes 2} \oplus (V_2)^{\otimes 2} \oplus 2(V_1 \otimes V_2),$$

and this in turn forces $M_4(G, V) \geq 1 + 1 + 2^2 = 6$.

To prove 2), we use the second interpretation (1.1.5) of $M_4(G, V)$. If $M_4(G, V) = 2$, then $\mathrm{End}(V)$ is the sum of two distinct irreducible representations of $G$. But under the bigger group $GL(V)$, $\mathrm{End}(V)$ is the sum of two representations of $GL(V)$, namely

$$\mathrm{End}(V) = \mathrm{End}^0(V) \oplus \mathbb{1} = \mathrm{Lie}(SL(V)) \oplus \mathbb{1}.$$

(The two summands are inequivalent irreducible representations of $GL(V)$, but we will not use this fact.) Because $M_4(G, V) = 2$, this must be the decomposition of $\mathrm{End}(V)$ as the sum of two distinct irreducible representations of $G$. In particular, $\mathrm{Lie}(SL(V))$ is $G$-irreducible.

The derived group $G^{\mathrm{der}}$ lies in $SL(V)$, so $\mathrm{Lie}(G^{\mathrm{der}})$ lies in $\mathrm{Lie}(SL(V))$. As $G^{\mathrm{der}}$ is a normal subgroup of $G$, $\mathrm{Lie}(G^{\mathrm{der}})$ is a $G$-stable submodule of $\mathrm{Lie}(SL(V))$. So by the $G$-irreducibility of $\mathrm{Lie}(SL(V))$, either $\mathrm{Lie}(G^{\mathrm{der}}) = \mathrm{Lie}(SL(V))$, or $\mathrm{Lie}(G^{\mathrm{der}}) = 0$. In the first case, $(G^{\mathrm{der}})^0 = SL(V)$, and so $G \supset SL(V)$. Thus if in addition $G \cap$ scalars is finite, $G^0$ is $SL(V)$.

In the second case, $G^{\mathrm{der}}$ is finite. For any fixed element $\gamma$ in $G(\mathbb{C})$, the morphism from $G^0$ to $G^{\mathrm{der}}$ defined by $g \mapsto g\gamma g^{-1}\gamma^{-1}$ is therefore the constant map $g \mapsto e$.

Therefore $G^0$ lies in $Z(G)$. As $G$ acts irreducibly on $V$, its center $Z(G)$ lies in the $\mathbb{G}_m$ of scalars. But $G^0 \subset Z(G)$, so $G^0$ lies in the $\mathbb{G}_m$ of scalars. Therefore $G^0 \subset G \cap$ scalars, whence $G/(G \cap \text{scalars})$ is finite. So if in addition $G \cap$ scalars is finite, then $G$ is finite.

To prove 3), use the first interpretation (1.1.4) of $M_4(G, V)$. If $M_4(G, V) = 3$, then $V^{\otimes 2}$ is the sum of three distinct irreducible representations of $G$. Under $GL(V)$, we first decompose

$$V^{\otimes 2} = \text{Sym}^2(V) \oplus \Lambda^2(V).$$

As $O(V)$-modules, we have an isomorphism

$$\Lambda^2(V) \cong \text{Lie}(SO(V))$$

and the further decompostion

$$\text{Sym}^2(V) = \text{SphHarm}^2(V) \oplus \mathbb{1}.$$

Thus as $O(V)$-module, we have the three term decomposition

$$V^{\otimes 2} = \text{SphHarm}^2(V) \oplus \mathbb{1} \oplus \text{Lie}(SO(V)).$$

(For $\dim(V) \geq 2$, the three summands are distinct irreducible representations of $O(V)$. If $\dim(V)$ is neither 2 nor 4, they are distinct irreducible representations of $SO(V)$. For $n = 2$ (resp. $n = 4$), $\text{SphHarm}^2(V)$ (resp. $\text{Lie}(SO(V))$) is a reducible representation of $SO(V)$. We will not use these facts.)

If $M_4(G, V) = 3$, then

$$V^{\otimes 2} = \text{SphHarm}^2(V) \oplus \mathbb{1} \oplus \text{Lie}(SO(V))$$

must be the decomposition of of $V^{\otimes 2}$ as the sum of three distinct irreducible representations of $G$.

We now exploit the fact that $\text{Lie}(SO(V))$ is $G$-irreducible. Since $G \subset O(V)$, $G^0 \subset SO(V)$, so $\text{Lie}(G^0)$ is a $G$-stable submodule of $\text{Lie}(SO(V))$. By $G$-irreducibility, $\text{Lie}(G^0)$ is either $\text{Lie}(SO(V))$ or is zero. If $\text{Lie}(G^0) = \text{Lie}(SO(V))$, then $G^0$ is $SO(V)$ and $G$, being caught between $SO(V)$ and $\text{O}(V)$, is either $SO(V)$ or $\text{O}(V)$. If $\text{Lie}(G^0)$ is zero, then $G$ is finite.

If $\dim(V)$ is 2 or 4, we claim $G$ cannot lie in $SO(V)$. Indeed, for $\dim(V) = 2$, $SO(V)$ is $\mathbb{G}_m$, $\text{Lie}(SO(V))$ is $\mathbb{1}$ as $SO(V)$-module, and $\text{SphHarm}^2(V)$ is $SO(V)$-reducible, so if $G \subset SO(V)$ then $M_4(G, V) \geq 6$. If $\dim(V) = 4$, then $SO(4)$ is $(SL(2) \times SL(2)) / \pm(1,1)$, hence $\text{Lie}(SO(4))$ is $SO(4)$-reducible: so if $G \subset SO(V)$ then $M_4(G, V) \geq 4$.

To prove 4), we begin with the $GL(V)$-decomposition

$$V^{\otimes 2} = \text{Sym}^2(V) \oplus \Lambda^2(V).$$

As $\text{Sp}(V)$-modules, we have an isomorphism

$$\text{Lie}(\text{Sp}(V)) \cong \text{Sym}^2(V),$$

and the further (because $\dim(V) > 2$) decomposition

$$\Lambda^2(V) = (\Lambda^2(V)/\mathbb{1}) \oplus \mathbb{1}.$$

Thus as $\mathrm{Sp}(V)$-module we have a three term deomposition

$$V^{\otimes 2} = \mathrm{Lie}\,(\mathrm{Sp}(V)) \oplus (\Lambda^2(V)/\mathbb{1}) \oplus \mathbb{1}.$$

(The three summands are distinct irreducible representations of $\mathrm{Sp}\,(V)$, but we will not use this fact.) Exactly as in the $SO$ case above, we infer that $\mathrm{Lie}\,(\mathrm{Sp}\,(V))$ is $G$-irreducible. But $G \subset \mathrm{Sp}\,(V)$, so $\mathrm{Lie}\,(G^0)$ is a $G$-stable submodule of $\mathrm{Lie}\,(\mathrm{Sp}\,(V))$, and so either $\mathrm{Lie}\,(G^0) = \mathrm{Lie}\,(\mathrm{Sp}\,(V))$, or $\mathrm{Lie}\,(G^0)$ is zero. In the first case, $G$ is $\mathrm{Sp}\,(V)$, and in the second case $G$ is finite. □

### 1.2. Remarks

1.2.1. We should call attention to a striking result of Beukers, Brownawell, and Heckmann, [BBH, Theorems A5 and A7 together], which is similar in spirit to 1.1, though more difficult: if $G$ is a Zariski closed subgroup of $GL(V)$ which acts irreducibly on $\mathrm{Sym}^2\,(V)$, then either $G/(G \cap \text{scalars})$ is finite, or $G$ contains $SL(V)$, or $\dim\,(V)$ is even and $\mathrm{Sp}\,(V) \subset G \subset \mathrm{GSp}\,(V)$.

1.2.2. There are connected semisimple subgroups $G \subset GL(V)$ with $M_4\,(G, V) = 3$ other than $SO(V)$ (for $\dim\,(V) \geq 3$, but $\neq 4$) and $\mathrm{Sp}\,(V)$ (for $\dim\,(V) \geq 4$). The simplest examples are these. Take a $\mathbb{C}$-vector space $W$ of dimension $\ell + 1$. Then for $V$ either $\mathrm{Sym}^2\,(W)$, if $\ell \geq 2$, or $\Lambda^2\,(W)$, if $\ell \geq 4$, the image $G$ of $SL(W)$ in $GL(V)$ has $M_4\,(G, V) = 3$, but $V$ is not self-dual as a representation of $G$ (not self-dual because we excluded the case $\ell = 3$, $V = \Lambda^2\,(W)$). Here is a bad proof. In the Bourbaki notation [Bour-L8, page 188], $\mathrm{Sym}^2\,(W)$ is the highest weight module $\mathrm{E}\,(2\omega_1)$, and $\Lambda^2\,(W)$ is the highest weight module $E(\omega_2)$. We use the first interpretation (1.1.4) of the fourth absolute moment. We have

$$\mathrm{End}\,(E(2\omega_1)) = E(2\omega_1) \otimes E(2\omega_1)^\vee = E(2\omega_1) \otimes E(2\omega_\ell) \tag{1.2.2.1}$$

and

$$\mathrm{End}\,(E(\omega_2)) = E(\omega_2) \otimes E(\omega_2)^\vee = E(\omega_2) \otimes E(\omega_{\ell-1}). \tag{1.2.2.2}$$

Now End (any nontrivial representation of $SL(W)$) contain both the trivial representation $\mathbb{1}$ of $SL(W)$ and its adjoint representation $E(\omega_1 + \omega_\ell)$.

From looking at highest weights, we see that $\mathrm{End}\,(E(2\omega_1))$ contains $E(2\omega_1 + 2\omega_\ell)$, and we see that $\mathrm{End}\,(E(\omega_2))$ contains $E\,(\omega_2 + \omega_{\ell-1})$.

Thus we have a priori decompositions

$$\mathrm{End}\,(E(2\omega_1)) = \mathbb{1} \oplus E(\omega_1 + \omega_\ell) \oplus E(2\omega_1 + 2\omega_\ell) \oplus (?), \tag{1.2.2.3}$$

$$\mathrm{End}\,(E(\omega_2)) = \mathbb{1} \oplus E(\omega_1 + \omega_\ell) \oplus E(\omega_2 + \omega_{\ell-1}) \oplus (?). \tag{1.2.2.4}$$

To see that in both cases there is no (?) term, it suffices to check that the dimensions add up, an exercise in the Weyl dimension formula we leave to the reader.

1.2.3. Other examples are (the image of) $E_6$ in either of its 27-dimensional irreducible representations, or Spin(10) in either of its 16-dimensional spin representations: according to simpLie [MPR], these all have fourth absolute moment 3.

### 1.3. The case of *G* finite: the primitivity theorem

1.3.1. What about finite groups $G \subset GL(V)$ with $M_4(G, V) = 2$, or finite groups $G$ in $O(V)$ or $\mathrm{Sp}(V)$ with $M_4(G, V) = 3$?

PRIMITIVITY THEOREM 1.3.2. *Let $V$ be a $\mathbb{C}$-vector space of dimension $N \geq 2$, $G \subset GL(V)$ a finite subgroup of $GL(V)$. With the notations of the previous theorem, suppose that one of the following conditions 1), 2), or 3) holds.*

(1) $M_4(G, V) = 2$

(2) *$G$ lies in $O(V)$,* $\dim(V) \geq 3$, *and* $M_4(G, V) = 3$.

(3) *$G$ lies in* $\mathrm{Sp}(V)$, $\dim(V) \geq 4$, *and* $M_4(G, V) = 3$.

*Then $G$ is an (irreducible) primitive subgroup of $GL(V)$, i.e., there exists no proper subgroup $H$ of $G$ such that $V$ is induced from a representation of $H$.*

1.3.3. Before giving the proof, we recall the following well-known lemma.

LEMMA 1.3.4. *Let $G$ be a group, $H$ a subgroup of $G$ of finite index, and $A$ and $B$ two finite-dimensional $\mathbb{C}$-representations of $H$.*

1. *Denoting by* $^\vee$ *the dual (contragredient) representation, we have a canonical $G$-isomorphism*

$$\left(\mathrm{Ind}_H{}^G(A)\right)^\vee \cong \mathrm{Ind}_H{}^G(A^\vee).$$

2. *There is a canonical surjective $G$-morphism ("cup product")*

$$\left(\mathrm{Ind}_H{}^G(A)\right) \otimes_{\mathbb{C}} \left(\mathrm{Ind}_H{}^G(B)\right) \to \mathrm{Ind}_H{}^G(A \otimes_{\mathbb{C}} B).$$

*Proof of Lemma 1.3.4.* Assertion 1) is proven in [Ka-TLFM, 3.1.3]. For assertion 2), we view induction as Mackey induction, cf. [Ka-TLFM, 3.0.1.2]. Thus $\mathrm{Ind}_H{}^G(A)$ is $\mathrm{Hom}_{\text{left H-sets}}(G, A)$, with left $G$-action defined by $(L_g\varphi)(x) := \varphi(xg)$. We define a $\mathbb{C}$-bilinear map

$$\left(\mathrm{Ind}_H{}^G(A)\right) \otimes_{\mathbb{C}} \left(\mathrm{Ind}_H{}^G(B)\right) \to \mathrm{Ind}_H{}^G(A \otimes_{\mathbb{C}} B)$$

as follows. Given maps $\varphi : G \to A$ and $\psi : G \to B$ of left $H$-sets, we define their cup product $\varphi \otimes \psi : G \to A \otimes_{\mathbb{C}} B$ by

$$(\varphi \otimes \psi)(x) := \varphi(x) \otimes \psi(x).$$

It is immediate that $\varphi \otimes \psi$ is a map of left $H$-sets, and so the cup product construction $(\varphi, \psi) \mapsto \varphi \otimes \psi$ is a $\mathbb{C}$-linear map

$$\left(\mathrm{Ind}_H{}^G(A)\right) \otimes_{\mathbb{C}} \left(\mathrm{Ind}_H{}^G(B)\right) \to \mathrm{Ind}_H{}^G(A \otimes_{\mathbb{C}} B).$$

This map is easily checked to be $G$-equivariant and surjective. □

*Proof of Theorem 1.3.2.* Let $H$ be a subgroup of a finite group $G$, of finite index $d \geq 2$, and A be a finite-dimensional $\mathbb{C}$-representation of $H$, of dimension $a \geq 1$. We wish to compute a lower bound for $M_4(G, \mathrm{Ind}_H{}^G(A))$. To do this we attempt to decompose $\mathrm{Ind}_H{}^G(A) \otimes (\mathrm{Ind}_H{}^G(A))^\vee$ as a sum of $G$-modules. By the previous lemma, we have a $g$-isomorphism

$$\left(\mathrm{Ind}_H{}^G(A)\right)^\vee \cong \mathrm{Ind}_H{}^G(A^\vee),$$

and a surjective $G$-map

$$\mathrm{Ind}_H{}^G(A) \otimes \mathrm{Ind}_H{}^G(A^\vee) \to \mathrm{Ind}_H{}^G(A \otimes A^\vee).$$

Its source has dimension $d^2a^2$, while its target has lower dimension $da^2$, so this map has a nonzero kernel “Ker,” which is a $G$-module of dimension $(d^2 - d)a^2$. So we have a $G$-isomorphism

$$\mathrm{Ind}_H{}^G(A) \otimes \left(\mathrm{Ind}_H{}^G(A)\right)^\vee \cong \mathrm{Ker} \oplus \mathrm{Ind}_H{}^G(A \otimes A^\vee).$$

Now the $H$-module $A \otimes A^\vee = \mathrm{End}\,(A)$ itself has an $H$-decomposition

$$\mathrm{End}\,(A) \cong \mathrm{End}^0\,(A) \oplus \mathbb{1}_H,$$

as the sum of the endomorphisms of trace zero with the scalars. [Of course, if $A$ is one-dimensional, then $\mathrm{End}^0\,(A)$ vanishes.] Thus we have a $G$-decomposition

$$\mathrm{Ind}_H{}^G(A \otimes A^\vee) \cong \mathrm{Ind}_H{}^G(\mathrm{End}^0\,(A)) \oplus \mathrm{Ind}_H{}^G(\mathbb{1}_H),$$

Now the trivial representation $\mathbb{1}_G$ occurs once in $\mathrm{Ind}_H{}^G(\mathbb{1}_H)$, so we have a further decomposition

$$\mathrm{Ind}_H{}^G(\mathbb{1}_H) \cong \mathrm{Ind}_H{}^G(\mathbb{1}_H)/\mathbb{1}_G \oplus \mathbb{1}_G.$$

So all in all we have a four term $G$-decomposition

$$\mathrm{Ind}_H{}^G(A) \otimes \left(\mathrm{Ind}_H{}^G(A)\right)^\vee \cong \mathrm{Ker} \oplus \mathrm{Ind}_H{}^G(\mathrm{End}^0\,(A)) \oplus \mathrm{Ind}_H{}^G(\mathbb{1}_H)/\mathbb{1}_G \oplus \mathbb{1}_G,$$

in which the dimensions of the terms are respectively $(d^2 - d)a^2$, $d(a^2 - 1)$, $d - 1$, and 1. So we obtain the a priori estimate

$$M_4\left(G, \mathrm{Ind}_H{}^G(A)\right) \geq 4 \text{ if } \dim(A) \geq 2,$$

$$M_4\left(G, \mathrm{Ind}_H{}^G(A)\right) \geq 3 \text{ if } \dim(A) = 1.$$

Thus if $M_4(G, V) = 2$, then $G$ is a primitive subgroup of $GL(V)$.

Suppose now that $M_4(G, V) = 3$, and that $V$ is induced from a subgroup $H$ of $G$ of finite index $d \geq 2$, from an $H$-module $A$. Then $\dim(A) = 1$, and $\dim(V) = d$. Moreover, $\mathrm{Ind}_H{}^G(A) \otimes (\mathrm{Ind}_H{}^G(A))^\vee$ is the sum of three distinct irreducibles, of dimensions $d^2 - d, d - 1$, and 1.

If we further suppose that $G$ lies in either $O(V)$ or $\mathrm{Sp}(V)$, then $V \cong \mathrm{Ind}_H{}^G(A)$ is self-dual, so we have a $G$-isomorphism

$$\mathrm{Ind}_H{}^G(A) \otimes \left(\mathrm{Ind}_H{}^G(A)\right)^\vee \cong \mathrm{Ind}_H{}^G(A) \otimes \mathrm{Ind}_H{}^G(A) \cong V \otimes V.$$

If $G$ lies in $O(V)$, and $\dim(V) \geq 3$, then we have the $G$-decomposition

$$V \otimes V \cong \mathrm{SphHarm}^2(V) \oplus \Lambda^2(V) \oplus \mathbb{1}_G.$$

In this decomposition, the dimensions of the terms are respectively $d(d+1)/2 - 1, d(d-1)/2$, and 1. Since $M_4(G, V) = 3$, these three terms must be distinct irreducibles. Thus $V \otimes V \cong V \otimes V^\vee$ is simultaneously presented as the sum of three distinct irreducibles of dimensions $d^2 - d, d - 1$, and 1, and the sum of three distinct irreducibles of dimensions $d(d+1)/2 - 1, d(d-1)/2$, and 1. As $d \geq 2$, we have $d(d+1)/2 - 1 \geq d(d-1)/2$. Comparing the dimensions of the largest irreducible constituent in the two presentations, we find

$$d^2 - d = d(d+1)/2 - 1,$$

which forces $d = 1$ or 2, contradiction.

If $G$ lies in $\mathrm{Sp}(V)$, and $\dim(V) \geq 4$, the argument is similar. We have the $G$-decomposition

$$V \otimes V \cong \mathrm{Sym}^2(V) \oplus \Lambda^2(V)/\mathbb{1}_G \oplus \mathbb{1}_G.$$

into what must be three distinct irreducibles, of dimensions $d(d+1)/2$, $d(d-1)/2 - 1$, and 1. Exactly as above, we compare dimensions of the largest irreducible constituent in the two presentations. We find

$$d^2 - d = d(d+1)/2,$$

which forces $d = 3$, contradiction. □

*Remark* 1.3.5. In the primitivity theorem, when $V$ is either symplectic or orthogonal, we required $\dim(V) > 2$. This restriction is necessary, because there exist imprimitive finite groups $G$ in both $O(2)$ and in $\mathrm{Sp}(2) = SL(2)$ whose fourth moment is 3 in their given representations. Indeed, fix an integer $n \geq 1$ which is not a divisor of 4, and denote by $\zeta$ a primitive $n$'th root of unity. The dihedral group $D_{2n} \subset O(2)$ of order $2n$ (denoted $D_n$ in [C-R-MRT, page 22]), the group generated by $\mathrm{Diag}(\zeta, \zeta^{-1})$ and Antidiag (1, 1), is easily checked to have fourth moment 3 in its given representation. If we further require $n$ to be even, the generalized quaternion group $Q_{2n} \subset SL(2)$ of order $2n$ (denoted $Q_{n/2}$ in [C-R-MRT, page 23]), the group generated by $\mathrm{Diag}(\zeta, \zeta^{-1})$ and Antidiag (1,−1), is easily checked to have fourth moment 3 in its given representation.

TENSOR INDECOMPOSABILITY LEMMA 1.3.6. *Let $V$ be a $\mathbb{C}$-vector space of dimension $N \geq 2$, $G \subset GL(V)$ a finite subgroup of $GL(V)$. Suppose that $M_4(G, V) \leq 3$. Then $V$ is tensor-indecomposable in the following (strong) sense. There exists no expression of the $\mathbb{C}$-vector space $V$ as a tensor product*

$$V = V_1 \otimes V_2$$

*of $\mathbb{C}$-vector spaces $V_1$ and $Y$ in such a way that all three of the following conditions are satisfied:*

$\dim(V_1) \geq 2,$

$\dim(V_2) \geq 2,$

*every element $g$ in $G$, viewed as lying in $GL(V) = GL(V_1 \otimes V_2)$, can be written in the form $A \otimes B$ with $A$ in $GL(V_1)$ and with $B$ in $GL(V_2)$.*

*Proof.* If not, $G$ lies in the image "$GL(V_1) \otimes GL(V_2)$" of the product group $GL(V_1) \times GL(V_2)$ in $GL(V_1 \otimes V_2)$. So we have the trivial inequality

$$M_4(G, V) = M_4(G, V_1 \otimes V_2) \geq M_4(GL(V_1) \otimes GL(V_2), V_1 \otimes V_2).$$

But by definition

$$\begin{aligned} &M_4(GL(V_1) \otimes GL(V_2), V_1 \otimes V_2) \\ &= \dim(((V_1 \otimes V_2)^{\otimes 2} \otimes ((V_1 \otimes V_2)^{\vee})^{\otimes 2})^{GL(V_1)\times GL(V_2)}) \\ &= \dim(((V_1{}^{\otimes 2} \otimes (V_1{}^{\vee})^{\otimes 2}) \otimes (V_2{}^{\otimes 2} \otimes (V_2{}^{\vee})^{\otimes 2}))^{GL(V_1)\times GL(V_2)}) \\ &\geq \dim(((V_1^{\otimes 2} \otimes (V_1^{\vee})^{\otimes 2})^{GL(V_1)}) \otimes ((V_2^{\otimes 2} \otimes (V_2^{\vee})^{\otimes 2})^{GL(V_2)})) \\ &= M_4(GL(V_1), V_1) \times M_4(GL(V_2), V_2) \\ &\geq 2 \times 2 = 4. \end{aligned}$$

□

NORMAL SUBGROUP COROLLARY 1.3.7 [LARSEN-CHAR, 1.6]. *Let $V$ be a $\mathbb{C}$-vector space of dimension $N \geq 2$, $G \subset GL(V)$ a finite subgroup of $GL(V)$. Let $H$ be a proper normal subgroup of $G$. Suppose that one of the following conditions 1), 2), or 3) holds.*

(1) *$M_4(G, V) = 2$.*

(2) *$G$ lies in $O(V)$, $\dim(V) \geq 3$, and $M_4(G, V) = 3$.*

(3) *$G$ lies in $\mathrm{Sp}(V)$, $\dim(V) \geq 4$, and $M_4(G, V) = 3$.*

*Then either $H$ acts on $V$ as scalars and lies in the center $Z(G)$, or $V$ is $H$-irreducible.*

*Proof.* By the Primitivity Theorem 1.3.2, $G$ is primitive. So the restriction of $V$ to $H$ must be $H$-isotypical, as otherwise $V$ is induced. Say $V|H \cong nV_1$, for some irreducible representation $V_1$ of $H$. If $\dim(V_1) = 1$, then $H$ acts on $V$ as scalars. But $H \subset G \subset GL(V)$, so $H$ certainly lies in $Z(G)$. If $n = 1$, then $V = V_1$ is $H$-irreducible. It remains to show that the case where $\dim(V_1) \geq 2$ and $n \geq 2$

cannot arise. To see this, write the vector space $V$ as $X \otimes Y$ with $X := V_1$ and $Y := \mathrm{Hom}_H(V_1, V)$. Then $\dim(X)$ and $\dim(Y)$ are both at least 2, and, by [C-R-MRT, 51.7], every element of $g$ is of the form $A \otimes B$ with $A$ in $GL(X)$ and $B$ in $GL(Y)$. But this contradicts the Tensor Indecomposability Lemma 1.3.5. □

## 1.4. Criteria for $G$ to be big

1.4.1. We next combine these results with some classical results of Blichfeld and of Mitchell, and with recent results of Wales and Zalesskii, to give criteria which force $G$ to be big. Recall that an element $A$ in $GL(V)$ is called a pseudoreflection if $\mathrm{Ker}(A - 1)$ has codimension 1 in $V$. A pseudoreflection of order 2 is called a reflection. Given an integer $r$ with $1 \leq r < \dim(V)$, an element $A$ of $GL(V)$ is called quadratic of drop $r$ if its minimal polynomial is $(T - 1)(T - \lambda)$ for some nonzero $\lambda$, if $V/\mathrm{Ker}(A - 1)$ has dimension $r$, and if $A$ acts on this space as the scalar $\lambda$. Thus a quadratic element of drop 1 is precisely a pseudoreflection.

THEOREM 1.4.2. *Let $V$ be a $\mathbb{C}$-vector space of dimension $N \geq 2$, $G$ in $GL(V)$ a (not necessarily connected) Zariski closed reductive subgroup of $GL(V)$ with $M_4(G, V) = 2$. Fix an integer $r$ with $1 \leq r <$* $\dim(V)$. *If any of the following conditions is satisfied, then $G \supset SL(V)$.*

(1) *$G$ contains a unipotent element $A \neq 1$.*

(2) *$G$ contains a quadratic element $A$ of drop $r$ which has finite order $n \geq 6$.*

(3) *$G$ contains a quadratic element $A$ of drop $r$ which has finite order 4 or 5, and* $\dim(V) > 2r$.

(4) *$G$ contains a quadratic element $A$ of drop $r$ which has finite order 3, and* $\dim(V) > 4r$.

(5) *$G$ contains a reflection $A$, and* $\dim(V) > 8$.

*Proof.* Suppose we have already proven the theorem in the case when $G \cap$ scalars is finite. To treat the remaining case, when $G$ contains the scalars, we make use of the following elementary lemma. □

LEMMA 1.4.3. *Let $V$ be a $\mathbb{C}$-vector space of dimension $N \geq 2$, $G \subset GL(V)$ a (not necessarily connected) Zariski closed reductive subgroup of $GL(V)$ which contains the scalars $\mathbb{C}^\times$. For each integer $d \geq 1$, denote by $G_d \subset G$ the closed subgroup*

$$G_d := \{g \text{ in } G \mid \det(g)^d = 1\}.$$

*Then $G_d$ is reductive, and for every integer $n \geq 1$, we have*

$$M_{2n}(G_d, V) = M_{2n}(G, V).$$

*Proof of Lemma 1.4.3.* Since $G$ contains the scalars, every element of $G$ can be written as $\lambda g_1$ with $\lambda$ any chosen $n$'th root of $\det(g)$, and $g_1 := \lambda^{-1}g$ an element of $G_1$. So we have $G = \mathbb{G}_m G_d$ for every $d \geq 1$. So for every $n \geq 1$, $G$ and $G_d$ acting on $V^{\otimes n}(V^\vee)^{\otimes n}$ have the same image in $GL(V^{\otimes n}(V^\vee)^{\otimes n})$ (simply because the scalars in $GL(V)$ act trivially on $V^{\otimes n}(V^\vee)^{\otimes n}$). Therefore we have the asserted equality of moments. Moreover, $G$ being reductive, each $V^{\otimes n}(V^\vee)^{\otimes n}$ is a completely reducible representation of $G_d$. Each has a finite kernel (because its kernel in $GL(V)$ is the scalars), so $G_d$ is reductive. □

*Proof of Theorem 1.4.2, suite.* Thus if $G$ contains the scalars, each $G_d$ is reductive, $G_d \cap$ scalars is finite, and $G_d$ has fourth moment 2. So we already know the theorem for $G_d$. In all of the cases 1) through 5), the given element $A$ in $G$ lies in some $G_d$. So $G_d \supset SL(V)$, and we are done.

It remains to treat the case in which $G \cap$ scalars is finite. By Larsen's theorem together with the primitivity theorem, either $G^0 = SL(V)$, or $G$ is a finite irreducible primitive subgroup of $GL(V)$. Suppose that $G$ is a finite irreducible primitive subgroup of $GL(V)$. We will show that each of the conditions 1) through 5) leads to a contradiction

For assertion 1), the contradiction is obvious: a nontrivial unipotent element is of infinite order.

Assertion 2) contradicts Blichfeld's "60° theorem" [Blich-FCG, paragraph 70, Theorem 8, page 96], applied to that power of $A$ whose only eigenvalues are 1 and $\exp(2\pi i/n)$: in a finite irreducible primitive subgroup $G$ of $GL(N, \mathbb{C})$, if an element $g$ in $G$ has an eigenvalue $\alpha$ such that every other eigenvalue of $g$ is within 60° of $\alpha$ (on either side, including the endpoints), then $g$ is a scalar.

Assertion 3) in the case $n = 5$ (resp. $n = 4$) contradicts a result of Zalesskii [Zal, 11.2] (resp. Wales [Wales, Thm. 1],) applied to $A$: if a finite irreducible primitive subgroup $G$ of $GL(N, \mathbb{C})$ contains a quadratic element of drop $r$ and order 5 (resp. order 4), then $\dim(V) = 2r$.

Assertion 4) contradicts a result of Wales [Wales, section 5], applied to $A$: if a finite irreducible primitive subgroup $G$ of $GL(N, \mathbb{C})$ contains a quadratic element of drop $r$ and order 3, then $\dim(V) \leq 4r$.

Assertion 5) contradicts the following theorem, the first (and essential) part of which was proved by Mitchell nearly a century ago. □

THEOREM 1.4.4 (MITCHELL). *Let $V$ be a $\mathbb{C}$-vector space of dimension $N > 8$, $G \subset GL(V)$ a finite irreducible primitive subgroup of $GL(V) \cong GL(N, \mathbb{C})$ which contains a reflection $A$. Let $\Gamma \subset G$ denote the normal subgroup of $G$ generated by all the reflections in $G$. Then we have:*

(1) *$\Gamma$ is (conjugate in $GL(V)$ to) the group $S_{N+1}$, viewed as a subgroup of $GL(N, \mathbb{C})$ by its "permutation of coordinates" action on the hyperplane* $\mathrm{Aug}_N$ *in $\mathbb{C}^{N+1}$ consisting of those vectors whose coordinates sum to zero.*

(2) *$G$ is the product of $\Gamma$ with the group $G \cap$ (scalars).*

(3) $M_4(G, V) > 3$.

*Proof.* By a theorem of Mitchell [Mi], if $N > 8$, and if $G$ is a finite irreducible primitive subgroup of $GL(V) \cong GL(\mathrm{Aug}_N)$ which contains a reflection, then the image of $G$ in the projective group $\mathrm{PGL}(\mathrm{Aug}_N) = GL(\mathrm{Aug}_N)/\mathbb{C}^\times$ is the image in that group of the symmetric group $S_{N+1}$.

We first exhibit an $S_{N+1}$ inside $G$. For this, we argue as follows. We have our reflection $A$ in $G$. Its image in $S_{N+1}$, and indeed the image in $S_{N+1}$ of any reflection in $G$, is a transposition. Renumbering, we may suppose $A \mapsto (1, 2)$. As all transpositions in $S_{N+1}$ are $S_{N+1}$-conjugate, for each $i$ with $1 \leq i \leq N$, there is a $G$-conjugate $A_i$ of $A$ which maps to the transposition $\sigma_i := (i, i+1)$. Now $A_i$ is itself a reflection, being a conjugate of the reflection $A$. We claim it is the unique reflection in $G$ which maps to $\sigma_i$. Indeed, any element in $G$ which maps to $\sigma_i$ is of the form $\lambda A_i$ for some invertible scalar $\lambda$; but $\lambda A_i$ has $\lambda$ as eigenvalue with multiplicity $N - 1 > 1$, so $\lambda A_i$ can be a reflection only if $\lambda = 1$. We next claim that the subgroup $H$ of $G$ generated by the $A_i$ maps isomorphically to $S_{N+1}$. We know $H$ maps onto $S_{N+1}$ (because $S_{N+1}$ is generated by the $\sigma_i$), so it suffices to show that the order of $H$ divides $(N+1)!$ For this, it suffices to show that $H$ is a quotient of $S_{N+1}$. We know [Bour-L4, pages 12 and 27] that $S_{N+1}$ is generated by elements $s_i$, $1 \leq i \leq N$, subject to the Coxeter relations

$$(s_i s_j)^{m(i,j)} = 1,$$

where

$$m(i,i) = 1,$$
$$m(i,j) = 2 \text{ if } |i-j| \geq 2,$$
$$m(i,j) = 3 \text{ if } |i-j| = 1.$$

(If we map $s_i$ to $\sigma_i$, we get the required isomorphism with $S_{N+1}$.) So it suffices to show that the $A_i$ satisfy these relations. Each $A_i$ is a reflection, so of order 2. For any $i$ and $j$, the subspace

$$\mathrm{Ker}\,(A_i - 1) \cap \mathrm{Ker}\,(A_j - 1)$$

of $V$ has codimension at most 2, and the product $A_i A_j$ fixes each element of this subspace. Therefore its power $(A_i A_j)^{m(i,j)}$ also fixes each element of this subspace. But $(A_i A_j)^{m(i,j)}$ maps to $(\sigma_i \sigma_j)^{m(i,j)} = 1$ in $S_{N+1}$, and hence $(A_i A_j)^{m(i,j)}$ is a scalar $\lambda$. As this scalar $\lambda$ fixes every vector in a subspace of codimension at most 2, we must have $\lambda = 1$.

We next observe that $H = \Gamma$, i.e., that $H$ contains every reflection $A$ in $G$. For the image of $A$ in the projective group is a transposition, so $A = \lambda h$ for some scalar $\lambda$ and some transposition $h$ in $H$. But such an $h$ is a reflection in $GL(\mathrm{Aug}_N)$. Thus both $h$ and $\lambda h$ are reflections, which forces $\lambda = 1$. This proves 1).

Since $H = \Gamma$ maps isomorphically to the image $S_{N+1} \cong G/G \cap (\text{scalars})$ of $G$ in $\mathrm{PGL}(\mathrm{Aug}_N)$, $G$ is generated by $\Gamma$ and by the central subgroup $G \cap (\text{scalars})$, and $\Gamma \cap (\text{scalars}) = \{1\}$. This proves 2).

To prove 3), notice that the scalars in $GL(V)$ act trivially on the tensor spaces $V^{\otimes n} \otimes (V^\vee)^{\otimes n}$ for every $n$, in particular for $n = 2$. So the action of

$G = \Gamma \times G \cap$ (scalars) on $V^n \otimes (V^\vee)^{\otimes n}$ factors through the action of $\Gamma$. Thus we have

$$M_{2n}(G, V) = M_{2n}(\Gamma, V) = M_{2n}(S_{N+1}, \mathrm{Aug}_N).$$

So it remains only to prove the following lemma. □

LEMMA 1.4.5. *For any $N \geq 4$, we have $M_4(S_{N+1}, \mathrm{Aug}_N) > 3$.*

*Remark 1.4.5.1.* We will see later (2.4.3) that, in fact, we have $M_4(S_{N+1}, \mathrm{Aug}_N) = 4$ for $N \geq 3$, but we do not need this finer result here.

*Proof of Lemma 1.4.5.* Aug := $\mathrm{Aug}_N$ is an orthogonal representation of $S_{N+1}$, so we have an $S_{N+1}$-decomposition

$$(\mathrm{Aug})^{\otimes 2} \cong \mathbb{1} \oplus \Lambda^2(\mathrm{Aug}) \oplus \mathrm{SphHarm}^2(\mathrm{Aug}),$$

and thus an a priori inequality $M_4(S_{N+1}, \mathrm{Aug}_N) \geq 3$, with equality if and only if the following condition (1.4.5.2) holds:

(1.4.5.2) $\mathbb{1}$, $\Lambda^2(\mathrm{Aug})$, and $\mathrm{SphHarm}^2(\mathrm{Aug})$ are three inequivalent irreducible representations of $S_{N+1}$. □

The dimensions of these three representations are 1, $N(N-1)/2$, and $N(N+1)/2 - 1$ respectively. Because $N \geq 4$, none of these dimensions is $N$. So if (1.4.5.2) holds, then the irreducible representation Aug does not occur in $(\mathrm{Aug})^{\otimes 2}$, or equivalently (Aug being self-dual), $\mathbb{1}$ does not occur in $(\mathrm{Aug})^{\otimes 3}$, or equivalently

$$\int_{S_{N+1}} \mathrm{Trace}\,(g|\,\mathrm{Aug})^3 = 0.$$

But in fact we have

$$\int_{S_{N+1}} \mathrm{Trace}\,(g|\,\mathrm{Aug})^3 > 0,$$

as the following argument shows. The representation Aug being irreducible and nontrivial, we have

$$\int_{S_{N+1}} \mathrm{Trace}\,(g|\,\mathrm{Aug}) = 0.$$

For $g$ in $S_{N+1}$, let us denote by $\mathrm{Fix}(g)$ the number of fixed points of $g$, viewed as a perrmutation of $\{1, \ldots, N+1\}$. Then

$$\mathrm{Trace}\,(g|\,\mathrm{Aug}) = \mathrm{Fix}\,(g) - 1.$$

So we get

$$\int_{S_{N+1}} (\mathrm{Fix}\,(g) - 1) = 0.$$

Now break up $S_{N+1}$ as the disjoint union $\mathrm{Fix}_{\geq 2} \amalg \mathrm{Fix}_{=1} \amalg \mathrm{Fix}_{=0}$, according to the number of fixed points. Then we may rewrite the above vanishing as

$$\int_{\mathrm{Fix}_{\geq 2}} (\mathrm{Fix}\,(g) - 1) - \int_{\mathrm{Fix}_{=0}} (1) = 0.$$

At the same time, we have

$$\int_{S_{N+1}} \mathrm{Trace}\,(g \mid \mathrm{Aug})^3 = \int_{\mathrm{Fix}_{\geq 2}} (\mathrm{Fix}\,(g) - 1)^3 - \int_{\mathrm{Fix}_{=0}} (1).$$

At every point of $\mathrm{Fix}_{\geq 2}$, we have

$$(\mathrm{Fix}\,(g) - 1)^3 \geq \mathrm{Fix}\,(g) - 1,$$

with strict inequality on the nonempty set $\mathrm{Fix}_{\geq 3}$. Thus we have

$$\int_{S_{N+1}} \mathrm{Trace}\,(g \mid \mathrm{Aug})^3 > \int_{S_{N+1}} \mathrm{Trace}\,(g \mid \mathrm{Aug}) = 0.$$

Therefore (1.4.5.2) does not hold, i.e., we have $M_4(S_{N+1}, \mathrm{Aug}_N) > 3$. This proves both Lemma 1.4.5 and Theorem 1.4.4.

Using Theorem 1.4.4, we also get a result in the orthogonal case.

THEOREM 1.4.6. *Let $V$ be a $\mathbb{C}$-vector space of dimension $N > 8$ equipped with a nondegenerate quadratic form. Let $G \subset O(V)$ be a (not necessarily connected) Zariski closed reductive subgroup of $O(V)$ with $M_4(G, V) = 3$. If $G$ contains a reflection, then $G = O(V)$.*

*Proof.* Theorem 1.4.4 rules out the possibility that $G$ is a finite irreducible primitive subgroup of $GL(V)$. So $G$ is either $SO(V)$ or $O(V)$. But $SO(V)$ does not contain a reflection. □

For the sake of completeness, let us also record the immediate consequence of Larsen's theorem (1.1.6) in the symplectic case.

THEOREM 1.4.7. *Let $V$ be a $\mathbb{C}$-vector space of dimension $N \geq 4$ equipped with a nondegenerate alternating form. Suppose that $G \subset \mathrm{Sp}(V)$ is a (not necessarily connected) Zariski closed reductive subgroup of $\mathrm{Sp}(V)$ with $M_4(G, V) = 3$. If $G$ contains a unipotent element $A \neq 1$, then $G = \mathrm{Sp}(V)$.*

*Proof.* By Theorem 1.1.6, $G$ is either $\mathrm{Sp}(V)$ or it is finite. Since $A$ has infinite order, $G$ is not finite. □

## 1.5. Examples of finite G: the Weil–Shale case

1.5.1. We begin with some examples of finite groups $G \subset GL(V)$ with $M_4(G, V) = 2$, pointed out to me by Deligne. Let $q$ be a power of an odd prime

$p$, i.e., q is the cardinality of a finite field $\mathbb{F}_q$ of odd characteristic $p$. Fix an integer $n \geq 1$, and a 2n-dimensional $\mathbb{F}_q$-vector space $F$, endowed with a nondegenerate symplectic form $< , >$. The Heisenberg group $\mathrm{Heis}_{2n}(\mathbb{F}_q)$ is the central extension of $F$ by $\mathbb{F}_q$ defined as the set of pairs ($\lambda$ in $\mathbb{F}_q$, $f$ in $F$), with group operation

$$(\lambda, f)(\mu, g) := (\lambda+\mu + <f, g>, f + g).$$

The symplectic group $\mathrm{Sp}(F)$ acts on $\mathrm{Heis}_{2n}(\mathbb{F}_q)$, $\gamma$ in $\mathrm{Sp}(F)$ acting by

$$\gamma(\lambda, f) := (\lambda, \gamma(f)).$$

The irreducible $\mathbb{C}$-representations of the group $\mathrm{Heis}_{2n}(\mathbb{F}_q)$ are well-known. There are $q^{2n}$ one-dimensional representations, those trivial on the center. For each of the $q - 1$ nontrivial $\mathbb{C}^\times$-valued characters $\psi$ of the center, there is precisely one irreducible representation with central character $\psi$, say $V_\psi$, which has dimension $q^n$. Because the action of $\mathrm{Sp}(F)$ on $\mathrm{Heis}_{2n}(\mathbb{F}_q)$ is trivial on the center, the action of $\mathrm{Heis}_{2n}(\mathbb{F}q)$ on $V_\psi$ extends to a projective representation of the semidirect product group $\mathrm{Heis}_{2n}(\mathbb{F}_q) \ltimes \mathrm{Sp}(F)$ on $V_\psi$. Because we are over a finite field, this projective representation in turn extends to a linear representation of $\mathrm{Heis}_{2n}(\mathbb{F}_q) \ltimes \mathrm{Sp}(F)$ on $V_\psi$, the Weil-Shale representation.

1.5.2. We claim that for any nontrivial character $\psi$ of the center, we have

(1.5.2.1) $$M_4(\mathrm{Heis}_{2n}(\mathbb{F}_q) \ltimes \mathrm{Sp}(F), V_\psi) = 2.$$

To see this, it suffices to work over the complex numbers. We fix a choice of the nontrivial character $\psi$, and denote by

$$\chi : \mathrm{Heis}_{2n}(\mathbb{F}_q) \ltimes \mathrm{Sp}(F) \to \mathbb{C}$$

the character of $V_\psi$:

$$\chi((\lambda, f, \gamma)) := \mathrm{Trace}\,((\lambda, f, \gamma) \mid V_\psi).$$

According to Howe [Howe, Prop. 2, (i), page 290], $\chi$ is supported on those conjugacy classes which meet (the center $Z$ of $\mathrm{Heis}_{2n}(\mathbb{F}_q)) \ltimes \mathrm{Sp}(F)$, where it is given by

(1.5.2.2) $$|\chi((\lambda, 0, \gamma))|^2 = q^{\dim(\mathrm{Ker}(\gamma-1)\text{ in }F)}.$$

Moreover, an element $(\lambda, f, \gamma)$ in $\mathrm{Heis}_{2n}(\mathbb{F}_q) \ltimes \mathrm{Sp}(F)$ is conjugate to an element of $Z \ltimes \mathrm{Sp}(F)$ if and only if it is conjugate to $(\lambda, 0, \gamma)$, and this happens if and only if $f$ lies in $\mathrm{Image}\,(\gamma - 1)$, cf. [Howe, page 294, first paragraph]. Thus we have

(1.5.2.3) $$|\chi((\lambda, f, \gamma))|^2 = q^{\dim(\mathrm{Ker}(\gamma-1))}, \text{ if } f \in \mathrm{Image}\,(\gamma - 1),$$
$$|\chi((\lambda, f, \gamma))|^2 = 0, \text{ if not.}$$

1.5.3. Using this explicit formula, we find a striking relation between the absolute moments of $\mathrm{Heis}_{2n}(\mathbb{F}_q)\ltimes\mathrm{Sp}(F)$ on $V_\psi$ and the absolute moments of its subgroup $\mathrm{Sp}(F)$ on $V_\psi$. For any integer $k \geq 1$, we have

$$M_{2k+2}(\mathrm{Heis}_{2n}(\mathbb{F}_q)\ltimes\mathrm{Sp}(F), V_\psi) = M_{2k}(\mathrm{Sp}(F), V_\psi). \tag{1.5.3.1}$$

To see this, we use the fact that $\dim(\mathrm{Ker}(\gamma - 1)) + \dim(\mathrm{Im}(\gamma - 1)) = \dim(F)$, and simply compute:

$$\begin{aligned}
&\#(\mathrm{Heis}_{2n}(\mathbb{F}_q)\ltimes\mathrm{Sp}(F)) \times M_{2k+2}(\mathrm{Heis}_{2n}(\mathbb{F}_q)\ltimes\mathrm{Sp}(F), V_\psi)\\
&:= \Sigma_{(\lambda, f, \gamma)}|\chi((\lambda, f, \gamma))|^{2k+2}\\
&= \Sigma_{(\lambda,0,\gamma)}\Sigma_{f \,\mathrm{in}\, \mathrm{Im}(\gamma-1)}|\chi((\lambda, f, \gamma))|^{2k+2}\\
&= \Sigma_{(\lambda,0,\gamma)}\, q^{\dim(\mathrm{Im}(\gamma-1))} \times |q^{\dim(\mathrm{Ker}(\gamma-1))}|^{k+1}\\
&= \Sigma_{\gamma \,\mathrm{in}\, \mathrm{Sp}(F)}\, q^{1+\dim(F)} \times |q^{\dim(\mathrm{Ker}(\gamma-1)}|^{k}\\
&= \Sigma_{\gamma \,\mathrm{in}\, \mathrm{Sp}(F)}\, q^{1+\dim(F)} \times |\chi((0, 0, \gamma))|^{2k}\\
&= q^{1+\dim(F)} \times \#(\mathrm{Sp}(F)) \times M_{2k}(\mathrm{Sp}(F), V_\psi)\\
&= \#(\mathrm{Heis}_{2n}(\mathbb{F}_q)\ltimes\mathrm{Sp}(F)) \times M_{2k}(\mathrm{Sp}(F), V_\psi).
\end{aligned}$$

So in particular we have

$$M_4(\mathrm{Heis}_{2n}(\mathbb{F}_q)\ltimes\mathrm{Sp}(F), V_\psi) = M_2(\mathrm{Sp}(F), V_\psi). \tag{1.5.3.2}$$

1.5.4. The formula (1.5.2.2) $|\chi((0, 0, \gamma))|^2 = q^{\dim(\mathrm{Ker}(\gamma-1))} = \#$ (fixed points of $\gamma$ on $F$) means precisely that $\mathrm{End}(V_\psi)$ as $\mathrm{Sp}(F)$-module is isomorphic to the natural permutation representation of $\mathrm{Sp}(F)$ on the space of $\mathbb{C}$-valued functions on $F$. So

$$M_2(\mathrm{Sp}(F), V_\psi) = M_{1,0}(\mathrm{Sp}(F), \mathrm{Fct}(F, \mathbb{C})) \tag{1.5.4.1}$$

is the dimension of the space of $\mathrm{Sp}(F)$-invariant functions on $F$, which is in turn equal to the number of $\mathrm{Sp}(F)$-orbits in $F$, cf. [Ger, proof of Cor. 4.4, first paragraph, page 85]. But $\mathrm{Sp}(F)$ acts transitively on $F - \{0\}$, so there are just two orbits. Thus

$$\begin{aligned}M_4(\mathrm{Heis}_{2n}(\mathbb{F}_q)\ltimes\mathrm{Sp}(F), V_\psi) &= M_2(\mathrm{Sp}(F), V_\psi)\\ &= M_{1,0}(\mathrm{Sp}(F), \mathrm{Fct}(F, \mathbb{C})) = 2,\end{aligned} \tag{1.5.4.2}$$

as asserted.

### 1.6. Examples of finite G from the Atlas

1.6.1. A perusal of the Atlas [CCNPW-Atlas] gives some finite simple groups $G$ with a low dimensional irreducible representation $V$ for which we have $M_4(G, V) = 2$. Here are some of them. In the table below, we give (in Atlas notation) the simple group $G$, the character $\chi$ of the lowest dimensional such $V$, the dimension of $V$, and the expression of $|\chi|^2$ as the sum of two distinct irreducible characters.

| G | character $\chi$ of $V$ | dim $(V)$ | $\|\chi\|^2$ |
|---|---|---|---|
| $L_3(2) = L_2(7)$ | $\chi_2, \chi_3$ | 3 | $\mathbb{1} + \chi_6$ |
| $U_4(2) = S_4(3)$ | $\chi_2, \chi_3$ | 5 | $\mathbb{1} + \chi_{10}$ |
| $U_5(2)$ | $\chi_3, \chi_4$ | 11 | $\mathbb{1} + \chi_{16}$ |
| $^2F_4(2)'$ | $\chi_2, \chi_3$ | 26 | $\mathbb{1} + \chi_{15}$ |
| $M_{23}$ | $\chi_3, \chi_4$ | 45 | $\mathbb{1} + \chi_{17}$ |
| $M_{24}$ | $\chi_3, \chi_4$ | 45 | $\mathbb{1} + \chi_{19}$ |
| $J_4$ | $\chi_2, \chi_3$ | 1333 | $\mathbb{1} + \chi_{11}$ |

1.6.2. What about finite subgroups of $O(V)$ with $M_4(G, V) = 3$? Again the Atlas gives some examples of finite simple groups $G$ with a low dimensional irreducible orthogonal representation $V$ for which we have $M_4(G, V) = 3$. Here are some of them:

| G | character $\chi$ of $V$ | dim $(V)$ | $\chi^2$ |
|---|---|---|---|
| $U_4(2)$ | $\chi_4$ | 6 | $\mathbb{1} + \chi_7 + \chi_9$ |
| $S_6(2)$ | $\chi_2$ | 7 | $\mathbb{1} + \chi_4 + \chi_6$ |
| $S_4(5)$ | $\chi_2$ | 13 | $\mathbb{1} + \chi_7 + \chi_9$ |
| | $\chi_3$ | 13 | $\mathbb{1} + \chi_8 + \chi_9$ |
| $G_2(3)$ | $\chi_2$ | 14 | $\mathbb{1} + \chi_6 + \chi_7$ |
| McL | $\chi_2$ | 22 | $\mathbb{1} + \chi_3 + \chi_4$ |
| $U_6(2)$ | $\chi_2$ | 22 | $\mathbb{1} + \chi_3 + \chi_4$ |
| $CO_2$ | $\chi_2$ | 23 | $\mathbb{1} + \chi_3 + \chi_4$ |
| $Fi_{22}$ | $\chi_2$ | 78 | $\mathbb{1} + \chi_6 + \chi_7$ |
| $HN = F_{5+}$ | $\chi_2$ | 133 | $\mathbb{1} + \chi_6 + \chi_8$ |
| | $\chi_3$ | 133 | $\mathbb{1} + \chi_7 + \chi_8$ |
| Th | $\chi_2$ | 248 | $\mathbb{1} + \chi_6 + \chi_7$ |

1.6.3. What about finite subgroups of $\mathrm{Sp}(V)$ with $M_4(G, V) = 3$?

The Atlas gives a few cases of finite simple groups $G$ with a low dimensional irreducible symplectic representation $V$ for which we have $M_4(G, V) = 3$. [As Deligne and Ramakrishnan explained to me, "most" simple groups have no symplectic representations, cf. the article [Pra] of Prasad.] Here are two lonely examples:

| G | character of $V$ | dim $(V)$ | $\chi^2$ |
|---|---|---|---|
| $U_3(2)$ | $\chi_2$ | 6 | $\mathbb{1} + \chi_6 + \chi_7$ |
| $U_5(2)$ | $\chi_2$ | 10 | $\mathbb{1} + \chi_5 + \chi_6$ |

## 1.7. Questions

1.7.1. Given a connected algebraic group $G$ over $\mathbb{C}$ with Lie $(G)$ simple, what if any are the finite subgroups of $G$ which act irreducibly on Lie $(G)$?

1.7.2. Given a finite set of irreducible representations $\{V_i\}_i$ of such a $G$, what if any are the finite subgroups $\Gamma$ of $G$ which act irreducibly on every $V_i$? From the data $(G, \{V_i\}_i)$, how can one tell if any such $\Gamma$ will exist? For example, if $G$ is simple and simply connected, can we find such a $\Gamma$ if we take for $\{V_i\}_i$ all the fundamental representations of $G$. [For $SL(N)$, pick any even $m \geq 4$: then the subgroup $\Gamma_m \subset SL(N)$ consisting of all permutation-shaped matrices of determinant one with entries in $\mu_m$ is such a subgroup.] If we take for $\{V_i\}_i$ all the irreducible representations whose highest weight is the sum of at most two fundamental weights? [For $SL(N)$, the groups $\Gamma_m$ above fail here, already for $\mathrm{Sym}^2(\mathrm{std}_N) = E(2\omega_1)$. Indeed, the $\mathbb{C}$-span of the squares $(e_1)^2$ of the standard basis elements $e_i$ of $\mathbb{C}^N$ is a $\Gamma_m$-stable subspace of $\mathrm{Sym}^2(\mathrm{std}_N)$.]

1.7.3. Given a reductive, Zariski closed subgroup $G$ of $GL(V)$, can one classify the finite subgroups $\Gamma \subset G$ for which $M_4(\Gamma, V) = M_4(G, V)$?

1.7.4. Given $G$ as in 3) above, and an integer $k \geq 1$, let us say that a finite subgroup $\Gamma \subset G$ "spoofs" $G$ to order $k$ if we have

(1.7.4.1) $$M_{2\ell}(\Gamma, V) = M_{2\ell}(G, V) \text{ for all } 1 \leq \ell \leq k?$$

For a given $G$, what can we say about the set Spoof($G$) of integers $k \geq 1$ for which there exists a finite subgroup $\Gamma \subset G$ which spoofs $G$ to order $k$? This set may consist of all $k \geq 1$. Take for $G$ the diagonal subgroup of $GL(N)$, and, for each integer $m \geq 2$, take $\Gamma_m$ the finite subgroup of $G$ consisting of diagonal matrices with entries in $\mu_m$. Then $\Gamma_m$ spoofs $G$ to order $m - 1$. Or take $G$ itself to be finite, then $\Gamma = G$ spoofs $G$ to any order. Is it true that if $G^0$ is semisimple and nontrivial, then the set Spoof($G$) is finite.?

## Part II: Applications to the Monodromy of Lefschetz Pencils

### 2.1. Diophantine preliminaries

2.1.1. Let $k$ be a finite field of cardinality $q$ and characteristic $p$, $\ell$ a prime number other than $p$, $w$ a real number, $\iota$ an embedding of $\overline{\mathbb{Q}}_\ell$ into $\mathbb{C}$, $S/k$ a smooth, geometrically connected $k$-scheme of dimension $D \geq 1$, and $\mathcal{F}$ a lisse $\overline{\mathbb{Q}}_\ell$-sheaf on $S$ of rank $r \geq 1$ which is $\iota$-pure of integer weight $w$. Pick a geometric point $s$ in $S$, and define $V := \mathcal{F}_s$. Denote by

(2.1.1.1) $$\rho_{\mathcal{F}} : \pi_1(S, s) \to GL(V) = GL(\mathcal{F}_s) \cong GL(r, \bar{\mathbb{Q}}_\ell),$$

the $\ell$-adic representation that $\mathcal{F}$ "is." Denote by $G \subset GL(V)$ the Zariski closure of the image of ${\pi_1}^{\mathrm{geom}}(S, s) := \pi_1(S \otimes_k \bar{k}, s)$ under $\rho_{\mathcal{F}}$. Because $\mathcal{F}$ is $\iota$-pure of some weight, we know [De-Weil II, 1.3.8 and 3.4.3 (iii)] that $G$ is a (not necessarily connected) semisimple subgroup of $GL(V)$.

2.1.2. Denote by $\mathcal{F}^\vee$ the linear dual (contragredient representation) of $\mathcal{F}$, and by $\overline{\mathcal{F}} := \mathcal{F}^\vee(-w)$ the "complex conjugate" of $\mathcal{F}$; the sheaves $\mathcal{F}$ and $\overline{\mathcal{F}}$ have, via $\iota$, complex conjugate local trace functions.

2.1.3. Our first task is to give a diophantine calculation of the absolute moments $M_{2n}(G, V)$, $n \geq 1$, in terms of moments $S_{2n}$ of the local trace function of $\mathcal{F}$. For each finite extension field $E/k$, define the real number $S_{2n}(E, \mathcal{F})$ by

(2.1.3.1) $$S_{2n}(E, \mathcal{F}) := (\#E)^{-\dim(S)-nw} \Sigma_{x \text{ in } S(E)} |\iota(\text{Trace}(\text{Frob}_{E,x}|\mathcal{F}))|^{2n}.$$

LEMMA 2.1.4. *Hypotheses and notations as in 2.1.1–3 above, for each $n \geq 1$ we have the limit formula*

$$M_{2n}(G, V) = \text{limsup}_{E/k \text{ finite}} S_{2n}(E, \mathcal{F}).$$

*Proof.* The moment $M_{2n}(G, V)$ is the dimension of the space of $G$-invariants, or equivalently of $\pi_1{}^{\text{geom}}(S, s)$-invariants, in $(V \otimes V^\vee)^{\otimes n}$, i.e., it is the dimension of $H^0(S \otimes_k \bar{k}, (\mathcal{F} \otimes \overline{\mathcal{F}})^{\otimes n})$. So, by Poincare duality, we have

$$M_{2n}(G, V) = \dim H_c^{2\dim(S)}\left(S \otimes_k \bar{k}, (\mathcal{F} \otimes \overline{\mathcal{F}})^{\otimes n}\right).$$

Because $\mathcal{F}$ is pure of weight $w$, $(\mathcal{F} \otimes \overline{\mathcal{F}})^{\otimes n}$ is $\iota$-pure of weight $2nw$, so this last cohomology group is $\iota$-pure of weight $2nw + 2\dim(S)$. So the endomorphism $A := \text{Frob}_k/q^{\text{wn}+\dim(S)}$ acting on it has, via $\iota$, all its eigenvalues on the unit circle. By a standard compactness argument (cf. [Ka-SE, 2.2.2.1]), we recover the dimension of the cohomology group by the limsup formula

$$\begin{aligned}
&\dim H_c^{2\dim(S)}(S \otimes_k \bar{k}, (\mathcal{F} \otimes \overline{\mathcal{F}})^{\otimes n} \\
&= \text{limsup}_m |\iota\left(\text{Trace}\left(A^m \left|H_c^{2\dim(S)}\left(S \otimes_k \bar{k}, (\mathcal{F} \otimes \overline{\mathcal{F}})^{\otimes n}\right)\right)\right)\right| \\
&= \text{limsup}_{E/k \text{ finite}} \\
&(\#E)^{-\dim(S)-nw}|\iota\left(\text{Trace}\left(\text{Frob}_E \left|H_c^{2\dim(S)}\left(S \otimes_k \bar{k}, (\mathcal{F} \otimes \overline{\mathcal{F}})^{\otimes n}\right)\right)\right)\right|.
\end{aligned}$$

By [De-Weil II, 3.3.4], the lower cohomology groups $H_c^j$, $j < 2\dim(S)$, are $\iota$-mixed of strictly lower weight, so we get $M_{2n}(G, V)$ as the limsup, over $E/k$ finite, of the quantities

$$(\#E)^{-\dim(S)-\text{nw}} \Sigma_j \left|\iota\left(\Sigma_j (-1)^j \text{ Trace }\left(\text{Frob}_E|H_c^j\left(S \otimes_k \bar{k}, (\mathcal{F} \otimes \overline{\mathcal{F}})^{\otimes n}\right)\right)\right)\right|.$$

By the Lefschetz Trace Formula, this last quantity is precisely $S_{2n}(E, \mathcal{F})$. □

FIRST VARIANT LEMMA 2.1.5. *Hypotheses and notations as in Lemma 2.1.4, suppose we are given in addition a $\overline{\mathbb{Q}}_\ell$-valued function $\varphi(E, x)$ on the set of pairs*

*(a finite extension field $E/k$, a point $x$ in $S(E)$)*

*such that there exists a positive real constant C for which we have the estimate*

$$|\iota(\varphi(E,x))| \le C(\#E)^{w-1/2}.$$

*For each finite extension $E/k$, define the approximate moment $\tilde{S}_{2n}(E,\mathcal{F})$ by*

$$\tilde{S}_{2n}(E,\mathcal{F}) := (\#E)^{-\dim(S)-nw}\Sigma_{x \text{ in } S(E)}|\iota(\mathrm{Trace}\,(\mathrm{Frob}_{E,x} \mid \mathcal{F}) + \varphi(E,x))|^{2n}.$$

*Then we have the limit formula*

$$M_{2n}(G,V) = \mathrm{limsup}_{E/k \text{ finite}}\,\tilde{S}_{2n}(E,\mathcal{F}).$$

*Proof.* One checks easily that $\tilde{S}_{2n}(E,\mathcal{F}) - S_{2n}(E,\mathcal{F}) \to 0$ as $\#E$ grows. □

SECOND VARIANT LEMMA 2.1.6. *Hypotheses and notations as in Lemma 2.1.5, suppose that S is an open subscheme of a smooth, geometrically connected k-scheme $T/k$ (necessarily of the same dimension D). Suppose that we are given a $\overline{\mathbb{Q}}_\ell$-valued function $\tau(E,x)$ on the set of pairs*

$$(\textit{a finite extension field } E/k,\ \textit{a point } x \textit{ in } T(E)),$$

*such that whenever x lies in S(E), we have*

$$\tau(E,x) = \mathrm{Trace}\,(\mathrm{Frob}_{E,x} \mid \mathcal{F}) + \varphi(E,x).$$

*For each finite extension $E/k$, define the mock moment $T_{2n}(E,\mathcal{F})$ by*

$$T_{2n}(E,\mathcal{F}) := (\#E)^{-\dim(S)-\mathrm{nw}}\Sigma_{x \text{ in } T(E)}|\iota(\tau(E,x))|^{2n}.$$

*Then we have the inequality*

$$M_{2n}(G,V) \le \lim\sup_{E/k \text{ finite}} T_{2n}(E,\mathcal{F}).$$

*Proof.* Obvious from the previous result and the observation that for each $E/k$ we have

$$\tilde{S}_{2n}(E,\mathcal{F}) \le T_{2n}(E,\mathcal{F})$$

simply because we obtain $T_{2n}(E,\mathcal{F})$ by adding positive quantities to $\tilde{S}_{2n}(E,\mathcal{F})$. □

## 2.2. Universal families of hypersurface sections

2.2.1. Recall that $k$ is a finite field, and $X/k$ is a projective, smooth, geometrical variety of dimension $n+1 \ge 1$, given with a projective embedding $X \subset \mathbb{P}$. We denote by $\mathrm{PHyp}_d/k$ the projective space of degree $d$ hypersurfaces in $\mathbb{P}$, and by

(2.2.1.1) $$\mathrm{Good}_X\mathrm{PHyp}_d \subset \mathrm{PHyp}_d$$

the dense open set consisting of those degree $d$ hypersurfaces $H$ which are transverse to $X$, i.e., such that the scheme-theoretic intersection $X \cap H$ is smooth and of codimension one in $X$. Over $\mathrm{Good}_X\mathrm{PHyp}_d$ we have the universal family of all smooth, degree $d$ hypersurface sections of $X$, say

$$\pi : \mathrm{Univ}_d \to \mathrm{Good}_X\mathrm{PHyp}_d, \tag{2.2.1.2}$$

whose fibre over a degree $d$ hypersurface $H$ in $\mathbb{P}$ is $X \cap H$.

2.2.2. For any finite extension $E/k$, and any point $H$ in $\mathrm{Good}_X\mathrm{PHyp}_d(E)$, the weak Lefschetz theorem tells us that the restriction map

$$H^i(X \otimes_k \bar{k}, \overline{\mathbb{Q}}_\ell) \to H^i((X \otimes_k \bar{k}) \cap H, \overline{\mathbb{Q}}_\ell)$$

is an isomorphism for $i < n$, and injective for $i = n$. By Poincare duality, the Gysin map

$$H^i((X \otimes_k \bar{k}) \cap H, \overline{\mathbb{Q}}_\ell) \to H^{i+2}(X \otimes_k \bar{k}, \overline{\mathbb{Q}}_\ell)(1)$$

is an isomorphism for $i > n$, and surjective for $i = n$. Thanks to the hard Lefschetz theorem, we know that, for $i = n$, the kernel of the Gysin map is a subspace

$$Ev^n((X \otimes_k \bar{k}) \cap H, \overline{\mathbb{Q}}_\ell) \subset H^n((X \otimes_k \bar{k}) \cap H, \overline{\mathbb{Q}}_\ell)$$

on which the cup-product remains nondegenerate, and which maps isomorphically to the quotient $H^n((X \otimes_k \bar{k}) \cap H, \overline{\mathbb{Q}}_\ell)/H^n(X \otimes_k \bar{k}, \overline{\mathbb{Q}}_\ell)$.

2.2.3. Over the space $\mathrm{Good}_X\mathrm{PHyp}_d$, there is a lisse $\overline{\mathbb{Q}}_\ell$-sheaf $\mathcal{F}_d$, such that for any finite extension $E/k$, and any $E$-valued point $H$ of $\mathrm{Good}_X\mathrm{PHyp}_d$, the stalk of $\mathcal{F}_d$ at $H$ is $Ev^n((X \otimes_k \bar{k}) \cap H, \overline{\mathbb{Q}}_\ell)$. The sheaf $\mathcal{F}_d$ is pure of weight $n$, and carries a cup-product autoduality toward $\overline{\mathbb{Q}}_\ell(-n)$. The autoduality is symplectic if $n$ is odd, and orthogonal if $n$ is even. For fixed $X$ but variable $d$, the rank $N_d$ of $\mathcal{F}_d$ is a polynomial in $d$ of degree $n+1$, of the form $\deg(X)d^{n+1}$ + lower terms.

THEOREM 2.2.4. *Suppose that $n \geq 2$ is even, that $d \geq 3$, and that $N_d > 8$. Then the geometric monodromy group $G_d$ of the lisse sheaf $\mathcal{F}_d$ is the full orthogonal group $O(N_d)$.*

*Proof.* The group $G_d$ is a priori a Zariski closed subgroup of $O(N_d)$. We first recall that $G_d$, indeed its subgroup $\rho_{\mathcal{F}_d}(\pi_1{}^{\mathrm{geom}}(\mathrm{Good}_X\mathrm{PHyp}_d))$, contains a reflection.

Take a sufficiently general line $L$ in $\mathrm{PHyp}_d$. Over its intersection $L - \Delta$ with $\mathrm{Good}_X\mathrm{PHyp}_d$, we get a Lefschetz pencil of smooth hypersurface sections of degree $d$ of $X$. Denote by

$$i : L - \Delta \to \mathrm{Good}_X\mathrm{PHyp}_d$$

the inclusion. We have the inequality

$$\#\Delta(\bar{k}) \geq 1 \text{ if } N_d \neq 0,$$

because $Ev^n((X \otimes_k \bar{k}) \cap H, \overline{\mathbb{Q}}_\ell)$ is spanned by the images, using all possible "chemins," of the vanishing cycles, one at each point of $\Delta(\bar{k})$, cf. [De-Weil II, 4.2.4 and 4.3.9]. (So long as char $(k)$ is not 2, we can choose a single chemin for each vanishing cycle, and we have the inequality $\#\Delta(\bar{k}) \geq N_d$, cf. [SGA 7, Expose XVIII, 6.6 and 6.6.1]).

By the Picard-Lefschetz formula [SGA 7, Exposé XV, 3.4], each of the $\#\Delta(\bar{k})$ local monodromies in a Lefschetz pencil is a reflection. Thus $\pi_1{}^{\mathrm{geom}}(L - \Delta)$ contains elements which act on $i^*\mathcal{F}_d$ as reflections, and their images in $\pi_1{}^{\mathrm{geom}}(\mathrm{Good}_X\mathrm{PHyp}_d)$ act as reflections on $\mathcal{F}_d$.

In view of Theorem 1.4.6, it suffices to show that, denoting by $V_d$ the representation of $G_d$ given by $\mathcal{F}_d$, we have $M_4(G_d, V_d) = 3$. Since $G_d$ lies in $O(N_d)$ and $N_d > 1$, we have the *a* priori inequality

$$M_4(G_d, V_d) \geq M_4(O(N_d), \mathrm{std}) = 3.$$

So the desired conclusion results from the following theorem. □

THEOREM 2.2.5. *Suppose that $n \geq 1$ and $d \geq 3$. Then $M_4(G_d, V_d) \leq 3$. If $n = 0$ and $d \geq 3$, we have $M_4(G_d, V_d) \leq 4$.*

*Proof.* Denote by $\mathrm{Hyp}_d/k$ the affine space over $k$ which is the affine cone of the projective space $\mathrm{PHyp}_d/k$. For any $k$-algebra $A$, the $A$-valued points of $\mathrm{Hyp}_d$ are the elements of $H^0(\mathbb{P}, \mathcal{O}(d)) \otimes_k A$. The natural projection map

$$\overline{\pi} : \mathrm{Hyp}_d - \{0\} \to \mathrm{PHyp}_d$$

is a (Zariski locally trivial) $\mathbb{G}_m$-bundle. We denote by

$$\mathrm{Good}_X\mathrm{Hyp}_d \subset \mathrm{Hyp}_d - \{0\}$$

the dense open set which is the inverse image of $\mathrm{Good}_X\mathrm{PHyp}_d$, and by

$$\pi : \mathrm{Good}_X\mathrm{Hyp}_d \to \mathrm{Good}_X\mathrm{PHyp}_d$$

its projection. Thus we have a cartesian diagram

$$\begin{array}{ccc} \mathrm{Good}_X\mathrm{Hyp}_d & \subset & \mathrm{Hyp}_d - \{0\} \\ \overline{\pi}\downarrow & & \pi\downarrow \\ \mathrm{Good}_X\mathrm{PHyp}_d & \subset & \mathrm{PHyp}_d \end{array}$$

We form the lisse sheaf $\pi^*\mathcal{F}_d$ on $\mathrm{Good}_X\mathrm{Hyp}_d$. By [Ka-La-FGCFT, Lemma 2, part (2)], for any geometric point $\xi$ of $\mathrm{Good}_X\mathrm{Hyp}_d$, the map

$$\pi_* : \pi_1{}^{\mathrm{geom}}(\mathrm{Good}_X\mathrm{Hyp}_d, \xi) \to \pi_1{}^{\mathrm{geom}}(\mathrm{Good}_X\mathrm{PHyp}_d, \pi(\xi))$$

is surjective. So we recover $G_d$ as the Zariski closure of the image of $\pi_1{}^{\mathrm{geom}}$ $(\mathrm{Good}_X\mathrm{Hyp}_d)$ acting on $\pi^*\mathcal{F}_d$.

The advantage is that the base space is now a dense open set of an affine space, namely $\mathrm{Hyp}_d$. We will now apply the diophantine method explained above, to the sheaf $\pi^*\mathcal{F}_d$ on the dense open set $\mathrm{Good}_X\mathrm{Hyp}_d$ of $\mathrm{Hyp}_d$.

Let $E/k$ be a finite extension field, and $H$ an $E$-valued point of $\mathrm{Good}_X\mathrm{Hyp}_d$. Then the stalk of $\pi^*\mathcal{F}_d$ at $H$ is $Ev^n((X \otimes_k \overline{k}) \cap (H = 0), \overline{\mathbb{Q}}_\ell)$. □

KEY LEMMA 2.2.6. *Given $X/k$ as above, denote by $\Sigma(X \otimes_k \overline{k}, \overline{\mathbb{Q}}_\ell)$ the sum of the $\overline{\mathbb{Q}}_\ell$-Betti numbers. Then for any finite extension field $E/k$, and for any $E$-valued point $H$ of* $\mathrm{Good}_X\mathrm{Hyp}_d$, *putting $Y := X \cap (H = 0)$, we have the estimate*

$$\begin{aligned}&|\mathrm{Trace}\,(\mathrm{Frob}_{E,H} \mid \pi^*\mathcal{F}_d) - (-1)^n(\#Y(E) - \#X(E)/\#E)|\\ &\leq \Sigma(X \otimes_k \overline{k}, \overline{\mathbb{Q}}_\ell)(\#E)^{(n-1)/2}.\end{aligned}$$

*Proof.* Use the Lefschetz Trace Formula on $Y$ to write $\#Y(E)$ as a sum of three terms:

$$\begin{aligned}\#Y(E) = {} & \Sigma_{i\leq n-1}(-1)^i\mathrm{Trace}\,(\mathrm{Frob}_E|H^i(Y \otimes_E \overline{k}, \overline{\mathbb{Q}}_\ell))\\ &+(-1)^n\mathrm{Trace}\,(\mathrm{Frob}_E|H^n(Y \otimes_E \overline{k}, \overline{\mathbb{Q}}_\ell))\\ &+\Sigma_{i\geq n+1}(-1)^i\mathrm{Trace}\,(\mathrm{Frob}_E|H^i(Y \otimes_E \overline{k}, \overline{\mathbb{Q}}_\ell)).\end{aligned}$$

Use the same formula to write $\#X(E)/\#E$ as the sum of three terms:

$$\begin{aligned}\#X(E)/\#E = {} & \Sigma_{i\leq n+1}(-1)^i\mathrm{Trace}\,(\mathrm{Frob}_E|H^i(X \otimes_k \overline{k}, \overline{\mathbb{Q}}_\ell)(1))\\ &+(-1)^{n+2}\mathrm{Trace}\,(\mathrm{Frob}_E|H^{n+2}(X \otimes_k \overline{k}, \overline{\mathbb{Q}}_\ell)(1))\\ &++\Sigma_{i\geq n+3}(-1)^i\mathrm{Trace}\,(\mathrm{Frob}_E|H^i(X \otimes_k k, \overline{\mathbb{Q}}_\ell)(1)).\end{aligned}$$

By the Poincare dual of the weak Lefschetz theorem, the third terms in the two expressions are equal. The difference of the second terms is precisely

$$(-1)^n\mathrm{Trace}\,(\mathrm{Frob}_E \mid Ev^n((X \otimes_k \overline{k}) \cap (H = 0), \overline{\mathbb{Q}}_\ell)),$$

i.e., it is $(-1)^n\mathrm{Trace}\,(\mathrm{Frob}_{E,H} \mid \pi^*\mathcal{F}_d)$. The difference of the first terms is

$$\begin{aligned}&\Sigma_{i\leq n-1}(-1)^i\mathrm{Trace}\,(\mathrm{Frob}_E|H^i(Y \otimes_E \overline{k}, \overline{\mathbb{Q}}_\ell))\\ &\quad-\Sigma_{i\leq n+1}(-1)^i\mathrm{Trace}\,(\mathrm{Frob}_E|H^i(X \otimes_k \overline{k}, \overline{\mathbb{Q}}_\ell)(1)).\end{aligned}$$

By Deligne's Weil I, each cohomology group occurring here is pure of some weight $\leq n-1$, so we get the asserted estimate with the constant

$$\Sigma_{i\leq n-1}h^i(Y \otimes_E \overline{k}, \overline{\mathbb{Q}}_\ell) + \Sigma_{i\leq n+1}h^i(X \otimes_k \overline{k}, \overline{\mathbb{Q}}_\ell).$$

Using weak Lefschetz, this is equal to

$$= \Sigma_{i\leq n-1}h^i(X \otimes_k \overline{k}, \overline{\mathbb{Q}}_\ell) + \Sigma_{i\leq n+1}h^i(X \otimes_k \overline{k}, \overline{\mathbb{Q}}_\ell).$$

Using Poincare duality on $X$, this in turn is equal to

$$= \Sigma_{i \geq n+3} h^i(X \otimes_k \overline{k}, \overline{\mathbb{Q}}_\ell) + \Sigma_{i \leq n+1} h^i(X \otimes_k \overline{k}, \overline{\mathbb{Q}}_\ell)$$
$$\leq \Sigma_i h^i(X \otimes_k \overline{k}, \overline{\mathbb{Q}}_\ell) := \Sigma(X \otimes_k \overline{k}, \overline{\mathbb{Q}}_\ell). \quad \square$$

For any finite extension field $E/k$, and for any $E$-valued point $H$ of $\mathrm{Hyp}_d$, we define

$$\tau(E, H) := (-1)^n((-1)^n(\#(X \cap (H = 0))(E) - \#X(E)/\#E).$$

Notice that $\tau$ takes values in $\mathbb{Q}$.

We then define the mock moment $T_4(E, \pi^*\mathcal{F}_d)$ by

$$T_4(E, \pi^*\mathcal{F}_d) := (\#E)^{-\dim(\mathrm{Hyp}_d)-2n} \Sigma_{H \text{ in } \mathrm{Hyp}_d(E)} \mid \tau(E, H)|^4$$
$$= (\#E)^{-\dim(\mathrm{Hyp}_d)-2n} \Sigma_{H \text{ in } \mathrm{Hyp}_d(E)} (\#(X \cap (H = 0))(E) - \#X(E)/\#E)^4.$$

(Because $\tau$ takes values in $\mathbb{Q}$, there is no need for the $\iota$ which figured in the general definition, where $\tau$ was allowed to be $\overline{\mathbb{Q}}_\ell$-valued.)

In view of the Second Variant Lemma 2.1.6, Theorem 2.2.5 now results from the following theorem:

THEOREM 2.2.7. *Let $X/k$ be as above, of dimension $n + 1 \geq 1$. If $n \geq 1$, then for any $d \geq 3$, we have the estimate*

$$|T_4(E, \pi^*\mathcal{F}_d) - 3| = O((\#E)^{-1/2}).$$

*If $n = 0$, then for any $d \geq 3$, we have the estimate*

$$|T_4(E, \pi^*\mathcal{F}_d) - 4| = O((\#E)^{-1/2}).$$

*Proof.* Fix a finite field extension $E/k$ with $\#E \geq 6$. We will use an exponential sum method to calculate $T_4(E, \pi^*\mathcal{F}_d)$ in closed form. Fix a nontrivial $\mathbb{C}^\times$-valued additive character $\psi$ of $E$. View the ambient $\mathbb{P} = \mathbb{P}^m$ as the space of lines in $\mathbb{A}^{m+1}$. For each point $x$ in $\mathbb{P}^m(E)$, choose a point $\tilde{x}$ in $\mathbb{A}^{m+1}(E) - \{0\}$ which lifts it. For any fixed $H$ in $\mathrm{Hyp}_d(E)$, the value $H(\tilde{x})$ depends upon the choice of $\tilde{x}$ lifting $x$, but only up to an $E^\times$-multiple. So the sum

$$\Sigma_{\lambda \text{ in } E^\times} \psi(\lambda H(\tilde{x}))$$

depends only on the original point $x$ in $\mathbb{P}(E)$. By the orthogonality relations for characters, we have

$$\Sigma_{\lambda \text{ in } E^\times} \psi(\lambda H(\tilde{x})) = -1 + \Sigma_{\lambda \text{ in } E} \psi(\lambda H(\tilde{x}))$$
$$= \#E - 1, \text{ if } H(x) = 0,$$
$$= -1, \text{ if not.}$$

So we get the identity

$$\Sigma_{x \text{ in } X(E)} \Sigma_{\lambda \text{ in } E^\times} \psi(\lambda H(\tilde{x})) = (\#E)(\#(X \cap (H=0))(E)) - \#X(E)$$
$$= (-1)^n (\#E)\tau(E, H).$$

This in turn gives the identity

$$(\#E)^{\dim(\mathrm{Hyp}_d)+2n+4}\, T_4(E, \pi^* \mathcal{F}_d)$$
$$= \Sigma_{H \text{ in } \mathrm{Hyp}_d(E)} (\Sigma_{x \text{ in } X(E)} \Sigma_{\lambda \text{ in } E^\times} \psi(\lambda H(\tilde{x})))^4.$$

We next open the inner sum and interchange orders of summation, to get

$$= \Sigma_{(x_i) \text{ in } X(E)^4} \Sigma_{(\lambda_i) \text{ in } (E^\times)^4} \Sigma_{H \text{ in } \mathrm{Hyp}_d(E)} \psi(\Sigma_{i=1 \text{ to } 4} \lambda_i H(\tilde{x}_i)).$$

The key observation is given by the following lemma. □

SINGLETON LEMMA 2.2.8. *Suppose $\#E \geq 4$. Given four (not necessarily distinct) points $x_1, x_2, x_3, x_4$ in $\mathbb{P}(E)$, suppose among them there is a singleton, i.e., a point which is not equal to any of the others. Then for any $(\lambda_i)$ in $(E^\times)^4$, we have the vanishing*

$$\Sigma_{H \text{ in } \mathrm{Hyp}_d(E)} \psi(\Sigma_{i=1 \text{ to } 4} \lambda_i H(\tilde{x}_i)) = 0.$$

Before proving this lemma, it will be convenient to give two other lemmas.

LEMMA 2.2.9. *If $\#E \geq 4$, then given four (not necessarily distinct) points $x_1, x_2, x_3, x_4$ in $\mathbb{P}(E)$, there exists an $E$-rational hyperplane $L$ in $\mathbb{P}$, i.e., a point $L$ in* PHyp$_1$*(E), such that all four points $x_i$ lie in the affine open set $\mathbb{P}^m[1/L]$.*

*Proof.* Say $\mathbb{P}$ is $\mathbb{P}^m$. In the dual projective space, the set of hyperplanes through a given point $x_i$ in $\mathbb{P}^m(E)$ form a $\mathbb{P}^{m-1}$, so there are precisely

$$((\#E)^m - 1)/(\#E - 1)$$

$E$-rational hyperplanes through $x_i$. So there are at least

$$((\#E)^{m+1} - 1)/(\#E - 1) - 4((\#E)^m - 1)/(\#E - 1)$$

$E$-rational hyperplanes which pass through none of the $x_i$. As $\#E$ is at least 4, this difference is strictly positive. □

EVALUATION LEMMA 2.2.10. *Let $E$ be a field, $m \geq 1$ and $d \geq 1$ integers. Denote by* Poly$_{\leq d}(E)$ *the $E$-vector space of $E$-rational polynomial functions on $\mathbb{A}^m$. For any integer $r \leq d+1$, and for any $r$ distinct points $x_i$, $i = 1$ to $r$, in $\mathbb{A}^m(E)$, the $E$-linear multi-evaluation map*

$$\mathrm{Poly}_{\leq d}(E) \to E^r$$
$$f \mapsto (f(x_1), \ldots, f(x_r))$$

*is surjective.*

*Proof.* The map being $E$-linear, its surjectivity map be checked over any extension field. Passing to a large enough such extension, we may add additional distinct points, so that our $x_i$ are the first $r$ of $d+1$ distinct points. It suffices to prove the lemma in the hardest case $r = d+1$ (then project onto the first $r$ coordinates in the target). To do this hardest case, we first treat the case $m = 1$. In this case, source and target have the same dimension, $d+1$, so it suffices that the map be injective. But its kernel consists of those polynomials in one variable of degree at most $d$, which have $d+1$ distinct zeroes. To do the general case, it suffices to find a linear form $T$ from $\mathbb{A}^m$ to $\mathbb{A}^1$ under which the $d+1$ points $x_i$ have $d+1$ distinct images. For then already polynomials of degree at most $d$ in $T$ will be a subspace of the source $\mathrm{Poly}_{\leq d}(E)$ which will map onto $E^r$. We can do this as soon as $\#E \geq \mathrm{Binom}(d+1, 2)$. Indeed, we are looking for a linear form $T$ with the property that for each of the Binom $(d+1, 2)$ pairs $(x_i, x_j)$ with $i < j$, we have $T(x_i) - T(x_j) \neq 0$. For each such pair, the set of $T$ for which $T(x_i) - T(x_j) = 0$ is a hyperplane in the dual space. So we need $T$ to not lie in the union of Binom $(d+1, 2)$ linear subspaces of codimension one. Since they all intersect in zero, their union has cardinality strictly less than Binom $(d+1, 2)(\#E)^{m-1}$. So as soon as $\#E \geq$ Binom $(d+1, 2)$, the desired $T$ exists. □

With these preliminaries out of the way, we can prove the Singleton Lemma 2.2.8. Because $\#E \geq 4$, we can find a non-zero linear form $L$ in $\mathrm{Hyp}_1(E)$ such that our four points $x_i$ all lie in $\mathbb{P}[1/L] \cong \mathbb{A}^m$. By means of the map $H \mapsto H/L^d$, we get an $E$-linear isomorphism

$$\mathrm{Hyp}_d(E) \cong \mathrm{Poly}_{\leq d}(E)$$

of $\mathrm{Hyp}_d(E)$ with the $E$-rational polynomial functions on $\mathbb{P}[1/L] \cong \mathbb{A}^m$ of degree at most $d$.

Moreover, for any $x$ in $\mathbb{P}[1/L](E)$, and any lifting $\tilde{x}$ in $\mathbb{A}^{m+1}(E)$, the two $E$-linear forms on $\mathrm{Hyp}_d(E)$,

$$H \mapsto H(\tilde{x})$$

and

$$H \to (H/L^d)(x),$$

are proportional. So whatever the four points $(x_i)$ in $\mathbb{P}[1/L](E) \cong \mathbb{A}^m(E)$, we can rewite the sum

$$\begin{aligned}&\Sigma_{H \text{ in } \mathrm{Hyp}_d(E)}\ \psi(\Sigma_{i=1 \text{ to } 4}\ \lambda_i H(\tilde{x}_i))\\ &\quad = \Sigma_{h \text{ in } \mathrm{Poly}_{\leq d}(E)}\ \psi(\Sigma_{i=1 \text{ to } 4}\ \lambda_i h(x_i)).\end{aligned}$$

Renumbering, we may suppose that $x_1$ is a singleton. We consider separately various cases (which, up to renumbering, cover all the cases when $x_1$ is a singleton).

If the four points are all distinct, then as $h$ runs over $\mathrm{Poly}_{\leq d}(E)$, the vector $(h(x_i))$ runs over $E^4$, and our sum becomes #(Ker of eval at $(x_i)$) times

$$\Sigma_{(t_i)\text{ in }E^4}\ \psi(\Sigma_{i=1\text{ to }4}\ \lambda_i t_i).$$

Since the vector $(\lambda_i)$ is nonzero, $(t_i) \mapsto \psi(\Sigma_{i=1\text{ to }4}\ \lambda_i t_i)$ is a nontrivial additive character of $E^4$, so the inner sum vanishes.

If the three remaining points are all equal, then as $h$ runs over $\mathrm{Poly}_{\leq d}(E)$, the vector $(h(x_1), h(x_2))$ runs over $E^2$, and our sum becomes #(Ker of eval at $(x_1, x_2)$) times

$$\Sigma_{(t_1,t_2)\text{ in }E^2}\ \psi(\lambda_1 t_1 + (\lambda_2 + \lambda_3 + \lambda_4)t_2).$$

Since the vector $(\lambda_1, \lambda_2 + \lambda_3 + \lambda_4)$ is nonzero, the sum again vanishes.

If the first three points are distinct, but $x_4 = x_3$, then our sum becomes #(Ker of eval at $(x_1, x_2, x_3)$) times

$$\Sigma_{(t_1,t_2,t_3)\text{ in }E^3}\ \psi(\lambda_1 t_1 + \lambda_2 t_2 + (\lambda_3 + \lambda_4)t_3).$$

Since the vector $(\lambda_1, \lambda_2, \lambda_3 + \lambda_4)$ is nonzero, the sum again vanishes.

In exactly the same way, we prove the following two elementary lemmas.

TWINNING LEMMA 2.2.11. *Suppose* $\#E \geq 2$. *Given two distinct points* $x_1, x_2$ *in* $\mathbb{P}(E)$, *put* $x_3 = x_1$, *and put* $x_4 = x_2$. *Then for* $(\lambda_i)$ *in* $(E^\times)^4$, *we have*

$$\begin{aligned}&\Sigma_{H\text{ in }\mathrm{Hyp}_d(E)}\ \psi(\Sigma_{i=1\text{ to }4}\ \lambda_i H(\tilde{x}_i))\\ &= \#\mathrm{Hyp}_d(E),\ \text{ if } \lambda_1 + \lambda_2 = \lambda_3 + \lambda_4 = 0,\\ &= 0, \quad \text{otherwise.}\end{aligned}$$

QUADRUPLES LEMMA 2.2.12. *Given a point* $x$ *in* $\mathbb{P}(E)$, *put* $x_i = x$ *for* $i = 1$ *to* 4. *Then for* $(\lambda_i)$ *in* $(E^\times)^4$, *we have*

$$\begin{aligned}&\Sigma_{H\text{ in }\mathrm{Hyp}_d(E)}\psi\ (\Sigma_{i=1\text{ to }4}\ \lambda_i H(\tilde{x}_i))\\ &= \#\mathrm{Hyp}_d(E),\ \text{if } \lambda_1 + \lambda_2 + \lambda_3 + \lambda_4 = 0,\\ &= 0, \quad \text{otherwise.}\end{aligned}$$

*Proof of Theorem 2.2.7:* suite. Recall that we have the identity

$$\begin{aligned}&(\#E)^{\dim(\mathrm{Hyp}_d)+2n+4}\ T_4(E, \pi^*\mathcal{F}_d)\\ &= \Sigma_{(x_i)\text{ in }X(E)^4}\ \Sigma_{(\lambda_i)\text{ in }(E^\times)^4}\ \Sigma_{H\text{ in }\mathrm{Hyp}_d(E)}\ \psi(\Sigma_{i=1\text{ to }4}\ \lambda_i H(\tilde{x}_i)).\end{aligned}$$

We now break up this sum by the coincidence pattern of the four-tuple $(x_1, x_2, x_3, x_4)$.

If there is any singleton, the entire inner sum vanishes.

If all the $x_i$ coincide, the inner sum is

$$\#\mathrm{Hyp}_d(E) \times \#\{(\lambda_i) \text{ in } (E^\times)^4 \text{ with } \lambda_1+\lambda_2+\lambda_3+\lambda_4=0\}.$$

This case occurs $\#X(E)$ times, one for each of the possible common values of the $x_i$.

If there are no singletons and exactly two among the $x_i$ are distinct, put $x := x_1$, and take for $y$ the other. Then the pattern is either $(x, x, y, y)$ or $(x, y, x, y)$ or $(x, y, y, x)$. In each case, the inner sum is

$$\begin{aligned}&\#\mathrm{Hyp}_d(E) \times \#\{(\lambda_i) \text{ in } (E^\times)^4 \text{ with } \lambda_1+\lambda_2=\lambda_3+\lambda_4=0\}\\ &\quad= \#\mathrm{Hyp}_d(E) \times (\#E-1)^2.\end{aligned}$$

This case occurs $3(\#X(E))(\#X(E)-1)$ times, 3 for the possible repeat pattern, $\#X(E)$ for the choice of $x_1$, $\#X(E)-1$ for the choice of $y \neq x_1$.

So all in all, we get a closed formula

$$\begin{aligned}&(\#E)^{\dim(\mathrm{Hyp}_d)+2n+4}\, T_4(E, \pi^*\mathcal{F}_d)\\ &\quad= 3(\#X(E))(\#X(E)-1)(\#\mathrm{Hyp}_d(E))(\#E-1)^2\\ &\quad+ (\#X(E))(\#\mathrm{Hyp}_d(E))(\#\{(\lambda_i) \text{ in } (E^\times)^4 \text{ with } \lambda_1+\lambda_2+\lambda_3+\lambda_4=0\}).\end{aligned}$$

Dividing through by $\#\mathrm{Hyp}_d(E) = (\#E)^{\dim(\mathrm{Hyp}_d)}$, we get

$$\begin{aligned}&(\#E)^{2n+4}\, T4(E, \pi^*\mathcal{F}_d)\\ &\quad= 3(\#X(E))(\#X(E)-1)(\#E-1)^2\\ &\quad+ (\#X(E))(\#\{(\lambda_i) \text{ in } (E^\times)^4 \text{ with } \lambda_1+\lambda_2=\lambda_3+\lambda_4=0\}).\end{aligned}$$

LEMMA 2.2.13. *We have the identity*

$$\begin{aligned}&\#\{(\lambda_i) \text{ in } (E^\times)^4 \text{ with } \lambda_1+\lambda_2+\lambda_3+\lambda_4=0\}\\ &\quad= (\#E-1)^3 - ((\#E-1)^2 - (\#E-1)).\end{aligned}$$

*Proof of Lemma* 2.2.13. View the set in question as the subset of $(E^\times)^3$ where $\lambda_1+\lambda_2+\lambda_3 \neq 0$ (solve for $\lambda_4$). Its complement in $(E^\times)^3$ is the subset of $(E^\times)^2$ where $\lambda_1+\lambda_2 \neq 0$ (solve for $\lambda_3$). The complement in $(E^\times)^2$ of this last set is the set of pairs $(\lambda, -\lambda)$. □

So now we have the identity

$$\begin{aligned}&(\#E)^{2n+4}\, T_4^{|}(E, \pi^*\mathcal{F}_d)\\ &\quad= 3(\#X(E))(\#X(E)-1)(\#E-1)^2\\ &\quad+ (\#X(E))((\#E-1)^3 - ((\#E-1)^2 - (\#E-1))).\end{aligned}$$

Dividing through, we get

$$
\begin{aligned}
&T_4(E, \pi^*\mathcal{F}_d)\\
&\quad = 3(\#X(E)/(\#E)^{n+1})(\#X(E)/(\#E)^{n+1} - 1/(\#E)^{n+1})(1 - 1/\#E)^2\\
&\quad + (\#X(E)/(\#E)^{n+1})((\#E - 1)^3 - ((\#E - 1)^2 - (\#E - 1)))/(\#E)^{n+3}.
\end{aligned}
$$

By Lang-Weil, we have

$$|(\#X(E)/(\#E)^{n+1} - 1| = O((\#E)^{-1/2}).$$

So the first term is $3 + O((\#E)^{-1/2})$. If $n = 0$, the second term is $1 + O((\#E)^{-1/2})$, while if $n \geq 1$ the second term is $O((\#E)^{-1})$. This concludes the proof of Theorem 2.2.7, and, with it, the proofs of Theorems 2.2.5 and 2.2.4.

2.2.14. We now give a supplement to Theorem 2.2.4, by combining our results with those of Deligne [De-Weil II, 4.4.1, 4.4.2$^s$, and 4.4.9]. This supplement will itself be supplemented in 2.3.6.

THEOREM 2.2.15 (SUPPLEMENT TO THEOREM 2.2.4). *Suppose that $n \geq 2$ is even, and that $d \geq 3$.*

(1) *If $N_d$ is 1, 3, 4, or 5, or if $N_d \geq 9$, then the geometric monodromy group $G_d$ of the lisse sheaf $\mathcal{F}_d$ is the full orthogonal group $O(N_d)$.*

(2) *If $N_d$ is 6, 7, or 8, then $G_d$ is either the full orthogonal group $O(N_d)$, or $G_d$ is the Weyl group of the root system $E_\alpha$, $\alpha := N_d$, in its standard $N_d$-dimensional representation as a Weyl group.*

(3) *If $N_d = 2$, then $G_d$ is the symmetric group $S_3$ in the representation* $\mathrm{Aug}_2$.

*Proof.* According to [De-Weil II, 4.4.1, 4.4.2$^s$, and 4.4.9], if $N_d \geq 1$, $G_d$ is either the full orthogonal group $O(N_d)$, or it is a finite reflection group. Moreover, the only finite reflection groups that arise are the Weyl groups of root systems of type $A_\alpha$ for $\alpha \geq 1$, $D_\alpha$ for $\alpha \geq 4$, or $E_\alpha$ for $\alpha = 6$, 7, or 8, in their standard $\alpha$-dimensional representations.

We have shown (Theorem 2.2.5) that for any $d \geq 3$, we have $M_4(G_d, V_d) \leq 3$. Suppose first that $G_d$ is finite, and that $N_d \geq 3$.

We cannot have the Weyl group of $A_\alpha$ for any $\alpha \geq 3$, in its standard representation, i.e., we cannnot have he group $S_{\alpha+1}$ in the representation $\mathrm{Aug}_\alpha$, because $M_4(S_{\alpha+1}, \mathrm{Aug}_\alpha) > 3$ for $\alpha \geq 3$. Indeed, for $\alpha \geq 4$ this is proven in Lemma 1.4.5, and for $\alpha = 3$ it is an elementary calculation we leave to the reader (or the reader can observe that $A_3 = D_3$, and see the discussion of $D_\alpha$ just below).

We can rule out having the Weyl group of $D_\alpha$ for any $\alpha \geq 3$, in its standard representation, as follows. By Theorem 1.3.2, $G_d$ is primitive. But the standard representation of the Weyl group of $D_\alpha$ is induced (in the Bourbaki notations

[Bour-L6, Planche IV, page 257], the lines spanned by the $\varepsilon_j$ are permuted among themselves).

So the only surviving finite group cases with $N_d \geq 3$ are the Weyl groups of $E_6$, $E_7$, and $E_8$ in their standard representations.

If $N_d = 2$, then $G_d$ must be finite, because it is a semisimple subgroup of $O(2)$. The only possibility is the Weyl group of $A_2$, i.e., $S_3$ in the representation $\mathrm{Aug}_2$.

If $N_d = 1$, then $O(1) = \{\pm 1\} = S_2$ in $\mathrm{Aug}_1$, so there is only one possibility. □

*Remark* 2.2.16. The Weyl groups of type $E$ in their standard Weyl group representations all have fourth moment 3. The Weyl group of $E_6$ occurs as the monodromy group attached to the universal family of smooth cubic surfaces in $\mathbb{P}^3$. (Since a smooth cubic surface has middle Betti number 7, and all its cohomology is algebraic, we have a case with $d = 3$, $N_d = 6$, and $G_d$ finite, so necessarily the Weyl group of $E_6$, cf. also [Beau].) We do not know if the Weyl groups of $E_7$ or of $E_8$ can occur as the monodromy group of the universal family of smooth hypersurface sections of degree $d \geq 3$ of some projective smooth $X$. (These groups certainly occur as the monodromy of suitable families of del Pezzo surfaces, but those families are not of the required form.)

*Remark* 2.2.17. In Theorems 2.2.4 and 2.2.15, the hypothesis that $d$ be at least 3 is absolutely essential. Indeed, fix an even integer $n \geq 0$, take for $X$ a smooth quadric hypersurface in $\mathbb{P}^{n+2}$, and consider the universal family of smooth, degree $d = 2$ hypersurface sections of $X$. Each member of this family is a smooth complete intersection of multi-degree (2, 2) in $\mathbb{P}^{n+2}$, so has middle betti number $n + 4$, and all cohomology algebraic. This family has $N_d = n + 3$, and its finite $G_d$ is the Weyl group of $D_{n+3}$. (Indeed, if $n = 0$ the two possibilities coincide. If $n \geq 2$, the only other possibility is $S_{n+4}$ in $\mathrm{Aug}_{n+3}$, or, if $n = 4$, the Weyl group of $E_7$ in its standard Weyl group representation, both of which are primitive. But by [Reid], cf. [Beau, page 16], the monodromy for the universal family of smooth complete intersections of multi-degree (2, 2) in $\mathbb{P}^{n+2}$ is the Weyl group of $D_{n+3}$. So our $G_d$ is a subgroup of the Weyl group of $D_{n+3}$. In particular, our $G_d$ is imprimitive.)

### 2.3. Higher moments

2.3.1. The same ideas used in proving Theorem 2.2.5 allow one to prove the following estimate for higher moments.

THEOREM 2.3.2. *Suppose that $n \geq 1$ and $d \geq 3$. For any integer $b \geq 1$ with $2b \leq d + 1$, we have the estimate*

$$M_{2b}(G_d, V_d) \leq (2b)!! := \Pi_{j=1 \text{ to } b}(2j - 1).$$

*Proof.* We proceed as in the proof of Theorem 2.2.7. We define the mock moment $T_{2b}(E, \pi^*\mathcal{F}_d)$ by

$$T_{2b}(E, \pi^*\mathcal{F}_d) := (\#E)^{-\dim(\mathrm{Hyp}_d)-bn} \Sigma_{H \text{ in } \mathrm{Hyp}_d(E)} |\tau(E, H)|^{2b}$$
$$= (\#E)^{-\dim(\mathrm{Hyp}_d)-bn} \Sigma_{H \text{ in } \mathrm{Hyp}_d(E)} (\#(X \cap (H = 0))(E) - \#X(E)/\#E)^{2b}.$$

It suffices to show that

$$|T_{2b}(E, \pi^*\mathcal{F}_d) - (2b)!!| = O((\#E)^{-1/2}).$$

Exactly as in the discussion of $T_4$, we find for $T_{2b}$ the identity

$$(\#E)^{\dim(\mathrm{Hyp}_d)+bn+2b} \ T_{2b}(E, \pi^*\mathcal{F}_d)$$
$$= \Sigma_{H \text{ in } \mathrm{Hyp}_d(E)} \ (\Sigma_{x \text{ in } X(E)} \ \Sigma_{\lambda \text{ in } E^\times} \ \psi(\lambda H(\tilde{x})))^{2b}.$$

We next open the inner sum and interchange orders of summation, to get

$$= \Sigma_{(x_i) \text{ in } X(E)^{2b}} \ \Sigma_{(\lambda_i) \text{ in } (E^\times)^{2b}} \ \Sigma_{H \text{ in } \mathrm{Hyp}_d(E)} \ \psi(\Sigma_{i=1 \text{ to } 2b} \ \lambda_i H(\tilde{x}_i)).$$

We next break up this sum according to the coincidence pattern of the $2b$ not necessarily distinct points $x_1, \ldots, x_{2b}$ in $X(E)$.

The coincidence pattern among the $x_i$ gives a partition $\mathcal{P}$ of the set $\{1, 2, \ldots, 2b\}$ into $\#\mathcal{P}$ disjoint nonempty subsets $S_\alpha : x_i = x_j$ if and only if $i$ and $j$ lie in the same $S_\alpha$.

Fix a point $(x_i)$ in $X(E)^{2b}$ with partition $\mathcal{P}$. Exactly as in the proof of Theorem 2.2.7, the innermost sum vanishes unless, for each $S_\alpha$ in $\mathcal{P}$, we have $\Sigma_{i \text{ in } S_\alpha} \lambda_i = 0$, in which case the innermost sum is equal to $(\#E)^{\dim(\mathrm{Hyp}_d)}$. So the inner double sum is equal to

$$(\#E)^{\dim(\mathrm{Hyp}_d)} \Pi_{\alpha \text{ in } \mathcal{P}} \#\{(\lambda_i)_{i \text{ in} S_\alpha} \text{with } \lambda_i \text{ in } E^\times \text{and } \Sigma_{i \text{ in } S_\alpha} \lambda_i = 0\}.$$

This visibly vanishes if some $S_\alpha$ is a singleton. More generally, consider the sequence of integer polynomials $P_r(X)$, $r \geq 1$, defined inductively by

$$P_1(X) = 0,$$
$$P_r(X) = X^{r-1} - P_{r-1}(X),$$

i.e.,

$$P_r(X) = X^{r-1} - X^{r-2} + X^{r-3} \ldots + (-1)^{r-2} X.$$

We have the elementary identity

$$\#\{(\lambda_i)_{i \text{ in } S_\alpha} \text{ with } \lambda_i \text{ in } E^\times \text{ and } \Sigma_{i, \text{in } S_\alpha} \lambda_i = 0\} = P_{\#S_\alpha}(\#E).$$

So the innermost double sum is

$$(\#E)^{\dim(\mathrm{Hyp}_d)} \Pi_{\alpha \text{ in } \mathcal{P}} P_{\#S_\alpha}(\#E).$$

This vanishes if any $S_\alpha$ is a singleton, otherwise it is given by a polynomial in $\#E$ of the form

$$(\#E)^{\dim(\mathrm{Hyp}_d)} \Pi_{\alpha \text{ in } \mathcal{P}} (\#E)^{\#S_\alpha - 1} + \text{lower terms}$$
$$= (\#E)^{\dim(\mathrm{Hyp}_d)+2b-\#\mathcal{P}} + \text{lower terms.}$$

The number of points $(x_i)$ in $X(E)^{2b}$ with given partition $\mathcal{P}$ is

$$\Pi_{j=0 \text{ to } \#\mathcal{P}-1}(\# X(E) - j) = \# X(E)^{\#\mathcal{P}} + \text{lower terms}$$
$$= (\# E)^{(n+1)\#\mathcal{P}} + O((\# E)^{(n+1)\#\mathcal{P}-1/2}).$$

As we have seen above, partitions with a singleton do not contribute.

For each partition $\mathcal{P}$ without singletons, the total contribution of all points with that coincidence pattern is thus the product

$$((\#E)^{\dim(\mathrm{Hyp}_d)+2b-\#\mathcal{P}} + \text{lower terms}) \times ((\#E)^{(n+1)\#\mathcal{P}} + O((\#E)^{(n+1)\#\mathcal{P}-1/2}))$$
$$= (\#E)^{\dim(\mathrm{Hyp}_d)+2b+n\#\mathcal{P}}(1 + +O(\#E)^{-1/2}).$$

So the terms of biggest size $(\#E)^{\dim(\mathrm{Hyp}_d)+2b+nb}$ come from those $\mathcal{P}$ without singletons having exact b members, and there are exactly (2b)!! such partitions. □

2.3.3. The relevance of Theorem 2.3.2 is this. Recall (cf. [Weyl, Theorem (2.9.A), page 53 and Theorem (6.1.A), page 167], [ABP, Appendix I, pages 322–326]) that for $O(V)$ or $\mathrm{Sp}(V)$, the invariants in the dual of any even tensor power $V^{\otimes 2b}$, $b \geq 1$, are the $\mathbb{C}$ -span of the "complete contractions," i.e., the linear forms on $V^{\otimes 2b}$ obtained by choosing a partition $\mathcal{P}$ of the index set $\{1, 2, \ldots, 2b\}$ into $b$ disjoint sets $S_\alpha$ of pairs, say $S_\alpha = \{i_\alpha, j_\alpha\}$ with $i_\alpha < j_\alpha$, and mapping

$$v_1 \otimes v_2 \otimes \ldots \otimes v_{2b} \to \Pi_{\alpha \text{ in } \mathcal{P}} < v_{i_\alpha}, v_{j_\alpha} >.$$

There are (2b)!! such complete contractions. If $\dim(V) \geq 2b$, they are linearly independent (cf. [Weyl, section 5 of Chapter $V$, pages 147–149]). So for any $N \geq 2b$, we have

$$M_{2b}(O(N), \text{ std}) = (2b)!!,$$

and for any even $N \geq 2b$, we have

$$M_{2b}(\mathrm{Sp}(N), \text{ std}) = (2b)!!,$$

(cf. [Larsen-Normal], [Dia-Sha]).

COROLLARY 2.3.4. *Suppose $n \geq 1$, and $d \geq 3$. For each $b \geq 1$ with*

$$2b \leq \mathrm{Max}(N_d, d+1),$$

*we have the equality*

$$M_{2b}(G_d, V_d) = (2b)!!.$$

*Proof.* Suppose $n$ is odd. Since $G_d$ is a subgroup of $\mathrm{Sp}(N_d) = \mathrm{Sp}(V_d)$, we have the $a$ priori inequality

$$M_{2b}(G_d, V_d) \geq M_{2b}(\mathrm{Sp}(N_d), \mathrm{std}).$$

If $2b \leq N_d$, we have

$$M_{2b}(\mathrm{Sp}(N), \mathrm{std}) = (2b)!!,$$

as explained in (2.3.3) above. So we find

$$M_{2b}(G_d, V_d) \geq (2b)!!.$$

If in addition $d \geq 3$ and $d + 1 \geq 2b$, we have the reverse inequality from Theorem 2.3.2. For the proof in the case of even $n$, simply replace $\mathrm{Sp}(N_d)$ by $O(N_d)$ in the above argument. □

2.3.5. We now use these estimates for higher moments to eliminate more possibilities of finite monodromy in our universal families.

THEOREM 2.3.6 (SUPPLEMENT TO THEOREM 2.2.15). *Suppose that $n \geq 2$ is even, that $d \geq 5$, and that $N_d \geq 3$. If $N_d \neq 8$, or if $d \geq 7$, then the geometric monodromy group $G_d$ of the lisse sheaf $\mathcal{F}_d$ is the full orthogonal group $O(N_d)$.*

*Proof.* Unless $N_d$ is 6, 7, or 8, the desired conclusion is given by 2.2.15.

If $N_d$ is 6, then $G_d$ is either $O(6)$ or it is $W(E_6)$, the Weyl group of $E_6$, in its standard reflection representation $\mathrm{std}_6$. According to the the computer program GAP [GAP], the sixth moment of $W(E_6)$ in $\mathrm{std}_6$ is given by

$$M_6(W(E_6), \mathrm{std}_6) = 16.$$

But if $d \geq 5$, then by 2.3.2, we have $M_6(G_d, V_d) \leq 6!! = 15$. So we cannot have $W(E_6)$ if $d \geq 5$.

If $N_d$ is 7, then $G_d$ is either $O(7)$ or it is $W(E_7)$, the Weyl group of $E_7$, in its standard reflection representation $\mathrm{std}_7$. According to GAP [GAP], the sixth moment of $W(E_7)$ in $\mathrm{std}_7$ is given by

$$M_6(W(E_7), \mathrm{std}_7) = 16.$$

But if $d \geq 5$, then by 2.3.2, we have $M_6(G_d, V_d) \leq 6!! = 15$. So we cannot have $W(E_7)$ if $d \geq 5$.

If $N_d = 8$, then $G_d$ is either $O(8)$ or it is $W(E_8)$, the Weyl group of $E_8$, in its standard reflection representation $\mathrm{std}_8$. According to GAP [GAP], the eighth moment of $W(E_8)$ in $\mathrm{std}_8$ is given by

$$M_8(W(E_8),\ \mathrm{std}_8) = 106.$$

But if $d \geq 7$, then by 2.3.2 we have $M_8(G_d, V_d) \leq 8!! = 105$. So we cannot have $W(E_8)$ if $d \geq 7$. □

### 2.4. Remarks on Theorem 2.2.4

2.4.1. We have stated Theorem 2.2.4 in terms of the universal family of smooth hypersurface sections of degree $d$. It results from Bertini's theorem [Ka-ACT, 3.11.1] that we also get the same $G_d$ for any sufficiently general Lefschetz pencil of hypersurface sections of degree $d$.

2.4.2. We have given a diophantine proof of Theorem 2.2.4, based on having a finite ground field. It follows, by standard spreading out techniques, that the same theorem is valid, for either the universal family of smooth hypersurface sections of degree $d$, or for a sufficiently general Lefschetz pencil thereof, over any field $k$ in which $\ell$ is invertible. When $k$ is $\mathbb{C}$, we have integral cohomology theory

$$X \mapsto H^*(X(\mathbb{C})^{\mathrm{an}}, \mathbb{Z}),$$

so $\mathcal{F}_d$ has a natural $\mathbb{Z}$-form, and we can speak of the integral monodromy group. In some cases, this finer invariant is known, cf. [Beau].

2.4.3. In the case $n = 0$, if we take $X$ to be $\mathbb{P}^1$, then $G_d$ is a subgroup of the symmetric group $S_d$, and $V_d$ is just the representation $\mathrm{Aug}_{d-1}$. (Of course, $G_d$ is equal to $S_d$, thanks to Abel, but we will not use this fact here, cf. 2.4.4 just below.) Since we have proven that

$$M_4(G_d, V_d = \mathrm{Aug}_{d-1}) \leq 4,$$

it follows that for the larger group $S_d$ we have

$$M_4(S_d, \mathrm{Aug}_{d-1}) \leq 4.$$

On the other hand, we have already proven (1.4.5) that

$$M_4(S_d, \mathrm{Aug}_{d-1}) > 3 \text{ for } d \geq 5.$$

Since in any case the moments are integers, we have

$$M_4(S_d, \mathrm{Aug}_{d-1}) = 4 \text{ for } d \geq 5.$$

(One can check by hand that $M_4(S_4, \mathrm{Aug}_3) = 4$, but that $M_4(S_3, \mathrm{Aug}_2) = 3$.)

2.4.4. In the case $n = 0$, $X \subset \mathbb{P}$ any smooth, geometrically connected, projective curve, we can see that $G_d$, the geometric monodromy group of $\mathcal{F}_d$, is the full symmetric group $S_{N_d+1}$ as follows. Since $G_d$ is a priori a subgroup of $S_{N_d+1}$, it suffices to exhibit a pullback of $\mathcal{F}_d$ whose geometric monodromy group is $S_{N_d+1}$. Any Lefschetz pencil of degree $d$ hypersurface sections on $X$ will do this. Indeed, such a pencil gives a finite flat map $f : X \to \mathbb{P}^1$ which is finite etale of degree

$$\deg(f) = \deg(\mathcal{O}_X(d)) = d \times \deg(X) = 1 + N_d$$

over a dense open set $\mathbb{P}^1 - S$, inclusion denoted

$$j : \mathbb{P}^1 - S \to \mathbb{P}^1,$$

such that for each geometric point $s$ in $S$, the geometric fibre $f^{-1}(s)$ consists of $\deg(f) - 1$ distinct points. The pullback to $\mathbb{P}^1 - S$ of the sheaf $\mathcal{F}_d$ is $j^*(f_*\overline{\mathbb{Q}}_\ell/\overline{\mathbb{Q}}_\ell)$. We must show that $j^*(f_*\overline{\mathbb{Q}}_\ell)$ has geometric monodromy group $S_{\deg(f)}$. From the commutative diagram

$$\begin{array}{ccccc} & & & k & \\ & X - f^{-1}(S) & & \subset X & \\ \tilde{f} & \downarrow & & \downarrow & f \\ & \mathbb{P}^1 - S & & \subset \mathbb{P}^1 & \\ & & & j & \end{array}$$

we see that $f_*\overline{\mathbb{Q}}_\ell = f_*k_*\overline{\mathbb{Q}}_\ell = j_*\tilde{f}_*\overline{\mathbb{Q}}_\ell = j_*j^*f_*\overline{\mathbb{Q}}_\ell$. From the equality $f_*\overline{\mathbb{Q}}_\ell = j_*j^*f_*\overline{\mathbb{Q}}_\ell$, we see that the local monodromy of $j^*(f_*\overline{\mathbb{Q}}_\ell)$ at each point of $S$ has a fixed space of codimension one, so is a reflection. The monodromy group of $j^*(f_*\overline{\mathbb{Q}}_\ell)$ is a subgroup of $S_{\deg(f)}$ which is transitive (the total space $X - f^{-1}(S)$ is geometrically connected) and generated by reflections (all the conjugates of the local monodromies at all the points of $S$), hence is the whole group $S_{\deg(f)}$.

### 2.5. A $p$-adic approach to ruling out finite monodromy for universal families of hypersurface sections

2.5.1. In the case of odd fibre dimension $n$, we know [De-Weil II, 4.4.1] that any Lefschetz pencil has monodromy group which is Zariski dense in the full symplectic group. The moment technique gives a variant proof, valid for the universal family (and then by Bertini for any sufficiently general Lefschtz pencil) of hypersurface sections of degree $d \geq 3$ such that $N_d \geq 4$. Indeed, the fourth moment is 3, so $G_d$ is either $\mathrm{Sp}(N_d)$ or it is finite. But $G_d$ cannot be finite, because in odd fibre dimension the local monodromies in a Lefschetz pencil are unipotent pseudoreflections (and so of infinite order).

2.5.2. In our discussion so far, we have made essential use of the Picard-Lefschetz formula [SGA 7, Exposé XV, 3.4], to know that $G_d$ contains a reflection in the case of even fibre dimension $n$, and, a unipotent pseudoreflection in the case of odd fibre dimension.

2.5.3. Suppose we did not know the Picard-Lefschetz formula, but did know all the results of [De-Weil II], an admittedly unlikely but nonetheless logically possible situation. In that case, a result of Koblitz [Kob, Lemma 4, page 132, and Theorem 1, page 139] leads to a $p$-adic proof that, given $X/k$ as above of dimension $n+1 \geq 2$, then for all $d$ sufficiently large, the group $G_d$ is not finite. Once $G_d$ is not finite for a given $d \geq 3$ with $N_d \geq 3$, we know from Larsen's Alternative that $G_d$ is $\mathrm{Sp}(N_d)$ if n is odd, and that $G_d$ is either $SO(N_d)$ or $O(N_d)$ if $n$ is even. We do not know how to prove, in the case of even fibre dimension, that the $SO$ case cannot occur, without appealing to the Picard-Lefschetz formula!

2.5.4. We now explain the $p$-adic proof that if $X/k$ as above has dimension $n+1 \geq 2$, then for $d$ sufficiently large, the group $G_d$ is not finite.

2.5.4.1. We know that $G_d$ is an irreducible subgroup of $GL(V_d)$. If $G_d$ is finite, then any element $A$ of the ambient $GL(V_d)$ which normalizes $G_d$ has some power a scalar. For the group $\mathrm{Aut}(G_d)$ is itself finite, so a power of $A$, acting by conjugation on $G_d$, will act trivially, i.e., a power of $A$ will commute with $G_d$, which, $G_d$ being irreducible, makes that power a scalar. This applies to the image in $GL(V_d)$ of any Frobenius element in $\pi_1(\mathrm{Good}_X\mathrm{PHyp}_d)$. So if $G_d$ is finite, then for any finite extension field $E/k$, and any $H$ in $\mathrm{Good}_X\mathrm{PHyp}_d(\mathrm{E})$, we find that a power of $\mathrm{Frob}_E$ acting on $Ev^n((X \otimes_k \bar{k}) \cap H, \overline{\mathbb{Q}}_\ell)$ is a scalar. Moreover, we know that $\mathrm{Frob}_E/(\#E)^{n/2}$ lies in either Sp or $O$, so has determinant $\pm 1$. Since $\mathrm{Frob}_E/(\#E)^{n/2}$ has a power which is a scalar, that scalar must be a root of unity. Thus every eigenvalue of $\mathrm{Frob}_E$ acting on $Ev^n((X \otimes_k \bar{k}) \cap H, \overline{\mathbb{Q}}_\ell)$ is of the form (a root of unity) $\times$ $(\#E)^{n/2}$, so in particular of the form $p \times$ (an algebraic integer).

2.5.4.2. On the other hand, we know that the characteristic polynomial of $\mathrm{Frob}_E$ on $H^i((X \otimes_k \bar{k}) \cap H, \overline{\mathbb{Q}}_\ell)$ or on $H^i(X \otimes_k \bar{k}, \overline{\mathbb{Q}}_\ell)$ has $\mathbb{Z}$-coefficients. By the hard Lefschetz theorem on $X$, for $i > n$, all eigenvalues of $\mathrm{Frob}_E$ on $H^i(X \otimes_k \bar{k}) \cap H, \overline{\mathbb{Q}}_\ell)$ are also of the form $p \times$ (an algebraic integer). So we get a congruence mod $p$ for the zeta function of $X \cap H/E$, viewed as an element of $1 + T\mathbb{Z}[[T]]$:

$$\mathrm{Zeta}(X \cap H/E, T)$$
$$\equiv \Pi_{i=0 \text{ to } n} \det(1 - T\,\mathrm{Frob}_E | H^i((X \otimes_k \bar{k}) \cap H, \overline{\mathbb{Q}}_\ell)^{(-1)^{i+1}}.$$

Using the weak Lefschetz theorem, this last product is equal to the product

$$(\Pi_{i=0 \text{ to } n} \det(1 - T\,\mathrm{Frob}_E | H^i(X \otimes_k \bar{k}, \overline{\mathbb{Q}}_\ell)^{(-1)^{i+1}})$$
$$\times \det(1 - T\,\mathrm{Frob}_E | Ev^n(X \otimes_k \bar{k}) \cap H, \overline{\mathbb{Q}}_\ell)^{(-1)^{n+1}}.$$

If $G_d$ is finite, then the second term is 1 mod $p$. So we get a congruence formula for $\mathrm{Zeta}(X \cap H/E, T)$ which shows that its reduction mod $p$ is a rational function whose degree as a rational function depends only on $X$. Indeed, if we denote by $\sigma_i$

the degree of the reduction mod $p$ of the integer polynomial

$$\det(1 - T\,\mathrm{Frob}_k | H^i(X \otimes_k \bar{k}, \overline{\mathbb{Q}}_\ell),$$

then Zeta $(X \cap H/E, T)$ mod $p$ has degree $\sigma(X) := \Sigma_{i=0\,\mathrm{to}\,n}(-1)^{i+1}\sigma_i$, for every finite extension $E/k$, and every point $H$ in $\mathrm{Good}_X\mathrm{PHyp}_d(E)$.

2.5.4.3. We now explain how this last conclusion leads to a contradiction for large $d$. By the congruence formula [SGA 7, Part II, Exposé XXII, 3.1] for the zeta function, we have the mod $p$ congruence

$$\begin{aligned}&\mathrm{Zeta}\,(X \cap H/E, T)\\ &\quad\equiv \Pi_{i=0\,\mathrm{to}\,n} \det(1 - T\,\mathrm{Frob}_E | H^i(X \cap H, \mathcal{O}_{X\cap H}))^{(-1)^{i+1}}.\end{aligned}$$

For $d$ sufficiently large, the restriction map

$$H^i((X, \mathcal{O}_X) \to H^i(X \cap H, \mathcal{O}_{X\cap H}))$$

is an isomorphism for $i < n$, and is injective for $i = n$ (i.e., for large $d$ we have vanishing of $H^i(X, \mathcal{O}_X(-d))$ for $i \leq n$). So we can factor this mod $p$ product as

$$\begin{aligned}&(\Pi_{i=0\,\mathrm{to}\,n} \det(1 - T\,\mathrm{Frob}_E | H^i(X, \mathcal{O}_X))^{(-1)^{i+1}})\\ &\quad\times \det(1 - T\,\mathrm{Frob}_E | H^n(X \cap H, \mathcal{O}_{X\cap H})/H^n(X, \mathcal{O}_X))^{(-1)^{n+1}}.\end{aligned}$$

The degree of the first factor depends only on $X$. Indeed, if we denote by $\tau_i$ the degree of the mod $p$ polynomial

$$\det(1 - T\,\mathrm{Frob}_k | H^i(X, \mathcal{O}_X)),$$

this degree is $\tau(X) := \Sigma_{i=0\,\mathrm{to}\,n}(-1)^{i+1}\tau_i$. So if $G_d$ is finite, then we conclude that the mod $p$ polynomial

$$\det(1 - T\,\mathrm{Frob}_E | H^n(X \cap H, \mathcal{O}_{X\cap H})/H^n(X, \mathcal{O}_X))$$

has degree $(-1)^n(\sigma(X) - \tau(X))$, for every finite extension $E/k$, and every point $H$ in $\mathrm{Good}_X\mathrm{PHyp}_d(E)$.

2.5.4.4. Thanks to Koblitz [Kob, Lemma 4, page 132, and Theorem 1, page 139], for $d$ sufficiently large, there is a dense open set of $\mathrm{Good}_X\mathrm{PHyp}_d$ on which the degree of the mod $p$ polynomial

$$\det(1 - T\,\mathrm{Frob}_E | H^n(X \cap H, \mathcal{O}_{X\cap H})/H^n(X, \mathcal{O}_X))$$

is constant, say $F(d)$, and $F(d)$ goes to infinity with $d$. More precisely, Koblitz shows that there is a $\mathbb{Q}$-polynomial $P_X(T)$ of degree $n+1$, of the form

$\deg(X)T^{n+1}/(n+1)! +$ lower terms, such that $F(d) \geq P_X(d)$. So for $d$ large enough that the following three conditions hold:

$$d \geq 3,$$
$$H^i(X, \mathcal{O}_X(-d)) = 0 \text{ for } i \leq n,$$
$$F(d) > (-1)^n(\sigma(X) - \tau(X)),$$

$G_d$ is not finite.

---

## REFERENCES


[ABP] M. Atiyah, R. Bott, and V. K. Patodi, On the heat equation and the index theorem, *Invent. Math.* **19** (1973), 279–330.

[Beau] A. Beauville, Le groupe de monodromie des familles universelles d'hypersurfaces et d'intersections complétes *Complex analysis and algebraic geometry, Springer Lecture Notes in Math,* vol. 1194, 1986.

[BBH] F. Beukers, D. Brownawell, and G. Heckman, Siegel Normality, *Ann. Math.* **127** (1988), 279–308.

[Blich-FCG] H. F. Blichfeldt, *Finite Collineation Groups,* Univ. Chicago Press, 1917.

[Bour-L4] N. Bourbaki, *Groupes et algebres de lie,* Masson, Paris, 1981, Ch. 4–6.

[Bour-L8] ———, *Groupes et algebres de lie,* Paris, 1975, Ch. 7–8, Diffusion CCLS.

[CCNPW-Atlas] J. H. Conway, R. T. Curtis, S. P. Norton, R. A. Parker, and R. A. Wilson, *Atlas of finite groups. Maximal subgroups and ordinary characters for simple groups. With computational assistance from J. G. Thackray,* Oxford University Press, 1985.

[C-R-MRT] C. W. Curtis, and I. Reiner, *Methods of Representation Theory with Applications To Finite Groups and Orders,* vol. 1, John Wiley and Sons, New York, 1981.

[De-Weil II] P. Deligne, La conjecture de Weil. II, *Pub. Math. I.H.E.S.* **52** (1981), 313–428.

[Dia-Sha] Persi Diaconis, and Mehrdad Shahshahani, On the eigenvalues of random matrices: Studies in applied probability. *J. Appl. Probab.* **31A** (1994), 49–62.

[GAP] Lehrstuhl D für Mathematik, RWTH Aachen, GAP, computer program, available from http://www-gap.dcs.st-and.ac.uk/~gap

[Ger] P. Gerardin, Weil representations associated to finite fields, *J. Alg.* **46** (1977), 54–101.

[Howe] R. Howe, On the character of Weil's representation, *Trans. A. M. S.* **177** (1973), 287–298.

[Ka-ACT] N. Katz, Affine cohomological transforms, perversity and monodromy, *JAMS* **6:** (1993), 149–222.

[Ka-LFM] ———, $L$-functions and monodromy: Four lectures on Weil. II, *Adv. Math.* **160** (2001), 81–132.

[Ka-SE] ———, *Sommes exponentielles,* rédigé par G. Laumon, *Asterisque* **79** (1980), xx.

[Ka-TLFM] ———, *Twisted L-functions and monodromy, Annals of Math. Study,* vol. 150, Princeton University Press, 2001.

[Ka-La-FGCFT] N. Katz, and S. Lang, Finiteness theorems in geometric classfield theory, with an appendix by Kenneth A. Ribet, *Enseign. Math.* **27: 3–4** (1981), 285–319.

[Kob] Neal Koblitz, $p$-adic variation of the zeta-function over families of varieties defined over finite fields. *Compositio Math.* **31:2** (1975), 119–218.

[Lar-Char] Michael Larsen, A characterization of classical groups by invariant theory, preprint.

[Lar-Normal] ———, The normal distribution as a limit of generalized Sato-Tate measures, preprint.

[Mi] H. H. Mitchell, Determination of all primitive collineation groups in more than four variables which contain homologies, *Amer. J. Math.* **36** (1914), 1–12.

[MPR] R. V. Moody, J. Patera, and D. W. Rand, simpLie, Version 2.1, Macintosh Software for Representations of Simple Lie Algebras, December, 2000, available from http://www.crm.umontreal.ca/~rand/simpLie.html

[Pra] Dipendra Prasad, On the self-dual representations of finite groups of Lie type, *J. Alg.* **210** (1998), 298–310.

[Reid] Miles Reid, The intersection of two or more quadrics, Ph.D. thesis, Cambridge University, 1972.

[SGA] A. Grothendieck et al., *Séminaire de Géométrie Algébrique du Bois-Marie,* SGA 1, SGA 4 Parts I, II, and III, SGA $4^1/_2$, SGA 5, SGA 7 Parts I and II, *Springer Lecture Notes in Math.* 224, 269–270–305, 569, 589, 288–340, 1971 to 1977.

[Wales] D. Wales, Quasiprimitive linear groups with quadratic elements, *J. Alg.* **245** (2001), 584–606.

[Weyl] Hermann Weyl, *The Classical Groups. Their Invariants and Representations.* Princeton University Press, 1939, pp. xii, 302.

[Zal] A. E. Zalesskii, Linear groups, *Russian Math Surveys* **36:5** (1981), 63–128.

## CHAPTER 19

# ON THE HOLOMORPHY OF CERTAIN $L$-FUNCTIONS

By Henry H. Kim and Freydoon Shahidi

---

*To Joseph A. Shalika*

**1. Introduction.** In this note we continue our study of automorphic $L$-functions using the Langlands-Shahidi method. In [Ki-Sh], we proved that symmetric cube $L$-functions of non-monomial cuspidal representations of $GL_2$ are entire using our method.

In this note we study the general case but when $m \geq 2$. Here $m$ is the number of automorphic $L$-functions which appear in the constant term of corresponding Eisenstein series. (See (3.1).) Moreover, we assume that the inducing cuspidal representation of maximal Levi $M$ has at least one supercuspidal component. The second author [Sh1] determined the unitary dual for such places and we can therefore use our method. We prove, under a standard conjecture on the normalized local intertwining operators, that $L(s, \sigma, r_i)$, $i \geq 3$, are all entire, since they come from non-self conjugate maximal parabolic subgroups. We then prove that if the second $L$-function has a pole at $s = 1$, then the first one is entire (Theorem 6.2).

We give several examples. The first one is that of classical groups. For example, let $\sigma$ be a cuspidal representation of $GL_k(\mathbb{A})$ and $\tau$ a generic cuspidal representation of $Sp_{2l}(\mathbb{A})$. Our result then implies that if $\sigma \otimes \tau$ has one supercuspidal component and $L(s, \sigma, \wedge^2)$ has a pole at $s = 1$, then $L(s, \sigma \times \tau)$ is entire. This clearly agrees with the parametrization problem. Here $\mathbb{A}$ is the ring of adeles of our number field $F$.

As our second example, we show that exterior cube $L$-functions for cuspidal representations of $PGL_6(\mathbb{A})$ are entire, if the cuspidal representations of $PGL_6(\mathbb{A})$ each have at least one supercuspidal component. These are degree 20 $L$-functions.

Finally, under our assumption on normalized local intertwining operators (Assumption (A) in Section 3), we prove that the following three $L$-functions are entire if the corresponding cuspidal representations each have one supercuspidal component;

(1) the degree 56 standard $L$-function attached to generic cuspidal representations of the adelic points of the adjoint group of type $E_7$,

(2) the degree 32 $L$-function $L(s, \pi, r)$, where $\pi$ is a generic cuspidal representation of $PSO_{12}(\mathbb{A})$ and $r$ is the half-spin representation of $Spin(12, \mathbb{C})$, and

---

Research of the first author supported in part by NSF grant DMS9988672, DMS9729992 (at IAS), and by Clay Mathematics Institute; research of the second author supported in part by NSF grant DMS9970156 and by Clay Mathematics Institute.

(3) the $L$-function $L(s, \pi, r)$, where $\pi$ is a generic cuspidal representation of $SO_7(\mathbb{A})$ and $r$ is the 14-dimensional representation of $Sp_6(\mathbb{C})$, the so called spherical harmonic representation.

We refer to [B-F-G], [G1-G3], [G-R], [G-R-S], and [G-PS-R] for integral representations for these $L$-functions.

*Acknowledgments.* It is our great pleasure to dedicate this paper to Joseph Shalika on the occasion of his 60th birthday. Being Shalika's first student, the second author is particularly indebted to him, since it was through his mentorship and guidance that he started his career in automorphic forms many years ago.

**2. Notation.** Throughout this paper, we refer to [Sh2] for more details and unexplained notations. Let $F$ be a number field and $\mathbb{A} = \mathbb{A}_F$ be its ring of adeles. Let $\mathbf{G}$ be a quasi-split connected reductive group over $F$ and let $G = \mathbf{G}(\mathbb{A}_F)$. Let $\mathbf{B} = \mathbf{TU}$ be a Borel subgroup, where $\mathbf{T}$ is a maximal torus and $\mathbf{U}$ is the unipotent radical of $\mathbf{B}$. Let $\mathbf{P}$ be a standard parabolic subgroup of $\mathbf{G}$ containing $\mathbf{B}$. Let $\mathbf{P} = \mathbf{MN}$ be a Levi decomposition, where $\mathbf{N} \subset \mathbf{U}$. Assume $\mathbf{T} \subset \mathbf{M}$ and let $\mathbf{A}$ be the split component of the center of $\mathbf{M}$. Hence $\mathbf{A} \subset \mathbf{T}$.

Let $v$ be a place of $F$ and let $F_v$ be the completion of $F$ with respect to $v$. All the groups above may be considered as objects over $F_v$. If $\mathbf{H}$ is one such, we write $H_v = \mathbf{H}(F_v)$.

*L-groups.* Let $\psi(\mathbf{G}) = (X^*, X_*, \Sigma^*, \Sigma_*)$ be the root datum, where $X^*$ and $X_*$ are the character group and the cocharacter group of $\mathbf{T}$ respectively. Moreover, $\Sigma^*$ and $\Sigma_*$ are the roots and the coroots of $\mathbf{T}$ in $\mathbf{U}$, respectively. Define a complex group $\hat{G}$ (connected and reductive) such that $\psi(\hat{G}) = \psi(\mathbf{G})^\vee = (X_*, X^*, \Sigma_*, \Sigma^*)$. Let $W = W(\bar{F}/F)$ be the Weil group. It acts on $\psi(\mathbf{G})$ and thus on $\psi(\hat{G})$ and on $\hat{G}$. Let ${}^LG = \hat{G} \rtimes W$ be the $L$-group of $\mathbf{G}$. Similarly define ${}^LM$ for $\mathbf{M}$. There is a natural definition of ${}^LN$, $L$-group of $\mathbf{N}$, such that ${}^LP = {}^LM\,{}^LN$ is a parabolic subgroup of ${}^LG$, ${}^LT \subset {}^LM$ and ${}^LN \subset {}^LU$ [Bo]. Let ${}^L\mathfrak{n}$ be the Lie algebra of ${}^LN$. Then ${}^LM$ acts by adjoint action on ${}^L\mathfrak{n}$. Denote this action by $r$.

From now on, we will assume $\mathbf{P}$ is maximal. This is equivalent to saying that either $\dim(\mathbf{A}/\mathbf{A} \cap Z(\mathbf{G})) = 1$, or that there exists a unique simple root $\alpha$ of $\mathbf{A}_0 \subset \mathbf{T}$, maximal split torus of $\mathbf{T}$, in $\mathbf{N}$. Let $\rho_{\mathbf{P}}$ be half the sum of roots generating the Lie algebra of $\mathbf{N}$, and let $\tilde{\alpha} = \langle \rho_{\mathbf{P}}, \alpha \rangle^{-1} \rho_{\mathbf{P}}$. We decompose $r = \oplus_{i=1}^m r_i$ to its irreducible constituents on ${}^L\mathfrak{n} = V = \oplus_{i=1}^m V_i$ and index them according to an upper central series of ${}^L\mathfrak{n}$. More precisely, $m$ is equal to the nilpotence class of ${}^LN$ (or ${}^L\mathfrak{n}$), and each $r_i$ corresponds to a factor in the series with $r_m$ corresponding to the action on the center $V_m$ of ${}^L\mathfrak{n}$, and then increasing up on the series. Notice that

$$V_i = \left\{ X_{\beta^\vee} \in {}^L\mathfrak{n} | \langle \tilde{\alpha}, \beta \rangle = i \right\}$$

and therefore the order is that of the increasing eigenvalues of the action of ${}^LA$ on ${}^L\mathfrak{n}$.

**3. Automorphic *L*-functions.** Let $\pi = \otimes_v \pi_v$ be a cuspidal representation of $M = \mathbf{M}(\mathbb{A}_F)$. For almost all $v$, $\pi_v$ is $\mathbf{M}(O_v)$-spherical and will be parametrized by a semisimple conjugacy class $\{t_v\} \subset {}^LM_v$, where ${}^LM_v$ is the $L$-group of $\mathbf{M}$, as a group over $F_v$. Recall that

$$L(s, \pi_v, r_{i,v}) = det\left(1 - r_{i,v}(t_v)q_v^{-s}\right)^{-1}.$$

Here $r_{i,v} : {}^LM_v \longrightarrow {}^LM \longrightarrow Aut(\mathcal{H}(r_i))$. If $S$ is a finite set of places of $F$ such that $\pi_v$ is $\mathbf{M}(O_v)$-spherical for $v \notin S$, we then let

$$L_S(s, \pi, r_i) = \prod_{v \notin S} L(s, \pi_v, r_{i,v}).$$

*Problem.* (1) Define $L(s, \pi_v, r_{i,v})$ and $\epsilon(s, \pi_v, r_{i,v}, \psi_v)$ for $v \in S$;

(2) Let $L(s, \pi, r_i) = \prod_v L(s, \pi_v, r_{i,v})$, and $\epsilon(s, \pi, r_i) = \prod_v \epsilon(s, \pi_v, r_{i,v}, \psi_v)$. Show that each $L(s, \pi, r_i)$ has an analytic continuation to the whole complex plane and satisfies a functional equation $L(s, \pi, r_i) = \epsilon(s, \pi, r_i)L(1 - s, \pi, \tilde{r}_i)$;

(3) Show that $L(s, \pi, r_i)$ has a finite number of poles.

Suppose $\pi$ is globally generic, i.e., has a non-zero Fourier coefficient with respect to a generic character of $\mathbf{U_M}(F)\backslash\mathbf{U_M}(\mathbb{A}_F)$. Then (1) and (2) are solved in [Sh1]. Local factors are defined canonically such that archimedean factors and those who have Iwahori fixed vectors are Artin factors. (3) is solved partially in [Sh2].

We would like to address the problem (3) in some generality whenever one local component $\pi_{v_0}$ is supercuspidal, and show that in some fairly general setting, poles are only at $s = 0, 1$. This has been fairly hard up to now.

*Example.* Let $\mathbf{G} = GL_{n+p}$, $\mathbf{M} = GL_n \times GL_p$ and let $\pi = \pi' \otimes \pi''$, where $\pi'$ (resp. $\pi''$) is a cuspidal representation of $GL_n(\mathbb{A})$ (resp. $GL_p(\mathbb{A})$). Let $t'_v = diag(\alpha'_{1,v}, \ldots, \alpha'_{n,v})$ and $t''_v = diag(\alpha''_{1,v}, \ldots, \alpha''_{p,v})$. We obtain $m = 1$ and

$$L(s, \pi_v, r_1) = L(s, \pi'_v \times \widetilde{\pi''_v}) = \prod_{1 \leq i \leq n, 1 \leq j \leq p} (1 - \alpha'_{i,v}{\alpha''_{j,v}}^{-1} q_v^{-s})^{-1},$$

the Rankin-Selberg product $L$-function. The conditions (1), (2) and (3) are all proved in this case: Local factors are the same as those defined by Rankin-Selberg method [Sh4]. At archimedean places they are given by Langlands' parametrization. At non-archimedean places they are the same as those from the parametrization a lá Harris-Taylor [Ha-Ta] and [He]. Functional equations are proved in [Sh3]. Recall that $\phi_v : W_{F_v} \longrightarrow {}^LM_v \longrightarrow Aut(\mathcal{H}(r_{i,v}))$ gives $r_{i,v} \circ \phi_v$ as a representation of $W_{F_v}$, to which one can attach Artin factors.

Recall the global intertwining operator and its normalization ([Sh1], [Ki1])

$$(3.1)\quad M(s,\pi)f = \prod_{i=1}^{m} \varepsilon(is,\pi,\tilde r_i)^{-1} L(is,\pi,\tilde r_i) L(1+is,\pi,\tilde r_i)^{-1} \otimes_v N(s\tilde\alpha,\pi_v,w)f_v.$$

In this paper, we assume the following [Ki1].

Assumption (A). *The local normalized intertwining operators $N(s\tilde\alpha,\pi_v,w_0)$ are all holomorphic and non-zero for $Re(s)\geq \frac{1}{2}$.*

This assumption was proved for the split classical groups $Sp_{2n}$ and $SO_{2n+1}$ in [Ki2]. Similar arguments work for the split group $SO_{2n}$. We will prove it for certain other cases later.

Next, let $E_\chi(s,\tilde\varphi,e,P)$ be the $\chi$–Fourier coefficient of the Eisenstein series attached to $\pi$ and $P$ as defined in [Sh2, Sh4]. Here $\tilde\varphi$ is the function attached to $f$ coming from a cusp form $\varphi$ on $M$. We record equation (3.4) of [Sh2] as:

Proposition 3.1. *One has*

$$E_\chi(s,\tilde\varphi,e,P) = \prod_{v\in S} W_{v,s}(e)\cdot \prod_{i=1}^{m} L_S(1+is,\pi,\tilde r_i)^{-1},$$

*where $W_{v,s}$ is the Whittaker function attached to $f_v$ by equation (3.2) of [Sh2].*

**4. Non-self conjugate cases.** Let $W=W(\mathbf{A}_0)=N_{\mathbf{G}}(\mathbf{A}_0)/\mathbf{A}_0$, $\mathbf{M}=\mathbf{M}_\theta$. Let $W_{\mathbf{M}}=N_{\mathbf{M}}(\mathbf{A}_0)/\mathbf{A}_0$. Let $\tilde w_l\in W$ (resp. $\tilde w_{l,\theta}\in W_{\mathbf{M}}$) be the longest element in $W$ (resp. $W_{\mathbf{M}}$). Let $\tilde w_0=\tilde w_l\tilde w_{l,\theta}$. Let $\alpha$ be the unique simple root of $\mathbf{A}_0$ in $\mathbf{N}$, i.e., $\Delta-\theta=\{\alpha\}$, where $\Delta$ is the set of simple roots of $\mathbf{A}_0$ in $\mathbf{U}$ and $\theta$ is the set of simple roots in $\mathbf{M}$. Then $P=P_\theta$ is self-conjugate or self-associate if and only if $\tilde w_0(\alpha)=\tilde w_l(\alpha)=-\alpha$ if and only if $\tilde w_0(\theta)=\theta$. Non-self conjugate maximal parabolic subgroups exist only in groups of type $A_n$, $D_n$ ($n$ odd), and $E_6$. Note that if $P$ is non-self conjugate, then in (3.1), $m=1$, except for one case in $E_6$ (the case $E_6-2$ in [Sh2]) for which $m=2$.

Recall Langlands' theory of Eisenstein series.

Proposition 4.1 [Ki1, Proposition 2.1]. *If $w_0(\pi)\not\simeq\pi$, then the global intertwining operator (3.1) is holomorphic for $Re(s)\geq 0$.*

Hence if $P$ is non-self conjugate, together with Assumption (A),

$$(4.1)\qquad \prod_{i=1}^{m}\frac{L(is,\pi,r_i)}{L(1+is,\pi,r_i)}$$

is holomorphic for $Re(s)\geq\frac{1}{2}$. By induction and the functional equation, we have:

THEOREM 4.2 [Ki1]. *Under Assumption (A), $L(s,\pi,r_1)$ is entire whenever* **P** *is not self conjugate.*

*Proof.* As we noted above, except for case $E_6-2$ [Sh2], $m=1$. Hence $L(s,\pi,r_1)L(1+s,\pi,r_1)^{-1}$ is holomorphic for $Re(s)\geq\frac{1}{2}$. Starting with $Re(s)$ large, and using induction, we see that $L(s,\pi,r_1)$ is holomorphic for $Re(s)\geq\frac{1}{2}$ since $L(s,\pi,r_1)$ is in fact holomorphic for $Re(s)$ large. By the functional equation, $L(s,\pi,r_1)$ is entire.

Now, in the case $E_6-2$, $m=2$ and we see that $L(s,\pi,r_1)L(2s,\pi,r_2)$ is holomorphic for $Re(s)\geq\frac{1}{2}$. However $L(s,\pi,r_2)$ is a unitary Hecke $L$-function, and it has no zeros for $Re(s)\geq 1$. Hence $L(s,\pi,r_1)$ is holomorphic for $Re(s)\geq\frac{1}{2}$ and is entire by the functional equation. □

PROPOSITION 4.3. *If* **P** *is non-self conjugate, then $L(s,\pi,r_1)$ has no zeros for $Re(s)\geq 1$, except in the case $E_6-2$. In this case, $L(s,\pi,r_1)$ has no zeros for $Re(s)>1$.*

*Proof.* By Proposition 4.1, the Eisenstein series $E(s,\tilde{\varphi},e,P)$ is holomorphic for $Re(s)\geq 0$ and so is $E_\chi(s,\tilde{\varphi},e,P)$. Note that Whittaker functions and $L$-functions are non-zero. Hence our assertion is immediate from Proposition 3.1 if $m=1$. In the case $E_6-2$, Proposition 3.1 implies that $L(1+s,\pi,r_1)$ $L(1+2s,\pi,r_2)$ has no zeros for $Re(s)\geq 0$. But $L(s,\pi,r_2)$ has no poles for $Re(s)>1$, as it is a Hecke $L$-function attached to a unitary character and hence our assertion follows. □

**5. The general case.** When **G** is exceptional, $m\geq 3$ is possible. In fact, for $\mathbf{G}=E_8$, $m=6$ also happens. But for $i\geq 3$, each $L$-function $L(s,\pi,r_i)$ already appears in one of the non-self conjugate cases and is therefore entire. Moreover, using the holomorphy of the non-constant Fourier coefficients of non-self conjugate cases, i.e., Proposition 3.1, we can remove all $L(s,\pi,r_i)$ for $i\geq 3$ from (4.1). Let us make this more precise. We begin with the main induction step of [Sh1] (cf. Propositions 4.1, 4.2, 5.1, Theorem 3.5 of [Sh1]).

PROPOSITION 5.1. *Let $F$ be a number field (resp., a local field of characteristic zero). Let* **G** *be a quasisplit connected reductive group over $F$. Let $\mathbf{P}=\mathbf{MN}$ be a standard maximal parabolic subgroup of* **G** *with respect to a $F$–Borel subgroup* **B** *as before. Let $\pi$ be a globally generic cuspidal representation of $M=\mathbf{M}(\mathbb{A}_F)$ (irreducible admissible generic representation of $M=\mathbf{M}(F)$, respectively). Let $r=\oplus_{i=1}^m r_i$ be the adjoint action of ${}^LM$ on ${}^L\mathfrak{n}$ as before. Then for each $i$, $2\leq i\leq m$, there exists a quasisplit connected reductive $F$–group $\mathbf{G}_i$, a maximal $F$–parabolic subgroup $\mathbf{P}_i=\mathbf{M}_i\mathbf{N}_i$ of $\mathbf{G}_i$, a globally generic cuspidal representation $\pi'$ of $M_i=\mathbf{M}_i(\mathbb{A}_F)$ (an irreducible admissible generic representation $\pi'$ of*

*$M_i = \mathbf{M}_i(F)$, respectively) such that, if the adjoint action $r'$ of ${}^LM_i$ on ${}^L\mathfrak{n}_i$ decomposes as $r' = \oplus_{j=1}^{m'} r'_j$, then*

$$L(s, \pi, r_i) = L(s, \pi', r'_1) \tag{5.1}$$

*and*

$$\varepsilon(s, \pi, r_i) = \varepsilon(s, \pi', r'_1) \tag{5.2}$$

*($\varepsilon(s, \pi, r_i, \psi_F) = \varepsilon(s, \pi', r'_1, \psi_F)$, respectively). Moreover $m' < m$ and if the data outside $S$ is unramified for $(\mathbf{G}, \mathbf{M}, \pi)$, then the same is true for each $(\mathbf{G}_i, \mathbf{M}_i, \pi')$.*

LEMMA 5.2. *If $i \geq 3$, then $L_S(s, \pi, r_i)$ is holomorphic for $Re(s) > \frac{1}{2}$. If Assumption (A) is satisfied, then the completed L-functions $L(s, \pi, r_i)$ are entire.*

*Proof.* By the above proposition, there exists a quasisplit group $\mathbf{G}_i$, a maximal parabolic subgroup $\mathbf{P}_i = \mathbf{M}_i\mathbf{N}_i$, and a cuspidal representation $\pi'$ of $\mathbf{M}_i(\mathbb{A})$ such that if the adjoint action $r'$ of ${}^LM_i$ on ${}^L\mathfrak{n}_i$ decomposes as $r' = \oplus_{j=1}^{m'} r'_j$, then

$$L_S(s, \pi, r_i) = L_S(s, \pi', r'_1).$$

Note that except for $r_3$ of the case $(E_8 - 1)$ of [Sh3], each $(\mathbf{G}_i, \mathbf{M}_i)$, $3 \leq i \leq m$, may be chosen to have length one with $\mathbf{P}_i$ non-self conjugate in $\mathbf{G}_i$. Then applying the same argument as in the discussion after Theorem 3.12 of [Ki1], we see that $L_S(s, \sigma, r_i)$ is holomorphic for $Re(s) > 0$ for $i \geq 3$, except for $r_3$ of the case $(E_8 - 1)$. (This is an inductive argument on $Re(s)$ using the constant term which is in the same vein as in Lemma 5.7 of [Sh2], but this time using the holomorphy of $M(s, \pi)$ for $Re(s) > 0$ as in [Ki1], instead of finiteness of poles used in [Sh2].) The representation $r_3$ of $(E_8 - 1)$ appears as $r'_1$ in the case $(E_6 - 2)$. In this case we showed (see the remark after Theorem 3.12 in [Ki1]) that $L_S(s, \pi', r'_1)$ is holomorphic for $Re(s) > \frac{1}{2}$. Hence $L_S(s, \pi, r_3)$ is holomorphic for $Re(s) > \frac{1}{2}$.

As for the completed $L$-functions, use the fact that $L(s, \pi, r_i) = L(s, \pi', r'_1)$ and Theorem 4.2. □

PROPOSITION 5.3. *If $i \geq 3$, $L_S(s, \pi, r_i)$ has no zeros for $Re(s) > 1$.*

*Proof.* As in the proof of Lemma 5.2, each $(\mathbf{G}_i, \mathbf{M}_i)$, $3 \leq i \leq m$, may be chosen to have $\mathbf{P}_i$ non-self conjugate in $\mathbf{G}_i$. Hence our result follows from Proposition 4.3. □

In many cases, if $w_0(\pi) \not\simeq \pi$, then $w'_0(\pi') \not\simeq \pi'$ in Proposition 5.1 with $i = 2$. (It is not always true that $w_0(\pi) \not\simeq \pi$ implies $w'_0(\pi') \not\simeq \pi'$. For example, let $\mathbf{G} = GSp_{2n}$, $\mathbf{M} = GL_n \times GL_1$. Let $\sigma$ be a cuspidal representation of $GL_n$ with central character $\omega_\sigma$. Then $w_0(\sigma \otimes \chi) = \tilde{\sigma} \otimes (\omega_\sigma \chi)$. But $\pi' = \sigma$. Hence if $\sigma \simeq \tilde{\sigma}$ and $\omega_\sigma \neq 1$, then $w'_0(\pi') \simeq \pi'$, but $w_0(\sigma \otimes \chi) \not\simeq \sigma \otimes \chi$.) When this is the case, we can prove that under Assumption (A), $L(s,\pi,r_i)$ is entire and has no zeros for $Re(s)\geq 1$ by induction.

**6. Main theorem.** From now on we will assume that $m\geq 2$, $\mathbf{P}$ is self-conjugate, and $w_0(\pi)\simeq\pi$. Suppose $\pi_{v_0}$ is supercuspidal for some finite place $v_0$.

Consider $I(s\tilde{\alpha},\pi_{v_0})$. Since $w_0(\pi_{v_0})\simeq\pi_{v_0}$, there exists a unique $s_0\geq 0$ such that $I(s,\pi_{v_0})$ is reducible at $s=s_0$ and irreducible at all other points. It is proved in [Sh1] that $s_0\in\{0,\frac{1}{2},1\}$;

(1) $s_0=0$. In this case, $I(s\tilde{\alpha},\pi_{v_0})$ is always irreducible and never unitary for $s>0$.

(2) $s_0=\frac{1}{2}$. This happens if and only if $L(s,\pi_{v_0},r_2)$ has a pole at $s=0$. In this case, $I(s\tilde{\alpha},\pi_{v_0})$ is always irreducible and never unitary for $s>\frac{1}{2}$.

(3) $s_0=1$. This happens if and only if $L(s,\pi_{v_0},r_1)$ has a pole at $s=0$. In this case, $I(s\tilde{\alpha},\pi_{v_0})$ is always irreducible and never unitary for $s>1$.

LEMMA 6.1. *Suppose $L(s,\pi_{v_0},r_1)$ does not have a pole at $s=0$. Then $L(s,\pi,r_1)$ is holomorphic, except possibly at $s=\frac{1}{2}$.*

*Proof.* For every pole $s_0$ of $L(s,\pi,r_1)$, we obtain a unitarizable Langlands' quotient for $I(s_0\tilde{\alpha},\pi_{v_0})$. By assumption, we are either in Case (1) or (2) discussed above and therefore $s_0\leq 1/2$. Consequently $L(s,\pi,r_1)$ is holomorphic for $Re(s)>\frac{1}{2}$. We now apply the functional equation. □

THEOREM 6.2. *Assume $m\geq 2$. Suppose $\pi_{v_0}$ is supercuspidal for some $v_0$. If $L(s,\pi,r_2)$ has a pole at $s=1$, then $L(s,\pi,r_1)$ is entire.*

In order to prove Theorem 6.2, we first need the following lemma.

LEMMA 6.3. *Suppose $L(s,\pi,r_2)$ has a pole at $s=1$. Then $L(s,\pi_{v_0},r_2)$ has a pole at $s=0$ and therefore $L(s,\pi_{v_0},r_1)$ is holomorphic at $s=0$.*

*Proof.* By induction, there exists $(\mathbf{G}',\mathbf{M}')$ and $\pi'$ a cuspidal representation of $\mathbf{M}'(\mathbb{A}_F)$ such that

$$L(s,\pi,r_2)=L(s,\pi',r_1'),\qquad L(s,\pi_{v_0},r_2)=L(s,\pi'_{v_0},r_1').$$

Here note that since $\pi_{v_0}$ is supercuspidal, then so is $\pi'_{v_0}$. Since $L(s,\pi,r_2)$ has a pole at $s=1$, $L(s,\pi',r_1')$ has a pole at $s=1$. This means that $I(s\tilde{\alpha}',\pi'_{v_0})$ is reducible at $s=1$. Consequently $L(s,\pi'_{v_0},r_1')$ must have a pole at $s=0$, hence the same is true for $L(s,\pi_{v_0},r_2)$. □

COROLLARY (OF THE PROOF). *If the edge of complementary series for* $I(s\tilde{\alpha}', \pi'_{v_0})$ *is at 1, then the edge for* $I(s\tilde{\alpha}, \pi_{v_0})$ *is at* $1/2$.

*Proof of the Theorem 6.2.* A pole for $L(s, \pi, r_1)$ at $s = \frac{1}{2}$ will give a double pole for Eisenstein series which is not possible. To prove the holomorphy elsewhere, we note that the condition of Lemma 6.1 is now satisfied by Lemma 6.3. □

*Remark.* One can replace supercuspidal $\pi_{v_0}$ by any irreducible unitary representation, for which $\frac{1}{2}$ is the edge of complementary series.

*Remark.* Theorem 6.2 is expected to be true for arbitrary $\pi$. In fact, one expects that the product $L(s, \pi, r_1)L(s, \pi, r_2)$ has at most a simple pole at $s = 1$ for any $\pi$.

## 7. Examples.

**7.1. Classical groups.** Let $\mathbf{G}_{n+p}$ be a quasisplit classical group of rank $n + p$. Let $\mathbf{M} = GL_n \times \mathbf{G}_p$. Then $m = 2$ and $r_1 = \rho_n \otimes \rho_p$, where $\rho_n$ is the standard representation of $GL_n(\mathbb{C})$ and $\rho_p$ is the standard representation of ${}^L G_p$;

$$r_2 = \begin{cases} Sym^2(\rho_n), & \mathbf{G} = \text{odd orthogonal group} \\ \wedge^2(\rho_n), & \mathbf{G} = \text{even orthogonal or symplectic group} \\ \text{Asai or Asai}\otimes\eta, & \mathbf{G} = U(n+p, n+p) \text{ or } U(n+p+1, n+p), \text{ resp.,} \end{cases}$$

where $\eta$ is the corresponding class field character (cf. [Sh2]).

Conjecture 7.1 of [Sh1], which demands the holomorphy of local $L$-functions attached to tempered data for $Re(s) > 0$, is true in these cases [Ca-Sh]. Suppose that $\pi$ contains a supercuspidal component. Then since Assumption (A) is valid for split classical groups [Ki2], we have:

PROPOSITION 7.1.1. *Let $\pi$ be a cuspidal representation of $GL_n(\mathbb{A})$ and assume $\sigma$ is a generic cuspidal representation of $G_r(\mathbb{A})$. Suppose $\pi \otimes \sigma$ has a supercuspidal component.*

(1) *If $\mathbf{G}_r$ = SO(even) or Sp, and $\pi$ comes from SO(odd), i.e., $L(s, \pi, \Lambda^2)$ has a pole at $s = 1$ (cf. [Sh5]), then $L(s, \pi \times \sigma)$ is entire.*

(2) *If $\mathbf{G}_r$ = SO(odd), and $\pi$ comes from SO(even) (cf. [Sh5]), i.e., $L(s, \pi, Sym^2)$ has a pole at $s = 1$, then $L(s, \pi \times \sigma)$ is entire.*

**7.2. Exceptional groups.** (1) Let $\mathbf{G}$ be the adjoint group of type $E_6$. Let $\mathbf{M} = \mathbf{M}_\theta$, where $\theta = \{\alpha_1, \alpha_3, \alpha_4, \alpha_5, \alpha_6\}$. Let $\pi$ be a cuspidal representation of $PGL_6(\mathbb{A})$. Extend $\pi$ to a cuspidal representation of $\mathbf{M}(\mathbb{A})$, trivial on $GL_1$. Then $r_1$ is the third exterior power representation $\wedge^3$ of $SL_6(\mathbb{C})$, i.e., $\delta_3$. Here $dim\ r_1 = 20$, while $r_2$ is the 1-dimensional trivial representation. Note that if $\pi_v$ is an unramified component with Satake parameter $diag(\beta_1, \ldots, \beta_6)$, then

$$L\left(s, \pi_v, \wedge^3\right) = \prod_{1 \leq i < j < k \leq 6} \left(1 - \beta_i \beta_j \beta_k q_v^{-s}\right)^{-1}.$$

Moreover, observe that $L(s, \pi, r_2)$ has a pole at $s = 1$. In this case, Conjecture 7.1 of [Sh1] is valid. We prove:

LEMMA 7.2.1. *Assumption (A) is valid.*

*Proof.* We proceed as in [Ki2, Proposition 3.4]. If $\pi_v$ is tempered, then $N(s, \pi_v, w_0)$ is holomorphic and non-zero for $Re(s) > 0$, since Conjecture 7.1 of [Sh1] is valid.

If $\pi_v$ is non-tempered, we write $I(s, \pi_v)$ as in [Ki1, p. 481], i.e.,

$$I(s, \pi_v) = I(s\tilde{\alpha} + \Lambda_0, \pi_0) = Ind_{M_0(F_v)N_0(F_v)}^{G(F_v)} \pi_0 \otimes q^{\langle s\tilde{\alpha} + H_{P_0}() \rangle},$$

where $\pi_0$ is a tempered representation of $M_0(F_v)$ and $P_0 = M_0 N_0$ is another parabolic subgroup of $G$. We can identify the normalized operator $N(s, \pi_v, w_0)$ with the normalized operator $N(s\tilde{\alpha} + \Lambda_0, \pi_0, \tilde{w}_0)$, which is a product of rank-one operators attached to tempered representations (cf. [Zh, Proposition 1]).

Suppose $\pi_v$ is not of the form $Ind\, \mu|\;|^r \otimes \sigma \otimes \mu|\;|^{-r}$, where $0 < r < \frac{1}{2}$, $\mu$ is a unitary character of $F_v^\times$, and $\sigma$ is a tempered representation of $GL_4(F_v)$. Then all the rank-one operators are operators for a parabolic subgroup whose Levi subgroup has a derived group isomorphic to $SL_k \times SL_l$ inside a group of type $A_{k+l-1}$.

By [M-W2, Proposition I.10] one knows that each rank-one operator for $GL_k \times GL_l$ is holomorphic for $Re(s) > -1$. Hence by identifying roots of $G$ with respect to a parabolic subgroup, with those of $G$ with respect to the maximal torus, it is enough to check $\langle s\tilde{\alpha} + \Lambda_0, \beta^\vee \rangle > -1$ for all positive roots $\beta$ if $Re(s) \geq \frac{1}{2}$.

Here in the notation of Bourbaki, $\tilde{\alpha} = \alpha_1 + 2\alpha_2 + 2\alpha_3 + 3\alpha_4 + 2\alpha_5 + \alpha_6$; $\Lambda_0 = r_1\alpha_1 + (r_1 + r_2)\alpha_3 + (r_1 + r_2 + r_3)\alpha_4 + (r_1 + r_2)\alpha_5 + r_1\alpha_6$, where $\frac{1}{2} > r_1 \geq r_2 \geq r_3 \geq 0$. Hence

$$\begin{aligned} s\tilde{\alpha} + \Lambda_0 &= (s + r_1)\alpha_1 + 2s\alpha_2 + (2s + r_1 + r_2)\alpha_3 + (3s + r_1 + r_2 + r_3)\alpha_4 \\ &\quad + (2s + r_1 + r_2)\alpha_5 + (s + r_1)\alpha_6. \end{aligned}$$

We observe that the least value of $Re(\langle s\tilde{\alpha} + \Lambda_0, \beta^\vee \rangle)$ is $Re(s) - (r_1 + r_2 + r_3)$ which is larger than $-1$, if $Re(s) \geq \frac{1}{2}$. Consequently, $N(s\tilde{\alpha} + \Lambda_0, \pi_0, \tilde{w}_0)$ is holomorphic for $Re(s) \geq \frac{1}{2}$. By Zhang's lemma (cf. [Ki2, Lemma 1.7]), it is non-zero as well.

This completes the proof except for the case of a Levi subgroup of type $A_3$ inside a group of type $D_4$. In this case, $r_2 = r_3 = 0$. Hence $s\tilde{\alpha} + \Lambda_0$ is in the closure of the positive Weyl chamber for $Re(s) \geq \frac{1}{2}$. In this case, $N(s\tilde{\alpha} + \Lambda_0, \pi_0, \tilde{w}_0)$ is holomorphic over the same region [Ki1, Proposition 2.4]. □

The following proposition is now a consequence of Theorem 6.2 and Lemma 7.2.1.

PROPOSITION 7.2.2. *Let $\pi$ be a cuspidal representation of $PGL_6(\mathbb{A})$ which has at least one supercuspidal component. Then the exterior cube L-function $L(s, \pi, \wedge^3)$ is entire.*

(2) Let $E_7^{ad}$ be the adjoint group of type $E_7$ and let $\pi$ be a generic cuspidal representation of $E_7^{ad}(\mathbb{A})$. Let $r_1$ be the degree 56 standard representation of $E_7^{sc}(\mathbb{C})$.

THEOREM 7.2.3. *Assume $\pi$ has a supercuspidal component and assume the validity of Assumption (A) (which holds; for example, if ramified ones are tempered). Then $L(s, \pi, r_1)$ is entire.*

*Proof.* This is case (*xxxii*) in [La]. We take $\mathbf{G} = E_8$ and consider

$$\mathbf{M} = \left(E_7^{sc} \times GL_1\right)/\{\pm 1\}.$$

Take $\pi$ as a cuspidal representation of $E_7^{sc}(\mathbb{A})$, trivial on $\{\pm 1\}$, and extend it to all of $M$, trivially on $GL_1$. Here $m = 2$ and the second $L$-function is the Hecke $L$-function attached to the trivial character which has a pole at $s = 1$. Observe that

$${}^LM = \left(E_7^{sc}(\mathbb{C}) \times GL_1(\mathbb{C})\right)/\{\pm 1\},$$

where $E_7^{sc}$ is the simply connected group of type $E_7$. Apply Theorem 6.2. □

(3) Let $\pi$ be a generic cuspidal representation of $PSO_{12}(\mathbb{A})$. Let $r_1$ be the degree 32 half-spin representation of $Spin(12, \mathbb{C})$.

THEOREM 7.2.4. *Assume $\pi$ has a supercuspidal component and assume the validity of Assumption (A) (e.g., if ramified ones are tempered). Then $L(s, \pi, r_1)$ is entire.*

*Proof.* This is case (*xxvi*) in [La]. We take $\mathbf{G} = E_7^{ad}$ and consider

$$\mathbf{M} = (HS(12) \times GL_1)/\{\pm 1\} = GHS(12).$$

Here $HS(12)$ is the half-spin group $Spin(12)/\{1, z\}$, where $z = H_{\alpha_2}(-1)H_{\alpha_5}(-1)H_{\alpha_7}(-1)$, i.e., the image of $Spin(12) \subset E_7^{sc}$ in $E_7^{ad}$ (cf. [C]). Since $PGHS(12) = PHS(12) \cong PSO(12) = PGSO(12)$, we can extend $\pi$ to all of $M$, trivially on $GL_1$. Here $m = 2$ and the second $L$-function is the Hecke $L$-function attached to the trivial character which has a pole at $s = 1$. Observe that ${}^LM = GSpin(12, \mathbb{C})$. Apply Theorem 6.2. □

(4) Let $\mathbf{G} = F_4$ be the (simply-connected) group of type $F_4$. Consider a generic cuspidal representation $\pi$ of $SO_7(\mathbb{A})$. Let $r$ be the 14-dimensional irreducible representation of $Sp_6(\mathbb{C})$, the so called spherical harmonic representation.

THEOREM 7.2.5. *Assume $\pi$ has a supercuspidal component and assume the validity of Assumption (A) (e.g., if ramified ones are tempered). Then $L(s, \pi, r)$ is entire.*

*Proof.* This is the case *(xviii)* in [La]. Take $\mathbf{G} = F_4$ and consider

$$\mathbf{M} = (Spin(7) \times GL_1)/\{\pm 1\}.$$

Lift $\pi$ to a cuspidal representation of $Spin_7(\mathbb{A})$ and extend it to all of $M$, trivially on $GL_1$. Here $m = 2$ and the second $L$-function is the Hecke $L$-function attached to the trivial character which has a pole at $s = 1$. Apply Theorem 6.2. □

DEPT. OF MATH.
UNIVERSITY OF TORONTO
TORONTO, ONTARIO M5S 3G3, CANADA
*E-mail:* henrykim@math.toronto.edu

DEPT. OF MATH.
PURDUE UNIVERSITY
WEST LAFAYETTE, IN 47906
*E-mail:* shahidi@math.purdue.edu

REFERENCES

[B-F-G] D. Bump, S. Friedberg, and D. Ginzburg, A Rankin–Selbert integral using the automorphic minimal representation of $SO_7$, preprint.

[C] R. Carter, Finite Groups of Lie Type, Conjugacy Classes and Complex Characters, Wiley Classics Library, John Wiley & Sons, 1993.

[Ca-Sh] W. Casselman and F. Shahidi, On irreducibility of standard modules for generic representations, *Ann. Sci. École Norm. Sup.* **31** (1998), 561–589.

[G1] D. Ginzburg, On standard $L$-functions for $E_6$ and $E_7$, *J. Reine Angew. Math.* **465** (1995), 101–131.

[G2] ———, On spin $L$-functions for orthogonal groups, *Duke Math. J.* **77** (1995), 753–780.

[G3] ———, $L$-functions for $SO_n \times GL_k$, *J. Reine Angew. Math.* **405** (1990), 156–180.

[G-R] D. Ginzburg and S. Rallis, The exterior cube $L$-function for GL(6), *Compositio Math.* **123** (2000), 243–272.

[G-R-S] D. Ginzburg, S. Rallis, and D. Soudry, $L$-functions for symplectic groups, *Bull. Soc. Math. France* **126** (1998), 181–244.

[G-PS-R] D. Ginzburg, I. Piatetski–Shapiro, and S. Rallis, $L$-functions for the orthogonal group, *Mem. Amer. Math. Soc.* **128** (1997, no. 611).

[Ha-Ta] M. Harris and R. Taylor, On the geometry and cohomology of some simple Shimura varieties, *Ann. of Math. Stud.,* vol. 151, Princeton Univ. Press, 2001.

[He] G. Henniart, Une Preuve simple des conjectures de Langlands pour $GL(n)$ sur un corps $p$–adique, *Inv. Math.* **139** (2000), 439–455.

[Ki1] H. Kim, Langlands-Shahidi method and poles of automorphic $L$-functions: Application to exterior square $L$-functions, *Can. J. Math.* **51** (1999), 835–849.

[Ki2] ________, Langlands-Shahidi method and poles of automorphic $L$-functions, II, *Israel J. Math* **117** (2000), 261–284.

[Ki-Sh] H. Kim and F. Shahidi, Symmetric cube $L$-functions for $GL_2$ are entire, *Ann. of Math.* **150** (1999), 645–662.

[La] R. P. Langlands, Euler Products, Yale University Press, 1971.

[M-W1] C. Moeglin and J.-L. Waldspurger, Spectral Decomposition and Eisenstein series, une paraphrase de l'Ecriture, Cambridge Tracts in Mathematics, vol. 113 Cambridge University Press, 1995.

[M-W2] ________, Le spectre résiduel de $GL(n)$, *Ann. Scient. Eć. Norm. Sup.* **22** (1989), 605–674.

[Sh1] F. Shahidi, A proof of Langlands conjecture on Plancherel measures; Complementary series for $p$-adic groups, *Annals of Math.* **132** (1990), 273–330.

[Sh2] ________, On the Ramanujan conjecture and finiteness of poles for certain $L$-functions, *Ann. of Math.* **127** (1988), 547–584.

[Sh3] ________, On certain $L$-functions, *Amer. J. Math* **103** (1981), 297–355.

[Sh4] ________, Fourier transforms of intertwining operators and Plancherel measures for $GL(n)$, *Amer. J. of Math.* **106** (1984), 67–111.

[Sh5] ________, Twisted endoscopy and reducibility of induced representations for $p$-adic groups, **66, No. 1** (1992), 1–41.

[Sh6] ________, On multiplicativity of local factors, In: Festschrift in honor of I.I. Piatetski-Shapiro, Part II, *Israel Math. Conf. Proc.* 3 (1990), Weizmann, Jerusalem, 279–289.

[Zh1] Y. Zhang, The holomorphy and nonvanishing of normalized intertwining operators, *Pac. J. Math.* **180** (1997), 386–398.

[Zh2] ________, $L$-packets and reducibilities, *J. Reine Angew. Math.* **510** (1999), 83–102.

## CHAPTER 20

# RICHARDSON VARIETIES IN THE GRASSMANNIAN

By Victor Kreiman and V. Lakshmibai

---

*Dedicated to Professor J. Shalika on his sixtieth birthday*

*Abstract.* The Richardson variety $X_w^v$ is defined to be the intersection of the Schubert variety $X_w$ and the opposite Schubert variety $X^v$. For $X_w^v$ in the Grassmannian, we obtain a standard monomial basis for the homogeneous coordinate ring of $X_w^v$. We use this basis first to prove the vanishing of $H^i(X_w^v, L^m), i > 0, m \geq 0$, where $L$ is the restriction to $X_w^v$ of the ample generator of the Picard group of the Grassmannian; then to determine a basis for the tangent space and a criterion for smoothness for $X_w^v$ at any $T$-fixed point $e_\tau$; and finally to derive a recursive formula for the multiplicity of $X_w^v$ at any $T$-fixed point $e_\tau$. Using the recursive formula, we show that the multiplicity of $X_w^v$ at $e_\tau$ is the product of the multiplicity of $X_w$ at $e_\tau$ and the multiplicity of $X^v$ at $e_\tau$. This result allows us to generalize the Rosenthal-Zelevinsky determinantal formula for multiplicities at $T$-fixed points of Schubert varieties to the case of Richardson varieties.

**0. Introduction.** Let $G$ denote a semisimple, simply connected, algebraic group defined over an algebraically closed field $K$ of arbitrary characteristic. Let us fix a maximal torus $T$ and a Borel subgroup $B$ containing $T$. Let $W$ be the Weyl group ($N(T)/T$, $N(T)$ being the normalizer of $T$). Let $Q$ be a parabolic subgroup of $G$ containing $B$, and $W_Q$, the Weyl group of $Q$. For the action of $G$ on $G/Q$ given by left multiplication, the $T$-fixed points are precisely the cosets $e_w := wQ$ in $G/Q$. For $w \in W/W_Q$, let $X_w$ denote the *Schubert variety* (the Zariski closure of the $B$-orbit $Be_w$ in $G/Q$ through the $T$-fixed point $e_w$), endowed with the canonical structure of a closed, reduced subscheme of $G/Q$. Let $B^-$ denote the Borel subgroup of $G$ opposite to $B$ (it is the unique Borel subgroup of $G$ with the property $B \cap B^- = T$). For $v \in W/W_Q$, let $X^v$ denote the *opposite Schubert variety*, the Zariski closure of the $B^-$-orbit $B^-e_v$ in $G/Q$.

Schubert and opposite Schubert varieties play an important role in the study of the generalized flag variety $G/Q$, especially, the algebraic-geometric and representation-theoretic aspects of $G/Q$. A more general class of subvarieties in $G/Q$ is the class of Richardson varieties; these are varieties of the form $X_w^v := X_w \cap X^v$, the intersection of the Schubert variety $X_w$ with opposite Schubert variety $X^v$. Such varieties were first considered by Richardson in (cf. [22]), who shows that such intersections are reduced and irreducible. Recently, Richardson varieties have shown up in several contexts: such double coset intersections $BwB \cap B^-xB$ first appear in [11], [12], [22], [23]. Very recently, Richardson varieties have also

Research of the second author supported in part by NSF grant DMS 9971295.

appeared in the context of K-theory of flag varieties ([3], [14]). They also show up in the construction of certain degenerations of Schubert varieties (cf. [3]).

In this paper, we present results for Richardson varieties in the Grassmannian variety. Let $G_{d,n}$ be the Grassmannian variety of $d$-dimensional subspaces of $K^n$, and $p : G_{d,n} \hookrightarrow \mathbb{P}^N (= \mathbb{P}(\wedge^d K^n))$, the Plücker embedding (note that $G_{d,n}$ may be identified with $G/P$, $G = SL_n(K)$, $P$ a suitable maximal parabolic subgroup of $G$). Let $X := X_w \cap X^v$ be a Richardson variety in $G_{d,n}$. We first present a standard monomial theory for $X$ (cf. Theorem 3.3.2). Standard monomial theory (SMT) consists in constructing an explicit basis for the homogeneous coordinate ring of $X$. SMT for Schubert varieties was first developed by the second author together with Musili and Seshadri in a series of papers, culminating in [16], where it is established for all classical groups. Further results concerning certain exceptional and Kac-Moody groups led to conjectural formulations of a general SMT, see [17]. These conjectures were then proved by Littelmann, who introduced new combinatorial and algebraic tools: the path model of representations of any Kac-Moody group, and Lusztig's Frobenius map for quantum groups at roots of unity (see [18, 19]); recently, in collaboration with Littelmann (cf. [14]), the second author has extended the results of [19] to Richardson varieties in $G/B$, for any semisimple $G$. Further, in collaboration with Brion (cf. [4]), the second author has also given a purely geometric construction of standard monomial basis for Richardson varieties in $G/B$, for any semisimple $G$; this construction in loc. cit. is done using certain flat family with generic fiber $\cong$ diag $(X^v_w) \subset X^v_w \times X^v_w$, and the special fiber $\cong \cup_{v \leq x \leq w} X^v_x \times X^x_w$.

If one is concerned with just Richardson varieties in the Grassmannian, one could develop a SMT in the same spirit as in [21] using just the Plücker coordinates, and one doesn't need to use any quantum group theory nor does one need the technicalities of [4]. Thus we give a self-contained presentation of SMT for unions of Richardson varieties in the Grassmannian. We should remark that Richardson varieties in the Grassmannian are also studied in [26], where these varieties are called *skew Schubert varieties*, and standard monomial bases for these varieties also appear in loc. cit. (Some discussion of these varieties also appears in [10].) As a consequence of our results for unions of Richardson varieties, we deduce the vanishing of $H^i(X, L^m)$, $i \geq 1$, $m \geq 0$, $L$, being the restriction to $X$ of $\mathcal{O}_{\mathbb{P}^N}(1)$ (cf. Theorem 5.0.6); again, this result may be deduced using the theory of Frobenius-splitting (cf. [20]), while our approach uses just the classical Pieri formula. Using the standard monomial basis, we then determine the tangent space and also the multiplicity at any $T$-fixed point $e_\tau$ on $X$. We first give a recursive formula for the multiplicity of $X$ at $e_\tau$ (cf. Theorem 7.6.2). Using the recursive formula, we derive a formula for the multiplicity of $X$ at $e_\tau$ as being the product of the multiplicities at $e_\tau$ of $X_w$ and $X^v$ (as above, $X = X_w \cap X^v$) (cf. Theorem 7.6.4). Using the product formula, we get a generalization of Rosenthal-Zelevinsky determinantal formula (cf. [24]) for the multiplicities at singular points of Schubert varieties to the case of Richardson varieties (cf. Theorem 7.7.3). It should be mentioned that the multiplicities of Schubert varieties at $T$-fixed points determine their multiplicities at all other points, because of the $B$-action; but this does not extend to Richardson varieties, since Richardson varieties have only a $T$-action. Thus even though, certain smoothness criteria at $T$-fixed points on a Richardson variety are given in Corollaries 6.7.3 and 7.6.5, the problem of the determination of singular loci of Richardson varieties still remains open.

In §1, we present basic generalities on the Grassmannian variety and the Plücker embedding. In §2, we define Schubert varieties, opposite Schubert varieties, and the more general Richardson varieties in the Grassmannian and give some of their basic properties. We then develop a standard monomial theory for a Richardson variety $X_w^v$ in the Grassmannian in §3 and extend this to a standard monomial theory for unions and nonempty intersections of Richardson varieties in the Grassmannian in §4. Using the standard monomial theory, we obtain our main results in the three subsequent sections. In §5, we prove the vanishing of $H^i(X_w^v, L^m)$, $i > 0$, $m \geq 0$, where $L$ is the restriction to $X_w^v$ of the ample generator of the Picard group of the Grassmannian. In §6, we determine a basis for the tangent space and a criterion for smoothness for $X_w^v$ at any $T$-fixed point $e_\tau$. Finally, in §7, we derive several formulas for the multiplicity of $X_w^v$ at any $T$-fixed point $e_\tau$.

*Acknowledgments.* We are thankful to the referee for many valuable comments and suggestions, especially for the alternate proof of Theorem 7.6.4.

**1. The Grassmannian variety $G_{d,n}$.** Let $K$ be the base field, which we assume to be algebraically closed of arbitrary characteristic. Let $d$ be such that $1 \leq d < n$. The *Grassmannian* $G_{d,n}$ is the set of all $d$-dimensional subspaces of $K^n$. Let $U$ be an element of $G_{d,n}$ and $\{a_1, \ldots a_d\}$ a basis of $U$, where each $a_j$ is a vector of the form

$$a_j = \begin{pmatrix} a_{1j} \\ a_{2j} \\ \vdots \\ a_{nj} \end{pmatrix}, \quad \text{with } a_{ij} \in K.$$

Thus, the basis $\{a_1, \cdots, a_d\}$ gives rise to an $n \times d$ matrix $A = (a_{ij})$ of rank $d$, whose columns are the vectors $a_1, \cdots, a_d$.

We have a canonical embedding

$$p : G_{d,n} \hookrightarrow \mathbb{P}(\wedge^d K^n),\ U \mapsto [a_1 \wedge \cdots \wedge a_d]$$

called the *Plücker embedding*. It is well known that $p$ is a closed immersion; thus $G_{d,n}$ acquires the structure of a projective variety. Let

$$I_{d,n} = \{\underline{i} = (i_1, \ldots, i_d) \in \mathbb{N}^d \ : \ 1 \leq i_1 < \cdots < i_d \leq n\}.$$

Then the projective coordinates (*Plücker coordinates*) of points in $\mathbb{P}(\wedge^d K^n)$ may be indexed by $I_{d,n}$; for $\underline{i} \in I_{d,n}$, we shall denote the $\underline{i}$-th component of $p$ by $p_{\underline{i}}$, or

$p_{i_1,\cdots,i_d}$. If a point $U$ in $G_{d,n}$ is represented by the $n \times d$ matrix $A$ as above, then $p_{i_1,\cdots,i_d}(U) = \det(A_{i_1,\ldots,i_d})$, where $A_{i_1,\ldots,i_d}$ denotes the $d \times d$ submatrix whose rows are the rows of $A$ with indices $i_1, \ldots, i_d$, in this order.

For $\underline{i} \in I_{d,n}$ consider the point $e_{\underline{i}}$ of $G_{d,n}$ represented by the $n \times d$ matrix whose entries are all 0, except the ones in the $i_j$-th row and $j$-th column, for each $1 \leq j \leq d$, which are equal to 1. Clearly, for $\underline{i}, \underline{j} \in I_{d,n}$,

$$p_{\underline{i}}(e_{\underline{j}}) = \begin{cases} 1, & \text{if } \underline{i} = \underline{j}; \\ 0, & \text{otherwise.} \end{cases}$$

We define a partial order $\geq$ on $I_{d,n}$ in the following manner: if $\underline{i} = (i_1, \ldots, i_d)$ and $\underline{j} = (j_1, \ldots, j_d)$, then $\underline{i} \geq \underline{j} \Leftrightarrow i_t \geq j_t, \forall t$. The following well known theorem gives the defining relations of $G_{d,n}$ as a closed subvariety of $\mathbb{P}(\wedge^d K^n)$ (cf. [9]; see [13] for details):

THEOREM 1.0.1. *The Grassmannian $G_{d,n} \subset \mathbb{P}(\wedge^d K^n)$ consists of the zeroes in $\mathbb{P}(\wedge^d K^n)$ of quadratic polynomials of the form*

$$p_{\underline{i}} p_{\underline{j}} - \sum \pm p_\alpha p_\beta$$

*for all $\underline{i}, \underline{j} \in I_{d,n}$, $\underline{i}, \underline{j}$ non-comparable, where $\alpha$, $\beta$ run over a certain subset of $I_{d,n}$ such that $\alpha >$ both $\underline{i}$ and $\underline{j}$, and $\beta <$ both $\underline{i}$ and $\underline{j}$.*

**1.1. Identification of $G/P_d$ with $G_{d,n}$.** Let $G = SL_n(K)$. Let $P_d$ be the maximal parabolic subgroup

$$P_d = \left\{ A \in G \;\middle|\; A = \begin{pmatrix} * & * \\ 0_{(n-d)\times d} & * \end{pmatrix} \right\}.$$

For the natural action of $G$ on $\mathbb{P}(\wedge^d K^n)$, we have, the isotropy at $[e_1 \wedge \cdots \wedge e_d]$ is $P_d$ while the orbit through $[e_1 \wedge \cdots \wedge e_d]$ is $G_{d,n}$. Thus we obtain a surjective morphism $\pi : G \to G_{d,n}$, $g \mapsto g \cdot a$, where $a = [e_1 \wedge \cdots \wedge e_d]$. Further, the differential $d\pi_e : \text{Lie}G \to T(G_{d,n})_a$ (= the tangent space to $G_{d,n}$ at $a$) is easily seen to be surjective. Hence we obtain an identification $f_d : G/P_d \cong G_{d,n}$ (cf. [1], Proposition 6.7).

**1.2. Weyl group and Root System.** Let $G$ and $P_d$ be as above. Let $T$ be the subgroup of diagonal matrices in $G$, $B$ the subgroup of upper triangular matrices in $G$, and $B^-$ the subgroup of lower triangular matrices in $G$. Let $W$ be the Weyl group of $G$ relative to $T$, and $W_{P_d}$ the Weyl group of $P_d$. Note that $W = S_n$, the group of permutations of a set of $n$ elements, and that $W_{P_d} = S_d \times S_{n-d}$. For a permutation $w$ in $S_n$, $l(w)$ will denote the usual length function. Note also that $I_{d,n}$ can be identified with $W/W_{P_d}$. In the sequel, we shall identify $I_{d,n}$ with the set of "minimal representatives" of $W/W_{P_d}$ in $S_n$; to be very precise, a $d$-tuple $\underline{i} \in I_{d,n}$ will be identified with the element $(i_1, \ldots, i_d, j_1, \ldots, j_{n-d}) \in S_n$, where

$\{j_1, \ldots, j_{n-d}\}$ is the complement of $\{i_1, \ldots, i_d\}$ in $\{1, \ldots, n\}$ arranged in increasing order. We denote the set of such minimal representatives of $S_n$ by $W^{P_d}$.

Let $R$ denote the root system of $G$ relative to $T$, and $R^+$ the set of positive roots relative to $B$. Let $R_{P_d}$ denote the root system of $P_d$, and $R^+_{P_d}$ the set of positive roots.

**2. Schubert, opposite Schubert, and Richardson varieties in $G_{d,n}$.** For $1 \leq t \leq n$, let $V_t$ be the subspace of $K^n$ spanned by $\{e_1, \ldots, e_t\}$, and let $V^t$ be the subspace spanned by $\{e_n, \ldots, e_{n-t+1}\}$. For each $\underline{i} \in I_{d,n}$, the *Schubert variety* $X_{\underline{i}}$ and *opposite Schubert variety* $X^{\underline{i}}$ associated to $\underline{i}$ are defined to be

$$X_{\underline{i}} = \{U \in G_{d,n} \mid \dim(U \cap V_{i_t}) \geq t \ , \ 1 \leq t \leq d\},$$

$$X^{\underline{i}} = \{U \in G_{d,n} \mid \dim(U \cap V^{n-i_{(d-t+1)}+1}) \geq t \ , \ 1 \leq t \leq d\}.$$

For $\underline{i}, \underline{j} \in I_{d,n}$, the *Richardson variety* $X_{\underline{i}}^{\underline{j}}$ is defined to be $X_{\underline{i}} \cap X^{\underline{j}}$. For $\underline{i}, \underline{j}, \underline{e}, \underline{f} \in I_{d,n}$, where $\underline{e} = (1, \ldots, d)$ and $\underline{f} = (n+1-d, \ldots, n)$, note that $G_{d,n} = X_{\underline{f}}^{\underline{e}}$, $X_{\underline{i}} = X_{\underline{i}}^{\underline{e}}$, and $X^{\underline{i}} = X_{\underline{f}}^{\underline{i}}$.

For the action of $G$ on $\mathbb{P}(\wedge^d K^n)$, the $T$-fixed points are precisely the points corresponding to the $T$-eigenvectors in $\wedge^d K^n$. Now

$$\wedge^d K^n = \bigoplus_{\underline{i} \in I_{d,n}} K e_{\underline{i}}, \quad \text{as } T\text{-modules},$$

where for $\underline{i} = (i_1, \ldots, i_d)$, $e_{\underline{i}} = e_{i_1} \wedge \cdots \wedge e_{i_d}$. Thus the $T$-fixed points in $\mathbb{P}(\wedge^d K^n)$ are precisely $[e_{\underline{i}}]$, $\underline{i} \in I_{d,n}$, and these points, obviously, belong to $G_{d,n}$. Further, the Schubert variety $X_{\underline{i}}$ associated to $\underline{i}$ is simply the Zariski closure of the $B$-orbit $B[e_{\underline{i}}]$ through the $T$-fixed point $[e_{\underline{i}}]$ (with the canonical reduced structure), $B$ being as in §1.2. The opposite Schubert variety $X^{\underline{i}}$ is the Zariski closure of the $B^-$-orbit $B^-[e_{\underline{i}}]$ through the $T$-fixed point $[e_{\underline{i}}]$ (with the canonical reduced structure), $B^-$ being as in §1.2.

**2.1. Bruhat decomposition.** Let $V = K^n$. Let $\underline{i} \in I_{d,n}$. Let $C_{\underline{i}} = B[e_{\underline{i}}]$ be the *Schubert cell* and $C^{\underline{i}} = B^-[e_{\underline{i}}]$ the *opposite Schubert cell* associated to $\underline{i}$. The $C_{\underline{i}}$'s provide a cell decomposition of $G_{d,n}$, as do the $C^{\underline{i}}$'s. Let $X = V \oplus \cdots \oplus V$ (d times). Let

$$\pi : X \to \wedge^d V, (u_1, \ldots, u_d) \mapsto u_1 \wedge \ldots \wedge u_d,$$

and

$$p : \wedge^d V \setminus \{0\} \to \mathbb{P}(\wedge^d V), u_1 \wedge \ldots \wedge u_d \mapsto [u_1 \wedge \ldots \wedge u_d].$$

Let $v_{\underline{i}}$ denote the point $(e_{i_1}, \ldots, e_{i_d}) \in X$.

Identifying $X$ with $M_{n \times d}$, $v_{\underline{i}}$ gets identified with the $n \times d$ matrix whose entries are all zero except the ones in the $i_j$-th row and $j$-th column, $1 \leq j \leq d$, which are equal to 1. We have

$$B \cdot v_{\underline{i}} = \left\{ A \in M_{n\times d} \mid x_{ij} = 0, i > i_j, \text{ and } \prod_t x_{i_t t} \neq 0 \right\},$$

$$B^- \cdot v_{\underline{i}} = \left\{ A \in M_{n\times d} \mid x_{ij} = 0, i < i_j, \text{ and } \prod_t x_{i_t t} \neq 0 \right\}.$$

Denoting $\overline{B \cdot v_{\underline{i}}}$ by $D_{\underline{i}}$, we have $D_{\underline{i}} = \{A \in M_{n\times d} \mid x_{ij} = 0, i > i_j\}$. Further, $\pi(B \cdot v_{\underline{i}}) = p^{-1}(C_{\underline{i}})$, $\pi(D_{\underline{i}}) = \widehat{X_{\underline{i}}}$, the cone over $X_{\underline{i}}$. Denoting $\overline{B^- \cdot v_{\underline{i}}}$ by $D^{\underline{i}}$, we have $D^{\underline{i}} = \{A \in M_{n\times d} \mid x_{ij} = 0, i < i_j\}$. Further, $\pi(B^- \cdot v_{\underline{i}}) = p^{-1}(C^{\underline{i}})$, $\pi(D^{\underline{i}}) = \widehat{X^{\underline{i}}}$, the cone over $X^{\underline{i}}$. From this, we obtain:

THEOREM 2.1.1. (1) *Bruhat Decomposition:* $X_{\underline{j}} = \dot{\cup}_{\underline{i} \leq \underline{j}} Be_{\underline{i}}$, $X^{\underline{j}} = \dot{\cup}_{\underline{i} \geq \underline{j}} B^- e_{\underline{i}}$.

(2) $X_{\underline{i}} \subseteq X_{\underline{j}}$ *if and only if* $\underline{i} \leq \underline{j}$.

(3) $X^{\underline{i}} \subseteq X^{\underline{j}}$ *if and only if* $\underline{i} \geq \underline{j}$.

COROLLARY 2.1.2. (1) $X_{\underline{j}}^{\underline{k}}$ *is nonempty* $\iff$ $\underline{j} \geq \underline{k}$; *further, when* $X_{\underline{j}}^{\underline{k}}$ *is nonempty, it is reduced and irreducible of dimension* $l(w) - l(v)$, *where* $w$ *(resp.* $v$*) is the permutation in* $S_n$ *representing* $\underline{j}$ *(resp.* $\underline{k}$*) as in §1.2.*

(2) $p_{\underline{j}}|_{X_{\underline{i}}^{\underline{k}}} \neq 0 \iff \underline{i} \geq \underline{j} \geq \underline{k}$.

*Proof.* (1) Follows from [22]. The criterion for $X_{\underline{j}}^{\underline{k}}$ to be nonempty, the irreducibility, and the dimension formula are also proved in [6].
(2) From Bruhat decomposition, we have $p_{\underline{j}}|_{X_{\underline{i}}} \neq 0 \iff e_{\underline{j}} \in X_{\underline{i}}$; we also have $p_{\underline{j}}|_{X^{\underline{k}}} \neq 0 \iff e_{\underline{j}} \in X^{\underline{k}}$. Thus $p_{\underline{j}}|_{X_{\underline{i}}^{\underline{k}}} \neq 0 \iff e_{\underline{j}} \in X_{\underline{i}}^{\underline{k}}$. Again from Bruhat decomposition, we have $e_{\underline{j}} \in X_{\underline{i}}^{\underline{k}} \iff \underline{i} \geq \underline{j} \geq \underline{k}$. The result follows from this. □

For the remainder of this paper, we will assume that all our Richardson varieties are nonempty.

*Remark* 2.1.3. In view of Theorem 2.1.1, we have $X_{\underline{i}} \subseteq X_{\underline{j}}$ if and only if $\underline{i} \leq \underline{j}$. Thus, under the set-theoretic bijection between the set of Schubert varieties and the set $I_{d,n}$, the partial order on the set of Schubert varieties given by inclusion induces the partial order $\geq$ on $I_{d,n}$.

**2.2. More results on Richardson varieties.**

LEMMA 2.2.1. *Let* $X \subseteq G_{d,n}$ *be closed and* $B$*-stable (resp.* $B^-$*-stable). Then* $X$ *is a union of Schubert varieties (resp. opposite Schubert varieties).*

The proof is obvious.

LEMMA 2.2.2. *Let $X_1$, $X_2$ be two Richardson varieties in $G_{d,n}$ with nonempty intersection. Then $X_1 \cap X_2$ is a Richardson variety (set-theoretically).*

*Proof.* We first give the proof when $X_1$ and $X_2$ are both Schubert varieties. Let $X_1 = X_{\tau_1}, X_2 = X_{\tau_2}$, where $\tau_1 = (a_1, \ldots, a_d), \tau_2 = (b_1, \ldots, b_d)$. By Lemma 2.2.1, $X_1 \cap X_2 = \cup X_{w_i}$, where $w_i < \tau_1, w_i < \tau_2$. Let $c_j = \min\{a_j, b_j\}$, $1 \leq j \leq d$, and $\tau = (c_1, \ldots, c_d)$. Then, clearly $\tau \in I_{d,n}$, and $\tau < \tau_i, i = 1, 2$. We have $w_i \leq \tau$, and hence $X_1 \cap X_2 = X_\tau$.

The proof when $X_1$ and $X_2$ are opposite Schubert varieties is similar. The result for Richardson varieties follows immediately from the result for Schubert varieties and the result for opposite Schubert varieties. □

*Remark* 2.2.3. Explicitly, in terms of the distributive lattice structure of $I_{d,n}$, we have that $X_{w_1}^{v_1} \cap X_{w_2}^{v_2} = X_{w_1 \wedge w_2}^{v_1 \vee v_2}$ (set-theoretically), where $w_1 \wedge w_2$ is the *meet* of $w_1$ and $w_2$ (the largest element of $W^{P_d}$ which is less than both $w_1$ and $w_2$) and $v_1 \vee v_2$ is the *join* of $v_1$ and $v_2$ (the smallest element of $W^{P_d}$ which is greater than both $v_1$ and $v_2$). The fact that $X_1 \cap X_2$ is reduced follows from [20]; we will also provide a proof in Theorem 4.3.1.

## 3. Standard monomial theory for Richardson varieties.

**3.1. Standard monomials.** Let $R_0$ be the homogeneous coordinate ring of $G_{d,n}$ for the Plücker embedding, and for $w, v \in I_{d,n}$, let $R_w^v$ be the homogeneous coordinate ring of the Richardson variety $X_w^v$. In this section, we present a standard monomial theory for $X_w^v$ in the same spirit as in [21]. As mentioned in the introduction, standard monomial theory consists in constructing an explicit basis for $R_w^v$.

*Definition* 3.1.1. A monomial $f = p_{\tau_1} \cdots p_{\tau_m}$ is said to be *standard* if

$$(*) \qquad \tau_1 \geq \cdots \geq \tau_m.$$

Such a monomial is said to be *standard on $X_w^v$*, if in addition to condition (*), we have $w \geq \tau_1$ and $\tau_m \geq v$.

*Remark* 3.1.2. Note that in the presence of condition (*), the standardness of $f$ on $X_w^v$ is equivalent to the condition that $f|_{X_w^v} \neq 0$. Thus given a standard monomial $f$, we have $f|_{X_w^v}$ is either 0 or remains standard on $X_w^v$.

**3.2. Linear independence of standard monomials.**

THEOREM 3.2.1. *The standard monomials on $X_w^v$ of degree $m$ are linearly independent in $R_w^v$.*

*Proof.* We proceed by induction on $\dim X_w^v$.

If $\dim X_w^v = 0$, then $w = v$, $p_w^m$ is the only standard monomial on $X_w^v$ of degree $m$, and the result is obvious. Let $\dim X_w^v > 0$. Let

$$(*) \qquad 0 = \sum_{i=1}^{r} c_i F_i, \quad c_i \in K^*,$$

be a linear relation of standard monomials $F_i$ of degree $m$. Let $F_i = p_{w_{i1}} \cdots p_{w_{im}}$. Suppose that $w_{i1} < w$ for some $i$. For simplicity, assume that $w_{11} < w$, and $w_{11}$ is a minimal element of $\{w_{j1} \mid w_{j1} < w\}$. Let us denote $w_{11}$ by $\varphi$. Then for $i \geq 2$, $F_i|_{X_\varphi^v}$ is either 0, or is standard on $X_\varphi^v$. Hence restricting $(*)$ to $X_\varphi^v$, we obtain a nontrivial standard sum on $X_\varphi^v$ being zero, which is not possible (by induction hypothesis). Hence we conclude that $w_{i1} = w$ for all $i$, $1 \leq i \leq m$. Canceling $p_w$, we obtain a linear relation among standard monomials on $X_w^v$ of degree $m-1$. Using induction on $m$, the required result follows. □

### 3.3. Generation by standard monomials.

THEOREM 3.3.1. *Let $F = p_{w_1} \cdots p_{w_m}$ be any monomial in the Plücker coordinates of degree $m$. Then $F$ is a linear combination of standard monomials of degree $m$.*

*Proof.* For $F = p_{w_1} \cdots p_{w_m}$, define

$$N_F = l(w_1)N^{m-1} + l(w_2)N^{m-2} + \ldots l(w_m),$$

where $N \gg 0$, say $N > d(n-d)$ $(= \dim G_{d,n})$ and $l(w) = \dim X_w$. If $F$ is standard, there is nothing to prove. Let $t$ be the first violation of standardness, i.e. $p_{w_1} \cdots p_{w_{t-1}}$ is standard, but $p_{w_1} \cdots p_{w_t}$ is not. Hence $w_{t-1} \not\geq w_t$, and using the quadratic relations (cf. Theorem 1.0.1)

$$(*) \qquad p_{w_{t-1}} p_{w_t} = \sum_{\alpha,\beta} \pm p_\alpha p_\beta,$$

$F$ can be expressed as $F = \sum F_i$, with $N_{F_i} > N_F$ (since $\alpha > w_{t-1}$ for all $\alpha$ on the right hand side of $(*)$). Now the required result is obtained by decreasing induction on $N_F$ (the starting point of induction, i.e. the case when $N_F$ is the largest, corresponds to standard monomial $F = p_\theta^m$, where $\theta = (n+1-d, n+2-d, \cdots, n)$, in which case $F$ is clearly standard). □

Combining Theorems 3.2.1 and 3.3.1, we obtain:

THEOREM 3.3.2. *Standard monomials on $X_w^v$ of degree $m$ give a basis for $R_w^v$ of degree $m$.*

As a consequence of Theorem 3.3.2 (or also Theorem 1.0.1), we have a qualitative description of a typical quadratic relation on a Richardson variety $X_w^v$ as given by the following:

PROPOSITION 3.3.3. *Let $w, \tau, \varphi, v \in I_{d,n}$, $w > \tau, \varphi$ and $\tau, \varphi > v$. Further let $\tau, \varphi$ be non-comparable (so that $p_\tau p_\varphi$ is a non-standard degree 2 monomial on $X_w^v$). Let*

$$(*) \qquad p_\tau p_\varphi = \sum_{\alpha,\beta} c_{\alpha,\beta} p_\alpha p_\beta, \quad c_{\alpha,\beta} \in k^*$$

*be the expression for $p_\tau p_\varphi$ as a sum of standard monomials on $X_w^v$. Then for every $\alpha, \beta$ on the right hand side we have, $\alpha >$ both $\tau$ and $\varphi$, and $\beta <$ both $\tau$ and $\varphi$.*

Such a relation as in (*) is called a *straightening relation.*

**3.4. Equations defining Richardson varieties in the Grassmannian.** Let $w, v \in I_{d,n}$, with $w \geq v$. Let $\pi_w^v$ be the map $R_0 \to R_w^v$ (the restriction map). Let $\ker \pi_w^v = J_w^v$. Let $Z_w^v = \{$all standard monomials $F \mid F$ contains some $p_\varphi$ for some $w \not\geq \varphi$ or $\varphi \not\geq v\}$. We shall now give a set of generators for $J_w^v$ in terms of Plücker coordinates.

LEMMA 3.4.1. *Let $I_w^v = (p_\varphi, w \not\geq \varphi$ or $\varphi \not\geq v)$ (ideal in $R_0$). Then $Z_w^v$ is a basis for $I_w^v$.*

*Proof.* Let $F \in I_w^v$. Then writing $F$ as a linear combination of standard monomials

$$F = \sum a_i F_i + \sum b_j G_j,$$

where in the first sum each $F_i$ contains some $p_\tau$, with $w \not\geq \tau$ or $\tau \not\geq v$, and in the second sum each $G_j$ contains only coordinates of the form $p_\tau$, with $w \geq \tau \geq v$. This implies that $\sum a_i F_i \in I_w^v$, and hence we obtain

$$\sum b_j G_j \in I_w^v.$$

This now implies that considered as an element of $R_w^v$, $\sum b_j G_j$ is equal to 0 (note that $I_w^v \subset J_w^v$). Now the linear independence of standard monomials on $X_w^v$ implies that $b_j = 0$ for all $j$. The required result now follows. □

PROPOSITION 3.4.2. *Let $w, v \in I_{d,n}$ with $w \geq v$. Then $R_w^v = R_0/I_w^v$.*

*Proof.* We have, $R_w^v = R_0/J_w^v$ (where $J_w^v$ is as above). We shall now show that the inclusion $I_w^v \subset J_w^v$ is in fact an equality. Let $F \in R_0$. Writing $F$ as a linear combination of standard monomials

$$F = \sum a_i F_i + \sum b_j G_j,$$

where in the first sum each $F_i$ contains some term $p_\tau$, with $w \not\geq \tau$ or $\tau \not\geq v$, and in the second sum each $G_j$ contains only coordinates $p_\tau$, with $w \geq \tau \geq v$, we have, $\sum a_i F_i \in I_w^v$, and hence we obtain

$F \in J_w^v$
$\Longleftrightarrow \sum b_j G_j \in J_w^v$ (since $\sum a_i F_i \in I_w^v$, and $I_w^v \subset J_w^v$)
$\Longleftrightarrow \pi_w^v(F)\ (= \sum b_j G_j)$ is zero
$\Longleftrightarrow \sum b_j G_j$ (= a sum of standard monomials on $X_w^v$) is zero on $X_w^v$
$\Longleftrightarrow b_j = 0$ for all $j$ (in view of the linear independence of standard monomials on $X_w^v$)
$\Longleftrightarrow F = \sum a_i F_i$
$\Longleftrightarrow F \in I_w^v$ .

Hence we obtain $J_w^v = I_w^v$. □

*Equations defining Richardson varieties.* Let $w, v \in I_{d,n}$, with $w \geq v$. By Lemma 3.4.1 and Proposition 3.4.2, we have that the kernel of $(R_0)_1 \to (R_w^v)_1$ has a basis given by $\{p_\tau \mid w \not\geq \tau \text{ or } \tau \not\geq v\}$, and that the ideal $J_w^v$ (= the kernel of the restriction map $R_0 \to R_w^v$) is generated by $\{p_\tau \mid w \not\geq \tau \text{ or } \tau \not\geq v\}$. Hence $J_w^v$ is generated by the kernel of $(R_0)_1 \to (R_w^v)_1$. Thus we obtain that $X_w^v$ is scheme-theoretically (even at the cone level) the intersection of $G_{d,n}$ with all hyperplanes in $\mathbb{P}(\wedge^d k^n)$ containing $X_w^v$. Further, as a closed subvariety of $G_{d,n}$, $X_w^v$ is defined (scheme-theoretically) by the vanishing of $\{p_\tau \mid w \not\geq \tau \text{ or } \tau \not\geq v\}$.

**4. Standard monomial theory for a union of Richardson varieties.** In this section, we prove results similar to Theorems 3.2.1 and 3.3.2 for a union of Richardson varieties.

Let $X_i$ be Richardson varieties in $G_{d,n}$. Let $X = \cup X_i$.

*Definition* 4.0.3. A monomial $F$ in the Plücker coordinates is *standard* on the union $X = \cup X_i$ if it is standard on some $X_i$.

**4.1. Linear independence of standard monomials on $X = \cup X_i$.**

THEOREM 4.1.1. *Monomials standard on $X = \cup X_i$ are linearly independent.*

*Proof.* If possible, let

$$(*) \qquad 0 = \sum_{i=1}^{r} a_i F_i,\ a_i \in K^*$$

be a nontrivial relation among standard monomials on $X$. Suppose $F_1$ is standard on $X_j$. Then restricting $(*)$ to $X_j$, we obtain a nontrivial relation among standard monomials on $X_j$, which is a contradiction (note that for any $i$, $F_i|_{X_j}$ is either 0 or remains standard on $X_j$; further, $F_1|_{X_j}$ is non-zero). □

**4.2. Standard monomial basis.**

THEOREM 4.2.1. *Let $X = \cup_{i=1}^r X_{w_i}^{v_i}$, and $S$ the homogeneous coordinate ring of $X$. Then the standard monomials on $X$ give a basis for $S$.*

*Proof.* For $w, v \in I_{d,n}$ with $w \geq v$, let $I_w^v$ be as in Lemma 3.4.1. Let us denote $I_t = I_{w_t}^{v_t}$, $X_t = X_{w_t}^{v_t}$, $1 \leq t \leq r$. We have $R_{w_t}^{v_t} = R_0/I_t$ (cf. Proposition 3.4.2). Let $S = R_0/I$. Then $I = \cap I_t$ (note that being the intersection of radical ideals, $I$ is also a radical ideal, and hence the set-theoretic equality $X = \cup X_i$ is also scheme-theoretic). A typical element in $R_0/I$ may be written as $\pi(f)$, for some $f \in R_0$, where $\pi$ is the canonical projection $R_0 \to R_0/I$. Let us write $f$ as a sum of standard monomials

$$f = \sum a_j G_j + \sum b_l H_l,$$

where each $G_j$ contains some $p_{\tau_j}$ such that $w_i \not\geq \tau_j$ or $\tau_j \not\geq v_i$, for $1 \leq i \leq r$; and for each $H_l$, there is some $i_l$, with $1 \leq i_l \leq r$, such that $H_l$ is made up entirely of $p_\tau$'s with $w_{i_l} \geq \tau \geq v_{i_l}$. We have $\pi(f) = \sum b_l H_l$ (since $\sum a_j G_j \in I$). Thus we obtain that $S$ (as a vector space) is generated by monomials standard on $X$. This together with the linear independence of standard monomials on $X$ implies the required result. □

### 4.3. Consequences.

THEOREM 4.3.1. *Let $X_1$, $X_2$ be two Richardson varieties in $G_{d,n}$. Then*
*(1) $X_1 \cup X_2$ is reduced.*
*(2) If $X_1 \cap X_2 \neq \emptyset$, then $X_1 \cap X_2$ is reduced.*

*Proof.* (1) Assertion is obvious.

(2) Let $X_1 = X_{w_1}^{v_1}$, $X_2 = X_{w_2}^{v_2}$, $I_1 = I_{w_1}^{v_1}$, and $I_2 = I_{w_2}^{v_2}$. Let $A$ be the homogeneous coordinate ring of $X_1 \cap X_2$. Let $A = R_0/I$. Then $I = I_1 + I_2$. Let $F \in I$. Then by Lemma 3.4.1 and Proposition 3.4.2, in the expression for $F$ as a linear combination of standard monomials

$$F = \sum a_j F_j,$$

each $F_j$ contains some $p_\tau$, where either $((w_1 \text{ or } w_2) \not\geq \tau)$ or $(\tau \not\geq (v_1 \text{ or } v_2))$. Let $X_1 \cap X_2 = X_\mu^\nu$ set theoretically, where $\mu = w_1 \wedge w_2$ and $\nu = v_1 \vee v_2$ (cf. Remark 2.2.3). If $B = R_0/\sqrt{I}$, then by Lemma 3.4.1 and Proposition 3.4.2, under $\pi : R_0 \to B$, $\ker \pi$ consists of all $f$ such that $f = \sum c_k f_k$, $f_k$ being standard monomials such that each $f_k$ contains some $p_\varphi$, where $\mu \not\geq \varphi$ or $\varphi \not\geq \nu$. Hence either $((w_1 \text{ or } w_2) \not\geq \varphi)$ or $(\varphi \not\geq (v_1 \text{ or } v_2))$. Hence $\sqrt{I} = I$, and the required result follows from this. □

*Definition 4.3.2.* Let $w > v$. Define $\partial^+ X_w^v := \cup_{w > w' \geq v} X_{w'}^v$, and $\partial^- X_w^v := \cup_{w \geq v' > v} X_w^{v'}$.

THEOREM 4.3.3. *(Pieri's formulas) Let $w > v$.*
*(1) $X_w^v \cap \{p_w = 0\} = \partial^+ X_w^v$, scheme theoretically.*
*(2) $X_w^v \cap \{p_v = 0\} = \partial^- X_w^v$, scheme theoretically.*

*Proof.* Let $X = \partial^+ X_w^v$, and let $A$ be the homogeneous coordinate ring of $X$. Let $A = R_w^v/I$. Clearly, $(p_w) \subseteq I$, $(p_w)$ being the principal ideal in $R_w^v$ generated by $p_w$. Let $f \in I$. Writing $f$ as

$$f = \sum b_i G_i + \sum c_j H_j,$$

where each $G_i$ is a standard monomial in $R_w^v$ starting with $p_w$ and each $H_j$ is a standard monomial in $R_w^v$ starting with $p_{\theta_{j1}}$, where $\theta_{j1} < w$, we have, $\sum b_i G_i \in I$. This now implies $\sum c_j H_j$ is zero on $\partial^+ X_w^v$. But now $\sum c_j H_j$ being a sum of standard monomials on $\partial^+ X_w^v$, we have by Theorem 4.1.1, $c_j = 0$, for all $j$. Thus we obtain $f = \sum b_i G_i$, and hence $f \in (p_w)$. This implies $I = (p_w)$. Hence we obtain $A = R_w^v/(p_w)$, and (1) follows from this. The proof of (2) is similar. □

**5. Vanishing theorems.** Let $X$ be a union of Richardson varieties. Let $S(X, m)$ be the set of standard monomials on $X$ of degree $m$, and $s(X, m)$ the cardinality of $S(X, m)$. If $X{=}X_w^v$ for some $w$, $v$, then $S(X, m)$ and $s(X, m)$ will also be denoted by just $S(w, v, m)$, respectively $s(w, v, m)$.

LEMMA 5.0.4. (1) *Let $Y = Y_1 \cup Y_2$, where $Y_1$ and $Y_2$ are unions of Richardson varieties such that $Y_1 \cap Y_2 \neq \emptyset$. Then*

$$s(Y, m) = s(Y_1, m) + s(Y_2, m) - s(Y_1 \cap Y_2, m).$$

(2) *Let $w > v$. Then*

$$\begin{aligned} s(w, v, m) &= s(w, v, m-1) + s\left(\partial^+ X_w^v, m\right) \\ &= s(w, v, m-1) + s\left(\partial^- X_w^v, m\right). \end{aligned}$$

(1) *and* (2) *are easy consequences of the results of the previous section.*

Let $X$ be a closed subvariety of $G_{d,n}$. Let $L = p^*(\mathcal{O}_{\mathbb{P}}(1))$, where $\mathbb{P} = \mathbb{P}(\wedge^d K^n)$, and $p : X \hookrightarrow \mathbb{P}$ is the Plücker embedding restricted to $X$.

PROPOSITION 5.0.5. *Let $r$ be an integer $\leq d(n-d)$. Suppose that all Richardson varieties $X$ in $G_{d,n}$ of dimension at most $r$ satisfy the following two conditions:*

(1) *$H^i(X, L^m) = 0$, for $i \geq 1$, $m \geq 0$.*

(2) *The set $S(X, m)$ is a basis for $H^0(X, L^m)$, $m \geq 0$.*

*Then any union of Richardson varieties of dimension at most $r$ which have nonempty intersection, and any nonempty intersection of Richardson varieties, satisfy* (1) *and* (2).

*Proof.* The proof for intersections of Richardson varieties is clear, since any nonempty intersection of Richardson varieties is itself a Richardson variety (cf. Lemma 2.2.2 and Theorem 4.3.1).

We will prove the result for unions by induction on $r$. Let $S_r$ denote the set of Richardson varieties $X$ in $G_{d,n}$ of dimension at most $r$. Let $Y = \cup_{j=1}^t X_j$, $X_j \in S_r$.

Let $Y_1 = \cup_{j=1}^{t-1} X_j$, and $Y_2 = X_t$. Consider the exact sequence

$$0 \to \mathcal{O}_Y \to \mathcal{O}_{Y_1} \oplus \mathcal{O}_{Y_2} \to \mathcal{O}_{Y_1 \cap Y_2} \to 0,$$

where $\mathcal{O}_Y \to \mathcal{O}_{Y_1} \oplus \mathcal{O}_{Y_2}$ is the map $f \mapsto (f|_{Y_1}, f|_{Y_2})$ and $\mathcal{O}_{Y_1} \oplus \mathcal{O}_{Y_2} \to \mathcal{O}_{Y_1 \cap Y_2}$ is the map $(f, g) \mapsto (f - g)|_{Y_1 \cap Y_2}$. Tensoring with $L^m$, we obtain the long exact sequence

$$\begin{aligned} \to H^{i-1}(Y_1 \cap Y_2, L^m) \to H^i(Y, L^m) \to H^i(Y_1, L^m) \oplus H^i(Y_2, L^m) \\ \to H^i(Y_1 \cap Y_2, L^m) \to \end{aligned}$$

Now $Y_1 \cap Y_2$ is reduced (cf. Theorem 4.3.1) and $Y_1 \cap Y_2 \in S_{r-1}$. Hence, by the induction hypothesis (1) and (2) hold for $Y_1 \cap Y_2$. In particular, if $m \geq 0$, then (2) implies that the map $H^0(Y_1, L^m) \oplus H^0(Y_2, L^m) \to H^0(Y_1 \cap Y_2, L^m)$ is surjective. Hence we obtain that the sequence

$$0 \to H^0(Y, L^m) \to H^0(Y_1, L^m) \oplus H^0(Y_2, L^m) \to H^0(Y_1 \cap Y_2, L^m) \to 0$$

is exact. This implies $H^0(Y_1 \cap Y_2, L^m) \to H^1(Y, L^m)$ is the zero map; we have, $H^1(Y, L^m) \to H^1(Y_1, L^m) \oplus H^1(Y_2, L^m)$ is also the zero map (since by induction $H^1(Y_1, L^m) = 0 = H^1(Y_2, L^m)$). Hence we obtain $H^1(Y, L^m) = 0,\ m \geq 0$, and for $i \geq 2$, the assertion that $H^i(Y, L^m) = 0,\ m \geq 0$ follows from the long exact cohomology sequence above (and induction hypothesis). This proves the assertion (1) for $Y$.

To prove assertion (2) for $Y$, we observe

$$\begin{aligned} h^0(Y, L^m) &= h^0(Y_1, L^m) + h^0(Y_2, L^m) - h^0(Y_1 \cap Y_2, L^m) \\ &= s(Y_1, m) + s(Y_2, m) - s(Y_1 \cap Y_2, m). \end{aligned}$$

Hence Lemma 5.0.4 implies that

$$h^0(Y, L^m) = s(Y, L^m).$$

This together with linear independence of standard monomials on $Y$ proves assertion (2) for $Y$. □

THEOREM 5.0.6. *Let $X$ be a Richardson variety in $G_{d,n}$. Then*
(a) *$H^i(X, L^m) = 0$ for $i \geq 1$, $m \geq 0$.*
(b) *$S(X, m)$ is a basis for $H^0(X, L^m)$, $m \geq 0$.*

*Proof.* We prove the result by induction on $m$, and $\dim X$.

If $\dim X = 0$, $X$ is just a point, and the result is obvious. Assume now that $\dim X \geq 1$. Let $X = X_w^v$, $w > v$. Let $Y = \partial^+ X_w^v$. Then by Pieri's formula (cf. §4.3.3), we have,

$$Y = X(\tau) \cap \{p_\tau = 0\} \quad \text{(scheme theoretically).}$$

Hence the sequence

$$0 \to \mathcal{O}_X(-1) \to \mathcal{O}_X \to \mathcal{O}_Y \to 0$$

is exact. Tensoring it with $L^m$, and writing the cohomology exact sequence, we obtain the long exact cohomology sequence

$$\cdots \to H^{i-1}(Y, L^m) \to H^i(X, L^{m-1}) \to H^i(X, L^m) \to H^i(Y, L^m) \to \cdots.$$

Let $m \geq 0$, $i \geq 2$. Then the induction hypothesis on $\dim X$ implies (in view of Proposition 5.0.5) that $H^i(Y, L^m) = 0$, $i \geq 1$. Hence we obtain that the sequence $0 \to H^i(X, L^{m-1}) \to H^i(X, L^m)$, $i \geq 2$, is exact. If $i = 1$, again the induction hypothesis implies the surjectivity of $H^0(X, L^m) \to H^0(Y, L^m)$. This in turn implies that the map $H^0(Y, L^m) \to H^1(X, L^{m-1})$ is the zero map, and hence we obtain that the sequence $0 \to H^1(X, L^{m-1}) \to H^1(X, L^m)$ is exact. Thus we obtain that $0 \to H^i(X, L^{m-1}) \to H^i(X, L^m), m \geq 0, i \geq 1$ is exact. But $H^i(X, L^m) = 0$, $m \gg 0, i \geq 1$ (cf. [25]). Hence we obtain

(1) $$H^i(X, L^m) = 0 \text{ for } i \geq 1,\ m \geq 0,$$

and

(2) $$h^0(X, L^m) = h^0(X, L^{m-1}) + h^0(Y, L^m).$$

In particular, assertion (a) follows from (1). The induction hypothesis on $m$ implies that $h^0(X, L^{m-1}) = s(X, m-1)$. On the other hand, the induction hypothesis on $\dim X$ implies (in view of Proposition 5.0.5) that $h^0(Y, L^m) = s(Y, m)$. Hence we obtain

(3) $$h^0(X, L^m) = s(X, m-1) + s(Y, m).$$

Now (3) together with Lemma 5.0.4, (2) implies $h^0(X, L^m) = s(X, m)$. Hence (b) follows in view of the linear independence of standard monomials on $X_w^v$ (cf. Theorem 3.2.1). □

COROLLARY 5.0.7. *We have*
(1) $R_w^v = \oplus_{m \in \mathbb{Z}^+} H^0(X_w^v, L^m)$, $w \geq v$.
(2) $\dim H^0(\partial^+ X_w^v, L^m) = \dim H^0(\partial^- X_w^v, L^m)$, $w > v$, $m \geq 0$.

*Proof.* Assertion 1 follows immediately from Theorems 3.3.2 and 5.0.6(b). Assertion 2 follows from Lemma 5.0.4, Theorem 5.0.5(2), and Theorem 5.0.6(b). □

## 6. Tangent Space and smoothness.

**6.1. The Zariski Tangent Space.** Let $x$ be a point on a variety $X$. Let $\mathfrak{m}_x$ be the maximal ideal of the local ring $\mathcal{O}_{X,x}$ with residue field $K(x)(= \mathcal{O}_{X,x}/\mathfrak{m}_x)$. Note that $K(x) = K$ (since $K$ is algebraically closed). Recall that the Zariski tangent space to $X$ at $x$ is defined as

$$\begin{aligned} T_x(X) &= \mathrm{Der}_K(\mathcal{O}_{X,x}, K(x)) \\ &= \{D : \mathcal{O}_{X,x} \to K(x),\ K\text{-linear such that } D(ab) = D(a)b + aD(b)\} \end{aligned}$$

(here $K(x)$ is regarded as an $\mathcal{O}_{X,x}$-module). It can be seen easily that $T_x(X)$ is canonically isomorphic to $\mathrm{Hom}_{K\text{-mod}}(\mathfrak{m}_x/\mathfrak{m}_x^2, K)$.

**6.2. Smooth and non-smooth points.** A point $x$ on a variety $X$ is said to be a *simple* or *smooth* or *nonsingular point of* $X$ if $\mathcal{O}_{X,x}$ is a regular local ring. A point $x$ which is not simple is called a *multiple* or *non-smooth* or *singular point* of $X$. The set $\mathrm{Sing}\, X = \{x \in X \mid x \text{ is a singular point}\}$ is called the *singular locus of* $X$. A variety $X$ is said to be *smooth* if $\mathrm{Sing} X = \emptyset$. We recall the well known:

THEOREM 6.2.1. *Let $x \in X$. Then $\dim_K T_x(X) \geq \dim \mathcal{O}_{X,x}$ with equality if and only if $x$ is a simple point of $X$.*

**6.3. The Space $T^v_{w,\tau}$.** Let $G, T, B, P_d, W, R, W_{P_d}, R_{P_d}$ etc., be as in §1.2. We shall henceforth denote $P_d$ by just $P$. For $\alpha \in R$, let $X_\alpha$ be the element of the Chevalley basis for $\mathfrak{g}$ ($=\mathrm{Lie}G$), corresponding to $\alpha$. We follow [2] for denoting elements of $R$, $R^+$ etc.

For $w \geq \tau \geq v$, let $T^v_{w,\tau}$ be the Zariski tangent space to $X^v_w$ at $e_\tau$. Let $w_0$ be the element of largest length in $W$. Now the tangent space to $G$ at $e_{id}$ is $\mathfrak{g}$, and hence the tangent space to $G/P$ at $e_{id}$ is $\oplus_{\beta\in R^+\setminus R^+_P} \mathfrak{g}_{-\beta}$. For $\tau \in W$, identifying $G/P$ with $G/{}^\tau P$ (where ${}^\tau P = \tau P \tau^{-1}$) via the map $gP \mapsto (n_\tau g n_\tau^{-1})^\tau P$, $n_\tau$ being a fixed lift of $\tau$ in $N_G(T)$, we have, the tangent space to $G/P$ at $e_\tau$ is $\oplus_{\beta\in\tau(R^+)\setminus\tau(R^+_P)} \mathfrak{g}_{-\beta}$, i.e.,

$$T^{\mathrm{id}}_{w_0,\tau} = \bigoplus_{\beta\in\tau(R^+)\setminus\tau(R^+_P)} \mathfrak{g}_{-\beta}.$$

Set

$$N^v_{w,\tau} = \left\{\beta \in \tau(R^+) \setminus \tau(R^+_P) \mid X_{-\beta} \in T^v_{w,\tau}\right\}.$$

Since $T^v_{w,\tau}$ is a $T$-stable subspace of $T^v_{w_0,\tau}$, we have

$$T^v_{w,\tau} = \text{ the span of } \left\{X_{-\beta},\ \beta \in N^v_{w,\tau}\right\}.$$

**6.4. Certain Canonical Vectors in $T^v_{w,\tau}$.** For a root $\alpha \in R^+ \setminus R^+_P$, let $Z_\alpha$ denote the $SL(2)$-copy in $G$ corresponding to $\alpha$; note that $Z_\alpha$ is simply the subgroup of $G$ generated by $U_\alpha$ and $U_{-\alpha}$. Given $x \in W^P$, precisely one of $\{U_\alpha, U_{-\alpha}\}$ fixes the point $e_x$. Thus $Z_\alpha \cdot e_x$ is a $T$-stable curve in $G/P$ (note that $Z_\alpha \cdot e_x \cong \mathbb{P}^1$), and conversely any $T$-stable curve in $G/P$ is of this form (cf. [5]). Now a $T$-stable curve $Z_\alpha \cdot e_x$ is contained in a Richardson variety $X^v_w$ if and only if $e_x, e_{s_\alpha x}$ are both in $X^v_w$.

LEMMA 6.4.1. *Let* $w, \tau, v \in W^P$, $w \geq \tau \geq v$. *Let* $\beta \in \tau(R^+ \setminus R_P^+)$. *If* $w \geq s_\beta\tau \geq v\,(\mathrm{mod}\, W_P)$, *then* $X_{-\beta} \in T_{w,\tau}^v$.

(Note that $s_\beta\tau$ need not be in $W^P$.)

*Proof.* The hypothesis that $w \geq s_\beta\tau \geq v(\mathrm{mod}\, W_P)$ implies that the curve $Z_\beta \cdot e_\tau$ is contained in $X_w^v$. Now the tangent space to $Z_\beta \cdot e_\tau$ at $e_\tau$ is the one-dimensional span of $X_{-\beta}$. The required result now follows. □

We shall show in Theorem 6.7.2 that $w, \tau, v$ being as above, $T_{w,\tau}^v$ is precisely the span of $\{X_{-\beta},\ \beta \in \tau(R^+ \setminus R_P^+) \mid w \geq s_\beta\tau \geq v\,(\mathrm{mod}\, W_P)\}$.

**6.5. A Canonical affine neighborhood of a $T$-fixed point.** Let $\tau \in W$. Let $U_\tau^-$ be the unipotent subgroup of $G$ generated by the root subgroups $U_{-\beta},\ \beta \in \tau(R^+)$ (note that $U_\tau^-$ is the unipotent part of the Borel subgroup ${}^\tau B^-$, opposite to ${}^\tau B\ (= \tau B \tau^{-1})$). We have

$$U_{-\beta} \cong \mathbb{G}_a,\quad U_\tau^- \cong \prod_{\beta \in \tau(R^+)} U_{-\beta}.$$

Now, $U_\tau^-$ acts on $G/P$ by left multiplication. The isotropy subgroup in $U_\tau^-$ at $e_\tau$ is $\prod_{\beta\in\tau(R_P^+)} U_{-\beta}$. Thus $U_\tau^- e_\tau \cong \prod_{\beta\in\tau(R^+\setminus R_P^+)} U_{-\beta}$. In this way, $U_\tau^- e_\tau$ gets identified with $\mathbb{A}^N$, where $N = \#(R^+ \setminus R_P^+)$. We shall denote the induced coordinate system on $U_\tau^- e_\tau$ by $\{x_{-\beta},\ \beta \in \tau(R^+ \setminus R_P^+)\}$. In the sequel, we shall denote $U_\tau^- e_\tau$ by $\mathcal{O}_\tau^-$ also. Thus we obtain that $\mathcal{O}_\tau^-$ is an affine neighborhood of $e_\tau$ in $G/P$.

**6.6. The affine variety $Y_{w,\tau}^v$.** For $w, \tau, v \in W$, $w \geq \tau \geq v$, let us denote $Y_{w,\tau}^v := \mathcal{O}_\tau^- \cap X_{w,\tau}^v$. It is a nonempty affine open subvariety of $X_w^v$, and a closed subvariety of the affine space $\mathcal{O}_\tau^-$.

Note that $L$, the ample generator of $\mathrm{Pic}(G/P)$, is the line bundle corresponding to the Plücker embedding, and $H^0(G/P, L) = (\wedge^d K^n)^*$, which has a basis given by the Plücker coordinates $\{p_\theta,\ \theta \in I_{d,n}\}$. Note also that the affine ring $\mathcal{O}_\tau^-$ may be identified as the homogeneous localization $(R_0)_{(p_\tau)}$, $R_0$ being as in §3.1. We shall denote $p_\theta/p_\tau$ by $f_{\theta,\tau}$. Let $I_{w,\tau}^v$ be the ideal defining $Y_{w,\tau}^v$ as a closed subvariety of $\mathcal{O}_\tau^-$. Then $I_{w,\tau}^v$ is generated by $\{f_{\theta,\tau} \mid w \not\geq \theta \text{ or } \theta \not\geq v\}$.

**6.7. Basis for tangent space and criterion for smoothness of $X_w^v$ at $e_\tau$.** Let $Y$ be an affine variety in $\mathbb{A}^n$, and let $I(Y)$ be the ideal defining $Y$ in $\mathbb{A}^n$. Let $I(Y)$ be generated by $\{f_1, f_2, \ldots, f_r\}$. Let $J$ be the Jacobian matrix $(\frac{\partial f_i}{\partial x_j})$. We have (cf. Theorem 6.2.1) the dimension of the tangent space to $Y$ at a point $P$ is greater than or equal to the dimension of $Y$, with equality if and only if $P$ is a smooth point; equivalently, rank $J_P \leq \mathrm{codim}_{\mathbb{A}^n} Y$ with equality if and only if $P$ is a smooth point of $Y$ (here $J_P$ denotes $J$ evaluated at $P$).

Let $w, \tau, v \in W$, $w \geq \tau \geq v$. The problem of determining whether or not $e_\tau$ is a smooth point of $X_w^v$ is equivalent to determining whether or not $e_\tau$ is a smooth point of $Y_{w,\tau}^v$ (since $Y_{w,\tau}^v$ is an open neighborhood of $e_\tau$ in $X_w^v$). In view of Jacobian criterion, the problem is reduced to computing $(\partial f_{\theta,\tau}/\partial x_{-\beta})_{e_\tau}$, $w \not\geq \theta$ or $\theta \not\geq v$ (the Jacobian matrix evaluated at $e_\tau$). To carry out this computation, we first observe the following:

Let $V$ be the $G$-module $H^0(G/P, L)\ (= (\wedge^d K^n)^*)$. Now $V$ is also a $\mathfrak{g}$-module. Given $X$ in $\mathfrak{g}$, we identify $X$ with the corresponding right invariant vector field $D_X$ on $G$. Thus we have $D_X p_\theta = X p_\theta$, and we note that

$$(\partial f_{\theta,\tau}/\partial x_{-\beta})(e_\tau) = X_{-\beta}\, p_\theta(e_\tau),\ \ \beta \in \tau(R^+ \setminus R_P^+),$$

where the left hand side denotes the partial derivative evaluated at $e_\tau$.

We make the following three observations:

(1) For $\theta, \mu \in W^P$, $p_\theta(e_\mu) \neq 0 \iff \theta = \mu$, where, recall that for $\theta = (i_1 \cdots i_d) \in W^P$, $e_\theta$ denotes the vector $e_{i_1} \wedge \cdots \wedge e_{i_d}$ in $\wedge^d K^n$, and $p_\theta$ denotes the Plücker coordinate associated to $\theta$.

(2) Let $X_\alpha$ be the element of the Chevalley basis of $\mathfrak{g}$, corresponding to $\alpha \in R$. If $X_\alpha p_\mu \neq 0$, $\mu \in W^P$, then $X_\alpha p_\mu = \pm p_{s_\alpha \mu}$, where $s_\alpha$ is the reflection corresponding to the root $\alpha$.

(3) For $\alpha \neq \beta$, if $X_\alpha p_\mu$, $X_\beta p_\mu$ are non-zero, then $X_\alpha p_\mu \neq X_\beta p_\mu$.

The first remark is obvious, since $\{p_\theta \mid \theta \in W^P\}$ is the basis of $(\wedge^P K^n)^*$ $(= H^0(G/P, L))$, dual to the basis $\{e_\varphi,\ \varphi \in I_{d,n}\}$ of $\wedge^d K^n$. The second remark is a consequence of $SL_2$ theory, using the following facts:

(a) $|\langle \chi, \alpha^* \rangle| = |\frac{2(\chi,\alpha)}{(\alpha,\alpha)}| = 0$ or $1$, $\chi$ being the weight of $p_\mu$.

(b) $p_\mu$ is the lowest weight vector for the Borel subgroup ${}^\mu B = \mu B \mu^{-1}$.

The third remark follows from weight considerations (note that if $X_\alpha p_\mu \neq 0$, then $X_\alpha p_\mu$ is a weight vector (for the $T$-action) of weight $\chi + \alpha$, $\chi$ being the weight of $p_\mu$).

THEOREM 6.7.1. *Let $w, \tau, v \in W^P$, $w \geq \tau \geq v$. Then*

$$\dim T_{w,\tau}^v = \#\{\gamma \in \tau(R^+ \setminus R_P^+) \mid w \geq s_\gamma \tau \geq v \ (\mathrm{mod}\ W_P)\}.$$

*Proof.* By Lemma 3.4.1 and Proposition 3.4.2, we have, $I_{w,\tau}^v$ is generated by $\{f_{\theta,\tau} \mid w \not\geq \theta \text{ or } \theta \not\geq v\}$. Denoting the affine coordinates on $\mathcal{O}_\tau^-$ by $x_{-\beta}$, $\beta \in \tau(R^+ \setminus R_P^+)$, we have the evaluations of $\frac{\partial f_{\theta,\tau}}{\partial x_\beta}$ and $X_\beta p_\theta$ at $e_\tau$ coincide. Let $J_w^v$ denote the Jacobian matrix of $Y_{w,\tau}^v$ (considered as a subvariety of the affine space $\mathcal{O}_\tau^-$). We shall index the rows of $J_w^v$ by $\{f_{\theta,\tau} \mid w \not\geq \theta \text{ or } \theta \not\geq v\}$ and the columns by $x_{-\beta}$, $\beta \in \tau(R^+ \setminus R_P^+)$. Let $J_w^v(\tau)$ denote $J_w^v$ evaluated at $e_\tau$. Now in view of (1) and

(2) above, the $(f_{\theta,\tau}, x_{-\beta})$-th entry in $J_w^v(\tau)$ is non-zero if and only if $X_\beta p_\theta = \pm p_\tau$. Hence in view of (3) above, we obtain that in each row of $J_w^v(\tau)$, there is at most one non-zero entry. Hence $\operatorname{rank} J_w^v(\tau) =$ the number of non-zero columns of $J_w^v(\tau)$. Now, $X_\beta p_\theta = \pm p_\tau$ if and only if $\theta \equiv s_\beta \tau \pmod{W_P}$. Thus the column of $J_w^v(\tau)$ indexed by $x_{-\beta}$ is non-zero if and only if $w \not\geq s_\beta\tau \pmod{W_P}$ or $s_\beta\tau \not\geq v \pmod{W_P}$. Hence $\operatorname{rank} J_w^v(\tau) = \#\{\gamma \in \tau(R^+ \setminus R_P^+) \mid w \not\geq s_\gamma\tau \pmod{W_P} \text{ or } s_\gamma\tau \not\geq v \pmod{W_P}\}$ and thus we obtain

$$\dim T_{w,\tau}^v = \#\{\gamma \in \tau(R^+ \setminus R_P^+) \mid w \geq s_\gamma\tau \geq v \pmod{W_P}\}. \qquad \square$$

THEOREM 6.7.2. *Let $w, \tau, v$ be as in Theorem 6.7.1. Then $\{X_{-\beta},\ \beta \in \tau(R^+ \setminus R_P^+) \mid w \geq s_\beta\tau \geq v \pmod{W_P}\}$ is a basis for $T_{w,\tau}^v$.*

*Proof.* Let $\beta \in \tau(R^+ \setminus R_P^+)$ be such that $w \geq s_\beta\tau \geq v \pmod{W_P}$. We have (by Lemma 6.4.1), $X_{-\beta} \in T_{w,\tau}^v$. On the other hand, by Theorem 6.7.1, $\dim T_{w,\tau}^v = \#\{\beta \in \tau(R^+ \setminus R_P^+) \mid w \geq s_\beta\tau \geq v \pmod{W_P}\}$. The result follows from this. $\square$

COROLLARY 6.7.3. *$X_w^v$ is smooth at $e_\tau$ if and only if $l(w) - l(v) = \#\{\alpha \in R^+ \setminus R_P^+ \mid w \geq \tau s_\alpha \geq v \pmod{W_P}\}$.*

*Proof.* We have, $X_w^v$ is smooth at $e_\tau$ if and only if $\dim T_{w,\tau}^v = \dim X_w^v$, and the result follows in view of Corollary 2.1.2 and Theorem 6.7.1 (note that if $\beta = \tau(\alpha)$, then $s_\beta\tau = \tau s_\alpha \pmod{W_P}$). $\square$

## 7. Multiplicity at a singular point.

**7.1. Multiplicity of an Algebraic variety at a point.** Let $B$ be a graded, affine $K$-algebra such that $B_1$ generates $B$ (as a $K$-algebra). Let $X = \operatorname{Proj}(B)$. The function $h_B(m)$ (or $h_X(m)$) $= \dim_K B_m$, $m \in \mathbb{Z}$ is called the *Hilbert function* of $B$ (or $X$). There exists a polynomial $P_B(x)$ (or $P_X(x)$) $\in \mathbb{Q}[x]$, called the *Hilbert polynomial* of $B$ (or $X$), such that $f_B(m) = P_B(m)$ for $m \gg 0$. Let $r$ denote the degree of $P_B(x)$. Then $r = \dim(X)$, and the leading coefficient of $P_B(x)$ is of the form $c_B/r!$, where $c_B \in \mathbb{N}$. The integer $c_B$ is called the *degree of* $X$, and denoted $\deg(X)$ (see [7] for details). In the sequel we shall also denote $\deg(X)$ by $\deg(B)$.

Let $X$ be an algebraic variety, and let $P \in X$. Let $A = \mathcal{O}_{X,P}$ be the stalk at $P$ and $\mathfrak{m}$ the unique maximal ideal of the local ring $A$. Then the *tangent cone* to $X$ at $P$, denoted $TC_P(X)$, is $\operatorname{Spec}(\operatorname{gr}(A, \mathfrak{m}))$, where $\operatorname{gr}(A, \mathfrak{m}) = \oplus_{j=0}^\infty \mathfrak{m}^j/\mathfrak{m}^{j+1}$. The *multiplicity* of $X$ at $P$, denoted $\operatorname{mult}_P(X)$, is $\deg(\operatorname{Proj}(\operatorname{gr}(A, \mathfrak{m})))$. (If $X \subset K^n$ is an affine closed subvariety, and $m_P \subset K[X]$ is the maximal ideal corresponding to $P \in X$, then $\operatorname{gr}(K[X], m_P) = \operatorname{gr}(A, \mathfrak{m})$.)

**7.2. Evaluation of Plücker coordinates on $U_\tau^- e_\tau$.** Let $X = X_w^v$. Consider a $\tau \in W^P$ such that $w \geq \tau \geq v$.

I. Let us first consider the case $\tau = \mathrm{id}$. We identify $U^-e_{\mathrm{id}}$ with

$$\left\{ \begin{pmatrix} & \mathrm{Id}_{d\times d} & \\ x_{d+1\,1} & \cdots & x_{d+1\,d} \\ \vdots & & \vdots \\ x_{n\,1} & \cdots & x_{n\,d} \end{pmatrix}, \quad x_{ij} \in k, \quad d+1 \le i \le n, 1 \le j \le d \right\}.$$

Let $A$ be the affine algebra of $U^-e_{\mathrm{id}}$. Let us identify $A$ with the polynomial algebra $k[x_{-\beta}, \beta \in R^+ \setminus R_P^+]$. To be very precise, we have $R^+ \setminus R_P^+ = \{\epsilon_j - \epsilon_i, 1 \le j \le d, d+1 \le i \le n\}$; given $\beta \in R^+ \setminus R_P^+$, say $\beta = \epsilon_j - \epsilon_i$, we identify $x_{-\beta}$ with $x_{ij}$. Hence we obtain that the expression for $f_{\theta,\mathrm{id}}$ in the local coordinates $x_{-\beta}$'s is homogeneous.

*Example* 7.2.1. Consider $G_{2,4}$. Then

$$U^-e_{\mathrm{id}} = \left\{ \begin{pmatrix} 1 & 0 \\ 0 & 1 \\ x_{31} & x_{32} \\ x_{41} & x_{42} \end{pmatrix}, \quad x_{ij} \in k \right\}.$$

On $U^-e_{\mathrm{id}}$, we have $p_{12} = 1$, $p_{13} = x_{32}$, $p_{14} = x_{42}$, $p_{23} = x_{31}$, $p_{24} = x_{41}$, $p_{34} = x_{31}x_{42} - x_{41}x_{32}$.

Thus a Plücker coordinate is homogeneous in the local coordinates $x_{ij}$, $d+1 \le i \le n$, $1 \le j \le d$.

II. Let now $\tau$ be any other element in $W^P$, say $\tau = (a_1, \ldots, a_n)$. Then $U_\tau^- e_\tau$ consists of $\{N_{d,n}\}$, where $N_{d,n}$ is obtained from $\binom{\mathrm{Id}}{X}_{n\times d}$ (with notations as above) by permuting the rows by $\tau^{-1}$. (Note that $U_\tau^- e_\tau = \tau U^- e_{\mathrm{id}}$.)

*Example* 7.2.2. Consider $G_{2,4}$, and let $\tau = (2314)$. Then $\tau^{-1} = (3124)$, and

$$U_\tau^- e_\tau = \left\{ \begin{pmatrix} x_{31} & x_{32} \\ 1 & 0 \\ 0 & 1 \\ x_{41} & x_{42} \end{pmatrix}, \quad x_{ij} \in k \right\}.$$

We have on $U_\tau^- e_\tau$, $p_{12} = -x_{32}$, $p_{13} = x_{31}$, $p_{14} = x_{31}x_{42} - x_{41}x_{32}$, $p_{23} = 1$, $p_{24} = x_{42}$, $p_{34} = -x_{41}$.

As in the case $\tau = \mathrm{id}$, we find that for $\theta \in W^P$, $f_{\theta,\tau} := p_\theta|_{U_\tau^- e_\tau}$ is homogeneous in local coordinates. In fact we have:

PROPOSITION 7.2.3. *Let $\theta \in W^P$. We have a natural isomorphism*

$$k[x_{-\beta}, \beta \in R^+ \setminus R_P^+] \cong k[x_{-\tau(\beta)}, \beta \in R^+ \setminus R_P^+],$$

*given by*

$$f_{\theta,id} \mapsto f_{\tau\theta,\tau}.$$

The proof is immediate from the above identifications of $U^-e_{\mathrm{id}}$ and $U_\tau^-e_\tau$. As a consequence, we have:

COROLLARY 7.2.4. *Let $\theta \in W^P$. Then the polynomial expression for $f_{\theta,\tau}$ in the local coordinates $\{x_{-\tau(\beta)}, \beta \in R^+ \setminus R_P^+\}$ is homogeneous.*

**7.3. The algebra $A_{w,\tau}^v$.** As above, we identify $A_\tau$, the affine algebra of $U_\tau^-e_\tau$ with the polynomial algebra $K[x_{-\beta}, \beta \in \tau(R^+ \setminus R_P^+)]$. Let $A_{w,\tau}^v = A_\tau/I_{w,\tau}^v$, where $I_{w\tau}^v$ is the ideal of elements of $A_\tau$ that vanish on $X_w^v \cap U_\tau^-e_\tau$.

Now $I(X_w^v)$, the ideal of $X_w^v$ in $G/P$, is generated by $\{p_\theta, \theta \in W^P \mid w \not\geq \theta \text{ or } \theta \not\geq v\}$. Hence we obtain (cf. Corollary 7.2.4) that $I_{w,\tau}^v$ is homogeneous. Hence we get

$$(*) \qquad \operatorname{gr}(A_{w,\tau}^v, M_{w,\tau}^v) = A_{w,\tau}^v,$$

where $M_{w,\tau}^v$ is the maximal ideal of $A_{w,\tau}^v$ corresponding to $e_\tau$. In particular, denoting the image of $x_{-\beta}$ under the canonical map $A_\tau \to A_{w,\tau}^v$ by just $x_{-\beta}$, the set $\{x_{-\beta} \mid \beta \in \tau(R^+ \setminus R_P^+)\}$ generates $A_{w,\tau}^v$. Let $R_w^v$ be the homogeneous coordinate ring of $X_w^v$ (for the Plücker embedding), $Y_{w,\tau}^v = X_w^v \cap U_\tau^-e_\tau$. Then $K[Y_{w,\tau}^v] = A_{w,\tau}^v$ gets identified with the homogeneous localization $(R_w^v)_{(p_\tau)}$, i.e. the subring of $(R_w^v)_{p_\tau}$ (the localization of $R_w^v$ with respect to $p_\tau$) generated by the elements

$$\left\{\frac{p_\theta}{p_\tau}, \theta \in W^P, w \geq \theta \geq v\right\}.$$

**7.4. The integer $\deg_\tau(\theta)$.** Let $\theta \in W^P$. We define $\deg_\tau(\theta)$ by

$$\deg_\tau(\theta) := \deg f_{\theta,\tau}$$

(note that $f_{\theta,\tau}$ is homogeneous, cf. Corollary 7.2.4). In fact, we have an explicit expression for $\deg_\tau(\theta)$, as follows (cf. [13]):

PROPOSITION 7.4.1. *Let $\theta \in W^P$. Let $\tau = (a_1, \ldots, a_n)$, $\theta = (b_1, \ldots, b_n)$. Let $r = \#\{a_1, \ldots, a_d\} \cap \{b_1, \ldots, b_d\}$. Then $\deg_\tau(\theta) = d - r$.*

**7.5. A basis for the Tangent Cone.** Let $Z_\tau = \{\theta \in W^P \mid \text{either } \theta \geq \tau \text{ or } \tau \geq \theta\}$.

THEOREM 7.5.1. *With notations as above, given $r \in \mathbb{Z}^+$,*

$$\left\{ f_{\theta_1,\tau} \cdots f_{\theta_m,\tau} \mid w \geq \theta_1 \geq \ldots \geq \theta_m \geq v,\ \theta_i \in Z_\tau,\ \sum_{i=1}^m \deg_\tau(\theta_i) = r \right\}$$

*is a basis for $(M_{w,\tau}^v)^r/(M_{w,\tau}^v)^{r+1}$.*

*Proof.* For $F = p_{\theta_1} \cdots p_{\theta_m}$, let $\deg F$ denote the degree of $f_{\theta_1,\tau} \cdots f_{\theta_m,\tau}$. Let

$$A_r = \{F = p_{\theta_1} \cdots p_{\theta_m},\ w \geq \theta_i \geq v \mid \deg F = r\}.$$

Then in view of the relation (*) in §7.3, we have, $A_r$ generates $(M_{w,\tau}^v)^r/(M_{w,\tau}^v)^{r+1}$. Let $F \in A_r$, say $F = p_{\tau_1} \cdots p_{\tau_m}$. From the results in §3, we know that $p_{\tau_1} \cdots p_{\tau_m}$

is a linear combination of standard monomials $p_{\theta_1} \cdots p_{\theta_m}$, $w \geq \theta_i \geq v$. We *claim* that in each $p_{\theta_1} \cdots p_{\theta_m}$, $\theta_i \in Z_\tau$, for all $i$. Suppose that for some $i$, $\theta_i \notin Z_\tau$. This means $\theta_i$ and $\tau$ are not comparable. Then using the fact that $f_{\tau,\tau} = 1$, on $Y^v_{w,\tau}$, we replace $p_{\theta_i}$ by $p_{\theta_i} p_\tau$ in $p_{\theta_1} \cdots p_{\theta_m}$. We now use the straightening relation (cf. Proposition 3.3.3) $p_{\theta_i} p_\tau = \sum c_{\alpha,\beta} p_\alpha p_\beta$ on $X^v_w$, where in each term $p_\alpha p_\beta$ on the right hand side, we have $\alpha < w$, and $\alpha >$ both $\theta_i$ and $\tau$, and $\beta <$ both $\theta_i$ and $\tau$; in particular, we have, in each term $p_\alpha p_\beta$ on the right hand side, $\alpha$, $\beta$ belong to $Z_\tau$. We now proceed as in the proof of Theorem 3.3.1 to conclude that on $Y^v_{w,\tau}$, $F$ is a linear combination of standard monomials in $p_\theta$'s, $\theta \in Z_\tau$ which proves the claim.

Clearly, $\{f_{\theta_1,\tau} \ldots f_{\theta_m,\tau} \mid w \geq \theta_1 \geq \ldots \geq \theta_m \geq v,\ \theta_i \in Z_\tau\}$ is linearly independent in view of Theorem 3.2.1 (since $p^l_\tau f_{\theta_1,\tau} \cdots f_{\theta_m,\tau} = p^{l-m}_\tau p_{\theta_1} \cdots p_{\theta_m}$ for $l \geq m$, and the monomial on the right hand side is standard since $\theta_i \in Z_\tau$). □

### 7.6. Recursive formulas for $\mathrm{mult}_\tau X^v_w$.

*Definition* 7.6.1. If $w > \tau \geq v$, define $\partial^{v,+}_{w,\tau} := \{w' \in W^P \mid w > w' \geq \tau \geq v,\ l(w') = l(w) - 1\}$. If $w \geq \tau > v$, define $\partial^{v,-}_{w,\tau} := \{v' \in W^P \mid w \geq \tau \geq v' > v, l(v') = l(v) + 1\}$.

THEOREM 7.6.2. (1) *Suppose $w > \tau \geq v$. Then*

$$(\mathrm{mult}_\tau X^v_w) \deg_\tau w = \sum_{w' \in \partial^{v,+}_{w,\tau}} \mathrm{mult}_\tau X^v_{w'}.$$

(2) *Suppose $w \geq \tau > v$. Then*

$$(\mathrm{mult}_\tau X^v_w) \deg_\tau v = \sum_{v' \in \partial^{v,-}_{w,\tau}} \mathrm{mult}_\tau X^{v'}_w.$$

(3) $\mathrm{mult}_\tau X^\tau_\tau = 1$.

*Proof.* Since $X^\tau_\tau$ is a single point, (3) is trivial. We will prove (1); the proof of (2) is similar.

Let $H_\tau = \cup_{w' \in \partial^{v,+}_{w,\tau}} X^v_{w'}$. Let $\varphi^v_w(r)$ (resp. $\varphi_{H_\tau}(r)$) be the Hilbert function for the tangent cone of $X^v_w$ (resp. $H_\tau$) at $e(\tau)$, i.e.

$$\varphi^v_w(r) = \dim((M^v_{w,\tau})^r/(M^v_{w,\tau})^{r+1}).$$

Let

$$\mathcal{B}^v_{w,\tau}(r) = \Big\{p_{\tau_1} \ldots p_{\tau_m},\ \tau_i \in Z_\tau \mid (1)\, w \geq \tau_1 \geq \cdots \geq \tau_m \geq v, (2) \sum \deg_\tau(\tau_i) = r\Big\}.$$

Let

$$\mathcal{B}_1 = \{p_{\tau_1} \ldots p_{\tau_m} \in \mathcal{B}^v_{w,\tau}(r) \mid \tau_1 = w\},$$

$$\mathcal{B}_2 = \{p_{\tau_1} \ldots p_{\tau_m} \in \mathcal{B}^v_{w,\tau}(r) \mid \tau_1 < w\}.$$

We have $\mathcal{B}^v_{w,\tau}(r) = \mathcal{B}_1 \dot{\cup} \mathcal{B}_2$. Hence denoting $\deg_\tau(w)$ by $d$, we obtain

$$\varphi^v_w(r+d) = \varphi^v_w(r) + \varphi_{H_\tau}(r+d).$$

Taking $r \gg 0$ and comparing the coefficients of $r^{u-1}$, where $u = \dim X^v_w$, we obtain the result. □

COROLLARY 7.6.3. *Let $w > \tau > v$. Then*

$$(\text{mult}_\tau X^v_w)(\deg_\tau w + \deg_\tau v) = \sum_{w' \in \partial^{v,+}_{w,\tau}} \text{mult}_\tau X^v_{w'} + \sum_{v' \in \partial^{v,-}_{w,\tau}} \text{mult}_\tau X^{v'}_w.$$

THEOREM 7.6.4. *Let $w \geq \tau \geq v$. Then* $\text{mult}_\tau X^v_w = (\text{mult}_\tau X_w) \cdot (\text{mult}_\tau X^v)$.

*Proof.* We proceed by induction on $\dim X^v_w$.

If $\dim X^v_w = 0$, then $w = \tau = v$. In this case, by Theorem 7.6.2 (3), we have that $\text{mult}_\tau X^\tau_\tau = 1$. Since $e_\tau \in Be_\tau \subseteq X_w$, and $Be_\tau$ is an affine space open in $X_w$, $e_\tau$ is a smooth point of $X_w$, i.e. $\text{mult}_\tau X_w = 1$. Similarly, $\text{mult}_\tau X^v = 1$.

Next suppose that $\dim X^v_w > 0$, and $w > \tau \geq v$. By Theorem 7.6.2 (1),

$$\begin{aligned}
\text{mult}_\tau X^v_w &= \frac{1}{\deg_\tau w} \sum_{w' \in \partial^{v,+}_{w,\tau}} \text{mult}_\tau X^v_{w'} \\
&= \frac{1}{\deg_\tau w} \sum_{w' \in \partial^{v,+}_{w,\tau}} \text{mult}_\tau X_{w'} \cdot \text{mult}_\tau X^v \\
&= \left( \frac{1}{\deg_\tau w} \sum_{w' \in \partial^{v,+}_{w,\tau}} \text{mult}_\tau X_{w'} \right) \cdot \text{mult}_\tau X^v \\
&= \left( \frac{1}{\deg_\tau w} \sum_{w' \in \partial^{v,+}_{w,\tau}} \text{mult}_\tau X^{\text{id}}_{w'} \right) \cdot \text{mult}_\tau X^v \\
&= \left( \text{mult}_\tau X^{\text{id}}_w \right) \cdot \text{mult}_\tau X^v = \text{mult}_\tau X_w \cdot \text{mult}_\tau X^v.
\end{aligned}$$

The case of $\dim X^v_w > 0$ and $w = \tau > v$ is proven similarly. □

COROLLARY 7.6.5. *Let $w \geq \tau \geq v$. Then $X^v_w$ is smooth at $e_\tau$ if and only if both $X_w$ and $X^v$ are smooth at $e_\tau$.*

*Remark* 7.6.6. The following alternate proof of Theorem 7.6.4 is due to the referee, and we thank the referee for the same.

> *Identify $\mathcal{O}^-_\tau$ with the affine space $\mathbb{A}^N$ where $N = d(n-d)$, by the coordinate functions defined in §6.5. Then $X_w \cap \mathcal{O}^-_\tau$ and $X^v \cap \mathcal{O}^-_\tau$ are closed subvarieties of that affine space, both invariant under scalar*

*multiplication (e.g. by Corollary 7.2.4). Moreover, $X_w$ and $X^v$ intersect properly along the irreducible subvariety $X_w^v$; in addition, the Schubert cells $C_w$ and $C^v$ intersect transversally (by* [22]*).*

*Now let $Y$ and $Z$ be subvarieties of $\mathbb{A}^N$, both invariant under scalar multiplication, and intersecting properly. Assume in addition that they intersect transversally along a dense open subset of $Y \cap Z$. Then*

$$\mathrm{mult}_{\mathbf{o}}(Y \cap Z) = \mathrm{mult}_{\mathbf{o}}(Y) \cdot \mathrm{mult}_{\mathbf{o}}(Z)$$

*where $\mathbf{o}$ is the origin of $\mathbb{A}^N$.*

*To see this, let $\mathbb{P}(Y)$, $\mathbb{P}(Z)$ be the closed subvarieties of $\mathbb{P}(\mathbb{A}^N) = \mathbb{P}^{N-1}$ associated with $Y$, $Z$. Then $\mathrm{mult}_{\mathbf{o}}(Y)$ equals the degree $\deg(\mathbb{P}(Y))$, and likewise for $Z$, $Y \cap Z$. Now*

$$\deg(\mathbb{P}(Y \cap Z)) = \deg(\mathbb{P}(Y) \cap \mathbb{P}(Z)) = \deg(\mathbb{P}(Y)) \cdot \deg(\mathbb{P}(Z))$$

*by the assumptions and the Bezout theorem (see [8], Proposition 8.4 and Example 8.1.11).*

It has also been pointed out by the referee that the above alternate proof in fact holds for Richardson varieties in a minuscule $G/P$, since the intersections of Schubert and opposite Schubert varieties with the opposite cell are again invariant under scalar multiplication (the result analogous to Corollary 7.2.4 for a minuscule $G/P$ follows from the results in [15]). Recall that for $G$ a semisimple algebraic group and $P$ a maximal parabolic subgroup of $G$, $G/P$ is said to be *minuscule* if the associated fundamental weight $\omega$ of $P$ satisfies

$$(\omega, \beta^*)(= 2(\omega, \beta)/(\omega, \beta)) \leq 1$$

for all positive roots $\beta$, where ( , ) denotes a $W$-invariant inner product on $X(T)$.

**7.7. Determinantal formula for mult $X_w^v$.** In this section, we extend the Rosenthal-Zelevinsky determinantal formula (cf. [24]) for the multiplicity of a Schubert variety at a $T$-fixed point to the case of Richardson varieties. We use the convention that the binomial coefficient $\binom{a}{b} = 0$ if $b < 0$.

THEOREM 7.7.1. *(Rosenthal-Zelevinsky) Let $w = (i_1, \ldots, i_d)$ and $\tau = (\tau_1, \ldots, \tau_d)$ be such that $w \geq \tau$. Then*

$$\mathrm{mult}_\tau X_w = (-1)^{\kappa_1 + \cdots + \kappa_d} \begin{vmatrix} \binom{i_1}{-\kappa_1} & \cdots & \binom{i_d}{-\kappa_d} \\ \binom{i_1}{1-\kappa_1} & \cdots & \binom{i_d}{1-\kappa_d} \\ \vdots & & \vdots \\ \binom{i_1}{d-1-\kappa_1} & \cdots & \binom{i_d}{d-1-\kappa_d} \end{vmatrix},$$

*where $\kappa_q := \#\{\tau_p \mid \tau_p > i_q\}$, for $q = 1, \ldots, d$.*

LEMMA 7.7.2. *$\operatorname{mult}_\tau X^v = \operatorname{mult}_{w_0\tau} X_{w_0 v}$, where $w_0 = (n+1-d, \ldots, n)$.*

*Proof.* Fix a lift $n_0$ in $N(T)$ of $w_0$. The map $f : X^v \to n_0 X^v$ given by left multiplication is an isomorphism of algebraic varieties. We have $f(e_\tau) = e_{w_0\tau}$, and $n_0 X^v = n_0 \overline{B^- e_v} = n_0 \overline{n_0 B n_0 e_v} = \overline{B n_0 e_v} = \overline{B e_{w_0 v}} = X_{w_0 v}$. □

THEOREM 7.7.3. *Let $w = (i_1, \ldots, i_d)$, $\tau = (\tau_1, \ldots, \tau_d)$, and $v = (j_1, \ldots, j_d)$ be such that $w \geq \tau \geq v$. Then*

$$\operatorname{mult}_\tau X_w^v = (-1)^c \left| \begin{pmatrix} \binom{i_1}{-\kappa_1} & \cdots & \binom{i_d}{-\kappa_d} \\ \binom{i_1}{1-\kappa_1} & \cdots & \binom{i_d}{1-\kappa_d} \\ \vdots & & \vdots \\ \binom{i_1}{d-1-\kappa_1} & \cdots & \binom{i_d}{d-1-\kappa_d} \end{pmatrix} \cdot \begin{pmatrix} \binom{n+1-j_d}{-\gamma_d} & \cdots & \binom{n+1-j_1}{-\gamma_1} \\ \binom{n+1-j_d}{1-\gamma_d} & \cdots & \binom{n+1-j_1}{1-\gamma_1} \\ \vdots & & \vdots \\ \binom{n+1-j_d}{d-1-\gamma_d} & \cdots & \binom{n+1-j_1}{d-1-\gamma_1} \end{pmatrix} \right|,$$

*where $\kappa_q := \#\{\tau_p \mid \tau_p > i_q\}$, for $q = 1, \ldots, d$, and $\gamma_q := \#\{\tau_p \mid \tau_p < j_q\}$, for $q = 1, \ldots, d$, and $c = \kappa_1 + \cdots + \kappa_d + \gamma_1 + \cdots + \gamma_d$.*

*Proof.* Follows immediately from Theorems 7.6.4, 7.7.1, and Lemma 7.7.2, in view of the fact that $w_0\tau = (n+1-\tau_d, \ldots, n+1-\tau_1)$ and $w_0 v = (n+1-j_d, \ldots, n+1-j_1)$. □

DEPARTMENT OF MATHEMATICS, NORTHEASTERN UNIVERSITY, BOSTON, MA 02115
*E-mail:* vkreiman@lynx.neu.edu

DEPARTMENT OF MATHEMATICS, NORTHEASTERN UNIVERSITY, BOSTON, MA 02115
*E-mail:* lakshmibai@neu.edu

## REFERENCES

[1] A. Borel, Linear algebraic groups, GTM, 126, Second edition, Springer-Verlag, New York, 1991.
[2] N. Bourbaki, Groupes et Algèbres de Lie, Hermann, Paris, 1968, chapitres 4, 5 et 6.
[3] M. Brion, Positivity in the Grothendieck group of complex flag varieties, preprint, *math.* AG/0105254, 2001.
[4] M. Brion and V. Lakshmibai, A geometric approach to standard monomial theory, preprint *math.* AG/0111054, 2001.
[5] J. Carrell, On the smooth points of a Schubert variety, CMS Conference Proceedings, vol 16, 15–24, Proceedings of the conference on "Representations of Groups: Lie, Algebraic, Finite, and Quantum," *Banff, Alberta,* 1994, pp. 15–14.

[6] V. Deodhar, On some geometric aspects of Bruhat orderings, I – A finer decomposition of Bruhat cells, *Inv., Math.* (1985), 499–511.

[7] D. Eisenbud, Commutative algebra with a view toward Algebraic Geometry, Springer-Verlag, GTM, 150.

[8] W. Fulton, Intersection Theory, 2nd ed., Springer-Verlag, 1998.

[9] W. V. D. Hodge, Some enumerative results in the theory of forms, *Proc. Camb. Phil. Soc.* **39** (1943), 22–30.

[10] W. V. D. Hodge and D. Pedoe, Methods of Algebraic Geometry, vols. 1 and 2, Cambridge University Press, 1953.

[11] D. Kazhdan and G. Lusztig, Representations of Coxeter groups and Hecke algebras, *Inv. Math.,* **53** (1979), 165–184.

[12] D. Kazhdan and G. Lusztig, Schubert varieties and Poincaré duality, *Proc. Symp. Pure. Math., A.M.S.* **36** (1980), 185–203.

[13] V. Lakshmibai and N. Gonciulea, Flag Varieties, Hermann, Éditeurs Des Sciences Et Des Arts, 2001.

[14] V. Lakshmibai and P. Littelmann, Equivariant K-theory and Richardson varieties, preprint, *math.* AG/0201075, 2002.

[15] V. Lakshmibai and J. Weyman, Multiplicities of points on a Schubert variety in a minuscule G/P, *Advances in Math.* **84** (1990), 179–208.

[16] V. Lakshmibai and C.S. Seshadri, Geometry of G/P-V, *J. Algebra* **100** (1986), 462–557.

[17] ———, Standard monomial theory, Proc. Hyderabad Conference on Algebraic Groups (S. Ramanan et al., eds.), Manoj Prakashan, Madras (1991), 279–323.

[18] P. Littelmann, A Littlewood—Richardson formula for symmetrizable Kac-Moody algebras. *Invent. Math.* **116** (1994) 329–346.

[19] P. Littelmann, Contracting modules and standard monomial theory. *JAMS* **11** (1998) 551–567.

[20] V. B. Mehta and A. Ramanathan Frobenius splitting and cohomology vanishing for Schubert varieties, *Ann. Math.* **122** (1985), 27–40.

[21] C. Musili, Postulation formula for Schubert varieties, *J. Indian Math. Soc.* **36** (1972), 143–171.

[22] R. W. Richardson, Intersections of double cosets in algebraic groups, *Indag. Math.* **3** (1992), 69–77.

[23] R. W. Richardson, G. Rörle, and R. Steinberg, Parabolic subgroups with abelian unipotent radical, *Inv. Math.* **110** (1992), 649–671.

[24] J. Rosenthal and A. Zelevinsky, Multiplicities of points on Schubert varieties in Grassmannians, *J. Algebraic Combin.* **13** (2001), 213–218.

[25] J.P. Serre, Faisceaux algèbriques cohérentsés, *Ann. Math.* **61** (1955), 197–278.

[26] R. Stanley, Some conbinatorial aspects of the Schubert calculus, *Combinatoire et Représentation du Groupe Symétrique, Lecture Notes in Mathematics,* Vol. 579, Springer-Verlag, 1977.

CHAPTER 21

# TYPES AND COVERS FOR SL(2)

By PHILIP KUTZKO

---

Let $F$ be a $p$-adic field and let $G$ be the set of $F$-points of a connected reductive group over $F$. Then it is a fundamental result of Bernstein that the category of smooth complex representations $\mathfrak{R}(G)$ of $G$ decomposes as a product of certain subcategories $\mathfrak{R}^{\mathfrak{s}}(G)$. Further, one knows that each of these categories is equivalent to the category of left $R$-modules of some ring with identity $R$. In many cases, the ring $R$ may be chosen to be a certain convolution algebra $\mathcal{H}(G, \lambda)$ associated to an irreducible representation $\lambda$ of a compact subgroup $J$ of $G$. In this case, one says that $(J, \lambda)$ is a *type* for $\mathfrak{R}^{\mathfrak{s}}(G)$ in $G$ or, simply, an $\mathfrak{s}$-*type*. It should be noted that the equivalence of categories $\mathfrak{R}^{\mathfrak{s}}(G) \cong \mathcal{H}(G, \lambda)$-Mod is quite explicit in this case and that the algebra $\mathcal{H}(G, \lambda)$ is, in all known cases, isomorphic to an affine Hecke algebra (with possibly unequal parameters). Thus, one may hope to transfer questions about the representation theory of $G$ to questions concerning the module theory of affine Hecke algebras, a theory that has been much studied [Lu].

Let $P = LN$ be an $F$-parabolic subgroup of $G$ and write $\iota = \iota_P : \mathfrak{R}(L) \to \mathfrak{R}(G)$ for the functor of normalized parabolic induction. Then $\iota$ maps Bernstein subcategories of $L$ to Bernstein subcategories of $G$ and, in certain cases, this is captured by a homomorphism of rings $j : \mathcal{H}(L, \lambda_L) \to \mathcal{H}(G, \lambda)$ for appropriate types $(J_L, \lambda_L)$, $(J, \lambda)$ in $L$ and $G$ respectively. When this happens, we say that $(J, \lambda)$ is a *G-cover* for $(J_L, \lambda_L)$.

If, in particular, $P$ is a maximal (proper) parabolic subgroup and if the Bernstein subcategory in $\mathfrak{R}(L)$ is one whose irreducible objects are supercuspidal, then one expects the resulting algebra $\mathcal{H}(G, \lambda)$ to be isomorphic to an affine Hecke algebra with two generators. In theory, then, a study of the module theory and harmonic analysis of these algebras should yield information about the representation theory and harmonic analysis of the functor $\iota_P$, information which, in light of the work of Langlands, Shahidi, and others has arithmetical implications [Sh].

We have recently, together with Aubert and Morris, undertaken such a study and have, among other things, given an analog of the Plancherel formula for affine Hecke algebras with two generators. (See also [Mat].) In case $G = SL(2, F)$ this formula, when combined with results of [BK] and [BHK], is sufficient to determine Plancherel measure for $G$ given that one has constructed types and covers for this group. The purpose of this paper is to provide these constructions.

---

Research partially supported by NSF grant DMS-9503140

This paper is organized as follows. In §1 we give a description of the Bernstein subcategories of $\mathfrak{R}(G)$ and note that there are two different kinds of subcategories: those whose irreducible objects are supercuspidal and those which arise by parabolic induction from the subgroup $L$ of diagonal matrices. In the former case, the existence of a type follows from the fact that irreducible supercuspidal representations of $G$ may be constructed by induction from compact, open subgroups. In the latter case, types may be constructed as covers; this is done in §2. In §3 we describe the algebras $\mathcal{H}(G, \lambda)$ and give isomorphisms to affine Hecke algebras with at most two generators. In §4 we use the isomorphisms given in §3 to transfer representations of $G$ to modules over our affine Hecke algebras.

We note that the results here have been known to us for some time and that there are now more general (though less explicit) results available for the groups $SL(N, F)$ [GR]. We are publishing them now in the hope that, apart from the applications described above, these results may serve as an introduction to the method of types and covers.

It should be noted that Shalika was among the first to exploit the existence of compact, open subgroups in the study of the representation theory and harmonic analysis of $p$-adic groups. My own work and the work of many others has been greatly influenced by a reading of Shalika's papers in this field. It is a pleasure to acknowledge this influence on the occasion of his sixtieth birthday.

**1. Background.** Let $F$ be a $p$-adic field with ring of integers $\mathcal{O}$ and maximal ideal $P$. Write $\nu$ for the usual (additive) valuation and fix $\varpi \in F$ with $\nu(\varpi) = 1$. Set $G = SL(2, F)$ and write $L, N, \bar{N}$ respectively for the diagonal, upper unipotent, and lower unipotent subgroups of $G$. Set $B = LN, \ \bar{B} = L\bar{N}$. It is a consequence of a general result of Bernstein [BD] that the category $\mathfrak{R}(G)$ of smooth complex representations of $G$ decomposes into a product of full subcategories

$$\mathfrak{R}(G) \cong \prod_{\mathfrak{s} \in \mathcal{B}(G)} \mathfrak{R}^{\mathfrak{s}}(G)$$

where $\mathcal{B}(G)$ is an indexing set whose elements we now describe for the case at hand.

(1) *Supercuspidal elements*

For each irreducible supercuspidal representation $\sigma$ of $G$ we write $\mathfrak{s}(\sigma)$ for the equivalence class of $\sigma$ in $\mathfrak{R}(G)$ . We let $\mathfrak{R}^{\mathfrak{s}(\sigma)}(G)$ be the full subcategory of $\mathfrak{R}(G)$ whose objects are isomorphic to sums of copies of $\sigma$.

(2) *Induced elements*

We identify $L$ with $F^\times$ via $a \to \begin{bmatrix} a & \\ & a^{-1} \end{bmatrix}$, $a \in F^\times$ and we note that under this identification, $\mathcal{O}^\times$ corresponds to the maximal compact subgroup $L^0$ of $L$. We write $\mathbf{X}_F, \ \mathbf{X}_{\mathcal{O}}$ for the groups of (not necessarily unitary) characters of $F^\times, \ \mathcal{O}^\times$ respectively, and, given $\chi \in \mathbf{X}_{\mathcal{O}}$, we let $\mathfrak{R}^\chi(L)$ be the full subcategory $\mathfrak{R}(L)$ whose objects $(\pi, V)$ have the property that $\pi(x)v = \chi(x)v, \ x \in L^0, \ v \in V$. Then one may easily verify that the category $\mathfrak{R}(L)$ decomposes as a product of its subcategories $\mathfrak{R}^{\chi}(L)$:

$$\mathfrak{R}(L) \cong \Pi_{\chi \in \mathbf{X}_{\mathcal{O}}} \mathfrak{R}^{\chi}(L).$$

Indeed, this is just the Bernstein decomposition of $\mathfrak{R}(L)$.

Given $\chi \in \mathbf{X}_{\mathcal{O}}$ we set $\mathfrak{s}(\chi) = \{\chi, \chi^{-1}\}$ and we let $\mathfrak{R}^{\mathfrak{s}(\chi)}(G)$ be the full subcategory of $\mathfrak{R}(G)$ whose objects are subrepresentations of representations of the form $\iota_B(\nu)$ where $\nu$ is an object in $\mathfrak{R}^{\chi}(L)$. (Here, $\iota_B : \mathfrak{R}(L) \to \mathfrak{R}(G)$ is the functor of normalized induction.) We note that $\mathfrak{R}^{\mathfrak{s}(\chi)}(G)$ depends only on the set $\mathfrak{s}(\chi)$.

The set $\mathcal{B}(G)$ is then just the union of the sets $\{\mathfrak{s}(\sigma)\}$, $\{\mathfrak{s}(\chi)\}$, $\sigma$, $\chi$ as above.

Let $J$ be a compact open subgroup of $G$, let $(\lambda, W)$ be a smooth irreducible representation of $J$ and write $(\hat{\lambda}, \hat{W})$ for the contragredient representation. Fixing once and for all Haar measure on $G$ one then has associated to the pair $(J, \lambda)$ the convolution (Hecke) algebra $\mathcal{H}(G, \lambda)$ of $\mathrm{End}_{\mathbb{C}}(\hat{W})$-valued functions $f$ on $G$ which satisfy $f(hxk) = \hat{\lambda}(h) f(x) \hat{\lambda}(k)$, $x \in G$, $h, k \in J$. Further, given a smooth $G$-representation $(\pi, V)$ there is a natural left $\mathcal{H}(G, \lambda)$-module structure on the space $V_\lambda = \mathrm{Hom}_J(W, V)$. (See §2 of [BK].) Given $\mathfrak{s} \in \mathcal{B}(G)$ we say that the pair $(J, \lambda)$ as above is a *type* for $\mathfrak{s}$ (or an $\mathfrak{s}$-*type*) if the map $(\pi, V) \to V_\lambda$ induces an equivalence of categories $\mathfrak{R}^{\mathfrak{s}}(G) \cong \mathcal{H}(G, \lambda)$-Mod.

In the case that $\mathfrak{s} = \mathfrak{s}(\sigma)$ is a supercuspidal element the existence of a type for $\mathfrak{s}$ is a consequence of the fact [KS] that there exists an irreducible representation $\lambda$ of a compact, open subgroup $J$ of $G$ such that

$$\sigma \cong \mathrm{ind}_J^G \lambda.$$

(Here, $\mathrm{ind}_J^G : \mathfrak{R}(J) \to \mathfrak{R}(G)$ is the functor of compact induction.)

In fact $(J, \lambda)$ above is an $\mathfrak{s}$-type and one notes that $\mathcal{H}(G, \lambda)$ is just the trivial $\mathbb{C}$-algebra in this case and that $\sigma$ corresponds to the trivial $\mathcal{H}(G, \lambda)$-module under the map $V \to V_\lambda$.

## 2. Types and covers.

**2.1.** Our goal is now to construct $\mathfrak{s}$-types in case $\mathfrak{s} = \mathfrak{s}(\chi)$ is an induced element. We will construct these types using the method of *covers* [BK]. To begin with, it is worth noting that for a character $\chi \in \mathbf{X}_{\mathcal{O}}$, the pair $(L^0, \chi)$ is a $\chi$-type in $L$; that is, $(L^0, \chi)$ is a type for the subcategory $\mathfrak{R}^{\chi}(L)$ of $\mathfrak{R}(L)$. Next, recall that a $G$-*cover* for $(L^0, \chi)$ is a pair $(J, \lambda)$ where $J$ is a compact open subgroup of $G$ and $\lambda$ is an irreducible representation of $J$ (necessarily one dimensional in this case), this pair having the following properties:

(1) $J = (J \cap \bar{N}) L^0 (J \cap N)$.

(2) $J \cap \bar{N}$, $J \cap N < \ker \lambda$; $\lambda|_{L^0} = \chi$.

(3) There are positive integers $n_1, n_2$ and invertible elements $f_1, f_2 \in \mathcal{H}(G, \lambda)$ such that $f_1, f_2$ are supported respectively on the double cosets $J\Pi^{n_1}J$, $J\Pi^{-n_2}J$ where $\Pi = \begin{bmatrix} \varpi & \\ & \varpi^{-1} \end{bmatrix}$.

(One checks that these properties are specializations to the present context of the defining properties of a cover given in 8.1 of [BK].) It follows from 8.3 of [BK] that a $G$-cover of $(L^0, \chi)$ is an $\mathfrak{s}(\chi)$ type; we construct these covers as follows:

For $\chi \in \mathbf{X}_{\mathcal{O}}$ define the integer $\mathrm{sw}(\chi)$ by setting $\mathrm{sw}(\chi) = 1$ if $(1 + P) < \ker \chi$ and letting $\mathrm{sw}(\chi)$ be the smallest integer $n$ so that $(1 + P^n) < \ker \chi$ otherwise. Now define the subgroup $J = J_\chi$ by

$$J = \left\{[c_{ij}] \in G \mid c_{11}, c_{22} \in \mathcal{O}^\times,\ c_{12} \in \mathcal{O},\ c_{21} \in P^{\mathrm{sw}(\chi)}\right\}$$

and define a function $\lambda = \lambda_\chi$ on $J$ by

$$\lambda([c_{ij}]) = \chi(c_{11}).$$

Then we will prove

PROPOSITION. *The pair $(J_\chi, \lambda_\chi)$ is a $G$-cover for $(L^0, \chi)$, $\chi \in \mathbf{X}_{\mathcal{O}}$.*

*Remark.* One checks that $\lambda$ is in fact a character and that properties (1) and (2) above hold for $(J, \lambda)$. We will show in the next two sections that property (3) holds as well.

**2.2.** We continue with the notation of the previous section. In addition we set $K = SL(2, \mathcal{O}_F)$ and we let $I$ be the Iwahori subgroup

$$I = \{[c_{ij}] \in K \mid c_{21} \in P\}$$

Then one has $K = I \cup IwI$ where $w = \begin{bmatrix} 0 & 1 \\ -1 & 0 \end{bmatrix}$.

Recall that an element $x \in G$ is said to *intertwine* $\lambda$ if $\lambda(xkx^{-1}) = \lambda(k)$, $k \in J \cap x^{-1}Jx$. We denote by $\mathcal{I} = \mathcal{I}_G(\lambda)$ the set of elements in $G$ which intertwine $\lambda$. Recall also that

$$\mathcal{H}(G, \lambda) = \oplus_{x \in \mathcal{I}} \mathcal{H}_x(G, \lambda)$$

where $\mathcal{H}_x(G, \lambda)$ is the space of functions in $\mathcal{H}(G, \lambda)$ which are supported on the double coset $JxJ$. Since $\lambda$ is one dimensional it follows that $\mathcal{H}_x(G, \lambda)$ is one-dimensional; indeed, $\mathcal{H}_x(G, \lambda)$ is spanned by the function $g_x$ where $g_x$ is supported on $JxJ$ and is given there by the formula $f(hxk) = \lambda(h^{-1})\lambda(k^{-1})$, $h, k \in J$. (Recall that the contragredient of $\lambda$ is just $\lambda^{-1}$ in this case.). We will need the following lemma.

LEMMA. (1) *We have $(I\Pi^n I) \cap \mathcal{I} = J\Pi^n J$, $n \in \mathbb{Z}$.*
(2) *If $\chi^2 \neq 1$ then $(J\bar{N}J) \cap \mathcal{I} = J$.*

*Proof.* (1) If $\mathrm{sw}(\chi) \leq 1$ there is nothing to prove. Suppose then that $\mathrm{sw}(\chi) \geq 2$ and assume without loss of generality that $n \geq 0$. (One notes that $\mathcal{I}$ is stable under $x \to x^{-1}$.) Then $\Pi^{-n}(I \cap \bar{N})\Pi^n \subset I$ and so $I\Pi^n I = J(I \cap \bar{N})\Pi^n I = J\Pi^n I$. Thus we only need to show that the element $a = \Pi^n \begin{bmatrix} 1 & 0 \\ y & 1 \end{bmatrix}$ does not intertwine

$\lambda$ if $1 \le \nu(y) \le \mathrm{sw}(\chi) - 1$. To this end, we fix $y$ with $1 \le \nu(y) \le \mathrm{sw}(\chi) - 1$, pick $z \in P^{\mathrm{sw}(\chi)-1}$ with $\chi(1+z) \ne 1$ and set

$$c(y) = \begin{bmatrix} 1 & 0 \\ y & 1 \end{bmatrix}, \quad b = \begin{bmatrix} 1 & 0 \\ -zy(1+z)^{-1} & 1 \end{bmatrix} \begin{bmatrix} 1 & -zy^{-1} \\ 0 & 1 \end{bmatrix}.$$

Then $b \in J$ and one computes that $aba^{-1} = \begin{bmatrix} 1+z & -zy^{-1}\Pi^{2n} \\ 0 & (1+z)^{-1} \end{bmatrix}$. Thus $aba^{-1} \in J$ and we have $\lambda(aba^{-1}) = \chi(1+z) \ne 1 = \chi(b)$. It follows that $a$ does not intertwine $\lambda$ as was to be shown.

(2) We may now assume that $\nu_F(y) \le 0$. Pick $a \in \mathcal{O}_F^\times$ with $\chi(a^2) \ne 1$ and set

$$x_1 = \begin{bmatrix} a & (a-a^{-1})y^{-1} \\ 0 & a^{-1} \end{bmatrix}, \quad x_2 = \begin{bmatrix} a^{-1} & (a-a^{-1})y^{-1} \\ 0 & a \end{bmatrix}.$$

Then $x_i \in J$, $i = 1, 2$ and $c(y)x_1c(y)^{-1} = x_2$. Since $\lambda(x_1) = \chi(a) \ne \chi(a^{-1}) = \lambda(x_2)$, we see that $c(y)$ does not intertwine $\lambda$. □

We may now prove

PROPOSITION. *Suppose that $\chi^2 \ne 1$. Then $(J, \lambda)$ is a cover for $(L^0, \chi)$.*

*Proof.* We have to verify condition (3) of the previous section. First, we note that the elements $\Pi$, $\Pi^{-1}$ intertwine $\lambda$; we set $f_1 = g_\Pi$, $f_2 = g_{\Pi^{-1}}$. Now the convolution $f_1 * f_2$ is supported on $(J\Pi J\Pi^{-1}J) \cap \mathcal{I}$ and since $\Pi J \Pi^{-1} \subset J\bar{N}J$, it follows from our lemma that $f_1 * f_2$ is supported on $J$. A direct calculation shows that $f_1 * f_2(1) \ne 0$. (See Lemma 3.2 below.) It follows that $f_1 * f_2$ is a non-zero scalar multiple of $g_1$ which is in turn a scalar multiple of the identity element in $\mathcal{H}(G, \lambda)$. Thus both $f_1$, $f_2$ are invertible as was to be shown. □

**2.3.** Continuing with the notation of the previous section we now assume that $\chi^2 = 1$.

LEMMA. *We have*

$$\mathcal{I}_K(\lambda) = J \cup JwJ$$

*Proof.* We have $K = I \cup IwI$ and we claim that $IwI = JwJ$. To see this we note that

$$I = (I \cap B)(I \cap \bar{N}) = (I \cap \bar{N})(I \cap B)$$

and also that

$$w^{-1}(I \cap \bar{N})w, \; w(I \cap \bar{N})w^{-1} \subset I \cap B.$$

It follows that $IwI = (I \cap B)(I \cap \bar{N})wI = (I \cap B)wI$ and similarly that $(I \cap B)wI = (I \cap B)w(I \cap B)$. Since $I \cap B \subset J$ we have $IwI \subset JwJ$ whence our claim, the other containment being obvious. Thus $K = I \cup JwJ$. But by the lemma in the preceding section, $\mathcal{I} \cap I = J$ and so $I_K \subset J \cup JwJ$. On the other hand, $w$ certainly intertwines $\lambda$ whence our lemma. □

COROLLARY. *The element $g_w$ is invertible in $\mathcal{H}(G, \lambda)$.*

*Proof.* The element $g_w * g_w$ is supported on $K \cap \mathcal{I}$. By our lemma, we have that $g_w * g_w = ag_w + bg_1$ for scalars $a, b$. One checks that $g_w * g_w(1) \neq 0$ whence $b \neq 0$. (See Proposition 3.2 below.) Since $g_1$ is a non-zero scalar multiple of the identity element of $\mathcal{H}(G, \lambda)$, it follows that $g_w$ satisfies a quadratic equation with non-zero constant term and is thus invertible. □

PROPOSITION. *$(J, \lambda)$ is a $G$-cover for $(L^0, \chi)$ in case $\chi^2 = 1$.*

*Proof.* Let $\alpha = \begin{bmatrix} 0 & 1 \\ \varpi^{\mathrm{sw}(\chi)} & 0 \end{bmatrix}$. Then conjugation by $\alpha$ induces an outer automorphism of $G$ that takes $J$ to $J$ and fixes $\lambda$. This automorphism thus induces an algebra automorphism $\tilde{\alpha}$ of $\mathcal{H}(G, \lambda)$ that takes $g_w$ to $g_{w'}$ where $w' = \alpha w \alpha^{-1} = \begin{bmatrix} 0 & -\varpi^{-\mathrm{sw}(\chi)} \\ \varpi^{\mathrm{sw}(\chi)} & 0 \end{bmatrix}$. Thus $g_{w'}$ is invertible in $\mathcal{H}(G, \lambda)$. It follows that $g_w * g_{w'}$ is invertible in $\mathcal{H}(G, \lambda)$ and is supported on $JwJw'J$. One checks that $JwJw'J = Jww'J = J\Pi^{\mathrm{sw}(\chi)}J$ so that $g_w * g_{w'}$ is invertible in $\mathcal{H}(G, \lambda)$ and supported on $J\Pi^{\mathrm{sw}(\chi)}J$. Now conjugation by $\alpha$ takes $\Pi$ to $\Pi^{-1}$ and so $\tilde{\alpha}(g_w * g_{w'})$ is invertible in $\mathcal{H}(G, \lambda)$ and supported on $J\Pi^{-\mathrm{sw}(\chi)}J$.

Thus we may satisfy condition (3) in §2.1 by setting $f_1 = g_w * g_{w'}$, $f_2 = \tilde{\alpha}(g_w * g_{w'})$. This proves our proposition and completes our proof of Proposition 21. □

## 3. The algebras $\mathcal{H}(G, \lambda)$.

**3.1.** We now apply Theorem 11.4 of [BK] to give a preliminary description of the algebra $\mathcal{H}(G, \lambda)$. To begin we, we note that the algebra $\mathcal{H}(L, \chi)$ has a particularly simple structure. Indeed, if we let $h$ be the function in $\mathcal{H}(L, \chi)$ which is supported on $L^0\Pi$ and defined there by $h(k\Pi) = \bar{\chi}(k)$, $k \in L^0$, then we have $\mathcal{H}(L, \chi) = \mathbb{C}[h, h^{-1}]$. Let $h_\Pi = \frac{1}{q}g_\Pi$ and define a map of algebras $j^+ : \mathbb{C}[h] \to \mathcal{H}(G, \lambda)$ be setting $j^+(h) = h_\Pi$. Then it is a consequence of Theorem 7.2 of [BK] that $j^+$ extends to an injective map of algebras $j : \mathcal{H}(L, \chi) \to \mathcal{H}(G, \lambda)$. (We will explain the factor $\frac{1}{q}$ in §4 below.) Set $\mathcal{B} = j(\mathcal{H}(L, \chi))$, $\mathcal{K} = \mathcal{H}(K, \lambda)$. Then we may apply Theorem 11.4 of [BK] to Theorem 2.1 above to obtain:

PROPOSITION. *The map $(\phi, f) \to \phi * f$ induces an isomorphism of left $\mathcal{B}$-modules*

$$\mathcal{B} \otimes_{\mathbb{C}} \mathcal{K} \cong \mathcal{H}(G, \lambda).$$

In case $\chi^2 \neq 1$, this proposition is sufficient to completely determine the structure of $\mathcal{H}(G, \lambda)$:

COROLLARY. *Suppose that $\chi^2 \neq 1$. Then the map $j$ above is an isomorphism.*

*Proof.* As we have seen, $K = I \cup JwJ$ and we have also seen that $\mathcal{I} \cap I = J$. On the other hand, $w$ does not intertwine $\chi$ since $\chi^2 \neq 1$. It follows that $\mathcal{K}$ is one dimensional whence our result. □

**3.2.** Our goal is now to give an explicit description of the algebra $\mathcal{H}(G, \lambda)$ in case $\chi^2 = 1$. (We note in this case that $\chi = \chi^{-1}$ so that we may replace $\lambda^{-1}$ by $\lambda$ in the definition of $\mathcal{H}(G, \lambda)$; see §1.) To facilitate calculations we fix Haar measure on $G$ such that $J$ has measure one. In addition, we set

$$w_1 = w = \begin{bmatrix} 0 & 1 \\ -1 & 0 \end{bmatrix}, \quad w_2 = \begin{bmatrix} 0 & -\varpi^{-1} \\ \varpi & 0 \end{bmatrix}.$$

PROPOSITION. *If $\chi$ is the trivial character then we have*

$$g_{w_i} * g_{w_i} = (q-1)g_{w_i} + qg_1, \quad i = 1, 2.$$

*Otherwise we have*

$$g_{w_i} * g_{w_i} = \chi(-1)qg_1 \quad \textit{if} \quad sw(\chi) = 1$$

*while*

$$g_{w_1} * g_{w_1} = \chi(-1)q^{sw(\chi)}g_1, \quad g_{w_2} * g_{w_2} = \chi(-1)q^{sw(\chi)-2}g_1 \quad \textit{if} \quad sw(\chi) \geq 2.$$

*Proof.* We begin with a lemma: □

LEMMA. *Let $a, b \in G$. Then*
(1) *We have*

$$g_a * g_b(x) = [J : J \cap aJa^{-1}] \int_J \lambda(k^{-1}) g_b(a^{-1}kx)dk, \; x \in G.$$

(2) *Write $c(u) = \begin{bmatrix} 1 & 0 \\ u & 1 \end{bmatrix}$, $u \in F$ and fix Haar measure on $F$ so that $\mathcal{O}$ has measure one. Then*

$$g_{w_1} * g_b(x) = q^{\mathrm{sw}(\chi)} \int_{\mathcal{O}} g_b\big(c(u)w_1^{-1}x\big)du \; x \in G.$$

*Proof of lemma.* By definition, we have

$$g_a * g_b(x) = \int_G g_a(y)g_b(y^{-1}x)\,dy.$$

Now we use the fact that the integrand is supported on $JaJ$ and is right invariant under $J$ together with the fact that the map $k \to ka$ induces a bijection $J/(J \cap aJa^{-1}) \to JaJ/J$ to conclude that

$$\int_G g_a(y) g_b(y^{-1}x)\,dy = [J : J \cap aJa^{-1}] \int_J g_a(ka) g_b(a^{-1}k^{-1}x)\,dk.$$

(We note that $G$ is unimodular.)

The first part of our lemma now follows on making the substitution $k \to k^{-1}$ in the last integral and noting that $g_a(k^{-1}a) = \lambda(k^{-1}),\ k \in J$.

Now set $a = w_1$. Then $J \cap w_1 J w_1^{-1} = \{[c_{ij}] \in J \mid c_{12} \in P^{\mathrm{sw}(\chi)}\}$ and so $[J : J \cap w_1 J w_1^{-1}] = q^{\mathrm{sw}(\chi)}$. The integrand in the first part of our lemma is left invariant under $J \cap \bar{B}$ and since $J = (J \cap \bar{B})(J \cap N)$ and $\lambda$ is trivial on $N$, we see that

$$g_{w_1} * g_b(x) = q^{\mathrm{sw}(\chi)} \int_{\mathcal{O}} g_b\left(w_1^{-1} \begin{bmatrix} 1 & -u \\ 0 & 1 \end{bmatrix} x\right) du.$$

The second part of our lemma now follows on noting that $w_1^{-1} \begin{bmatrix} 1 & -u \\ 0 & 1 \end{bmatrix} = c(u) w_1^{-1}$. □

*Proof of proposition.* We know that $g_{w_1} * g_{w_1} = a_1 g_{w_1} + b_1 g_{w_1}$ for some scalars $a_1, b_1$ and clearly $a_1 = g_{w_1} * g_{w_1}(w_1)$, $b_1 = g_{w_1} * g_{w_1}(1)$. Now $[J : J \cap w_1 J w_1^{-1}] = q^{\mathrm{sw}(\chi)}$ and so by the first part of our lemma,

$$b_1 = q^{\mathrm{sw}(\chi)} \int_J \lambda(k^{-1}) g_{w_1}\left(w_1^{-1} k\right) dk.$$

But $w_1^{-1} = -w_1$ and so the integrand is just $\chi(-1)$. Thus $b_1 = \chi(-1) q^{\mathrm{sw}(\chi)}$ as was to be shown.

By the second part of our lemma we have

$$a_1 = q^{\mathrm{sw}(\chi)} \int_{\mathcal{O}} g_{w_1}(c(u))\,du.$$

Now one checks that $c(u) \in J w_1 J$ if and only if $u \in \mathcal{O}^\times$. If $\chi = 1$ then we see that $a_1 = q\mu(\mathcal{O}^\times) = q - 1$. Thus our proposition is proved in this case. Now suppose that $\chi \neq 1$. Then for $u \in \mathcal{O}^\times$, we have

$$c(u) = -\begin{bmatrix} u^{-1} & 1 \\ 0 & u \end{bmatrix} w_1 \begin{bmatrix} 1 & u^{-1} \\ 0 & 1 \end{bmatrix}$$

whence $g_{w_1}(c(u)) = \chi(-u^{-1})$. Therefore, $\int_{\mathcal{O}} g_{w_1}(c(u)) du = \int_{\mathcal{O}^\times} \chi(-u^{-1}) du = 0$ since $\chi$ is non-trivial and so $a_1 = 0$ as was to be shown in this case.

We now turn to the computation of $g_{w_2} * g_{w_2}$. Let

$$K_2 = \left\{[c_{ij}] \in G \mid c_{11}, c_{22} \in \mathcal{O},\ c_{12} \in P^{-1},\ c_{21} \in P\right\};$$

that is, $K_2 = \beta K \beta^{-1}$ for $\beta = \begin{bmatrix} 0 & 1 \\ \varpi & 0 \end{bmatrix}$. Then $J < K_2$; we set $\mathcal{K}_2 = \mathcal{H}(K_2, \lambda)$. Now it follows from 11.5 and 11.6 of [BK] that $\dim_{\mathbb{C}} \mathcal{K}_2 = 2$. Since $w_2$ intertwines $\lambda$, we see that $\{g_1, g_{w_2}\}$ is a basis for $\mathcal{K}_2$; in particular, $g_{w_2} * g_{w_2} = a_2 g_{w_2} + b_2 g_1$ for scalars $a_2, b_2$. Now if $\mathrm{sw}(\chi) = 1$, then $\beta J \beta^{-1} = J$ and conjugation by $\beta$ fixes $\lambda$. Thus conjugation by $\beta$ induces an algebra isomorphism $\mathcal{K} \cong \mathcal{K}_2$ and this isomorphism clearly takes $g_{w_1}$ to $g_{w_2}$. It follows that $a_2 = a_1,\ b_2 = b_1$ as was shown. We now may take $\mathrm{sw}(\chi) \geq 2$.

In this case, we note that $w_2 J w_2 \subset I$ so that $J w_2 J w_2 J \subset I$. Since, $g_{w_2} * g_{w_2}$ is supported on $J w_2 J w_2 J \cap \mathcal{I}$, it follows from the first part of Lemma 2.1 that $g_{w_2} * g_{w_2}$ is supported on $J$; that is, $a_2 = 0$. We have $b_2 = g_{w_2} * g_{w_2}(1)$ and by the first part of our lemma, $g_{w_2} * g_{w_2}(1) = \chi(-1)[J : J \cap w_2 J w_2^{-1}]$. Since $J \cap w_2 J w_2^{-1} = \{[c_{ij}] \in J \mid c_{12} \in P^{\mathrm{sw}(\chi)-2}\}$ we see that $[J : J \cap w_2 J w_2^{-1}] = q^{\mathrm{sw}(\chi)-2}$, whence our result. □

**3.3.** We continue with the notation of the preceding section and define functions $h_j \in \mathcal{H}(G, \lambda),\ j = 1.2$ by setting $h_j = \epsilon^{2j-1} |g_{w_j} * g_{w_j}|^{\frac{1}{2}} g_{w_j}$ where $\epsilon = 1$ if $\chi(1) = 1,\ \epsilon = i$ if $\chi(1) = -1$.

LEMMA. *We have* $h_1 * h_2 = h_\Pi$

*Proof.* We work first with $g_{w_1} * g_{w_2}$ and note that this function is supported on $J w_1 J w_2 J \cap \mathcal{I} \subset I w_1 I w_2 I \cap \mathcal{I}$. Now $I = (I \cap \bar{N})(I \cap B$ and $w_1(I \cap \bar{N}) w_1^{-1},\ w_2^{-1}(I \cap B) w_2 \subset I$; it follows that $I w_1 I w_2 I = I \Pi I$. This being so, it follows from Lemma 3.1 that $g_{w_1} * g_{w_2}$ is supported on $J \Pi J$; that is, $g_{w_1} * g_{w_2} = (g_{w_1} * g_{w_2}(1)) g_\Pi$. Now by the second part of Lemma 3.2 we have $g_{w_1} * g_{w_2}(1) = q^{\mathrm{sw}(\chi)} \int_{\mathcal{O}} g_{w_2}(c(u) w_2) du$ and we see that the integrand is 0 unless $c(u) \in (J w_2 J w_2^{-1}) \cap J$ in which case the integrand is identically one. A direct calculation shows that if $\mathrm{sw}(\chi) = 1$ then $c(u) \in (J w_2 J w_2^{-1}) \cap J$ if and only if $u \in P$, while, if $\mathrm{sw}(\chi) \geq 2$, then $c(u) \in (J w_2 J w_2^{-1}) \cap J$ if and only if $u \in P^2$. We conclude that $g_{w_1} * g_{w_2} = g_\Pi$ if $\mathrm{sw}(\chi) = 1$ while $g_{w_1} * g_{w_2} = q^{\mathrm{sw}(\chi)-2} g_\Pi$ if $\mathrm{sw}(\chi) \geq 2$. Our lemma now follows from the definition of the functions $h_i$ given above together with the calculation of the constants $g_{w_i} * g_{w_i}(1)$ given in the proof of Proposition 3.2. □

PROPOSITION. *Fix a real number* $l \geq 1$ *and let* $\mathcal{H}(l)$ *be the algebra with identity* $1_{\mathcal{H}(l)}$ *and two generators* $s_1, s_2$ *subject only to the relations*

$$s_i^2 = (l^{\frac{1}{2}} - l^{-\frac{1}{2}}) s_i + 1_{\mathcal{H}(l)}, \quad i = 1, 2.$$

*Then the map* $s_i \to h_i,\ i = 1, 2$ *induces an isomorphism of algebras*

$$\Phi : \mathcal{H}(l) \cong \mathcal{H}(G, \lambda)$$

*where* $l = q$ *if* $\chi$ *is trivial and* $l = 1$ *otherwise.*

*Proof.* The case that $\chi$ is trivial is well known and its proof will be omitted. (See, e.g., [Bo].) We assume now that $\chi$ is non-trivial. Then $\mathcal{H}(l)$ is just the group algebra on the infinite dihedral group with generators $s_1, s_2$ and we have shown that $h_i^2 = g_1,\ i = 1, 2$. Since $g_1$ is the identity element for $\mathcal{H}(G, \lambda)$ the map $s_i \to h_i$ certainly induces a homomorphism of algebras $\Phi : \mathcal{H}(l) \to \mathcal{H}(G, \lambda)$. Now the reduced words on $s_1, s_2$ form a $\mathbb{C}$-basis for $\mathcal{H}(l)$. But by Proposition 5, the set of elements $h_\Pi^n, h_\Pi^n h_1,\ n \in \mathbb{Z}$ form a $\mathbb{C}$-basis for $\mathcal{H}(G, \lambda)$ and, by our lemma, these are exactly the set of reduced words on $h_1, h_2$. Thus $\Phi$ maps a basis in $\mathcal{H}(l)$ onto a basis in $\mathcal{H}(G, \lambda)$ so that $\Phi$ is an isomorphism as was shown. □

**4. Parabolic induction.** We now turn to the question of computing the functor $\iota_B : \mathfrak{R}^\chi(L) \to \mathfrak{R}^\mathfrak{s}(G),\ \mathfrak{s} = \mathfrak{s}(\chi)$, using the method of types and covers. The basic tool here is Corollary 8.4 of [BK].

There is a unique embedding of algebras $t = t_B : \mathcal{H}(L, \chi) \to \mathcal{H}(G, \lambda)$ so that the following diagram commutes:

$$\begin{array}{ccc} \mathfrak{R}^\chi(L) & \xrightarrow{\mathrm{Ind}_B} & \mathfrak{R}^\mathfrak{s}(G) \\ \downarrow & & \downarrow \\ \mathcal{H}(L, \chi)\text{-Mod} & \xrightarrow[t_*]{} & \mathcal{H}(G, \lambda)\text{-Mod}. \end{array}$$

Here, the vertical arrows are given by the functors $X \to X_\chi,\ V \to V_\lambda$ respectively, the functor $\mathrm{Ind}_B^G$ is that of smooth (non-normalized) parabolic induction while the functor $t_*$ is given on objects by $t_*(M) = \hom_{\mathcal{H}(L,\chi)}(\mathcal{H}(G, \lambda), M)$ where $\mathcal{H}(G, \lambda)$ is viewed a left $\mathcal{H}(L, \chi)$-module via $(f, g) \to t(f)g,\ f \in \mathcal{H}(L, \chi),\ g \in \mathcal{H}(G, \lambda)$. One may thus say that the map $t_B$ implements parabolic induction at the level of Hecke algebras.

The map $t$ is given explicitly (§7 of [BK]); in our case, it is just the map determined by $t(h) = \delta g_\Pi$ where $\delta$ is the module of the action of $\Pi$ on $N$. One computes $\delta = [N(\mathcal{O}) : \Pi N(\mathcal{O})\Pi^{-1}]^{-1} = \frac{1}{q^2}$. Thus $t$ is determined by $t(h) = \frac{1}{q^2} g_\Pi$.

The functor $\iota_B$ of normalized parabolic induction is given by $\iota_B((\xi, X)) = \mathrm{Ind}_B^G((\xi \otimes \delta^{\frac{1}{2}}, X)$ for any smooth representation $(\xi, X)$ of $L$. It follows easily from the above commutative diagram that the map $j : \mathcal{H}(L, \chi) \to \mathcal{H}(G, \lambda)$ given in §3.1 implements normalized parabolic induction at the level of Hecke algebras; that is, that the following diagram commutes:

$$\begin{array}{ccc} \mathfrak{R}^\chi(L) & \xrightarrow{\iota_B} & \mathfrak{R}^\mathfrak{s}(G) \\ \downarrow & & \downarrow \\ \mathcal{H}(L, \chi)\text{-Mod} & \xrightarrow[j_*]{} & \mathcal{H}(G, \lambda)\text{-Mod}. \end{array}$$

Now in case $\chi^2 = 1$, Proposition 3.3 gives us another commutative diagram. To be precise, let $d = s_1 s_2$, set $\mathcal{D} = \mathbb{C}[d, d^{-1}] \subset \mathcal{H}(l)$ and define an algebra isomorphism $\Phi_l : \mathcal{D} \to \mathcal{H}(L, \chi)$ by $\Phi_L(d) = h$. Then $j(\Phi_L(d)) = \Phi(d)$ whence the following diagram commutes:

$$\begin{array}{ccc} \mathcal{H}(L,\chi)\text{-Mod} & \xrightarrow{j_*} & \mathcal{H}(G,\lambda)\text{-Mod} \\ \Phi_L^* \downarrow & & \downarrow \Phi^* \\ \mathcal{D}\text{-Mod} & \xrightarrow[\hom_{\mathcal{D}}(\mathcal{H}(l),\ -\ )]{} & \mathcal{H}(l)\text{-Mod}. \end{array}$$

If, on the other hand, $\chi^2 \neq 1$ then by Corollary 3.1 we have $\mathcal{H}(G,\lambda) = \mathbb{C}[h_\Pi, h_\Pi^{-1}]$, and so we have an isomorphism $\Psi : \mathcal{D} \to \mathcal{H}(G,\lambda)$ given by $\Psi(d) = h_\Pi$. (We think of $d$ as an indeterminate in case $\chi^2 \neq 1$.) We then have the commutative diagram

$$\begin{array}{ccc} \mathcal{H}(L,\chi)\text{-Mod} & \xrightarrow{j_*} & \mathcal{H}(G,\lambda)\text{-Mod} \\ \Phi_L^* \downarrow & & \downarrow \Psi^* \\ \mathcal{D}\text{-Mod} & = & \mathcal{D}\text{-Mod}. \end{array}$$

The following terminology will be useful.

*Definition.* Let $\chi \in \mathbf{X}_{\mathcal{O}}$.

(1) If $\chi^2 = 1$ then we say that the $\mathcal{H}(l)$-module $M$ corresponds to the smooth $G$-representation $(\pi, V)$ if $M = \Phi^*(V_\lambda)$.

(2) If $\chi^2 \neq 1$ then we say that the $\mathcal{D}$-module $M$ corresponds to the smooth $G$-representation $(\pi, V)$ if $M = \Psi^*(V_\lambda)$.

**4.1.** We consider the implications of §4.1 for the case of a principal series representation of $G$. To this end, we fix a character $\chi \in \mathbf{X}_{\mathcal{O}}$ and for $t \in \mathbb{C}$ we define the character $\chi_t$ on $F^\times$ by

$$\chi_t|_{\mathcal{O}^\times} = \chi; \quad \chi_t(\varpi) = q^{-t}.$$

We think of $\chi_t$ as giving us a one-dimensional representation $(\chi_t, \mathbb{C}_t)$ of $L$ and we set $(\pi_t, V_t) = \iota_B(\chi_t, \mathbb{C}_t)$. Now $\Phi_L^*(C_t)$ is just $\mathbb{C}_t$ and one checks that the structure of $\mathbb{C}_t$ as a left $\mathcal{D}$ module is given by $d \cdot z = q^{-t}z,\ z \in \mathbb{C}_t$. It therefore follows from our discussion in §4.1 that

PROPOSITION. *Fix $\chi \in \mathbf{X}_{\mathcal{O}}$, if $\chi^2 = 1$, set $l = q$ if $\chi$ is trivial, and $l = 1$ if $\chi$ is non-trivial. Then*

*(1) If $\chi^2 = 1$ then the representation $(\pi_t, V_t)$ corresponds to the left $\mathcal{H}(l)$-module $M_t = \hom_{\mathcal{D}}(\mathcal{H}(l), \mathbb{C}_t)$.*

*(2) If $\chi^2 \neq 1$ then the representation $(\pi_t, V_t)$ corresponds to the left $\mathcal{D}$-module $C_t$.*

*Remark.* The above proposition effectively transfers almost all questions concerning the representation theory and harmonic analysis of the categories $\mathfrak{R}^{\mathfrak{s}}(G)$

to analogous questions about the module theory and harmonic analysis of the algebras $\mathcal{H}(l)$, $\mathcal{D}$. For example, given enough information about the module theory of $\mathcal{H}(l)$, $\mathcal{D}$ one may determine the parameters $t$ for which the representation $\pi_t$ is reducible and the parameters $t_1, t_2$ for which $\pi_{t_1} = \pi_{t_2}$. Similarly, given enough information about the harmonic analysis of $\mathcal{H}(l)$, $\mathcal{D}$ one may determine which of the representations $\pi_t$ is unitarizable and one may determine the Plancherel measure on that part of the reduced dual of $G$ whose smooth vectors are objects in $\mathfrak{R}^{\mathfrak{s}(\chi)}(G)$. The module theory and harmonic analysis of $\mathcal{D}$ is well known; that of the algebras $\mathcal{H}(l)$ is the subject of forthcoming work with Aubert and Morris.

DEPARTMENT OF MATHEMATICS, UNIVERSITY OF IOWA, IOWA CITY IA 54420, USA
*E-mail:* pkutzko@blue.weeg.uiowa.edu

---

REFERENCES

[BD] J.-N. Bernstein (rédigé par P. Deligne), Le "centre" de Bernstein, *Représentations des groupes réductifs sur un corps local,* Hermann, Paris, 1984, pp. 1–32.

[Bo] Nicholas Bourbaki, Groupes et algèbres de Lie, Ch.IV, V & VI, Hermann, Paris, 1968.

[BHK] C. J. Bushnell, G. Henniart, and P. C. Kutzko, Towards and explicit Plancherel formula for reductive p-adic groups, preprint.

[BK] C. J. Bushnell and P. C. Kutzko, Smooth representations of reductive p-adic groups: structure theory via types, *Proc. London Math. Soc.* **77** (1998), 582–634.

[Cs] W. Casselman, Introduction to the theory of admissible representations of p-*adic reductive groups,* unpublished notes.

[KS] P. C. Kutzko and Paul J. Sally, Jr., All supercuspidal representations of $SL_l$ are induced, *Proc. Conference on Reperesntations of Reductive Groups,* Park City, Utah, 1983.

[GR] David Goldberg and Alan Roche, preprint.

[Lu] G. Lusztig, Classification of unipotent reprsentations of simple $p$-adic groups, *Internat. Math. Res. Notices* **11** (1995).

[Mat] H. Matsumoto, Analyse harmonique dans les systèmes de Tits bornologiques de type affines, Springer Lecture Notes no. 590, Springer-Verlag, 1977.

[Sh] F. Shahidi, Twisted endoscopy and reducibility of induced representations for $p$-adic groups, *Duke Math. J.* **66** (1992), 1–41.

CHAPTER 22

# BEYOND ENDOSCOPY

By Robert P. Langlands

Ya tutarsa – Nasreddin Hoca

---

*Dedicated to Joseph Shalika on the occasion of his sixtieth birthday*

**Informal reference.** There is available at http://SunSITE.UBC.CA/Digital MathArchive/Langlands the text of a lecture *Endoscopy and beyond* that can serve as an introduction to this paper. It has the advantage of being informal, but there are misprints and some suggestions towards the end are red herrings. The present paper may well turn out to have the same defects!

*Acknowledgments.* I would like to thank James Arthur, who once again guided me through the subtleties of weighted orbital integrals, Erez Lapid and Peter Sarnak for useful conversations related to the material of this paper and Werner Hoffmann for his comments on [H] and on Appendix C and D.

**0.1. Functoriality and related matters.** The notion of $L$-group and the principle of functoriality appeared in [L] and were explained at more length in [Cor] and elsewhere. The principle of functoriality, which is now widely believed but is very far from being established in general, can be roughly stated as follows:

(I) *If $H$ and $G$ are two reductive groups over the global field $F$ and the group $G$ is quasi-split then to each homomorphism*

$$\phi: \quad {}^LH \longrightarrow {}^LG$$

*there is associated a* transfer *of automorphic representations of $H$ to automorphic representations of $G$.*

A second problem that arose some time after functoriality is that of associating to an automorphic representation $\pi$, now on the group $G$, an algebraic subgroup ${}^\lambda H_\pi$ of ${}^LG$ that would at best be defined up to conjugacy, although even that might often fail, and would have the following property. (I use the notation ${}^\lambda H$ to stress that we are dealing with a subgroup of the $L$-group ${}^LG$ that may not itself be an $L$-group, but is close to one. Although there is not yet a group $H$ attached to ${}^\lambda H$, I use, for simplicity, in the next statement and subsequently, the notation $m_H(\rho)$ or $m_{H_\pi}(\rho)$ rather than $m_{{}^\lambda H}(\rho)$ or $m_{{}^\lambda H_\pi}(\rho)$.)

(II) *If $\rho$ is a representation of ${}^LG$ then the multiplicity $m_H(\rho)$ of the trivial representation of ${}^\lambda H_\pi$ in the restriction of $\rho$ to ${}^\lambda H_\pi$ is the order $m_\pi(\rho)$ of the pole of $L(s,\pi,\rho)$ at $s=1$.*

Once again, this is not intended as an absolutely precise statement.

**0.2. Some touchstones.** There are three. The first two form a part of functoriality. The third does not. It is a question raised by a theorem of Deligne-Serre [DS]. I take for expository purposes the ground field $F$ to be an arbitrary number field (of finite degree).

(T1) *Take $H$ to be $GL(2)$, $G$ to be $GL(m+1)$, and $\phi$ to be the $m^{\text{th}}$ symmetric power representation.*

(T2) *Take $H$ to be the group consisting of a single element and $G$ to be $GL(2)$. Then ${}^LH$ is a Galois group and problem* (I) *is that of associating an automorphic form to a two-dimensional Galois representation.*

(T3) *Take $G$ to be $GL(2)$ and $\pi$ to be an automorphic representation such that every infinite place $v$ of the $\pi_v$ is associated to a two-dimensional representation not merely of the Weil group but of the Galois group over $F_v$. Show that $H_\pi$ is finite.*

A positive solution of the first problem has as consequence the Ramanujan-Petersson conjecture and the Selberg conjecture in their strongest forms; the Artin conjecture follows from the second. As is well known, all these problems have been partially solved; some striking results for the first problem are very recent. For various reasons, the partial solutions all leave from a methodological point of view a great deal to be desired. Although none of these problems refer to the existence of ${}^\lambda H_\pi$, I am now inclined to the view that the key to the solution of the first two and of functoriality in general lies in the problem (II), whose ultimate formulation will include functoriality. Moreover, as I shall observe at the very end of the paper, the problem (T3) can be approached in the same spirit.

I by no means want to suggest that I believe the solution to (II) is imminent. What I want to suggest rather, and to establish on the basis of the concrete calculations in this paper, is that reflecting on the problem of attacking (II) with the help of the trace formula, in my opinion, the only analytic tool of any substantial promise available for either (I) or (II), one is led to concrete problems in analytic number theory. They are difficult; but an often successful strategy, even though slow and usually inglorious, for breaching an otherwise unassailable mathematical problem is to reduce some aspect of it to a concrete, accessible form on which at least small inroads can be made and some experience acquired. The calculations, tentative as they are, described in the third part of this paper are intended as a first step in this direction for problems (I) and (II). I concentrate on (T2), for which $G$ is $GL(2)$ and on $\pi$ for which ${}^\lambda H_\pi$ is finite. The same approach applied to (T1) would entail dealing with $GL(m+1)$ and $\pi$ for which ${}^\lambda H$ was the image of $GL(2)$ under the $m$th symmetric power. This would require the use of the trace formula for $GL(m+1)$, much more sophisticated than that for $GL(2)$ although perhaps not completely inaccessible to numerical investigation for very small $m$.

**Part I: Formal structure.**

**1.1. The group $^\lambda H_\pi$.** We might take (II) as a definition of $^\lambda H_\pi$, but there are several difficulties. It is, first of all, perhaps best not to try to define $^\lambda H_\pi$ for all $\pi$. Arthur in his study of the trace formula has been led to a classification of automorphic representations that, in spite of its apparent reliance on objects whose existence is not established, can, in fact, in the context of the trace formula, usually be formulated in decidable terms. The classification is above all a separation into representations that are of *Ramanujan type* and those that are not. It is of conceptual significance that one expects to prove ultimately that the representations of Ramanujan type are exactly those that satisfy the general form of the Ramanujan conjecture, but that is not essential to the classification. The point is that a given trace formula will give a sum over both types of automorphic representation but the contribution to the formula of the representations that are not of Ramanujan type will be expressible in terms of traces from groups of lower dimension, so that the remainder can be regarded as the sum over the representations of Ramanujan type. We shall see a simple application of this principle to $GL(2)$. If $\pi$ is not of Ramanujan type, it will be natural to define $^\lambda H_\pi$ as the product $^\lambda H_{\pi'} \times S$ of a group $^\lambda H_{\pi'}$ defined by an ancillary $\pi'$ of Ramanujan type with an image $S$ of $SL(2, \mathbb{C})$, but this is a matter for which any great concern would be premature.

The other difficulties are more severe. The first is that even though we may expect that when $\pi$ is of Ramanujan type, the functions $L(s, \pi, \rho)$ are analytic on $\mathrm{Re}(s) \geq 1$, except perhaps for a finite number of poles on $\mathrm{Re}(s) = 1$, we are in no position to prove it. So an alternative definition of $m_\pi(\rho)$ is called for, even though, as must be stressed, the definition need at first only be used operationally—as a guide to the construction of various algebraic and analytic expressions whose meaning will be clear and unambiguous.

There are two more difficulties: given $\pi$ (implicitly of Ramanujan type) why should there exist an $^\lambda H$ (implicitly a reductive, but often not a connected, group) such that

$$m_H(\rho) = m_\pi(\rho)$$

for all $\rho$. Even if there is such an $^\lambda H$, why should it be unique, or rather why should its conjugacy class under $\hat{G}$ be unique? Recall that the $L$-group is the semi-direct product of its connected component $\hat{G}$ with the Galois group $\mathrm{Gal}(K/F)$ of a finite Galois extension of $F$ that has to be allowed to be arbitrarily large, so that the $L$-group is really an inverse sequence of groups with a common connected component. It normally suffices, however, to fix a $K$ large enough for the purposes at hand.

The second of these difficulties is easily resolved. The conjugacy class may not be unique and there may be several groups to be denoted $^\lambda H_\pi$. This is related to the multiplicity problem for automorphic representations. It will, however, be important to establish that if the function $\rho \to m_H(\rho)$ is given, then there are only finitely many possibilities for the conjugacy class of ${}^\lambda H$. Jean-Pierre Wintenberger has pointed out to me that as a consequence of a theorem of Larsen-Pink [LP] the group ${}^\lambda H$ is uniquely determined by the numbers $m_H(\rho)$ if ${}^L G$ is $GL(n, \mathbb{C})$, thus if $G$ is $GL(n)$ over $F$ and the Galois extension of $F$ used to define the $L$-group is $F$ itself. (There are certain supplementary conditions to be taken into account even in this case.)

In so far as the condition that the function $m_\pi$ be an $m_H$ is a linear condition—thus $m_\pi(\rho) = \operatorname{tr} \pi(f^\rho)$, where $f^\rho$ is some kind of generalized function on $G(\mathbb{A}_F)$—the existence of ${}^\lambda H_\pi$ is something to be verified by the trace formula. In the simplest of cases, there would be a linear form

$$\sum \alpha_\rho m_\pi(\rho), \qquad \alpha_\rho = \alpha_\rho^H, \tag{1}$$

which is 0 if ${}^\lambda H_\pi$ is not conjugate to a given ${}^\lambda H$ but is 1 if it is. The trace formula will, with any luck, yield an expression for the sum over all $\pi$ with appropriate multiplicities of (1) and will thus select exactly those $\pi$ attached to ${}^\lambda H$, but a similar sum that selected exactly, perhaps with multiplicity, those $\pi$ such that ${}^\lambda H_\pi$ lies in a given ${}^\lambda H$ would be better. Thus $\sum \alpha_\rho m_\pi(\rho)$ is to be 0 if none of the possible ${}^\lambda H_\pi$ is conjugate to a subgroup of ${}^\lambda H$ but is otherwise to be $\beta_\pi^H \neq 0$, where $\beta_\pi^H$ depends only on the collection of possible ${}^\lambda H_\pi$ and is to be 1 if ${}^\lambda H_\pi = {}^\lambda H$.

If we admit the possibility that there is a second group ${}^\lambda H'$ such that $m_{H'}(\rho) = m_H(\rho)$ for all $\rho$, then we see that we are demanding too much from the form (1). We might rather introduce a partial ordering on the collection of ${}^\lambda H$, writing

$$\lambda_{H'} \prec_{LP} \lambda_H,$$

if $m_{H'}(\rho) \geq m_H(\rho)$ for all $\rho$. Then we could at best hope that (1) would be different from 0 only if ${}^\lambda H_\pi \prec_{LP} {}^\lambda H$, and that it would be 1 if ${}^\lambda H_\pi \sim_{LP} {}^\lambda H$, thus if $m_{H_\pi}(\rho) = m_H(\rho)$ for all $\rho$. We would then, for each ${}^\lambda H_\pi$, try to obtain from the trace formula an expression for

$$\sum_{{}^\lambda H_\pi \prec_{LP} {}^\lambda H} \sum_\rho \alpha_\rho^H m_\pi(\rho). \tag{2}$$

It is best, however, to admit frankly that the first of the two difficulties, which amounts to understanding the conditions on the linear form $\rho \to m(\rho)$ that guarantee it is given by a subgroup ${}^\lambda H$ and to showing that $m_\pi$ satisfies these conditions, is a very serious problem that is not broached here. I content myself with a basic example or two that suggest it is prudent to keep an open mind about the properties to be possessed by (1) and about the final structure of the arguments. So (1) and (2) are at best provisional approximations to what is to be investigated.

**1.2. A simple observation.** Not only is the $L$-group an inverse sequence but so is, implicitly, each ${}^\lambda H$. If the occasion arises to distinguish the group in the sequence that lies in ${}^LG^K = \hat{G} \rtimes \mathrm{Gal}(K/F)$, we denote it ${}^\lambda H^K$. If $K \subset K'$, there is a surjective map

$$ {}^\lambda H^{K'} \to {}^\lambda H^K. $$

Among the representations $\rho$ are those that factor through the projection of ${}^LG$ on the Galois group, $\mathrm{Gal}(K/F)$. Since $L(s, \pi, \rho)$ is, for such a representation, equal to the Artin $L$-function $L(s, \rho)$, the number $m_\pi(\rho) = m_{H_\pi}(\rho)$ is just the multiplicity with which the trivial representation occurs in $\rho$. If $\mathfrak{H}$ is the image of ${}^\lambda H_\pi$ in $\mathfrak{G} = \mathrm{Gal}(K/F)$, it is also $m_{\mathfrak{H}}(\rho)$, calculated with respect to $\mathfrak{G}$. This is clearly possible for all $\rho$ only if $\mathfrak{H} = \mathfrak{G}$. Thus if ${}^\lambda H_\pi$ exists it will have to be such that its projection on $\mathrm{Gal}(K/F)$ is the full group. We shall implicitly assume throughout the paper that any group ${}^\lambda H$ appearing has this property.

**1.3. Calculation of $m_H(\rho)$ in some simple cases.** In the second part of the paper, I shall consider only the group $G = GL(2)$, and only over the base field $\mathbb{Q}$. I have not reflected on any other cases. I shall also often consider only $\pi$ whose central character is trivial, so that $\pi$ is an automorphic representation of $PGL(2)$. Then $m_\pi(\rho)$ will not change when $\rho$ is multiplied by any one-dimensional representation of $GL(2, \mathbb{C})$, and ${}^\lambda H_\pi$ will lie in $SL(2, \mathbb{C})$ or, to be more precise, in the family $\{SL(2, \mathbb{C}) \times \mathrm{Gal}(K/\mathbb{Q})\}$. It is instructive to compute $m_H(\rho)$ for a few ${}^\lambda H^K$ in $SL(2, \mathbb{C}) \times \mathrm{Gal}(K/\mathbb{Q})$ and a few $\rho$. We may as well confine ourselves to the standard symmetric powers $\sigma_m$, $m = 1, 2, \ldots$ of dimension $m + 1$ and to their tensor products with irreducible Galois representations $\tau$.

If ${}^\lambda H \subset SL(2, \mathbb{C}) \times \mathrm{Gal}(K/\mathbb{Q})$, the multiplicity $m_H(\sigma_1)$ is 2 if the projection of ${}^\lambda H$ on the first factor is $\{1\}$ and is 0 otherwise. Thus if we confine ourselves to groups ${}^\lambda H$ that project onto $\mathrm{Gal}(K/\mathbb{Q})$, then

$$ \text{(A)} \qquad a_1 m_H(\rho_1), \quad a_1 = \frac{1}{2}, \quad \rho_1 = \sigma_1, $$

is 1 if ${}^\lambda H = \{1\} \times \mathrm{Gal}(K/\mathbb{Q})$ and 0 otherwise. On the other hand,

$$ \text{(B)} \qquad a_1 m_{H'}(\rho_1), \quad a_1 = 1, \quad \rho_1 = \det, $$

is 1 for all subgroups ${}^\lambda H'$ of ${}^\lambda H = SL(2, \mathbb{C}) \times \mathrm{Gal}(K/\mathbb{Q})$ but 0 for groups that are not contained in ${}^\lambda H$. When and if the occasion arises for a precise reference, we denote the groups in these two cases by ${}^\lambda H_A$ and ${}^\lambda H_B$.

In general, as in (1) and (2), given ${}^\lambda H$, we would like to find a collection $\rho_1, \ldots, \rho_n$ of representations and a collection $a_1, \ldots, a_n$ of real numbers such that

$$ \sum_k a_k m_{H'}(\rho_k) = 1, $$

if ${}^\lambda H' \subset {}^\lambda H$ and 0 if it is not. We will normally want to consider only ${}^\lambda H$ and ${}^\lambda H'$ defined with respect to a given $K$. To make clear to which group given collections are associated I sometimes write as before $\rho_k = \rho_k^H$, $a_k = a_k^H$.

If the kernel of the projection of ${}^{\lambda}H$ to $\mathrm{Gal}(K/\mathbb{Q})$ is infinite, it is either $SL(2,\mathbb{C})$, a trivial case already treated, or contains the group

$$\hat{H} = \left\{ \begin{pmatrix} a & 0 \\ 0 & a^{-1} \end{pmatrix} \mid a \in \mathbb{C}^{\times} \right\}$$

as a normal subgroup of index 1 or 2. The group of outer automorphisms of $\hat{H}$, through which the action of ${}^{\lambda}H$ on $\hat{H}$ factors, is of order two and the image of ${}^{\lambda}H$ in it may or may not be trivial. If it is trivial, then ${}^{\lambda}H = \hat{H} \times \mathrm{Gal}(K/\mathbb{Q})$ and $m_H(\sigma_m \otimes \tau)$ is 1 if $m$ is even and $\tau$ is trivial and otherwise 0. We take

(C) $$a_1 = 1 \quad \rho_1 = \sigma_2,$$

and denote the pertinent group by ${}^{\lambda}H_C$.

If the image of ${}^{\lambda}H$ in the group of outer automorphisms, identified with $\mathbb{Z}_2$, is not trivial, the map ${}^{\lambda}H \to \mathbb{Z}_2$ may or may not factor through the Galois group. If it does not, then $\hat{H}\backslash{}^{\lambda}H$ is isomorphic to $\mathbb{Z}_2 \times \mathrm{Gal}(K/\mathbb{Q})$ and ${}^{\lambda}H$ contains the normalizer of $\hat{H}$ in $SL(2,\mathbb{C})$. Moreover $m_H(\sigma_m \otimes \tau) = 0$ unless $m \equiv 0 \pmod 4$ and $\tau$ is trivial, when it is 1. If the map ${}^{\lambda}H \to \mathbb{Z}_2$ factors through the Galois group, then $\hat{H}\backslash H$ is isomorphic to $\mathrm{Gal}(K/\mathbb{Q})$ and $m_H(\sigma_m \otimes \tau)$ is 1, if and only if $m \equiv 0 \pmod 4$ and $\tau$ is trivial or $m \equiv 2 \pmod 4$ is even and $\tau$ is the one-dimensional representation $\tau_0$ of $\mathrm{Gal}(K/\mathbb{Q})$ obtained by projecting onto the group $\mathbb{Z}_2$ and then taking the nontrivial character of this group, which is of order two. Otherwise $m_H(\sigma_m \otimes \tau)$ is 0. We take in these two cases:

(D) $$a_1 = 1, \quad \rho_1 = \sigma_4;$$

(E) $$a_1 = 1, \quad \rho_1 = \sigma_2 \otimes \tau_0.$$

The two groups will of course be denoted by ${}^{\lambda}H_D$ and ${}^{\lambda}H_E$.

If ${}^{\lambda}H'$ and ${}^{\lambda}H$ are each one of the five groups just described, then

$$\sum_k a_k^H m_{H'}(\rho_k^H)$$

is different from 0 only if ${}^{\lambda}H'$ is conjugate to a subgroup of ${}^{\lambda}H$ and is 1 if ${}^{\lambda}H' = {}^{\lambda}H$. Observe as well that in each of these cases, $m_H(\sigma_m \otimes \tau)$ depends only on $\tau$ and on $m$ modulo 4.

The only remaining possibility is that ${}^{\lambda}H$ projects to a finite nontrivial subgroup in $SL(2,\mathbb{C})$. The projection is either abelian, dihedral, tetrahedral, octahedral, or icosahedral. For the last three cases, the numbers $m_H(\sigma_m)$ are calculated for $m = 1, \ldots, 30$ to be the following.

Tetrahedral: 0, 0, 0, 0, 0, 1, 0, 1, 0, 0, 0, 2, 0, 1, 0, 1, 0, 2, 0, 2, 0, 1, 0, 3, 0, 2, 0, 2, 0, 3;

Octahedral: 0, 0, 0, 0, 0, 0, 0, 1, 0, 0, 0, 1, 0, 0, 0, 1, 0, 1, 0, 1, 0, 0, 0, 2, 0, 1, 0, 1, 0, 1;

Icosahedral: 0, 0, 0, 0, 0, 0, 0, 0, 0, 0, 0, 1, 0, 0, 0, 0, 0, 0, 0, 1, 0, 0, 0, 1, 0, 0, 0, 0, 0, 1.

As a consequence, if we take $K$ to be $\mathbb{Q}$ and let ${}^\lambda H_T$, ${}^\lambda H_O$, and ${}^\lambda H_I$ be the three subgroups of $SL(2,\mathbb{C})$ corresponding to the regular solids, and if we set

$$a_1^T = 1, \quad a_2^T = -1, \quad \rho_1^T = \sigma_6, \quad \rho_2^T = \sigma_2,$$
$$a_1^O = 1, \quad a_2^O = -1, \quad \rho_1^O = \sigma_8, \quad \rho_2^O = \sigma_4,$$
$$a_1^I = 1, \quad a_2^I = -1, \quad \rho_1^I = \sigma_{12}, \quad \rho_2^I = \sigma_8,$$

then, for ${}^\lambda H'$ infinite or equal to one of the same three groups,

$$\sum_k a_k^H m_{H'}(\rho_k^H)$$

is 0 if ${}^\lambda H'$ is not conjugate to a subgroup of ${}^\lambda H$, and is 1 if ${}^\lambda H' = {}^\lambda H$.

On the other hand, if the projection on $SL(2,\mathbb{C})$ is abelian of order $l$, then $m_H(\sigma_m)$ is the number $N$ of integers in $\{m, m-2, \ldots, -m\}$ divisible by $l$, and if it is dihedral with center of order $l \geq 3$, then $m_H(\sigma)$ is $N/2$ if $m$ is odd and $(N+1)/2$ if $m \equiv 0 \pmod 4$ and $(N-1)/2$ if $m \equiv 2 \pmod 4$. Suppose, for example, that it is dihedral with center of order 6. Then $N = 3$ for $m = 6$ and $N = 1$ for $m = 2$. Thus

$$a_1^T m_H(\rho_1^T) - a_2^T m_H(\rho_2^T) = 1 \neq 0,$$

but the group $H$ is not contained in the tetrahedral group. If we try to exclude the group $H$ by adding other representations to the sequence $\rho_1^T$, $\rho_2^T$, for example, $\sigma_{10}$, then we will introduce other groups, like the abelian group of order 10, that should be, but will not be, subgroups of the tetrahedral group. So we are still hoping for too much from the form (1). It looks as though we will have to accept in (2) groups that are not subgroups of the tetrahedral group, but that are finite dihedral groups or abelian. Since ${}^\lambda H_\pi$ is abelian only if $\pi$ is associated to Eisenstein series, we can envisage treating them by first treating the infinite dihedral groups along the lines of (1) and (2), and then treating dihedral ${}^\lambda H$ as subgroups of the $L$-group of the group defined by the elements of norm 1 in a quadratic extension. This is clumsier than one might hope. On the other hand, we would be using these arguments in combination with the trace formula, in which there is always an implicit upper bound on the ramification of the $\pi$ that occur. Since $\pi$ with large finite ${}^\lambda H_\pi$ would, in all likelihood, necessarily have large ramification, we can imagine that these two contrary influences might allow us to remove the unwanted groups from (2).

Suppose the group ${}^\lambda H^{\mathbb{Q}} = {}^\lambda H = {}^\lambda H_\pi$ is defined and finite for $K = \mathbb{Q}$. Then for an extension $K$ the projection of the group ${}^\lambda H^K$ on $SL(2,\mathbb{C})$ will be ${}^\lambda H^{\mathbb{Q}}$ and

$${}^\lambda H^K \subset {}^\lambda H^{\mathbb{Q}} \times \mathrm{Gal}(K/\mathbb{Q}).$$

There are two possibilities: there exists a $K$ such that the projection of ${}^\lambda H^K$ onto $\mathrm{Gal}(K/\mathbb{Q})$ is an isomorphism, or there does not, so that the kernel is never trivial. If our definitions are correct, it should be possible to decide which from the behavior of the $m_H(\rho)$ as $K$ and $\rho$ vary.

Take as an example the case that ${}^\lambda H^{\mathbb{Q}}$ is a cyclic group of odd prime order $l$, a possibility that will certainly arise. Then ${}^\lambda H^K$ will be a subgroup of $\mathbb{Z}/l\mathbb{Z} \times \mathrm{Gal}(K/\mathbb{Q})$. If it is a proper subgroup, then its projection to $\mathrm{Gal}(K/\mathbb{Q})$ is an isomorphism. If it is not the case to be considered, then it is the full product. In both cases, $m_H(\sigma_l) = 2$, $m_H(\sigma_{l-2}) = 0$, and $m_H(\rho_a) = 2$ if

$$\rho_a = \sigma_l - \sigma_{l-2}$$

is defined as a virtual representation.

The numbers

$$l-2, l-4, \ldots, 1, -1, \ldots, 2-l$$

run over all the nonzero residues of $l$, so that every nontrivial character of $\mathbb{Z}/l\mathbb{Z} = {}^\lambda H^{\mathbb{Q}}$ appears exactly once in the restriction of the representation $\sigma_{l-2}$ to ${}^\lambda H^{\mathbb{Q}}$. Suppose that $\tau$ is a character of the Galois group of order $l$ and consider the representation,

$$\rho_b = \sigma_{l-2} \otimes \tau.$$

If ${}^\lambda H$ is the full group ${}^\lambda H^{\mathbb{Q}} \times \mathrm{Gal}(K/\mathbb{Q})$, then $m_H(\rho_b) = 0$ because $\rho_b$ does not contain the trivial representation of ${}^\lambda H$. If, on the other hand, it is not the full group and $\tau$ factors through $\mathrm{Gal}(K/\mathbb{Q}) \simeq {}^\lambda H^K \to {}^\lambda H^{\mathbb{Q}}$, then it contains the trivial representation exactly once and $m_H(\rho_b) = 1$. Thus

$$\frac{1}{2} m_H(\rho_a) - m_H(\rho_b) \neq 0 \tag{3}$$

if ${}^\lambda H^K$ is the full group, but can be 0 if it is not.

The question with which we began is very difficult, but an obvious hypothesis lies at hand.

**1.4. A splitting hypothesis.** Suppose that for some automorphic representation $\pi$ of Ramanujan type the group ${}^\lambda H_\pi = {}^\lambda H_\pi^K \subset \hat{G} \rtimes \mathrm{Gal}(K/F)$, whose existence is only hypothetical, were finite. Then I expect—and there is no reason to believe that I am alone—that for a perhaps larger extension $L$ and the group ${}^\lambda H_\pi^L$ in $\hat{G} \rtimes \mathrm{Gal}(L/F)$, the projection of ${}^\lambda H_\pi^L$ to $\mathrm{Gal}(L/F)$ will be an isomorphism and that this will then continue to be true for all Galois extensions of $F$ that contain $L$. Moreover if $\pi$ is unramified outside a finite set $S$ it is natural to suppose that $L$ can also be taken unramified outside of $S$ and of a degree that is bounded by an integer determined by the order of the intersection of ${}^\lambda H_\pi$ with $\hat{G}$. Thus $L$ could be chosen among one of a finite number of fields.

In general, even when ${}^\lambda H_\pi \cap \hat{G}$ is not finite, we can expect that for some sufficiently large $L$, the group ${}^\lambda H_\pi^L \cap \hat{G}$ will be connected and that $L$ can be taken unramified where $\pi$ is unramified and of a degree over $K$ bounded by an integer determined by the number of connected components of ${}^\lambda H_\pi^K \cap \hat{G}$. The observations at the end of the previous section indicate what, at least from the point of view of this paper, the proof of the hypothesis will entail a special case: it must be shown that the expression (3), which we still do not know how we might calculate, is 0 for at least one of the finitely many cyclic extensions of $\mathbb{Q}$ of order $p$ unramified outside a finite set that depends on the original $\pi$. One might expect that the general hypothesis, or rather each case of it, reduces to similar statements.

**1.5. Alternative definition of $m_\pi$.** The integers $m_\pi(\rho)$ have been defined by residues of the logarithmic derivatives of automorphic $L$-functions at a point $s = 1$ outside the region at which they are known to be absolutely convergent. So it is not clear how this definition might be implemented. Since these integers have been introduced in the hope of broaching the problem of functoriality and thus that of analytic continuation, an alternative definition has to be found that better lends itself to harmonic analysis and to numerical investigation. For this purpose, I recall some familiar basic principles of analytic number theory. Since the extension of the principles and the definitions to other number fields will be patent, I confine myself for simplicity to the rationals.

If $c > 0$ is sufficiently large and $X > 0$, then

$$-\frac{1}{2\pi i}\int_{c-i\infty}^{c+i\infty}\frac{L'}{L}(s,\pi,\rho)X^s\frac{ds}{s} \tag{4}$$

is equal to

$$\frac{1}{2\pi i}\sum_p\sum \ln(p)\int_{c-i\infty}^{c+i\infty}\frac{\operatorname{tr}\left(\rho\left(A(\pi_p)^k\right)\right)}{p^{ks}}X^s\frac{ds}{s}. \tag{5}$$

This expansion shows that the integral (4) converges at least conditionally. Those terms of (5) for which $X < p^k$ are 0, as is shown by moving the contour to the right. The finite number of terms for which $X > p^k$ are calculated by moving the integral to the left as a residue at $s = 0$. So (5) is equal to

$$\sum_{p^k<X}\ln(p)\operatorname{tr}\left(\rho\left(A(\pi_p)^k\right)\right)$$

On the other hand, if the $L$-function can be analytically continued to a region containing the closed half-plane $\mathrm{Re}(s) \geq 1$, where it has no poles except for a finite number at points $1 + i\rho_l$, $l = 1, \dots, n$, and if its behavior in $\mathrm{Im}(s)$ permits a deformation of the contour of integration in (4) to a contour $C$ that except for small semi-circles skirting these points on the left runs directly from $1 - i\infty$ to $1 + i\infty$ on the line $\mathrm{Re}(s) = 1$, then (4) is (morally) equal to

$$\sum_l \frac{m_{1+i\rho_l}}{1+i\rho_l}X^{1+i\rho_l} + o(X).$$

As a consequence

$$(6)\qquad m_\pi(\rho) = m_1 = \lim_{M\to\infty,\, X\to\infty} \frac{1}{M}\int_X^{X+M} \frac{\sum_{p^k<Y} \ln(p)\,\mathrm{tr}\,\big(\rho\big(A(\pi_p)^k\big)\big)}{Y}dY.$$

If, for whatever reason, we know that the only possible pole is at 1, then this may be simplified to

$$(7)\qquad m_\pi(\rho) = \lim_{X\to\infty} \frac{\sum_{p^k<X} \ln(p)\,\mathrm{tr}\,\big(\rho\big(A(\pi_p)^k\big)\big)}{X}.$$

The possible appearance of other poles and thus the introduction of $M$ are simply nuisances that we could well do without.

For summation over primes, the sums [Lan]

$$\vartheta(X) = \sum_{p<X} \ln(p)$$

are the analogues of the sums over all positive integers

$$\sum_{1\le n<X} 1.$$

In particular, $\vartheta(X) = X + o(X)$. Moreover,

$$\psi(X) = \sum_{p^k<X} \ln(p) = \vartheta(X) + o(X).$$

Since it is expected that for $\pi$ of Ramanujan type the eigenvalues of $\rho(A(\pi_p))$ all have absolute value equal to 1, it is therefore not unreasonable in a tentative treatment to replace (6) and (7), both nothing but possible definitions, by

$$(8)\qquad m_\pi(\rho) = \lim_{M\to\infty,\, X\to\infty} \frac{1}{M}\int_X^{X+M} \frac{\sum_{p<Y} \ln(p)\,\mathrm{tr}(\rho(A(\pi_p)))}{Y}dY$$

and by

$$(9)\qquad m_\pi(\rho) = \lim_{X\to\infty} \frac{\sum_{p<X} \ln(p)\,\mathrm{tr}(\rho(A(\pi_p)))}{X}.$$

We want to see to what extent these definitions can be given real content and how.

We could modify (5) by replacing the denominator $s$ by $s(s+1)$. The residues at $s = 1 + i\rho_l$ become $1/(1+i\rho_l)(2+i\rho_l)$ and the residue at $s = 0$ is replaced by residues at $s = 0$ and $s = 1$. The result is that

$$(6')\qquad m_\pi(\rho) = \lim_{M\to\infty,\, X\to\infty} \frac{2}{M}\int_X^{X+M} \frac{\sum_{p^k<X} \ln(p)(1-p/X)\,\mathrm{tr}\,\big(\rho\big(A(\pi_p)^k\big)\big)}{X}dX,$$

or in the favorable case that there is only a pole at $s = 1$,

$$(7')\qquad m_\pi(\rho) = \lim_{X\to\infty} \frac{2\sum_{p^k<X} \ln(p)(1-p/X)\,\mathrm{tr}\,\big(\rho\big(A(\pi_p)^k\big)\big)}{X}.$$

The two formulas (8) and (9) can be similarly modified. Some of the experiments have been made using (7′) on the somewhat doubtful and certainly untested assumption that this improves convergence.

**1.6. The role of the trace formula.** As we have already stressed, in the general theory of automorphic forms, it is usually unwise to attempt to calculate directly any invariant associated to individual automorphic representations. Rather one calculates the sum—often weighted as, for example, in endoscopy—of the invariants over all automorphic representations of one group and compares them with an analogous sum for a second group, establishing by a term-by-term comparison their equality. For present purposes, what we might hope to calculate from the trace formula is

$$\sum_{\pi} \mu_\pi m_\pi(\rho) \prod_{v \in S} \operatorname{tr}(\pi_v(f_v)). \tag{10}$$

(We have to expect that it will at first be unknown whether the $m_\pi(\rho)$ are integers. To show that they are integers comparisons like those envisaged in (15) will very likely be necessary.) The finite-dimensional complex-analytic representation $\rho$ of ${}^LG$ is arbitrary. The set $S$ is a finite set of places of the base field $F$, including all archimedean places and all places where the group $G$ is not quasi-split and split over an unramified extension, and $f_v$ is a suitable function on $G(F_v)$. Implicitly we also fix a hyperspecial maximal compact subgroup at each place outside of $S$. The coefficient $\mu_\pi$ is usually a multiplicity; the sum is over automorphic representations of Ramanujan type unramified outside of $S$, ultimately perhaps only over the cuspidal ones, although it is best not to try without more experience to anticipate exactly what will be most useful—or the exact nature of $\mu_\pi$.

For the base field $F = \mathbb{Q}$, at this stage an adequate representative of the general case, to arrive at (10) we choose, for each given prime $p \notin S$, $f_q, q \notin S$, and $q \neq p$ to be the unit element of the Hecke algebra at $q$ and we choose $f_p$ in the Hecke algebra to be such that

$$\operatorname{tr}(\pi_p(f_p)) = \operatorname{tr}(\rho(A(\pi_p))) \tag{11}$$

if $\pi_p$ is unramified. Then we take $f^p(g) = \prod_v f_v(g_v)$, where $f_v$, $v \in S$, is given in (10). If $R$ is the representation of $G$ on the space of cuspidal automorphic forms of Ramanujan type and if we can get away with (9), then (10) is equal to

$$\lim_{X\to\infty} \sum_{\pi} \mu_\pi \frac{\sum_{p<X} \ln(p)\operatorname{tr}(R(f^p))}{X}. \tag{12}$$

If we use (7′) then (12) is replaced by

$$2\lim_{X\to\infty} \sum_{\pi} \mu_\pi \frac{\sum_{p<X} \ln(p)(1-p/X)\operatorname{tr}(R(f^p))}{X}. \tag{12′}$$

Not only is it unclear at this stage whether it is the representation on the space of cuspidal automorphic forms that is most appropriate or whether it might be better to include some noncuspidal representations, but it is also unclear whether it is best to take the ordinary trace or the stable trace. Such questions are premature. The important questions are whether we can hope to prove that the limit of (12) exists and whether we can find a useful, concrete expression for it.

We shall address some very particular cases of this question in the second part of this paper. Grant for the moment that we have such a representation for representations $\rho_k$, $1 \leq k \leq n$. Then for any coefficients $a_k$ we also have an expression for

$$\sum_{\pi} \mu_\pi \sum_{k=1}^{n} a_k m_\pi(\rho_k) \prod_{v \in S} \operatorname{tr}(\pi_v(f_v))$$

If we could find $a_k$ such that

$$\sum_{k} a_k m_\pi(\rho_k) \tag{13}$$

is equal to 1 if and only if ${}^\lambda H_\pi$ is IP-dominated by a given group ${}^\lambda H$ and is otherwise 0, then we would have an expression for

$$\sum_{{}^\lambda H_\pi \prec {}^\lambda H} \mu_\pi \prod_{v \in S} \operatorname{tr}(\pi_v(f_v)), \tag{14}$$

the sum being over automorphic representations of $G$ unramified outside of $S$, principally over cuspidal but perhaps with some noncuspidal terms present as well. The multiplicities $\mu_\pi$ could be ordinary multiplicities, but they will more likely be stable multiplicities and may even depend on ${}^\lambda H_\pi$. As we observed, we will have to content ourselves with satisfying the conditions imposed on (13) approximately; some of the representations for which it is not zero may have to be dealt with separately by an iterative procedure or the argument modified.

The existence of coefficients $a_k$ for which (13) has the desired properties, exactly or approximately, is an algebraic question that I have not broached except for $GL(2)$. The group $GL(2)$ has a center, so that the representation of $GL(2, \mathbb{A})$ on the space of cusp forms is not the direct sum of irreducible representations. To achieve this it is necessary, as usual, to consider the cusp forms transforming under a given character of $Z_+ = \mathbb{R}^+$ and there is no good reason at this stage not to suppose that this character is the trivial character. So we treat the representation on the space of functions on $GL(2, \mathbb{Q})Z_+ \backslash GL(2, \mathbb{A})$. Then $f_\infty$ will be a smooth function of compact support on $Z_+ \backslash GL(2, \mathbb{R})$ and the only $\pi$ to be considered are those for which $\pi_\infty$ is trivial on $Z_+$. This implies that the central character of $\pi$ is trivial on $\mathbb{R}^+$. In addition, if we suppose that $S$ consists of the infinite place alone—this is an assumption to be made purely for convenience as it removes inessential complications from the preliminary algebra and from the experiments—then we conclude that the only $\pi$ to be considered are those whose central character is trivial on $\mathbb{R}^+$ and unramified and thus trivial. Since the central character controls the group $\det(H_\pi)$, this means that we are taking only $\pi$ with ${}^\lambda H_\pi \subset SL(2,\mathbb{C})$, or, more precisely, ${}^\lambda H_\pi \subset SL(2,\mathbb{C}) \times \mathrm{Gal}(K\backslash\mathbb{Q})$. These are the very simple groups that we considered in a previous section and for which we are in a position to find—in so far as they are available—the coefficients of (13).

**1.7. Comparison.** If we managed by a combination of the trace formula with various limiting processes to obtain a formula for (14), then we would want to compare it with the trace formula on ${}^\lambda H$ itself, except that ${}^\lambda H$ may not be an $L$-group, for it may not be defined by a semidirect product. When, however, the kernel of ${}^\lambda H^K \to \mathrm{Gal}(K/\mathbb{Q})$ is connected, it is possible as a consequence of, for example, Prop. 4 of [L1] to imbed the center $\hat{Z}$ of $\hat{H}$, the connected component of the identity in ${}^\lambda H$, in the connected dual $\hat{T}$ of a product of tori, $T = \prod_i K_i^\times$, where each $K_i$ is a field over $F$, and to imbed it in such a way that ${}^L\tilde{H}$, the quotient of the semidirect product $\hat{T} \rtimes {}^\lambda H$ by the diagonally imbedded $\hat{Z}$ becomes an $L$-group. (The $L$-group may have to be defined by the Weil group and not by the Galois group, but that is of no import.) Notice that the Galois group $\mathrm{Gal}(K/\mathbb{Q})$ acts on $\hat{T}$, so that ${}^\lambda H$ does as well. Maps $\phi$ into ${}^\lambda H$ may be identified with maps into ${}^L\tilde{H}$ that correspond to automorphic representations of $\tilde{H}$ whose central character is prescribed by the structure of ${}^\lambda H$. They can presumably be identified in the context of the trace formula.

Then, to make use of (14), we would have to introduce a transfer $f \to f^H$ from functions on $G(\mathbb{A}_F)$ to functions on $H(\mathbb{A}_F)$ (if ${}^\lambda H = {}^L H$ is an $L$-group but to functions on $\tilde{H}(\mathbb{A}_F)$ in the general case) and compare (14) with

$$\sum \prod_{v\in S} \mathrm{tr}\left(\pi'_v(f_v^H)\right), \tag{15}$$

the sum being over automorphic representations of $H$ of, say, Ramanujan type unramified outside of $S$, so that there will also be a formula for (15) which is to be compared with that for (14). The difference between *IP*-domination and inclusion will undoubtedly complicate this comparison.

There is no reason not to admit the possibility that (14) is replaced by a sum over groups ${}^\lambda H$,

$$\sum_{{}^\lambda H} \sum_{{}^\lambda H_\pi \prec {}^\lambda H} \mu_\pi \prod_{v\in S} \mathrm{tr}(\pi_v(f_v)). \tag{14'}$$

Then (15) would be replaced by a similar sum (15′).

It is perhaps well to underline explicitly the differences between the comparison envisaged here and endoscopic comparison. For endoscopy the transfer $f \to f^H$ is defined in terms of a correspondence between conjugacy classes. In general, the transfer $f \to f^H$, which is defined locally, will be much less simple. There will already be much more knowledge of local harmonic analysis, especially of irreducible characters, implicit in its definition. Secondly, there will be difficult *analytic* problems to overcome in taking the limit of the trace formula on $G$. Thirdly, the groups ${}^{\lambda}H$ that occur are essentially arbitrary subgroups of ${}^{L}G$, not just those defined by endoscopic conditions.

**1.8. Further concrete cases.** I consider $GL(2)$ and icosahedral representations but in two different ways. The ground field $F$ may as well be taken to be $\mathbb{Q}$. Suppose $K/\mathbb{Q}$ is a Galois extension and $\mathrm{Gal}(K/\mathbb{Q})$ admits an imbedding $\tau$ in $GL(2,\mathbb{C})$ as an icosahedral representation. Thus $\mathrm{Gal}(K/\mathbb{Q})$ is an extension of the icosahedral group by $\mathbb{Z}_2$. Take ${}^{L}G = {}^{L}G^{K}$ and consider $\rho = \sigma_1 \otimes \tilde{\tau}$, where $\tilde{\tau}$ is the contragedient of $\tau$. If $m_H(\rho) \neq 0$, then $\sigma_1$ and $\tau$ define the same representation of ${}^{\lambda}H$. Therefore the kernels of ${}^{\lambda}H \to \mathrm{Gal}(K/\mathbb{Q})$ and ${}^{\lambda}H \to GL(2,\mathbb{C})$ are the same and thus $\{1\}$. So the projection of ${}^{\lambda}H$ to $\mathrm{Gal}(K/\mathbb{Q})$ is an isomorphism; ${}^{\lambda}H$ is an $L$-group, that attached to the group $H = \{1\}$; and $\sigma_1$ restricted to ${}^{\lambda}H$ is $\tau$, or rather the composition of $\tau$ with the isomorphism ${}^{\lambda}H \to \mathrm{Gal}(K/\mathbb{Q})$.

Thus we can expect that $m_\pi(\rho) \neq 0$ if and only if $\pi = \pi(\tau)$ is the automorphic representation attached to $\tau$ by functoriality. To compare (14), provided we can find such a formula, and (15), we will need to define the local transfer $f_v \to f_v^H$ by means of the characters of $\pi_v(\tau)$.

On the other hand, define the $L$-group ${}^{L}G$ to be ${}^{L}G^{\mathbb{Q}}$ and take $\rho = \sigma_{12} - \sigma_8$. We have seen that $m_H(\rho)$ is nonzero only if ${}^{\lambda}H$ is a subgroup of the icosahedral group or perhaps a finite abelian or dihedral group that can be treated independently. Then $m_\pi(\rho)$ will be nonzero only if ${}^{\lambda}H_\pi$ is such a subgroup. There will be many such $\pi$ and although (15) will not have to take them all into account, it will have to contain a sum over all icosahedral extensions unramified outside a given set of places.

So the first approach has at least one advantage: it singles out a unique $\pi$. It may have another. Numerical experiments involving $\sigma_1$ are manageable. Those for $\sigma_m$ quickly become impossible as $m$ grows. Even $m = 3$ is very slow. On the other hand, the first approach alone cannot, so far as I can see, assure us that if ${}^{\lambda}H_\pi \subset {}^{L}G^{\mathbb{Q}}$ is an icosahedral group, then $\pi$ is associated to an icosahedral representation of the Galois group. No matter what $\tau$ we choose, it necessarily overlooks $\pi$ for which this is false.

**1.9. A cautionary example.** Take the group $G$ to be $GL(1)$ over $\mathbb{Q}$ and take ${}^{\lambda}H$ to be the finite group of order $m$ in ${}^{L}G = \mathbb{C}^\times$. If $\rho$ is the representation $z \to z^m$, then $m_{H'}(\rho) = 1$ if ${}^{\lambda}H'$ lies in ${}^{\lambda}H$ and is otherwise 0. Let $S$ be, as usual, a finite set of places containing the infinite places. In order to have a discrete spectrum under the action of $G(\mathbb{A})$, we consider functions, thus automorphic forms, on $\mathbb{R}^+\mathbb{Q}^\times\backslash I$, $I$ being the group of idèles. This is the space $R^+G(\mathbb{Q})\backslash G(\mathbb{A})$. The function $f = \prod_v f_v$ will be such that $f_\infty$ is in fact a function on $\mathbb{R}^+\backslash\mathbb{R}^\times = \{\pm 1\}$. The function $f_p$ is the characteristic function of the set of integral $\gamma$ with $|\gamma| = p^{-m}$.

If we take the measure on $\mathbb{R}^+\backslash G(\mathbb{A})$ to be a product measure, with the measure of $G(\mathbb{Z}_p)$ and of $\mathbb{R}^+\backslash\mathbb{R}^+$ equal to 1, then $\mu(\mathbb{R}^+G(\mathbb{Q})\backslash G(\mathbb{A}))$ is equal to 1 and

$$\mathrm{tr}\, R(f) = \sum_{\pi} \mathrm{tr}\,\pi(f) = \sum_{\gamma\in\mathbb{Q}^\times} f(\gamma). \tag{16}$$

The element $\gamma$ must be equal to $ap^m$, where

$$a = \pm \prod_{q \in S} q^{\alpha_q}. \tag{17}$$

Thus the expression (16) is equal to $g(p^m)$, where $g$ is the function on $\prod_{q\in S} \mathbb{Z}_q$ given by $g(x) = \sum f(ax)$, the sum being over all $a$ of the form (17).

Thus (12) is

$$\lim_{X\to\infty} \frac{\sum_{p<X} \ln(p) g(p^m)}{X},$$

which is equal to

$$\frac{\sum_{x \bmod M} g(x^m)}{\varphi(M)},$$

where $M$ is a positive integer that is divisible only by primes in $S$ and that depends on the collection of functions $f_q$, $q \in S$, each of them being smooth. The number $\varphi(M)$ is the order of the multiplicative group of $\mathbb{Z}/M\mathbb{Z}$. In terms of $f$, this is

$$\frac{\int_{\mathbb{R}^+\mathbb{Q}_S I_S^m} f(x)dx}{\int_{\mathbb{R}^+\mathbb{Q}_S I_S^m} dx}. \tag{18}$$

where $\mathbb{Q}_S$ is the set of nonzero rational numbers that are units outside of $S$ and $I_S$ is the product $\prod_{v\in S} \mathbb{Q}_v$, the first regarded as a subgroup of the second.

The expression (18) is certainly in an appropriate form and is equal to

$$\sum_\chi \chi(f),$$

where $\chi$ runs over all characters of $\mathbb{R}^+\mathbb{Q}_S I_S$ of order dividing $m$. This, however, is pretty much the point from which we began. We are still left, as in class-field theory, with the problem of showing that these characters can be deduced from characters of the Galois group. Thus we cannot expect that the trace formula will spare us the arithmetical investigations. It will, at best, make it clear what these must be.

## Part II: Preliminary analysis.

**2.1. Measures and orbital integrals.** In this part of the paper, we shall review the trace formula for $GL(2)$, the only group with which we are seriously concerned at present, and examine the possibility of obtaining an expression for (14) or (15′). It would be worthwhile to undertake a similar study of the trace formula for other groups. If the general trace formula admits a similar analysis and transformation, it will be an encouraging sign.

To obtain expressions that can then be used for numerical purposes, we have to be clear about the conventions. As we already observed, we shall consider automorphic forms on $G(\mathbb{Q})Z_+\backslash G(\mathbb{A})$, $\mathbb{A} = \mathbb{A}_\mathbb{Q}$, and $G = GL(2)$. The functions whose trace is to be calculated are functions on $Z_+\backslash G(\mathbb{A})$ and are taken to be products $f(g)=\prod_v f_v(g_v)$. The measure on $Z_+\backslash G(\mathbb{A})$ is to be a product measure as is the measure on $Z_+\backslash G_\gamma(\mathbb{A})$ if $\gamma$ is regular and semisimple. The group $G_\gamma$ is then defined by the multiplicative group of a ring $E_\gamma$, the centralizer of $\gamma$ in the ring of $2\times 2$ matrices. At a nonarchimedean place $p$, the subgroup $G_\gamma(\mathbb{Z}_p)$ has a natural definition and we normalize the local measures by the conditions:

$$\mu(G_\gamma(\mathbb{Z}_p))=1, \quad \mu(G(\mathbb{Z}_p))=1.$$

At infinity, the choice of measure on $Z_+\backslash G(\mathbb{R})$ is not important, nor is that on $Z_+\backslash G_\gamma(\mathbb{R})$. It is not necessary to be explicit about the first, but it is best to be explicit about the second.

(a) *Elliptic torus.* Here I mean that the torus is elliptic at infinity and thus that $E=E_\gamma$ is an imaginary quadratic extension. I assume, for simplicity, that it is neither $\mathbb{Q}(\sqrt{-2})$ nor $\mathbb{Q}(\sqrt{-3})$. An element in $G_\gamma(\mathbb{R})$ is given by its eigenvalues, $\sigma e^{i\theta}$ and $\sigma e^{-i\theta}$. The value of $\sigma>0$ is irrelevant and I take the measure to be $d\theta$. The volume of

$$Z_+G_\gamma(\mathbb{Q})\backslash G_\gamma(\mathbb{A})=Z_+E^\times\backslash I_E \tag{19}$$

is the class number $C_E$ times the measure of

$$\pm Z_+\backslash\mathbb{C}^\times\times\prod_p G_\gamma(\mathbb{Z}_p),$$

which, according to the conventions chosen, is the measure of $\pm Z_+\backslash\mathbb{C}^\times$ or

$$\int_0^\pi d\theta=\pi.$$

(b) *Split torus.* Once again, the torus is only to be split at infinity, so that $E_\gamma$ is a real quadratic field. If the eigenvalues of an element $\delta$ are $\alpha$ and $\beta$, set

$$r=\alpha+\beta,$$

$$N=4\alpha\beta,$$

$$\frac{r}{\sqrt{|N|}}=\frac{1}{2}\left(\operatorname{sgn}\alpha\sqrt{\frac{|\alpha|}{|\beta|}}+\operatorname{sgn}\beta\sqrt{\frac{|\beta|}{|\alpha|}}\right)=\pm\frac{1}{2}(\lambda\pm\lambda^{-1}), \tag{20}$$

$$\lambda=\sqrt{\frac{|\alpha|}{|\beta|}},\quad \sigma=\sqrt{|\alpha\beta|},$$

$$\alpha=\pm\sigma\lambda,\quad \beta=\pm\frac{\sigma}{\lambda}.$$

The value of $\sigma$ is irrelevant and I take the measure to be $\frac{d\lambda}{\lambda}$. Notice that

$$d\left(\frac{r}{\sqrt{|N|}}\right)=\frac{1}{2}(1\mp\lambda^{-2})d\lambda=\frac{1}{2}\left(1-\frac{\beta}{\alpha}\right)d\lambda. \tag{21}$$

The upper sign is that of $N$. The parameters $r = \operatorname{tr}\delta$ and $N = 4\det\delta$ can also be defined when the torus is elliptic at infinity or globally or at any other place. When the torus is elliptic at infinity,

$$(22)\qquad d\left(\frac{r}{\sqrt{|N|}}\right) = d\cos\theta = \frac{i}{2}(1-\lambda^{-2})d\lambda, \quad \lambda = e^{i\theta}.$$

The fundamental unit $\epsilon$ can be taken to be the unit with the smallest absolute value $|\epsilon| > 1$. Thus $\ln|\epsilon|$ is the regulator as it appears in [C]. The measure of the quotient (19) is now the class number times the measure of

$$(23)\qquad \pm\mathbb{R}^+\backslash\mathbb{R}^\times \times \mathbb{R}^\times/\{\epsilon^k | k \in \mathbb{Z}\}.$$

Since $\pm\mathbb{R}^+\backslash\mathbb{R}^\times \times \mathbb{R}^\times$ can be identified with $\mathbb{R}^\times$ by projecting on the first factor, the measure of (23) is $2\ln|\epsilon|$, the measure of (19) is, in the notation of [C], $2h(D)R(D)$ if $D$ is the discriminant of the field $E_\gamma$.

There is a very small point to which attention has to be paid when computing with the trace formula. Locally there are two measures to be normalized, that on $G_\gamma(\mathbb{Q}_v)\backslash G(\mathbb{Q}_v)$ and that on $Z_+\backslash G_\gamma(\mathbb{R})$ or $G_\gamma(\mathbb{Q}_p)$. They appear in two ways in the measure on $G(\mathbb{Q}_v)$: once when fixing it, as $d\delta d\bar{g}$, by the measure on the subgroup $G_\gamma(\mathbb{Q}_p)$ (or $Z_+\backslash G_\gamma(\mathbb{R})$) and the measure on the quotient space $G_\gamma(\mathbb{Q}_v)\backslash G(\mathbb{Q}_v)$; and once, as in the Weyl integration formula, when fixing the measure on

$$\{g^{-1}\delta g \mid \delta \in G_\gamma(\mathbb{Q}_v),\ g \in G(\mathbb{Q}_v)\}$$

by means of the map

$$(24)\qquad (\delta, g) \to g^{-1}\delta g, \quad G_\gamma(\mathbb{Q}_v) \times (G_\gamma(\mathbb{Q}_v)\backslash G(\mathbb{Q}_v)) \to G(\mathbb{Q}_v).$$

Since (24) is a double covering, the measure to be used in the Weyl integration formula is

$$\frac{1}{2}\prod_\alpha |1-\alpha(\delta)| d\delta d\bar{g},$$

the product over $\alpha$ being a product over the two roots of the torus.

If $m$ is a nonnegative integer, let $T_p^m$ be the characteristic function of

$$\{X \in \operatorname{Mat}(\mathbb{Z}_p) | \quad |\det X| = p^{-m}\},$$

where $\operatorname{Mat}(\mathbb{Z}_p)$ is the algebra of $2\times 2$-matrices over $\mathbb{Z}_p$. If $\rho = \sigma_m$, then $T_p^m/p^{m/2}$ is the function $f_p$ of (11). In other words,

$$\operatorname{tr}\pi_p\left(T_p^m\right) = p^{m/2}\sum_{k=0}^{m}\alpha^{m-k}(\pi_p)\beta^k(\pi_p)$$

if $\pi_p$ is unramified and $\alpha(\pi_p)$ and $\beta(\pi_p)$ are the eigenvalues of $A(\pi_p)$. I recall the standard calculation.

Take $\pi_p$ to be the usual induced representation, so that the vector fixed by $GL(2,\mathbb{Z}_p)$ is the function

$$\phi(ntk) = |a|^{-s_1+1/2}|b|^{-s_2-1/2}, \qquad t = \begin{pmatrix} a & 0 \\ 0 & b \end{pmatrix},$$

and $\{\alpha(\pi_p), \beta(\pi_p)\} = \{p^{s_1}, p^{s_2}\}$. Then

$$\int \phi(g)T_p^m(g)dg = \sum_{k=0}^{m} p^{(m-k)s_1+ks_2}p^{(2k-m)/2}\int_{|x|\le p^{m-k}} dx = p^{m/2}\sum_{k=0}^{m} p^{(m-k)s_1+ks_2}.$$

We shall need the orbital integrals of the functions $T_p^m/p^{m/2}$ for all $m$, but $m=0$ is particularly important as it is the unit element of the Hecke algebra. The pertinent calculations can be found in [JL] but there is no harm in repeating them here. If $\gamma$ is a regular semisimple element in $G(\mathbb{Q}_p)$, set

$$U^m(\gamma) = \int_{G_\gamma(\mathbb{Q}_p)\backslash G(\mathbb{Q}_p)} T_p^m(g^{-1}\gamma g). \tag{25}$$

Denote the two eigenvalues of $\gamma$ by $\gamma_1$ and $\gamma_2$ and extend the usual norm on $\mathbb{Q}_p$ to $\mathbb{Q}_p(\gamma_1,\gamma_2)$ or to $E_p = E_\gamma \otimes \mathbb{Q}_p$, which we identify, taking $\gamma_1 = \gamma$. The ring of integral elements in $E_p$ is of the form $\mathbb{Z}_p \oplus \mathbb{Z}_p\Delta$. If $\bar\Delta$ is the conjugate of $\Delta$, so that $\Delta + \bar\Delta = \mathrm{tr}(\Delta)$, set $\delta_\gamma = p|\Delta - \bar\Delta|$. Let $\gamma_1 - \gamma_2 = b(\Delta - \bar\Delta)$ with $|b| = p^{-k}$, $k = k_\gamma$.

Lemma 1. *$U^m(\gamma)$ is 0 unless $\gamma_1$ and $\gamma_2$ are integral and $|\gamma_1\gamma_2| = p^{-m}$, when it is given by the following formulas.*

(a) *If $\gamma$ is split then (25) is*

$$p^k = \frac{1}{|\gamma_1 - \gamma_2|}.$$

(b) *If $\gamma$ is not split and $E_\gamma$ is unramified then* (2) *is*

$$p^k\frac{p+1}{p-1} - \frac{2}{p-1}.$$

(c) *If $\gamma$ is not split and $E_\gamma$ is ramified then* (25) *is*

$$\frac{p^{k+1}}{p-1} - \frac{1}{p-1}.$$

The proof is familiar and easy. As the lemma is basic to our calculations, I repeat it. The value of the characteristic function $T_p^m(g^{-1}\gamma g)$ is 1 if and only if $g^{-1}\gamma g$ takes the lattice $L_0 = \mathbb{Z}_p \oplus \mathbb{Z}_p$ into itself and has determinant with absolute value $p^{-m}$, thus only if it stabilizes the lattice and $|\det(\gamma)| = p^{-m}$. Thus, assuming this last condition, if and only if $\gamma$ stabilizes $L = gL_0$. Knowing $L$ is equivalent to knowing $g$ modulo $G(\mathbb{Z}_p)$ on the right. Multiplying $g$ on the left by an element of $G_\gamma(\mathbb{Q}_p) = E_p^\times$ is equivalent to multiplying $L$ by the same element.

If $E_p$ is split, then we can normalize $L$ up to such a multiplication by demanding that

$$L \cap \{(0, z) \mid z \in \mathbb{Q}_p\} = \{(0, z) \mid z \in \mathbb{Z}_p\}$$

and that its projection onto the first factor is $Z_p$. Then the $x$ such that $(1, x)$ lies in $L$ are determined modulo $Z_p$ by $L$. Multiplying by

$$\begin{pmatrix} \alpha & 0 \\ 0 & \beta \end{pmatrix}, \quad |\alpha| = |\beta| = 1,$$

we replace $(1, x)$ by $(1, \beta x/\alpha)$, so that only the absolute value $|x|$ counts. The measure in $G_\gamma(\mathbb{Q}_p)\backslash G(\mathbb{Q}_p)$ of the set of $g$ giving the lattice $L$ is the index in $G_\gamma(\mathbb{Z}_p)$ of the stabilizer of $L$. This is just the number of $y$ modulo $\mathbb{Z}_p$ with the same absolute value as $x$ (or with $|y| \le 1$ if $|x| \le 1$), thus the number of lattices that can be obtained from the given one by multiplying by an element of $G_\gamma(\mathbb{Z}_p)$. The condition that $L$ be fixed by $\gamma = (\gamma_1, \gamma_2)$ is that $\gamma_1$ and $\gamma_2$ be integral and that

$$(\gamma_1, \gamma_2 x) = \gamma_1(1, x) + (0, (\gamma_2 - \gamma_1)x)$$

lie in $L$, thus that $(\gamma_1 - \gamma_2)x$ be integral. We conclude finally that (25) is equal to $1/|\gamma_1 - \gamma_2|$.

The argument is the same in the remaining cases. Identifying $G(\mathbb{Q}_p)$ with the automorphisms of the vector space $E_p$, we identify the quotient $G(\mathbb{Q}_p)/G(\mathbb{Z}_p)$ with the lattices in $E_p$. Modulo the action of $G_\gamma(\mathbb{Q}_p)$, these can be put in the form $\mathbb{Z}_p + \mathbb{Z}_p p^j \Delta$, $j \ge 0$. Such a lattice is fixed by $\gamma$ if and only if $k \ge j$. In the unramified case, the stabilizer of the lattice in $G_\gamma(\mathbb{Z}_p)$ has index 1 if $j = 1$ and index $p^j(1 + 1/p)$ otherwise. So (25) is equal to

$$1 + \sum_{j=1}^{k} p^j \left(1 + \frac{1}{p}\right) = p^k \frac{p+1}{p-1} - \frac{2}{p-1},$$

as asserted by (b). If $E_p$ is ramified, the stabilizer has index 1 if $j = 0$ and index $p^j$ otherwise. So (25) is now equal to

$$1 + \sum_{j=1}^{k} p^j = \frac{p^{k+1}}{p-1} - \frac{1}{p-1}.$$

This is (c).

This lemma provided us with the orbital integrals that we need outside of $S$. The discussion inside of $S$ is quite different. Since we are going to take, for the present purposes, $S = \{\infty\}$, I confine myself to this case. The same principles apply in all cases. Over the field $\mathbb{R} = \mathbb{Q}_\infty$, the necessary information is in the discussion of HCS-families in Chap. 6 of [L2], although it is not elegantly expressed. Let

$\mathrm{ch}(\gamma) = (4\,\mathrm{Nm}(\gamma), \mathrm{tr}(\gamma))$. For any $\gamma$ in $GL(2, \mathbb{R})$,

$$\text{(26)} \qquad \int f_\infty(g^{-1}\gamma g)dg = \psi(\mathrm{ch}(\gamma)) = \psi'_\infty(\mathrm{ch}(\gamma)) + \psi''_\infty(\mathrm{ch}(\gamma))\frac{|\,\mathrm{Nm}\,\gamma|^{1/2}}{|\gamma_1 - \gamma_2|},$$

where $\psi'_\infty$ and $\psi''_\infty$ depend on $f_\infty$. The second is a smooth function on the plane with the $y$-axis removed. The first is 0 outside the parabola $y^2 - x \leq 0$, but inside and up to the boundary of this parabola, it is a smooth function of $x$ and $y^2 - x$. The functions $\psi'$ and $\psi''$ are not uniquely determined. Since we have taken $f$ positively homogeneous, the function $\psi$ is positively homogeneous, $\psi(\lambda^2 N, \lambda r) = \psi(N, r)$ for $\lambda > 0$. Thus it is determined by the two functions $\psi(\pm 1, r)$ on the line. The function $\psi_- = \psi(-1, r)$ is smooth; the function $\psi_+ = \psi(1, r)$ may not be. They are both compactly supported.

If $\theta(\gamma) = \theta(\mathrm{ch}(\gamma))$ is any positively homogeneous class function on $G(\mathbb{R})$, the Weyl integration formula and formulas (21) and (22) give

$$\text{(27)} \qquad \int_{Z_+\backslash G(\mathbb{R})} \theta(g)f(g)dg = \frac{1}{2}\sum \int_{Z_+\backslash T(R)} \theta(\mathrm{ch}(\gamma))\psi(\mathrm{ch}(\gamma))\left|1 - \frac{\alpha}{\beta}\right|\left|1 - \frac{\beta}{\alpha}\right|\frac{d\lambda}{|\lambda|}$$
$$= 4\sum \int_{-\infty}^{\infty} \psi_\pm(r)\theta(\pm 1, r)\sqrt{|r^2 \mp 1|}dr$$

because

$$\left|1 - \frac{\alpha}{\beta}\right|\left|1 - \frac{\beta}{\alpha}\right|\frac{d\lambda}{|\lambda|} = 2\left|1 - \frac{\alpha}{\beta}\right| |\lambda| \frac{dr}{\sqrt{|N|}},$$
$$\lambda\left(1 - \frac{\alpha}{\beta}\right) = \lambda \mp \lambda^{-1},$$
$$\frac{r^2}{N} - 1 = \pm\frac{1}{4}(\lambda \pm \lambda^{-1})^2 - 1 = \pm\frac{(\lambda \mp \lambda^{-1})^2}{4},$$

and

$$2|\lambda \mp \lambda^{-1}|\frac{dr}{\sqrt{|N|}} = 4\sqrt{\left|\frac{r^2}{N} - 1\right|}\frac{dr}{\sqrt{|N|}}.$$

The sums in (27) are over the two tori and then, in the last line, over the two possible signs. The elliptic torus corresponds to the region $-1 < r < 1$, $N = 1$; the split torus to the rest. The factor $1/2$ is removed in the passage from the first to the second line of (27) because the map $\gamma \to \mathrm{ch}(\gamma)$ from each of the tori to the plane is also a double covering.

The formula (27) is applicable if $\theta$ is a one-dimensional representation of $G(\mathbb{R})$, in particular, if it is identically equal to 1, and then (27) yields

$$\text{(28)} \qquad \mathrm{tr}(\theta(f)) = 4\sum \int_{-\infty}^{\infty} \psi_\pm(r)\sqrt{|r^2 \mp 1|}dr.$$

Another possibility is to take $\theta$ to be the character of the representation $\pi_\chi$ unitarily induced from a character

$$\chi : \begin{pmatrix} \alpha & 0 \\ 0 & \beta \end{pmatrix} \to \operatorname{sgn} \alpha^k \operatorname{sgn} \beta^l$$

of the diagonal matrices. Only the parities of $k$ and $l$ matter. The character is 0 on the elliptic elements, where $N > 0$ and $r^2 < N$. Otherwise it is constant on the four sets determined by fixing the signs of $N$ and $r$, where it is given by

$$(\operatorname{sgn} N + 1) \frac{\operatorname{sgn}(r)}{\sqrt{|1 - \alpha/\beta||1 - \beta/\alpha|}} \tag{29.a}$$

if $k \neq l$ and by

$$2 \frac{\operatorname{sgn}(N)^l}{\sqrt{|1 - \alpha/\beta||1 - \beta/\alpha|}} \tag{29.b}$$

otherwise. The eigenvalues $\alpha$ and $\beta$ of $\gamma$ with $\operatorname{ch}(\gamma) = (N, r)$ are of course one-half the roots of $x^2 - 2rx + N = 0$. Since

$$\frac{|\gamma_1 - \gamma_2|^2}{|\operatorname{Nm} \gamma|} = |1 - \alpha/\beta||1 - \beta/\alpha| = |\lambda|^2 |1 \mp \lambda^{-2}|^2 = 4 \left| \frac{r^2}{N} - 1 \right|,$$

we conclude that $\psi_\pm$, although not necessarily bounded, are integrable functions of $r$ and that $\operatorname{tr}(\pi_\chi(f))$ is given by

$$4 \left\{ \int_{-\infty}^{-1} \epsilon_{+-} \psi_+(r) + \int_1^\infty \epsilon_{++} \psi_+(r) + \int_{-\infty}^0 \epsilon_{--} \psi_-(r) + \int_0^\infty \epsilon_{-+} \psi_-(r) \right\}, \tag{30}$$

where the constants $\epsilon_{\pm\pm}$, which are $\pm 1$ or 0, are to be chosen as prescribed by (29).

**2.2. Calculating with the trace formula.** Rather than refer to Arthur's general trace formula, as I should if I were intent on preparing for the general case, I prefer to appeal to the formula on pp. 516–517 of [JL] with which I am more at ease and to which the reader is requested to refer. There are eight terms in that formula, but for a base field of characteristic zero the term (iii) is absent. We shall also only consider, for reasons already given, automorphic representations whose central character is trivial on $\mathbb{R}^+$. The formula of [JL] gives the sum of the traces of $\pi(f)$ over all automorphic representations occurring discretely in $L^2(Z_+ G(\mathbb{Q}) \backslash G(\mathbb{A}))$. So we need to subtract those representations that are not of Ramanujan type. For $G = GL(2)$, these are the one-dimensional representations. Their traces will be subtracted from the term (ii) of [JL] and the difference will be more important than (ii) itself. We refer to the difference as the elliptic term. It is the most difficult and will be discussed—not treated—in §2.5.

The principal question that concerns us is whether there are any possible developments in analytic number theory that might enable us to find an explicit expression for (12). The numbers $\mu_\pi$ are here equal to 1 (or, for those $\pi$ that are absent from the sum, 0). Lacking all experience, I fell back on the obvious and made explicit calculations. For $\rho = \sigma_m \otimes \tau$ they are feasible and not all too slow for $m = 1$. For $m = 2, 3$ something can still be done, but for higher $m$, at least with my inefficient programs, they are too slow to provide any useful information. On the other hand, as the first problem of §1.8 demonstrates, calculations for $m = 1$ are of considerable interest provided that we take the tensor product of $\sigma_m$ with a general $\tau$ or even just a $\tau$ of icosahedral type. Although taking such a tensor product demands a simultaneous study of icosahedral representations or other Galois representations, there is no reason not to expect that the important features of the problem are not already present for $\tau$ trivial and that they persist. Of course, there may be accidental features, but these the wise student should recognize and resolutely ignore. The addition of the Galois representation will add to the labor but should not put additional demands on raw computer power, only on the skill of the programmer. So I confine myself to trivial $\tau$ and, by and large, to $m = 1$. Although it is important for theoretical purposes to envisage taking $S$ arbitrarily large, computations and theory for larger $S$ should not differ essentially from the case that $S$ consists of the infinite place alone, although there will be many more terms in the trace formula to be taken into account.

The sum over $r \in \mathbb{Z}$ that occurs in the elliptic term will be replaced by sums over $r$ satisfying a congruence condition. This will entail that whatever behavior we find for $S = \{\infty\}$ should remain valid when congruence conditions are imposed. Such an assertion, which implies a greater theoretical regularity that may make the proofs easier to come by, has to be tested further, but these are the principles that justify confining myself at first to $m = 1$ and to representations unramified at all finite places. We know, of course, a good deal about such representations. In particular, there are none of Galois type, but this is an accidental circumstance that we will use to verify that the programs are functioning well, but that will be otherwise irrelevant to our conclusions.

The representations are to be unramified at all finite places; so the central character $\eta$ of [JL] is trivial. Since we will also examine, at least briefly, some $m > 1$, I do not fix $m$ to be 1. The representation $\tau$ will be, however, trivial. The trace formula replaces the expression (12) by a sum of seven terms, corresponding to its seven terms. The function $\Phi$ of [JL] is now being denoted $f$, $f^p$, or even $f^{p,m}$, and

$$f^{p,m}(g) = f_\infty(g_\infty) f_p^m(g_p) \prod_{q \neq p} f_q(g_q),$$

in which $f_\infty$ is a variable function, $f_p^m$ depends on $m$, but all the other $f_q$ do not depend on $m$. Thus the function $\Phi$ does not satisfy the conditions of [JL]; it does not transform according to a character of the center $Z(\mathbb{A})$ of $G(\mathbb{A})$, and the resulting trace formula is different, but not very different. In (i) there is a sum over the scalar matrices. In (ii) and (iv) there are sums over the full tori, not just over the tori divided by the center. In (v) there is also a sum over the scalar matrices, the $n_0$ defining

$\theta(s, f_v)$ being replaced by $zn_0$. In principle, (vi), (vii), and (viii) are different, but because $f_q$ is a spherical function for all $q$, the sum over $(\mu, \nu)$ implicit in these expressions reduces to the single term $\mu = \nu = 1$.

Since we are in a situation where (7) is appropriate and (6) unnecessary, the contribution of the first term of the trace formula to (12) is given by

$$\text{(TF.1)} \qquad \sum_{Z(\mathbb{Q})} \frac{\mu(Z_+G(\mathbb{Q})\backslash G(\mathbb{A}))}{X} \sum \ln(p) f^m(z).$$

Since

$$f^m(z) = \frac{f_\infty(z)}{p^{m/2}},$$

if $z = \pm p^{m/2}$ and 0 otherwise, the limit that appears in (12) or (12′) will be 0.

The second term is the elliptic term to be treated in the next section. None of the terms (iv), (v), and (viii) of [JL], is invariant on its own, so that some recombination of these terms is necessary. The terms (vi) and (vii) can, however, be treated directly.

I begin with (vi), which yields a contribution that is not in general 0. Since $f_q$ is a spherical function for all $q$, the only pair $(\mu, \nu)$ that contributes to (vi) or to (vii) is the pair of trivial characters and $\rho(\cdot, s)$, denoted $\xi_s$ in this paper to avoid a conflict of notation, is the global (or local) representation unitarily induced from the representation

$$\begin{pmatrix} \alpha & x \\ 0 & \beta \end{pmatrix} \to |\alpha|^{\frac{s}{2}} |\beta|^{-\frac{s}{2}}$$

of the adelic superdiagonal matrices. It is, moreover, easily verified that $M(0)$ is the operator $-I$. Thus the contribution of (vi) to (12) is

$$\text{(TF.2)} \qquad \frac{1}{4X} \sum \ln(p)\, \mathrm{tr}(\xi_0(f_\infty))\, \mathrm{tr}\left(\xi_0\left(f_p^m\right)\right).$$

Since $\mathrm{tr}(\xi_0(f_p^m)) = m + 1$, the limit as $X \to \infty$ is

$$\text{(31)} \qquad \frac{m+1}{4} \mathrm{tr}(\xi_0(f_\infty)).$$

From (30) we conclude that for $m = 1$, this is

$$\text{(32)} \qquad 2\left\{\int_{-\infty}^{-1} \psi_+(r)dr + \int_1^{\infty} \psi_+(r)dr + \int_{-\infty}^{\infty} \psi_-(r)dr\right\}.$$

Apart from the elliptic term, this will be the only nonzero contribution to the limit. Since the standard automorphic $L$-function $L(s, \pi, \sigma_1)$ does not have poles on $\mathrm{Re}(s) = 1$, we expect that (32) will be cancelled by the elliptic term. This is accidental and will not be for us, even numerically, the principal feature of the elliptic term.

The function $m(s)$ that appears in (vii) is

$$\pi\frac{\Gamma((1-s)/2)}{\Gamma((1+s)/2)}\frac{\zeta(1-s)}{\zeta(1+s)}.$$

Thus

$$\frac{m'(s)}{m(s)}=-\frac{1}{2}\frac{\Gamma'((1-s)/2)}{\Gamma((1-s)/2)}-\frac{1}{2}\frac{\Gamma'((1+s)/2)}{\Gamma((1+s)/2)}-\frac{\zeta'(1-s)}{\zeta(1-s)}-\frac{\zeta'(1+s)}{\zeta(1+s)}. \tag{33}$$

It is to be multiplied by the product of

$$\mathrm{tr}(\xi_s(f_\infty)) \tag{34}$$

and $\mathrm{tr}(\xi_s(f_p^m)$. The first of these two functions, as a function on $(-i\infty, i\infty)$, is the Fourier transform of a smooth function of compact support. The second is equal to

$$p^{im\frac{s}{2}}+p^{i(m-2)\frac{s}{2}}+\cdots+p^{i(2-m)\frac{s}{2}}+p^{-im\frac{s}{2}}. \tag{35}$$

The estimates of §48 and §77 of [Lan] assure us that the product of (33) and (34) is an $L^1$-function on $(-i\infty, i\infty)$. From the Riemann-Lebesgue lemma we then conclude that, for odd $m$, the integral of the product of (33), (34), and (35) over that line approaches 0 as $p$ approaches infinity. So, for $m$ odd, (vii) does not contribute to the limit in (12) or (12′).

**2.3. The noninvariant terms.** Both $\omega(\gamma, f_v)$ and $\omega_1(\gamma, f_v)$ are 0 unless there is a matrix

$$n=\begin{pmatrix}1 & x\\ 0 & 1\end{pmatrix}, \quad x\in F_v,$$

and a matrix $k$ in the maximal compact subgroup of $GL(2,\mathbb{Q}_v)$ such that the element $k^{-1}n^{-1}\gamma nk$ lies in the support of $f_v$. Since $f_\infty$ is fixed at present, this means that the two eigenvalues $\alpha$ and $\beta$ of $\gamma$ are units away from $p$ and that there is a fixed $\delta>0$ such that $\delta<|\alpha/\beta|_\infty<1/\delta$. From the product formula we conclude that $\delta<|\alpha/\beta|_p<1/\delta$. Since $\alpha\beta=\alpha^2(\beta/\alpha)=\pm p^m$ if $\omega(\gamma, f_v)$ or $\omega_1(\gamma, f_v)$ is not 0, we conclude that (iv) is 0 for all but a finite number of $p$ if $m$ is odd and thus does not contribute to (12) or (12′). If $m$ is even, there are only a finite number of $\gamma$ that yield a nonzero contribution to (iv). Indeed, such $\gamma$ have to be of the form

$$\gamma=\begin{pmatrix}\pm p^k & 0\\ 0 & \pm p^l\end{pmatrix}, \quad k+l=m.$$

Since $\alpha/\beta=\pm p^{k-m}$ is bounded in absolute value, for all but a finite number of $p$ only

$$\gamma\begin{pmatrix}\pm p^{m/2} & 0\\ 0 & \pm p^{m/2}\end{pmatrix}$$

contribute. Since $\gamma$ is not central, the signs must be different.

In the new form of (v), $\theta(s, f_v)$ depends upon a nonzero scalar $z$,

$$\theta_z(s, f_v) = \iint f_v\left(k_v^{-1} a_v^{-1} z n_0 a_v k_v\right) \left|\frac{\alpha_v}{\beta_v}\right|^{-1-s} da_v dk_v.$$

So the only contribution to (v) will be from $z = \pm p^{m/2}$ and it will only occur for even $m$.

At finite places $q$, the operator $R'(\mu_q, \nu_q, s)$ that occurs in (viii) annihilates the vector fixed by $G(\mathbb{Z}_q)$. So, with our assumptions, (viii) reduces to

$$\frac{1}{4\pi}\int_{-i\infty}^{i\infty} \operatorname{tr}(R^{-1}(s)R'(s)\xi_s(f_\infty))\left(\sum_{k=0}^{m} p^{i(m-2k)s}\right) d|s|$$

in which $R$ is the local intertwining operator at infinity normalized as in [JL] and in which it is implicit that $\mu_\infty = \nu_\infty = 1$. According to the estimates of [A],

$$|\operatorname{tr}(R^{-1}(s)R'(s)\xi_s(f_\infty))| = O\left(\frac{1}{s^2}\right), s \to \infty.$$

Thus we can once again apply the Riemann-Lebesgue lemma to conclude that, for $m$ odd, there will be no contribution to the limits (12) or (12′) from (viii).

**2.4. The case of even *m*.** There are four sections devoted to even $m$, by and large to $m = 2$, or to weighted orbital integrals: §2.4, §4.3, and Appendices B and C. They are not used in this paper and are best omitted on a first reading. The formulas for even $m$ are given for almost no other purpose than to make clear that for odd $m$ many significant simplifications occur. Since the formulas are not elegant and are applied neither theoretically nor numerically, I very much fear that errors may have slipped in and advise the reader to be cautious.

For even $m$, there are several contributions in addition to those from (ii) that survive when we take the limit in $X$. Since the term $p^0$ occurs in (35), the expression (vii) contributes

$$\frac{1}{4\pi}\int_{-i\infty}^{i\infty} \frac{m'(s)}{m(s)} \operatorname{tr}(\xi_s(f_\infty)) d|s| \tag{36}$$

to (12) or (12′). From (viii) we have

$$\frac{1}{4\pi}\int_{-i\infty}^{i\infty} \operatorname{tr}(R^{-1}(s)R'(s)\xi_s(f_\infty)) d|s| \tag{37}$$

To treat (36), or at least part of it, we deform the contour from $\operatorname{Re}(s) = 0$ to $\operatorname{Re}(s) > 0$ or to $\operatorname{Re}(s) < 0$, as the usual estimates permit [Lan], expand

$$-\frac{\zeta'(1-s)}{\zeta(1-s)} - \frac{\zeta'(1+s)}{\zeta(1+s)}$$

as

$$\sum_q \sum_{n>0} \frac{\ln q}{q^{n(1-s)}} + \sum_q \sum_{n>0} \frac{\ln q}{q^{n(1+s)}} \tag{38}$$

and integrate term by term, deforming the contours of the individual integrals back to $\mathrm{Re}(s) = 0$. In fact, because of the pole of

$$\frac{\zeta'(1 \pm s)}{\zeta(1 \pm s)}$$

at $s = 0$, we have first to move the contour to the right and then, for the contribution from $\zeta'(1-s)/\zeta(1-s)$, move it back to the left. The result is that we pick up a supplementary contribution $-\operatorname{tr}(\xi_0(f_\infty))/2$.

Since the character of $\xi_s$ is the function

$$\frac{|\alpha/\beta|^{s/2} + |\beta/\alpha|^{s/2}}{\sqrt{|1-\alpha/\beta||1-\beta/\alpha|}},$$

the calculation that led to (30) shows that

$$\operatorname{tr}(\xi_s(f_\infty)) = 2\int (\lambda^{s/2} + \lambda^{-s/2})\psi_\pm(r)dr, \tag{39}$$

where the integral is to be taken over the set of $(\pm 1, r)$ with the interval

$$\{(1,r)\} \mid -1 \le r \le 1\}$$

removed. This may be rewritten as

$$2\int_{-\infty}^{\infty} |t|^s \{|t - t^{-1}|\psi_+(t + 1/t) + |t + t^{-1}|\psi_-(t - 1/t)\}\frac{dt}{|t|},$$

so that, for $s$ purely imaginary, $\operatorname{tr}(\xi_s(f_\infty))$ is the Fourier transform of

$$2\{|e^x - e^{-x}|\psi_+(e^x + e^{-x}) + |e^x + e^{-x}|\psi_-(e^x - e^{-x})\}. \tag{40}$$

As a result, the contribution of (38) to (36) is

(41)

$$\sum_q \sum_{n>0} \frac{\ln q}{q^n}\{|q^n - q^{-n}|\psi_+(q^n + q^{-n}) + |q^n + q^{-n}|\psi_-(q^n - q^{-n})\} - \frac{\operatorname{tr}(\xi_0(f_\infty))}{2},$$

in which the terms for large $q$ or large $n$ are 0. Since this expression occurs for every $p$, it remains in the average, as in part a sum of atomic measures that may well be finally cancelled by a contribution from the elliptic term, but it is hard to see at present how this will occur!

Although the local normalization of the intertwining operators to $R$ used in [JL] is necessary if the products and sums appearing in the trace formula are to converge, or at least if the global contribution (vii), which entails no study of local harmonic analysis, is to be clearly separated from the contributions (viii) for which the primary difficulty lies in the local harmonic analysis. Nonetheless it is best that, having separated (38) from (33) to obtain a term that could be analyzed more easily, we combine what remains of (36) with (37) so that we can more readily appeal to known results on weighted orbital integrals.

Since

$$-\frac{1}{2}\frac{\Gamma'((1-s)/2)}{\Gamma((1-s)/2)}-\frac{1}{2}\frac{\Gamma'((1+s)/2)}{\Gamma((1+s)/2)} \tag{42}$$

is the logarithmic derivative of

$$\frac{\pi^{(s-1)/2}\Gamma((1-s)/2)}{\pi^{(s+1)/2}\Gamma((1+s)/2)},$$

the combination of the two terms amounts to multiplying the unnormalized operator

$$J_s : \phi \to J_s\phi, \quad J_s\phi(g)=\int_{\mathbb{R}}\phi(\bar{n}(x)g)dx \tag{43}$$

on the space of the induced representation $\xi_s$ by

$$\frac{\pi^{(s-1)/2}\Gamma((1-s)/2)}{\pi^{s/2}\Gamma(s/2)}. \tag{44}$$

I set

$$n(x)=\begin{pmatrix}1 & x\\ 0 & 1\end{pmatrix}, \quad \bar{n}(x)=\begin{pmatrix}1 & 0\\ x & 1\end{pmatrix}.$$

The choice of measure is irrelevant, because a logarithmic derivative is to be taken. Moreover there is a slight difference between (43) and the intertwining operator of [JL], but this difference too disappears when the logarithmic derivative is taken. So we use (43), which is the definition used in [H].

The logarithmic derivatives of both (44) and $J_s$ now have a pole at $s=0$, the poles cancelling, so that, when we replace the sum of (37) and the contribution of (42) to (36) by the integrals of the logarithmic derivative of (43) and of (44), the contour of integration has to be deformed whenever we want to discuss them separately. Hoffmann prefers to avoid 0 by skirting it to the right. I follow his convention. So if $C$ is the new contour, we are left with two terms,

$$\frac{1}{4\pi i}\int_C\left\{-\frac{1}{2}\frac{\Gamma'((1-s)/2)}{\Gamma((1-s)/2)}-\frac{1}{2}\frac{\Gamma'(s/2)}{\Gamma(s/2)}\right\}\operatorname{tr}\xi_s(f_\infty)ds \tag{45}$$

and

$$\frac{1}{4\pi i}\int_C\operatorname{tr}\left(J_s^{-1}J_s'\xi_s(f_\infty)\right)ds. \tag{46}$$

The contribution (46) is not invariant and must be paired with terms from (iv) and (v) to obtain an invariant distribution, the only kind that is useful in our context, for it is the only kind expressible in terms of $\psi$ alone.

The two expressions will be, however, ultimately combined. Indeed, there is a danger in discussing them separately. We need an explicit expression for the sum as

$$\frac{1}{4\pi i}\int_{-i\infty}^{i\infty}\Omega(s)\operatorname{tr}\xi_s(f_\infty)ds.$$

Since $\operatorname{tr}\xi_s(f_\infty)$ is even but otherwise essentially arbitrary, the function (or distribution) $\Omega$ will be unique if it is assumed even. The integrands of (36) and (37) are even, so that if we stay with them it is easier to use parity to monitor the manipulations. On the other hand, the factor multiplying the trace in (45) is not even; nor is the integrand of (46). Since Hoffmann's results for (46) are in a form that is not only transparent but also symmetric, and since we can easily put (45) in symmetric form, we can readily restore the symmetry, the only cost being the replacement of (45) by a somewhat lengthier expression, in which there is one surprise, the final term in the following formula. If we avoid 0 by a small semi-circle of radius $\epsilon$ then (45) becomes, up to a term of order $O(\epsilon)$

$$\frac{1}{4\pi i}\int_{-i\infty}^{-i\epsilon}+\int_{i\epsilon}^{i\infty}\left\{-\frac{1}{2}\frac{\Gamma'((1-s)/2)}{\Gamma((1-s)/2)}-\frac{1}{2}\frac{\Gamma'(s/2)}{\Gamma(s/2)}\right\}\operatorname{tr}\xi_s(f_\infty)ds+\frac{\operatorname{tr}\xi_0(f_\infty)}{4}.$$

The first factor in the integrand may be symmetrized, so that the singularity $1/s$ at $s=0$ disappears, and then $\epsilon$ allowed to go to 0. The result is the sum of

$$\frac{1}{16\pi i}\int_{-i\infty}^{i\infty}\left\{-\frac{\Gamma'((1-s)/2)}{\Gamma((1-s)/2)}-\frac{\Gamma'(s/2)}{\Gamma(s/2)}-\frac{\Gamma'((1+s)/2)}{\Gamma((1+s)/2)}-\frac{\Gamma'(-s/2)}{\Gamma(-s/2)}\right\}\operatorname{tr}\xi_s(f_\infty)ds,$$

or better, since

$$\frac{\Gamma'(s/2)}{\Gamma(s/2)}=\frac{\Gamma'(1+s/2)}{\Gamma(1+s/2)}-\frac{2}{s},$$

of

(47) $$\frac{-1}{16\pi i}\int\left\{\frac{\Gamma'((1-s)/2)}{\Gamma((1-s)/2)}+\frac{\Gamma'(1+s/2)}{\Gamma(1+s/2)}+\frac{\Gamma'((1+s)/2)}{\Gamma((1+s)/2)}+\frac{\Gamma'(1-s/2)}{\Gamma(1-s/2)}\right\}\operatorname{tr}\xi_s(f_\infty)$$

and

(48) $$\frac{\operatorname{tr}\xi_0(f_\infty)}{4}.$$

When considering (iv) and (v), we suppose that $p\neq 2$ since, as we observed, the values for a particular $p$ have no influence on the limit. We may also suppose that $\gamma=\pm p^{m/2}\delta$, where $\delta$ is a matrix with eigenvalues $\pm 1$. The signs are equal for (v) and different for (iv).

We begin with (iv). (There appears to be a factor of $1/2$ missing in (iv). It was lost on passing from p. 530 to p. 531 of [JL], but is included below.) We invert the order of summation and discard all terms that do not contribute to the average, so that it becomes a sum over just two $\gamma$ followed by a sum over the places of $\mathbb{Q}$.

If $v$ is a finite place $q$ different from $p$, then

$$n(x)^{-1}\gamma n(x) = \pm p^{m/2}\begin{pmatrix}1 & 2x\\ 0 & -1\end{pmatrix}$$

is integral in $\mathbb{Q}_v$ if and only if $2x$ is integral. Consequently, $\omega_1(\gamma, f_q)$, $q \neq p$, is 0 except for $q = 2$, but for $q = 2$,

$$\omega_1(\gamma, f_2) = -\frac{\ln(2^2)}{2} = -\ln 2.$$

Moreover,

$$\omega(\gamma, f_q) = \begin{cases}1, & q \neq 2, p;\\ 2, & q = 2.\end{cases}$$

On the other hand,

$$\omega(\gamma, f_p) = 1,$$

if $p \neq 2$. Finally

$$\omega_1(\gamma, f_p) = -\frac{\int_{1<|x|\le p^{m/2}} \ln|x|^2 dx}{p^{m/2}} = -\left(1-\frac{1}{p}\right)\frac{\ln p}{p^{m/2}}\sum_{j=1}^{m/2} 2jp^j.$$

The integral is taken in $\mathbb{Q}_p$.

Thus the sum over $v$ in (iv) reduces to three terms, those for $v = \infty$, $v = 2$, and $v = p$. Since our emphasis is on $f_\infty$, the only variable part of $f$, the first plays a different role than the last two. It is invariant only in combination with (46). The first two are already invariant as functions of $f_\infty$.

Before continuing, we give the values of the three constants $c$, $\lambda_0$, and $\lambda_{-1}$ appearing in the trace formula as given in [JL]. First of all, $\lambda_{-1} = 1$ and $\lambda_0$ is Euler's constant. The constant $c$ is the ratio between two measures, the numerator being the measure introduced in §2.1 and used to define the operators

$$R(f) = \int_{Z_+\backslash G(\mathbb{A})} f(g)dg$$

appearing in the trace formula. (So the symbol $R$ has two different roles. It is not the only symbol of the paper whose meaning depends on the context.) The denominator being that given locally and globally as $d(ank) = dadndk$, $g = ank$ being the Iwasawa decomposition. Thus both measures are product measures, so that $c = \prod c_v$. If we choose, as we implicitly do, the measures $da$ and $dn$ so that $A(\mathbb{Z}_q)$ and $N(\mathbb{Z}_q)$ have measure 1 for all $q$, then $c_q = 1$ at all finite places. On the other hand, we have not been explicit about the measure on $Z_+\backslash G(\mathbb{R})$. There was no need for it. We may as well suppose that it is taken to be $dadndk$, where $a$ now belongs to $Z_+\backslash A(\mathbb{R})$. Then $c = c_\infty = 1$.

The measure on $Z_+\backslash A(\mathbb{R})$ has already been fixed, but the choice of measures on $N_\infty$ and $K_\infty$ do not enter the formulas explicitly. We have

$$\omega(\gamma, f_\infty) = \psi(\mathrm{ch}(\gamma)).$$

Thus the contribution from (iv) is the sum of two terms. The first

$$-\frac{1}{2}\sum_{\gamma}\omega_1(\gamma, f_\infty)\prod_{q\neq\infty}\omega(\gamma, f_q), \tag{49}$$

in which only two $\gamma$ appear,

$$\gamma = \pm p^{m/2}\begin{pmatrix}1 & 0\\ 0 & -1\end{pmatrix},$$

is to be combined with (46). Since $\omega(\gamma, f_q)$ is just the orbital integral of $f_q$, it is calculated by Lemma 1 and, as $|\gamma_1 - \gamma_2| = |2p^{m/2}|$ for the $\gamma$ in question,

$$\prod_{q\neq\infty}\omega(\gamma, f_q) = \frac{1}{p^{m/2}}2p^{m/2} = 2. \tag{50}$$

The second stands alone and is

$$\frac{1}{2}\psi_-(0)\left\{\ln 2 + \left(1-\frac{1}{p}\right)\frac{\ln p}{p^{m/2}}\sum_{j=1}^{m/2}2jp^j\right\}, \tag{51}$$

an expression that is about $\ln p$ in size. Its occurrence is certainly unexpected, as it is not bounded in $p$, so that the elliptic term will have to contain something that compensates for it. The source of this atomic contribution to the elliptic term—if it is present—should not be hard to find, but I have not yet searched for it.

We verify immediately that

$$\theta_z(0, f_v) = \int_{\mathbb{Q}_v}\int_{K_v} f_v\big(\pm p^{m/2}k_v^{-1}n(x)k_v\big)dk_v dx, \quad z = \pm p^{m/2},$$

if $v = q$ is finite. If $q \neq p$, this is equal to 1. If $q = p$, it is equal to 1 because $p^{m/2}x$ is integral for $|x| \leq p^{m/2}$. Since (the notation is that of [JL, p. 194])

$$L(1) = \pi^{-1/2}\Gamma\left(\frac{1}{2}\right) = 1$$

and $f_\infty$ is positively homogeneous, the first term of (v) contributes

$$\sum_{\pm}\lambda_0\theta_{\pm 1}(0, f_\infty) = \lambda_0\sum\int_{\mathbb{R}}\int_{K_\infty} f_\infty(\pm k^{-1}n(x)k)dkdx. \tag{52}$$

to the average.

The expression (52) can be calculated easily in terms of $\psi_+$. We take

$$\gamma = \pm\begin{pmatrix}e^t & 0\\ 0 & e^{-t}\end{pmatrix},$$

and let $t$ approach 0. Then $\mathrm{ch}(\gamma) = (N, r) = (4, \pm 2\cosh t)$, $r/\sqrt{|N|}$ approaches $\pm 1$, and, as a simple change of variables shows,

$$(53)\quad 2\left(\sqrt{\frac{r^2}{N}} - 1\right)\psi_+(r) = 2|\sinh t| \int_{\mathbb{R}}\int_{K_\infty} f_\infty(k^{-1}\gamma n((1-e^{-2t})x)k)dxdk$$

approaches the integral of (52). According to (26), the limit of (53) is $\psi''_\infty(4, \pm 2)$. It is nonzero only when $\psi_+$ is singular at 1 or $-1$.

The derivative

$$(54)\quad \theta'_z(0, f_v) = -\frac{\ln q}{q-1}\int f_v(k^{-1}zn(x)k)dxdk + \int f_v(k^{-1}zn(x)k)\ln|x|dxdk$$

if $v = q$ is nonarchimedean. If it is archimedean, then

$$(55)\quad \theta'_z(0, f_v) = \kappa\int f_\infty(k^{-1}zn(x)k)dxdk + \int f_\infty(k^{-1}zn(x)k)\ln|x|dxdk,$$

where

$$\kappa = -\frac{\pi^{-1/2}\Gamma(1/2)}{2}\ln\pi + \pi^{-1/2}\Gamma'(1/2) = -\frac{\lambda_0}{2} - \frac{\ln\pi}{2} - \ln 2,$$

a result of

$$\Gamma'\left(\frac{1}{2}\right) = (-\lambda_0 - 2\ln 2)\sqrt{\pi}, \quad \text{(cf. [N, p. 15])}.$$

The expression (54) is deceptive. If $v \neq p$ and $z = \pm p^{m/2}$, then $\theta_z(s, f_q)$ is identically 1 and its derivative 0. If $q = p$, then

$$\theta_z(s, f_q) = \frac{1}{p^{m/2}L_q(1+s, 1)}\int_{|\beta|\leq p^{m/2}} |\beta|^{1+s}\frac{d\beta}{|\beta|} = p^{ms/2},$$

so that

$$\theta'_z(0, f_p) = \frac{m\ln p}{2}.$$

The sum in (v) is a double sum, over $\gamma = \pm p^{m/2}$ and over $v$. Only $v = \infty$ and $v = p$ yield a contribution different from 0. The first will be combined with (52) to give the sum of

$$(56)\quad \kappa_1\sum\int_{\mathbb{R}}\int_{K_\infty} f_\infty(\pm k^{-1}n(x)k)dx, \quad \kappa_1 = \frac{\lambda_0}{2} - \frac{\ln\pi}{2} - \ln 2$$

and the noninvariant expression

$$(57)\quad \sum\int f_\infty(k^{-1}zn(x)k)\ln|x|dxdk,$$

which will have to be combined with (46). The second is

$$\sum \frac{m \ln p}{2} \int_{\mathbb{R}} \int_{K_\infty} f_\infty(\pm k^{-1} n(x) k) dx, \tag{58}$$

in which two disagreeable features appear: the logarithm of $p$ that cannot possibly have an average and the integral that is expressible only in terms of the singularities of $\psi_+$ at $\pm 1$. So there is no question of the logarithmic terms in (51) and (58) cancelling.

**2.5. The elliptic term.** The sum in the expression (ii) from [JL] is over the global regular elliptic elements $\gamma$, each $\gamma$ being determined by its trace, which we have denoted $r$ and by 4 times its determinant, $N = 4\det(\gamma)$. Only $\gamma$ for which $r$ is integral and $N = \pm 4p^m$ appear. The eigenvalues of $\gamma$ are

$$\frac{r}{2} \pm \frac{\sqrt{r^2 - N}}{2}.$$

Their difference is $\pm\sqrt{r^2 - N}$. Thus $\gamma$ will be elliptic if and only if $r^2 - N$ is not a square. We write $r^2 - N = s^2 D$, where $D$ is a fundamental discriminant, thus $D \equiv 0, 1 \pmod 4$. Both $D$ and $s$ are understood to be functions of $r$ and $N$. If $r^2 = N$, then $D$ is taken to be 0; if it is a square, then $D = 1$. Note that the symbol $s$ appears in the paper in two quite different ways: here and elsewhere, as an integer by whose square we divide to obtain the fundamental discriminant; previously and also below, as a variable parametrizing the characters of $\mathbb{R}^+$.

I claim that

$$\sum_{f|s} f \prod_{q|f} \left(1 - \frac{\left(\frac{D}{q}\right)}{q}\right), \tag{59}$$

in which $(\frac{D}{p})$ is the Kronecker symbol ([C]), is equal to the product of $U^m(\gamma)$ taken at $p$ with the product over $q \neq p$ of $U^1(\gamma)$. By multiplicativity, it is enough to consider

$$1 + \sum_{j=1}^{k} q^j \left(1 - \frac{\left(\frac{D}{q}\right)}{q}\right)$$

for each prime $q$. If $(\frac{D}{q}) = 1$, this is $1 + q^k - 1$, but if $(\frac{D}{q}) = -1$, it is

$$1 + (q^k - 1)\frac{q+1}{q-1} = q^k \frac{q+1}{q-1} - \frac{2}{q-1}.$$

Finally, if $(\frac{D}{q}) = 0$, it is

$$\sum_{j=0}^{k} q^j = \frac{q^{k+1}}{q-1} - \frac{1}{q-1}.$$

So we have only to appeal to Lemma 1.

Let $\mu_D$ be the volume $\mu(Z_+G_\gamma(\mathbb{Q})\backslash G_\gamma(\mathbb{A}))$ if $D \neq 1$. Then the uncorrected elliptic term is the sum

$$\sum_{N=\pm 4p^m} \sum_r \mu_D \frac{\psi(N,r)}{p^{m/2}} \sum_{f|s} f \prod_{q|f} \left(1 - \frac{\left(\frac{D}{q}\right)}{q}\right), \tag{60}$$

in which the function $\psi$ continues to be defined as in (26). The factor $1/2$ in (ii) has been removed because each $r$ accounts for two $\gamma$. Because of the presence of the term $\psi(N,r)$, the sum is finite, the number of terms being of order $\sqrt{|N|}$. The terms with $D = 0, 1$ are excluded because they do not correspond to regular elliptic $\gamma$. Moreover $p$ is fixed for the moment.

We now make use of formulas from [C] (§5.3.3 and §5.6.2—the general form of the second formula is stated incorrectly but we do not need the general form) for $\mu_D$. If $n$ is a positive integer and $x$ a real number, define the function $\varphi(x,n)$ by the following formulas.

$x < 0$:

$$\varphi(x,n) = \pi \operatorname{erfc}\left(\frac{n\sqrt{\pi}}{\sqrt{|x|}}\right) + \frac{\sqrt{|x|}}{n}\exp(-\pi n^2/|x|).$$

$x > 0$:

$$\varphi(x,n) = \frac{\sqrt{x}}{n} \operatorname{erfc}\left(\frac{n\sqrt{\pi}}{\sqrt{|x|}}\right) + \mathrm{E}_1\left(\frac{\pi n^2}{x}\right),$$

where $\mathrm{E}_1$ is defined to be the function

$$-\gamma - \ln(x) + \sum_{k\geq 1}(-1)^{k-1}\frac{x^k}{k!k},$$

$\gamma$ being Euler's constant. Then, on making use of the formulas in §2.1 for $\mu_D$ in terms of the class number, we obtain

$$\mu_D = \sum_{n=1}^{\infty}\left(\frac{D}{n}\right)\varphi(D,n). \tag{61}$$

The series (61) is absolutely convergent and we substitute it in (60) to obtain

$$2\sum_n \sum_f \sum_{\{(r,N)\,|\,f|s\}} f\left(\frac{D}{n}\right)\varphi(D,n)\frac{\psi(N,r)}{\sqrt{|N|}}\prod_{q|f}\left(1 - \frac{\left(\frac{D}{q}\right)}{q}\right). \tag{62}$$

Recall that $N$ assumes only two values $\pm 4p^m$, but that $r$ runs over all integers except the very few for which $D = 0, 1$. By homogeneity, we may replace $\psi(N,r)$

by $\psi_\pm(x_r)$, where for brevity of notation I set $x_r = r/\sqrt{|N|}$. I rewrite (62) as

$$(63)\quad 2\sum_f \sum_{\{n\,|\,(n,f)=1\}} \sum_{\{(r,N)\,|\,f|s\}} \sum_{f'} f\left(\frac{D}{nf'}\right)\varphi(D,nf')\frac{\psi_\pm(x_r)}{\sqrt{|N|}}\prod_{q|f}\left(1-\frac{\left(\frac{D}{q}\right)}{q}\right).$$

The sum over $f'$ is over all positive numbers all of whose prime divisors are prime divisors of $f$. The sum over $(r, N)$ is over those pairs for which $f|s$, $s$ continuing to be defined by $r^2 - N = s^2 D$. In principle, we want to examine the individual terms

$$(64)\qquad 2\sum_{f|s}\sum_{f'} f\left(\frac{D}{nf'}\right)\varphi(D,nf')\frac{\psi_\pm(x_r)}{\sqrt{|N|}}\prod_{q|f}\left(1-\frac{\left(\frac{D}{q}\right)}{q}\right),$$

the outer sum being a sum over $r$ and the two possible $N$, but we must first subtract the contribution (28) from the trivial representation. So we have to express it too as a sum over $n$ and $f$.

The contribution from the trivial representation $\theta$ is the product of (28) with

$$\operatorname{tr}\theta\left(f_p^m\right) = \sum_{k=0}^{m} p^{(m-2k)/2} = p^{m/2}\frac{1-p^{-m}}{1-p^{-1}} = \frac{\sqrt{|N|}}{2}\frac{1-p^{-m}}{1-p^{-1}}.$$

So it is

$$(65)\qquad 2\sqrt{|N|}\frac{1-p^{-m}}{1-p^{-1}}\sum\int \psi_\pm(x)\sqrt{|x^2\mp 1|}\,dx,$$

the sum being over the set $\{+,-\}$. To see how this is to be expressed as a sum over $n$ and $f$, we observe that $\varphi(D,n)$ behaves for large $|D|$ like $\sqrt{|D|}/n = \sqrt{|r^2-N|}/sn$, so that, for a rough analysis, (63) may be replaced by

$$(66)\quad 2\sqrt{|N|}\sum_f \sum_{\{n\,|\,(n,f)=1\}} \sum_{f|s}\sum_{f'} \frac{f}{snf'}\left(\frac{D}{nf'}\right)\psi_\pm(x_r)\frac{\sqrt{|x_r^2\mp 1|}}{\sqrt{|N|}}\prod_{q|f}\left(1-\frac{\left(\frac{D}{q}\right)}{q}\right).$$

Suppose we replace each of the factors

$$(67)\qquad \sum_{f'}\frac{f}{snf'}\left(\frac{D}{nf'}\right)\prod_{q|f}\left(1-\frac{\left(\frac{D}{q}\right)}{q}\right) = \frac{f}{sn}\left(\frac{D}{n}\right)$$

by a number $\epsilon_{n,f}(N)$, an approximation to its average value on intervals long with respect to $n$ but short with respect to $\sqrt{|N|}$. Then (63) is replaced by

$$(68)\qquad 2\sqrt{|N|}\sum_f \sum_{\{n\,|\,(n,f)=1\}} \sum_{\pm}\epsilon_{n,f}(N)\int \psi_\pm(x)\sqrt{|x^2\mp 1|}\,dx.$$

The inner sum is over the two possible values of $N$.

The exact sense in which $\epsilon_{n,f}(N)$ is an approximation to the average is not important, provided the choice works, but we do need to show that

$$\sum_{n,f} \epsilon_{n,f}(N) = \frac{1}{1-p^{-1}} + O(|N|^{-1}), \tag{69}$$

$$= \frac{1-p^{-m}-1}{1-p^{-1}} + O(|N|^{-1}),$$

so that (65) is equal to (68) and the difference between (63) and (65) has some chance of being $o(|N|^{1/2})$. For the purposes of further examination, we write this difference as the sum over $n$ and $f$, $\gcd(n, f) = 1$, of

$$2\left\{\sum f\left(\frac{D}{nf'}\right)\varphi(D, nf')\frac{\psi_\pm(x_r)}{\sqrt{|N|}}\Phi - \sqrt{|N|}\sum_{\pm}\epsilon_{n,f}(N)\int\psi_\pm(x)\sqrt{|x^2 \mp 1|}dx\right\}, \tag{70}$$

with

$$\Phi = \Phi_f = \prod_{q|f}\left(1 - \frac{\left(\frac{D}{q}\right)}{q}\right).$$

The first sum in (70) is over $r$, $f'$, and $\pm$. (Notice that the sum (70) has a simpler mathematical structure than (60), especially for $f = 1$ for then the sum over $f'$ is absent. The only element that varies irregularly with $r$ is the square $s^2$ dividing $r^2 - N$.)

I now explain how we choose $\epsilon_{n,f}(N)$. Let $t$ be an integer large with respect to $\sqrt{|N|}$ and divisible by a multiplicatively very large square. The average of (68) is to be first calculated on $[0, t)$. The divisibility of $r^2 - N$ by $s^2$ is then decided for all $s$ up to a certain point by the residue of $r$ modulo $t$. Whether $(r^2 - N)/s^2$ is then divisible by further squares is not, but it is except for squares that are only divisible by very large primes. There will be very few $r$ for which this occurs. Otherwise $f$ divides $s$ if and only if $f^2$ divides $r^2 - N$ with a remainder congruent to 0 or 1 modulo 4. Then, for $(n, f) = 1$,

$$\left(\frac{(r^2-N)f^2/s^2}{n}\right) = \left(\frac{D}{n}\right).$$

Thus $\epsilon_{n,f}(N)$ will be an approximation to the average value of

$$\frac{f}{sn}\left(\frac{(r^2-N)f^2/s^2}{n}\right). \tag{71}$$

If $(n, f)$ were not 1, these expressions would be 0, and it is useful to set $\epsilon_{n,f}(N) = 0$ if $\gcd(n, f) \neq 1$. The calculation of these factors is long and tedious, but their values are needed for the numerical experiments, and (69) is a confirmation of the correctness of the calculation. So I present the calculation in an appendix.

## Part III: Numerical experiments.

**3.1. A first test.** We observed that (32) was, apart from the elliptic term, the only nonzero contribution to the limit. Since $L(s, \pi, \sigma_1)$ is regular and nonzero on $\mathrm{Re}(s) = 1$ for all cuspidal automorphic $\pi$, we expect that the limit (12′) is 0. For $m = 1$, we have calculated explicitly all contributions to the limit (12′) except for the difference between the elliptic term (60) and the contribution (65) from the one-dimensional representations. So we have to show that this difference, or rather its average in the sense of (12′) over $p < X$, cancels the simple, but in our context fundamental, distribution (32). A first test is numeric.

Both (60) and (65) are distributions, even measures, on the pair $(\psi_+, \psi_-)$. The first is a sum of atomic measures. The second is absolutely continuous with respect to Lebesgue measure. So their difference and the average over $p$ is also a measure, symmetric with respect to $r \to -r$. I divide the interval from $-3$ to 3 on each of the lines $N = \pm 1$ into 60 equal parts of length 0.1 and calculate numerically for each $p$ the measure of each interval. In the unlikely event that a point common to two intervals has nonzero mass, I assign half of this mass to each of the two intervals. Then I average over the first $n$ primes in the sense of (12′). The result should be approaching $-0.2$ on each of the intervals except those on $N = +1$ between $-1$ and 1, where it should approach 0. From Table 3.1, in which the first two columns refer to the average over the first 200 primes, the second to that over the first 3600 primes, and the third to that over the first 9400, we see that the average is almost immediately approximately correct at least for the intervals closer to 0, that it does seem to converge to the correct values, but that the convergence is slow, sometimes even doubtfully slow.

Thanks to the symmetry, only the results for the intervals from $-3$ to 0 need be given. In each set of two columns, the numbers in the first column are for intervals of $r$ with $N = -1$, and those in the second for $N = 1$. Once the results get within about 0.007 of the expected values they cease to improve. I assume they would with better programming.

**3.2. A rough estimate.** For $m$ odd and in particular for $m = 1$, the elliptic term, or more precisely the difference between the elliptic term and the contribution from the one-dimensional representations, is a formidable expression, with which it is difficult, probably very difficult, to deal. The limit of the average is nevertheless expected to exist and is, moreover, expected to be, even if $S$ contains finite places, a linear combination of the distributions (30), of which there are three, because $\epsilon_{--} = \epsilon_{-+}$. The coefficients will depend on the functions $f_q$, $q \neq \infty$, $q \in S$, thus on congruence conditions. So there is a great deal of uniformity present in the limit, and it is fair to assume that it will influence the structure of the proofs.

On the other hand, the average of the difference, with the elliptic term expressed as in (62) and (65), decomposed with the help of (69), will be a sum over three parameters, $r$, $p$, and $n$. More precisely, the last sum is over $n$ and $f$, but the

Table 3.1

| | | | | | |
|---|---|---|---|---|---|
| −0.686858, | −0.010181, | −0.232848, | −0.181406, | −0.207348, | −0.213738, |
| 0.186493, | −0.509315, | −0.143169, | −0.267028, | −0.160988, | −0.214750, |
| −0.291132, | −0.231973, | −0.268148, | −0.267438, | −0.226005, | −0.252182, |
| −0.199118, | −0.079383, | −0.183245, | −0.132645, | −0.202296, | −0.171024, |
| −1.025438, | −0.527494, | −0.233874, | −0.247813, | −0.248718, | −0.231158, |
| 0.017161, | −0.245755, | −0.271121, | −0.200476, | −0.211613, | −0.186097, |
| 0.057466, | −0.073425, | −0.243888, | −0.170394, | −0.227328, | −0.194990, |
| −0.604006, | −0.449603, | −0.106058, | −0.211854, | −0.123694, | −0.198008, |
| −0.147666, | −0.198848, | −0.244547, | −0.154175, | −0.267187, | −0.169350, |
| −0.232995, | −0.460777, | −0.199048, | −0.265850, | −0.163095, | −0.238796, |
| −0.352068, | 0.088846, | −0.154301, | −0.183977, | −0.186853, | −0.186285, |
| −0.183918, | −0.399292, | −0.319741, | −0.250550, | −0.273583, | −0.235420, |
| −0.218394, | −0.137239, | −0.112422, | −0.158514, | −0.140592, | −0.170782, |
| −0.331184, | −0.328770, | −0.223462, | −0.234503, | −0.237538, | −0.201874, |
| −0.330528, | −0.277603, | −0.236737, | −0.227831, | −0.191458, | −0.218459, |
| −0.107266, | −0.126031, | −0.195398, | −0.185583, | −0.213405, | −0.202951, |
| −0.138815, | −0.188041, | −0.211230, | −0.181449, | −0.197189, | −0.192290, |
| −0.388114, | −0.267641, | −0.236432, | −0.223636, | −0.239041, | −0.204733, |
| −0.285824, | −0.179338, | −0.159782, | −0.179327, | −0.176213, | −0.194733, |
| −0.147042, | −0.182515, | −0.213875, | −0.195378, | −0.207439, | −0.192548, |
| −0.137437, | −0.008915, | −0.225177, | −0.013033, | −0.211749, | −0.000857, |
| −0.413068, | −0.056893, | −0.220921, | 0.007333, | −0.201721, | 0.003851, |
| −0.080076, | −0.004867, | −0.169043, | −0.005863, | −0.182254, | −0.007391, |
| −0.270411, | −0.037313, | −0.235472, | −0.020661, | −0.224686, | −0.006618, |
| −0.282038, | −0.001461, | −0.188183, | 0.019859, | −0.193986, | 0.005356, |
| −0.232331, | −0.095297, | −0.217932, | −0.011204, | −0.224461, | 0.004303, |
| −0.125913, | 0.028871, | −0.208961, | −0.020619, | −0.194989, | −0.011606, |
| −0.238424, | −0.026239, | −0.177729, | 0.004167, | −0.197464, | −0.006211, |
| −0.175674, | −0.020565, | −0.215916, | 0.008770, | −0.192947, | −0.007468, |
| −0.249100, | 0.015408, | −0.184689, | −0.009332, | −0.198297, | 0.006565. |

additional sum over $f$ may be little more than an unfortunate complication, whose implications are limited, of the sum over $n$. At the moment, I am not concerned with it. There are also sums over $\pm$ and $f'$ that occur simultaneously with the sum over $r$ and are understood to be part of it.

The sum over $r$ has a simple structure, except for the dependence on $s$. The use of the logarithmic derivatives that leads us to an average over $p$ with the factor $\ln p$ is alarming as any incautious move puts us dangerously close to the mathematics of the Riemann hypothesis, but there is nothing to be done about it. The structure suggested by functoriality and the $L$-group imposes the use of the logarithmic derivative on us, and any attempt to avoid it for specious (in the sense of MacAulay) technical advantages is likely to lead us away from our goal, not toward it. Although in the middle of the nineteenth century the word *specious* did not yet have its present thoroughly pejorative sense, it did evoke doubt; so a few words of explanation are in order. The only tool presently in sight for passing from $\pi$ to ${}^{\lambda}H$ are the functions $m_\pi$ and $m_H$, which are linear in $\rho$. We are already familiar, thanks to basic results for the Artin $L$-functions, with the importance of the linearity in $\rho$ of the order of the pole of the $L$-functions at $s = 1$. The linearity in $\rho$ is naturally accompanied by a linearity in $\pi$. The functions $m_\pi$ not only incorporate the structural advantages suggested by functoriality that will be of great importance when we pass to groups of large dimension but also are fully adapted to the trace formula, provided we take them as defined by the logarithm of the $L$-function $L(s, \pi, \rho)$ or its derivative. On the other hand, we have none of the necessary analytic experience. We are faced with sums and limits in which we do not know what is large, what is small, what converges, what does not, and we desperately need insight. If some experience and some feeling for the analysis can be acquired by a modification of the problem in special cases in which the sums over primes that appear in the logarithmic derivative are replaced by easier sums over integers, then common sense suggests that we start there. Moreover, we do want to discover something about the behaviour of automorphic $L$-functions near $s = 1$, although if we are careful we should not otherwise find ourselves inside the critical strip.

By passing from (60) to (62) we remove the class number, an almost intractable factor, but at the cost of the additional sum over $n$ and $f$. Although the contribution from the one-dimensional representations is not at first expressed as a sum over $n$ and $f$, we observed in §2.5 that there was a natural way so to express it, so that the difference becomes the sum over $n$ and $f$ of (70).

If it turned out that for each $n$ and $f$, the sum over $r$ and $p$ behaved well, then we would, it seems to me, have a much better chance of dealing with the elliptic term. More precisely, it would be a real windfall if the average of (70) approached a limit for each $n$ and $f$, and if the sum over $n$ and $f$ followed by the average could be replaced by the average followed by the sum. The most important observation of this paper is that preliminary numerical investigations suggest that the average of (70) does indeed have a regular behavior, but there are no windfalls. Since my experience as a programmer is limited and mistakes are easy to make, either outright blunders or a careless analysis of possible systematic errors in what are necessarily approximate calculations, I very much hope that others will find the results sufficiently curious to be worthy of their attention. Not only should my conclusions be examined again and more extensively, but, apart from any theoretical efforts, higher $m$, especially $m = 2, 3$, need to be considered as does the effect of congruence conditions or of characters of the Galois group.

Interchanging the order of summation and the passage to the limit is another matter. In the summation there are three ranges: $n$ substantially smaller than $\sqrt{|N|}$; $n$ about equal to $\sqrt{|N|}$; $n$ substantially larger than $\sqrt{|N|}$. We can expect that the interchange picks out the first range. The function $\varphi(x, n)$ is such that we can expect the last range to contribute nothing. This leaves the intermediate range, which may very well contribute but about which nothing is said in this paper, whose tentative explorations, instructive though they are, stop short of all difficult analytic problems. Since $\varphi(x, n)$ is a function of $n/\sqrt{x}$, the factor $\varphi(D, n)$ in (62) can almost be treated as a constant when $n \sim \sqrt{N}$. Thus, in so far as $D$ is just $r^2 - N$, the pertinent expression in the intermediate range is pretty much

$$\sum_{-cn \le r \le cn} \left(\frac{r^2 - N}{n}\right) \psi_\pm \left(\frac{r}{n}\right).$$

More extensive investigations, which I have not yet undertaken, would examine, at least numerically but also theoretically if this is possible, the sum over the intermediate range in this light as well as the validity of the separation into three ranges. Is it possible to hope that the average over $p < X$ of the previous expression will have features like those described in §3.3 for the average over the first range? Can the separation be made cleanly so that any contributions from intermediate domains on the marches of the three ranges are small?

Although we persuaded ourselves that (70) might very well be $o(|N|^{1/2})$, so that it is smaller than the two expressions of which it is a difference, we made no effort to see what size it might be. Its average over $p$ is intended to have a limit, thus, in particular, to be $O(1)$, but that does not prevent violent oscillations in the individual terms. Besides, the existence of a limit may be too much to expect. The numerical results described later in this section suggest that (70) is $O(\ln^2 |N|)$, but I have not yet even been able to show that it is $O(\ln^c |N|)$ for some exponent $c$. Before coming to the experiments, I describe briefly the difficulties that I met in trying to estimate (70) directly. I have not yet made a serious attempt to overcome them.

Recall first that $\psi_\pm$ in (70) are zero outside some interval $[c_1, c_2]$, so that $r$ need be summed only over $c_1\sqrt{|N|} \le r \le c_2\sqrt{|N|}$. To simplify the—in any case rough—analysis, I suppose that both $\psi_\pm$ are bounded; thus I ignore the possible singularity of $\psi_+$ at $r = \pm 1$. Observe that, for large $N$,

$$\varphi(x, n) = \frac{\sqrt{|x|}}{n} + A + B \ln |x| + O(|x|^{-1/2}),$$

where $A$ and $B$ are well-determined constants that depend only on the sign of $x$. The constant implicit in the error term depends on $n$. Since

$$\sum_{c_1\sqrt{|N|} \le r \le c_2\sqrt{|N|}} \frac{1}{\sqrt{|N|}} = O(1),$$

we can replace $\varphi(x, n)$ by $\sqrt{|x|}/n$ at a cost that is $O(\ln |N|)$, a price that we are willing to pay.

Thus, at that level of precision, we can make the same modifications as led from (63) to (66) and replace (70) by twice the sum over $\pm$ of the difference

$$(72)\quad \sum_r \frac{f}{sn}\left(\frac{D}{n}\right)\psi_\pm(x_r)\sqrt{|x_r^2 \mp 1|}\Phi - \sqrt{|N|}\epsilon_{n,f}(N)\int \psi_\pm(x)\sqrt{|x^2 \mp 1|}dx.$$

In (72) there is no longer a sum over $f'$ and no need to sum over $\pm$, as we can simply fix the sign.

To simplify further, I take $n = 1$. In so far as there is any real argument in the following discussion, it can easily be extended to an arbitrary $n$. This is just a matter of imposing further congruence conditions on $r$ modulo primes dividing $2n$. For similar reasons, I also take $f = 1$. Then (72) becomes

$$\sum_r \psi_\pm(x_r)\frac{\sqrt{|D|}}{\sqrt{|N|}} - \sqrt{|N|}\epsilon_{1,1}(N)\int \psi_\pm(x)\sqrt{x^2 \mp 1}dx.$$

It might be better to take $\psi_\pm$ to be the characteristic function of an interval and to attack this fairly simple expression directly. I tried a different approach.

I indicate explicitly the dependence of $s$ on $r$ by setting $s = s_r$ and then write the first term of the difference, with $n$ now equal to 1, as

$$(73)\quad \sum_s \frac{1}{s}\sum_{s_r=s}\psi_\pm(x_r)\sqrt{|x_r^2 \mp 1|}.$$

We then compare

$$(74)\quad \sum_{s_r=s}\psi_\pm(x_r)\sqrt{|x_r^2 \mp 1|}$$

with

$$(75)\quad \frac{\sqrt{|N|}}{s^2}\int_{-\infty}^{\infty}\psi_\pm(x)\sqrt{|x^2 \mp 1|}g_s(x)dx,$$

where $g_s(x)$ is constant on each interval $[ks^2/\sqrt{|N|}, (k+1)s^2/\sqrt{|N|})$ and equal to the number $C_s(k)$ of integral points $r$ in $[ks^2, (k+1)s^2)$ such that $r^2 - N$ divided by $s^2$ is a fundamental discriminant. The sum (73) is compared with

$$(76)\quad \sum_s \frac{\sqrt{|N|}}{s^3}\int_{-\infty}^{\infty}\psi_\pm(x)\sqrt{|x^2 \mp 1|}g_s(x)dx.$$

We have to compare (76) not only with (73) but also with the second term of (72), which is, for $n = f = 1$,

$$(77)\quad \sqrt{|N|}\epsilon_{1,1}(N)\int \psi_\pm(x)\sqrt{|x^2 \mp 1|}dx.$$

I truncate both (73) and (76) at $s \le M = |N|^{1/4}$. An integer $r$ contributes to the number $C_s(k)$ only if $s^2$ divides $r^2 - N$. This already fixes $r$ up to a number of possibilities modulo $s^2$ bounded by $2^{\#(s)}$, where $\#(s)$ is the number of prime divisors of $s$. Thus the truncation of (76) leads to an error whose order is no larger than

$$\sum_{s>M} \frac{\sqrt{|N|}2^{\#(s)}}{s^3} = O(\ln^{2c-1}|N|),$$

as in Lemma B.1 of Appendix B. As observed there, the constant $c$ may very well be 1.

For $s > M$, the number of $r$ in $[c_1\sqrt{|N|}, c_2\sqrt{|N|})$ such that $s^2$ divides $r^2 - N$ is $O(2^{\#(s)})$, because $\sqrt{|N|}/s^2$ is bounded by 1. Thus, according to Lemma B.3, the error entailed by the truncation of (73) is of order no worse than

$$\sum_{C\sqrt{|N|}>s>M} \frac{2^{\#(s)}}{s} = O(\ln^{2c}|N|).$$

To estimate the difference between (74) and (75), we regard

$$\psi_\pm(x_r)\sqrt{|x_r^2 \mp 1|} = \frac{\sqrt{|N|}}{s^2}\psi_\pm(x_r)\sqrt{|x_r^2 \mp 1|}\frac{s^2}{\sqrt{|N|}},$$

$ks^2 \leq r < (k+1)s^2$, as an approximation to

$$\frac{\sqrt{|N|}}{s^2}\int_{\frac{ks^2}{\sqrt{|N|}}}^{\frac{(k+1)s^2}{\sqrt{|N|}}} \psi_\pm(x)\sqrt{|x^2 \mp 1|}dx.$$

Besides difficulties around $x = \pm 1$—where $\psi_+$ may not be bounded, much less smooth—the approximation will be good to within

$$\frac{\sqrt{|N|}}{s^2}O\left(\left(\frac{s^2}{\sqrt{|N|}}\right)^2\right) = O\left(\frac{s^2}{\sqrt{|N|}}\right).$$

Multiplying by $1/s$ and summing up to $M$, we obtain as an estimate for the truncated difference between (73) and (76)

$$\frac{1}{\sqrt{|N|}}O\left(\sum_{s\leq M} 2^{\#(s)}s\right),$$

which is estimated according to Corollary B.4 as $O(\ln^{2c}|N|)$.

For a given $N$ and a given natural number $s$, the condition that $(r^2 - N)/s^2$ be integral but divisible by the square of no odd prime dividing $s$ and that $(r^2 - N)/4$ have some specified residue modulo 4, or any given higher power of 2 is a condition on $r$ modulo $4s^4$ or some multiple of this by a power of 2, so that it makes sense to speak of the average number $\alpha'(N, s)$ of such $r$. The number $\alpha'(N, s)$ is $O(2^{\#(s)}/s^2)$. If $q$ is odd and prime to $s$ then the average number of $r$ for which, in addition, $r^2 - N$

is not divisible by $q^2$ is

$$1-\frac{1+\left(\frac{N}{q}\right)}{q^2}.$$

Thus the average number of $r$ for which $r^2-N$ divided by $s^2$ is a fundamental discriminant can be defined as

$$\alpha(N,s)=\alpha'(N,s)\prod_{\gcd(q,2s)=1}\left(1-\frac{1+\left(\frac{N}{q}\right)}{q^2}\right).$$

As in the first appendix,

$$\epsilon_{1,1}(N)=\sum_{s=1}^{\infty}\frac{\alpha(N,s)}{s}.$$

It remains to compare (75) with

$$\sqrt{|N|}\alpha(N,s)\int_{-\infty}^{\infty}\psi_{\pm}(x)\sqrt{|x^2\mp 1|}dx \tag{78}$$

remembering that their difference is to be divided by $s$ and then summed over $s$, although by Lemma B.1, the sum can be truncated at $s\le M$. I had difficulties with the estimates that I have not yet been able to overcome. I describe them.

Let $\bar{g}_s$ be the average of $g_s$ on some interval $[-C,C]$ large enough to contain in its interior the support of $\psi_{\pm}$. The difference between (75) and (78) divided by $s$ is the sum of two terms. First of all,

$$\frac{\sqrt{|N|}}{s^3}\int_{-C}^{C}\psi_{\pm}(x)\sqrt{|x^2\mp 1|}(g_s(x)-\bar{g}_s(x))dx; \tag{79}$$

and secondly,

$$\frac{\sqrt{|N|}}{s^3}\int_{-C}^{C}\psi_{\pm}(x)\sqrt{|x^2\mp 1|}(\bar{g}_s(x)-s^2\alpha(N,s))dx. \tag{80}$$

The first should be smallest when $\psi_{\pm}$ is very flat; the second when its mean is 0. So it appears they are to be estimated separately.

First of all, to calculate $C_s(k)$ and thus $g_s$, we have to examine the $O(2^{\#(s)})$ integers $r$ in the pertinent interval such that $s^2$ divides $r^2-N$. For simplicity, rather than work with $g_s$ and $\bar{g}_s$, I work with the contributions to $C_s(k)$ from a single residue class $\bar{r}$ modulo $s^2$, but without changing the notation. As a result, the estimates obtained will have to be multiplied by the familiar factor $2^{\#(s)}$. Moreover, the definition of $\alpha(N,s)$ will have to be modified according to the same principle.

If $r$ lies in $[-C\sqrt{|N|},C\sqrt{|N|}]$ and has residue $\bar{r}$, then we attach to $r$ the set $p_1,p_2,\dots,p_l$ such that $s^2p_i^2$ divides $r^2-N$ and is congruent to 0 or 1 modulo 4. Then $s(r)$ is divisible by $sp_1\dots p_l$ and $s^2p_1^2\dots p_l^2\le(C^2+1)|N|$. So there are only a finite number of sets $\{p_1,\dots,p_l\}$ that arise. Let $A(p_1,\dots,p_l)$ be the set of $k$ such that $r \in [ks^2, (k+1)s^2) \subset [-C\sqrt{|N|}, C\sqrt{|N|}]$ with the given residue $\bar{r}$ has $s(r)$ divisible by $sp_1 \dots p_l$ and by no prime but those in $\{p_1, \dots, p_l\}$. Let $|A|$ be the total number of elements in all the $A(p_1, \dots, p_l), l \geq 0$. Then $|A| - 2C\sqrt{|N|}/s^2$ is $O(1)$ and $1/|A| = s^2/2C\sqrt{|N|} + O(1/|A|^2)$. Let $A(+)$ be the union of $A(p_1, \dots, p_l)$, $l > 0$.

Set $\Psi_\pm(k)$ equal to the integral over the interval $[ks^2/\sqrt{|N|}, (k+1)s^2/\sqrt{|N|}]$ of $\psi_\pm(x)\sqrt{|x^2 \pm 1|}$. Then with our new conventions, the integral in (79) becomes

$$\sum_{k \in A()} \Psi_\pm(k)\left(1 - \sum_{i \in A()} 1/|A|\right) - \sum_{k \in A(+)} \sum_{i \in A()} \Psi_\pm(k)/|A| + O(s^2/\sqrt{|N|}).$$

Thanks to (B.10) we may ignore the error term. The main term is

$$\frac{1}{|A|} \sum_{k \in A()} \sum_{i \in A(+)} (\Psi_\pm(k) - \Psi_\pm(i)). \tag{81}$$

Each term $\Psi_\pm(k)$ that appears in (81) is assigned not only to a $k \in A()$ but also to an $i \in A(+)$, say $i \in A(p_1, \dots, p_l)$. We can change the assignation and thus rearrange the sum by decomposing the integers into intervals $I_m = [ms^2 p_1^2 \dots p_l^2, (m+1)s^2 p_1^2 \dots p_l^2)$, choosing for each of these intervals an $i'$ in it such that $i' \equiv i \pmod{s^2 p_1^2 \dots p_l^2}$ and assigning $\Psi_\pm(k)$ to $k$ and to that $i'$ lying in the same interval $I_m$ as $k$. For this to be effective, we introduce sets $B(p_1, \dots, p_l)$, defined as the set of $k$ such that the $r \in [ks^2, (k+1)s^2)$ with the given residue $\bar{r}$ modulo $s$ has $s(r)$ divisible by $sp_1 \dots p_l$. Then $i'$ necessarily lies in $B(p_1, \dots, p_l)$, although it may not lie in $A(p_1, \dots, p_l)$. Then the union $B(+)$ of all the $B(p_1, \dots, p_l)$, $l > 0$, is again $A(+)$ but these sets are no longer disjoint. The number of times $Q(k, i')$ that $k$ is assigned to a given $i'$ is clearly $O(\sqrt{|N|}/s^2 p_1^2 \dots p_l^2)$.

If we change notation, replacing $i'$ by $i$, the sum (81) becomes

$$\frac{1}{|A|} \sum_{i \in A(+)} \sum_{k \in A()} Q(k, i)(\Psi_\pm(k) - \Psi_\pm(i)). \tag{82}$$

If $\psi_\pm$ is continuously differentiable and if $i \in A(p_1, \dots, p_l)$ and $Q(k, i) \neq 0$, then

$$\Psi_\pm(k) - \Psi_\pm(i) = O\left(\frac{s^2}{\sqrt{|N|}}\right) O\left(\frac{s^2 \prod_{j=1}^l p_j^2}{\sqrt{|N|}}\right),$$

the first factor coming from the length of the interval, the second from the difference of the functions $\psi_\pm$ on the two intervals. Since the number of elements in $B(p_1, \dots, p_l)$ is $O(2^{l'}\sqrt{|N|}/s^2 p_1^2 \dots p_l^2)$, $l'$ being the number of $p_j$, $1 \leq j \leq l$, that do not divide $s$, (81) is estimated as

$$\frac{1}{|A|} \sum_{l>0} \sum_{p_1, \dots, p_l} 2^{l'} O\left(\frac{s^2}{\sqrt{|N|}}\right) O\left(\frac{s^2 \prod_{j=1}^l p_j^2}{\sqrt{|N|}}\right) O\left(\frac{\sqrt{|N|}}{s^2 p_1^2 \dots p_l^2}\right)^2 O(p_1^2 \dots p_l^2),$$

the final factor being the number of intervals of length $s^2$ in an interval $I_m$. This expression is

$$O\left(\frac{s^2}{\sqrt{|N|}}\right) O\left(\sum_{l>0} \sum_{p_1,\dots,p_l} 2^{l'}\right), \tag{83}$$

which multiplied by $\sqrt{|N|}/s^3$ yields

$$O\left(\frac{1}{s}\right) O\left(\sum_{l>0} \sum_{p_1,\dots,p_l} 2^{l'}\right). \tag{84}$$

Were it not for the second factor, we could appeal to (B.10). Even though this factor is a finite sum because $s^2p_1^2 \dots Pl^2 \le (C^2+1)\sqrt{|N|}$, it is far too large to be useful. It is likely to have been very wasteful to estimate the terms in (82) individually. We can after all expect that if $0 \le \bar{i} < s^2p_1^2 \dots p_l^2$ is the residue of $i$ in $A(p_1, \dots, p_l)$ then $\bar{i}/s^2p_1^2 \dots p_l^2$ is distributed fairly uniformly over $[0, 1)$ as $p_1, \dots, p_l$ vary, but at the moment I do not know how to establish or to use this. So the poor estimate (83) is one obstacle to establishing a reasonable estimate for (72).

As in the analysis of (79), we may calculate, with an error that is easily estimated as $O(s^2/\sqrt{|N|})$, $\bar{g}_s$ as $|A()|/|A|$ or, better, $s^2|A()|/2C\sqrt{|N|}$. It is clear that

$$|A()| = |B()| - \sum_{p_1} |B(p_1)| + \sum_{p_1,p_2} |B(p_1, p_2)| - + \dots, \tag{85}$$

in which the sum is over $s^2p_1^2 \dots p_l^2 \le C^2 + 1$. Each set $B(p_1, \dots, p_l)$ corresponds to an interval $[ks^2p_1^2 \dots p_l^2, (k+1)s^2p_1^2 \dots p_l^2)$ and it is implicit in the definition that this interval must meet $[-C\sqrt{|N|}, C\sqrt{|N|}]$. It is not, however, necessary that it be contained in the larger interval. Then

$$|B(p_1, \dots, p_l)| = \alpha(p_1, \dots, p_l)\left(\frac{2C\sqrt{N}}{s^2p_1^2 \dots p_l^2} + \epsilon(p_1, \dots, p_l)\right). \tag{86}$$

Here

$$\alpha(p_1, \dots, p_l) = \prod_{j=1}^{l} \alpha(p_j).$$

If $p$ is odd, $\alpha(p)$, which is 0, 1, or 2, is the number of solutions of $r^2 - N \equiv 0 \pmod{s^2p^2}$, $(r^2 - N)/s^2p^2 \equiv 0, 1 \pmod 4$ with the condition that the residue of $r$ modulo $s^2$ is $\bar{r}$. If $p = 2$, it is $1/4$ the number of solutions of the same conditions but with $r$ taken modulo $4s^2p^2$. Because those intervals $[ks^2p_1^2 \dots p_l^2,$ $(k+1)s^2p_1^2 \dots p_l^2)$ that lie partly inside $[-C\sqrt{|N|}, C\sqrt{|N|}]$ and partly outside may or not belong to $B(p_1, \dots, p_l)$ the number $\epsilon(p_1, \dots, p_l)$ lies between $-1$ and 1 if no $p_j$ is 2. Otherwise it lies between $-4$ and 4.

We can also calculate the modified $\alpha(N, s)$ as

$$\sum_{p_1,\dots,p_l} (-1)^l \frac{\alpha(p_1,\dots,p_l)}{s^2 p_1^2 \dots p_l^2}.$$

We conclude from (85) and (86) that, apart from an error that we can allow ourselves, the difference $\bar{g}_s - s^2\alpha$ is

$$\sum_{l\geq 0} \sum_{p_1,\dots,p_l} (-1)^l \alpha(p_1,\dots,p_l)\epsilon(p_1,\dots,p_l)\frac{s^2}{2C\sqrt{|N|}}.$$

So once again, we have to deal with

$$\sum_s \frac{1}{s}\left\{\sum_{l\geq 0} \sum_{p_1,\dots,p_l} (-1)^l \alpha(p_1,\dots,p_l)\epsilon(p_1,\dots,p_l)\right\}. \tag{87}$$

The expression in parentheses in (87) depends strongly on $s$ and is, once again, apparently far too large, a coarse estimate suggesting that the inner sum is of magnitude

$$\sum_{p_1^2\dots p_l^2 \leq (C^2+1)\sqrt{|N|}/s^2} 2^{l'}, \tag{88}$$

where $l'$ is once again the number of $j$, $1 \leq j \leq l$ such that $p_j$ does not divide $s$. Perhaps we have to take into account that the signs of the factors $\epsilon(p_1,\dots,p_l)$ vary and cancel each other. I have not tried to do this.

Our estimate of (70) is unsatisfactory, so that at this stage it is useful to examine it numerically. The numerical results that I now describe suggest strongly that all estimates that look, for one reason or another, weak are indeed so and that (70) is $O(\ln^2|N|)$. The experimental results, too, leave a good deal to be desired, partly because it is impossible to detect slowly growing coefficients but also because it is inconvenient (for me with my limited programming skills) to work with integers greater than $2^{31} = 2147483648$. For example, when testing the divisibility properties of $r^2 - N$ by $s^2$, it is inconvenient to take $s$ greater than $2^{15}$. Since we can work with remainders when taking squares, we can let $r$ be as large as $2^{31}$. Nonethless, if we do not want to take more time with the programming and do not want the machine to be too long with the calculations, there are limits on the accuracy with which we can calculate the $s = s_r$ appearing in (72). We can calculate a large divisor of $s$, for example, the largest prime divisor that is the product of powers $q^a = q^{a_q}$ of the first $Q$ primes, where $Q$ is at our disposition and where $q^{a_q}$ is at most $2^{15}$. The same limitations apply to the calculation of $\epsilon_{n,f}(N)$ and, in particular, of $\epsilon_{1,1}(N)$. So we can only approximate (72), the approximation depending also on $Q$.

In Table 3.2.A, which has three parts, we give three approximations not to the difference itself but to the difference divided by $\ln p$. Each is for $n = f = 1$ and for three different primes of quite different sizes, the 6000th, $p = 59369$, the 60000th, $p = 746777$, and the 600000th, $p = 8960467$. The three approximations are for

Table 3.2.A: Part 1: $p = 59369$

| | | | | | |
|---|---|---|---|---|---|
| 0.356779, | −0.136681, | 0.358137, | −0.135542, | 0.358815, | −0.136301, |
| −0.305870, | 0.089564, | −0.304567, | 0.090622, | −0.303917, | 0.089916, |
| 0.102864, | 0.114616, | 0.104113, | 0.115592, | 0.104736, | 0.114941, |
| −0.108330, | 0.054990, | −0.107135, | 0.055883, | −0.106538, | 0.055287, |
| −0.027594, | 0.071878, | −0.026452, | 0.072683, | −0.025881, | 0.072146, |
| −0.212283, | 0.083968, | −0.211192, | 0.084682, | −0.210647, | 0.084206, |
| 0.117788, | −0.003163, | 0.118829, | −0.002547, | 0.119348, | −0.002958, |
| 0.091523, | −0.015066, | 0.092514, | −0.014557, | 0.093010, | −0.014897, |
| 0.020256, | −0.084660, | 0.021200, | −0.084275, | 0.021671, | −0.084532, |
| −0.252761, | −0.016231, | −0.251863, | −0.016025, | −0.251414, | −0.016162, |
| 0.133049, | 0.067864, | 0.133903, | 0.068064, | 0.134330, | 0.067930, |
| 0.088015, | 0.014307, | 0.088828, | 0.014663, | 0.089234, | 0.014425, |
| −0.081067, | −0.030958, | −0.080293, | −0.030509, | −0.079906, | −0.030808, |
| 0.017027, | 0.076392, | 0.017766, | 0.076908, | 0.018135, | 0.076564, |
| 0.121633, | 0.025750, | 0.122340, | 0.026318, | 0.122693, | 0.025939, |
| −0.081617, | 0.053260, | −0.080938, | 0.053867, | −0.080599, | 0.053463, |
| −0.002066, | −0.126718, | −0.001409, | −0.126081, | −0.001082, | −0.126505, |
| 0.068478, | 0.082597, | 0.069116, | 0.083256, | 0.069435, | 0.082816, |
| −0.004951, | 0.124929, | −0.004325, | 0.125601, | −0.004012, | 0.125153, |
| −0.239656, | 0.099643, | −0.239035, | 0.100322, | −0.238725, | 0.099869. |

$Q = 80, 160, 320$. They give not (72) itself, but the measure implicit in it, thus the mass with respect to the measure of the twenty intervals of length 0.1 between $-2$ and 0, a point mass falling between exactly at the point separating two intervals being assigned half to one and half to the other interval. All these masses are divided by $\ln p$. For the smallest of the three primes, all approximations give similar results. For the largest of the primes, even the best two are only close to another. For numbers with any claim to precision, either a larger value of $Q$ or a larger bound on the powers of the primes would be necessary. Nevertheless, the change in the numbers with increasing $Q$ is far, far less than suggested by (84) and (88).

In each part of the table one of the three primes is considered. Each part has three double columns, each of them corresponding to one value of $Q$. For a given $Q$, the first element of the double column is the measure for $\psi_-$ and the second for $\psi_+$. The interval in the first row is $[-2, -1.9]$ and in the last is $[-0.1, 0]$. Notice that the mass divided by $\ln p$ does not seem to grow much or to decrease much but does behave irregularly. Thus the mass itself at first glance seems to be about $O(\ln p)$, but, as already suggested, this is not the correct conclusion.

To exhibit the fluctuating character of these numbers, a similar table for the 6001st prime $p = 746791$ is included as Table 3.2.B, but I only give the results for $Q = 320$. Once again, they come in pairs, for $N = -1$ and $N = 1$, but there are

Table 3.2.A: Part 2: $p = 746777$

| | | | | | |
|---|---|---|---|---|---|
| −0.219263, | 0.068058, | −0.215435, | 0.071714, | −0.208291, | 0.070964, |
| 0.143721, | 0.001004, | 0.146339, | 0.004404, | 0.153195, | 0.003706, |
| −0.020184, | 0.035014, | −0.016663, | −0.065354, | −0.010093, | −0.065998, |
| −0.091281, | −0.098202, | −0.087911, | −0.095335, | −0.081622, | −0.095923, |
| 0.003985, | −0.017991, | 0.007207, | −0.015405, | 0.013220, | −0.015936, |
| 0.087754, | 0.076422, | −0.042187, | 0.078715, | −0.036444, | 0.078245, |
| 0.180775, | −0.060344, | 0.183710, | −0.058364, | 0.189187, | −0.058771, |
| 0.107415, | 0.038662, | 0.110212, | 0.013086, | 0.115430, | 0.012750, |
| −0.058412, | −0.094389, | −0.167936, | −0.093154, | −0.162967, | −0.093407, |
| −0.073063, | 0.007405, | −0.070531, | 0.008066, | −0.065803, | 0.007930, |
| 0.070096, | 0.045556, | 0.072506, | 0.046198, | 0.077003, | 0.046066, |
| −0.023146, | 0.060006, | −0.020854, | 0.061151, | −0.016574, | 0.060916, |
| 0.129606, | −0.071048, | 0.084723, | −0.117606, | 0.088799, | −0.117902, |
| −0.044026, | 0.110108, | −0.041942, | 0.111766, | −0.038054, | 0.111426, |
| −0.015731, | −0.062830, | −0.013737, | −0.061007, | −0.010015, | −0.061381, |
| −0.180290, | 0.079119, | −0.258895, | 0.081069, | −0.255319, | 0.080669, |
| 0.077034, | −0.078929, | 0.078885, | −0.076884, | 0.082340, | −0.077304, |
| −0.011454, | 0.060534, | −0.009653, | −0.008355, | −0.006292, | −0.008789, |
| −0.003079, | 0.082049, | −0.001312, | 0.084207, | 0.001986, | 0.083764, |
| 0.031046, | 0.121403, | −0.041293, | 0.123583, | −0.038028, | 0.123136. |

two columns, the first for the interval from $-2$ to $-1$ and the second for the interval from $-1$ to 0. Table 3.2.B can be compared with Table 3.2.A, Part 1, to see the change on moving from one prime to the next.

As a further test, I took the largest of the absolute values of the masses of the 2 times 20 intervals for the 1000th, the 2000th, and so on, up to the 100000th prime and divided it by $\ln p$. The one hundred numbers so obtained appear as Table 3.2.C, which is to be read like a normal text, from left to right and then from top to bottom. In the calculations, the integer $Q$ was taken to be 160, but doubling this has only a slight effect. Although at first glance, there is no obvious sign in the table of any increase, a plot of the numbers, as in Diagram 3.2.A, suggests that they do increase and rather dramatically. (Unfortunately, it was not always convenient to insert the tables and the diagrams at the points where they are discussed in the text.)

On the other hand, if we continue up to the 600000th prime we obtain the results of Diagram 3.2.B, where once again $Q = 160$ and where once again doubling $Q$ leads to essentially the same scattering with only a slight displacement of the points. So Diagram 3.2.A is misleading and there is no dramatic rise! A second, more careful glance at the diagram suggests, however, that a slow movement of the points upward, perhaps compatible with the $O(\ln^2 |N|) = O(\ln^2 p)$ hypothesis, is not out of the question. We will return to this point when we have more and different

Table 3.2.A: Part 3: $p = 8960467$

| | | | | | |
|---|---|---|---|---|---|
| 0.065614, | −0.026226, | 0.077929, | −0.016149, | 0.086276, | −0.021717, |
| −0.099652, | −0.151652, | −0.087834, | −0.142283, | −0.079825, | −0.147459, |
| −0.148913, | 0.172582, | −0.137585, | 0.181226, | −0.129909, | 0.176450, |
| 0.403084, | −0.036168, | 0.413927, | −0.028269, | 0.421275, | −0.032633, |
| −0.156494, | 0.060108, | −0.146127, | 0.067236, | −0.139102, | 0.063297, |
| 0.016583, | 0.038520, | 0.026483, | 0.044839, | 0.033191, | 0.041348, |
| 0.273885, | 0.195733, | 0.283327, | 0.201189, | 0.289726, | 0.198174, |
| −0.074761, | −0.041762, | −0.065764, | −0.037253, | −0.059667, | −0.039744, |
| 0.092875, | 0.010855, | 0.101440, | 0.014259, | 0.107244, | 0.012378, |
| −0.126962, | 0.015473, | −0.118812, | 0.017294, | −0.113290, | 0.016288, |
| 0.065242, | 0.009336, | 0.072994, | 0.011104, | 0.078248, | 0.010127, |
| 0.258894, | 0.023944, | 0.266270, | 0.027098, | 0.271269, | 0.025356, |
| −0.120552, | −0.028011, | −0.113526, | −0.024038, | −0.108765, | −0.026233, |
| −0.059059, | 0.034596, | −0.052355, | 0.039166, | −0.047812, | 0.036641, |
| −0.106739, | −0.015397, | −0.100324, | −0.010374, | −0.095977, | −0.013149, |
| 0.093762, | −0.044505, | 0.099926, | −0.039132, | 0.104104, | −0.042101, |
| −0.101929, | 0.219005, | −0.095973, | 0.224641, | −0.091937, | 0.221527, |
| 0.101531, | 0.011202, | 0.107326, | 0.017028, | 0.111253, | 0.013809, |
| 0.036660, | −0.033043, | 0.042345, | −0.027093, | 0.046198, | −0.030380, |
| −0.010312, | −0.008676, | −0.004683, | −0.002666, | −0.000869, | −0.005986. |

data at our disposition. As a convenient comparison, Diagram 3.2.C superposes the points of Diagram 3.2.B on the graph of the curve $0.4\ln(1000x\ln(1000x))/15$, $1 \le x \le 600$. The diagram confirms, to the extent it can, the hypothesis.

**3.3. Some suggestive phenomena.** The previous section does not establish beyond doubt that (70) is $O(\ln^2 p)$ or even the slightly weaker hypothesis that for some integer $l$ the expression (70) is $O(\ln^l p)$. We now consider fixing $n$ and $f$

Table 3.2.B

| | | | |
|---|---|---|---|
| 0.065424, | 0.139319, | 0.009127, | 0.020117, |
| −0.339535, | 0.095697, | 0.244114, | 0.028861, |
| 0.064939, | −0.125936, | −0.242039, | 0.013350, |
| 0.227577, | 0.215971, | 0.068047, | 0.008164, |
| −0.077371, | −0.069531, | 0.011311, | 0.034411, |
| −0.210577, | −0.056532, | 0.159543, | −0.064707, |
| 0.373104, | −0.083462, | −0.002606, | 0.046361, |
| −0.279405, | 0.042905, | 0.147926, | 0.197334, |
| 0.176183, | 0.114329, | −0.041558, | 0.238247, |
| −0.049009, | −0.016793, | −0.180851, | −0.082724. |

Table 3.2.C

| | | | | |
|---|---|---|---|---|
| 0.358120, | 0.872623, | 0.258640, | 0.295414, | 0.345120, |
| 0.358137, | 0.310121, | 0.427307, | 0.392101, | 0.461449, |
| 0.301074, | 0.242229, | 0.353707, | 0.498970, | 0.449767, |
| 0.405747, | 0.256198, | 0.461453, | 0.381769, | 0.347241, |
| 0.317558, | 0.345492, | 0.324273, | 0.559732, | 0.305104, |
| 0.246601, | 0.355806, | 0.287550, | 0.435331, | 0.400707, |
| 0.275095, | 0.324584, | 0.376984, | 0.427550, | 0.321304, |
| 0.319035, | 0.306974, | 0.494958, | 0.301518, | 0.393844, |
| 0.394138, | 0.252000, | 0.429559, | 0.365034, | 0.407917, |
| 0.359968, | 0.458391, | 0.338244, | 0.312106, | 0.300587, |
| 0.291630, | 0.489896, | 0.327670, | 0.405218, | 0.209386, |
| 0.227849, | 0.481018, | 0.556393, | 0.322056, | 0.258895, |
| 0.361781, | 0.383069, | 0.374638, | 0.337790, | 0.287852, |
| 0.441601, | 0.695974, | 0.321117, | 0.627571, | 0.324480, |
| 0.391816, | 0.830769, | 0.615896, | 0.358815, | 0.291243, |
| 0.644122, | 0.228597, | 0.557525, | 0.313941, | 0.440433, |
| 0.343996, | 0.864512, | 0.356637, | 0.678889, | 0.582523, |
| 0.314871, | 0.329813, | 0.398283, | 0.385383, | 0.645377, |
| 0.314966, | 0.470168, | 0.331259, | 0.298338, | 0.479059, |
| 0.302799, | 0.579901, | 0.365380, | 0.457965, | 0.388941. |

and taking the average of (70), in the sense of (12′) over the primes up to $X$. If $X = x \ln x$, then, under the hypothesis that (70) is $O(\ln p)$, the order of the average will be majorized by a constant times

$$\frac{\sum_{n<x} \ln^2(n \ln n)}{x \ln x}. \tag{89}$$

This is approximately

$$\frac{\int_2^x \ln^2(t \ln t)}{x \ln x} \sim \ln x \sim \ln(x \ln x) = \ln X.$$

If the order were $\ln^l p$, then (89) would be majorized by a constant times $\ln^l X$. The average is a measure $\nu_{n,f,X}$, which we may also consider as a distribution on the set of possible $\psi_\pm$. Suppose

$$\nu_{n,f,X} = \alpha_{n,f} + \beta_{n,f} \ln X + o_{n,f}(1),$$

where $\alpha_{n,f}$ and $\beta_{n,f}$ are two measures or distributions. Then interchanging the order of the sum up to $X$ and the sum over $n$, we find that we are to take the limit of

$$\sum_{n,f} \alpha_{n,f} + \sum_{n,f} \beta_{n,f} \ln X + \sum_{n,f} o_{n,f}(1). \tag{90}$$

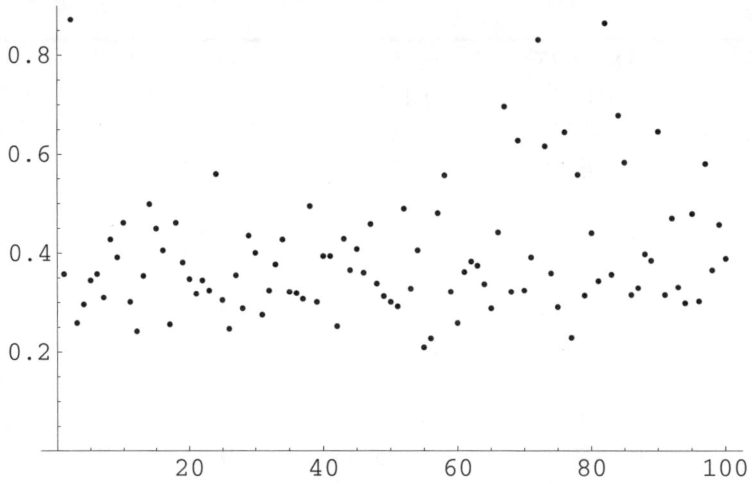


*Diagram 3.2.A*

If there were no contributions from the other two ranges and if the third sum was itself $o(1)$, then the sum

$$\sum_{n,f} \beta_{n,f} \tag{91}$$

would have to be 0, and the sum

$$\sum_{n,f} \alpha_{n,f} \tag{92}$$

the limit for which we are looking, thus the contribution from the first range of summation where $n$ is smaller than $\sqrt{|N|}$. There is, however, no good reason to

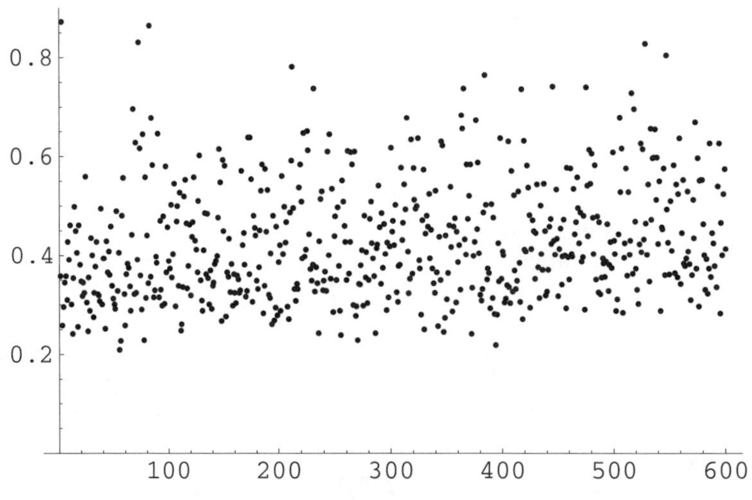


*Diagram 3.2.B*

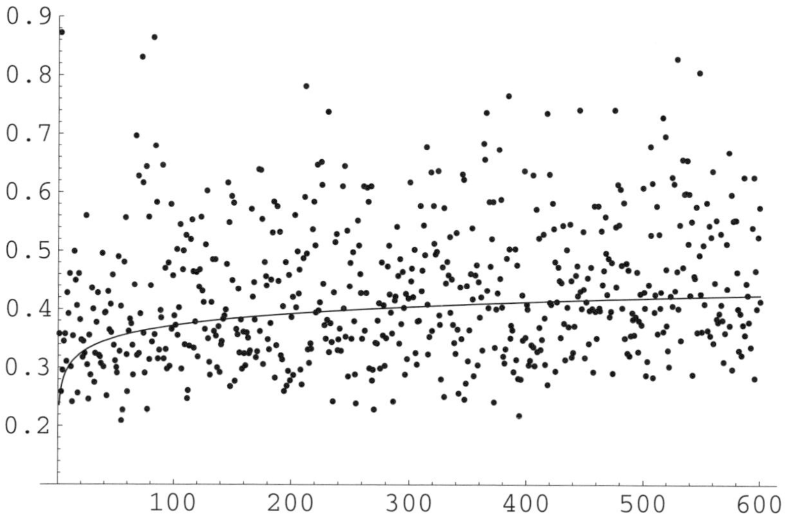


*Diagram 3.2.C*

expect that (91) is 0. It may be cancelled by a contribution from the intermediate range.

We can certainly envisage polynomials of higher degree in (90). For such asymptotic behavior to make sense, it is best that the change in $\ln^l X$ be $o(1)$, as $X$ changes from $n \ln n$ to $(n+1)\ln(n+1)$, thus, in essence, as we pass from one prime to the next. Since

$$(n+1)\ln(n+1) = (n+1)\ln n + O(1) = n \ln n \left(1 + O\left(\frac{1}{n}\right)\right),$$

we have

$$\ln((n+1)\ln(n+1)) = \ln(n \ln n) + O\left(\frac{1}{n}\right).$$

I examined the behavior of the average of the sum (70) for $n = 1, 3, 5, 15$ and $f = 1$, treating it again as a measure on the two lines $N = \pm 1$ and plotting the average, in the sense of (12′), over the first $1000k$ primes for $1 \leq k \leq 60$ against $1000k \ln(1000k)$. The results are given at the end of the paper in Diagrams 3.3.A to 3.3.D. The results are not so simple as (90), although they do make it clear that the average behaves regularly and is naturally expressed as a quadratic function of $\ln(X)$, so that $\nu_{f,n,X}$ would be a quadratic function of $\ln X$ with a small remainder and there would be another sum in (91) that would have to vanish. (It is perfectly clear to me that these suggestions are far-fetched. I feel, nevertheless, that they are worth pursuing.) I divided the interval $[-3, 0]$ into six intervals of length 0.5, each column of each diagram contains the six graphs for the six intervals, the first column for $N = -1$ and the second for $N = 1$. They are close to linear as (90) suggests, but not exactly linear.

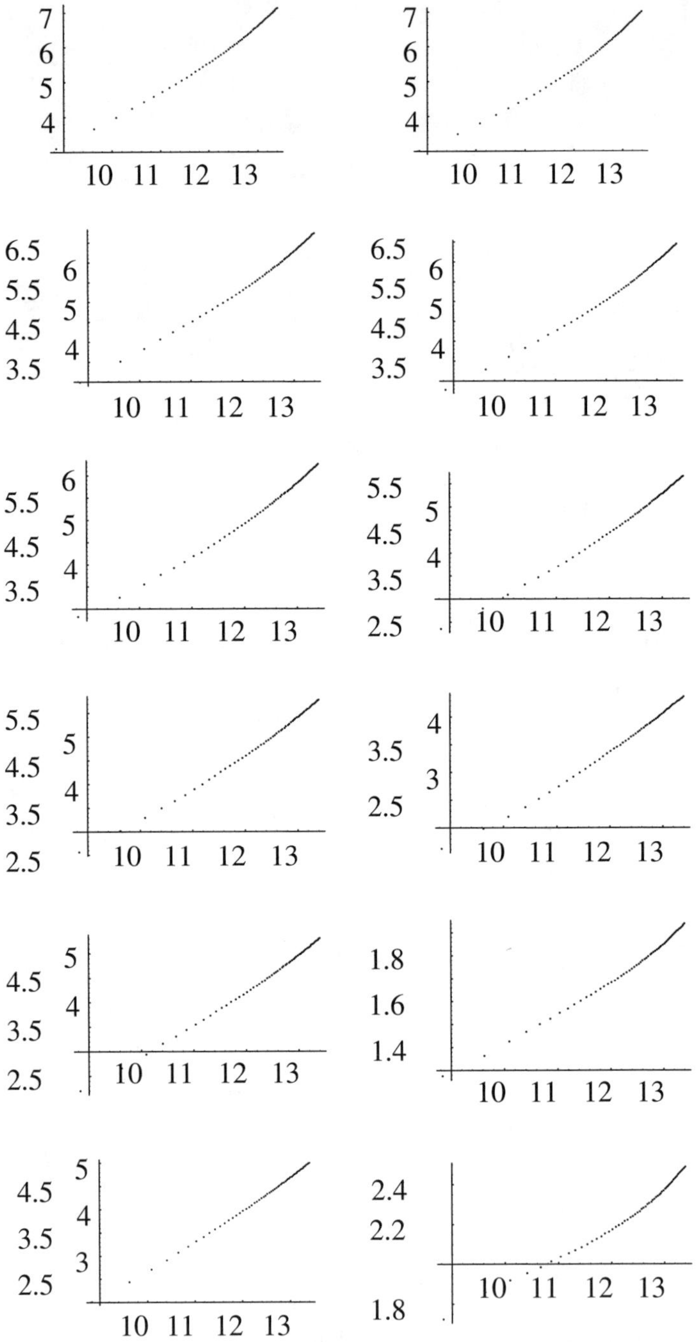
7
6
5
4
10 11 12 13
7
6
5
4
10 11 12 13
6.5
5.5
4.5
3.5
6
5
4
10 11 12 13
6.5
5.5
4.5
3.5
6
5
4
10 11 12 13
5.5
4.5
3.5
6
5
4
10 11 12 13
5.5
4.5
3.5
2.5
5
4
10 11 12 13
5.5
4.5
3.5
2.5
5
4
10 11 12 13
4
3.5
3
2.5
10 11 12 13
4.5
3.5
2.5
5
4
10 11 12 13
1.8
1.6
1.4
10 11 12 13
4.5
3.5
2.5
5
4
3
10 11 12 13
2.4
2.2
1.8
10 11 12 13

***Diagram 3.3.A***

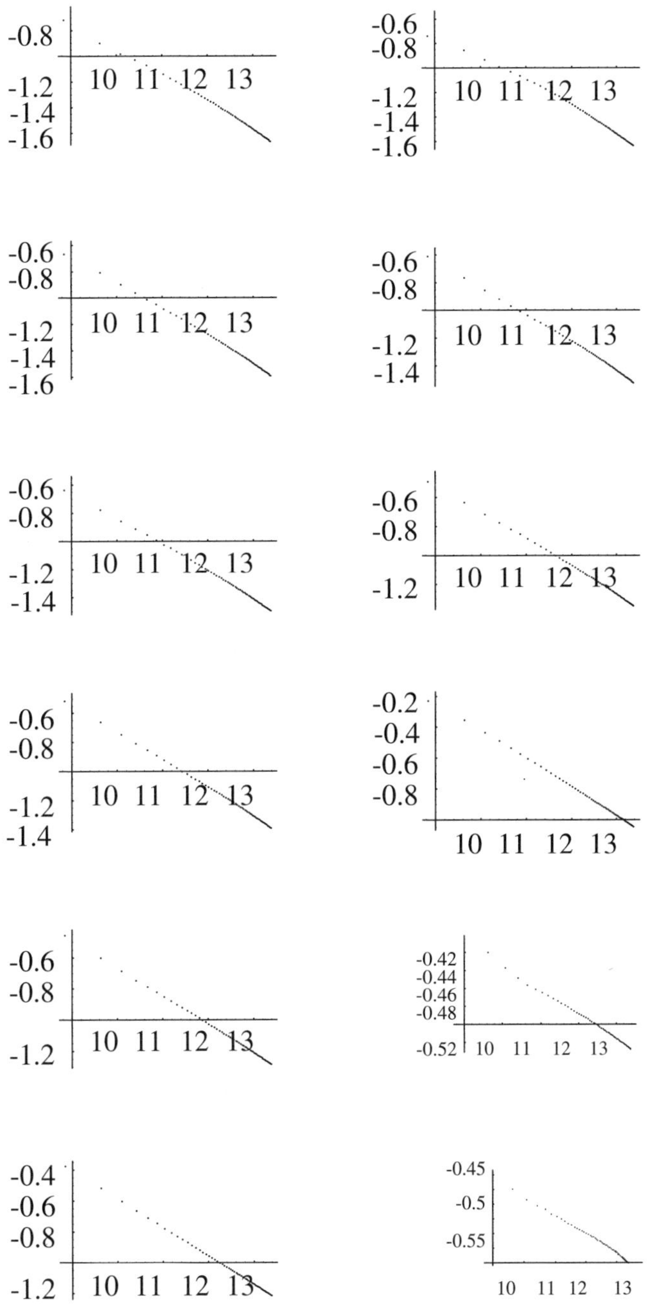
-0.8
-1.2
-1.4
-1.6
10 11 12 13
-0.6
-0.8
-1.2
-1.4
-1.6
10 11 12 13
-0.6
-0.8
-1.2
-1.4
-1.6
10 11 12 13
-0.6
-0.8
-1.2
-1.4
10 11 12 13
-0.6
-0.8
-1.2
-1.4
10 11 12 13
-0.6
-0.8
-1.2
10 11 12 13
-0.6
-0.8
-1.2
-1.4
10 11 12 13
-0.2
-0.4
-0.6
-0.8
10 11 12 13
-0.6
-0.8
-1.2
10 11 12 13
-0.42
-0.44
-0.46
-0.48
-0.52
10 11 12 13
-0.4
-0.6
-0.8
-1.2
10 11 12 13
-0.45
-0.5
-0.55
10 11 12 13

***Diagram 3.3.B***

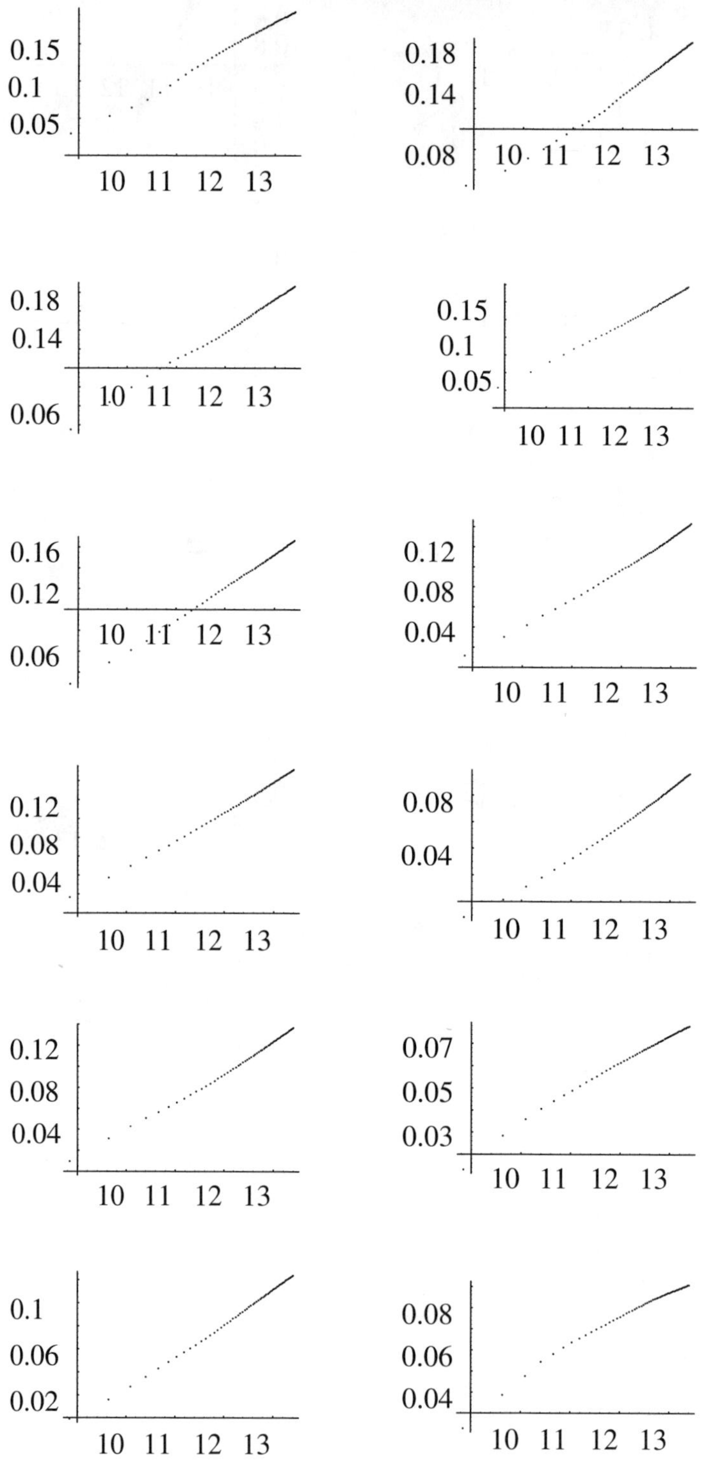
0.15
0.1
0.05
10 11 12 13
0.18
0.14
0.08
10 11 12 13
0.18
0.14
0.06
10 11 12 13
0.15
0.1
0.05
10 11 12 13
0.16
0.12
0.06
10 11 12 13
0.12
0.08
0.04
10 11 12 13
0.12
0.08
0.04
10 11 12 13
0.08
0.04
10 11 12 13
0.12
0.08
0.04
10 11 12 13
0.07
0.05
0.03
10 11 12 13
0.1
0.06
0.02
10 11 12 13
0.08
0.06
0.04
10 11 12 13

***Diagram 3.3.C***

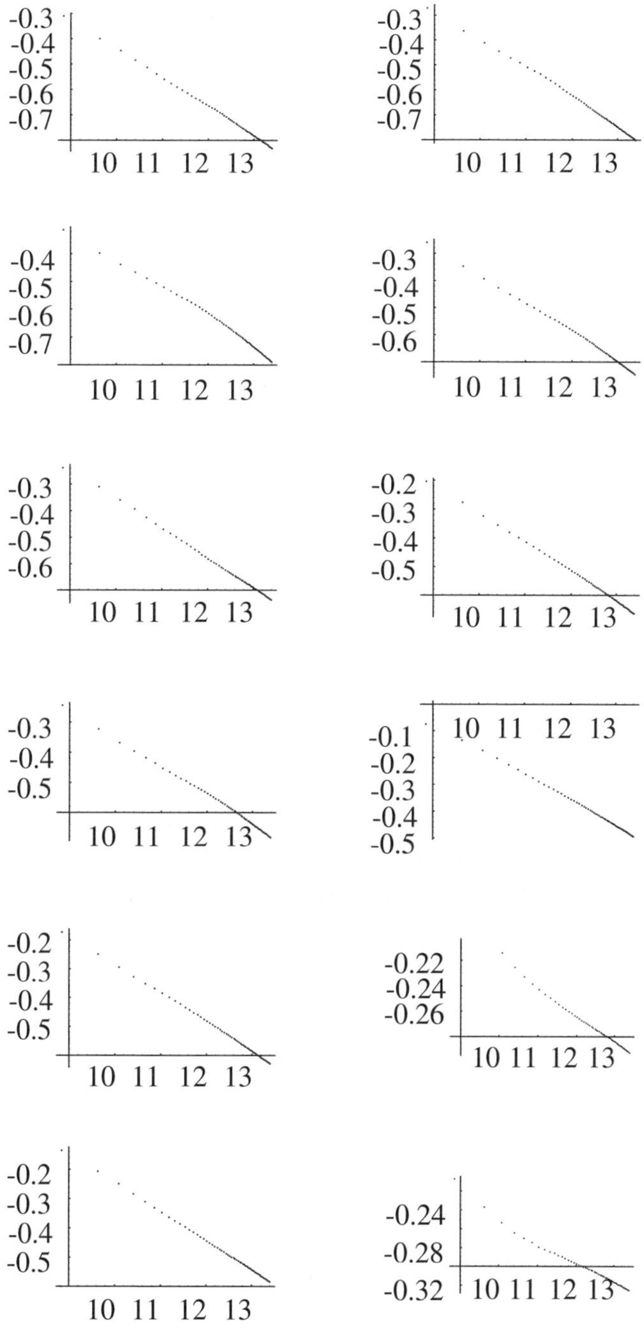
-0.3 -0.4 -0.5 -0.6 -0.7
10 11 12 13
-0.3 -0.4 -0.5 -0.6 -0.7
10 11 12 13
-0.4 -0.5 -0.6 -0.7
10 11 12 13
-0.3 -0.4 -0.5 -0.6
10 11 12 13
-0.3 -0.4 -0.5 -0.6
10 11 12 13
-0.2 -0.3 -0.4 -0.5
10 11 12 13
-0.3 -0.4 -0.5
10 11 12 13
10 11 12 13
-0.1 -0.2 -0.3 -0.4 -0.5
-0.2 -0.3 -0.4 -0.5
10 11 12 13
-0.22 -0.24 -0.26
10 11 12 13
-0.2 -0.3 -0.4 -0.5
10 11 12 13
-0.24 -0.28 -0.32
10 11 12 13

*Diagram 3.3.D*

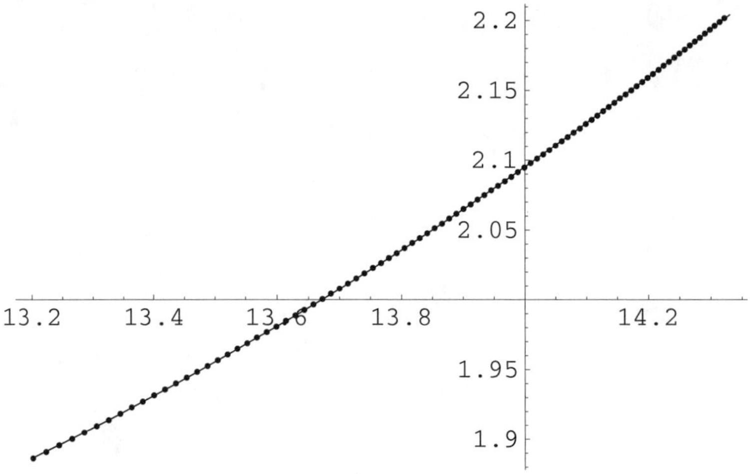


*Diagram 3.3.E*

So I redid the experiments for $\det N > 0$ and $n = 1$ on the intervals in $[-1, 0]$ for primes up to 140000, using a slightly better approximation to the integral

$$\int \sqrt{1 - x^2} dx$$

over the two intervals, but continuing to use only 320 primes to compute the various factors. Since $\ln(k \ln k)$ is 13.4 for $k = 60000$, 14.32 for $k = 140000$, and 15.7 for $k = 600000$, not much is gained by taking even more primes. The two resulting curves, but only for $50000 \leq k \leq 140000$, together with quadratic approximations to them are shown in Diagrams 3.3.E, for the first interval, and 3.3.F, for the second.

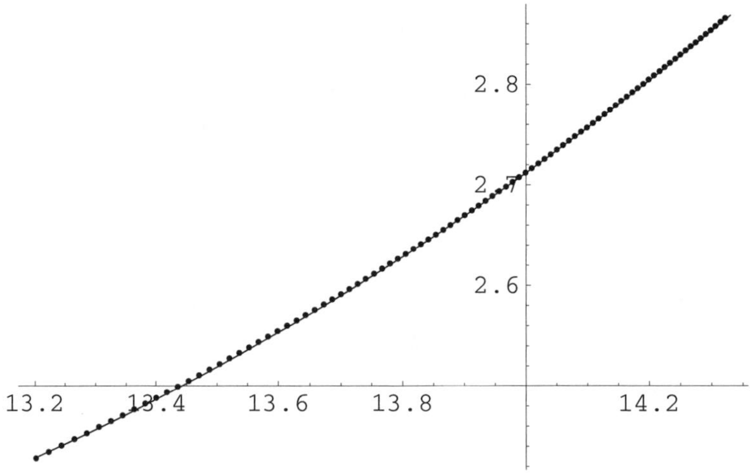


*Diagram 3.3.F*

The quadratic approximations are

$$-1.37552 + 0.24677x + 0.06329(x - 13.5)^2$$

for the first interval and

$$-1.97565 + 0.33297x + 0.10656(x - 13.5)^2$$

for the second. The quadratic term looks to be definitely present. There may even be terms of higher order, but there appears to be little question that we are dealing with a function that as a function of $\ln X$ is essentially polynomial. Thus the natural parameter is $\ln X$ and not some power of $X$.

**Part IV: Supplementary remarks.**

**4.1. Quaternion algebras.** There is some advantage in treating quaternion algebras, as similar results are to be expected, but only the terms (i) and (ii) appear in the trace formula. The disadvantage, especially for numerical purposes, is that some ramification has to be admitted immediately. Apart from that, the only formal difference in the elliptic term is that the discriminant $D$ is subject to the condition that $(\frac{D}{q}) = 1$ for those $q$ that ramify in the quaternion algebra. Moreover, if the algebra is ramified at infinity then only $D < 0$ are allowed.

**4.2. Transfer from elliptic tori.** The representation $\sigma_2$ is of course the representation

$$X \to AXA^t \tag{93}$$

on the space of symmetric matrices. Thus if a reductive subgroup ${}^\lambda H^{\mathbb{Q}}$ of ${}^L G^{\mathbb{Q}} = GL(2, \mathbb{C})$ is not abelian but has a fixed vector in the representation, it is contained in an orthogonal group. Observe that the condition of §1.3 may no longer be fulfilled: the group ${}^\lambda H$ may not lie in $SL(2, \mathbb{C}) \times \mathrm{Gal}(K/F)$. If ${}^\lambda H^{\mathbb{Q}}$ is the first term of an inverse system ${}^\lambda H$ in the system ${}^L G$, then ${}^\lambda H^{\mathbb{Q}}$ is contained in the usual image in ${}^L G^{\mathbb{Q}}$ of the $L$-group of an elliptic torus. Thus, if we take the $\rho$ implicit in (12′) to be $\sigma_2$, then we can expect to single out in the limit those cuspidal representations $\pi$ that are transfers from elliptic tori. They will, however, have an additional property. If the torus is associated to the quadratic extension $E$ with associated character $\chi_E$, then $\chi_E$ will be the central character of $\pi$. Since we can, in the context of the trace formula, fix the central character of the representations $\pi$ to be considered in any way we like, we can in fact single out those representations that are transfers from a given elliptic torus. Then the sum in (14′) will be a sum over a single torus.

If we want an arbitrary central character, then we have to replace (93) by the tensor product of $\sigma_4$ with $\det^{-2}$. Thus the sum in (14′) will be an infinite sum, over all elliptic tori. Moreover, there will in all likelihood be no choice but to let the transfer $f \to f^H$ reflect the reality of the situation. It will have to be defined by the condition that

$$\operatorname{tr}\theta(f^H) = \operatorname{tr}\Theta(f)$$

if $\Theta$ is the transfer of the character $\theta$. These transfers are certainly known to exist, but the relation between the characters of $\theta$ and $\Theta$ remains obscure. So the definition of $f^H$, which is to be made locally, is by no means clear.

If the base field is $\mathbb{Q}$, we cannot take $f_v$ to be unramified at all finite places, because $f_v^H$ would then necessarily be 0 at those places where the quadratic field defining the torus $H$ was ramified. So for experimental purposes, some ramification in $f$ has to be admitted.

If we consider only representations trivial on $Z_+$, then (14′) will be

$$\sum_H \sum_\theta \operatorname{tr}\theta(f^H), \tag{94}$$

with those $\theta$ that lead to noncuspidal representations excluded. Since they can be taken care of separately, it is best to include them. Then (94) can be written as

$$\sum_H \mu(Z_+ H(\mathbb{Q})\backslash H(\mathbb{A})) \sum_{\gamma\in H(\mathbb{Q})} f^H(\gamma). \tag{95}$$

Although this sum appears infinite, it will not be, because $f^H$ will necessarily be 0 for those $H$ that ramify where $f_v$ is unramifed. The sum (95) is very much like the elliptic term of the trace formula, except that the $\gamma$ in the center appear more than once.

The transfer $\theta \to \Theta$ is well understood at infinity. There, at least, the inverse tranfer $f \to f^H$ differs in an important way from endoscopic transfer. Endoscopic transfer is local in the sense that the support of (the orbital integrals of) $f^H$ is, in the stable sense, the same as the support of (those of) $f^H$. In contrast, even if the orbital integrals of $f_\infty$ are supported on hyperbolic elements, $f_\infty^H$ may be nonzero for tori elliptic at infinity. This does not prevent a comparison between (14′) and (17′), but does suggest that it may have a number of novel elements not present for endoscopy.

The first, simplest test offers itself for the representations unramified everywhere. Since every quadratic extension of $\mathbb{Q}$ is ramified somewhere, there are no unramified representations arising from elliptic tori. Thus the limit (12′) should be 0 for $\rho = \sigma_m$, $m = 2$. This is even less obvious than for $m = 1$ and everything will depend on the elliptic contribution to the trace formula. It must cancel all the others. I have made no attempt to understand numerically how this might function, but it would be very useful to do so. A distillation that separates the different kinds of contribution in the elliptic term may be necessary. It would then be useful to understand clearly the orders of magnitude of these contributions.

As a convenient reference for myself, and for anyone else who might be inclined to pursue the matter, I apply the formulas of Appendix B to the conclusions of §2.4 to obtain a list of all the contributions to be cancelled. As it stands, the list has no structure and the terms no meaning. Until they do, §4.3 has to be treated with scepticism.

**4.3. Contributions for even *m*.** I consider all contributions but the elliptic. The first is made up of (31) from the term (ii), corrected by the last term in (41) and by (48) to yield

$$\text{(a)}\qquad \frac{m}{4}\operatorname{tr}(\xi_0(f_\infty)).$$

The second is the sum of atomic measures in (41):

$$\text{(b)}\qquad \sum_q\sum_{n>0}\{|q^n-q^{-n}|\psi_+(q^n+q^{-n})+|q^n+q^{-n}|\psi_-(q^n-q^{-n})\}.$$

The third arises from (51), which is equal to

$$\psi_-(0)\left\{\ln 2+m\ln p\left(1+O\left(\frac{1}{p}\right)\right)\right\}$$

and whose average is

$$\text{(c)}\qquad \psi_-(0)(\ln 2+m\ln X).$$

As was already suggested, this means that for $m=2$ the analogue of (91) will not be 0, but will have to cancel, among other things, (c), at least when there is no ramification.

The contributions from (56) and (48) yield together, in the notation of Appendix C,

$$\sum_\pm\left(\kappa_1+\frac{m\ln p}{2}\right)\hat{f}_\infty(a(1,\pm1)),$$

or when averaged

$$\text{(d)}\qquad \sum_\pm\left(\kappa_1+\frac{m\ln X}{2}\right)\hat{f}_\infty(a(1,\pm1)).$$

I offer no guarantee for the constants in (c) and (d).

All that remains are the terms resulting from the combination of (49) and (57) with (46) and an application of Hoffmann's formula. There is, first of all, the contribution from (C.13) (which must be multiplied by $1/2$)

$$\text{(e)}\qquad -\frac{1}{2}\sum\int_{-\infty}^{\infty}\frac{e^{-|x|}}{1+e^{-|x|}}\hat{f}_\infty(a)dx,$$

where the sum is over the arbitrary sign before the matrix

$$a=\pm\begin{pmatrix}e^x & 0\\ 0 & -e^{-x}\end{pmatrix},$$

and, from (C.17) and (C.18),

$$\text{(f)} \qquad \frac{1}{2}\sum\int_{-\infty}^{\infty}\left(\frac{e^{-|x|}}{1-e^{-|x|}}-\frac{1}{|x|}\right)\hat{f}_\infty(a)dx,$$

in which

$$a = a(x) = \pm\begin{pmatrix} e^x & 0 \\ 0 & e^{-x}\end{pmatrix},$$

the sum being again over the sign, and

$$\text{(g)} \qquad -\frac{1}{2}\int_{-\infty}^{\infty}\ln|x|\frac{d\hat{f}_\infty}{dx}(a)dx,$$

which according to the formula of Appendix D is equal to

$$\int_{-i\infty}^{i\infty}(\ln|s|+\lambda_0)\operatorname{tr}\xi_s(f_\infty).$$

From (47) we have

$$\text{(h)} \quad \frac{1}{16\pi i}\int_{-i\infty}^{i\infty}\left\{-\frac{\Gamma'((1-s)/2)}{\Gamma((1-s)/2)}-\frac{\Gamma'(s/2)}{\Gamma(s/2)}-\frac{\Gamma'((1+s)/2)}{\Gamma((1+s)/2)}-\frac{\Gamma'(-s/2)}{\Gamma(-s/2)}\right\}\operatorname{tr}\xi_s(f_\infty)ds.$$

Finally, from (D.19) there is the completely different contribution

$$\text{(i)} \qquad -\frac{1}{2}\sum_{k=0}^{\infty}(\pm 1)^{k-1}\Theta_{\pi_k}(f).$$

The usual formulas [N, §72] for the logarithmic derivative of the $\Gamma$-function suggest that there should be cancellation among (f), (g), and (h). The Fourier transform of $\xi_s(f_\infty)$ is, however, a function on all four components of the group of diagonal matrices, each component determined by the signs in

$$a = a(x) = \begin{pmatrix} \pm e^x & 0 \\ 0 & \pm e^{-x}\end{pmatrix}.$$

So any cancellation between (h) and (f) would also have to involve (e). I am not familiar with any formula that relates (e) to the $\Gamma$-function and have not searched for one.

**4.4. The third touchstone.** The problem (T3) is, on the face of it, different than the first two, but may be amenable to the same kind of arguments. If the base field is $\mathbb{Q}$, the pertinent representations of $GL(2,\mathbb{R})$ are those obtained by induction from the representations

$$\begin{pmatrix} a & x \\ 0 & b\end{pmatrix} \to (\operatorname{sgn} a)^k(\operatorname{sgn} b)^l\left|\frac{a}{b}\right|^{s/2}, \qquad k,l=\pm 1.$$

We can try to isolate them by a function $f_\infty$ such that $\operatorname{tr}\pi(f_\infty)$ is 0 if $\pi$ lies in the discrete series and $\operatorname{tr}\xi_s^{k,l}(f_\infty)$ is independent of $k$, $l$, but, as a function of $s$, is an approximation to the $\delta$-function at $s = 0$. This means that $\hat{f}_\infty$ is concentrated on $a$ with positive eigenvalues and that it is approaching the function identically equal to 1. Thus $\psi_-$ will be 0 and $\psi_+$ will be 0 for $x < -1$. For $x > 1$, it will be approaching

$$\psi_+(x) = \frac{1}{e^t - e^{-t}} = \frac{1}{\sqrt{|x^2 - 1|}}, \qquad r = e^t + e^{-t}, \qquad x = \frac{r}{2}.$$

What will happen on the range $-1 < x < 1$ remains to be worked out.

Since the approximation at infinity would be occurring while $f_q$ remained fixed at the other places, the sum over $r$ in the elliptic term of the trace formula would be a sum over a fixed lattice—the lattice of integral $r$ if $f_q$ were the unit element of the Hecke algebra everywhere. So the problems that arise look to be different than those for (T2): the limits to be taken are of a different nature. They are perhaps easier, perhaps more difficult; but I have not examined the matter. I have also not examined the role of the other terms in the trace formula.

**4.5. General groups.** Is there an obvious obstacle to extending the considerations of this paper to general groups? Recall that the structure of the trace formula is the equality of a spectral side and a geometric side. The principal term of the spectral side is the sum over the representations occurring discretely in $L^2(G(\mathbb{Q})\backslash G(\mathbb{A}))$ of $\operatorname{tr}\pi(f)$. As for $GL(2)$, we will expect that an inductive procedure will be necessary to remove the contributions from representations that are not of Ramanujan type. This will leave

$$\sum_{\pi}^{R} \operatorname{tr}\pi(f)$$

in which to substitute appropriate $f$ before passing to the limit.

On the geometric side, there will also be a main term, the sum over the elliptic elements. For $GL(k)$ an elliptic element $\gamma$ corresponds to a monic polynomial

$$x^k + a_1x^{k-1} + \dots a_{k-1}x + a_k.$$

For $GL(2)$, $a_1 = -r$, $a_2 = N/4$. Of course, for $\gamma$ to be regular certain degenerate sequences $a_1, a_2, \dots, a_k$ will have to be excluded. For $GL(2)$, not only is $N \neq 0$ but $r^2 - N \neq 0$. In addition, split $\gamma$ are excluded. We should like to say that for a general group, an elliptic element is defined, after the exclusion of singular or partially split elements, by the values of a similar sequence $a_1, a_2, \dots, a_k$. If the group is semisimple and simply connected, these could be the characters of the representations with highest weight $\lambda_i$, $(\lambda_i, \alpha_j) = \delta_{i,j}$, but only if we deal not with conjugacy classes in the usual sense but with stable conjugacy classes, as is perfectly reasonable if we first stabilize the trace formula. For groups that are not semisimple or not simply connected, something can surely be arranged. So we can expect in general a sum over a lattice, analogous either to the lattice of integral $(r, N)$, or, if we recognize that the values of the rational characters of $G$ on those $\gamma$ that yield a contribution different from 0 will be determined up to a finite number of possibilities by $f$, over an analogue of the lattice of $r$. As for $GL(2)$, it will be appropriate to allow a fixed denominator or to impose congruence conditions.

The limits of the remaining terms, either on the spectral side or on the geometric side, we can hope to treat by induction. So the question arises during these preliminary reflections whether the terms in the sum over the lattice have the same structural features as for $GL(2)$. If so, and if there is a procedure for passing rigorously to the limit in the sum over $p < X$, either one in the spirit of the remarks in Part III or some quite different method, then we can continue to hope that the constructs of this paper have some general validity.

There are several factors in the sum: the volume $\mu_\gamma$ of $G_\gamma(\mathbb{Q})\backslash G_\gamma(\mathbb{A})$; the orbital integral at infinity, a function of $a_1, \dots, a_k$ and the analogue of $\psi$; the orbital integrals at the finite number of finite primes in $S$ that give congruence conditions and conditions on the denominators; the orbital integrals at the primes outside of $S$. The latter accounts for the contribution

$$\sum_{f|s} f \prod_{q|f} \left(1 - \frac{\left(\frac{D}{q}\right)}{q}\right) \tag{96}$$

of (59).

The usual calculations of the volume of $T(\mathbb{Q})\backslash T(\mathbb{A})$ (see Ono's appendix to [W]) show that it is expressible as the value of an $L$-function at $s = 1$ so that it will be given by an expression similar to (61). There will be changes. In particular, the $L$-function will be a product of nonabelian Artin $L$-functions. For $GL(k)$ the Kronecker symbols $(\frac{D}{n})$ will be replaced by an expression determined by the behavior of $x^k + a_1x^{k-1} + \cdots \equiv 0$ in the local fields defined by this equation and associated to the primes dividing $n$. This behavior is periodic in $a_1, \dots, a_k$ with period given by some bounded power of the primes dividing $n$, so that the nature of the contribution of $\mu_\gamma$ to the numerical analysis appears to be unchanged. For other groups the relation between the coefficients $a_1, a_2, \dots$ and the stable conjugacy class will be less simple, but the principle is the same.

The contribution of the orbital integrals for places outside $S$ will not be so simple as that given by Lemma 1. It has still to be examined, but it will have similar features. Lemma 1 expresses, among other things, a simple form of the Shalika germ expansion, and it may very well be that this structural feature of orbital integrals will be pertinent to the general analysis. It is reassuring for those who have struggled with the fundamental lemma and other aspects of orbital integrals to see that the arithmetic structure of the orbital integrals of functions in the Hecke algebra, especially of the unit element, may have an even deeper signifance than yet appreciated.

It remains, however, to be seen whether anything serious along these lines can be accomplished!

**Appendix A: Calculation of $\epsilon_{n,f}(N)$.** Both $n$ and $f$ are products of prime powers, $n = \prod q^a$ and $f = \prod q^b$. Thanks to the Chinese remainder theorem,

$$\epsilon_{n,f}(N) = \prod_q \epsilon_{q^a,q^b}(N).$$

It will suffice to show that

$$\text{(A.1)} \qquad \begin{aligned} \sum_{a,b=0}^{\infty} \epsilon_{q^a,q^b}(N) &= 1, & q \neq p, \\ &= \frac{1}{1-p^{-1}} + O(|N|^{-1/2}), & q = p. \end{aligned}$$

When $q$ is fixed, we set for brevity $\epsilon_{q^a,q^b}(N) = \Lambda_{a,b}$. It will be more convenient to define $\Lambda_{a,b,c}$, $c \geq b$, as the product of the average value of

$$\left(\frac{(r^2-N)q^{2b}/q^{2c}}{q^{2a}}\right)$$

on the set of $r$ for which $q^{2c}$ is the highest even power of $q$ dividing $r^2 - N$ with a remainder congruent to 0 or 1 modulo 4 with the density of the set, and to calculate $\Lambda_{a,b}$ as

$$\text{(A.2)} \qquad \sum_{c \geq b} \frac{q^b}{q^{a+c}} \Lambda_{a,b,c}.$$

That the $\Lambda_{a,b,c}$ are at least as natural to calculate as the $\Lambda_{a,b}$ suggests that rather than expressing the elliptic term as a sum over $f$ and $n$ as in the experiments to be described, one might want to express it as a sum over $f$, $n$, and $s$. This would mean that $a$,$d = c - b$ and $c$ were as good a choice of parameters as $a$, $b$, and $c$, or that (71) could be replaced by

$$\text{(A.3)} \qquad \frac{1}{gn}\left(\frac{(r^2-N)/g^2}{n}\right), \quad g = \frac{s}{f}.$$

A direct analytic attack on the problems leads to (A.3) and not to (71).

Suppose first that $q$ is odd and not equal to $p$. Then $N$ is prime to $q$. If $t$ is a high power of $q$, then the density of $r$ modulo $t$ such that $r^2 - N$ is divisible by $q^c$ is

$$\left(1 + \left(\frac{N}{q}\right)\right) q^{-c},$$

if $c > 0$. Thus the density $\mu_c$ of $r$ such that it is divisible by $q^{2c}$ and not by $q^{2c+2}$ is

$$\text{(A.4)}\qquad \begin{array}{ll} 1-\dfrac{1}{q^2}\left(1+\left(\dfrac{N}{q}\right)\right), & c=0 \\[2ex] \dfrac{1}{q^{2c}}\left(1+\left(\dfrac{N}{q}\right)\right)\left(1-\dfrac{1}{q^2}\right), & c>0. \end{array}$$

For positive even $a$, it is the density $\nu_c$ of $r$ such that $r^2 - N$ is divisible by $q^{2c}$ and not by $q^{2c+1}$ that is pertinent. This is

$$\text{(A.5)}\qquad \begin{array}{ll} 1-\dfrac{1}{q}\left(1+\left(\dfrac{N}{q}\right)\right), & c=0 \\[2ex] \dfrac{1}{q^{2c}}\left(1+\left(\dfrac{N}{q}\right)\right)\left(1-\dfrac{1}{q}\right), & c>0. \end{array}$$

When $c > 0$, if $r^2 = N + uq^c$, then

$$(r + vq^c)^2 \equiv N + (u + 2v)q^c \pmod{q^{c+1}}.$$

Thus, the average value of

$$\left(\frac{(r^2 - N)/q^{2c}}{q^a}\right)$$

on those $r$ for which $r^2 - N$ is divisible by $q^{2c}$ and not by $q^{2c+2}$ is 0 if $a$ is odd. For $c = 0$ and $a$ odd, we have a simple lemma that shows that the average is $-1/q$.

Lemma 2. *The sum $A$ of*

$$\left(\frac{r^2 - N}{q}\right)$$

*over $r$ modulo $q$ is $-1$.*

Since the number of solutions of

$$\text{(A.6)}\qquad y^2 = x^2 - N$$

for a given value of $x$ is

$$\left(\frac{x^2 - N}{q}\right) + 1,$$

the number $A + q$ is just the number of points on the rational curve (A.6) modulo $q$ whose coordinates are finite. The lemma follows.

The value of all $\Lambda_{a,b,c}$ and $\Lambda_{a,b}$ can now be calculated. For $a = b = 0$, $\Lambda_{0,0,c} = \mu_c$ and

$$\begin{aligned}\Lambda_{0,0} &= \sum \frac{1}{q^c}\mu_c \\ &= 1 - \frac{1}{q^2}\left(1 + \left(\frac{N}{q}\right)\right) + \sum_{c=1}^{\infty} \frac{1}{q^{3c}}\left(1 + \left(\frac{N}{q}\right)\right)\left(1 - \frac{1}{q^2}\right) \\ &= 1 - \left(1 + \left(\frac{N}{q}\right)\right)\left\{\frac{1}{q^2} - \frac{1}{q^2}\frac{q^2 - 1}{q^3 - 1}\right\} \\ &= 1 - \left(1 + \left(\frac{N}{q}\right)\right)\frac{q - 1}{q^3 - 1}.\end{aligned}$$

If $a > 0$ and $b > 0$ then $\Lambda_{a,b,c} = 0$ and $\Lambda_{a,b} = 0$. For $b > 0$,

$$\begin{aligned}\Lambda_{0,b} &= \sum_{c=b}^{\infty} \frac{q^b}{q^c}\Lambda_{0,b,c} \\ &= \sum_{c=b}^{\infty} \frac{q^b}{q^c}\mu_c \\ &= \left(1 - \frac{1}{q^2}\right)\left(1 + \left(\frac{N}{q}\right)\right)\sum_{c=b}^{\infty} \frac{q^b}{q^{3c}} \\ &= \left(1 - \frac{1}{q^2}\right)\left(1 + \left(\frac{N}{q}\right)\right)\frac{1}{q^{2b}}\frac{1}{1 - q^{-3}}.\end{aligned}$$

Thus

$$\sum_{b=1}^{\infty} \Lambda_{0,b} = \frac{1}{q^2}\left(1 + \left(\frac{N}{q}\right)\right)\frac{1}{1 - q^{-3}} = \frac{q}{q^3 - 1}\left(1 + \left(\frac{N}{q}\right)\right). \tag{A.7}$$

If $a > 0$ is even,

$$\begin{aligned}\Lambda_{a,0} &= \frac{1}{q^a}\sum_{c=0}^{\infty} \frac{\nu_c}{q^c} \\ &= \frac{1}{q^a}\left\{1 - \frac{1}{q}\left(1 + \left(\frac{N}{q}\right)\right) + \frac{1}{q^3 - 1}\left(1 + \left(\frac{N}{q}\right)\right)\left(1 - \frac{1}{q}\right)\right\}.\end{aligned}$$

The sum of this over all positive even integers is

$$\begin{aligned}\sum_{a=1}^{\infty} \Lambda_{2a,0} &= \frac{1}{q^2 - 1}\left\{1 - \frac{1}{q}\left(1 + \left(\frac{N}{q}\right)\right) + \frac{1}{q^3 - 1}\left(1 + \left(\frac{N}{q}\right)\right)\left(1 - \frac{1}{q}\right)\right\} \\ &= \frac{1}{q^2 - 1} - \frac{1}{q^3 - 1}\left(1 + \left(\frac{N}{q}\right)\right).\end{aligned}$$

If $a$ is odd

$$\Lambda_{a,0} = -\frac{1}{q^{a+1}}.$$

Thus

$$\sum_{a=0}^{\infty} \Lambda_{2a+1} = -\frac{1}{q^2}\frac{1}{1-\frac{1}{q^2}} = -\frac{1}{q^2-1}.$$

Examining the previous calculations, we conclude that

$$\sum_{a,b=0}^{\infty} \Lambda_{a,b} = 1.$$

We now consider $q = 2 \neq p$, calculating first of all for each $r$ the highest even power $2^{2c}$ of 2 that divides $r^2 - N$ with a remainder congruent to 0 or 1 modulo 4. We begin by observing that 4 divides $r^2 - N$ if and only if $r = 2t$ is even and then

$$\frac{r^2 - N}{4} = t^2 - M, \quad M = \pm p^m,$$

which is congruent to 0 modulo 4, if and only if $t$ is odd and $(\frac{-1}{M}) = 1$, and to 1, if and only if $t$ is even and $(\frac{-1}{M}) = -1$. In the first of these two cases, $c > 0$; in the second $c = 1$. Otherwise $c = 0$.

There are thus two ways in which $c$ can be 0. Either $r$ is odd or $r$ is even. Since $r^2 - N$ is odd, if and only if $r$ is odd, and is then congruent to $1 - N$ modulo 8,

$$\Lambda_{0,0,0} = \frac{1}{2} + \frac{1}{4} = \frac{3}{4},$$

$$\Lambda_{a,0,0} = \frac{1}{2}\left(\frac{1-N}{2}\right) = \frac{1}{2}\left(\frac{5}{2}\right) = -\frac{1}{2}, \quad a > 0, \quad a \text{ odd},$$

$$\Lambda_{a,0,0} = \frac{1}{2}, \qquad a > 0, \quad a \text{ even}.$$

If $(\frac{-1}{M}) = -1$ and $c > 0$, then $c$ is necessarily 1. Thus for such $M$,

$$\Lambda_{a,b,c} = 0, \quad c > 1.$$

Moreover, recalling that the Kronecker symbol $(\frac{n}{2})$ is 0 for $n$ even, 1 for $n \equiv 1, 7 \pmod 8$, and $-1$ for $n \equiv 3, 5 \pmod 8$ and that $t^2 - M$ is odd only for $t$ even and then takes on the values $-M, 4 - M$ modulo 8 with equal frequency, we see that for the same $M$,

$$\Lambda_{0,b,1} = \frac{1}{4} = \frac{1}{4} - \frac{1}{16}\left(1 + \left(\frac{-1}{M}\right)\right), \quad 1 \geq b \geq 0,$$

$$\Lambda_{a,0,1} = 0, \quad a > 0, \quad a \text{ odd},$$

$$\Lambda_{a,0,1} = \frac{1}{4}, \quad a > 0, \quad a \text{ even}.$$

Now suppose that $(\frac{-1}{M}) = 1$ and $c > 0$. Then, as observed, 4 divides $t^2 - M$ if and only if $t = 2u + 1$. The integer $u^2 + u$ is necessarily even and for any even $v$ and any $d \geq 2$, $u^2 + u \equiv v \pmod{2^d}$ has exactly two solutions modulo $2^d$. In particular, $u \equiv 0, 1, 2, 3 \pmod 4$ yield respectively $u^2 + u \equiv 0, 2, 2, 0 \pmod 4$. Since

$$\text{(A.8)} \qquad \frac{t^2 - M}{4} = u^2 + u + \frac{1 - M}{4},$$

and we conclude that $c = 1$ for $1/2$ of the possible values of $u$ and $c = 2$ for the other half when $M \equiv 5 \pmod 8$. Thus, in this case,

$$\Lambda_{0,b,1} = \frac{1}{8} = \frac{1}{4} - \frac{1}{16}\left(1 + \left(\frac{-1}{M}\right)\right), \quad 1 \geq b \geq 0,$$

$$\Lambda_{0,b,2} = \frac{1}{8} = \frac{1}{32}\left(1 + \left(\frac{-1}{M}\right)\right)\left(1 - \left(\frac{M}{2}\right)\right), \quad 2 \geq b \geq 0,$$

$$\Lambda_{a,0,1} = 0, \quad \Lambda_{a,0,2} = 0, \quad a > 0, \quad a \text{ odd},$$

$$\Lambda_{a,0,1} = 0, \quad \Lambda_{a,0,2} = \frac{1}{8}. \quad a > 0, \quad a \text{ even}.$$

These numbers are to be incorporated with the factor

$$\frac{1}{4}\left(1 + \left(\frac{-1}{M}\right)\right)\left(1 - \left(\frac{M}{2}\right)\right)$$

in so far as it is not already present.

For $M \equiv 1 \pmod 8$, (A.8) can be any even number and the density of $u$ for which it can be divided by $2^{2d}$, $d \geq 0$, to give a number congruent to 0, 1 modulo 4 is $1/2$ if $d = 0$ and $1/2^{2d}$ if $d > 0$. Since $d$ will be $c - 2$, this is $1/2^{2c-4}$. On the other hand, the density is multiplied by $1/4$ when we pass from $u$ to $r$, so that

$$\begin{aligned}
\Lambda_{0,b,1} &= \frac{1}{8} = \frac{1}{4} - \frac{1}{16}\left(1+\left(\frac{-1}{M}\right)\right), && 1 \geq b \geq 0,\\
\Lambda_{a,0,1} &= \Lambda_{a,0,2} = 0, && a > 0, \quad a \text{ odd},\\
\Lambda_{a,0,1} &= \Lambda_{a,0,2} = 0, && a > 0, \quad a \text{ even},\\
\Lambda_{0,b,2} &= \frac{1}{16}, && 2 \geq b \geq 0,\\
\Lambda_{0,b,c} &= \frac{1}{2^{2c-2}}\left(1-\frac{1}{4}\right), && c \geq b \geq 0, \quad c > 2,\\
\Lambda_{a,0,c} &= 0, && a > 0, \quad a \text{ odd}, \quad c > 2,\\
\Lambda_{a,0,c} &= \frac{1}{2^{2c-1}}, && a > 0, \quad a \text{ even}, \quad c > 2.
\end{aligned}$$

These numbers are to be incorporated with the factor

$$\frac{1}{4}\left(1+\left(\frac{-1}{M}\right)\right)\left(1+\left(\frac{M}{2}\right)\right).$$

Then $\Lambda_{0,0}$ is the sum of

$$\frac{7}{8} - \frac{1}{32}\left(1+\left(\frac{-1}{M}\right)\right) + \frac{1}{2^7}\left(1+\left(\frac{-1}{M}\right)\right)\left(1-\left(\frac{M}{2}\right)\right) \tag{A.9$'$}$$

and

$$\left\{\frac{1}{2^8} + \frac{3}{2^4}\sum_{c=3}^{\infty}\frac{1}{2^{3c-2}}\right\}\left(1+\left(\frac{-1}{M}\right)\right)\left(1+\left(\frac{M}{2}\right)\right)$$

or

$$\Lambda'_{0,0} = \frac{1}{2^8}\left(1+\frac{3}{7}\right)\left(1+\left(\frac{-1}{M}\right)\right)\left(1+\left(\frac{M}{2}\right)\right). \tag{A.9$''$}$$

For $b > 0$,

$$\Lambda_{0,b} = \sum_{c=b}^{\infty}\frac{2^b}{2^c}\Lambda_{0,b,c}.$$

Thus $\Lambda_{0,1}$ is

$$\frac{1}{4} - \frac{1}{16}\left(1+\left(\frac{-1}{M}\right)\right) + \frac{1}{64}\left(1+\left(\frac{-1}{M}\right)\right)\left(1-\left(\frac{M}{2}\right)\right)$$
$$+ \frac{1}{2^7}\left(1+\frac{3}{7}\right)\left(1+\left(\frac{-1}{M}\right)\right)\left(1+\left(\frac{M}{2}\right)\right),$$

while

$$\Lambda_{0,2} = \frac{1}{32}\left(1+\left(\frac{-1}{M}\right)\right)\left(1-\left(\frac{M}{2}\right)\right) + \frac{1}{2^6}\left(1+\frac{3}{7}\right)\left(1+\left(\frac{-1}{M}\right)\right)\left(1+\left(\frac{M}{2}\right)\right)$$

and

$$\Lambda_{0,b} = \frac{3}{7}\frac{1}{2^{2b-1}}\left(1+\left(\frac{-1}{M}\right)\right)\left(1+\left(\frac{M}{2}\right)\right), \qquad b > 2,$$

because

$$\frac{3}{16}\sum_{c=b}^{\infty}\frac{2^b}{2^{3c-2}} = \frac{3}{7}\frac{1}{2^{2b-1}}.$$

Thus

$$\sum_{b>2}\Lambda_{0,b} = \frac{1}{2^3}\frac{1}{7}\left(1+\left(\frac{-1}{M}\right)\right)\left(1+\left(\frac{M}{2}\right)\right)$$

and

$$\Lambda'_{0,0} + \sum_{b>0}\Lambda_{0,b}$$

is equal to the sum of

(A.10′)
$$\frac{1}{4} - \frac{1}{16}\left(1+\left(\frac{-1}{M}\right)\right)$$

and

(A.10″)
$$\frac{3}{64}\left(1+\left(\frac{-1}{M}\right)\right)\left(1-\left(\frac{M}{2}\right)\right)$$

and

(A.10‴)
$$\left\{\frac{5}{2^7} + \frac{1}{7}\frac{1}{2^3}\right\}\left(1+\left(\frac{-1}{M}\right)\right)\left(1+\left(\frac{M}{2}\right)\right),$$

because

$$\frac{5}{7}\left(\frac{1}{2^7}+\frac{1}{2^6}+\frac{1}{2^5}\right) + \frac{1}{7}\frac{1}{2^3} = \frac{5}{2^7} + \frac{1}{7}\frac{1}{2^3}.$$

Finally

$$\Lambda_{a,0} = \sum_{c=0}^{\infty}\frac{1}{2^{a+c}}\Lambda_{a,0,c}, \quad a > 0.$$

I express it as a sum of three terms, the first of which is

$$\Lambda'_{a,0} = -\frac{1}{2^{a+1}}, \quad a \text{ odd},$$

or

$$\Lambda'_{a,0} = \frac{1}{2^{a+1}} + \frac{1}{2^{a+4}}\left(1 - \left(\frac{-1}{M}\right)\right), \quad a \text{ even}.$$

The other two, $\Lambda''_{a,0}$ and $\Lambda'''_{a,0}$, will be multiples of $(1+(\frac{-1}{M}))(1+(\frac{M}{2}))$ and $(1+(\frac{-1}{M}))(1-(\frac{M}{2}))$, respectively. Observe that

$$\text{(A.11)} \qquad \sum_{a=1}^{\infty} \Lambda'_{a,0} = -\frac{1}{3} + \frac{1}{6} + \frac{1}{48}\left(1 - \left(\frac{-1}{M}\right)\right)$$
$$= -\frac{1}{16}\left(1 - \left(\frac{-1}{M}\right)\right) - \frac{1}{12}\left(1 + \left(\frac{-1}{M}\right)\right).$$

Since

$$\text{(A.12)} \qquad \frac{1}{8} = \frac{1}{16}\left(1 + \left(\frac{-1}{M}\right)\right) + \frac{1}{16}\left(1 - \left(\frac{-1}{M}\right)\right),$$

we can conclude at least that $\sum \Lambda_{a,b} - 1$ is a multiple of $(1+(\frac{-1}{M}))$.

The terms that involve $(1+(\frac{-1}{M}))$ alone without a second factor $(1 \pm (\frac{M}{2}))$ come from (A.9′), (A.10′), (A.11), and (A.12).

$$\text{(A.13)} \qquad \left\{-\frac{1}{32} - \frac{1}{16} + \frac{1}{16} - \frac{1}{12}\right\}\left(1 + \left(\frac{-1}{M}\right)\right).$$

Since all other terms involve the second factor, I multiply (A.13) by

$$\frac{1}{2}\left(1 + \left(\frac{M}{2}\right)\right) + \frac{1}{2}\left(1 - \left(\frac{M}{2}\right)\right).$$

To establish the first equality of (A.1) for $q = 2$, we have to show that the coefficients of the two expressions $(1+(\frac{-1}{M}))(1 \pm (\frac{M}{2}))$ add up to 0.

The remaining terms that involve the product $(1+(\frac{-1}{M}))(1-(\frac{M}{2}))$ come from (A.9′), (A.10″), and

$$\sum_{a>0} \frac{1}{2^{a+2}} \Lambda''_{a,0} = \sum_{a>0} \frac{1}{2^{2a+7}}\left(1 + \left(\frac{-1}{M}\right)\right)\left(1 - \left(\frac{M}{2}\right)\right)$$
$$= \frac{1}{2^7}\frac{1}{3}\left(1 + \left(\frac{-1}{M}\right)\right)\left(1 - \left(\frac{M}{2}\right)\right).$$

They multiply it by the factor

$$\frac{1}{2^7} + \frac{3}{64} + \frac{1}{2^7}\frac{1}{3}.$$

The sum of this factor and 1/2 of that of (A.13) is

$$\frac{1}{32} - \frac{1}{24} + \frac{1}{2^5}\frac{1}{3} = 0.$$

Since

$$\sum_{a>0} \Lambda'''_{a,0} = \frac{1}{4}\left(1+\left(\frac{-1}{M}\right)\right)\left(1+\left(\frac{M}{2}\right)\right) \sum_{a=2d>0} \sum_{c=3}^{\infty} \frac{1}{2^{a+3c-1}}$$
$$= \frac{1}{2^7}\frac{1}{3}\frac{1}{7}\left(1+\left(\frac{-1}{M}\right)\right)\left(1+\left(\frac{M}{2}\right)\right),$$

the terms involving the factor $(1+(\frac{-1}{M}))(1+(\frac{M}{2}))$ yield

$$\left\{\frac{5}{2^7}+\frac{1}{7}\frac{1}{2^3}+\frac{1}{2^7}\frac{1}{3}\frac{1}{7}\right\}\left(1+\left(\frac{-1}{M}\right)\right)\left(1+\left(\frac{M}{2}\right)\right)$$
$$= \frac{11}{2^6 3}\left(1+\left(\frac{-1}{M}\right)\right)\left(1+\left(\frac{M}{2}\right)\right). \tag{A.14}$$

Since

$$-\frac{1}{32}-\frac{1}{12}=-\frac{11}{2^5 3},$$

the term (A.14) cancels the contribution from (A.13).

I treat the second equality of (A.1) only for $q$ odd as this suffices for our purposes. We calculate $\Lambda_{a,b}$ using (A.2). Since

$$\sum_{a\geq 0}\sum_{c\geq b}\sum_{c\geq m/2} \frac{q^b}{q^{a+c}} O\left(\frac{1}{q^c}\right) = \left\{\sum_{a\geq 0}\sum_{d\geq 0}\frac{1}{q^{a+d}}\right\}\sum_{c\geq m/2} O\left(\frac{1}{q^c}\right),$$

we need not use the exact value of $\Lambda_{a,b,c}$ for $2c \geq m$. We need only approximate it uniformly within $O(\frac{1}{q^c})$.

For $2c < m$, the density of $r$ for which $r^2 - N$ is exactly divisible by $q^{2c}$ is $(1-1/q)/q^c$. For $2c \geq m$, it is $O(1/q^c)$. Thus, as an approximation,

$$\Lambda_{0,0} \sim \left(1-\frac{1}{q}\right)\sum_{c=0}^{\infty}\frac{1}{q^{2c}} = \frac{q}{q+1}.$$

Moreover, again as an approximation,

$$\Lambda_{0,b} \sim \left(1-\frac{1}{q}\right)\sum_{c=b}^{\infty}\frac{q^b}{q^{2c}} = \frac{q}{q+1}\frac{1}{q^b}, \quad b>0,$$

so that

$$\sum_{b>0}\Lambda_{0,b} \sim \frac{q}{q+1}\frac{1}{q-1}.$$

If $2c < m$ and $r = q^c t$, $(q,t)=1$, then

$$\frac{r^2-N}{q^2c} \equiv t^2 \pmod q$$

and

$$\left(\frac{(r^2 - N)/q^{2c}}{q}\right) = 1.$$

Thus, the approximation is

$$\Lambda_{a,0} \sim \left(1 - \frac{1}{q}\right) \sum_{c=0}^{\infty} \frac{1}{q^{a+2c}} = \frac{1}{q^a} \frac{q}{q+1},$$

and

$$\sum_{a>0} \Lambda_{a,0} \sim \frac{q}{q+1} \frac{1}{q-1}.$$

Finally

$$\Lambda_{0,0} + \sum_{b>0} \Lambda_{0,b} + \sum_{a>0} \Lambda_{a,0} \sim \frac{1}{1 - q^{-1}}.$$

**Appendix B: Some estimates.** I collect here a few simple estimates needed in Section 3.2. They are provisional and made without any effort to search the literature. To simplify the notation, take $N$ to be positive and $M = N^{1/4}$. If $s$ is a positive integer, let $\#(s)$ be the number of distinct prime divisors of $s$.

LEMMA B.1. *There is a constant $c \geq 1$ such that*

$$\sqrt{N} \sum_{s>M} \frac{2^{\#(s)}}{s^3} = O(\ln^{2c-1} N).$$

There is a chance that the constant $c$ is 1. It is even very likely, but I make no effort to prove it here. The analysis would certainly be more difficult. For the lemma as stated, it is sufficient to use the well-known Tchebychef estimate [HW, p. 10] for the $n$-th prime number $p(n) \asymp n \ln 2n$. (Following [HW], I use the notation $p(n) \asymp n \ln 2n$ to mean that $C_1 n \ln 2n \leq p(n) \leq C_2 n \ln 2n$, with positive constants $C_1$ and $C_2$.) I have used $n \ln 2n$ rather than $n \ln n$ only to avoid dividing by $\ln 1 = 0$. To verify the lemma with $c = 1$ would undoubtedly entail the use the prime number theorem, thus the asymptotic relation $p(n) \sim n/\ln 2n \sim n/\ln n$, and a different, more incisive treatment of the sums that appear.

Let $q(n)$ be the $n^{\text{th}}$ element of the sequence of prime powers $\{2, 3, 4, 5, 7, 8, 9, \ldots\}$, and $\sigma(x)$ the number of prime powers less than $x$. I observe first that the Tchebychef estimate $\pi(x) \asymp x \ln x$ implies that $\sigma(x) \asymp x \ln x$ as well, and thus that $q(n) \asymp n \ln n \asymp n \ln 2n$.

Indeed,

$$\sigma(x) = \pi(x) + \pi(x^{1/2}) + \ldots \pi(x^{1/D}) + O(1), \quad D = [\ln x],$$

and

$$\sum_{j=2}^{D} \pi(x^{1/j}) \le C \int_{t=1}^{D} \frac{x^{1/t}}{\ln x^{1/t}} dt \le C \int_{t=1}^{\ln x} \frac{x^{1/t}}{\ln x^{1/t}} dt,$$

because $y/\ln y$ is an increasing function for $y \ge e$. The integral is

$$\begin{aligned}\frac{1}{\ln x}\int_1^{\ln x} t e^{\ln x/t} dt &= \frac{1}{\ln x}\int_{1/\ln x}^{1} e^{t\ln x}\frac{dt}{t^3} \\ &= \ln x \int_1^{\ln x} e^t \frac{dt}{t^3} \\ &\le \ln x \int_1^{\ln x/2} e^t \frac{dt}{t^3} + \frac{8}{\ln^2 x}\int_{\ln x/2}^{\ln x} e^t dt = O\left(\frac{x}{\ln^2 x}\right).\end{aligned}$$

Thus $\sigma(x) \asymp \pi(x)$.

To prove the lemma we write $s$ as $s = p_1^{a_1} \dots p_l^{a_l}$, where all the primes $p_1, \dots, p_l$, are different. At first, take $p_1 < p_2 < \dots < p_l$. The expression of the lemma may be written as

$$\sqrt{N}\left\{\sum_{l>0}\sum_{p_1,p_2,\dots,p_l} \frac{2^l}{p_1^{3a_1}\dots p_l^{3a_l}}\right\}.$$

There is certainly a sequence $1' < \dots < k', k' \le l$ such that $p_{1'}^{a_{1'}} \dots p_{k'}^{a_{k'}} > M$ while $p_{1'}^{a_{1'}} \dots \hat{p}_{i'}^{a_{i'}} \dots p_{k'}^{a_{k'}} \le M$ for any $i'$, $1 \le i' \le k'$. The notation signifies that $p_{i'}^{a_{i'}}$ is removed from the product. Thus the expression of the lemma is bounded by

$$\sqrt{N}\left\{\sum_{k>0}\sum_{p_1,p_2,\dots,p_k} \frac{2^k}{p_1^{a_1^3}\dots p_k^{a_k^3}}\left(\sum_t \frac{2^{\#(t)}}{t^3}\right)\right\},$$

where $t$ is allowed to run over all integers prime to $p_1, \dots, p_k$, but where $p_1 < \dots < p_k$, $p_1^{a_1} \dots p_k^{a_k} > M$, and $p_1^{a_1} \dots \hat{p}_i^{a_i} \dots p_k^{a_k} \le M$.

I next allow $p_1, \dots, p_k$ to appear in any order, so that I have to divide by $k!$. It is still the case, however, that $p_1^{a_1} \dots p_k^{a_k} > M$ and that $p_1^{a_1} \dots p_{k-1}^{a_{k-1}} \le M$. Since

$$\sum_t \frac{2^{\#(t)}}{t^3} \le \prod_p \left(1 + \frac{2}{p^3} + \frac{2}{p^6} + \dots\right)$$

is finite, we may drop it from the expression and consider

$$\text{(B.1)} \qquad \begin{aligned}&\sqrt{N}\sum_{k=1}^{\infty}\frac{2^k}{k!}\sum_{\substack{p_1^{a_1}\dots p_k^{a_k} > M \\ p_1^{a_1}\dots p_{k-1}^{a_{k-1}} \le M}} \frac{1}{p_1^{3a_1}\dots p_k^{3a_k}} \\ &= \sqrt{N}\sum_{k=1}^{\infty}\frac{2^k}{k!}\sum_{\substack{q_1\dots q_k > M \\ q_1\dots q_{k-1} \le M}} \frac{1}{q_1^3\dots q_k^3},\end{aligned}$$

where $q_1, \dots q_k$ are prime powers. It is this sum that is to be estimated. In it,

$$q_k > A = \frac{M}{q_1 \dots q_{k-1}} \geq 1.$$

So in general, as a first step, we need to estimate, for any $A \geq 1$,

$$\sum_{q>A} \frac{1}{q^3}. \tag{B.2}$$

We apply the Tchebychef estimate. Thus, if $C$ is taken to be an appropriate positive constant independent of $A$, (B.2) is majorized by a constant times

$$\begin{aligned}\sum_{n>C\frac{A}{\ln 2A}} \frac{1}{n^3 \ln^3 2n} &\leq \frac{1}{\ln^3(CA/\ln 2A)} \sum_{n>C\frac{A}{\ln 2A}} \frac{1}{n^3} \\ &= O\left(\frac{1}{(A/\ln 2A)^2} \frac{1}{\ln^3(2A/\ln 2A)}\right) \\ &= O\left(\frac{1}{A^2 \ln 2A}\right).\end{aligned}$$

Although the argument itself is doubtful for small $A$, especially if $C$ is also small, the conclusion is not.

As a result, (B.1) is bounded by

$$C\sqrt{N} \sum_{k=1}^{\infty} \frac{2^k}{k!} \sum_{q_1 \dots q_{k-1} \leq M} \frac{1}{q_1^3 \dots q_{k-1}^3} \frac{q_1^2 \dots q_{k-1}^2}{M^2} \frac{1}{\ln(2M/q_1 \dots q_{k-1})},$$

with perhaps a new constant $C$. Since $M^2 = \sqrt{N}$, this is

$$C \sum_k \frac{2^k}{k!} \sum_{q_1 \dots q_{k-1} \leq M} \frac{1}{q_1 \dots q_{k-1}} \frac{1}{\ln(2M/q_1 \dots q_{k-1})}. \tag{B.3}$$

To complete the proof of Lemma B.1, we shall use another lemma.

LEMMA B.2. *If $A \geq 1$, then*

$$\sum_{q \leq A} \frac{1}{q \ln(2A/q)} \leq c \frac{\ln \ln A}{\ln 2A}, \tag{B.4}$$

*the sum running over prime powers.*

The constant of this lemma is the constant that appears in Lemma B.1. So it is Lemma B.2 that will have to be improved.

Before proving the lemma, we complete the proof of Lemma B.1. Set $A = M/p_1 \dots p_{k-2}$. Then

$$\sum_{q_1 \dots q_{k-1} \le M} \frac{1}{q_1 \dots q_{k-1}} \frac{1}{\ln(2M/q_1 \dots q_{k-1})} \tag{B.5}$$

may be rewritten as

$$\sum_{q_1 \dots q_{k-2} \le M} \frac{1}{q_1 \dots q_{k-2}} \sum_{q_{k-1} \le A} \frac{1}{q_{k-1}} \frac{1}{\ln(2A/q_{k-1})},$$

which, by Lemma B.2, is at most

$$c \sum_{q_1 \dots q_{k-2} \le M} \frac{1}{q_1 \dots q_{k-2}} \frac{\ln \ln A}{\ln 2A} \le c \ln \ln M \sum_{q_1 \dots q_{k-2} \le M} \frac{1}{q_1 \dots q_{k-2}} \frac{1}{\ln 2A}.$$

It is clear that (B.5) is $O((c \ln \ln M)^{k-1}/\ln 2M)$ for $k = 1$, and this estimate now follows readily by induction for all $k$ uniformly in $k$. As a result (B.3) is

$$O\left(\sum_{k=1}^{\infty} \frac{2^k (c \ln \ln M)^{k-1}}{k! \ln M}\right) = O\left(\frac{e^{2c \ln \ln M}}{\ln \ln M \ln M}\right) = O(\ln^{2c-1} M),$$

where we have discarded a $\ln \ln M$ in the denominator that is of no help.

If we are willing to accept a very large constant $c$ in (B.4), then we can replace $\ln 2A/q$ in the denominator by $\ln CA/\ln p$, where $C$ is any given constant greater than 1 or by $CA/n \ln 2n$, if $q = q(n)$ is the $n$th prime power and $C$ is chosen sufficiently large in comparison to the constant in the Tchebychef inequality. We can also replace the $p(n)$ in the denominator by $n \ln 2n$. Thus, at the cost of adding some terms, we may replace the sum (B.4) by

$$\sum_{n \ln 2n \le C'A} \frac{1}{n \ln 2n} \frac{1}{\ln(CA/n \ln 2n)}. \tag{B.6}$$

There is no harm in supposing that $C' = 1$, Clearly, we can also demand that the sum run over $n \ln 2n \ge C_1$, where $C_1$ is a fixed arbitrary constant, because the sum

$$\sum_{n \ln 2n \le C_1} \frac{1}{n \ln 2n} \frac{1}{\ln(CA/n \ln 2n)}$$

is certainly $O(1/\ln A)$. Set

$$\frac{CA}{n \ln 2n} = A^{1-\alpha}, \qquad \alpha = e^{-a}.$$

If $C_1 \ge C$, $\alpha = \alpha(n) \ge 0$. Moreover, as we have agreed to exclude the initial terms of the original sum, $\alpha < 1$ and $a > 0$. If $\beta$ is some fixed number less than 1, then

$$\sum_{\alpha(n) \le \beta} \frac{1}{n \ln 2n} \frac{1}{\ln(CA/n \ln 2n)} \le C_2 \frac{1}{\ln 2A} \sum_{\alpha(n) \le \beta} \frac{1}{n \ln 2n} \le C_3 \frac{\ln \ln A}{\ln 2A}.$$

So we may sum over $\alpha(n) > \beta$ or $a = a(n) < b$, $b = -\ln\beta$. We now confine ourselves to this range.

In addition $(1-\alpha)\ln A \geq \ln C$, so that $(1-\alpha) \geq C_4/\ln A$ and

$$a \geq C_5/\ln A.$$

Let $\epsilon > 0$ and set $b(k) = b(1+\epsilon/\ln A)^{-k}$. I shall decompose the sum into sums over the intervals $b(k+1) \leq a(n) < b(k)$, for all those $k$ such that $b(k+2) \geq C_5/\ln A$, and into one last interval $C_5/\ln A \leq a(n) < b(k)$, where $k$ is the first integer such that $b(k+2) < C_5/\ln A$. I shall denote these intervals by $I$ and use the Hardy-Wright notation to indicate uniformity with respect to $I$.

Notice first that

$$A^{\alpha(n+1)-\alpha(n)} = \frac{(n+1)\ln 2(n+1)}{n\ln 2n} = 1 + O\left(\frac{1}{n}\right) = 1 + O\left(\frac{1}{\ln 2n}\right).$$

Thus $\alpha(n+1) - \alpha(n) \leq C_6/\ln^2 A$ when $\alpha(n) > \beta$. As a result, on the same range $a(n) - a(n+1) \leq C_7/\ln^2 A$. Moreover,

$$b(k) - b(k+1) \geq b(k)\frac{\epsilon}{\ln A} > \frac{C_5\epsilon}{\ln^2 A}.$$

Thus each of these intervals contains at least two terms of our sum provided that $C_5\epsilon > 2C_7$, as we assume. Moreover, if $a'$ and $a$ lie in the same interval, then $a'/a \asymp 1$ and $(1-\alpha')/(1-\alpha) \asymp 1$, so that

$$\frac{\ln(CA/n'\ln 2n')}{\ln(CA/2n\ln 2n)} \asymp 1,$$

where $n' = n(\alpha')$ and $n = n(\alpha)$ are not necessarily integers.

We conclude first of all that, for any point $a_I$ in $I$,

$$\sum_{a(n)\in I} \frac{1}{n\ln 2n}\,\frac{1}{\ln(CA/n\ln 2n)} \asymp \frac{1}{a_I \ln 2A} \sum_{a(n)\in I} \frac{1}{n\ln 2n}$$

and that

$$\int_I \frac{1}{a}da \asymp \frac{1}{a_I}\int_I da.$$

So, if we can show that

$$\sum_{a(n)\in I} \frac{1}{n\ln 2n} \asymp \int_I da, \tag{B.7}$$

the lemma will follow, because

$$\sum_I \int_I \frac{1}{a}da = \int_{C_5/\ln A}^{b} \frac{1}{a}da = O(\ln\ln A).$$

Since

$$\frac{(n+1)\ln 2(n+1)}{n\ln 2n} = O\left(\left(1+\frac{1}{n}\right)^2\right),$$

the sum in (B.7) may be replaced by the integral with respect to $dn$ from $n_1$ to $n_2$ if $a_2 = a(n_2 - 1)$ and $a_1 = a(n_1)$ are the first and last points in the interval associated to integers. The integral is equal to

$$\int_{n_1}^{n_2} \frac{1}{n\ln 2n} dn = \ln\ln 2n_2 - \ln\ln 2n_1.$$

We show that the right-hand side is equivalent in the sense of Hardy-Wright to $a_2 - a_1$ or, what is the same on the range in question, to $\alpha_1 - \alpha_2$. Thus all three are of comparable magnitudes uniformly in $I$. Since $a_2 - a_1$ is equivalent, again in the sense of Hardy-Wright, to the length of $I$, the relation (B.7) will follow.

Since $n\ln 2n = CA^\alpha$, $\ln n + \ln\ln 2n = \ln C + \alpha \ln A$,

$$\ln n + \ln\ln 2n = \ln n + \left(1 + \frac{\ln\ln 2n}{\ln n}\right),$$

and $\alpha = \alpha(n)$ is bounded below by $-\ln b$, we infer that $\ln n \asymp \ln A$. Moreover

$$\text{(B.8)}\qquad \begin{aligned} \ln\ln n + \left(1+\frac{\ln\ln n}{\ln 2n}\right) &= \ln\left(\alpha\ln A\left(1+\frac{\ln C}{\alpha\ln A}\right)\right) \\ &= \ln\alpha + \ln\ln A + \ln\left(1+\frac{\ln C}{\alpha \ln A}\right). \end{aligned}$$

Since a difference between the values of a continuously differentiable function at two values of the argument is equal to the difference of the arguments times the derivative at some intermediate point,

$$\text{(B.9)}\qquad \begin{aligned} \ln\left(1+\frac{\ln C}{\alpha_2\ln A}\right) - \ln\left(1+\frac{\ln C}{\alpha_1 \ln A}\right) &= O\left(\frac{\ln C}{\ln A}(\alpha_1-\alpha_2)\right) \\ &= O\left(\frac{1}{\ln A}(a_2 - a_1)\right). \end{aligned}$$

The expression

$$\ln\left(1+\frac{\ln\ln 2n}{\ln n}\right) = \ln\left(1+\frac{\ln X}{X-\ln 2}\right), \qquad X = \ln 2n.$$

So the difference

$$\ln\left(1+\frac{\ln_2 \ln 2n_2}{\ln n_2}\right) - \ln\left(1+\frac{\ln_1\ln 2n_1}{\ln n_1}\right) = O\left(\frac{\ln\ln A}{\ln^2 A}(\ln n_2 - \ln n_1)\right).$$

Since

$$\ln\ln n_2 - \ln\ln n_1 \asymp \frac{1}{\ln A}(\ln n_2 - \ln n_1),$$

we conclude from (B.8) and (B.9)that

$$\ln\ln 2n_2 - \ln\ln 2n_1 \asymp \ln\ln n_2 - \ln\ln n_1 \asymp \alpha_2 - \alpha_1.$$

The next lemma is similar to Lemma B.1.

LEMMA B.3. *There is a positive constant $c \geq 1$ such that for any positive constant $C$,*

$$\sum_{C\sqrt{N}>s>M} \frac{2^{\#(s)}}{s} = O(\ln^{2c} N).$$

It is again very likely that $c$ may be taken equal to 1, but once again our proof will squander a good deal of the force even of the Tchebychef inequality.

I have stated the lemma in the way it will be used, but the constant $C$ is clearly neither here nor there. Moreover, we prove the stronger statement

$$\sum_{s\leq\sqrt{N}} \frac{2^{\#(s)}}{s} = O(\ln^{2c} N). \tag{B.10}$$

Thus the lower bound on $s$ in the sum is unnecessary. We take $A = \sqrt{N}$ and write $s = p_1 \dots p_l t$, where $t$ is prime to $p_1, \dots, p_l$ and where $p|t$ implies that $p^2|t$. So the left side of (B.10) is majorized by

$$\left(\sum_{l\geq 0} \sum_{p_1\dots p_l<A} \frac{2^l}{p_1 \dots p_l}\right) \prod_p \left(1 + \frac{2}{p^2} + \frac{2}{p^3} + \dots\right).$$

The product is a constant factor and can be dropped for purposes of the estimation. So we are left with

$$\sum_{l\geq 0} \sum_{p_1\dots p_l<A} \frac{2^l}{p_1 \dots p_l} = \sum_{l\geq 0} \sum_{p_1\dots p_l<A} \frac{2^l}{l!} \frac{1}{p_1 \dots p_l}, \tag{B.11}$$

the difference between the left and the right sides being that the first is over $p_1 < \dots < p_l$, whereas in the second the primes are different but the order arbitrary.

It is clear that

$$\sum_{p<A} \frac{1}{p} = O\left(\sum_{n\ln 2n<CA} \frac{1}{n\ln 2n}\right) = O\left(\ln\ln\left(\frac{A}{\ln A}\right)\right) = O(\ln\ln A).$$

Thus,

$$\sum_{p_1\dots p_l<A} \frac{1}{p_1 \dots p_l} \leq \left(\sum_{p<A} \frac{1}{p}\right)^l \leq (c\ln\ln A)^l,$$

uniformly in $l$. The estimate (B.10) follows from (B.11).

Applying Lemma B.3 with $N$ replaced by $\sqrt{N}$ we obtain

COROLLARY B.4. *There is a constant $c \geq 1$ such that*

$$\frac{1}{\sqrt{N}} \sum_{s \leq M} 2^{\#(s)} s = O(\ln^{2c} N).$$

**Appendix C: Weighted orbital integrals.** This is largely a matter of recollecting results from [H] and earlier papers, amply acknowledged in [H]. More must be said than would be necessary had the author, W. Hoffmann, not assumed that his groups were connected, for, like many groups that arise in the arithmetic theory of automorphic forms, $Z_+ \backslash GL(2, \mathbb{R})$ is unfortunately disconnected, but there is no real difficulty and I shall be as brief as possible. The goal of §2.4 and §4.3, for which we need these results, is just to make clear what terms in addition to the elliptic term contribute to the limit (12′) when $m$ is even and how. We first establish the relation between the notation of this paper and that of [H], as well as the connection between $\omega_1(\gamma, f_\infty)$ and $\theta'_z(0, f_\infty)$, or rather, on referring to (55), between $\omega_1(\gamma, f_\infty)$ and

$$\int f_\infty(k^{-1} z n(x) k) \ln |x| dx dk. \tag{C.1}$$

Let

$$\gamma = \begin{pmatrix} \alpha & 0 \\ 0 & \beta \end{pmatrix}.$$

According to its definition in [JL],

$$\begin{aligned} \omega_1(\gamma, f_\infty) &= -\iint f_\infty(k^{-1} n^{-1}(x) \gamma n(x) k) \ln(1 + x^2) dx dk \\ &= -\iint f_\infty(k^{-1} \gamma n((1 - \beta/\alpha)x) k) \ln(1 + x^2) dx dk, \end{aligned}$$

which is equal to

$$-\frac{1}{|1 - \beta/\alpha|} \iint f_\infty(k^{-1} \gamma n(x) k) \{\ln((1 - \beta/\alpha)^2 + x^2) - \ln(1 - \beta/\alpha)^2\} dx dk.$$

Thus

$$|1 - \beta/\alpha| \omega_1(\gamma, f_\infty) - \ln(1 - \beta/\alpha)^2 \omega(\gamma, f_\infty) \tag{C.2}$$

approaches $-2$ times (C.1) as $\alpha$ and $\beta$ approach $z$. So we shall be able to deduce a convenient expression for (C.1) from Hoffmann's formulas, which are valid for $\alpha\beta > 0$. Since the singularity of $|1 - \beta/\alpha| \omega_1(\gamma, f_\infty)$ at $\alpha = \beta$ is only logarithmic, we may multiply it in (C.2) by any smooth function that assumes the value 1 for $\alpha = \beta$.

Because

$$\gamma = z \begin{pmatrix} 1 & 0 \\ 0 & -1 \end{pmatrix}$$

does not lie in the connected component of $Z_+\backslash GL(2,\mathbb{R})$, Hoffmann's arguments do not apply directly to $\omega_1(\gamma, f_\infty)$ for this $\gamma$.

When comparing the notation of this paper with that of Hoffmann, it is best to replace, without comment, all of Hoffmann's group elements by their inverses. Otherwise the conventions are not those of number-theorists and not those of this paper. For him maximal compact subgroups operate on the left, and parabolic and discrete groups on the right.

The group $P$ of Hoffmann is for us the group of upper-triangular matrices, $\bar{P}$ the group of lower-triangular matrices, and $M$ is the quotient of the group of diagonal matrices by $Z_+$ and has as Lie algebra $\mathfrak{a}_{\mathbb{R}}$. His map $\lambda_P$, which is determined by the weight in the noninvariant orbital integral defining $\omega_1$, we take to be

$$\begin{pmatrix} a & 0 \\ 0 & b \end{pmatrix} \to a - b,$$

and the $\lambda$ defining his $\sigma$ to be $s/2$ times $\lambda_P$. In addition, his $d\lambda$ is $ds/2$. Then, as a result of the transfer of the parabolic subgroup to the right in [H], Hoffmann's $v(n(x))$ is $\ln(1+x^2)$ and is, as he observes, positive. (What with signs and factors of 2, there is considerable room for error when attempting to reconcile conventions from various sources.) Since

$$D_G(m) = \left(1 - \frac{\beta}{\alpha}\right)\left(1 - \frac{\alpha}{\beta}\right), \quad m = \gamma,$$

we conclude that

$$J_M(m, f_\infty) = -\frac{|\alpha - \beta|}{|\alpha\beta|^{\frac{1}{2}}}\omega_1(\gamma, f_\infty).$$

So we may replace $|1 - \beta/\alpha|\omega_1(\gamma, f_\infty)$ in (C.2) by $-J_M(m, f_\infty)$. Here and elsewhere in this appendix I freely use the symbol $m$ as it is used by Hoffmann. Elsewhere in the paper, the symbol $m$ is reserved for the degree of the symmetric power.

Before entering into further comparisons between our notation and that of Hoffmann, I review my understanding of his conventions about the measure on $M$ and on its dual. He takes the two measures to be dual with respect to the Fourier transform. So when they both appear, the normalization is immaterial. On the other hand, only one may appear; moreover, there is a second choice, that of $\lambda_P$, which is fixed by the weighting factor $v$. Hoffmann's $I_P$ is a linear combination of $J_M(m, f)$ and an integral over the dual $\hat{M}$. $J_M(m, f)$ depends directly on $\lambda_P$ but not on the two Haar measures. There is a further dependence on the measure on $M\backslash G$, but this dependence is the same in every pertinent expression in his paper and can be ignored. The integral over the dual depends directly on the measure on $\hat{M}$ and directly on the measure on $M$ because of the presence of $\pi_{P,\sigma}(f)$, which depends directly on the measure on $G$, thus on the measures on $M$ and $M\backslash G$; because of the derivative $\delta_P$, it depends directly on $\lambda_P$ as well. Since the measures on $M$ and its dual are inversely proportional, the dependence on the two measures is cancelled and both terms of the sum depend on $\lambda_P$ alone.

This must therefore be the case for the right side of the formula in his Theorem 1 as well. In the second term, the integral over $\hat{M}$, this is clear, because $\Theta_{\pi_\sigma}$ depends directly on the measure on $M$ and $\Omega_{P,\Sigma}$ depends depends directly on $\lambda_P$. In the first term, however, the only dependence is through $\Omega_\pi(f)$ and is a direct dependence on the measure on $M$. If the theorem is to be valid, this measure must be defined directly in terms of the form $\lambda_P$. This Hoffmann does in a straightforward manner. I refer to his paper for more precision. For the group $SL(2,\mathbb{R}) = Z_+\backslash G^+$, with $G^+ = \{g \in GL(2,\mathbb{R}) \mid \det(g) > 0\}$ and with our parameters, $s$, for the characters of $M$ and $t$ for $a = a(x)$ as in §4.3, the measures are $d\sigma = d|s|/2$ and $da = dx$, which is also the measure $d\lambda/\lambda$ of §2.1.

The collection $\hat{M}$ of unitary representations of $M$ has four connected components, corresponding to the four choices of $k, l = 0, 1$,

$$\sigma : \gamma \to \operatorname{sgn}(\alpha)^k \operatorname{sgn}(\beta)^l \left|\frac{\alpha}{\beta}\right|^{s/2},$$

with $s$ purely imaginary. Although $J_s$ and $\operatorname{tr}(J_s^{-1}J_s'\xi_s(f_\infty))$ were defined in §2.3 only for $k = l = 0$, they are defined for all choices of $k$ and $l$ and Hoffmann's $-J_P(\sigma, f_\infty)$ is nothing but $2\operatorname{tr}(J_s^{-1}J_s'\xi_s^{k,l}(f_\infty))$, an expression in which all implicit dependence on $k$ and $l$ is not indicated. Earlier in the paper, $\xi_s^{0,0}$ appeared simply as $\xi_s$. The factor 2 is a result of the relation $\lambda = s\lambda_P/2$.

Recalling that $D_M(m) = 1$, we consider

$$\text{(C.3)}\quad J_M(m, f_\infty) + \frac{1}{8\pi i}\sum \int_C \operatorname{sgn}(\alpha)^k \operatorname{sgn}(\beta)^l \left|\frac{\alpha}{\beta}\right|^{-s/2} \operatorname{tr}\left(J_s^{-1}J_s'\xi_s^{k,l}(f_\infty)\right)ds.$$

The sum before the integration is over the four possible choices for the pair $(k, l)$. If $f$ is supported on $G^+$ and if $\det(m) > 0$, then the integrand does not change when $k, l$ are replaced modulo 2 by $k+1, l+1$. So the sum over $l$ can be dropped, $l$ can be taken to be 0 and the 8 becomes 4. So (C.3) would reproduce Hoffmann's definition if we were concerned with $G^+$ alone.

We will, in general, be summing (C.3) over $\pm m$, so that the total contribution from the integrals for $k \neq l$ will be 0 and for $k = l$ the $8\pi i$ in the denominator will be replaced by $4\pi i$. Moreover replacing $k = l = 0$ by $k = l = 1$ has the effect of replacing $\xi_s(g)$ by $\operatorname{sgn}(\det(g))\xi_s(g)$ and has no effect on $J_s$. For the contribution from (iv), we shall be concerned with $\alpha = -\beta$ and, for such an $m$, $\operatorname{sgn}\alpha^k \operatorname{sgn}\beta^k$ is 1 for $k = 0$ and $-1$ for $k = 1$. The sum of (C.3) over $\pm m$ therefore reduces to

$$\text{(C.4)}\qquad J_M(m, f_\infty) + J_M(-m, f_\infty) + \frac{1}{2\pi i}\int_C \operatorname{tr}\left(J_s^{-1}J_s'\xi_s(f_\infty^-)\right)ds,$$

where $f_\infty^-$ is the product of $f_\infty$ with the characteristic function of the component of $Z_+\backslash GL(2,\mathbb{R})$ defined by $\det(g) = -1$. The analogous $f_\infty^+$ will appear below. For the $m$ in question, the factor $|D_G(m)|^{1/2}$ is equal to 2. This is the factor coming from $\omega(\gamma, f_2)$. Thus (C.4) is twice the negative of the sum of the contribution to the limit (12′) of (iv), in which there is yet another minus sign, and of that part of (viii) associated to $f_\infty^-$.

The expression (C.3) has no meaning for the $\gamma$ that are pertinent in the contribution of (v) to the limit (12′) for even symmetric powers, namely for $\alpha = \beta$. We may, however, consider it for $\alpha$ unequal but close to $\beta$. Once again we consider the sum over $\pm m$. Then only the terms with $k = l$ remain. Since $\operatorname{sgn}\alpha$ will be equal to $\operatorname{sgn}\beta$, we obtain

$$J_M(m, f_\infty) + J_M(-m, f_\infty) + \frac{1}{2\pi i}\int_C \left|\frac{\alpha}{\beta}\right|^{-s/2} \operatorname{tr}\left(J_s^{-1}J_s'\xi_s(f_\infty^+)\right)ds. \tag{C.5}$$

We add to this

$$\ln(1-\beta/\alpha)^2\{\omega(\gamma, f_\infty) + \omega(-\gamma, f_\infty)\}, \quad \gamma = m.$$

Since the second term in (C.3) is well behaved as $\alpha \to \beta$, the result will have a limit as $\alpha$ and $\beta$ approach a common value $z$ because the integrals themselves will have a limit. The limit is

$$2\sum_{j=0}^{1}\int f_\infty(k^{-1}(-1)^j zn(x)k)\ln|x|dxdk + \frac{1}{2\pi i}\int_C \operatorname{tr}\left(J_s^{-1}J_s'\xi_s(f_\infty^+)\right)ds. \tag{C.6}$$

This is twice the contribution of (57) and of that part of (viii) associated to $f_\infty^+$ to the limit (12′),

Although the results of Hoffmann cannot be applied directly to the general form of (C.3) or (C.4), they can be applied to (C.5). In fact, the material necessary for extending his arguments is available, although not all in print. The principal ingredients are the differential equation for the weighted orbital integrals and an analysis of their asymptotic behavior. The first is available in general [A1] and the second will appear in the course of time in a paper by the same author. Since irreducible representations of $Z_+\backslash GL(2,\mathbb{R})$ are obtained by decomposing—into at most two irreducible constituents—representations induced from its connected component $SL(2,\mathbb{R})$, the Plancherel measure of the larger group is, at least for the discrete series, the same as that of the smaller one. So I feel free to apply Hoffmann's results to (C.3) and (C.4) as well, taking care that the measures used are compatible on restriction to functions supported on $G^+$ with his.

For any diagonal matrix $m$ with diagonal entries of different absolute value, Hoffmann ([H],Th.1) finds—at least for $f$ supported on $G^+$—that $I_P(m, f_\infty)$ is equal to

$$-\frac{|\alpha-\beta|}{|\alpha\beta|^{\frac{1}{2}}}\sum_\pi \Theta_{\check{\pi}}(m)\Theta_\pi(f) + \frac{1}{8\pi i}\sum_{k,l}\int_{-i\infty}^{i\infty}\Omega(m,s)\operatorname{tr}\xi_s^{k,l}(f_\infty)ds, \tag{C.7}$$

where

$$\Omega(m,s) = \eta_{k,l}(m,s) + \eta_{l,k}(m,-s)$$

and

$$\text{(C.8)} \qquad \eta_{k,l}(m,s) = \operatorname{sgn}\alpha^k \operatorname{sgn}\beta^l e^{ts} \begin{cases} \sum_{n=1}^{\infty} \frac{(\alpha/\beta)^{-n}}{n-s}, & t > 0 \\ \sum_{n=0}^{\infty} \frac{(\alpha/\beta)^{n}}{n+s} + \frac{\pi(-1)^{k+l}}{\sin(\pi s)}, & t < 0, \end{cases}$$

if

$$m = m(t) = \begin{pmatrix} \alpha & 0 \\ 0 & \beta \end{pmatrix} = \begin{pmatrix} \pm e^t & 0 \\ 0 & \pm e^{-t} \end{pmatrix},$$

the two signs being chosen independently. The factor $\lambda_P(H_\alpha)/2$ that appears in [H] is 1.

There are two observations to be made. First of all, $\Omega$ depends not only on $m$ and $s$, but also on $k$ and $l$, which determine the character of $M_I$. Secondly, $\Omega(m,s)$ is, for a given $m$, symmetric in $s$ and, despite appearances, does not have a singularity at $s = 0$, so that the contour of integration can pass through that point.

For our purposes, it is best to represent (C.7) in terms of the Fourier transform of $\xi_s^{k,l}(f_\infty)$. We begin with the case that $\det m$ is negative, for the passage to the limit $|\alpha| = |\beta|$ is then more direct. We refer to the first term in (C.7) as the elliptic contribution and to the second as the hyperbolic contribution. If $\det m$ is negative, then the character of a discrete-series representation vanishes at $m$, $\Theta_{\tilde{\pi}}(m) = 0$. So the elliptic contribution is 0.

The character of the representation $\xi_s^{k,l}$ is 0 on the elliptic elements of $GL(2,\mathbb{R})$, but on a hyperbolic element

$$\text{(C.9)} \qquad a = a(x) = \epsilon \begin{pmatrix} e^x & 0 \\ 0 & \delta e^{-x} \end{pmatrix}, \qquad \delta, \epsilon = \pm 1,$$

it is equal to

$$\text{(C.10)} \qquad \epsilon^{k+l} \frac{\delta^l e^{sx} + \delta^k e^{-sx}}{\sqrt{|1-\alpha/\beta||1-\beta/\alpha|}},$$

where the signs are that appearing in the matrix. Since

$$r = \epsilon(e^x \pm e^{-x}),$$

the numbers $e^x$ and $e^{-x}$ can of course be recovered from $r$ and the sign. The measure $d\lambda/\lambda$ is in this new notation $dx$. If

$$\hat{f}_\infty(a) = \sqrt{|1-\alpha/\beta||1-\beta/\alpha|} \int_{M\backslash GL(2,\mathbb{R})} f_\infty(g^{-1}ag)dg,$$

then, by the Weyl integration formula,

$$\operatorname{tr} \xi_s^{k,l}(f_\infty) = \sum \int_{-\infty}^{\infty} \epsilon^{k+l} \delta^l e^{sx} \hat{f}_\infty(a) dx,$$

where $a$ is given by (C.9) and there is a sum over the two free signs in $a$. Thus $\mathrm{tr}(\xi_s^{k,l}(f_\infty))$ is expressed as the Fourier transform of the functions $\hat{f}_\infty(a)$, although the formula (C.10) and the calculations that led to (30) allow us to express this immediately as an integral of the two functions $\psi_\pm$. It is, however, too soon to pass to the variable $r$.

What we want to do is to express the hyperbolic contribution to (C.7), for $|\alpha| \neq |\beta|$, in terms not of $\mathrm{tr}\,\xi_s^{k,l}(f_\infty)$ but in terms of its Fourier transform, then to pass to $\alpha = -\beta$, and at this point and for this particular choice to express the result in terms of $\psi_\pm$. I stop short of this final transformation.

Since we shall be taking the limit $t \to 0$, it suffices to take $t > 0$. Since the signs of $\alpha$ and $\beta$ are supposed different, the function $\eta(m, s)$ is the Fourier transform of the function that is

$$\operatorname{sgn}\alpha^k \operatorname{sgn}\beta^l \sum_{n=1}^{\infty}(-1)^n e^{-n(t+x)} = -\operatorname{sgn}\alpha^k \operatorname{sgn}\beta^l \frac{e^{-(t+x)}}{1+e^{-(t+x)}}$$

for $x > t$ and 0 for $x < t$. Thus, the hyperbolic contribution is

$$-\frac{1}{4}\sum_{k,l}\sum \int_t^\infty \operatorname{sgn}\alpha^k \operatorname{sgn}\beta^l \epsilon^{k+l}\delta^l \frac{e^{-(t+x)}}{1+e^{-(t+x)}}\hat{f}_\infty(a)dx,$$

in which the inner sum is over the free signs in $a$. The effect of the sum over $k$ and $l$ together with the factor $1/4$ is to remove all terms of the inner sum except the one for which $\epsilon = \alpha$ and $\delta\epsilon = \beta$, as we could have predicted. Thus the signs of $a$ are those of $m$ and the hyperbolic contribution is

$$-\sum \int_t^\infty \frac{e^{-(t+x)}}{1+e^{-(t+x)}}\hat{f}_\infty(a)dx. \tag{C.11}$$

The limit as $t \to 0$ can be taken without further ado and gives

$$-\sum \int_0^\infty \frac{e^{-x}}{1+e^{-x}}\hat{f}_\infty(a)dx, \tag{C.12}$$

where $a$ has eigenvalues of opposite sign. Which is positive and which is negative does not matter because of the summation over the two possible opposing signs. When we take $\eta(m, -s)$ into account as well, we obtain in addition

$$-\sum \int_{-\infty}^0 \frac{e^{-x}}{1+e^{-x}}\hat{f}_\infty(a)dx, \tag{C.12$'$}$$

The two are to be added together. Since we take the sum of $I_P(m, f_\infty)$ and $I_P(-m, f_\infty)$, it is probably best to represent it as the sum of (C.12) (together with (C.12′)),

$$-\sum \int_{-\infty}^\infty \frac{e^{-|x|}}{1+e^{-|x|}}\hat{f}_\infty(a)dx, \quad a = a(x). \tag{C.13}$$

If $\det m$ is positive, then, the sign no longer appearing, (C.11) is replaced by

$$\sum \int_t^\infty \frac{e^{-(t+x)}}{1 - e^{-(t+x)}} \hat{f}_\infty(a)dx, \tag{C.14}$$

where, of course, the signs of $a$ are those of $m$. When we need to be explicit, we denote by $a(x, \epsilon)$ the diagonal matrix with eigenvalues $\epsilon e^x$ and $\epsilon e^{-x}$, $\epsilon$ being $\pm 1$. For the passage to the limit, we replace (C.14) by the sum of

$$\sum \int_t^\infty \left( \frac{e^{-(t+x)}}{1 - e^{-(t+x)}} - \frac{1}{t+x} \right) \hat{f}_\infty(a)dx, \tag{C.15}$$

whose limit is obtained by setting $t = 0$, and

(C.16)

$$\frac{1}{2} \sum \int_t^\infty \frac{1}{t+x} \hat{f}_\infty(a)dx = -\sum \hat{f}_\infty(\pm m) \ln(2t) - \int_t^\infty \ln(t+x) \frac{d\hat{f}_\infty}{dx}(a)dx,$$

where we have integrated by parts. (The formulas here are variants of those to be found in [H], especially Lemma 6. They are not necessarily more useful.) Once again, there will be similar terms arising from $\eta(m, -s)$. The first term is an even function of $x$ and will thus contribute

$$-2 \sum \hat{f}_\infty(\pm m) \ln(2t).$$

Since $1 - \beta/\alpha \sim 2t$, we are to add to this

$$\ln(4t^2)\omega(\gamma, f_\infty) = 2\ln(2t)\hat{f}_\infty(m), \qquad \gamma = m,$$

because in spite of our notation, taken as it is from a variety of sources, $\hat{f}(m) = \omega(\gamma, f_\infty)$.

So the limit as $t \to 0$ of the sum over $m$ and $-m$ is the sum of

$$\sum \int_{-\infty}^\infty \left( \frac{e^{-|x|}}{1 - e^{-|x|}} - \frac{1}{|x|} \right) \hat{f}_\infty(a)dx \tag{C.17}$$

and

$$-\sum \int_{-\infty}^\infty \ln|x| \operatorname{sgn} x \frac{d\hat{f}_\infty}{dx}(a)dx. \tag{C.18}$$

In both (C.17) and (C.18) there is a sum over $a$ and $-a$, as in (C.13).

For the elliptic contribution, we recall that from the formula for the discrete-series character with parameter $k \geq 0$, as found, for example, in [K]

$$-\frac{|\alpha - \beta|}{|\alpha\beta|^{\frac{1}{2}}} \Theta_{\tilde{\pi}}(m) = -(\pm 1)^{k-1} e^{-kt}, \quad m = a(t, \pm 1), \ t > 0.$$

This has a limit as $t \to 0$. It is $-(\pm 1)^{k-1}$. Since

$$-\sum_{k=0}^\infty (\pm 1)^{k-1} \Theta_{\pi_k}(f) \tag{C.19}$$

is absolutely convergent, we can provisionally take (C.19) as the contribution of the elliptic term of Hoffmann's formula. The contribution (C.19) does not appear to be expressible as an integral of the pair of functions $\psi_\pm$ against a measure. So for the moment I prefer to leave it as it stands.

**Appendix D: A Fourier transform.** The Fourier transform of the distribution

$$h \to \int_0^\infty \ln x \frac{dh}{dx}(x)dx \tag{D.1}$$

is calculated by treating the distribution as minus the derivative with respect to the purely imaginary Fourier transform variable $s$ of

$$\lim_{\epsilon\to 0} \frac{d}{dt} \int_0^\infty x^t e^{-\epsilon x} h(x)dx$$

for $s = 0$. The Fourier transform of the distribution without either the derivative or the limit is calculated directly as

$$\int_0^\infty x^t e^{-\epsilon x} e^{sx} dx = (\epsilon - s)^{-1-t}\Gamma(t+1),$$

where $s$ is purely imaginary. Differentiating, setting $t = 0$, and multiplying by $-s$, we obtain

$$\frac{s}{\epsilon - s}\Gamma(1)\ln(\epsilon + s) - \Gamma'(1)\frac{s}{\epsilon - s}.$$

Careful attention to the real content of this formal argument reveals that $\ln(\epsilon + s)$ is to be chosen between $-\pi/2$ and $\pi/2$. Letting $\epsilon$ approach 0, this becomes

$$-\ln s + \Gamma'(1), \tag{D.2}$$

where $\ln s$ is $\ln|s| + \frac{\pi}{2}\operatorname{sgn} s$. The symmetric form of (D.1) is

$$\int_{-\infty}^\infty \ln|x| \operatorname{sgn} x \frac{dh}{dx}(x)dx \tag{D.1$'$}$$

and the symmetric form of (D.2) is $-2\ln|s| + 2\Gamma'(1)$. Recall from [N, p. 15] that $\Gamma'(1) = -\lambda_0$ is the negative of Euler's constant.

REFERENCES

[A] J. Arthur, On the Fourier transforms of weighted orbital integrals, *J. reine angew. Math.* **451** (1994), 163–217.

[A1] ———, Some tempered distributions on groups of real rank one, *Ann. of Math.* **100** (1974), 553–584; The local behaviour of weighted orbital integrals, *Duke Math. J* **56** (1988), 223–293.

[C] H. Cohen, A course in computational number theory, Springer Verlag, 1996.

[Cor] A. Borel and W. Casselman, Automorphic forms, representations and $L$-functions, *Proc. Symposia in Pure Math., Amer. Math. Soc.*, 1979.

[DS] P. Deligne and J.-P. Serre, Formes modulaires de poids 1, *Ann. Sc. Éc. Norm. Sup.*(4) **7** (1974), 507–530.
[HW] G. H. Hardy and E. M. Wright, An Introduction to the Theory of Numbers, Oxford at the Clarendon Press.
[H] W. Hoffmann, The Fourier transforms of weighted orbital integrals on semisimple groups of real rank one, *J. reine angew. Math.* **489** (1997), 53–97.
[JL] H. Jacquet and R. P. Langlands, Automorphic forms on GL(2), SLM 114, Springer Verlag, 1970.
[K] A. W. Knapp, Representation theory of semisimple groups, Princeton University Press, 1986.
[Lan] E. Landau, Primzahlen, Chelsea Publishing Co.
[L] R. P. Langlands, Problems in the theory of automorphic forms, Lectures in Modern Analysis and Applications, vol. III, SLM 170, 1970, pp. 18–61.
[L1] ———, Stable conjugacy: definitions and lemmas, *Can. J. Math* **31** (1979), 700–725.
[L2] ———, Base change for GL (2), Ann. Math Studies, vol. 96, Princeton University Press, 1980.
[LP] M. Larsen and R. Pink, Determining representations from invariant dimensions, *Invent. Math.* **102** (1990), 377–398.
[N] N. Nielsen, Die Gammafunktion, Chelsea Publishing Co.
[W] A. Weil, Adèles and algebraic groups, Birkhäuser, 1982.

*Added in proof.* As the reader was cautioned, much of the material of the paper was provisional: the analysis rough and the numerical experiments preliminary. Although further reflection confirms so far the general conclusions, details will have to be modified. There is little point in precise explanations until arguments or experiments have reached a more mature stage. There are only three observations to make. First of all, the $O(\ln^l(p))$ hypothesis for the size of (70) appears more than doubtful. On the other hand, the averages continue to be behave, with better code, as described, but better, for the quadratic term in $\ln p$ seems to have coefficient 0. Finally formula (81) is not correct as it stands.

CHAPTER 23

# AN ANALOGUE OF A CONJECTURE OF MAZUR: A QUESTION IN DIOPHANTINE APPROXIMATION ON TORI

By Dipendra Prasad

*To Professor Shalika, with admiration*

*Abstract.* B. Mazur has considered the question of density in the Euclidean topology of the set of $\mathbb{Q}$-rational points on a variety $X$ defined over $\mathbb{Q}$, in particular for Abelian varieties. In this paper we consider the question of closures of the image of finitely generated subgroups of $T(\mathbb{Q})$ in $\Gamma\backslash T(\mathbb{R})$ where $T$ is a torus defined over $\mathbb{Q}$, $\Gamma$ an arithmetic subgroup such that $\Gamma\backslash T(\mathbb{R})$ is compact. Assuming Schanuel's conjecture, we prove that the closures correspond to *algebraic* sub-tori of $T$.

Let $V$ be a smooth algebraic variety over $\mathbb{Q}$. The set $V(\mathbb{R})$ acquires a topological structure from the Euclidean topology of $\mathbb{R}$. It is known that $V(\mathbb{R})$ has finitely many connected components. If $V(\mathbb{Q})$ is Zariski dense in $V$, it was conjectured by B. Mazur, cf. [M1] and [M2], that the closure of $V(\mathbb{Q})$ in $V(\mathbb{R})$ is a finite union of connected components of $V(\mathbb{R})$. This conjecture was shown to be false in this generality for an elliptic surface by Colliot-Thélène, Skorobogatov, and Swinnerton-Dyer, who have proposed a slightly reformulated conjecture, cf. [CSS]. However, the present evidence seems to suggest that the following special case of Mazur's conjecture is true.

*Conjecture 1 (Mazur's conjecture for Abelian varieties).* Let $A$ be an abelian variety over $\mathbb{Q}$, and $G$ a subgroup of $A(\mathbb{Q})$. Then the closure of $G$ in the Euclidean topology of $A(\mathbb{R})$ contains $B(\mathbb{R})^0$ as a subgroup of finite index for a certain abelian subvariety $B$ defined over $\mathbb{Q}$.

The following theorem of M. Waldschmidt [W1] is the best result known toward Mazur's conjecture.

Theorem 1 (Waldschmidt). *If $A$ is a simple abelian variety over $\mathbb{Q}$ of dimension $d$ if the rank of $A(\mathbb{Q})$ is $\geq d^2 - d + 1$, then the closure of $A(\mathbb{Q})$ in the Euclidean topology contains $A(\mathbb{R})^0$.*

**1. The conjecture about tori.** In this section we propose the following analogue of Mazur's conjecture for tori. We begin by recalling certain standard definitions.

Manuscript received October 17, 2002.

Let $S$ be a torus defined over $\mathbb{Q}$, i.e., let $S$ be a commutative linear algebraic group over $\mathbb{Q}$ which becomes isomorphic to $\mathbb{G}_m^n$ over the algebraic closure $\overline{\mathbb{Q}}$ of $\mathbb{Q}$ for a certain integer $n \geq 0$. The torus $S$ is called isotropic over $\mathbb{Q}$ if there exists a homomorphism of algebraic groups $S \to \mathbb{G}_m$ defined over $\mathbb{Q}$. If $S$ is not isotropic over $\mathbb{Q}$, it is called anisotropic.

Given a linear algebraic group over $\mathbb{Q}$ such as $S$, it makes sense to talk of arithmetic subgroups $\Gamma$ of $S(\mathbb{Q})$. Any two arithmetic subgroups $\Gamma_1$ and $\Gamma_2$ are commensurable, i.e., $\Gamma_1 \cap \Gamma_2$ is of finite index in both $\Gamma_1$ and $\Gamma_2$. It is a consequence of Dirichlet unit theorem (or the general theorem due to Borel and Harish-Chandra) that if $S$ is an anisotropic torus over $\mathbb{Q}$, then $\Gamma \backslash S(\mathbb{R})$ is a compact abelian group for $\Gamma$ any arithmetic subgroup of $S(\mathbb{Q})$. The connected component of identity of $\Gamma \backslash S(\mathbb{R})$ is a torus in the usual sense of the word, i.e., a topological group isomorphic to $(S^1)^n$.

Unlike most other algebraic groups, tori have the property that there is a unique *maximal* arithmetic group. This unique maximal arithmetic subgroup is the subgroup $\Gamma$ of $T(\mathbb{Q})$ defined as follows:

$$\Gamma = \left\{ \gamma \in T(\mathbb{Q}) \;\middle|\; \begin{array}{l} \chi(\gamma) = \text{ a unit in the ring of integers of } \bar{\mathbb{Q}}^* \\ \text{for all characters } \chi : T \to \mathbb{G}_m \end{array} \right\}.$$

We now make the following analogue of Mazur's conjecture for tori.

*Conjecture* 2. Let $S$ be an anisotropic algebraic torus defined over $\mathbb{Q}$. Let $F$ be a finitely generated subgroup of $S(\mathbb{Q})$. For $\Gamma$ any arithmetic subgroup of $S(\mathbb{Q})$, the connected component of identity of the closure of the image of $F$ in $\Gamma \backslash S(\mathbb{R})$ equals connected component of identity of $\Gamma_T \backslash T(\mathbb{R})$ for a certain subtorus $T$ of $S$ defined over $\mathbb{Q}$ with $\Gamma_T = \Gamma \cap T(\mathbb{Q})$. Equivalently, the closure of the image of $F$ is dense in the Euclidean topology in the identity component of $\Gamma \backslash S(\mathbb{R})$ if and only if any subgroup $G$ of $S(\mathbb{Q})$ surjecting on the image of $F$ in $\Gamma \backslash S(\mathbb{R})$ is Zariski dense in $S$.

*Remark* 1. To any finitely generated subgroup $F$ of $S(\mathbb{Q})$, there is a natural subtorus $T$ of $S$ over $\mathbb{Q}$ defined as the (connected component of the identity of the) kernel of the group of characters

$$X_F = \{\chi : S \to \mathbb{G}_m | \chi(a) = \text{ a unit in the ring of integers of } \bar{\mathbb{Q}}^* \text{ for all } a \in F\}.$$

By embedding an anisotropic torus in a product of norm 1 tori (as in Lemma 3 below), it is easy to see that the torus $T$ defined here is the same as that which appears in the conjecture above. We are sloppy here, as well as in other places in the paper, about making such assertions only up to connected components.

*Remark* 2. Let the finitely generated group $F$ be generated by $\{a_1, a_2, \ldots, a_n\}$. Since the product of compact subgroups is compact, it is clear that the closure of the image of $F$ in $\Gamma \backslash S(\mathbb{R})$, $\Gamma$ an arithmetic subgroup of $S(\mathbb{R})$, is the product of the closures of cyclic groups generated by $a_i$. Hence it suffices to prove the conjecture for cyclic groups $F$. By embedding an anisotropic torus in a product of norm 1 tori (as in Lemma 3 below), we are further reduced to proving the conjecture for a cyclic subgroup in a product of norm 1 tori.

*Remark* 3. The conjecture above can also be formulated as the closure of a finitely generated subgroup in $S(\mathbb{Q})$ containing an arithmetic subgroup in $S(\mathbb{R})$. We note that since an arithmetic group in an anisotropic torus over $\mathbb{Q}$, which is split over $\mathbb{R}$, is Zariski dense, so will this finitely generated subgroup. Thus in this sense, our conjecture has a different flavor than Mazur's conjecture. We note that there is also a conjecture, different from the one formulated here, due to Waldschmidt [W3, Conjecture 3.5 of Chapter 3] about the closure in the Euclidean topology of a finitely generated subgroup of $S(\mathbb{Q})$.

*Remark* 4. In this paper we will often be using without explicit mention the trivial remark that the connected component of identity of the closure in either Euclidean or Zariski topology of a finitely generated subgroup $F$ (of a topological group or an Algebraic group) or a subgroup $G$ of $F$ of finite index is the same.

*Example* 1. The simplest case of our conjecture is when $T$ is a torus over $\mathbb{Q}$ such that $T(\mathbb{R})$ itself is compact. In this case we can take $T(\mathbb{Z})$ to be the trivial subgroup of $T(\mathbb{Q})$. We will thus be comparing the closures in Euclidean topology and Zariski topology of a finitely generated subgroup of $T(\mathbb{Q})$. That the two closures are the same follows from the easily proven assertion that any continuous homomorphism from the compact group $T(\mathbb{R})$ to $S^1$ is the restriction to $T(\mathbb{R})$ of an algebraic character (defined over $\mathbb{C}$) from $T(\mathbb{C})$ to $\mathbb{C}^*$.

*Example* 2. Let $k$ be a totally real cubic extension of $\mathbb{Q}$. Let $T = k^1$ denote the group of norm 1 elements of $k$. So $T(\mathbb{Q}) = k^1$, and the group of units of (the ring of integers of) $k$ of norm 1 can be taken to be an arithmetic subgroup of $T(\mathbb{Q})$. We have,

$$\begin{aligned} T(\mathbb{R}) &= [k \otimes \mathbb{R}]^1 \\ &= [\mathbb{R} \oplus \mathbb{R} \oplus \mathbb{R}]^1 \\ &\cong \mathbb{R}^* \times \mathbb{R}^*. \end{aligned}$$

We note that $(x, y) \to (\log x, \log y)$ gives an isomorphism of the product of two copies of positive reals (under multiplication) with real numbers (under addition). We have thus a homomorphism from $T(\mathbb{R})$ to $\mathbb{R} \times \mathbb{R}$ such that the image of the group of units of norm 1 is a discrete cocompact subgroup with quotient $S^1 \times S^1$.

It is easy to see that the torus $T$ of norm 1 elements of a cubic field has no nontrivial subtorus defined over $\mathbb{Q}$. Our conjecture in this case will therefore say that any element of $T(\mathbb{Q})$ which is not of finite order will generate a dense subgroup of $T(\mathbb{Z})\backslash T(\mathbb{R})$. We make this more concrete.

Observe that if $\Lambda \subset \mathbb{R}^2$ is a lattice in $\mathbb{R}^2$ and $v$ is a vector in $\mathbb{R}^2$, then integral multiples of $v$ is dense in $\Lambda\backslash\mathbb{R}^2$ if and only if there does not exist a nonzero integer $q$, and a lattice point $\lambda \in \Lambda$ such that $qv + \lambda$ is real multiple of a vector in $\Lambda$. Suppose if possible, $qv + \lambda = r\lambda_1$ for an integer $q$, and a real number $r$. By looking at the coordinates of the vectors on the two sides of the equality, it is easily seen that it suffices to prove that if $(\log|x_1|, \log|x_2|)$ is a real multiple of $(\log|\epsilon_1|, \log|\epsilon_2|)$, then it is a rational multiple of $(\log|\epsilon_1|, \log|\epsilon_2|)$. Here $x_1$ and $x_2$ are the images of an element (corresponding to $qv + r$) in $k^1$ under two fixed embeddings into the reals, and $\epsilon_1$ and $\epsilon_2$ are the images of a unit element in $k^1$ under the same embeddings into the reals. Even this simple-minded question seems beyond present knowledge. However, we note that this will be a consequence of the following well-known conjecture, cf. [W2].

*Conjecture* 3. (4 Exponential Conjecture due to Schneider, Lang, and Ramachandra) Let $M$ be a $2 \times 2$ matrix consisting of a logarithm of algebraic numbers. Assume that the rows of the matrix are linearly independent over $\mathbb{Q}$, and also that the columns of the matrix are linearly independent over $\mathbb{Q}$, then the determinant of $M$ is nonzero.

The following theorem is a step toward the proof of the 4 exponential conjecture.

THEOREM 2 (LANG AND RAMACHANDRA). *Let $M$ be a $2 \times 3$ matrix consisting of a logarithm of algebraic numbers. Assume that the rows of the matrix are linearly independent over $\mathbb{Q}$, and also that the columns of the matrix are linearly independent over $\mathbb{Q}$. Then the rank of $M$ is 2.*

**2. Some lemmas about Tori.** In this section we collect together some elementary lemmas about tori. We will be considering closures of finitely generated subgroups in the Euclidean and Zariski topologies.

LEMMA 1. (a) *For a discrete subgroup $\Lambda \subset \mathbb{R}^n$ with $\mathbb{R}^n/\Lambda$ compact, the integral multiples of a point $x \in \mathbb{R}^n$ are dense inside $\mathbb{R}^n/\Lambda$ if and only if no nontrivial continuous homomorphism of $\mathbb{R}^n/\Lambda$ to $S^1$ takes $x$ to the identity element of $S^1$.*

(b) *The integral multiples of $x \in \mathbb{R}^n$ are dense inside $\mathbb{R}^n/\Lambda$ if and only if $rx + \lambda$ does not belong to $\Lambda_1 \otimes_{\mathbb{Z}} \mathbb{R}$ for any subgroup $\Lambda_1$ of $\Lambda$ with rank ${}_{\mathbb{Z}}\Lambda_1 < n$, any nonzero integer $r$, and any element $\lambda \in \Lambda$.*

(c) *The integral multiples of $x = (x_1, \ldots, x_n) \in \mathbb{R}^n$ are dense inside $\mathbb{R}^n/\Lambda$ if and only if for $\ell_1 = (\ell_{11}, \ldots, \ell_{1n}), \ldots, \ell_{n-1} = (\ell_{n-1,1}, \ldots, \ell_{n-1,n})$, belonging to*

*$\Lambda$ and generating a rank $(n-1)$ subgroup of $\Lambda$, the matrix*

$$\begin{pmatrix} rx_1+\lambda_1 & rx_2+\lambda_2 & \cdots & rx_n+\lambda_n \\ \ell_{11} & \ell_{12} & \cdot & \ell_{1n} \\ \cdot & \cdot & \cdot\cdot & \cdot \\ \cdot & \cdot & \cdot & \cdot \\ \cdot & \cdot & \cdot\cdot & \cdot \\ \ell_{n-1,1} & \ell_{n-1,2} & \cdots & \ell_{n-1,n} \end{pmatrix}$$

*is nonsingular, i.e., the determinant is nonzero, for any nonzero integer $r$, and any $\lambda=(\lambda_1,\dots,\lambda_n)\in\Lambda$.*

Now we have a lemma about density of the abelian group generated by a point on a torus in the Zariski topology. In this lemma for a number field $K$, we will be looking at the torus $T=R_{K/\mathbb{Q}}\mathbb{G}_m$ defined over $\mathbb{Q}$ to be the algebraic group whose group of rational points over any $\mathbb{Q}$-algebra $A$ is $T(A)=(K\otimes_{\mathbb{Q}}A)^*$; in particular $T(\mathbb{Q})=K^*$.

LEMMA 2. (a) *An element $x\in T(\mathbb{Q})=K^*$ generates $K$ (i.e., $K$ is the smallest field extension of $\mathbb{Q}$ containing $x$) if and only if all its conjugates (i.e., the image of $x$ under all the distinct embeddings of $K$ in $\mathbb{C}$) $\{x_1,\dots,x_n\}$ are distinct.*

(b) *An element $x\in T(\mathbb{Q})=K^*$ lies in no proper algebraic subgroup defined over $\mathbb{Q}$ if and only if the abelian subgroup generated by $\{x_1,\dots,x_n\}$ is free abelian of rank $n$.*

*Proof.* Part $(a)$ is clear. For part $(b)$ note that any algebraic character of $T(\mathbb{Q})=K^*$ (defined over the algebraic closure) is defined by $z\to z_1^{d_1}\cdot z_2^{d_2}\cdots z_n^{d_n}$, where $z_i$ denotes the image of $z$ under the various embeddings of $K$ into $\mathbb{C}$. Since for an element belonging to a proper algebraic subgroup of $T$, there is a character of $T$ trivial on that element, therefore if $x$ belongs to a proper algebraic subgroup, the subgroup generated by $\{x_1,\dots,x_n\}$ will not be free.

Conversely, if the subgroup generated by $\{x_1,\dots,x_n\}$ is not free abelian, $x$ belongs to the kernel of a nontrivial character $\chi$ of $T$ defined over $\overline{\mathbb{Q}}$. Hence $x\in T(\mathbb{Q})$ lies in $S(\overline{\mathbb{Q}})$ for an algebraic subgroup $S$ of $T$. By Galois conjugation, $x\in S^\sigma(\overline{\mathbb{Q}})$ for all Galois conjugates of $S$. Hence $x\in\cap(S^\sigma)(\overline{\mathbb{Q}})$. However, $A=\cap(S^\sigma)$ is an algebraic group defined over $\mathbb{Q}$. Hence $x$ belongs to $A(\mathbb{Q})$ for $A$ a proper algebraic subgroup of $T$. □

LEMMA 3. *For any anisotropic torus $T$ over $\mathbb{Q}$, there are field extensions $K_1,\dots,K_d$ of $\mathbb{Q}$, such that if $S$ denotes the product of the norm 1 tori associated to $K_i$, then there is an embedding of $T$ into $S$.*

*Proof.* As is well known, there is an equivalence of categories between tori over $\mathbb{Q}$ and finitely generated free $\mathbb{Z}$-module with an action of the Galois group $\mathrm{Gal}(\bar{\mathbb{Q}}/\mathbb{Q})$ of the algebraic closure $\bar{\mathbb{Q}}$ of $\mathbb{Q}$. The equivalence is given by associating to any torus $T$, its character group $X^*(T)$. Choose a $\mathbb{Z}$-basis, say $\{e_1, \ldots, e_d\}$ of the character group of $T$. Suppose that $H_i$ is the subgroup of $\mathrm{Gal}(\bar{\mathbb{Q}}/\mathbb{Q})$ which stabilizes the vector $e_i$. The mapping $g \to g \cdot e_i$ from $\mathrm{Gal}(\bar{\mathbb{Q}}/\mathbb{Q})$ to $X^*(T)$ gives a mapping from $\mathbb{Z}[G/H_i]$ to $X^*(T)$. Summing over $i$, we get a surjective map from $\sum_i \mathbb{Z}[G/H_i]$ to $X^*(T)$. This gives an embedding from $T$ to $\prod_i R_{K_i/\mathbb{Q}}\mathbb{G}_m$ where $K_i$ is the fixed field of $H_i$. Since $T$ is anisotropic, the image of $T$ lands inside the product of norm 1 tori. □

**3. Schanuel's conjecture implies conjecture 2.** In this section we prove that our conjecture 2 about closures in Euclidean topology of finitely generated subgroups in general tori is a consequence of Schanuel's conjecture.

We should, however, add that although our approach in this paper is via Schanuel's conjecture, there is a possibility that there may be a simpler proof for conjecture 2, just by using the more primitive methods of Geometry of Numbers, as we are dealing with a rather specific number theoretic context.

We begin with the statement of Schanuel's conjecture, which is one of the most outstanding problems in transcendental number theory. In the statement of this conjecture, as well as everywhere else in the paper, one means by $\log A$, for a complex number $A$ to be *any* complex number $B$ such that $\exp(B) = A$.

*Conjecture 4 (Schanuel's Conjecture).* If $\alpha_1, \ldots, \alpha_n$ are algebraic numbers such that $\log\alpha_1, \ldots, \log\alpha_n$ are linearly independent over $\mathbb{Q}$, then $\log\alpha_1, \ldots, \log\alpha_n$ are algebraically independent over $\mathbb{Q}$.

We will need the following lemma about number fields.

LEMMA 4. *Let $L$ be a number field that is Galois over $\mathbb{Q}$. Enumerate the elements of the Galois group $G$ of $L$ over $\mathbb{Q}$ as $\sigma_1 = 1, \sigma_2, \ldots, \sigma_d$. For an element $z$ of $L^*$, denote the various (not necessarily distinct) Galois conjugates of $z$ by $z_1 = z, z_2 = \sigma_2(z), \ldots, z_d = \sigma_d(z)$. Let $x$ be an element $L^*$. Then there is a nonzero integer $m$ and a unit $\epsilon$ in (the ring of integers of) $L$, such that whenever $x_1^{n_1}x_2^{n_2}\ldots x_d^{n_d}$ is a unit in (the ring of integers of) $L$ for a $d$-tuple $(n_1, n_2, \ldots, n_d)$ inside $\mathbb{Z}^d$, $(x^m\epsilon)_1^{n_1}(x^m\epsilon)_2^{n_2}\ldots(x^m\epsilon)_d^{n_d} = 1$. The integer $m$ can be taken to be the order of the class group of $L$ times the degree of $L$ over $\mathbb{Q}$, hence can be chosen to be independent of $x$.*

*Proof.* Write the (fractional) ideal generated by $x$ as a product of prime ideals:

$$(x) = \wp_1^{m_1}\cdots\wp_r^{m_r}.$$

(We will assume that if a certain prime $\wp_i$ occurs in the above decomposition, so does any Galois conjugate of it, with exponent perhaps 0.) Let $k$ be an integer such that each of the ideals $\wp_i^k$ is a principal ideal generated by, say $\varpi_i$. If $H_i$ denotes the subgroup of the Galois group that fixes the prime ideal $\wp_i$, then the elements of $H_i$ will take $\varpi_i$ into itself up to a unit: $h(\varpi_i) = h_i \cdot \varpi_i$. Clearly $h \to h_i$ is a 1-cocycle on $H_i$ with values in the group of units of $L^*$. Since $H_i$ is a finite group, a finite power of the cocycle becomes a coboundary, i.e., there is a positive integer $r$, and a unit $\nu_i$ such that

$$h(\varpi_i^r) = h_i^r \cdot \varpi_i^r = h(\nu_i)\nu_i^{-1}\varpi_i^r.$$

It follows that $\nu_i^{-1}\varpi_i^r$ is invariant under $H_i$. Thus we can choose generators $\pi_i$ of the principal ideals $\wp_i^{rk}$ in such a way that if an element of the Galois group takes one such ideal into another such, then the same holds for the generators (and not just only up to units). From the equality of ideals, $(x^{rk}) = (\pi_1^{m_1}\pi_2^{m_2}\cdots\pi_r^{m_r})$, there is a unit $\epsilon$ with, $x^{rk}\epsilon = \pi_1^{m_1}\pi_2^{m_2}\cdots\pi_r^{m_r}$. Now observe that if a product of certain elements of $L^*$ with any two *distinct* elements coprime (such as $\pi_i$'s) is a unit, then the product is in fact 1 (and is in some sense the "empty product"). From this it follows that $(x^{kr}\epsilon)_1^{n_1}(x^{kr}\epsilon)_2^{n_2}\cdots(x^{kr}\epsilon)_d^{n_d} = 1$. Finally, the proof given here works with the choice of $m = rk$ to be the product of the order of the class group of $L$ and the degree of $L$ over $\mathbb{Q}$. □

COROLLARY 1 (OF THE PROOF). *With the notation as in the lemma, given elements $\{x^{(1)}, x^{(2)}, \ldots, x^{(n)}\}$ in $L^*$, there is an integer $m$ and units $\epsilon_i$ in the ring of integers of $L$ such that for $y^{(i)} = (x^{(i)})^m\epsilon_i$, the subgroup of $L^*$ generated by $\sigma_j(y^{(i)})$ does not contain any unit of the ring of integers of $L$ other than 1.*

THEOREM 3. *Schanuel's conjecture implies conjecture 2.*

*Proof.* As already noted in Remark 2, it suffices to prove Conjecture 2 for cyclic subgroups. Furthermore, because of Lemma 3, we can assume that the anisotropic torus $S$ is the product of the norm 1 tori: $S = \prod_{i=1}^m (R_{K_i/\mathbb{Q}}\mathbb{G}_m)^1$. We will further assume (by enlarging the anisotropic torus which does not affect the conclusion regarding closures) that the fields $K_i$ are Galois over $\mathbb{Q}$, and by taking the compositum, we assume that the fields $K_i$ are the same, say $L$, a Galois extension of degree $d+1$ over $\mathbb{Q}$, which we will assume to be totally real. The case when $L$ has complex places is very similar, although notationally more cumbersome.

For an element $x^{(i)} \in L$, we let $x_j^{(i)}$, $j \in \{1, 2, \ldots, d+1\}$, denote the various Galois conjugates of $x^{(i)}$.

Let $x = (x^{(1)}, x^{(2)}, \ldots, x^{(m)}) \in S = \prod_{i=1}^m (R_{L/\mathbb{Q}}\mathbb{G}_m)^1$. Replacing $x^{(i)}$ by $y^{(i)} = (x^{(i)})^m\epsilon_i$ as in Corollary 1, we can assume that $y = (y^{(1)}, y^{(2)}, \ldots, y^{(m)})$ is such that the group generated by the various Galois conjugates $y_j^{(i)}$ intersects the units in the ring of integers of $L$ in identity alone.

We note that a general algebraic character of $L^*$ is given by $z \to \prod_j (z_j)^{n_j}$. Denote by $A$ the group of characters $\chi$ of $S = \prod_{i=1}^m (R_{L/\mathbb{Q}}\mathbb{G}_m)^1$ such that $\chi(y) = 1$.

Let the rank of the abelian group $A$ be $dm - k$. Therefore the subgroup of $L^*$ generated by $\{y_j^{(i)}\}$ is a free abelian group of rank $k$.

We will use the homomorphism with finite kernel:

$$\begin{array}{ccc} [L \otimes \mathbb{R}]^1 & \to & \mathbb{R}^d \\ (x_1, \dots, x_{d+1}) & \to & (\log|x_1|, \dots, \log|x_d|), \end{array}$$

under which (by Dirichlet unit theorem) the group of units of the ring of integers of $L$ of norm 1, $\mathcal{O}_L^{*1}$ goes to a lattice $\Lambda$ in $\mathbb{R}^d$ with $\mathbb{R}^d/\Lambda$ compact.

Taking the direct sum of this homomorphism $m$ number of times, we get a homomorphism from $S(\mathbb{R})$ to $\mathbb{R}^{dm}/\Lambda^m$, whose kernel is an arithmetic group in $S(\mathbb{R})$, to be denoted by $S(\mathbb{Z})$. We will denote the image of $y = (y^{(1)}, y^{(2)}, \dots, y^{(m)})$ in $\mathbb{R}^{dm}$ as $(\log(|y_1^{(1)}|), \dots, \log(|y_d^{(1)}|), \dots, \log(|y_1^{(m)}|), \dots, \log(|y_d^{(m)}|))$. By Lemma 1$(c)$, to prove this theorem it suffices to prove that the rank of the matrix

$$A = \begin{pmatrix} a_{11} & \cdots & a_{1d} & \cdots & a_{m1} & \cdots & a_{md} \\ \ell(1)_{11} & \cdots & \ell(1)_{1d} & \cdots & \ell(1)_{m1} & \cdots & \ell(1)_{md} \\ \cdot & \cdots & \cdot & \cdots & \cdot & \cdots & \cdot \\ \cdot & \cdots & \cdot & \cdots & \cdot & \cdots & \cdot \\ \cdot & \cdots & \cdot & \cdots & \cdot & \cdots & \cdot \\ \ell(k-1)_{11} & \cdots & \ell(k-1)_{1d} & \cdots & \ell(k-1)_{m1} & \cdots & \ell(k-1)_{md} \end{pmatrix}$$

is $k$ where $(a_{11}, \dots, a_{1d}, \dots, a_{m1}, \dots, a_{md}) = (r\log(|y_1^{(1)}|) + \ell_{11}, \dots, r\log(|y_d^{(1)}|) + \ell_{1d}, \dots, r\log(|y_1^{(m)}|) + \ell_{m1}, \dots, r\log(|y_d^{(m)}|)) + \ell_{md}$, $r$ is a nonzero integer, $(\ell_{11}, \dots, \ell_{1,d}, \dots, \ell_{m1}, \dots, \ell_{md})$ is a vector in $\Lambda^m$, and where the 2nd to $k$th rows of this matrix represent $(k-1)$ $\mathbb{Z}$-linearly independent vectors in $\Lambda^m$.

Since the rank of the matrix

$$B = \begin{pmatrix} \ell(1)_{11} & \cdots & \ell(1)_{1d} & \cdots & \ell(1)_{m1} & \cdots & \ell(1)_{md} \\ \cdot & \cdots & \cdot & \cdots & \cdot & & \cdots \\ \cdot & \cdots & \cdot & \cdots & \cdot & & \cdots \\ \cdot & \cdots & \cdot & \cdots & \cdot & & \cdots \\ \ell(k-1)_{11} & \cdots & \ell(k-1)_{1d} & \cdots & \ell(k-1)_{m1} & \cdots & \ell(k-1)_{md} \end{pmatrix}$$

is $(k-1)$, there is a $(k-1) \times (k-1)$ submatrix with nonzero determinant. After re-indexing the coordinates in $\mathbb{R}^{dm}$, we assume that the first $(k-1) \times (k-1)$ submatrix of $B$ has rank $(k-1)$, i.e., has nonzero determinant.

Since the rank of the group generated by $\{y_j^{(i)}\}$ is $k$, there is at least one index, say $y_{j_0}^{(i_0)}$, such that no power of it belongs to the group generated by the $y's$ corresponding to the first $(k-1)$ entries in the first row of $A$ (after re-indexing introduced above). Denote these $y's$ as $y_1, y_2, \dots, y_{k-1}$, and the corresponding $\ell's$ as $\ell_1, \ell_2, \dots, \ell_{k-1}$. Also, denote $y_{j_0}^{(i_0)}$ as $y_k$.

Let $C$ be the $k \times k$ submatrix of $A$ comprised of the first $(k-1)$ columns of $A$, and the $k$th column corresponding to $y_{j_0}^{(i_0)}$. We want to prove that $\det(C) \neq 0$.

Clearly $\det(C) = \sum_{i=1}^{k} [r\log(|y_i|) + \ell_i] \det(L_i)$, where $L_i$ is a $(k-1) \times (k-1)$ matrix consisting of log of units in $L$, and $\ell_i$ are also log of units in $L$.

It follows from Schanuel's conjecture and our hypothesis that no (nonzero) power of $y_k$ belongs to the group generated by $y_i, i = 1, \ldots, k-1$, that $\log(|y_k|)$ is algebraically independent over the subfield of $\mathbb{C}$ generated by log of algebraic units and the $\log(|y_i|), i = 1, \ldots, k-1$.

By our assumption, the first $(k-1) \times (k-1)$ submatrix of $B$ has nonzero determinant, which is $\det(L_k)$, hence $\det(C) = \sum[r \log(|y_i|) + \ell_i] \det(L_i)$ is nonzero (by algebraic independence of the $k$th term from the rest). □

**4. Counter-example to a more general question.** It is natural to ask if an analogue of Conjecture 2 can be made more generally. The general question is about the algebraicity of the connected component of identity of the closure in Euclidean topology of a finitely generated subgroup of $\mathbb{C}^{*n}$ with algebraic coordinates, where $\mathbb{C}^{*n}$ is considered as the $2n$-dimensional torus defined over $\mathbb{R}$ as the Weil restriction of scalars $R_{\mathbb{C}/\mathbb{R}}\mathbb{G}_m^n$. For example, can one drop the condition on the torus $S$ in Conjecture 2 being anisotropic over $\mathbb{Q}$, and instead of taking $S(\mathbb{Z})$, which is a cocompact discrete subgroup of $S(\mathbb{R})$ if $S$ is anisotropic, take any cocompact discrete subgroup $\Gamma$ of $S(\mathbb{R})$ contained in $S(\mathbb{Q})$? A simple counter-example shows that this is not possible, shattering any hope for a simple answer to the general question above.

To construct the counter-example, take $S = \mathbb{G}_m^3$, the 3-dimensional split torus over $\mathbb{Q}$. The principle behind the counter-example is the well-known observation that although the determinant of a skew-symmetric $n \times n$ matrix consisting of logarithm of algebraic numbers is 0 if $n$ is odd, the rows and columns could be linearly independent over $\mathbb{Q}$, such as for the matrix:

$$\begin{pmatrix} 0 & \log 2 & \log 3 \\ -\log 2 & 0 & \log 5 \\ -\log 3 & -\log 5 & 0 \end{pmatrix}.$$

Since the determinant of the following matrix is nonzero,

$$\begin{pmatrix} 0 & \log 2 & \log 3 \\ -\log 2 & 0 & \log 5 \\ \log 7 & 0 & \log 2 \end{pmatrix},$$

it follows that the subgroup of $\mathbb{R}^{*3}$ generated by the elements, $x = (1, 2, 3)$, $y = (1/2, 1, 5)$, $z = (7, 1, 2)$, is a discrete cocompact subgroup $\Gamma$ of $\mathbb{R}^{*3}$. However, the closure of the image of the cyclic group generated by the image of $w = (3, 5, 1)$ in $\Gamma\backslash\mathbb{R}^{*3}$ is a 2-dimensional topological torus (this follows from Lemma, 1(b)), which does not arise from an algebraic subtorus of $\mathbb{R}^{*3}$, as follows from Lemma 2(b).

*Remark.* The connected component of identity of the closure of a finitely generated subgroup of algebraic numbers of $\mathbb{C}^*$ is $\{1, \mathbb{R}^+, \mathbb{S}^1, \mathbb{C}^*\}$. We refer to Theorem 1.10, p. 56 of [W3] for a proof of this assuming Schanuel's conjecture. (The subtlety lies in proving that algebraic points cannot be dense on a closed connected subgroup of $\mathbb{C}^*$ besides those mentioned above.)

**5. Non-abelian analogue.** It seems very natural to extend the scope of Conjecture 2 by replacing the anisotropic torus $T$ by a general algebraic group $G$ over $\mathbb{Q}$.

We recall that by a theorem due to Borel and Harish-Chandra, for a reductive algebraic group $G$ over $\mathbb{Q}$, which is anisotropic over $\mathbb{Q}$, $G(\mathbb{Z})\backslash G(\mathbb{R})$ is compact. We would like to suggest the analogue of Conjecture 2 to assert that the closure of the image in $G(\mathbb{Z})\backslash G(\mathbb{R})$ of a finitely generated subgroup $F$ of $G(\bar{\mathbb{Q}}_{\mathbb{R}})$ (where $\bar{\mathbb{Q}}_{\mathbb{R}}$ is the subfield of algebraic numbers in $\mathbb{R}$) is of the form $\Gamma_H\backslash H(\mathbb{R})$ for an algebraic subgroup $H$ of $G$ defined over $\mathbb{Q}$ with $\Gamma_H = H(\mathbb{R}) \cap G(\mathbb{Z})$. (Note that if the image of a subgroup $H(\mathbb{R})$ in $G(\mathbb{Z})\backslash G(\mathbb{R})$ is closed, hence compact, then $\Gamma_H = H(\mathbb{R}) \cap G(\mathbb{Z})$ is a cocompact lattice in $H(\mathbb{R})$, and hence if $H$ is algebraic, it is defined over $\mathbb{Q}$ by the Borel density theorem.) Notice that we have not proved even for a torus (even after assuming Schanuel's conjecture), a theorem in this generality as we have always restricted ourselves to finitely generated subgroups $F$ of the torus which are contained in the group of $\mathbb{Q}$-rational points. This seems to have been necessary for the proof of Conjecture 2 given here.

We remark that our suggested analogue contains a consequence of M. Ratner's theorem (the proof of the so-called Raghunathan conjecture) as observed by Dani and Raghunathan, cf. Cor. 4.9 in [V] in a very special case. It states that if a semi-simple group G over $\mathbb{R}$ (with $G(\mathbb{R})$ noncompact) has two distinct $\mathbb{Q}$ structures, with corresponding lattices $\Gamma_1$ and $\Gamma_2$, then $\Gamma_1 \cdot \Gamma_2$ is dense in $G(\mathbb{R})$ (in the Euclidean topology).

*Acknowledgment.* This note was conceived several years back. It is being published now in the hope of stimulating further research. The author thanks B. Mazur, Gopal Prasad, T. N. Venkataramana and M. Waldschmidt for encouragement and helpful remarks.

HARISH-CHANDRA RESEARCH INSTITUTE, CHHATNAG ROAD, JHUSI, ALLAHABAD 211019, INDIA
*E-mail:* dprasad@mri.ernet.in

---

## REFERENCES

[CSS] J.-L.Colliot-Thélène, A. N. Skorobogatov, and P. Swinnerton-Dyer, Double fibres and double covers: Paucity of rational points, *Acta Arith.* **79** (1997), 113–135.

[M1] B. Mazur, Topology of rational points, *Experimental Mathematics* **1** (1992), 35–45.

[M2] B. Mazur, Open problems regarding rational points on curves and varieties; Galois representations in arithmetic algebraic geometry (Durham, 1996), 239–265, *London Math. Soc. Lecture Note Ser., 254, Cambridge Univ. Press*, Cambridge, (1998).

[V] V. Vatsal, Uniform distribution of Heegner points, *Invent. Math.* **148** (2002), 1–46.

[W1] M. Waldschmidt, Densité des points rationnels sur un groupe algébrique, *Experimental Mathematics*, **3** (1994), 329–352.

[W2] M. Waldschmidt, Diophantine approximation on linear algebraic groups, *Grundlehren der mathematischen Wissenschaften,* vol. 326, Springer Verlag (2000).

[W3] M. Waldschmidt, Topologie des Points Rationnels, lecture notes from a course at P. et M. Curie 94/95 available at www.mathp6.jussieu.fr/ miw/TPR.html (1998).

CHAPTER 24

# EXISTENCE OF RAMANUJAN PRIMES FOR GL(3)

By DINAKAR RAMAKRISHNAN

---

*To Joe Shalika with admiration*

**Introduction.** Let $\pi$ be a cusp form on GL $(n)/\mathbb{Q}$, i.e., a cuspidal automophic representation of GL $(n, \mathbb{A})$, where $\mathbb{A}$ denotes the adele ring of $\mathbb{Q}$. We will say that a prime $p$ is a Ramanujan prime for $\pi$ iff the corresponding $\pi_p$ is tempered. The local component $\pi_p$ will necessarily be unramified for almost all $p$, determined by an unordered $n$-tuple $\{\alpha_{1,p}, \alpha_{2,p}, \ldots, \alpha_{n,p}\}$ of nonzero complex numbers, often represented by the corresponding diagonal matrix $A_p(\pi)$ in GL $(n, \mathbb{C})$, unique up to permutation of the diagonal entries. The $L$-factor of $\pi$ at $p$ is given by

$$L(s, \pi_p) = \det\left(I - A_p(\pi)p^{-s}\right)^{-1} = \prod_{j=1}^{n}\left(1 - \alpha_{j,p}p^{-s}\right)^{-1}.$$

As $\pi$ is unitary, $\pi_p$ is tempered (in the unramified case) iff each $\alpha_{j,p}$ is of absolute value 1. It was shown in [Ra] that for n $= 2$, the set $\mathcal{R}(\pi)$ of Ramanujan primes for $\pi$ is of lower density at least $9/10$. When one applies in addition the deep recent results of H. Kim and F. Shahidi [KSh] on the symmetric cube and the symmetric fourth power liftings for GL (2), the lower bound improves from $9/10$ to $34/35$ (loc. cit.), which is $0.971428\ldots$.

For $n > 2$ there is a dearth of results for general $\pi$, though for cusp forms of regular algebraic infinity type, assumed for $n > 3$ to have a discrete series component at a finite place, one knows by [Pic] for $n = 3$, which relies on the works of J. Rogawski, et al, and by the work of Clozel [C$\ell$] for $n > 3$ (see also the nontrivial refinement due to Harris and Taylor [HaT]) that all the unramified primes are Ramanujan primes for $\pi$. (Of course for $n = 2$, a regular algebraic cusp form is necessarily holomorphic of weight $k \geq 2$, and it is known, by Eichler-Shimura for $k = 2$ and by Deligne for $k > 2$, that every prime is a Ramanujan prime in this case; ditto for $k = 1$ by the work of Deligne and Serre.) One is interested in knowing whether there exists even one Ramanujan prime for general $\pi$ on GL $(n)$. Thanks to the work of Kim and Shahidi, one sees the importance of knowing a positive answer to such a question.

The main result of this note is the following:

THEOREM A. *Let $\pi$ be a cusp form on GL $(3)/\mathbb{Q}$. Then there exist infinitely many Ramanujan primes for $\pi$.*

Let us now explain the main issues behind its proof. One can show (see Section 1) that at any prime $p$ where $\pi$ is unramified, if the coefficient $a_p(\pi)$ is bounded in absolute value by 1, then $\pi_p$ is tempered. A general result proved in [Ra] for GL $(n)$ implies that for any real number $b > 1$, the set of $p$ where $|a_p(\pi)| \leq b$ is infinite, even of lower Dirichlet density $\geq 1 - \frac{1}{b^2}$. But this gives us nothing for $b = 1$. Our aim here is to show that for infinitely many primes $p$, $a_p(\pi)$ is indeed bounded in absolute value by 1. The key idea is to exploit the adjoint $L$-function (whose definition makes sense for $\pi$ on GL $(n)$ for any $n$):

$$L(s, \pi; Ad) = \frac{L(s, \pi \times \overline{\pi})}{\zeta(s)}, \tag{0.1}$$

where $L(s, \pi \times \overline{\pi})$ is the Rankin-Selberg $L$-function of the pair $(\pi, \overline{\pi})$. (As usual, $\overline{\pi}$ signifies the complex conjugate of $\pi$, which, by the unitarity of $\pi$, is the same as the contragredient of $\pi$.) One knows (see [HRa], Lemma $a$ of Section 2) that $L(s, \pi \times \overline{\pi})$ is of positive type, i.e., the Dirichlet series defined by its logarithm has non-negative coefficients. The proof of Theorem A relies on the following:

PROPOSITION B. *Let $\pi$ be a cusp form on* GL $(n)/\mathbb{Q}$. *Then for any finite set $S$ of primes containing infinity, the incomplete adjoint $L$-function $L^S(s, \pi, Ad)$ is not of positive type.*

The proof given here of this proposition, and hence of Theorem A, will work over any number field $F$ having no real zeros in the interval (0, 1). In the case of real zeros one has to proceed differently.

*Acknowledgments.* When I was at Hopkins as an Assistant Professor during 1983–85, I learnt a lot from Joe Shalika about the $L$-functions of GL $(n)$. It is a pleasure to dedicate this article to him. Thanks are due to Jeff Lagarias and Freydoon Shahidi for making comments on an earlier version of this article, which led to an improvement of the exposition, and to the NSF for financial support through the grant DMS–0100372.

**1. Why Proposition B implies Theorem A.** Let $\pi \simeq \pi_\infty \otimes (\otimes'_p \pi_p)$ be a cuspidal automorphic representation of $\mathrm{GL}(3, \mathbb{A}) = \mathrm{GL}(3, \mathbb{R}) \times \mathrm{GL}(3, \mathbb{A}_f)$. Let $S_0$ be the set of primes $p$ where $\pi_p$ is unramified and tempered, and let $S_1$ be the finite set of primes where $\pi_p$ is ramified. Put

$$S = S_0 \cup S_1 \cup \{\infty\}. \tag{1.1}$$

For any $L$-function with an Euler product $\prod_v L_v(s)$ over $\mathbb{Q}$, put

$$L^S(s) = \prod_{p \notin S} L_p(s), \tag{1.2}$$

which we call the incomplete Euler product relative to, or outside, $S$.

Pick any $p$ outside $S$ (if there is one) and consider the Langlands class

$$A_p = A_p(\pi) = \{\alpha_{1,p}, \alpha_{2,p}, \alpha_{3,p}\}. \tag{1.3}$$

As $\pi_p$ is by assumption nontempered, there is a nonzero real number $t$ and a complex number $u$ of absolute value 1 such that, after possibly renumbering the $\alpha_{j,p}$,

$$\alpha_{1,p} = up^t.$$

On the other hand, by the unitarity of $\pi_p$, we must have

$$\{\overline{\alpha}_{1,p}, \overline{\alpha}_{2,p}, \overline{\alpha}_{3,p}\} = \left\{\alpha_{1,p}^{-1}, \alpha_{2,p}^{-1}, \alpha_{3,p}^{-1}\right\}.$$

This then implies that

$$A_p = \{up^t, up^{-t}, w\}, \tag{1.4}$$

for some complex number $w$ of absolute value 1. We may, and we will, assume that $t$ is positive. Put

$$u^{-1}w = e^{i\theta}, \tag{1.6}$$

for some $\theta \in [0, 2\pi) \subset \mathbb{R}$.

So we have

(1.6)
$$|a_p|^2 = (p^t + p^{-t} + \cos\theta)^2 + \sin^2\theta = 3 + p^{2t} + p^{-2t} + 2\cos\theta(p^t + p^{-t}).$$

Now let us look at the adjoint $L$-function. By definition,

$$L^S(s, \pi; Ad) = \frac{\prod_{p\notin S} L(s, \pi_p\overline{\pi}_p)}{\prod_{p\notin S}(1 - p^{-s})^{-1}}. \tag{1.7}$$

So for any $p$ outside $S$, the Langlands class of the Adjoint $L$-function is

$$A_p(\pi; Ad) = A_p \otimes \overline{A}_p - \{1\}.$$

Applying (1.4) and (1.5), we get

$$A_p(\pi; Ad) = \{p^{2t}, p^{-2t}, 1, 1, u\overline{w}p^t, u\overline{w}p^{-t}, \overline{u}wp^t, \overline{u}wp^{-t}\} \tag{1.8}$$

and

$$a_p(\pi; Ad) = \mathrm{tr}\,(A_p(\pi; Ad)) = 2 + p^{2t} + p^{-2t} + 2\cos\theta(p^t + p^{-t}).$$

Consequently,

$$\log L^S(s, \pi; Ad) = \sum_{p\notin S}\sum_{m\geq 1} \frac{a_{p^m}(\pi; Ad)}{p^{ms}}, \tag{1.9}$$

where (by (1.8) and (1.6))

(1.10). $$a_{p^m}(\pi; Ad) = 2 + p^{2mt} + p^{-2mt} + 2\cos m\theta(p^{mt} + p^{-mt}).$$

Since

$$p^{mt} + p^{-mt} \geq 2,$$

and since

$$a_{p^m}(\pi; Ad) = (p^{mt} + p^{-mt})((p^{mt} + p^{-mt}) + 2\cos m\theta)),$$

we get the following:

LEMMA 1.11. *Let $\pi$ be a cusp form on* GL(3)/$\mathbb{Q}$ *and $S$ the set of primes containing $\infty$, the primes where $\pi$ is ramified and the Ramanujan primes for $\pi$. Then $L^S(s, \pi; Ad)$ is of positive type.*

But if $S_0$, and hence $S$, is finite, this lemma contradicts the conclusion of Proposition B. Hence the set of Ramanujan primes for $\pi$ must be infinite, once we accept Proposition B.

**2. Proof of Proposition B.** In this section $\pi$ will be a unitary, cuspidal automorphic representation of GL$(n, \mathbb{A})$. At each place $v$, the local factor of $L(s, \pi; Ad)$ is given by

$$L(s, \pi_v; Ad) = \frac{L(s, \pi_v \times \overline{\pi}_v)}{\zeta_v(s)}, \tag{2.1}$$

where $\zeta_v(s)$ is $(1 - p^{-s})^{-1}$ if $v$ is a finite place defined by a prime $p$, and it equals $\pi^{-s/2}\Gamma(s/2)$ if $v$ is the archimedean place. By convention, $\zeta(s)$ is the product of $\zeta_v(s)$ over all the *finite* $v$, while all the other automorphic $L$-functions occurring in this paper will also involve the archimedean factors.

The $L$-group of GL$(n)$ is GL$(n, \mathbb{C})$, and the Euler factor $L(s, \pi; Ad)$ is associated to the representation

$$Ad : \mathrm{GL}(n, \mathbb{C}) \to \mathrm{GL}(n^2 - 1, \mathbb{C}), \tag{2.2}$$

given by composing the natural projection of GL$(n, \mathbb{C})$ onto PGL$(n, \mathbb{C})$ with the $(n^2 - 1)$-dimensional Adjoint representation of PGL$(n, \mathbb{C})$, whence the notation $Ad$. In any case, we have for every $v$:

$$L(s, \pi_v; Ad) \neq 0, \ \forall s \in \mathbb{C}. \tag{2.3}$$

One way to see this will be to use the local Langlands correspondence, established recently by Harris-Taylor [HaT] and Henniart [He], associating to each $\pi_v$ an $n$-dimensional representation $\sigma_v$ of $W_{F_v} \times \mathrm{SL}(2, \mathbb{C})$ (resp. $W_{F_v}$) for $v$ finite (resp. infinite). (Here $W_{F_v}$ denotes as usual the Weil group of $F_v$.) Since this correspondence preserves the local factors of pairs and matches the central character of $\pi_v$ with the determinant of $\sigma_v$, one gets in particular

$$L(s, \pi_v; Ad) = L(s, Ad(\sigma_v)),$$

where $Ad(\sigma_v)$ denotes $\sigma_v \otimes \sigma_v^\vee \ominus 1$, which is a genuine representation because the trivial representation occurs in $\sigma_v \otimes \sigma_v^\vee \simeq \mathrm{End}(\sigma_v)$. It is well known that for any representation $\tau_v$ of $W_{F_v} \times \mathrm{SL}(2, \mathbb{C})$, such as $Ad(\sigma_v)$, the associated $L$-factor has no zeros.

Now let $S$ be any finite set of primes containing $\infty$ and the primes where $\pi$ is ramified. Put

$$L^S(s, \pi; Ad) = \prod_{v \notin S} L(s, \pi_v; Ad). \tag{2.4}$$

Suppose $L^S(s, \pi; Ad)$ is of positive type. Then by definition, its logarithm defines a Dirichlet series with positive coefficients, absolutely convergent in a right half plane. By the theory of Landau, this Dirichlet series converges on $(\beta, \infty)$, where $\beta$ is the largest real number where $\log L^S(s, \pi; Ad)$ diverges. But such a point of divergence must be a pole, and not a zero, of $L^S(s, \pi; Ad)$ because its logarithm is positive in $(\beta, \infty)$.

LEMMA 2.5. *Let $\beta$ be the smallest real number such that $L^S(s, \pi; Ad)$ converges for all real $s > \beta$. Then*

$$\beta < 0.$$

*Proof of the Lemma 2.5.* By the standard properties of the Rankin-Selberg $L$-functions ([JPSS], [JS], [Sh1-3], [MW]—see also [BRa]), $L^S(s, \pi \times \overline{\pi})$ is invertible for $\Re(s) > 1$ and admits a meromorphic continuation to the whole $s$-plane with a unique simple pole at $s = 1$. The same properties hold of course for $\zeta^S(s)$; so $L^S(s, \pi; Ad)$ has no pole in $\Re(s) \geq 1$. In other words we have $\beta < 1$. Moreover, one knows that $\zeta^S(s)$ is nonzero on $(0, 1) \subset \mathbb{R}$. Hence we have

$$\beta \leq 0.$$

Now let us look at the point $s = 0$. By definition,

$$L^S(s, \pi; Ad) = \frac{L^\infty(s, \pi \times \overline{\pi})}{\zeta(s) \prod_{v \in S - \{\infty\}} L(s, \pi_v; Ad)}. \qquad \square$$

The numerator on the right has no pole at $s = 0$, and $\zeta(s)$ does not vanish at $s = 0$. Moreover, as we noted above, the local factors $L(s, \pi_v; Ad)$ have no zeros. Consequently, $L^S(s, \pi; Ad)$ has no pole at $s = 0$, and this proves the lemma.

LEMMA 2.6. *$L(s, \pi_\infty; Ad)$ has a pole at $s = 0$.*

*Proof of the Lemma 2.6.* There exist complex numbers $z_1, z_2, z_3$ such that

$$L(s, \pi_\infty) = \prod_{j=1}^{3} \Gamma_{\mathbb{R}}(s + z_j + \delta_j), \tag{2.7}$$

with $\delta_j \in \{\pm 1\}$, $\forall j$, and

$$\Gamma_{\mathbb{R}}(s) = \pi^{-s/2}\Gamma(s/2).$$

By the unitarity of $\pi_\infty$, we see that either all the $z_j$ have absolute value 1, in which case $\pi_\infty$ is tempered, or exactly one of the $z_j$, say $z_1$, has absolute value 1, and moreover,

$$z_2 = u + t, \qquad z_3 = u - t,$$

for some positive real number $t$ and a complex number $u$ of absolute value 1. In either case we see that the set

$$B(\pi_\infty; Ad) := \{z_1, z_2, z_3\} \cup \{\overline{z}_1, \overline{z}_2, \overline{z}_3\} - \{0\} \tag{2.8}$$

contains 0. The standard yoga of Langlands $L$-functions furnishes the identity

$$L(s, \pi_\infty; Ad) = \prod_{z \in B(\pi_\infty; Ad)} \Gamma_{\mathbb{R}}(s + z). \tag{2.9}$$

Recall that $\Gamma_{\mathbb{R}}(s)$ never vanishes and has simple poles at the even non-positive integers, in particular at $s = 0$. Since $B(\pi_\infty; Ad)$ contains 0, $\Gamma_{\mathbb{R}}(s)$ is a factor of $L(s, \pi_\infty; Ad)$. We must then have

$$-\mathrm{ord}_{s=0} L(s, \pi_\infty; Ad) \geq 1,$$

as asserted. □

*Proof of Proposition B (continued).* As the local factors $L(s, \pi_v; Ad)$ never vanish, and since $S$ is finite by assumption, the function

$$L_{S-\{\infty\}}(\pi; Ad) := \prod_{p \in S-\{\infty\}} L(s, \pi_p; Ad)$$

is nonzero at $s = 0$. Hence by Lemma 2.6,

$$-\mathrm{ord}_{s=0} L_S(s, \pi; Ad) \geq 1. \tag{2.10}$$

But we know that the full adjoint $L$-function $L(s, \pi; Ad)$ has no pole at $s = 1$, nor at $s = 0$ by the functional equation. So all this forces the following:

$$L^S(0, \pi; Ad) = 0,$$

which contradicts Lemma 2.5. The only unsupported assumption we made was that $L^S(s, \pi; Ad)$ is of positive type, which must be wrong if $S$ is finite. We are done. □

Note that the proof uses the base field $\mathbb{Q}$ in order to use the crucial property of the Riemann zeta function that it does not vanish in the real interval (0, 1). For general number fields $F$, the Dedekind zeta function $\zeta_F(s)$ should not have any such real zero either, save possibly at $s = 1/2$. Clearly, Theorem A will follow for any $F$ for which Proposition B can be established. One way to get around the difficulty for general $F$ would be to prove a priori that the adjoint $L$-function, which has been studied from different points of view by D. Ginzburg, Y. Flicker, H. Jacquet, S. Rallis, F. Shahidi, and D. Zagier, has no pole in $(0, 1)$, which is, to our knowledge, unknown. To elaborate a little, a particular version of the trace formula due to H. Jacquet and D. Zagier [JZ] suggests that the divisibility of $L(s, \pi \times \overline{\pi})$ by $\zeta_F(s)$ for all cuspidal automorphic representations $\pi$ of $\mathrm{GL}(n, \mathbb{A}_F)$ with trivial central character is equivalent to the divisibility of $\zeta_K(s)$ by $\zeta_F(s)$ for all commutative cubic algebras $K$ over $F$. Since the latter is known for $n = 3$, one hopes that the former holds. This divisibility has been investigated by relating it to an Eisenstein series on $G_2$ by D. Jiang and S. Rallis [JiR], and the desired result could be close to being established in the $n = 3$ case.

253-37 CALTECH, PASADENA, CA 91125
*E-mail:* dinakar@its.caltech.edu

## REFERENCES

[BRa] L. Barthel and D. Ramakrishnan, A non-vanishing result for twists of $L$-functions of GL$(n)$, *Duke Math. Journal* **74**, no. 3, (1994), 681–700.

[C$\ell$] L. Clozel, Représentations galoisiennes associées aux représentations automorphes autoduales de GL$(n)$, *IHES Publications Math.* **73** (1991), 97–145.

[HaT] M. Harris and R. Taylor, The geometry and cohomology of some simple Shimura varieties, with an appendix by V. Berkovich, *Annals of Math.* Studies **151**, Princeton, 2001.

[He] G. Henniart, Une preuve simple des conjectures de Langlands pour GL$(n)$ sur un corps $p$-adique, *Inventiones Math.* **139** no. 2 (2000), 439–455.

[HRa] J. Hoffstein and D. Ramakrishnan, Siegel zeros and cusp forms, *International Math. Research Notices* (IMRN) no. 6 (1995), 279–308.

[JPSS] H. Jacquet, I. Piatetski-Shapiro, and J. Shalika, Rankin-Selberg convolutions, *Amer. J. Math.* **105** (1983), 367–464.

[JS] H. Jacquet and J. A. Shalika, On Euler products and the classification of automorphic forms. I & II, *Amer. J of Math.* **103** (1981), 499–558 & 777–815.

[JZ] H. Jacquet and D. Zagier, Eisenstein series and the Selberg trace formula. II, *Transactions of the AMS* **300** no. 1 (1987), 1–48.

[JiR] D. Jiang and S. Rallis, Fourier coefficients of the Eisenstein series of the exceptional group of type $G_2$, *Pacific Journal of Math.* **181**, no. 2 (1997), 281–314.

[KSh] H. Kim and F. Shahidi, Cuspidality of symmetric powers with applications, *Duke Math. Journal* **112** no. 1 (2002), 177–197.

[MW] C. Moeglin and J.-L. Waldspurger, Poles des fonctions $L$ de paires pour GL$(N)$, Appendice, Ann. Sci. École Norm. **22** Sup. (4) (1989), 667–674.

[Pic] *Zeta functions of Picard modular surfaces*, ed. by R. P. Langlands and D. Ramakrishnan, CRM Publications, Montréal, 1992.

[Ra] D. Ramakrishnan, On the coefficients of cusp forms, *Math Research Letters* **4** nos. 2–3 (1997), 295–307.

[Sh1] F. Shahidi, On certain $L$-functions, *American Journal of Math.* **103** (1981), 297–355.

[Sh2] ———, On the Ramanujan conjecture and the finiteness of poles for certain $L$-functions, *Ann. of Math.* **127** no. 2 (1988), 547–584.

[Sh3] ———, A proof of the Langlands conjecture on Plancherel measures; Complementary series for $p$-adic groups, *Ann. of Math.* **132** (1990), 273–330.

CHAPTER 25

# NONVANISHING OF $L$-FUNCTIONS ON $\Re(s) = 1$

By PETER SARNAK

---

*To Joe Shalika on his 60th birthday*

*Abstract.* In [Ja-Sh], Jacquet and Shalika use the spectral theory of Eisenstein series to establish a new result concerning the nonvanishing of $L$-functions on $\Re(s) = 1$. Specifically they show that the standard $L$-function $L(s, \pi)$ of an automorphic cusp form $\pi$ on $GL_m$ is nonzero for $\Re(s) = 1$. We analyze this method, make it effective, and also compare it with the more standard methods. This note is based on the letter [Sa1].

**1. Review of de la Vallée Poussin's method.** A celebrated result of Hadamard and de la Vallée Poussin is the Prime Number Theorem. Their proof involved showing that the Riemann zeta function $\zeta(s)$ is not zero for $\Re(s) = 1$. In fact these two results turn out to be equivalent. De la Vallée Poussin (1899) extended this method to give a zero-free region for $\zeta(s)$ of the form

$$\zeta(s) \neq 0 \text{ for } \sigma \geq 1 - \frac{c}{\log(|t| + 2)} . \tag{1.1}$$

Here $c$ is an absolute positive constant and $s = \sigma + it$. We will call a zero-free region of the type (1), a standard zero-free region.

De la Vallée Poussin's method is based on the construction of an auxillary $L$-function, $D(s)$ with positive coefficients. $D(s)$ should be analytic in $\Re(s) > 1$, have a pole at $s = 1$ of order, say, $k$, and if $L(\sigma + it_0) = 0$, (here $L(s)$ stands for a generic $L$-function whose nonvanishing we seek to establish.) $D(s)$ should vanish to order at least $k$ at $s = \sigma$. (Since the coefficients of $D(s)$ are positive, the Euler product for $D(s)$ converges absolutely for $\Re(s) > 1$.) This is enough to ensure that $L(1 + it_0) \neq 0$ and if the order of vanishing at $\sigma$ is bigger than $k$ then one obtains an effective standard zero-free region for $L(s)$. To arrange for $D(s)$ to have positive coefficients one often uses a positive definite function on an appropriate group. For example $L(s, \pi \times \tilde{\pi})$, where $\pi$ is any unitary isobaric representation of $GL_m$ (see, for example, [Ho-Ra] for definitions and examples) has this property. See also [De] for such positive definite functions on other groups.

In the most basic case of $\zeta(s)$ and $t_0 \in \mathbb{R}$ with $|t_0| \geq 2$, one can take

$$\begin{aligned} D(s) &= \zeta^3(s)\,\zeta^2(s + it_0)\,\zeta^2(s - it_0)\,\zeta(s + 2it_0)\,\zeta(s - 2it_0) \\ &= L(s, \Pi \times \widetilde{\Pi}) . \end{aligned} \tag{1.2}$$

Here $\Pi = \mathbf{1} \boxplus \alpha^{-it_0} \boxplus \alpha^{it_0}$ is an isobaric representation of $GL_3$ and $\alpha$ the principal quasi-character of $\mathbb{A}_{\mathbb{Q}}^*$. $D(s)$ has a pole of order 3 at $s = 1$ and if $\zeta(\sigma + it_0) = 0$

then $D(s)$ will have a zero of order 4 at $s = \sigma$. Hence, by a standard function theoretic argument (see [Ho-Ra], for example) we have that $\zeta(\sigma + it_0) \neq 0$ for $\sigma \geq 1 - \frac{c}{\log |t_0|}$. This establishes the standard zero-free region (1) for $\zeta(s)$.

We note that (1.1) is not the best zero-free region that is known for $\zeta(s)$. Vinogradov [Vi] and his school have developed sophisticated techniques that lead to zero-free regions of the type; $\zeta(s) \neq 0$ for $\sigma \geq 1 - c_\alpha/(\log(|t| + 2))^\alpha$ for $\alpha > \frac{2}{3}$ and $c_\alpha > 0$.

Another well-known example of nonvanishing is that of a Dirichlet $L$-function $L(s, \chi)$, with $\chi$ a quadratic character of conductor $q$. For $D(s)$ we can take $\zeta(s)L(s, \chi)$ (or, if one prefers, $(\zeta(s)L(s, \chi))^2 = L(s, (\mathbf{1} \boxplus \chi) \times (\widetilde{\mathbf{1} \boxplus \chi}))$. In this case the order of zero at $s = 1$ is equal to the order of pole. Hence $L(1, \chi) \neq 0$ (see Landau's Lemma [Da], p. 34) but this does not yield a standard zero-free region near $s = 1$ for $L(s, \chi)$—i.e., in terms of the conductor. (That is to say, $L(\sigma, \chi) \neq 0$ for $\sigma \geq 1 - \frac{c}{\log q}$ and some $c > 0$.) In fact no such zero-free region is known for $L(s, \chi)$—this being the notorious problem of the exceptional, or "Landau-Siegel," zero. In this note we will only be concerned with zero-free regions for a fixed $L$-function (i.e., what is called the $t$-aspect). For a recent discussion of the exceptional zero problem in general, see the paper [Ho-Ra].

The de la Vallée Poussin method generalizes to automorphic $L$-functions. Let $K$ be a number field, $m \geq 1$, and let $\pi$ be an automorphic cusp form on $GL_m(\mathbb{A}_k)$. The standard (finite part) $L$-function associated with $\pi$, namely, $L(s, \pi)$, has an analytic continuation to $\mathbb{C}$ and a functional equation $s \longrightarrow 1 - s, \pi \longrightarrow \tilde{\pi}$ [Go-Ja]. Also well known by now are the analytic properties (i.e., continuation and functional equation) of the Rankin-Selberg $L$-functions $L(s, \pi \times \pi')$, where $\pi$ and $\pi'$ are cusp forms on $GL_m(\mathbb{A}_k)$ and $GL_{m'}(\mathbb{A}_k)$, respectively. This follows from [Ja -PS-Sh], [Sh1], and [Mo-Wa].

Apply de la Vallée Poussin's method with

$$(1.3) \qquad \begin{aligned} D(s) &= \zeta(s)\, L^2(s, \pi \times \tilde{\pi}) L^2(s + it_0, \pi)\, L^2(s - it_0, \tilde{\pi}) \,\cdot \\ &\quad L(s + 2it_0, \pi \times \pi)\, L(s - 2it_0, \tilde{\pi} \times \tilde{\pi}) \\ &= L(s, \Pi \times \tilde{\Pi}), \end{aligned}$$

where $\Pi := \mathbf{1} \boxplus \pi \otimes \alpha^{it_0} \boxplus \tilde{\pi} \otimes \alpha^{-it_0}$ (we are tacitly assuming that we have normalized $\pi$ so that $\tilde{\pi} \neq \pi \otimes \alpha^{\pm 2it_0}$).

This yields a standard zero-free region for $L(s, \pi)$—that is, a zero-free region as in (1) but with

$$(1.4) \qquad c = c(\pi).$$

As mentioned in the abstract, the nonvanishing of such an $L(s, \pi)$ for $\sigma = 1$ was first established (before the Rankin-Selberg theory was developed) by Jacquet-Shalika who used the method discussed in Section 2. The advantage of de la Vallée Poussin's method here is that it yields a standard zero-free region (1.4). According to the general functoriality conjectures of Langlands, any automorphic $L$-function should be a (finite) product of such standard $L$-functions. If so, we would be in good shape at least in that the zero-free region for the general $L$-function would be of the same quality as for $\zeta(s)$.

One can apply de la Vallée Poussin's method to certain Rankin-Selberg $L$-functions. If $\pi \neq \pi'$, take for $D$,

$$(1.5) \quad D(s) = L(s, \pi \times \tilde{\pi})\, L(s, \pi' \times \tilde{\pi}')\, L(s + it_0, \pi \times \pi')\, L(s - it_0, \tilde{\pi} \times \tilde{\pi}').$$

From this it follows that $L(1 + it_0, \pi \times \pi') \neq 0$. This nonvanishing result was established in this way by Ogg [Og] for $m = 2 = m'$. The general case of any $m$ and $m'$ was first proven by Shahidi [Sh1] using the Eisenstein series method discussed in Section 2. If $\pi$ and $\pi'$ are self-dual, Moreno [Mo] established a standard zero-free region for $L(s, \pi \times \pi')$. To see this, one can take $D(s) = L(s, \Pi \times \widetilde{\Pi})$ where

$$(1.6) \quad \Pi = \pi \boxplus \pi \otimes \alpha^{it_0} \boxplus \pi \otimes \alpha^{-it_0} \boxplus \pi' \boxplus \pi' \otimes \alpha^{it_0} \boxplus \pi' \otimes \alpha^{-it_0}.$$

In this case, $D(s)$ has a pole of order 6 at $s = 1$ and a zero of order 8 at $s = \sigma$ if $L(\sigma + it, \pi \times \pi') = 0$.

These $L$-functions $L(s, \pi)$ and $L(s, \pi \times \pi')$ and any others that can be expressed in terms of these by known cases of functoriality are more or less all that can be handled by de la Vallée Poussin's method. We turn now to the Eisenstein series method.

**2. Nonvanishing via Eisenstein series.** This method is based on the spectral theory of locally homogeneous spaces and in particular Eisenstein series. In as much as this method works effectively in all cases where the methods of Section 1 apply as well as in many other cases, it must, at least at the present time, be considered a principal method. There have been a number of suggestions as well as evidence for useful spectral interpretations of the zeroes of $L$-functions [Od], [Ka-Sa], [Co], [Za]. However, it does not seem to be widely appreciated that the spectral interpretation of the zeroes of $L$-functions through poles of Eisenstein series (i.e., as resonances) has already proven to be very powerful.

To illustrate this, consider the symmetric power $L$-functions $L(s, \pi, \mathrm{sym}^k)$, where $\pi$ is a cusp form on $GL_2(\mathbb{A}_{\mathbb{Q}})$ with trivial central character and $k \geq 1$ (see [Sh2] for definitions). The recently established functorial lifts, $\mathrm{sym}^3 : \mathrm{GL}_2 \longrightarrow \mathrm{GL}_4$ and $\mathrm{sym}^4 : \mathrm{GL}_2 \longrightarrow \mathrm{GL}_5$, due to Kim and Shahidi [K-S1], [Ki] (see also [He]), allow one to study $L(s, \pi, \mathrm{sym}^k)$ for $1 \leq k \leq 8$. By decomposing $L(s, \mathrm{sym}^i \pi \times \mathrm{sym}^j \pi)$ $1 \leq i \leq 4$, $1 \leq j \leq 4$ into a product of primitive $L$-functions (see [K - S2]) and using (6) of Section 1, one sees that for $1 \leq k \leq 8$, $L(s, \pi, \mathrm{sym}^k)$ satisfies a standard zero-free region. However, the next symmetric power $L(s, \pi, \mathrm{sym}^9)$ falls outside of the range of this approach. It is not known whether the Euler product for $L(s, \pi, \mathrm{sym}^9)$ converges absolutely for $\Re(s) > 1$. In particular $L(s, \pi, \mathrm{sym}^9)$ might even have zeroes in $\Re(s) > 1$! Thus an application of de la Vallée Poussin's method appears to be problematic. Concerning the absolute convergence, if we use the best bounds known towards the Ramanujan Conjectures [Ki-Sa] one sees that $L(s,\pi,\mathrm{sym}^9)$ converges absolutely for $\Re(s) > \frac{71}{64}$. Given the above comments, it is remarkable that the theory of Eisenstein series on $E_8$ together with the Langlands-Shahidi (see [Mi] for a recent summary and outline) method allows one to show that (see [K-S2]): $L(s,\pi,\mathrm{sym}^9)$ is meromorphic in the plane and is analytic and nonvanishing in $\Re(s) \geq 1$, except possibly for a finite number of simple zeros or poles in $\left[1, \frac{71}{64}\right]$. The same arguments give the analyticity and nonvanishing in $\{s \,|\, \Re(s) \geq 1\} \backslash [1,\infty)$, for a quite general class of $L$-functions—see [Ge-Sh]. Given the success of this technique the question arises as to what zero-free region it yields. The proof of nonvanishing, though very simple (even magical), is quite indirect and it is not clear how to make it effective so as to yield zero-free regions. We show below that in the simplest case of $\zeta(s)$, one can with some effort make the proof effective, and it gives zero-free regions which are almost as good as the standard zero-free regions—see (53) below. At the end of this section we indicate how one might proceed in the general case.

In order to deal with $\zeta(s)$ we consider the Eisenstein series for the modular quotient $X$ of the upper half-plane $\mathbb{H}$, that is, $X = SL(2,\mathbb{Z})\backslash\mathbb{H}$. It is instructive in this analysis to consider more general Fuchsian groups $\Gamma \leq SL(2,\mathbb{R})$ for this allows us to separate the arithmetic and analytic features. So assume that $\Gamma\backslash\mathbb{H}$ has one cusp at infinity and it is normalized so that the stabilizer of infinity is $\Gamma_\infty = \{\pm\left(\begin{smallmatrix}1 & m\\ 0 & 1\end{smallmatrix}\right) \,|m \in \mathbb{Z}\}$.

The corresponding Eisenstein series is defined by

$$E_\Gamma(z,s) = \sum_{\gamma\in\Gamma_\infty\backslash\Gamma} (y(\gamma z))^s\,, \text{ for } \Re(s) > 1 \tag{2.1}$$

and $z = x + iy \in \mathbb{H}$.

The spectral theory of Eisenstein series due to Selberg in this setting [Se1], and Langlands in general [La], asserts that $E_\Gamma(z,s)$ is meromorphic in $s$. Moreover, it is analytic in $\Re(s) \geq \frac{1}{2}$ except possibly for simple poles in $(\frac{1}{2}, 1]$. These general properties when applied to $\Gamma = SL(2,\mathbb{Z})$ imply that $\zeta(s) \neq 0$ for $\Re(s) = 1$.

One can formulate this in a very explicit way (see (2.15) below), which I learned from a lecture of Selberg's (at Stanford, $\pm$ 1980). Let $\Gamma_\infty\backslash\Gamma/\Gamma_\infty$ be a set of representatives for these double cosets for such a $\Gamma \leq PSL(2,\mathbb{R})$. The set of $c$ appearing in $\left[\begin{smallmatrix}a & b\\ c & d\end{smallmatrix}\right] \in \Gamma_\infty\backslash\Gamma/\Gamma_\infty$ can be chosen so that $c \geq 0$ and $c = 0$ corresponds to the identity coset. They form a discrete subset in $[0,\infty)$. For $m \in \mathbb{Z}$ and $c > 0$ as above set

$$\tau_{m,\Gamma}(c) = \sum_{\left[\begin{smallmatrix}a & b\\ c & d\end{smallmatrix}\right]\in\Gamma_\infty\backslash\Gamma/\Gamma_\infty} e\left(\frac{md}{c}\right). \tag{2.2}$$

Here $e(z) = e^{2\pi i z}$ and the sum is easily seen to be well defined. A standard calculation [Ku] shows that the $m-^{\text{th}}$ coefficient of $E_\Gamma(z,s)$, $\int_0^1 E_\Gamma(z,s)\,e(-mx)\,dx$, is given by

$$y^s + \left(\sum_{c>o} \frac{\tau_{0,\Gamma}(c)}{c^{2s}}\right) \frac{\pi^{1/2}\Gamma(s-\frac{1}{2})}{\Gamma(s)}\, y^{1-s}, \quad \text{if } m = 0 \tag{2.3}$$
$$:= y^s + \phi_\Gamma(s)\, y^{1-s},$$

and

$$= \left(\sum_{c>0} \frac{\tau_{m,\Gamma}(c)}{c^{2s}}\right) \frac{2\pi^s |m|^{s-\frac{1}{2}} y^{1/2} K_{s-\frac{1}{2}}(2\pi |m| y)}{\Gamma(s)}, \quad \text{if } m \neq 0\,. \tag{2.4}$$

Using (2.4) and the theory of Eisenstein series we can meromorphically continue the functions $D_m(s)$, where

$$D_m(s) := \sum_{c>0} \frac{\tau_{m,\Gamma}(c)}{c^{2s}}\,. \tag{2.5}$$

In order to give growth bounds on $D_m(s)$ for $\Re(s) \geq \frac{1}{2}$, we use the technique in [Go-Sa]. For $m > 0$ set

$$P_m(z, s) = \sum_{\gamma \in \Gamma\infty \backslash \Gamma} (y(\gamma z))^s\, e(m\gamma z)\,. \tag{2.6}$$

This series converges absolutely for $\Re(s) > 1$ and for $\Re(s) \geq \frac{3}{2}$, we have the bound

$$P_m(z, s) \ll y^{1-\sigma} \quad \text{for} \quad y \geq \frac{1}{2}. \tag{2.7}$$

Hence for $\sigma \geq \frac{1}{2}$ we may form

$$\begin{aligned} &\langle E_\Gamma(\cdot, s),\, P_m(\cdot\,, \bar{s} + 1)\rangle \\ &= \int_{\Gamma\backslash\mathbb{H}} E_\Gamma(z, s)\, \overline{P_m(z, \bar{s} + 1)}\, \frac{dx\, dy}{y^2} \\ &= a\, b^s\, \frac{\Gamma(2s)}{\Gamma(s)\,\Gamma(s+1)}\, D_m(s), \end{aligned} \tag{2.8}$$

for suitable constants $a$ and $b$ (depending on $m$).

We assume that $E_\Gamma(z, s)$ has no poles in $(\frac{1}{2}\,, 1)$. For a given $\Gamma$ this can be checked, and if such poles are present, they will enter in the asymptotics below in an explicit way. From (2.8) and the properties of Eisenstein series we see that $D_m(s)$ is holomorphic in $\Re(s) \geq \frac{1}{2}$ (there being no pole at $s = 1$ since $\langle 1, P_m\rangle = 0$). Also from (2.8) and Stirling's series we see that for $\Re(s) = \frac{1}{2}$ and $|t|$ large,

$$|D_m(s)| \sim |t|^{1/2} |< E(\cdot\,, s)\,, P_m(\cdot\,, \bar{s} + 1)\rangle|\,. \tag{2.9}$$

For $T$ large and $T \leq t \leq T + 1$ write

$$P_m\left(\frac{3}{2} + it\right) = P_m\left(\frac{3}{2} + iT\right) + \frac{1}{i}\int_T^t P_m'\left(\frac{3}{2} + i\,\tau\right) d\tau.$$

Hence

$$\int_T^{T+1} \left| \left\langle E\left(\frac{1}{2}+it\right), P_m\left(\frac{3}{2}-it\right)\right\rangle \right| dt \tag{2.10}$$

$$\leq \int_T^{T+1} \left\{ \left| \left\langle E\left(\frac{1}{2}+it\right), P_m\left(\frac{3}{2}-iT\right)\right\rangle \right| + \int_T^t \left| \left\langle E\left(\frac{1}{2}+it\right), P_m'\left(\frac{3}{2}-i\tau\right)\right\rangle \right| d\tau \right\} dt.$$

The spectral theory of $L^2(\Gamma\backslash\mathbb{H})$ and, in particular, Bessel's inequality, yields

$$\int_{-\infty}^{\infty} \left| \left\langle E\left(\frac{1}{2}+it\right), P_m'\left(\frac{3}{2}-i\tau\right)\right\rangle \right|^2 dt \leq \left\| P_m'\left(\frac{3}{2}-i\tau\right)\right\|_2^2 \ll 1\,. \tag{2.11}$$

Putting this in (2.10) and applying Cauchy's inequality yields,

$$\int_T^{T+1} \left| \left\langle E\left(\frac{1}{2}+it\right), P_m\left(\frac{3}{2}-it\right)\right\rangle \right| dt \ll 1\,.$$

Combined with (2.9) we get

$$\int_T^{T+1} \left| D_m\left(\frac{1}{2}+it\right)\right| dt \ll T^{1/2}\,. \tag{2.12}$$

Equipped with (2.12) we return to (2.5) and apply Perron's formula. For $x > 0$

$$\sum_{c\leq x} \left(1-\frac{c}{x}\right) \tau_{m,\Gamma}(c) = \frac{1}{2\pi i}\int_{\Re(s)=2} D_m(s)\, x^{2s}\, \frac{ds}{s(2s+1)}\,. \tag{2.13}$$

Now shift the contour in the last integral to $\Re(s) = \frac{1}{2}$. The estimate (2.12) easily justifies this shift and, moreover, we don't pick up any poles. Thus

$$\sum_{c\leq x} \left(1-\frac{c}{x}\right) \tau_{m,\Gamma}(c) = \frac{x}{2\pi}\int_{-\infty}^{\infty} \frac{D_m\left(\frac{1}{2}+it\right) e^{it\log x}}{\left(\frac{1}{2}+it\right)(1+2it)}\, dt. \tag{2.14}$$

(2.12) ensures that the integral in (2.14) is absolutely convergent. Hence by the Riemann-Lebesgue lemma we conclude that as $x \longrightarrow \infty$

$$\sum_{c\leq x} \left(1-\frac{c}{x}\right) \tau_{m,\Gamma}(c) = o(x)\,. \tag{2.15}$$

This general result holds for any $\Gamma$ for which $E(z, s)$ has no poles in $(\frac{1}{2}, 1)$. In particular, it holds if $\lambda_1\,(\Gamma\backslash\mathbb{H}) \geq \frac{1}{4}$ where $\lambda_1$ is the second smallest eigenvalue of the Laplacian on $\Gamma\backslash\mathbb{H}$. For $SL(2,\mathbb{Z})\backslash\mathbb{H}$ as well as other low genus Riemann surfaces one can use geometric methods, see for example [S2, p. 34], to show that $\lambda_1 \geq \frac{1}{4}$. Also, for $\Gamma = SL(2,\mathbb{Z})$ the $c$'s run over integers and the sums $\tau_{m,SL(2,\mathbb{Z})}(c)$ are

Ramanujan sums that may be evaluated explicitly in terms of the Mobius function $\mu$.

$$\tau_{m,SL(2,\mathbb{Z})}(c) = \sum_{\substack{d|m \\ d|c}} \mu\left(\frac{c}{d}\right) d \,. \tag{2.16}$$

Thus for $m = 1$ and $\Gamma = SL(2, \mathbb{Z})$, (15) asserts that

$$\sum_{c \le x} \left(1 - \frac{c}{x}\right) \mu(c) = o(x) \,. \tag{2.17}$$

Now (2.17) is elementarily equivalent to the prime number theorem. Thus the above spectral analysis provides us with a nonarithmetic setting in which (2.15), a form of the prime number theorem, is valid in a family (the Beurling theory of generalized primes [Be] provides another such setting). However the standard zero-free region for $\zeta(s)$ does not persist in the family and is apparently a rigid feature. To see this, note that a zero-free region is equivalent to a rate of decay in (2.17) or (2.15) and that this in turn is equivalent to a pole-free region in $\beta < \frac{1}{2}$ for the poles $\rho = \beta + i\gamma$ of $D_m(s)$. From (2.8) this amounts to such pole-free regions for $E_\Gamma(z, s)$, and according to the theory of Eisenstein series these poles occur at poles of $\phi_\Gamma(s)$. We apply the theory [P-S1], [P-S2], and see also [Wo1] and [Wo2], concerning the behavior of such poles under deformations of $\Gamma$. A consequence of the theory is that if the critical values of certain $L$-functions are nonzero, then the corresponding eigenvalues of the Laplacian are dissolved into poles of Eisenstein series. For example, using the recent nonvanishing results of Luo [Lu] for such critical values of Rankin-Selberg $L$-functions and assuming a suitable form of a standard multiplicity bound conjecture for the eigenvalues of the Laplacian on a congruence quotient of $\mathbb{H}$, we have that for the generic $\Gamma$:

(2.18) There is $c_\Gamma > 0$ such that the number of poles $\rho = \beta + i\gamma$ of $\phi_\Gamma$ (s) with $|\gamma| \le T$ and $0 \le \beta < \dfrac{1}{2}$ is at least $c_\Gamma T^2$.

On the other hand Selberg [Se2] has shown that for any $\Gamma$

$$\sum_{\substack{\rho \\ |\gamma| \le T \\ \beta < \frac{1}{2}}} \left(\frac{1}{2} - \beta\right) = O(T \log T) \,. \tag{2.19}$$

Combining (2.18) and (2.19) we see that for generic $\Gamma$ there is a sequence of poles $\rho_j = \beta_j + i\gamma_j$ of $\phi_\Gamma(s)$ with

$$\frac{1}{2} > \beta_j > \frac{1}{2} - \frac{c'_\Gamma \log |\gamma_j|}{|\gamma_j|} \,. \tag{2.20}$$

In the case of $\Gamma = SL(2, \mathbb{Z})$, according to (2.2), (2.3), and (2.16) we have

$$\phi_{SL(2,\mathbb{Z})}(s) = \frac{\sqrt{\pi}\,\Gamma\left(s - \frac{1}{2}\right)}{\Gamma(s)} \frac{\zeta(2s-1)}{\zeta(2s)} \tag{2.21}$$

and the $m-^{\text{th}}$ coefficient of $E(z,s)$ is

$$\frac{2\pi^s |m|^{s-1/2}\, y^{1/2}\, K_{s-\frac{1}{2}}(2\pi |m| y)}{\Gamma(s)\,\zeta(2s)}\,\sigma_{1-2s}(|m|), \tag{2.22}$$

where

$$\sigma_s(m) = \sum_{d|m} d^s\,. \tag{2.23}$$

Hence the standard zero-free region for $\zeta(s)$ implies the pole-free region for $SL_2(\mathbb{Z})$ of the form:

$$\beta \leq \frac{1}{2} - \frac{c}{\log(|\gamma|+2)}\,. \tag{2.24}$$

Thus while the analogue of the nonvanishing of $\zeta(s)$ on $\sigma = 1$ is valid for general $\Gamma$, the pole-free region (2.24) is not. In particular this shows that one cannot apply this Eisenstein series method of showing nonvanishing of $L$-functions on $\sigma = 1$ to get zero-free regions at least if all one uses is the general spectral theory.

The key to effectivizing the nonvanishing proof in which the Fourier coefficients of $E(\cdot\,, s)$ along unipotents of parabolics are ratios of $L$-functions (as in the Langlands-Shahidi method, see [Ge-Sh]) is to exploit the inhomogeneous form of the Maass-Selberg-Langlands relation. For the rest we stick to the case $\Gamma = SL(2,\mathbb{Z})$. It is perhaps worth pointing out the simple magical and standard derivation of the nonvanishing of $\zeta(s)$ for $\sigma = 1$ using $E(z,s)$. $E(z,s)$ is analytic for $\Re(s) = \frac{1}{2}$, thus if we look at (2.22) with $m \neq 0$, we see that such a coefficient is analytic for $\Re(s) = \frac{1}{2}$. Clearly, consideration of the denominator shows that $\zeta(1+2it_0)$ cannot be zero.

To continue our quantitative analysis, we will use freely the analytic properties of $\zeta(s)$—that is, the location of poles and the functional equation (all of which are essentially known for the general $L$-function that can be continued by the theory of Eisenstein series). We also use freely the asymptotic properties of the Whittaker functions $K_s(y)$ as well as the Gamma function.

The Maass-Selberg relation for $SL(2,\mathbb{Z})$ reads:

$$\begin{aligned}\int_X \left| E_A\left(z, \frac{1}{2}+it\right)\right|^2 \frac{dxdy}{y^2} \\ = 2\log A - \frac{\phi'}{\phi}\left(\frac{1}{2}+it\right) + \frac{\bar\phi\left(\frac{1}{2}+it\right)A^{2it} - \phi\left(\frac{1}{2}+it\right)A^{-2it}}{2it},\end{aligned} \tag{2.25}$$

where $A \geq 1$ and where for $z \in \mathcal{F}$ (the standard fundamental domain),

$$E_A(z,s) = \begin{cases} E(z,s), & \text{for } y \leq A \\ E(z,s) - y^s - \phi(s)y^{1-s}, & \text{for } y > A. \end{cases} \tag{2.26}$$

Normalizing the denominator in $\phi(s)$—(see (2.21))—gives

$$\int_X |\zeta(1+2it)|^2 \left|E_A\left(z, \frac{1}{2}+it\right)\right|^2 \frac{dx\,dy}{y^2} \tag{2.27}$$

$$= |\zeta(1+2it)|^2 \left| 2\log A - \frac{\phi'}{\phi}\left(\frac{1}{2}+it\right) + \frac{\overline{\phi\left(\frac{1}{2}+it\right)}A^{2it} - \phi\left(\frac{1}{2}+it\right)A^{-2it}}{2it}\right|.$$

From (2.22), Parseval's inequality and the shape of $\mathcal{F}$ we have that

$$\text{LHS of (2.27)} \gg \sum_{m=1}^{\infty} \int_1^{\infty} \left|\frac{K_{it}(2\pi|m|y)\,\sigma_{-2it}(m)}{\Gamma\left(\frac{1}{2}+it\right)}\right|^2 \frac{dy}{y}. \tag{2.28}$$

On the other hand, (2.21) together with Stirling for $\frac{\Gamma'}{\Gamma}(s)$ and the functional equation for $\zeta(s)$ shows that for $t \geq 2$

$$\left|\zeta(1+2it)\frac{\phi'}{\phi}\left(\frac{1}{2}+it\right)\right| \ll \log t + |\zeta(1+2it)| + \left|\zeta'(1+2it)\right|. \tag{2.29}$$

It is elementary (see p. 49 of [Ti] and such upper bounds for $L$-functions can be derived generally) that for $t \geq 2$

$$\begin{aligned} |\zeta(1+2it)| &\ll \log t \\ |\zeta'(1+2it)| &\ll (\log t)^2. \end{aligned} \tag{2.30}$$

Hence

$$\left|\zeta(1+2it)\frac{\phi'}{\phi}\left(\frac{1}{2}+it\right)\right| \ll (\log t)^2. \tag{2.31}$$

Assuming that $|\zeta(1+2it)| \leq 1$, which we can in estimating this quantity from below, we have from (2.28) and (2.29) that

$$\sum_{m=1}^{\infty} |\sigma_{-2it}(m)|^2 \int_m^{\infty} \left|\frac{K_{it}(2\pi y)}{\Gamma\left(\frac{1}{2}+it\right)}\right|^2 \frac{dy}{y} \ll |\zeta(1+2it)|\,(\log t)^2. \tag{2.32}$$

Using the asymptotics [E, p. 87–88] we have that for $y < \frac{t}{4}$

$$K_{it}(y) \sim \frac{e^{\frac{-\pi}{4}t}\sqrt{2\pi}}{\sqrt[4]{t^2-y^2}} \sin\left[\frac{\pi}{4} + t h(y/t)\right], \tag{2.33}$$

with $h$ a fixed smooth function.

Hence

$$\int_{t/8}^{t/4} \left| \frac{K_{it}(2\pi y)}{\Gamma\left(\frac{1}{2}+it\right)} \right|^2 \frac{dy}{y} \gg \frac{1}{t}, \tag{2.34}$$

and so for $m \leq t/8$ we have

$$\int_{m}^{\infty} \left| \frac{K_{it}(2\pi y)}{\Gamma\left(\frac{1}{2}+it\right)} \right|^2 \frac{dy}{y} \gg \frac{1}{t}. \tag{2.35}$$

Applying (2.32) with $m = 1$ and using (2.35) and $|\sigma_{-2it}(1)| = 1$, we obtain

$$|\zeta(1+2it)| \gg \frac{1}{t(\log t)^2}. \tag{2.36}$$

We can do better by using many $m$'s in (2.32). For example, note that for $m = p$, a prime

$$\left|\sigma_{-2it}(p) - \sigma_{-2it}(p^2)\right| = 1.$$

Hence

$$\left|\sigma_{-2it}(p^2)\right|^2 + \left|\sigma_{-2it}(p)\right|^2 \gg 1. \tag{2.37}$$

Thus

$$\sum_{m \leq t/8} \left|\sigma_{-2it}(m)\right|^2 \gg \sum_{p \leq \frac{\sqrt{t}}{8}} 1.$$

The last sum is by elementary means (Chebyshev) $\gg \sqrt{t}/\log t$. Combining this with (2.32) and (2.35) yields

$$|\zeta(1+2it)| \gg \frac{1}{(\log t)^3 \sqrt{t}}. \tag{2.38}$$

To further improve this, we examine intervals of primes $p$ where $|1 + p^{it2}|$ is small, i.e., where $\sigma_{+2it}(p)$ is small. To proceed we need more flexibility on the range of $m$'s in (2.32). Set $\eta = t^{-\delta}$ with $\delta > 0$ (e.g., $\delta = 1$ will work) and consider instead of (2.27), the quantity

$$I = \int_{\eta}^{\infty} \int_{0}^{1} |\zeta(1+2it)|^2 \left| E_A\left(z, \frac{1}{2}+it\right) \right|^2 \frac{dxdy}{y^2}. \tag{2.39}$$

If $N(z, \eta) = |\{\gamma \in \Gamma_\infty \backslash \Gamma \,|\, y(\gamma z) \geq \eta\}|$ then it is not difficult to see ([Iw], p. 54) that for $\eta \leq 1$

$$N(z, \eta) \ll \frac{1}{\eta}. \tag{2.40}$$

Hence

$$I = \int_{\mathcal{F}} N(z,\eta) \left| E_A\left(z, \frac{1}{2} + it\right)\right|^2 |\zeta(1+2it)|^2 \frac{dxdy}{y^2} \tag{2.41}$$
$$\ll \frac{1}{\eta} \int_{\mathcal{F}} |\zeta(1+2it)|^2 \left| E_A\left(z, \frac{1}{2} + it\right)\right|^2 \frac{dxdy}{y^2}.$$

By (2.27) and (2.29), this gives

$$I \ll \frac{1}{\eta} |\zeta(1+2it)| \left[(\log t)^2 + 2\log A\right]. \tag{2.42}$$

Now if $\frac{1}{A} = \eta$, then we see that the nonzero Fourier coefficients of $E_A(z,s)$ coincide with those of $E(z,s)$ for $y \geq \eta$. So as in the discussion following (2.27) we deduce that

$$I \gg \sum_{m=1}^{\infty} |\sigma_{-2it}(m)|^2 \int_{\eta m}^{\infty} \left| \frac{K_{it}(2\pi y)}{\Gamma\left(\frac{1}{2} + it\right)}\right|^2 \frac{dy}{y} \tag{2.43}$$
$$\gg \frac{1}{t} \sum_{m \leq t/4\eta} |\sigma_{-2it}(m)|^2 .$$

On the other hand, for $A = 1/\eta$, $\log A = O(\log t)$ so that (2.42) and (2.43) yield

$$\frac{\eta}{t} \sum_{m \leq t/4\eta} |\sigma_{-2it}(m)|^2 \ll (\log t)^2 |\zeta(1+2it)|. \tag{2.44}$$

This gives us the flexibility we need. If $\delta = 1$, i.e., $\eta = t^{-1}$, we have

$$\frac{1}{t^2} \sum_{\frac{t^2}{8} \leq m \leq \frac{t^2}{4}} |\sigma_{-2it}(m)|^2 \ll |\zeta(1+2it)|(\log t)^2 . \tag{2.45}$$

To give a lower bound for the left hand side of (2.45) we restrict $m$ to primes $N \leq p \leq 2N$ with $N = \frac{t^2}{8}$. For integers $m$ with

$$t \log N \leq 2\pi m \leq t \log(2N), \tag{2.46}$$

let $I_m$ be the interval of $p$ given by

$$|2t \log p - 2\pi m + \pi| < \frac{1}{100}. \tag{2.47}$$

Note that for $p \notin I_m$

$$|\sigma_{2it}(p)| \gg 1. \tag{2.48}$$

The length $I_m$ satisfies

$$|I_m| \leq \frac{N}{50t}. \tag{2.49}$$

A well-known application of sieve theory (which is independent of the zeta function!), see, for example, [Bo-Da], who use the large sieve, asserts that for $N,\ M \geq 2$

$$\pi(M+N) - \pi(N) \leq \frac{3M}{\log M}, \tag{2.50}$$

where $\pi(x) = \sum_{p \leq x} 1$.

Hence the number of primes in $I_m$ is at most

$$\frac{3N}{50t\ \log(N/50t)} \leq \frac{3N}{25t\ \log N}. \tag{2.51}$$

The number of intervals $I_m$ is less than $t$ according to the set up (2.46). Thus the total number of primes in $\cup_m I_m$ is at most

$$\frac{3N}{25\ \log N}. \tag{2.52}$$

Again, elementary arguments show that the number of primes $p$ satisfying $N \leq p \leq 2N$ is at least $\frac{N}{4 \log N}$. Hence there are at least $\frac{N}{8 \log N}$ primes $N \leq p \leq 2N$ which are not in any $I_m$. Thus from (2.45) and (2.47) it follows that

$$|\zeta(1+2it)| \gg \frac{1}{(\log t)^3}. \tag{2.53}$$

This is the effective nonvanishing of $\zeta$ on $\sigma = 1$ that we sought to establish using $E(z,s)$. It leads immediately to a zero-free region of the type $\zeta(s) \neq 0$ for $\Re(s) > 1 - \frac{c}{(\log t)^5}$. This is not quite the standard zero-free region but it is of the same general quality.

Note that once we arrive at (2.45) we are in a similar position to a proof of the nonvanishing of $\zeta(s)$ on $\sigma = 1$ due to Ingham [In]. He uses the identity

$$\sum_{n=1}^{\infty} |\sigma_{it_0}(n)|^2\, n^{-s} = \frac{\zeta^2(s)\,\zeta(s+it_0)\,\zeta(s-it_0)}{\zeta(2s)}. \tag{2.54}$$

Indeed, Balasubramanian and Ramachandra [Ba-Ra] in deriving zero-free regions from (2.54), use similar sieving arguments to those used after (2.45). See also the comments by Heath-Brown on p. 68 of [Ti]. Our point is that we arrive at (2.45) in a geometric way using the Maass-Selberg relation, and hence our analysis can be generalized to the Langlands-Shahidi setting.

In that setting one would probably (at least at first) go as far as (2.36) in the above argument. That is to consider only one nonzero Fourier coefficient $E_\chi(s)$ (in the notation of [Ge-Sh]) of the Eisenstein series. This will require understanding the asymptotic behaviour in $t$ of the archimedian local Whittaker function $W_{it}(e)$. Mckee [Mc] has made the first steps in this direction. We expect that these ideas will lead to effective lower bounds in general of the type

$$|L_{\text{finite}}\,(1+it,\pi,r)| \gg |t|^{-\delta} \tag{2.55}$$

for a constant $\delta$ depending on the group on which the Eisenstein series lives. This falls short of a standard zero-free region, but it would be more than enough to establish the conjecture in [Ge-Sh].

*Acknowledgement.* I would like to thank S. Gelbart and F. Shahidi for their comments on an earlier version of this note.

---

REFERENCES

[Ba-Ra] R. Balasubramanian and K. Ramachandra, The place of an identity of Ramanujan in prime number theory, *Proc. Ind. Acad. Sc.*, Sect. A **83** (1976), 156–165.

[Be] A. Beurling, Analyse de la loi asymptotique de la distribution des nombres premiers généralisés, *Acta Math.* **68** (1937), 255–291.

[Bo-Da] E. Bombieri and H. Davenport, On the large sieve method, *Number Theory and Analysis*, Plenum, New York, 1969, pp. 9–22.

[Co] A. Connes, Trace formula in noncommutative geometry and zeros of the Riemann zeta function, *Selecta Math.* **5** (1999), 29–106.

[Da] H. Davenport, Multiplicative number theory, *Springer-Verlag, Grad. Texts* **74** (1980).

[De] P. Deligne, La Conjecture de Weil II, *Pub. Math.* IHES, no. 52, (1980).

[E] A. Erdélyi et al, *Higher Transcendental Functions,* vol. 2, McGraw-Hill, 1953.

[Ge-Sh] S. Gelbart and F. Shahidi, Boundedness of automorphic $L$-Functions in vertical strips, *J. Amer. Math. Soc.* **14** no. 1 (2001), 79–107.

[Go-Sa] D. Goldfeld and P. Sarnak, Sums of Kloosterman Sums, *Invent. Math.* **71** (1983), 471–542.

[Go-Ja] R. Godement and H. Jacquet, Zeta functions of simple algebras, *Lecture Notes in Math.,* vol. **260** Springer-Verlag, 1972.

[He] G. Henniart, Progrés récents en fonctorialité de Langlands, Seminare Bourbaki, June 2001.

[Ho-Ra] J. Hoffstein and D. Ramakrishnan, Siegel zeros and cusp forms, *IMRN*, no. 6, (1995), 279–308.

[In] A. Ingham, Note on Riemann's $\zeta$-function and Dirichlet's $L$-functions, *J. London. Math. Soc.* **5** (1930), 107–112.

[Iw] H. Iwaniec, Introduction to the Spectral Theory of Automorphic Forms, *Bib. Revista Mat. Iber.* (1995).

[J-PS-S] H. Jacquet, I. Piatetski-Shapiro, and J. Shalika, Rankin-Selberg Convolutions, *Amer. J. Math.* **105** (1983), 367–464.

[Ja-Sh] H. Jacquet and J. Shalika, A nonvanishing theorem for zeta-functions of $GL_n$, *Invent. Math.* **38** (1976), 1–16.

[Ka-Sa] N. Katz and P. Sarnak, Zeros of zeta functions and symmetry, *BAMS* **36** (1999), 1–26.

[Ki] H. Kim, Functoriality for the exterior square of $GL_4$ and symmetric fourth of $GL_2$, (to appear in *JAMS*).

[K-S1] H. Kim and F. Shahidi, Functorial products for $GL_2 \times GL_3$ and the symmetric cube for $GL_2$, *C.R. ACAD. SC.*, Paris, Math 331, **8** (2000), 599–604.

[K-S2] ———, Cuspidality of symmetric powers with applications, *Duke Math. J.* **112** (2002), 177–197.

[Ki-Sa] H. Kim and P. Sarnak, Refined estimates towards the Ramanujan and Selberg conjectures, Appendix to [Ki], 2001.

[Ku] T. Kubota, *Elementary Theory of Eisenstein Series,* Kodansha Ltd., 1973.

[La] R. Langlands, On the functional equations satisfied by Eisenstein series, *S.L.N. in Math.* **544** (1976).

[Lu] W. Luo, Nonvanishing of $L$-values and Weyl's Law, *Ann. Math.*, **154**, no. 2 (2001), 477–502.

[Mc] M. Mckee, Ph.D. thesis, Princeton, 2002.

[Mi] S. D. Miller, A summary of the Langlands-Shahidi method of constructing $L$-functions, www.math.rutgers.edu/~sdmiller/l-functions.

[Mo-Wa] C. Moeglin and L. Waldspurger, Le spectre residual de $GL(n)$, *Ann. Sci. Ecole Norm* **22**, Sup (4) (1989), 605–674.

[Mo] C. Moreno, An analytic proof of the Strong Multiplicity One Theorem, *Amer. J. Math.* **107** (1985), 163–206.

[Og] A. Ogg, On a convolution of $L$-series, *Invent. Math.* **7** (1969), 297–312.

[Od] A. Odlyzko, History of Hilbert-Polya conjecture, www.dtc.vmn.edu/ odlyzko/polya/.

[P-S1] R. Phillips and P. Sarnak, The Weyl theorem and deformations of discrete groups, *C.P.A.M.*, **38** (1985), 853–866.

[P-S2] ———, On cusp forms for cofinite subgroups of $PSL(2, \mathbb{R})$, *Inv. Math.* **80** (1985), 339–364.

[Sh1] F. Shahidi, On certain $L$-functions, *Amer. J. Math.* **103**, no. 2, (1981).

[Sh2] ———, Automorphic $L$-functions: A survey, *Automorphic Forms, Shimura Varieties and L-functions,* (L. Clozel and J. S. Milne, eds.) *Perspectives in Math.*, A.P. (1990), 415–437.

[Sa1] P. Sarnak, letter to S. Gelbart and F. Shahidi, January 2001.

[Sa2] ———, Some Applications of modular forms, Cambridge Tracts in Math., vol. 99, (1990).

[Se1] A. Selberg, Harmonic Analysis, *Collected Works,* vol. I, Springer-Verlag, 1989, 626–674.

[Se2] ———, Remarks on poles of Eisenstein series, *Collected Works,* vol. II, 1991, 15–47.

[Ti] E. Titchmarsh, *The Theory of the Riemann Zeta Function,* 2nd ed., Oxford Press, 1986.

[Vi] I. M. Vinogradov, A new estimate for the function $\zeta(1 + it)$, *Izs. Akad Nauk SSSR Ser. Mat* **22** (1958), 161–164.

[Wo1] S. Wolpert, Spectral limits for hyperbolic surfaces I, II, *Invent. Math.* **108** (1992), 67–89, 91–129.

[Wo1] ———, Disappearance of cusp forms in special families, *Ann. Math.* **139** (1994), 239–291.

[Za] D. Zagier, Eisenstein series and the Riemann zeta-function, *Automprphic Forms Representation Theory and Arithmetic,* Tata Institute, Springer, 1981, 275–301.

CHAPTER 26

# RATIONAL POINTS AND AUTOMORPHIC FORMS

By JOSEPH A. SHALIKA, RAMIN TAKLOO-BIGHASH, and YURI TSCHINKEL

*Abstract.* We study the distribution of rational points of bounded height on certain equivariant compactifications of anisotropic inner forms of semi-simple groups.

**1. Introduction.** Let $\mathbf{x} \in \mathbb{P}^n(\mathbb{Q})$ be a $\mathbb{Q}$-rational point in the projective space of dimension $n$ with coordinates $\mathbf{x} = (x_0 : x_1 : \cdots : x_n)$, such that

$$(x_0, x_1, \ldots, x_n) \in \mathbb{Z}^{n+1}_{\text{prim}},$$

that is, the set of primitive $(n+1)$-tuples of integers. Define a height function

$$H(\mathbf{x}) := \max_j(|x_j|).$$

Of course, we could replace this norm by any other norm on $\mathbb{R}^{n+1}$, for example, $\sqrt{x_0^2 + \cdots + x_n^2}$. Generally, for any number field $F$ and $\mathbf{x} \in \mathbb{P}^n(F)$ we can define

$$H(\mathbf{x}) := \prod_{v \in \text{Val}(F)} \max_j (|x_j|_v),$$

where the product is over all valuations of $F$. By the product formula, this does not depend on a particular choice of homogeneous coordinates for $\mathbf{x}$. Clearly, the number

$$N(\mathbb{P}^n, B) := \#\{\mathbf{x} \in \mathbb{P}^n(F) \mid H(\mathbf{x}) \leq B\}$$

is finite, for any $B > 0$. In 1964, Schanuel computed its asymptotic behavior, as $B \to \infty$,

$$N(\mathbb{P}^n, B) = \mathrm{c}(n, F, H) \cdot B^{n+1}(1 + o(1)),$$

where $\mathrm{c}(n, F, H)$ is an explicit constant (see [35]).

Let $X$ be an algebraic variety over a $F$ and $\mu : X \longrightarrow \mathbb{P}^n$ a projective embedding. Then $H \circ \mu$ defines a height function on the set of $F$-rational points $X(F)$ (more conceptually, the height is defined by means of an adelic metrization $\mathcal{L} = (L, \|\cdot\|_{\mathbb{A}})$ of the line bundle $L := \mu^*(\mathcal{O}(1))$). We obtain an induced counting function

$$N(X, \mathcal{L}, B) := \#\{\mathbf{x} \in X(F) \mid H \circ \mu(\mathbf{x}) \leq B\}.$$

One of the main themes of modern arithmetic geometry and number theory is the study of distribution properties of rational points on algebraic varieties. In particular, one is interested in understanding the asymptotic distribution of rational points of bounded height.

All theoretical and numerical evidence available so far indicates that one should expect an asymptotic expression of the form

$$N(X, \mathcal{L}, B) = \mathsf{c} \cdot B^{\mathsf{a}} \log(B)^{\mathsf{b}-1}(1 + o(1)),$$

for some $\mathsf{a} \in \mathbb{Q}$, $\mathsf{b} \in \frac{1}{2}\mathbb{Z}$, and a positive real $\mathsf{c}$. In 1987, Manin had initiated a program aimed at interpreting the constants $\mathsf{a}$, $\mathsf{b}$, and $\mathsf{c}$ in terms of intrinsic algebro-geometric and arithmetic invariants of $X$. The main observation was that $\mathsf{a}$ and $\mathsf{b}$ should depend only on the class of the embedding line bundle $L$ in the Picard group $\mathrm{Pic}(X)$ of the variety $X$, more precisely its position with respect to the anticanonical class $[-K_X]$ and the cone of effective divisors $\Lambda_{\mathrm{eff}}(X) \subset \mathrm{Pic}(X)_{\mathbb{R}}$. The constant $\mathsf{c}$, on the other hand, should reflect the dependence of the asymptotic expression on finer structures (like the choice of a norm in the definition of the height and $p$-adic densities).

Of course, it may happen that $X$ has no rational points at all, or that $X(F)$ is entirely contained in a proper Zariski closed subset. In these cases, it is hopeless to try to read off the geometry of $X$ from the asymptotics of rational points. We will therefore assume that $X(F)$ is Zariski dense. In general, it is not so easy to produce examples of interesting varieties with a Zariski dense set of rational points (unless, of course, $X$ admits an action of an algebraic group with a Zariski dense orbit). For example, $X$ could be a flag variety or an abelian variety. It is expected that the density of points (at least after a finite extension of the groundfield) holds for *Fano* varieties (that is, varieties with ample anticanonical class $[-K_X]$). This question is still open even in dimension 3 (see [20]). Here is a version of Manin's conjecture:

*Conjecture* 1.1. Let $X$ be an algebraic variety over a number field $F$ such that its anticanonical class $[-K_X]$ is ample and $X(F)$ is Zariski dense. Then there exists a Zariski open subset $U \subset X$ such that

$$N(U, -\mathcal{K}_X, B) = \mathsf{c}(\mathcal{K}_X) \cdot B(\log B)^{\mathsf{b}(X)-1}(1 + o(1))$$

for $B \to \infty$, where $-\mathcal{K}_X$ is a (metrized) anticanonical line bundle, $\mathsf{b}(X)$ is the rank of the Picard group $\mathrm{Pic}(X)$ and $\mathsf{c}(\mathcal{K}_X)$ a nonzero constant.

*Remark* 1.2. The restriction to Zariski open subsets is necessary since $X$ may contain *accumulating* subvarieties (the asymptotics of rational points on these subvarieties will dominate the asymptotics of the complement). The constant $\mathsf{c}(\mathcal{K}_X)$ has an interpretation as a Tamagawa number (defined by Peyre in [33]). Finally, there is a similar description for arbitrary ample line bundles, proposed in [3], resp. [8].

Conjecture 1.1 and its refinements have been proved for the following classes of varieties:

- smooth complete intersections of small degree in $\mathbb{P}^n$ (circle method);
- generalized flag varieties [17];
- toric varieties [5], [6];

- horospherical varieties [40];
- equivariant compactifications of $\mathbb{G}_a^n$ [11];
- bi-equivariant compactifications of unipotent groups [38], [39].

We expect that Manin's conjecture (and its refinements) should hold for equivariant compactifications of *all* linear algebraic groups G and their homogeneous spaces G/H. We provide further evidence for this expectation by outlining a proof of the above conjecture for certain smooth equivariant compactifications of $\mathbb{Q}$-anisotropic semi-simple $\mathbb{Q}$-groups of adjoint type (for complete proofs see [37]).

This work focuses on the interplay between arithmetic geometry and automorphic forms. Though the main problem is inspired by Manin's conjecture in arithmetic geometry, our tools and techniques, which are naturally suited to the current context, are from the theory of automorphic forms and representations of $p$-adic groups. Our approach is inspired by the work of Batyrev and Tschinkel on compactifications of anisotropic tori [4] and the work of Godement and Jacquet on central simple algebras [19]. We are currently working on a generalization of our results to higher rank, where the presence of the Eisenstein series makes the problem even more interesting from the analytic point of view.

Finally, we would like to mention related work of Duke, Rudnick, and Sarnak [14], Eskin, McMullen [15], Eskin, Mozes, and Shah [16] on asymptotics of *integral* points of bounded height on homogeneous varieties. Their theorems neither imply our results nor follow from them.

*Acknowledgments.* We have greatly benefited from conversations with Arthur and Sarnak. The second author wishes to thank the Clay Mathematics Institute for partial support of this project. The third author was partially supported by the NSA, NSF and the Clay Mathematics Institute.

**2. Methods and results.** Let $F$ be a number field and D a central simple algebra of rank $m$ over $F$. Let $\Lambda$ be a lattice in D. Denote by $\mathrm{Val}(F)$ the set of all valuations and by $S_\infty$ the subset of archimedean valuations of $F$. For each $v \in \mathrm{Val}(F)$, we put $\mathrm{D}_v = \mathrm{D} \otimes_F F_v$ and, for $v \notin S_\infty$, $\Lambda_v = \Lambda \otimes_{\mathcal{O}_F} \mathcal{O}_v$. For almost all $v$, $\Lambda_v$ is a maximal order in $\mathrm{D}_v$. We proceed to define a family of norms $\|\cdot\|_{\Lambda_v}$ on $\mathrm{D}_v$, one for each place $v$ of $F$.

- *nonarchimedean $v$:* Choose a basis $\{\xi_1^v, \ldots, \xi_k^v\}$ for $\Lambda_v$. For $g \in \mathrm{D}_v$, write $g = \sum_i c_i(g)\xi_i^v$ and set

$$\|g\|_v = \|g\|_{\Lambda_v} := \max_{i=1,\ldots,k}\{|c_i(g)|_v\}.$$

  It is easy to see that this norm is independent of the choice of basis.

- *archimedean $v$:* Fix a Banach space norm $\|\cdot\|_v = \|\cdot\|_{\mathrm{D}_v}$ on the finite-dimensional real (or complex) vector space $\mathrm{D}_v = \mathrm{D} \otimes_F F_v$.

Clearly, for $c \in F_v$ and $g \in \mathrm{D}_v$, we have

$$\|cg\|_v = |c|_v \cdot \|g\|_v.$$

Consequently, for $c \in F^\times$ and $g \in \mathrm{D}$, we have

$$\prod_{v \in \mathrm{Val}(F)} \|cg\|_v = \prod_{v \in \mathrm{Val}(F)} \|g\|_v, \tag{2.1}$$

by the product formula. For an adelic point $g = (g_v)_v \in \mathrm{D}(\mathbb{A})$ define the *global height function:*

$$H(g) := \prod_{v \in \mathrm{Val}(F)} H_v(g) = \prod_{v \in \mathrm{Val}(F)} \|g_v\|_v.$$

By the product formula, $H$ is well defined on the projective group $\mathrm{D}(F)^\times/F^\times$. Moreover, $H$ is invariant under the right and left action of a compact open subgroup

$$\mathrm{K}_0 = \prod_{v \notin S_\infty} \mathrm{K}_{0,v} \subset \mathrm{G}(\mathbb{A}_{\mathrm{fin}})$$

(if we fix an integral model for G then $\mathrm{K}_{0,v} = \mathrm{G}(\mathcal{O}_v)$, for almost all $v$). It will be convenient to assume that the Haar measure $dg$ is such that $\mathrm{vol}\,(\mathrm{K}_0) = 1$.

From now on, we let G be an $F$-anisotropic inner form of a split semi-simple group $\tilde{\mathrm{G}}$ of adjoint type over a number field $F$. Let

$$\varrho_F : \mathrm{G} \longrightarrow \mathrm{D}^\times$$

be an $F$-group morphism from G to the multiplicative group of a central simple algebra over $F$ of rank $m$. Extending scalars to a finite Galois extension $E/F$ over which both G and D are split, we obtain a homomorphism

$$\varrho_E : \tilde{\mathrm{G}}(E) \longrightarrow \mathrm{GL}_m(E).$$

This homomorphism is obtained from an algebraic representation

$$\varrho : \tilde{\mathrm{G}} \longrightarrow \mathrm{GL}_m,$$

defined over $F$.

*Remark* 2.1. Conversely, from any algebraic representation $\varrho : \tilde{\mathrm{G}} \longrightarrow \mathrm{GL}_m$ over $F$ we obtain a group homomorphism

$$\varrho_E : \tilde{\mathrm{G}}(E) \longrightarrow \mathrm{GL}_m(E),$$

which induces a map

$$\varrho_E^* : \mathrm{Z}^1(\mathrm{Gal}\,(E/F), \tilde{\mathrm{G}}(E)) \longrightarrow \mathrm{Z}^1(\mathrm{Gal}\,(E/F), \mathrm{PGL}_m(E)).$$

Let $c \in \mathrm{Z}^1(\mathrm{Gal}\,(E/F), \tilde{\mathrm{G}}(E))$ be the cocycle that defines the inner form G. Then $\varrho_E^*(c)$ defines a central simple algebra $\mathrm{D} \subset \mathrm{Mat}_m(E)$. It is easy to verify that $\varrho_E$

descends to a morphism of $F$-groups

$$\varrho_F : \mathrm{G} \to \mathrm{D}^\times.$$

Thus we can use $\varrho_F$ to pull back the height function from $\mathrm{D}^\times$ to G. We are interested in the asymptotics of

$$N(\varrho, B) := \#\{\gamma \in \mathrm{G}(F) \mid H(\varrho_F(\gamma)) \leq B\},$$

as $B \to \infty$. To put this in geometric perspective, the pair $(\mathrm{G}, \varrho_F)$ defines an equivariant compactification $X$ of G and a G-linearized ample line bundle on $X$ (and vice versa). Thus we are counting rational points on a Zariski open subset $\mathrm{G} \subset X$, with respect to some adelically metrized line bundle (depending on $\varrho$). Below we will see that for appropriate choices of $\varrho$ the asymptotic formula for $N(\varrho, B)$ matches precisely Manin's prediction.

Our main technical assumption (used in the computation (2.5)) is the following:

*Assumption* 2.2. The representation $\varrho_F$ is absolutely irreducible.

In order to state our theorem we need to introduce some notation. Fix a Borel subgroup B with maximal split torus T in $\tilde{\mathrm{G}}$ and denote by $\mathfrak{X}^*(\mathrm{T})$ the character group of T. Let $\Phi$ be the root system of $(\tilde{\mathrm{G}}, \mathrm{T})$, and $\Delta = \{\alpha_1, \dots, \alpha_r\}$ the set of simple roots. Also let $2\rho_{\mathrm{G}} = \sum_{\alpha \in \Phi^+} \alpha$. Since $\tilde{\mathrm{G}}$ is of adjoint type it is immediate that there are one-parameter subgroups $\{\hat{\alpha}_1, \dots, \hat{\alpha}_r\}$ of T such that

$$< \hat{\alpha}_j, \alpha_i >= -\delta_{ij}.$$

Let $\varrho = \varrho_\lambda$ be the irreducible algebraic representation of $\tilde{\mathrm{G}}$ associated with a dominant weight $\lambda$. Let $\chi_\lambda$ be the character of T associated with $\lambda$. Since $\lambda$ is dominant and $\tilde{\mathrm{G}}$ is of adjoint type, there exist nonnegative integers $k_1(\varrho), \dots, k_r(\varrho)$ such that

$$\chi_\lambda(t) = \prod_{i=1}^{r} \alpha_i(t)^{k_i(\varrho)}.$$

The numbers $k_i(\varrho)$, $1 \leq i \leq r$, are all non zero if the representation $\varrho$ is nontrivial. Set then

$$\mathsf{a}_\varrho := \max_{j=1,\dots,r} \frac{1- < \hat{\alpha}_j, 2\rho_{\mathrm{G}} >}{k_j(\varrho)}, \quad \text{and} \quad \mathsf{b}_\varrho := \#\left\{ j \mid \frac{1- < \hat{\alpha}_j, 2\rho_{\mathrm{G}} >}{k_j(\varrho)} = \mathsf{a}_\varrho \right\}.$$

*Remark* 2.3. The anticanonical embedding of the *wonderful* compactification $X$ of G of de Concini-Procesi is associated with the weight $\kappa = 2\rho_{\mathrm{G}} + \sum_{i=1}^r \alpha_i$. In particular, $X$ is Fano (see [9],[13] for more details).

It is not hard to see that if $\varrho = \varrho_\kappa$, then $\mathsf{a}_\varrho = 1$ and $\mathsf{b}_\varrho = r$. We set

$$\mathsf{c}_\kappa := \lim_{s\to 1}(s-1)^r \int_{\mathrm{G}(\mathbb{A})} H(\varrho_F(g))^{-s}\, dg,$$

(where $dg$ is a suitably normalized Haar measure on $\mathrm{G}(\mathbb{A})$). By (2.5), the limit exists and is a positive real number. Our main theorem is the following:

THEOREM 2.4. *We have*

$$N(\varrho_\kappa, B) = \frac{\mathsf{c}_\kappa}{(r-1)!}\cdot B(\log B)^{r-1}(1+o(1)),$$

*as* $B \to \infty$.

We note that this theorem implies Manin's conjecture and its refinement due to Peyre for the wonderful compactification of G as above. We have also proved analogous results for arbitrary irreducible representations $\varrho$ (in other words, for height functions associated with arbitrary ample line bundles on the wonderful compactification of G).

We will now sketch the proof (in the case $\varrho = \varrho_\kappa$). Using Tauberian theorems one deduces the asymptotic properties of $N(\varrho, B)$ from the analytic properties of the height zeta function

$$\mathcal{Z}(s,\varrho) = \sum_{\gamma\in \mathrm{G}(F)} H(\varrho_F(\gamma))^{-s}.$$

Actually, we will use the function

$$\mathcal{Z}(s,\varrho,g) = \sum_{\gamma\in \mathrm{G}(F)} H(\varrho_F(\gamma g))^{-s}.$$

For $\Re(s) \gg 0$, the right side converges (uniformly on compacts) to a function which is holomorphic in $s$ and continuous in $g$ on $\mathbb{C}\times \mathrm{G}(\mathbb{A})$. Since G is $F$-anisotropic, $\mathrm{G}(F)\backslash\mathrm{G}(\mathbb{A})$ is compact, and

$$\mathcal{Z} \in \mathsf{L}^2(\mathrm{G}(F)\backslash\mathrm{G}(\mathbb{A}))^{\mathrm{K}_0}$$

(recall that $H$ is bi-invariant under $\mathrm{K}_0$). Again since G is anisotropic, we have

$$\mathsf{L}^2(\mathrm{G}(F)\backslash\mathrm{G}(\mathbb{A})) = \Big(\hat{\bigoplus_\pi}\mathcal{H}_\pi\Big)\bigoplus\Big(\bigoplus_\chi \mathbb{C}_\chi\Big), \tag{2.2}$$

as a direct sum of irreducible subspaces. Here the first direct sum is over infinite-dimensional representations of $\mathrm{G}(\mathbb{A})$ and the second direct sum is a sum over all automorphic characters of $\mathrm{G}(\mathbb{A})$. Consequently,

$$\mathsf{L}^2(\mathrm{G}(F)\backslash\mathrm{G}(\mathbb{A}))^{\mathrm{K}_0} = \Big(\hat{\bigoplus_\pi}\mathcal{H}_\pi^{\mathrm{K}_0}\Big)\bigoplus\Big(\bigoplus_\chi \mathbb{C}_\chi\Big), \tag{2.3}$$

a sum over representations containing a $K_0$-fixed vector (in particular, the sum over characters is finite). For each infinite-dimensional $\pi$ occurring in (2.3) we choose an orthonormal basis $\mathcal{B}_\pi = \{\phi^\pi_\alpha\}_\alpha$ for $\mathcal{H}^{K_0}_\pi$. We have next the "Poisson formula:"

$$(2.4)\quad \mathcal{Z}(s,\varrho,g) = \sum_\pi \sum_{\phi\in\mathcal{B}_\pi} \langle \mathcal{Z}(s,\varrho,g),\phi(g)\rangle\phi(g) + \sum_\chi \langle \mathcal{Z}(s,\varrho,g),\chi(g)\rangle\chi(g).$$

Here the series on the right converges normally to $\mathcal{Z}(\varrho, g)$ for $\Re(s) \gg 0$. We will establish a meromorphic continuation of the right side of (2.4), leading to a proof of the main theorem.

A key result is the computation of the individual inner products $\langle\mathcal{Z},\phi\rangle$. After the usual unfolding it turns out that each of these is an Euler product with an explicit regularization. In particular, the pole of highest order of $\mathcal{Z}(s,\varrho,g)$ (or the main term in the asymptotic expression of $N(\varrho, B)$) is contributed by the trivial representation:

$$\int_{G(\mathbb{A})} H(\varrho_F(g))^{-s}dg = \prod_{v\in \mathrm{Val}(F)} \int_{G(F_v)} H_v(\varrho_F(g_v))^{-s}dg_v.$$

Local integrals of such type can be computed explicitly at almost all places (see [11]). They are reminiscent of Igusa's local zeta functions and their modern generalizations: "motivic" integrals of Batyrev, Kontsevich, and Denef-Loeser (see [2], [12]). In our case, we have

$$(2.5)\qquad \int_{G(\mathbb{A})} H(\varrho_F(g))^{-s}dg = \prod_{j=1}^r \zeta_F(k_js + <\hat{\alpha}_j, 2\rho_G>)\cdot h_F(s,\varrho)$$

(where $h_F(s,\varrho)$ is a holomorphic function for $\Re(s) > 1-\epsilon$ and some $\epsilon > 0$ and $h_F(1,\varrho)\in\mathbb{R}_{>0}$).

Next we prove that each remaining term is holomorphic around $\Re(s) = 1$. In general, we have

$$\begin{aligned}\langle\mathcal{Z},\phi\rangle &= \int_{G(F)\backslash G(\mathbb{A})} \mathcal{Z}(s,\varrho,g)\overline{\phi(g)}\,dg\\ &= \int_{G(\mathbb{A})} H(\varrho_F(g))^{-s}\overline{\phi(g)}\,dg\\ &= \int_{G(\mathbb{A})} H(\varrho_F(g))^{-s}\int_{K_0}\overline{\phi(kg)}\,dk\,dg.\end{aligned}$$

Next we follow an argument by Godement and Jacquet in [19]. Without loss of generality we can assume that

$$K_0 = \prod_{v\notin S} K_v \times K_0^S,$$

for a finite set of places $S$. Here for $v\notin S$, $K_v$ is a maximal special open compact subgroup in $G(F_v)$. After enlarging $S$ to contain all the places where G is not split, we can assume that $K_v = G(\mathcal{O}_v)$. In particular, for $v\notin S$ the local representations

$\pi_v$ are spherical. Thus we have a normalized local spherical function $\varphi_v$ associated to $\pi_v$. We have assumed that each $\phi$ is right $\mathrm{K}_0$-invariant. In conclusion,

$$\langle \mathcal{Z}, \phi \rangle = \prod_{v \notin S} \int_{\mathrm{G}(F_v)} \varphi_v(g_v) H_v(\varrho_F(g_v))^{-s}\, dg_v$$
$$\times \int_{\mathrm{G}(\mathbb{A}_S)} H(\varrho_F(\eta(g_S)))^{-s} \int_{\mathrm{K}_0^S} \phi(k\eta(g_S))\, dk\, dg_S.$$

(Here $\eta : \mathrm{G}(\mathbb{A}_S) \to \mathrm{G}(\mathbb{A})$ is the natural inclusion map.) The second factor is relatively easy to deal with. Our main concern here is with the first factor. To proceed we need to invoke some fairly deep results from the representation theory of reductive groups to find nontrivial bounds on spherical functions. Depending on the semi-simple rank of G, there are two cases to consider:

*Case 1: semi-simple rank* 1. In this situation, G is an inner form of $\mathrm{PGL}_2$—that is, the projective group of a quaternion algebra. By the Jacquet-Langlands correspondence [22], there is an irreducible cuspidal automorphic representation $\pi' = \otimes_v \pi'_v$ of $\mathrm{PGL}_2$ such that for $v \notin S$, we have $\pi_v = \pi'_v$. In particular, in order to obtain nontrivial bounds on spherical functions of infinite-dimensional representations, we need to examine local components of cuspidal representations of $\mathrm{GL}_2$ with trivial central character. Here the estimate we need follows from a classical result of Gelbart and Jacquet who established the symmetric square lifting from $\mathrm{GL}_2$ to $\mathrm{GL}_3$ [18], combined with a result of Jacquet and Shalika (see [23]). We also note the recent beautiful results of Kim and Shahidi towards sharper bounds in the Ramanujan-Petersson conjecture [24].

*Case 2: semi-simple rank* $> 1$. First we use a strong approximation argument to show that for $v \notin S$, the representation $\pi_v$ is not one-dimensional, unless $\pi$ itself is one-dimensional (a similar argument appears in the work of Clozel and Ullmo [10]). Then we apply a recent result of Oh [32] giving bounds for all nontrivial spherical matrix coefficients of the unitary dual of semi-simple groups of rank at least 2, which can be considered as a quantitative version of property (T) for these groups. Moreover, the bounds are uniform over all primes $p$, which is crucial in our applications. Let us mention that a (weaker) form of Oh's results can be deduced from [21] for $\mathrm{Sp}_{2n}(\mathbb{Q}_p)$ and for $\mathrm{GL}_n(\mathbb{Q}_p)$ from the known classification of the unitary dual (see [41]).

Putting everything together, we obtain that there exists an $\epsilon > 0$ such that for all nontrivial representations and all $\phi$ the inner product $\langle \mathcal{Z}, \phi \rangle$ is holomorphic for $\Re(s) > 1 - \epsilon$.

Finally, to prove the convergence of the right side (for appropriate $s$), we integrate by parts (with respect to the Laplacian $\Delta$ on the compact Riemannian manifold associated with $\mathrm{G}(\mathbb{A})$ and $\mathrm{K}_0$), and combine $\mathsf{L}^\infty$-estimates for $\Delta$-eigenfunctions with standard facts about spectral zeta functions of compact manifolds.

*Remark* 2.5. Similar arguments lead to a proof of equidistribution of rational points of bounded anticanonical height with respect to the Tamagawa measure associated with the metrization of $-K_X$.

---

REFERENCES


[1] J. Arthur and L. Clozel, Simple algebras, base change, and the advanced theory of the trace formula, *Annals of Mathematics Studies,* vol. **120,** Princeton University Press, 1989.

[2] V. V. Batyrev, Non-archimedean integrals and stringy Euler numbers of log terminal pairs, *Journ. of EMS* **1** (1999), 5–33.

[3] V. V. Batyrev and Yu. I. Manin, Sur le nombre de points rationnels de hauteur bornée des variétés algébriques, *Math. Ann.* **286** (1990), 27–43.

[4] V. V. Batyrev, Yu. Tschinkel, Rational points of bounded height on compactifications of anisotropic tori, *Intern. Math. Res. Notices* **12** (1995), 591–635.

[5] ———, Manin's conjecture for toric varieties, *Journ. Algebraic Geometry* **7** no. 1 (1998), 15–53.

[6] ———, Height zeta functions of toric varieties, *Algebraic Geometry* 5, (Manin's Festschrift), *Journ. Math. Sci.* **82** no. 1 (1998), 3220–3239.

[7] ———, Rational points on some Fano cubic bundles, *C.R. Acad. Sci., Paris* **323** Ser I (1996), 41–46.

[8] ———, Tamagawa numbers of polarized algebraic varieties, in Nombre et répartition des points de hauteur bornée Asterisque **251** (1998), 299–340.

[9] M. Brion and P. Polo, Large Schubert varieties, *Journ. Repr. Th.* **4** (2000), 97–126.

[10] L. Clozel and E. Ullmo, Equidistribution des points de Hecke, preprint (2001).

[11] A. Chambert-Loir and Yu. Tschinkel, On the distribution of points of bounded height on equivariant compactifications of vector groups, *Inventiones Math.* **48** no. 2 (2002), 421–452.

[12] J. Denef and F. Loeser, Germs of arcs on singular algebraic varieties and motivic integration, *Inventiones Math.* **135** (1999), 201–232.

[13] C. De Concini and C. Procesi, Complete symmetric varieties, LMN **996** (1983), 1–44.

[14] W. Duke, Z. Rudnick and P. Sarnak, Density of integer points on affine homogeneous spaces, *Duke Math. Journ.* **71** (1993), 181–209.

[15] A. Eskin and C. McMullen, Mixing, counting and equidistribution in Lie group, *Duke Math. Journ.* **71** (1993), 143–180.

[16] A. Eskin, S. Mozes and N. Shah, Unipotent flows and counting lattice points in homogeneous varieties, *Annals of Math.* **143** (1996), 253–299.

[17] J. Franke, Yu. I. Manin and Yu. Tschinkel, Rational points of bounded height on Fano varieties, *Inventiones Math.* **95** no. 2 (1989), 421–435.

[18] S. Gelbart and H. Jacquet, A relation between automorphic representations of GL(2) and GL(3), *Ann. Sci. École Norm. Sup.* **(4) 11** (1978), 471–542.

[19] R. Godement and H. Jacquet, Zeta functions of simple algebras, LNM **260** (1972).

[20] J. Harris and Yu. Tschinkel, Rational points on quartics, *Duke Math. Journ.* **104** (2000), 477–500.

[21] R. Howe, On a notion of rank for unitary representations of the classical groups, Harmonic analysis and group representations, 223–331, Liguori, Naples, (1982).

[22] H. Jacquet and R. Langlands Automorphic forms on GL(2), LNM **114,** *Springer-Verlag, Berlin-New York,* (1970).

[23] H. Jacquet and J. Shalika, On Euler products and the classification of automorphic forms, I and II, *Amer. Journ. of Math.* **103** no. 3 and 4 (1981), 499–558 and 777–815.

[24] H. Kim and F. Shahidi, Functorial products for $GL_2 \times GL_3$ and functorial symmetric cube for $GL_2$, *C. R. Acad. Sci. Paris Sér. I Math.* **331** no. 8 (2000), 599–604.

[25] ———, Cuspidality of symmetric powers with applications, *Duke Math. Journ.* **112** (2002), 177–197.

[26] J.-P. Labesse and J. Schwermer, On liftings and cusp cohomology of arithmetic groups, *Inventiones Math.* **83** no. 2 (1986), 383–401.

[27] R. Langlands and D. Ramakrishnan, The zeta functions of Picard modular surfaces, CRM Publications, (1992).

[28] W. Luo, Z. Rudnick and P. Sarnak, On the generalized Ramanujan conjecture for GL($n$), Automorphic forms, automorphic representations, and arithmetic (Fort Worth, TX, 1996), 301–310, *Proc. Sympos. Pure Math.,* **66,** Part 2, AMS, Providence, RI, (1999).

[29] I. G. Macdonald, Special functions on a group of $p$-adic type, Publ. Ramanujan *Inst. for Adv. Studies,* (1971).

[30] Moeglin and Waldspurger, Le spectre résiduel de GL($n$), *Ann. Sci. Ecole Norm. Sup.* **(4) 22** no. 4 (1989), 605–674.

[31] H. Oh, Tempered subgroups and representations with minimal decay of matrix coefficients, *Bull. Soc. Math.* France **126** no. 3 (1998), 355–380.

[32] ———, Uniform pointwise bounds for matrix coefficients of unitary representations and applications to Kazhdan's constants, *Duke Math. Journ.* **113** (2002), 133–192.

[33] E. Peyre, Hauteurs et mesures de Tamagawa sur les variétés de Fano, *Duke Math. Journ.* **79** (1995), 101–218.

[34] J. Rogawski, Automorphic representations of unitary groups in three variables, *Ann. of Math. Studies* **123,** Princeton Univ. Press, (1990).

[35] J.-P. Serre, Lectures on the Mordell-Weil theorem, Aspects of Mathematics, Vieweg, Braunschweig, (1997).

[36] S. Schanuel, On heights in number fields, *Bull. Amer. Math. Soc.* **70** (1964), 262–263.

[37] J. Shalika, R. Takloo-Bighash and Yu. Tschinkel, Distribution of rational points on equivariant compactifications of reductive groups I, math.princeton.edu/ytschink/publications.html, (2001)

[38] J. Shalika and Yu. Tschinkel, Height zeta functions of equivariant compactifications of the Heisenberg group, *Contribution to Automorphic Forms, Geometry, and Number Theory* (H. Hida, D. Ramakrishnan, and F. Shahidi, eds.), The Johns Hopkins University Press, Baltimore, MD, 2004, pp. 743–771.

[39] ———, Height zeta functions of equivariant compactifications of unipotent groups, in preparation.

[40] M. Strauch and Yu. Tschinkel, Height zeta functions of toric bundles over flag varieties, *Selecta Math.* **5(3)** (1999), 325–396.

[41] M. Tadić, Topology of unitary dual of non-Archimedean GL($n$), *Duke Math. Journ.* **55** (1987), 385–422.

CHAPTER 27

# HEIGHT ZETA FUNCTIONS OF EQUIVARIANT COMPACTIFICATIONS OF THE HEISENBERG GROUP

By Joseph A. Shalika and Yuri Tschinkel

---

*Abstract.* We study analytic properties of height zeta functions of equivariant compactifications of the Heisenberg group.

**Introduction.** Let $\mathrm{G} = \mathrm{G}_3$ be the three-dimensional Heisenberg group:

$$\mathrm{G} = \left\{ g = g(x, z, y) = \begin{pmatrix} 1 & x & z \\ 0 & 1 & y \\ 0 & 0 & 1 \end{pmatrix} \right\}.$$

Let $X$ be a projective equivariant compactification of G (for example, $X = \mathbb{P}^3$). Thus $X$ is a projective algebraic variety over $\mathbb{Q}$, equipped with a (left) action of G (and containing G as a dense Zariski open subset). Such varieties can be constructed as follows: consider a $\mathbb{Q}$-rational algebraic representation $\rho : \mathrm{G} \to \mathrm{PGL}_{n+1}$ and take $X \subset \mathbb{P}^n$ to be the Zariski closure of an orbit (with trivial stabilizer). This closure need not be smooth (or even normal). Applying G-equivariant resolution of singularities and passing to a desingularization, we may assume that $X$ is smooth and that the boundary $D = X \backslash \mathrm{G}$ consists of geometrically irreducible components $D = \cup_{\alpha \in \mathcal{A}} D_\alpha$, intersecting transversally. In this chapter, we will always assume that $X$ is a bi-equivariant compactification, that is, $X$ carries a left and right G-action, extending the left and right action of G on itself. Equivalently, $X$ is an equivariant compactification of the homogeneous space $\mathrm{G} \times \mathrm{G}/\mathrm{G}$.

Let $L$ be a very ample line bundle on $X$. It defines an embedding of $X$ into some projective space $\mathbb{P}^n$. Let $\mathcal{L} = (L, \|\cdot\|_{\mathbb{A}})$ be a (smooth adelic) metrization of $L$ and

$$H_{\mathcal{L}} : X(\mathbb{Q}) \to \mathbb{R}_{>0}$$

the associated (exponential) height. Concretely, fix a basis $\{f_j\}_{j=0,\dots,n}$ in the vector space of global sections of $L$ and put

$$H_{\mathcal{L}}(x) := \prod_p \max_j(|f_j(x)|_p) \cdot \left( \sum_{j=0}^n f_j(x)^2 \right)^{1/2}.$$

We are interested in the asymptotics of

$$N(B) = N(\mathcal{L}, B) := \{\gamma \in \mathrm{G}(\mathbb{Q}) \mid H_{\mathcal{L}}(\gamma) \le B\}$$

as $B \to \infty$.

The main result of this chapter is the determination of the asymptotic behavior of $N(B)$ for arbitrary bi-equivariant compactifications $X$ of G and arbitrary projective embeddings.

To describe this asymptotic behavior it is necessary to introduce some geometric notions. Denote by $\mathrm{Pic}(X)$ the Picard group of $X$. For smooth equivariant compactifications of unipotent groups, $\mathrm{Pic}(X)$ is freely generated by the classes of $D_\alpha$ (with $\alpha \in \mathcal{A}$). We will use these classes as a basis. In this basis, the (closed) cone of effective divisors $\Lambda_{\mathrm{eff}}(X) \subset \mathrm{Pic}(X)_{\mathbb{R}}$ consists of classes

$$[L] = (l_\alpha) = \sum_{\alpha \in \mathcal{A}} l_\alpha [D_\alpha] \in \mathrm{Pic}(X)_{\mathbb{R}},$$

with $l_\alpha \geq 0$ for all $\alpha$. Let $\mathcal{L} = (L, \|\cdot\|_{\mathbb{A}})$ be a metrized line bundle on $X$ such that its class $[L]$ is contained in the interior of $\Lambda_{\mathrm{eff}}(X)$. Conjecturally, at least for varieties with sufficiently positive anticanonical class, asymptotics of rational points of bounded height are related to the location of (the class of) $L$ in $\mathrm{Pic}(X)$ with respect to the anticanonical class $[-K_X] = \kappa = (\kappa_\alpha)$ and the cone $\Lambda_{\mathrm{eff}}(X)$ (see [4], [10], and [20]). In the special case of G-compactifications $X$ as above and $[L] = (l_\alpha)$, define:

- $a(L) := \inf\{a \mid a[L] + [K_X] \in \Lambda_{\mathrm{eff}}(X)\} = \max_\alpha(\kappa_\alpha / l_\alpha)$;
- $b(L) := \#\{\alpha \mid \kappa_\alpha = a(L) l_\alpha\}$;
- $\mathcal{C}(L) := \{\alpha \mid \kappa_\alpha \neq a(L) l_\alpha\}$;
- $c(L) := \prod_{\alpha \notin \mathcal{C}(L)} l_\alpha^{-1}$.

Let

$$\mathcal{Z}(s, \mathcal{L}) := \sum_{\gamma \in \mathrm{G}(\mathbb{Q})} H_{\mathcal{L}}(\gamma)^{-s},$$

be the height zeta function (the series converges a priori to a holomorphic function for ample $\mathcal{L}$ and $\Re(s) \gg 0$). The Tauberian theorems relate the asymptotics of $N(\mathcal{L}, B)$ to analytic properties of $\mathcal{Z}(s, \mathcal{L})$.

THEOREM. *Let $X$ be a smooth projective bi-equivariant compactification of the Heisenberg group* G *and $\mathcal{L} = (L, \|\cdot\|_{\mathbb{A}})$ a line bundle (equipped with a smooth adelic metrization) such that its class $[L] \in \mathrm{Pic}(X)$ is contained in the interior of the cone of effective divisors $\Lambda_{\mathrm{eff}}(X)$. Then*

$$\mathcal{Z}(s, \mathcal{L}) = \frac{c(L)\tau(\mathcal{L})}{(s - a(L))^{b(L)}} + \frac{h(s)}{(s - a(L))^{b(L)-1}},$$

*where $h(s)$ is a holomorphic function (for $\Re(s) > a(L) - \epsilon$, some $\epsilon > 0$) and $\tau(\mathcal{L})$ is a positive real number. Consequently,*

$$N(\mathcal{L}, B) \sim \frac{c(L)\tau(\mathcal{L})}{a(L)(b(L) - 1)!} B^{a(L)} \log(B)^{b(L)-1}$$

*as $B \to \infty$.*

*Remark* 0.1. The constant $\tau(-\mathcal{K}_X)$ is the Tamagawa number associated to the metrization of the anticanonical line bundle (see [20]). For arbitrary polarizations $\tau(\mathcal{L})$ has been defined in [4].

This chapter is structured as follows: in Section 1 we describe the relevant geometric invariants of equivariant compactifications of unipotent groups. In Section 2 we introduce the height pairing

$$H = \prod_p H_p \cdot H_\infty \; : \; \mathrm{Pic}(X)_{\mathbb{C}} \times \mathrm{G}(\mathbb{A}) \to \mathbb{C}$$

between the complexified Picard group and the adelic points of G, generalizing the usual height, and the height zeta function

$$\mathcal{Z}(\mathbf{s}, g) := \sum_{\gamma \in \mathrm{G}(\mathbb{Q})} H(\mathbf{s}, \gamma g)^{-1}. \tag{0.1}$$

By the projectivity of $X$, the series converges to a function that is continuous and bounded in $g$ and holomorphic in $\mathbf{s}$, for $\Re(\mathbf{s})$ contained in some (shifted) cone $\Lambda \subset \mathrm{Pic}(X)_{\mathbb{R}}$. Our goal is to obtain a meromorphic continuation of $\mathcal{Z}(\mathbf{s}, g)$ to the tube domain T over an open neighborhood of $[-K_X] = \kappa \in \mathrm{Pic}(X)_{\mathbb{R}}$ and to identify the poles.

The bi-equivariance of $X$ implies that $H$ is invariant under the action *on both sides* of a compact open subgroup K of the finite adeles $\mathrm{G}(\mathbb{A}_{\mathrm{fin}})$. Moreover, $H_\infty$ is smooth. We observe that

$$\mathcal{Z} \in \mathsf{L}^2(\mathrm{G}(\mathbb{Q})\backslash\mathrm{G}(\mathbb{A}))^{\mathrm{K}}.$$

Next, we have, for $\Re(\mathbf{s})$ contained in some shifted cone in $\mathrm{Pic}(X)_{\mathbb{R}}$, an identity in $\mathsf{L}^2(\mathrm{G}(\mathbb{Q})\backslash\mathrm{G}(\mathbb{A}))$ (Fourier expansion):

$$\mathcal{Z}(\mathbf{s}, g) = \sum_{\varrho} \mathcal{Z}_\varrho(\mathbf{s}, g), \tag{0.2}$$

where the sum is over all irreducible unitary representations $(\varrho, \mathcal{H}_\varrho)$ of $\mathrm{G}(\mathbb{A})$, occuring in the right regular representation of $\mathrm{G}(\mathbb{A})$ in $\mathsf{L}^2(\mathrm{G}(\mathbb{Q})\backslash\mathrm{G}(\mathbb{A}))$ and having K-fixed vectors. We recall the relevant results from representation theory in Section 3.

We will establish the above identity as an identity of continuous functions by analyzing the individual terms on the right. Thus we need to use the (well-known) theory of irreducible unitary representations of the Heisenberg group. We will see that for $L = -K_X$ the pole of highest order of the height zeta function is supplied by the trivial representation. This need not be the case for other line bundles. Depending on the geometry of $X$, it can happen that infinitely many nontrivial representations contribute to the leading pole of $\mathcal{Z}(s, \mathcal{L})$. In such cases the coefficient at the pole of highest order is an infinite (convergent) sum of Euler products.

To analyze the contributions in (0.2) from the various representations, we need to compute local height integrals. For example, for the trivial representation, we need to compute the integral

$$\int_{\mathrm{G}(\mathbb{Q}_p)} H_p(\mathbf{s}, g_p)^{-1} dg_p$$

for almost all $p$ (see Section 4). This has been done in [6] for equivariant compactifications of additive groups $\mathbf{G}_a^n$; the same approach applies here. We regard the height integrals as geometric versions of Igusa's integrals. They are closely related to "motivic" integrals of Batyrev, Kontsevich, Denef, and Loeser (see [9], [14], and [18]).

The above integral is in fact equal to:

$$p^{-\dim(X)}\left(\sum_{A\subseteq\mathcal{A}} \#D_A^0(\mathbf{F}_p)\prod_{\alpha\in A}\frac{p-1}{p^{s_\alpha-\kappa_\alpha+1}-1}\right), \tag{0.3}$$

where

$$D_\emptyset := \mathrm{G}, \quad D_A := \cap_{\alpha\in A} D_\alpha, \quad D_A^0 := D_A \setminus \cup_{A'\supsetneq A} D_{A'},$$

and $\mathbf{F}_p$ is the finite field $\mathbb{Z}/p\mathbb{Z}$. The resulting Euler product has a pole of order rk Pic$(X)$ at $\mathbf{s} = \kappa$ and also the expected leading coefficient at this pole.

The bi-K-invariance of the height insures us that the trivial representation is "isolated" (c.f. especially Proposition 4.9). Using "motivic" integration as above, we prove that *each* of the terms on the right side in (0.2) admits a meromorphic continuation. We will identify the poles of $\mathcal{Z}_\varrho$ for nontrivial representations: for $\mathbf{s} \in \mathrm{T}$, they are contained in the real hyperplanes $s_\alpha = \kappa_\alpha$, and the order of the pole at $\mathbf{s} = \kappa$ is strictly smaller than rk Pic$(X)$. Finally, it will suffice to prove the convergence of the series (0.2), for $\mathbf{s}$ in the appropriate domain. This is done in Section 4.

This chapter is part of a program initiated in [10] to relate asymptotics of rational points of bounded height to geometric invariants. It continues the work of Chambert-Loir and the second author on compactifications of additive groups [6]. Many statements are direct generalizations from that paper. Here we explore the interplay between the theory of infinite-dimensional representations of adelic groups and the theory of height zeta functions of algebraic varieties. The main theorem holds for bi-equivariant compactifications of arbitrary unipotent groups. We decided to explain in detail, in a somewhat expository fashion, our approach in the simplest possible case of the Heisenberg group over $\mathbb{Q}$ and to postpone the treatment of the general case to a subsequent publication. We have also included the example of $\mathbb{P}^3$ in which most of the technicalities are absent.

*Acknowledgments.* The second author was partially supported by the NSA, NSF and the Clay Foundation.

## 1. Geometry.

*Notations* 1.1. Let $X$ be a smooth projective algebraic variety. We denote by $\mathrm{Pic}(X)$ the Picard group, by $\Lambda_{\mathrm{eff}}(X)$ the (closed) cone of effective divisors and by $K_X$ the canonical class of $X$. If $X$ admits an action by a group G, we write $\mathrm{Pic}^{\mathrm{G}}(X)$ for the group of (classes of) G-linearized line bundles on $X$.

*Definition* 1.2. Let $X$ be a smooth projective algebraic variety. Assume that $\Lambda_{\mathrm{eff}}(X)$ is a finitely generated polyhedral cone. Let $L$ be a line bundle such that its class $[L]$ is contained in the interior of $\Lambda_{\mathrm{eff}}(X)$. Define

$$a(L) = \inf\{a \mid a[L] + [K_X] \in \Lambda_{\mathrm{eff}}(X)\}$$

and $b(L)$ as the codimension of the face of $\Lambda_{\mathrm{eff}}(X)$ containing $a(L)[L] + [K_X]$.

*Notations* 1.3. Let G be a linear algebraic group over a number field $F$. An algebraic variety $X$ (over $F$) will be called a *good* compactification of G if:

- $X$ is smooth and projective;
- $X$ contains G as a dense Zariski open subset and the action of G on itself (by left translations) extends to $X$;
- the boundary $X \setminus \mathrm{G}$ is a union of smooth geometrically irreducible divisors intersecting transversally (a divisor with strict normal crossings).

*Remark* 1.4. Equivariant resolution of singularities (over a field of characteristic zero) implies that for *any* equivariant compactification $X$ there exists an equivariant desingularization (a composition of equivariant blowups) $\rho : \tilde{X} \to X$, such that $\tilde{X}$ is a good compactification. By the functoriality of heights, the counting problem for a metrized line bundle $\mathcal{L}$ on $X$ can then be transferred to a counting problem for $\rho^*(\mathcal{L})$ on $\tilde{X}$. Thus it suffices to prove the theorem for good compactifications (the answer, of course, does not depend on the chosen desingularization).

Proposition 1.5. *Let $X$ be a good compactification of a unipotent algebraic group* G. *Let $D := X \setminus \mathrm{G}$ be the boundary and $\{D_\alpha\}_{\alpha \in \mathcal{A}}$ the set of its irreducible components. Then:*

- $\mathrm{Pic}^{\mathrm{G}}(X)_{\mathbb{Q}} = \mathrm{Pic}\,(X)_{\mathbb{Q}}$;
- $\mathrm{Pic}(X)$ *is freely generated by the classes* $[D_\alpha]$;
- $\Lambda_{\mathrm{eff}}(X) = \oplus_\alpha \mathbb{R}_{\geq 0}[D_\alpha]$;
- $[-K_X] = \sum_\alpha \kappa_\alpha [D_\alpha]$ *with* $\kappa_\alpha \geq 2$ *for all* $\alpha \in \mathcal{A}$.

*Proof.* Analogous to the proofs in Section 2 of [12]. In particular, it suffices to assume that $X$ carries only a one-sided action of G. □

*Notations* 1.6. Introduce coordinates on $\mathrm{Pic}(X)$ using the basis $\{D_\alpha\}_{\alpha \in \mathcal{A}}$: a vector $\mathbf{s} = (s_\alpha)$ corresponds to $\sum_\alpha s_\alpha D_\alpha$.

COROLLARY 1.7. *The divisor of every nonconstant function $f \in F[\mathrm{G}]$ can be written as*

$$\mathrm{div}(f) = E(f) - \sum_{\alpha} d_\alpha(f) D_\alpha,$$

*where $E(f)$ is the unique irreducible component of $\{f = 0\}$ in G and $d_\alpha(f) \geq 0$ for all $\alpha$. Moreover, there is at least one $\alpha \in \mathcal{A}$ such that $d_\alpha(f) > 0$.*

## 2. Height zeta function.

*Notations* 2.1. For a number field $F$, we denote by $\mathrm{Val}(F)$ the set of all places of $F$, by $S_\infty$ the set of archimedean and by $S_{\mathrm{fin}}$ the set of nonarchimedean places. For any finite set $S$ of places containing $S_\infty$, we denote by $\mathfrak{o}_S$ the ring of $S$-integers. We denote by $\mathbb{A}$ (resp. $\mathbb{A}_{\mathrm{fin}}$) the ring of adeles (resp. finite adeles).

*Definition* 2.2. Let $X$ be a smooth projective algebraic variety over a number field $F$. A smooth adelic metrization of a line bundle $L$ on $X$ is a family $\| \cdot \|_{\mathbb{A}}$ of $v$-adic norms $\| \cdot \|_v$ on $L \otimes_F F_v$, for all $v \in \mathrm{Val}(F)$, such that:

- for $v \in S_\infty$, the norm $\| \cdot \|_v$ is $\mathrm{C}^\infty$;
- for $v \in S_{\mathrm{fin}}$, the norm of every local section of $L$ is locally constant in the $v$-adic topology;
- there exist a finite set $S \subset \mathrm{Val}(F)$, a flat projective scheme (an integral model) $\mathcal{X}$ over $\mathrm{Spec}(\mathfrak{o}_S)$ with generic fiber $X$ together with a line bundle $\mathcal{L}$ on $\mathcal{X}$, such that for all $v \notin S$, the $v$-adic metric is given by the integral model.

PROPOSITION 2.3. *Let* G *be a unipotent algebraic group defined over a number field $F$ and $X$ a good bi-equivariant compactification of* G. *Then there exists a height pairing*

$$H = \prod_{v \in \mathrm{Val}(F)} H_v \, : \, \mathrm{Pic}(X)_{\mathbb{C}} \times \mathrm{G}(\mathbb{A}) \to \mathbb{C}$$

*such that:*

- *for all $[L] \in \mathrm{Pic}(X)$, the restriction of $H$ to $[L] \times \mathrm{G}(F)$ is a height corresponding to some smooth adelic metrization of $L$;*
- *the pairing is exponential in the* $\mathrm{Pic}(X)$ *component:*

$$H_v(\mathbf{s} + \mathbf{s}', g) = H_v(\mathbf{s}, g) H_v(\mathbf{s}', g),$$

  *for all $\mathbf{s}, \mathbf{s}' \in \mathrm{Pic}(X)_{\mathbb{C}}$, all $g \in \mathrm{G}(\mathbb{A})$ and all $v \in \mathrm{Val}(F)$;*
- *there exists a compact open subgroup (depending on $H$)*

$$\mathrm{K} = \mathrm{K}(H) = \prod_v \mathrm{K}_v \subset \mathrm{G}(\mathbb{A}_{\mathrm{fin}})$$

  *such that, for all $v \in S_{\mathrm{fin}}$, one has $H_v(\mathbf{s}, kgk') = H_v(\mathbf{s}, g)$ for all $\mathbf{s} \in \mathrm{Pic}(X)_{\mathbb{C}}$, $k, k' \in \mathrm{K}_v$, and $g \in \mathrm{G}(F_v)$.*

*Proof.* For $\mathrm{G} = \mathbf{G}_a^n$ the proposition is proved in [6], Lemma 3.2. The same proof applies to any unipotent group. □

*Notations* 2.4. For $\delta \in \mathbb{R}$, we denote by $\mathsf{T}_\delta \subset \mathrm{Pic}(X)_{\mathbb{C}}$ the tube domain $\Re(s_\alpha) - \kappa_\alpha > \delta$ (for all $\alpha \in \mathcal{A}$).

*Definition* 2.5. The height zeta function on $\mathrm{Pic}(X)_{\mathbb{C}} \times \mathrm{G}(\mathbb{A})$ is defined as

$$\mathcal{Z}(\mathbf{s}, g) = \sum_{\gamma \in \mathrm{G}(F)} H(\mathbf{s}, \gamma g)^{-1}.$$

PROPOSITION 2.6. *There exists a $\delta > 0$ such that, for all $\mathbf{s} \in \mathsf{T}_\delta$ and all $g \in \mathrm{G}(\mathbb{A})$, the series defining the height zeta function $\mathcal{Z}(\mathbf{s}, g)$ converges normally (for $g$ and $\mathbf{s}$ contained in compacts in $\mathrm{G}(\mathbb{A})$, resp. $\mathsf{T}_\delta$) to a function that is holomorphic in $\mathbf{s}$ and continuous in $g$.*

*Proof.* The proof is essentially analogous to the proof of Proposition 4.5 in [6] (and follows from the projectivity of $X$). □

COROLLARY 2.7. *For $\mathbf{s} \in \mathsf{T}_\delta$, one has an identity in $\mathsf{L}^2(\mathrm{G}(F)\backslash\mathrm{G}(\mathbb{A}))$, as above:*

$$\mathcal{Z}(\mathbf{s}, g) = \sum_{\varrho} \mathcal{Z}_\varrho(\mathbf{s}, g). \tag{2.1}$$

*The sum is over all irreducible unitary representations $\varrho$ of $\mathrm{G}(\mathbb{A})$ occuring $\mathsf{L}^2(\mathrm{G}(F)\backslash\mathrm{G}(\mathbb{A}))$ and having a $\mathrm{K}$-fixed vector (cf. Proposition 3.3).*

## 3. Representations.

**3.1.** From now on, for the sake of simplicity, we suppose $F = \mathbb{Q}$. Denote by $\mathrm{Z} = \mathbf{G}_a$ the one-dimensional center and by $\mathrm{G}^{\mathrm{ab}} = \mathrm{G}/\mathrm{Z} = \mathbf{G}_a^2$ the abelianization of G. Let $\mathrm{U} \subset \mathrm{G}$ be the subgroup

$$\mathrm{U} := \{u \in \mathrm{G} \mid u = (0, z, y)\}$$

and

$$\mathrm{W} := \{w \in \mathrm{G} \mid w = (x, 0, 0)\}.$$

We have $\mathrm{G} = \mathrm{W} \cdot \mathrm{U} = \mathrm{U} \cdot \mathrm{W}$. We may assume that the compact open subgroup

$$\mathrm{K} = \prod_p \mathrm{K}_p \subset \mathrm{G}(\mathbb{A}_{\mathrm{fin}})$$

of Proposition 2.3 is given by

$$\mathrm{K} = \prod_{p \notin S_H} \mathrm{G}(\mathbb{Z}_p) \cdot \prod_{p \in S_H} \mathrm{G}(p^{n_p}\mathbb{Z}_p), \tag{3.1}$$

where $S_H$ is a finite set of primes and the $n_p$ are positive integers. We denote by $\mathrm{K}^{\mathrm{ab}}$, $\mathrm{K}_{\mathrm{Z}}$, etc., the corresponding compact subgroups of the (finite) adeles of $\mathrm{G}^{\mathrm{ab}}$, Z, U, W, respectively, and put

$$\mathsf{n}(\mathrm{K}) = \prod_{p \in S_H} p^{n_p}.$$

We denote by $dg = \prod_p dg_p \cdot dg_\infty$ the Haar measure on $\mathrm{G}(\mathbb{A})$, where we have set $dg_p = dx_p dy_p dz_p$ with the normalization $\int_{\mathbb{Z}_p} dx_p = 1$, etc., (similarly at the real place). We write $du_p = dz_p dy_p$ (resp. $du_\infty$, $du$) for the Haar measure on $\mathrm{U}(\mathbb{Q}_p)$ (resp. $\mathrm{U}(\mathbb{R})$, $\mathrm{U}(\mathbb{A})$). We let $dk_p$ be the Haar measure on $\mathrm{K}_p$ obtained by restriction of $dg_p$ to $\mathrm{K}_p$. Further, our normalization of measures implies that $\int_{\mathrm{K}_p} dk_p = 1$. As usual, a choice of a measure on the local (or global) points of G and of a subgroup $\mathrm{H} \subset \mathrm{G}$ determines a unique measure on the local (resp. global) points of the homogeneous space G/H.

LEMMA 3.2. *One has:*

- $\mathrm{G}(\mathbb{Z}_p) = (\mathrm{G}(\mathbb{Z}_p) \cap \mathrm{U}(\mathbb{Q}_p)) \cdot (\mathrm{G}(\mathbb{Z}_p) \cap \mathrm{W}(\mathbb{Q}_p))$;
- $\mathrm{U}(\mathbb{Q}_p) \cdot \mathrm{W}(\mathbb{Z}_p)$ *is a subgroup of* $\mathrm{G}(\mathbb{Q}_p)$;
- $\mathrm{G}(\mathbb{A}) = \mathrm{G}(\mathbb{Q}) \cdot \mathrm{G}(\mathbb{R}) \cdot \mathrm{K}$;
- *there exists a subgroup* $\Gamma \subset \mathrm{G}(\mathbb{Z})$ *(of finite index) such that*

$$\mathrm{G}(\mathbb{Q})\backslash\mathrm{G}(\mathbb{A})/\mathrm{K} = \Gamma\backslash\mathrm{G}(\mathbb{R});$$

- *the quotient* $\Gamma\backslash\mathrm{G}(\mathbb{R})$ *is compact.*

These statements are well known and easily verified.

We now recall the well known representation theory of the Heisenberg group in an adele setting [13]. Denote by $\varrho$ the right regular representation of $\mathrm{G}(\mathbb{A})$ on the Hilbert space

$$\mathcal{H} := \mathsf{L}^2(\mathrm{G}(\mathbb{Q})\backslash\mathrm{G}(\mathbb{A})).$$

Consider the action of the compact group $\mathrm{Z}(\mathbb{A})/\mathrm{Z}(\mathbb{Q})$ on $\mathcal{H}$ (recall that $\mathrm{Z} = \mathbf{G}_a$). By the Peter-Weyl theorem, we obtain a decomposition

$$\mathcal{H} = \oplus \mathcal{H}_\psi$$

and corresponding representations $(\varrho_\psi, \mathcal{H}_\psi)$ of $\mathrm{G}(\mathbb{A})$. Here

$$\mathcal{H}_\psi := \{\varphi \in \mathcal{H} \mid \varrho(z)(\varphi)(g) = \psi(z)\varphi(g)\}$$

and $\psi$ runs over the set of (unitary) characters of $\mathrm{Z}(\mathbb{A})$, which are trivial on $\mathrm{Z}(\mathbb{Q})$. For nontrivial $\psi$, the corresponding representation $(\varrho_\psi, \mathcal{H}_\psi)$ of $\mathrm{G}(\mathbb{A})$ is nontrivial, irreducible and unitary. On the other hand, when $\psi$ is the trivial character, the corresponding representation $\varrho_0$ decomposes further as a direct sum of one-dimensional representations $\varrho_\eta$:

$$\mathcal{H}_0 = \oplus_\eta \mathcal{H}_\eta.$$

Here $\eta$ runs once over all (unitary) characters of the group $\mathrm{G}^{\mathrm{ab}}(\mathbb{Q})\backslash\mathrm{G}^{\mathrm{ab}}(\mathbb{A})$. It is convenient to consider $\eta$ as a function on $\mathrm{G}(\mathbb{A})$, trivial on the $\mathrm{Z}(\mathbb{A})$-cosets. Precisely, let $\psi_1 = \prod_p \psi_{1,p} \cdot \psi_{1,\infty}$ be the Tate-character (which has exponent zero at each finite prime, see [27] and [28]). For $\mathbf{a} = (a_1, a_2) \in \mathbb{A} \oplus \mathbb{A}$, consider the corresponding linear form on

$$\mathrm{G}^{\mathrm{ab}}(\mathbb{A}) = \mathbb{A} \oplus \mathbb{A}$$

given by

$$g(x, z, y) \mapsto a_1 x + a_2 y,$$

and denote by $\eta = \eta_{\mathbf{a}}$ ($\mathbf{a} = (a_1, a_2)$) the corresponding adelic character

$$\eta \, : \, g(x, z, y) \mapsto \psi_1(a_1 x + a_2 y)$$

of $\mathrm{G}(\mathbb{A})$. For $a \in \mathbb{A}$, we will denote by $\psi_a$ the adelic character of $\mathrm{Z}(\mathbb{A})$ given by

$$z \mapsto \psi_1(az).$$

As in Section 2, the starting point of our analysis of the height zeta function is the spectral decomposition of $\mathcal{H}$. A more detailed version of Corollary 2.7 is the following proposition.

PROPOSITION 3.3. *There exists a $\delta > 0$ such that, for all $\mathbf{s} \in \mathsf{T}_\delta$, one has an identity of $\mathsf{L}^2$-functions*

$$\mathcal{Z}(\mathbf{s}, g) = \mathcal{Z}_0(\mathbf{s}, g) + \mathcal{Z}_1(\mathbf{s}, g) + \mathcal{Z}_2(\mathbf{s}, g), \tag{3.2}$$

*where*

$$\mathcal{Z}_0(\mathbf{s}, \mathit{id}) = \int_{\mathrm{G}(\mathbb{A})} H(\mathbf{s}, g)^{-1} dg, \tag{3.3}$$

$$\mathcal{Z}_1(\mathbf{s}, g) = \sum_\eta \eta(g) \cdot \mathcal{Z}(\mathbf{s}, \eta), \tag{3.4}$$

*and*

$$\mathcal{Z}_2(\mathbf{s}, g) = \sum_\psi \sum_{\omega^\psi} \omega^\psi(g) \cdot \mathcal{Z}(\mathbf{s}, \omega^\psi). \tag{3.5}$$

*Here we have set*

$$\mathcal{Z}(\mathbf{s},\eta) := \langle \mathcal{Z}(\mathbf{s},\cdot),\eta\rangle = \int_{\mathrm{G}(\mathbb{A})} H(\mathbf{s},g)^{-1}\overline{\eta}(g)dg,$$

$$\mathcal{Z}(\mathbf{s},\omega^{\psi}) := \langle \mathcal{Z}(s,g),\omega^{\psi}\rangle = \int_{\mathrm{G}(\mathbb{Q})\backslash\mathrm{G}(\mathbb{A})} \mathcal{Z}(\mathbf{s},g)\overline{\omega}^{\psi}(g)dg$$

$$= \int_{\mathrm{G}(\mathbb{A})} H(\mathbf{s},g)^{-1}\overline{\omega}^{\psi}(g)dg,$$

*$\eta$ ranges over all nontrivial characters of*

$$\mathrm{G}^{\mathrm{ab}}(\mathbb{Q})\cdot\mathrm{K}^{\mathrm{ab}}\backslash\mathrm{G}^{\mathrm{ab}}(\mathbb{A}),$$

*$\psi$ ranges over all nontrivial characters of*

$$\mathrm{Z}(\mathbb{Q})\cdot\mathrm{K}_{\mathrm{Z}}\backslash\mathrm{Z}(\mathbb{A}),$$

*and $\omega^{\psi}$ ranges over a fixed orthonormal basis of $\mathcal{H}_{\psi}^{\mathrm{K}}$ (for each $\psi$).*

*In particular, for $\eta=\eta_{\mathbf{a}}$ and $\psi=\psi_a$ occuring in this decomposition, we have*

$$a_1,a_2,a\in\frac{1}{\mathsf{n}(\mathrm{K})}\mathbb{Z}.$$

*Proof.* We use the (right) K-invariance of the height for the last statement (for $\eta$). For $\psi$ see also Lemma 3.11 as well as Proposition 2.6. □

*Remark* 3.4. The desired meromorphic properties of $\mathcal{Z}_0$ and $\mathcal{Z}_1$ have, in fact, already been established in [6]. The height integrals are computed as in the abelian case, and the convergence of the series $\mathcal{Z}_1$ is proved in the same way as in [6]. In particular, (3.4) is an identity of continuous functions. The novelty here is the treatment of $\mathcal{Z}_2$.

We now proceed to describe the various standard models of infinite-dimensional representations of the Heisenberg group.

**3.5. Locally.** Let $\psi=\psi_p$ (resp. $\psi=\psi_\infty$) be a *local* nontrivial character of $\mathbb{Q}_p$ (resp. $\mathbb{R}$). Extend $\psi$ to $\mathrm{U}(\mathbb{Q}_p)$ by setting

$$\psi((0,z,y))=\psi(z).$$

The one-dimensional representation of $\mathrm{U}(\mathbb{Q}_p)$ thus obtained induces a representation $\pi_\psi=\pi_{\psi,p}$ of $\mathrm{G}(\mathbb{Q}_p)$. The representation $\pi_\psi$ acts on the Hilbert space of measurable functions

$$\phi:\mathrm{G}(\mathbb{Q}_p)\to\mathbb{C},$$

which satisfy the conditions:

- $\phi(ug) = \psi(u)\phi(g)$, for all $u \in \mathrm{U}(\mathbb{Q}_p)$ and $g \in \mathrm{G}(\mathbb{Q}_p)$;
- $\|\phi\|^2 := \int_{\mathrm{U}(\mathbb{Q}_p)\backslash\mathrm{G}(\mathbb{Q}_p)} |\phi(g)|^2 dg < \infty$.

The action is given by

$$\pi_\psi(g')\phi(g) = \phi(gg'), \quad g' \in \mathrm{G}(\mathbb{Q}_p).$$

On the other hand, we have a representation $\pi'_\psi = \pi'_{\psi,p}$ (the oscillator representation) on

$$\mathsf{L}^2(\mathrm{W}(\mathbb{Q}_p)) = \mathsf{L}^2(\mathbb{Q}_p),$$

where the action of $\mathrm{G}(\mathbb{Q}_p)$ on a function $\varphi \in \mathsf{L}^2(\mathbb{Q}_p)$ is given by

$$\pi'_\psi(g(x',0,0))\varphi(x) = \varphi(x+x') \tag{3.6}$$

$$\pi'_\psi(g(0,0,y))\varphi(x) = \psi(y\cdot x)\varphi(x)$$

$$\pi'_\psi(g(0,z,0))\varphi(x) = \psi(z)\varphi(x).$$

It is easy to see that the representations $\pi_\psi$ and $\pi'_\psi$ are unitarily equivalent. We will identify the unitary representations $\pi_\psi$ and $\pi'_\psi$ in what follows.

**Globally.** In the adelic situation, to each nontrivial character $\psi$ of $\mathrm{Z}(\mathbb{A})$ we can associate a representation $\pi_\psi$ of $\mathrm{G}(\mathbb{A})$, where $\pi_\psi = \otimes_p \pi_{\psi,p} \otimes \pi_{\psi,\infty}$, and the action on $\mathsf{L}^2(\mathrm{U}(\mathbb{A})\backslash\mathrm{G}(\mathbb{A})) = \mathsf{L}^2(\mathbb{A})$ is given by the formulas (3.6) (with $\psi_p$ replaced by $\psi$). The representations $\pi_\psi$ and $\varrho_\psi$ are equivalent irreducible unitary representations of $\mathrm{G}(\mathbb{A})$. We will recall the explicit intertwining map between $\pi_\psi$ and $\varrho_\psi$ (c.f. Lemma 3.11).

We also recall that the space $\mathcal{S}(\mathbb{A}) \subset \mathsf{L}^2(\mathbb{A})$ of Schwartz-Bruhat functions coincides with the space of smooth vectors of $\pi_\psi$ (for the real place, see the appendix in [7]) and note that $\mathsf{L}^2(\mathbb{Q}_p)^{\mathrm{K}_p} = \mathcal{S}(\mathbb{Q}_p)^{\mathrm{K}_p}$.

For a character $\psi(z) = \psi_\infty(z) = e^{2\pi i a z}$ (with $a \neq 0$), consider the following operators on the subspace of Schwartz functions $\mathcal{S}(\mathbb{R}) \subset \mathsf{L}^2(\mathbb{R})$:

$$\mathfrak{d}^+_\psi\varphi(x) = \frac{d}{dx}\varphi(x)$$

$$\mathfrak{d}^-_\psi\varphi(x) = 2\pi i a x\varphi(x)$$

$$\Delta_\psi = (\mathfrak{d}^+_\psi)^2 + (\mathfrak{d}^-_\psi)^2.$$

We have

$$\Delta_\psi\varphi(x) = \varphi''(x) - (2\pi a x)^2\varphi(x)$$

(harmonic oscillator). The eigenvalues of $\Delta_\psi$ are given by

$$\lambda^\psi_n = -2\pi(2n+1)|a|$$

(with $n = 0, 1, 2, \dots$). They have multiplicity one. Denote by $h_n^\psi(x)$ the $n$-th Hermite polynomial:

$$h_0^\psi(x) = 1$$
$$h_1^\psi(x) = 4\pi|a|x$$
$$h_2^\psi(x) = -4\pi|a|(1 - 4\pi|a|x^2)$$

and, in general,

$$\frac{d^n}{dx^n} e^{-2\pi|a|x^2} = (-1)^n h_n^\psi(x) e^{-2\pi|a|x^2}.$$

The (essentially unique) eigenfunction $\varphi_n^\psi$ corresponding to $\lambda_n^\psi$ is given by

$$\varphi_n^\psi := c_n e^{-\pi|a|x^2} h_n^\psi(x).$$

Here we choose the constants $c_n$, so that the $\mathsf{L}^2$-norm of $\varphi_n^\psi$ is 1.

LEMMA 3.6. *The set* $\mathcal{B}_\infty(\pi'_\psi) := \{\varphi_n^\psi\}$ *is a complete orthonormal basis of* $\mathsf{L}^2(\mathbb{R})$.

*Proof.* For details see, for example, [5], Chapter 13, or [8]. □

**3.7.**

*Notations* 3.8. For $\eta = \eta_{\mathbf{a}}$ with $\mathbf{a} = (a_1, a_2)$ and $a_1, a_2 \in \frac{1}{\mathsf{n}(\mathrm{K})}\mathbb{Z}$, denote by $S_\eta$ the set of primes $p$ dividing either $\mathsf{n}(\mathrm{K})a_1$ or $\mathsf{n}(\mathrm{K})a_2$. Similarly, for $\psi = \psi_a$ with $a \in \frac{1}{\mathsf{n}(\mathrm{K})}\mathbb{Z}$, denote by $S_\psi$ the set of primes dividing $\mathsf{n}(\mathrm{K})a$.

LEMMA 3.9. *Let* $\psi = \psi_a$ *be a nontrivial character of* $\mathrm{Z}(\mathbb{Q})\backslash\mathrm{Z}(\mathbb{A})$ *and* $\varrho_\psi = \otimes_p \varrho_{\psi,p} \otimes \varrho_{\psi,\infty}$ *the corresponding infinite-dimensional automorphic representation. Suppose* $\varrho_\psi$ *contains a* K*-fixed vector (for* K *as in (3.1)). Then:*

- $a \in \frac{1}{\mathsf{n}(\mathrm{K})}\mathbb{Z}$ *(for* $\mathsf{n}(\mathrm{K}) = \prod_{p \in S_H} p^{n_p}$*);*
- $\dim \varrho_{\psi,p}^{\mathrm{K}_p} = 1$ *for* $p \notin S_\psi$*;*
- $\dim \varrho_{\psi,p}^{\mathrm{K}_p} = |\mathsf{n}(\mathrm{K})^2 a|_p^{-1}$ *for* $p \in S_\psi$*, provided* $p^{n_p} \cdot \mathsf{n}(\mathrm{K}) \in \mathbb{Z}_p$.

*Proof.* We need only use the explicit form of the representation $\pi_{\psi,p}$ given in (3.6). Suppose first that $\pi_{\psi,p}$ has a nonzero $\mathrm{K}_p$-fixed vector $\varphi$. Taking $z \in p^{n_p} \cdot \mathbb{Z}_p$, we get

$$\psi_p(p^{n_p} r) = \psi_{1,p}(a p^{n_p} r) = 1$$

for all $r \in \mathbb{Z}_p$. Since the exponent of $\psi_{1,p}$ is zero, we have

$$a \cdot p^{n_p} \in \mathbb{Z}_p,$$

from which the first assertion follows.

Let us assume then that $p^{n_p} \cdot \mathsf{n}(\mathrm{K}) \in \mathbb{Z}_p$. Then the space of $\mathrm{K}_p$-fixed vectors $\varphi$ in $\mathsf{L}^2(\mathbb{Q}_p)$ is precisely the set of $\varphi$ satisfying

- $\varphi(u + p^{n_p} r_1) = \varphi(u)$;
- $\varphi(u) = \psi(p^{n_p} r_2 u)\varphi(u)$

for all $r_1, r_2 \in \mathbb{Z}_p$, $u \in \mathbb{Q}_p$. The first identity implies that $\varphi$ is a continuous function and the second that $\mathrm{Supp}(\varphi) \subset a^{-1} p^{-n_p} \cdot \mathbb{Z}_p$. The second and the third assertions of the lemma follow at once. □

*Notations* 3.10. Let $V_{\psi,p}$ be the space of the induced representation of $\pi_{\psi,p}$. Denote by $V^{\infty}_{\psi,p}$ the space of smooth vectors in $V_{\psi,p}$. Thus $V^{\infty}_{\psi,p}$ is the set of all $v \in V_{\psi,p}$ fixed by some open compact subgroup of $\mathrm{G}(\mathbb{Q}_p)$. Note that $V^{\infty}_{\psi,p}$ is stable under the action of $\mathrm{G}(\mathbb{Q}_p)$. Note also that in the explicit realization of $\pi_{\psi,p}$ given in (3.6), $\mathsf{L}^2(\mathbb{Q}_p)^{\infty} = \mathcal{S}(\mathbb{Q}_p)$ (see the proof of Lemma 3.9)

For $\varphi \in \mathcal{S}(\mathbb{A})$ define the theta-distribution

$$\Theta(\varphi) := \sum_{x \in \mathbb{Q}} \varphi(x).$$

Clearly, $\Theta$ is a $\mathrm{G}(\mathbb{Q})$-invariant linear functional on $\mathcal{S}(\mathbb{A})$. This gives a map

$$\begin{aligned} j_\psi &: \mathcal{S}(\mathbb{A}) \to \mathsf{L}^2(\mathrm{G}(\mathbb{Q})\backslash \mathrm{G}(\mathbb{A})) \\ j_\psi(\varphi)(g) &= \Theta(\pi_\psi(g)\varphi). \end{aligned}$$

LEMMA 3.11. *The map $j_\psi$ extends to an isometry*

$$j_\psi : \mathsf{L}^2(\mathbb{A}) \xrightarrow{\sim} \mathcal{H}_\psi \subset \mathsf{L}^2(\mathrm{G}(\mathbb{Q})\backslash \mathrm{G}(\mathbb{A})),$$

*intertwining $\pi_\psi$ and $\varrho_\psi$. Moreover,*

$$j_\psi : \mathsf{L}^2(\mathbb{A})^{\mathrm{K}} \xrightarrow{\sim} \mathcal{H}_\psi^{\mathrm{K}}.$$

Let us recall the definition of a restricted algebraic tensor product: for all primes $p$, let $V_p$ be a (pre-unitary) representation space for $\mathrm{G}(\mathbb{Q}_p)$. Let $(e_p)_p$ be a family of vectors $e_p \in V_p$, defined for all primes $p$ outside a finite set $S_0$. Suppose that, for almost all $p$, $e_p$ is fixed by $\mathrm{K}_p$. We will also assume that the norm of $e_p$ is equal to 1. Let $S$ be a finite set of primes containing $S_0$. A *pure* tensor is a vector, $v = v_S \otimes e^S$, where $e^S = \otimes_{p \notin S} e_p$ and $v_S$ is a pure tensor in the finite tensor product $\otimes_{p \in S} V_p$. The restricted algebraic tensor product $V = \otimes_p V_p$ is generated by finite linear combinations of pure tensors (see [16] for more details).

*Example* 3.12. Consider the representation $\pi_\psi$ of $\mathrm{G}(\mathbb{A})$ on the Schwartz-Bruhat space $\mathcal{S}(\mathbb{A}_{\mathrm{fin}}) = \otimes_p \mathcal{S}(\mathbb{Q}_p)$ and the corresponding representation $\pi_{\psi_p}$ of $\mathrm{G}(\mathbb{Q}_p)$ on $\mathcal{S}(\mathbb{Q}_p)$. In this case, for all primes $p \notin S_\psi$, $e_p$ is unique (up to scalars) and may be taken to be the characteristic function of $\mathbb{Z}_p$. We have $j_\psi(\mathcal{S}(\mathbb{A}_{\text{fin}}) \otimes \mathcal{S}(\mathbb{R})) = \mathcal{H}_\psi^{\text{smooth}}$ (by [7]).

We now fix an orthonormal basis $\mathcal{B}_{\text{fin}}(\pi_\psi)$ for the space $\mathcal{S}(\mathbb{A}_{\text{fin}})^{\mathrm{K}}$ as follows. We let $\mathcal{B}_{\text{fin}}(\pi_\psi) = \otimes_p \mathcal{B}_p(\pi_{\psi_p})$, where, for $p \in S_0 = S = S_\psi$, $\mathcal{B}_p(\pi_{\psi_p})$ is any fixed orthonormal basis for $\mathcal{S}(\mathbb{Q}_p)^{\mathrm{K}_p}$ and, for $p \notin S$, $\mathcal{B}_p(\pi_{\psi_p}) = e_p$. Thus any $\varphi \in \mathcal{B}_{\text{fin}}(\pi_\psi)$ has the form

$$\varphi = \varphi_S \otimes e^S,$$

with $e^S = \otimes_{p \in S} e_p$, as above. We have then the following lemma:

LEMMA 3.13. *The set*

$$\mathcal{B}(\varrho_\psi) := j_\psi(\mathcal{B}_{\text{fin}}(\pi_\psi) \otimes \mathcal{B}_\infty(\pi_\psi))$$

*is a complete orthonormal basis of* $\mathcal{H}_\psi^{\mathrm{K}}$. *The number of elements* $\omega \in \mathcal{B}(\varrho_\psi)$ *(c.f. Lemma 3.9) with given eigenvalue* $\lambda_n^\psi$ *is* $|\mathsf{n}(\mathrm{K})^2 a|$ *if* $a \in \frac{1}{\mathsf{n}(\mathrm{K})}\mathbb{Z}$ *(and zero otherwise).*

*Definition* 3.14. Suppose $p \notin S_\psi$. The normalized spherical function $f_p$ on $\mathrm{G}(\mathbb{Q}_p)$ is defined by

$$f_p(g_p) := \langle \pi_{\psi_p}(g_p) e_p, e_p \rangle.$$

Here $\langle \cdot, \cdot \rangle$ is the standard inner product on $\mathsf{L}^2(\mathbb{Q}_p)$.

LEMMA 3.15 (FACTORIZATION). *For* $\omega \in \mathcal{B}(\varrho_\psi)$ *and* $S = S_\psi \cup \{\infty\}$, *we have an identity*

$$\int_{\mathrm{K}^S} \omega(k^S g) dk^S = \prod_{p \notin S} f_p(g_p) \cdot \omega(g_S).$$

*Here* $\mathrm{K}^S = \prod_{p \notin S_\psi} \mathrm{K}_p$, $g = g^S \cdot g_S$, *with* $g^S$ *(resp.* $g_S$*) in* $\mathrm{G}(\mathbb{A}^S)$ *(resp.* $\mathrm{G}(\mathbb{A}_S)$*).*

*Proof.* Define a linear form $\mu$ on $V = \mathcal{S}(\mathbb{A})$ by setting

$$\mu(\varphi) := \int_{\mathrm{K}^S} j(\varphi)(k^S) dk^S,$$

(where $\varphi \in \mathcal{S}(\mathbb{A})$). Set

$$V^S := \otimes_{p \notin S} \mathcal{S}(\mathbb{Q}_p)$$

and

$$V_S := \otimes_{p \in S_\psi} \mathcal{S}(\mathbb{Q}_p) \otimes \mathcal{S}(\mathbb{R}),$$

so that $V = V_S \otimes V^S$. Then from Lemma 3.9 we have, for $\varphi^S \in V^S$, with $\pi_\psi^S = \otimes_{p \notin S} \pi_{\psi,p}$, an equality of the form

$$\int_{\mathrm{K}^S} \pi_\psi^S(k^S)\varphi^S dk^S = \nu^S(\varphi^S) \cdot e^S$$

for a unique linear form $\nu^S$ on $V^S$. Note that $\nu^S(\varphi^S) = \langle \varphi^S, e^S \rangle$, for $\varphi^S \in V^S$. Now we have, for $\varphi$ of the form $\varphi = \varphi_S \otimes \varphi^S$, with $\varphi_S \in V_S$ and $\varphi^S \in V^S$,

$$\mu\left(\varphi_S \otimes \pi_\psi^S(k^S)\varphi^S\right) = \mu(\varphi_S \otimes \varphi^S),$$

from which it follows at once that

$$\mu(\varphi_S \otimes \varphi^S) = \mu_S(\varphi_S) \cdot \nu^S(\varphi^S),$$

for some linear form $\mu_S$ on $V_S$. From this we obtain in turn, for $\varphi = \varphi_S \otimes e^S$, the identity

$$\begin{aligned} \textstyle\int_{\mathrm{K}^S} j(\varphi)(k^S g)dk^S &= \mu(\pi_\psi(g)\varphi) \\ &= \mu_S(\pi_{\psi,S}(g_S)\varphi_S) \cdot \nu^S\left(\pi_\psi^S(g^S)e_S\right) \\ &= \mu_S(\pi_{\psi,S}(g_S)\varphi_S) \cdot \textstyle\prod_{p \notin S} f_p(g_p) \end{aligned}$$

for $g \in \mathrm{G}(\mathbb{A})$. Here $\pi_{\psi,S} = \otimes_{p \in S} \pi_{\psi,p}$. Taking $\omega = j(\varphi)$, with $\varphi = \varphi_S \otimes e^S$, $\varphi_S \in V_S$ as above, we arrive next at the equality

$$\int_{\mathrm{K}^S} \omega(k^S g)dk^S = \omega'(g_S) \cdot \prod_{p \notin S} f_p(g_p),$$

for some function $\omega'$ on $\mathrm{G}(\mathbb{A}_S)$. Finally, if $g = g_S \in \mathrm{G}(\mathbb{A}_S)$, we obtain from the last expression

$$\omega'(g_S) = \int_{\mathrm{K}^S} \omega(k^S g_S)dk^S = \int_{\mathrm{K}^S} \omega(g_S k^S)dk^S = \omega(g_S),$$

since, in fact, $\omega$ is K-invariant on the right. This completes the proof of the lemma. □

COROLLARY 3.16. *Let $\psi = \psi_a$ be as above (with $a \in \mathbb{Q}^\times$) and $\varrho_\psi$ the associated irreducible unitary automorphic representation of* $\mathrm{G}(\mathbb{A})$. *Suppose that $\varrho_\psi$ has a* K*-fixed vector. Then, for $S = S_\psi \cup \{\infty\}$, all $\omega \in \mathcal{B}(\varrho_\psi)$, all primes $p \notin S_\psi$ and all (integrable) functions $H$ on* $\mathrm{G}(\mathbb{A})$ *such that*

$$H_p(k_p g_p) = H_p(g_p k_p) = H_p(g_p),$$

*for all* $k_p \in \mathrm{K}_p$ *and* $g_p \in \mathrm{G}(\mathbb{Q}_p)$, *one has*

(3.7)
$$\int_{\mathrm{G}(\mathbb{A})} H(g)\omega(g)dg = \prod_{p \notin S} \int_{\mathrm{G}(\mathbb{Q}_p)} H_p(g_p) f_p(g_p) dg_p \cdot \int_{\mathrm{G}(\mathbb{A}_S)} H(g_S)\omega_S(g_S)dg_S,$$

*where* $\omega_S$ *is the restriction of* $\omega$ *to* $\mathrm{G}(\mathbb{A}_S)$.

LEMMA 3.17. *For all* $\psi$ *and all* $p \notin S_\psi$ *one has, for* $H_p$ *as above,*

$$\int_{\mathrm{G}(\mathbb{Q}_p)} H_p(g_p) f_p(g_p) dg_p = \int_{\mathrm{U}(\mathbb{Q}_p)} H_p(u_p)\psi_p(u_p)du_p.$$

*Proof.* Suppose $p \notin S_\psi$. Let $\chi_p$ be the characteristic function of $\mathrm{K}_p$. Define a function $\tilde{\psi}_p$ on $\mathrm{G}(\mathbb{Q}_p)$ by setting

$$\tilde{\psi}_p(g_p) := \int_{\mathrm{U}(\mathbb{Q}_p)} \chi_p(u_p g_p)\overline{\psi}_p(u_p)du_p$$

(with $g_p \in \mathrm{G}(\mathbb{Q}_p)$). Clearly, $\tilde{\psi}_p$ belongs to the space $V_{\psi,p}$ of the induced representation $\pi_{\psi,p}$; moreover, $\tilde{\psi}_p$ is $\mathrm{K}_p$-invariant (on the right).

Next we have, with our normalization of Haar measures,

$$\tilde{\psi}_p(g_p) = \psi_p(u_p)$$

provided $g_p = u_p k_p$, with $u_p \in \mathrm{U}(\mathbb{Q}_p)$, $k_p \in \mathrm{K}_p$, and zero otherwise. In particular,

$$|\tilde{\psi}_p(g_p)|^2 = \int_{\mathrm{U}(\mathbb{Q}_p)} \chi_p(u_p g_p)du_p,$$

from which it follows that

$$\|\tilde{\psi}_p\|^2 = \int_{\mathrm{U}(\mathbb{Q}_p)\backslash \mathrm{G}(\mathbb{Q}_p)} |\tilde{\psi}_p(g_p)|^2 d^*g_p = \int_{\mathrm{G}(\mathbb{Q}_p)} \chi_p(g_p)dg_p = \int_{\mathrm{K}_p} dg_p = 1.$$

(Here $d^*g_p$ is normalized so that $dg_p = du_p d^*g_p$ as in Section 3.1.) Next, for $\nu \in V^\infty_{\psi,p}$, we have, with $\pi_p = \pi_{\psi,p}$,

$$\int_{\mathrm{K}_p} \pi_p(k_p)\nu dk_p = \mu(\nu)\tilde{\psi}_p,$$

for a unique linear form $\mu$ on $V^\infty_{\psi,p}$. Note that $\mu(\nu) = \langle \nu, \tilde{\psi}_p\rangle$. We have then, using $\tilde{\psi}_p(e) = 1$,

$$f_p(g_p) = \langle \pi_p(g_p)\tilde{\psi}_p, \tilde{\psi}_p\rangle = \int_{\mathrm{K}_p} \tilde{\psi}_p(k_p g_p)dk_p.$$

To complete the proof, we note first, from the left $\mathrm{K}_p$-invariance of $H_p$, that

$$\int_{\mathrm{G}(\mathbb{Q}_p)} H_p(g_p) f_p(g_p) dg_p = \int_{\mathrm{G}(\mathbb{Q}_p)} H_p(g_p)\tilde{\psi}_p(g_p)dg_p.$$

In turn, the last integral is

$$= \int_{\mathrm{U}(\mathbb{Q}_p)} \overline{\psi}_p(u_p) \int_{\mathrm{G}(\mathbb{Q}_p)} H_p(g_p)\chi_p(u_p g_p) dg_p$$

$$= \int_{\mathrm{U}(\mathbb{Q}_p)} \psi_p(u_p) \int_{\mathrm{G}(\mathbb{Q}_p)} H_p(u_p g_p)\chi_p(g_p) dg_p$$

$$= \int_{\mathrm{U}(\mathbb{Q}_p)} H_p(u_p)\psi_p(u_p) du_p,$$

the last equality from the right $\mathrm{K}_p$-invariance of $H_p$. □

**4. Euler products.** In this section we show that each summand in the $\mathsf{L}^2$-expansion of the height zeta function in Proposition 3.3 is regularized by an explicit Euler product. First we record the integrability of local heights:

LEMMA 4.1. *For all compacts $\mathsf{K} \subset \mathsf{T}_{-1}$ and all primes $p$, there exists a constant $\mathsf{c}_p(\mathsf{K})$ such that, for all $\mathbf{s} \in \mathsf{K}$, one has:*

$$\int_{\mathrm{G}(\mathbb{Q}_p)} |H_p(\mathbf{s}, g_p)^{-1}| dg_p \le \mathsf{c}_p(\mathsf{K}).$$

*Moreover, for all $\partial$ in the universal enveloping algebra $\mathfrak{U}(\mathfrak{g})$ of* G *and all compacts $\mathsf{K} \subset \mathsf{T}_{-1}$, there exists a constant $\mathsf{c}(\mathsf{K}, \partial)$ such that, for all $\mathbf{s} \in \mathsf{K}$,*

$$\int_{\mathrm{G}(\mathbb{R})} |\partial H_\infty(\mathbf{s}, g_\infty)^{-1}| dg_\infty \le \mathsf{c}(\mathsf{K}, \partial).$$

*Proof.* This is Lemma 8.2 and Proposition 8.4 of [6]. □

*Notations* 4.2. Denote by $S_X$ the set of all primes such that one of the following holds:

- $p$ is 2 or 3;
- $\mathrm{K}_p \neq \mathrm{G}(\mathbb{Z}_p)$;
- over $\mathbb{Z}_p$, the union $\cup_\alpha D_\alpha$ is not a union of smooth relative divisors with strict normal crossings.

*Remark* 4.3. For all $p \notin S_X$, the height $H_p$ is invariant with respect to the right and left $\mathrm{G}(\mathbb{Z}_p)$-action.

PROPOSITION 4.4. *For all primes $p \notin S_X$ and all $\mathbf{s} \in \mathsf{T}_{-1}$, one has*

$$\int_{\mathrm{G}(\mathbb{Q}_p)} H(\mathbf{s}, g_p)^{-1} dg_p = p^{-3} \left( \sum_{A \subseteq \mathcal{A}} \# D_A^0(\mathbf{F}_p) \prod_{\alpha \in A} \frac{p-1}{p^{s_\alpha - \kappa_\alpha + 1} - 1} \right),$$

*where $X = \sqcup D_A^0$ is the stratification of $X$ by locally closed subvarieties as in the Introduction and $\mathbf{F}_p = \mathbb{Z}/p\mathbb{Z}$ is the finite field with $p$ elements.*

*Proof.* This is Theorem 9.1 in [6]. The proof proceeds as follows: for $p \notin S_X$ there is a *good* model $\mathcal{X}$ of $X$ over $\mathbb{Z}_p$: all boundary components $D_\alpha$ (and G) are defined over $\mathbb{Z}_p$ and form a strict normal crossing divisor. We can consider the reduction map

$$\mathrm{red} \ : \ X(\mathbb{Q}_p) = X(\mathbb{Z}_p) \to X(\mathbf{F}_p) = \sqcup_{A \subset \mathcal{A}} D_A^0(\mathbf{F}_p).$$

The main observation is that, in a neighborhood of the preimage in $X(\mathbb{Q}_p)$ of the point $\tilde{x}_v \subset D_A^0(\mathbf{F}_p)$, one can introduce local $p$-adic analytic coordinates $\{x_\alpha\}_{\alpha=1,\dots,n}$ such that

$$H_p(\mathbf{s}, g) = \prod_{\alpha \in A} |x_\alpha|_p^{s_\alpha}.$$

Now it suffices to keep track of the change of the measure $dg_p$:

$$dg_p = \prod_{\alpha \notin A} dx_\alpha \cdot \prod_{\alpha \in A} |x_\alpha|_p^{k_\alpha} dx_\alpha,$$

where $dx_\alpha$ are standard Haar measures on $\mathbb{Q}_p$. The integrals obtained are elementary:

$$\int_{\mathrm{red}^{-1}(\tilde{x}_p)} H_p(\mathbf{s}, g_p)^{-1} dg_p = \prod_{\alpha \notin A} \int_{p\mathbb{Z}_p} dx_\alpha \cdot \prod_{\alpha \in A} \int_{p\mathbb{Z}_p} p^{-(s_\alpha - k_\alpha) v_p(x_\alpha)} dx_\alpha$$

(where $v_p(x) = \log_p(|x|_p)$ is the ordinal of $x$ at $p$). Summing over all $\tilde{x}_p \in X(\mathbf{F}_p)$, we obtain the proof (see [6] for more details.) □

COROLLARY 4.5. *For all primes p one has the identity*

$$\int_{\mathrm{G}(\mathbb{Q}_p)} H_p(\mathbf{s}, g_p)^{-1} dg_p = \prod_{\alpha \in \mathcal{A}} \zeta_p(s_\alpha - \kappa_\alpha + 1) \cdot f_{0,p}(\mathbf{s}),$$

*where $f_{0,p}(\mathbf{s})$ is a holomorphic function in $\mathsf{T}_{-1+\epsilon}$. Moreover, there exist a $\delta > 0$ and a function $f_0(\mathbf{s}, g)$, which is holomorphic in $\mathsf{T}_{-\delta}$ and continuous in $g \in \mathrm{G}(\mathbb{A})$, such that*

$$\mathcal{Z}_0(\mathbf{s}, g) = f_0(\mathbf{s}, g) \cdot \prod_{\alpha \in \mathcal{A}} \zeta(s_\alpha - \kappa_\alpha + 1);$$

*moreover,*

$$\lim_{\mathbf{s} \to \kappa} \mathcal{Z}_0(\mathbf{s}, e) \cdot \prod_{\alpha \in \mathcal{A}} (s_\alpha - \kappa_\alpha) = \tau(\mathcal{K}_X) \neq 0,$$

*where $\tau(\mathcal{K}_X)$ is the Tamagawa number defined in* [20].

*Proof.* Apply Corollary 9.6 in [6]. □

*Notations* 4.6. Let $\mathbf{a} = (a_1, a_2) \in \mathbb{Q}^2$ and let $f_{\mathbf{a}}$ be the $\mathbb{Q}$-rational linear form

$$(x, y) \mapsto a_1 x + a_2 y.$$

The linear form $f_{\mathbf{a}}$ defines an adelic character $\eta = \eta_{\mathbf{a}}$ of $\mathrm{G}(\mathbb{A})$:

$$\eta(g(x, z, y)) = \psi_1(a_1 x + a_2 y),$$

where again $\psi_1$ is the Tate-character of $\mathbb{A}/\mathbb{Q}$. Write

$$\mathrm{div}(\eta) = E(\eta) - \sum_{\alpha \in \mathcal{A}} d_\alpha(\eta) D_\alpha$$

for the divisor of the function $f_{\mathbf{a}}$ on the compactification $X$ (by Corollary 1.7, $d_\alpha \geq 0$ for all $\alpha \in \mathcal{A}$ and $d_\alpha > 0$ for at least one $\alpha \in \mathcal{A}$). Denote by

$$\mathcal{A}_0(\eta) = \{\alpha \mid d_\alpha(\eta) = 0\}.$$

Let $V \subset X$ be the induced equivariant compactification of $\mathrm{U} \subset \mathrm{G}$.

*Assumption* 4.7. From now on we will assume that the boundary $V \setminus \mathrm{U}$ is a strict normal crossing divisor whose components are obtained by intersecting the boundary components of $X$ with $V$:

$$V \setminus \mathrm{U} = \cup_{\alpha \in \mathcal{A}^V} D_\alpha^V = \cup_{\alpha \in \mathcal{A}} D_\alpha \cap V,$$

(with $\mathcal{A}^V \subseteq \mathcal{A}$).

*Remark* 4.8. The general case can be reduced to this situation by (equivariant) resolution of singularities.

By Lemma 7.3 of [6], we have

$$-K_V = \sum_{\alpha \in \mathcal{A}^V} \kappa_\alpha^V D_\alpha^V,$$

with $\kappa_\alpha^V \leq \kappa_\alpha$ (for all $\alpha$) and equality holding for $\alpha$ in a *proper* subset of $\mathcal{A}$. Denote by $f_a$ the $\mathbb{Q}$-rational linear form on the center $\mathrm{Z}$ of $\mathrm{G}$

$$z \mapsto a \cdot z.$$

The linear form $f_a$ defines an adelic character $\psi = \psi_a$ of $\mathrm{U}(\mathbb{A})/\mathrm{U}(\mathbb{Q})$:

$$\psi_a(g(0, z, y)) = \psi_1(az).$$

Write

$$\mathrm{div}(\psi) = E(\psi) - \sum_{\alpha \in \mathcal{A}^V} d_\alpha(\psi) D_\alpha$$

for the divisor of the function $f_a$ on $V$ and denote by

$$\mathcal{A}_0(\psi) = \{\alpha \mid d_\alpha(\psi) = 0\}.$$

We note that both $\mathcal{A}_0(\eta)$ and $\mathcal{A}_0(\psi)$ are *proper* subsets of $\mathcal{A}$. A precise formulation of the statement that the trivial representation of $\mathrm{G}(\mathbb{A})$ is "isolated" in the automorphic spectrum is the following proposition.

PROPOSITION 4.9. *Let $\eta = \eta_{\mathbf{a}}$ and $\psi = \psi_a$ be the nontrivial adelic characters occuring in Proposition 3.3 ($a_1, a_2, a \in \frac{1}{\mathsf{n}(\mathsf{K})}\mathbb{Z}$). For any $\epsilon > 0$ there exist a constant $\mathsf{c}(\epsilon)$ and holomorphic bounded functions $\phi_\eta(\mathbf{a}, \cdot)$, $\varphi_\psi(a, \cdot)$ on $\mathsf{T}_{-1/2+\epsilon}$ such that, for any $\mathbf{s} \in \mathsf{T}_0$, one has*

$$\prod_{p \notin S_\eta} \int_{\mathrm{G}(\mathbb{Q}_p)} H_p(\mathbf{s}, g_p)^{-1} \overline{\eta}_p(g_p) dg_p = \phi_\eta(\mathbf{a}, \mathbf{s}) \prod_{\alpha \in \mathcal{A}_0(\eta)} \zeta^{S_\eta}(s_\alpha - \kappa_\alpha + 1);$$

$$\prod_{p \notin S_\psi} \int_{\mathrm{U}(\mathbb{Q}_p)} H_p(\mathbf{s}, u_p)^{-1} \overline{\psi}_p(u_p) du_p = \varphi_\psi(a, \mathbf{s}) \prod_{\alpha \in \mathcal{A}_0(\psi)} \zeta^{S_\psi}(s_\alpha - \kappa_\alpha + 1),$$

*where $\zeta^S(s) = \prod_{p \notin S}(1 - p^{-s})^{-1}$ is the incomplete Riemann zeta function. Moreover,*

$$|\phi_\eta(\mathbf{a}, \mathbf{s})| \le \mathsf{c}(\epsilon),$$
$$|\varphi_\psi(a, \mathbf{s})| \le \mathsf{c}(\epsilon).$$

*Proof.* The integrals can be computed as in Proposition 4.4. They are regularized by the indicated products of (partial) zeta functions. The remaining Euler products are expressions involving the number of $\mathbf{F}_p$-points for boundary strata (and their intersections with div($\eta$), resp. div($\psi$)). In particular, they are uniformly bounded on compacts in $\mathsf{T}_{-1/2+\epsilon}$. For details we refer to [6], Proposition 5.5 (which follows from Proposition 10.2, loc. cit.). □

COROLLARY 4.10. *In particular,* each *term in the sums $\mathcal{Z}_1(\mathbf{s}, g)$ and $\mathcal{Z}_2(\mathbf{s}, g)$ has a meromorphic continuation to the domain $\mathsf{T}_{-1/2}$.*

LEMMA 4.11. *For any $\epsilon > 0$ and any compact $\mathsf{K}$ in $\mathsf{T}_{-1/2+\epsilon}$, there exist constants $\mathsf{c}(\mathsf{K})$ and $\mathsf{n}' = \mathsf{n}'(\mathsf{K}) > 0$, such that*

$$|\prod_{p \in S_\eta} \int_{\mathrm{G}(\mathbb{Q}_p)} H_p(\mathbf{s}, g_p)^{-1} dg_p| \le \mathsf{c}(\mathsf{K}) \cdot (1 + \|\mathbf{a}\|)^{\mathsf{n}'}$$
$$|\prod_{p \in S_\psi} \int_{\mathrm{G}(\mathbb{Q}_p)} H_p(\mathbf{s}, g_p)^{-1} dg_p| \le \mathsf{c}(\mathsf{K}) \cdot (1 + |a|)^{\mathsf{n}'}$$

*for all $\mathbf{s} \in \mathsf{K}$.*

*Proof.* For $p \in S_X$ we use the bound from Lemma 4.1. For $p \in S_\eta \setminus S_X$ (resp. $S_\psi \setminus S_X$) we apply Proposition 4.4: there is a constant $\mathsf{c} > 0$ (depending only on $X$ and $\mathsf{K}$) such that

$$\left| \int_{\mathrm{G}(\mathbb{Q}_p)} H_p(\mathbf{s}, g_p)^{-1} dg_p \right| \le \left(1 + \frac{\mathsf{c}}{\sqrt{p}}\right)$$

for all $p$. Using the bound

$$\prod_{p|b} \left(1 + \frac{\mathsf{c}}{\sqrt{p}}\right) \le |b|^{\mathsf{n}'}$$

(for $b = a \cdot \mathsf{n}(\mathrm{K})$ and some $\mathsf{n}' = \mathsf{n}'(\mathrm{K}) > 0$), we conclude the proof. □

PROPOSITION 4.12. *For any* $\mathsf{n} > 0$ *and any compact* $\mathsf{K} \subset \mathsf{T}_{-1/2+\epsilon}$, *there exists a constant* $\mathsf{c}(\mathsf{K}, \mathsf{n})$ *such that, for any* $\mathbf{s} \in \mathsf{K}$, *and any* $\mathbf{a} = (a_1, a_2)$ *and* $a$ *as above, one has the estimates*

$$| \int_{G(\mathbb{R})} H_\infty(\mathbf{s}, g_\infty)^{-1} \overline{\eta}_\infty(g_\infty) dg_\infty | \le \mathsf{c}(\mathsf{K}, \mathsf{n})(1 + \|\mathbf{a}\|)^{-\mathsf{n}},$$

$$| \int_{G(\mathbb{A}_S)} H_S(\mathbf{s}, g_S)^{-1} \overline{\omega}_S(g_S) dg_S | \le \mathsf{c}(\mathsf{K}, \mathsf{n})(1 + |\lambda|)^{-\mathsf{n}}(1 + |a|)^{\mathsf{n}'},$$

*where* $\mathsf{n}' = \mathsf{n}'(\mathrm{K})$ *is the bound from Lemma 4.11,* $S = S_\psi \cup \{\infty\}$, $\lambda = \lambda(\omega)$ *is the eigenvalue of* $\omega_S \in \mathcal{B}_S(\varrho_\psi)$ *(with respect to the elliptic operator* $\Delta$*).*

*Proof.* We use Lemma 4.1 and integration by parts. For $\eta$ we apply the operator $\partial = \partial_x^2 + \partial_y^2$ (as in [6]) and for $\psi$ the elliptic operator $\Delta = \partial_x^2 + \partial_y^2 + \partial_z^2$ (and use the eigenfunction property of $\omega_S$, or, what amounts to the same, of $\omega \in \mathcal{B}(\varrho_\psi)$). More precisely, the second integral is majorized by

$$|\lambda|^{-\mathsf{m}} \cdot | \int_{G(\mathbb{A}_S)} \Delta^{\mathsf{m}} H_S(\mathbf{s}, g_S)^{-1} dg_S | \cdot \sup_{g_S \in G(\mathbb{A}_S)} |\omega_S(g_S)|.$$

Using the class number one property

$$G(\mathbb{A}) = G(\mathbb{Q}) \cdot G(\mathbb{R}) \cdot \mathrm{K}$$

and the invariance of $\omega$ under $G(\mathbb{Q})$ and K, we obtain the estimates

$$\sup_{g_S \in G(\mathbb{A}_S)} |\omega_S(g_S)| \le \sup_{g \in G(\mathbb{A})} |\omega(g)| = \sup_{g \in \Gamma \backslash G(\mathbb{R})} |\omega_\infty(g)|. \tag{4.1}$$

Further we have

$$\sup_{g \in \Gamma \backslash G(\mathbb{R})} |\omega_\infty(g)| \ll |\lambda|^{\mathsf{m}'} \cdot \|\omega\|_{\mathsf{L}^2(\Gamma \backslash G(\mathbb{R}))} = |\lambda|^{\mathsf{m}'} \cdot \|\omega\|_{\mathsf{L}^2(G(\mathbb{Q}) \backslash G(\mathbb{A}))} = |\lambda|^{\mathsf{m}'} \tag{4.2}$$

for some constant $\mathsf{m}'$ (see [11], [23], p. 22, and [25] for the comparison between the $\mathsf{L}^2$ and the $\mathsf{L}^\infty$ norms of an eigenfunction of an elliptic operator on a compact manifold and other applications of this inequality). The rest of the proof follows at once from Lemma 4.1 and Lemma 4.11. (Notice that the implied constants, including $\mathsf{m}'$, in the above inequalities depend only on the choice of K.) □

Before continuing to the proof of the main theorem, we discuss the individual inner products

$$\mathcal{Z}(\mathbf{s}, \eta) = \langle \mathcal{Z}(\mathbf{s}, \cdot), \eta \rangle,$$

$$\mathcal{Z}(\mathbf{s}, \omega^\psi) = \langle \mathcal{Z}(\mathbf{s}, \cdot), \omega^\psi \rangle.$$

Let us first set

$$\zeta_\eta(\mathbf{s}) = \prod_{\alpha\in\mathcal{A}_0(\eta)} \zeta(s_\alpha - \kappa_\alpha + 1),$$

$$\zeta_\psi(\mathbf{s}) = \prod_{\alpha\in\mathcal{A}_0(\psi)} \zeta(s_\alpha - \kappa_\alpha + 1).$$

We have then the following corollary.

COROLLARY 4.13. *The functions*

$$\zeta_\eta(\mathbf{s})^{-1} \cdot \mathcal{Z}(\mathbf{s}, \eta)$$

*and*

$$\zeta_\psi(\mathbf{s})^{-1} \cdot \mathcal{Z}(\mathbf{s}, \omega^\psi),$$

*initially defined for* $\mathbf{s} \in \mathsf{T}_\delta$ *(cf. 2.4), have an analytic continuation to the domain* $\mathsf{T}_{-1/2+\epsilon}$ *(for all* $\epsilon > 0$*).*

*Proof.* We will consider the function $\zeta_\psi(\mathbf{s})^{-1} \cdot \mathcal{Z}(\mathbf{s}, \omega^\psi)$ and leave the first case to the reader. We start by observing that, for $\mathbf{s} \in \mathsf{T}_\delta$, we have

$$\int_{\mathrm{G}(\mathbb{A})} |H(\mathbf{s}, g)|^{-1} dg < \infty \tag{4.3}$$

(this follows from Proposition 2.4 together with the compactness of $\mathrm{G}(\mathbb{Q})\backslash\mathrm{G}(\mathbb{A})$). Consequently,

$$\mathcal{Z}(\mathbf{s}, \overline{\omega}^\psi) = \int_{\mathrm{G}(\mathbb{A})} H(\mathbf{s}, g)^{-1} \omega^\psi(g) dg,$$

again for $\mathbf{s} \in \mathsf{T}_\delta$. Using the left-K, and in particular, the left $\mathrm{K}^S$-invariance of $H$, we have, for $\mathbf{s} \in \mathsf{T}_\delta$,

$$\mathcal{Z}(\mathbf{s}, \overline{\omega}^\psi) = \int_{\mathrm{G}(\mathbb{A})} H(\mathbf{s}, g)^{-1} \int_{\mathrm{K}^S} \omega^\psi(k^S g) dk^S dg.$$

Then, from Lemma 3.15 (factorization), we have (with $\mathbf{s}$ in the same domain),

$$\mathcal{Z}(\mathbf{s}, \overline{\omega}^\psi) = \int_{\mathrm{G}(\mathbb{A}_S)} H(\mathbf{s}, g_S)^{-1} \omega^\psi(g_S) dg_S \cdot \int_{\mathrm{G}(\mathbb{A}^S)} H(\mathbf{s}, g^S)^{-1} f^S(g^S) dg^S, \tag{4.4}$$

where we have set

$$f^S(g^S) := \prod_{p\notin S} f_p(g_p)$$

(recall that $S = S_\psi \cup \{\infty\}$). Both integrals above are convergent for $\mathbf{s} \in \mathsf{T}_\delta$ by (4.3). By Lemma 4.1, the first integral on the right in (4.4) is absolutely convergent for

$\mathbf{s} \in \mathsf{T}_{-1}$. Next it follows from Proposition 4.9 that the second integral above actually converges for $\mathbf{s} \in \mathsf{T}_0$. Moreover, we have for that integral the product expression

$$\prod_{p \notin S_\psi} \int_{\mathrm{G}(\mathbb{Q}_p)} H_p(\mathbf{s}, g_p)^{-1} f_p(g_p) dg_p.$$

As we have noted in Proposition 4.9, the infinite product is convergent to a holomorphic function, for $\mathbf{s} \in \mathsf{T}_0$. Further, we then have for this infinite product the expression

$$\varphi_\psi(a, \mathbf{s}) \cdot \prod_{\alpha \in \mathcal{A}_0(\psi)} \prod_{p \in S_\psi} \zeta(s_\alpha - \kappa_\alpha + 1) \cdot \zeta_\psi(\mathbf{s}),$$

for $\mathbf{s} \in \mathsf{T}_0$. It follows, again from Proposition 4.9, that

$$\zeta_\psi(\mathbf{s})^{-1} \cdot \mathcal{Z}(\mathbf{s}, \overline{\omega}^\psi)$$

can be continued holomorphically to the domain $\mathsf{T}_{-1/2+\epsilon}$. (Note that we have used the meromorphic continuation of $\zeta(s)$ to $\Re(s) > 1/2 + \epsilon$.) □

Moreover, we have the following lemma:

LEMMA 4.14. *For any $\epsilon, \mathsf{n} > 0$ and any compact $\mathsf{K} \subset \mathsf{T}_{-1/2+\epsilon}$, there is a constant $\mathsf{c}(\mathsf{K}, \mathsf{n})$ and an integer $\mathsf{n}' > 0$ such that, for any $\mathbf{s} \in \mathsf{K}$ and $a$ as above ($\psi = \psi_a$), we have*

$$|\zeta_\psi(\mathbf{s})^{-1} \cdot \mathcal{Z}(\mathbf{s}, \overline{\omega}^\psi)| \le \mathsf{c}(\mathsf{K}, \mathsf{n})(1 + \|\lambda|)^{-\mathsf{n}}(1 + |a|)^{\mathsf{n}'}.$$

*Proof.* We have from the preceding (proof of Corollary 4.13)

$$\begin{aligned}
&\zeta_\psi(\mathbf{s})^{-1} \cdot \mathcal{Z}(\mathbf{s}, \overline{\omega}^\psi) \\
&\quad = \varphi_\psi(a, \mathbf{s}) \cdot \int_{\mathrm{G}(\mathbb{A}_S)} H(\mathbf{s}, g_S)^{-1} \omega(g_S) dg_S \cdot \prod_{\alpha \in \mathcal{A}_0(\psi)} \prod_{p \in S_\psi} \zeta_p(s_\alpha - \kappa_\alpha + 1),
\end{aligned}$$

for $\mathbf{s} \in \mathsf{T}_{-1/2+\epsilon}$. Our conclusion follows from Proposition 4.9 and Proposition 4.12 and, for example, the elementary inequality

$$\prod_{p|b} \left(1 + \frac{1}{\sqrt{p}}\right) \le |b|^{\mathsf{n}'}$$

applied to $b = a\mathsf{n}(\mathrm{K})$ (for some $\mathsf{n}' > 0$, independent of $a$). □

THEOREM 4.15. *The height zeta function $\mathcal{Z}(\mathbf{s})$ is holomorphic for $\mathbf{s} \in \mathsf{T}_0$. Moreover,*

$$\prod_{\alpha \in \mathcal{A}} (s_\alpha - \kappa_\alpha) \cdot \mathcal{Z}(\mathbf{s})$$

*admits a holomorphic continuation to* $\mathsf{T}_{-\delta}$ *(for some* $\delta > 0$*) and*

$$\lim_{\mathbf{s}\to\kappa} \prod_{\alpha\in\mathcal{A}} (s_\alpha - \kappa_\alpha) \cdot \mathcal{Z}(\mathbf{s}) = \tau(\mathcal{K}_X).$$

*Proof.* Set

$$\mathsf{z}(\mathbf{s}) := \prod_{\alpha\in\mathcal{A}} (s_\alpha - \kappa_\alpha).$$

We prove first that both series

$$\sum_{\eta\neq\mathbf{1}} \mathsf{z}(\mathbf{s}) \cdot \mathcal{Z}(\mathbf{s}, \eta) \cdot \eta(g) \tag{4.5}$$

and

$$\sum_{\psi\neq 1} \sum_{\omega^\psi\in\mathcal{B}(\varrho_\psi)} \mathsf{z}(\mathbf{s}) \cdot \mathcal{Z}(\mathbf{s}, \omega^\psi) \cdot \omega^\psi(g) \tag{4.6}$$

are normally convergent for $\mathbf{s}$ in a compact subset of $\mathsf{T}_{-1/2+\epsilon}$ and $g$ in a compact subset of $\mathsf{G}(\mathbb{A})$. We note that, by Proposition 4.12, the products

$$\mathsf{z}(\mathbf{s})\mathcal{Z}(\mathbf{s}, \eta) \quad \text{and} \quad \mathsf{z}(\mathbf{s})\mathcal{Z}(\mathbf{s}, \omega^\psi)$$

are defined for $\mathbf{s} \in \mathsf{T}_{-1/2+\epsilon}$. We will prove our assertion for the second series; the proof for the first is entirely similar.

We have a map from the set of nontrivial characters $\{\psi\}$ of $\mathbb{A}/\mathbb{Q}$ to the set of subsets of $\mathcal{A}$ given by

$$\psi \mapsto \mathcal{A}_0(\psi).$$

It suffices to prove our assertion for each subseries $\mathcal{Z}_A$ of $\mathcal{Z}_2$ corresponding to $\psi$ with $\mathcal{A}_0(\psi) = A$ (for $A \subset \mathcal{A}$). From Lemma 4.14 we have a uniform majoration (for real $\mathbf{s}$)

$$\mathsf{z}(\mathbf{s}) \cdot \mathcal{Z}(\mathbf{s}, \omega^\psi) \ll \mathsf{z}(\mathbf{s}) \cdot \zeta_\psi(\mathbf{s}) \cdot (1 + |\lambda|)^{-\mathsf{n}} \cdot (1 + |a|)^{\mathsf{n}'}.$$

By definition, the function $\zeta_\psi$ is the same for all for $\psi$ occuring in $\mathcal{Z}_A$. It remains then to prove the assertion for the series

$$\sum_{\psi} \sum_{\omega^\psi\in\mathcal{B}(\varrho_\psi)} |\lambda|^{-\mathsf{n}+\mathsf{m}'} |a|^{\mathsf{n}'}, \tag{4.7}$$

where we have used the estimates (4.1) and (4.2) (and the sum is over all characters $\psi$ occuring in $\mathcal{Z}_A$). We recall that $\lambda = \lambda(\omega^\psi)$ is the $\Delta$-eigenvalue of $\omega^\psi$ and $\psi = \psi_a$. We also recall (Lemma 3.13) that (with $S = S_\psi$)

$$\omega^\psi = j(\varphi_S \otimes e^S \otimes \varphi_n^\psi),$$

for $n = 0, 1, 2, \ldots$, where $\varphi_S$ varies over an orthonormal basis of $\mathcal{S}(\mathbb{A}_S)^{\mathsf{K}_S}$. Thus our series (4.7) is bounded from above by

$$\sum_{a\in\mathbb{Z}, a\neq 0} \sum_{n} |\lambda_n|^{-\mathsf{n}} |a|^{\mathsf{n}'+1} \cdot \mathsf{n}(\mathsf{K})^2$$

(see Lemma 3.9). Our claim now follows upon remarking that

$$\lambda_n = (-2\pi(n+1)|a| - 4\pi^2 a^2).$$

At this point we may conclude that the series (4.5) and (4.7) converge as stated. It now follows that, for $\mathbf{s} \in \mathsf{T}_\delta$,

$$\mathcal{Z}(\mathbf{s}, g) = \mathcal{Z}_0(\mathbf{s}, g) + \mathcal{Z}_1(\mathbf{s}, g) + \mathcal{Z}_2(\mathbf{s}, g),$$

as an equality of continuous functions on $\mathrm{G}(\mathbb{A})$. In particular, we have

$$\mathsf{z}(\mathbf{s})\mathcal{Z}(\mathbf{s}) = \mathsf{z}(\mathbf{s})(\mathcal{Z}_0(\mathbf{s}, id) + \mathcal{Z}_1(\mathbf{s}, id) + \mathcal{Z}_2(\mathbf{s}, id)), \tag{4.8}$$

again for $\mathbf{s} \in \mathsf{T}_\delta$. Finally, we obtain, from (4.8), Corollary 4.5, and the preceding, the meromorphic continuation of $\mathcal{Z}(\mathbf{s})$ to the domain $\mathsf{T}_{-1/2+\epsilon}$.

Further, since for nontrivial $\psi$ the set $\mathcal{A}_0(\psi)$ is a *proper* subset of $\mathcal{A}$, we also see that the function

$$\mathsf{z}(\mathbf{s})(\mathcal{Z}_1(\mathbf{s}, id) + \mathcal{Z}_2(\mathbf{s}, id))$$

vanishes for $\mathbf{s} = \kappa$. Thus we have finally

$$\mathsf{z}(\mathbf{s})\mathcal{Z}(\mathbf{s}, id)|_{\mathbf{s}=\kappa} = \mathsf{z}(\mathbf{s})\mathcal{Z}_0(\mathbf{s}, id)|_{\mathbf{s}=\kappa}.$$

Applying Corollary 4.5 we conclude the proof. □

*Remark* 4.16. Theorem 4.15 implies that for each $L$ in the interior of $\Lambda_{\mathrm{eff}}(X)$ the (one-parameter) height zeta function $\mathcal{Z}(s, \mathcal{L})$ is holomorphic for $\Re(s) > a(L)$ and admits a meromorphic continuation to $\Re(s) > a(L) - \epsilon$ (for some $\epsilon > 0$) with an isolated pole at $s = a(L)$ of order *at most* $b(L)$. The proof that the order is exactly $b(L)$ and that the leading coefficient of $\mathcal{Z}(s, \mathcal{L})$ at this pole is $c(L) \cdot \tau(\mathcal{L})$ is analogous to the proof of the corresponding statement for height zeta functions of equivariant compactifications of additive groups (see Section 7 in [6]).

**5. Example: $\mathbb{P}^3$.** A standard bi-equivariant compactification of the Heisenberg group G is the three-dimensional projective space $X = \mathbb{P}^3$. The boundary $D = X \backslash \mathrm{G}$ consists of a single irreducible divisor (the hyperplane section). The class of this divisor generates the Picard group $\mathrm{Pic}(X)$. The anticanonical class $-[K_X] = 4[D]$ and the cone of effective divisors $\Lambda_{\mathrm{eff}}(X) = \mathbb{R}_{\geq 0}[D]$. The height pairing is given by

$$H(s, g) := \prod_p H_p(s, g_p) \cdot H_\infty(s, g_\infty), \tag{5.1}$$

where $g \in \mathrm{G}(\mathbb{A})$,

$$H_p(s, g_p) = \max\{1, |x|_p, |y|_p, |z|_p\}^s \tag{5.2}$$

and

$$H_\infty(s, g_\infty) = (1 + x^2 + y^2 + z^2)^{s/2}. \tag{5.3}$$

The heights $H_p$ are invariant with respect to the action of $\mathrm{G}(\mathbb{Z}_p)$ (on both sides). We are interested in the analytic properties of the height zeta function

$$\mathcal{Z}(s, g) = \sum_{\gamma=(x,z,y)\in\mathbb{Q}^3} H(s, \gamma g)^{-1}. \tag{5.4}$$

As above, we consider the Fourier expansion of $\mathcal{Z}(s, g)$. Each term in this expansion will be regularized by an explicit Euler product of height integrals. We need to compute these height integrals at good primes and estimate them at bad primes and at the real place.

LEMMA 5.1. *For $\Re(s) > 4$, one has*

$$\int_{\mathrm{G}(\mathbb{A}_{\mathrm{fin}})} H(s, g)^{-1} dg = \frac{\zeta(s-3)}{\zeta(s)}.$$

LEMMA 5.2. *For $\Re(s) > 3$ and all $p \notin S_\eta$, one has*

$$\int_{\mathrm{G}(\mathbb{Q}_p)} H_p(s, g_p)^{-1} \overline{\eta}_{\mathbf{a}}(g_p) dg_p = \zeta_p^{-1}(s).$$

*Proof.* Both lemmas may be proved by direct computation using the definition of $H_p$ in (5.2). □

LEMMA 5.3. *For $\Re(s) > 3$, all $\psi = \psi_a$ and all $p \notin S_\psi$, one has*

$$\int_{\mathrm{G}(\mathbb{Q}_p)} H_p(s, g_p)^{-1} f_p(g_p) dg_p = \int_{\mathrm{U}(\mathbb{Q}_p)} H_p(s, u_p)^{-1} \overline{\psi}_a(u_p) du_p = \zeta_p^{-1}(s)$$

*(where $f_p$ is the local normalized spherical function).*

*Proof.* Direct computation. Note that the second integral is similar to the integral in Lemma 5.2 for the variety $\mathbb{P}^2 \subset \mathbb{P}^3$ (the induced equivariant compactification of U). □

LEMMA 5.4. *For all $\epsilon > 0$, $\mathsf{n} > 0$ and all compacts $\mathsf{K}$ in the domain $\Re(s) > 3 + \epsilon$, there exists a constant $\mathsf{c}(\mathsf{n}, \mathsf{K})$ such that, for all $s \in \mathsf{K}$ and all $\eta = \eta_{\mathbf{a}}$ (with $\mathbf{a} \in \mathbb{Z}^2$), the finite product*

$$\left| \prod_{p\in S_\eta} \int_{\mathrm{G}(\mathbb{Q}_p)} H_p(s, g_p)^{-1} \overline{\eta}(g_p) dg_p \int_{\mathrm{G}(\mathbb{R})} H_\infty(s, g_\infty)^{-1} \overline{\eta}(g_\infty) dg_\infty \right|$$

*is bounded by*

$$\mathsf{c}(\mathsf{n}, \mathsf{K})(1 + |a_1| + |a_2|)^{-\mathsf{n}}.$$

*Proof.* We replace $\eta$ by 1, $H_p(s, g_p)$ by $H_p(\Re(s), g)$ and obtain

$$\left|\int_{\mathrm{G}(\mathbb{Q}_p)} H_p(s, g_p)^{-1} dg_p\right| \leq \frac{1}{1-p^{-\epsilon}}.$$

For $a \in \mathbb{Z}$, we have

$$\prod_{p|a}(1+p^{-\epsilon}) \leq (1+|a|)^{\mathsf{n}'}$$

(for some positive integer $\mathsf{n}'$). Using the definition of $H_\infty$:

$$\left|\int_{\mathbb{R}^3}(1+x^2+y^2+z^2)^{-s/2} e^{-2\pi i(a_1 x + a_2 y)} dx dy dz\right| < \mathsf{c}(\mathsf{n}, \mathsf{K})(1+|a_1|+|a_2|)^{-\mathsf{n}}$$

for all $\mathsf{n}$ (this is an easy consequence of integration by parts). □

LEMMA 5.5. *For all $\epsilon > 0$, $\mathsf{n} > 0$ and all compacts $\mathsf{K}$ in the domain $\Re(s) > 3+\epsilon$, there exists a constant $\mathsf{c}(\mathsf{n}, \mathsf{K})$ such that, for all $s \in \mathsf{K}$, all $\psi = \psi_a$ (with $a \in \mathbb{Z}, a \neq 0$), $S = S_\psi \cup \{\infty\}$, and all $\omega_S \in \mathcal{B}_S(\varrho_\psi)$, the expression*

$$\left|\int_{\mathrm{G}(\mathbb{A}_S)} H(s, g_S)^{-1}\overline{\omega}_S(g_S) dg_S\right|$$

*is bounded by*

$$\mathsf{c}(\mathsf{K}, \mathsf{n})|an|^{-\mathsf{n}}$$

*(where the real component of $j_\psi^{-1}(\omega_S)$ is equal to $c_n\varphi_n^\psi$, cf. Lemma 3.13).*

*Proof.* Let $\lambda$ be the $\Delta$-eigenvalue of $\omega$. Here

$$\Delta = \partial_x^2 + \partial_y^2 + \partial_z^2$$

is an elliptic differential operator on $\mathrm{G}(\mathbb{Z})\backslash\mathrm{G}(\mathbb{R})$, and $\partial_x$ (resp. $\partial_y$, $\partial_z$) is the invariant vector field corresponding to $g(x, 0, 0)$ (resp. $g(0, 0, y)$ and $g(0, z, 0)$). On each subspace

$$\mathcal{H}_\psi^{\mathrm{K}} \subset \mathsf{L}^2(\mathrm{G}(\mathbb{Q})\backslash\mathrm{G}(\mathbb{A}))^{\mathrm{K}} = \mathsf{L}^2(\mathrm{G}(\mathbb{Z})\backslash\mathrm{G}(\mathbb{R})),$$

we have

$$\partial_z\omega = (2\pi i a)\omega$$

(here we used the $\pi_\psi$ realization). It follows that

$$\partial_z^2\omega = -4\pi^2 a^2 \cdot \omega,$$

and

$$\Delta\omega = (\lambda_n^\psi - 4\pi^2 a^2)\omega,$$

where $\lambda_n^\psi = -2\pi(2n+1)|a|$ is the $\Delta_\psi$-eigenvalue of $\varphi_n^\psi$, the real component of $j_\psi^{-1}(\omega_S)$.

After these preparations, we can assume that $s$ is real. Using repeated integration by parts, we find the following estimate for the above integral:

$$\lambda^{-\mathrm{n}} \cdot \|\omega\|_{\mathrm{L}^\infty} \cdot \prod_{p|a} \int_{\mathrm{G}(\mathbb{Q}_p)} H_p(s, g_p)^{-1} dg_p \cdot \int_{\mathbb{R}^3} \Delta^{\mathrm{n}}(1+x^2+y^2+z^2)^{-s/2} dxdydz.$$

Here we have again used the estimates (4.1) and (4.2). Continuing, we estimate the finite product of $p$-adic integrals as in the proof of Lemma 5.4. Finally, we find from Lemma 4.1 that the integral

$$\int_{\mathbb{R}^3} \Delta^{\mathrm{n}}(1+x^2+y^2+z^2)^{-s/2} dxdydz$$

is convergent for $s \in \mathsf{K}$ and, further, is bounded on the same region. □

PROPOSITION 5.6. *The height zeta function $\mathcal{Z}(s)$ defined in (5.4)*
- *is holomorphic for $\Re(s) > 4$;*
- *admits a meromorphic continuation to $\Re(s) > 3 + \epsilon$ (for any $\epsilon > 0$); and*
- *has a simple pole in this domain at $s = 4$ with residue*

$$\zeta(4)^{-1} \int_{\mathbb{R}^3} (1+x^2+y^2+z^2)^{-2} dxdydz = \frac{\pi^2}{\zeta(4)}.$$

*Proof.* Using the estimates of Lemma 5.5, and (4.1) and (4.2), we see, as in the proof of Theorem 4.15, that the series for $\mathcal{Z}_2(s, g)$ in Proposition 3.3 is normally convergent for $s$ in a compact set $\mathsf{K}$ of $\Re(s) > 3$ and for $g \in \mathrm{G}(\mathbb{A})$. It now follows (as in the proof of Theorem 4.15) that, for $\Re(s) > 4$,

$$\mathcal{Z}(s, g) = \mathcal{Z}_0(s, g) + \mathcal{Z}_1(s, g) + \mathcal{Z}_2(s, g),$$

as an equality of continuous functions on $\mathrm{G}(\mathbb{A})$. In particular,

$$\mathcal{Z}(s, id) = \mathcal{Z}_0(s, id) + \mathcal{Z}_1(s, id) + \mathcal{Z}_2(s, id), \tag{5.5}$$

again for $\Re(s) > 4$. We obtain then, as in the proof of Theorem 4.15, the meromorphic continuation of $\mathcal{Z}(s)$ to $\Re(s) > 3 + \epsilon$ (see esp. Lemma 5.1 for $\mathcal{Z}_0$). Finally,

$$(s-4)\mathcal{Z}(s)|_{s=4} = (s-4)\mathcal{Z}_0(s, id)|_{s=4} = \frac{\pi^2}{\zeta(4)}.$$ □

---

## REFERENCES

[1] V. V. Batyrev, Non-archimedean integrals and stringy Hodge numbers of log-terminal pairs, *Journ. EMS* **1** (1999), 5–33.

[2] V. V. Batyrev and Yu. I. Manin, Sur le nombre de points rationnels de hauteur bornée des variétés algébriques, *Math. Ann.* **286** (1990), 27–43.

[3] V. V. Batyrev and Yu. Tschinkel, Manin's conjecture for toric varieties, *Journ. Algebraic Geometry* **7**, no. 1 (1998), 15–53.

[4] ———, Tamagawa numbers of polarized algebraic varieties, in *Nombre et répartition des points de hauteur bornée, Asterisque* **251** (1998), 299–340.

[5] D. Bohm, Quantum theory, *Dover Publ. Inc.*, New York, 1989.

[6] A. Chambert-Loir and Yu. Tschinkel, On the distribution of points of bounded height on equivariant compactifications of vector groups, *Invent. Math.* **148** no. 2 (2002), 421–452.

[7] L. Corwin and F. P. Greenleaf, Representations of nilpotent groups and their applications 1. *Cambridge Studies in Adv. Math.* **18** (1990).

[8] R. Courant and D. Hilbert, *Methods of Mathematical Physics*, vol. 1., Wiley and Sons, 1989.

[9] J. Denef and F. Loeser, Germs of arcs on singular algebraic varieties and motivic integration, *Invent. Math.* **135** (1999), 2001–2232.

[10] J. Franke, Yu. I. Manin and Yu. Tschinkel, Rational points of bounded height on Fano varieties, *Invent. Math.* **95** no. 2 (1989), 421–435.

[11] G. B. Folland, Introduction to partial differential equations, *Princeton Univ.* Press, New Jersey, 1995.

[12] B. Hassett and Yu. Tschinkel, Geometry of equivariant compactifications of $\mathrm{G}_a^n$, *Internat. Math. Res. Notices* **22** (1999), 1211–1230.

[13] R. Howe, Θ-series and invariant theory, *Proc. Symp. Pure Math.* **33** (1979), 275–285.

[14] J.-I. Igusa, An introduction to the theory of local zeta functions, AMS/IP Studies in Advanced Mathematics, 14. AMS, International Press, Cambridge, MA, 2000.

[15] ———, Forms of higher degree, Tata Lectures on Mathematics and Physics, 59. Bombay; by the Narosa Publishing House, New Delhi, 1978.

[16] H. Jacquet and R. Langlands, Automorphic forms on GL(2), *SLN* **114** Springer Verlag, Berlin-Heidelberg-New York, 1970.

[17] S. Lang and A. Weil, Number of points of varieties in finite fields, *Amer. Journ. Math.* **76** (1954), 819–827.

[18] E. Looijenga, Motivic measures, *Séminaire Bourbaki* **874** (2000).

[19] C. Moore, Decomposition of unitary representations defined by discrete subgroups of nilpotent groups, *Ann. of Math.* **82** (1965), 146–182.

[20] E. Peyre, Hauteurs et mesures de Tamagawa sur les variétés de Fano, *Duke Math. Journ.* **79** (1995), 101–218.

[21] ———, Nombre et répartition des points de hauteur bornée, Astérisque **251** (1998).

[22] ———, Torseurs universels et méthode du cercle, "Rational points on algebraic varieties", *Progr. Math.* **199**, Birkhäuser, 2001, 221–274.

[23] D. Ramakrishnan, Modularity of Rankin-Selberg $L$-series and multiplicity one for $SL(2)$, *Annals of Math.* **152** (2000), 45–111.

[24] M. Rosenlicht, Some basic theorems on algebraic groups, *American Journ. of Math.* **78** (1956), 401–443.

[25] C. Sogge, Concerning the $L^p$-norm of spectral clusters of second order elliptic operators on compact manifolds, *Journ. Funct. Anal.* **77** (1988), 123–138.

[26] M. Strauch and Yu. Tschinkel, Height zeta functions of toric bundles over flag varieties, *Selecta Math.* **5,** no. 3 (1999), 325–396.

[27] J. T. Tate, Fourier analysis in number fields, and Hecke's zeta-functions, Algebraic Number Theory (Proc. Instructional Conf., Brighton, 1965), Thompson, Washington, D.C., 1967, 305–347.

[28] A. Weil, Basic Number Theory, Springer-Verlag, Berlin, 1995.

[29] ———, Adeles and algebraic groups, *Progr. Math.* **23,** Birkhäuser, 1982.

CHAPTER 28

# ON HIGHEST WHITTAKER MODELS AND INTEGRAL STRUCTURES

By Marie-France Vignéras

*Abstract.* We show that the integral functions in a highest Whittaker model of an irreducible integral $\overline{\mathbf{Q}}_\ell$-representation of a $p$-adic reductive connected group form an integral structure.

**Introduction.** This work is motivated by a question of E. Urban (March 2001) for the group $Sp(4)$. The fact that the integral Whittaker functions form an integral structure is an ingredient at the nonarchimedean places for deducing congruences between Eisenstein series and cuspidal automorphic forms from congruences between special values of $L$-functions using the theory of Langlands-Shahidi. Many fundamental and deep theorems in the theory of Whittaker models and of $L$-functions attached to automorphic representations of reductive groups with arithmetical applications are due to Joseph Shalika and his collaborators, or inspired by him. Whittaker models and their generalizations as the Shalika models have become a basic tool to study automorphic representations and they may become soon a basic tool for studying congruences between them.

Let $(F, G, \ell)$ be the triple formed by a local nonarchimedean field $F$ of residual characteristic $p$, the group $G$ of rational points of a reductive connected $F$-group, a prime number $\ell$ different from $p$. We denote by $\overline{\mathbf{Q}}_\ell$ an algebraic closure of the field $\mathbf{Q}_\ell$ of $\ell$-adic numbers, $\overline{\mathbf{Z}}_\ell$ the ring of its integers, $\Lambda$ the maximal ideal, $\overline{\mathbf{F}}_\ell = \overline{\mathbf{Z}}_\ell/\Lambda\overline{\mathbf{Z}}_\ell$ the residual field (an algebraic closure of the finite field $\mathbf{F}_\ell$ with $\ell$-elements), $\mathrm{Mod}_{\overline{\mathbf{Q}}_\ell} G$ the category of $\overline{\mathbf{Q}}_\ell$-representations of $G$, $\mathrm{Irr}_{\overline{\mathbf{Q}}_\ell} G$ the subset of irreducible representations. All representations $(\pi, V)$ of $G$ are smooth: the stabilizer of any vector $v \in V$ is open. The dimension of a representation of $G$ is usually infinite.

However, a reductive $p$-adic group tries very hard to behave like a finite group. A striking example of this principle is the strong Brauer-Nesbitt theorem:

Theorem 1. *Let $(\pi, V)$ be a $\overline{\mathbf{Q}}_\ell$-representation of $G$ of finite length, which contains a $G$-stable free $\overline{\mathbf{Z}}_\ell$-submodule $L$. Then the $\overline{\mathbf{Z}}_\ell G$-module $L$ is finitely generated, $L/\Lambda L$ has finite length and the semi-simplification of $L/\Lambda L$ is independent of the choice of $L$.*

This is a stronger version of the Brauer-Nesbitt theorem in [V2, II.5.11.b] because the hypotheses (loc. cit.) contained the property that the $\overline{\mathbf{Z}}_\ell G$-module $L$

Manuscript received September 24, 2001, revised July 7, 2002.

is finitely generated and $\overline{\mathbf{Z}}_\ell$-free. Here we prove that the $\overline{\mathbf{Z}}_\ell$-freeness of $L$ implies that $L$ is $\overline{\mathbf{Z}}_\ell G$-finitely generated.

A representation $(\pi, V) \in \mathrm{Mod}_{\overline{\mathbf{Q}}_\ell} G$ is called *integral* when the vector space $V$ contains a $G$-stable free $\overline{\mathbf{Z}}_\ell$-submodule $L$ containing a $\overline{\mathbf{Q}}_\ell$-basis, and $L$ is called an *integral structure*.

There is not yet a standard notation for the Whittaker models. Our notation is the following. A Whittaker $\overline{\mathbf{Q}}_\ell$-representation of $G$ is associated to a pair $(Y, \mu)$ where $Y$ is a nilpotent element of $\mathrm{Lie}\, G$ and $\mu$ is a cocharacter of $G$ related by $\mathrm{Ad}\,\mu(x)Y = x^{-2}Y$ for all $x \in F^*$. The Whittaker $\overline{\mathbf{Q}}_\ell$-representation of $G$ is an induced representation $\mathrm{Ind}_N^G \Omega$, where $N$ is the unipotent subgroup of $G$ defined by the cocharacter $\mu$ and $\Omega$ is an admissible irreducible representation (character or an infinite dimensional metaplectic representation) of $N$ defined by the nilpotent element $Y$ [MW]. The contragredient $(N, \tilde{\Omega})$ of $(N, \Omega)$ is associated to $(-Y, \mu)$. When $Y = 0$, $\Omega$ is the trivial character of $N$. When $Y$ is regular, i.e., the dimension $d(Y)$ of the nilpotent orbit $\mathcal{O} = \mathrm{Ad}\, G, Y$ is maximal among the dimensions of the nilpotent orbits of $\mathrm{Lie}\, G$, $N$ is a maximal unipotent subgroup and $\Omega$ is a generic character of $N$; the corresponding Whittaker $\overline{\mathbf{Q}}_\ell$-representation of $G$ is called generic. We need the assumption that the characteristic of $F$ is 0 and $p \neq 2$ in order to refer to [MW]. It is clear that a generic Whittaker $\overline{\mathbf{Q}}_\ell$-representation of $G$ can be defined without any assumption on $F$.

Let $\pi \in \mathrm{Mod}_{\overline{\mathbf{Q}}_\ell} G$, which may fail to be irreducible. A Whittaker model of $\pi$ associated to $(Y, \mu)$ is a subrepresentation of $\mathrm{Ind}_N^G \Omega$ isomorphic to $\pi$, if there exists one. If $\pi$ has a model in a generic Whittaker $\overline{\mathbf{Q}}_\ell$-representation of $G$, then $\pi$ is called generic and the model is called a generic Whittaker model. The "highest Whittaker models" of $\pi$ are the Whittaker models of $\pi$ associated to $(Y, \mu)$ when the nilpotent orbit $\mathcal{O}$ is maximal among the nilpotent orbits of $\mathrm{Lie}\, G$ associated to the Whittaker models of $\pi$, when $\pi$ has a Whittaker model. When $\pi$ is irreducible and generic, the generic Whittaker models are the highest Whittaker models of $\pi$.

When $\pi$ is irreducible, the characteristic of $F$ is 0 and $p \neq 2$, a Whittaker model with our definition is called a degenerate Whittaker model in [MW]; the set of Whittaker models of $\pi$ is not empty [MW].

We relate now the Whittaker models with the integral structures. The representation $\Omega$ has a natural integral structure $L_\Omega$ but the induction does not respect integral structures: in general, the $\overline{\mathbf{Z}}_\ell G$-submodule $\mathrm{Ind}_H^G L_\Omega$ is not $\overline{\mathbf{Z}}_\ell$-free and does not generate $\mathrm{Ind}_N^G \Omega$, and the Whittaker representation $\mathrm{Ind}_H^G \Omega$ is not integral. However, we have the following remarkable property.

THEOREM 2. *Let $\pi \in \mathrm{Mod}_{\overline{\mathbf{Q}}_\ell} G$ admissible and let $V \subset \mathrm{Ind}_N^G \Omega$ be a highest Whittaker model of $\pi$. Then the two following properties (1) and (2) are equivalent:*

(1) *$\pi$ is integral.*

(2) *The functions in $V$ with values in $L_\Omega$ form an integral structure of $\pi$.*

*Under the restriction on $(F, \pi)$, the characteristic of $F$ is 0 and $p \neq 2$, $\pi$ is irreducible.*

*When $V$ is a generic Whittaker model of $\pi$, the equivalence is true without restriction on $(F, \pi)$.*

As (2) implies clearly (1), the key point is to show that (1) implies (2). We prove that (1) implies (2) iff any element $v$ of $V$ has a denominator, i.e., the values of a multiple of $v$ belong to $L_\Omega$ (II.5), and we give two general criteria A, B for this property (II.6 and II.7).

Criterion A given in (II.6) is that $(\pi, V)$ contains an integral structure $L$ such that the $\Omega$-coinvariant $p_\Omega L$ is $\overline{\mathbf{Z}}_\ell N$-finitely generated. This is an integral version of the fact that the $\Omega$-coinvariant $p_\Omega V$ is finite dimensional (Moeglin and Waldspurger) when $V$ is a highest Whittaker model of $\pi \in \mathrm{Irr}_{\overline{\mathbf{Q}}_\ell} G$ attached to $(N, \Omega)$. To explain the method due to Rodier, let us suppose that $\Omega$ is a character. One approximates $(N, \Omega)$ by characters $\chi_n$ of open compact pro-p-subgroups $K_n$ of $G$. The key point is to prove that the projection $p_\Omega$ on the $(N, \Omega)$-coinvariants restricts to an isomorphism $e_n V \simeq p_\Omega V$, where $e_n$ is the projector on the $(K_n, \chi_n)$-invariants, when $n$ is big enough. Recall that $V$ is admissible, hence $p_\Omega V$ is finite dimensional. The tool to prove the isomorphism is the expansion of the trace of $\pi$ around 1. As $e_n L$ is a lattice of $e_n V$ for any integral structure $L$ of $(\pi, V)$, criterion A is satisfied if $p_\Omega$ restricts to an isomorphism $e_n L \simeq p_\Omega L$. This is proved in Section III by a careful analysis of the proof of [MW].

Compact induction behaves well for integral structures. A compact Whittaker representation $\mathrm{ind}_H^G \Omega$ is integral and $\mathrm{ind}_H^G L_\Omega$ is an integral structure. The Whittaker representation $\mathrm{Ind}_N^G \Omega$ is the contragredient of the compact Whittaker representation $\mathrm{ind}_N^G \tilde{\Omega}$, where $\tilde{\Omega}$ is the contragredient of $\Omega$ because $\Omega$ is admissible and $N$ unimodular. The criterion B given in (II.7) is a property of the $K$-invariants of $\mathrm{ind}_N^G \tilde{L}_\Omega$ as a right module for the Hecke algebra of $(G, K)$ when $K$ is an open compact subgroup of $G$. It is an integral version of a finiteness theorem: the component of $\mathrm{ind}_N^G \tilde{\Omega}$ in any Bernstein block is finitely generated. In the generic case and without restriction on $(F, \pi)$, this has been recently proved by Bushnell and Henniart [BH 7.1]. Their proof is well adapted to criterion B and one can, after some simplifications, obtain that a generic compact Whittaker representation satisfies criterion B. This is done in Section IV.

For a generic irreducible representation with the restriction on $F$, we get two very different proofs of the Theorem 2, using criteria A and B. For $GL(n, F)$ with no restriction on $F$, when the representation is also cuspidal, a third proof was known and showed that modulo homotheties, the Kirillov model is the unique integral structure [V4]. The Kirillov integral model was used for $GL(2, F)$ to prove that the semi-simple local Langlands correspondence modulo $\ell$ is uniquely defined by equalities between $\epsilon$ factors [V6]. The characterization of the local Langlands correspondence modulo $\ell$ in the general case $n > 2$ by $L$ and $\epsilon$ factors remains open. Probably the case $n = 3$ is accessible.

As noticed by Jacquet and Shalika for $GL(n, F)$, the Whittaker models of representations induced from tempered irreducible representations are useful. Being aware of future applications, we did not consider only integral models of irreducible representations. The criteria A, B, as well as Theorem 1 and the generic case of Theorem 2 are given for representations that may fail to be irreducible.

In the appendix, we compare, for a representation $V$ of $G$ over an algebraically closed field $R$ of characteristic $\neq p$, the three properties:

(i) the right $\mathcal{H}_R(G, K)$-modules $V^K$ are finitely generated for the open compact subgroups $K$ of $G$;

(ii) the components of $V$ in the blocks of $\mathrm{Mod}_R\, G$ are finitely generated;

(iii) the irreducible quotients of $V$ have finite multiplicity.

The criterion A is an integral version of (iii), the criterion B is an integral version of (i). The property (i) is equivalent (ii) in the complex case [BH] and it is clear that (ii) implies (iii) but is not equivalent. We give a proof of the equivalence between (i) and (ii) in the modular case, and in the complex case we give certain properties of $V$ and of its Jacquet functors implying the equivalence between (ii) and (iii). For instance, the complex representation of $GL(2, F)$ compactly induced from a character of a maximal (split or not split) torus and its coinvariants by a unipotent subgroup satisfy this properties. This representation introduced by Waldspurger and studied also by Tunnel, plays a role in the arithmetic theory of automorphic forms.

*Acknowledgments.* I thank the Institute for Advanced Study for its invitation during the spring term 2001, where this work started and was completed in the best possible conditions. I thank Guy Henniart and Steve Rallis for discussions on Gelfand-Graev-Whittaker models. I thank also the C.N.R.S. for the delegation that allows me to come here and to do research full-time for one year.

**I. Proof of the strong Brauer-Nesbitt theorem.** Let $(\pi, V)$ be a finite length $\overline{\mathbf{Q}}_\ell$-representation of $G$ which contains a $G$-stable free $\overline{\mathbf{Z}}_\ell$-submodule $L$. We will prove that the $\overline{\mathbf{Z}}_\ell G$-module $L$ is finitely generated. The rest of the theorem follows from the Brauer-Nesbitt theorem proved in [V2, II.5.11.b].

The proof uses an unrefined theory of types for $G$ as in [V2, II.5.11.b]. One can take either the mottes [V1] or the more sophisticated Moy-Prasad types.

The subrepresentation $\pi'$ of $\pi$ generated by $L$ has finite length and we may suppose that $\pi = \pi'$ is generated by $L$.

One may replace $\overline{\mathbf{Z}}_\ell$ by the ring of integers $O_E$ of a finite extension $E$ of $\mathbf{Q}_\ell$ as in [V II.4.7]. What is important is that $O_E$ is a principal local ring. Let $p_E$ be a generator of the maximal ideal, let $k_E := O_E/p_E O_E$ be the residual field.

The theory of unrefined types shows that $L/p_E L \in \mathrm{Mod}_{k_E}\, G$ has finite length because it contains only finitely many unrefined minimal types modulo $G$-conjugation ([V II.5.11.a], where the condition $O_E G$-finitely generated is useless).

Let $m$ be the length of $L/p_E L$. We will prove the $\overline{\mathbf{Z}}_\ell G$-module $L$ is generated by $m$ elements.

We cannot conclude immediately because the free $O_E$-module $L$ is usually not of finite rank. As $\ell \neq p$, the open compact pro-p-subgroups $K$ of $G$ form a fundamental system of neighborhoods of 1. The finite length $\overline{\mathbf{Q}}_\ell$-representation $(\pi, V)$ of $G$ is admissible: for any open compact pro-p-subgroup $K$ of $G$, the $E$-dimension of the vector space $V^K$ is finite. The $O_E$-modules $L^K$ are free of finite rank. By smoothness we have $L = \cup_K L^K$.

The $k_E G$-module $L/p_E L$ is generated by $m$ elements $w_1, \ldots, w_m$. We lift these elements arbitrarily to $v_1, \ldots, v_m$ in $L$ and we consider the $O_E G$-submodule $L'$ of $L$ that they generate. As $O_E$ is principal and $L$ is $O_E$-free, the $O_E$-submodule $L'$ of $L$ is $O_E$-free. We have by construction

$$L = L' + p_E L.$$

The $O_E$-modules $L'^K$, $L^K$ are free of finite rank and $L^K = L'^K + p_E L^K$. The theory of invariants for free modules of finite rank over a principal ring implies that $L'^K = L^K$. As $L = \cup_K L^K$, $L' = \cup_K L'^K$, we deduce $L = L'$. Thus Theorem 1 is proved.

**II. Integral structures in induced representations (criteria A and B).** The framework of this section is very general, $R$ is a commutative ring and $G$ is a locally profinite group that contains an open compact subgroup $C$ of pro-order invertible in $R$, such that $G/C$ is countable. The criteria A and B are given in (II.6) and (II.7). The proofs are given at the end of the section.

**II.1.** We fix the notations:

$\mathrm{Mod}_R$ is the category of $R$-modules;

$\mathrm{Mod}_R\, G$ is the category of smooth representations of $G$ on $R$-modules;

$\mathrm{Irr}_R\, G$ is the subset of irreducible representations;

$H$ is a closed subgroup of $G$;

$O_E$ is a principal ring with quotient field $E$;

$(\Omega, W) \in \mathrm{Mod}_E\, H$ of countable dimension;

$\mathrm{Ind}_H^G(\Omega, W) \in \mathrm{Mod}_E\, G$ is the space of functions $f : G \to W$ right invariant by some open compact subgroup $K_f$, with functional equation $f(hg) = \Omega(h) f(g)$ for $h, g \in H, G$, with the action of $G$ by right translations;

$\mathrm{ind}_H^G(\Omega, W) \in \mathrm{Mod}_E\, G$ is the compactly induced representation, subrepresentation of $\mathrm{Ind}_H^G(\Omega, W)$ on the functions $f$ with compact support modulo $H$.

We often forget the module $V$ or the action $\pi$ in the notation $(\pi, V)$ of a representation.

The induced representation $\operatorname{Ind}_H^G \Omega$ and the compactly induced representation $\operatorname{ind}_H^G \Omega$ can be equal even when $G$ is not compact modulo $H$. There are two typical examples with $\operatorname{ind}_H^G \Omega = \operatorname{Ind}_H^G \Omega$:

– a metaplectic representation [MVW I.3, I.6]: $G$ is a $p$-adic Heisenberg group of center $Z$, $H$ is a maximal commutative subgroup of $G$, $\ell \neq p$ a prime number, $E$ is the field generated over $\overline{\mathbf{Q}}_\ell$ by the roots of 1 of order any power of $p$ (the ring of integers $O_E$ is principal), $\Omega$ is an $E$-character of $H$ nontrivial on $Z$.

– a cuspidal representation [V5]: $G$ is a $p$-adic connected reductive group, $H$ is the normalizer in $G$ of a maximal parahoric subgroup $K$, $R$ is an algebraically closed field of characteristic $\neq p$, $\Omega \in \operatorname{Irr}_R H$ such that $\Omega|_K$ contains the inflation of a cuspidal irreducible representation of the quotient $K/K_p$ (a finite connected reductive group).

A representation $(\pi, V) \in \operatorname{Mod}_E G$ is called $O_E$-integral when it contains an $O_E$-integral structure $L$, i.e., a $G$-stable $O_E$-free submodule $L$ that contains an $E$-basis of $V$.

**II.2.** Let $L$ be an $O_E$-integral structure of a representation $(\pi, V) \in \operatorname{Mod}_E G$ and let $(\pi', V')$ be a subrepresentation of $(\pi, V)$. Then $L' := L \cap V$ is an $O_E$-integral structure of $(\pi', V')$.

This is a basic fact with an easy proof: clearly $L'$ is $G$-stable; as $O_E$ is principal and the $O_E$-module $L$ is free, the $O_E$-submodule $L' \subset L$ is free; if $(v_i)_{i\in I}$ is a basis of $V'$ then for each $i \in I$ there exists $a_i \in O_E$ such that $v_i a_i \in L$ hence $v_i a_i \in L'$. Therefore $L'$ is an $O_E$-integral structure of $V'$.

In contrast with (II.2): *a quotient of an integral representation is not always integral.* A counter-example is given after (II.3).

We suppose in this section that $(\Omega, W) \in \operatorname{Irr}_E H$ is *$O_E$-integral with an $O_E G$-finitely generated, $O_E$-integral structure $L_W$.* Are the induced representations $\operatorname{Ind}_H^G \Omega$ and $\operatorname{ind}_H^G \Omega$ integral? In general, the induced representation without condition on the support is not integral by (II.2) because $\operatorname{Ind}_H^G \Omega$ may contain a nonintegral irreducible representation. This contrasts with the compactly induced representation $\operatorname{ind}_H^G \Omega$, which is integral.

PROPOSITION II.3. $\operatorname{ind}_H^G L_W$ *is an $O_E$-integral structure of* $\operatorname{ind}_H^G(\Omega, W)$.

The integral representation $\operatorname{ind}_H^G \Omega$ may have nonintegral quotients: $\operatorname{ind}_1^{\mathbf{Q}_p^*} 1_E$ is integral but there are characters of $\mathbf{Q}_p^*$ with values not contained in $O_E^*$.

The $O_E$-module $\operatorname{Ind}_H^G L_W$ is clearly $G$-stable. But when $\operatorname{ind}_H^G \Omega \neq \operatorname{Ind}_H^G \Omega$, the $O_E$-module $\operatorname{Ind}_H^G L_W$ is not free and does not contain a basis of $\operatorname{Ind}_H^G W$. Hence the following property is particularly nice:

PROPOSITION II.4. *For any admissible subrepresentation* $(\pi, V)$ *of* $\operatorname{Ind}_H^G(\Omega, W)$, *the $O_E$-module* $V \cap \operatorname{Ind}_H^G L_W$ *is free or zero.*

Clearly $V \cap \operatorname{Ind}_H^G L_W$ is $G$-stable, hence $V \cap \operatorname{Ind}_H^G L_W$ is an $O_E$-integral structure of $(\pi, V)$ if and only if any element of $V$ has a nonzero multiple in $\operatorname{Ind}_H^G L_W$.

*Denominators.* Let $(\pi, V) \subset \operatorname{Ind}_H^G(\Omega, W)$. We say that $v \in V$ has a denominator if there exists $a \in O_E$ nonzero with $av \in \operatorname{Ind}_H^G L_W$. We say that $V$ has a bounded denominator if there exists $a \in O_E$ nonzero and an $E$-basis of $V$ with $av \in \operatorname{Ind}_H^G L_W$ for all $v$ in the basis.

Two $O_E G$-finitely generated, $O_E$-integral structures $L_W, L'_W$ of $(\Omega, W) \in \operatorname{Irr}_E H$ are commensurable:

$$aL_W \subset L'_W \subset bL_W, \quad \text{for some } a, b \in O_E$$

and the definition of a denominator or of a bounded denominator does not depend on the choice of $L_W$. Any element of $V$ has a denominator iff every element in a set of generators of $V$ has a denominator. If $(\pi, V)$ is finitely generated, any element of $V$ has a denominator iff $V$ has a bounded denominator; this is false if $(\pi, V)$ is not finitely generated. From (II.4) we deduce:

COROLLARY II.5. *Let* $(\pi, V) \in \operatorname{Mod}_E G$ *admissible contained in* $\operatorname{Ind}_H^G(\Omega, W)$. *Then any element of* $V$ *has a denominator iff* $V \cap \operatorname{Ind}_H^G L_W$ *is an* $O_E$*-integral structure of* $(\pi, V)$.

We give two criteria A in (II.6), B in (II.7) for this property.

**II.6.** Criterion A uses the $H$-equivariant projection

$$p_\Omega : V \to V_\Omega$$

on the $\Omega$-coinvariants $V_\Omega$ of $(\pi, V) \in \operatorname{Mod}_E G$; by definition $V_\Omega$ is the maximal semi-simple $\Omega$-isotypic quotient of the restriction of $(\pi, V)$ to $H$.

CRITERION A. *Let* $(\pi, V) \in \operatorname{Mod}_E G$ *contained in* $\operatorname{Ind}_H^G(\Omega, W)$. *If* $(\pi, V)$ *contains an* $O_E$*-integral model* $L$ *such that the* $O_E H$*-module* $p_\Omega L$ *is finitely generated, then* $V$ *has a bounded denominator.*

Criterion A is equivalent to: $V_\Omega$ is isomorphic to a finite sum $\oplus^{m(\pi)}\Omega$ and $p_\Omega L$ is an $O_E$-structure of $V_\Omega$. This is clear except may be the $O_E$-freeness of $p_\Omega L$ that results from the fact that $O_E$ is principal and that a multiple of $p_\Omega L$ is contained in the $O_E$-integral structure of $V_\Omega$ defined by $L_W$. By adjunction

$$m(\pi) = \dim_E \operatorname{Hom}_{EG}\left(\pi, \operatorname{Ind}_H^G \Omega\right).$$

Criterion A is an integral version of the finite multiplicity of $\pi$ in $\operatorname{Ind}_H^G \Omega$.

In the section III, for $(F, G, \ell)$ as in the introduction under the restriction on $F$ given in the Theorem 2, we will prove that any highest Whittaker model of

$(\pi, V) \in \mathrm{Irr}_{\overline{\mathbf{Q}}_\ell} G$ satisfies the criterion A. As $(\pi, V)$ is admissible, it follows that (1) implies (2) in Theorem 2.

**II.7.** Criterion B uses the Hecke algebras. One denotes by $\simeq_R$ an isomorphism of $R$-modules. For any open compact subgroup $K$ of $G$, one defines the Hecke $R$-algebra of $(G, K)$,

$$\mathcal{H}_R(G, K) := \mathrm{End}_{RG}\, R[K\backslash G] \simeq_R R[K\backslash G/K].$$

For $g \in G$, the $RG$-endomorphism of $R[K\backslash G]$ sending the characteristic function of $K$ to the characteristic function of $KgK$ identifies with the natural image $[KgK]$ of $KgK$ in $R[K\backslash G/K]$. The set $V^K$ of $K$-invariants of $(\pi, V) \in \mathrm{Mod}_R\, G$, has a natural structure of right $\mathcal{H}_R(G, K)$-module, which satisfies for any $v \in V^K$, $g \in G$:

$$v * [KgK] = \sum_h \pi(h)^{-1} v, \tag{1}$$

where $KgK = \cup_h Kh$ (disjoint union).

CRITERION B. *We suppose that the $\mathcal{H}_{O_E}(G, K)$-module $(\mathrm{ind}_H^G L_W)^K$ is finitely generated for all $K$ in a separated decreasing sequence of open compact subgroups of $G$ of pro-order invertible in $O_E$. Let $(\pi, V) \in \mathrm{Mod}_E\, G$ be a quotient of $\mathrm{ind}_H^G(\Omega, W)$. Then $(\pi, V)$ is $O_E$-integral iff the image of $\mathrm{ind}_H^G L_W$ in $V$ is an $O_E$-integral structure of $(\pi, V)$.*

Criterion B does not depend on the choice of $L_W$. There is no restriction on $(\pi, V)$. Its application to the integral structures of subrepresentations of $\mathrm{Ind}_H^G$ $(\Omega, W)$ is obtained by using the contragredient (II.8.3); for the contragredient, we need to restrict to admissible representations.

Criterion B implies that the $\mathcal{H}_E(G, K)$-modules $(\mathrm{ind}_H^G \Omega)^K$ are finitely generated. This implies that for any admissible representation$(\pi, V) \in \mathrm{Mod}_E\, G$,

$$m_K(\pi) := \dim_E \mathrm{Hom}_{\mathcal{H}_E(G,K)}\left(\left(\mathrm{ind}_H^G \Omega\right)^K, \pi^K\right) < \infty.$$

The converse is false in general, the finite multiplicity $m_K(\pi) < \infty$ for all $(\pi, V) \in \mathrm{Mod}_E\, G$ does not implies that the $\mathcal{H}_E(G, K)$-modules $(\mathrm{ind}_H^G \Omega)^K$ are finitely generated.

For $(F, G, \ell)$ as in the introduction, we will prove in (IV.2.1) that any generic compact Whittaker $\overline{\mathbf{Q}}_\ell$-representation of $G$ satisfies the Criterion B. Therefore (1) implies (2) in the Theorem 2 for any generic admissible representation, without restriction on $F$.

**II.8.** We recall some general properties of the contragredient. The contragredient $(\tilde{\pi}, \tilde{V}) \in \mathrm{Mod}_R\, G$ of an $R$-representation $(\pi, V) \in \mathrm{Mod}_R\, G$ of $G$ is given the natural action of $G$ on the smooth linear forms of $V$ [V2, I.4.12]. The representation $(\pi, V)$ is called reflexive when $(\pi, V)$ is the contragredient of $(\tilde{\pi}, \tilde{V})$.

The contragredient$\tilde{}$: $\mathrm{Mod}_E\, G \to \mathrm{Mod}_E\, G$ is exact [V2, I.4.18] and relates the induced representation to the compactly induced representation

$$\left(\mathrm{ind}_H^G\, \Omega\right)\tilde{} \simeq \mathrm{Ind}_H^G(\Omega\tilde{} \otimes \delta_H),$$

where $\delta_H$ is the module of $H$ [V2, I.5.11].

Admissible representations of $\mathrm{Mod}_E\, G$ are reflexive and conversely [V2, I.4.18]. Note that the induced representations $\mathrm{Ind}_H^G\, \Omega$, $\mathrm{ind}_H^G\, \Omega$ are not admissible in general. To apply the Criterion B to a subrepresentation $(\pi, V)$ of $\mathrm{Ind}_H^G(\Omega, W)$, we suppose $(\pi, V)$ *and* $(\Omega, W)$ *admissible so* that:

$$\left(\mathrm{ind}_H^G\, \tilde{\Omega} \otimes \delta_H^{-1}\right)\tilde{} \simeq \mathrm{Ind}_H^G\, \Omega$$

and $(\tilde{\pi}, \tilde{V})$ *is a quotient of* $\mathrm{ind}_H^G(\tilde{\Omega} \otimes \delta_H^{-1}, \tilde{W})$.

The assertion on the quotient results from a property (II.8.1) of the following isomorphism [V2, I.4.13]: Let $V_1, V_2 \in \mathrm{Mod}_R\, G$, then there is an isomorphism

$$\mathrm{Hom}_{RG}(V_1, \tilde{V}_2) \simeq \mathrm{Hom}_{RG}(V_2, \tilde{V}_1)$$

sending $f \in \mathrm{Hom}_{RG}(V_1, \tilde{V}_2)$ to $\phi \in \mathrm{Hom}_{RG}(V_2, \tilde{V}_1)$ such that

$$< f(v_1), v_2 >=< v_1, \phi(v_2) > \quad \text{for all } v_1 \in V_1, v_2 \in V_2$$

(for $\tilde{v} \in \tilde{V}, v \in V$, one denotes $\tilde{v}(v)$ by $< \tilde{v}, v >$ or by $< v, \tilde{v} >$).

*Claim* II.8.1. Suppose that $R$ is a field. If $\phi$ is surjective then $f$ is injective; if $V_1$ is admissible then the converse is true.

An integral $O_E$-structure $L$ (II.1) of an admissible representation $(\pi, V) \in \mathrm{Mod}_E\, G$ is an admissible integral $O_E$-structure in the sense of [V2, I.9.1-2] and conversely. The contragredient $\tilde{L}$ of $L$ in $\mathrm{Mod}_{O_E}\, G$ is an $O_E$-structure of $(\tilde{\pi}, \tilde{V})$ [V2, I.9.7], and $L$ is reflexive in $\mathrm{Mod}_{O_E}\, G$, i.e., the contragredient of $\tilde{L}$ is equal to $L$. These results are false without the admissibility.

The values of the module $\delta_H$ are units in $O_E$ hence $\tilde{L}_W \subset \tilde{W}$ is stable by the action of $\tilde{\Omega} \otimes \delta_H^{-1}$. The $O_E$-module $\tilde{L}_W$ is an $O_E$-integral structure of $(\tilde{\Omega} \otimes \delta_H^{-1}, \tilde{W})$. The space of $\mathrm{ind}_H^G(\tilde{\Omega} \otimes \delta_H^{-1}, \tilde{L}_W) \in \mathrm{Mod}_{O_E}\, G$ is the $O_E$-module of functions $f \in \mathrm{ind}_H^G(\tilde{\Omega} \otimes \delta_H^{-1}, \tilde{W})$ with values in $\tilde{L}_W$.

$\mathrm{ind}_H^G(\tilde{\Omega} \otimes \delta_H^{-1}, \tilde{L}_W)$ *is an* $O_E$*-integral structure of* $\mathrm{ind}_H^G(\tilde{\Omega} \otimes \delta_H^{-1}, \tilde{W})$ by (II.3).

$(\mathrm{Ind}_H^G\, \Omega, \mathrm{Ind}_H^G\, L_W)$ *is the contragredient of* $\mathrm{ind}_H^G(\tilde{\Omega} \otimes \delta_H^{-1}, \tilde{L}_W)$ by [V2, I.5.11].

But $\mathrm{Ind}_H^G\, L_W$ is not an $O_E$-integral structure of $\mathrm{Ind}_H^G(\Omega, W)$ in general.

We deduce:

Let $(\pi, V) \in \mathrm{Mod}_E\, G$ admissible, $O_E$-integral, and contained in $\mathrm{Ind}_H^G(\Omega, W)$. Then $(\tilde{\pi}, \tilde{V}) \in \mathrm{Mod}_E\, G$ is admissible, $O_E$-integral, and a quotient of $\mathrm{ind}_H^G(\tilde{\Omega} \otimes \delta_H^{-1}, \tilde{W})$.

The image $L'$ of $\mathrm{ind}_H^G(\tilde{\Omega} \otimes \delta_H^{-1}, \tilde{L}_W)$ in $\tilde{V}$ is always nonzero. When $L'$ is an $O_E$-integral structure of $(\tilde{\pi}, \tilde{V})$, then $\tilde{L}'$ is an $O_E$-integral structure of $(\pi, V)$.

PROPOSITION II.8.2. *Let $(\Omega, W) \in \mathrm{Irr}_E\, H$ admissible and let $(\pi, V) \in \mathrm{Mod}_E\, G$ admissible contained in $\mathrm{Ind}_H^G(\Omega, W)$ and $O_E$-integral. The following properties are equivalent:*

– *$L := V \cap \mathrm{Ind}_H^G L_W$ contains an $E$-basis of $V$,*

– *the image $L'$ of $\mathrm{ind}_H^G(\tilde{\Omega} \otimes \delta_H^{-1}, \tilde{L}_W)$ in $\tilde{V}$ is $O_E$-free.*

– *$L, L'$* are *$O_E$-integral structures of $(\pi, V), (\tilde{\pi}, \tilde{V})$, contragredient of each other.*

*Remarks:* (i) When $\pi$ is irreducible, the first property is equivalent to $L \neq 0$.

(ii) When $L'$ is $O_E G$-finitely generated, the second property is satisfied because a multiple of $L'$ is contained in an $O_E$-integral structure of $(\tilde{\pi}, \tilde{V})$ and $O_E$ is principal.

With Criterion B (II.7) we deduce:

COROLLARY II.8.3. *Suppose that the $\mathcal{H}_{O_E}(G, K)$-module $\mathrm{ind}_H^G(\tilde{\Omega} \otimes \delta_H^{-1}, \tilde{L}_W)^K$ is finitely generated for all $K$ as in (II.7). Let $(\pi, V) \subset \mathrm{Ind}_H^G(\Omega, W)$ admissible. Then $(\pi, V)$ is $O_E$-integral iff $V \cap \mathrm{Ind}_H^G L_W$ is an $O_E$-integral structure of $(\pi, V)$.*

**Proofs of II.3, II.4, II.6, II.7, II.8.**

*Proof of* II.3. Let $K$ be an arbitrary open compact subgroup of $G$ of pro-order invertible in $O_E$. We have the Mackey relations [V2, I.5.5]:

$$\text{(II.3.1)} \qquad \left(\mathrm{ind}_H^G W\right)^K = \oplus_{HgK}\, \mathrm{ind}_H^{HgK} W, \quad \mathrm{ind}_H^{HgK} W \simeq_R W^{H \cap gKg^{-1}}.$$

The hypotheses on $G, H, W$ insure that the dimension of $\mathrm{ind}_H^G W = \cup_K (\mathrm{ind}_H^G W)^K$ is countable. The relations (II.3.1) are valid for any $O_E$-representation of $H$. We apply them to $L_W \in \mathrm{Mod}_{O_E} H$. As $O_E$ is principal and $L_W$ is an $O_E$-free module that generates $W$, the $O_E$-module $L_W^{H \cap gKg^{-1}}$ is free and generates $W^{H \cap gKg^{-1}}$. We deduce that the $O_E$-module $(\mathrm{ind}_H^G L_W)^K$ is free and contains a basis of $(\mathrm{ind}_H^G W)^K$.

As $K$ is arbitrary, this implies that $\mathrm{ind}_H^G L_W$ contains a basis of the vector space $\mathrm{ind}_H^G W$ and is free as an $O_E$-module, by the characterization of free modules on a principal commutative ring [V2, I.9.2 or I.C.4]. □

*Proof of* II.4. Let $(e_i)_{i\in I}$ be an $O_E$-basis of $(\operatorname{ind}_H^G L_W)^K$. We have

$$\left(\operatorname{Ind}_H^G W\right)^K = \prod_{i\in I} Ee_i, \quad \left(\operatorname{Ind}_H^G L_W\right)^K = \prod_{i\in I} O_E e_i.$$

We suppose, as we may, $\operatorname{Ind}_H^G W \neq \operatorname{ind}_H^G W$; the set $I$ is infinite and countable. The $E$-dimension $N$ of $V^K$ if finite because $V$ is admissible. Let $(v_j)_{1\le j\le N}$ be an $E$-basis of $V^K$. We write $v_j = \sum_{i\in I} x_{j,i}e_i$ with $x_{j,i} \in E$, and the support of the map $i \to x_{j,i}$ is finite. We can extract a square submatrix $A = (x_{j,i})$ for $i = i_1, \dots, i_N$ and $1 \le j \le N$ of nonzero determinant; the projection $p : V^K \to \oplus_{1\le k\le N} Ee_{i_k}$ is an isomorphism. The projection $p$ restricts to an injective $O_E$-homomorphism

$$V^K \cap \left(\operatorname{Ind}_H^G L_W\right)^K = \left(V \cap \operatorname{Ind}_H^G L_W\right)^K \to \oplus_{1\le k\le N} O_E e_{i_k}.$$

As $O_E$ is principal, the $O_E$-submodule $p(V \cap \operatorname{Ind}_H^G L_W)^K$ of $\oplus_{1\le k\le N} O_E e_{i_k}$ is $O_E$-free or zero. This is true for all $K$ and we deduce that $V \cap \operatorname{Ind}_H^G L_W$ is $O_E$-free or zero as in the proof of II.3. □

*Proof of* II.6. The value at 1 defines an $H$-equivariant nonzero linear form $V \to W$, and hence factorizes through $p_\Omega V$. There exists an $H$-equivariant linear map $q : p_\Omega V \to W$ such that $v(1) = q \circ p_\Omega(v)$ for all $v \in V$. As $V_\Omega$ is semi-simple, $q$ splits and we can suppose that $q$ corresponds to the first projection $\oplus^{m(\pi)} W \to W$.

By hypothesis $p_\Omega(L)$ is $O_E H$-finitely generated, the same is true for its image by the $H$-equivariant linear map $q$, therefore there exists $a \in O_E$ such that $a(q \circ p_\Omega)L \subset L_W$. Let $(v, g) \in L \times G$ arbitrary. We have $v(g) = gv(1) = q \circ p_\Omega(gv)$ and $gv \in L$, hence $av(g) \in L_W$, that is, $aL \subset \operatorname{Ind}_H^G L_W$. As $L$ contains an $E$-basis of $V$, $V$ has a bounded denominator. □

*Proof of* II.7. We suppose that $(\pi, V)$ is $O_E$-integral. We want to prove that the image $L$ of $\operatorname{ind}_H^G L_W$ in $V$ is an $O_E$-integral structure of $(\pi, V)$. Clearly $L$ is $G$-stable and generates the $E$-vector space $V$. The only property that needs some argument is the $O_E$-freeness of $L$. As in the proofs of (II.3) and of (II.6) it is equivalent to prove that $L^K$ is contained in a $O_E$-free module for all $K$, as in the Criterion B, with $V^K \neq 0$. This results from the fact that the right $\mathcal{H}_{O_E}(G, K)$-module $L^K$ is finitely generated, being the quotient of $(\operatorname{ind}_H^G L_W)^K$ ( as $p \neq \ell$, the $K$-invariant functor is exact), hence a multiple of $L^K$ is contained in an $O_E$-structure of $(\pi, V)$, and $O_E$ is principal. □

*Proof of* II.8.1. $f$ is not injective iff there exist $v_1 \in V_1$ nonzero such that $f(v_1) = 0$, i.e., $< \phi(v_2), v_1 >= 0$ for any $v_2 \in V_2$. Let $K$ be an open compact subgroup of $G$ of pro-order invertible in $R$ such that $v_1 \in V_1^K$. Then $(\tilde{V}_1)^K$ is the linear dual of $V_1^K$, and as we supposed that $R$ is a field, there exists a linear form of $V_1^K$ that does not vanish on $v_1$. Hence $\phi$ is not surjective.

$\phi$ is not surjective iff there exists $K$, as above, such that $\phi(V_2)^K$ is not the linear dual of $V_1^K$. Suppose $V_1$ admissible. The vector space $V_1^K$ is finite dimensional and $\phi(V_2)^K$ is not the linear dual of $V_1^K$ iff there exists $v_1 \in V_1^K$ nonzero such that $\phi(V_2)^K$ vanish on $v_1$. Hence $f$ is not injective. □

*Proof of* II.8.2. (a) By definition $L = V \cap \mathrm{Ind}_H^G L_W$ and $L'$ is the image in $\tilde{V}$ of $\mathrm{ind}_H^G \tilde{L}_W$ (we supressed $\Omega$, $\tilde{\Omega} \otimes \delta_H^{-1}$ to simplify).

An element $v \in \mathrm{Ind}_H^G W$ belongs to $\mathrm{Ind}_H^G L_W$ iff $< v, \phi > \in O_E$ for all $\phi \in \mathrm{ind}_H^G(\tilde{L}_W)$, because $\mathrm{Ind}_H^G(\Omega, L_W)$ is the contragredient of $\mathrm{ind}_H^G(\tilde{\Omega} \otimes \delta_H^{-1}, \tilde{L}_W)$. An element $\phi \in \mathrm{ind}_H^G(\tilde{W})$ acts on $V$ via the quotient map $\mathrm{ind}_H^G(\tilde{W}) \to \tilde{V}$.

We deduce that $L$ is the set of $v \in V$ such that $< v, \phi > \in O_E$ for all $\phi \in \mathrm{ind}_H^G(\tilde{L}_W)$ and $< L', L > \subset O_E$.

(b) Suppose that $L'$ is $O_E$-free. Then $L'$ is an $O_E$-integral structure of $(\tilde{\pi}, \tilde{V})$. Its contragredient $\tilde{L}'$ is equal to $L$ by the above description of $L$. Hence $L = \tilde{L}'$ is an $O_E$-integral structure of $(\pi, V)$.

(c) Suppose that $L$ contains an $E$-basis of $V$, that is, by (II.4), $L$ is an $O_E$-integral structure of $(\pi, V)$. Its contragredient $\tilde{L}$ is an $O_E$-integral structure of $\tilde{V}$. By the last formula in (a), $L' \subset \tilde{L}$ hence $L'$ is $O_E$-free because $O_E$ is principal. From (b) we deduce that $L'$ is the $O_E$-integral structure of $(\tilde{\pi}, \tilde{V})$ contragredient to $L$. □

**III. Integral highest Whittaker model.** Let $(F, G, \ell)$ be as in the introduction with the restriction: the characteristic of $F$ is zero and $p \neq 2$.

We define a Whittaker data and a Whittaker representation following [MW]. We choose:

(a) A continuous homomorphism $\phi : F \to \mathbf{C}^*$, trivial on $O_F$ but not on $p_F^{-1} O_F$.

(b) A nondegenerate $\mathrm{Ad}\, G$-invariant bilinear form $B : \mathcal{G} \times \mathcal{G} \to F$ on the Lie algebra $\mathcal{G}$ of $G$.

(c) An exponential $\exp : \mathcal{V}(0) \to V(1)$, which is a bijective $G$-equivariant homeomorphism defined on an $\mathrm{Ad}\, G$-invariant open closed subset $\mathcal{V}(0)$ of $\mathcal{G}$ containing the nilpotent elements with image an $G$-invariant open closed subset $V(1)$ of $G$, with inverse a logarithm $\log : V(1) \to \mathcal{V}(0)$.

(d) A nilpotent element $Y$ of $\mathcal{G}$ of orbit $\mathcal{O} = \mathrm{Ad}\, G.Y$.

(e) A cocharacter $\mu : F^* \to G$ of $G$ defining via the adjoint action a grading of the Lie algebra $\mathcal{G} = \oplus_{i \in \mathbf{Z}} \mathcal{G}_i$,

$$\mathcal{G}_i := \{X \in \mathcal{G} \mid \mathrm{Ad}\, \mu(s).X = s^i X \ \text{ for all } s \in \mu(F^*)\}$$

such that $Y \in \mathcal{G}_{-2}$. Set $\mathcal{G}_{\geq ?} := \oplus_{i \geq ?} \mathcal{G}_i$ and $?_i := \mathcal{G}_i \cap ?$.

Clearly the grading is finite, $[\mathcal{G}_i, \mathcal{G}_j] \subset \mathcal{G}_{i+j}$ and $B(\mathcal{G}_i, \mathcal{G}_j) = 0$ if $i + j \neq 0$. The centralizer $\mathcal{G}^Y := \{Z \in \mathcal{G} \mid [Y, Z] = 0\}$ of $Y$ in $\mathcal{G}$ satisfies $B(Y, \mathcal{G}^Y) = 0$ [MW, p. 438]. There is a unique $\mu(F^*)$-invariant decomposition

$$\mathcal{G} = \mathcal{M} \oplus \mathcal{G}^Y$$

and $\mathcal{M} = \oplus_{i\in\mathbf{Z}}\mathcal{M}_i, \quad \mathcal{G}^Y = \oplus_{i\in\mathbf{Z}}\mathcal{G}_i^Y$. The skew bilinear form

$$B_Y(X, Z) := B(Y, [X, Z]) : \mathcal{G} \times \mathcal{G} \to F$$

has a radical $\{Z \in \mathcal{G} \mid B(Y, [X, Z]) = 0$ for all $X \in \mathcal{G}\}$ equal to $\mathcal{G}^Y$. Therefore $B_Y$ induces a duality between $\mathcal{M}_i$ and $\mathcal{M}_{i+2}$ for all $i \in \mathbf{Z}$ and a symplectic form on $\mathcal{M}_1$. The dimension of $\mathcal{M}_1$ is an even integer $2m_1$.

(f) An $O_F$-lattice $\mathcal{M}_1(O_F)$ of $\mathcal{M}_1$, which is self-dual for $B_Y$:

$$\mathcal{M}_1(O_F) = \{m \in \mathcal{M}_1 \mid B_Y(m, \mathcal{M}_1(O_F)) \subset O_F\}.$$

The group $N := \exp \mathcal{G}_{\geq 1}$ is unipotent and depends only on the choice of $\mu$ and exp. We consider the open subgroup $H$ of $N$ and the character $\chi$ of $H$ defined by:

$$H := \exp\left(\mathcal{M}_1(O_F) \oplus \mathcal{G}_1^Y \oplus \mathcal{G}_{i\geq 2}\right), \quad \chi(\exp X) := \phi(B(Y, X)).$$

Clearly $H = N$ iff $\mathcal{M}_1 = 0$, and $\chi(\exp X) = \chi(\exp X_2)$, where $X_2$ is the component of $X$ in $\mathcal{M}_2$. The character $\chi$ does not change if $(\phi, B)$ is replaced by $(\phi_a, a^{-1}B)$, where $\phi_a(x) := \phi(ax)$ with $a \in O_F^*$.

*Definition* III.1. We call $(\phi, B, \exp, Y, \mathcal{O}, \mu, \mathcal{M}_1(O_F))$ a Whittaker data of $G$ and

$$\operatorname{Ind}_H^G \chi = \operatorname{Ind}_N^G \Omega, \quad \text{where } \Omega := \operatorname{Ind}_H^N \chi$$

a Whittaker representation of $G$.

When $H \neq N$ the representation $\Omega$ is a metaplectic representation of the Heisenberg group $H/\operatorname{Ker}\chi$. The representation $\Omega$ is irreducible and admissible [MVW chapter 2, I.6 (3)] and its isomorphism class does not depend on the choice of $\mathcal{M}_1(O_F)$. The isomorphism class of the Whittaker representation depends only on $(Y, \mu)$ when $(\phi, B, \exp)$ are fixed, and does not change if $(Y, \mu)$ is replaced by a $G$-conjugate.

The complex field $\mathbf{C}$ appears only in the definition of the nontrivial additive character $\phi$ of $F$. The same definitions can be given over any field (or even a commutative ring) $R$, which contains roots of 1 of any $p$-power order.

We define the highest Whittaker models of $(\pi, V) \in \operatorname{Irr}_{\overline{\mathbf{Q}}_\ell} G$ as in the introduction. When $V \subset \operatorname{Ind}_H^G \chi$ is a highest Whittaker model of $\pi$, we want to show that the projection on the $(H, \chi)$-coinvariant vectors

$$p_\chi : V \to V_\chi$$

behaves well with integral structures.

THEOREM III.2. *Let* $(\pi, V) \in \mathrm{Irr}_{\overline{\mathbf{Q}}_\ell} G$ *integral with* $V \subset \mathrm{Ind}_H^G \chi$ *a highest Whittaker model. Let L be a* $\overline{\mathbf{Z}}_\ell$*-integral structure of* $(\pi, V)$. *Then* $p_\chi L$ *is a* $\overline{\mathbf{Z}}_\ell$*-free module.*

As $p_\chi V$ is a finite dimensional $\overline{\mathbf{Q}}_\ell$-space by Moeglin and Waldspurger, and as $p_\chi L$ is a $\overline{\mathbf{Z}}_\ell$-integral structure (a lattice) of $p_\chi V$ by the Theorem III.2, $(\pi, V)$ satisfies the criterion A, modulo the fact that $\overline{\mathbf{Z}}_\ell$ is not a principal ring. But we may replace $(\overline{\mathbf{Z}}_\ell, \overline{\mathbf{Q}}_\ell)$ by $(O_E, E)$, where $O_E$ is the ring of integers of a finite extension $E/\mathbf{Q}_\ell(\mu_{p^\infty})$ such that $\pi$ is defined over $E$, where $\mu_{p^\infty}$ is the group of roots of 1 of any order of $p$ in $\overline{\mathbf{Q}}_\ell$. The extension $\mathbf{Q}_\ell(\mu_{p^\infty})/Q_\ell$ is infinite and unramified hence $O_E$ is principal.

Therefore (III.2) implies the Theorem 2 of the introduction under the restrictions on $(F, \pi)$. The theorem (III.2) results from (III.4.6) and the remark following (III.4.3). The rest of the section III is devoted to the proof of (III.2).

The fundamental idea due to Rodier is to approximate the character $\chi$ of $H$ by characters $\chi_n$ of open compact subgroups $K_n$ with the property that the projections $e_n$ on the $(K_n, \chi_n)$-invariant vectors approximate the projection $p_\chi$ on the $(H, \chi)$-coinvariant vectors in the following sense: when $n$ is big enough, $p_\chi$ restricts to an isomorphism $e_n V \to p_\chi V$. We want to prove the same thing for an integral structure $L$ instead of $V$. There is not much to add to the original proof for $V$, only another technical computation (III.4.1), and this is the purpose of this section.

**III.3.** We recall the construction of the geometric approximation $(K_n, \chi_n)$ of $(H, \chi)$ following [MW I.2 (2), I.4 (1), I.9, I.13] (our $\chi_n$ is not the character $\chi_n$ of [MW]). Set $t := \mu(p_F)$. We choose a lattice $\mathcal{L}$ of $\mathcal{G}$ such that $[\mathcal{L}, \mathcal{L}] \subset \mathcal{L}$ and we complete $\mathcal{M}_1(O_F)$ to a self-dual lattice $\mathcal{M}(O_F) = \oplus_i \mathcal{M}_i(O_F)$ of $\mathcal{M}$. The $O_F$-module

$$\mathcal{L}' := \mathcal{M}(O_F) \oplus \oplus_{i \in \mathbf{Z}} \left(\mathcal{L} \cap \mathcal{G}_i^Y\right)$$

is an $O_F$-lattice of $\mathcal{G}$. For a big enough fixed integer $A$ and a fixed integer $c \geq A$, we set for all $n \geq A$

$$G_n := \exp\left(p_F^n \mathcal{L}'\right), \quad A_n := \exp\left(p_F^{[n/2]+c} (\mathcal{L} \cap \mathcal{G}_1)^Y\right), \quad K_n := t^{-n} (G_n A_n) t^n$$

$$\xi_n(\exp X) = \chi_n(t^{-n} \exp(Z_1) \exp(X) t^n) := \phi\big(p_F^{-2n} B(Y, X)\big)$$

for all $X \in p_F^n \mathcal{L}'$, $Z_1 \in p_F^{[n/2]+c} (\mathcal{L} \cap \mathcal{G}_1)^Y$, where $[n/2]$ is the smallest integer $\leq n/2$. The particular form of $K_n$ will be explained soon.

We set $N' := \exp(\mathcal{G}_1^Y \oplus \mathcal{G}_{i \geq 2})$. The Campbell-Hausdorff formula shows that $N'$ is a normal subgroup of $H$. The closed subgroup $C$ of $H$ generated by $\exp(\mathcal{M}_1(O_F))$ is compact and $H = CN'$. The character $\chi$ of $H$ is trivial on $C$. The sequence $(K_n, \chi_n)_{n \geq A}$ is an approximation of $(H, \chi)$ in the following sense:

$K_n = (K_n \cap P^-)(K_n \cap H) = (K_n \cap H)(K_n \cap P^-)$ [MW I.4] where $P^-$ is the stabilizer in $G$ of $\oplus_{i<0}\mathcal{G}_i$, the sequence of groups $K_n \cap P^-$ is decreasing with trivial intersection, the sequence of groups $K_n \cap H = C(K_n \cap N')$ is increasing with union $H$, the restriction of $\chi_n$ to $K_n \cap P^-$ is trivial and $\chi_n = \chi$ on $K_n \cap H$.

The sequence of open compact subgroups $G_n$ of $G$ is decreasing with trivial intersection, and $\xi_n$ is a character of $G_n$. A basic property of $(G_n, \xi_n)$ is [MW I.6]:

*Claim* III.3.1. For any integers $A \leq m \leq n$, the group $G_n$ is normal in $G_m$ and the stabilizer of $\xi_n$ in $G_m$ is equal to $G_n \exp(p_F^m \mathcal{L}^Y)$.

We introduce now an admissible representation $(\pi, V) \in \mathrm{Mod}_{\overline{\mathbf{Q}}_\ell} G$. Let $I_n$ be the projection of $V$ on its $(G_n, \xi_n)$-invariant vectors. The dimension of the $\overline{\mathbf{Q}}_\ell$-vector space $I_n V$ is finite. The profinite group $\exp(p_F^{c+[n/2]} \mathcal{L}^Y)$ acts on $I_n V$ by (III.3.1). The action is trivial iff the trace $\mathrm{tr}_{I_n V}\, u$ of the action of any element $u \in \exp(p_F^{c+[n/2]} \mathcal{L}^Y)$ is equal to $\dim I_n V$.

Suppose that $(\pi, V)$ is irreducible hence admissible. When $n$ is big enough, $\mathrm{tr}_{I_n V}\, u$ can be computed using the expansion of the trace $\mathrm{tr}\,\pi$ of $\pi$ around 1. The computation simplifies when the nilpotent orbit $\mathcal{O}$ is maximal among the nilpotent orbits with a nonzero coefficient. When $\mathcal{O}$ satisfies this property we say that $\mathcal{O}$ is maximal for $\mathrm{tr}\,\pi$. Then we have [MW I.13]:

*Claim* III.3.2. Let $(\pi, V) \in \mathrm{Irr}_{\overline{\mathbf{Q}}_\ell} G$. When $\mathcal{O}$ is maximal for $\mathrm{tr}\,\pi$ and when $n$ is big enough, the action of $\exp(p_F^{c+[n/2]} \mathcal{L}^Y)$ on $I_n V$ is trivial.

For two integers $n, m \geq A$, we denote by $I_{n,m} : I_n V \to I_m V$ the restriction to $I_n V$ of $I_m t^{m-n}$. In particular

$$I_{n+1,n} = I_n t^{-1} : I_{n+1} V \to I_n V,$$

$$I_{n,n+1} = I_{n+1} t : I_n V \to I_{n+1} V.$$

The property [MW I.15]: "*Let* $(\pi, V) \in \mathrm{Irr}_{\overline{\mathbf{Q}}_\ell} G$. *When* $\mathcal{O}$ *is maximal for* $\mathrm{tr}\,\pi$ *and when n is big enough,* $I_{n+1,n} I_{n,n+1} I_n$ *is a nonzero multiple of* $I_n$" is used to prove that the nilpotent orbits maximal for $\mathrm{tr}_\pi$ are those maximal for the Whittaker models [MW I.16] and that the dimension of the $(H, \chi)$-coinvariants of $(V, \pi)$ is equal to the coefficient attached to $\mathcal{O}$ in the expansion of $\mathrm{tr}_\pi$ [MW I.17].

**III.4.** We give variants of this property that will be the key to prove (III.2).

LEMMA III.4.1. *Let* $(\pi, V) \in \mathrm{Mod}_{\overline{\mathbf{Q}}_\ell} G$ *such that the action of* $\exp(p_F^{c+[n/2]} \mathcal{L}^Y)$ *on* $I_n V$ *is trivial when n is big enough. Then, when* $n \geq n_o$ *is big enough, there exist integers* $b(n), b'(n) \geq 0$ *such that*

$$I_{n+1,n} I_{n,n+1} I_n = p^{b(n)} I_n, \quad I_{n,n+1} I_{n+1,n} I_{n+1} = p^{b'(n)} I_{n+1}.$$

We will prove in (III.4.2) that $b(n) = b'(n)$.

*Proof of* III.4.1. To simplify we set $gv := \pi(g)v$ for $g \in G, v \in V$.

(a) It is proved in [MW I.15] under the hypothesis that $\pi$ is irreducible but without using this property, that for any $w_n \in I_n V$, $I_{n+1,n} I_{n,n+1} w_n$ is the product of a power of $p$ and of a sum

$$\sum_h \xi_{n+1}(h^{-1}) t^{-1} h t w_n,$$

where $h \in G_{n+1}/(G_{n+1} \cap t G_n t^{-1})$ and $t^{-1}ht$ stabilizes $\xi_n$. The number of terms of the sum is a power of $p$. It is claimed in [MW I.15] that each term of the sum is equal to $w_n$ when $n$ is big enough; we deduce that there exists an integer $b(n)$ such that $I_{n+1,n} I_{n,n+1} I_n = p^{b(n)} I_n$. We give a proof of the claim because the same method is used for the second equality of the lemma. Let $h \in G_{n+1}$ such that $t^{-1}ht$ stabilizes $\xi_n$. The definition of $G_n$ shows that if $n$ is big enough, the group $t^{-1} G_{n+1} t$ is contained in $G_{n+1-a}$ for some integer $a$ such that $c + [n/2] \leq n + 1 - a$. There exists $g \in G_n$ and $y \in \exp p_F^{n+1-a} \mathcal{L}^Y$ such that $t^{-1}ht = gy$ by (III.3.1). By (III.3.2) $y$ acts trivially on $I_n V$ hence $t^{-1} h t w_n = g w_n = \xi_n(g) w_n$. Denote by $X_2$ the component of $\log g$ in $\mathcal{M}_2$. Then

$$\xi_n(g) = \phi\big(p_F^{-2n} B(Y, X_2)\big) = \phi\big(p_F^{-2n-2} B(Y, \mathrm{Ad}\, t . X_2)\big) = \xi_{n+1}(h).$$

Hence each term in the sum is equal to $w_n$.

(b) We prove the second equality with the same method. For all $n \geq A$, we choose on $G_n$ the Haar measure normalized by $\mathrm{vol}\, G_n = 1$. By definition, $I_{n,n+1} I_{n+1,n} I_{n+1} = I_{n+1} t I_n t^{-1} I_{n+1}$ is equal to

$$\int_{G_{n+1}} \int_{G_n} \int_{G_{n+1}} \xi_{n+1}(g')^{-1} \xi_n(h)^{-1} \xi_{n+1}(g)^{-1} g' t h t^{-1} g \, dg' \, dh \, dg.$$

When $h \in G_n \cap t^{-1} G_{n+1} t$, the action of $\xi_n(h)^{-1} t h t^{-1}$ on $I_{n+1} V$ is trivial because $\xi_n(h) = \xi_{n+1}(tht^{-1})$ as in (a). The volume of $G_n \cap t^{-1} G_{n+1} t$ is a power of $p$. The triple integral is the product of this volume and of:

$$\sum_{h \in G_n/(G_n \cap t^{-1} G_{n+1} t)} \xi_n(h)^{-1} \int_{G_{n+1}} \int_{G_{n+1}} \xi_{n+1}(g')^{-1} \xi_{n+1}(g)^{-1} g' t h t^{-1} g \, dg' \, dg.$$

The group $t G_n t^{-1}$ normalizes $G_{n+1}$, because $t G_n t^{-1}$ is contained in $G_{n-a}$ and $n - a \geq A$ when $n$ is big enough. After the change of variables $y = (tht^{-1})^{-1} g' t h t^{-1}$ and $x = yg$ in $G_{n+1}$ we get

$$\sum_{h \in G_n/(G_n \cap t^{-1} G_{n+1} t)} \xi_n(h)^{-1} t h t^{-1} \int_{G_{n+1}} \int_{G_{n+1}} \xi_{n+1}(tht^{-1} y (tht^{-1})^{-1} y^{-1} x)^{-1} x \, dx \, dy,$$

which is equal to the product of a power of $p$ and of

$$J := \sum_h \xi_n(h)^{-1} t h t^{-1} \int_{G_{n+1}} \xi_{n+1}(x)^{-1} x \, dx = \sum_h \xi_n(h)^{-1} t h t^{-1} I_{n+1},$$

where $h \in G_n/(G_n \cap t^{-1}G_{n+1}t)$ and $tht^{-1}$ stabilizes $\xi_{n+1}$. The number of $h$ is a power of $p$. Let $w_{n+1} \in I_{n+1}V$. We have

$$Jw_{n+1} = \sum_h \xi_n(h)^{-1} tht^{-1} w_{n+1}$$

for $h$ as above. As in (a), one shows that each term of the sum is equal to $w_{n+1}$. Let $h \in G_n$ such that $tht^{-1}$ stabilizes $\xi_{n+1}$. As in (a), $tht^{-1} \in G_{n-a}$ and the stabilizer of $\xi_{n+1}$ in $G_{n-a}$ is $G_{n+1} \exp p_F^{n-a}\mathcal{L}^Y$ with $n-a > c + [(n+1)/2]$ when $n$ is big enough. Hence $tht^{-1} = gy$ for some $g \in G_{n+1}$ and the action of $y$ on $I_{n+1}V$ is trivial. Hence $tht^{-1}w_{n+1} = gw_{n+1} = \xi_{n+1}(g)w_{n+1}$. Denote by $X_2$ the component of $\log g$ in $\mathcal{G}_2$. Then

$$\xi_{n+1}(g) = \phi\big(p_F^{-2n-2}B(Y, X_2)\big) = \phi\big(p_F^{-2n}B(Y, \operatorname{Ad} t^{-1}.X_2)\big) = \xi_n(h).$$

Hence each term in the sum is equal to $w_{n+1}$. We deduce that there exists an integer $b'(n)$ such that $I_{n,n+1}I_{n+1,n}I_{n+1} = p^{b'(n)}I_{n+1}$. The lemma is proved. □

For the application that we have in mind, we replace the projection $I_n$ on the $(G_n, \xi_n)$-invariant vectors by the projection $e_n$ on the $(K_n, \chi_n)$-invariant vectors in the Lemma III.4.1, and we prove $b(n) = b'(n)$.

Stabilization Lemma III.4.2. *Let $(\pi, V) \in \operatorname{Mod}_{\overline{\mathbf{Q}}_\ell} G$ such that the action of $\exp(p_F^{c+[n/2]}\mathcal{L}^Y)$ on $I_nV$ is trivial when $n$ is big enough. Then, when $n \geq n_o$ is big enough, there exists an integer $b(n) \geq 0$ such that*

$$e_n e_{n+1} e_n = p^{b(n)} e_n, \quad e_{n+1} e_n e_{n+1} = p^{b(n)} e_{n+1}.$$

*In particular, $e_{n+1}$ induces an isomorphism $e_nV \simeq e_{n+1}V$ of inverse $p^{-b(n)}e_n$ restricted to $e_{n+1}V$.*

*Proof of* III.4.2. Suppose that $n$ is big enough. By (III.3) $K_n = t^{-n}A_nt^n\, t^{-n}G_nt^n$, as $t^{-n}A_nt^n$ acts trivially on $t^{-n}I_nV$ and as $\chi_n(t^{-n}gt^n) = \xi_n(g)$ for all $g \in G_n$, we have

$$I_n = t^n e_n t^{-n}.$$

The action of $t$ on $V$ is invertible hence $I_nV = t^ne_nV$. We have

$$I_{n+1,n} = I_nt^{-1} = t^ne_nt^{-n-1} : t^{n+1}e_{n+1}V \to t^ne_nV,$$

$$I_{n,n+1} = I_{n+1}t = t^{n+1}e_{n+1}t^{-n} : t^ne_nV \to t^{n+1}e_{n+1}V,$$

$$I_{n+1,n}I_{n,n+1} = t^ne_ne_{n+1}t^{-n} : t^ne_nV \to t^ne_nV,$$

$$I_{n,n+1}I_{n+1,n} = t^{n+1}e_{n+1}e_nt^{-n-1} : t^{n+1}e_{n+1}V \to t^{n+1}e_{n+1}V,$$

$$I_{n+1,n}I_{n,n+1}I_n = t^ne_ne_{n+1}e_nt^{-n} : V \to t^ne_nV,$$

$$I_{n,n+1}I_{n+1,n}I_{n+1} = t^{n+1}e_{n+1}e_ne_{n+1}t^{-n-1} : V \to t^{n+1}e_{n+1}V.$$

The equalities in III.4.1 are equivalent to

$$e_n e_{n+1} e_n = p^{b(n)} e_n, \quad e_{n+1} e_n e_{n+1} = p^{b'(n)} e_{n+1}.$$

We compute $e_n e_{n+1} e_n e_{n+1}$ in two different ways using the two equalities. We get $p^{b(n)} e_n e_{n+1} = p^{b'(n)} e_n e_{n+1}$. The first equality implies $e_n e_{n+1} \neq 0$, hence $b(n) = b'(n)$. The equalities in (III.4.2) are proved.

Let $v_n \in e_n V$. The first equality gives $e_n e_{n+1} v_n = p^{b(n)} v_n$. In particular $e_{n+1}$ is injective on $e_n V$. For $v_{n+1} \in e_{n+1} V$ the second equality gives $e_{n+1} e_n v_{n+1} = p^{b(n)} v_{n+1}$. In particular $e_{n+1} e_n V = e_{n+1} V$. Hence $e_{n+1}$ induces an isomorphism $e_n V \to e_{n+1} V$. By the first equality $p^{-b(n)} e_n e_{n+1} v_n = v_n$, by the second equality $e_{n+1} p^{-b(n)} e_n v_{n+1} = v_{n+1}$. Hence $p^{-b(n)} e_n$ induces the inverse isomorphism $e_{n+1} V \to e_n V$. □

STABILIZATION PROPERTY III.4.3. *We say that the stabilization property holds for $(H, \chi)$ in $(\pi, V) \in \mathrm{Mod}_{\overline{\mathbf{Q}}_\ell} G$ when: for all big enough integers $n \geq n_o$, there exists an integer $b(n)$ such that $e_{n+1}$ restricted to $e_n V$ is an isomorphism $e_n V \simeq e_{n+1} V$ of inverse $p^{-b(n)} e_n$ restricted to $e_{n+1} V$.*

*Remark.* When $(\pi, V) \in \mathrm{Irr}_{\overline{\mathbf{Q}}_\ell} G$ and $V \subset \mathrm{Ind}_H^G \chi$ is a highest Whittaker model, then the stabilization property III.4.3 holds for $(H, \chi)$ in $(\pi, V)$ by (III.3.2) and (III.4.2).

We consider finally the projections $\varepsilon_n$ on the $(K_n \cap H, \chi|_{K_n \cap H})$-invariant vectors.

LEMMA III.4.4. *The stabilization property for $(H, \chi)$ in $(\pi, V) \in \mathrm{Mod}_{\overline{\mathbf{Q}}_\ell} G$ implies for any big enough integers $n \geq m \geq n_o$:*

(a) *$\varepsilon_n = e_n$ on $e_m V$ and $\varepsilon_n$ restricted to $e_m V$ is an isomorphism $e_m V \to e_n V$,*

(b) *if $(\pi, V)$ has an integral structure $L$, we can replace $V$ by $L$ in (a).*

*Proof of* III.4.4. $\varepsilon_n v = e_n v$ for any $v \in V$, which is invariant by $K_n \cap P^-$ because $K_n = (K_n \cap P^-)(K_n \cap H)$ and $\chi_n$ is trivial on $K_n \cap P^-$ and equal to $\chi$ on $K_n \cap H$. In particular $\varepsilon_n v_m = e_n v_m$ for any $v_m \in e_m V$ because the sequence of groups $K_n \cap P^-$ is decreasing and $\chi_m$ is trivial on $K_m \cap P^-$. The stabilization property implies that $\varepsilon_{m+1}$ restricted to $e_m V$ is an isomorphism $e_m V \simeq e_{m+1} V$. By induction, $\varepsilon_n \circ \ldots \circ \varepsilon_{m+1}$ restricted to $e_m V$ is an isomorphism $e_m V \simeq e_n V$. The open compact groups $K_n \cap H$ form an increasing sequence, hence for any $n \geq m$ and $m$ big enough, $\varepsilon_n = \varepsilon_n \circ \ldots \circ \varepsilon_m$. We proved (a).

If $(\pi, V)$ has an integral structure $L$, $e_{n+1}$ and $p^{-b(n)} e_n$ give by restriction isomorphisms $e_n L \simeq e_{n+1} L$, which are inverse of each other, because the $K_n$ are pro-p-groups, $p \neq \ell$, and $e_n L = L \cap e_n V$. The arguments given in the proof (a) are valid when $V$ is replaced by $L$. □

As the open compact groups $K_n \cap H$ form an increasing sequence of union $H$, the projections $\varepsilon_n$ on the $(K_n \cap H, \chi|_{K_n \cap H})$-invariant vectors approximate the projection $p_\chi$ on the $(H, \chi)$-invariants in the following sense:

(III.4.5) $$p_\chi \varepsilon_n = p_\chi, \quad \operatorname{Ker} p_\chi = \cup_{n \geq m} \operatorname{Ker} \varepsilon_n$$

for any integer $m$.

PROPOSITION III.4.6. *The stabilization property (III.4.3) for $(H, \chi)$ in $(\pi, V) \in \mathrm{Mod}_{\overline{\mathbf{Q}}_\ell} G$ implies for a big enough integer $m \geq n_o$:*

(1) *$p_\chi$ restricted to $e_m V$ is an isomorphism $e_m V \simeq p_\chi V$,*

(2) *if $(\pi, V)$ is integral with integral structure $L$, $p_\chi e_m L \simeq p_\chi L$ is a lattice of $p_\chi V$.*

The property (1) is a reformulation of [MW I.14] when $(\pi, V) \in \mathrm{Irr}_{\overline{\mathbf{Q}}_\ell} G$ and $V \subset \mathrm{Ind}_H^G \chi$ is a highest Whittaker model.

*Proof of* III.4.6. (a) Injectivity of $p_\chi$ restricted to $e_m V$. Apply (III.4.5), (III.4.4), and the injectivity of $\varepsilon_n$ restricted to $e_m V$ for all $n \geq m \geq n_o$.

(b) Surjectivity of $p_\chi$ restricted to $e_m V$. We have $V = \cup_{n \geq m} V^{K_n \cap P^-}$ and by (III.4.4), and its proof:

$$p_\chi(V^{K_n \cap P^-}) = p_\chi \varepsilon_n(V^{K_n \cap P^-}) = p_\chi e_n(V^{K_n \cap P^-}) \subset p_\chi e_n V = p_\chi \varepsilon_n e_m V = p_\chi e_m V.$$

Hence $p_\chi V = p_\chi e_m V$.

(c) $p_\chi e_m L = p_\chi L$. The arguments of (b) apply to $L$ instead of $V$.

(d) $e_m L$ is a lattice of $e_m V$; this remains true when one applies the isomorphism $p_\chi$. □

## IV. Integral generic compact Whittaker representation

*Notation* IV.1. Let $(F, G)$ be as in the introduction and let $R$ be a commutative ring that contains roots of the unity of any power of $p$. The characteristic of $R$ is automatically different from $p$. We choose in $G$ a maximal split $F$-torus $T$ (the group of rational points a maximal split $F$-torus) and a minimal parabolic $F$-group $B = TU$ that contains $T$ and of unipotent radical $U$. We denote by $Z$ the centralizer of $T$ in $G$ (not the center of $G$), and by $\overline{B} = T\overline{U}$ the opposite of $B$ in $G$. We denote by $\Phi$, $\Phi^{red}$, $\Delta$, $\Phi^+$, $\Phi^{+,red}$ the set of roots of $(G, T)$ in $\mathrm{Lie}\, U$, of reduced roots, of simple positive roots, of positive roots, of positive reduced roots with respect to $B$. Let $U_{(\alpha)}$ be the unipotent subgroup of $U$ normalized by $Z$ with Lie algebra $\mathcal{U}_\alpha + \mathcal{U}_{2\alpha}$ for any root $\alpha \in \Phi$ (when $2\alpha$ is not a root, $\mathcal{U}_{2\alpha} = 0$ and $U_{(2\alpha)} = \{1\}$).

*Definition* IV.1.1. A character $\phi : U \to R^*$ is nondegenerate if the restrictions $\phi_{(\alpha)}$ of $\phi$ to $U_{(\alpha)}$ satisfy the two following properties (1) and (2):

1. $\phi_{(\alpha)}$ is trivial for any $\alpha \in \Phi^+ - \Delta$.

The character $\phi$ satisfying (1) identifies to a character of the direct product

$$\prod_{\alpha\in\Delta} U_{(\alpha)}/U_{(2\alpha)} \to R^*.$$

2. The kernel $\operatorname{Ker}\phi_{(\alpha)}$ of $\phi_{(\alpha)}$ is an open compact subgroup of $U_{(\alpha)}$ for all $\alpha \in \Delta$.

In particular (2) implies that $\phi_{(\alpha)}$ is nontrivial for all $\alpha \in \Delta$.

*Remarks* IV.1.2. (1) When $G$ is anisotropic, $U = \{1\}$ is the trivial group, the regular representation of $G$ on the $R$-module $C_c^\infty(G; R)$ of locally constant functions $f : G \to R$ with compact support is the compact generic Whittaker $R$-representation of $G$.

(2) When $G$ is split, the property (1) of (IV.1.1) is true except in some exceptional cases [Borel Tits *Ann. Math.* **97** (1973), 449–571, see page 519 4.3], and the property (2) of (IV.1.1) is equivalent to: $\phi_{(\alpha)}$ is nontrivial for all $\alpha \in \Delta$.

(3) The set of nondegenerate characters of $U$ is stable by the natural action of $Z$, because $Z$ normalizes $U_{(\alpha)}$ for all roots $\alpha \in \Phi$.

**IV.2.** We choose an open compact subgroup $K_o$ of $G$ such that

$$G = BK_o$$

and a normal subgroup $K$ of $K_o$ of finite index, normalized by

$$T^+ := \{t \in T \mid |\alpha(t)| \le 1 \text{ for all } \alpha \in \Delta\},$$

with an Iwahori decomposition

$$K = (K \cap \overline{U})(K \cap Z)(K \cap U) = (K \cap U)(K \cap Z)(K \cap \overline{U}).$$

$$K \cap U = \prod_{\alpha\in\Phi^{+,red}} K \cap U_{(\alpha)}, \quad K \cap \overline{U} = \prod_{\alpha\in\Phi^{+,red}} K \cap U_{(-\alpha)}.$$

The theory of Bruhat-Tits gives a subgroup $K_o$ of $G$ and a decreasing separated sequence of subgroups $K$ of $G$ satisfying these properties.

THEOREM IV.2.1. *The right $\mathcal{H}_R(G, K)$-module $(\operatorname{ind}_U^G \phi)^K$ is finitely generated for any nondegenerate character $\phi : U \to R^*$.*

This implies that a generic compact Whittaker representation satisfies the Criterion B of (II.7). The rest of this section is devoted to the proof of the theorem.

When $R = \mathbf{C}$ is the field of complex numbers, this is a theorem of Bushnell and Henniart [BH 7.1].

The theorem follows from a geometric property (IV.2.2) and a computation (IV.2.3). This proof is valid over any $R$ and does not use the theorem over **C**, and is a variant of the proof of [BH].

The support of $(\mathrm{ind}_U^G \phi)^K$ is

$$G(U, \phi, K) = \{g \in G \mid gKg^{-1} \cap U \subset \mathrm{Ker}\,\phi\} \tag{1}$$

by the Mackey decomposition of $(\mathrm{ind}_U^G \phi)^K$ (proof of (II.3)). This means the following :

- for $g \in G(U, \phi, K)$ there exists a function $\phi_{UgK} \in (\mathrm{ind}_U^G \phi)^K$ with support $UgK$ and value 1 at $g$,
- the functions $\phi_{UgK}$ for the $(U, K)$-cosets $UgK$ of $G(U, \phi, K)$ form a basis of the $R$-module $(\mathrm{ind}_U^G \phi)^K$ over $R$.

*Claim* IV.2.2. The support $G(U, \phi, K)$ of $(\mathrm{ind}_U^G \phi)^K$ satisfies the geometric property: $G(U, \phi, K)$ is a finite union of $UzT_+K_o$ with $z \in Z \cap G(U, \phi, K)$.

*Claim* IV.2.3. The right action of the Hecke algebra $\mathcal{H}_R(G, K)$ on $(\mathrm{ind}_U^G \phi)^K$ satisfies:

(a) We have for $x \in G(U, \phi, K)$ and $k_o \in K_o$:

$$\phi_{UxK} * [Kk_oK] = \phi_{Uxk_oK}$$

(b) We have for $z \in Z \cap G(U, \phi, K)$ and $t_+ \in T_+$:

$$\phi_{UzK} * [Kt_+K] = \phi_{Uzt_+K}.$$

Clearly, the Theorem IV.2.1 follows from the claims (IV.2.2) and (IV.2.3).

**IV.3.** The geometric property (IV.2.2) results from a known fact: when $X_{(\alpha)}$ is a group in the Bruhat-Tits filtration of $U_{(\alpha)}$ for $\alpha \in \Delta$ [T 1.4.2], we have the equality of semi-groups (deduced from [T 1.2 (1), 1.4.2]):

$$T(X, X) = T_+, \tag{IV.3.1}$$

where $T(X, X) := \{t \in T \mid tX_{(\alpha)}t^{-1} \subset X_{(\alpha)} \text{ for all } \alpha \in \Delta\}$.

For (IV.2.2) it is enough to know that for any $\alpha \in \Delta$ there exists an open compact subgroup $X_{(\alpha)}$ of $U_{(\alpha)}$ such that (IV.3.1) is true. We give a variant of (IV.3.1) when $T$ is replaced by $Z$ and the $X_{(\alpha)}$ are replaced by pairs $(K_{(\alpha)}, C_{(\alpha)})$ of open compact subgroups of $U_{(\alpha)}$ with $K_{(\alpha)}$ normalized by $T_+$ for any $\alpha \in \Delta$, and $T(X, X)$ is replaced by

$$Z(K, C) := \{z \in Z \mid zK_{(\alpha)}z^{-1} \subset C_{(\alpha)} \text{ for all } \alpha \in \Delta\}.$$

*Claim* IV.3.2. $Z(K, C) = Z_o T_+$ for some compact subset $Z_o$ of $Z(K, C)$.

The proof of the variant (IV.3.2) uses the particular case (IV.3.1) and the fact that $T_+$ contains the maximal compact open subgroup $T^o$ of $T$ with semi-group quotient $T/T^o \simeq \mathbf{N}^d$ where $\mathbf{N}$ is the set of natural integers and $d > 0$ an integer. One reduces (IV.3.2) to the combinatorial finiteness property:

*Claim* IV.3.3. Any non-empty subset $Y$ of $\mathbf{N}^d$ saturated under addition by $\mathbf{N}^d$ is a finite union of $y + \mathbf{N}^d$ for $y \in Y$.

The proof is elementary. When $d = 1$, we choose the minimum element $y$ of $Y$. Then $Y = y + \mathbf{N}$. By induction on $d$, we suppose that the property is true for $d - 1$. Let us call "minimal" an element $y$ of $Y$ such that $z + \mathbf{N}^d \subset y + \mathbf{N}^d$ and $z \in Y$ implies $z = y$. Then $Y$ is the union of $y + \mathbf{N}^d$ for $y \in Y$ minimal. The property is equivalent to the finiteness of minimum elements. If $y, z \in Y$ are minimum and distinct, then some component of $z$ is strictly smaller than some component of $y$. We are reduced to prove that the set $M(i, m)$ of minimum elements of $Y$ with a given $i$-th component $m \in \mathbf{N}$ is finite, for any $1 \leq i \leq d$ and any $m \in \mathbf{N}$. Suppose that $M(i, m)$ is not empty and let $Y_{i,m}$ be the union of $y + \mathbf{N}^d$ for $y \in M(i, m)$. Via the components different from $i$, the set of elements of $Y_{i,m}$ with $i$-th component $m$, identifies with a non-empty subset of $Y(i, m) \subset \mathbf{N}^{d-1}$ saturated under under addition by $\mathbf{N}^{d-1}$. Under this identification $M(i, m)$ becomes the set of minimum elements of $Y(i, m)$. By induction hypothesis, the set $M(i, m)$ is finite.

**IV.3.4.** We explain how (IV.3.1) and (IV.3.3) imply (IV.3.2).

(1) We replace $Z$ by $T$. There exists an open compact subgroup $Z_o$ of $Z$ that normalizes $K_{(\alpha)}$ for any $\alpha \in \Delta$. There exists a finite set of $z_k \in Z$ such that

$$Z = \cup_k z_k T Z_o$$

because the quotient $Z/T$ is compact. The subset $C_{k,(\alpha)} := z_k^{-1} C_{(\alpha)} z_k$ of $U_{(\alpha)}$ is open and compact. Let $t \in T$, $z_o \in Z_o$. Then $z_k t z_o \in Z(K, C)$ iff $t$ belongs to $T(K, C_k)$ where

$$T(K, C) := \{t \in T \mid t K_{(\alpha)} t^{-1} \subset C_{(\alpha)} \text{ for all } \alpha \in \Delta\}.$$

Hence $Z(K, C) = \cup_k z_k T(K, C_k) Z_o$. The set $T(K, C)$ is stable by multiplication by $T_+$ because the $K_{(\alpha)}$ are normalized by $T_+$. Hence the property (IV.3.2) is true if for any $(K, C)$ iff $T(K, C) = T_o T_+$ for some compact $T_o \subset T(K, C)$ for any $(K, C)$. When $T(K, C)$ satisfies this property we say simply that $T(K, C)$ is compact modulo $T_+$.

(2) Change of $(K, C)$ by $(K', C')$. The conjugation by $t \in T$ respects the property of being an open compact subgroup of $T$ or of being an open compact subgroup of $T$ normalized by $T_+$. Let $t_1, t_2 \in T$. Then $(t_1^{-1} K t_1, t_2 C t_2^{-1})$ satisfies the same hypotheses than $(K, C)$. An element $t \in T$ satisfies $t t_1^{-1} K_{(\alpha)} t_1 t^{-1} \subset t_2 C_{(\alpha)} t_2^{-1}$

iff $x := t(t_1t_2)^{-1}$ satisfies $xK_{(\alpha)}x^{-1} \subset C_{(\alpha)}$. In other terms,

$$T(K,C) = T(t_1^{-1}Kt_1, t_2Ct_2^{-1})(t_1t_2)^{-1}. \tag{2a}$$

We deduce that $T(t_1^{-1}Kt_1, t_2Ct_2^{-1})$ is compact modulo $T_+$ iff the same is true for $T(K,C)$.

Let $(K', C')$ satisfying the same hypotheses than $(K, C)$. For $Y = K, C$ and $\alpha \in \Delta$, there exists $t_+ \in T^+$ such that

$$t_+Y_{(\alpha)}t_+^{-1} \subset Y'_{(\alpha)} \subset t_+^{-1}Y_{(\alpha)}t_+.$$

We can choose $t_+$ independent of the finite set of $\alpha \in \Delta$. The inclusions $K'_{(\alpha)} \subset t_+^{-1}K_{(\alpha)}t_+$, $t_+C_{(\alpha)}t_+^{-1} \subset C'_{(\alpha)}$ imply $T(t_+^{-1}Kt_+, t_+Ct_+^{-1}) \subset T(K', C')$. By symmetry and by (2a), we obtain:

$$T(K', C')t_+^2 \subset T(K,C) \subset T(K', C')t_+^{-2}. \tag{2b}$$

(3) Choosing $(K', C') = (X, X)$ and applying (IV.3.1) we deduce from (2a) and (2b) that there exists $t_+ \in T_+$ such that $T_+t_+^4 \subset T(t_+^{-1}Kt_+, t_+Ct_+^{-1}) \subset T_+$. Using the remark following (2a) and that $t_+^4 \in T_+$, we deduced that $T(K,C)$ is compact modulo $T_+$ for all $(K, C)$ iff this is true when

$$T_+t_+ \subset T(K,C) \subset T_+$$

for some $t_+ \in T_+$. The image of these inclusions under the natural projection $T \to T/T^o$ followed by an isomorphism $T/T^o \simeq \mathbf{N}^d$ is

$$a + \mathbf{N}^d \subset Y \subset \mathbf{N}^d,$$

where $(Y, a)$ is the image of $(T(K,C), t_+)$ in $\mathbf{N}^d$. We have $Y + \mathbf{N}^d \subset Y$ because $T(K,C)$ is stable by multiplication by $T^+$. By (IV.3.3), $Y$ is a finite union of $y + \mathbf{N}^d$ with $y \in Y$. We deduce that $T(K,C) = T_oT_+$ is compact modulo $T^+$.

The claim (IV.3.2) is proved.

**IV.3.5.** We explain how the geometric property (IV.2.2) can be deduced from (IV.3.2). We start from the decomposition $G = UZK_o$. As $K$ is normal in $K_o$, the support $G(U, \phi, K)$ of $\operatorname{ind}_U^G \phi$ described in (IV.2.1) (1) is a union of double $(U, K_o)$-cosets. Hence $G(U, \phi, K) = U(Z \cap G(U, \phi, K))K_o$. We have

$$Z \cap G(U, \phi, K) = \{z \in Z \mid z(K \cap U)z^{-1} \subset \operatorname{Ker} \phi\}.$$

because $zKz^{-1} \cap U = z(K \cap U)z^{-1}$ as $z \in Z$ normalizes $U$. As $\phi_{(\alpha)}$ is trivial for all positive non simple roots $\alpha \in \Phi^+ - \Delta$ by hypothesis (IV.1.1), and as $z \in Z$ normalizes $U_{(\alpha)}$ for all roots $\alpha \in \Phi$, the decomposition of $K \cap U$ implies that

$$Z \cap G(U, \phi, K) = \{z \in Z \mid z(K \cap U_{(\alpha)})z^{-1} \subset \operatorname{Ker} \phi_{(\alpha)} \text{ for all } \alpha \in \Delta\}.$$

By hypothesis (IV.1.1), $\operatorname{Ker} \phi_{(\alpha)}$ is an open compact subgroup of $U_{(\alpha)}$ for all $\alpha \in \Delta$. The open compact subgroups $K \cap U_{(\alpha)}$ of $U_{(\alpha)}$ are normalized by $T_+$. Hence by

(IV.3.2) $Z \cap G(U, \phi, K)$ is compact modulo $T^+$. Therefore $G(U, \phi, K)$ is a finite union of $UzK_o$ with $z \in Z$. The geometric property (IV.2.2) is proved.

**IV.3.6.** We check the computations of (IV.2.3). The first one (a) follows from the formula (II.7) (1) and from the fact that $K$ is normal in $K_o$ hence $Kk_oK = k_oK = Kk_o$ and $UxKk_oK = Uxk_oK$ for any $k_o \in K_o, x \in G$. We check now the second one (b). Any element $t_+ \in T_+$ satisfies the relations

$$t_+(K \cap U)t_+^{-1} \subset K \cap U, \ \ t_+(K \cap Z)t_+^{-1} = K \cap Z, \ \ t_+^{-1}(K \cap \overline{U})t_+ \subset K \cap \overline{U}.$$

These relations and the Iwahori decomposition of $K$ imply

(a) $t_+K = (K \cap Z\overline{U})t_+K$,

(b) $Kt_+ = Kt_+(K \cap ZU)$,

(c) $Kt_+K = \cup_{u^-} Kt_+u^-$(disjoint) with $K \cap U^- = \cup_{u^-} t_+^{-1}(K \cap U^-)t_+u^-$ (disjoint),

(d) $UzKt_+K = Uz(K \cap Z\overline{U})t_+K = Uzt_+K$ for any $z \in Z$ ($z$ normalizes $U \cap K$).

By (d) the support of $f := \phi_{UzK} * [Kt_+K]$ is contained in $Uzt_+K$. Hence $f = f(zt_+)\phi_{Uzt_+K}$. We want to prove $f(zt^+) = 1$. We have using (c):

$$f(zt_+) = \sum_{u^-} \phi_{UzK}\big(zt_+(t_+u^-)^{-1}\big) = \sum_{u^-} \phi_{UzK}\big(zt_+u^{--1}t_+^{-1}\big)$$

for $u^-$ as in (c). Only the $u^-$ with $zt_+u^{--1}t_+^{-1} \in UzK$ give a nonzero contribution. As $z$ normalises $U$, we can forget it and the condition is $u^{--1} \in t_+^{-1}UKt_+$ which means $u^- \in t_+^{-1}(K \cap U^-)t_+$ because $UK \subset B(K \cap U^-)$. With (c), only one term contributes and $f(zt^+) = 1$.

**Appendix.** Let $(F, G, \ell)$ be as in the introduction and let $R$ be any algebraically closed field of characteristic $\ell$. The aim of this appendix is to compare three properties of a representation $(\rho, V) \in \mathrm{Mod}_R\, G$:

(i) The $\mathcal{H}_R(G, K)$-module $V^K$ is finitely generated for all $K$ in a separated decreasing sequence of open compact pro-p-subgroups of $G$.

(ii) $(\rho, V)$ is finitely generated in each block of $\mathrm{Mod}_R\, G$.

(iii) For any irreducible $R$-representation $\pi$, the quotient multiplicity $\dim_R \mathrm{Hom}_{RG}(\rho, \pi)$ is finite.

*Example.* $G = GL(2, F)$, $H$ is a maximal torus (split or not split), $\Omega : H \to R^*$ a character. The representation $\rho = \mathrm{ind}_H^G \Omega$ was originally considered by Waldspurger in his work on modular forms of half integral weight leading to a proof of nonvanishing of values of $L$ functions of automorphic cuspidal representations for $GL(2)$ at the center of the critical strip. We call it a *Waldspurger representation*.

THEOREM.
- *(i) is equivalent to (ii).*
- *(ii) implies (iii).*
- *(iii) implies (ii) for a complex Waldspurger representation.*

*Remarks.* (1) The finite quotient multiplicity of $\rho \in \mathrm{Mod}_R G$ is equivalent to the finite multiplicity of the contragredient $\tilde{\rho}$ : for all $\pi \in \mathrm{Irr}_R G$, the multiplicity $\dim_R \mathrm{Hom}_{RG}(\pi, \tilde{\rho})$ is finite. To prove this, one uses that the contragredient is an involution on $\mathrm{Irr}_R G$ and the isomorphism (see II.8): $\mathrm{Hom}_{RG}(\pi, \tilde{\rho}) \simeq \mathrm{Hom}_{RG}(\rho, \tilde{\pi})$.

(2) When $G$ is noncompact, their are infinitely many irreducible representations in a block, their direct sum is not finitely generated but satisfies the finite quotient multiplicity.

(3) When $R$ is the field of complex numbers, the equivalence between (i) and (ii) is proved in [BH].

(4) The category $\mathrm{Mod}_R G$ is a product of blocks. Each block has a level $r \in \mathbf{Q}$ and there are finitely many blocks of a given level [V2, II.5.8, II.5.9] and [V3, III.6]

(5) By the theory of Bernstein, in the complex case, the cuspidal blocks are well understood and the blocks are related with the cuspidal blocks of the Levi subgroups $M$ of the parabolic subgroups of $G$. The groups $M$ are the $F$-points of a reductive connected group, just as $G$, always with a noncompact center when $M \neq G$.

*Proof (i) $\Leftrightarrow$ (ii).* We need some preliminaries on the theory of Moy-Prasad minimal unrefined $R$-types. There are finitely many blocks of a given level $r \in \mathbf{Q}$. We denote by $\mathrm{Mod}_R G(r)$ their sum. The Moy-Prasad minimal unrefined types of level $r$ contained in $V \in \mathrm{Mod}_{\mathbf{C}} G$ generate the component $V(r)$ of $V$ in $\mathrm{Mod}_R G(r)$. There are only finitely many Moy-Prasad minimal unrefined types of a given level $r$, modulo $G$-conjugation [V2, II.5.5]. *For each level $r$, there exists $K(r)$ such that $V(r)$ is generated by $V(r)^{K(r)}$, this is also true for a smaller $K \subset K(r)$.* Note that $V$ is generated by $V^K$ for some $K$ iff $V$ has only finitely many non zero components in the blocks of $G$. The letter $K$ or $K(r)$ always stands for an open compact pro-$p$-subgroup of $G$. The properties (i), (ii) are respectively equivalent to: *For any level $r \in \mathbf{Q}$,*

(i)′ *the $\mathcal{H}_R(G, K)$-module $V(r)^K$ is finitely generated for some $K \subset K(r)$.*

(ii)′ *$V(r)$ finitely generated.*

We prove that (i)′ and (ii)′ are equivalent. We have $V^K = e_K V$ where $e_K \in \mathcal{H}_R(G)$ is an idempotent such that the Hecke algebra $\mathcal{H}_R(G, K)$ identifies to the subalgebra $e_K\mathcal{H}_R(G)e_K$ of the global Hecke algebra $\mathcal{H}_R(G)$, using that $K$ is a pro-$p$-group [V2, I.3.2]. Let $(v_i)_{i\in I}$ be elements of $V^K$. The two relations

$$V^K = \sum_{i\in I}\mathcal{H}_R(G,K)v_i, \quad \mathcal{H}_R(G)V^K = \sum_{i\in I}\mathcal{H}_R(G)v_i$$

are equivalent. Take $V = V(r)$ then $\mathcal{H}_R(G)V(r)^K = V(r)$ for any $K \subset K(r)$; we deduce from this the equivalence of (i)′ and (ii)′. □

**Comparaison between (ii) and (iii).** It is clear that the finite generation in each block implies the finite quotient multiplicity because each irreducible representation is admissible. The converse is not true in general. We will describe certain properties which imply that the converse is true for complex representations.

We consider first a cuspidal block $\mathcal{B} \subset \mathrm{Mod}_{\mathbf{C}}\, G$. We recall some known facts [BDK]. As for a torus (IV.3), the compact subgroups of $G$ generate a normal subgroup $G^o$ with quotient isomorphic to $\mathbf{Z}^d$ where $d$ is the rank of the maximal central split torus $T$ of $G$. The unipotent subgroups of $G$ are contained in $G^o$. If $Z$ is the center of $G$ (and not the centralizer of $T$ as in the chapter IV), the quotient $G/G^oZ$ is finite. Let $\pi \in \mathcal{B}$ irreducible. The restriction

$$\pi|_{G^o} = \oplus\sigma_i, \quad \sigma_i \in \mathrm{Irr}_{\mathbf{C}}\, G^o,$$

of $\pi$ to $G^o$ is semi-simple of finite length, and the irreducible representations in $\mathcal{B}$ are the representations of $G$ with the same restriction to $G^o$. Each $\sigma_i$ is the unique irreducible representation in a block of $\mathrm{Mod}_{\mathbf{C}}\, G^o$. We denote by $\mathcal{B}^o$ the sum of the blocks containing the $\sigma_i$. For $V \in \mathrm{Mod}_{\mathbf{C}}\, G$, the restriction of $V$ to $G^o$ belongs to $\mathcal{B}^o$ iff $V$ belongs to $\mathcal{B}$. There are infinitely many irreducible non isomorphic cuspidal representations in $\mathcal{B}$ iff $d > 0$. The abelian subcategory $\mathcal{B}_\omega$ of representations in $\mathcal{B}$ with a central character $\omega$ contains only finitely many irreducible representations modulo isomorphism.

The categories $\mathcal{B}^o$ and $\mathcal{B}_\omega$ are semi-simple. In these categories, the properties finitely generated, finite length, finite multiplicity, finite quotient multiplicity are trivially equivalent.

For any representation $V = \mathrm{ind}_{G^o}^G W \in \mathcal{B}$ compactly induced from $W \in \mathcal{B}^o$, the property: $V$ has finite quotient multiplicity is equivalent to the same property for $W$ using that the functor $\mathrm{ind}_{G^o}^G$ is the left adjoint of the restriction from $G$ to $G^o$. It implies that $W$ is finitely generated hence $V$ is finitely generated. By transitivity of the compact induction, this is also true for any $V \in \mathcal{B}$ compactly induced from a closed subgroup $H$ of $G^o$. Any complex irreducible representation of a closed subgroup $H$ of $G$ has a central character because the cardinal of $\mathbf{C}$ is strictly bigger than the cardinal of $G$, hence $V = \mathrm{ind}_H^G W$ has a central character when $Z \subset H$. We summarize:

*Let $H$ be a closed subgroup of $G$ with $H \subset G^o$ or $Z \subset H$ and let $\Omega \in \mathrm{Irr}_{\mathbf{C}}\, H$. Then the cuspidal irreducible quotients of $\mathrm{ind}_H^G\,\Omega$ have finite multiplicity if and only if $\mathrm{ind}_H^G\,\Omega$ is finitely generated in any cuspidal block.*

*Remarks.* (1) This applies to all the representations used to give models of irreducible representations in the theory of automorphic forms related with $L$-functions, that I am aware of. For the Whittaker representations, $H$ is nilpotent hence $H \subset G^o$. For the Waldspurger representations, $H$ contains the center $Z$ of $G$.

(2) There are of course other properties of $(H, \Omega)$ implying the same property for $\mathrm{ind}_H^G \Omega$. A variant that we will use for the component of a Waldspurger representation in a non cuspidal block is: $H = G^o Z'$ where $Z'$ is a closed subgroup acting on $\Omega \in \mathrm{Mod}_\mathbb{C} H$ by a character.

**Reduction to a cuspidal block.** We consider now a noncuspidal block $\mathcal{B}$ of $\mathrm{Mod}_\mathbb{C} G$. There exists a pair $(P, \mathcal{B}_M)$ where $P = MN$ is a parabolic subgroup of $G$ with unipotent radical $N$ and Levi subgroup $M$ and $\mathcal{B}_M$ is a cuspidal block of $M$, unique modulo association, such that the normalized functor of $N$-coinvariants, called the Jacquet functor, $r_P^G : \mathcal{B} \to \sum \mathcal{B}_M$ restricted to $\mathcal{B}$ is exact and faithful [R] Corollary 2.4 of image contained in the finite sum $\sum \mathcal{B}_M$ of the blocks of $\mathrm{Mod}_\mathbb{C} M$ conjugate to $\mathcal{B}_M$ by the normalizer of $M$ in $G$. We need all of them, at the level of blocks $r_P^G(\mathcal{B}) = \sum \mathcal{B}_M$. Let $(\pi, V) \in \mathcal{B}$. We claim:

*$(\pi, V)$ is finitely generated iff $r_P^G(\pi, V)$ is finitely generated.*

*$(\pi, V)$ has finite quotient multiplicity iff $r_P^G(\pi, V)$ has finite quotient multiplicity.*

$r_P^G(\pi, V)$ is finitely generated iff $r_P^G(\pi, V)$ is finitely generated in each cuspidal block because the sum $\sum \mathcal{B}_M$ is finite. The computation of the Jacquet functors of the representations used for models in the theory of automorphic forms is a well known and basic question, originally considered by Rodier, Casselman, and Shalika for the generic Whittaker representation.

The proof of the claim is easy. Finitely generated: if because of exactness and faithfulness of $r_P^G$, any subset $(v_i)$ of $V$ which lifts a set of generators of $r_P^G(\pi, V)$ generates $(\pi, V)$. Iff because $G/P$ is compact, a finite set $(v_i)$ of generators of $(\pi, V)$ is fixed by an open compact subgroup $K$, $G = \cup_j P k_j K$ (finite union), the finite set $(k_j v_i)$ generates $r_P^G(\pi, V)$.

Finite quotient multiplicity: $r_P^G$ is the left adjoint of the normalized parabolic induction $i_P^G$, so $\mathrm{Hom}_{\mathbb{C}G}(\pi, i_P^G \tau) \simeq \mathrm{Hom}_{\mathbb{C}M}(r_P^G \pi, \tau)$ for all $\tau \in \mathrm{Irr}_\mathbb{C} M$. As $i_P^G \tau$ has finite length, the finite quotient multiplicity for $\pi$ implies the finite quotient multiplicity for $r_P^G \pi$ (one does not need to suppose $\pi \in \mathcal{B}$).

Conversely, the faithfulness of $r_P^G$ on $\mathcal{B}$ implies that $r_P^G \rho \neq 0$ for any irreducible representation $\rho$ which is a quotient of $\pi \in \mathcal{B}$; as $r_P^G \rho$ has finite length it has an irreducible quotient $\tau$; by adjunction $\rho$ is contained in $i_P^G \tau$ and $\dim_\mathbb{C} \mathrm{Hom}_{\mathbb{C}G}(\pi, \rho) \leq \dim_\mathbb{C} \mathrm{Hom}_{\mathbb{C}G}(r_P^G \pi, \tau)$. Hence the finite quotient multiplicity for $r_P^G \pi$ implies the finite quotient multiplicity for $\pi$.

*Example.* Let $G = GL(2, F)$ and $B = TN$ is the upper triangular subgroup with unipotent radical $N$ and $T$ the diagonal subgroup. Let $V \in \mathrm{Mod}_\mathbb{C} G$. Then

(iii) implies (ii) for the noncuspidal part of $V$ iff (iii) implies (ii) for the $N$-coinvariants $V_N$. We need to analyze $V_N$. We take the example of a complex Waldspurger representation $\mathrm{ind}_H^G \Omega$ defined at the beginning of the appendix.

First case: $H = T$. We have $G = B \cup BsN$ where $s = \begin{pmatrix} 0 & 1 \\ 1 & 0 \end{pmatrix}$ and a $\mathbf{C}N$-equivariant exact sequence:

(1) $$0 \to \mathrm{ind}_T^{BsN} \Omega \to \mathrm{ind}_T^G \Omega \to \mathrm{ind}_T^B \Omega \to 0.$$

The functor of $N$-coinvariants is exact and $(\mathrm{ind}_T^G \Omega)_N$ can be computed using (2) and (3) below. We have

(2) $$(\mathrm{ind}_T^B \Omega)_N \simeq \Omega$$

by the linear form $f \to \int_N f(n)\,dn$ for $f \in \mathrm{ind}_T^B \Omega$ and a Haar measure $dn$ on $N$. We can neglect the character $\Omega$ for the properties (ii) and (iii). We compute $(\mathrm{ind}_T^{BsN} \Omega)_N$. The linear map $f(bsn) \to \phi(b) := \int_N f(bsn)dn$ for $b \in B$, followed by the restriction to $N$ identifies $(\mathrm{ind}_T^{BsN} \Omega)_N$ with the space $C_c^\infty(N; \mathbf{C})$. The action of $t \in T$ on $\phi \in C_c^\infty(N; \mathbf{C})$ is

$$(t * \phi)(n') = \int_N f(n'snt)\,dn = \Omega(sts) \int_N f(n''s\,t^{-1}nt)dn = \Omega\delta_B(sts)\phi(n'')$$

where $\delta_B$ is the module of $B$ and $n'' := (sts)^{-1}n'sts$ for $n' \in N$. We have

(3) $$\left(\mathrm{ind}_T^{BsN} \Omega\right)_N \simeq (\Omega\delta_B \otimes \rho) \circ s,$$

where $\rho$ is the natural action of $T$ on $C_c^\infty(N; \mathbf{C})$ by $(t.\phi)(n) = \phi(t^{-1}nt)$. For the properties (ii) and (iii) we can neglect the character $\Omega\delta_B$ and $s$. As $T$ has two orbits in $N$, the trivial element of stabilizer $T$ and the nontrivial elements of stabilizer the center $Z$ of $G$, we have a $T$-equivariant exact sequence

(4) $$0 \to \mathrm{ind}_Z^T 1 \to \rho \to 1 \to 0.$$

For (ii) and (iii) we can neglect the trivial character, and we are reduced to examine $\mathrm{ind}_Z^T 1$. The blocks of $\mathrm{Mod}_{\mathbf{C}} T$ are parametrized by the characters $\chi^o$ of the maximal compact subgroup $T^o$ of $T$, and the component of $\mathrm{ind}_Z^T 1$ in the block parametrized by $\chi^o$ is the cyclic representation $\mathrm{ind}_{ZT^o}^T \chi_o$ if $\chi^o$ is trivial on $Z \cap T^o$ and 0 otherwise. We deduce that the Waldspurger representation $\mathrm{ind}_T^G \Omega$ is finitely generated in the non cuspidal blocks of $G$.

Second case: $H$ nonsplit. Modulo conjugation, $H$ is contained in one of the two maximal, compact modulo the center $Z$, subgroups of $G$

$$C_1 := KZ, \quad C_2 := ZI \cup ZIt,$$

where $K = GL(2, O_F)$, $I$ is the standard Iwahori subgroup normalized by $t := \begin{pmatrix} 0 & 1 \\ p_F & 0 \end{pmatrix}$. We suppose $H \subset C$ where $C = C_1$ or $C_2$. Using $G = CTN$ and the transitivity of the compact induction, we compute:

$$\left(\operatorname{ind}_H^G \Omega\right)_N \simeq \operatorname{ind}_{C\cap T}^T(\tau_{C\cap N}), \tag{5}$$

with $\tau_{C\cap N}$ equal to the $C \cap N$-coinvariants of $\tau = \operatorname{ind}_H^C \Omega$. As $C \cap T = T^o Z$ and $Z$ acts on $\tau_{C\cap N}$ by multiplication by a character. We deduce from the cuspidal case seen above, that the Waldspurger representation $\operatorname{ind}_H^G \Omega$ are finitely generated in the non cuspidal blocks if and only if the noncuspidal quotients have finite multiplicity.

---

## REFERENCES


[BDK] Bernstein, Deligne, and Kazhdan, *Le Centre de Bernstein,* in Bernstein, Deligne, Kazhdong, Vignéras, *Representations des groups reductifs sur un corps local,* Hemann, Paris, 1984.

[BH] Colin Bushnell and Guy Henniart, Generalized Whittaker models and the Bernstein centre, preprint, May 2001.

[MW] Colette Moeglin and Jean-Loup Waldspurger, Modéles de Whittaker dégénérés pour des groupes $p$-adiques. *Math. Z.* **196,** no. 3 (1987), 427–452.

[MVW] Colette Moeglin, Marie-France Vignéras, and Jean-Loup Waldspurger, *Correspondances de Howe sur un corps p-adique. Lecture Notes in Mathematics,* vol. 1291, Springer-Verlag, Berlin, 1987.

[R] Alain Roche, Parabolic induction and the Bernstein decomposition, *Compositio. Math.* **134** (2002), 113–133.

[V1] Marie-France Vignéras, $\ell$-principe de Brauer pour un groupe de Lie $p$-adique, $p \neq \ell$, *Math. Nachr.* **159** (1992), 37–45.

[V2] ———, Représentations $l$-modulaires d'un groupe réductif $p$-adique avec $\ell \neq p$, *Progress in Mathematics,* vol. 137, Birkhäuser, Boston, 1996.

[V3] ———, Induced representations of $p$-adic reductive groups, *Sel. Math.,* new series 4 (1998), 549–623.

[V4] ———, Integral Kirillov model, *C. R. Acad. Sci. Paris Serie 1 Math.* **326**, no.4 (1998), 411–416.

[V5] ———, Irreducible modular representations of a reductive $p$-adic group and simple modules for Hecke algebras, ECM3, Barcelone, 2000.

[V6] ———, Congruence modulo $\ell$ between $\epsilon$ factors for cuspidal representations of GL(2), *Journal de Théorie des Nombres de Bordeaux* **12** (2000), 571–580.

CHAPTER 29

# REPRÉSENTATIONS DE RÉDUCTION UNIPOTENTE POUR SO(2N+1): QUELQUES CONSÉQUENCES D'UN ARTICLE DE LUSZTIG

By JEAN-LOUP WALDSPURGER

**0. Introduction.** Cet article fait suite à l'article [MW] de C. Moeglin et l'auteur. Rappelons brièvement quel en était le sujet. Soient $p$ un nombre premier, $F$ une extension finie de $\mathbb{Q}_p$, $n$ un entier $\geq 1$. On suppose $p$ "grand." On note $\mathbf{G}_{iso}$ et $\mathbf{G}_{an}$ les deux formes possibles sur $F$ du groupe spécial orthogonal d'une forme quadratique de dimension $2n+1$; $\mathbf{G}_{iso}$ est déployé et $\mathbf{G}_{an}$ est une forme intérieure non déployée de $\mathbf{G}_{iso}$. Quand on n'a pas besoin de distinguer les deux formes, on les note uniformément $\mathbf{G}_\sharp$. On pose $G_\sharp = \mathbf{G}_\sharp(F)$. Soit $\pi$ une représentation admissible irréductible de $G_\sharp$ dans un espace vectoriel complexe $E$. Soit $K$ un sous-groupe parahorique de $G_\sharp$, notons $K^u$ son radical pro-$p$-unipotent et $E^{K^u}$ le sous-espace des éléments de $E$ invariants par $K^u$. De $\pi$ se déduit une représentation de $K/K^u$ dans $E^{K^u}$, notons-la $res_K(\pi)$. Le quotient $K/K^u$ est le groupe des points sur le corps résiduel $\mathbb{F}_q$ de $F$ d'un groupe algébrique réductif connexe défini sur $\mathbb{F}_q$. On connaît la notion de représentation irréductible unipotente d'un tel groupe. On dit alors que $\pi$ est de réduction unipotente s'il existe $K$ comme ci-dessus tel que $E^{K^u}$ soit non nul et tel que $res_K(\pi)$ contienne une représentation irréductible unipotente de ce groupe. On note $Irr_u^{G_\sharp}$ l'ensemble des classes d'isomorphie de représentations de $G_\sharp$, admissibles, irréductibles et de réduction unipotente. On note $Irr_{utemp}^{G_\sharp}$ le sous-ensemble des classes d'isomorphie des représentations qui sont de plus tempérées. Si $\pi \in Irr_{utemp}^{G_\sharp}$, notons $\Theta_\pi$ son caractère, que l'on considère comme une distribution sur $G_\sharp$. Le but de l'article [MW] est de décrire les combinaisons linéaires $\Sigma c_i \Theta_{\pi_i}$, où les $\pi_i$ appartiennent à $Irr_{utemp}^{G_\sharp}$, qui sont des distributions stablement invariantes.

Notons $W_F$ le groupe de Weil de $F$. Considérons un couple $(\psi, \epsilon)$, où:

$$\psi : W_F \times SL(2, \mathbb{C}) \to Sp(2n, \mathbb{C})$$

est un homomorphisme tel que $\psi_{|W_F}$ est semi-simple et non ramifié et $\psi_{|SL(2,\mathbb{C})}$ est algébrique;

$$\epsilon : Z_{Sp(2n,\mathbb{C})}(\psi)/Z_{Sp(2n,\mathbb{C})}(\psi)^0 \to \{\pm 1\}$$

est un caractère, où $Z_{Sp(2n,\mathbb{C})}(\psi)$ est le commutant dans $Sp(2n, \mathbb{C})$ de l'image de $\psi$ et $Z_{Sp(2n,\mathbb{C})}(\psi)^0$ est sa composante neutre.

Manuscript received Juillet 2001

Notons $\hat{Z}$ le centre de $Sp(2n, \mathbb{C})$. On note $\epsilon_{|\hat{Z}}$ le caractère de $\hat{Z}$ composé de $\epsilon$ et de l'homomorphisme naturel:

$$\hat{Z} \to Z_{Sp(2n,\mathbb{C})}(\psi)/Z_{Sp(2n,\mathbb{C})}(\psi)^0.$$

Identifions à $\{\pm 1\}$ le groupe des caractères de $\hat{Z}$. On note $\underline{\Psi}_u^{G_\sharp}$ l'ensemble des classes de conjugaison par $Sp(2n, \mathbb{C})$ des couples $(\psi, \epsilon)$ comme ci-dessus tels que:

$$\epsilon_{|\hat{Z}} = \begin{cases} 1, & \text{si } \sharp = iso, \\ -1, & \text{si } \sharp = an. \end{cases}$$

Lusztig a construit une bijection de $\underline{\Psi}_u^{G_\sharp}$ sur $Irr_u^{G_\sharp}$, cf. [L1]; l'existence de cette bijection était prédite par une conjecture de Langlands, raffinée par Deligne et Lusztig. Dans [MW], nous avons besoin de renseignements supplémentaires, dans les trois directions suivantes.

(1) Notons $\underline{\Psi}_{utemp}^{G_\sharp}$ le sous-ensemble des (classes de conjugaison de) $(\psi, \epsilon) \in \underline{\Psi}_u^{G_\sharp}$ tels que l'image de $W_F$ par $\psi_{|W_F}$ soit un sous-groupe relativement compact de $Sp(2n, \mathbb{C})$. On veut construire une bijection

$$\begin{aligned} \underline{\Psi}_{utemp}^{G_\sharp} &\to Irr_{utemp}^{G_\sharp} \\ (\psi, \epsilon) &\mapsto \pi_{\psi,\epsilon}. \end{aligned}$$

(2) On veut déterminer les combinaisons linéaires d'éléments de $\underline{\Psi}_{utemp}^{G_\sharp}$ dont les images par la bijection précédente soient des représentations (virtuelles) elliptiques au sens d'Arthur [Ar].

(3) Soient $(\psi, \epsilon) \in \underline{\Psi}_{utemp}^{G_\sharp}$ et $K$ un sous-groupe parahorique de $G_\sharp$. Supposons que $\pi_{\psi,\epsilon}$ soit composante d'une représentation (virtuelle) elliptique. On veut calculer, ou du moins obtenir assez de renseignements sur la représentation $res_K(\pi_{\psi,\epsilon})$.

Nous résolvons ces problèmes à l'aide des méthodes introduites par Lusztig. Remplaçons la bijection qu'il a construite par sa composée avec l'involution $\mathbf{D}^{G_\sharp}$ de Zelevinsky-Aubert-Schneider-Stuhler. On note:

$$\begin{aligned} \underline{\Psi}_u^{G_\sharp} &\to Irr_u^{G_\sharp} \\ (\psi, \epsilon) &\mapsto \pi_{\psi,\epsilon} \end{aligned}$$

cette application composée. On prouve alors, pour $(\psi, \epsilon) \in \underline{\Psi}_u^{G_\sharp}$:

- $\pi_{\psi,\epsilon}$ est de la série discrète si et seulement si l'image de $\psi$ n'est contenue dans aucun sous-groupe parabolique propre de $Sp(2n, \mathbb{C})$ (proposition 4.2);

- $\pi_{\psi,\epsilon}$ est tempérée si et seulement si $(\psi, \epsilon) \in \underline{\Psi}_{utemp}^{G_\sharp}$ (proposition 4.3); cela résout le problème (1).

Notre bijection est compatible à la classification de Langlands. Soit $(\psi, \epsilon) \in \underline{\Psi}_u^{G_\sharp}$. Notons *Frob* un élément de Frobenius de $W_F$. Alors $z = \psi_{|W_F}(Frob)$ est un élément semi-simple de $Sp(2n, \mathbb{C})$. On lui associe un sous-groupe parabolique $\hat{Q}$ et un sous-groupe de Lévi $\hat{L}$ de $\hat{Q}$: les algèbres de Lie correspondantes sont les sommes des sous-espaces de l'algèbre de Lie $\mathfrak{sp}(2n, \mathbb{C})$, propres pour $Ad(z)$, de valeurs propres de modules $\leq 1$, resp. $= 1$. On introduit dualement un sous-groupe parabolique $\mathbf{Q}$ de $\mathbf{G}_\sharp$ et un sous-groupe de Lévi $\mathbf{L}$ de $\mathbf{Q}$. En fait, $\psi$ prend ses valeurs dans $\hat{L}$ et une généralisation facile de la bijection ci-dessus permet de définir une représentation $\pi^L_{\psi,\epsilon}$ de $L$. Elle vérifie les conditions requises pour que la représentation induite $Ind_Q^{G_\sharp}(\pi^L_{\psi,\epsilon})$ possède un unique quotient irréductible, le quotient de Langlands. On montre que ce quotient est égal à $\pi_{\psi,\epsilon}$ (théorème 4.4).

La solution du problème (2) ci-dessus est exprimée en termes combinatoires par le théorème 4.7.

Pour ce qui est du problème (3), nous revenons à la bijection initiale de Lusztig, c'est-à-dire que nous calculons $res_K \circ \mathbf{D}^{G_\sharp}(\pi_{\psi,\epsilon})$. Les représentations des groupes finis qui interviennent sont paramétrisées par des représentations de groupes de Weyl. Nous calculons $res_K \circ \mathbf{D}^{G_\sharp}(\pi_{\psi,\epsilon})$ en termes de cette paramétrisation (proposition 5.3).

Décrivons brièvement le contenu de chacun des paragraphes de l'article. Le premier rappelle les constructions de [L1] et [L2]. On les précise un peu car Lusztig laisse indéterminés certains choix. Expliquons ce point. Notons $\mu_{nr}$ le caractère non ramifié d'ordre 2 de $W_F$, prolongeons-le trivialement à $W_F \times SL(2, \mathbb{C})$. Soit $\pi$ un élément cuspidal de $Irr_u^{G_\sharp}$. Les constructions de Lusztig définissent deux couples $(\psi', \epsilon'), (\psi'', \epsilon'') \in \underline{\Psi}_u^{G_\sharp}$, candidats à paramétriser $\pi$. Ils se déduisent l'un de l'autre par torsion: $\psi' = \psi'' \otimes \mu_{nr}, \epsilon' = \epsilon''$. Pour ce que fait Lusztig, le choix d'un élément de ce couple est indifférent. Pour ce que nous faisons dans [MW], à savoir regrouper les représentations en paquets, le choix ne peut plus être arbitraire, c'est pourquoi nous devons le préciser.

On doit étudier comment les constructions de Lusztig se comportent par induction. Le problème se décompose en deux, comme les constructions elles-mêmes. Une partie "géométrique" est traitée au paragraphe 2, à l'aide des résultats de [L3] et [L4]. La démonstration de la proposition principale 2.11 s'inspire de celle du théorème 6.2 de [KL]. Une deuxième partie, "algébrique," est traitée au paragraphe 3. Elle repose sur les constructions de [L2] et des résultats de Bushnell et Kutzko. Les problèmes (1) et (2) sont ensuite résolus au paragraphe 4, comme de simples conséquences des deux paragraphes précédents.

Le caractère explicite des constructions rappelées au paragraphe 1 permet, par un argument de déformation (remplacer $q$ par 1), de résoudre le problème (3), plus exactement, comme on l'a dit ci-dessus, ce problème modifié par l'involution $\mathbf{D}^{G_\sharp}$. Cela est fait au paragraphe 5, où on utilise largement des constructions combinatoires de [MW].

Le sixième paragraphe revient sur la question des choix évoquée plus haut. Le problème est en fait équivalent au suivant, concernant les groupes orthogonaux pairs sur les corps finis. Une représentation unipotente cuspidale du groupe spécial orthogonal étant donnée, comment distinguer ses deux prolongements au groupe orthogonal? Nous donnons une façon de le faire en 6.3. Nous montrons ensuite que cette définition coïncide avec celle de [MW]. Cette dernière repose sur une conjecture que nous admettons pour prouver cette coïncidence. Mais, pour le présent article, nous n'avons pas besoin de cette conjecture.

L'article se termine par un index des principales notations.

On a choisi de ne traiter que le cas des groupes orthogonaux impairs, alors que l'article [L1] s'applique à tout groupe adjoint dont la forme intérieure quasi-déployée est déployée. Cela pour plusieurs raisons. D'abord parce que, pour les applications à notre article [MW], ce cas suffit. Ensuite parce que travailler dans une situation concrète nous semble permettre davantage de précision. Enfin parce que le cadre des groupes adjoints n'est pas non plus d'une généralité maximale. Mais, bien que nous ne traitions qu'un groupe particulier, les méthodes que nous employons sont pour l'essentiel générales (à l'exception de la démonstration du lemme 2.13, où nous utilisons le plongement naturel de $\mathbf{G}_\sharp$ dans $\mathbf{GL}(2n+1)$). Elles doivent pouvoir s'appliquer à tout groupe réductif connexe. Le fait de se limiter aux groupes orthogonaux impairs a toutefois un inconvénient. Pour de nombreux raisonnements utilisant l'induction parabolique, on doit supposer connus les résultats pour les sous-groupes de Lévi de sous-groupes paraboliques de $\mathbf{G}_\sharp$. Or ces Lévi ne sont pas des groupes orthogonaux impairs, mais des produits de tels groupes avec des groupes linéaires. Nous avons choisi d'admettre sans démonstration les résultats correspondants pour ces groupes linéaires. Cela nous paraît légitime car il n'y a pas de doute que nos méthodes s'appliquent à ces groupes (et même se simplifient dans ce cas). La théorie des représentations de ces groupes est d'ailleurs suffisamment connue grâce à Bernstein et Zelevinsky, cela n'aurait pas grand intérêt de la refaire.

Il s'est avéré que, pendant que j'écrivais cet article, Lusztig écrivait l'article [L10]. Quelques mois plus tard, il publiait [L11]. Ces articles contiennent les résultats de nos paragraphes 2 et 3 et du début du paragraphe 4. Ils se placent dans un cadre beaucoup plus général que le nôtre. Je remercie d'ailleurs Lusztig de m'avoir signalé une définition incorrecte dans une première version du présent article.

## 1. Paramétrisation.

**1.1.** Rappelons quelques notations de l'article I. On note $F$ un corps local non archimédien de caractéristique nulle, $\mathfrak{o}_F$ son anneau des entiers, $\varpi_F$ une uniformisante, $\mathbb{F}_q$ le corps résiduel, qui est fini et a $q$ éléments. On note $p$ la caractéristique de $\mathbb{F}_q$. On suppose $p \neq 2$.

Les groupes algébriques définis sur $F$ ou $\mathbb{F}_q$ seront désignés par des lettres majuscules grasses et leurs groupes de points sur $F$ ou $\mathbb{F}_q$ par les lettres majuscules ordinaires correspondantes ($H = \mathbf{H}(F)$). Interviendront aussi des groupes algébriques complexes. Leurs algèbres de Lie seront désignées par des lettres gothiques minuscules: $\mathfrak{h}$ est l'algèbre de Lie de $H$. En tout cas, pour un groupe algébrique $\mathbf{H}$ défini sur un corps quelconque, on note $\mathbf{H}^0$ sa composante neutre.

Soient $V$ un espace vectoriel de dimension finie sur $F$, muni d'une forme quadratique $Q$ non dégénérée. On note $\mathbf{G}$ le groupe spécial orthogonal de $(V, Q)$ et $\mathbf{G}^{\pm}$ le groupe orthogonal. Eventuellement, on précisera ces notations en ajoutant un indice $V$: $\mathbf{G}_V$, $\mathbf{G}_V^{\pm}$. On dispose d'un homomorphisme:

$$sp : G \to \{\pm 1\},$$

composé de la norme spinorielle:

$$G \to F^\times/F^{\times 2}$$

et de l'homomorphisme:

$$\begin{array}{rcl} F^\times/F^{\times 2} & \to & \{\pm 1\} \\ x & \mapsto & (-1)^{v_F(x)} \end{array}$$

où $v_F$ est bien sûr la valuation. On note $G_{sp=1}$ le noyau de $sp$.

Si $L \subseteq V$ est un $\mathfrak{o}_F$-réseau, on pose:

$$\tilde{L} = \{v \in V;\ \forall v' \in V,\ \ Q(v, v') \in \mathfrak{o}_F\}.$$

On dit que $L$ est presque autodual si:

$$\tilde{L} \supseteq L \supseteq \varpi_F \tilde{L}.$$

Soit $L_. = (L_0, \ldots, L_r)$ une chaîne de réseaux presque autoduaux. Cela signifie que chaque $L_i$ est un réseau presque autodual et que l'on a les inclusions:

$$\tilde{L}_0 \supseteq L_0 \supset L_1 \supset \ldots \supset L_r \supseteq \varpi_F \tilde{L}_r.$$

On note $K(L_.)$ le sous-groupe des éléments de $G_{sp=1}$ qui stabilisent chaque réseau $L_i$. On note $K(L_.)^u$ son plus grand sous-groupe distingué pro-p-nilpotent. Les espaces:

$$l'_r = L_r/\varpi_F \tilde{L}_r \text{ et } l''_0 = \tilde{L}_0/L_0$$

sont munis de formes quadratiques non dégénérées à valeurs dans $\mathbb{F}_q$. On a l'isomorphisme:

$$K(L_.)/K(L_.)^u \simeq GL(L_0/L_1) \times \ldots \times GL(L_{r-1}/L_r) \times G_{l'_r} \times G_{l''_0}.$$

On parlera indifféremment de représentations de $G$ ou de $G$-modules. C'est-à-dire que, une représentation étant un couple (ou une classe d'isomorphie de couple) $(\pi, E)$, où $E$ est un espace vectoriel complexe et $\pi$ un homomorphisme de $G$ dans $GL(E)$, on désignera ce couple comme "la représentation $\pi$" ou bien comme "le $G$-module $E$." On fera de même pour les représentations des divers groupes et algèbres qui interviendront.

On note $Irr^G$ l'ensemble des classes d'isomorphie de représentations lisses irréductibles de $G$. Soit $(\pi, E)$ une telle représentation. On dit que $\pi$ est de réduction unipotente s'il existe une chaîne de réseaux presque autoduaux $L_.$ de $V$ telle que:

- le sous-espace des invariants $E^{K(L_.)^u}$ est non nul;
- la représentation déduite de $\pi$ de $K(L_.)/K(L_.)^u$ dans cet espace est somme de représentations irréductibles unipotentes (on sait définir cette notion car $K(L_.)/K(L_.)^u$ est le groupe des points sur $\mathbb{F}_q$ d'un groupe réductif connexe).

On note $Irr_u^G$ l'ensemble des classes d'isomorphie de représentations lisses irréductibles de $G$, de réduction unipotente.

On fixe pour tout l'article un entier $n \geq 1$ et deux espaces $V_{iso}$ et $V_{an}$ de dimension $2n+1$ sur $F$, munis de formes quadratiques non dégénérées $Q_{iso}$ et $Q_{an}$ telles que:

- $(-1)^n det(Q_{iso}) \in F^{\times 2}$, $(-1)^n det(Q_{an}) \in F^{\times 2}$;
- le noyau anisotrope de $Q_{iso}$, resp. $Q_{an}$, est de dimension 1, resp. 3.

Quand on n'aura pas besoin de distinguer $V_{iso}$ de $V_{an}$, on notera souvent $V_\sharp$ l'un quelconque de ces espaces. On note $\mathbf{G}_\sharp$ le groupe spécial orthogonal de $(V_\sharp, Q_\sharp)$.

Le but du premier chapitre est d'expliciter les constructions de Lusztig qui paramétrisent $Irr_u^{G_\sharp}$ en termes d'objets vivant dans le L-groupe $Sp(2n, \mathbb{C})$ de $\mathbf{G}_\sharp$.

**1.2.** Soit $\sharp = iso$ ou $an$. On note $\Theta_\sharp$ l'ensemble des couples $(R', R'')$ d'entiers $\geq 0$ tels que:

- $R'^2 + R' + R''^2 \leq n$,
- $R''$ est pair si $\sharp = iso$, $R''$ est impair si $\sharp = an$.

Soit $\theta = (R', R'') \in \Theta_\sharp$, posons $N = n - R'^2 - R' - R''^2$. Fixons dans $V_\sharp$ une chaîne de réseaux presque autoduaux $L_. = (L_0, \ldots, L_N)$, telle qu'en posant:

$$l'_N = L_N/\varpi_F \tilde{L}_N, \quad l''_0 = \tilde{L}_0/L_0,$$

on ait:

$$dim_{\mathbb{F}_q}(l'_N) = 2R'^2 + 2R' + 1, \quad dim_{\mathbb{F}_q}(l''_0) = 2R''^2.$$

Posons $K_\theta = K(L_.)$. C'est un sous-groupe parahorique de $G_\sharp$ et l'on a:

$$K_\theta/K_\theta^u \simeq (\mathbb{F}_q^\times)^N \times G_{l'_N} \times G_{l''_0}.$$

Chacun des groupes $G_{l'_N}$ et $G_{l''_0}$ possède, à isomorphisme près, une unique représentation irréductible cuspidale et unipotente, cf. I.2.6, I.2.9. Nous la notons $\pi'_{R'}$, resp. $\pi''_{R''}$. Considérons $\pi'_{R'} \otimes \pi''_{R''}$ comme une représentation de $K_\theta/K_\theta^u$, triviale sur $(\mathbb{F}_q^\times)^N$. Notons $\pi_\theta$ la représentation de $K_\theta$ composée de la représentation précédente et de la projection de $K_\theta$ sur $K_\theta/K_\theta^u$. Enfin, notons $Irr_\theta^{G_\sharp}$ l'ensemble des classes d'isomorphie de représentations admissibles irréductibles de $G_\sharp$ dont la restriction à $K_\theta$ contient $\pi_\theta$. Alors $Irr_u^{G_\sharp}$ est réunion disjointe des $Irr_\theta^{G_\sharp}$, quand $\theta$ parcourt $\Theta_\sharp$.

On note $Rep_\theta^{G_\sharp}$ la catégorie des représentations lisses de longueur finie de $G_\sharp$ dont tous les sous-quotients irréductibles appartiennent à $Irr_\theta^{G_\sharp}$.

**1.3.** Soient $\sharp = iso$ ou $an$ et $\theta = (R', R'') \in \Theta_\sharp$. Introduisons un espace $E_\theta$ dans lequel se réalise $\pi_\theta$. Notons $\mathcal{H}_\theta$ l'espace des fonctions $f : G_\sharp \to End(E_\theta)$, à support compact, telles que:

$$f(kgk') = \pi_\theta(k)f(g)\pi_\theta(k')$$

pour tous $g \in G_\sharp$, $k$, $k' \in K_\theta$. Cet espace est muni d'une structure d'algèbre pour laquelle l'unité est la fonction $\mathbf{1}$ à support dans $K_\theta$ et telle que $\mathbf{1}(k) = \pi_\theta(k)$ pour tout $k \in K_\theta$. Dans la suite, ne sont considérés que des $\mathcal{H}_\theta$-modules dans lesquels $\mathbf{1}$ agit par l'identité.

Si $E$ est un $G_\sharp$-module appartenant à $Rep_\theta^{G_\sharp}$, l'espace $Hom_{K_\theta}(E_\theta, E)$ est muni d'une action de $\mathcal{H}_\theta$, cf. [BK] 2.7.

*Remarque.* On utilise ici, pour simplifier, le fait que $\pi_\theta$ est isomorphe à sa contragrédiente.

D'après [M] 4.13, [MP] théorème 6.14 et [L1 ] 1.6, l'application $E \mapsto Hom_{K_\theta}(E_\theta, E)$ est une équivalence de catégories entre $Rep_\theta^{G_\sharp}$ et la catégorie des $\mathcal{H}_\theta$-modules de longueur finie. Nous allons décrire $\mathcal{H}_\theta$.

Fixons, ainsi qu'il est loisible:

- un sous-espace $U$ de $V_\sharp$;
- des éléments $v_i$ de $V_\sharp$, pour $i \in \{\pm 1, \ldots, \pm N\}$;

de sorte que les conditions suivantes soient vérifiées:

- pour tous $i$, $j \in \{\pm 1, \ldots, \pm N\}$,

$$Q_\sharp(v_i, v_j) = \begin{cases} 0, & si\ i + j \neq 0, \\ 1, & si\ i + j = 0; \end{cases}$$

- la famille $(v_i)$ est une base de l'orthogonal $U^\perp$ de $U$ dans $V_\sharp$;
- pour tout $i \in \{0, \ldots, N\}$, $L_i$ est somme directe de $L_0 \cap U$ et du réseau de $U^\perp$ engendré par:

$$v_1, \ldots, v_N, v_{-N}, \ldots, v_{-i-1}, \varpi_F v_{-i}, \ldots, \varpi_F v_{-1}.$$

Notons $K_\theta^\pm$ le normalisateur de $K_\theta$ dans $G_\sharp$, et $K_\theta^- = K_\theta^\pm \setminus K_\theta$. Si $R'' > 0$, on a:

$$K_\theta^\pm / K_\theta^u \simeq (\mathbb{F}_q^\times)^N \times G_{l'_N} \times G_{l''_0}^\pm,$$

(cf. I.1.2). On a fixé en I.2.9 un prolongement $\tilde{\pi}''_{R''}$ de $\pi''_{R''}$ à $G^{\pm}_{l''_0}$. On a besoin de fixer un prolongement $\tilde{\pi}_\theta$ de $\pi_\theta$ à $K^{\pm}_\theta$. Dans l'espoir de rendre les choses plus compréhensibles, on reporte au paragraphe 6.3 la définition précise de ce prolongement. On note $\omega$ l'élément de $\mathcal{H}_\theta$ à support dans $K^-_\theta$, tel que $\omega(k) = \tilde{\pi}_\theta(k)$ pour tout $k \in K^-_\theta$. Si $R'' = 0$ et $N > 0$, $K_\theta$ est encore d'indice 2 dans $K^{\pm}_\theta$. Notons $s_\omega$ l'élément de $G_\sharp$ qui agit par $-1$ sur $U$ et tel que:

$$s_\omega(v_i) = v_i, \text{ pour tout } i \in \{\pm 2, \dots, \pm N\},$$

$$s_\omega(v_1) = \varpi_F v_{-1}, \ s_\omega(v_{-1}) = \varpi_F^{-1} v_1.$$

Alors $s_\omega \in K^-_\theta$. On note $\omega$ l'unique élément de $\mathcal{H}_\theta$, à support dans $K^-_\theta$ et tel que $\omega(s_\omega)$ soit l'identité de $E_\theta$. En tout cas, on a $\omega^2 = \mathbf{1}$.

Si $N = R'' = 0$, on a $\mathcal{H}_\theta = \mathbb{C}$.

Si $N = 0$ et $R'' > 0$, on a $\mathcal{H}_\theta = \mathbb{C}[\omega]$.

Supposons $N > 0$. Introduisons les éléments suivants $s_i$ de $G^{\pm}_\sharp$, pour $i \in \{0, \dots, N\}$. Ils fixent tout élément de $U$ ainsi que les $v_j$, sauf ceux indiqués ci-après:

- pour $i \in \{1, \dots, N-1\}$, $s_i$ échange $v_i$ et $v_{i+1}$, ainsi que $v_{-i}$ et $v_{-i-1}$;
- $s_N$ échange $v_N$ et $v_{-N}$;
- si $R'' \neq 0$, $s_0(v_1) = \varpi_F v_{-1}$, $s_0(v_{-1}) = \varpi_F^{-1} v_1$;
- si $R'' = 0$ et $N \geq 2$, $s_0(v_1) = \varpi_F v_{-2}$, $s_0(v_2) = \varpi_F v_{-1}$, $s_0(v_{-2}) = \varpi_F^{-1} v_1$, $s_0(v_{-1}) = \varpi_F^{-1} v_2$;
- si $R'' = 0$ et $N = 1$, $s_0(v_1) = \varpi_F^2 v_{-1}$, $s_0(v_{-1}) = \varpi_F^{-2} v_1$.

Rappelons que l'on note $G_{\sharp, sp=1}$ le noyau de l'homomorphisme $sp$. Notons $N_{G^{\pm}_\sharp}(K_\theta)$ le normalisateur de $K_\theta$ dans $G^{\pm}_\sharp$. L'intersection:

$$(s_i N_{G^{\pm}_\sharp}(K_\theta)) \cap G_{\sharp, sp=1}$$

est formée d'une seule classe à droite modulo $K_\theta$. Fixons $s'_i \in N_{G^{\pm}_\sharp}(K_\theta)$ tel que $s_i s'_i$ appartienne à cette classe. Il existe alors dans $\mathcal{H}_\theta$ un unique élément $S_i$, à support dans $K_\theta s_i s'_i K_\theta$, vérifiant l'équation:

(1) $$(S_i + \mathbf{1})(S_i - q^{L(i)}\mathbf{1}) = 0,$$

où:

$$L(1) = L(2) = \dots = L(N-1) = 1, \quad L(N) = 2R' + 1,$$

$$L(0) = \begin{cases} 2R'', & si\ R'' > 0, \\ 1, & si\ R'' = 0. \end{cases}$$

Notons $\mathcal{H}'_\theta$ la sous-algèbre engendrée par les $S_i$ pour $i \in \{0, \dots, N\}$. On sait que les relations vérifiées par les $S_i$ sont engendrées par les relations (1) et par les relations suivantes:

- pour $i, j \in \{0, \ldots, N\}$ tels que $i \neq j$ et $s_i s_j$ est d'ordre fini $m(i, j)$ dans $G_\sharp$,

$$S_i S_j S_i \ldots = S_j S_i S_j \ldots$$

où chaque produit a $m(i, j)$ termes.

On a l'égalité:

$$\mathcal{H}_\theta = \mathbb{C}[\omega] \otimes_{\mathbb{C}} \mathcal{H}'_\theta.$$

La structure d'algèbre est le produit direct si $R'' > 0$, le produit semi-direct pour lequel $\omega$ échange $S_0$ et $S_1$ si $R'' = 0$.

On aura besoin des précisions suivantes concernant les $S_i$. Dans le cas où $i \in \{1, \ldots, N-1\}$, le calcul de $S_i$ se ramène aisément au cas du groupe $GL_2(F)$. On voit que l'on peut supposer $s'_i = 1$ et que $S_i(s_i)$ est l'identité de $E_\theta$. La détermination de $S_0$ et $S_N$ est plus délicate. On a déjà fixé un prolongement $\tilde{\pi}_\theta$ de $\pi_\theta$ à $K_\theta^\pm$. On le prolonge encore en une représentation de $N_{G_\sharp^\pm}(K_\theta)$, notée encore $\tilde{\pi}_\theta$. Ce prolongement est bien déterminé de la façon suivante. Notons $\xi$ l'élément de $G_\sharp^\pm$ qui agit par $-1$ sur $U$ et par l'identité sur $U^\perp$. Alors $\tilde{\pi}_\theta(\xi)$ est l'identité de $E_\theta$. Alors, pour $i \in \{0, N\}$, il existe $\nu_i \in \mathbb{C}^\times$ tel que:

$$S_i(s_i s'_i) = \nu_i \tilde{\pi}_\theta(s'_i).$$

PROPOSITION. *On a les égalités:*

$$\nu_0 = q^{-R''^2+R''}, \quad \nu_N = (-1)^{R'} q^{-R'^2}.$$

*Cette proposition sera démontrée en 6.11.*

**1.4.** Soient $N$, $A$, $B$ trois entiers tels que $N \geq 1$, $A > B \geq 0$. Introduisons, à la suite de Lusztig, l'algèbre de Hecke affine $\mathcal{H}(N; A, B)$, de type $C_N$, de paramètres $A$ et $B$. Dans les notations de [L2], elle correspond au système de racines $(X_N, Y_N, R_N, \check{R}_N, \Pi_N)$ et aux fonctions $\lambda$, $\lambda^*$ suivants:

- $X_N = Y_N = \mathbb{Z}^N$; on note $(e_i)_{i=1,\ldots,N}$, resp. $(\check{e}_i)_{i=1,\ldots,N}$, la base canonique de $X_N$, resp. $Y_N$;
- $R_N = \{\pm e_i \pm e_j;\ 1 \leq i < j \leq N\} \cup \{\pm e_i;\ 1 \leq i \leq N\}$;
- $\check{R}_N = \{\pm \check{e}_i \pm \check{e}_j;\ 1 \leq i < j \leq N\} \cap \{\pm 2\check{e}_i;\ 1 \leq i \leq N\}$;
- $\Pi_N = \{\alpha_1, \ldots, \alpha_N\}$, où $\alpha_i = e_i - e_{i+1}$ pour $i = 1, \ldots, N-1$, $\alpha_N = e_N$;
- $\lambda : \Pi_N \to \mathbb{N}$ est définie par $\lambda(\alpha_i) = 1$, pour $i = 1, \ldots, N-1$, $\lambda(\alpha_N) = A$;
- $\lambda^*(\alpha_N) = B$.

Notons $W_N$ le groupe de Weyl du système de racines ci-dessus. C'est un groupe de type $C_N$ qui est engendré par les symétries élémentaires $w_i$, pour $i = 1, \ldots, N$, associées aux éléments de $\Pi_N$. On note $W_N^{aff}$ le produit semi-direct de $X_N$ et $W_N$. C'est un groupe de Weyl affine qui est engendré par les éléments $w_i$, pour $i = 0, \dots, N$, où $w_0$ est défini ainsi: notons $\sigma_0'$ l'élément de $W_N$ qui envoie $e_1$ sur $-e_1$ et fixe $e_i$ pour tout $i \geq 2$; alors $w_0 = e_1\sigma_0'$.

Introduisons une indéterminée $v$ et l'algèbre $\mathcal{A} = \mathbb{C}[v, v^{-1}]$. Alors $\mathcal{H}(N; A, B)$ est la $\mathcal{A}$-algèbre engendrée par des éléments $T_i$ pour $i \in \{0, \dots, N\}$, soumis aux relations suivantes:

- pour $i, j \in \{0, \dots, N\}$, avec $i \neq j$,

$$T_iT_j = T_jT_i, \text{ si } |i - j| \neq 1,$$

$$T_iT_jT_i = T_jT_iT_j, \text{ si } |i - j| = 1 \text{ et } \{i, j\} \cap \{0, N\} = \emptyset,$$

$$T_iT_jT_iT_j = T_jT_iT_jT_i, \text{ si } |i - j| = 1,\ N \neq 1 \text{ et } \{i, j\} \cap \{0, N\} \neq \emptyset;$$

- pour $i \in \{1, \dots, N - 1\}$, $(T_i + 1)(T_i - v^2) = 0$;
- $(T_0 + 1)(T_0 - v^{2B}) = 0$, $(T_N + 1)(T_N - v^{2A}) = 0$.

A tout $w \in W_N^{aff}$, on peut associer un élément $T(w)$ de $\mathcal{H}(N; A, B)$ de sorte que $T(w_i) = T_i$ pour tout $i \in \{0, \dots, N\}$ et $T(w'w'') = T(w')T(w'')$ si $\ell(w'w'') = \ell(w') + \ell(w'')$, où $\ell$ est la fonction longueur.

Lusztig donne une autre présentation de l'algèbre $\mathcal{H}(N; A, B)$. Notons $\mathcal{H}(N; A)$ la sous-algèbre engendrée par les $T_i$ pour $i \in \{1, \dots, N\}$. Notons $\mathcal{A}[X_N]$ la $\mathcal{A}$-algèbre du groupe $X_N$ et, pour $x \in X_N$, $\xi_x$ l'élément de $\mathcal{A}[X_N]$ associé. Alors:

$$\mathcal{H}(N; A, B) = \mathcal{A}[X_N] \otimes_{\mathcal{A}} \mathcal{H}(N; A),$$

la structure d'algèbre étant définie par les relations suivantes:

- pour $x \in X_N$ et $i \in \{1, \dots, N - 1\}$,

$$\xi_x T_i - T_i \xi_{w_i(x)} = (v^2 - 1)\frac{\xi_x - \xi_{w_i(x)}}{1 - \xi_{-\alpha_i}};$$

- pour $x \in X_N$,

$$\xi_x T_N - T_N \xi_{w_N(x)} = \left((v^{2A} - 1) + \xi_{-\alpha_N}(v^{A+B} - v^{A-B})\right)\frac{\xi_x - \xi_{w_N(x)}}{1 - \xi_{-2\alpha_N}}.$$

On passe d'une présentation à l'autre par les formules suivantes:

- pour $i \in \{1, \dots, N\}$,

$$\xi_{e_1+\dots+e_i} = \left(v^{-2N-A-B+i+1}T_0T_1 \dots T_NT_{N-1} \dots T_i\right)^i;$$

- $T_0 = v^{2N+A+B-2}\xi_{e_1}T(\sigma_0')^{-1}$.

Pour unifier les notations, pour $A$ et $B$ comme précédemment, on pose $\mathcal{H}(0; A, B) = \mathcal{A}$.

Par spécialisation de $v$ en $q^{1/2}$ (la racine carrée positive de $q$), on déduit de l'algèbre $\mathcal{H}(N; A, B)$ une $\mathbb{C}$-algèbre notée $\mathcal{H}_q(N; A, B)$.

**1.5.** Soient $\sharp = iso$ ou $an$ et $\theta = (R', R'') \in \Theta_\sharp$. Posons $N = n - R'^2 - R' - R''^2$, $A = sup(2R' + 1, 2R'')$, $B = inf(2R' + 1, 2R'')$. Supposons $N > 0$. On vérifie qu'un homomorphisme:

$$h : \mathcal{H}_q(N; A, B) \to \mathcal{H}_\theta$$

est bien défini par les formules suivantes, pour $i \in \{0, \ldots, N\}$:

- si $R' \geq R'' > 0$, $h(T_i) = S_i$;
- si $R' \geq R'' = 0$, $h(T_i) = S_i$ si $i \neq 0$, $h(T_0) = \omega$;
- si $R'' > R'$, $h(T_i) = S_{N-i}$.

L'homomorphisme $h$ est injectif, d'image $\mathcal{H}'_\theta$ si $R'' > 0$, $\mathcal{H}_\theta$ si $R'' = 0$. On en déduit un isomorphisme:

$$\mathcal{H}_\theta \simeq \begin{cases} \mathcal{H}_q(N; A, B), & si\ R'' = 0, \\ \mathbb{C}[\omega] \otimes_{\mathbb{C}} \mathcal{H}_q(N; A, B), & si\ R'' > 0, \end{cases}$$

ce dernier produit étant direct.

Bien sûr, cela reste vrai si $N = 0$.

**1.6.** Soient $N$, $A$, $B$ comme en 1.4. Le groupe $W_N$ agit sur $\mathcal{A}[X_N]$. Notons $\mathcal{Z} = \mathcal{A}[X_N]^{W_N}$ la sous-algèbre des invariants et $\mathcal{F}$ le corps des fractions de $\mathcal{Z}$. L'algèbre $\mathcal{Z}$ est le centre de $\mathcal{H}(N; A, B)$. Définissons les éléments de $\mathcal{A}[X_N] \otimes_{\mathcal{Z}} \mathcal{F}$ suivants:

- pour $i \in \{1, \ldots, N-1\}$, $\mathcal{G}(i) = \frac{\xi_{\alpha_i} v^2 - 1}{\xi_{\alpha_i} - 1}$,
- $\mathcal{G}(N) = \frac{\left(\xi_{\alpha_N} v^{A+B} - 1\right)\left(\xi_{\alpha_N} v^{A-B} + 1\right)}{\xi_{2\alpha_N} - 1}$;

et les éléments de $\mathcal{H}(N; A, B) \otimes_{\mathcal{Z}} \mathcal{F}$ suivants, pour $i \in \{1, \ldots, N\}$:

$$\tau(w_i) = -1 + (T_i + 1)\mathcal{G}(i)^{-1}.$$

Pour tout $w \in W_N$, on peut définir $\tau(w)$ de sorte que $\tau(w'w'') = \tau(w')\tau(w'')$ pour tous $w'$, $w'' \in W_N$. Pour tous $x \in X_N$ et $w \in W_N$, on a la relation:

$$\tau(w)\xi_x = \xi_{w(x)}\tau(w),$$

et $\mathcal{H}(N; A, B) \otimes_{\mathcal{Z}} \mathcal{F}$ est engendrée par $\mathcal{A}[X_N] \otimes_{\mathcal{Z}} \mathcal{F}$ et les $\tau(w)$, pour $w \in W_N$.

Le groupe $\mathfrak{S}_N$ est naturellement un sous-groupe de $W_N$. On note $\mathcal{H}^{\mathfrak{S}}(N)$ la sous-algèbre de $\mathcal{H}(N; A, B)$ engendrée par $\mathcal{A}[X_N]$ et les $T(w)$, pour $w \in \mathfrak{S}_N$. Elle est indépendante de $A$ et $B$. C'est une algèbre affine étendue de type $A_{N-1}$. L'algèbre $\mathcal{H}^{\mathfrak{S}}(N) \otimes_{\mathcal{Z}} \mathcal{F}$ est engendrée par $\mathcal{A}[X_N] \otimes_{\mathcal{Z}} \mathcal{F}$ et les $\tau(w)$, pour $w \in \mathfrak{S}_N$.

**1.7.** On conserve les mêmes hypothèses. Considérons:

- un élément $\underline{s} = (s_1, \ldots, s_N; v_0) \in (Y_N \otimes_{\mathbb{Z}} \mathbb{C}^\times) \times \mathbb{C}^\times \simeq (\mathbb{C}^\times)^{N+1}$;
- une décomposition:

$$N = N^+ + N_1 + \ldots + N_a + N^-$$

en entiers tels que $N^+ \geq 0$, $N^- \geq 0$, $N_j \geq 1$ pour tout $j \in \{1, \ldots, a\}$;

- un signe $\zeta \in \{\pm 1\}$.

A la décomposition de $N$ est associée la décomposition en intervalles:

(1) $\{1, \ldots, N\}$

$$= \{1, \ldots, N^+\} \cup \{N^+ + 1, \ldots, N^+ + N_1\} \cup \ldots \cup \{N - N^- + 1, \ldots, N\}.$$

Pour tout $z \in \mathbb{C}^\times$, notons $< z >$ le sous-groupe de $\mathbb{C}^\times$ engendré par $z$. Supposons vérifiées les conditions suivantes:

- pour tout $i \in \{1, \ldots, N\}$,

$$s_i \in \zeta < v_0 > \Leftrightarrow i \in \{1, \ldots, N^+\},$$

$$-s_i \in \zeta < v_0 > \Leftrightarrow i \in \{N - N^- + 1, \ldots, N\};$$

- pour tous $i,\ j \in \{N^+ + 1, \ldots, N - N^-\}$,

si $i$ et $j$ appartiennent au même intervalle (cf. (1)), $s_i s_j^{-1} \in < v_0^2 >$,

si $i$ et $j$ n'appartiennent pas au même intervalle, $s_i s_j^{-1} \notin < v_0^2 >$ et $s_i s_j \notin < v_0^2 >$. Posons:

$$\mathcal{H}' = \mathcal{H}(N^+; A, B) \otimes_{\mathcal{A}} \mathcal{H}^{\mathfrak{S}}(N_1) \otimes_{\mathcal{A}} \ldots \otimes_{\mathcal{A}} \mathcal{H}^{\mathfrak{S}}(N_a) \otimes_{\mathcal{A}} \mathcal{H}(N^-; A, B).$$

De la décomposition (1) sont issus:

- un plongement $W' \to W_N$, où

$$W' = W_{N^+} \times \mathfrak{S}_{N_1} \times \ldots \times \mathfrak{S}_{N_a} \times W_{N^-};$$

- un isomorphisme

$$X_N \simeq X_{N^+} \oplus X_{N_1} \oplus \ldots \oplus X_{N_a} \oplus X_{N^-};$$

puis un plongement $\mathcal{A}[X_N] \to \mathcal{H}'$. On note $\mathcal{Z}' = \mathcal{A}[X_N]^{W'}$ la sous-algèbre des invariants par $W'$. C'est le centre de $\mathcal{H}'$.

Notons $\ell_{\underline{s}} : \mathcal{A}[X_N] \to \mathbb{C}$ l'évaluation au point $\underline{s}$: $\ell_{\underline{s}}(v) = v_0$, $\ell_{\underline{s}}(\xi_{e_i}) = s_i$ pour tout $i \in \{1, \ldots, N\}$. On note encore $\ell_{\underline{s}}$ sa restriction à $\mathcal{Z}$ ou $\mathcal{Z}'$.

PROPOSITION. *Il existe une équivalence entre la catégorie des $\mathcal{H}'$-modules de dimension finie sur lesquels $\mathcal{Z}'$ agit par l'homomorphisme $\ell_{\underline{s}}$ et celle des $\mathcal{H}(N; A, B)$-modules de dimension finie sur lesquels $\mathcal{Z}$ agit par l'homomorphisme $\ell_{\underline{s}}$.*

Cf. [L2] paragraphe 8. Rappelons la construction que donne Lusztig de cette équivalence. Notons $\mathcal{F}'$ le corps des fractions de $\mathcal{Z}'$. Remarquons que $\mathcal{F}$ se plonge dans $\mathcal{F}'$. En appliquant à chaque facteur de $\mathcal{H}'$ les constructions de 1.4 et 1.6, on dispose pour tout $w \in W'$ d'un élément de $\mathcal{H}' \otimes_{\mathcal{Z}'} \mathcal{F}'$, que nous noterons $\tau'(w)$. Notons $\mathcal{U}$ le sous-ensemble des $w \in W_N$ qui sont de longueur minimale dans leur classe $W'w$. Posons

$$\mathrm{u} = |\mathcal{U}| = |W' \backslash W_N|.$$

On définit un homomorphisme:

$$\begin{array}{rcl} m : \mathcal{H}(N; A, B) \otimes_{\mathcal{Z}} \mathcal{F} & \rightarrow & M_{\mathrm{u}}(\mathcal{H}' \otimes_{\mathcal{Z}'} \mathcal{F}') \\ h & \mapsto & (m(h)_{w',w''})_{w',w'' \in \mathcal{U}} \end{array}$$

de la façon suivante:

- pour $x \in X_N$ et $w', w'' \in \mathcal{U}$,

$$m(\xi_x)_{w',w''} = \begin{cases} 0, & si\ w' \neq w'', \\ \xi_{w'(x)}, & si\ w' = w''; \end{cases}$$

- pour $w \in W_N$ et $w', w'' \in \mathcal{U}$,

$$m(\tau(w))_{w',w''} = \begin{cases} 0, & si\ w'ww''^{-1} \notin W', \\ \tau'(w'ww''^{-1}), & si\ w'ww''^{-1} \in W'. \end{cases}$$

Notons $\mathcal{Z}_{(\underline{s})} \subseteq \mathcal{F}$, resp. $\mathcal{Z}'_{(\underline{s})} \subseteq \mathcal{F}'$, le localisé de $\mathcal{Z}$, resp. $\mathcal{Z}'$, relativement à l'idéal $\mathcal{Z} \cap Ker(\ell_{\underline{s}})$, resp. $\mathcal{Z}' \cap Ker(\ell_{\underline{s}})$, et $\bar{\mathcal{Z}}_{(\underline{s})}$, resp. $\bar{\mathcal{Z}}'_{(\underline{s})}$, son complété. En modifiant légèrement la preuve de [L2], on montre que l'image par $m$ de $\mathcal{H}(N; A, B) \otimes_{\mathcal{Z}} \mathcal{Z}_{(\underline{s})}$ est incluse dans $M_{\mathrm{u}}(\mathcal{H}' \otimes_{\mathcal{Z}'} \mathcal{Z}'_{(\underline{s})})$ et que $m$ se prolonge en un isomorphisme:

$$\mathcal{H}(N; A, B) \otimes_{\mathcal{Z}} \bar{\mathcal{Z}}_{(\underline{s})} \simeq M_{\mathrm{u}}(\mathcal{H}' \otimes_{\mathcal{Z}'} \mathcal{Z}'_{(\underline{s})}).$$

Cet isomorphisme identifie $\bar{\mathcal{Z}}_{(\underline{s})}$ à $\bar{\mathcal{Z}}'_{(\underline{s})}$ plongé diagonalement dans l'algèbre de droite. Soit $E'$ un $\mathcal{H}'$-module de dimension finie sur lequel $\mathcal{Z}'$ agit par l'homomorphisme $\ell_{\underline{s}}$. Alors $E'$ s'étend en un $\mathcal{H}' \otimes_{\mathcal{Z}'} \mathcal{Z}'_{(\underline{s})}$-module. L'espace $(E')^{\otimes \mathrm{u}}$ est un $M_{\mathrm{u}}(\mathcal{H}' \otimes_{\mathcal{Z}'} \mathcal{Z}'_{(\underline{s})})$-module, donc un $\mathcal{H}(N; A, B) \otimes_{\mathcal{Z}} \bar{\mathcal{Z}}_{(\underline{s})}$-module et, par restriction, un $\mathcal{H}(N; A, B)$-module. L'algèbre $\mathcal{Z}$ y agit par l'homomorphisme $\ell_{\underline{s}}$. Le foncteur $E' \mapsto (E')^{\oplus \mathrm{u}}$ est l'équivalence de catégories annoncée.

Décrivons une variante de la construction ci-dessus. On définit un plongement:

$$j' : \mathcal{H}' \otimes_{\mathcal{Z}'} \mathcal{F}' \rightarrow \mathcal{H}(N; A, B) \otimes_{\mathcal{Z}} \mathcal{F}$$

de la façon suivante:

- pour $x \in X_N$, $j'(\xi_x) = \xi_x$;
- pour $w \in W'$, $j'(\tau'(w)) = \tau(w)$.

Définissons le sous-espace:

$$J'_{(\underline{s})} = \sum_{w \in \mathcal{U}} j'(\mathcal{H}' \otimes_{\mathcal{Z}'} \mathcal{Z}'_{(\underline{s})})\tau(w)$$

de $\mathcal{H}(N; A, B) \otimes_{\mathcal{Z}} \mathcal{F}$. Il est stable par multiplication à gauche par $j'(\mathcal{H}')$. Il est aussi stable par multiplication à droite par $\mathcal{H}(N; A, B)$. En effet, soient $h \in \mathcal{H}$ et

$$x = \sum_{w \in \mathcal{U}} j'(h'_w)\tau(w) \in J'_{(\underline{s})},$$

avec $h'_w \in \mathcal{H}' \otimes_{\mathcal{Z}'} \mathcal{Z}'_{(\underline{s})}$ pour tout $w \in \mathcal{U}$. On calcule:

$$xh = \sum_{w \in \mathcal{U}} \left( \sum_{w' \in \mathcal{U}} j'\left(h'_w m(h)_{w',w}\right) \right) \tau(w).$$

Or, d'après les résultats ci-dessus, $m(h)_{w',w} \in \mathcal{H}' \otimes_{\mathcal{Z}'} \mathcal{Z}'_{(\underline{s})}$ pour tous $w', w \in \mathcal{U}$. Donc $xh \in J'_{(\underline{s})}$.

Soit $E'$ un $\mathcal{H}'$-module de dimension finie sur lequel $\mathcal{Z}'$ agit par l'homomorphisme $\ell_{\underline{s}}$. Posons:

$$E = Hom_{\mathcal{H}'}\left(J'_{(\underline{s})}, E'\right),$$

i.e., $E$ est l'ensemble des applications linéaires $f : J'_{(\underline{s})} \to E'$ telles que $f(j'(h')x) = h' f(x)$ pour tous $x \in J'_{(\underline{s})}$ et $h \in \mathcal{H}'$. Puisque $\mathcal{H}(N; A, B)$ agit à droite sur $J'_{(\underline{s})}$, $E$ est muni d'une structure de $\mathcal{H}(N; A, B)$-module. On vérifie que $E$ est isomorphe au module $(E')^{\oplus \mathfrak{u}}$ construit ci-dessus.

*Remarque.* On a l'égalité:

$$J'_{(\underline{s})} = j'\left(\mathcal{H}' \otimes_{\mathcal{Z}'} \mathcal{Z}'_{(\underline{s})}\right)\mathcal{H}(N; A, B).$$

En effet, puisque $J'_{(\underline{s})}$ est stable par multiplication à gauche par $j'(\mathcal{H}' \otimes_{\mathcal{Z}'} \mathcal{Z}'_{(\underline{s})})$ et à droite par $\mathcal{H}(N; A, B)$, le membre de droite ci-dessus est contenu dans $J'_{(\underline{s})}$. Pour démontrer l'inclusion opposée, il suffit de prouver que pour tout $w \in \mathcal{U}$,

$$\tau(w) \in j'(\mathcal{Z}'_{(\underline{s})})\mathcal{H}(N; A, B). \tag{2}$$

Pour tout $\alpha \in R_N$, posons:

- si $\alpha$ est longue, $\mathcal{G}_d(\alpha) = \xi_{-\alpha} v^2 - 1$,
- si $\alpha$ est courte, $\mathcal{G}_d(\alpha) = (\xi_{-\alpha} v^{A+B} - 1)(\xi_{-\alpha} v^{A-B} + 1)$.

La base $\Pi_N$ définit un ordre sur $R_N$. Pour $w \in W_N$, posons:

$$\mathcal{G}_d(w) = \prod_{\substack{\alpha > 0 \\ w^{-1}(\alpha) < 0}} \mathcal{G}_d(\alpha).$$

On vérifie sur les définitions que $\mathcal{G}_d(w)\tau(w) \in \mathcal{H}(N; A, B)$. Mais, si $w \in \mathcal{U}$, on vérifie que $\mathcal{G}_d(w)^{-1} \in j'(\mathcal{Z}'_{(s)})\mathcal{A}[X_N]$. La relation (2) en résulte.

**1.8.** Soient $N$ un entier $\geq 0$ et $C$ un entier $\geq 1$. On note $\mathcal{S}$ l'algèbre symétrique de $(X_N \otimes_{\mathbb{Z}} \mathbb{C}) \oplus \mathbb{C}$. C'est une algèbre de polynômes en des coordonnées que nous noterons de façon évidente $x_1, \ldots, x_N$, $r$. Le groupe $W_N$ agit sur $\mathcal{S}$, en fixant $r$. On note $\bar{\mathcal{H}}(N; C)$ la $\mathcal{S}$-algèbre engendrée par des éléments $t_w$, pour $w \in W_N$, soumis aux relations suivantes:

- pour $w$, $w' \in W_N$, $t_{ww'} = t_w t_{w'}$;
- pour tout $i \in \{1, \ldots, N\}$, posons $t_i = t_{w_i}$; alors, pour toute $f \in \mathcal{S}$,

$$t_i f - w_i(f) t_i = \begin{cases} 2r\frac{f - w_i(f)}{\alpha_i}, & si\ i < N, \\ Cr\frac{f - w_i(f)}{\alpha_i}, & si\ i = N. \end{cases}$$

*Remarque.* Si $N = 0$, on a simplement $\bar{\mathcal{H}}(N; C) = \mathbb{C}[r]$.

Notons $\mathcal{M}$ l'algèbre des fonctions méromorphes sur $(Y_N \otimes_{\mathbb{Z}} \mathbb{C}) \oplus \mathbb{C}$ et $\mathcal{M}\bar{\mathcal{H}}(N; C)$ l'analogue de l'algèbre $\bar{\mathcal{H}}(N; C)$ quand $\mathcal{S}$ est remplacé par $\mathcal{M}$. De l'injection naturelle $\mathcal{S} \subseteq \mathcal{M}$ se déduit une injection:

$$\bar{\mathcal{H}}(N; C) \subseteq \mathcal{M}\bar{\mathcal{H}}(N; C).$$

Définissons les éléments de $\mathcal{M}$ suivants:

- pour $i \in \{1, \ldots, N - 1\}$, $\bar{\mathcal{G}}(i) = 1 + \frac{2r}{\alpha_i}$,
- $\bar{\mathcal{G}}(N) = 1 + \frac{Cr}{\alpha_N}$;

et les éléments de $\mathcal{M}\bar{\mathcal{H}}(N; C)$ suivants:

- pour $i \in \{1, \ldots, N\}$, $\bar{\tau}(w_i) = -1 + (t_i + 1)\bar{\mathcal{G}}(i)^{-1}$.

Alors, pour tout $w \in W_N$, on peut définir $\bar{\tau}(w)$ de sorte que $\bar{\tau}(w'w'') = \bar{\tau}(w')\bar{\tau}(w'')$ pour tous $w'$, $w'' \in W_N$. Pour tous $f \in \mathcal{M}$, $w \in W_N$, on a la relation $\bar{\tau}(w)f = w(f)\bar{\tau}(w)$. L'algèbre $\mathcal{M}\bar{\mathcal{H}}(N; C)$ est engendrée par $\mathcal{M}$ et les $\bar{\tau}(w)$, pour $w \in W_N$.

Notons $\mathcal{Z}_{\mathcal{S}} = \mathcal{S}^{W_N}$ l'algèbre des invariants par $W_N$. C'est le centre de $\bar{\mathcal{H}}(N; C)$. Pour tout $\underline{\sigma} = (\sigma_1, \ldots, \sigma_N; r_0) \in (Y \otimes_{\mathbb{Z}} \mathbb{C}) \oplus \mathbb{C}$, notons $\ell_{\underline{\sigma}} : \mathcal{S} \to \mathbb{C}$ l'évaluation au point $\underline{\sigma}$, et $W_N(\underline{\sigma}) = \{w(\underline{\sigma});\ w \in W_N\}$. Remarquons que pour $w \in W_N$, les restrictions à $\mathcal{Z}_{\mathcal{S}}$ de $\ell_{\underline{\sigma}}$ et $\ell_{w(\underline{\sigma})}$ coïncident. Notons $\mathcal{M}_{W_N(\underline{\sigma})}$ la sous-algèbre des éléments de $\mathcal{M}$ qui sont holomorphes en tout point de $W_N(\underline{\sigma})$. On définit

$\mathcal{M}_{W_N(\underline{\sigma})}\bar{\mathcal{H}}(N;C)$ comme on a défini $\bar{\mathcal{H}}(N;C)$: on remplace $\mathcal{S}$ par $\mathcal{M}_{W_N(\underline{\sigma})}$. On a les inclusions:

$$\bar{\mathcal{H}}(N;C) \subseteq \mathcal{M}_{W_N(\underline{\sigma})}\bar{\mathcal{H}}(N;C) \subseteq \mathcal{M}\bar{\mathcal{H}}(N;C).$$

Soient $\underline{\sigma}$ comme ci-dessus et $E$ un $\bar{\mathcal{H}}(N;C)$-module. Supposons que $\mathcal{Z}_{\mathcal{S}}$ agisse dans $E$ par l'homomorphisme $\ell_{\underline{\sigma}}$. Alors l'action de $\bar{\mathcal{H}}(N;C)$ dans $E$ s'étend en une action de $\mathcal{M}_{W_N(\underline{\sigma})}\bar{\mathcal{H}}(N;C)$. On doit pour cela définir une action de $\mathcal{M}_{W_N(\underline{\sigma})}$ dans $E$. Posons $I = \mathcal{Z}_{\mathcal{S}} \cap Ker(\ell_{\underline{\sigma}})$. Pour tout $\underline{\sigma}' \in W_N(\underline{\sigma})$, introduisons l'anneau topologique $\bar{\mathcal{S}}_{(\underline{\sigma}')}$, complété du localisé de $\mathcal{S}$ en $\underline{\sigma}'$, et notons $\bar{I}_{(\underline{\sigma}')}$ la clôture de $I\bar{\mathcal{S}}_{(\underline{\sigma}')}$ dans $\bar{\mathcal{S}}_{(\underline{\sigma}')}$. On a un isomorphisme:

$$I\mathcal{S}\backslash\mathcal{S} \stackrel{\sim}{\to} \oplus_{\underline{\sigma}'\in W_N(\underline{\sigma})}\bar{I}_{(\underline{\sigma}')}\backslash\bar{\mathcal{S}}_{(\underline{\sigma}')},$$

et une application naturelle:

$$\mathcal{M}_{W_N(\underline{\sigma})} \to \oplus_{\underline{\sigma}'\in W_N(\underline{\sigma})}\bar{\mathcal{S}}_{(\underline{\sigma}')}$$

(le développement en série au voisinage de chaque $\underline{\sigma}'$). D'où un homomorphisme $\mathcal{M}_{W_N(\underline{\sigma})} \to I\mathcal{S}\backslash\mathcal{S}$. Puisque $\mathcal{S}$ agit dans $E$ via son quotient $I\mathcal{S}\backslash\mathcal{S}$, on déduit de cet homomorphisme l'action cherchée de $\mathcal{M}_{W_N(\underline{\sigma})}$.

**1.9.** Soient $N$, $A$, $B$ comme en 1.4 et $\epsilon \in \{\pm 1\}$. Posons $C = A + \epsilon B$. On définit un homomorphisme:

$$gr_\epsilon : \mathcal{H}(N; A, B) \otimes_{\mathcal{Z}} \mathcal{F} \to \mathcal{M}\bar{\mathcal{H}}(N;C)$$

de la façon suivante:

- pour tout $i \in \{1, \ldots, N\}$, $gr_\epsilon(\xi_{e_i}) = \epsilon exp(x_i)$;
- pour tout $w \in W_N$, $gr_\epsilon(\tau(w)) = \bar{\tau}(w)$.

Soit $\underline{\sigma} = (\sigma_1, \ldots, \sigma_N; r_0) \in (Y_N \otimes_{\mathbb{Z}} \mathbb{C}) \oplus \mathbb{C}$. Supposons $r_0 \notin 2\pi i\mathbb{Q}$ (où $i$ est ici une racine carrée de $-1$) et $\frac{\sigma_j}{r_0} \in \mathbb{Z}$ pour tout $j \in \{1, \ldots, N\}$. Alors Lusztig montre que l'image de $\mathcal{H}(N; A, B)$ par $gr_\epsilon$ est incluse dans $\mathcal{M}_{W_N(\underline{\sigma})}\bar{\mathcal{H}}(N;C)$. Soit $E$ un $\bar{\mathcal{H}}(N;C)$-module. Supposons que $\mathcal{Z}_{\mathcal{S}}$ y agisse par l'homomorphisme $\ell_{\underline{\sigma}}$. Alors $E$ est aussi un $\mathcal{M}_{W_N(\underline{\sigma})}\bar{\mathcal{H}}(N;C)$-module (cf. 1.8). Via $gr_\epsilon$, c'est donc un $\mathcal{H}(N; A, B)$-module. L'algèbre $\mathcal{Z}$ y agit par l'homomorphisme $\ell_{\underline{s}}$, où:

$$\underline{s} = (\epsilon exp(\sigma_1), \ldots, \epsilon exp(\sigma_N); exp(r_0)).$$

Le foncteur ainsi défini est une équivalence entre la catégorie des $\bar{\mathcal{H}}(N;C)$-modules de longueur finie sur lesquels $\mathcal{Z}_{\mathcal{S}}$ agit par $\ell_{\underline{\sigma}}$ et la catégorie des $\mathcal{H}(N; A, B)$-modules de longueur finie sur lesquels $\mathcal{Z}$ agit par $\ell_{\underline{s}}$.

**1.10.** Soient $d$ un entier pair $\geq 2$ et $\hat{V}$ un espace de dimension $d$ sur $\mathbb{C}$, muni d'une forme symplectique. On note $\hat{G}$ son groupe symplectique et $\hat{\mathfrak{g}}$ l'algèbre de

Lie de ce groupe. Considérons un quadruplet $(\sigma, r_0, y, \epsilon)$, où:

- $\sigma$ est un élément semi-simple de $\hat{\mathfrak{g}}$;
- $r_0 \in \mathbb{C}, r_0 \neq 0$;
- $y$ est un élément nilpotent de $\hat{\mathfrak{g}}$;
- $\epsilon$ est un caractère du groupe de composantes:

$$Z_{\hat{G}}(\sigma, y)/Z_{\hat{G}}(\sigma, y)^0$$

où $Z_{\hat{G}}(\sigma, y)$ est l'intersection des commutants dans $\hat{G}$ de $\sigma$ et $y$, et $Z_{\hat{G}}(\sigma, y)^0$ est sa composante neutre; ces données vérifiant la relation:

$$[\sigma, y] = 2r_0 y.$$

On sait que l'on peut décomposer $\sigma$ en $\sigma_L + \sigma^L$, où $\sigma_L$ commute à $\sigma$ et $y$ et $(r_0^{-1}\sigma^L, y)$ peut être complété en un $sl_2$-triplet. La décomposition est unique à conjugaison près par $Z_{\hat{G}}(\sigma, y)$. Notons $\hat{V}_0$ l'espace propre pour $\sigma_L$ associé à la valeur propre 0, et $\hat{V}_\perp$ son orthogonal. Notons $\hat{G}_0$, resp. $\hat{G}_\perp$, le groupe symplectique de $\hat{V}_0$, resp. $\hat{V}_\perp$, et $\hat{\mathfrak{g}}_0$, resp. $\hat{\mathfrak{g}}_\perp$, leurs algèbres de Lie. L'élément $y$ se décompose en $y_0 + y_\perp$, où $y_0 \in \hat{\mathfrak{g}}_0$ et $y_\perp \in \hat{\mathfrak{g}}_\perp$. On a l'isomorphisme:

$$Z_{\hat{G}}(\sigma, y)/Z_{\hat{G}}(\sigma, y)^0 \simeq Z_{\hat{G}_0}(y_0)/Z_{\hat{G}_0}(y_0)^0,$$

qui permet de considérer $\epsilon$ comme un caractère de ce dernier groupe. Par la correspondance de Springer généralisée (cf. [L5] paragraphe 12), à $(y_0, \epsilon)$ sont associés un entier $k \in \mathbb{N}$ tel que $k(k+1) \leq dim_{\mathbb{C}}(\hat{V}_0)$ et un couple cuspidal $(y_c, \epsilon_c)$. Notons $\hat{V}_c$ un espace de dimension $k(k+1)$ sur $\mathbb{C}$, muni d'une forme symplectique, $\hat{G}_c$ son groupe symplectique et $\hat{\mathfrak{g}}_c$ l'algèbre de Lie de ce groupe. Le terme $y_c$ est un élément nilpotent de $\hat{G}_c$ et $\epsilon_c$ est un caractère de $Z_{\hat{G}_c}(y_c)/Z_{\hat{G}_c}(y_c)^0$. Plus précisément, $y_c$ est paramétrisé par la partition:

$$(2k, 2(k-1), \ldots, 2).$$

Le groupe $Z_{\hat{G}_c}(y_c)/Z_{\hat{G}_c}(y_c)^0$ est isomorphe à $(\mathbb{Z}/2\mathbb{Z})^k$; il a une base naturelle $\{f_2, \ldots, f_{2k}\}$ sur $\mathbb{Z}/2\mathbb{Z}$, indexée par les termes de la partition. On a l'égalité $\epsilon_c(f_{2i}) = (-1)^i$ pour tout $i \in \{1, \ldots, k\}$.

Comme on le verra plus loin (cf. 2.4), il en résulte que l'on peut plonger $\hat{V}_c$ dans $\hat{V}_0$, de sorte qu'en notant $\hat{V}_1$ son orthogonal dans $\hat{V}$, $\hat{V}_c$ et $\hat{V}_1$ soient stables par $\sigma$ et, si $\sigma_c$ désigne la restriction de $\sigma$ à $\hat{V}_c$, on ait la relation:

$$[\sigma_c, y_c] = 2r_0 y_c.$$

Posons:

$$N = (d - k(k+1))/2, \ \ C = 2k + 1.$$

Notons $\sigma_1$ la restriction de $\sigma$ à $\hat{V}_1$. Sa classe de conjugaison dans le groupe symplectique de $\hat{V}_1$ est déterminée par la collection de ses valeurs propres, que l'on peut voir comme un élément $(\sigma_1, \ldots, \sigma_N)$ de $Y_N \otimes_{\mathbb{C}} \mathbb{C}$, bien déterminé modulo l'action de $W_N$. Posons:

$$\underline{\sigma} = (\sigma_1, \ldots, \sigma_N; r_0).$$

Au quadruplet $(\sigma, r_0, y, \epsilon)$, Lusztig associe en [L4] un module irréductible $\mathbb{P}(\sigma, r_0, y, \epsilon)$ sur l'algèbre $\bar{\mathcal{H}}(N; C)$. Le centre $\mathcal{Z}_S$ agit sur ce module par l'homomorphisme $\ell_{\underline{\sigma}}$.

*Remarque.* Si $k \geq 1$, c'est directement la construction de Lusztig. Si $k = 0$, l'algèbre qu'il construit n'est pas à première vue la nôtre: pour lui, $C = 2$. Mais son système de racines n'est pas non plus le nôtre: pour lui, $\alpha_N$ est multiplié par 2. Ces deux modifications se compensent.

**1.11.** Soient $\hat{V}$ un espace de dimension $2n$ sur $\mathbb{C}$, muni d'une forme symplectique. On utilise les notations du paragraphe précédent. Considérons un triplet $(s, y, \epsilon)$, où:

- $s$ est un élément semi-simple de $\hat{G}$;
- $y$ est un élément nilpotent de $\hat{\mathfrak{g}}$;
- $\epsilon$ est un caractère de $Z_{\hat{G}}(s, y)/Z_{\hat{G}}(s, y)^0$, avec une notation évidente;

ces données vérifiant la relation:

$$Ad(s)(y) = qy.$$

On peut trouver:

- une décomposition orthogonale

$$\hat{V} = \hat{V}^+ \oplus \hat{V}^- \oplus (\oplus_{j=1,\ldots,a} \hat{V}_{\pm j});$$

- pour tout $j \in \{1, \ldots, a\}$, une décomposition en deux lagrangiens $\hat{V}_{\pm j} = \hat{V}_j \oplus \hat{V}_{-j}$ et un nombre complexe $z_j$;

de sorte que:

- chacun des espaces $\hat{V}^+$, $\hat{V}^-$, $\hat{V}_{\pm j}$, $\hat{V}_j$, $\hat{V}_{-j}$ soit stable par $s$;
- les valeurs propres de $s$ dans $\hat{V}^+$, resp. $\hat{V}^-$, $\hat{V}_j$, $\hat{V}_{-j}$, appartiennent à $< q^{1/2} >$, resp. $- < q^{1/2} >$, $z_j < q >$, $z_j^{-1} < q >$;
- pour tout $j \in \{1, \ldots, a\}$, $z_j \notin \pm < q^{1/2} >$;
- pour tous $j, k \in \{1, \ldots, a\}$, avec $j \neq k$, $z_j z_k^{-1} \notin < q >$, $z_j z_k \notin < q >$.

Notons $\hat{G}^+$, resp. $\hat{G}^-$, le groupe symplectique de $\hat{V}^+$, resp. $\hat{V}^-$, et $\hat{\mathfrak{g}}^+$, resp. $\hat{\mathfrak{g}}^-$, son algèbre de Lie. Le produit $\hat{G}^+ \times \hat{G}^- \times \prod_{j=1,\ldots,a} GL(\hat{V}_j)$ est naturellement un sous-groupe de $\hat{G}$, et $s$ appartient à ce sous-groupe. De même, $y$ appartient à $\hat{\mathfrak{g}}^+ \oplus \hat{\mathfrak{g}}^- \oplus (\oplus_{j=1,\ldots,a}\mathfrak{gl}(\hat{V}_j))$. Ecrivons de façon évidente $s = s^+s^- \prod_{j=1,\ldots,a} s_j$, $y = y^+ + y^- + \sum_{j=1,\ldots,a} y_j$. On a l'isomorphisme:

$$Z_{\hat{G}}(s,y)/Z_{\hat{G}}(s,y)^0 \simeq Z_{\hat{G}^+}(s^+,y^+)/Z_{\hat{G}^+}(s^+,y^+)^0 \times Z_{\hat{G}^-}(s^-,y^-)/Z_{\hat{G}^-}(s^-,y^-)^0.$$

Le caractère $\epsilon$ se décompose ainsi en un couple $(\epsilon^+, \epsilon^-)$.

Soit $\eta \in \{\pm 1\}$, que l'on identifie à un signe $\pm$. Il existe un unique élément $\sigma^\eta$ de l'algèbre de Lie $\hat{\mathfrak{g}}^\eta$ tel que:

- ses valeurs propres appartiennent à $\frac{log(q)}{2}\mathbb{Z}$;
- $exp(\sigma^\eta) = \eta s^\eta$.

On a l'égalité:

$$[\sigma^\eta, y^\eta] = log(q) y^\eta.$$

Appliquons la construction de 1.10 à $\hat{V}^\eta$ et au quadruplet $(\sigma^\eta, \frac{log(q)}{2}, y^\eta, \epsilon^\eta)$. On en déduit un entier $k^\eta$ et un module irréductible sur l'algèbre $\bar{\mathcal{H}}(N^\eta; 2k^\eta + 1)$, où

$$N^\eta = \frac{1}{2}(dim_{\mathbb{C}}(\hat{V}^\eta) - k^\eta(k^\eta + 1)).$$

On en déduit aussi un élément $\underline{\sigma}^\eta$ de $(Y_{N^\eta} \otimes_{\mathbb{Z}} \mathbb{C}) \oplus \mathbb{C}$.

Définissons deux entiers $R', R'' \in \mathbb{N}$ et un signe $\zeta \in \{\pm 1\}$ de la façon suivante:

- si $k^+ = k^-$, $R' = k^+ = k^-$, $R'' = 0$, $\zeta = (-1)^{R'}$;
- si $k^+ > k^-$ et $k^+ \equiv k^- \, mod\, 2\mathbb{Z}$, $R' = \frac{k^+ + k^-}{2}$, $R'' = \frac{k^+ - k^-}{2}$, $\zeta = 1$;
- si $k^+ < k^-$ et $k^+ \equiv k^- \, mod\, 2\mathbb{Z}$, $R' = \frac{k^+ + k^-}{2}$, $R'' = \frac{k^- - k^+}{2}$, $\zeta = -1$;
- si $k^+ > k^-$ et $k^+ \not\equiv k^- \, mod\, 2\mathbb{Z}$, $R' = \frac{k^+ - k^- - 1}{2}$, $R'' = \frac{k^+ + k^- + 1}{2}$, $\zeta = 1$;
- si $k^+ < k^-$ et $k^+ \not\equiv k^- \, mod\, 2\mathbb{Z}$, $R' = \frac{k^- - k^+ - 1}{2}$, $R'' = \frac{k^+ + k^- + 1}{2}$, $\zeta = -1$.

Remarquons que:

$$R'^2 + R' + R''^2 = \frac{k^+(k^+ + 1) + k^-(k^- + 1)}{2}.$$

Posons:

$$A = sup(2R' + 1, 2R''), \quad B = inf(2R' + 1, 2R'').$$

Soit $\eta \in \{\pm 1\}$. On a l'égalité $2k^\eta + 1 = A + \eta\zeta B$. Appliquons la construction de 1.9 à $N^\eta$, $A$, $B$, $\epsilon = \eta\zeta$ et $\underline{\sigma}^\eta$. Du module irréductible ci-dessus pour l'algèbre $\bar{\mathcal{H}}(N^\eta; 2k^\eta + 1)$ se déduit un module irréductible sur l'algèbre $\mathcal{H}(N^\eta; A, B)$.

Des constructions analogues à celles de 1.9 et 1.10 sont valables pour les algèbres de Hecke relatives au groupe linéaire. Nous ne les détaillerons pas, cf. [L2] et [L4]. Le résultat est que, pour tout $j \in \{1, \ldots, a\}$, on peut associer aux données $(\zeta s_j, y_j)$ un module irréductible sur l'algèbre $\mathcal{H}^{\mathfrak{S}}(N_j)$, où $N_j = dim_{\mathbb{C}}(\hat{V}_j)$.

Posons $N = n - R'^2 - R' - R''^2$. On a la décomposition:

$$N = N^+ + N_1 + \ldots + N_a + N^-$$

et on a construit ci-dessus des représentations irréductibles des algèbres $\mathcal{H}(N^+; A, B)$, $\mathcal{H}(N^-; A, B)$ et $\mathcal{H}^{\mathfrak{S}}(N_j)$ pour tout $j \in \{1, \ldots, a\}$. On en déduit par tensorisation une représentation de $\mathcal{H}'$, où:

$$\mathcal{H}' = \mathcal{H}(N^+; A, B) \otimes_{\mathcal{A}} \mathcal{H}^{\mathfrak{S}}(N_1) \otimes_{\mathcal{A}} \ldots \otimes_{\mathcal{A}} \mathcal{H}^{\mathfrak{S}}(N_a) \otimes_{\mathcal{A}} \mathcal{H}(N^-; A, B).$$

On peut appliquer la construction de 1.7 à cette situation et en déduire une représentation irréductible de $\mathcal{H}(N; A, B)$. Par construction, $v$ agit dans cette représentation par multiplication par $q^{1/2}$. Il s'agit donc d'une représentation de $\mathcal{H}_q(N; A, B)$.

Posons $\sharp = iso$ si $R''$ est pair, $\sharp = an$ si $R''$ est impair. Le couple $\theta = (R', R'')$ appartient à $\Theta_\sharp$. Si $R'' = 0$, on a $\mathcal{H}_\theta = \mathcal{H}_q(N; A, B)$ et la représentation précédente s'identifie à une représentation de $\mathcal{H}_\theta$. Si $R'' > 0$, on a:

$$\mathcal{H}_\theta = \mathbb{C}[\omega] \otimes_{\mathbb{C}} \mathcal{H}_q(N; A, B).$$

On prolonge la représentation précédente de $\mathcal{H}_q(N; A, B)$ en une représentation irréductible de $\mathcal{H}_\theta$ dans laquelle $\omega$ agit par multiplication par $(-1)^{R'}\zeta$.

Du $\mathcal{H}_\theta$-module ainsi construit, on déduit un $G_\sharp$-module irréductible, noté $\mathbb{P}(s, y, \epsilon)$, qui appartient à $Irr_u^{G_\sharp}$. En résumé, au triplet $(s, y, \epsilon)$, on a associé un élément $\mathbb{P}(s, y, \epsilon)$ de $Irr_u^{G_{iso}} \cup Irr_u^{G_{an}}$. Lusztig a prouvé que cette application se quotientait en une bijection entre l'ensemble des classes de conjugaison par $\hat{G}$ de triplets $(s, y, \epsilon)$ et l'ensemble $Irr_u^{G_{iso}} \cup Irr_u^{G_{an}}$.

**1.12.** Soit $(s, y, \epsilon)$ un triplet vérifiant les conditions du paragraphe précédent. On lui a associé un terme $\sharp$ qu'il est utile de préciser. Le groupe $Z(\hat{G}) \simeq \{\pm 1\}$ s'envoie naturellement dans $Z_{\hat{G}}(s, y)/Z_{\hat{G}}(s, y)^0$, de façon en général non injective. Donc de $\epsilon$ se déduit un caractère de $Z(\hat{G})$ que l'on note improprement $\epsilon_{|Z(\hat{G})}$. Identifions à $\{\pm 1\}$ le groupe des caractères de $Z(\hat{G})$.

LEMME. *On a les relations:*

$$\sharp = \begin{cases} iso, & si\ \epsilon_{|Z(\hat{G})} = 1; \\ an, & si\ \epsilon_{|Z(\hat{G})} = -1. \end{cases}$$

*Preuve.* En 1.11, on a associé à $(s, y, \epsilon)$ des objets $\hat{G}^+, s^+, y^+, \epsilon^+, \hat{G}^-, s^-, y^-, \epsilon^-$. En 1.10, on leur a associé d'autres objets que nous noterons de façon évidente $\hat{G}_0^+, y_0^+, \hat{G}_0^-, y_0^-$. On a l'isomorphisme:

$$Z_{\hat{G}}(s, y)/Z_{\hat{G}}(s, y)^0 \simeq \left(Z_{\hat{G}_0^+}(y_0^+)/Z_{\hat{G}_0^+}(y_0^+)^0\right) \times \left(Z_{\hat{G}_0^-}(y_0^-)/Z_{\hat{G}_0^-}(y_0^-)^0\right).$$

Supposons pour simplifier $\hat{G}_0^+$ et $\hat{G}_0^-$ non triviaux (le raisonnement s'adapte aisément au cas général). En identifiant à $\{\pm 1\}$ les trois groupes des caractères de $Z(\hat{G})$, $Z(\hat{G}_0^+)$ et $Z(\hat{G}_0^-)$, on a l'égalité:

$$\epsilon_{|Z(\hat{G})} = \epsilon^+_{|Z(\hat{G}_0^+)}\epsilon^-_{|Z(\hat{G}_0^-)}.$$

Les entiers $k^+$, resp. $k^-$, sont associés par la correspondance de Springer généralisée aux couples $(y_0^+, \epsilon^+)$, resp. $(y_0^-, \epsilon^-)$. On a calculé ces entiers en [W] XI.3. On voit que pour tout signe $\eta = \pm$,

$$k^\eta \equiv \begin{cases} 0 \text{ ou } 3 \bmod 4\mathbb{Z}, & \text{si } \epsilon^\eta_{|Z(\hat{G}_0^\eta)} = 1, \\ 1 \text{ ou } 2 \bmod 4\mathbb{Z}, & \text{si } \epsilon^\eta_{|Z(\hat{G}_0^\eta)} = -1. \end{cases}$$

Il résulte alors de la définition de $R''$ (cf.1.11) que:

$R''$ est pair si $\epsilon_{|Z(\hat{G})} = 1$,

$R''$ est impair si $\epsilon_{|Z(\hat{G})} = -1$.

Par définition, $\sharp = iso$ si $R''$ est pair, $\sharp = an$ si $R''$ est impair. D'où le lemme.

## 2. Modules standard.

**2.1.** Soient $d$ un entier pair $\geq 1$, $\hat{V}$ un espace de dimension $d$ sur $\mathbb{C}$ muni d'une forme symplectique, $\hat{G}$ son groupe symplectique, $k$ un entier $\geq 0$ tel que $k(k+1) \leq d$. Posons:

$$N = (d - k(k+1))/2, \quad \bar{\mathcal{H}} = \bar{\mathcal{H}}(N; 2k+1).$$

Fixons un sous-groupe parabolique $\hat{P}$ de $\hat{G}$ et une décomposition de Lévi $\hat{P} = \hat{M}\hat{U}$ de sorte que:

$$\hat{M} \simeq (\mathbb{C}^\times)^N \times \hat{G}_c,$$

où $\hat{G}_c$ est un groupe symplectique de rang $k(k+1)/2$. On note $\hat{T}$ le plus grand tore central dans $\hat{M}$. On note par les lettres gothiques minuscules les algèbres de Lie correspondantes: $\hat{\mathfrak{g}}$, $\hat{\mathfrak{g}}_c$, $\hat{\mathfrak{p}}$ etc ... Soit $\mathcal{C}$ l'unique classe de conjugaison nilpotente dans $\hat{\mathfrak{g}}_c$ (ou $\hat{\mathfrak{m}}$) paramétrisée par la partition $(2k, 2k-2, \ldots, 2)$. Fixons $y_c \in \mathcal{C}$. Le caractère $\epsilon_c$ de $Z_{\hat{G}_c}(y_c)/Z_{\hat{G}_c}(y_c)^0$ décrit en 1.10 définit un système local de rang 1 sur $\mathcal{C}$ que l'on note $\mathcal{L}$. Il est $\hat{M}$-équivariant.

Soit $y$ un élément nilpotent de $\hat{\mathfrak{g}}$. On pose:

$$\mathcal{P}_y = \{g\hat{P} \in \hat{G}/\hat{P};\ Ad(g^{-1})y \in \mathcal{C} + \hat{\mathfrak{u}}\}.$$

Considérons le diagramme:

$$\begin{array}{ccc} & \{g\hat{T}\hat{U} \in \hat{G}/\hat{T}\hat{U};\ Ad(g^{-1})y \in \mathcal{C} + \hat{\mathfrak{u}}\} & \\ f_1 \swarrow & & \searrow f_2 \\ \mathcal{P}_y & & \mathcal{C} \end{array}$$

($f_1$ est la projection évidente; $f_2$ envoie $g\hat{T}\hat{U}$ sur la projection sur $\mathcal{C}$ de $Ad(g^{-1})y$). Parce que $\mathcal{L}$ est $\hat{M}$-équivariant, il existe un unique système local sur $\mathcal{P}_y$, que l'on note encore $\mathcal{L}$, de sorte que l'on ait l'égalité: $f_1^*\mathcal{L} = f_2^*\mathcal{L}$.

*Remarque.* Dans la suite, on déduira de $\mathcal{L}$ des systèmes locaux sur diverses variétés par des procédés du même genre que celui ci-dessus. On n'explicitera pas les définitions. Les systèmes locaux obtenus seront tous notés $\mathcal{L}$.

On pose $\hat{G}_{\mathbb{C}} = \hat{G} \times \mathbb{C}^\times$ et, de même, $\hat{P}_{\mathbb{C}} = \hat{P} \times \mathbb{C}^\times$ etc... Le groupe $\hat{G}_{\mathbb{C}}$ agit sur $\hat{\mathfrak{g}}$ par:

$$(g, \lambda)x = \lambda^{-2} Ad(g)x.$$

Posons:

$$\hat{A}(y) = Z_{\hat{G}_{\mathbb{C}}}(y)^0, \quad \bar{A}(y) = Z_{\hat{G}_{\mathbb{C}}}(y)/Z_{\hat{G}_{\mathbb{C}}}(y)^0.$$

Remarquons que l'injection $Z_{\hat{G}}(y) \to Z_{\hat{G}_{\mathbb{C}}}(y)$ induit un isomorphisme:

$$Z_{\hat{G}}(y)/Z_{\hat{G}}(y)^0 \simeq \bar{A}(y).$$

Le groupe $Z_{\hat{G}_{\mathbb{C}}}(y)$ agit sur $\mathcal{P}_y$, par $((g, \lambda), g'\hat{P}) \mapsto gg'\hat{P}$, et $\mathcal{L}$ est équivariant pour cette action. On définit l'espace d'homologie équivariante:

$$H^{\hat{A}(y)}(\mathcal{P}_y, \mathcal{L}) = \oplus_{j \in \mathbb{N}} H_{2j}^{\hat{A}(y)}(\mathcal{P}_y, \mathcal{L})$$

([L3]1.1; la définition sera rappelée brièvement en 2.3; les groupes de degré impair sont nuls, cf. [L3] proposition 8.6). Cet espace est muni d'une action de l'algèbre de cohomologie $H_{\hat{A}(y)}$ et d'une action du groupe $\bar{A}(y)$. Lusztig définit une action de $\bar{\mathcal{H}}$ sur $H^{\hat{A}(y)}(\mathcal{P}_y, \mathcal{L})$, qui commute aux actions précédentes ([L3] paragraphe 8, et ci-dessous 2.5).

L'algèbre de Lie $\hat{\mathfrak{a}}(y)$ de $\hat{A}(y)$ est incluse dans $\hat{\mathfrak{g}}_{\mathbb{C}} = \hat{\mathfrak{g}} \oplus \mathbb{C}$. Soit $(\sigma, r_0)$ un élément semi-simple de $\hat{\mathfrak{a}}(y)$. On a $\sigma \in \hat{\mathfrak{g}}$, $r_0 \in \mathbb{C}$ et $[\sigma, y] = 2r_0 y$. Posons:

$$\bar{A}(\sigma, y) = Z_{\hat{G}}(\sigma, y)/Z_{\hat{G}}(\sigma, y)^0.$$

C'est un groupe abélien. De $(\sigma, r_0)$ se déduit un homomorphisme d'algèbres:

$$\chi_{\sigma, r_0} : H_{\hat{A}(y)} \to \mathbb{C}$$

(cf. ci-dessous 2.4). On en déduit par tensorisation un espace:

$$H^{\hat{A}(y)}(\mathcal{P}_y, \mathcal{L})_{\sigma, r_0} = \mathbb{C} \otimes_{H_{\hat{A}(y)}} H^{\hat{A}(y)}(\mathcal{P}_y, \mathcal{L}).$$

Le groupe $\bar{A}(\sigma, y)$ agit sur cet espace, lequel se décompose selon les caractères de ce groupe. On note $\mathcal{E}_k(\sigma, r_0, y)$ l'ensemble des caractères qui interviennent et, pour tout $\epsilon \in \mathcal{E}_k(\sigma, r_0, y)$, $\mathbb{M}(\sigma, r_0, y, \epsilon)$ le sous-espace correspondant. Chaque espace $\mathbb{M}(\sigma, r_0, y, \epsilon)$ est un $\bar{\mathcal{H}}$-module, dont la classe d'isomorphie ne dépend que de $r_0$ et de la classe de conjugaison par $\hat{G}$ du triplet $(\sigma, r_0, \epsilon)$. Les $\bar{\mathcal{H}}$-modules $\mathbb{M}(\sigma, r_0, y, \epsilon)$

sont appelés par Lusztig des modules standard. Ils ne sont pas irréductibles en général.

**2.2.** Pour tout élément semi-simple $(\sigma, r_0)$ de $\hat{\mathfrak{g}}_{\mathbb{C}}$, posons:

$$\hat{\mathfrak{g}}^{\sigma,r_0} = \{y \in \hat{\mathfrak{g}};\ [\sigma, y] = 2r_0 y\}.$$

Supposons $r_0 \neq 0$. Alors le groupe $Z_{\hat{G}}(\sigma)$ agit dans $\hat{\mathfrak{g}}^{\sigma,r_0}$ et il n'y a dans cet espace qu'un nombre fini d'orbites pour cette action. Si $y \in \hat{\mathfrak{g}}^{\sigma,r_0}$, on note $\mathfrak{o}_y^{\sigma,r_0}$ son orbite pour cette action et $\bar{\mathfrak{o}}_y^{\sigma,r_0}$ la clôture de cette orbite. La finitude du nombre d'orbites entraîne qu'il existe une et une seule orbite dense dans $\hat{\mathfrak{g}}^{\sigma,r_0}$.

Décrivons le lien entre les constructions de 2.1 et celles de 1.10. Soit $(\sigma, r_0, y, \epsilon)$ comme en 1.10. On a dit que Lusztig associait à ce quadruplet un $\bar{\mathcal{H}}$-module irréductible $\mathbb{P}(\sigma, r_0, y, \epsilon)$. Supposons que l'entier $k$ de 2.1 soit égal à celui associé en 1.10 à $(\sigma, r_0, y, \epsilon)$. Alors $\epsilon \in \mathcal{E}_k(\sigma, r_0, y)$ et, dans le groupe de Grothendieck des $\bar{\mathcal{H}}$-modules de dimension finie, on a une égalité:

$$\mathbb{P}(\sigma, r_0, y, \epsilon) = \mathbb{M}(\sigma, r_0, y, \epsilon) + \sum_{y'} \sum_{\epsilon' \in \mathcal{E}_k(\sigma, r_0, y')} c(y', \epsilon')\mathbb{M}(\sigma, r_0, y', \epsilon'),$$

où les $c(y', \epsilon')$ sont des entiers relatifs et $y'$ parcourt un ensemble fini d'éléments nilpotents de $\hat{\mathfrak{g}}$ tels que:

- $y' \in \hat{\mathfrak{g}}^{\sigma,r_0}$,
- $\mathfrak{o}_y^{\sigma,r_0} \subset \bar{\mathfrak{o}}_{y'}^{\sigma,r_0}$ et $\mathfrak{o}_y^{\sigma,r_0} \neq \mathfrak{o}_{y'}^{\sigma,r_0}$,

(cf. [L4] corollaire 10.7). En particulier, si $\mathfrak{o}_y^{\sigma,r_0}$ est l'orbite dense dans $\hat{\mathfrak{g}}^{\sigma,r_0}$, alors $\mathbb{M}(\sigma, r_0, y, \epsilon)$ est irréductible et égal à $\mathbb{P}(\sigma, r_0, y, \epsilon)$. En général, la formule ci-dessus est "triangulaire" et s'inverse en une formule analogue exprimant $\mathbb{M}(\sigma, r_0, y, \epsilon)$ en fonction des $\mathbb{P}(\sigma, r_0, y', \epsilon')$. Il en résulte que $\mathbb{P}(\sigma, r_0, y, \epsilon)$ est un sous-quotient irréductible de $\mathbb{M}(\sigma, r_0, y, \epsilon)$.

**2.3.** Rappelons brièvement les définitions des groupes d'homologie et de cohomologie équivariante. Si $L$ est un groupe agissant à droite sur un ensemble $A$ et à gauche sur un ensemble $B$, on note $A \times_L B$ le quotient de $A \times B$ par la relation d'équivalence $(a\ell, b) \sim (a, \ell b)$ pour tous $a \in A, b \in B, \ell \in L$. Si $B'$ est un autre ensemble muni d'une action à gauche de $L$ et si $f : B \to B'$ est une application entrelaçant les actions de $L$, on note:

$$id_A \times_L f : A \times_L B \to A \times_L B'$$

l'application quotient de $id_A \times f$, où $id_A$ est l'identité de $A$. Dans ce qui suit, les variétés considérées sont des variétés algébriques complexes, les groupes sont des groupes algébriques complexes, les actions des groupes sur les variétés sont algébriques. Soient $L$ un groupe, $X$ une variété munie d'une action de $L$, $\mathcal{F}$ un système local $L$-équivariant sur $X$, $\mathcal{F}^*$ le système dual. Un entier $j \geq 0$ étant fixé, on choisit une variété $\Gamma$ lisse irréductible sur laquelle $L$ agit librement, telle que

$H^i(\Gamma) = \{0\}$ pour $i \in \{1, \ldots, m\}$, $m$ étant un entier assez grand. Des systèmes locaux $\mathcal{F}$ et $\mathcal{F}^*$ se déduisent des systèmes locaux encore notés $\mathcal{F}$ et $\mathcal{F}^*$ sur la variété $\Gamma \times_L X$. On pose:

$$H_L^j(X, \mathcal{F}) = H^j(\Gamma \times_L X, \mathcal{F}),\ H_j^L(X, \mathcal{F}) = H_c^{2\delta - j}(\Gamma \times_L X, \mathcal{F}^*)^*,$$

où $\delta = dim(\Gamma \times_L X)$ et le dernier $^*$ signifie "dual." Dans le cas particulier où $X = \{.\}$, c'est-à-dire $X$ est réduit à un point, et $\mathcal{F} = \mathbb{C}$, on pose simplement:

$$H_L^j = H_L^j(\{.\}, \mathbb{C}),\ H_j^L = H_j^L(\{.\}, \mathbb{C}).$$

Remarquons que la condition "$m$ assez grand" dépend de $j$. On ignorera cette difficulté en fixant $\Gamma$. Strictement parlant, nos constructions ne seront correctes que pour un ensemble borné de $j$. Peu importe puisqu'on peut choisir $\Gamma$ tel que cet ensemble soit aussi grand qu'on le veut. On fixe donc une variété $\Gamma$ lisse irréductible, sur laquelle $\hat{G}_C$ agit librement, "assez acyclique."

**2.4.** Fixons un sous-groupe de Borel $\hat{P}_0$ de $\hat{P}$ et un sous-tore maximal $\hat{T}_0$ de $\hat{P}_0 \cap \hat{M}$. Notons $W$, resp. $W^{\hat{M}}$, le groupe de Weyl de $\hat{G}$, resp. $\hat{M}$, relativement à $\hat{T}_0$. Posons:

$$W(\hat{M}) = \{w \in W; w\hat{M}w^{-1} = \hat{M}\}.$$

C'est un sous-groupe de $W$ et $W^{\hat{M}}$ est un sous-groupe distingué de $W(\hat{M})$. Remarquons que le quotient $W(\hat{M})/W^{\hat{M}}$ agit naturellement sur $\hat{T}$.

Fixons un homomorphisme $\phi_c : SL(2, \mathbb{C}) \to \hat{M}$ tel que sa dérivée $d\phi_c$ envoie $\left(\begin{smallmatrix} 0 & 1 \\ 0 & 0 \end{smallmatrix}\right)$ sur $y_c \in \mathcal{C}$. Posons:

$$\hat{Z}(\phi_c) = \{(m, \lambda) \in \hat{M}_{\mathbb{C}}; m\phi_c(\gamma)m^{-1} = \phi_c\left(\begin{pmatrix} \lambda & 0 \\ 0 & \lambda^{-1} \end{pmatrix}\gamma\begin{pmatrix} \lambda^{-1} & 0 \\ 0 & \lambda \end{pmatrix}\right)$$
$$\text{pour tout } \gamma \in SL(2, \mathbb{C})\}.$$

Notons $\hat{A}_c$ la composante neutre de $\hat{Z}(\phi_c)$ et $\bar{A}_c = \hat{Z}(\phi_c)/\hat{A}_c$. On vérifie que $\hat{Z}(\phi_c)$ est abélien. Soit $(m, \lambda) \in \hat{A}_c$. Posons $m_T = m\phi_c(\left(\begin{smallmatrix} \lambda^{-1} & 0 \\ 0 & \lambda \end{smallmatrix}\right))$. Alors $m_T \in \hat{T}$. L'application $(m, \lambda) \mapsto (m_T, \lambda)$ est un isomorphisme de $\hat{A}_c$ sur $\hat{T}_{\mathbb{C}}$. L'isomorphisme dérivé de $\hat{\mathfrak{a}}_c$ sur $\hat{\mathfrak{t}}_{\mathbb{C}}$ sera noté $(\sigma, r_0) \mapsto (\sigma_T, r_0)$, avec $\sigma_T = \sigma - d\phi_c(\left(\begin{smallmatrix} r_0 & 0 \\ 0 & -r_0 \end{smallmatrix}\right))$. Remarquons que, pour deux éléments $(\sigma, r_0)$, $(\sigma', r_0') \in \hat{\mathfrak{a}}_c$, avec $r_0' = r_0$, $\sigma$ et $\sigma'$ sont conjugués par un élément de $\hat{G}$ si et seulement si $\sigma_T$ et $\sigma_T'$ le sont. On a défini $Y_N$ et $W_N$ en 1.4. On peut identifier $Y_N$ au groupe des cocaractères $X_*(\hat{T})$ et $W_N$ au groupe $W(\hat{M})/W^{\hat{M}}$, de sorte que:

- l'action de $W_N$ sur $Y_N$ s'identifie à l'action naturelle de $W(\hat{M})/W^{\hat{M}}$ sur $X_*(\hat{T})$;
- les éléments de $\check{R}_N \subseteq Y_N$ s'identifient à des cocaractères positifs relativement à $\hat{P}$.

Le $\mathbb{Z}$-module $Y_N \oplus \mathbb{Z}$ s'identifie à $X_*(\hat{T}_{\mathbb{C}})$, donc à $X_*(\hat{A}_c)$. Il résulte alors de [L3] 1.11 que l'on a l'isomorphisme $H_{\hat{A}_c} \simeq \mathcal{S}$.

Soit $y$ un élément nilpotent de $\hat{\mathfrak{g}}$. Notons $\mathcal{Z}$ la variété qui classifie les classes de conjugaison semi-simples dans l'algèbre de Lie $\hat{\mathfrak{a}}(y)$ de $\hat{A}(y)$, pour la conjugaison par $\hat{A}(y)$. L'algèbre $H_{\hat{A}(y)}$ n'est autre que l'algèbre des fonctions régulières sur $\mathcal{Z}$ ([L3] 1.11). Tout élément semi-simple $(\sigma, r_0) \in \hat{\mathfrak{a}}(y)$ détermine un homomorphisme d'évaluation:

$$\chi_{\sigma,r_0} : H_{\hat{A}(y)} \to \mathbb{C}.$$

On pose:

$$H^{\hat{A}(y)}(\mathcal{P}_y, \mathcal{L})_{\sigma,r_0} = \mathbb{C} \otimes_{H_{\hat{A}(y)}} H^{\hat{A}(y)}(\mathcal{P}_y, \mathcal{L}).$$

Fixons $(\sigma, r_0)$. Notons $\hat{D}$ le plus petit sous-tore de $\hat{A}(y)$ dont l'algèbre de Lie contient $(\sigma, r_0)$ et $\mathcal{P}_y^{\hat{D}}$ le sous-ensemble des invariants par $\hat{D}$ dans $\mathcal{P}_y$. On a encore un homomorphisme d'évaluation:

$$\chi_{\sigma,r_0} : H_{\hat{D}} \to \mathbb{C}.$$

De l'homomorphisme naturel $H^{\hat{A}(y)}(\mathcal{P}_y, \mathcal{L}) \to H^{\hat{D}}(\mathcal{P}_y, \mathcal{L})$ se déduit un isomorphisme:

$$H^{\hat{A}(y)}(\mathcal{P}_y, \mathcal{L})_{\sigma,r_0} = \mathbb{C} \otimes_{H_{\hat{D}}} H^{\hat{D}}(\mathcal{P}_y, \mathcal{L})$$

([L3], 7.5). D'après [L4], proposition 4.4, de l'homomorphisme $H^{\hat{D}}(\mathcal{P}_y^{\hat{D}}, \mathcal{L}) \to H^{\hat{D}}(\mathcal{P}_y, \mathcal{L})$ déduit de l'injection $\mathcal{P}_y^{\hat{D}} \to \mathcal{P}_y$, se déduit un isomorphisme:

$$H^{\hat{A}(y)}(\mathcal{P}_y, \mathcal{L})_{\sigma,r_0} = \mathbb{C} \otimes_{H_{\hat{D}}} H^{\hat{D}}(\mathcal{P}_y^{\hat{D}}, \mathcal{L}).$$

Supposons $H^{\hat{A}(y)}(\mathcal{P}_y, \mathcal{L})_{\sigma,r_0} \neq \{0\}$. Alors $\mathcal{P}_y^{\hat{D}}$ est non vide. A fortiori, il existe $g \in \hat{G}$ tel que $Ad(g^{-1})y \in \mathcal{C} + \hat{\mathfrak{u}}$ et $Ad(g^{-1})\sigma \in \hat{\mathfrak{p}}$. Quitte à conjuguer $\sigma$ et $y$, on peut supposer $g = 1$. Quitte à conjuguer encore $\sigma$ et $y$, on peut supposer $\sigma \in \hat{\mathfrak{m}}$, puis $y \in y_c + \hat{\mathfrak{u}}$. L'équation $[\sigma, y] = 2r_0 y$ entraîne $[\sigma, y_c] = 2r_0 y_c$. Quitte à conjuguer encore $\sigma$ et $y$, on peut donc supposer $(\sigma, r_0) \in \hat{\mathfrak{a}}_c$.

*Remarque.* Cela explique une construction de 1.10: en supposant $(\sigma, r_0) \in \hat{\mathfrak{a}}_c$, les éléments $\sigma_c$, resp. $\sigma_1$, de 1.10 ne sont autres que $d\phi_c(\left(\begin{smallmatrix} r_0 & 0 \\ 0 & -r_0 \end{smallmatrix}\right))$, resp. $\sigma_T$.

**2.5.** Jusqu'en 2.11, on fixe un élément nilpotent $y \in \hat{\mathfrak{g}}$. On simplifie les notations en posant $\mathcal{P} = \mathcal{P}_y$, $\hat{A} = \hat{A}(y)$, $\bar{A} = \bar{A}(y)$. Remarquons qu'une action de $\bar{\mathcal{H}}$ dans un espace $E$ est déterminée par:

- une action de $\mathcal{S}$ dans $E$, via l'injection naturelle $\mathcal{S} \to \bar{\mathcal{H}}$;
- une action de $W_N$ dans $E$, via l'homomorphisme $w \mapsto t_w$ de $W_N$ dans $\bar{\mathcal{H}}$.

Rappelons les définitions des actions de $\mathcal{S}$ et $W_N$ dans $H^{\hat{A}}(\mathcal{P}, \mathcal{L})$. Posons:

$$\dot{\mathfrak{g}} = \{(x, g\hat{P}) \in \hat{\mathfrak{g}} \times \hat{G}/\hat{P};\ Ad(g^{-1})x \in \mathcal{C} + \hat{\mathfrak{t}} + \hat{\mathfrak{u}}\}.$$

Le groupe $\hat{G}_{\mathbb{C}}$ agit sur $\dot{\mathfrak{g}}$ par:

$$(g', \lambda)(x, g\hat{P}) = (\lambda^{-2} Ad(g')x, g'g\hat{P}).$$

On a un isomorphisme:

$$H_{\hat{G}_{\mathbb{C}}}(\dot{\mathfrak{g}}) \simeq \mathcal{S}, \tag{1}$$

qui est le composé des isomorphismes (2) à (7) ci-dessous. L'application:

$$\begin{array}{ccc} \hat{G}_{\mathbb{C}} \times_{\hat{P}_{\mathbb{C}}} (\mathcal{C} + \hat{\mathfrak{t}} + \hat{\mathfrak{u}}) & \to & \dot{\mathfrak{g}} \\ ((g, \lambda), x) & \mapsto & (\lambda^{-2} Ad(g)x, g\hat{P}) \end{array}$$

est un isomorphisme. Donc:

$$\Gamma \times_{\hat{P}_{\mathbb{C}}} (\mathcal{C} + \hat{\mathfrak{t}} + \hat{\mathfrak{u}}) \simeq \Gamma \times_{\hat{G}_{\mathbb{C}}} \dot{\mathfrak{g}},$$

et:

$$H_{\hat{G}_{\mathbb{C}}}(\dot{\mathfrak{g}}) \simeq H_{\hat{P}_{\mathbb{C}}}(\mathcal{C} + \hat{\mathfrak{t}} + \hat{\mathfrak{u}}). \tag{2}$$

D'après [L3] 1.4 (h):

$$H_{\hat{P}_{\mathbb{C}}}(\mathcal{C} + \hat{\mathfrak{t}} + \hat{\mathfrak{u}}) \simeq H_{\hat{M}_{\mathbb{C}}}(\mathcal{C} + \hat{\mathfrak{t}} + \hat{\mathfrak{u}}). \tag{3}$$

De la projection $p_{\mathcal{C}} : \mathcal{C} + \hat{\mathfrak{t}} + \hat{\mathfrak{u}} \to \mathcal{C}$ se déduit un isomorphisme:

$$p_{\mathcal{C}}^* : H_{\hat{M}_{\mathbb{C}}}(\mathcal{C}) \simeq H_{\hat{M}_{\mathbb{C}}}(\mathcal{C} + \hat{\mathfrak{t}} + \hat{\mathfrak{u}}). \tag{4}$$

Posons $\hat{Z}_c = Z_{\hat{M}_{\mathbb{C}}}(y_c)$. Puisque $\mathcal{C} \simeq \hat{M}_{\mathbb{C}} \times_{\hat{Z}_c} \{y_c\}$, on a $\Gamma \times_{\hat{M}_{\mathbb{C}}} \mathcal{C} \simeq \Gamma \times_{\hat{Z}_c} \{y_c\}$ et:

$$H_{\hat{M}_{\mathbb{C}}}(\mathcal{C}) \simeq H_{\hat{Z}_c}. \tag{5}$$

Mais $\hat{Z}(\phi_c)$ est un sous-groupe réductif maximal de $\hat{Z}_c$. D'après [L3] 1.4 (h) et 1.12 (a), on a donc:

$$H_{\hat{Z}_c} \simeq H_{\hat{Z}(\phi_c)} \simeq H_{\hat{A}_c}. \tag{6}$$

Enfin, comme on l'a dit en 2.4,

$$H_{\hat{A}_c} \simeq \mathcal{S}. \tag{7}$$

Posons $\tilde{\mathcal{P}} = \hat{G}_{\mathbb{C}} \times_{\hat{A}} \mathcal{P}$. De $\mathcal{L}$ se déduit un système local sur $\tilde{\mathcal{P}}$, encore noté $\mathcal{L}$ selon nos conventions. Puisque $\Gamma \times_{\hat{G}_{\mathbb{C}}} \tilde{\mathcal{P}} \simeq \Gamma \times_{\hat{A}} \mathcal{P}$, on a:

$$H^{\hat{A}}(\mathcal{P}, \mathcal{L}) \simeq H^{\hat{G}_{\mathbb{C}}}(\tilde{\mathcal{P}}, \mathcal{L}). \tag{8}$$

Définissons l'application:

$$\begin{array}{rccc} h : & \tilde{\mathcal{P}} & \to & \dot{\mathfrak{g}} \\ & (g, \lambda, g'\hat{P}) & \mapsto & (\lambda^{-2} Ad(g)y, gg'\hat{P}). \end{array}$$

On en déduit un homomorphisme:

$$h^* : H_{\hat{G}_{\mathbb{C}}}(\dot{\mathfrak{g}}) \to H_{\hat{G}_{\mathbb{C}}}(\tilde{\mathcal{P}}).$$

On dispose d'un accouplement, issu du cup-produit:

$$H_{\hat{G}_{\mathbb{C}}}(\tilde{\mathcal{P}}) \times H^{\hat{G}_{\mathbb{C}}}(\tilde{\mathcal{P}}, \mathcal{L}) \to H^{\hat{G}_{\mathbb{C}}}(\tilde{\mathcal{P}}, \mathcal{L}).$$

Via $h^*$, on en déduit un accouplement:

$$H_{\hat{G}_{\mathbb{C}}}(\dot{\mathfrak{g}}) \times H^{\hat{G}_{\mathbb{C}}}(\tilde{\mathcal{P}}, \mathcal{L}) \to H^{\hat{G}_{\mathbb{C}}}(\tilde{\mathcal{P}}, \mathcal{L}). \tag{9}$$

Via les isomorphismes (1) et (8), c'est la définition de l'action de $\mathcal{S}$ sur $H^{\hat{A}}(\mathcal{P}, \mathcal{L})$ ([L3] 8.2). On peut transcrire cette définition d'une façon plus maniable pour nous. Posons:

$$\tilde{\mathcal{P}}_{\hat{P}} = \{(g, \lambda, g'\hat{P}) \in \tilde{\mathcal{P}};\ gg'\hat{P} = \hat{P}\},$$
$$\tilde{\mathcal{V}} = \{(g, \lambda, g'\hat{P}) \in \tilde{\mathcal{P}}_{\hat{P}};\ \lambda^{-2}Ad(g)y \in y_c + \hat{\mathfrak{u}}\}.$$

On a:

$$\tilde{\mathcal{P}} = \hat{G}_{\mathbb{C}} \times_{\hat{P}_{\mathbb{C}}} \tilde{\mathcal{P}}_{\hat{P}},\ \ \tilde{\mathcal{P}}_{\hat{P}} = \hat{M}_{\mathbb{C}} \times_{\hat{Z}_c} \tilde{\mathcal{V}}.$$

Parallèlement aux isomorphismes (2) à (6), on a les isomorphismes:

$$\begin{aligned} H^{\hat{G}_{\mathbb{C}}}(\tilde{\mathcal{P}}, \mathcal{L}) &\simeq H^{\hat{P}_{\mathbb{C}}}(\tilde{\mathcal{P}}_{\hat{P}}, \mathcal{L}) \simeq H^{\hat{M}_{\mathbb{C}}}(\tilde{\mathcal{P}}_{\hat{P}}, \mathcal{L}) \simeq H^{\hat{Z}_c}(\tilde{\mathcal{V}}, \mathcal{L}) \\ &\simeq H^{\hat{Z}(\phi_c)}(\tilde{\mathcal{V}}, \mathcal{L}) \simeq H^{\hat{A}_c}(\tilde{\mathcal{V}}, \mathcal{L})^{\bar{A}_c}, \end{aligned} \tag{10}$$

où le dernier exposant signifie que l'on prend les invariants par le groupe $\bar{A}_c$, cf. [L3] 1.9 (a) pour le dernier isomorphisme. Il est facile de suivre la transformation de l'accouplement (9) via les isomorphismes (2) à (6) et (10). On obtient la description suivante de l'action de $\mathcal{S}$ sur $H^{\hat{A}}(\mathcal{P}, \mathcal{L})$: identifions $\mathcal{S}$ à $H_{\hat{A}_c}$ via (7), et $H^{\hat{A}}(\mathcal{P}, \mathcal{L})$ à $H^{\hat{A}_c}(\tilde{\mathcal{V}}, \mathcal{L})^{\bar{A}_c}$ via (8) et (10); alors l'action de $\mathcal{S}$ sur $H^{\hat{A}}(\mathcal{P}, \mathcal{L})$ s'identifie à l'action naturelle de $H_{\hat{A}_c}$ sur $H^{\hat{A}_c}(\tilde{\mathcal{V}}, \mathcal{L})^{\bar{A}_c}$.

Considérons la projection naturelle:

$$p_{\hat{\mathfrak{g}}} : \dot{\mathfrak{g}} \to \hat{\mathfrak{g}}.$$

Posons $K = p_{\hat{\mathfrak{g}}!}\mathcal{L}^*$. Il s'agit d'un objet de la catégorie dérivée des complexes $\hat{G}_{\mathbb{C}}$-équivariants sur $\hat{\mathfrak{g}}$. En fait, $K$ est un faisceau pervers et c'est le prolongement d'intersection d'un faisceau lisse défini sur la variété:

$$\mathcal{Y} = \{Ad(g)(\sigma + y_c);\ g \in \hat{G},\ \sigma \in \hat{\mathfrak{t}},\ Z_{\hat{G}}(\sigma) = \hat{M}\}.$$

D'autre part, fixons un élément nilpotent $y_0 \in \hat{\mathfrak{g}}$ tel qu'en notant $\mathfrak{o}(y_0)$ son orbite pour l'action adjointe de $\hat{G}$, l'intersection $\mathfrak{o}(y_0) \cap (\mathcal{C} + \hat{\mathfrak{u}})$ soit dense dans $\mathcal{C} + \hat{\mathfrak{u}}$. La fibre $(\mathcal{H}^0K)_{y_0}$ du faisceau de cohomologie $\mathcal{H}^0K$ est de dimension 1 ([L5], théorème 9.2.c). Lusztig a défini une action de $W_N$ sur $K$ ([L3], paragraphe 3).

Elle est caractérisée par les propriétés suivantes. Identifions $W_N$ à un ensemble de représentants dans le normalisateur de $\hat{M}$ dans $\hat{G}$. Pour tout $w \in W_N$, le système local $Ad(w)^*\mathcal{L}$ sur $\mathcal{C}$ est isomorphe à $\mathcal{L}$. Fixons un isomorphisme $a_w : \mathcal{L} \to Ad(w)^*\mathcal{L}$. Soit $\sigma$ un élément de $\hat{\mathfrak{t}}$ tel que $Z_{\hat{G}}(\sigma) = \hat{M}$. Posons $x = \sigma + y_c$. La fibre de $p_{\hat{\mathfrak{g}}}$ au-dessus de $x$ est l'ensemble des sous-groupes paraboliques de $\hat{G}$ de sous-groupe de Lévi $\hat{M}$, que l'on identifie à $W_N$ par $w \mapsto w\hat{P}w^{-1}$. La fibre $K_x$ de $K$ en $x$ est naturellement isomorphe à:

$$\oplus_{w \in W_N} \mathcal{L}_{Ad(w)^{-1}y_c}.$$

Pour tout $w$, l'isomorphisme $a_w$ identifie $\mathcal{L}_{Ad(w)^{-1}y_c}$ à $\mathcal{L}_{y_c}$. En fixant un isomorphisme $\mathcal{L}_{y_c} \simeq \mathbb{C}$, on identifie donc $K_x$ à $\mathbb{C}^{W_N}$. En [L5], théorème 9.2, Lusztig montre que l'on peut choisir d'une unique façon la famille d'isomorphismes $(a_w)_{w \in W_N}$, de telle sorte que, pour tout $w \in W_N$, il existe un unique automorphisme $\underline{\rho}(w)$ de $K$ vérifiant:

- pour tout $(z_{w'})_{w' \in W_N} \in \mathbb{C}^{W_N} \simeq K_x$, $\underline{\rho}(w)(z_{w'})_{w' \in W_N} = (z_{w'w})_{w' \in W_N}$;
- $\underline{\rho}(w)$ agit trivialement sur $(\mathcal{H}^0 K)_{y_0}$.

De plus, l'application $w \mapsto \underline{\rho}(w)$ est un homomorphisme de $W_N$ dans le groupe des automorphismes de $K$. On a ainsi défini une action de $W_N$ sur $K$.

Considérons le diagramme évident:

$$\begin{array}{ccccc} \Gamma \times_{\hat{G}_{\mathbb{C}}} \dot{\mathfrak{g}} & \leftarrow & \Gamma \times \dot{\mathfrak{g}} & \rightarrow & \dot{\mathfrak{g}} \\ p \downarrow & & \downarrow & & \downarrow p_{\hat{\mathfrak{g}}}. \\ \Gamma \times_{\hat{G}_{\mathbb{C}}} \hat{\mathfrak{g}} & \leftarrow & \Gamma \times \hat{\mathfrak{g}} & \stackrel{\alpha}{\rightarrow} & \hat{\mathfrak{g}} \end{array}$$

Parce que $\alpha$ est lisse, à fibres connexes, le faisceau $K$ se remonte en un faisceau sur $\Gamma \times \hat{\mathfrak{g}}$. Il est $\hat{G}_{\mathbb{C}}$-équivariant et se descend en un faisceau sur $\Gamma \times_{\hat{G}_{\mathbb{C}}} \hat{\mathfrak{g}}$. Parce que les carrés du diagramme sont cartésiens, ce faisceau n'est autre que $p_!\mathcal{L}^*$. Il est encore pervers, à un décalage près, et a même anneau d'endomorphismes que $K$ ([BBD], proposition 4.2.5). D'où une action de $W_N$ sur $p_!\mathcal{L}^*$.

Considérons le diagramme:

$$\begin{array}{ccc} \Gamma \times_{\hat{A}} \mathcal{P} & \stackrel{\dot{h}}{\rightarrow} & \Gamma \times_{\hat{G}_{\mathbb{C}}} \dot{\mathfrak{g}} \\ \downarrow p_y & & \downarrow p \\ \Gamma \times_{\hat{A}} \{y\} & \stackrel{h}{\rightarrow} & \Gamma \times_{\hat{G}_{\mathbb{C}}} \hat{\mathfrak{g}} \end{array}$$

(les flèches $p_y$ et $h$ sont évidentes; $\dot{h}$ est définie par $\dot{h}(\gamma, g\hat{P}) = (\gamma, y, g\hat{P})$). Il est cartésien, donc $h^* p_!\mathcal{L}^* = p_{y!}\mathcal{L}^*$. Par fonctorialité par $h^*$, on obtient une action de $W_N$ sur $p_{y!}\mathcal{L}^*$. Soit $j$ un entier $\geq 0$. On a:

$$H_j^{\hat{A}}(\mathcal{P}, \mathcal{L}) = H_c^{2\delta-j}(\Gamma \times_{\hat{A}} \mathcal{P}, \mathcal{L}^*)^* = H_c^{2\delta-j}(\Gamma \times_{\hat{A}} \{y\}, p_{y!}\mathcal{L}^*)^*,$$

où $\delta = dim(\Gamma \times_{\hat{A}} \mathcal{P})$. De l'action de $W_N$ sur $p_{y!}\mathcal{L}^*$ se déduit une action de $W_N$ sur le dernier espace ci-dessus, donc sur $H_j^{\hat{A}}(\mathcal{P}, \mathcal{L})$. C'est l'action cherchée.

L'action que l'on vient de définir de $\bar{\mathcal{H}}$ sur $H^{\hat{A}}(\mathcal{P}, \mathcal{L})$ commute à l'action naturelle de $H_{\hat{A}}$ ([L3] 8.3). Pour $(\sigma, r_0) \in \hat{\mathfrak{a}}$, l'espace $H^{\hat{A}}(\mathcal{P}, \mathcal{L})_{\sigma,r_0}$ défini en 2.4 est donc muni d'une structure de $\bar{\mathcal{H}}$-module.

*Remarque.* Si $(\pi, E)$ est une représentation de dimension finie de $\bar{\mathcal{H}}$, on note $\mathcal{E}xp(\pi)$, ou $\mathcal{E}xp(E)$, l'ensemble des $\underline{\sigma} = (\sigma, r_0) \in (Y_N \otimes_{\mathbb{Z}} \mathbb{C}) \oplus \mathbb{C}$ tels qu'il existe $e \in E, e \neq 0$, vérifiant:

$$\pi(f)e = \ell_{\underline{\sigma}}(f)e$$

pour tout $f \in \mathcal{S}$, cf. 1.8 pour les notations. Soit $(\sigma, r_0) \in \hat{\mathfrak{a}}$, supposons $H^{\hat{A}}(\mathcal{P}, \mathcal{L})_{\sigma,r_0} \neq \{0\}$ et soit $(\sigma_1, r_1) \in \mathcal{E}xp(H^{\hat{A}}(\mathcal{P}, \mathcal{L})_{\sigma,r_0})$. Soit $(\sigma_1', r_1) \in \hat{\mathfrak{a}}_c$ tel que $\sigma_{1,T}' = \sigma_1$, cf. 2.4 pour cette notation. On montre alors que $(\sigma_1', r_1)$ et $(\sigma, r_0)$ appartiennent à la même orbite pour l'action adjointe de $\hat{G}_{\mathbb{C}}$, cf. [L3] preuve du théorème 8.17 (c). En particulier $r_1 = r_0$.

**2.6.** Soient $\hat{Q}$ un sous-groupe parabolique de $\hat{G}$ et $\hat{Q} = \hat{L}\hat{N}$ une décomposition de Lévi. On suppose $\hat{P}_0 \subseteq \hat{Q}$, $\hat{T}_0 \subseteq \hat{L}$, on note $W^{\hat{L}}$ le groupe de Weyl de $\hat{L}$ relativement à $\hat{T}_0$. C'est un sous-groupe de $W$.

LEMME. *Supposons que $\hat{Q}$ ne contient pas $\hat{P}$ et que y appartient à $\hat{\mathfrak{l}}$. Alors* $H^{\hat{A}}(\mathcal{P}, \mathcal{L}) = \{0\}$.

*Preuve.* Considérons la variété $\mathcal{Z}$ qui classifie les classes de conjugaison semi-simples dans l'algèbre de Lie $\hat{\mathfrak{a}}$ de $\hat{A}$, pour la conjugaison par $\hat{A}$. On sait que l'algèbre $H_{\hat{A}}$ est l'algèbre des fonctions régulières sur $\mathcal{Z}$. De plus, $H^{\hat{A}}(\mathcal{P}, \mathcal{L})$ est un module projectif de type fini sur $H_{\hat{A}}$ ([L3], proposition 8.6). Il suffit pour démontrer l'énoncé de trouver un ouvert dense $\mathcal{Z}' \subseteq \mathcal{Z}$ tel que, pour tout élément semi-simple $(\sigma, r_0) \in \hat{\mathfrak{a}}$ dont la classe appartient à $\mathcal{Z}'$, on ait $H^{\hat{A}}(\mathcal{P}, \mathcal{L})_{\sigma,r_0} = \{0\}$.

Notons $\hat{T}_L$ le plus grand tore central dans $\hat{L}$. Il s'identifie, par $t \mapsto (t, 1)$, à un sous-tore de $\hat{A}$. Fixons un sous-tore maximal $\hat{T}_A$ de $\hat{A}$ contenant $\hat{T}_L$. On a $\hat{T}_A \subseteq \hat{L}_{\mathbb{C}}$. Notons $\hat{\mathfrak{t}}_A'$ l'ensemble des $(t, r_0) \in \hat{\mathfrak{t}}_A$ tels que:

- $(t, r_0)$ est régulier dans $\hat{\mathfrak{a}}$;

(1) l'application:

$$\begin{aligned} \hat{\mathfrak{n}} &\to \hat{\mathfrak{n}} \\ x &\mapsto Ad(t)x \end{aligned}$$

est bijective.

Parce que $\hat{\mathfrak{t}}_L$ est inclus dans $\hat{\mathfrak{t}}_A$, $\hat{\mathfrak{t}}_A'$ est un ouvert dense dans $\hat{\mathfrak{t}}_A$. Il nous suffit de prouver que $H^{\hat{A}}(\mathcal{P}, \mathcal{L})_{t,r_0} = \{0\}$ pour tout $(t, r_0) \in \hat{\mathfrak{t}}_A'$. Fixons donc $(t, r_0) \in \hat{\mathfrak{t}}_A'$. Notons $\hat{D}$ le plus petit sous-tore de $\hat{A}$ dont l'algèbre de Lie contient $(t, r_0)$ et $\mathcal{P}^{\hat{D}}$

le sous-ensemble des invariants par $\hat{D}$ dans $\mathcal{P}$. On a l'isomorphisme:

(2) $$H^{\hat{A}}(\mathcal{P}, \mathcal{L})_{t,r_0} \simeq \mathbb{C} \otimes_{H_{\hat{D}}} H^{\hat{D}}(\mathcal{P}^{\hat{D}}, \mathcal{L})$$

(cf. 2.4). Fixons un ensemble de représentants $W'$ de l'ensemble de doubles classes $W^{\hat{L}} \backslash W / W^{\hat{M}}$. Pour tout $w \in W'$, posons:

$$\mathcal{P}_w = \{g\hat{P} \in \mathcal{P}; g \in \hat{L}w\hat{P}\}.$$

Quand $w$ varie dans $W'$, ces sous-ensembles sont disjoints. Montrons que:

(3) $$\mathcal{P}^{\hat{D}} = \cup_{w \in W'} \mathcal{P}_w^{\hat{D}}.$$

Fixons, ainsi qu'il est loisible, un élément $x \in \hat{L}$ tel que $Ad(x)t \in \hat{\mathfrak{t}}_0$, posons $t_0 = Ad(x)t$. Notons $\hat{U}_0$ le radical unipotent de $\hat{P}_0$ et $\hat{U}'$ celui du sous-groupe parabolique de $\hat{G}$, de Lévi $\hat{M}$, opposé à $\hat{P}$. Soit $g\hat{P} \in \mathcal{P}^{\hat{D}}$. On sait qu'il existe $w \in W$ et $u \in \hat{U}_0 \cap w\hat{U}'w^{-1}$ tels que $g\hat{P} = x^{-1}uw\hat{P}$. On a l'égalité:

$$\hat{U}_0 \cap w\hat{U}'w^{-1} = (\hat{N} \cap w\hat{U}'w^{-1})(\hat{U}_0 \cap \hat{L} \cap w\hat{U}'w^{-1}).$$

Fixons donc $w \in W$, $u_1 \in \hat{N} \cap w\hat{U}'w^{-1}$ et $u_2 \in \hat{U}_0 \cap \hat{L} \cap w\hat{U}'w^{-1}$, de sorte que $g\hat{P} = x^{-1}u_1u_2w\hat{P}$. Puisque $g\hat{P} \in \mathcal{P}^{\hat{D}}$, on a $t \in Ad(g)\hat{\mathfrak{p}}$, ou encore $Ad(u_2^{-1})Ad(u_1^{-1})t_0 \in Ad(w)\hat{\mathfrak{p}}$. Parce que $t_0 \in \hat{\mathfrak{t}}_0$, on a $Ad(u^{-1})t_0 - t_0 \in \hat{\mathfrak{n}} \cap Ad(w)\hat{\mathfrak{u}}'$ pour tout $u \in \hat{N} \cap w\hat{U}'w^{-1}$ et, grâce à l'hypothèse (1), on voit que:

(4) l'application:

$$\begin{array}{ccc} \hat{N} \cap w\hat{U}'w^{-1} & \to & \hat{\mathfrak{n}} \cap Ad(w)\hat{\mathfrak{u}}' \\ u & \mapsto & Ad(u^{-1})t_0 - t_0 \end{array}$$

est bijective.

Notons $y_1$ l'image de $u_1$ par cette application. Alors:

$$Ad(u_2^{-1})t_0 + Ad(u_2^{-1})y_1 \in Ad(w)\hat{\mathfrak{p}}.$$

Mais le premier terme ci-dessus appartient à $\hat{\mathfrak{l}} \cap Ad(w)\hat{\mathfrak{u}}'$, le deuxième à $\hat{\mathfrak{n}} \cap Ad(w)\hat{\mathfrak{u}}'$. La relation ci-dessus entraîne que ces deux termes sont nuls. Or $Ad(u_2^{-1})$ est un automorphisme de $\hat{\mathfrak{n}} \cap Ad(w)\hat{\mathfrak{u}}'$. Donc $y_1 = 0$ et, grâce à (4), $u_1 = 1$. Alors $g\hat{P} \in \hat{L}w\hat{P}$. On peut dans cette relation remplacer $w$ par un élément de $W'$. Alors $g\hat{P} \in \mathcal{P}_w^{\hat{D}}$. Cela démontre (3).

Soit $w \in W'$. Montrons que:

(5) $$\text{si } \mathcal{P}_w \neq \emptyset, \text{ alors } w\hat{M}w^{-1} \subseteq \hat{L}.$$

Supposons $\mathcal{P}_w \neq \emptyset$, soit $h \in \hat{L}$ tel que $hw\hat{P} \in \mathcal{P}$. Alors $Ad(w^{-1}h^{-1})y \in \mathcal{C} + \hat{\mathfrak{u}}$. Ecrivons:

$$Ad(h^{-1})y = Ad(w)(y_c' + y''),$$

avec $y'_c \in \mathcal{C}$ et $y'' \in \hat{\mathfrak{u}}$. Soit $(t_n)_{n\geq 1}$ une suite d'éléments de $\hat{T}$ telle que $lim_{n\to\infty} Ad(t_n)y'' = 0$. Pour tout $n$, posons $t'_n = wt_nw^{-1}$. Alors:

$$lim_{n\to\infty} Ad(t'_n h^{-1})y = Ad(w)y'_c.$$

Mais, pour tout $n$, $t'_n \in \hat{T}_0 \subseteq \hat{L}$, donc $Ad(t'_n h^{-1})y \in \hat{\mathfrak{l}}$. Il en résulte que $Ad(w)y'_c \in \hat{\mathfrak{l}}$. Alors, le sous-tore $(w\hat{T}w^{-1})\hat{T}_L$ de $\hat{T}_0$ est contenu dans $Z_{\hat{G}}(Ad(w)y'_c)$. Or $w\hat{T}w^{-1}$ est un sous-tore maximal de ce centralisateur. Donc $\hat{T}_L \subseteq w\hat{T}w^{-1}$, puis $\hat{L} \supseteq w\hat{M}w^{-1}$. Cela démontre (5).

On a supposé que $\hat{Q}$ ne contenait pas $\hat{P}$. De la forme particulière du groupe $\hat{P}$ résulte que $w\hat{M}w^{-1}$ n'est inclus dans $\hat{L}$ pour aucun $w \in W'$. D'après (3) et (5), on a donc $\mathcal{P}^{\hat{D}} = \emptyset$. Alors l'espace de droite de (2) est nul, donc aussi $H^{\hat{A}}(\mathcal{P}, \mathcal{L})_{t,r_0}$. Cela achève la démonstration.

**2.7.** On conserve les notations du paragraphe précédent, on suppose maintenant et jusqu'en 2.11 que $\hat{Q}$ contient $\hat{P}$ et $y \in \hat{\mathfrak{l}}$. En remplaçant $\hat{G}$ par $\hat{L}$ et $\hat{P}$ par $\hat{L} \cap \hat{P}$ dans les définitions, on définit la variété $\mathcal{P}^{\hat{L}}$, le groupe $\hat{A}^{\hat{L}}$, que l'on note pour simplifier $\hat{B}$, et le groupe d'homologie $H^{\hat{B}}(\mathcal{P}^{\hat{L}}, \mathcal{L})$. Il est muni d'une action d'une algèbre $\bar{\mathcal{H}}^{\hat{L}}$ analogue à $\bar{\mathcal{H}}$. Il est facile d'identifier $\bar{\mathcal{H}}^{\hat{L}}$. En effet, posons:

$$W_N^{\hat{L}} = \{w \in W^{\hat{L}}; w\hat{M}w^{-1} = \hat{M}\}/W^{\hat{M}}.$$

C'est un sous-groupe de $W_N$ et $\bar{\mathcal{H}}^{\hat{L}}$ est la sous-algèbre de $\bar{\mathcal{H}}$ engendrée par $\mathcal{S}$ et les éléments $t_w$ de $\bar{\mathcal{H}}$, pour $w \in W_N^{\hat{L}}$.

Puisque $\hat{B}$ est un sous-groupe de $\hat{A}$, on dispose de deux homomorphismes:

$$H_{\hat{A}} \to H_{\hat{B}}, \quad H^{\hat{A}}(\mathcal{P}, \mathcal{L}) \to H^{\hat{B}}(\mathcal{P}, \mathcal{L}),$$

dont on déduit un homomorphisme de $H_{\hat{B}}$-modules:

$$H_{\hat{B}} \otimes_{H_{\hat{A}}} H^{\hat{A}}(\mathcal{P}, \mathcal{L}) \to H^{\hat{B}}(\mathcal{P}, \mathcal{L}).$$

C'est un isomorphisme ([L3], propositions 7.5 et 8.6). De l'action de $\bar{\mathcal{H}}$ sur $H^{\hat{A}}(\mathcal{P}, \mathcal{L})$ se déduit une action de $\bar{\mathcal{H}}$ sur $H^{\hat{B}}(\mathcal{P}, \mathcal{L})$.

De l'injection naturelle $i : \mathcal{P}^{\hat{L}} \to \mathcal{P}$ se déduit un homomorphisme de $H_{\hat{B}}$-modules:

$$i_! : H^{\hat{B}}(\mathcal{P}^{\hat{L}}, \mathcal{L}) \to H^{\hat{B}}(\mathcal{P}, \mathcal{L})$$

([L3], 1.4).

Lemme. *L'application $i_!$ est un homomorphisme de $\mathcal{S}$-modules.*

*Preuve.* Posons:

$$\tilde{\mathcal{P}}_B = \hat{G}_{\mathbb{C}} \times_{\hat{B}} \mathcal{P}, \quad \tilde{\mathcal{V}}_B = \{(g, \lambda, g'\hat{P}) \in \tilde{\mathcal{P}}_B; gg'\hat{P} = \hat{P},\ \lambda^{-2}Ad(g)y \in y_c + \hat{\mathfrak{u}}\},$$

et définissons $\tilde{\mathcal{V}}$ comme en 2.5. Comme en 2.5 (10), on a des isomorphismes:

$$H^{\hat{B}}(\mathcal{P},\mathcal{L}) \simeq H^{\hat{G}}(\tilde{\mathcal{P}}_B,\mathcal{L}) \simeq H^{\hat{A}_c}(\tilde{\mathcal{V}}_B,\mathcal{L})^{\bar{A}_c}. \tag{1}$$

De l'application naturelle $\tilde{\mathcal{V}}_B \to \tilde{\mathcal{V}}$ se déduit un homomorphisme:

$$H^{\hat{A}_c}(\tilde{\mathcal{V}},\mathcal{L})^{\bar{A}_c} \to H^{\hat{A}_c}(\tilde{\mathcal{V}}_B,\mathcal{L})^{\bar{A}_c}.$$

Le diagramme suivant est commutatif:

$$\begin{array}{ccc} H^{\hat{A}}(\mathcal{P},\mathcal{L}) \simeq & H^{\hat{A}_c}(\tilde{\mathcal{V}},\mathcal{L})^{\bar{A}_c} \\ \downarrow & \downarrow \\ H^{\hat{B}}(\mathcal{P},\mathcal{L}) \simeq & H^{\hat{A}_c}(\tilde{\mathcal{V}}_B,\mathcal{L})^{\bar{A}_c} \end{array}.$$

De la description de 2.5 résulte que, quand on identifie $H^{\hat{B}}(\mathcal{P},\mathcal{L})$ à $H^{\hat{A}_c}(\tilde{\mathcal{V}}_B,\mathcal{L})^{\bar{A}_c}$ et $\mathcal{S}$ à $H_{\hat{A}_c}$, l'action de $\mathcal{S}$ sur $H^{\hat{B}}(\mathcal{P},\mathcal{L})$ s'identifie à l'action naturelle de $H_{\hat{A}_c}$ sur $H^{\hat{A}_c}(\tilde{\mathcal{V}}_B,\mathcal{L})^{\bar{A}_c}$. Posons:

$$\tilde{\mathcal{P}}^{\hat{L}} = \hat{L}_{\mathbb{C}} \times_{\hat{B}} \mathcal{P}^{\hat{L}}, \quad \tilde{\mathcal{V}}^{\hat{L}} = \{(g,\lambda,g'(\hat{L}\cap\hat{P})) \in \tilde{\mathcal{P}}^{\hat{L}};$$
$$gg'(\hat{L}\cap\hat{P}) = \hat{L}\cap\hat{P},\ \lambda^{-2}Ad(g)y \in y_c + \hat{\mathfrak{l}}\cap\hat{\mathfrak{u}}\}.$$

Comme en 2.5 (10), on a des isomorphismes:

$$H^{\hat{B}}(\mathcal{P}^{\hat{L}},\mathcal{L}) \simeq H^{\hat{L}_{\mathbb{C}}}(\tilde{\mathcal{P}}^{\hat{L}},\mathcal{L}) \simeq H^{\hat{A}_c}(\tilde{\mathcal{V}}^{\hat{L}},\mathcal{L})^{\bar{A}_c}. \tag{2}$$

Posons:

$$\tilde{\mathcal{V}}_B^1 = \{(g,\lambda,g'\hat{P}) \in \tilde{\mathcal{V}}_B; g \in \hat{Q}\}.$$

Notons $\tilde{i} : \tilde{\mathcal{V}}_B^1 \to \tilde{\mathcal{V}}_B$ l'injection naturelle. Remarquons que l'application:

$$\begin{array}{ccc} \hat{N} \times \tilde{\mathcal{V}}^{\hat{L}} & \to & \tilde{\mathcal{V}}_B^1 \\ (n,g,\lambda,g'(\hat{L}\cap\hat{P})) & \to & (ng,\lambda,g'\hat{P}) \end{array}$$

est un isomorphisme. On dispose donc d'une suite d'homomorphismes:

$$H^{\hat{A}_c}(\tilde{\mathcal{V}}^{\hat{L}},\mathcal{L}) \simeq H^{\hat{A}_c}(\hat{N}\times\tilde{\mathcal{V}}^{\hat{L}},\mathcal{L}) \simeq H^{\hat{A}_c}(\tilde{\mathcal{V}}_B^1,\mathcal{L}) \xrightarrow{\tilde{i}_!} H^{\hat{A}_c}(\tilde{\mathcal{V}}_B,\mathcal{L}). \tag{3}$$

En dévissant la construction des isomorphismes (1) et (2), on vérifie que le diagramme ci-dessous est commutatif:

$$\begin{array}{ccc} H^{\hat{B}}(\mathcal{P}^{\hat{L}},\mathcal{L}) & \xrightarrow{\sim} & H^{\hat{A}_c}(\tilde{\mathcal{V}}^{\hat{L}},\mathcal{L})^{\bar{A}_c} \\ \downarrow i_! & & \downarrow \\ H^{\hat{B}}(\mathcal{P},\mathcal{L}) & \xrightarrow{\sim} & H^{\hat{A}_c}(\tilde{\mathcal{V}}_B,\mathcal{L})^{\bar{A}_c} \end{array}$$

(les flèches horizontales sont celles de (1) et (2), la flèche verticale de droite est celle de (3)). Dire que $i_!$ entrelace les actions de $\mathcal{S}$ revient à dire que la flèche de droite entrelace les actions de $H_{\hat{A}_c}$. C'est clair.

**2.8.** Notons $\mathfrak{z}_{\hat{N}}(y)$ le commutant de $y$ dans $\hat{\mathfrak{n}}$. Pour tout élément $(\sigma, r_0)$ de l'algèbre de Lie $\hat{\mathfrak{b}}$, l'opérateur $ad(\sigma)$ conserve $\hat{\mathfrak{n}}$ et $\mathfrak{z}_{\hat{N}}(y)$. Posons:

$$\Delta(\sigma, r_0) = det(ad(\sigma) - 2r_0|\hat{\mathfrak{n}})\, det(ad(\sigma)|\mathfrak{z}_{\hat{N}}(y))\, det(ad(\sigma)|\hat{\mathfrak{n}})^{-1}.$$

Cela définit une fonction rationnelle sur $\hat{\mathfrak{b}}$, invariante par conjugaison par $\hat{B}$, autrement dit un élément du corps des fractions de $H_{\hat{B}}$.

Rappelons que pour tout élément semi-simple $(\sigma, r_0)$ de $\hat{\mathfrak{b}}$, on peut trouver un homomorphisme $\phi : SL(2, \mathbb{C}) \to \hat{L}$ et un élément semi-simple $z$ de $\hat{\mathfrak{l}}$ tels que:

- $z$ commute à l'image de $\phi$;
- $d\phi(\begin{smallmatrix}0 & 1\\ 0 & 0\end{smallmatrix}) = y$;
- $\sigma = z + d\phi(\begin{smallmatrix}r_0 & 0\\ 0 & -r_0\end{smallmatrix})$.

Le couple $(z, \phi)$ est bien déterminé à conjugaison près par l'intersection des commutants de $\sigma$ et de $y$ dans $\hat{L}$. Avec ces notations, nous dirons que $(\sigma, r_0)$ contracte $\hat{\mathfrak{q}}$ si $r_0 > 0$ et si toutes les valeurs propres de $ad(z)$ dans $\hat{\mathfrak{n}}$ ont une partie réelle $\leq 0$.

LEMME. (i) *La fonction $\Delta$ est régulière sur $\hat{\mathfrak{b}}$, i.e. $\Delta \in H_{\hat{B}}$.*

(ii) *Si $(\sigma, r_0)$ est un élément semi-simple de $\hat{\mathfrak{b}}$ qui contracte $\hat{\mathfrak{q}}$, alors $\Delta(\sigma, r_0) \neq 0$.*

*Preuve.* Fixons un homomorphisme $\phi : SL(2, \mathbb{C}) \to \hat{L}$ tel que:

$$d\phi \begin{pmatrix} 0 & 1 \\ 0 & 0 \end{pmatrix} = y,$$

et un sous-tore maximal $\hat{T}_\phi$ du commutant dans $\hat{L}$ de l'image de $\phi$. Le produit $\hat{T}_\phi \times SL(2, \mathbb{C})$ agit naturellement dans $\hat{\mathfrak{n}}$: $\hat{T}_\phi$ agit par $Ad$, $SL(2, \mathbb{C})$ par $Ad \circ \phi$. Décomposons $\hat{\mathfrak{n}}$ en somme directe de sous-espaces irréductibles pour cette action:

$$\hat{\mathfrak{n}} = \oplus_{s \in S} \hat{\mathfrak{n}}_s.$$

Soit $s \in S$, posons $d_s = dim_{\mathbb{C}}(\hat{\mathfrak{n}}_s)$, $D_s = \{-d_s + 1, -d_s + 3, \ldots, d_s - 3, d_s - 1\}$. Il existe une base $(e_{s,\delta})_{\delta \in D_s}$ de $\hat{\mathfrak{n}}_s$ telle que, pour tout $\lambda \in \mathbb{C}$, on ait:

$$ad \circ d\phi(\begin{pmatrix} \lambda & 0 \\ 0 & -\lambda \end{pmatrix}) e_{s,\delta} = \delta\lambda e_{s,\delta}.$$

Fixons une telle base. Alors la famille $(e_{s,d_s-1})_{s \in S}$ est une base de $\mathfrak{z}_{\hat{N}}(y)$. D'autre part, pour tout $s \in S$, il existe $\alpha_s \in X^*(\hat{T}_\phi)$ tel que, pour tout $z \in \hat{\mathfrak{t}}_\phi$ et tout $e \in \hat{\mathfrak{n}}_s$, on ait $ad(z)e = \alpha_s(z)e$, où l'on identifie $\alpha_s$ à une forme linéaire sur $\hat{\mathfrak{t}}_\phi$.

Il nous suffit de considérer les éléments $(\sigma, r_0) \in \hat{\mathfrak{b}}$ tels qu'il existe $z \in \hat{\mathfrak{t}}_\phi$ de sorte que $\sigma = z + d\phi(\left(\begin{smallmatrix} r_0 & 0 \\ 0 & -r_0 \end{smallmatrix}\right))$. Pour un tel élément et pour $s \in S$, posons:

$$\Delta_s(\sigma, r_0) = det(ad(\sigma) - 2r_0|\hat{\mathfrak{n}}_s)\, det(ad(\sigma)|\mathfrak{z}_{\hat{N}}(y) \cap \hat{\mathfrak{n}}_s) det(ad(\sigma)|\hat{\mathfrak{n}}_s)^{-1}.$$

On calcule:

$$\Delta_s(\sigma, r_0) = \left(\prod_{\delta \in D_s} (\alpha_s(z) + (\delta - 2)r_0)\right)(\alpha_s(z) + (d_s - 1)r_0)\left(\prod_{\delta \in D_s} (\alpha_s(z) + \delta r_0)\right)^{-1},$$

$$= \alpha_s(z) - (d_s + 1)r_0.$$

Donc $\Delta_s$ est une fonction régulière. Si $(\sigma, r_0)$ contracte $\hat{\mathfrak{q}}$, on a $r_0 > 0$ et $Re(\alpha_s(z)) \leq 0$, donc $\Delta_s(\sigma, r_0) \neq 0$. Puisque:

$$\Delta(\sigma, r_0) = \prod_{s \in S} \Delta_s(\sigma, r_0),$$

le lemme s'ensuit.

**2.9.** Notons $w_{max}$ et $w^{\hat{L}}_{max}$ les éléments de plus grande longueur de $W$ et $W^{\hat{L}}$. Posons $w_0 = w^{\hat{L}}_{max} w_{max}$. On a $w_0^2 = 1$. On identifie $w_0$ à un représentant dans $\hat{G}$. Remarquons que $w_0$ normalise $\hat{M}, \hat{L}, \hat{L} \cap \hat{P}, (\mathcal{C} + \hat{\mathfrak{u}}) \cap \hat{L}$. On supposera, ainsi qu'il est loisible, que $Ad(w_0)$ agit trivialement sur le sous-groupe $\hat{G}_c$ de $\hat{M}$, a fortiori sur $\mathcal{C}$. Il agit donc sur $\hat{Z}(\phi_c)$, $\hat{A}_c$ et $\bar{A}_c$. Son action sur $\bar{A}_c$ est triviale. Posons:

$$\mathcal{P}^{w_0} = \{g\hat{P} \in \mathcal{P}; g \in \hat{Q} w_0 \hat{P}\}.$$

L'injection naturelle $j : \mathcal{P}^{w_0} \to \mathcal{P}$ est un plongement ouvert. Soit $g \in \hat{Q} w_0 \hat{P}$. Ecrivons $g = xhw_0 z$, avec $x \in \hat{N}$, $h \in \hat{L}$ et $z \in \hat{P}$. L'élément $x$ et la classe $h(\hat{L} \cap \hat{P})$ sont uniquement déterminés. Alors $g\hat{P} \in \mathcal{P}$ si et seulement si $x \in Z_{\hat{N}}(y)$ et $h(\hat{L} \cap \hat{P}) \in \mathcal{P}^{\hat{L}}$. L'application:

$$\begin{aligned} \mathcal{P}^{w_0} &\to Z_{\hat{N}}(y) \times \mathcal{P}^{\hat{L}} \\ g\hat{P} &\mapsto (x, h(\hat{L} \cap \hat{P})) \end{aligned}$$

est un isomorphisme. On note $\pi : \mathcal{P}^{w_0} \to \mathcal{P}^{\hat{L}}$ le composé de cet isomorphisme et de la projection naturelle de $Z_{\hat{N}}(y) \times \mathcal{P}^{\hat{L}}$ sur $\mathcal{P}^{\hat{L}}$.

Le groupe $\hat{B}$ agit sur $\mathcal{P}$, $\mathcal{P}^{w_0}$ et $\mathcal{P}^{\hat{L}}$. Les applications $j$ et $\pi$ sont équivariantes pour ces actions. D'après [L3] 1.4, de $j$ et $\pi$ se déduisent:

- un homomorphisme $j^* : H^{\hat{B}}(\mathcal{P}, \mathcal{L}) \to H^{\hat{B}}(\mathcal{P}^{w_0}, \mathcal{L})$;
- un isomorphisme $\pi^* : H^{\hat{B}}(\mathcal{P}^{\hat{L}}, \mathcal{L}) \xrightarrow{\sim} H^{\hat{B}}(\mathcal{P}^{w_0}, \mathcal{L})$.

On note $\rho$, resp. $\rho^{\hat{L}}$, l'action de $\bar{\mathcal{H}}$, resp. $\bar{\mathcal{H}}^{\hat{L}}$, sur $H^{\hat{B}}(\mathcal{P}, \mathcal{L})$, resp.$H^{\hat{B}}(\mathcal{P}^{\hat{L}}, \mathcal{L})$. D'autre part, l'élément $\Delta$ de $H_{\hat{B}}$ définit un endomorphisme de chacun de ces espaces, que l'on note encore $\Delta$. Pour $w \in W_N$, resp. $w \in W_N^{\hat{L}}$, l'application $(\pi^*)^{-1} \circ j^* \circ \rho(t_w) \circ i_!$, resp. $\Delta \circ \rho^{\hat{L}}(t_w)$, est un endomorphisme de $H^{\hat{B}}(\mathcal{P}^{\hat{L}}, \mathcal{L})$.

PROPOSITION. (i) *Pour tout* $w \in W_N \setminus w_0 W_N^{\hat{L}}$, $(\pi^*)^{-1} \circ j^* \circ \rho(t_w) \circ i_! = 0$.
(ii) *Pour tout* $w \in W_N^{\hat{L}}$, $(\pi^*)^{-1} \circ j^* \circ \rho(t_{w_0 w}) \circ i_! = \Delta \circ \rho^{\hat{L}}(t_w)$.

*Preuve.* Introduisons les variétés:

$$\dot{\mathfrak{q}} = \{(x, g\hat{P}) \in \hat{\mathfrak{q}} \times \hat{G}/\hat{P}; Ad(g^{-1})x \in \mathcal{C} + \hat{\mathfrak{t}} + \hat{\mathfrak{u}}\},$$
$$\dot{\mathfrak{q}}^1 = \{(x, g\hat{P}) \in \dot{\mathfrak{q}}; g \in \hat{Q}\},$$
$$\dot{\mathfrak{q}}^{w_0} = \{(x, g\hat{P}) \in \dot{\mathfrak{q}}; g \in \hat{Q} w_0 \hat{P}\}.$$

Le groupe $\hat{L}_{\mathbb{C}}$ agit sur chacune d'elles par $(h, \lambda)(x, g\hat{P}) = (\lambda^{-2}Ad(h)x, hg\hat{P})$. On définit la variété $\dot{\mathfrak{l}}$ comme on a défini $\dot{\mathfrak{g}}$, en remplaçant $\hat{G}$ par $\hat{L}$.

L'application:

$$\begin{array}{ccc} \hat{\mathfrak{n}} \times \dot{\mathfrak{l}} & \rightarrow & \dot{\mathfrak{q}}^1 \\ (v, x, h(\hat{L} \cap \hat{P})) & \mapsto & (x + v, h\hat{P}) \end{array} \tag{1}$$

est un isomorphisme. Elle est équivariante pour les actions de $\hat{L}_{\mathbb{C}}$, si l'on munit $\hat{\mathfrak{n}} \times \dot{\mathfrak{l}}$ de l'action de $\hat{L}_{\mathbb{C}}$ ainsi définie: $(h, \lambda)(v, x, h'(\hat{L} \cap \hat{P})) = (\lambda^{-2}Ad(h)v, \lambda^{-2}Ad(h)x, hh'(\hat{L} \cap \hat{P}))$.

L'application:

$$\begin{array}{ccc} \hat{N} \times \dot{\mathfrak{l}} & \rightarrow & \dot{\mathfrak{q}}^{w_0} \\ (n, x, h(\hat{L} \cap \hat{P})) & \mapsto & (Ad(n)x, nhw_0\hat{P}) \end{array} \tag{2}$$

est un isomorphisme. Elle est équivariante pour les actions de $\hat{L}_{\mathbb{C}}$, si l'on munit $\hat{N} \times \dot{\mathfrak{l}}$ de l'action de $\hat{L}_{\mathbb{C}}$ ainsi définnie: $(h, \lambda)(n, x, h'(\hat{L} \cap \hat{P})) = (hnh^{-1}, \lambda^{-2}Ad(h)x, hh'(\hat{L} \cap \hat{P}))$. Remarquons que $\hat{\mathfrak{n}} \times \dot{\mathfrak{l}}$ et $\hat{N} \times \dot{\mathfrak{l}}$ sont isomorphes en tant que variétés, mais pas en tant que variétés munies d'une action de $\hat{L}_{\mathbb{C}}$.

Introduisons les variétés:

$$\mathcal{Y} = \{(x, g\hat{P}) \in \dot{\mathfrak{q}}; x \in y + \hat{\mathfrak{n}}\}, \quad \mathcal{Y}^1 = \mathcal{Y} \cap \dot{\mathfrak{q}}^1, \quad \mathcal{Y}^{w_0} = \mathcal{Y} \cap \dot{\mathfrak{q}}^{w_0}.$$

L'action sur $\dot{\mathfrak{q}}$ du sous-groupe $\hat{B}$ de $\hat{L}_{\mathbb{C}}$ conserve chacune d'elles. Rappelons que $\mathcal{P}^{\hat{L}}$ se plonge dans $\dot{\mathfrak{l}}$ par $h(\hat{L} \cap \hat{P}) \mapsto (y, h(\hat{L} \cap \hat{P}))$. Alors les isomorphismes (1) et (2) se restreignent en des isomorphismes:

$$\hat{\mathfrak{n}} \times \mathcal{P}^{\hat{L}} \xrightarrow{\sim} \mathcal{Y}^1, \tag{3}$$

$$\hat{N} \times \mathcal{P}^{\hat{L}} \xrightarrow{\sim} \mathcal{Y}^{w_0}. \tag{4}$$

On note $\pi^1 : \mathcal{Y}^1 \rightarrow \mathcal{P}^{\hat{L}}$, resp. $\pi^{w_0} : \mathcal{Y}^{w_0} \rightarrow \mathcal{P}^{\hat{L}}$, les composés des inverses de ces isomorphismes et des projections évidentes de $\hat{\mathfrak{n}} \times \mathcal{P}^{\hat{L}}$, resp. $\hat{N} \times \mathcal{P}^{\hat{L}}$, sur $\mathcal{P}^{\hat{L}}$. De ces applications se déduisent des isomorphismes:

$$\pi^{1*} : H^{\hat{B}}(\mathcal{P}^{\hat{L}}, \mathcal{L}) \xrightarrow{\sim} H^{\hat{B}}(\mathcal{Y}^1, \mathcal{L}),$$
$$\pi^{w_0*} : H^{\hat{B}}(\mathcal{P}^{\hat{L}}, \mathcal{L}) \xrightarrow{\sim} H^{\hat{B}}(\mathcal{Y}^{w_0}, \mathcal{L}).$$

On a des applications naturelles:

$$\mathcal{Y}^1 \xrightarrow{i_{\mathcal{Y}}} \mathcal{Y} \xleftarrow{j_{\mathcal{Y}}} \mathcal{Y}^{w_0}.$$

L'application $i_{\mathcal{Y}}$ est une immersion fermée et $j_{\mathcal{Y}}$ est une immersion ouverte. On en déduit des applications:

$$i_{\mathcal{Y}!} : H^{\hat{B}}(\mathcal{Y}^1, \mathcal{L}) \to H^{\hat{B}}(\mathcal{Y}, \mathcal{L}),$$
$$j_{\mathcal{Y}}^* : H^{\hat{B}}(\mathcal{Y}, \mathcal{L}) \to H^{\hat{B}}(\mathcal{Y}^{w_0}, \mathcal{L}).$$

Considérons le diagramme évident:

$$\begin{array}{ccccc} \Gamma \times_{\hat{B}} \mathcal{Y} & \xrightarrow{\dot{k}} & \Gamma \times_{\hat{L}_{\mathbb{C}}} \dot{\mathfrak{q}} & \xrightarrow{\dot{h}_{\mathfrak{q}}} & \Gamma \times_{\hat{G}_{\mathbb{C}}} \dot{\mathfrak{g}} \\ \downarrow p_{\mathcal{Y}} & & \downarrow \tilde{p} & & \downarrow p \\ \Gamma \times_{\hat{B}} (y + \hat{\mathfrak{n}}) & \xrightarrow{k} & \Gamma \times_{\hat{L}_{\mathbb{C}}} \hat{\mathfrak{q}} & \xrightarrow{h_{\mathfrak{q}}} & \Gamma \times_{\hat{G}_{\mathbb{C}}} \hat{\mathfrak{g}} \end{array}$$

Chacun des carrés est cartésien, donc $h_{\mathfrak{q}}^* p_! \mathcal{L}^* = \tilde{p}_! \mathcal{L}^*$, $(h_{\mathfrak{q}} \circ k)^* p_! \mathcal{L}^* = p_{\mathcal{Y}!} \mathcal{L}^*$. Par fonctorialité par $h_{\mathfrak{q}}^*$, resp. $(h_{\mathfrak{q}} \circ k)^*$, l'action de $W_N$ sur $p_! \mathcal{L}^*$ définie en 2.5 se transporte en une action de $W_N$ sur $\tilde{p}_! \mathcal{L}^*$, resp. $p_{\mathcal{Y}!} \mathcal{L}^*$. Comme en 2.5, de cette action sur $p_{\mathcal{Y}!} \mathcal{L}^*$ se déduit une action de $W_N$ sur $H^{\hat{B}}(\mathcal{Y}, \mathcal{L})$, que nous identifions à une action $\rho_{\mathcal{Y}}$ du sous-groupe $\{t_w; w \in W_N\}$ de $\bar{\mathcal{H}}$. Pour $w \in W_N$, resp. $w \in W_N^{\hat{L}}$, les applications suivantes sont des endomorphismes de $H^{\hat{B}}(\mathcal{P}^{\hat{L}}, \mathcal{L})$: $(\pi^{w_0*})^{-1} \circ j_{\mathcal{Y}}^* \circ \rho_{\mathcal{Y}}(t_w) \circ i_{\mathcal{Y}!} \circ \pi^{1*}$, resp. $\rho^{\hat{L}}(t_w)$. Nous prouverons ci-dessous les deux relations:

(5) pour tout $w \in W_N \setminus w_0 W_N^{\hat{L}}$, $(\pi^{w_0*})^{-1} \circ j_{\mathcal{Y}}^* \circ \rho_{\mathcal{Y}}(t_w) \circ i_{\mathcal{Y}!} \circ \pi^{1*} = 0$;

(6) pour tout $w \in W_N^{\hat{L}}$, $(\pi^{w_0*})^{-1} \circ j_{\mathcal{Y}}^* \circ \rho_{\mathcal{Y}}(t_{w_0 w}) \circ i_{\mathcal{Y}!} \circ \pi^{1*} = \rho^{\hat{L}}(t_w)$.

Admettons-les et démontrons la proposition. Considérons le diagramme:

$$\begin{array}{ccccc} \mathcal{Y}^1 & \xrightarrow{i_{\mathcal{Y}}} & \mathcal{Y} & \xleftarrow{j_{\mathcal{Y}}} & \mathcal{Y}^{w_0} \\ \uparrow \alpha^1 & & \uparrow \alpha & & \uparrow \alpha^{w_0} \\ \mathcal{P}^{\hat{L}} & \xrightarrow{i} & \mathcal{P} & \xleftarrow{j} & \mathcal{P}^{w_0} \end{array}$$

où $\alpha$ et $\alpha^{w_0}$ sont définis par $g\hat{P} \mapsto (y, g\hat{P})$ et $\alpha^1$ l'est par $h(\hat{L} \cap \hat{P}) \mapsto (y, h\hat{P})$. Les carrés sont cartésiens et $\alpha$, $\alpha^1$, $\alpha^{w_0}$ sont des immersions fermées. On en déduit un diagramme:

$$\begin{array}{ccccc} H^{\hat{B}}(\mathcal{Y}^1, \mathcal{L}) & \xrightarrow{i_{\mathcal{Y}!}} & H^{\hat{B}}(\mathcal{Y}, \mathcal{L}) & \xrightarrow{j_{\mathcal{Y}}^*} & H^{\hat{B}}(\mathcal{Y}^{w_0}, \mathcal{L}) \\ \uparrow \alpha_!^1 & & \uparrow \alpha_! & & \uparrow \alpha_!^{w_0} \\ H^{\hat{B}}(\mathcal{P}^{\hat{L}}, \mathcal{L}) & \xrightarrow{i_!} & H^{\hat{B}}(\mathcal{P}, \mathcal{L}) & \xrightarrow{j^*} & H^{\hat{B}}(\mathcal{P}^{w_0}, \mathcal{L}) \end{array}$$

Il résulte des définitions que ses carrés sont commutatifs. Il résulte aussi des définitions que, pour tout $w \in W_N$, on a l'égalité $\rho_{\mathcal{Y}}(t_w) \circ \alpha_! = \alpha_! \circ \rho(t_w)$. Introduisons les endomorphismes de $H^{\hat{B}}(\mathcal{P}^{\hat{L}}, \mathcal{L})$ suivants:

$$D^1 = (\pi^{1*})^{-1} \circ \alpha_!^1, \quad D^{w_0} = (\pi^{w_0*})^{-1} \circ \alpha_!^{w_0} \circ \pi^*.$$

Pour $w \in W_N$, l'égalité:

$$(7) \quad D^{w_0} \circ (\pi^{1*})^{-1} \circ j^* \circ \rho(t_w) \circ i_! = (\pi^{w_0*})^{-1} \circ j_{\mathcal{Y}}^* \circ \rho_{\mathcal{Y}}(t_w) \circ i_{\mathcal{Y}!} \circ \pi^{1*} \circ D^1$$

résulte formellement des propriétés ci-dessus. Les endomorphismes $D^1$ et $D^{w_0}$ se calculent aisément. Grâce à (3), les applications $\alpha^1$ et $\pi^1$ s'identifient aux applications évidentes:

$$\mathcal{P}^{\hat{L}} \xrightarrow{\alpha^1} \hat{\mathfrak{n}} \times \mathcal{P}^{\hat{L}} \xrightarrow{\pi^1} \mathcal{P}^{\hat{L}}.$$

D'après [L3] 1.10, $D^1$ est donc la multiplication par un élément de $H_{\hat{B}}$, que nous notons encore $D^1$. En un point $(\sigma, r_0) \in \hat{\mathfrak{b}}$, $D^1(\sigma, r_0)$ est le déterminant de l'action de $(\sigma, r_0)$ sur $\hat{\mathfrak{n}}$. On se rappelle que $\hat{B}$ agit sur $\hat{\mathfrak{n}}$ par $(h, \lambda)v = \lambda^{-2}Ad(h)v$. Donc:

$$D^1(\sigma, r_0) = det(ad(\sigma) - 2r_0|\hat{\mathfrak{n}}).$$

Grâce à (4), les applications $\pi$, $\alpha^{w_0}$ et $\pi^{w_0}$ s'identifient aux applications évidentes:

$$\mathcal{P}^{\hat{L}} \xleftarrow{\pi} Z_{\hat{N}}(y) \times \mathcal{P}^{\hat{L}} \xrightarrow{\alpha^{w_0}} \hat{N} \times \mathcal{P}^{\hat{L}} \xrightarrow{\pi^{w_0}} \mathcal{P}^{\hat{L}}.$$

Par l'exponentielle, on peut remplacer $\hat{N}$ et $Z_{\hat{N}}(y)$ par leurs algèbres de Lie $\hat{\mathfrak{n}}$ et $\mathfrak{z}_{\hat{N}}(y)$. En identifiant $\hat{\mathfrak{n}}/\mathfrak{z}_{\hat{N}}(y)$ à un supplémentaire $\hat{B}$-invariant de $\mathfrak{z}_{\hat{N}}(y)$ dans $\hat{\mathfrak{n}}$, on peut remplacer la suite d'applications ci-dessus par:

$$\mathcal{P}^{\hat{L}} \xleftarrow{\pi} \mathfrak{z}_{\hat{N}}(y) \times \mathcal{P}^{\hat{L}} \xrightarrow{\alpha^{w_0}} (\hat{\mathfrak{n}}/\mathfrak{z}_{\hat{N}}(y)) \times \mathfrak{z}_{\hat{N}}(y) \times \mathcal{P}^{\hat{L}} \xrightarrow{\beta} \mathfrak{z}_{\hat{N}}(y) \times \mathcal{P}^{\hat{L}} \xrightarrow{\pi} \mathcal{P}^{\hat{L}}.$$

Comme ci-dessus, $(\beta^*)^{-1} \circ \alpha_!^{w_0}$ est la multiplication par un élément de $H_{\hat{B}}$, que nous notons encore $D^{w_0}$. En un point $(\sigma, r_0) \in \hat{\mathfrak{b}}$, $D^{w_0}(\sigma, r_0)$ est le déterminant de l'action de $(\sigma, r_0)$ sur $\hat{\mathfrak{n}}/\mathfrak{z}_{\hat{N}}(y)$. On se rappelle que $\hat{B}$ agit sur $\hat{N}$ par $(h, \lambda)n = hnh^{-1}$. Donc:

$$D^{w_0}(\sigma, r_0) = det(ad(\sigma)|\hat{\mathfrak{n}})\, det(ad(\sigma)|\mathfrak{z}_{\hat{N}}(y))^{-1}.$$

Puisque notre endomorphisme originel $D^{w_0}$ est égal à $(\pi^*)^{-1} \circ (\beta^*)^{-1} \circ \alpha_!^{w_0} \circ \pi^*$, il est aussi égal à la multiplication par le $D^{w_0}$ ci-dessus.

On a l'égalité $D^1 = \Delta D^{w_0}$. Le $H_{\hat{B}}$-module $H^{\hat{B}}(\mathcal{P}^{\hat{L}}, \mathcal{L})$ étant projectif ([L3] proposition 8.6), on peut diviser l'égalité (7) par $D^{w_0}$ et on obtient l'égalité:

$$(\pi^{1*})^{-1} \circ j^* \circ \rho(t_w) \circ i_! = \Delta \circ (\pi^{w_0 *})^{-1} \circ j_{\mathcal{Y}}^* \circ \rho_{\mathcal{Y}}(t_w) \circ i_{\mathcal{Y}!} \circ \pi^{1*}.$$

La proposition résulte alors de (5) et (6), qu'il nous reste à démontrer.

Considérons le diagramme commutatif:

(8)

$$\begin{array}{ccccc}
\Gamma \times_{\hat{B}} (\hat{\mathfrak{n}} \times \mathcal{P}^{\hat{L}}) = \Gamma \times_{\hat{B}} \mathcal{Y}^1 & \overset{id_\Gamma \times_{\hat{B}} i_{\mathcal{Y}}}{\longrightarrow} & \Gamma \times_{\hat{B}} \mathcal{Y} & \overset{id_\Gamma \times_{\hat{B}} j_{\mathcal{Y}}}{\longleftarrow} & \Gamma \times_{\hat{B}} \mathcal{Y}^{w_0} = \Gamma \times_{\hat{B}} (\hat{N} \times \mathcal{P}^{\hat{L}}) \\
\downarrow id_\Gamma \times_{\hat{B}} \pi^1 & & \downarrow p_{\mathcal{Y}} & & \downarrow id_\Gamma \times_{\hat{B}} \pi^{w_0} \\
\Gamma \times_{\hat{B}} \mathcal{P}^{\hat{L}} & & \Gamma \times_{\hat{B}} (y + \hat{\mathfrak{n}}) & & \Gamma \times_{\hat{B}} \mathcal{P}^{\hat{L}} \\
& \searrow p_y^{\hat{L}} & \downarrow s & p_y^{\hat{L}} \swarrow & \\
& & \Gamma \times_{\hat{B}} \{y\} & &
\end{array}$$

Parce que $\pi^1$ et $\pi^{w_0}$ sont des fibrations en espaces vectoriels, on a $(id_\Gamma \times_{\hat{B}} \pi^1)_! \mathcal{L}^* = \mathcal{L}^*[-2\delta]$, $(id_\Gamma \times_{\hat{B}} \pi^{w_0})_! \mathcal{L}^* = \mathcal{L}^*[-2\delta]$, où $\delta = dim_{\mathbb{C}}(\hat{\mathfrak{n}}) = dim_{\mathbb{C}}(\hat{N})$. Parce que $i_{\mathcal{Y}}$ est une immersion fermée et $j_{\mathcal{Y}}$ une immersion ouverte, on dispose d'homomorphismes naturels:

$$(id_\Gamma \times_{\hat{B}} j_{\mathcal{Y}})_! \mathcal{L}^* = (id_\Gamma \times_{\hat{B}} j_{\mathcal{Y}})_! (id_\Gamma \times_{\hat{B}} j_{\mathcal{Y}})^! \mathcal{L}^* \rightarrow \mathcal{L}^*$$
$$\rightarrow (id_\Gamma \times_{\hat{B}} i_{\mathcal{Y}})_* (id_\Gamma \times_{\hat{B}} i_{\mathcal{Y}})^* \mathcal{L}^* = (id_\Gamma \times_{\hat{B}} i_{\mathcal{Y}})_! \mathcal{L}^*.$$

En appliquant $(s \circ p_{\mathcal{Y}})_!$, on en déduit des homomorphismes:

$$p_{y!}^{\hat{L}} (id_\Gamma \times_{\hat{B}} \pi^{w_0})_! \mathcal{L}^* = p_{y!}^{\hat{L}} \mathcal{L}^*[-2\delta] \overset{\epsilon^{w_0}}{\longrightarrow} s_! p_{\mathcal{Y}!} \mathcal{L}^* \overset{\epsilon^1}{\longrightarrow} p_{y!}^{\hat{L}} \mathcal{L}^*[-2\delta]$$
$$= p_{y!}^{\hat{L}} (id_\Gamma \times_{\hat{B}} \pi^1)_! \mathcal{L}^*.$$

On a défini plus haut une action de $W_N$ sur $p_{\mathcal{Y}!}\mathcal{L}^*$. Par fonctorialité par $s_!$, on en déduit une action de $W_N$ sur $s_! p_{\mathcal{Y}!} \mathcal{L}^*$, que l'on note $\underline{\rho}_{\mathcal{Y}}$. Soit $w \in W_N$. Alors $\epsilon^1 \circ \underline{\rho}_{\mathcal{Y}}(w^{-1}) \circ \epsilon^{w_0}$ est un endomorphisme de $p_{y!}^{\hat{L}} \mathcal{L}^*[-2\delta]$. On en déduit pour tout entier $j \geq 0$ un endomorphisme de $H_c^j(\Gamma \times_{\hat{B}} \{y\}, p_{y!}^{\hat{L}} \mathcal{L}^*)$, i.e. de $H_c^j(\Gamma \times_{\hat{B}} \mathcal{P}^{\hat{L}}, \mathcal{L}^*)$. Par dualité, on en déduit pour tout entier $j \geq 0$ un endomorphisme de $H_j^{\hat{B}}(\mathcal{P}^{\hat{L}}, \mathcal{L})$. Il résulte des définitions que cet endomorphisme est égal à $(\pi^{w_0 *})^{-1} \circ j_{\mathcal{Y}}^* \circ \rho_{\mathcal{Y}}(t_w) \circ i_{\mathcal{Y}!} \circ \pi^{1*}$. De la même façon, en remplaçant $\hat{G}$ par $\hat{L}$ dans les définitions de 2.5, on dispose d'une action de $W_N^{\hat{L}}$ sur $p_{y!}^{\hat{L}} \mathcal{L}^*$, ou, si l'on préfère, sur $p_{y!}^{\hat{L}} \mathcal{L}^*[-2\delta]$. Notons $\underline{\rho}^{\hat{L}}$ cette action. Pour $w \in W_N^{\hat{L}}$, l'endomorphisme $\rho^{\hat{L}}(t_w)$ se déduit de $\underline{\rho}^{\hat{L}}(w^{-1})$ par le même procédé que ci-dessus. Alors (5) et (6) résultent des assertions suivantes concernant des endomorphismes de $p_{y!}^{\hat{L}} \mathcal{L}^*[-2\delta]$:

(9) pour tout $w \in W_N \setminus W_N^{\hat{L}} w_0$, $\epsilon^1 \circ \underline{\rho}_{\mathcal{Y}}(w) \circ \epsilon^{w_0} = 0$;

(10) pour tout $w \in W_N^{\hat{L}}$, $\epsilon^1 \circ \underline{\rho}_{\mathcal{Y}}(w w_0) \circ \epsilon^{w_0} = \underline{\rho}^{\hat{L}}(w)$.

Considérons le diagramme commutatif:

(11)

$$\begin{array}{ccccc}
\Gamma \times_{\hat{L}_{\mathbb{C}}} (\hat{\mathfrak{n}} \times \dot{\mathfrak{l}}) = \Gamma \times_{\hat{L}_{\mathbb{C}}} \dot{\mathfrak{q}}^1 & \stackrel{\tilde{i}}{\longrightarrow} & \Gamma \times_{\hat{L}_{\mathbb{C}}} \dot{\mathfrak{q}} & \stackrel{\tilde{j}}{\longleftarrow} & \Gamma \times_{\hat{L}_{\mathbb{C}}} \dot{\mathfrak{q}}^{w_0} = \Gamma \times_{\hat{L}_{\mathbb{C}}} (\hat{N} \times \dot{\mathfrak{l}}) \\
\downarrow \tilde{\pi}^1 & & \downarrow \tilde{p} & & \downarrow \tilde{\pi}^{w_0} \\
\Gamma \times_{\hat{L}_{\mathbb{C}}} \dot{\mathfrak{l}} & & \Gamma \times_{\hat{L}_{\mathbb{C}}} \hat{\mathfrak{q}} & & \Gamma \times_{\hat{L}_{\mathbb{C}}} \dot{\mathfrak{l}} \\
 & \searrow p^{\hat{L}} & \downarrow \tilde{s} & p^{\hat{L}} \swarrow & \\
 & & \Gamma \times_{\hat{L}_{\mathbb{C}}} \hat{\mathfrak{l}} & &
\end{array}$$

Les mêmes procédés que ci-dessus permettent de définir:

- des endomorphismes:

$$p_!^{\hat{L}} \mathcal{L}^*[-2\delta] \stackrel{\tilde{\epsilon}^{w_0}}{\longrightarrow} \tilde{s}_! \tilde{p}_! \mathcal{L}^* \stackrel{\tilde{\epsilon}^1}{\longrightarrow} p_!^{\hat{L}} \mathcal{L}^*[-2\delta];$$

- une action $\underline{\tilde{\rho}}$ de $W_N$ sur $\tilde{s}_! \tilde{p}_! \mathcal{L}^*$;
- une action $\underline{\tilde{\rho}}^{\hat{L}}$ de $W_N^{\hat{L}}$ sur $p_!^{\hat{L}} \mathcal{L}^*[-2\delta]$.

Or le diagramme (8) est l'image réciproque, en un sens évident, du diagramme (11) par l'application naturelle:

$$\Gamma \times_{\hat{B}} \{y\} \stackrel{h^{\hat{L}}}{\longrightarrow} \Gamma \times_{\hat{L}_{\mathbb{C}}} \hat{\mathfrak{l}}.$$

Et les objets $\epsilon^{w_0}$, $\epsilon^1$, $\underline{\rho}_y$, $\underline{\rho}^{\hat{L}}$ se déduisent par fonctorialité par $h^{\hat{L}*}$ des objets $\tilde{\epsilon}^{w_0}$, $\tilde{\epsilon}^1$, $\underline{\tilde{\rho}}$, $\underline{\tilde{\rho}}^{\hat{L}}$. Alors (9) et (10) résultent des assertions suivantes:

(12) $\qquad$ pour tout $w \in W_N \setminus W_N^{\hat{L}} w_0$, $\tilde{\epsilon}^1 \circ \underline{\tilde{\rho}}(w) \circ \tilde{\epsilon}^{w_0} = 0$;

(13) $\qquad$ pour tout $w \in W_N^{\hat{L}}$, $\tilde{\epsilon}^1 \circ \underline{\tilde{\rho}}(w w_0) \circ \tilde{\epsilon}^{w_0} = \underline{\tilde{\rho}}^{\hat{L}}(w)$.

Rappelons que $p_!^{\hat{L}} \mathcal{L}^*$ est un faisceau pervers et que c'est le prolongement d'intersection d'un faisceau lisse sur la variété:

$$\Gamma \times_{\hat{L}_{\mathbb{C}}} \{Ad(x)(\sigma + y_c);\ x \in \hat{L},\ \sigma \in \hat{\mathfrak{t}},\ Z_{\hat{L}}(\sigma) = \hat{M}\}.$$

On peut fixer $\gamma \in \Gamma$ et $s \in \hat{\mathfrak{t}}$ tel que $Z_{\hat{G}}(\sigma) = \hat{M}$ et se contenter de vérifier (12) et (13) sur la fibre de $p_!^{\hat{L}} \mathcal{L}^*[-2\delta]$ au-dessus de $(\gamma, \sigma + y_c)$. L'application:

$$\begin{array}{rcl}
\hat{N} \times W_N & \rightarrow & \Gamma \times_{\hat{L}_{\mathbb{C}}} \dot{\mathfrak{q}} \\
(n, w) & \mapsto & (\gamma, Ad(n)(\sigma + y_c), nw\hat{P})
\end{array}$$

définit un isomorphisme de $\hat{N} \times W_N$ sur la fibre de $\tilde{s} \circ \tilde{p}$ au-dessus de $(\gamma, \sigma + y_c)$. Les fibres de $p^{\hat{L}} \circ \tilde{\pi}^1$, resp. $p^{\hat{L}} \circ \tilde{\pi}^{w_0}$, s'identifient à $\hat{N} \times W_N^{\hat{L}}$, resp. $\hat{N} \times W_N^{\hat{L}} w_0$, les applications $\tilde{i}$ et $\tilde{j}$ étant les injections évidentes. Les fibres de $\tilde{s}_! \tilde{p}_! \mathcal{L}^*$, resp. $p_!^{\hat{L}} \mathcal{L}^*[-2\delta]$, au-dessus de $(\gamma, \sigma + y_c)$ s'identifient donc à $\mathbb{C}[-2\delta]^{W_N}$, resp.

$\mathbb{C}[-2\delta]^{W_N^{\hat{L}}}$. Les applications:

$$\tilde{\epsilon}^1 : \mathbb{C}[-2\delta]^{W_N} \to \mathbb{C}[-2\delta]^{W_N^{\hat{L}}},$$
$$\tilde{\epsilon}^{w_0} : \mathbb{C}[-2\delta]^{W_N^{\hat{L}}} \to \mathbb{C}[-2\delta]^{W_N},$$

se décrivent ainsi:

$$\tilde{\epsilon}^1((z_w)_{w\in W_N}) = (z_w)_{w\in W_N^{\hat{L}}},$$
$$\tilde{\epsilon}^{w_0}((z_w)_{w\in W_N^{\hat{L}}}) = (z'_w)_{w\in W_N},$$

où $z'_w = 0$ si $w \notin W_N^{\hat{L}} w_0$, $z'_{ww_0} = z_w$ si $w \in W_N^{\hat{L}}$. Pour $w \in W_N$, resp. $w \in W_N^{\hat{L}}$, on a décrit en 2.5 les actions $\underline{\tilde{\rho}}(w)$, resp. $\underline{\tilde{\rho}}^{\hat{L}}(w)$. On a:

$$\underline{\tilde{\rho}}(w)((z_{w'})_{w'\in W_N}) = (z_{w'w})_{w'\in W_N},$$
$$\text{resp. } \underline{\tilde{\rho}}^{\hat{L}}(w)((z_{w'})_{w'\in W_N}) = (z_{w'w})_{w'\in W_N^{\hat{L}}}.$$

Alors (12) et (13) deviennent évidents. Cela achève la démonstration.

**2.10.** De l'automorphisme $Ad(w_0)$ de $\hat{L}$ se déduit un automorphisme de $\bar{\mathcal{H}}^{\hat{L}}$, encore noté $Ad(w_0)$:

- pour $x \in X_N \simeq X^*(\hat{T})$, on a $Ad(w_0)(\xi_x) = \xi_{w_0(x)}$;
- pour $w \in W_N^{\hat{L}}$, on a $Ad(w_0)(t_w) = t_{w_0 w w_0^{-1}}$.

Si $\rho^{\hat{L}}$ est une action de $\bar{\mathcal{H}}^{\hat{L}}$ dans un espace complexe, $\rho^{\hat{L}} \circ Ad(w_0)$ est encore une action de $\bar{\mathcal{H}}^{\hat{L}}$ dans le même espace.

On note $\rho$ l'action de $\bar{\mathcal{H}}$ sur $H^{\hat{B}}(\mathcal{P}, \mathcal{L})$ (cf. 2.7) et $\rho_{|\bar{\mathcal{H}}^{\hat{L}}}$ sa restriction à $\bar{\mathcal{H}}^{\hat{L}}$. On note $\rho^{\hat{L}}$ l'action de $\bar{\mathcal{H}}^{\hat{L}}$ sur $H^{\hat{B}}(\mathcal{P}^{\hat{L}}, \mathcal{L})$.

LEMME. (i) *L'homomorphisme:*

$$(\pi^*)^{-1} \circ j^* : H^{\hat{B}}(\mathcal{P}, \mathcal{L}) \to H^{\hat{B}}(\mathcal{P}^{\hat{L}}, \mathcal{L})$$

*entrelace les actions* $\rho_{|\bar{\mathcal{H}}^{\hat{L}}}$ *et* $\rho^{\hat{L}} \circ Ad(w_0)$.

(ii) *L'homomorphisme:*

$$i_! : H^{\hat{B}}(\mathcal{P}^{\hat{L}}, \mathcal{L}) \to H^{\hat{B}}(\mathcal{P}, \mathcal{L})$$

*entrelace les actions* $\rho^{\hat{L}}$ *et* $\rho_{|\bar{\mathcal{H}}^{\hat{L}}}$.

*Preuve.* Pour tout élément semi-simple $(\sigma, r_0) \in \hat{\mathfrak{b}}$, on dispose d'un homomorphisme d'évaluation $\chi_{\sigma,r_0} : H_{\hat{B}} \to \mathbb{C}$. Si $\mathcal{X}$ est un $H_{\hat{B}}$-module, on pose $\mathcal{X}_{\sigma,r_0} = \mathbb{C} \otimes_{H_{\hat{B}}} \mathcal{X}$, la tensorisation se faisant via $\chi_{\sigma,r_0}$. De même, si $\alpha : \mathcal{X} \to \mathcal{Y}$ est un homomorphisme de $H_{\hat{B}}$-modules, on note $\alpha_{\sigma,r_0} : \mathcal{X}_{\sigma,r_0} \to \mathcal{Y}_{\sigma,r_0}$ l'homomorphisme spécialisé. Posons pour simplifier:

$$\mathcal{X} = H^{\hat{B}}(\mathcal{P}, \mathcal{L}), \ \mathcal{Y} = H^{\hat{B}}(\mathcal{P}^{\hat{L}}, \mathcal{L}), \ \alpha = (\pi^*)^{-1} \circ j^*.$$

Fixons un ensemble de représentants $\mathcal{U}$ de l'ensemble de classes $W_N^{\hat{L}} \backslash W_N$. Définissons une application:

$$\begin{aligned} \tilde{\alpha} : \mathcal{X} &\to \mathcal{Y}^{\mathcal{U}} \\ h &\mapsto (\alpha \circ \rho(t_{w_0 w})(h))_{w \in \mathcal{U}}. \end{aligned}$$

C'est un homomorphisme de $H_{\hat{B}}$-modules. Il vérifie la propriété:

(1) si $(\sigma, r_0)$ est un élément semi-simple de $\hat{\mathfrak{b}}$ tel que $\Delta(\sigma, r_0) \neq 0$, alors $\tilde{\alpha}_{\sigma,r_0}$ est surjectif.

En effet, soit $(h_w)_{w \in \mathcal{U}} \in \mathcal{Y}^{\mathcal{U}}_{\sigma,r_0}$. Posons:

(2) $$h = \Delta(\sigma, r_0)^{-1} \sum_{w \in \mathcal{U}} (\rho(t_{w^{-1}}) \circ i_!)_{\sigma,r_0}(h_w).$$

Alors $\tilde{\alpha}_{\sigma,r_0}(h) = (h'_w)_{w \in \mathcal{U}}$, où, pour tout $w \in \mathcal{U}$,

$$h'_w = \Delta(\sigma, r_0)^{-1} \sum_{w' \in \mathcal{U}} (\alpha \circ \rho(t_{w_0 w w'^{-1}}) \circ i_!)_{\sigma,r_0}(h_{w'}).$$

Soient $w, w' \in \mathcal{U}$. Si $w \neq w'$, on a $w_0 w w'^{-1} \notin w_0 W_N^{\hat{L}}$ et $\alpha \circ \rho(t_{w_0 w w'^{-1}}) \circ i_! = 0$. Si $w = w'$, $(\alpha \circ \rho(t_{w_0 w w'^{-1}}) \circ i_!)_{\sigma,r_0} = \Delta(\sigma, r_0)$. Alors $h'_w = h_w$ pour tout $w \in \mathcal{U}$, et cela démontre (1).

D'autre part:

(3) pour tout élément semi-simple $(\sigma, r_0)$ de $\hat{\mathfrak{b}}$, on a $dim_{\mathbb{C}}(\mathcal{X}_{\sigma,r_0}) = dim_{\mathbb{C}}(\mathcal{Y}^{\mathcal{U}}_{\sigma,r_0})$. Parce que $\mathcal{X}$ et $\mathcal{Y}$ sont des $H_{\hat{B}}$-modules projectifs de type fini, ces dimensions ne dépendent pas du point $(\sigma, r_0)$. On peut supposer que $(\sigma, r_0)$ est régulier dans $\hat{\mathfrak{b}}$ et que l'application $Ad(\sigma)$ définit un automorphisme de $\hat{\mathfrak{n}}$. Notons $\hat{D}$ le plus petit sous-tore de $\hat{B}$ dont l'algèbre de Lie contient $(\sigma, r_0)$ et notons $\mathcal{P}^{\hat{D}}$, resp. $\mathcal{P}^{\hat{L},\hat{D}}$, le sous-ensemble des invariants par $\hat{D}$ dans $\mathcal{P}$, resp. $\mathcal{P}^{\hat{L}}$. Reprenons la démonstration du lemme 2.6. On a les isomorphismes:

$$\mathcal{Y}_{\sigma,r_0} \simeq \mathbb{C} \otimes_{H_{\hat{D}}} H^{\hat{D}}(\mathcal{P}^{\hat{L},\hat{D}}, \mathcal{L}), \ \mathcal{X}_{\sigma,r_0} \simeq \mathbb{C} \otimes_{H_{\hat{D}}} H^{\hat{D}}(\mathcal{P}^{\hat{D}}, \mathcal{L}),$$

la tensorisation se faisant via l'homomorphisme $H_{\hat{D}} \to \mathbb{C}$ d'évaluation en $(\sigma, r_0)$. Remarquons que $\mathcal{U}$ est un système de représentants de l'ensemble des doubles classes $w \in W^{\hat{L}} \backslash W / W^{\hat{M}}$ telles que $w \hat{M} w^{-1} \subseteq \hat{L}$. Grâce à 2.6 (3) et (5), et avec les notations de ce paragraphe, on a:

$$\mathcal{P}^{\hat{D}} = \bigcup_{w \in \mathcal{U}} \mathcal{P}^{\hat{D}}_w.$$

C'est une décomposition en union disjointe et chaque sous-variété $\mathcal{P}^{\hat{D}}_w$ est fermée dans $\mathcal{P}^{\hat{D}}$ (car $\hat{L} w \hat{P}$ est fermé dans $\hat{G}$). Donc:

$$\mathcal{X}_{\sigma,r_0} \simeq \oplus_{w \in \mathcal{U}} \big(\mathbb{C} \otimes_{H_{\hat{D}}} H^{\hat{D}}(\mathcal{P}^{\hat{D}}_w, \mathcal{L})\big).$$

Enfin, pour tout $w \in \mathcal{U}$, l'application $h(\hat{L} \cap \hat{P}) \mapsto hw\hat{P}$ est un isomorphisme de $\mathcal{P}^{\hat{L},\hat{D}}$ sur $\mathcal{P}_w^{\hat{D}}$. Donc:

$$\mathcal{X}_{\sigma,r_0} \simeq \left(\mathbb{C} \otimes_{H_{\hat{D}}} H^{\hat{D}}(\mathcal{P}^{\hat{L},\hat{D}}, \mathcal{L})\right)^{\mathcal{U}} \simeq \mathcal{Y}^{\mathcal{U}}_{\sigma,r_0}.$$

Cela démontre (3).

De (1) et (3) résulte bien sûr:

(4) si $(\sigma, r_0)$ est un élément semi-simple de $\hat{\mathfrak{b}}$ tel que $\Delta(\sigma, r_0) \neq 0$, alors $\tilde{\alpha}_{\sigma,r_0}$ est bijectif.

Fixons un tel élément $(\sigma, r_0)$. Puisque $\tilde{\alpha}_{\sigma,r_0}$ est bijectif, il a un inverse, qui est donné par la formule (2). Supposons que $w_0 \in \mathcal{U}$, ainsi qu'il est loisible. Soit $h \in \mathcal{X}_{\sigma,r_0}$, posons $\tilde{\alpha}_{\sigma,r_0}(h) = (h_w)_{w\in\mathcal{U}}$. On a $h_{w_0} = \alpha_{\sigma,r_0}(h)$. Soit $w \in W_N^{\hat{L}}$. Grâce à (2), on a:

$$(\alpha \circ \rho(t_w))_{\sigma,r_0}(h) = \Delta(\sigma, r_0)^{-1} \sum_{w'\in\mathcal{U}} (\alpha \circ \rho(t_{ww'^{-1}}) \circ i_!)_{\sigma,r_0}(h_{w'}).$$

Si $w' \in \mathcal{U}$ et $w' \neq w_0$, on a $ww'^{-1} \notin w_0 W_N^{\hat{L}}$ et $\alpha \circ \rho(t_{ww'^{-1}}) \circ i_! = 0$. Si $w' = w_0$, on a $ww'^{-1} = w_0 w_0 w w_0$ et:

$$\alpha \circ \rho(t_{ww'^{-1}}) \circ i_! = \Delta \circ \rho^{\hat{L}}(t_{w_0 w w_0}).$$

D'où les égalités:

$$(5) \qquad (\alpha \circ \rho(t_w))_{\sigma,r_0}(h) = \rho^{\hat{L}}(t_{w_0 w w_0})_{\sigma,r_0}(h_{w_0}) = \left(\rho^{\hat{L}}(t_{w_0 w w_0}) \circ \alpha\right)_{\sigma,r_0}(h).$$

Soit maintenant $x \in X_N$. Grâce à (2), on a:

$$(\alpha \circ \rho(\xi_x))_{\sigma,r_0}(h) = \Delta(\sigma, r_0)^{-1} \sum_{w\in\mathcal{U}} (\alpha \circ \rho(\xi_x t_{w^{-1}}) \circ i_!)_{\sigma,r_0}(h_w).$$

Soit $w \in \mathcal{U}$. Dans l'algèbre $\bar{\mathcal{H}}$, on a une égalité:

$$\xi_x t_{w^{-1}} = t_{w^{-1}} \xi_{w(x)} + \sum_{w'} t_{w'} \varphi_{w',x},$$

où:

- $w'$ parcourt l'ensemble des éléments de $W_N$ de longueur $\ell(w') < \ell(w)$;
- pour tout tel $w'$, $\varphi_{w',x} \in \mathcal{S}$.

Grâce au lemme 2.7, on a donc:

$$\alpha \circ \rho(\xi_x t_{w^{-1}}) \circ i_! = \alpha \circ \rho(t_{w^{-1}}) \circ i_! \rho^{\hat{L}}(\xi_{w(x)}) + \sum_{w'} \alpha \circ \rho(t_{w'}) \circ i_! \circ \rho^{\hat{L}}(\varphi_{w',x}).$$

Pour un élément $w'$ intervenant dans cette dernière somme, on a $\ell(w') < \ell(w^{-1}) \leq \ell(w_0)$. Donc $w' \notin w_0 W_N^{\hat{L}}$ et $\alpha \circ \rho(t_{w'}) \circ i_! = 0$. Si $w \neq w_0$, on a aussi $\alpha \circ \rho(t_{w^{-1}}) \circ i_! = 0$. Si $w = w_0$, alors $\alpha \circ \rho(t_{w^{-1}}) \circ i_! = \Delta$. Finalement:

$$(6) \qquad (\alpha \circ \rho(\xi_x))_{\sigma,r_0}(h) = \rho^{\hat{L}}(\xi_{w(x)})_{\sigma,r_0}(h_{w_0}) = (\rho^{\hat{L}}(\xi_{w(x)}) \circ \alpha)_{\sigma,r_0}(h).$$

Les relations (5) et (6) montrent que $\alpha_{\sigma,r_0}$ entrelace les représentations de $\bar{\mathcal{H}}^{\hat{L}}$ dans $\mathcal{X}_{\sigma,r_0}$ et $\mathcal{Y}_{\sigma,r_0}$. Parce que $\mathcal{X}$ et $\mathcal{Y}$ sont des $H_{\hat{B}}$-modules projectifs, donc sans torsion, il en résulte que $\alpha$ lui-même est un entrelacement. Cela démontre la première assertion du lemme. La démonstration de la seconde assertion est analogue, on la laisse au lecteur. Notons d'ailleurs que l'on a déjà fait la moitié du travail en 2.7.

**2.11.** Soit $(\rho, E)$ un $\bar{\mathcal{H}}^{\hat{L}}$-module de dimension finie. On note $Ind_{\bar{\mathcal{H}}^{\hat{L}}}^{\bar{\mathcal{H}}}(E)$ l'espace des fonctions $f : \bar{\mathcal{H}} \to E$ telles que $f(h'h) = \rho(h')f(h)$ pour tous $h \in \bar{\mathcal{H}}$, $h' \in \bar{\mathcal{H}}^{\hat{L}}$. De l'action de $\bar{\mathcal{H}}$ sur lui-même par multiplication à droite se déduit une action de $\bar{\mathcal{H}}$ sur $Ind_{\bar{\mathcal{H}}^{\hat{L}}}^{\bar{\mathcal{H}}}(E)$. On définit aussi l'espace $\bar{\mathcal{H}} \otimes_{\bar{\mathcal{H}}^{\hat{L}}} E$, quotient de $\bar{\mathcal{H}} \otimes_{\mathbb{C}} E$ par le sous-espace engendré par les éléments $hh' \otimes e - h \otimes \rho(h')e$, pour $h \in \bar{\mathcal{H}}$, $h' \in \bar{\mathcal{H}}^{\hat{L}}$, $e \in E$. De l'action de $\bar{\mathcal{H}}$ sur lui-même par multiplication à gauche se déduit une action de $\bar{\mathcal{H}}$ sur $\bar{\mathcal{H}} \otimes_{\bar{\mathcal{H}}^{\hat{L}}} E$.

D'autre part, on note $Ad(w_0) \circ E$ le $\bar{\mathcal{H}}^{\hat{L}}$-module ainsi défini: en tant qu'espace vectoriel complexe, $Ad(w_0) \circ E = E$; l'action de $\bar{\mathcal{H}}^{\hat{L}}$ sur $Ad(w_0) \circ E$ est $\rho \circ Ad(w_0)$ (cf. 2.10).

Soit $(\sigma, r_0)$ un élément semi-simple de l'algèbre de Lie $\hat{\mathfrak{b}}$. On a défini en 2.1 un ensemble $\mathcal{E}_k(\sigma, r_0, y)$ de caractères du groupe $\bar{A}(\sigma, y)$ et, pour $\epsilon \in \mathcal{E}_k(\sigma, r_0, y)$, un $\bar{\mathcal{H}}$-module $\mathbb{M}(\sigma, r_0, y, \epsilon)$. Des constructions analogues s'appliquent en remplaçant $\hat{G}$ par $\hat{L}$: on définit un ensemble $\mathcal{E}_k^{\hat{L}}(\sigma, r_0, y)$ de caractères de $\bar{A}^{\hat{L}}(\sigma, y)$ et, pour tout $\epsilon^{\hat{L}}$ dans cet ensemble, un $\bar{\mathcal{H}}^{\hat{L}}$-module $\mathbb{M}^{\hat{L}}(\sigma, r_0, y, \epsilon^{\hat{L}})$.

Remarquons que, de l'injection naturelle $Z_{\hat{L}}(\sigma, y) \to Z_{\hat{G}}(\sigma, y)$, se déduit une injection:

$$\bar{A}^{\hat{L}}(\sigma, y) \to \bar{A}(\sigma, y).$$

Pour tout caractère $\epsilon$ de $\bar{A}(\sigma, y)$, on note $\epsilon_{|\bar{A}^{\hat{L}}(\sigma,y)}$ le composé de $\epsilon$ et de cette injection.

Proposition. *Soit $(\sigma, r_0)$ un élément semi-simple de l'algèbre de Lie $\hat{\mathfrak{b}}$.*

(i) *Pour tout caractère $\epsilon$ de $\bar{A}(\sigma, y)$, $\epsilon$ appartient à $\mathcal{E}_k(\sigma, r_0, y)$ si et seulement si $\epsilon_{|\bar{A}^{\hat{L}}(\sigma,y)}$ appartient à $\mathcal{E}_k^{\hat{L}}(\sigma, r_0, y)$.*

(ii) *Supposons que $(\sigma, r_0)$ contracte $\hat{\mathfrak{q}}$ (cf. 2.8), soit $\epsilon^{\hat{L}} \in \mathcal{E}_k^{\hat{L}}\sigma, r_0, y)$. Alors les trois $\bar{\mathcal{H}}$-modules suivants sont isomorphes:*

$$Ind_{\bar{\mathcal{H}}^{\hat{L}}}^{\bar{\mathcal{H}}}\left(Ad(w_0) \circ \mathbb{M}^{\hat{L}}\left(\sigma, r_0, y, \epsilon^{\hat{L}}\right)\right), \quad \bar{\mathcal{H}} \otimes_{\bar{\mathcal{H}}^{\hat{L}}} \mathbb{M}^{\hat{L}}\left(\sigma, r_0, y, \epsilon^{\hat{L}}\right), \quad \oplus_\epsilon \mathbb{M}(\sigma, r_0, y, \epsilon),$$

*où l'on somme sur les caractères $\epsilon$ de $\bar{A}(\sigma, y)$ tels que $\epsilon_{|\bar{A}^{\hat{L}}(\sigma,y)} = \epsilon^{\hat{L}}$.*

*Preuve.* Supposons d'abord $(\sigma, r_0) = (0, 0)$. Dans ce cas, $\bar{A}(0, y) = \bar{A}(y)$ (cf.2.1) et, pour un caractère $\epsilon$ de ce groupe, $\epsilon$ appartient à $\mathcal{E}_k(0, 0, y)$ si et seulement si $k$ est l'entier associé au couple $(y, \epsilon)$ par la correspondance de Springer généralisée (cf. [L5]). Il résulte des propriétés relatives à l'induction de cette correspondance ([L5], théorème 8.3) que cet entier se conserve par induction, i.e., $\epsilon \in \mathcal{E}_k(0,0,y) \iff \epsilon_{|\bar{A}^{\hat{L}}(0,y)} \in \mathcal{E}_k^{\hat{L}}(0,0,y)$. Passons au cas général. On a un diagramme commutatif d'homomorphismes injectifs:

$$\begin{array}{ccc} \bar{A}^{\hat{L}}(\sigma,y) & \rightarrow & \bar{A}(\sigma,y) \\ \downarrow & & \downarrow \\ \bar{A}^{\hat{L}}(y) & \rightarrow & \bar{A}(y) \end{array}$$

Soit $\epsilon$ un caractère de $\bar{A}(\sigma,y)$, fixons un caractère $\epsilon'$ de $\bar{A}(y)$ tel que $\epsilon'_{|\bar{A}(\sigma,y)} = \epsilon$, avec une notation évidente. On a alors les équivalences:

$$\epsilon \in \mathcal{E}_k(\sigma,r_0,y) \iff \epsilon' \in \mathcal{E}_k(0,0,y) \iff \epsilon'_{|\bar{A}^{\hat{L}}(y)} \in \mathcal{E}_k^{\hat{L}}(0,0,y)$$

$$\iff \epsilon_{|\bar{A}^{\hat{L}}(\sigma,y)} \in \mathcal{E}_k^{\hat{L}}(\sigma,r_0,y).$$

La deuxième équivalence est le cas particulier déjà traité. Les deux autres résultent de [L4] propositions 8.16 et 8.17. Cela démontre (i).

Plaçons-nous sous les hypothèses de (ii). Posons pour simplifier:

$$\mathcal{X} = H^{\hat{B}}(\mathcal{P},\mathcal{L}),\ \mathcal{Y} = H^{\hat{B}}(\mathcal{P}^{\hat{L}},\mathcal{L}),\ \alpha = (\pi^*)^{-1}\circ j^*,\ \beta = i_!.$$

D'après le lemme précédent, les applications:

$$\mathcal{Y} \overset{\beta}{\rightarrow} \mathcal{X} \overset{\alpha}{\rightarrow} Ad(w_0)\circ\mathcal{Y}$$

sont des homomorphismes de $\bar{\mathcal{H}}^{\hat{L}}$-modules. On en déduit des homomorphismes de $\bar{\mathcal{H}}$-modules:

$$\bar{\mathcal{H}} \otimes_{\bar{\mathcal{H}}^{\hat{L}}} \mathcal{Y} \overset{\tilde{\beta}}{\rightarrow} \mathcal{X} \overset{\tilde{\alpha}}{\rightarrow} Ind_{\bar{\mathcal{H}}^{\hat{L}}}^{\bar{\mathcal{H}}}(Ad(w_0)\circ\mathcal{Y})$$

par les formules suivantes, où $\rho$ désigne l'action de $\bar{\mathcal{H}}$ sur $\mathcal{X}$:

- pour $\varphi \in \bar{\mathcal{H}}$ et $h' \in \mathcal{Y}$, $\tilde{\beta}(\varphi \otimes h') = \rho(\varphi)\beta(h')$;
- pour $h \in \mathcal{X}$ et $\varphi \in \bar{\mathcal{H}}$, $\tilde{\alpha}(h)(\varphi) = (\alpha \circ \rho(\varphi))(h)$.

En définissant $\mathcal{U}$ comme dans la preuve précédente, on a des isomorphismes d'espaces vectoriels:

$$\begin{array}{ccc} Ind_{\bar{\mathcal{H}}^{\hat{L}}}^{\bar{\mathcal{H}}}(Ad(w_0)\circ\mathcal{Y}) & \rightarrow & \mathcal{Y}^{\mathcal{U}} \\ f & \mapsto & (f(t_{w_0 w}))_{w\in\mathcal{U}} \\ \mathcal{Y}^{\mathcal{U}} & \rightarrow & \bar{\mathcal{H}} \otimes_{\bar{\mathcal{H}}^{\hat{L}}} \mathcal{Y} \\ (h_w)_{w\in\mathcal{U}} & \mapsto & \sum_{w\in\mathcal{U}} t_{w^{-1}} \otimes h_w. \end{array}$$

Il résulte alors de 2.10 (4) et du lemme 2.8 (ii) que l'application spécialisée $\tilde{\alpha}_{\sigma,r_0}$ est un isomorphisme. De même, on a vu au cours de la preuve précédente que l'application définie par 2.10 (2) était un isomorphisme. Il en est donc de même de $\tilde{\beta}_{\sigma,r_0}$. Le groupe $\bar{A}^{\hat{L}}(\sigma,y)$ agit naturellement sur $\mathcal{X}_{\sigma,r_0}$, $Ind_{\bar{\mathcal{H}}^{\hat{L}}}^{\bar{\mathcal{H}}}(Ad(w_0)\circ\mathcal{Y}_{\sigma,r_0})$ et

$\bar{\mathcal{H}} \otimes_{\bar{\mathcal{H}}^{\hat{L}}} \mathcal{Y}_{\sigma,r_0}$. Par construction, $\tilde{\alpha}_{\sigma,r_0}$ et $\tilde{\beta}_{\sigma,r_0}$ entrelacent ces actions. Donc $\tilde{\alpha}_{\sigma,r_0}$ et $\tilde{\beta}_{\sigma,r_0}$ se restreignent en des isomorphismes entre les parties isotypiques de type $\epsilon^{\hat{L}}$ de leurs espaces de départ et d'arrivée. Or ces parties isotypiques sont les modules de l'énoncé.

**2.12.** Avec les notations de 1.4, posons:

$$\check{\Pi}_N = \{\check{e}_1 - \check{e}_2, \ldots, \check{e}_{N-1} - \check{e}_N, 2\check{e}_N\}.$$

On note ${}^+C(Y_N)$ le sous-cône ouvert de $Y_N \otimes_{\mathbb{Z}} \mathbb{R}$ engendré par $\check{\Pi}_N$ et ${}^+\bar{C}(Y_N)$ sa clôture. On définit aussi le cône positif aigu:

$$C^+(Y_N) = \{\mu \in Y_N \otimes_{\mathbb{Z}} \mathbb{R}; \forall \alpha \in \Pi_N, < \alpha, \mu >> 0\},$$

et sa clôture $\bar{C}^+(Y_N)$. Plus généralement, soit $J$ un sous-ensemble de $\{1, \ldots, N\}$. On note:

- ${}^+C^J$ le sous-cône des $\mu \in Y_N \otimes_{\mathbb{Z}} \mathbb{R}$ qui s'écrivent $\mu = \sum_{j \in J} \mu_j \check{\alpha}_j$, avec des $\alpha_j > 0$;
- $C_J^+$ le sous-cône des $\mu \in Y_N \otimes_{\mathbb{Z}} \mathbb{R}$ tels que, pour tout $j \in \{1, \ldots, N\}$, on ait $< \alpha_j, \mu >= 0$ si $j \in J$ et $< \alpha_j, \mu >> 0$ si $j \notin J$;
- ${}^+\bar{C}^J$ et $\bar{C}_J^+$ les clôtures de ${}^+C^J$ et $C_J^+$;
- $W_N^J$ le sous-groupe de $W_N$ engendré par les symétries élémentaires $w_j$ pour $j \in J$;
- $W_{N,J}$ l'ensemble des $w \in W_N$ de longueur minimale dans leur classe $W_N^J w$.

On munit l'espace $Y_N \otimes_{\mathbb{Z}} \mathbb{R}$ de l'ordre partiel pour lequel:

$$\mu_1 \leq \mu_2 \iff \mu_2 - \mu_1 \in {}^+\bar{C}(Y_N).$$

D'après le lemme de Langlands, pour tout $\mu \in Y_N \otimes_{\mathbb{Z}} \mathbb{R}$, il existe un unique ensemble $J(\mu) \subseteq \{1, \ldots, N\}$ tel que l'on puisse écrire $\mu = \mu_+ - \mu_-$, avec $\mu_+ \in \bar{C}_{J(\mu)}^+$ et $\mu_- \in {}^+C^{J(\mu)}$. L'application:

$$\begin{array}{ccc} Y_N \otimes_{\mathbb{Z}} \mathbb{R} \to & \bar{C}^+(Y_N) \\ \mu & \mapsto \mu_+ \end{array}$$

est croissante. De plus, pour tout $\mu \in Y_N \otimes_{\mathbb{Z}} \mathbb{R}$ et tout $w \in W_{N,J(\mu)}$, on a l'inégalité $(w^{-1}(\mu))_+ \leq \mu_+$.

Soit $(\pi, E)$ une représentation de dimension finie de $\bar{\mathcal{H}}$. Nous dirons que $\pi$ est anti-tempérée, resp. anti-discrète, si et seulement si tout exposant $(\sigma, r_0) \in \mathcal{E}xp(\pi)$ (cf. 2.5) vérifie les relations:

- $r_0 \in \mathbb{R}$ et $r_0 > 0$;
- $\sigma \in {}^+\bar{C}(Y_N)$, resp. $\sigma \in {}^+C(Y_N)$.

**2.13.** Soient $\sigma$, resp. $y$, un élément semi-simple, resp. nilpotent, de $\hat{\mathfrak{g}}$, et $r_0$ un réel $> 0$. On suppose:

- $[\sigma, y] = 2r_0 y$;

(1) $$H^{\hat{A}(y)}(\mathcal{P}_y, \mathcal{L})_{\sigma,r_0} \neq \{0\}.$$

On sait que l'on peut trouver un homomorphisme $\phi : SL(2, \mathbb{C}) \to \hat{G}$ et un élément semi-simple $z$ de $\hat{\mathfrak{g}}$ tels que:

- $z$ commute à l'image de $\phi$;
- $d\phi\begin{pmatrix} 0 & 1 \\ 0 & 0 \end{pmatrix} = y$;
- $\sigma = z + d\phi\begin{pmatrix} r_0 & 0 \\ 0 & -r_0 \end{pmatrix}$.

Fixons de telles données et fixons un sous-tore maximal $\hat{T}_L$ du centralisateur dans $\hat{G}$ de l'image de $\phi$, tel que $z \in \hat{\mathfrak{t}}_L$. Notons $\hat{L}$ le commutant de $\hat{T}_L$ dans $\hat{G}$. C'est un groupe de Lévi de $\hat{G}$, c'est-à-dire un sous-groupe de Lévi d'un sous-groupe parabolique de $\hat{G}$.

*Remarque.* On a $y \in \hat{\mathfrak{l}}$. On vérifie que si $\hat{L}_1$ est un groupe de Lévi de $\hat{G}$ tel que $y \in \hat{\mathfrak{l}}_1$, alors il existe $g \in Z_{\hat{G}}(y)^0$ tel que $g\hat{L}g^{-1} \subseteq \hat{L}_1$. En particulier $\hat{L}$ est minimal dans l'ensemble des groupes de Lévi $\hat{L}_1$ tels que $y \in \hat{\mathfrak{l}}_1$.

Le tore $\hat{T}_L$ est le plus grand tore central dans $\hat{L}$. Pour toute racine $\alpha$ de $\hat{T}_L$ dans $\hat{\mathfrak{g}}$, notons $\hat{\mathfrak{g}}_\alpha$ le sous-espace correspondant, et identifions $\alpha$ à une forme linéaire sur $\hat{\mathfrak{t}}_L$. Notons $\hat{Q}_{temp}$ le sous-groupe parabolique de $\hat{G}$ dont l'algèbre de Lie est la somme de $\hat{\mathfrak{l}}$ et des $\hat{\mathfrak{g}}_\alpha$ pour les $\alpha$ tels que $Re(\alpha(z)) \leq 0$. Fixons un sous-groupe parabolique $\hat{Q}$ de $\hat{G}$, de sous-groupe de Lévi $\hat{L}$, tel que $\hat{Q} \subseteq \hat{Q}_{temp}$. On a fixé en 2.4 un sous-groupe de Borel $\hat{P}_0$ et un sous-tore maximal $\hat{T}_0$ de $\hat{P}_0$. Quitte à conjuguer $\sigma$ et $y$, on peut supposer $\hat{P}_0 \subseteq \hat{Q}$ et $\hat{T}_0 \subseteq \hat{L}$. D'après notre hypothèse (1) et la proposition 2.6, on a alors $\hat{P} \subseteq \hat{Q}$, $\hat{M} \subseteq \hat{L}$ et l'inclusion $\hat{\mathfrak{t}}_L \subseteq \hat{\mathfrak{t}} \simeq Y_N \otimes_{\mathbb{Z}} \mathbb{C}$. Il existe un unique sous-ensemble $J \subseteq \{1, \ldots, N\}$ tel que, par cette inclusion, $\hat{\mathfrak{t}}_L$ s'identifie à l'ensemble des $\mu \in Y_N \otimes_{\mathbb{Z}} \mathbb{C}$ tels que $< \alpha_j, \mu >= 0$ pour tout $j \in J$.

L'élément $z$ de $\hat{\mathfrak{t}}_L$ s'identifie à un élément de $Y_N \otimes_{\mathbb{Z}} \mathbb{C}$. On note $Re(z)$ sa partie réelle dans $Y_N \otimes_{\mathbb{Z}} \mathbb{R}$ et on pose:

$$\tau(\sigma, y) = -z.$$

Par construction, $\tau(\sigma, y) \in \bar{C}_J^+ \subseteq \bar{C}^+(Y_N)$. On vérifie que cet élément $\tau(\sigma, y)$ ne dépend pas des choix effectués dans la construction ci-dessus.

On renvoie à 2.2 pour les notations utilisées ci-dessous.

LEMME. *Soient $\sigma$, $y$, $y'$ tels que les couples $(\sigma, y)$ et $(\sigma, y')$ vérifient tous deux les hypothèses précédentes. Supposons $\mathfrak{o}_y^{\sigma,r_0} \subseteq \bar{\mathfrak{o}}_{y'}^{\sigma,r_0}$. Alors $\tau(\sigma, y) \geq \tau(\sigma, y')$. Supposons de plus $\mathfrak{o}_y^{\sigma,r_0} \neq \mathfrak{o}_{y'}^{\sigma,r_0}$. Alors $\tau(\sigma, y) > \tau(\sigma, y')$.*

*Preuve.* De même que l'on a identifié $\hat{\mathfrak{t}}$ à $Y_N \otimes_{\mathbb{Z}} \mathbb{C}$, on peut identifier $\hat{\mathfrak{t}}_0$ à $Y_{d/2} \otimes_{\mathbb{Z}} \mathbb{C}$. Le plongement de $\hat{\mathfrak{t}}$ dans $\hat{\mathfrak{t}}_0$ s'identifie à un plongement:

$$Y_N \otimes_{\mathbb{Z}} \mathbb{C} \to Y_{d/2} \otimes_{\mathbb{Z}} \mathbb{C}, \tag{2}$$

dont on peut supposer qu'il se déduit de l'injection naturelle $\{1, \dots, N\} \to \{1, \dots, d/2\}$. De même que l'on a défini $\tau(\sigma, y)$, on peut définir $\tau_0(\sigma, y) \in \bar{C}^+(Y_{d/2})$. Cette construction ne fait pas intervenir l'entier $k$, en particulier n'utilise pas l'hypothèse (1) qui ne nous a servi que pour plonger $\hat{\mathfrak{t}}_L$ dans $\hat{\mathfrak{t}}$. Mais par construction, $\tau(\sigma, y) = \tau_0(\sigma, y)$. De plus, par le plongement (2), l'ordre sur $Y_N \otimes_{\mathbb{Z}} \mathbb{R}$ s'identifie à la restriction de l'ordre sur $Y_{d/2} \otimes_{\mathbb{Z}} \mathbb{R}$. On peut donc pour notre problème remplacer $\tau(\sigma, y)$ et $\tau(\sigma, y')$ par $\tau_0(\sigma, y)$ et $\tau_0(\sigma, y')$, et $Y_N \otimes_{\mathbb{Z}} \mathbb{R}$ par $Y_{d/2} \otimes_{\mathbb{Z}} \mathbb{R}$, autrement dit supposer $k = 0$.

Débarrassées de $k$, nos définitions s'étendent à tout groupe réductif connexe sur $\mathbb{C}$. Montrons que l'analogue de l'énoncé pour le groupe $GL(\hat{V})$ entraîne notre énoncé pour le groupe $\hat{G}$. On a un plongement naturel $\hat{G} \to GL(\hat{V})$. On peut fixer un sous-groupe de Borel $\tilde{P}$ de $GL(\hat{V})$ et un sous-tore maximal $\tilde{T}$ de $\tilde{P}$ de sorte que $\hat{P} = \tilde{P} \cap \hat{G}$, $\hat{T} = \tilde{T} \cap \hat{G}$ (rappelons que l'on suppose $k = 0$, donc $\hat{P} = \hat{P}_0$, $\hat{T} = \hat{T}_0$). Posons $\tilde{Y} = X_*(\tilde{T})$. On définit les cônes $C^+(\tilde{Y})$, ${}^+C(\tilde{Y})$, etc... et un ordre sur $\tilde{Y} \otimes_{\mathbb{Z}} \mathbb{R}$. L'espace $Y_N \otimes_{\mathbb{Z}} \mathbb{R}$ se plonge dans $\tilde{Y} \otimes_{\mathbb{Z}} \mathbb{R}$ et l'ordre sur $Y_N \otimes_{\mathbb{Z}} \mathbb{R}$ n'est autre que la restriction de l'ordre sur $\tilde{Y} \otimes_{\mathbb{Z}} \mathbb{R}$. De même que l'on a défini $\tau(\sigma, y) \in \bar{C}^+(Y_N)$, on définit un élément $\tilde{\tau}(\sigma, y) \in \bar{C}^+(\tilde{Y})$. Mais, via le plongement précédent, on a l'égalité $\tau(\sigma, y) = \tilde{\tau}(\sigma, y)$. De même que l'on a défini la $Z_{\hat{G}}(\sigma)$-orbite $\mathfrak{o}_y^{\sigma, r_0}$, on définit la $Z_{GL(\hat{V})}(\sigma)$-orbite $\tilde{\mathfrak{o}}_y^{\sigma, r_0}$. Il est clair que si $\mathfrak{o}_y^{\sigma, r_0} \subseteq \bar{\mathfrak{o}}_{y'}^{\sigma, r_0}$, alors $\tilde{\mathfrak{o}}_y^{\sigma, r_0} \subseteq \bar{\tilde{\mathfrak{o}}}_{y'}^{\sigma, r_0}$. Alors, si la première assertion de l'énoncé est vraie pour le groupe $GL(\hat{V})$, elle l'est aussi pour $\hat{G}$. Pour appliquer le même raisonnement à la deuxième assertion, on doit démontrer la propriété moins immédiate:

$$\text{si } \mathfrak{o}_y^{\sigma, r_0} \neq \mathfrak{o}_{y'}^{\sigma, r_0}, \text{ alors } \tilde{\mathfrak{o}}_y^{\sigma, r_0} \neq \tilde{\mathfrak{o}}_{y'}^{\sigma, r_0}. \tag{3}$$

Fixons $z$ et $\phi$ comme dans les constructions précédant l'énoncé, avec $d\phi(\left(\begin{smallmatrix} 0 & 1 \\ 0 & 0 \end{smallmatrix}\right)) = y$, et de même $z'$ et $\phi'$ avec $d\phi'(\left(\begin{smallmatrix} 0 & 1 \\ 0 & 0 \end{smallmatrix}\right)) = y'$. Supposons $\tilde{\mathfrak{o}}_y^{\sigma, r_0} = \tilde{\mathfrak{o}}_{y'}^{\sigma, r_0}$. Fixons $\tilde{h} \in Z_{GL(\hat{V})}(\sigma)$ tel que $y' = Ad(\tilde{h})y$. Les données $Ad(\tilde{h})z$ et $Ad(\tilde{h})\phi$ vérifient, pour le groupe $GL(\hat{V})$, les mêmes conditions que $z'$ et $\phi'$. Il en résulte (cf. [L6] proposition 3.3) qu'il existe $\tilde{h}' \in Z_{GL(\hat{V})}(\sigma)$ tel que $Ad(\tilde{h}'\tilde{h})z = z'$ et $Ad(\tilde{h}'\tilde{h})\phi = \phi'$.

On a des isomorphismes:

$$Z_{\hat{G}}(\sigma) \simeq GL(N_1, \mathbb{C}) \times \cdots \times GL(N_a, \mathbb{C}) \times \hat{G}^0,$$

$$Z_{GL(\hat{V})}(\sigma) \simeq (GL(N_1, \mathbb{C}) \times GL(N_1, \mathbb{C})) \times \cdots \times (GL(N_a, \mathbb{C}) \times GL(N_a, \mathbb{C})) \times GL(\hat{V}_0),$$

où $\hat{V}_0$ est un sous-espace non dégénéré de $\hat{V}$ et $\hat{G}_0$ est son groupe symplectique. Le plongement $Z_{\hat{G}}(\sigma) \to Z_{GL(\hat{V})}(\sigma)$ est le produit du plongement naturel $\hat{G}_0 \to$

$GL(\hat{V}_0)$ et des plongements:

$$\begin{array}{ccc} GL(N_i,\mathbb{C}) & \to & GL(N_i,\mathbb{C}) \times GL(N_i,\mathbb{C}) \\ g & \mapsto & (g, {}^t g^{-1}). \end{array}$$

Les deux éléments $z, z'$ appartiennent à l'algèbre de Lie $\mathfrak{z}_{\hat{G}}(\sigma)$ et sont semi-simples. Ils sont dans la même orbite pour l'action de $Z_{GL(\hat{V})}(\sigma)$. De la description ci-dessus résulte qu'ils sont aussi dans la même orbite pour l'action de $Z_{\hat{G}}(\sigma)$. Soit donc $h \in Z_{\hat{G}}(\sigma)$ tel que $Ad(h)z = z'$. Puisque $Ad(h)\sigma = \sigma$, on a aussi $Ad(h)d\phi((\begin{smallmatrix} r_0 & 0 \\ 0 & -r_0\end{smallmatrix})) = Ad(h)d\phi'((\begin{smallmatrix} r_0 & 0 \\ 0 & -r_0\end{smallmatrix}))$. Notons $\hat{H}'$ le commutant de $z'$ dans $\hat{G}$. Alors $Ad(h)\phi$ et $\phi'$ sont deux $SL(2)$-triplets à valeurs dans $\hat{H}'$, tels que les valeurs de leurs dérivées en $(\begin{smallmatrix} r_0 & 0 \\ 0 & -r_0\end{smallmatrix})$ soient égales. Il en résulte que ces $SL(2)$-triplets sont dans la même orbite pour l'action de $\hat{H}'$, cf. [C] proposition 5.6.4. Soit donc $h' \in \hat{H}'$ tel que $Ad(h'h)\phi = \phi'$. On a alors $Ad(h'h)d\phi((\begin{smallmatrix} r_0 & 0 \\ 0 & -r_0\end{smallmatrix})) = d\phi'((\begin{smallmatrix} r_0 & 0 \\ 0 & -r_0\end{smallmatrix}))$, i.e., $Ad(h'h)\sigma = \sigma$, donc $h'h \in Z_{\hat{G}}(\sigma)$. On a aussi:

$$Ad(h'h)y = Ad(h'h)\,d\phi\left(\begin{pmatrix} 0 & 1 \\ 0 & 0 \end{pmatrix}\right) = d\phi'\left(\begin{pmatrix} 0 & 1 \\ 0 & 0 \end{pmatrix}\right) = y'.$$

Donc $\mathfrak{o}_y^{\sigma,r_0} = \mathfrak{o}_{y'}^{\sigma,r_0}$. Cela démontre (3).

On est maintenant ramené au cas du groupe $GL(\hat{V})$. Les objets se décrivent de façon élémentaire. On identifie $\tilde{Y} \otimes_{\mathbb{Z}} \mathbb{C}$ à $\mathbb{C}^d$, $\tilde{Y} \otimes_{\mathbb{Z}} \mathbb{R}$ à $\mathbb{R}^d$ et $\bar{C}^+(\tilde{Y})$ au cône des $\underline{x} = (x_1, \ldots, x_d) \in \mathbb{R}^d$ tels que $x_1 \geq x_2 \geq \cdots \geq x_d$. Pour $\underline{x} = (x_1, \ldots, x_d)$, $\underline{x}' = (x_1', \ldots, x_d') \in \mathbb{R}^d$, on a $\underline{x} \leq \underline{x}'$ si et seulement si:

$$(d-i)(x_1 + \cdots + x_i) - i(x_{i+1} + \cdots + x_d) \leq (d-i)(x_1' + \cdots + x_i') - i(x_{i+1}' + \cdots + x_d')$$

pour tout $i \in \{1, \ldots, d-1\}$. Appelons segment un sous-ensemble de $\mathbb{C}$ de la forme:

$$\Delta = \{a, a+1, \ldots, a+\ell-1\},$$

où $a \in \mathbb{C}$ et $\ell \in \mathbb{N}$. Avec ces notations, on pose $\ell(\Delta) = \ell$ et $c(\Delta) = a + \frac{\ell-1}{2}$ si $\ell \neq 0$. Appelons multi-ensemble de segments une famille finie $(\Delta_k)_{k\in K}$ de segments. On identifie deux familles qui ne diffèrent que par leur indexation, ainsi que deux familles qui ne diffèrent que par des segments vides. Notons $\mathcal{D}(\sigma)$ l'ensemble des multi-ensembles de segments $(\Delta_k)_{k\in K}$ tels que, pour tout $x \in \mathbb{C}$, le nombre d'éléments de l'ensemble $\{k \in K; x \in \Delta_k\}$ soit égal à la dimension de l'espace propre de $\sigma$ associé à la valeur propre $2xr_0$. On définit un ordre dans $\mathcal{D}(\sigma)$. Il est engendré par la relation élémentaire suivante. Soient $\underline{\Delta} = (\Delta_k)_{k\in K} \in \mathcal{D}(\sigma)$ et $k, k' \in K$, avec $k \neq k'$. Supposons:

- $\Delta_k \not\subseteq \Delta_{k'}$ et $\Delta_{k'} \not\subseteq \Delta_k$;
- $\Delta_k \cup \Delta_{k'}$ et $\Delta_k \cap \Delta_{k'}$ sont des segments.

Notons $\underline{\Delta}'$ la famille réunion, en un sens évident, de $(\Delta_{k''})_{k'' \in K \setminus \{k,k'\}}$ et de $(\Delta_k \cup \Delta_{k'}, \Delta_k \cap \Delta_{k'})$. Alors $\underline{\Delta} < \underline{\Delta}'$.

Pour $\underline{\Delta} = (\Delta_k)_{k \in K} \in \mathcal{D}(\sigma)$, on définit $\tau(\underline{\Delta}) = (x_1, \ldots, x_d) \in \bar{C}^+(\tilde{Y})$ par la propriété: pour tout $x \in \mathbb{R}$, on a l'égalité:

$$|\{i \in \{1, \ldots, d\}; x_i = x\}| = \sum \ell(\Delta_k),$$

où l'on somme sur les $k \in K$ tels que $\ell(\Delta_k) \neq 0$ et $Re(c(\Delta_k)) = x$.

Il résulte de [Z] qu'il existe une bijection $\mathfrak{o} \mapsto \underline{\Delta}(\mathfrak{o})$ définie sur l'ensemble des $Z_{GL(\hat{V})}(\sigma)$-orbites dans l'espace:

$$\mathfrak{gl}(\hat{V})^{\sigma, r_0} = \{y'' \in \mathfrak{gl}(\hat{V}); [\sigma, y''] = 2r_0 y''\},$$

à valeurs dans $\mathcal{D}(\sigma)$, telle que:

- si $\mathfrak{o} \subseteq \bar{\mathfrak{o}}'$, alors $\underline{\Delta}(\mathfrak{o}) \leq \underline{\Delta}(\mathfrak{o}')$;
- si $\mathfrak{o}$ est l'orbite de $y'' \in \mathfrak{gl}(\hat{V})^{\sigma, r_0}$, alors $\tau(\sigma, y'') = \tau(\underline{\Delta}(\mathfrak{o}))$.

Pour achever la démonstration, il reste à montrer que si $\underline{\Delta}, \underline{\Delta}' \in \mathcal{D}(\sigma)$ vérifient $\underline{\Delta} < \underline{\Delta}'$, alors $\tau(\underline{\Delta}) > \tau(\underline{\Delta}')$. On peut supposer que $\underline{\Delta}'$ se déduit de $\underline{\Delta}$ par une opération élémentaire décrite ci-dessus. La preuve est alors facile. Cela achève la démonstration.

**2.14.** On fixe jusqu'à la fin du paragraphe 2 deux éléments $\sigma$ et $y$ de $\hat{\mathfrak{g}}$ et un réel $r_0 > 0$. On suppose vérifiées les hypothèses du paragraphe précédent, c'est-à-dire:

- $\sigma$ est semi-simple et $y$ est nilpotent;
- $[\sigma, y] = 2r_0 y$;
- $H^{\hat{A}(y)}(\mathcal{P}_y, \mathcal{L})_{\sigma, r_0} \neq \{0\}$.

Cette dernière hypothèse est équivalente à: $\mathcal{E}_k(\sigma, r_0, y) \neq \emptyset$.

COROLLAIRE. *Supposons $\tau(\sigma, r_0) = 0$. Alors, pour tout $\epsilon \in \mathcal{E}_k(\sigma, r_0, y)$, le $\bar{\mathcal{H}}$-module $\mathbb{M}(\sigma, r_0, y, \epsilon)$ est irréductible et égal à $\mathbb{P}(\sigma, r_0, y, \epsilon)$.*

*Preuve.* D'après le lemme précédent, il n'existe pas de $y' \in \hat{\mathfrak{g}}^{\sigma, r_0}$ tel que $\mathfrak{o}_y^{\sigma, r_0} \subset \bar{\mathfrak{o}}_{y'}^{\sigma, r_0}$ et $\mathfrak{o}_y^{\sigma, r_0} \neq \mathfrak{o}_{y'}^{\sigma, r_0}$. Il reste à appliquer l'égalité de 2.2.

**2.15.**

LEMME. *Supposons que $y$ n'est contenu dans aucune sous-algèbre de Lévi d'une sous-algèbre parabolique propre de $\hat{\mathfrak{g}}$. Alors, pour tout $\epsilon \in \mathcal{E}_k(\sigma, r_0, y)$, le $\bar{\mathcal{H}}$-module $\mathbb{M}(\sigma, r_0, y, \epsilon)$ est anti-discret.*

*Preuve.* Fixons $\epsilon \in \mathcal{E}_k(\sigma, r_0, y)$. On doit prouver que, si $(\mu, r_0) \in \mathcal{E}xp(\mathbb{M}(\sigma, r_0, y, \epsilon))$, alors $Re(\mu) \in {}^+C(Y_N)$. En 2.4, on a identifié $\hat{\mathfrak{a}}_c$ et $(Y_N \otimes_{\mathbb{Z}} \mathbb{C}) \oplus \mathbb{C}$ par une application $(s, \lambda) \mapsto (s_T, \lambda)$. Soit $(s, r_0) \in \hat{\mathfrak{a}}_c$, posons $\mu = s_T$. On a défini en 1.8 l'homomorphisme d'évaluation $\ell_{\mu,r_0} : \mathcal{S} \to \mathbb{C}$. Supposons $(\mu, r_0) \in \mathcal{E}xp(\mathbb{M}(\sigma, r_0, y, \epsilon))$. Par définition:

$$\mathbb{C} \otimes_{\mathcal{S}} \mathbb{M}(\sigma, r_0, y, \epsilon) \neq \{0\},$$

le produit tensoriel se faisant via $\ell_{\mu,r_0}$. Par construction, l'espace précédent est quotient de:

$$\mathbb{C} \otimes_{\mathcal{S}} H^{\hat{A}(y)}(\mathcal{P}_y, \mathcal{L}).$$

En 2.4 et 2.5, on a identifié $\mathcal{S}$ à $H_{\hat{A}_c}$ et $H^{\hat{A}(y)}(\mathcal{P}_y, \mathcal{L})$ à un espace $H^{\hat{A}_c}(\tilde{\mathcal{V}}, \mathcal{L})^{\bar{A}_c}$. L'homomorphisme $\ell_{\mu,r_0}$ s'identifie à l'homomorphisme d'évaluation $\chi_{s,r_0} : H_{\hat{A}_c} \to \mathbb{C}$. Alors l'espace ci-dessus est quotient de:

$$\mathbb{C} \otimes_{H_{\hat{A}_c}} H^{\hat{A}_c}(\tilde{\mathcal{V}}, \mathcal{L}), \tag{1}$$

le produit tensoriel se faisant via $\chi_{s,r_0}$. Cet espace est donc non nul.

Notons $\hat{D}$ le plus petit sous-tore de $\hat{A}_c$ dont l'algèbre de Lie contient $(s, r_0)$ et notons $\tilde{\mathcal{V}}^{\hat{D}}$ le sous-ensemble des invariants par $\hat{D}$ dans $\tilde{\mathcal{V}}$. D'après [L4] proposition 4.4(a), l'espace (1) est égal à: $\mathbb{C} \otimes_{H_{\hat{A}_c}} H^{\hat{A}_c}(\tilde{\mathcal{V}}^{\hat{D}}, \mathcal{L})$. Cet espace étant non nul, $\tilde{\mathcal{V}}^{\hat{D}}$ est non vide. Notons $\mathfrak{o}_y$ l'orbite de $y$ pour l'action adjointe de $\hat{G}$. La projection:

$$\begin{array}{rl} \tilde{\mathcal{V}} & \to (y_c + \hat{\mathfrak{u}}) \cap \mathfrak{o}_y \\ (g, \lambda, g'\hat{P}) \mapsto & \lambda^{-2} Ad(g) y \end{array}$$

entrelace les actions de $\hat{D}$. Puisque $\tilde{\mathcal{V}}^{\hat{D}}$ est non vide, l'ensemble des invariants $((y_c + \hat{\mathfrak{u}}) \cap \mathfrak{o}_y)^{\hat{D}}$ ne l'est pas non plus. Quitte à conjuguer $y$, on peut supposer qu'il appartient à cet ensemble. On a donc: $y \in y_c + \hat{\mathfrak{u}}$, $[s, y] = 2r_0 y$. Comme en 2.13, il existe un homomorphisme $\phi : SL(2, \mathbb{C}) \to \hat{G}$ tel que:

$$d\phi \begin{pmatrix} 0 & 1 \\ 0 & 0 \end{pmatrix} = y, \quad d\phi \begin{pmatrix} r_0 & 0 \\ 0 & -r_0 \end{pmatrix} = s;$$

(l'élément $z$ de 2.13 est nul car le Lévi $\hat{L}$ de ce paragraphe est $\hat{G}$ tout entier d'après l'hypothèse sur $y$). Soient $\omega \in X^*(\hat{T})$ un poids dominant relativement à $\hat{P}$ et $\hat{P}_\omega$ le sous-groupe parabolique propre maximal de $\hat{G}$ associé à $\omega$. On a $\hat{P}_\omega \supseteq \hat{P}$. Il existe une représentation algébrique $\rho$ de $\hat{G}$ dans un espace $E$ et un élément $e \in E \setminus \{0\}$ tels que:

- le stabilisateur dans $\hat{G}$ de la droite $\mathbb{C}e$ est $\hat{P}_\omega$;
- pour tout $x \in \hat{\mathfrak{p}}_\omega$, $d\rho(x)e = \omega(x)e$;

($\omega$ définit un caractère de $\hat{P}_\omega$ puis, par dérivation, une forme linéaire sur $\hat{\mathfrak{p}}_\omega$). Posons $\rho' = \rho \circ \phi$. On a:

$$d\rho' \begin{pmatrix} 0 & 1 \\ 0 & 0 \end{pmatrix} e = 0, \; d\rho' \begin{pmatrix} r_0 & 0 \\ 0 & -r_0 \end{pmatrix} e = < \omega, s > e.$$

De la théorie des représentations de $SL(2, \mathbb{C})$ résulte que $< \omega, s > \geq 0$. Si $< \omega, s >= 0$, il résulte de la même théorie que $d\rho'(x)e = 0$ pour tout $x \in \mathfrak{sl}(2, \mathbb{C})$. Alors l'image de $\phi$ est contenue dans $\hat{P}_\omega$, donc dans un sous-groupe de Lévi de $\hat{P}_\omega$. L'élément $y$ appartient à l'algèbre de Lie de ce sous-groupe, contrairement à l'hypothèse. Donc $< \omega, s >> 0$. Cela est vrai pour tout poids dominant $\omega$. Cela revient à dire que $s_T$, c'est-à-dire $\mu$, est réel et appartient à ${}^+C(Y_N)$. Cela achève la démonstration.

**2.16.** Comme en 2.13, on associe à $\sigma$, $y$ et $r_0$ des groupes $\hat{Q}$, $\hat{L}$, un sous-ensemble $J \subseteq \{1, \dots, N\}$ et un élément $\tau(\sigma, y) \in \bar{C}_J^+$. On a défini en 2.12 l'ensemble $W_{N,J}$.

Lemme. *Soit* $\epsilon \in \mathcal{E}_k(\sigma, r_0, y)$.

(i) *Pour tout* $(\mu, r_0) \in \mathcal{E}xp(\mathbb{M}(\sigma, r_0, y, \epsilon))$, *il existe* $w \in W_{N,J}$ *et* $\mu^J \in {}^+C^J$ *tels que* $Re(\mu) = w^{-1}(-\tau(\sigma, y) + \mu^J)$.

(ii) *Il existe* $(\mu, r_0) \in \mathcal{E}xp(\mathbb{M}(\sigma, r_0, y, \epsilon))$ *et* $\mu^J \in {}^+C^J$ *tels que* $Re(\mu) = -\tau(\sigma, y) + \mu^J$.

*Preuve.* Posons $\epsilon^{\hat{L}} = \epsilon_{|\hat{A}^{\hat{L}}(\sigma,y)}$. Le couple $(\sigma, r_0)$ appartient à l'algèbre de Lie $\hat{\mathfrak{a}}^{\hat{L}}(y)$ et, par construction, il contracte $\hat{\mathfrak{q}}$. Il résulte de la proposition 2.11 que:

(1) $\mathbb{M}(\sigma, r_0, y, \epsilon)$ est sous-module de $Ind_{\bar{\mathcal{H}}^{\hat{L}}}^{\bar{\mathcal{H}}}(Ad(w_0) \circ \mathbb{M}^{\hat{L}}(\sigma, r_0, y, \epsilon^{\hat{L}}))$;

(2) $\mathbb{M}(\sigma, r_0, y, \epsilon)$ est quotient de $\bar{\mathcal{H}} \otimes_{\bar{\mathcal{H}}^{\hat{L}}} \mathbb{M}^{\hat{L}}(\sigma, r_0, y, \epsilon^{\hat{L}})$.

Reprenons les notations de 2.13 qui nous ont servi à définir $\tau(\sigma, y)$. Considérons le triplet $\sigma - z, r_0, y$. Le lemme 2.15, que nous avons démontré pour le groupe $\hat{G}$, se généralise au groupe $\hat{L}$: il faut traiter le cas d'un groupe linéaire, ce qui se fait de la même façon; évidemment, pour un tel groupe, il faut imposer dans l'énoncé que le "caractère central" est unitaire. Le triplet ci-dessus vérifie les hypothèses de ce lemme, relativement au groupe $\hat{L}$. Donc le $\bar{\mathcal{H}}^{\hat{L}}$-module $\mathbb{M}^{\hat{L}}(\sigma - z, r_0, y, \epsilon^{\hat{L}})$ est anti-discret. Si $(\mu, r_0)$ est l'un de ses exposants, on a $Re(\mu) \in {}^+C^J$. D'autre part, il est clair que l'ensemble des exposants de $\mathbb{M}^{\hat{L}}(\sigma, r_0, y, \epsilon^{\hat{L}})$ est l'ensemble des $(\mu + z, r_0)$, pour $(\mu, r_0)$ parcourant $\mathcal{E}xp(\mathbb{M}^{\hat{L}}(\sigma - z, r_0, y, \epsilon^{\hat{L}}))$. L'image de l'application:

$$\begin{array}{ccc} \mathbb{M}^{\hat{L}}\left(\sigma, r_0, y, \epsilon^{\hat{L}}\right) & \to & \bar{\mathcal{H}} \otimes_{\bar{\mathcal{H}}^{\hat{L}}} \mathbb{M}^{\hat{L}}\left(\sigma, r_0, y, \epsilon^{\hat{L}}\right) \\ h & \mapsto & \mathbf{1} \otimes h \end{array}$$

(où $\mathbf{1} = t_1$ est l'unité de $\bar{\mathcal{H}}$) engendre $\bar{\mathcal{H}} \otimes_{\bar{\mathcal{H}}^{\hat{L}}} \mathbb{M}^{\hat{L}}(\sigma, r_0, y, \epsilon^{\hat{L}})$ sur $\bar{\mathcal{H}}$. Sa projection est donc non nulle dans tout quotient non nul de ce module. Grâce à (2), l'intersection:

$$\mathcal{E}xp(\mathbb{M}(\sigma, r_0, y, \epsilon)) \cap \mathcal{E}xp(\mathbb{M}^{\hat{L}}(\sigma, r_0, y, \epsilon^{\hat{L}}))$$

est donc non vide. En se rappelant que $\tau(\sigma, y) = -Re(z)$, on voit que tout élément de cette intersection vérifie la condition du (ii) de l'énoncé.

Soit $(\rho, E)$ un $\bar{\mathcal{H}}^{\hat{L}}$-module. Alors on a l'égalité:

$$\mathcal{E}xp\big(Ind_{\bar{\mathcal{H}}^{\hat{L}}}^{\bar{\mathcal{H}}}(E)\big) = \cup_{w \in W_{N,J}} w^{-1}(\mathcal{E}xp(E)) \tag{3}$$

l'action de $W_N$ sur $Y_N \otimes_{\mathbb{Z}} \mathbb{C}$ étant prolongée trivialement à $(Y_N \otimes_{\mathbb{Z}} \mathbb{C}) \oplus \mathbb{C}$. En effet, munissons $W_{N,J}$ d'un ordre total raffinant l'ordre de Bruhat. Pour $w \in W_{N,J}$, notons $\mathcal{F}_w$ le sous-espace des $f \in Ind_{\bar{\mathcal{H}}^{\hat{L}}}^{\bar{\mathcal{H}}}(E)$ tels que $f(t_{w'}) = 0$ pour tout $w' \in W_{N,J}, w' < w$. C'est un sous-$\mathcal{S}$-module de $Ind_{\bar{\mathcal{H}}^{\hat{L}}}^{\bar{\mathcal{H}}}(E)$ et la famille $(\mathcal{F}_w)_{w \in W_{N,J}}$ forme une filtration décroissante. Notons $\oplus_{w \in W_{N,J}} \mathcal{F}_w^{gr}$ le gradué associé. Pour $w \in W_{N,J}$, l'application:

$$\begin{array}{ccc} \mathcal{F}_w & \to & E \\ f & \mapsto & f(t_w) \end{array}$$

se quotiente en un isomorphisme d'espaces vectoriels $\mathcal{F}_w^{gr} \simeq E$. Pour $f \in \mathcal{F}_w$ et $x \in X_N$, on a la relation:

$$f(t_w \xi_x) = f(\xi_{w(x)} t_w) = \rho(\xi_{w(x)}) f(t_w).$$

On en déduit que $\mathcal{E}xp(\mathcal{F}_w^{gr}) = w^{-1}(\mathcal{E}xp(E))$. L'ensemble des exposants de $Ind_{\bar{\mathcal{H}}^{\hat{L}}}^{\bar{\mathcal{H}}}(E)$ étant le même que celui de son gradué, on obtient (3).

D'autre part, il est clair que:

$$\mathcal{E}xp(Ad(w_0) \circ \mathbb{M}^{\hat{L}}(\sigma, r_0, y, \epsilon^{\hat{L}})) = w_0(\mathcal{E}xp(\mathbb{M}^{\hat{L}}(\sigma, r_0, y, \epsilon^{\hat{L}}))).$$

De (1) et (3) se déduit l'égalité:

$$\mathcal{E}xp(\mathbb{M}(\sigma, r_0, y, \epsilon)) \subseteq \cup_{w \in W_{N,J}} w^{-1}(\mathcal{E}xp(\mathbb{M}^{\hat{L}}(\sigma, r_0, y, \epsilon^{\hat{L}}))).$$

Grâce aux propriétés déjà démontrées des exposants de $\mathbb{M}^{\hat{L}}(\sigma, r_0, y, \epsilon^{\hat{L}})$, on en déduit (i).

**2.17.** On conserve les mêmes notations.

LEMME. *Soit $\epsilon \in \mathcal{E}_k(\sigma, r_0, y)$. Alors il existe $(\mu, r_0) \in \mathcal{E}xp(\mathbb{P}(\sigma, r_0, y, \epsilon))$ et $\mu^J \in {}^+C^J$ tels que $Re(\mu) = -\tau(\sigma, y) + \mu^J$.*

*Preuve.* Soient $(\mu, r_0) \in \mathcal{E}xp(\mathbb{M}(\sigma, r_0, y, \epsilon))$ et $\mu^J \in {}^+C^J$ tels que $Re(\mu) = -\tau(\sigma, y) + \mu^J$. Supposons $(\mu, r_0) \notin \mathcal{E}xp(\mathbb{P}(\sigma, r_0, y, \epsilon))$. Alors, d'après 2.2, il existe $y' \in \hat{\mathfrak{g}}^{\sigma, r_0}$ et $\epsilon' \in \mathcal{E}_k(\sigma, r_0, y')$ tels que:

- $\mathfrak{o}_y^{\sigma, r_0} \subset \bar{\mathfrak{o}}_{y'}^{\sigma, r_0}$ et $\mathfrak{o}_y^{\sigma, r_0} \neq \mathfrak{o}_{y'}^{\sigma, r_0}$;
- $(\mu, r_0) \in \mathcal{E}xp(\mathbb{M}(\sigma, r_0, y', \epsilon'))$.

Fixons de tels $y', \epsilon'$. D'après le lemme 2.16 (i), appliqué à $\sigma, r_0, y', \epsilon'$, et les rappels de 2.12, on a $(-Re(\mu))_+ \leq \tau(\sigma, y')$. Par construction de $\mu$, on a $(-Re(\mu))_+ =$

$\tau(\sigma, y)$. Donc $\tau(\sigma, y) \leq \tau(\sigma, y')$. Cela contredit le lemme 2.13. Cette contradiction prouve que $(\mu, r_0) \in \mathcal{E}xp(\mathbb{P}(\sigma, r_0, y, \epsilon))$.

**2.18.**

Proposition. *Soit $\epsilon \in \mathcal{E}_k(\sigma, r_0, y)$. Alors le module $\mathbb{P}(\sigma, r_0, y, \epsilon)$ est anti-discret si et seulement si $y$ n'appartient à aucune sous-algèbre de Lévi d'une sous-algèbre parabolique propre de $\hat{\mathfrak{g}}$.*

*Preuve.* Si cette condition sur $y$ est vérifiée, on a $\tau(\sigma, y) = 0$, donc $\mathbb{P}(\sigma, r_0, y, \epsilon) = \mathbb{M}(\sigma, r_0, y, \epsilon)$ d'après le corollaire 2.14. Ce module est anti-discret d'après 2.15.

Inversement, supposons $\mathbb{P}(\sigma, r_0, y, \epsilon)$ anti-discret. Soit $(\mu, r_0)$ vérifiant les conditions du lemme 2.17. Puisque $\mathbb{P}(\sigma, r_0, y, \epsilon)$ est anti-discret, on a $Re(\mu) \in {}^+C(Y_N)$. Cela entraîne $\tau(\sigma, r_0) = 0$ et $J = \{1, \ldots, N\}$. Alors $\hat{L} = \hat{G}$, d'où la conclusion.

**2.19.**

Proposition. *Soit $\epsilon \in \mathcal{E}_k(\sigma, r_0, y)$. Alors le module $\mathbb{P}(\sigma, r_0, y, \epsilon)$ est anti-tempéré si et seulement si $\tau(\sigma, y) = 0$. S'il en est ainsi, on a l'égalité $\mathbb{M}(\sigma, r_0, y, \epsilon) = \mathbb{P}(\sigma, r_0, y, \epsilon)$.*

*Preuve.* Si $\tau(\sigma, y) = 0$, $\mathbb{M}(\sigma, r_0, y, \epsilon)$ est anti-tempéré: cela résulte du (i) du lemme 2.16 et des rappels de 2.12. La démonstration de la proposition est alors similaire à la démonstration précédente.

## 3. Exposants, foncteur de Jacquet, induction.

**3.1.** Soient $\mathbf{H}$ un groupe réductif connexe défini sur $F$. Soient $\mathbf{P}$ un sous-groupe parabolique de $\mathbf{H}$, de radical unipotent $\mathbf{U}$, et $\mathbf{M}$ un sous-groupe de Lévi de $\mathbf{P}$. Notons $\mathbf{T}_M$ le plus grand tore déployé central dans $\mathbf{M}$ et $X^*(\mathbf{T}_M)$, resp. $X_*(\mathbf{T}_M)$, le groupe des caractères de $\mathbf{T}_M$, resp. des cocaractères. Notons $\Pi_P$ l'ensemble des racines simples de $\mathbf{T}_M$ agissant dans l'algèbre de Lie $\mathfrak{u}$. C'est un sous-ensemble de $X^*(\mathbf{T}_M)$. Dans $X^*(\mathbf{T}_M) \otimes_{\mathbb{Z}} \mathbb{R}$, on définit le cône:

$$ {}^+C_P = \left\{ \sum_{\beta \in \Pi_P} c_\beta \beta;\ c_\beta > 0 \text{ pour tout } \beta \in \Pi_P \right\} $$

et sa clôture ${}^+\bar{C}_P$. De l'application:

$$ \begin{aligned} \mathbb{C}^\times &\to \quad \mathbb{R} \\ z &\mapsto log(|z|), \end{aligned} $$

se déduit une application, notée:

$$\begin{array}{ccc} X^*(\mathbf{T}_M)\otimes_{\mathbb{Z}}\mathbb{C}^\times & \to & X^*(\mathbf{T}_M)\otimes_{\mathbb{Z}}\mathbb{R} \\ \chi & \mapsto & log\,|\chi|. \end{array}$$

Le groupe $X_*(\mathbf{T}_M)$ se plonge dans $T_M$ par $x \mapsto x(\varpi_F)$. D'autre part, le groupe $X^*(\mathbf{T}_M)\otimes_{\mathbb{Z}}\mathbb{C}^\times$ s'identifie au groupe des caractères complexes de $X_*(\mathbf{T}_M)$.

Soit $\pi$ une représentation lisse de $H$ dans un espace $E$, de longueur finie. Par le foncteur de Jacquet, on en déduit une représentation $\pi_P$ de $M$ dans un espace $E_P$.

*Remarque.* Pour nous, ces modules sont "normalisés," i.e., si $pr : E \to E_P$ est la projection naturelle, on a l'égalité:

$$pr(\pi(m)e) = \delta_P(m)^{1/2}\pi_P(m)pr(e)$$

pour tous $m \in M, e \in E$.

Le groupe $T_M$, donc aussi son sous-groupe $X_*(\mathbf{T}_M)$, agit dans $E_P$. On note $\mathcal{E}xp_P(\pi)$, ou $\mathcal{E}xp_P(E)$, l'ensemble fini des $\chi \in X^*(\mathbf{T}_M)\otimes_{\mathbb{Z}}\mathbb{C}^\times$ tels qu'il existe $e \in E_P$, $e \neq 0$, vérifiant:

$$\pi_P(x(\varpi_F))e = \chi(x)e$$

pour tout $x \in X_*(\mathbf{T}_M)$.

Supposons $\pi$ irréductible et de caractère central unitaire. On sait bien qu'alors $\pi$ est tempérée, resp. de la série discrète, si et seulement s'il existe $\mathbf{P}$ de sorte que:

- $E_P \neq \{0\}$;
- $\pi_P$ est cuspidale;
- pour tout $\chi \in \mathcal{E}xp_P(\pi)$, $-log\,|\chi| \in {}^+\bar{C}_P$, resp. ${}^+C_P$.

**3.2.** Notons $Irr^H$ l'ensemble des classes d'isomorphie de représentations lisses irréductibles de $H$. D'après [Au], il existe une involution $\mathbf{D}^H$ de l'ensemble $Irr^H$ vérifiant les propriétés suivantes:

- si $\pi \in Irr^H$ est cuspidale, $\mathbf{D}^H(\pi) = \pi$; soient $\mathbf{P} = \mathbf{MU}$ un sous-groupe parabolique de $\mathbf{H}$ et $\bar{\mathbf{P}} = \mathbf{M}\bar{\mathbf{U}}$ le sous-groupe parabolique opposé;
- soient $\pi^M \in Irr^M$ et $\pi \in Irr^H$; alors la multiplicité de $\pi$ dans la semi-simplifiée de la représentation induite $Ind_P^H(\pi^M)$ est égale à celle de $\mathbf{D}^H(\pi)$ dans la semi-simplifiée de $Ind_P^H(\mathbf{D}^M(\pi^M))$ ou de $Ind_{\bar{P}}^H(\mathbf{D}^M(\pi^M))$;
- si $\pi \in Irr^H$, $\mathcal{E}xp_P(\mathbf{D}^H(\pi)) = \mathcal{E}xp_{\bar{P}}(\pi)$.

Remarquons que, dans le cas où $\mathbf{H}$ est un groupe $\mathbf{G}_\sharp$ comme en 1.1, les propriétés ci-dessus entraînent que $\mathbf{D}^{G_\sharp}$ conserve l'ensemble $Irr_u^{G_\sharp}$ ainsi que chacun des sous-ensembles $Irr_\theta^{G_\sharp}$ pour $\theta \in \Theta_\sharp$ (cf. 1.2).

Soit $(\pi, E) \in Irr^H$, supposons le caractère central de $\pi$ unitaire. Nous dirons que $\pi$ est anti-tempérée, resp. anti-discrète, si et seulement si $\mathbf{D}^H(\pi)$ est tempérée, resp. de la série discrète. Il revient au même de dire qu'il existe $\mathbf{P}$ tel que:

- $E_P \neq \{0\}$;
- $\pi_P$ est cuspidale;
- pour tout $\chi \in \mathcal{E}xp_P(\pi)$, $log|\chi| \in {}^+\bar{C}_P$, resp. ${}^+C_P$.

**3.3.** Soient $\sharp = iso$ ou $an$ et $\theta = (R', R'') \in \Theta_\sharp$. En 1.5, on a associé à $\theta$ un triplet $N$, $A$, $B$. On a défini en 2.12 l'ensemble de coracines $\check{\Pi}_N$ ainsi que les sous-cônes ${}^+C(Y_N)$ et ${}^+\bar{C}(Y_N)$ de $Y_N \otimes_{\mathbb{Z}} \mathbb{R}$.

Reprenons les notations de 1.2. Supposons d'abord $R' \geq R''$. On note $\mathbf{P}$ le sous-groupe parabolique de $\mathbf{G}_\sharp$ qui stabilise le drapeau de sous-espaces de $V_\sharp$:

$$Fv_{-1}, Fv_{-1} \oplus Fv_{-2}, \ldots, Fv_{-1} \oplus \ldots \oplus Fv_{-N}.$$

On note $\mathbf{M}$ le sous-groupe de Lévi de $\mathbf{P}$ qui stabilise chaque droite $Fv_i$, pour $i \in \{\pm 1, \ldots, \pm N\}$. Pour tout $i \in \{1, \ldots, N\}$, notons $x_i$ l'élément de $X_*(\mathbf{T}_M)$ tel que, pour tout $z \in F^\times$ et tout $\ell \in \{\pm 1, \ldots, \pm N\}$,

$$x_i(z)(v_\ell) = \begin{cases} v_\ell, & \text{si } \ell \notin \{\pm i\}, \\ z^{-1}v_i, & \text{si } \ell = i, \\ zv_{-i}, & \text{si } \ell = -i. \end{cases}$$

L'application qui, pour tout $i$, envoie $e_i$ sur $x_i$ est un isomorphisme de $X_N$ sur $X_*(\mathbf{T}_M)$. On note $j : X^*(\mathbf{T}_M) \to Y_N$ l'isomorphisme dual. On note encore:

$$j : X^*(\mathbf{T}_M) \otimes_{\mathbb{Z}} \mathbb{C}^\times \to Y_N \otimes_{\mathbb{Z}} \mathbb{C}^\times, \;\; j : X^*(\mathbf{T}_M) \otimes_{\mathbb{Z}} \mathbb{R} \to Y_N \otimes_{\mathbb{Z}} \mathbb{R},$$

les isomorphismes qui s'en déduisent. Remarquons que pour tout $\beta \in \Pi_P$, on a $j(\beta) \in \check{\Pi}_N$ ou $2j(\beta) \in \check{\Pi}_N$. A fortiori $j({}^+C_P) = {}^+C(Y_N)$.

Supposons maintenant $R'' > R'$. Pour tout $i \in \{1, \ldots, N\}$, posons $v'_i = v_{N+1-i}$, $v'_{-i} = v_{-N-1+i}$. On définit alors $\mathbf{P}$, $\mathbf{M}$, $x_i$ pour $i \in \{1, \ldots, N\}$ et l'isomorphisme $j$ de la même façon que ci-dessus, à ceci près que l'on remplace chaque vecteur $v_i$ par $v'_i$. L'isomorphisme $j$ a les mêmes propriétés que ci-dessus.

Remarquons que $Y_N \otimes_{\mathbb{Z}} \mathbb{C}^\times$ s'identifie au groupe des caractères de $X_N$. Soit $(\pi, E)$ une représentation de dimension finie de l'algèbre $\mathcal{H}_q(N; A, B)$. On note $\mathcal{E}xp(\pi)$, ou $\mathcal{E}xp(E)$, l'ensemble des $s \in Y_N \otimes_{\mathbb{Z}} \mathbb{C}^\times$ tels qu'il existe $e \in E$, $e \neq 0$, vérifiant:

$$\pi(\xi_x)e = s(x)e$$

pour tout $x \in X_N$.

Fixons $\zeta \in \{\pm 1\}$. Si $R'' = 0$, on suppose $\zeta = (-1)^{R'}$. Soit $(\pi, E)$ une représentation de dimension finie de l'algèbre $\mathcal{H}_q(N; A, B)$. Si $R'' > 0$, on l'étend en une représentation, encore notée $(\pi, E)$, de l'algèbre:

$$\mathbb{C}[\omega] \otimes_{\mathbb{C}} \mathcal{H}_q(N; A, B)$$

dans laquelle $\omega$ agit par multiplication par $(-1)^{R'}\zeta$. Par 1.5, $(\pi, E)$ s'identifie à une représentation de $\mathcal{H}_\theta$. On en déduit par 1.3 une représentation de longueur finie $(\tilde{\pi}, \tilde{E})$ de $G_\sharp$. Définissons l'élément:

$$\underline{s}(\zeta) = (\zeta, \ldots, \zeta)$$

de $Y_N \otimes_{\mathbb{Z}} \mathbb{C}^\times \simeq (\mathbb{C}^\times)^N$.

Proposition. *Sous les hypothèses ci-dessus, les propriétés suivantes sont vérifiées:*

(i) $\tilde{E}_P \neq \{0\}$;

(ii) $\tilde{\pi}_P$ *est cuspidale;*

(iii) $j(\mathcal{E}xp_P(\tilde{\pi})) = \underline{s}(\zeta)\mathcal{E}xp(\pi)$.

*Preuve.* En généralisant les définitions de 1.2 et 1.3 et en y remplaçant $G_\sharp$ par $M$, on définit l'ensemble de représentations $Irr_\theta^M$ et l'algèbre $\mathcal{H}_\theta^M$. Cette algèbre est commutative et il est connu que dans ce cas, toutes les représentations appartenant à $Irr_\theta^M$ sont cuspidales. Posons:

$$E_P = Hom_{K_\theta \cap M}(E_\theta, \tilde{E}_P).$$

C'est un $\mathcal{H}_\theta^M$-module. D'après [BK] théorème 7.9, de la projection $pr : \tilde{E} \to \tilde{E}_P$ se déduit un isomorphisme:

$$J : E \to E_P.$$

En particulier, ce dernier espace est non nul et (i) est vérifié.

Puisque $E_P$ est non nul, au moins un sous-quotient irréductible de $\tilde{\pi}_P$, notons-le $\sigma$, appartient à $Irr_\theta^M$. Alors $\sigma$ est cuspidale. Donc $\tilde{\pi}_P$ elle-même est cuspidale et (ii) est vérifiée. De plus, tous les sous-quotients irréductibles de $\tilde{\pi}_P$ sont de la forme $w(\sigma)$, où $w$ appartient au normalisateur $N_{G_\sharp}(M)$ de $M$ dans $G_\sharp$. Un tel $w(\sigma)$ appartient aussi à $Irr_\theta^M$. Chacun de ces sous-quotients contribue de façon non nulle à l'espace $E_P$. Il en résulte que $\mathcal{E}xp_P(\tilde{\pi})$ est égal à l'ensemble des $\chi \in X^*(\mathbf{T}_M) \otimes_{\mathbb{Z}} \mathbb{C}^\times$ tels qu'il existe $e \in E_P$, $e \neq 0$, de sorte que:

$$\tilde{\pi}_P(x(\varpi_F))e = \chi(x)e$$

pour tout $x \in X_*(\mathbf{T}_M)$, où on a encore noté $\tilde{\pi}_P$ l'action de $T_M$ sur $E_P$ naturellement déduite de son action sur $\tilde{E}_P$.

Notons $\pi^M$ l'action naturelle de $\mathcal{H}_\theta^M$ sur $E_P$. Pour tout $x \in X_*(\mathbf{T}_M)$, notons $f_x$ l'unique élément de $\mathcal{H}_\theta^M$, à support dans $x(\varpi_F)(K_\theta \cap M)$, tel que $f_x(x(\varpi_F))$ soit l'identité de $E_\theta$. L'opérateur $\tilde{\pi}_P(x(\varpi_F))$ ci-dessus est égal à $\pi^M(f_x)$. Bushnell et

Kutzko ont défini un homomorphisme:

$$t_u : \mathcal{H}_\theta^M \to \mathcal{H}_\theta,$$

et prouvé la relation:

$$\pi^M(f) \circ J = J \circ \pi(t_u(f))$$

pour tout $f \in \mathcal{H}_\theta^M$ ([BK], théorème 7.9). Donc $\mathcal{E}xp_P(\tilde{\pi})$ est l'ensemble des $\chi \in X^*(\mathbf{T}_M) \otimes_{\mathbb{Z}} \mathbb{C}^\times$ tels qu'il existe $e \in E$, $e \neq 0$, vérifiant:

$$\pi(t_u(f_x))e = \chi(x)e$$

pour tout $x \in X_*(\mathbf{T}_M)$. On va prouver la relation:

(1) pour tout $i \in \{1, \cdots, N\}$, $\pi(t_u(f_{x_1+\cdots+x_i})) = \zeta^i \pi(\xi_{e_1+\cdots+e_i})$.

Puisque les éléments $x_1 + \cdots + x_i$, pour $i \in \{1, \ldots, N\}$, engendrent $X_*(\mathbf{T}_M)$, l'assertion (iii) résulte de cette relation.

D'après [BK] paragraphe 7, pour un élément $x \in X_*(\mathbf{T}_M)$, dominant, c'est-à-dire tel que $< x, \check{x} >\geq 0$ pour tout $\check{x} \in^+ \bar{C}_P$, $t_u(f_x)$ est l'unique élément de $\mathcal{H}_\theta$, à support dans $K_\theta x(\varpi_F)K_\theta$, tel que

$$t_u(f_x)(x(\varpi_F)) = \delta_P(x(\varpi_F))^{1/2} id_\theta,$$

où $id_\theta$ est l'identité de $E_\theta$.

*Remarque.* Cette normalisation diffère de celle de [BK] parce que nous avons normalisé différemment nos modules de Jacquet.

Revenons aux constructions de 1.3. Soit $i \in \{1, \ldots, N\}$, posons $x_{\leq i} = x_1 + \cdots + x_i$. Supposons d'abord $R' \geq R'' > 0$. On a l'égalité:

$$x_{\leq i}(\varpi_F) = (s_0 s_1 \ldots s_N s_{N-1} \ldots s_i)^i.$$

Il existe donc $k \in K_\theta^\pm$ tel que:

$$x_{\leq i}(\varpi_F) = k(s_0 s_0' s_1 \ldots s_N s_N' s_{N-1} \ldots s_i)^i.$$

On vérifie que $sp\,(x_{\leq i}(\varpi_F)) = (-1)^i$, donc aussi $sp\,(k) = (-1)^i$. Posons:

$$\varphi_i = (\omega S_0 S_1 \ldots S_N S_{N-1} \ldots S_i)^i.$$

La décomposition ci-dessus étant “de longueur minimale”, on vérifie que $\varphi_i$ est à support dans $K_\theta x_{\leq i}(\varpi_F)K_\theta$ et que l'on a l'égalité:

$$\varphi_i(x_{\leq i}(\varpi_F)) = (\nu_0 \nu_N)^i id_\theta.$$

Grâce à la proposition 1.3, on a donc l'égalité:

$$\varphi_i(x_{\leq i}(\varpi_F)) = ((-1)^{R'} q^{-R'^2 - R''^2 + R''})^i id_\theta.$$

On en déduit l'égalité:

$$t_u(f_{x_{\leq i}}) = \delta_P(x_{\leq i}(\varpi_F))^{1/2}\big((-1)^{R'}q^{R'^2+R''^2-R''}\big)^i\varphi_i.$$

On calcule:

$$\delta_P(x_{\leq i}(\varpi_F))^{1/2}q^{i(R'^2+R''^2-R'')} = q^{\frac{1}{2}(-2N-A-B+i+1)}.$$

En identifiant $\mathcal{H}_\theta$ à $\mathbb{C}[\omega]\otimes_{\mathbb{C}}\mathcal{H}_q(N;A,B)$, les égalités ci-dessus entraînent:

$$t_u(f_{x_{\leq i}}) = (-1)^{iR'}\omega^i\xi_{e_1+\cdots+e_i}.$$

Puisque $\omega$ agit sur $E$ par multiplication par $(-1)^{R'}\zeta$, on obtient la relation (1).

Supposons maintenant $R'' > R'$. Un calcul analogue conduit à la même conclusion: la permutation que l'on a introduite dans la définition des $x_i$ compense celle qui intervient dans l'identification de $\mathcal{H}_\theta$ à $\mathbb{C}[\omega]\otimes_{\mathbb{C}}\mathcal{H}_q(N;A,B)$, cf. 1.5.

Supposons enfin $R' \geq R'' = 0$. On a cette fois:

$$x_{\leq i}(\varpi_F) = (s_\omega s_1\ldots s_N s_{N-1}\ldots s_i)^i.$$

Il existe donc $k \in K_\theta^{\pm}$ tel que:

$$x_{\leq i}(\varpi_F) = k(s_\omega s_1\ldots s_N s'_N s_{N-1}\ldots s_i)^i.$$

On a cette fois $sp\,(s_\omega) = -1$, d'où $sp\,(k) = 1$. On pose:

$$\varphi_i = (\omega S_1\ldots S_N S_{N-1}\ldots S_i)^i,$$

et le raisonnement se poursuit comme précédemment, grâce à la proposition 1.3. Cela achève la démonstration.

**3.4.** Soient $N$ un entier $\geq 0$ et $C$ un entier $\geq 1$. Soit $(\pi, E)$ une représentation de dimension finie de $\bar{\mathcal{H}}(N;C)$ (cf. 1.8). On a défini en 2.5 l'ensemble de ses exposants $\mathcal{E}xp(\pi)$, que l'on notera aussi $\mathcal{E}xp(E)$. On dit que $(\pi, E)$ est $q$-entière si les deux conditions suivantes sont vérifiées:

(1) $\pi(r)$ est la multiplication par $\frac{1}{2}\,log(q)$.

Il en résulte que pour tout $\underline{\sigma} = (\sigma, r_0) \in \mathcal{E}xp(\pi)$, on a $r_0 = \frac{1}{2}log\,(q)$. Par la projection $\underline{\sigma} \mapsto \sigma$, on identifie $\mathcal{E}xp(\pi)$ à un sous-ensemble de $Y_N \otimes_{\mathbb{Z}} \mathbb{C} \simeq \mathbb{C}^N$.

(2) Pour tout $\sigma = (\sigma_1, \ldots, \sigma_N) \in \mathcal{E}xp(\pi)$ et tout $j \in \{1, \ldots, N\}$, on a $\sigma_j \in \frac{1}{2}log\,(q)\mathbb{Z}$.

*Remarque.* Cette définition s'étend aux représentations de l'algèbre $\bar{\mathcal{H}}^{\mathfrak{S}}(N)$, ou même de $\mathcal{S}$.

Soient $\sharp = iso$ ou $an$, $\theta = (R', R'') \in \Theta_\sharp$ et $\zeta \in \{\pm 1\}$. Dans le cas où $R'' = 0$, on suppose $\zeta = (-1)^{R'}$. Posons:

$$N = n - R'^2 - R' - R''^2, \quad A = sup\,(2R'+1, 2R''),$$

$$B = inf(2R'+1, 2R''), \quad \mathcal{H} = \mathcal{H}_q(N;A,B).$$

Soient $N^+$ et $N^-$ deux entiers $\geq 0$ tels que $N = N^+ + N^-$. Pour $\eta \in \{\pm 1\}$, que l'on identifie à un signe $\pm$, posons:

$$\bar{\mathcal{H}}^\eta = \bar{\mathcal{H}}(N^\eta; A + \zeta\eta B), \quad \mathcal{H}^\eta = \mathcal{H}_q(N^\eta; A, B).$$

Soit $E^\eta$ un $\bar{\mathcal{H}}^\eta$-module de dimension finie. On suppose que $E^\eta$ est $q$-entier et que le centre de $\bar{\mathcal{H}}^\eta$ agit sur $E^\eta$ par un caractère.

La construction de 1.11 s'étend à cette situation. Par 1.9, appliqué à $\epsilon = \zeta\eta$, $E^\eta$ devient un $\mathcal{H}^\eta$-module. On déduit du $\mathcal{H}^+ \otimes_{\mathbb{C}} \mathcal{H}^-$-module $E^+ \otimes_{\mathbb{C}} E^-$ un $\mathcal{H}$-module par la construction de 1.7. On étend ce $\mathcal{H}$-module en un $\mathcal{H}_\theta$-module: dans le cas où $R'' > 0$, $\omega$ y agit par multiplication par $(-1)^{R'}\zeta$. Enfin, par l'équivalence de catégories de 1.3, on obtient un $G_\sharp$-module $\Phi(E^+, E^-)$. Le foncteur $\Phi$ ainsi défini est une équivalence entre des catégories convenables. En particulier, si, pour $\eta \in \{\pm 1\}$, $(E_i^\eta)_{i \in I^\eta}$ est une suite de Jordan-Hölder de $E^\eta$, alors:

$$\left(\Phi\left(E_i^+, E_j^-\right)\right)_{(i,j) \in I^+ \times I^-}$$

est une suite de Jordan-Hölder de $\Phi(E^+, E^-)$. De plus, $\Phi(E^+, E^-)$ est semi-simple si et seulement si $E^+$ et $E^-$ le sont tous deux.

Posons $W' = W_{N^+} \times W_{N^-}$. Comme en 1.7, de la décomposition:

$$\{1, \ldots, N\} = \{1, \ldots, N^+\} \cup \{N^+ + 1, \ldots, N\}$$

se déduit un plongement de $W'$ dans $W_N$. Notons $\mathcal{U}$ l'ensemble des éléments $w$ de $W_N$ qui sont de longueur minimale dans leur classe $W'w$.

Définissons une application:

$$\begin{aligned} exp^{\pm} &: (Y_{N^+} \otimes_{\mathbb{Z}} \mathbb{C}) \oplus (Y_{N^-} \otimes_{\mathbb{Z}} \mathbb{C}) \to Y_N \otimes_{\mathbb{Z}} \mathbb{C}^\times \\ &\left(\sigma_1^+, \ldots, \sigma_{N^+}^+\right) + \left(\sigma_1^-, \ldots, \sigma_{N^-}^-\right) \\ &\mapsto \left(exp(\sigma_1^+), \ldots, exp(\sigma_{N^+}^+), -exp(\sigma_1^-), \ldots, -exp(\sigma_{N^-}^-)\right). \end{aligned}$$

Comme en 3.3, on associe à $\theta$ un sous-groupe parabolique $\mathbf{P}$ de $\mathbf{G}_\sharp$ et une application $j$.

Lemme. *Pour $\eta \in \{\pm 1\}$, soit $E^\eta$ un $\bar{\mathcal{H}}^\eta$-module de dimension finie. On suppose que $E^\eta$ est $q$-entier et que le centre de $\bar{\mathcal{H}}^\eta$ agit sur $E^\eta$ par un caractère. Posons $\tilde{E} = \Phi(E^+, E^-)$. Alors on a l'égalité:*

$$\begin{aligned} j(\mathcal{E}xp_P(\tilde{E})) = \{&w^{-1}(exp^{\pm}(\sigma^+ \oplus \sigma^-));\ w \in \mathcal{U},\ \sigma^+ \in \mathcal{E}xp(E^+), \\ &\sigma^- \in \mathcal{E}xp(E^-)\}. \end{aligned}$$

*Preuve.* Soit $\eta \in \{\pm 1\}$. Comme on l'a dit, $E^\eta$ devient un $\mathcal{H}^\eta$-module par la construction de 1.9 appliquée à $\epsilon = \zeta\eta$. Notons plus précisément $\underline{E}^\eta$ ce $\mathcal{H}^\eta$-module. D'après 1.9,

$$\mathcal{E}xp(\underline{E}^\eta) = \{(\zeta\eta exp(\sigma_1), \ldots, \zeta\eta exp(\sigma_{N^\eta}));\ (\sigma_1, \ldots, \sigma_{N^\eta}) \in \mathcal{E}xp(E^\eta)\}.$$

En identifiant $(Y_{N^+} \otimes_{\mathbb{Z}} \mathbb{C}^\times) \times (Y_{N^-} \otimes_{\mathbb{Z}} \mathbb{C}^\times)$ à $Y_N \otimes_{\mathbb{Z}} \mathbb{C}^\times$ de façon évidente, on obtient:

$$\mathcal{E}xp(\underline{E}^+ \otimes_{\mathbb{C}} \underline{E}^-) = \{\underline{s}(\zeta)exp^{\pm}(\sigma^+ + \sigma^-);\ \sigma^+ \in \mathcal{E}xp(E^+),\ \sigma^- \in \mathcal{E}xp(E^-)\}.$$

Soit $\underline{E}$ le $\mathcal{H}$-module déduit de $\underline{E}^+ \otimes_{\mathbb{C}} \underline{E}^-$ par la construction de 1.7. Il résulte de cette construction que:

$$\mathcal{E}xp(\underline{E}) = \{w^{-1}s;\ w \in \mathcal{U},\ s \in \mathcal{E}xp(\underline{E}^+ \otimes_{\mathbb{C}} \underline{E}^-)\}.$$

Enfin, d'après la proposition 2.3,

$$j(\mathcal{E}xp_P(\tilde{E})) = \underline{s}(\zeta)\mathcal{E}xp(\underline{E}).$$

D'où la formule de l'énoncé.

**3.5.** On a défini en 2.12 les notions de $\bar{\mathcal{H}}^\eta$-module anti-tempéré, resp. anti-discret.

COROLLAIRE. *Sous les hypothèses du lemme précédent, le $G_\sharp$-module $\tilde{E}$ est anti-tempéré, resp. anti-discret, si et seulement si les modules $E^+$ et $E^-$ le sont tous deux.*

*Preuve.* Cela résulte du lemme précédent et de la propriété suivante. Soit $\sigma^+ \in Y_{N^+} \otimes_{\mathbb{Z}} \mathbb{R}$, $\sigma^- \in Y_{N^-} \otimes_{\mathbb{Z}} \mathbb{R}$. Alors $w^{-1}(\sigma^+ + \sigma^-) \in {}^+\bar{C}(Y_N)$, resp. ${}^+C(Y_N)$, pour tout $w \in \mathcal{U}$ si et seulement si $\sigma^+ \in {}^+\bar{C}(Y_{N^+})$ et $\sigma^- \in {}^-\bar{C}(Y_{N^-})$, resp. $\sigma^+ \in {}^+C(Y_{N^+})$ et $\sigma^- \in {}^+C(Y_{N^-})$.

**3.6.** Soit $(s, y, \epsilon)$ comme en 1.11. Utilisons les notations de ce paragraphe. Supposons $a = 0$. Pour tout signe $\eta$, on a défini en 2.13 un élément $\tau(\sigma^\eta, y^\eta) \in \bar{C}^+(Y_{N^\eta})$. Identifions $(Y_{N^+} \otimes_{\mathbb{Z}} \mathbb{R}) \oplus (Y_{N^-} \otimes_{\mathbb{Z}} \mathbb{R})$ à $Y_N \otimes_{\mathbb{Z}} \mathbb{R}$ de façon évidente. Notons $\tau(s, y)$ l'unique élément de $\bar{C}^+(Y_N)$ qui appartient à la $W_N$-orbite de $\tau(\sigma^+, y^+) \oplus \tau(\sigma^-, y^-)$. A $(s, y, \epsilon)$, on a associé en 1.11 des données $\sharp$ et $\theta = (R', R'') \in \Theta_\sharp$. On en déduit comme en 3.3 un sous-groupe parabolique $\mathbf{P} = \mathbf{MU}$ de $\mathbf{G}_\sharp$ et un isomorphisme:

$$j : X^*(\mathbf{T}_M) \otimes_{\mathbb{Z}} \mathbb{R} \to Y_N \otimes_{\mathbb{Z}} \mathbb{R}.$$

COROLLAIRE. *Sous ces hypothèses, il existe $\chi \in \mathcal{E}xp_P(\mathbb{P}(s, y, \epsilon))$ tel que $(-j(log|\chi|))_+ = \tau(s, y)$.*

Cf. 2.12 pour la définition de l'application $\mu \mapsto \mu_+$.

*Preuve.* Pour tout signe $\eta$, posons $E^\eta = \mathbb{P}(\sigma^\eta, \frac{log(q)}{2}, y^\eta, \epsilon^\eta)$. D'après 2.17, il existe un sous-ensemble $J^\eta \subseteq \{1, \ldots, N^\eta\}$ et un élément $\mu^\eta \in \mathcal{E}xp(E^\eta)$ de sorte que:

$$\tau(\sigma^\eta, y^\eta) \in C^+_{J^\eta}, \quad Re(\mu^\eta) \in -\tau(\sigma^\eta, y^\eta) + {}^+\bar{C}^{J^\eta}.$$

On fixe de tels éléments. Il existe un élément $w$ de l'ensemble $\mathcal{U}$ de 3.4 tel que $w^{-1}(\tau(\sigma^+, y^+) \oplus \tau(\sigma^-, y^-)) = \tau(s, y)$. Fixons-en un. Notons $J$ l'unique sous-ensemble de $\{1, \dots, N\}$ tel que $\tau(s, y) \in C^+_J$. L'ensemble ${}^+\bar{C}^{J^+} \oplus {}^+\bar{C}^{J^-}$ étant plongé dans $Y_N \otimes_{\mathbb{Z}} \mathbb{R}$ via les applications:

$${}^+\bar{C}^{J^+} \oplus {}^+\bar{C}^{J^-} \to (Y_{N^+} \otimes_{\mathbb{Z}} \mathbb{R}) \oplus ((Y_{N^-} \otimes_{\mathbb{Z}} \mathbb{R}) \to Y_N \otimes_{\mathbb{Z}} \mathbb{R},$$

on vérifie que l'on a l'inclusion:

$$w^{-1}({}^+\bar{C}^{J^+} \oplus {}^+\bar{C}^{J^-}) \subseteq {}^+\bar{C}^J.$$

Alors $w^{-1}(\mu^+ \oplus \mu^-) \in -\tau(s, y) + {}^+\bar{C}^J$, et $(-w^{-1}(\mu^+ \oplus \mu^-))_+ = \tau(s, y)$. On a l'égalité $\mathbb{P}(s, y, \epsilon) = \Phi(E^+, E^-)$. D'après le lemme 3.4, il existe $\chi \in \mathcal{E}xp_P(\mathbb{P}(s, y, \epsilon))$ tel que $j(log|\chi|) = w^{-1}(\mu^+ \oplus \mu^-)$. Ce $\chi$ vérifie la condition de l'énoncé.

**3.7.** Soient $C$ un entier $\geq 1$ et $N_1, \dots, N_b, N_0, N$ des entiers tels que $N_j \geq 1$ pour tout $j \in \{1, \dots, b\}$, $N_0 \geq 0$, $N = N_1 + \dots + N_b + N_0$. Posons:

$$\bar{\mathcal{H}} = \bar{\mathcal{H}}(N; C),$$

$$\bar{\mathcal{H}}^L = \bar{\mathcal{H}}^{\mathfrak{S}}(N_1) \otimes_{\mathbb{C}[r]} \dots \otimes_{\mathbb{C}[r]} \bar{\mathcal{H}}^{\mathfrak{S}}(N_b) \otimes_{\mathbb{C}[r]} \bar{\mathcal{H}}(N_0; C).$$

De la décomposition:

$$\{1, \dots, N\} = \{1, \dots, N_1\} \cup \dots \cup \{N_1 + \dots + N_{b-1} + 1, \dots, N_1 + \dots + N_b\}$$
$$\cup \{N_1 + \dots + N_b + 1, \dots, N\}$$

se déduit une identification de $\bar{\mathcal{H}}^L$ à une sous-algèbre de $\bar{\mathcal{H}}$. Soient $E_1$, resp. $E_2, \dots, E_b, E_0$, des modules de dimension finie sur les algèbres $\bar{\mathcal{H}}^{\mathfrak{S}}(N_1)$, resp. $\bar{\mathcal{H}}^{\mathfrak{S}}(N_2), \dots, \bar{\mathcal{H}}^{\mathfrak{S}}(N_b), \bar{\mathcal{H}}(N_0; C)$. Supposons que $r$ agisse sur tous ces modules par un même scalaire. Posons:

$$E^L = E_1 \otimes_{\mathbb{C}} \dots \otimes_{\mathbb{C}} E_b \otimes_{\mathbb{C}} E_0.$$

C'est un $\bar{\mathcal{H}}^L$-module. On définit comme en 2.11 le $\bar{\mathcal{H}}$-module induit $E = Ind_{\bar{\mathcal{H}}^L}^{\bar{\mathcal{H}}}(E^L)$.

Soient maintenant $\sharp = iso$ ou $an$, $\theta = (R', R'') \in \Theta_\sharp$ et $\zeta \in \{\pm 1\}$. Si $R'' = 0$, on suppose $\zeta = (-1)^{R'}$. On définit $N$, $A$ et $B$ comme en 2.4. Soient $N_1^+, \dots, N_{b^+}^+, N_1^-, \dots, N_{b^-}^-$ des entiers $\geq 1$ et $N_0^+$, $N_0^-$ des entiers $\geq 0$ tels que:

$$N = N_1^+ + \dots + N_{b^+}^+ + N_1^- + \dots + N_{b^-}^- + N_0^+ + N_0^-.$$

Soient $E_1^+, \dots, E_{b^+}^+, E_1^-, \dots, E_{b^-}^-, E_0^+, E_0^-$ des modules de dimension finie sur les algèbres $\bar{\mathcal{H}}^{\mathfrak{S}}(N_1^+)$, resp. $\dots, \bar{\mathcal{H}}^{\mathfrak{S}}(N_{b^+}^+), \bar{\mathcal{H}}^{\mathfrak{S}}(N_1^-), \dots, \bar{\mathcal{H}}^{\mathfrak{S}}(N_{b^-}^-), \bar{\mathcal{H}}(N_0^+; A + \zeta B)$,

$\bar{\mathcal{H}}(N_0^-; A - \zeta B)$. On suppose que tous ces modules sont $q$-entiers et que les centres des algèbres respectives y agissent par des caractères. Posons:

$$N_0 = N_0^+ + N_0^-, \ n_0 = N_0 + R'^2 + R' + R''^2,$$

soit $\mathbf{G}_{\sharp,0}$ un groupe de même type que $\mathbf{G}_\sharp$, mais de rang absolu $n_0$. De $E_0^+$ et $E_0^-$ se déduit un $G_{\sharp,0}$-module $\tilde{E}_0 = \Phi(E_0^+, E_0^-)$.

Pour $\eta \in \{\pm 1\}$, posons:

$$\begin{aligned} N^\eta &= N_1^\eta + \cdots + N_{b^\eta}^\eta + N_0^\eta, \ \bar{\mathcal{H}}^\eta = \bar{\mathcal{H}}(N^\eta; A + \zeta\eta B), \\ \bar{\mathcal{H}}^{\eta,L} &= \bar{\mathcal{H}}^{\mathfrak{S}}(N_1^\eta) \otimes_{\mathbb{C}[r]} \ldots \otimes_{\mathbb{C}[r]} \bar{\mathcal{H}}^{\mathfrak{S}}(N_{b^\eta}^\eta) \otimes_{\mathbb{C}[r]} \bar{\mathcal{H}}(N_0^\eta; A + \zeta\eta B), \\ E^{\eta,L} &= E_1^\eta \otimes_{\mathbb{C}} \ldots \otimes_{\mathbb{C}} E_{b^\eta}^\eta \otimes_{\mathbb{C}} E_0^\eta, \\ E^\eta &= Ind_{\bar{\mathcal{H}}^{\eta,L}}^{\bar{\mathcal{H}}^\eta}(E^{\eta,L}). \end{aligned}$$

De $E^+$ et $E^-$ se déduit un $G_\sharp$-module $\tilde{E} = \Phi(E^+, E^-)$.

Soient $\eta \in \{\pm 1\}$ et $i \in \{1, \ldots, b^\eta\}$. En adaptant les constructions des paragraphes précédents, du $\bar{\mathcal{H}}^{\mathfrak{S}}(N_i^\eta)$-module $E_i^\eta$ se déduit un $GL(N_i^\eta)$-module $\tilde{E}_i^\eta$. On doit seulement prendre garde aux signes: l'application $gr_\eta$ de 1.9 se restreint en un homomorphisme $\mathcal{H}^{\mathfrak{S}}(N_i^\eta) \to \mathcal{M}\bar{\mathcal{H}}^{\mathfrak{S}}(N_i^\eta)$, avec une notation évidente. Via cet homomorphisme, $E_i^\eta$ devient un $\mathcal{H}^{\mathfrak{S}}(N_i^\eta)$-module, dont on déduit le $GL(N_i^\eta)$-module cherché.

Notons $\tilde{\pi}$, resp. $\tilde{\pi}_0$, $\tilde{\pi}_i^\eta$, la représentation de $G_\sharp$ dans $\tilde{E}$, resp. de $G_{\sharp,0}$ dans $\tilde{E}_0$, resp. de $GL(N_i^\eta)$ dans $\tilde{E}_i^\eta$.

*Remarque.* Notons $\mu_{nr}$ l'unique caractère de $F^\times$, non ramifié et d'ordre 2. Alors le support cuspidal de $\tilde{\pi}_i^\eta$ est composé de caractères de la forme $|.|^{j/2}$ si $\eta = 1$, $\mu_{nr}|.|^{j/2}$, si $\eta = -1$, où $j \in \mathbb{Z}$.

Soit $\mathbf{Q} = \mathbf{LU}$ un sous-groupe parabolique de $\mathbf{G}_\sharp$ tel que:

$$\mathbf{L} = \mathbf{GL}(N_1^+) \times \ldots \times \mathbf{GL}(N_{b^+}^+) \times \mathbf{GL}(N_1^-) \times \cdots \times \mathbf{GL}(N_{b^-}^-) \times \mathbf{G}_{\sharp,0},$$

les facteurs **GL** étant rangés dans l'ordre évident, c'est-à-dire que **Q** stabilise un drapeau de sous-espaces de dimensions $N_1^+$, $N_1^+ + N_2^+$, etc ... On définit la représentation de $L$:

$$\tilde{\pi}^L = \tilde{\pi}_1^+ \times \cdots \times \tilde{\pi}_{b^+}^+ \times \tilde{\pi}_1^- \times \cdots \times \tilde{\pi}_{b^-}^- \times \tilde{\pi}_0.$$

PROPOSITION. *La représentation $\tilde{\pi}$ est isomorphe à la représentation induite $Ind_Q^{G_\sharp}(\tilde{\pi}^L)$.*

*Preuve.* Commençons par introduire quelques notations. On pose

$$\mathcal{H} = \mathcal{H}(N; A, B), \quad \mathcal{H}_0 = \mathcal{H}(N_0; A, B),$$

et, pour $\eta \in \{\pm 1\}$,

$$\mathcal{H}^\eta = \mathcal{H}\big(N^\eta; A, B\big), \quad \mathcal{H}_0^\eta = \mathcal{H}\big(N_0^\eta; A, B\big),$$
$$\mathcal{H}_i^\eta = \mathcal{H}^{\mathfrak{S}}\big(N_i^\eta\big), \text{ pour } i \in \{1, \ldots, b^\eta\}.$$

Tous les modules considérés sur ces différentes algèbres sont de dimension finie et $v$ y agit par multiplication par $q^{1/2}$.

Si $X^+$ est un $\mathcal{H}^+$-module et $X^-$ un $\mathcal{H}^-$-module, on en déduit par la construction de 1.7 un $\mathcal{H}$-module que nous noterons ici $\Gamma(X^+, X^-)$. Si $X$ est un $\mathcal{H}$-module, on le prolonge en un $\mathcal{H}_\theta$-module, où $\omega$ agit par multiplication par $(-1)^{R'}\zeta$ dans le cas où $R'' > 0$, et on en déduit une représentation de $G_\sharp$ que nous noterons $\tilde{\pi}[X]$. On utilise des notations analogues si l'on remplace $\mathcal{H}^+, \mathcal{H}^-$ etc... par $\mathcal{H}_0^+, \mathcal{H}_0^-$ etc...

Si $m$ est un entier $\geq 1$ et $(\sigma, X)$ un $\mathcal{H}^{\mathfrak{S}}(m)$-module, on note $\tilde{\pi}[X]$ la représentation de $GL(m)$ qui s'en déduit. Si de plus $z \in \mathbb{C}^\times$, on définit le $\mathcal{H}^{\mathfrak{S}}(m)$-module $(\sigma[z], X[z])$: on pose $X[z] = X$; pour $w \in \mathfrak{S}_m$, $\sigma[z](w) = \sigma(w)$; pour $i \in \{1, \ldots, m\}$, $\sigma[z](\xi_{e_i}) = z\sigma(\xi_{e_i})$. En particulier $\tilde{\pi}[X[-1]] = \tilde{\pi}[X] \otimes \mu_{nr}$.

On considère désormais $E_0^\eta$, $E^\eta$ et $E_i^\eta$, pour $\eta \in \{\pm 1\}$ et $i \in \{1, \ldots, b^\eta\}$, comme des $\mathcal{H}_0^\eta$, resp. $\mathcal{H}^\eta$, $\mathcal{H}_i^\eta$-modules grâce à 1.8. Insistons sur le fait que l'homomorphisme utilisé en 1.8 est $gr_{\zeta\eta}$ dans le cas des modules $E_0^\eta$ et $E^\eta$, mais $gr_\eta$ dans celui des modules $E_i^\eta$. Posons:

$$\mathcal{H}^{\eta,L} = \mathcal{H}_1^\eta \otimes_{\mathcal{A}} \ldots \otimes_{\mathcal{A}} \mathcal{H}_{b^\eta}^\eta \otimes_{\mathcal{A}} \mathcal{H}_0^\eta,$$

définissons le $\mathcal{H}^{\eta,L}$-module:

$$E^{\eta,L} = E_1^\eta[\zeta] \otimes_{\mathbb{C}} \ldots \otimes_{\mathbb{C}} E_{b^\eta}^\eta[\zeta] \otimes_{\mathbb{C}} E_0^\eta,$$

et le $\mathcal{H}^\eta$-module induit (en un sens évident):

$$Ind_{\mathcal{H}^{\eta,L}}^{\mathcal{H}^\eta}(E^{\eta,L}).$$

Il est isomorphe à $E^\eta$: la torsion par $\zeta$ compense le fait que les homomorphismes $gr_\epsilon$ utilisés ne sont pas tous associés au même $\epsilon$. Par définition, $\tilde{\pi} = \tilde{\pi}[\Gamma(E^+, E^-)]$, donc:

$$\tilde{\pi} = \tilde{\pi}[\Gamma(Ind_{\mathcal{H}^{+,L}}^{\mathcal{H}^+}(E^{+,L}), Ind_{\mathcal{H}^{-,L}}^{\mathcal{H}^-}(E^{-,L}))]. \tag{1}$$

Posons:

$$\mathcal{H}^L = \mathcal{H}_1^+ \otimes_{\mathcal{A}} \ldots \otimes_{\mathcal{A}} \mathcal{H}_{b^+}^+ \otimes_{\mathcal{A}} \mathcal{H}_1^- \otimes_{\mathcal{A}} \ldots \otimes_{\mathcal{A}} \mathcal{H}_{b^-}^- \otimes_{\mathcal{A}} \mathcal{H}_0.$$

Rappelons un résultat de Bushnell et Kutzko. Pour $\eta \in \{\pm 1\}$ et $i \in \{1, \ldots, b^\eta\}$, soit $X_i^\eta$ un $\mathcal{H}_i^\eta$-module. Soit $X_0$ un $\mathcal{H}_0$-module. Posons:

$$X^L = X_1^+[\zeta] \otimes_{\mathbb{C}} \cdots \otimes_{\mathbb{C}} X_{b^+}^+[\zeta] \otimes_{\mathbb{C}} X_1^-[\zeta] \otimes_{\mathbb{C}} \ldots \otimes_{\mathbb{C}} X_{b^-}^-[\zeta] \otimes_{\mathbb{C}} X_0,$$
$$\pi^L = \tilde{\pi}\big[X_1^+\big] \times \cdots \times \tilde{\pi}\big[X_{b^+}^+\big] \times \tilde{\pi}\big[X_1^-\big] \times \cdots \times \tilde{\pi}\big[X_{b^-}^-\big] \times \tilde{\pi}[X_0].$$

L'espace $X^L$ est un $\mathcal{H}^L$-module et $\pi^L$ est une représentation de $L$. Alors les deux représentations de $G_\sharp$:

$$Ind_Q^{G_\sharp}(\pi^L) \text{ et } \tilde{\pi}\left[Ind_{\mathcal{H}^L}^{\mathcal{H}}(X^L)\right]$$

sont isomorphes, cf. [BK] 8.4. La torsion par $\zeta$ est la même que celle intervenant dans la proposition 2.3 et s'introduit pour les mêmes raisons. Posons:

$$E^L = E_1^+[\zeta] \otimes_{\mathbb{C}} \ldots \otimes_{\mathbb{C}} E_{b^+}^+[\zeta] \otimes_{\mathbb{C}} E_1^-[\zeta] \otimes_{\mathbb{C}} \ldots \otimes_{\mathbb{C}} E_{b^-}^-[\zeta] \otimes_{\mathbb{C}} \Gamma(E_0^+, E_0^-).$$

Le résultat ci-dessus entraîne l'égalité:

(2) $$Ind_Q^{G_\sharp}(\tilde{\pi}^L) = \tilde{\pi}[Ind_{\mathcal{H}^L}^{\mathcal{H}}(E^L)].$$

En comparant (1) et (2), on voit qu'il nous suffit de démontrer l'isomorphisme:

(3) $$\Gamma\left(Ind_{\mathcal{H}^{+,L}}^{\mathcal{H}^+}(E^{+,L}), Ind_{\mathcal{H}^{-,L}}^{\mathcal{H}^-}(E^{-,L})\right) = Ind_{\mathcal{H}^L}^{\mathcal{H}}(E^L).$$

A ce point, $\zeta$ n'intervient plus que par une torsion sur tous les modules $E_i^\eta$. On peut supposer $\zeta = 1$.

Posons:

$$\mathcal{H}_0^{+-} = \mathcal{H}_0^+ \otimes_{\mathcal{A}} \mathcal{H}_0^-, \ \mathcal{H}^{+-} = \mathcal{H}^+ \otimes_{\mathcal{A}} \mathcal{H}^-,$$
$$\mathcal{K} = \mathcal{H}_1^+ \otimes_{\mathcal{A}} \ldots \otimes_{\mathcal{A}} \mathcal{H}_{b^+}^+ \otimes_{\mathcal{A}} \mathcal{H}_0^+ \otimes_{\mathcal{A}} \mathcal{H}_1^- \otimes_{\mathcal{A}} \ldots \otimes_{\mathcal{A}} \mathcal{H}_{b^-}^- \otimes_{\mathcal{A}} \mathcal{H}_0^-.$$

On note $\mathcal{Z}_0^{+-}$ le centre de $\mathcal{H}_0^{+-}$, $\mathcal{F}_0^{+-}$ son corps des fractions, on pose $\mathcal{H}_{0,\mathcal{F}}^{+-} = \mathcal{H}_0^{+-} \otimes_{\mathcal{Z}_0^{+-}} \mathcal{F}_0^{+-}$. On définit de façon similaire $\mathcal{H}_{\mathcal{F}}^{+-}$, $\mathcal{H}_{\mathcal{F}}$, $\mathcal{H}_{\mathcal{F}}^L$, $\mathcal{H}_{0,\mathcal{F}}$ et $\mathcal{K}_{\mathcal{F}}$. Pour $\eta \in \{\pm 1\}$, fixons $\underline{s}_0^\eta \in (Y_{N_0^\eta} \otimes_{\mathbb{Z}} \mathbb{C}^\times) \times \mathbb{C}^\times$ tel que le centre de $\mathcal{H}_0^\eta$ agisse sur $E_0^\eta$ par l'homomorphisme $\ell_{\underline{s}_0^\eta}$. Pour $i \in \{1, \ldots, b^\eta\}$, fixons de même $\underline{s}_i^\eta \in (Y_{N_i^\eta} \otimes_{\mathbb{Z}} \mathbb{C}^\times) \times \mathbb{C}^\times$ tel que le centre de $\mathcal{H}_i^\eta$ agisse sur $E_i^\eta$ par l'homomorphisme $\ell_{\underline{s}_i^\eta}$. Ces points définissent naturellement des idéaux dans les centres des algèbres $\mathcal{H}_0^{+-}$, $\mathcal{H}^{+-}$, $\mathcal{K}$. Par exemple, écrivons:

$$\underline{s}_0^\eta = \left(s_{0,1}^\eta, \ldots, s_{0,N_0^\eta}^\eta; q^{1/2}\right),$$

posons:

$$\underline{s}_0 = \left(s_{0,1}^+, \ldots, s_{0,N_0^+}^+, s_{0,1}^-, \ldots, s_{0,N_0^-}^-; q^{1/2}\right).$$

L'idéal en question de $\mathcal{Z}_0^{+-}$ est $\mathcal{Z}_0^{+-} \cap Ker(\ell_{\underline{s}_0})$. On note $\mathcal{Z}_{0,loc}^{+-}$ le localisé de $\mathcal{Z}_0^{+-}$ relativement à cet idéal et:

$$\mathcal{H}_{0,loc}^{+-} = \mathcal{H}_0^{+-} \otimes_{\mathcal{Z}_0^{+-}} \mathcal{Z}_{0,loc}^{+-}.$$

On définit de même $\mathcal{H}_{loc}^{+-}$ et $\mathcal{K}_{loc}$.

Considérons les décompositions en intervalles, croissantes en un sens évident:

$$\{1,\dots,N_0\} = \{1,\dots,N_0^+\} \cup \{N_0^+ + 1,\dots,N_0\},$$

$$\{1,\dots,N\} = \{1,\dots,N^+\} \cup \{N^+ + 1,\dots,N\},$$

(4) $$\{1,\dots,N\} = I_1^+ \cup \dots \cup I_{b^+}^+ \cup I_0^+ \cup I_1^- \cup \dots \cup I_{b^-}^- \cup I_0^-,$$

(5) $$\{1,\dots,N\} = J_1^+ \cup \dots \cup J_{b^+}^+ \cup J_1^- \cup \dots \cup J_{b^-}^- \cup J_0^+ \cup J_0^-,$$

où, pour $\eta \in \{\pm 1\}$ et $i \in \{0,\dots,b^\eta\}$, $|I_i^\eta| = |J_i^\eta| = N_i^\eta$. A l'aide des deux premières décompositions, on a défini en 1.7 des plongements:

$$j_0 : \mathcal{H}_{0,\mathcal{F}}^{+-} \to \mathcal{H}_{0,\mathcal{F}}, \quad j : \mathcal{H}_{\mathcal{F}}^{+-} \to \mathcal{H}_{\mathcal{F}}.$$

De même, à l'aide des décompositions (4) et (5), on définit des plongements:

$$j' : \mathcal{K}_{\mathcal{F}} \to \mathcal{H}_{\mathcal{F}}, \text{ resp. } j'' : \mathcal{K}_{\mathcal{F}} \to \mathcal{H}_{\mathcal{F}}.$$

Notons que, pour définir $j''$, on permute implicitement l'ordre des facteurs de $\mathcal{K}$. Introduisons le $\mathcal{K}$-module:

$$E^{\mathcal{K}} = E_1^+ \otimes_{\mathbb{C}} \dots \otimes_{\mathbb{C}} E_{b^+}^+ \otimes_{\mathbb{C}} E_0^+ \otimes_{\mathbb{C}} E_1^- \otimes_{\mathbb{C}} \dots \otimes_{\mathbb{C}} E_{b^-}^- \otimes_{\mathbb{C}} E_0^-.$$

Comme en 1.7, introduisons le sous-espace $j'(\mathcal{K}_{loc})\mathcal{H}$ de $\mathcal{H}_{\mathcal{F}}$. C'est un $\mathcal{H}$-module à droite et un $\mathcal{K}$-module à gauche, via le plongement $j'$. Posons:

$$E' = Hom_{\mathcal{K}}(j'(\mathcal{K}_{loc})\mathcal{H}, E^{\mathcal{K}}).$$

C'est un $\mathcal{H}$-module. Montrons que:

(6) $$\Gamma\big(Ind_{\mathcal{H}^{+,L}}^{\mathcal{H}^+}(E^{+,L}), Ind_{\mathcal{H}^{-,L}}^{\mathcal{H}^-}(E^{-,L})\big) \simeq E'.$$

Pour $\eta \in \{\pm 1\}$, posons $\mathcal{I}^\eta = Ind_{\mathcal{H}^{\eta,L}}^{\mathcal{H}^\eta}(E^{\eta,L})$. Par définition (cf. 1.7):

$$\Gamma(\mathcal{I}^+, \mathcal{I}^-) = Hom_{\mathcal{H}^{+-}}\big(j\big(\mathcal{H}_{loc}^{+-}\big)\mathcal{H}, \mathcal{I}^+ \otimes_{\mathbb{C}} \mathcal{I}^-\big).$$

L'algèbre $\mathcal{K}$ se plonge naturellement dans $\mathcal{H}^{+-}$ et l'on a:

$$\mathcal{I}^+ \otimes_{\mathbb{C}} \mathcal{I}^- = Hom_{\mathcal{K}}(\mathcal{H}^{+-}, E^{\mathcal{K}}).$$

Donc:

$$\Gamma(\mathcal{I}^+, \mathcal{I}^-) = Hom_{\mathcal{H}^{+-}}\big(j\big(\mathcal{H}_{loc}^{+-}\big)\mathcal{H}, Hom_{\mathcal{K}}(\mathcal{H}^{+-}, E^{\mathcal{K}})\big).$$

Le plongement de $\mathcal{K}$ dans $\mathcal{H}^{+-}$ s'étend en un plongement de $\mathcal{K}_{\mathcal{F}}$ dans $\mathcal{H}_{\mathcal{F}}^{+-}$. Le composé de $j$ et de ce plongement est égal à $j'$. On vérifie de plus que $j(\mathcal{H}_{loc}^{+-}) \subseteq j'(\mathcal{K}_{loc})\mathcal{H}$. On définit alors une application:

$$\begin{aligned} E' = Hom_{\mathcal{K}}(j'(\mathcal{K}_{loc})\mathcal{H}, E^{\mathcal{K}}) &\to \Gamma(\mathcal{I}^+, \mathcal{I}^-) \\ &= Hom_{\mathcal{H}^{+-}}\big(j\big(\mathcal{H}_{loc}^{+-}\big)\mathcal{H}, Hom_{\mathcal{K}}(\mathcal{H}^{+-}, E^{\mathcal{K}})\big) \\ e \mapsto \quad & \qquad f \end{aligned}$$

par $f(h_1)(h_2) = e(j(h_2)h_1)$ pour tous $h_1 \in j(\mathcal{H}_{loc}^{+-})\mathcal{H}, h_2 \in \mathcal{H}^{+-}$. Cette application est injective et est un morphisme de $\mathcal{H}$-modules. Elle est surjective car les deux modules ont même dimension sur $\mathbb{C}$, cf. 1.7. Cela démontre (6).

Introduisons maintenant le sous-espace $j''(\mathcal{K}_{loc})\mathcal{H}$ de $\mathcal{H}_{\mathcal{F}}$. C'est un $\mathcal{H}$-module à droite et un $\mathcal{K}$-module à gauche, via le plongement $j''$. Posons:

$$E'' = Hom_{\mathcal{K}}(j''(\mathcal{K}_{loc})\mathcal{H}, E^{\mathcal{K}}).$$

C'est un $\mathcal{H}$-module. Montrons que:

(7) $$Ind_{\mathcal{H}^L}^{\mathcal{H}}(E^L) \simeq E''.$$

Considérons la suite d'applications:

$$\begin{aligned}\mathcal{K} = \mathcal{H}_1^+ \otimes_{\mathcal{A}} \ldots \otimes_{\mathcal{A}} \mathcal{H}_{b^+}^+ \otimes_{\mathcal{A}} \mathcal{H}_0^+ \otimes_{\mathcal{A}} \mathcal{H}_1^- \otimes_{\mathcal{A}} \ldots \otimes_{\mathcal{A}} \mathcal{H}_{b^-}^- \otimes_{\mathcal{A}} \mathcal{H}_0^- \\ \to \mathcal{H}_1^+ \otimes_{\mathcal{A}} \ldots \otimes_{\mathcal{A}} \mathcal{H}_{b^+}^+ \otimes_{\mathcal{A}} \mathcal{H}_1^- \otimes_{\mathcal{A}} \ldots \otimes_{\mathcal{A}} \mathcal{H}_{b^-}^- \otimes_{\mathcal{A}} \mathcal{H}_0^+ \otimes_{\mathcal{A}} \mathcal{H}_0^- \to \mathcal{H}_{\mathcal{F}}^L.\end{aligned}$$

La première est la permutation des facteurs, la seconde est issue du plongement $j_0$. Leur composée se prolonge en un plongement:

$$j_0^L : \mathcal{K}_{\mathcal{F}} \to \mathcal{H}_{\mathcal{F}}^L.$$

D'après 1.7, on a:

$$\Gamma\left(E_0^+, E_0^-\right) = Hom_{\mathcal{H}_0^{+-}}\left(j_0\left(\mathcal{H}_{0,loc}^{+-}\right)\mathcal{H}_0, E_0^+ \otimes_{\mathbb{C}} E_0^-\right).$$

D'où:

$$\begin{aligned}E^L &= Hom_{\mathcal{K}}\left(j_0^L(\mathcal{K}_{loc})\mathcal{H}^L, E^{\mathcal{K}}\right), \\ Ind_{\mathcal{H}^L}^{\mathcal{H}}(E^L) &= Hom_{\mathcal{H}^L}\left(\mathcal{H}, Hom_{\mathcal{K}}\left(j_0^L(\mathcal{K}_{loc})\mathcal{H}^L, E^{\mathcal{K}}\right)\right).\end{aligned}$$

Le plongement de $\mathcal{H}^L$ dans $\mathcal{H}$ s'étend en un plongement de $\mathcal{H}_{\mathcal{F}}^L$ dans $\mathcal{H}_{\mathcal{F}}$. Le composé de $j_0^L$ et de ce plongement est égal à $j''$. On définit une application:

$$\begin{aligned}E'' = Hom_{\mathcal{K}}(j''(\mathcal{K}_{loc})\mathcal{H}, E^{\mathcal{K}}) &\to Ind_{\mathcal{H}^L}^{\mathcal{H}}(E^L) \\ &= Hom_{\mathcal{H}^L}\left(\mathcal{H}, Hom_{\mathcal{K}}\left(j_0^L(\mathcal{K}_{loc})\mathcal{H}^L, E^{\mathcal{K}}\right)\right) \\ e &\mapsto f\end{aligned}$$

par $f(h_1)(h_2) = e(h_2h_1)$ pour $h_1 \in \mathcal{H}$, $h_2 \in j_0^L(\mathcal{K}_{loc})\mathcal{H}^L$. Cette application est injective et est un morphisme de $\mathcal{H}$-modules. Elle est surjective car les deux modules ont même dimension sur $\mathbb{C}$, cf. 1.7. Cela démontre (7).

Notons $w$ la permutation de $\{1, \ldots, N\}$ telle que, pour tout $\eta \in \{\pm 1\}$ et tout $i \in \{0, \ldots, b^\eta\}$, on ait $w(I_i^\eta) = J_i^\eta$ (cf. (4) et (5)) et $w$ soit croissante sur $I_i^\eta$. Identifions $w$ à un élément de $W_N$. Il résulte des définitions que:

$$j''(h) = \tau(w)j'(h)\tau(w^{-1})$$

pour tout $h \in \mathcal{K}_{loc}$. Comme en 1.7, on montre que $\tau(w^{-1}) \in j'(\mathcal{K}_{loc})\mathcal{H}$. Donc:

$$j''(\mathcal{K}_{loc})\mathcal{H} = \tau(w)j'(\mathcal{K}_{loc})\mathcal{H}.$$

Alors l'application:

$$\begin{aligned} E'' &\to E' \\ e'' &\mapsto e' \end{aligned}$$

définie par $e'(h) = e''(\tau(w)h)$ pour tout $h \in j'(\mathcal{K}_{loc})\mathcal{H}$, est un isomorphisme. Les relations (6) et (7) démontrent (3), ce qui achève la démonstration.

**3.8.** Soient $s$ un élément semi-simple de $\hat{G}$ et $y$ un élément nilpotent de $\hat{\mathfrak{g}}$, tels que $Ad(s)y = qy$. Une partie des constructions de 1.11 s'applique au couple $(s, y)$, c'est-à-dire est indépendante de la donnée supplémentaire $\epsilon$ de ce paragraphe. En particulier, on définit un entier $a$. Nous supposons:

(1) $$a = 0.$$

On a une décomposition $\hat{V} = \hat{V}^+ \oplus \hat{V}^-$ et, pour tout signe $\eta$, deux éléments $\sigma^\eta$ et $y^\eta$ de $\hat{\mathfrak{g}}^\eta$.

Pour tout signe $\eta$, donnons-nous une décomposition orthogonale:

$$\hat{V}^\eta = \hat{V}_0^\eta \oplus \big(\oplus_{\ell=1,\dots,b^\eta} \hat{V}_{\pm\ell}^\eta\big),$$

et pour tout $\ell \in \{1, \dots, b^\eta\}$, une décomposition en deux lagrangiens $\hat{V}_{\pm\ell}^\eta = \hat{V}_\ell^\eta \oplus \hat{V}_{-\ell}^\eta$. On note $\hat{L}^\eta$ le sous-groupe des éléments de $\hat{G}^\eta$ qui conservent chacun de ces sous-espaces. On a:

$$\hat{L}^\eta = GL\big(\hat{V}_1^\eta\big) \times \cdots \times GL\big(\hat{V}_{b^\eta}^\eta\big) \times \hat{G}_0^\eta,$$

avec une notation évidente. Supposons que $\sigma^\eta$ et $y^\eta$ appartiennent à $\hat{\mathfrak{l}}^\eta$. Pour tout $\ell \in \{-b^\eta, \dots, b^\eta\}$, on note $\sigma_\ell^\eta$ et $y_\ell^\eta$ les restrictions de $\sigma^\eta$ et $y^\eta$ à $\hat{V}_\ell^\eta$. Donnons-nous de plus un caractère $\epsilon_0^\eta$ du groupe:

$$Z_{\hat{G}^\eta}\big(\sigma_0^\eta, y_0^\eta\big)/Z_{\hat{G}^\eta}\big(\sigma_0^\eta, y_0^\eta\big)^0.$$

Du quadruplet $(\sigma_0^\eta, \frac{log(q)}{2}, y_0^\eta, \epsilon_0^\eta)$ se déduisent:

- des entiers $k^\eta$ et $N_0^\eta = \frac{1}{2}(dim_{\mathbb{C}}(\hat{V}_0^\eta) - k^\eta(k^\eta + 1))$ (cf. 1.10);
- un module $E_0^\eta = \mathbb{M}(\sigma_0^\eta, \frac{log(q)}{2}, y_0^\eta, \epsilon_0^\eta)$ sur l'algèbre $\bar{\mathcal{H}}(N_0^\eta; 2k^\eta + 1)$ (cf. 2.1);
- un élément $\tau(\sigma_0^\eta, y_0^\eta) \in \overline{C}^+(Y_{N_0^\eta})$ (cf. 2.13).

Nous supposons:

(2) $$\tau\big(\sigma_0^\eta, y_0^\eta\big) = 0.$$

Pour $\ell \in \{1, \dots, b^\eta\}$, on pose $N_\ell^\eta = dim_{\mathbb{C}}(\hat{V}_\ell^\eta)$. En adaptant au groupe linéaire les constructions du paragraphe 2, on déduit du triplet $(\sigma_\ell^\eta, \frac{log(q)}{2}, y_\ell^\eta)$:

- un module $E_\ell^\eta = \mathbb{M}(\sigma_\ell^\eta, \frac{log(q)}{2}, y_\ell^\eta)$ sur l'algèbre $\bar{\mathcal{H}}^{\mathfrak{S}}(N_\ell^\eta)$;

- un élément $\tau(\sigma_\ell^\eta, y_\ell^\eta) \in T_{N_\ell^\eta} \otimes_{\mathbb{Z}} \mathbb{R}$, qui appartient à la fermeture de la chambre de Weyl positive pour l'ensemble de racines $\Pi_{N_\ell^\eta} \setminus \{\alpha_{N_\ell^\eta}\}$ (cf. 1.4).

Nous supposons:

(3)
$\tau\big(\sigma_\ell^\eta, y_\ell^\eta\big)$ est "central", i.e., est diagonal quand on identifie $Y_{N_\ell^\eta} \otimes_{\mathbb{Z}} \mathbb{R}$ à $\mathbb{R}^{N_\ell^\eta}$.

Du couple $(k^+, k^-)$ se déduisent comme en 1.11 un couple $(R', R'')$ et un signe $\zeta$. Soit $\sharp = iso$ ou $an$ tel que $(R', R'') \in \Theta_\sharp$. On a maintenant toutes les données requises pour appliquer les constructions de 3.7. On construit donc comme dans ce paragraphe un sous-groupe parabolique $\mathbf{Q} = \mathbf{LU}$ de $\mathbf{G}_\sharp$ et une représentation $\tilde{\pi}^L$ de $L$.

On a une injection naturelle:

$$\big(Z_{\hat{G}_0^+}(\sigma_0^+, y_0^+)/Z_{\hat{G}_0^+}(\sigma_0^+, y_0^+)^0\big) \times \big(Z_{\hat{G}_0^-}(\sigma_0^-, y_0^-)/Z_{\hat{G}_0^-}(\sigma_0^-, y_0^-)^0\big)$$
$$\to Z_{\hat{G}}(s, y)/Z_{\hat{G}}(s, y)^0.$$

Notons $E(\epsilon_0^+, \epsilon_0^-)$ l'ensemble des caractères de $Z_{\hat{G}}(s, y)/Z_{\hat{G}}(s, y)^0$ dont le composé avec cette injection soit égal à $(\epsilon_0^+, \epsilon_0^-)$.

Lemme. *Rappelons que l'on a imposé les hypothèses (1), (2) et (3).*

(i) *Pour tout* $\epsilon \in E(\epsilon_0^+, \epsilon_0^-)$, *la représentation de* $G_\sharp$ *dans* $\mathbb{P}(s, y, \epsilon)$ *est sous-quotient de la représentation induite* $Ind_Q^{G_\sharp}(\tilde{\pi}^L)$.

(ii) *Supposons que pour tout signe* $\eta$ *et tout* $\ell \in \{1, \dots, b^\eta\}$, *on ait l'égalité* $\tau(\sigma_\ell^\eta, y_\ell^\eta) = 0$. *Alors la représentation induite* $Ind_Q^{G_\sharp}(\tilde{\pi}^L)$ *est isomorphe à la représentation de* $G_\sharp$ *dans:*

$$\oplus_{\epsilon \in E(\epsilon_0^+, \epsilon_0^-)} \mathbb{P}(s, y, \epsilon).$$

*Preuve.* On reprend les notations de 3.7. Soit $\eta$ un signe. On peut effectuer certaines opérations sur nos objets de départ, ainsi que sur le couple $\mathbf{L}, \tilde{\pi}^L$. Les propriétés suivantes sont claires:

- pour $\ell, m \in \{1, \dots, b^\eta\}$ avec $\ell \neq m$, échanger les couples $(\hat{V}_\ell^\eta, \hat{V}_{-\ell}^\eta)$ et $(\hat{V}_m^\eta, \hat{V}_{-m}^\eta)$ revient à permuter les facteurs $\mathbf{GL}(N_\ell^\eta)$ et $\mathbf{GL}(N_m^\eta)$ de $\mathbf{L}$ ainsi que les représentations $\tilde{\pi}_\ell^\eta$ et $\tilde{\pi}_m^\eta$;
- pour $\ell \in \{1, \dots, b^\eta\}$, échanger $\hat{V}_\ell^\eta$ et $\hat{V}_{-\ell}^\eta$ revient à remplacer $\tilde{\pi}_\ell^\eta$ par sa contragrédiente.

De telles opérations ne modifient pas la semi-simplifiée de la représentation induite $Ind_Q^{\hat{G}_\sharp}(\tilde{\pi}^L)$. Quitte à en effectuer, on peut supposer vérifiée la condition suivante. Pour tout signe $\eta$ et tout $\ell \in \{1, \dots, b^\eta\}$, écrivons $\tau(\sigma_\ell^\eta, y_\ell^\eta) = (\tau_\ell^\eta, \dots, \tau_\ell^\eta) \in$

$\mathbb{R}^{N_\ell^\eta} \simeq Y_{N_\ell^\eta} \otimes_{\mathbb{Z}} \mathbb{R}$. On demande que pour chaque $\eta$, on ait:

$$\tau_1^\eta \geq \tau_2^\eta \geq \cdots \geq \tau_{b^\eta}^\eta \geq 0. \tag{4}$$

D'après la proposition 3.7, on peut remplacer $Ind_Q^{G_\sharp}(\tilde{\pi}^L)$ par la représentation de $G_\sharp$ dans le module $\tilde{E} = \Phi(E^+, E^-)$. Pour tout signe $\eta$, notons $E(\epsilon_0^\eta)$ l'ensemble des caractères de $Z_{\hat{G}^\eta}(\sigma^\eta, y^\eta)/Z_{\hat{G}^\eta}(\sigma^\eta, y^\eta)^0$ dont le composé avec l'injection:

$$Z_{\hat{G}_0^\eta}(\sigma_0^\eta, y_0^\eta)/Z_{\hat{G}_0^\eta}(\sigma_0^\eta, y_0^\eta)^0 \to Z_{\hat{G}^\eta}(\sigma^\eta, y^\eta)/Z_{\hat{G}^\eta}(\sigma^\eta, y^\eta)^0$$

soit égal à $\epsilon_0^\eta$. On a une bijection:

$$\begin{array}{ccc} E(\epsilon_0^+, \epsilon_0^-) & \to & E(\epsilon_0^+) \times E(\epsilon_0^-) \\ \epsilon & \mapsto & (\epsilon^+, \epsilon^-). \end{array}$$

Rappelons que, pour $\epsilon \in E(\epsilon_0^+, \epsilon_0^-)$, on a l'égalité:

$$\mathbb{P}(s, y, \epsilon) = \Phi\left(\mathbb{P}\left(\sigma^+, \frac{log(q)}{2}, y^+, \epsilon^+\right), \mathbb{P}\left(\sigma^-, \frac{log(q)}{2}, y^-, \epsilon^-\right)\right).$$

Les propriétés du foncteur $\Phi$ (cf. 3.4) nous ramènent à démontrer les propriétés suivantes, pour tout signe $\eta$:

(5) pour tout $\epsilon^\eta \in E(\epsilon_0^\eta)$, $\mathbb{P}\left(\sigma^\eta, \frac{log(q)}{2}, y^\eta, \epsilon^\eta\right)$ est sous-quotient de $E^\eta$;

(6) sous les hypothèses de (ii), $E^\eta$ est isomorphe à le somme directe:

$$\oplus_{\epsilon^\eta \in E(\epsilon_0^\eta)} \mathbb{P}\left(\sigma^\eta, \frac{log(q)}{2}, y^\eta, \epsilon^\eta\right).$$

Fixons $\eta$, notons $\hat{Q}^\eta$ le sous-groupe parabolique de $\hat{G}^\eta$ qui stabilise le drapeau;

$$\hat{V}_1^\eta \subseteq \hat{V}_1^\eta \oplus \hat{V}_2^\eta \subseteq \ldots \subseteq \hat{V}_1^\eta \oplus \ldots \oplus \hat{V}_{b^\eta}^\eta,$$

et introduisons un élément $w_0$ comme en 2.9, relatif au groupe $\hat{G}^\eta$ et à ce sous-groupe parabolique. Avec les notations de 2.11, on a l'égalité:

$$E^{\eta,L} = \mathbb{M}^{\hat{L}^\eta}\left(\sigma^\eta, \frac{log(q)}{2}, y^\eta, \epsilon_0^\eta\right).$$

Par simple transport de structure, il est clair que:

$$E^{\eta,L} = Ad(w_0) \circ \mathbb{M}^{\hat{L}^\eta}\left(Ad(w_0)(\sigma^\eta), \frac{log(q)}{2}, Ad(w_0)(y^\eta), \epsilon_0^\eta \circ Ad(w_0)\right).$$

Grâce à (4), $(Ad(w_0)(\sigma^\eta), \frac{log(q)}{2})$ contracte $\hat{\mathfrak{q}}$. La proposition 2.11 nous dit que $E^\eta$, qui est par définition $Ind_{\bar{\mathcal{H}}^{\eta,L}}^{\bar{\mathcal{H}}^\eta}(E^{\eta,L})$, est isomorphe à:

$$\oplus_{\epsilon^\eta \in E(\epsilon_0^\eta)} \mathbb{M}\left(\sigma^\eta, \frac{log(q)}{2}, y^\eta, \epsilon^\eta\right)$$

(on a utilisé le fait que ce dernier module ne dépend que de la classe de conjugaison de $(\sigma^\eta, y^\eta, \epsilon^\eta)$). Pour $\epsilon^\eta \in E(\epsilon_0^\eta)$, $\mathbb{P}(\sigma^\eta, \frac{log(q)}{2}, y^\eta, \epsilon^\eta)$ est sous-quotient de $\mathbb{M}(\sigma^\eta, \frac{log(q)}{2}, y^\eta, \epsilon^\eta)$, et (5) en résulte. Sous les hypothèses de (ii), l'élément $\tau(\sigma^\eta, y^\eta)$ défini en 2.13 est nul. D'après 3.14, on a l'égalité:

$$\mathbb{M}\left(\sigma^\eta, \frac{log(q)}{2}, y^\eta, \epsilon^\eta\right) = \mathbb{P}\left(\sigma^\eta, \frac{log(q)}{2}, y^\eta, \epsilon^\eta\right)$$

pour tout $\epsilon^\eta \in E(\epsilon_0^\eta)$ et (6) en résulte. Cela achève la démonstration.

**3.9.** Soit $(s, y, \epsilon)$ un triplet comme en 1.11, reprenons les notations de ce paragraphe. Notons $\pi$ la représentation de $G_\sharp$ dans $\mathbb{P}(s, y, \epsilon)$. Introduisons un groupe $\mathbf{G}_{\sharp,0}$, de même type que $\mathbf{G}_\sharp$, mais de rang absolu $n_0 = \frac{1}{2}(dim_{\mathbb{C}}(\hat{V}^+) + dim_{\mathbb{C}}(\hat{V}^-))$. Notons $\hat{G}_0$ le groupe symplectique de $\hat{V}^+ \oplus \hat{V}^-$, $s_0 = s^+s^-$, $y_0 = y^+ + y^-$, identifions $\epsilon$ à un caractère du groupe:

$$Z_{\hat{G}_0}(s_0, y_0)/Z_{\hat{G}_0}(s_0, y_0)^0.$$

Le triplet $(s_0, y_0, \epsilon)$ paramétrise un $G_{\sharp,0}$-module irréductible $\mathbb{P}(s_0, y_0, \epsilon)$. Notons $\pi_0$ la représentation de $G_{\sharp,0}$ dans ce module.

Soit $j \in \{1, \ldots, a\}$. Par des constructions analogues à celles du paragraphe 1 (cf. [L4]), le couple $(s_j, y_j)$ paramétrise une représentation irréductible $\pi_j$ du groupe linéaire $GL(N_j)$.

*Remarque.* La paramétrisation de [L4] n'est pas la plus usuelle: c'est la composée de celle-ci avec l'involution de 2.2. En particulier, supposons que la collection des valeurs propres de $s_j$ soit:

$$\left(z_j q^{(N_j-1)/2}, \ldots, z_j q^{(1-N_j)/2}\right),$$

et que $y_j$ soit un nilpotent régulier. Alors $\pi_j$ est le caractère $|det|^{\sigma_j}$ où $\sigma_j$ est tel que $q^{-\sigma_j} = z_j$.

Soit $\mathbf{Q} = \mathbf{LU}$ un sous-groupe parabolique de $\mathbf{G}_\sharp$ tel que:

$$\mathbf{L} = \mathbf{GL}(N_1) \times \cdots \times \mathbf{GL}(N_a) \times \mathbf{G}_{\sharp,0}.$$

Introduisons la représentation de $L$:

$$\pi^L = \pi_1 \times \cdots \times \pi_a \times \pi_0.$$

PROPOSITION. *Sous ces hypothèses, $\pi$ est isomorphe à la représentation induite $Ind_Q^{G_\sharp}(\pi^L)$.*

La démonstration est similaire à celle de la proposition 3.7.

Si $a \geq 1$, $\pi$ est une induite irréductible. Son image $\mathbf{D}^{G_\sharp}(\pi)$ par l'involution l'est aussi. Donc $\mathbf{D}^{G_\sharp}(\pi)$ n'est pas de la série discrète et ne peut pas intervenir dans l'une des représentations elliptiques construites par Arthur.

*Remarque.* Du point de vue de la fonctorialité, les énoncés 3.7 et 3.9 sont raisonnables. Cela légitime les choix de signes que nous avons effectués pour construire la paramétrisation $(s, y, \epsilon) \mapsto \mathbb{P}(s, y, \epsilon)$.

## 4. Paramétrisation de Langlands.

**4.1.** On revient aux hypothèses de 1.1. Comme en 1.11, on fixe un espace symplectique de dimension $2n$ sur $\mathbb{C}$, dont on note $\hat{G}$ le groupe symplectique. On note $W_F$ le groupe de Weil de $F$, qui est un sous-groupe de $Gal(\bar{F}/F)$. Soit $\sharp = iso$ ou $an$. Notons $\underline{\Psi}_u^{G_\sharp}$ l'ensemble des classes de conjugaison par $\hat{G}$ de couples $(\psi, \epsilon)$ tels que:

- $\psi : W_F \times SL(2, \mathbb{C}) \to \hat{G}$ est un homomorphisme tel que $\psi_{|W_F}$ soit semi-simple et non ramifié:
- $\epsilon : Z_{\hat{G}}(\psi)/Z_{\hat{G}}(\psi)^0 \to \{\pm 1\}$ est un caractère;
- $\epsilon_{|Z(\hat{G})} = \begin{cases} 1, & \text{si } \sharp = iso, \\ -1, & \text{si } \sharp = an. \end{cases}$

Expliquons les notations. On a noté $Z_{\hat{G}}(\psi)$ le commutant dans $\hat{G}$ de l'image de $\psi$, et $Z_{\hat{G}}(\psi)^0$ sa composante neutre. Il y a un homomorphisme naturel:

$$Z(\hat{G}) \to Z_{\hat{G}}(\psi)/Z_{\hat{G}}(\psi)^0,$$

en général non injectif. On note $\epsilon_{|Z(\hat{G})}$ le caractère composé de $\epsilon$ et de cet homomorphisme. Enfin $Z(\hat{G}) = \{\pm 1\}$, et on identifie son groupe de caractères à $\{\pm 1\}$.

Soit $(\psi, \epsilon) \in \underline{\Psi}_u^{G_\sharp}$ (plus exactement, soit $(\psi, \epsilon)$ un couple dont la classe de conjugaison appartient à $\underline{\Psi}_u^{G_\sharp}$). Soit *Frob* un élément de Frobenius de $W_F$. Posons:

$$s = \psi\left(Frob, \begin{pmatrix} q^{1/2} & 0 \\ 0 & q^{-1/2} \end{pmatrix}\right).$$

Remarquons que $s$ ne dépend pas du choix de $Frob$, puisque $\psi_{|W_F}$ est non ramifié. Posons:

$$y = d(\psi_{|SL(2,\mathbb{C})})\left(\begin{pmatrix} 0 & 1 \\ 0 & 0 \end{pmatrix}\right).$$

Il y a un isomorphisme naturel:

$$Z_{\hat{G}}(\psi)/Z_{\hat{G}}(\psi)^0 \simeq Z_{\hat{G}}(s, y)/Z_{\hat{G}}(s, y)^0,$$

et $\epsilon$ s'identifie à un caractère de ce dernier groupe. Alors le triplet $(s, y, \epsilon)$ vérifie les conditions de 1.11. On a associé à ce triplet un élément $\sharp' = iso$ ou $an$ et un $G_{\sharp'}$-module irréductible $\mathbb{P}(s, y, \epsilon)$. Grâce au lemme 1.12, on a $\sharp' = \sharp$. On note $\pi_{\psi,\epsilon}$ la représentation de $G_\sharp$ dans le module $\mathbf{D}^{G_\sharp}(\mathbb{P}(s, y, \epsilon))$, image de $\mathbb{P}(s, y, \epsilon)$ par l'involution de 3.2. L'énoncé suivant ne fait que reformuler le résultat de Lusztig [L1].

Proposition. *L'application:*

$$\begin{aligned}\underline{\Psi}_u^{G_\sharp} &\to Irr_u^{G_\sharp}\\ (\psi,\epsilon) &\mapsto \pi_{\psi,\epsilon}\end{aligned}$$

*est bijective.*

Cela résulte de 1.11.

**4.2.** Notons $\underline{\Psi}_{udisc}^{G_\sharp}$ le sous-ensemble des $(\psi,\epsilon)\in\underline{\Psi}_u^{G_\sharp}$ tels que l'image de $\psi$ ne soit contenue dans aucun sous-groupe parabolique propre de $\hat{G}$.

Proposition. *Soit* $(\psi,\epsilon)\in\underline{\Psi}_u^{G_\sharp}$. *Alors* $\pi_{\psi,\epsilon}$ *est de la série discrète si et seulement si* $(\psi,\epsilon)\in\underline{\Psi}_{udisc}^{G_\sharp}$.

*Preuve.* Supposons $\pi_{\psi,\epsilon}$ de la série discrète, soit $(s,y,\epsilon)$ le triplet associé à $(\psi,\epsilon)$ comme dans le paragraphe précédent. Utilisons les notations de 1.11 relatives à ce triplet. Puisque $\pi_{\psi,\epsilon}$ est de la série discrète, $\mathbb{P}(s,y,\epsilon)$ est anti-discret. Alors $a=0$ d'après 3.9. Pour tout signe $\eta$, le $\bar{\mathcal{H}}(N^\eta;2k^\eta+1)$-module $\mathbb{P}(\sigma^\eta,\frac{log(q)}{2},y^\eta,\epsilon^\eta)$ est anti-discret, d'après 3.5. D'après 2.18, $y^\eta$ n'appartient à aucune sous-algèbre de Lévi d'une sous-algèbre parabolique propre de $\hat{\mathfrak{g}}^\eta$. De ces conditions résulte qu'il n'existe pas de sous-groupe parabolique propre de $\hat{G}$ contenant l'image de $\psi$. Donc $(\psi,\epsilon)\in\underline{\Psi}_{udisc}^{G_\sharp}$.

Des arguments analogues prouvent la réciproque.

**4.3.** Notons $\underline{\Psi}_{utemp}^{G_\sharp}$ le sous-ensemble des $(\psi,\epsilon)\in\underline{\Psi}_u^{G_\sharp}$ tels que $\psi(W_F)$ soit un sous-groupe relativement compact de $\hat{G}$.

Proposition. *Soit* $(\psi,\epsilon)\in\underline{\Psi}_u^{G_\sharp}$. *Alors* $\pi_{\psi,\epsilon}$ *est tempérée si et seulement si* $(\psi,\epsilon)\in\underline{\Psi}_{utemp}^{G_\sharp}$.

*Preuve.* Supposons $\pi_{\psi,\epsilon}$ tempérée, soit $(s,y,\epsilon)$ le triplet associé à $(\psi,\epsilon)$ comme en 4.1. Utilisons les notations de 3.9. La représentation $\pi_{\psi,\epsilon}$ est isomorphe à la représentation induite:

$$Ind_{\bar{Q}}^{G_\sharp}(\mathbf{D}^{GL(N_1)}(\pi_1)\times\cdots\times\mathbf{D}^{GL(N_a)}(\pi_a)\times\mathbf{D}^{G_{0,\sharp}}(\pi_0)).$$

Or une représentation induite est tempérée si et seulement si la représentation que l'on induit l'est. Donc $\pi_1,\ldots,\pi_a$ et $\pi_0$ sont toutes anti-tempérées. On utilise ci-après les notations de 1.11. Soit un signe $\eta$. Puisque $\pi_0$ est anti-tempérée, le $\bar{\mathcal{H}}(N^\eta;2k^\eta+1)$-module $\mathbb{P}(\sigma^\eta,\frac{log(q)}{2},y^\eta,\epsilon^\eta)$ l'est aussi, d'après 3.5. Donc $\tau(\sigma^\eta,y^\eta)=0$, d'après 2.19. Par construction, l'élément $\sigma^\eta$ est "réel". Ces deux conditions entraînent l'existence d'un homomorphisme:

$$\phi^\eta: SL(2,\mathbb{C})\to\hat{G}^\eta$$

de sorte que:

$$d\phi^\eta \begin{pmatrix} 0 & 1 \\ 0 & 0 \end{pmatrix} = y^\eta \text{ et } d\phi^\eta \begin{pmatrix} \frac{log(q)}{2} & 0 \\ 0 & -\frac{log(q)}{2} \end{pmatrix} = \sigma^\eta.$$

Alors $s^\eta = \eta\phi^\eta(\begin{smallmatrix} q^{1/2} & 0 \\ 0 & q^{-1/2} \end{smallmatrix})$.

Pour tout groupe linéaire $GL(m)$, on peut démontrer un énoncé analogue à la proposition 2.19. D'ailleurs, la classification des représentations d'un tel groupe est assez connue. Soit $j \in \{1, \ldots, a\}$. Parce que $\pi_j$ est anti-tempérée, on montre comme ci-dessus qu'il existe un homomorphisme:

$$\phi_j : SL(2, \mathbb{C}) \to GL(\hat{V}_j)$$

et un nombre complexe $z_j$ de module 1 de sorte que:

$$d\phi_j \begin{pmatrix} 0 & 1 \\ 0 & 0 \end{pmatrix} = y_j, \; s_j = z_j\phi_j \begin{pmatrix} q^{1/2} & 0 \\ 0 & q^{-1/2} \end{pmatrix}.$$

Notons $z'$ l'élément semi-simple de $\hat{G}$ qui agit par multiplication par 1 sur $\hat{V}^+$, $-1$ sur $\hat{V}^-$, $z_j$ sur $\hat{V}_j$, $z_j^{-1}$ sur $\hat{V}_{-j}$ pour tout $j \in \{1, \ldots, a\}$. Notons $\phi'$ : $SL(2, \mathbb{C}) \to \hat{G}$ l'unique homomorphisme tel que, pour $x \in SL(2, \mathbb{C})$, $\phi'(x)$ respecte la décomposition:

$$\hat{V} = \hat{V}^+ \oplus \hat{V}^- \oplus \oplus_{j=1,\ldots,a}(\hat{V}_j \oplus \hat{V}_{-j}),$$

et agit comme $\phi^+(x)$ sur $\hat{V}^+$, comme $\phi^-(x)$ sur $\hat{V}^-$ et comme $\phi_j(x)$ sur $\hat{V}_j$ pour tout $j \in \{1, \ldots, a\}$. Alors:

$$d\phi' \begin{pmatrix} 0 & 1 \\ 0 & 0 \end{pmatrix} = y, \; s = z'\phi' \begin{pmatrix} q^{1/2} & 0 \\ 0 & q^{-1/2} \end{pmatrix},$$

et $z'$ commute à l'image de $\phi'$. Posons;

$$z = \psi_{|W_F}(Frob), \; \phi = \psi_{|SL(2\mathbb{C})}.$$

Le couple $(z, \phi)$ vérifie les mêmes conditions que $(z', \phi')$. De la théorie des $SL(2)$-triplets résulte que $z$ est conjugué à $z'$ dans $\hat{G}$. Toutes les valeurs propres de $z'$ sont de module 1, donc $z$ engendre un sous-groupe relativement compact de $\hat{G}$. Cela démontre que $(\psi, \epsilon) \in \underline{\Psi}_{utemp}^{G_\sharp}$.

Des arguments analogues prouvent la réciproque.

**4.4.** Soit $(\psi, \epsilon) \in \underline{\Psi}_u^{G_\sharp}$, posons $z = \psi_{|W_F}(Frob)$. On peut décomposer $z$ en un produit $\nu z_c$ de deux éléments semi-simples qui commutent, de sorte que toutes les valeurs propres de $\nu$, resp. $z_c$, agissant dans $\hat{V}$, soient réelles $> 0$, resp. de module 1. Cette décomposition est unique. Considérons l'ensemble réunion de $\{1\}$ et de l'ensemble des valeurs propres de $\nu$. On note ses éléments:

$$\nu_1 < \cdots < \nu_t < \nu_0 = 1 < \nu_{-t} < \cdots < \nu_{-1}$$

(on a $\nu_{-r} = \nu_r^{-1}$ pour tout $r$). Pour $r \in \{-t, \dots, t\}$, notons $\hat{V}_{.r}$ le sous-espace propre attaché à la valeur propre $\nu_r$, sauf dans le cas où $r = 0$ et 1 n'est pas valeur propre de $\nu$, auquel cas on pose $\hat{V}_{.0} = \{0\}$. On note $\hat{L}$ le sous-groupe des éléments de $\hat{G}$ qui conservent $\hat{V}_{.r}$ pour tout $r$. On note $\hat{Q}$ le sous-groupe des éléments de $\hat{G}$ qui conservent le drapeau de sous-espaces:

$$\hat{V}_{.1} \subset \hat{V}_{.1} \oplus \hat{V}_{.2} \subset \cdots \subset \hat{V}_{.1} \oplus \hat{V}_{.2} \oplus \ldots \oplus \hat{V}_{.t}.$$

C'est un sous-groupe parabolique de $\hat{G}$, de sous-groupe de Lévi $\hat{L}$. On a l'isomorphisme:

$$\hat{L} \simeq GL(\hat{V}_{.1}) \times \cdots \times GL(\hat{V}_{.t}) \times \hat{G}_{.0},$$

où $\hat{G}_{.0}$ est le groupe symplectique de $\hat{V}_{.0}$. Remarquons que $\psi$ prend ses valeurs dans $\hat{L}$ et s'identifie à un produit d'homomorphismes $\psi_1 \times \cdots \times \psi_t \times \psi_0$, où:

$$\psi_0 : W_F \times SL(2, \mathbb{C}) \to \hat{G}_{.0},$$

$$\psi_r : W_F \times SL(2, \mathbb{C}) \to GL(\hat{V}_{.r}), \ pour\ r \in \{1, \dots, t\}.$$

Soient $(s, y, \epsilon)$ le triplet associé à $(\psi, \epsilon)$ comme en 4.1 et $\theta = (R', R'')$ l'élément de $\Theta_\sharp$ associé en 1.11 à ce triplet. Si l'on se reporte aux définitions de 1.10 et 1.11, on voit que, parce que $y \in \hat{\mathfrak{l}}$, on a l'inégalité:

(1) $$dim_{\mathbb{C}}(\hat{V}_{.0}) \geq 2(R'^2 + R' + R''^2).$$

En particulier, si $\sharp = an$, $dim_{\mathbb{C}}(\hat{V}_{.0}) \geq 2$ puisque $R''$ est impair. Il existe donc dans $\mathbf{G}_\sharp$ un sous-groupe parabolique $\mathbf{Q}$ stabilisant un drapeau de sous-espaces isotropes de $V_\sharp$ de dimensions $d_{.1}, d_{.1} + d_{.2}, \cdots, d_{.1} + \cdots + d_{.t}$, où, pour tout $r \in \{-t, \dots, t\}$, on a posé $d_{.r} = dim_{\mathbb{C}}(\hat{V}_{.r})$. Fixons $\mathbf{Q}$ et un sous-groupe de Lévi $\mathbf{L}$ de $\mathbf{Q}$. On a un isomorphisme:

$$\mathbf{L} \simeq \mathbf{GL}(d_{.1}) \times \cdots \times \mathbf{GL}(d_{.t}) \times \mathbf{G}_{.0},$$

où $\mathbf{G}_{.0}$ est un groupe de même type que $\mathbf{G}_\sharp$, dont $\hat{G}_{.0}$ est le L-groupe.

Il y a un isomorphisme naturel:

$$Z_{\hat{G}_{.0}}(\psi_0)/Z_{\hat{G}_{.0}}(\psi_0)^0 \simeq Z_{\hat{G}}(\psi)/Z_{\hat{G}}(\psi)^0$$

et $\epsilon$ s'identifie à un caractère du premier quotient. Le couple $(\psi_0, \epsilon)$ appartient à $\underline{\Psi}_u^{G_{.0}}$, et même à $\underline{\Psi}_{utemp}^{G_{.0}}$. D'après la proposition 4.3, il paramétrise une représentation tempérée $\pi_{\psi_0,\epsilon}$ de $G_{.0}$.

Soit $r \in \{1, \dots, t\}$. Notons:

$$\psi_{r,temp} : W_F \times SL(2, \mathbb{C}) \to GL(\hat{V}_{.r})$$

l'unique homomorphisme tel que:

$$\psi_{r,temp|SL(2,\mathbb{C})} = \psi_{r|SL(2,\mathbb{C})},$$

$$\psi_{r,temp|W_F}(Frob) = \nu_j^{-1} \psi_{r|W_F}(Frob).$$

Les définitions et résultats de 4.1 et 4.3 s'étendent au cas d'un groupe linéaire. Pour un tel groupe, les caractères $\epsilon$ disparaissent car les centralisateurs sont connexes. Alors $\psi_{r,temp}$ paramétrise une représentation tempérée $\pi_{\psi_{r,temp}}$ de $GL(d_{.r})$.

On introduit la représentation tempérée de $L$:

$$\pi^L = \pi_{\psi_{1,temp}} \otimes \cdots \otimes \pi_{\psi_{t,temp}} \otimes \pi_{\psi_0,\epsilon}.$$

Notons $\mathbf{T}_L$, resp. $\hat{T}_L$, le plus grand tore central de $\mathbf{L}$, resp. $\hat{L}$. On a:

$$\nu \in X_*(\hat{T}_L) \otimes_{\mathbb{Z}} \mathbb{R}_+^\times = X^*(\mathbf{T}_L) \otimes_{\mathbb{Z}} \mathbb{R}_+^\times = X^*(\mathbf{L}) \otimes_{\mathbb{Z}} \mathbb{R}_+^\times,$$

avec des notations usuelles. Un tel élément paramétrise un caractère de $L$ à valeurs réelles $> 0$, que l'on note $\underline{\nu}$. On a choisi $\mathbf{Q}$ de sorte que le couple $(\underline{\nu}, \pi^L)$ vérifie les conditions requises pour que la représentation induite $Ind_Q^{G_\sharp}(\underline{\nu} \otimes \pi^L)$ possède un unique quotient irréductible, que l'on appelle son quotient de Langlands (cf. [S]).

THÉORÈME. *La représentation $\pi_{\psi,\epsilon}$ est égale au quotient de Langlands de la représentation induite $Ind_Q^{G_\sharp}(\underline{\nu} \otimes \pi^L)$.*

*Preuve.* On introduit le sous-groupe parabolique $\mathbf{P} = \mathbf{MU}$ de $\mathbf{G}_\sharp$ associé à $\theta$, cf. 3.3. On note $\mathbf{T}_M$ le plus grand tore central de $\mathbf{M}$. Grâce à l'inégalité (1), on peut supposer que $\mathbf{Q}$ contient $\mathbf{P}$ et $\mathbf{L}$ contient $\mathbf{M}$. Posons $N = n - R'^2 - R' - R''^2$. On a une suite d'applications:

$$X_*(\hat{T}_L) \otimes_{\mathbb{Z}} \mathbb{R} \simeq X^*(\mathbf{T}_L) \otimes_{\mathbb{Z}} \mathbb{R} \simeq X^*(\mathbf{L}) \otimes_{\mathbb{Z}} \mathbb{R} \to$$

$$X^*(\mathbf{M}) \otimes_{\mathbb{Z}} \mathbb{R} \simeq X^*(\mathbf{T}_M) \otimes_{\mathbb{Z}} \mathbb{R} \simeq Y_N \otimes_{\mathbb{Z}} \mathbb{R},$$

le dernier isomorphisme étant le $j$ de 3.3. Puisque $\nu \in X_*(\hat{T}_L) \otimes_{\mathbb{Z}} \mathbb{R}_+^\times$, on peut définir $log(\nu) \in X_*(\hat{T}_L) \otimes_{\mathbb{Z}} \mathbb{R}$, que l'on identifie à un élément de $Y_N \otimes_{\mathbb{Z}} \mathbb{R}$. On pose $\tau = -log(\nu)$. On a choisi $\mathbf{Q}$ de sorte que $\tau \in \bar{C}^+(Y_N)$. De même, pour $\chi \in X^*(\mathbf{T}_M) \otimes_{\mathbb{Z}} \mathbb{C}^\times$, on définit $log|\chi|$, que l'on identifie à un élément de $Y_N \otimes_{\mathbb{Z}} \mathbb{R}$.

Le couple $\theta$ que l'on a associé à $(\psi, \epsilon) \in \underline{\Psi}_u^{G_\sharp}$ est égal au couple associé de même à $(\psi_0, \epsilon) \in \underline{\Psi}_u^{G_{.0}}$. Il en résulte que $\pi^L$ est sous-quotient d'une induite d'une représentation cuspidale de $M$. Le quotient de Langlands de la représentation $Ind_Q^{G_\sharp}(\underline{\nu} \otimes \pi^L)$ se caractérise de la façon suivante: c'est l'unique représentation admissible irréductible $\pi$ de $G_\sharp$ telle que:

- $\pi$ est sous-quotient de $Ind_{\bar{Q}}^{G_\sharp}(\underline{\nu} \otimes \pi^L)$;
- il existe $\chi \in \mathcal{E}xp_{\bar{P}}(\pi)$ tel que $(-log|\chi|)_+ = \tau$.

Cette caractérisation résulte de la démonstration par Silberger de l'unicité du quotient de Langlands (cf. [S]). On va prouver que $\pi_{\psi,\epsilon}$ vérifie ces propriétés.

Pour tout $r \in \{-t, \ldots, t\}$, $s$ et $y$ laissent stables $\hat{V}_{.r}$. On note $s_{.r}$ et $y_{.r}$ leurs restrictions à $\hat{V}_{.r}$. Le triplet $(s_{.0}, y_{.0}, \epsilon)$ paramétrise un $G_{.0}$-module $\mathbb{P}(s_{.0}, y_{.0}, \epsilon)$.

Pour $r \in \{1, \dots, t\}$, le couple $(s_{.r}, y_{.r})$ paramétrise un $GL(d_{.r})$-module $\mathbb{P}(s_{.r}, y_{.r})$. Posons:

$$\mathbb{P} = \mathbb{P}(s_{.1}, y_{.1}) \otimes \cdots \otimes \mathbb{P}(s_{.t}, y_{.t}) \otimes \mathbb{P}(s_{.0}, y_{.0}, \epsilon).$$

La représentation $\underline{\nu} \otimes \pi^L$ est la représentation de $L$ dans le module $\mathbf{D}^L(\mathbb{P}^L)$. Les propriétés de l'involution rappelées en 3.2 nous ramènent à démontrer:

(2) $\mathbb{P}(s, y, \epsilon)$ est sous-quotient de $Ind_Q^{G_\sharp}(\mathbb{P}^L)$;

(3) il existe $\chi \in \mathcal{E}xp_P(\mathbb{P}(s, y, \epsilon))$ tel que $(-log|\chi|)_+ = \tau$.

On a une première décomposition $\hat{V} = \oplus_{r=-t,\dots,t} \hat{V}_{.r}$. En 1.11, on a attaché à $(s, y, \epsilon)$ une autre décomposition que, pour la distinguer de la précédente, on note ici:

$$\hat{V} = \hat{V}_{0.}^+ \oplus \hat{V}_{0.}^- \oplus (\oplus_{j=1,\dots,a}(\hat{V}_{j.} \oplus \hat{V}_{-j.})).$$

Ces deux décompositions sont compatibles, c'est-à-dire que si l'on pose, pour tout $r \in \{-t, \dots, t\}$:

$$\hat{V}_{0r}^+ = \hat{V}_{0.}^+ \cap \hat{V}_{.r}, \ \ \hat{V}_{0r}^- = \hat{V}_{0.}^- \cap \hat{V}_{.r},$$
$$\hat{V}_{jr} = \hat{V}_{j.} \cap \hat{V}_{.r}, \ \text{ pour tout } j \in \{\pm 1, \dots, \pm a\},$$

on a l'égalité:

$$\hat{V} = \oplus_{r=-t,\dots,t} \left(\hat{V}_{0r}^+ \oplus \hat{V}_{0r}^- \oplus (\oplus_{j=1,\dots,a}(\hat{V}_{jr} \oplus \hat{V}_{(-j)r}))\right).$$

On regroupera parfois les espaces $\hat{V}_{0.}^+$ et $\hat{V}_{0.}^-$ en posant:

$$\hat{V}_{0.} = \hat{V}_{0.}^+ \oplus \hat{V}_{0.}^-, \ \ \hat{V}_{0r} = \hat{V}_{0r}^+ \oplus \hat{V}_{0r}^-.$$

Les éléments $s$ et $y$ stabilisent tous ces sous-espaces. On note de façon évidente leurs restrictions à ces sous-espaces. Par exemple, $s_{0r}^+$, resp. $s_{jr}$, est la restriction de $s$ à $\hat{V}_{0r}^+$, resp. $\hat{V}_{jr}$. De même, on note par exemple $d_{0r}^+$, resp. $d_{jr}$, la dimension de $\hat{V}_{0r}^+$, resp. $\hat{V}_{jr}$.

Remarquons que la décomposition:

$$\hat{V}_{.0} = \hat{V}_{00}^+ \oplus \hat{V}_{00}^- \oplus \left(\oplus_{j=1,\dots,a}(\hat{V}_{j0} \oplus \hat{V}_{(-j)0})\right)$$

est celle attachée en 1.11 au triplet $(s_{.0}, y_{.0}, \epsilon)$. Fixons un sous-groupe parabolique $\mathbf{Q}_{.0}$ de $\mathbf{G}_{.0}$ de sous-groupe de Lévi:

$$\mathbf{L}_{.0} = \mathbf{GL}(d_{10}) \times \cdots \times \mathbf{GL}(d_{a0}) \times \mathbf{G}_{00},$$

où $\mathbf{G}_{00}$ est un groupe analogue à $\mathbf{G}_\sharp$, dont le L-groupe est le groupe symplectique $\hat{G}_{00}$ de l'espace $\hat{V}_{00}$. On a un isomorphisme naturel:

$$Z_{\hat{G}_{00}}(s_{00}, y_{00})/Z_{\hat{G}_{00}}(s_{00}, y_{00})^0 \simeq Z_{\hat{G}}(s, y)/Z_{\hat{G}}(s, y)^0,$$

et $\epsilon$ s'identifie à un caractère du premier quotient. Le triplet $(s_{00}, y_{00}, \epsilon)$ paramétrise un $G_{00}$-module $\mathbb{P}(s_{00}, y_{00}, \epsilon)$. De même, pour tout $j \in \{1, \dots, a\}$, le couple

$(s_{j0}, y_{j0})$ paramétrise un $GL(d_{j0})$-module $\mathbb{P}(s_{j0}, y_{j0})$. Posons:

$$\mathbb{P}^{L.0} = \mathbb{P}(s_{10}, y_{10}) \otimes \cdots \otimes \mathbb{P}(s_{a0}, y_{a0}) \otimes \mathbb{P}(s_{00}, y_{00}, \epsilon).$$

D'après la proposition 3.9, $\mathbb{P}(s_{.0}, y_{.0}, \epsilon)$ est isomorphe au module induit $Ind_{Q.0}^{G.0}(\mathbb{P}^{L.0})$.

Soit $r \in \{1, \ldots, t\}$. Introduisons un sous-groupe parabolique $\mathbf{Q}_{.r}$ de $\mathbf{GL}(d_{.r})$, de sous-groupe de Lévi:

$$\mathbf{L}_{.r} = \mathbf{GL}\left(d_{0r}^{+}\right) \times \mathbf{GL}\left(d_{0r}^{-}\right) \times \mathbf{GL}\left(d_{1r}\right) \times \mathbf{GL}\left(d_{(-1)r}\right)$$
$$\times \cdots \times \mathbf{GL}\left(d_{ar}\right) \times \mathbf{GL}\left(d_{(-a)r}\right).$$

Le couple $(s_{0r}^{+}, y_{0r}^{+})$, resp ... $(s_{(-a)r}, y_{(-a)r})$, paramétrise un $GL(d_{0r}^{+})$-module $\mathbb{P}(s_{0r}^{+}, y_{0r}^{+})$, resp ... un $GL(d_{(-a)r})$-module $\mathbb{P}(s_{(-a)r}, y_{(-a)r})$. Posons:

$$\mathbb{P}^{L.r} = \mathbb{P}\left(s_{0r}^{+}, y_{0r}^{+}\right) \otimes \cdots \otimes \mathbb{P}(s_{(-a)r}, y_{(-a)r}).$$

D'après l'analogue pour le groupe $GL(d_{.r})$ de la proposition 3.9, $\mathbb{P}(s_{.r}, y_{.r})$ est isomorphe au module induit $Ind_{Q.r}^{GL(d_{.r})}(\mathbb{P}^{L.r})$.

*Remarque.* Cette analogue de la proposition 3.9 revient essentiellement à dire que ce module induit est irréductible. Cela résulte d'un critère d'irréductibilité bien connu, dû à Bernstein et Zelevinsky: si $\chi_1$ et $\chi_2$ sont des caractères de $F^{\times}$ apparaissant dans le support cuspidal de deux facteurs distincts de $\mathbb{P}^{L.r}$, alors $\chi_1\chi_2^{-1} \notin \{|.|, |.|^{-1}\}$. On a même $\chi_1\chi_2^{-1} \notin \{|.|^m; m \in \mathbb{Z}\}$.

Soit $\mathbf{Q}_.$ l'unique sous-groupe parabolique de $\mathbf{G}_\sharp$, inclus dans $\mathbf{Q}$, et tel que:

$$\mathbf{Q}_. \cap \mathbf{L} = \mathbf{Q}_{.1} \times \cdots \times \mathbf{Q}_{.t} \times \mathbf{Q}_{.0}.$$

Posons:

$$\mathbf{L}_. = \mathbf{L}_{.1} \times \cdots \times \mathbf{L}_{.t} \times \mathbf{L}_{.0},$$
$$\mathbb{P}^{L.} = \mathbb{P}^{L.1} \otimes \cdots \otimes \mathbb{P}^{L.t} \otimes \mathbb{P}^{L.0}.$$

Le groupe $\mathbf{L}_.$ est un sous-groupe de Lévi de $\mathbf{Q}_.$ et $\mathbb{P}^{L.}$ est un $L_.$-module. Des isomorphismes précédents se déduit un isomorphisme:

$$Ind_{Q}^{G_\sharp}(\mathbb{P}^{L}) = Ind_{Q.}^{G_\sharp}(\mathbb{P}^{L.}). \tag{4}$$

Fixons maintenant un sous-groupe parabolique $\mathbf{Q}'$ de $\mathbf{G}_\sharp$, de sous-groupe de Lévi:

$$\mathbf{L}' = \mathbf{GL}(d_{1.}) \times \cdots \times \mathbf{GL}(d_{a.}) \times \mathbf{G}_{0.},$$

où $\mathbf{G}_{0.}$ est un groupe de même type que $\mathbf{G}_\sharp$, dont le L-groupe est le groupe symplectique $\hat{G}_{0.}$ de l'espace $\hat{V}_{0.}$. Le triplet $(s_{0.}, y_{0.}, \epsilon)$ paramétrise un $G_{0.}$-module $\mathbb{P}(s_{0.}, y_{0.}, \epsilon)$. Pour $j \in \{1, \ldots, a\}$, le couple $(s_{j.}, y_{j.})$ paramétrise un $GL(d_{j.})$-module $\mathbb{P}(s_{j.}, y_{j.})$. Posons:

$$\mathbb{P}^{L'} = \mathbb{P}(s_{1.}, y_{1.}) \otimes \cdots \otimes \mathbb{P}(s_{a.}, y_{a.}) \otimes \mathbb{P}(s_{0.}, y_{0.}, \epsilon).$$

D'après la proposition 3.9, on a l'isomorphisme:

$$\mathbb{P}(s, y, \epsilon) = Ind_{Q'}^{G_\sharp}(\mathbb{P}^{L'}). \tag{5}$$

Fixons un sous-groupe parabolique $\mathbf{Q}'_{0.}$ de $\mathbf{G}_{0.}$, de sous-groupe de Lévi:

$$\mathbf{L}'_{0.} = \mathbf{GL}(d_{01}^+) \times \cdots \times \mathbf{GL}(d_{0t}^+) \times \mathbf{GL}(d_{01}^-) \times \cdots \times \mathbf{GL}(d_{0t}^-) \times \mathbf{G}_{00}.$$

Introduisons le $L'_{0.}$-module:

$$\mathbb{P}^{L'_{0.}} = \mathbb{P}(s_{01}^+, y_{01}^+) \otimes \cdots \otimes \mathbb{P}(s_{0t}^+, y_{0t}^+) \otimes \mathbb{P}(s_{01}^-, y_{01}^-) \otimes \cdots \otimes \mathbb{P}(s_{0t}^-, y_{0t}^-)$$
$$\otimes \mathbb{P}(s_{00}, y_{00}, \epsilon).$$

Grâce au (i) du lemme 3.8, le $G_{0.}$-module $\mathbb{P}(s_{0.}, y_{0.}, \epsilon)$ est sous-quotient du module induit $Ind_{Q'_{0.}}^{G_{0.}}(\mathbb{P}^{L'_{0.}})$.

Soit $j \in \{1, \ldots, a\}$. Fixons un sous-groupe parabolique $\mathbf{Q}'_{j.}$ de $\mathbf{GL}(d_{j.})$, de sous-groupe de Lévi:

$$\mathbf{L}'_{j.} = \mathbf{GL}(d_{j1}) \times \cdots \times \mathbf{GL}(d_{jt}) \times \mathbf{GL}(d_{j0}) \times \mathbf{GL}(d_{j(-t)}) \times \cdots \times \mathbf{GL}(d_{j(-1)}).$$

Introduisons le $L'_{j.}$-module:

$$\mathbb{P}^{L'_{j.}} = \mathbb{P}(s_{j1}, y_{j1}) \otimes \cdots \otimes \mathbb{P}(s_{jt}, y_{jt}) \otimes \mathbb{P}(s_{j0}, y_{j0}) \otimes \mathbb{P}(s_{j(-t)}, y_{j(-t)}) \otimes \cdots \otimes$$
$$\mathbb{P}(s_{j(-1)}, y_{j(-1)}).$$

Un analogue du lemme 3.8 vaut pour les groupes linéaires. Alors le $GL(d_{j.})$-module $\mathbb{P}(s_{j.}, y_{j.})$ est sous-quotient du module induit $Ind_{Q'_{j.}}^{GL(d_{j.})}(\mathbb{P}^{L'_{j.}})$.

Soit $\mathbf{Q}'_{.}$ l'unique sous-groupe parabolique de $\mathbf{G}_\sharp$, inclus dans $\mathbf{Q}'$, tel que:

$$\mathbf{Q}'_{.} \cap \mathbf{L}' = \mathbf{Q}'_{1.} \times \cdots \times \mathbf{Q}'_{a.} \times \mathbf{Q}'_{0.}.$$

Posons:

$$\mathbf{L}'_{.} = \mathbf{L}'_{1.} \times \ldots \times \mathbf{L}'_{a.} \times \mathbf{L}'_{0.},$$
$$\mathbb{P}^{L'_{.}} = \mathbb{P}^{L'_{1.}} \otimes \cdots \otimes \mathbb{P}^{L'_{a.}} \otimes \mathbb{P}^{L'_{0.}}.$$

Le groupe $\mathbf{L}'_{.}$ est un sous-groupe de Lévi de $\mathbf{Q}'_{.}$ et $\mathbb{P}^{L'_{.}}$ est un $L'_{.}$-module. D'après les résultats ci-dessus, $\mathbb{P}(s, y, \epsilon)$ est sous-quotient du module induit $Ind_{Q'_{.}}^{G_\sharp}(\mathbb{P}^{L'_{.}})$. Mais les triplets $(\mathbf{Q}_{.}, \mathbf{L}_{.}, \mathbb{P}^{L_{.}})$ et $(\mathbf{Q}'_{.}, \mathbf{L}'_{.}, \mathbb{P}^{L'_{.}})$ sont conjugués par un élément de $G_\sharp$. En effet, une telle conjugaison nous permet:

- de permuter les facteurs **GL** de nos Lévi;
- de remplacer un facteur $\mathbf{GL}(d_{j(-r)})$ par $\mathbf{GL}(d_{(-j)r})$, à condition de remplacer $\mathbb{P}(s_{j(-r)}, y_{j(-r)})$ par $\mathbb{P}(s_{(-j)r}, y_{(-j)r})$ (l'opération duale consiste à échanger les espaces isotropes duaux $\hat{V}_{j(-r)}$ et $\hat{V}_{(-j)r}$.

Une succession de telles opérations réalise la conjugaison cherchée. Alors les modules induits:

$$Ind_{Q.}^{G_\sharp}(\mathbb{P}^{L.}) \text{ et } Ind_{Q'.}^{G_\sharp}(\mathbb{P}^{L'.})$$

ont mêmes sous-quotients irréductibles. Donc $\mathbb{P}(s, y, \epsilon)$ est sous-quotient du module induit $Ind_{Q.}^{G_\sharp}(\mathbb{P}^{L.})$ et (4) entraîne (2).

Si $m_1, \ldots, m_h, m$ sont des entiers $\geq 0$ tels que $m = m_1 + \cdots + m_h$ et si $c_1, \ldots, c_h$ sont des réels, on note:

$$c_1^{\oplus m_1} \oplus \cdots \oplus c_h^{\oplus m_h}$$

l'élément de $\mathbb{R}^m$ suivant:

$$(c_1, \ldots, c_1, \ldots, c_h, \ldots, c_h),$$

où chaque $c_i$ intervient $m_i$ fois.

Pour tout signe $\eta$, considérons le couple $(\sigma^\eta, y^\eta)$ défini en 1.11. On a défini en 2.13 l'élément $\tau(\sigma^\eta, y^\eta)$ de $Y_{N^\eta} \otimes_{\mathbb{Z}} \mathbb{R} \simeq \mathbb{R}^{N^\eta}$. On a l'égalité:

$$\tau(\sigma^\eta, y^\eta) = (-log(\nu_1))^{\oplus d_{01}^\eta} \oplus \cdots \oplus (-log(\nu_t))^{\oplus d_{0t}^\eta} \oplus 0^{\oplus N_{00}^\eta}.$$

On déduit de $\tau(\sigma^+, y^+)$ et $\tau(\sigma^-, y^-)$ un élément $\tau(s_{0.}, y_{0.}) \in Y_{N_{0.}} \otimes_{\mathbb{Z}} \mathbb{R}$, où $N_{0.} = N^+ + N^-$, cf. 3.6. On a l'égalité:

$$\tau(s_{0.}, y_{0.}) = (-log(\nu_1))^{\oplus d_{01}} \oplus \cdots \oplus (-log(\nu_t))^{\oplus d_{0t}} \oplus 0^{\oplus N_{00}},$$

où $N_{00} = N_{00}^+ + N_{00}^-$. Notons $J_0$ l'unique sous-ensemble de $\{1, \ldots, N_0\}$ tel que $\tau(s_{0.}, y_{0.}) \in {}^+C_{J_0}$. D'après le corollaire 3.6, il existe $\chi_0 \in \mathcal{E}xp_{P\cap G_{0.}}(\mathbb{P}(s_{0.}, y_{0.}, \epsilon))$ tel que:

$$-log|\chi_0| \in \tau(s_{0.}, y_{0.}) - {}^+\bar{C}^{J_0}.$$

On fixe un tel $\chi_0$.

Soit $j \in \{1, \ldots, a\}$. Un résultat analogue vaut pour le $GL(d_{j.})$-module $\mathbb{P}(s_{j.}, y_{j.})$. Il faut adapter les définitions au cas d'un système de racines $A_{d_{j.}-1}$. Le résultat est le suivant. Posons:

$$\tau(s_{j.}, y_{j.}) = (-log(\nu_1))^{\oplus d_{j1}} \oplus \ldots \oplus (-log(\nu_t))^{\oplus d_{jt}} \oplus 0^{\oplus d_{j0}} \oplus (log(\nu_t))^{\oplus d_{j(-t)}}$$
$$\oplus \cdots \oplus (log(\nu_1))^{\oplus d_{j(-1)}}.$$

Notons $J_j$ l'ensemble des $m \in \{1, \ldots, d_{j.} - 1\}$ tels que $< \alpha_m, \tau(s_{j.}, y_{j.}) >= 0$. Alors il existe $\chi_j \in \mathcal{E}xp_{P\cap GL(d_{j.})}(\mathbb{P}(s_{j.}, y_{j.}))$ tel que:

$$-log|\chi_j| \in \tau(s_{j.}, y_{j.}) - {}^+\bar{C}^{J_j}.$$

On fixe un tel $\chi_j$.

Posons:

$$I = \{1, \ldots, N\}, \ I_{N_{0.}} = \{N - N_{0.} + 1, \ldots, N\}$$

et, pour tout $j \in \{1, \ldots, a\}$:

$$I_j = \{d_{1.} + \cdots + d_{j-1.} + 1, \ldots, d_{1.} + \cdots + d_{j.}\}.$$

On identifie $\{1, \ldots, N_{0.}\}$ à $I_{N_{0.}}$ par l'unique bijection croissante et de même, pour tout $j \in \{1, \ldots, a\}$, on identifie $\{1, \ldots, d_{j.}\}$ à $I_j$. On en déduit une identification de $Y_N \otimes_{\mathbb{Z}} \mathbb{R}$ à:

$$(Y_{d_{1.}} \otimes_{\mathbb{Z}} \mathbb{R}) \oplus \cdots \oplus (Y_{d_{a.}} \otimes_{\mathbb{Z}} \mathbb{R}) \oplus (Y_{N_{0.}} \otimes_{\mathbb{Z}} \mathbb{R}).$$

Le groupe de Weyl $N_{G_\sharp}(M)/M$ s'identifie au sous-groupe $W_N$ du groupe des permutations de $I \cup (-I)$. Définissons le sous-groupe $W^{L'} = N_{L'}(M)/M$ de $W_N$ et l'ensemble $W_{L'}$ des éléments $w \in W_N$ de longueur minimale dans leur classe $W^{L'}w$. L'égalité (5) et le calcul habituel des modules de Jacquet entraînent que l'ensemble $\{log|\chi|; \chi \in \mathcal{E}xp_P(\mathbb{P}(s, y, \epsilon))\}$ est égal à celui des:

$$w^{-1}(log|\xi_1| \oplus \cdots \oplus log|\xi_a| \oplus log|\xi_0|),$$

où:

- $w$ parcourt $W_{L'}$;
- $\xi_1$, resp $\ldots$ $\xi_a, \xi_0$, parcourt $\mathcal{E}xp_{P\cap GL(d_{1.})}(\mathbb{P}(s_{1.}, y_{1.}))$, resp $\ldots$ $\mathcal{E}xp_{P\cap GL(d_{a.})}$ $(\mathbb{P}(s_{a.}, y_{a.}))$, $\mathcal{E}xp_{P\cap G_{0.}}(\mathbb{P}(s_{0.}, y_{0.}, \epsilon))$.

Le terme $\tau = -log(\nu)$ est égal à:

$$(-log(\nu_1))^{\oplus d_{.1}} \oplus \cdots \oplus (-log(\nu_t))^{\oplus d_{.t}} \oplus 0^{\oplus N_{.0}},$$

où $N_{.0} = N - \sum_{r=1,\ldots,t} d_{.r}$. Notons $J$ l'unique sous-ensemble de $I$ tel que $\tau \in C_J^+$. On vérifie qu'il existe $w \in W_{L'}$ tel que:

$$w^{-1}(\tau(s_{1.}, y_{1.}) \oplus \cdots \oplus \tau(s_{a.}, y_{a.}) \oplus \tau(s_{0.}, y_{0.})) = \tau,$$

$$w^{-1}({}^+\bar{C}^{J_1} \oplus \cdots \oplus {}^+\bar{C}^{J_a} \oplus {}^+\bar{C}^{J_0}) \subseteq {}^+\bar{C}^J.$$

Pour un tel $w$, soit $\chi \in \mathcal{E}xp_P(\mathbb{P}(s, y, \epsilon))$ tel que:

$$log|\chi| = w^{-1}(log|\chi_1| \oplus \cdots \oplus log|\chi_a| \oplus log|\chi_0|).$$

Alors $-log|\chi| \in \tau - {}^+\bar{C}^J$. L'assertion (3) en résulte. Cela achève la démonstration.

**4.5.** Soit $\mathbf{Q}$ un sous-groupe parabolique de $\mathbf{G}_\sharp$ de sous-groupe de Lévi:

$$\mathbf{L} = \mathbf{GL}(d_{.1}) \times \cdots \times \mathbf{GL}(d_{.t}) \times \mathbf{G}_{.0},$$

où $\mathbf{G}_{.0}$ est un groupe analogue à $\mathbf{G}_\sharp$. Soient $(\psi_0, \epsilon_0) \in \underline{\Psi}_{utemp}^{G_{.0}}$ et, pour tout $r \in \{1, \ldots, t\}$, $\psi_r \in \underline{\Psi}_{utemp}^{GL(d_{.r})}$ (l'ensemble analogue à $\underline{\Psi}_{utemp}^{G_\sharp}$ quand on remplace $G_\sharp$ par $GL(d_{.r})$). On associe à $(\psi_0, \epsilon_0)$ une représentation $\pi_{\psi_0,\epsilon_0}$ de $G_{.0}$. Pour tout

$r \in \{1, \ldots, t\}$, on associe à $\psi_r$ une représentation $\pi_{\psi_r}$ de $GL(d_{.r})$. On introduit la représentation de $L$:

$$\pi^L = \pi_{\psi_1} \otimes \ldots \otimes \pi_{\psi_t} \otimes \pi_{\psi_0, \epsilon_0}.$$

Soit $\hat{Q}$ un sous-groupe parabolique de $\hat{G}$ dual de $\mathbf{Q}$, de sous-groupe de Lévi:

$$\hat{L} = GL(d_{.1}, \mathbb{C}) \times \cdots \times GL(d_{.t}, \mathbb{C}) \times \hat{G}_{.0}.$$

On considère le produit $\psi_1 \times \cdots \times \psi_t \times \psi_0$ comme un homomorphisme de $W_F \times SL(2, \mathbb{C})$ dans $\hat{L}$. Par composition avec l'injection de $\hat{L}$ dans $\hat{G}$, on obtient un homomorphisme $\psi : W_F \times SL(2, \mathbb{C}) \to \hat{G}$.

On dispose de l'injection naturelle:

$$Z_{\hat{G}_{.0}}(\psi_0)/Z_{\hat{G}_{.0}}(\psi_0)^0 \to Z_{\hat{G}}(\psi)/Z_{\hat{G}}(\psi)^0.$$

Notons $E(\epsilon_0)$ l'ensemble des caractères du quotient de droite dont la restriction au quotient de gauche est $\epsilon_0$. Pour tout $\epsilon \in E(\epsilon_0)$, le couple $(\psi, \epsilon)$ appartient à $\underline{\Psi}^{G_\sharp}_{utemp}$ et paramétrise une représentation $\pi_{\psi,\epsilon}$ de $G_\sharp$.

PROPOSITION. *La représentation induite $Ind_Q^{G_\sharp}(\pi^L)$ est somme directe des représentations $\pi_{\psi,\epsilon}$, où $\epsilon$ décrit $E(\epsilon_0)$.*

*Preuve.* Puisque $\pi^L$ est tempérée, son induite est semi-simple, on peut se contenter de décrire ses sous-quotients irréductibles.

Introduisons une décomposition:

(1) $$\hat{V} = \oplus_{r=-t,\ldots,t} \hat{V}_{.r},$$

de sorte que:

- la décomposition moins fine:

$$\hat{V} = \hat{V}_{.0} \oplus \oplus_{r=1,\ldots,t}(\hat{V}_{.r} \oplus \hat{V}_{.-r})$$

soit orthogonale;

- pour tout $r \in \{1, \ldots, t\}$, $\hat{V}_{.r}$ et $\hat{V}_{.-r}$ soient totalement isotropes;
- $\hat{L}$ conserve la décomposition (1);
- $\hat{Q}$ soit le stabilisateur du drapeau de sous-espaces:

$$\hat{V}_{.1} \subseteq \hat{V}_{.1} \oplus \hat{V}_{.2} \subseteq \ldots \subseteq \hat{V}_{.1} \oplus \ldots \oplus \hat{V}_{.t}.$$

Posons:

$$s = \psi\left(Frob \times \begin{pmatrix} q^{1/2} & 0 \\ 0 & q^{-1/2} \end{pmatrix}\right), \quad y = d(\psi_{|SL(2,\mathbb{C})})\left(\begin{pmatrix} 0 & 1 \\ 0 & 0 \end{pmatrix}\right).$$

Reprenons les notations de la démonstration précédente. Posons:

$$\mathbb{P}^L = \mathbb{P}(s_{.1}, y_{.1}) \otimes \ldots \otimes \mathbb{P}(s_{.t}, y_{.t}) \otimes \mathbb{P}(s_{.0}, y_{.0}, \epsilon_0).$$

En utilisant les propriétés de l'involution de 3.2, on est ramené à prouver que les sous-quotients irréductibles du module induit $Ind_Q^{G_\sharp}(\mathbb{P}^L)$ sont les $\mathbb{P}(s, y, \epsilon)$, pour $\epsilon \in E(\epsilon_0)$, chacun d'eux intervenant avec multiplicité 1.

On peut reprendre les définitions de la démonstration précédente (le $\epsilon$ de celle-ci étant remplacé par $\epsilon_0$). On a alors:

$$Ind_Q^{G_\sharp}(\mathbb{P}^L) = Ind_{Q.}^{G_\sharp}(\mathbb{P}^{L.}),$$

et les modules induits:

$$Ind_{Q.}^{G_\sharp}(\mathbb{P}^{L.}) \text{ et } Ind_{Q'_.}^{G_\sharp}(\mathbb{P}^{L'_.})$$

ont mêmes semi-simplifiés. Il suffit d'étudier ce dernier module.

Nos données vérifient les hypothèses requises pour appliquer le (ii) du lemme 3.8. Il en résulte que le module induit $Ind_{Q'_{0.}}^{G_{0.}}(\mathbb{P}^{L'_{0.}})$ est isomorphe à:

$$\oplus_{\epsilon \in E(\epsilon_0)} \mathbb{P}(s_{0.}, y_{0.}, \epsilon).$$

Comme toujours, un résultat analogue vaut pour les groupes linéaires. On obtient que, pour tout $j \in \{1, \ldots, a\}$, les $GL(d_{j.})$-modules $Ind_{Q'_{j.}}^{GL(d_{j.})}(\mathbb{P}^{L'_{.j}})$ et $\mathbb{P}(s_{j.}, y_{j.})$ sont isomorphes.

De ces assertions résulte que le module $Ind_{Q'_.}^{G_\sharp}(\mathbb{P}^{L'_.})$ est isomorphe à:

$$\oplus_{\epsilon \in E(\epsilon_0)} Ind_{Q'}^{G_\sharp}(\mathbb{P}(s_{1.}, y_{1.}) \otimes \ldots \otimes \mathbb{P}(s_{a.}, y_{a.}) \otimes \mathbb{P}(s_{0.}, y_{0.}, \epsilon)).$$

D'après la proposition 3.8, ce dernier module n'est autre que:

$$\oplus_{\epsilon \in E(\epsilon_0)} \mathbb{P}(s, y, \epsilon).$$

Cela achève la démonstration.

**4.6.** Introduisons quelques notations. Soient $\mathfrak{o}$ une orbite nilpotente dans $\hat{\mathfrak{g}}$ et $y$ un élément de $\mathfrak{o}$. On peut décomposer $y$ en blocs de Jordan. On note:

- $mult_{\mathfrak{o}} : \mathbb{N} \setminus \{0\} \to \mathbb{N}$ la fonction telle que, pour tout $i \in \mathbb{N} \setminus \{0\}$, $mult_{\mathfrak{o}}(i)$ soit le nombre de blocs de Jordan de longueur $i$;
- $Jord(\mathfrak{o}) = \{i \in \mathbb{N} \setminus \{0\}; mult_{\mathfrak{o}}(i) \geq 1\}$;
- $Jord^{bp}(\mathfrak{o})$ le sous-ensemble des entiers pairs dans $Jord(\mathfrak{o})$.

Soit de plus $f : \mathbb{N} \setminus \{0\} \to \mathbb{N}$ une fonction. On pose:

$$Jord^{bp}_{f \geq 1}(\mathfrak{o}) = \{i \in Jord^{bp}(\mathfrak{o}); f(i) \geq 1\},$$

et, pour tout $h \in \mathbb{N}$,

$$Jord^{bp}_{f=h}(\mathfrak{o}) = \{i \in Jord^{bp}(\mathfrak{o}); f(i) = h\}.$$

Notons $\bar{\Psi}^{G_\sharp}_{uquad}$ l'ensemble des quintuplets $\underline{\mathfrak{o}} = (\mathfrak{o}, m^+, \epsilon^+, m^-, \epsilon^-)$ tels que:

- $\mathfrak{o}$ est une orbite nilpotente dans $\hat{\mathfrak{g}}$;
- $m^+, m^- : \mathbb{N} \setminus \{0\} \to \mathbb{N}$ sont des fonctions telles que $m^+ + m^- = mult_{\mathfrak{o}}$ et $m^+(i)$ et $m^-(i)$ sont pairs pour tout entier $i$ impair;
- $\epsilon^+$ et $\epsilon^-$ sont deux fonctions:

$$\epsilon^+ : Jord^{bp}_{m^+\geq 1}(\mathfrak{o}) \to \{\pm 1\},\ \epsilon^- : Jord^{bp}_{m^-\geq 1}(\mathfrak{o}) \to \{\pm 1\};$$

- $$\left(\prod \epsilon^+(i)\right)\left(\prod \epsilon^-(i)\right) = \begin{cases} 1, & si\ \sharp = iso, \\ -1, & si\ \sharp = an, \end{cases}$$

où les produits sont pris sur les $i \in Jord^{bp}(\mathfrak{o})$ tels que $m^+(i)$ est impair, resp. $m^-(i)$ est impair.

Notons $\underline{\Psi}^{G_\sharp}_{uquad}$ l'ensemble des $(\psi, \epsilon) \in \underline{\Psi}^{G_\sharp}_u$ tels que $\psi_{|W_F}$ soit somme de caractères d'ordre au plus 2 (ces caractères sont non ramifiés par définition de $\underline{\Psi}^{G_\sharp}_u$). Soit $(\psi, \epsilon) \in \underline{\Psi}^{G_\sharp}_{uquad}$. Associons-lui un triplet $(s, y, \epsilon)$ comme en 4.1. La décomposition de $\hat{V}$ attachée en 1.11 à ce triplet se réduit à $\hat{V} = \hat{V}^+ \oplus \hat{V}^-$. Les termes $s, y, \epsilon$ se décomposent en $s = s^+s^-$, $y = y^+ + y^-$, $\epsilon = (\epsilon^+, \epsilon^-)$. Notons $\mathfrak{o}$, resp. $\mathfrak{o}^+$, $\mathfrak{o}^-$, l'orbite de $y$ dans $\hat{\mathfrak{g}}$, resp. de $y^+$ dans $\hat{\mathfrak{g}}^+$, de $y^-$ dans $\hat{\mathfrak{g}}^-$. Posons $m^+ = mult_{\mathfrak{o}^+}$, $m^- = mult_{\mathfrak{o}^-}$. Soit $\eta \in \{\pm 1\}$, identifié à un signe. L'hypothèse sur $(\psi, \epsilon)$ entraîne qu'il existe un homomorphisme $\phi^\eta : SL(2, \mathbb{C}) \to \hat{G}^\eta$ tel que:

$$s^\eta = \eta\phi^\eta \begin{pmatrix} q^{1/2} & 0 \\ 0 & q^{-1/2} \end{pmatrix}, \quad y^\eta = d\phi^\eta \begin{pmatrix} 0 & 1 \\ 0 & 0 \end{pmatrix}.$$

Alors:

$$Z_{\hat{G}^\eta}(s^\eta, y^\eta)/Z_{\hat{G}^\eta}(s^\eta, y^\eta)^0 \simeq Z_{\hat{G}^\eta}(y^\eta)/Z_{\hat{G}^\eta}(y^\eta)^0.$$

Ce groupe s'identifie naturellement à $(\mathbb{Z}/2\mathbb{Z})^{Jord^{bp}(\mathfrak{o}^\eta)}$. Remarquons que $Jord^{bp}(\mathfrak{o}^\eta) = Jord^{bp}_{m^\eta\geq 1}(\mathfrak{o})$. Puisque $\epsilon^\eta$ est un caractère du groupe ci-dessus, il s'identifie à une fonction:

$$\epsilon^\eta : Jord^{bp}_{m^\eta\geq 1}(\mathfrak{o}) \to \{\pm 1\}.$$

On vérifie que le quintuplet $\underline{\mathfrak{o}} = (\mathfrak{o}, m^+, \epsilon^+, m^-, \epsilon^-)$ appartient à $\bar{\Psi}^{G_\sharp}_{uquad}$. L'application $(\psi, \epsilon) \mapsto \underline{\mathfrak{o}}$ ainsi définie est une bijection de $\underline{\Psi}^{G_\sharp}_{uquad}$ sur $\bar{\Psi}^{G_\sharp}_{uquad}$. Pour $\underline{\mathfrak{o}} \in \bar{\Psi}^{G_\sharp}_{uquad}$, on pose $\pi_{\underline{\mathfrak{o}}} = \pi_{\psi,\epsilon}$, où $(\psi, \epsilon)$ est l'image réciproque de $\underline{\mathfrak{o}}$ par l'application précédente. Remarquons que $\pi_{\underline{\mathfrak{o}}}$ est tempérée puisque $\underline{\Psi}^{G_\sharp}_{uquad} \subseteq \underline{\Psi}^{G_\sharp}_{utemp}$.

Il est facile de déterminer l'image de $\underline{\Psi}^{G_\sharp}_{udisc}$ par l'application précédente. C'est le sous-ensemble $\bar{\Psi}^{G_\sharp}_{udisc}$ des $\underline{\mathfrak{o}} = (\mathfrak{o}, m^+, \epsilon^+, m^-, \epsilon^-) \in \bar{\Psi}^{G_\sharp}_{uquad}$ tels que $m^+(i) \leq 1$ et $m^-(i) \leq 1$ pour tout entier $i \in \mathbb{N} \setminus \{0\}$. Remarquons que cette relation entraîne $m^+(i) = m^-(i) = 0$ pour tout entier $i$ impair, puisqu'alors $m^+(i)$ et $m^-(i)$ sont pairs. Donc $mult_{\mathfrak{o}}(i) = 0$ pour tout tel entier, i.e. $Jord^{bp}(\mathfrak{o}) = Jord(\mathfrak{o})$. L'application $\underline{\mathfrak{o}} \mapsto$

$\pi_{\underline{\mathfrak{o}}}$ se restreint en une bijection de $\bar{\Psi}^{G_\sharp}_{udisc}$ sur l'ensemble des classes d'isomorphie de représentations lisses irréductibles de $G_\sharp$, de réduction unipotente et de la série discrète.

**4.7.** Notons $\bar{Ell}^{G_\sharp}_u$ l'ensemble des quintuplets $\underline{\mathfrak{o}}_{ell} = (\mathfrak{o}, m^+, \epsilon^+_{ell}, m^-, \epsilon^-_{ell})$ tels que:

- $\mathfrak{o}$ est une orbite nilpotente dans $\hat{\mathfrak{g}}$;
- $Jord^{bp}(\mathfrak{o}) = Jord(\mathfrak{o})$;
- $m^+, m^- : \mathbb{N} \setminus \{0\} \to \mathbb{N}$ sont des fonctions telles que $m^+ + m^- = mult_{\mathfrak{o}}$ et $m^+(i) \leq 2$ et $m^-(i) \leq 2$ pour tout entier $i \geq 1$;
- $\epsilon^+_{ell}, \epsilon^-_{ell}$ sont deux fonctions:
$$\epsilon^+_{ell} : Jord^{bp}_{m^+=1}(\mathfrak{o}) \to \{\pm 1\}, \ \epsilon^-_{ell} : Jord^{bp}_{m^-=1}(\mathfrak{o}) \to \{\pm 1\};$$
- $\left(\prod \epsilon^+_{ell}(i)\right)\left(\prod \epsilon^-_{ell}(i)\right) = \begin{cases} 1, & \text{si } \sharp = iso, \\ -1, & \text{si } \sharp = an, \end{cases}$

où les produits sont pris respectivement sur $i \in Jord^{bp}_{m^+=1}(\mathfrak{o})$, $i \in Jord^{bp}_{m^-=1}(\mathfrak{o})$.

Pour un tel quintuplet, on pose:

$$t(\underline{\mathfrak{o}}_{ell}) = \left|Jord^{bp}_{m^+=2}(\mathfrak{o})\right| + \left|Jord^{bp}_{m^-=2}(\mathfrak{o})\right|$$

et on note $Irr(\underline{\mathfrak{o}}_{ell})$ le sous-ensemble des $\underline{\mathfrak{o}} = (\mathfrak{o}, m^+, \epsilon^+, m^-, \epsilon^-) \in \bar{\Psi}^{G_\sharp}_{uquad}$ tels que:

- les données $\mathfrak{o}$, $m^+$ et $m^-$ sont les mêmes que celles figurant dans $\underline{\mathfrak{o}}_{ell}$;
- pour tout signe $\eta$, la restriction de $\epsilon^\eta$ à $Jord^{bp}_{m^\eta=1}(\mathfrak{o})$ est égale à $\epsilon^\eta_{ell}$.

L'application qui, à un tel quintuplet $\underline{\mathfrak{o}}$, associe le couple des restrictions de $\epsilon^+$ à $Jord^{bp}_{m^+=2}(\mathfrak{o})$ et de $\epsilon^-$ à $Jord^{bp}_{m^-=2}(\mathfrak{o})$ est une bijection de $Irr(\underline{\mathfrak{o}}_{ell})$ sur l'ensemble:

(1) $$\{\pm 1\}^{Jord^{bp}_{m^+=2}(\mathfrak{o})} \times \{\pm 1\}^{Jord^{bp}_{m^-=2}(\mathfrak{o})}.$$

Donc:

$$|Irr(\underline{\mathfrak{o}}_{ell})| = 2^{t(\underline{\mathfrak{o}}_{ell})}.$$

Introduisons l'élément suivant de $\mathbb{C}(Irr^{G_\sharp}_{utemp})$:

$$\pi_{\underline{\mathfrak{o}}_{ell}} = 2^{-t(\underline{\mathfrak{o}}_{ell})} \sum_{\underline{\mathfrak{o}}=(\mathfrak{o},m^+,\epsilon^+,m^-,\epsilon^-)\in Irr(\underline{\mathfrak{o}}_{ell})} \left(\prod_{i\in Jord^{bp}_{m^+=2}(\mathfrak{o})} \epsilon^+(i)\right)\left(\prod_{i\in Jord^{bp}_{m^-=2}(\mathfrak{o})} \epsilon^-(i)\right) \pi_{\underline{\mathfrak{o}}}.$$

En [MW]1.7, à la suite d'Arthur, on a défini le sous-ensemble des représentations elliptiques $Ell_u^{G_\sharp} \subseteq \mathbb{C}(Irr_{utemp}^{G_\sharp})$. En fait, on avait dû choisir des signes, la normalisation des opérateurs d'entrelacement n'étant pas canonique. En [MW]1.7, ces signes avaient été choisis arbitrairement.

THÉORÈME. *On peut choisir ces signes de sorte que, pour tout* $\underline{\mathfrak{o}}_{ell} \in \bar{Ell}_u^{G_\sharp}$, $\pi_{\underline{\mathfrak{o}}_{ell}}$ *appartienne à* $Ell_u^{G_\sharp}$. *L'application* $\underline{\mathfrak{o}}_{ell} \mapsto \pi_{\underline{\mathfrak{o}}_{ell}}$ *est alors une bijection de* $\bar{Ell}_u^{G_\sharp}$ *sur* $Ell_u^{G_\sharp}$.

*Preuve.* On a expliqué en [MW]1.7 comment se construisaient les représentations elliptiques de $G_\sharp$. En modifiant légérement les notations de [MW]1.7, on part d'un sous-groupe parabolique $\mathbf{Q} = \mathbf{LN}$ de $\mathbf{G}_\sharp$ et d'une représentation irréductible $\delta$ de $L$, de la série discrète. On écrit:

$$\mathbf{L} = \mathbf{GL}(d_1) \times \cdots \times \mathbf{GL}(d_r) \times \mathbf{G}_0,$$
$$\delta = \delta_1 \otimes \cdots \otimes \delta_r \otimes \delta_0.$$

Puisqu'on ne s'intéresse qu'aux représentations elliptiques de réduction unipotente, on peut supposer $\delta_1, \ldots, \delta_r$ et $\delta_0$ de réduction unipotente. Alors $\delta_0$ est paramétrisée par $(\psi_0, \epsilon_0) \in \underline{\Psi}_{udisc}^{G_0}$. De même, pour $i \in \{1, \ldots, r\}$, $\delta_i$ est paramétrisée par $\psi_i \in \underline{\Psi}_{udisc}^{GL(d_i)}$. Pour un tel $\psi_i$, la représentation $\psi_{i|SL(2,\mathbb{C})}$ est isomorphe à la représentation $Sym^{d_i-1}$ de $SL(2, \mathbb{C})$ et $\psi_{i|W_F}(Frob)$ est central, égal à l'homothétie de rapport, disons, $z_i \in \mathbb{C}^\times$. L'homomorphisme $\psi_i$ est entièrement déterminé, à conjugaison près, par $z_i$. La condition $s = t$ de [MW]1.7 signifie que $\delta_i = \delta_i^*$ pour tout $i \in \{1, \ldots, r\}$, autrement dit que $z_i \in \{\pm 1\}$ pour tout $i \in \{1, \ldots, r\}$. On peut supposer qu'il existe un entier $r^+ \in \{0, \ldots, r\}$ de sorte que:

$$z_i = \begin{cases} 1, & pour\ i \in \{1, \ldots, r^+\}, \\ -1, & pour\ i \in \{r^+ + 1, \ldots, r\}. \end{cases}$$

Les autres conditions imposées en [MW]1.7 reviennent à dire que $|R_\delta| = 2^r$, autrement dit que la représentation $Ind_Q^{G_\sharp}(\delta)$ est de longueur $2^r$. La décomposition de cette représentation induite est calculée par la proposition 4.5: avec les notations de cette proposition, l'induite est la somme directe des représentations $\pi_{\psi,\epsilon}$ quand $\epsilon$ décrit $E(\epsilon_0)$. Pour tout tel $\epsilon$, $(\psi, \epsilon)$ appartient à $\underline{\Psi}_{uquad}^{G_\sharp}$ et correspond à un élément de $\bar{\Psi}_{uquad}^{G_\sharp}$. C'est un simple exercice de calculer l'ensemble $\mathcal{U}$ des $\underline{\mathfrak{o}} \in \bar{\Psi}_{uquad}^{G_\sharp}$ correspondant aux $(\psi, \epsilon)$ quand $\epsilon$ décrit $E(\epsilon_0)$. Soit $\underline{\mathfrak{o}}_0 = (\mathfrak{o}_0, m_0^+, \epsilon_0^+, m_0^-, \epsilon_0^-) \in \bar{\Psi}_{udisc}^{G_0}$ l'élément correspondant à $(\psi_0, \epsilon_0)$. Définissons des fonctions $m^+, m^- : \mathbb{N} \setminus \{0\} \to \mathbb{N}$ par:

$$m^+(i) = m_0^+(i) + 2|\{j \in \{1, \ldots, r^+\}; d_j = i\}|,$$
$$m^-(i) = m_0^-(i) + 2|\{j \in \{r^+ + 1, \ldots, r\}; d_j = i\}|,$$

pour tout $i \in \mathbb{N} \setminus \{0\}$. Il existe une unique orbite nilpotente $\mathfrak{o}$ dans $\hat{\mathfrak{g}}$ telle que $mult_{\mathfrak{o}} = m^+ + m^-$. Pour tout signe $\eta$, posons pour simplifier:

$$J^\eta = Jord^{bp}_{m^\eta \geq 1}(\mathfrak{o}), \quad J^\eta_0 = Jord^{bp}_{m^\eta_0 \geq 1}(\mathfrak{o}).$$

On a l'inclusion $J^\eta_0 \subseteq J^\eta$. Si $\epsilon^\eta$ est une fonction sur $J^\eta$, notons $res_0(\epsilon^\eta)$ sa restriction à $J^\eta_0$. Alors $\mathcal{U}$ est l'ensemble des quadruplets $\underline{\mathfrak{o}} = (\mathfrak{o}, m^+, \epsilon^+, m^-, \epsilon^-) \in \bar{\Psi}^{G_\sharp}_{uquad}$ tels que $\mathfrak{o}, m^+, m^-$ soient les termes définis ci-dessus et $\epsilon^+$,$\epsilon^-$ vérifient:

$$res_0(\epsilon^+) = \epsilon^+_0, \quad res_0(\epsilon^-) = \epsilon^-_0.$$

L'application qui, à un tel quadruplet, associe la restriction de $(\epsilon^+, \epsilon^-)$ à $(J^+ \setminus J^+_0) \times (J^- \setminus J^-_0)$ est une bijection de $\mathcal{U}$ sur:

$$\{\pm 1\}^{J^+ \setminus J^+_0} \times \{\pm 1\}^{J^- \setminus J^-_0}.$$

Notre condition $|\mathcal{U}| = 2^r$ se traduit par l'égalité:

$$|J^+ \setminus J^+_0| + |J^- \setminus J^-_0| = r.$$

Par construction, on a aussi:

$$\left(\sum_{i \geq 1} \frac{m^+(i) - m^+_0(i)}{2}\right) + \left(\sum_{i \geq 1} \frac{m^-(i) - m^-_0(i)}{2}\right) = r.$$

De plus, pour tout signe $\eta$ et tout $i \in J^\eta \setminus J^\eta_0$, on a $\frac{m^\eta(i) - m^\eta_0(i)}{2} \geq 1$. On se rappelle que $\underline{\mathfrak{o}}_0 \in \bar{\Psi}^{G_0}_{udisc}$, donc $Jord^{bp}(\mathfrak{o}_0) = Jord(\mathfrak{o}_0)$ et $m^\eta_0(i) \leq 1$ pour tout $\eta$ et tout $i$. La comparaison des égalités ci-dessus entraîne que pour tout signe $\eta$ et tout entier $i \geq 1$, on a les relations:

- si $i \notin J^\eta$, $m^\eta(i) = m^\eta_0(i) = 0$;
- si $i \in J^\eta \setminus J^\eta_0$, $m^\eta(i) = 2$ et $m^\eta_0(i) = 0$;
- si $i \in J^\eta_0$, $m^\eta(i) = m^\eta_0(i) = 1$.

Alors $Jord^{bp}(\mathfrak{o}) = Jord(\mathfrak{o})$, on a $m^+(i) \leq 2$ et $m^-(i) \leq 2$ pour tout entier $i \geq 1$, et on a les égalités:

$$Jord^{bp}_{m^+ = 1}(\mathfrak{o}) = Jord^{bp}_{m^+_0 \geq 1}(\mathfrak{o}_0), \quad Jord^{bp}_{m^- = 1}(\mathfrak{o}) = Jord^{bp}_{m^-_0 \geq 1}(\mathfrak{o}_0).$$

Posons $\underline{\mathfrak{o}}_{ell} = (\mathfrak{o}, m^+, \epsilon^+_0, m^-, \epsilon^-_0)$. Alors $\underline{\mathfrak{o}}_{ell} \in \bar{Ell}^{G_\sharp}_u$ et l'ensemble $\mathcal{U}$ n'est autre que $Irr(\underline{\mathfrak{o}}_{ell})$. Remarquons qu'on a l'égalité $r = t(\underline{\mathfrak{o}}_{ell})$.

D'après [MW]1.7, de nos données est issue une unique représentation elliptique $\pi$. Si $t(\underline{\mathfrak{o}}_{ell}) = 0$, on a $\underline{\mathfrak{o}}_{ell} = \underline{\mathfrak{o}}_0$ et $\pi = \pi_{\underline{\mathfrak{o}}_0} = \pi_{\underline{\mathfrak{o}}_{ell}}$. Supposons $t(\underline{\mathfrak{o}}_{ell}) \geq 1$. La représentation $\pi$ est de la forme:

$$\pi = 2^{-t(\underline{\mathfrak{o}}_{ell})} \sum_{\underline{\mathfrak{o}} \in Irr(\underline{\mathfrak{o}}_{ell})} c(\underline{\mathfrak{o}}) \pi_{\underline{\mathfrak{o}}}$$

où $c(\underline{\mathfrak{o}}) \in \{\pm 1\}$ pour tout $\underline{\mathfrak{o}} \in Irr(\underline{\mathfrak{o}}_{ell})$. Pour déterminer ces signes, introduisons sur l'espace $\mathbb{C}(Irr^{G_\sharp}_{utemp})$ le produit scalaire naïf pour lequel $Irr^{G_\sharp}_{utemp}$ forme une base orthonormée. Il résulte de [MW]1.7 que $\pi$ est orthogonale à toute représentation induite à partir d'un sous-groupe parabolique propre de $\mathbf{G}_\sharp$. Fixons un signe $\eta$ et un entier $i \geq 1$ tel que $m^\eta(i) = 2$. Supposons pour simplifier la rédaction $\eta = +$. Introduisons un groupe $\mathbf{G}'_\sharp$ analogue à $\mathbf{G}_\sharp$, de rang absolu $n - i$, et son L-groupe $\hat{G}'$. Définissons $m'^{+} : \mathbb{N} \setminus \{0\} \to \mathbb{N}$ par:

$$m'^{+}(i') = \begin{cases} m^{+}(i'), & \text{si } i' \neq i, \\ 0, & \text{si } i' = i. \end{cases}$$

Il existe une unique orbite nilpotente $\mathfrak{o}'$ dans $\hat{\mathfrak{g}}'$ telle que $mult_{\mathfrak{o}'} = m'^{+} + m^{-}$. On a les égalités:

$$Jord^{bp}_{m'^{+}=1}(\mathfrak{o}') = Jord^{bp}_{m_0^{+}\geq 1}(\mathfrak{o}_0), \quad Jord^{bp}_{m^{-}=1}(\mathfrak{o}') = Jord^{bp}_{m_0^{-}\geq 1}(\mathfrak{o}_0).$$

Soient:

$$\epsilon'^{+} : Jord^{bp}_{m'^{+}\geq 1}(\mathfrak{o}') \to \{\pm 1\}, \quad \epsilon^{-} : Jord^{bp}_{m^{-}\geq 1}(\mathfrak{o}') \to \{\pm 1\}$$

deux fonctions dont les restrictions à $Jord^{bp}_{m_0^{+}\geq 1}(\mathfrak{o}_0)$, resp. $Jord^{bp}_{m_0^{-}\geq 1}(\mathfrak{o}_0)$ soient égales à $\epsilon_0^{+}, \epsilon_0^{-}$. Le quintuplet $\underline{\mathfrak{o}}' = (\mathfrak{o}', m'^{+}, \epsilon'^{+}, m^{-}, \epsilon^{-})$ appartient à $\bar{\Psi}^{G'_\sharp}_{uquad}$ et paramétrise une représentation tempérée $\pi_{\underline{\mathfrak{o}}'}$ de $G'_\sharp$. Par construction de $m^{+}$, il existe un unique $j \in \{1, \ldots, r^{+}\}$ tel que $d_j = i$. Considérons le sous-groupe parabolique $\mathbf{Q}'$ de $\mathbf{G}_\sharp$, de sous-groupe de Lévi $\mathbf{L}' = \mathbf{GL}(d_j) \times \mathbf{G}'_\sharp$. Considérons la représentation $\pi' = \delta_j \otimes \pi_{\underline{\mathfrak{o}}'}$ de $L'$. On décompose la représentation induite:

(2) $$Ind^{G_\sharp}_{Q'}(\pi')$$

de la même façon que l'on a décomposé l'induite $Ind^{G_\sharp}_{Q}(\delta)$. Le résultat est le suivant. On a l'égalité:

$$Jord^{bp}_{m^{+}\geq 1}(\mathfrak{o}) = Jord^{bp}_{m'^{+}\geq 1}(\mathfrak{o}') \cup \{i\}.$$

Notons $\epsilon_1^{+}$ et $\epsilon_2^{+}$ les deux fonctions sur $Jord^{bp}_{m^{+}\geq 1}(\mathfrak{o})$, à valeurs dans $\{\pm 1\}$, dont les restrictions à $Jord^{bp}_{m'^{+}\geq 1}(\mathfrak{o}')$ sont égales à $\epsilon'^{+}$. Pour $\ell \in \{1, 2\}$, posons $\underline{\mathfrak{o}}_\ell = (\mathfrak{o}, m^{+}, \epsilon_\ell^{+}, m^{-}, \epsilon^{-})$. Alors l'induite (2) est la somme de $\pi_{\underline{\mathfrak{o}}_1}$ et de $\pi_{\underline{\mathfrak{o}}_2}$. Remarquons que $\underline{\mathfrak{o}}_1, \underline{\mathfrak{o}}_2 \in Irr(\underline{\mathfrak{o}}_{ell})$. Puisque $\pi$ est orthogonale à cette induite, on a l'égalité $c(\underline{\mathfrak{o}}_1) = -c(\underline{\mathfrak{o}}_2)$. En faisant varier $i, \epsilon'^{+}, \epsilon^{-}$, puis en faisant la même chose pour $\eta = -$, on obtient le résultat suivant. Identifions $Irr(\underline{\mathfrak{o}}_{ell})$ à l'ensemble (1). Soient $\underline{\mathfrak{o}}_1, \underline{\mathfrak{o}}_2$ deux éléments de $Irr(\underline{\mathfrak{o}}_{ell})$ dont toutes les composantes sont égales, sauf pour un indice, pour lequel ces composantes sont opposées. Alors $c(\underline{\mathfrak{o}}_1) = -c(\underline{\mathfrak{o}}_2)$. Cette propriété entraîne qu'il existe $c \in \{\pm 1\}$ tel que, pour tout

$\underline{\mathfrak{o}} = (\mathfrak{o}, m^+, \epsilon^+, m^-, \epsilon^-) \in Irr(\underline{\mathfrak{o}}_{ell})$, on a l'égalité:

$$c(\underline{\mathfrak{o}}) = c\left(\prod_{i \in Jord^{bp}_{m^+=2}(\mathfrak{o})} \epsilon^+(i)\right)\left(\prod_{i \in Jord^{bp}_{m^-=2}(\mathfrak{o})} \epsilon^-(i)\right).$$

Quitte à changer les signes des normalisations de [MW]1.7, on peut supposer $c = 1$. Alors $\pi = \pi_{\underline{\mathfrak{o}}_{ell}}$.

A toute représentation $\pi \in Ell_u^{G_\sharp}$, on vient d'associer un élément $\underline{\mathfrak{o}}_{ell} \in \bar{Ell}_u^{G_\sharp}$ tel que $\pi = \pi_{\underline{\mathfrak{o}}_{ell}}$. Il est facile de vérifier la bijectivité de cette application. Cela démontre le théorème.

## 5. Restriction aux normalisateurs des sous-groupes parahoriques.

**5.1.** Soient $d$ un entier pair $\geq 2$, $\hat{V}$ un espace de dimension $d$ sur $\mathbb{C}$ muni d'une forme symplectique, $\hat{G}$ son groupe symplectique. Soit $y$ un élément nilpotent de $\hat{\mathfrak{g}}$. Comme en 2.1, posons:

$$\bar{A}(y) = Z_{\hat{G}}(y)/Z_{\hat{G}}(y)^0.$$

Soit $\epsilon$ un caractère de ce groupe. La correspondance de Springer généralisée (cf. [L6]) associe au couple $(y, \epsilon)$:

- un entier $k \geq 0$ tel que $k(k+1) \leq d$:
- une représentation irréductible $\rho(y, \epsilon)$ du groupe $W_N$, où $N = (d - k(k+1))/2$ (on rappelle que $W_N$ est le groupe de Weyl de type $C_N$).

Si besoin est, nous noterons plus précisément $k(y, \epsilon)$ l'entier $k$.

Considérons les constructions de 2.1, relatives à cet entier $k$. L'algèbre de Lie $\hat{\mathfrak{a}}(y)$ contient en particulier le point $(\sigma, r_0) = (0, 0)$. On a défini un ensemble $\mathcal{E}_k(0, 0, y)$ de caractères du groupe $\bar{A}(0, y) = \bar{A}(y)$ et, pour tout $\epsilon' \in \mathcal{E}_k(0, 0, y)$, un $\bar{\mathcal{H}}$-module $\mathbb{M}(0, 0, y, \epsilon')$, où:

$$\bar{\mathcal{H}} = \bar{\mathcal{H}}(N; 2k+1).$$

Par définition de la correspondance de Springer généralisée, $\epsilon$ appartient à $\mathcal{E}_k(0, 0, y)$. On dispose donc du $\bar{\mathcal{H}}$-module $\mathbb{M}(0, 0, y, \epsilon)$. Rappelons que l'on a un plongement:

$$\begin{aligned} W_N &\rightarrow \bar{\mathcal{H}} \\ w &\mapsto t_w \end{aligned}$$

grâce auquel $\mathbb{M}(0, 0, y, \epsilon)$ apparaît comme un $W_N$-module. Nous noterons $\underline{\rho}(y, \epsilon)$ la représentation de $W_N$ dans ce module. Elle n'est pas irréductible en général. Grâce à [L7] 24.2.14, on a une décomposition:

$$\underline{\rho}(y, \epsilon) = \rho(y, \epsilon) \oplus (\oplus_{(y', \epsilon')} p_{y', \epsilon'; y, \epsilon}\, \rho(y', \epsilon')),$$

où les coefficients $p_{y',\epsilon';y,\epsilon}$ sont des entiers $\geq 0$ et $(y',\epsilon')$ parcourt un ensemble fini de couples tels que:

- $y'$ est un élément nilpotent de $\hat{\mathfrak{g}}$ tel que, si l'on note $\mathfrak{o}_y$, resp. $\mathfrak{o}_{y'}$, l'orbite de $y$, resp. $y'$, pour l'action adjointe de $\hat{G}$, on a $\mathfrak{o}_y \subset \bar{\mathfrak{o}}_{y'}$ et $\mathfrak{o}_y \neq \mathfrak{o}_{y'}$;
- $\epsilon'$ est un caractère de $\bar{A}(y')$ tel que $k(y',\epsilon') = k$.

**5.2.** Soit $(s, y, \epsilon)$ un triplet comme en 1.11. En utilisant les notations de ce paragraphe, nous supposons:

- $a = 0$;
- pour tout signe $\eta$, il existe un homomorphisme $\phi^\eta : SL(2,\mathbb{C}) \to \hat{G}^\eta$ tel que:

$$d\phi^\eta\left(\begin{pmatrix} \frac{log(q)}{2} & 0 \\ 0 & -\frac{log(q)}{2} \end{pmatrix}\right) = \sigma^\eta, \quad d\phi^\eta\left(\begin{pmatrix} 0 & 1 \\ 0 & 0 \end{pmatrix}\right) = y^\eta.$$

On introduit les entiers $N^+$, $N^-$, $N$, $k^+$, $k^-$, $R'$, $R''$, $A$, $B$ et le signe $\zeta$. On a associé à $(s, y, \epsilon)$ une représentation de l'algèbre $\mathcal{H}_\theta$, où $\theta = (R', R'')$. Notons $\pi$ cette représentation.

Considérons maintenant deux entiers $N', N'' \geq 0$ tels que $N' + N'' = N$. Introduisons les sous-algèbres $\mathcal{H}'_\theta$ et $\mathcal{H}''_\theta$ suivantes (cf. 1.3 pour les notations):

- $\mathcal{H}'_\theta$ est engendrée par $\{S_i; i = N''+1, \dots, N\}$;
- si $N'' = R'' = 0$, $\mathcal{H}''_\theta = \mathbb{C}$;
- si $N'' > 0$ et $R'' = 0$, $\mathcal{H}''_\theta$ est engendrée par $\omega$ et $\{S_i; i = 0, \dots, N''-1\}$;
- si $R'' > 0$, $\mathcal{H}''_\theta$ est engendrée par $\{S_i; i = 0, \dots, N''-1\}$.

L'algèbre $\mathcal{H}'_\theta$ est une algèbre de Hecke, de paramètre $q$, pour le groupe $W_{N'}$. Ses représentations irréductibles sont en bijection avec celles du groupe $W_{N'}$. Si $R'' > 0$, l'algèbre $\mathcal{H}''_\theta$ est aussi une algèbre de Hecke, de paramètre $q$, pour le groupe $W_{N''}$. Ses représentations irréductibles sont en bijection avec celles du groupe $W_{N''}$. Si $R'' = 0$ et $N'' > 0$, $\mathcal{H}''_\theta$ est le produit tensoriel semi-direct de $\mathbb{C}[\omega]$ et d'une algèbre de Hecke, de paramètre $q$, pour le groupe $W^D_{N''}$ (le groupe de Weyl de type $D_{N''}$). Comme il est expliqué en [MW]2.9, ses représentations irréductibles sont encore en bijection avec les représentations irréductibles de $W_{N''}$.

Notons $res(\pi)$ la restriction de $\pi$ à $\mathcal{H}'_\theta \otimes_{\mathbb{C}} \mathcal{H}''_\theta$. On peut considérer $res(\pi)$ comme un élément de $\mathbb{C}(\hat{W}_{N'}) \otimes_{\mathbb{C}} \mathbb{C}(\hat{W}_{N''})$.

Pour tout signe $\eta$, on a défini en 5.1 une représentation $\underline{\rho}(y^\eta, \epsilon^\eta)$ de $W_{N^\eta}$. On en déduit un élément:

$$\underline{\rho}(y^+,\epsilon^+) \otimes \underline{\rho}(y^-,\epsilon^-) \in \mathbb{C}(\hat{W}_{N^+}) \otimes_{\mathbb{C}} \mathbb{C}(\hat{W}_{N^-}).$$

Reprenons les constructions de [MW]3.9 et [MW]3.10. Définissons deux éléments $\hat{h}^+, \hat{h}^- \in \hat{h}$ de la façon suivante:

- si $R'' \leq R'$ et $\zeta = 1$, $\hat{h}^+ = (0,0)$, $\hat{h}^- = (0,1)$;
- si $R'' \leq R'$ et $\zeta = -1$, $\hat{h}^+ = (0,1)$, $\hat{h}^- = (0,0)$;
- si $R' < R''$ et $\zeta = 1$, $\hat{h}^+ = (0,0)$, $\hat{h}^- = (1,0)$;
- si $R' < R''$ et $\zeta = -1$, $\hat{h}^+ = (1,0)$, $\hat{h}^- = (0,0)$.

A l'aide du couple $\hat{h}^+, \hat{h}^-$, on a défini en [MW]3.10 une application:

$$\iota_{N^+,N^-} : \mathbb{C}(\hat{W}_{N^+}) \otimes_{\mathbb{C}} \mathbb{C}(\hat{W}_{N^-}) \to \mathbb{C}(\hat{\mathcal{W}}_N).$$

On a aussi défini en [MW]3.9 une application:

$$\rho^*_{N',N''} : \mathbb{C}(\hat{\mathcal{W}}_N) \to \mathbb{C}(\hat{W}_{N'}) \otimes_{\mathbb{C}} \mathbb{C}(\hat{W}_{N''}).$$

PROPOSITION. *On a l'égalité:*

$$res(\pi) = \rho^*_{N',N''} \circ \iota_{N^+,N^-}(\underline{\rho}(y^+,\epsilon^+) \otimes \underline{\rho}(y^-,\epsilon^-)).$$

*Preuve.* Supposons $R'' \leq R'$ et $\zeta = 1$. Posons $\mathcal{H} = \mathcal{H}(N; A, B)$ et introduisons les sous-$\mathcal{A}$-algèbres $\mathcal{H}'$ et $\mathcal{H}''$ de $\mathcal{H}$ suivantes:

- $\mathcal{H}'$ est engendrée par $\{T_i; i = N''+1, \dots, N\}$;
- $\mathcal{H}''$ est engendrée par $\{T_i; i = 0, \dots, N''-1\}$,

cf. 1.4 pour les notations. Il résulte de 1.5 et 1.11 que, pour notre problème, on peut remplacer $\mathcal{H}_\theta$ par $\mathcal{H}$, $\mathcal{H}'_\theta$ par $\mathcal{H}'$, $\mathcal{H}''_\theta$ par $\mathcal{H}''$ et $\pi$ par la représentation de $\mathcal{H}$ associée en 1.11 à $(s, y, \epsilon)$. C'est cette dernière représentation que nous noterons $\pi$ désormais.

Pour tout signe $\eta$, posons:

$$\mathcal{H}^\eta = \mathcal{H}(N^\eta; A, B), \quad \bar{\mathcal{H}}^\eta = \bar{\mathcal{H}}(N^\eta; A + \eta B).$$

Soit $z \in \mathbb{C}$. On dispose du $\bar{\mathcal{H}}^\eta$-module $\mathbb{M}(z\sigma^\eta, z\frac{log(q)}{2}, y^\eta, \epsilon^\eta)$. Notons $\bar{\pi}^\eta(z)$ la représentation de $\bar{\mathcal{H}}^\eta$ dans ce module. Pourvu que $z \notin \frac{4\pi i}{log(q)}\mathbb{Q}$, on en déduit une représentation $\pi^\eta(z)$ de $\mathcal{H}^\eta$ (cf. 1.9, où l'on prend $\epsilon = \eta$). Grâce à 1.7, on déduit de la représentation $\pi^+(z) \otimes \pi^-(z)$ de $\mathcal{H}^+ \otimes_{\mathcal{A}} \mathcal{H}^-$ une représentation $\pi(z)$ de $\mathcal{H}$. On la restreint ensuite en une représentation $res(\pi(z))$ de $\mathcal{H}' \otimes_{\mathcal{A}} \mathcal{H}''$. Par construction, $\pi = \pi(1)$ (remarquons que, pour tout signe $\eta$, on a l'égalité:

$$\mathbb{P}\left(\sigma^\eta, \frac{log(q)}{2}, y^\eta, \epsilon^\eta\right) = \mathbb{M}\left(\sigma^\eta, \frac{log(q)}{2}, y^\eta, \epsilon^\eta\right)$$

d'après le corollaire 2.13 et nos hypothèses).

On va montrer que $\pi(z)$ est continue en $z \in \mathbb{C} \setminus \frac{4\pi i}{log(q)}\mathbb{Q}$ et qu'elle a une limite quand $z$ tend vers 0. En notant $\pi(0)$ cette limite, on va calculer $res(\pi(0))$.

Remarquons que chaque représentation $res(\pi(z))$ s'identifie à une représentation de $W_{N'} \times W_{N''}$. Celle-ci varie continûment en $z$ et est donc constante. On a donc l'égalité $res(\pi) = res(\pi(0))$, les deux membres étant interprétés comme des éléments de $\mathbb{C}(\hat{W}_{N'}) \otimes_{\mathbb{C}} \mathbb{C}(\hat{W}_{N''})$. La proposition découlera donc des résultats que l'on va maintenant démontrer.

Soit $\eta \in \{\pm 1\}$, identifié à un signe. On reprend les constructions du paragraphe 2. Notons $\hat{D}$ le sous-tore de $\hat{G}^{\eta}_{\mathbb{C}}$, image de l'homomorphisme:

$$\begin{array}{ccc} \mathbb{C}^{\times} & \to & \hat{G}^{\eta}_{\mathbb{C}} \\ \lambda & \mapsto & \left(\phi^{\eta}\left(\begin{pmatrix} \lambda & 0 \\ 0 & \lambda^{-1} \end{pmatrix}\right), \lambda\right). \end{array}$$

Il est inclus dans $\hat{A}(y)$. Son algèbre de Lie est la droite $\{(z\sigma^{\eta}, z\frac{log(q)}{2}); z \in \mathbb{C}\}$, et l'algèbre d'homologie $H_{\hat{D}}$ est l'algèbre des polynômes sur cette droite, notons-la $\mathbb{C}[Z]$, où $Z$ est le polynôme valant $z$ au point $(z\sigma^{\eta}, z\frac{log(q)}{2})$. Considérons l'espace de cohomologie $H^{\hat{D}}(\mathcal{P}_y, \mathcal{L})$. On a l'isomorphisme:

$$H_{\hat{D}} \otimes_{H_{\hat{A}(y)}} H^{\hat{A}(y)}(\mathcal{P}_y, \mathcal{L}) \simeq H^{\hat{D}}(\mathcal{P}_y, \mathcal{L})$$

(cf. [L3] proposition 7.5). Donc $H^{\hat{D}}(\mathcal{P}_y, \mathcal{L})$ se retrouve muni de deux actions de $\bar{\mathcal{H}}^{\eta}$ et $\bar{A}(y)$. Notons $H^{\hat{D}}(\mathcal{P}_y, \mathcal{L})^{\epsilon^{\eta}}$ le sous-espace de $H^{\hat{D}}(\mathcal{P}_y, \mathcal{L})$ dans lequel $\bar{A}(y)$ agit par le caractère $\epsilon^{\eta}$. C'est un sous-$H_{\hat{D}}$-module. D'après l'isomorphisme ci-dessus, pour tout $z \in \mathbb{C}$, on a l'isomorphisme:

$$\mathbb{M}\left(z\sigma^{\eta}, z\frac{log(q)}{2}, y^{\eta}, \epsilon^{\eta}\right) = \mathbb{C} \otimes_{H_{\hat{D}}} H^{\hat{D}}(\mathcal{P}_y, \mathcal{L})^{\epsilon^{\eta}},$$

la tensorisation se faisant via l'homomorphisme:

$$\begin{array}{ccc} H_{\hat{D}} = \mathbb{C}[Z] & \to & \mathbb{C} \\ P(Z) & \mapsto & P(z). \end{array}$$

Mais $H^{\hat{D}}(\mathcal{P}_y, \mathcal{L})$ est un $H_{\hat{D}}$-module projectif de type fini ([L3] proposition 7.2). Il en est de même de $H^{\hat{D}}(\mathcal{P}_y, \mathcal{L})^{\epsilon^{\eta}}$, qui est un facteur direct du précédent. Tout $\mathbb{C}[Z]$-module projectif de type fini est libre. Grâce à l'isomorphisme ci-dessus, $\bar{\pi}^{\eta}(z)$ apparaît comme la spécialisation d'une représentation de $\bar{\mathcal{H}}^{\eta}$ algébrique en $z \in \mathbb{C}$. A fortiori, elle est continue. Pour $z = 0$, l'ensemble des exposants $\mathcal{E}xp(\bar{\pi}^{\eta}(0))$ est réduit à $\{(0, 0)\}$, cf. la remarque de 2.5. L'idéal d'augmentation $\mathcal{I}$ de la sous-algèbre $\mathcal{S}$ de $\bar{\mathcal{H}}^{\eta}$ agit donc de façon nilpotente sur $\mathbb{M}(0, 0, y^{\eta}, \epsilon^{\eta})$. Il agit trivialement sur le semi-simplifié de ce module. Notons $\bar{\pi}^{\eta}(0)^{ss}$ la représentation de $\bar{\mathcal{H}}^{\eta}$ dans ce semi-simplifié. Elle apparaît comme une représentation de $\bar{\mathcal{H}}^{\eta}/\bar{\mathcal{H}}^{\eta}\mathcal{I}$. Cette algèbre n'est autre que la $\mathbb{C}$-algèbre du groupe $W_{N^{\eta}}$. La représentation $\underline{\rho}(y^{\eta}, \epsilon^{\eta})$ du groupe $W_{N^{\eta}}$ s'interprète comme une représentation de cette algèbre. Elle est égale à $\bar{\pi}^{\eta}(0)^{ss}$, par définition de $\underline{\rho}(y^{\eta}, \epsilon^{\eta})$.

Fixons un élément $\underline{\lambda}^{\eta} = (\lambda_1^{\eta}, \ldots, \lambda_{N^{\eta}}^{\eta}; \frac{log(q)}{2})$ de $\mathcal{E}xp(\bar{\pi}^{\eta}(1))$. Alors $\mathcal{E}xp(\bar{\pi}^{\eta}(1)) \subseteq W_{N^{\eta}}(\underline{\lambda}^{\eta})$, cf. 1.8 pour la notation. Plus généralement, pour tout

$z \in \mathbb{C}$, $\mathcal{E}xp(\bar{\pi}^{\eta}(z)) \subseteq W_{N^\eta}(z\underline{\lambda}^{\eta})$. Soit $z_0 \in \mathbb{C}$. Supposons $z_0 = 0$ ou $z_0 \notin \frac{4\pi i}{log(q)}\mathbb{Q}$. L'image de $\mathcal{H}^{\eta}$ par l'homomorphisme $gr_{\eta}$ de 1.9 est incluse dans $\mathcal{M}_{W_{N^\eta}(z_0\underline{\lambda}^{\eta})}\bar{\mathcal{H}}^{\eta}$: si $z_0 \neq 0$, cela résulte de [L2], lemme 9.5 (c'est ce que l'on a utilisé en 1.9); si $z_0 = 0$, cela résulte de [L2] 9.7. Pour tout voisinage $\mathcal{V}$ de $W_{N^\eta}(z_0\underline{\lambda}^{\eta})$, invariant par $W_{N^\eta}$, notons $\mathcal{M}_{\mathcal{V}}$ la sous-algèbre des éléments de $\mathcal{M}$ qui sont holomorphes dans $\mathcal{V}$. On définit $\mathcal{M}_{\mathcal{V}}\bar{\mathcal{H}}^{\eta}$ comme on a défini $\bar{\mathcal{H}}^{\eta}$, en remplaçant $\mathcal{S}$ par $\mathcal{M}_{\mathcal{V}}$ dans les définitions. Puisque $\mathcal{H}^{\eta}$ est de type fini, si $\mathcal{V}$ est assez petit, l'image de $\mathcal{H}^{\eta}$ par $gr_{\eta}$ est incluse dans $\mathcal{M}_{\mathcal{V}}\bar{\mathcal{H}}^{\eta}$. Si $z$ est assez proche de $z_0$, $W_{N^\eta}(z\underline{\lambda}^{\eta})$ est inclus dans $\mathcal{V}$ et $\bar{\pi}^{\eta}(z)$ se prolonge en une représentation de $\mathcal{M}_{\mathcal{V}}\bar{\mathcal{H}}^{\eta}$. Ce prolongement est continu en $z$. Puisque $\pi^{\eta}(z)$ est la composée de cette représentation et de l'homomorphisme $gr_{\eta}$, $\pi^{\eta}(z)$ est continue au voisinage de $z_0$.

Notons $\pi^{\eta}(0)^{ss}$ la semi-simplifiée de $\pi^{\eta}(0)$. C'est la composée de $\bar{\pi}^{\eta}(0)^{ss}$ et de $gr_{\eta}$. Evidemment, $\pi^{\eta}(0)^{ss}(v)$ est l'identité. On utilise ci-dessous les termes $\xi_{e_i}$, $\tau(w_i)$, $\bar{\tau}(w_i)$, $\mathcal{G}(i)$, $\bar{\mathcal{G}}(i)$ définis en 1.4, 1.6, 1.8. Pour $i \in \{1, \ldots, N^{\eta}\}$, on a:

$$gr_{\eta}(\xi_{e_i}) = \eta exp(x_i).$$

Puisque $\bar{\pi}^{\eta}(0)^{ss}(x_i) = 0$, $\pi^{\eta}(0)^{ss}(\xi_{e_i})$ est l'homothétie de rapport $\eta$. On a l'égalité:

$$gr_{\eta}(\tau(w_i)) = \bar{\tau}(w_i),$$

d'où:

$$gr_{\eta}(T_i) = -1 + (t_i + 1)\bar{\mathcal{G}}(i)^{-1}gr_{\eta}(\mathcal{G}(i)). \tag{1}$$

Le terme $\bar{\mathcal{G}}(i)^{-1}gr_{\eta}(\mathcal{G}(i))$ est un élément de $\mathcal{M}$, holomorphe au point $\underline{0} = (0, 0) \in (Y_{N^\eta} \otimes_{\mathbb{Z}} \mathbb{C}) \oplus \mathbb{C}$. Montrons qu'il vaut 1 en ce point. Supposons d'abord $i \neq N^{\eta}$. Par définition:

$$\bar{\mathcal{G}}(i)^{-1}gr_{\eta}(\mathcal{G}(i)) = \frac{\alpha_i}{\alpha_i + 2r}\frac{exp(\alpha_i + 2r) - 1}{exp(\alpha_i) - 1}.$$

Chacun des termes:

$$\frac{\alpha_i}{exp(\alpha_i) - 1} \quad \text{et} \quad \frac{exp(\alpha_i + 2r) - 1}{\alpha_i + 2r}$$

est holomorphe en $\underline{0}$ et vaut 1 en ce point. Supposons $i = N^{\eta}$. Alors, en posant $\alpha = \alpha_{N^\eta}$, on a:

$$\begin{aligned}
&\bar{\mathcal{G}}(i)^{-1}gr_{\eta}(\mathcal{G}(i)) \\
&\quad = \frac{\alpha}{\alpha + (A + \eta B)r}\frac{(\eta exp(\alpha + (A + B)r) - 1)(\eta exp(\alpha + (A - B)r) + 1)}{exp(2\alpha) - 1}, \\
&\quad = \frac{\alpha}{\alpha + (A + \eta B)r}\frac{(exp(\alpha + (A + \eta B)r) - 1)(exp(\alpha + (A - \eta B)r) + 1)}{exp(2\alpha) - 1}.
\end{aligned}$$

Chacun des termes:

$$\frac{\alpha}{exp(2\alpha) - 1}, \quad \frac{exp(\alpha + (A + \eta B)r) - 1}{\alpha + (A + \eta B)r}, \quad exp(\alpha + (A - \eta B)r) + 1,$$

est holomorphe en $\underline{0}$. En ce point, ils valent respectivement $\frac{1}{2}$, 1 et 2. Cela démontre l'assertion. On en déduit que $\bar{\pi}^{\eta}(0)^{ss}(\bar{\mathcal{G}}(i)^{-1}gr_{\eta}(\mathcal{G}(i)))$ est l'identité, puis, grâce à (1), que:

$$\pi^{\eta}(0)^{ss}(T_i) = \bar{\pi}^{\eta}(0)^{ss}(t_i) = \underline{\rho}(y^{\eta}, \epsilon^{\eta})(w_i).$$

On peut formuler le calcul de $\pi^{\eta}(0)^{ss}$ à l'aide des groupes introduits en [MW]3.9. Soit $m \in \mathbb{N}$. Rappelons que $W_m$, resp. $\mathcal{W}_m$, est un sous-groupe du groupe des permutations de $\{\pm 1\} \times \{1, \dots, m\}$, resp. $\{\pm 1\} \times \{\pm 1\} \times \{1, \dots, m\}$. Soit $i \in \{1, \dots, m\}$ et $w \in W_m$. On définit deux éléments $\underline{e}_i$ et $\underline{w}$ de $\mathcal{W}_m$ par les formules suivantes, pour $\alpha, \beta \in \{\pm 1\}$ et $j \in \{1, \dots, m\}$:

$$\underline{e}_i(\alpha, \beta, j) = \begin{cases} (\alpha, \beta, j), & \text{si } j \neq i, \\ (-\alpha, -\beta, i), & \text{si } j = i; \end{cases}$$

$$\underline{w}(\alpha, \beta, j) = (\alpha', \beta, j'), \text{ où } w(\alpha, j) = (\alpha', j').$$

Pour tout groupe fini $X$, notons $\mathbb{C}[X]$ la $\mathbb{C}$-algèbre du groupe $X$. On vérifie que l'on peut définir un homomorphisme:

$$\delta_m : \mathcal{H}(m; A, B) \to \mathbb{C}[\mathcal{W}_m]$$

par les formules suivantes:

- $\delta_m(v) = 1$;
- pour $i \in \{1, \dots, m\}$, $\delta_m(\xi_{e_i}) = \underline{e}_i$;
- pour $w \in W_m$, $\delta_m(T_w) = \underline{w}$.

En [MW]3.9, on a défini une projection $p_m : \mathcal{W}_m \to W_m$ qui se prolonge en un homomorphisme $p_m : \mathbb{C}[\mathcal{W}_m] \to \mathbb{C}[W_m]$. Pour $\hat{h} \in \hat{H}$, on a défini un caractère $\chi_{\hat{h}}$ de $\mathcal{W}_m$, qui se prolonge en un homomorphisme $\chi_{\hat{h}} : \mathbb{C}[\mathcal{W}_m] \to \mathbb{C}$. On peut alors reformuler les calculs ci-dessus par l'égalité:

(2) $$\pi^{\eta}(0)^{ss} = ((\underline{\rho}(y^{\eta}, \epsilon^{\eta}) \circ p_{N^{\eta}}) \otimes \chi_{\hat{h}^{\eta}}) \circ \delta_{N^{\eta}},$$

où $\hat{h}^{+} = (0, 0)$ et $\hat{h}^{-} = (0, 1)$.

Pour $z \in \mathbb{C}$, posons:

$$\underline{s}(z) = \left( exp(z\lambda_1^{+}), \dots, exp(z\lambda_{N^{+}}^{+}), -exp(z\lambda_1^{-}), \dots, -exp(z\lambda_{N^{-}}^{-}); exp\left(z\frac{log(q)}{2}\right)\right).$$

C'est un élément de $(Y_N \otimes_{\mathbb{Z}} \mathbb{C}^{\times}) \times \mathbb{C}^{\times}$. Si $z \notin \frac{4\pi i}{log(q)}\mathbb{Q}$ ou si $z = 0$, on peut appliquer la construction de 1.7 au point $\underline{s}$ et à la représentation $\pi^{+}(z) \otimes \pi^{-}(z)$ de $\mathcal{H}' = \mathcal{H}_{N^{+}} \otimes_{\mathcal{A}} \mathcal{H}_{N^{-}}$. Fixons un tel point $z_0$. L'image de $\mathcal{H}$ par l'homomorphisme $m$ de 1.7 prend ses valeurs dans $M_{\mathrm{u}}(\mathcal{H}' \otimes_{\mathcal{Z}'} \mathcal{Z}'_{(\underline{s}(z_0))})$. Puisque $\mathcal{H}$ est de type fini, on peut fixer un voisinage de Zariski $\mathcal{V}$ de $\underline{s}(z_0)$ tel qu'en notant $\mathcal{Z}'_{\mathcal{V}}$ l'anneau des éléments du corps des fractions de $\mathcal{Z}'$ réguliers dans $\mathcal{V}$, l'image de $\mathcal{H}$ par $m$ soit incluse dans $M_{\mathrm{u}}(\mathcal{H}' \otimes_{\mathcal{Z}'} \mathcal{Z}'_{\mathcal{V}})$. Si $z$ est assez proche de $z_0$ (et $z \notin \frac{4\pi i}{log(q)}\mathbb{Q}$ ou $z = 0$), $\underline{s}(z)$

appartient à $\mathcal{V}$ et la représentation $(\pi^+(z) \otimes \pi^-(z))^{\oplus \mathrm{u}}$ de $M_{\mathrm{u}}(\mathcal{H}')$ se prolonge en une représentation de $M_{\mathrm{u}}(\mathcal{H}' \otimes_{\mathcal{Z}'} \mathcal{Z}'_{\mathcal{V}})$, qui est continue en $z$. Puisque $\pi(z)$ est la composée de cette représentation et de l'homomorphisme $m$, $\pi(z)$ est continue au voisinage de $z_0$.

On a un plongement naturel:

$$\mathcal{W}_{N^+} \times \mathcal{W}_{N^-} \to \mathcal{W}_N.$$

Posons:

$$\mathcal{C} = \mathbb{C}[\mathcal{W}_{N^+} \times \mathcal{W}_{N^-}] = \mathbb{C}[\mathcal{W}_{N^+}] \otimes_{\mathbb{C}} \mathbb{C}[\mathcal{W}_{N^-}].$$

En 1.7, on a introduit le sous-ensemble $\mathcal{U}$ des éléments $w \in W_N$ qui sont de longueur minimale dans leur classe $(W_{N^+} \times W_{N^-})w$. On vérifie que l'on peut définir un homomorphisme:

$$\begin{aligned} \mu : \mathbb{C}[\mathcal{W}_N] &\to \quad M_{\mathrm{u}}(\mathcal{C}) \\ f \quad &\mapsto (\mu(f)_{w',w''})_{w',w'' \in \mathcal{U}} \end{aligned}$$

par la formule suivante, pour $w \in \mathcal{W}_N$ et $w', w'' \in \mathcal{U}$:

$$\mu(w)_{w',w''} = \begin{cases} 0, & \text{si } \underline{w}' w \underline{w}''^{-1} \notin \mathcal{W}_{N^+} \times \mathcal{W}_{N^-}, \\ \underline{w}' w \underline{w}''^{-1}, & \text{si } \underline{w}' w \underline{w}''^{-1} \in \mathcal{W}_{N^+} \times \mathcal{W}_{N^-}. \end{cases}$$

Posons:

$$\underline{\rho} = ((\underline{\rho}(y^+, \epsilon^+) \circ p_{N^+}) \otimes \chi_{\hat{h}^+}) \otimes ((\underline{\rho}(y^-, \epsilon^-) \circ p_{N^-}) \otimes \chi_{\hat{h}^-}).$$

C'est une représentation de $\mathcal{C}$. On en déduit une représentation $\underline{\rho}^{\oplus \mathrm{u}}$ de $M_{\mathrm{u}}(\mathcal{C})$. Posons:

$$\underline{\pi} = \pi^+(0)^{ss} \otimes \pi^-(0)^{ss},$$

et notons $\pi(0)^{ss}$ la semi-simplifiée de $\pi(0)$. C'est la composée de $\underline{\rho}^{\oplus \mathrm{u}}$ et de l'homomorphisme $m$. On va démontrer l'égalité:

(3) $$\pi(0)^{ss} = \underline{\rho}^{\oplus \mathrm{u}} \circ \mu \circ \delta_N.$$

Par construction, il s'agit de prouver que, pour tous $h \in \mathcal{H}$, $w', w'' \in \mathcal{U}$, on a l'égalité:

(4) $$\underline{\pi}(m(h)_{w',w''}) = \underline{\rho}(\mu(\delta_N(h))_{w',w''}).$$

Il suffit de la démontrer pour $h$ parcourant un ensemble de générateurs de $\mathcal{H}$.

Soit $i \in \{1, \ldots, N\}$, vérifions (4) pour $h = \xi_{e_i}$. On voit que les deux membres sont nuls si $w' \neq w''$. Supposons $w' = w''$. Remarquons que $\mathcal{U}$ s'identifie à l'ensemble des permutations $w$ de $\{1, \ldots, N\}$ telles que $w^{-1}$ est croissante sur chacun des intervalles $\{1, \ldots, N^+\}$, $\{N^+ + 1, \ldots, N\}$. Alors:

$$\underline{\pi}(m(\xi_{e_i})_{w',w''}) = \underline{\pi}(\xi_{w'(e_i)}) = \underline{\pi}\big(\xi_{e_{w'(i)}}\big).$$

Grâce à (2), c'est égal à $\underline{\rho}(\underline{e}_{w'(i)})$. De même:

$$\underline{\rho}(\mu(\delta_N(\xi_{e_i}))_{w',w''}) = \underline{\rho}(\mu(\underline{e}_i)_{w',w''}) = \underline{\rho}(\underline{e}_{w'(i)}).$$

D'où (4) dans ce cas.

Soit $i \in \{1, \dots, N-1\}$, vérifions (4) pour $h = T_i$. On a:

$$T_i = -1 + (\tau(w_i) + 1)\mathcal{G}(i).$$

La forme des matrices $m(\tau(w_i))$ et $m(\mathcal{G}(i))$ entraîne que:

$$m(T_i)_{w',w''} = \begin{cases} 0, & \text{si } w' \neq w'' \text{ et } \\ & w'w_iw''^{-1} \notin W_{N^+} \times W_{N^-}, \\ -1 + m(\mathcal{G}(i))_{w',w''}, & \text{si } w' = w'' \text{ et } \\ & w'w_iw''^{-1} \notin W_{N^+} \times W_{N^-}, \\ m(\tau(w_i))_{w',w''} m(\mathcal{G}(i))_{w'',w''}, & \text{si } w' \neq w'' \text{ et } \\ & w'w_iw''^{-1} \in W_{N^+} \times W_{N^-}, \\ -1 + (m(\tau(w_i))_{w',w'} + 1) m(\mathcal{G}(i))_{w',w'}, & \text{si } w' = w'' \text{ et } \\ & w'w_iw''^{-1} \in W_{N^+} \times W_{N^-}. \end{cases}$$

Posons $j = w''(i)$, $k = w''(i+1)$. Supposons d'abord $j$, $k$ dans le même intervalle $\{1, \dots, N^+\}$ ou $\{N^+ + 1, \dots, N\}$. Alors $k = j + 1$ et $w'w_iw''^{-1} \in W_{N^+} \times W_{N^-}$ si et seulement si $w' = w''$. Les deuxième et troisième cas du tableau ci-dessus sont exclus. Supposons $w' = w''$. Alors $w'w_iw'^{-1} = w_j$, $m(\tau(w_i))_{w',w'} = \tau'(w_j)$, $m(\mathcal{G}(i))_{w',w'} = \mathcal{G}'(j)$, où, pour plus de précision, on note par des $'$ les éléments relatifs à l'algèbre $\mathcal{H}' = \mathcal{H}^+ \otimes_{\mathcal{A}} \mathcal{H}^-$. Alors:

$$m(T_i)_{w',w''} = -1 + (\tau'(w_j) + 1)\mathcal{G}'(j) = T'_j = T'(w'w_iw''^{-1}).$$

Supposons maintenant que $j$ et $k$ n'appartiennent pas au même intervalle. On a $w'ww''^{-1} \in W_{N^+} \times W_{N^-}$ si et seulement si $w' = w''w_i$. Le quatrième cas du tableau ci-dessus est exclu. Supposons $w' = w''$ (on est alors dans le deuxième cas). On a:

$$m(\mathcal{G}(i))_{w',w'} = \frac{\xi_{e_j - e_k} v^2 - 1}{\xi_{e_j - e_k} - 1}.$$

Parce que $j$ et $k$ ne sont pas dans le même intervalle, ce terme appartient à l'algèbre localisée $\mathcal{A} \otimes_{\mathcal{Z}'} \mathcal{Z}'_{(\underline{s}(0))}$. Plus précisément, $\underline{\pi}(\xi_{e_j})$ et $\underline{\pi}(\xi_{e_k})$ sont des homothéties de rapport des signes opposés. Donc $\underline{\pi}(m(\mathcal{G}(i))_{w',w'})$ est l'identité. Alors $\underline{\pi}(m(T_i)_{w',w''}) = 0$. Supposons maintenant $w' = w''w_j$. On est dans le troisième cas du tableau ci-dessus. Comme précédemment, $\underline{\pi}(m(\mathcal{G}(i))_{w'',w''})$ est l'identité. On a:

$$m(\tau(w_i))_{w',w''} = \tau'(w'w_iw''^{-1}) = \tau'(1) = T'(1) = T'(w'w_iw''^{-1}),$$

et $\underline{\pi}(m(T_i)_{w',w''}) = \underline{\pi}(T'(w'w_iw''^{-1}))$. En résumé, on a prouvé l'égalité:

$$\underline{\pi}(m(T_i)_{w',w''}) = \begin{cases} 0, & \text{si } w'w_iw''^{-1} \notin W_{N^+} \times W_{N^-}, \\ \underline{\pi}(T'(w'w_iw''^{-1})), & \text{si } w'w_iw''^{-1} \in W_{N^+} \times W_{N^-}. \end{cases}$$

Grâce à (2), on peut remplacer ce dernier terme $\underline{\pi}(T'(w'w_iw''^{-1}))$ par $\underline{\rho}(\underline{w}'\underline{w}_i\underline{w}''^{-1})$. L'égalité (4), pour $h = T_i$, en découle.

Vérifions enfin (4) pour $h = T_N$. On a le même tableau que précédemment. Maintenant $w'w_Nw''^{-1} \in W_{N^+} \times W_{N^-}$ si et seulement si $w' = w''$. Les deuxième et troisième cas du tableau sont exclus. Supposons $w' = w''$. On a nécessairement $w''(N) = N^+$ ou $w''(N) = N$. Supposons par exemple $w''(N) = N^+$. Alors:

$$m(\tau(w_N)_{w',w'}) = \tau(w'w_Nw'^{-1}) = \tau'(w'_{N^+}),$$

où $w'_{N^+}$ est l'élément de $W_{N^+}$ analogue de $w_N$. Et $m(\mathcal{G}(N))_{w',w'} = \mathcal{G}'(N^+)$. Alors:

$$m(T_N)_{w',w'} = -1 + (\tau'(w'_{N^+}) + 1)\mathcal{G}'(N^+) = T'_{N^+} = T'(w'w_Nw''^{-1}).$$

Le calcul se poursuit comme dans le cas $h = T_i, i < N$, et conduit encore à l'égalité (4). Cela achève la preuve de (3).

On veut calculer $res(\pi(0))$, identifiée à une représentation de $W_{N'} \times W_{N''}$. Remarquons que toute représentation de ce groupe étant semi-simple, on peut remplacer $res(\pi(0))$ par $res(\pi(0)^{ss})$. Pour plus de précision, notons $res_{\mathcal{H}}(\pi(0)^{ss})$ la représentation de $\mathcal{H}' \otimes_{\mathcal{A}} \mathcal{H}''$ et $res_W(\pi(0)^{ss})$ la représentation de $W_{N'} \times W_{N''}$ à laquelle elle s'identifie. On dispose de l'homomorphisme de spécialisation en $v = 1$:

$$Sp : \mathcal{H}' \otimes_{\mathcal{A}} \mathcal{H}'' \to \mathbb{C}[W_{N'}] \otimes_{\mathbb{C}} \mathbb{C}[W_{N''}],$$

et on a l'égalité $res_{\mathcal{H}}(\pi(0)^{ss}) = res_W(\pi(0)^{ss}) \circ Sp$. Notons $R$ l'homomorphisme tel que le diagramme suivant soit commutatif:

$$\begin{array}{ccc} \mathcal{H}' \otimes_{\mathcal{A}} \mathcal{H}'' & \to & \mathcal{H} \\ Sp \downarrow & & \downarrow \delta_N \\ \mathbb{C}[W_{N'}] \otimes_{\mathbb{C}} \mathbb{C}[W_{N''}] & \xrightarrow{R} & \mathbb{C}[\mathcal{W}_N] \end{array}$$

Il résulte de (3) que:

$$res_W(\pi(0)^{ss}) = \underline{\rho}^{\oplus \mathrm{u}} \circ \mu \circ R. \tag{5}$$

Calculons $R$. Notons $\{w'_i; i = 1, \dots, N'\}$, resp. $\{w''_i; i = 1, \dots, N''\}$, les générateurs habituels de $W_{N'}$, resp. $W_{N''}$. On a les égalités:

$$Sp(T_i) = w''_{N''-i}, \text{ pour } i \in \{0, \dots, N''-1\},$$

$$Sp(T_i) = w'_{i-N''}, \text{ pour } i \in \{N''+1, \dots, N\}.$$

Pour $i \in \{1, .., N\}$, $\delta_N(T_i) = \underline{w}_i$. Donc:

$$R(w'_i) = \underline{w}_{i+N''}, \text{ pour } i \in \{1, \dots, N'\},$$

$$R(w''_i) = \underline{w}_{N''-i}, \text{ pour } i \in \{1, \dots, N''-1\}.$$

Avec les notations de 1.4, on a l'égalité:

$$T_0 = v^{2N+A+B-2}\xi_{e_1}T(\sigma'_0)^{-1}.$$

D'où $R(w''_{N''}) = \delta_N(T_0) = \underline{e}_1\underline{\sigma}_0'^{-1}$. En [MW], 3.9, on a introduit un homomorphisme:

$$\rho_{N',N''} : W_{N'} \times W_{N''} \to \mathcal{W}_N.$$

On vérifie sur les formules ci-dessus qu'il existe un élément $w$ de $\mathcal{W}_N$ tel que $R$ soit l'homomorphisme d'algèbres déduit de l'homomorphisme $Ad(w) \circ \rho_{N',N''}$. Dans l'égalité (5), on peut aussi bien remplacer $R$ par l'homomorphisme d'algèbres déduit de $\rho_{N',N''}$. Interprétons maintenant $res_W(\pi(0)^{ss})$ comme un élément de $\mathbb{C}(\hat{W}_{N'}) \otimes_{\mathbb{C}} \mathbb{C}(\hat{W}_{N''})$. En se remémorant les définitions de [MW]3.9 et [MW]3.10, l'égalité (5) devient:

$$res_W(\pi(0)^{ss}) = \rho^*_{N',N''} \circ \iota_{N^+,N^-}(\underline{\rho}(y^+, \epsilon^+) \otimes \underline{\rho}(y^-, \epsilon^-)).$$

On a déjà dit que $res(\pi) = res_W(\pi(0)^{ss})$. On obtient la formule de l'énoncé.

On a supposé $R'' \leq R'$ et $\zeta = 1$. Dans le cas où $\zeta = -1$, la construction de $\pi$ (cf. 1.11) échange les rôles de $\bar{\mathcal{H}}^+$ et $\bar{\mathcal{H}}^-$. Cela conduit au même résultat que ci-dessus, où l'on échange $\hat{h}^+$ et $\hat{h}^-$. Dans le cas où $R' < R''$, c'est l'identification de $\mathcal{H}_\theta$ à $\mathcal{H}$ ou à $\mathbb{C}[\omega] \otimes_{\mathbb{C}} \mathcal{H}$ qui échange les rôles de $\mathcal{H}'$ et $\mathcal{H}''$ (cf. 1.4). Dans les calculs ci-dessus, on doit remplacer $\rho_{N',N''}$ par l'homomorphisme:

$$(6) \qquad \begin{array}{ccc} W_{N'} \times W_{N''} & \to & \mathcal{W}_N \\ (w', w'') & \mapsto & \rho_{N'',N'}(w'', w') \end{array}$$

Pour tout entier $m \in \mathbb{N}$, notons $\nu_m$ l'automorphisme de $\{\pm 1\} \times \{\pm 1\} \times \{1, \dots, m\}$ défini par la formule:

$$\nu_m(\alpha, \beta, i) = (\beta, \alpha, i),$$

pour $\alpha, \beta \in \{\pm 1\}$, $i \in \{1, \dots, m\}$. La conjugaison par $\nu_m$ conserve le groupe $\mathcal{W}_m$. On vérifie qu'il existe $w \in \mathcal{W}_N$ tel que le plongement (6) soit égal à $Ad(w\nu_N) \circ \rho_{N',N''}$. On peut aussi bien remplacer le plongement (6) par $Ad(\nu_N) \circ \rho_{N',N''}$. On obtient:

$$res(\pi) = \rho^*_{N',N''} \circ Ad(\nu_N)^* \circ \iota_{N^+,N^-}(\underline{\rho}(y^+, \epsilon^+) \otimes \underline{\rho}(y^-, \epsilon^-)),$$

où $\iota_{N^+,N^-}$ est défini comme dans le cas $R'' \leq R'$. Reprenons les notations de [MW]3.10. Soient $f' \in \mathbb{C}(\hat{W}_{m'})$, $f'' \in \mathbb{C}(\hat{W}_{m''})$, calculons:

$$Ad(\nu_m)^* \circ \iota_{m',m''}(f' \otimes f''),$$

l'application $\iota_{m',m''}$ étant définie à l'aide d'un couple $(\hat{h}', \hat{h}'') = ((a', b'), (a'', b''))$. L'automorphisme $Ad(\nu_m)$ de $\mathcal{W}_m$ conserve le sous-groupe $\mathcal{W}_{m'} \times \mathcal{W}_{m''}$ et agit sur celui-ci comme $Ad(\nu_{m'}) \times Ad(\nu_{m''})$. Définissons la fonction $f_0^\nu$ sur $\mathcal{W}_{m'} \times \mathcal{W}_{m''}$ par:

$$f_0^\nu(w', w'') = f_0(Ad(\nu_{m'})(w'), Ad(\nu_{m''})(w'')).$$

Alors $Ad(\nu_m)^* \circ \iota_{m',m''}(f' \otimes f'')$ est l'induite à $\mathcal{W}_m$ de $f_0^\nu$. On vérifie que:

$$p_{m'} \circ Ad(\nu_{m'}) = p_{m'}, \quad \underline{\chi}_{\hat{h}'} \circ Ad(\nu_{m'}) = \underline{\chi}_{\nu(\hat{h}')},$$

où $\nu(\hat{h}') = (b', a')$. On a bien sûr des formules analogues pour le facteur $\mathcal{W}_{m''}$. D'où l'égalité:

$$Ad(\nu_m)^* \circ \iota_{m',m''}(f' \otimes f'') = \iota^\nu_{m',m''}(f' \otimes f''),$$

où $\iota^\nu_{m',m''}$ est définie à l'aide du couple $(\nu(\hat{h}'), \nu(\hat{h}''))$. On obtient finalement que $res(\pi)$ est calculée par la même formule que dans le cas $R'' \leq R'$, le couple $(\hat{h}^+, \hat{h}^-)$ étant changé en $(\nu(\hat{h}^+), \nu(\hat{h}^-))$. Cela conduit aux formules de l'énoncé.

**5.3.** Soient $\sharp = iso$ ou $an$ et $(\psi, \epsilon) \in \underline{\Psi}^{G_\sharp}_{uquad}$. On lui associe un triplet $(s, y, \epsilon)$ comme en 4.1. Ce triplet vérifie les hypothèses de 5.2. On lui associe les entiers $N^+, N^-, R', R''$, le signe $\zeta$ et les représentations $\underline{\rho}(y^+, \epsilon^+)$ de $W_{N^+}$ et $\underline{\rho}(y^-, \epsilon^-)$ de $W_{N^-}$. Posons $\gamma = (R', \zeta R'', N^+, N^-)$. C'est un élément de l'ensemble $\Gamma$ de [MW]3.17. On a l'égalité:

$$\mathcal{R}(\gamma) = \mathbb{C}(\hat{W}_{N^+}) \otimes_{\mathbb{C}} \mathbb{C}(\hat{W}_{N^-}),$$

cf. [MW], 3.17. La représentation $\underline{\rho}(y^+, \epsilon^+) \otimes \underline{\rho}(y^-, \epsilon^-)$ appartient à cet ensemble. Puisque $\mathcal{R}(\gamma)$ est un sous-espace de l'espace $\mathcal{R}$ de [MW]3.16, on peut considérer cette représentation comme un élément de $\mathcal{R}$. On a défini:

- en [MW]3.16, un espace $\mathcal{R}^{par}$ et un isomorphisme $Rep : \mathcal{R} \to \mathcal{R}^{par}$;
- en [MW]3.18, un endomorphisme $\rho\iota$ de $\mathcal{R}$;
- en [MW]4.2, un homomorphisme $Res$: $\mathbb{C}(Irr_u^{G_\sharp}) \to \mathcal{R}^{par}$;
- en 3.2, une involution $\mathbf{D}^{G_\sharp}$ de $Irr_u^{G_\sharp}$.

*Remarque.* La définition de $Rep$ nécessite des choix de prolongements, cf. [MW] 2.9. Nous les fixons ici comme expliqué en 6.3.

PROPOSITION. *Avec les notations ci-dessus, on a l'égalité:*

$$Res \circ \mathbf{D}^{G_\sharp}(\pi_{\psi,\epsilon}) = Rep \circ \rho\iota(\underline{\rho}(y^+, \epsilon^+) \otimes \underline{\rho}(y^-, \epsilon^-)).$$

*Preuve.* Posons $\pi = \mathbf{D}^{G_\sharp}(\pi_{\psi,\epsilon})$. C'est la représentation de $G_\sharp$ dans le module $\mathbb{P}(s, y, \epsilon)$. Soient $(n', n'') \in D(n)$, i.e. $n', n'' \in \mathbb{N}$ tels que $n' + n'' = n$. Si $\sharp = an$, on suppose $n'' \geq 1$. Calculons la composante $Res^{par}_{n',n''}(\pi)$ de $Res^{par}(\pi)$ dans le sommand:

$$\mathbb{C}[SO(2n'+1)]_u \otimes_{\mathbb{C}} \mathbb{C}[O(2n'')_\sharp]_u$$

de $\mathcal{R}^{par}$. Choisissons un réseau presque autodual $L$ de $V_\sharp$ tel que:

$$dim_{\mathbb{F}_q}(L/\varpi_F\tilde{L}) = 2n' + 1, \quad dim_{\mathbb{F}_q}(\tilde{L}/L) = 2n'',$$

cf. 1.1 pour les notations. Alors:

$$K^{\pm}(L)/K(L)^u \simeq SO(2n'+1) \times O(2n'')_\sharp,$$

et $Res^{par}_{n',n''}(\pi)$ est la trace de la représentation de ce groupe dans le sous-espace des invariants $\mathbb{P}(s, y, \epsilon)^{K(L)^u}$. D'après [MP] théorème 3.5, cet espace n'est non nul que si $K(L)$ contient un conjugué du groupe $K_\theta$ de 1.2, i.e., si $n' \geq R'^2 + R'$ et $n'' \geq R''^2$. Supposons ces conditions vérifiées. On peut alors supposer que $L$ est le réseau $L_{n''-R''^2}$ de 1.2. Notons $\mathcal{H}^K_\theta$ la sous-algèbre des éléments de l'algèbre $\mathcal{H}_\theta$ à support dans $K^{\pm}(L)$. Avec les notations de 5.2, où l'on prend $N' = n' - R'^2 - R'$, $N'' = n'' - R''^2$, on a:

$$\mathcal{H}^K_\theta = \begin{cases} \mathcal{H}'_\theta \otimes_{\mathbb{C}} \mathcal{H}''_\theta, & \text{si } R'' = 0, \\ \mathcal{H}'_\theta \otimes_{\mathbb{C}} (\mathbb{C}[\omega] \otimes_{\mathbb{C}} \mathcal{H}''_\theta), & \text{si } R'' > 0. \end{cases}$$

Soit $E$ un sous-$K^{\pm}(L)$-module irréductible de $\mathbb{P}(s, y, \epsilon)^{K(L)^u}$. Toujours d'après [MP] théorème 3.5, l'espace $Hom_{K_\theta}(E_\theta, E)$ est non nul et c'est un $\mathcal{H}^K_\theta$-module irréductible. Il lui correspond deux représentations irréductibles $\rho'_E$ de $W_{N'}$ et $\rho''_E$ de $W_{N''}$, plus un signe $\nu_E$ dans le cas où $R'' > 0$, à savoir le signe par lequel agit $\omega$. Posons $\gamma_E = (R', 0, N', N'')$ si $R'' = 0$, $\gamma_E = (R', \nu_E R'', N', N'')$ si $R'' > 0$, $\rho_E = \rho'_E \otimes \rho''_E$. Alors $\rho_E \in \mathcal{R}(\gamma_E)$ et, par définition de l'application $Rep$, $Rep(\rho_E)$ est la trace de la représentation de $K^{\pm}(L)$ dans $E$. Le terme $Res^{par}_{n',n''}(\pi)$ est égal à la somme des $Rep(\rho_E)$ quand $E$ parcourt les sous-modules irréductibles de $\mathbb{P}(s, y, \epsilon)^{K(L)^u}$, comptés bien sûr avec leurs multiplicités.

Notons $\tilde{\pi}$ la représentation de $\mathcal{H}_\theta$ associée à $\pi$. Quand $E$ parcourt l'ensemble de sous-modules ci-dessus, les représentations associées de $\mathcal{H}^K_\theta$ parcourent l'ensemble des composantes irréductibles (comptées avec leurs multiplicités) de la restriction de $\tilde{\pi}$ à $\mathcal{H}^K_\theta$. Dans le cas où $R'' > 0$, $\tilde{\pi}(\omega)$ est la multiplication par $(-1)^{R'}\zeta$. Le signe $\nu_E$ est donc constant; en posant $\gamma_{N',N''} = (R', (-1)^{R'}\zeta R'', N', N'')$, on a toujours $\gamma_E = \gamma_{N',N''}$. D'autre part, la somme des $\rho'_E \otimes \rho''_E$ n'est autre que la représentation $res(\tilde{\pi})$ définie en 5.2. Cette représentation est calculée par la proposition 5.2. On obtient que $Res^{par}_{n',n''}(\pi)$ est égal à l'image par $Rep$ de l'élément:

$$\rho^*_{N',N''} \circ \iota_{N^+,N^-}(\underline{\rho}(y^+, \epsilon^+) \otimes \underline{\rho}(y^-, \epsilon^-))$$

de $\mathcal{R}(\gamma_{N',N''})$, l'application $\iota_{N^+,N^-}$ étant définie comme en 5.2. On vérifie sur les définitions que ce terme n'est autre que la composante dans $\mathcal{R}(\gamma_{N',N''})$ de:

$$\rho\iota(\underline{\rho}(y^+, \epsilon^+) \otimes \underline{\rho}(y^-, \epsilon^-)).$$

Pour obtenir $Res^{par}(\pi)$, il suffit de sommer sur $(n', n'') \in D(n)$. Cela revient à sommer les termes ci-dessus sur $(N', N'') \in D(n - R'^2 - R' - R''^2)$. On obtient alors l'égalité de l'énoncé.

## 6. Choix des prolongements.

**6.1.** Soient $r$ un entier $\geq 0$, $V$ un espace vectoriel sur $\mathbb{F}_q$, muni d'une forme quadratique non dégénérée $Q$. On note $d$ la dimension de $V$ sur $\mathbb{F}_q$. On suppose vérifiée l'une des hypothèses suivantes:

$$d = 2r^2 + 2r + 1;$$
$$d = 2r^2,\ r \text{ est pair} > 0 \text{ et } det(Q) \in \mathbb{F}_q^{\times 2};$$
$$d = 2r^2,\ r \text{ est impair et } -det(Q) \notin \mathbb{F}_q^{\times 2}.$$

On a introduit en [MW]2.6 et [MW]2.9 une représentation irréductible de $G_V$, unipotente et cuspidale. Notons-la $(\pi_r, E_r)$. On la prolonge, ainsi qu'il est loisible, en une représentation $\hat{\pi}_r$ de $G_V^{\pm}$ dans $E_r$, ce prolongement étant provisoirement arbitraire.

Soit $V_1$ un espace vectoriel de dimension 2 sur $\mathbb{F}_q$, muni d'une base $\{e, f\}$. On note $Q_1$ la forme quadratique sur $V_1$ dont la matrice dans cette base est $\begin{pmatrix} 0 & 1 \\ 1 & 0 \end{pmatrix}$. Posons $V' = V \oplus V_1$, que l'on munit de la forme $Q'$ somme directe orthogonale de $Q$ et $Q_1$. Notons $\mathbf{P}$ le sous-groupe parabolique de $\mathbf{G}_{V'}$ qui stabilise la droite $\mathbb{F}_q e$, et $\mathbf{U}$ son radical unipotent. Par la projection naturelle $P \to G_V$, on remonte $\pi_r$ en une représentation de $P$. Notons $\mathcal{H}$ l'algèbre d'entrelacement de la représentation $Ind_P^{G_{V'}}(\pi_r)$. Elle est de dimension 2 sur $\mathbb{C}$. Décrivons une base de $\mathcal{H}$. Notons:

$$\mathbf{1} : G_{V'} \to End(E_r)$$

la fonction à support dans $P$ et telle que $\mathbf{1}(p) = \pi_r(p)$ pour tout $p \in P$. Fixons un élément $s \in G_V^-$, notons $s_1$ l'élément de $G_{V_1}^-$ qui échange $e$ et $f$, et $s'$ l'élément de $G_V$ produit de $s$ et $s_1$. Notons:

$$\phi : G_{V'} \to End(E_r)$$

la fonction à support dans $PsP$ et telle que:

$$\phi(p_1 s' p_2) = \pi_r(p_1)\hat{\pi}_r(s)\pi_r(p_2)$$

pour tous $p_1, p_2 \in P$. Alors $\{\mathbf{1}, \phi\}$ est une base de $\mathcal{H}$. Le produit de $\mathcal{H}$ est normalisé ainsi:

(1) $$\varphi_1 * \varphi_2(g) = |P|^{-1} \sum_{h \in G_{V'}} \varphi_1(gh) \circ \varphi_2(h^{-1}).$$

Alors $\mathbf{1}$ est l'unité de $\mathcal{H}$. Il est connu qu'il existe un unique nombre complexe non nul, que l'on note $\nu_d$, tel qu'en posant $S = \nu_d \phi$, on ait l'égalité:

(2) $$(S + \mathbf{1})(S - q^L \mathbf{1}) = 0,$$

où:

$$L = \begin{cases} 2r + 1, & \text{si } d \text{ est impair,} \\ 2r, & \text{si } d \text{ est pair} \end{cases}$$

cf. [L1] 1.18 et [L3] 2.13.

**6.2.** Supposons $d$ impair. On a l'égalité $G_V^{\pm} = \{\pm 1\} \times G_V$. On fixe le prolongement $\hat{\pi}_r$ de sorte qu'il soit trivial sur $\{\pm 1\}$.

LEMME. *Sous ces hypothèses, on a l'égalité* $\nu_d = (-1)^r q^{-r^2}$.

C'est un résultat d'Asai, cf. [As] paragraphe 1.3 et lemme 1.5.12.

**6.3.** On suppose $d$ pair. Pour tout $v \in V$ tel que $Q(v) \neq 0$ (où $Q(v) = Q(v, v)$), notons $\sigma_v \in G_V^-$ la symétrie par rapport à l'hyperplan orthogonal à $v$. Si $v, v' \in V$ vérifient $Q(v) \neq 0$, $Q(v') \neq 0$ et $Q(v)Q(v')^{-1} \in \mathbb{F}_q^{\times 2}$, alors $\sigma_v$ et $\sigma_{v'}$ sont conjugués dans $G_V^{\pm}$. Fixons donc deux éléments $v_+, v_- \in V$ tels que:

$$Q(v_+) \in det(Q)\mathbb{F}_q^{\times 2}, \quad Q(v_-) \in \mathbb{F}_q^{\times} \setminus det(Q)\mathbb{F}_q^{\times 2}.$$

Posons:

$$\sigma_+ = \sigma_{v_+}, \quad \sigma_- = \sigma_{v_-},$$
$$t = trace(\hat{\pi}_r(\sigma_+)) + trace(\hat{\pi}_r(\sigma_-)).$$

Ce terme ne dépend pas des choix de $v_+, v_-$. Posons:

(1) $$\mu = \frac{(q-1)q^{r-1}(q^{r^2} + (-1)^{r+1})t}{2(q^{2r}-1)dim_{\mathbb{C}}(E_r)}.$$

Le résultat suivant est bien connu.

LEMME. *Le nombre* $\mu$ *appartient à* $\{\pm 1\}$ *et on a l'égalité* $\nu_d = \mu q^{-r^2+r}$.

*Preuve.* Calculons $\phi^2(1)$ et $\phi^2(s')$, où $\phi^2 = \phi * \phi$. En appliquant 6.1(1), on voit que:

(2) $$\phi^2(s') = \sum_{u \in U} \phi(s'us'^{-1}) \circ \phi(s').$$

considérons l'application de $U$ dans $V$ qui à $u \in U$ associe la projection orthogonale $\underline{u}$ de $u(f)$ sur $V$. Elle est bijective. Un calcul matriciel montre que, pour $u \in U$, les propriétés suivantes sont vérifiées:

$$s'us'^{-1} \in Ps'P \iff Q(\underline{u}) \neq 0;$$
$$\text{si } Q(\underline{u}) \neq 0, \ \phi(s'us'^{-1}) = \hat{\pi}_r(\sigma_{\underline{u}}).$$

Posons:

$$A = \sum_{v \in V, Q(v) \neq 0} \hat{\pi}_r(\sigma_v).$$

L'égalité (2) devient $\phi^2(s') = A \circ \hat{\pi}_r(s)$. On vérifie que $A$ commute à $\pi_r(g)$ pour tout $g \in G_V$. C'est donc une homothétie de rapport $trace(A) dim_{\mathbb{C}}(E_r)^{-1}$. Posons:

$$a_+ = |\{v \in V; Q(v) \in det(Q)\mathbb{F}_q^{\times 2}\}|,$$

$$a_- = |\{v \in V; Q(v) \in \mathbb{F}_q^{\times} \setminus det(Q)\mathbb{F}_q^{\times 2}\}|.$$

Alors:

$$trace(A) = a_+ \, trace(\hat{\pi}_r(\sigma_+)) + a_- \, trace(\hat{\pi}_r(\sigma_-)),$$

et:

(3) $$\phi^2(s') = \frac{a_+ \, trace(\hat{\pi}_r(\sigma_+)) + a_- \, trace(\hat{\pi}_r(\sigma_-))}{dim_{\mathbb{C}}(E_r)} \hat{\pi}_r(s).$$

Le calcul de $\phi^2(1)$ est immédiat:

(4) $$\phi^2(1) = |P|^{-1}|Ps'P| \, id = |U| \, id = |V| \, id,$$

où $id$ est l'identité de $E_r$.

L'égalité 6.1(2) est équivalente aux deux égalités:

(5) $$\nu_d^2 \phi^2(1) = q^L \, id, \quad \nu_d \phi^2(s') = (q^L - 1)\hat{\pi}_r(s).$$

De la première et de (4) résulte l'existence de $\mu_1 \in \{\pm 1\}$ tel que $\nu_d = \mu_1 q^{-r^2+r}$. Pour $\eta \in \{\pm 1\}$, identifié à un signe, on explicite aisément:

$$a_\eta = q^{r^2-1}(q^{r^2} + (-1)^{r+1})\frac{q-1}{2}.$$

De la deuxième égalité de (5) et de (3) résulte alors l'égalité $\mu_1 = \mu$.

On a fixé arbitrairement le prolongement $\hat{\pi}_r$. Si l'on change ce prolongement, c'est-à-dire si on le multiplie par le caractère $det$ de $G_V^{\pm}$, le nombre $\mu$ se change en son opposé. Le prolongement $\hat{\pi}_r$ que l'on a fixé en 1.3 est choisi de sorte que l'on ait l'égalité $\mu = 1$, ou encore $\nu_d = q^{-r^2+r}$.

**6.4.** On suppose encore $d$ pair. La conjecture [MW]2.11 affirme l'existence d'un certain prolongement $\tilde{\pi}_r$ de $\pi_r$.

PROPOSITION. *Supposons d pair et admettons la conjecture [MW]2.11. Alors les deux prolongements $\hat{\pi}_r$ et $\tilde{\pi}_r$ coïncident.*

La preuve occupe les paragraphes 6.5 à 6.9, dans lesquels on suppose $d$ pair. Définissons comme dans le paragraphe précédent des termes $t$ et $\mu$, mais relatifs au prolongement $\tilde{\pi}_r$. Tout revient à prouver que $\mu = 1$.

**6.5.** Bien que ce ne soit pas indispensable, nous utiliserons le lemme suivant, dont la démonstration est laissée au lecteur.

LEMME. *Soit* $D \in \mathbb{C}(X)$ *une fraction rationnelle. Supposons qu'il existe un entier* $N_0 > 0$ *tel que, pour tout* $N \geq N_0$, $D(q^N)$ *appartienne à* $\mathbb{Z}$. *Alors* $D \in \mathbb{Q}[X]$, *i.e., D est un polynôme à coefficients dans* $\mathbb{Q}$.

**6.6.** Les nombres $dim_{\mathbb{C}}(E_r)$ et $t$ qui interviennent dans nos formules dépendent de $r$ et $q$. L'entier $r$ étant fixé, $q$ peut être considéré comme variable: c'est une puissance quelconque d'un nombre premier quelconque $p \geq 3$. Cela a un sens de dire que $dim_{\mathbb{C}}(E_r)$ et $t$ dépendent polynomialement de $q$. C'est ce qu'affirment les lemmes suivants.

LEMME. *Il existe un polynôme* $D_r \in \mathbb{Q}[X]$ *tel que* $dim_{\mathbb{C}}(E_r) = D_r(q)$. *Si l'on note* $\delta_r$ *le degré de* $D_r$ *et* $\alpha_r$ *son coefficient dominant, on a les égalités:*

$$\delta_r = r^4 - \frac{2}{3}r^3 - \frac{r^2}{2} + \frac{r}{6};$$

$$\alpha_r = 2^{-r+1}.$$

*Preuve.* La dimension $dim_{\mathbb{C}}(E_r)$ est calculée par le théorème 8.2 de [L8]. Le symbole $\Lambda$ associé à $\pi_r$ est $\Lambda = (X, Y)$, où $Y = \emptyset$ et:

$$X = \{2r - 1, 2r - 2, \ldots, 0\}.$$

En explicitant les formules de [L8], on obtient le résultat, à ceci près que $D_r$ n'est a priori qu'une fraction rationnelle. Puisque $dim_{\mathbb{C}}(E_r)$ est forcément un entier, le lemme 6.5 montre que $D_r$ est bien un polynôme.

**6.7.**

LEMME. *Il existe un polynôme* $D_{r,t} \in \mathbb{Q}[X]$ *tel que* $t = D_{r,t}(q)$. *Notons* $\delta_{r,t}$ *son degré et* $\alpha_{r,t}$ *son coefficient dominant. On a les égalités:*

$$\delta_{r,t} = r^4 - \frac{2}{3}r^3 - \frac{3r^2}{2} + \frac{7r}{6}, \quad \alpha_{r,t} = 2^{2-r}.$$

Cela sera démontré en 6.9.

**6.8.** Admettons ce lemme et démontrons la proposition 6.4. D'après la définition 6.3 (1), et grâce aux lemmes précédents, $\mu$ est donné par le quotient de deux polynômes dont on connaît les degrés et les coefficients dominants. Puisqu'on sait que $\mu \in \{\pm 1\}$, les degrés de ces deux polynômes sont nécessairement égaux (ce qui résulte aussi du calcul de ces degrés) et $\mu$ est égal au rapport des coefficients dominants. On voit alors que $\mu = 1$.

**6.9.** Rappelons que par hypothèse, $Q$ est déployée si $r$ est pair, non déployée si $r$ est impair. On pose $\sharp = iso$ dans le premier cas, $\sharp = an$ dans le second. A tout symbole $\Lambda'' = (X'', Y'') \in \tilde{S}_{r^2,pair}$ (cf. [MW]2.1), on a associé en [MW]2.4 un couple $(r'', \rho)$, où $r''$ est un entier relatif tel que $|r''| \leq r$ et $\rho$ est une représentation irréductible de $W_{r^2-r''^2}$. A ce couple $(r'', \rho)$, on a associé en [MW]2.10 une fonction $k(r'', \rho)_\sharp$ sur $G_V^\pm$, que l'on notera aussi $k(\Lambda'')_\sharp$. L'élément $\Lambda$ de $\tilde{S}_{r^2,pair}$ associé à $\tilde{\pi}_r$ est le même que celui introduit en 6.6. Notons $\tilde{\mathcal{F}am}(\Lambda) \subseteq \tilde{D}_{r^2,pair}$ sa famille. D'après la conjecture [MW]2.11, que nous admettons, on a l'égalité:

$$(1) \qquad trace(\tilde{\pi}_r)(g) = 2^{-r} \sum_{\Lambda'' \in \tilde{\mathcal{F}am}(\Lambda)} (-1)^{<\Lambda,\Lambda''>} k(\Lambda'')_\sharp(g),$$

pour tout $g \in G_V^\pm$.

Soient $r''$ un entier tel que $|r''| \leq r$ et $\rho$ une représentation irréductible de $W_{r^2-r''^2}$. Soit $\eta \in \{\pm 1\}$, identifié à un signe. D'après [MW]2.10, on a les égalités:

- $k(r'', \rho)_\sharp(\sigma_\eta) = 0$, si $r''$ est pair;
- (2) $k(-r'', \rho)_\sharp(\sigma_\eta) = (-1)^r k(r'', \rho)_\sharp(\sigma_\eta)$.

Supposons $r'' \geq 0$ et $r''$ impair. D'après [MW]2.13, on a l'égalité:

$$k(r'', \rho)_\sharp(\sigma_\eta) = |W_{r^2-r''^2}|^{-1} \sum_{w \in W_{r^2-r''^2}} trace(\rho(w)) k(r'', w)_\sharp(\sigma_\eta).$$

Fixons $w \in W_{r^2-r''^2}$. On a calculé $k(r'', w)_\sharp$ en [MW]2.16, sous l'hypothèse que $w$ était elliptique. Cette hypothèse ne nous servait qu'à prouver que $k(r'', w)_\sharp(x)$ était nul si la partie semi-simple de $x$ n'était pas elliptique. Ici $\sigma_\eta$ est semi-simple elliptique et la proposition [MW]2.16 s'applique. Le (i) de cette proposition nous dit que $k(r'', w)_\sharp(\sigma_\eta) = 0$ si $r''^2 > 1$, i.e., si $r'' \neq 1$. Supposons $r'' = 1$. Notons $V_+$ l'espace propre pour $\sigma_\eta$ associé à la valeur propre 1. Il est de dimension $2r^2 - 1$. On dispose de la fonction $Q(1, w)$ sur $G_{V_+}$ (cf. [MW]2.14). Alors le (ii) de la proposition [MW]2.16 nous dit que:

$$k(1, w)_\sharp(\sigma_\eta) = (-1)^{r+1} Q(1, w)(1).$$

Introduisons le tore maximal $\mathbf{T}$ de $\mathbf{G}_{V_+}$ paramétrisé par la classe de conjugaison de $w$ et notons $R_T$ le caractère de Deligne-Lusztig associé à $\mathbf{T}$ et au caractère trivial de $T$. D'après [MW]2.14 et [L7] théorème 1.14, on a les égalités:

$$Q(1, w)(1) = trace(\phi | K^u(1, w)_1) = (-1)^{r+1} R_T(1)$$

(la formule du théorème de Lusztig fait intervenir le signe $(-1)^{rg}$, où $rg$ est le rang de $\mathbf{G}_{V_+}$; on a $rg = r^2 - 1$). D'après [C], théorème 7.5.1, on a l'égalité:

$$R_T(1) = \epsilon_{G_{V_+}} \epsilon_T |G_{V_+}|_{p'} |T|^{-1},$$

où $\epsilon_{G_{V_+}}$, resp. $\epsilon_T$, est le rang "déployé" de $\mathbf{G}_{V_+}$, resp. $\mathbf{T}$, et $|G_{V_+}|_{p'}$ est la partie première à $p$ de $|G_{V_+}|$. On a l'égalité $\epsilon_{G_{V_+}} = (-1)^{r+1}$. Faisons agir naturellement sur $\mathbb{R}^{r^2-1}$ l'élément $w$ de $W_{r^2-1}$. On a l'égalité bien connue:

$$\epsilon_T |T| = det(1 - qw | \mathbb{R}^{r^2-1}).$$

Pour tout entier $h \geq 0$, définissons le polynôme:

$$\theta(h, X) = \prod_{i=1,\dots,h} (X^i - 1).$$

On calcule:

$$|G_{V_+}|_{p'} = \theta(r^2 - 1, q^2).$$

D'où l'égalité:

$$k(1, w)_\sharp(\sigma_\eta) = (-1)^{r+1}\theta(r^2 - 1, q^2)\, det(1 - qw|\mathbb{R}^{r^2-1}).$$

Pour tout entier $m \in \mathbb{N}$ et toute représentation irréductible $\chi$ de $W_m$, posons:

$$\delta_\chi(q) = (-1)^m |W_m|^{-1} \sum_{w \in W_m} trace(\chi(m))\, det(1 - qw|\mathbb{R}^m)^{-1}.$$

De l'égalité précédente résulte l'égalité:

$$k(1, \rho)_\sharp(\sigma_\eta) = \theta(r^2 - 1, q^2)\delta_\rho(q). \tag{3}$$

Dans la situation ci-dessus où $\chi \in \hat{W}_m$, le terme $\delta_\chi(q)$ est calculé en [L9] lemme 2.4. Soit $(A, B)$ le symbole associé au couple $(0, \chi)$, avec:

$$A = \{a_1 > \cdots > a_{c+1}\}, \quad B = \{b_1 > \cdots > b_c\}.$$

Alors $\delta_\chi(q)$ est une fraction rationnelle en $q$, de degré:

$$\left(\sum_{i=1,\dots,c+1} a_i(2c + 1 - 2i)\right) + \left(\sum_{i=1,\dots,c} b_i(2c - 2i)\right) - \left(\sum_{i=1,\dots,c+1} a_i^2\right) - \left(\sum_{i=1,\dots,c} b_i^2\right) - \frac{2c^3}{3} + \frac{c^2}{2} + \frac{c}{6}, \tag{4}$$

et de coefficient dominant 1.

Soient maintenant $\Lambda'' = (X'', Y'') \in \tilde{\mathcal{F}am}(\Lambda)$, avec $|X''| + |Y''| = 2r$, et $(r'', \rho)$ le couple correspondant. La condition $r'' = 1$, resp. $r'' = -1$, est équivalente à $|X''| = r + 1$, $|Y''| = r - 1$, resp. $|X''| = r - 1$, $|Y''| = r + 1$. Supposons l'une de ces conditions vérifiées, posons $\sigma(\Lambda'') = (Y'', X'')$. Son couple associé est $(-r'', \rho)$. De plus,

$$< \Lambda, \sigma(\Lambda'') > \equiv < \Lambda, \Lambda'' > +r \; mod\, 2\mathbb{Z}.$$

En utilisant (2), on voit que la formule (1) se simplifie:

$$trace(\tilde{\pi}_r)(\sigma_\eta) = 2^{1-r} \sum_{\substack{\Lambda'' = (X'', Y'') \in \tilde{\mathcal{F}am}(\Lambda) \\ |X''| = r+1, |Y''| = r-1}} (-1)^{<\Lambda,\Lambda''>} k(\Lambda'')_\sharp(\sigma_\eta). \tag{5}$$

Soient donc $\Lambda'' = (X'', Y'') \in \tilde{\mathcal{F}am}(\Lambda)$, avec $|X''| = r+1$, $|Y''| = r-1$, et $(1, \rho)$ le couple correspondant. Pour calculer $\delta_\rho(q)$, on doit paramétriser $\rho$ par un symbole $(A, B)$ tel que $|A| = |B| + 1$. On peut prendre pour couple $(X'', Y^*)$, où, si $Y'' = \{y''_1 > \cdots > y''_{r-1}\}$, on a posé $Y^* = \{y''_1 + 1 > \cdots > y''_{r-1} + 1 > 0\}$. On vérifie grâce à (3) et (4) les propriétés suivantes:

- $k(\Lambda'')_\sharp(\sigma_\eta)$ est une fraction rationnelle en $q$ de degré $\leq r^4 - \frac{2r^3}{3} - \frac{3r^2}{2} + \frac{7r}{6}$, de coefficient dominant 1;
- cette inégalité est une égalité pour l'unique symbole $\Lambda'' = (X'', Y'')$ tel que:

$$X'' = \{2r-1, 2r-2, 2r-4, \ldots, 2, 0\}, \quad Y'' = \{2r-3, 2r-5, \ldots, 3, 1\}.$$

On vérifie que, pour ce symbole, $< \Lambda, \Lambda'' > \equiv 0 \, mod \, 2\mathbb{Z}$ (cf. [MW]2.1). De la formule (4) et des propriétés ci-dessus résultent les assertions du lemme 6.7, à ceci près qu'a priori, on obtient des fractioons rationnelles au lieu de polynômes. Mais $t$ est une somme de traces d'éléments dont le carré est l'identité. Donc $t \in \mathbb{Z}$. Le lemme 6.5 permet d'affirmer que les fractions rationnelles sont des polynômes. Cela achève la preuve de la proposition 6.4.

**6.10.** Démontrons maintenant la proposition 1.3. Les notations sont celles du paragraphe 1. Supposons $R'' \neq 0$. On a l'isomorphisme:

$$N_{G^{\pm}_\sharp}(K_\theta)/K^u_\theta \simeq (\mathbb{F}_q^\times)^N \times G^{\pm}_{l'_N} \times G^{\pm}_{l''_0}.$$

On a défini en 6.2 et 6.3 les prolongements $\tilde{\pi}'_{R'}$ de $\pi'_{R'}$ à $G^{\pm}_{l'_N}$, $\tilde{\pi}''_{R''}$ de $\pi''_{R''}$ à $G^{\pm}_{l''_0}$. Par produit tensoriel, on en déduit un prolongement de $\pi_\theta$ à $N_{G^{\pm}_\sharp}(K_\theta)$. On vérifie que c'est le prolongement $\tilde{\pi}_\theta$ défini en 1.3 (cela résulte simplement du fait connu que $\pi''_{R''}(-1)$ est l'identité).

Considérons la chaîne de réseaux:

$$\mathcal{L}_. = (L_0, L_1, \ldots, L_{N-1}),$$

posons $l'_{N-1} = L_{N-1}/\varpi_F \tilde{L}_{N-1}$. On a:

$$K(\mathcal{L}_.)/K(\mathcal{L}_.)^u \simeq (\mathbb{F}_q^\times)^{N-1} \times G_{l'_{N-1}} \times G_{l''_0}.$$

L'image de $K_\theta$ dans ce quotient est de la forme:

$$(\mathbb{F}_q^\times)^{N-1} \times P' \times G_{l''_0},$$

où $\mathbf{P}'$ est un sous-groupe parabolique de $\mathbf{G}_{l'_{N-1}}$. En appliquant les constructions de 6.1 à l'espace $l'_{N-1}$ et à ce parabolique $\mathbf{P}'$, on dispose de deux fonctions $\mathbf{1}', \phi'$ : $G_{l'_{N-1}} \to End(E'_{R'})$ (on a affecté d'un $'$ les notations de 6.1). Soit $f'$ l'une d'elles. Définissons une fonction:

$$\begin{array}{rcl} (\mathbb{F}_q^\times)^{N-1} \times G_{l'_{N-1}} \times G_{l''_0} & \to & End(E_\theta) \\ (z_1, \ldots, z_{N-1}; x', x'') & \mapsto & f'(x') \otimes \pi''_{R''}(x''). \end{array}$$

Remontons-la en une fonction sur $K(\mathcal{L}_\cdot)$ puis prolongeons-la à $G_\sharp$ tout entier par 0 en dehors de $K(\mathcal{L}_\cdot)$. Notons $\mathbf{1}$, resp. $\phi$, la fonction obtenue, dans le cas où $f' = \mathbf{1}'$, resp. $f' = \phi'$. La fonction $\mathbf{1}$ est bien celle notée ainsi en 1.3, à savoir l'unité de l'algèbre $\mathcal{H}_\theta$. Il résulte de la construction de $S_N$ que l'on a l'égalité $S_N = \nu_N \phi$. L'égalité (1) de 1.3 pour cette fonction $S_N$ est équivalente à l'identité:

$$(\nu_N \phi' + \mathbf{1}')(\nu_N \phi' - q^{L(N)}\mathbf{1}') = 0.$$

Autrement dit, $\nu_N$ est égal au terme noté $\nu_d$ en 6.1, relatif à l'espace $l'_{N-1}$. L'égalité $\nu_N = (-1)^{R'} q^{-R'^2}$ résulte du lemme 6.2 .

Le calcul de $\nu_0$ est analogue. On considère cette fois la chaîne de réseaux $\mathcal{L}_\cdot = (L_1, \dots, L_N)$, pour laquelle:

$$K(\mathcal{L}_\cdot)/K(\mathcal{L}_\cdot)^u \simeq (\mathbb{F}_q^\times)^{N-1} \times G_{l'_N} \times G_{l''_1},$$

où $l'_1 = \tilde{L}_1/L_1$. L'égalite $\nu_0 = q^{-R''^2+R''}$ résulte du lemme 6.3 appliqué à l'espace $l''_1$.

Dans le cas où $R'' = 0$, le calcul de $\nu_N$ s'effectue comme ci-dessus. Celui de $\nu_0$ se ramène à un calcul dans un groupe $GL(2)$. Cela achève la démonstration.

CNRS, Institut mathématiques de JUSSIEU
*E-mail:* waldspur@math.jussieu.fr

## REFERENCES

[Ar] J. Arthur, On elliptic tempered characters, *Acta Math.* **171** (1993), 73–138.

[As] T. Asai, The unipotent class functions on the symplectic groups and the odd orthogonal groups over finite fields, *Comm. in Algebra* **12** (1984), 617–645.

[Au] A.-M. Aubert, Dualité dans le groupe de Grothendieck de la catégorie des représentations lisses de longueur finie d'un groupe réductif p-adique, *Trans. AMS* **347** (1995), 2179–2189.

[BBD] A. Beilinson, J. Bernstein and P. Deligne, Faisceaux pervers, in Analyse et topologie sur les espaces singuliers, *Astérisque* **100** (1982).

[BK] C. Bushnell and P. Kutzko, Smooth representations of reductive $p$-adic groups: Structure theory via types, *Proc. London Math. Soc.* **77** (1998), 582–634.

[C] R. Carter, Finite groups of Lie type, conjugacy classes and complex characters, Wiley-Interscience Publ. 1993.

[KL] D. Kazhdan and G. Lusztig, Proof of the Deligne-Langlands conjecture for Hecke algebras, *Inventiones Math.* **87** (1987), 153–215.

[L1] G. Lusztig, Classification of unipotent representations of simple $p$-adic groups, *Internat. Math. Res. Notices* **11** (1995), 517–589.

[L2] ———, Affine Hecke algebras and their graded version, *J. Amer. Math. Soc.* **2** (1989), 599–635.

[L3] ———, Cuspidal local systems and graded algebras. I, *Publ. Math. IHES* **67** (1988), 145–202.

[L4] ———, Cuspidal local systems and graded algebras. II, *Can. Math. Soc.,* Conference Proc. **16** (1995).

[L5] ———, Intersection cohomology complexes on a reductive group, *Invent. Math.* **75** (1984), 205–272.

[L6] ———, Study of perverse sheaves arising from graded Lie algebras, *Adv. in Math.* **112** (1995), 147–217.

[L7] ———, Character sheaves. V, *Adv. in Math.* **61** (1986), 103–155.

[L8] ———, Irreductible representations of finite classical groups, *Inventiones Math.* **43** (1977), 125–175.
[L9] ———,Green functions and characters sheaves, *Annals of Math.* **131** (1990), 355–408.
[L10] ———, Cuspidal local systems and graded Hecke algebras. III, prépublication, 2001.
[L11] ———, Classification of unipotent representations of simple $p$-adic groups. II, prépublication, 2001.
[MW] C. Moeglin and J.-L. Waldspurger, Paquets stables de représentations tempérées et de réduction unipotente pour $SO(2n+1)$, *Invent. Math.* **152** (2003), 461–623.
[M] L. Morris, Level 0 $G$-types, *Compositio Math.* **118** (1999), 135–157.
[MP] A. Moy and G. Prasad, Jacquet functors and unrefined minimal K-types, *Comment. Math. Helvetici* **71** (1996), 98–121.
[S] A. Silberger, The Langlands quotient theorem for $p$-adic groups, *Math. Ann.* **236** (1978), 95–104.
[W] J.-L. Waldspurger, Intégrales orbitales nilpotentes et endoscopie pour les groupes classiques non ramifiés, *Astérisque* **269** (2001).
[Z] A. Zelevinsky, A p-adic analogue of the Kazhdan-Lusztig conjecture, *Funkt. Anal. Pril.* **15** (1981), 9–21.